METRIC INTERNATIO

STUDENT SOLUTIONS MANUAL

FOR

Calculus

Concepts and Contexts • 3E

Jeffery A. Cole

ANOKA RAMSEY COMMUNITY COLLEGE

Dan Clegg

PALOMAR COLLEGE

THOMSON

BROOKS/COLE

Australia • Canada • Mexico • Singapore • Spain
United Kingdom • United States

Printed in the United States of America
1 2 3 4 5 6 7 09 08 07 06 05

Printer: West Group

ISBN: 0-495-01427-3

Thomson Higher Education
10 Davis Drive
Belmont, CA 94002-30980
USA

Asia (including India)
Thomson Learning
5 Shenton Way
#01-01 UIC Building
Singapore 068808

Australia/New Zealand
Thomson Learning Australia
102 Dodds Street
Southbank, Victoria 3006
Australia

Canada
Thomson Nelson
1120 Birchmount Road
Toronto, Ontario M1K 5G4
Canada

UK/Europe/Middle East/Africa
Thomson Learning
High Holborn House
50–51 Bedford Row
London WC1R 4LR
United Kingdom

Latin America
Thomson Learning
Seneca, 53
Colonia Polanco
11560 Mexico
D.F. Mexico

Spain (including Portugal)
Thomson Paraninfo
Calle Magallanes, 25
28015 Madrid, Spain

☐ PREFACE

This *Student Solutions Manual* contains strategies for solving and solutions to selected exercises in the texts *Metric International Version Calculus: Concepts and Contexts,* Third Edition, and *Metric International Version Single Variable Calculus: Concepts and Contexts*, Third Edition, by James Stewart. It contains solutions to the odd-numbered exercises in each section, the review sections, the True-False Quizzes, and the Problem Solving sections, as well as solutions to all the exercises in the Concept Checks.

This manual is a text supplement and should be read along with the text. You should read all exercise solutions in this manual because many concept explanations are given and then used in subsequent solutions. All concepts necessary to solve a particular problem are not reviewed for every exercise. If you are having difficulty with a previously covered concept, refer back to the section where it was covered for more complete help.

A significant number of today's students are involved in various outside activities, and find it difficult, if not impossible, to attend all class sessions; this manual should help meet the needs of these students. In addition, it is my hope that this manual's solutions will enhance the understanding of all readers of the material and provide insights to solving other exercises.

Some nonstandard use of notation is used in order to save space. If you see a symbol which you don't recognize, refer to the table of Abbreviations and Symbols on page v.

I appreciate feedback concerning errors, solution correctness or style, and manual style. Any comments may be sent directly to me at the address below, at jeff.cole@anokaramsey.edu, or in care of the publisher: Thomson Brooks/Cole, 10 Davis Drive, Belmont, CA 94002.

I would like to thank Jim Stewart, for his guidance; Dan Clegg, of Palomar College, for his work in previous editions and on the multivariable chapters in this manual; Craig Chamberlain, of Palomar College, for his careful assistance with most of the new solutions; Brian Betsill, Kathi Townes, and Jenny Turney, of TECH-arts, for their production services; and Bob Pirtle and Stacy Green, of Thomson Brooks/Cole, for entrusting me with this project as well as for their patience and support.

I dedicate this book to my wife, Joan.

Jeffery A. Cole
Anoka Ramsey Community College
11200 Mississippi Blvd. NW
Coon Rapids, MN 55433

ABBREVIATIONS AND SYMBOLS

CD	concave downward
CU	concave upward
D	the domain of f
FDT	First Derivative Test
HA	horizontal asymptote(s)
I	interval of convergence
IP	inflection point(s)
R	radius of convergence
VA	vertical asymptote(s)
$\overset{\text{CAS}}{=}$	indicates the use of a computer algebra system.
$\overset{\text{H}}{=}$	indicates the use of l'Hospital's Rule.
$\overset{j}{=}$	indicates the use of Formula j in the Table of Integrals in the back endpapers.
$\overset{\text{s}}{=}$	indicates the use of the substitution $\{u = \sin x, du = \cos x\,dx\}$.
$\overset{\text{c}}{=}$	indicates the use of the substitution $\{u = \cos x, du = -\sin x\,dx\}$.

CONTENTS

3 Differentiation Rules 71

4 Applications of Differentiation 109

5 Integrals 165

1 □ FUNCTIONS AND MODELS

1.1 Four Ways to Represent a Function

In exercises requiring estimations or approximations, your answers may vary slightly from the answers given here.

1. (a) The point $(-1, -2)$ is on the graph of f, so $f(-1) = -2$.

(b) When $x = 2$, y is about 2.8, so $f(2) \approx 2.8$.

(c) $f(x) = 2$ is equivalent to $y = 2$. When $y = 2$, we have $x = -3$ and $x = 1$.

(d) Reasonable estimates for x when $y = 0$ are $x = -2.5$ and $x = 0.3$.

(e) The domain of f consists of all x-values on the graph of f. For this function, the domain is $-3 \le x \le 3$, or $[-3, 3]$. The range of f consists of all y-values on the graph of f. For this function, the range is $-2 \le y \le 3$, or $[-2, 3]$.

(f) As x increases from -1 to 3, y increases from -2 to 3. Thus, f is increasing on the interval $[-1, 3]$.

3. From Figure 1 in the text, the lowest point occurs at about $(t, a) = (12, -85)$. The highest point occurs at about $(17, 115)$. Thus, the range of the vertical ground acceleration is $-85 \le a \le 115$. In Figure 11, the range of the north-south acceleration is approximately $-325 \le a \le 485$. In Figure 12, the range of the east-west acceleration is approximately $-210 \le a \le 200$.

5. No, the curve is not the graph of a function because a vertical line intersects the curve more than once. Hence, the curve fails the Vertical Line Test.

7. Yes, the curve is the graph of a function because it passes the Vertical Line Test. The domain is $[-3, 2]$ and the range is $[-3, -2) \cup [-1, 3]$.

9. The person's weight increased to about 64 kg at age 20 and stayed fairly steady for 10 years. The person's weight dropped to about 52 kg for the next 5 years, then increased rapidly to about 68 kg. The next 30 years saw a gradual increase to 76 kg. Possible reasons for the drop in weight at 30 years of age: diet, exercise, health problems.

11. The water will cool down almost to freezing as the ice melts. Then, when the ice has melted, the water will slowly warm up to room temperature.

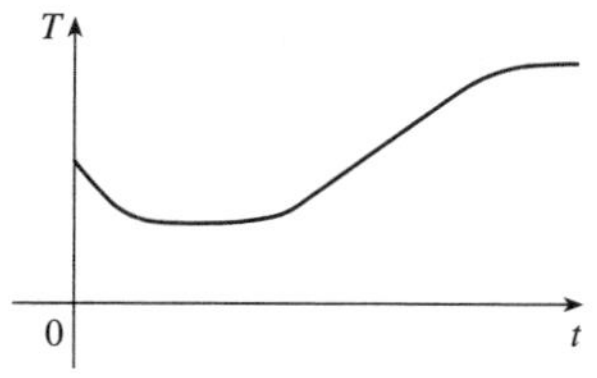

13. Of course, this graph depends strongly on the geographical location!

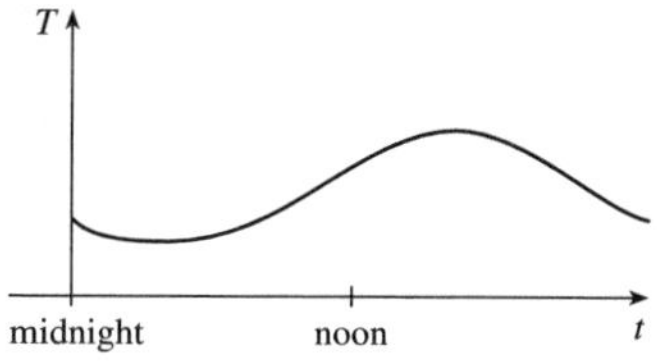

15. As the price increases, the amount sold decreases.

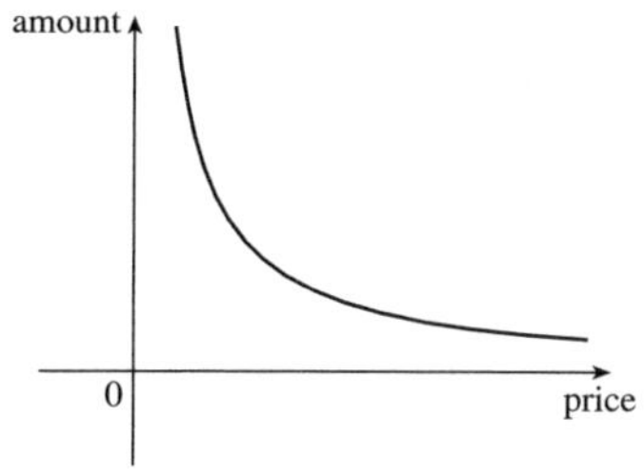

17.

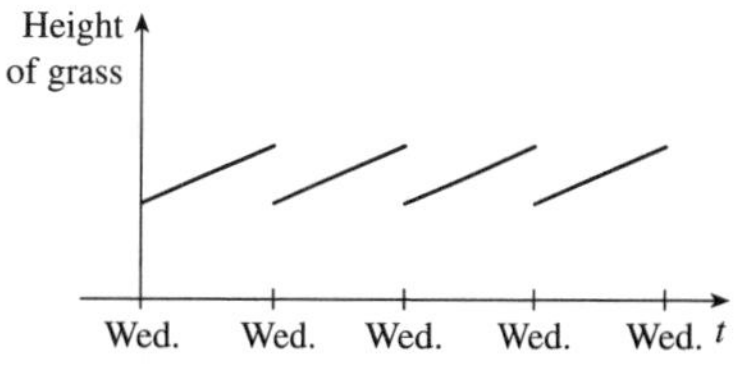

19. (a)

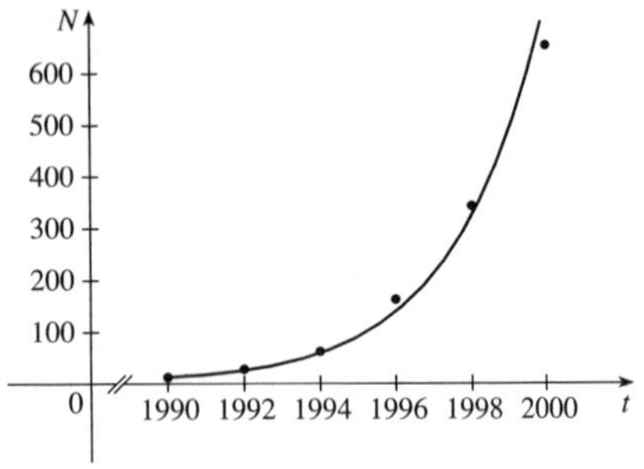

(b) From the graph, we estimate the number of cell-phone subscribers worldwide to be about 92 million in 1995 and 485 million in 1999.

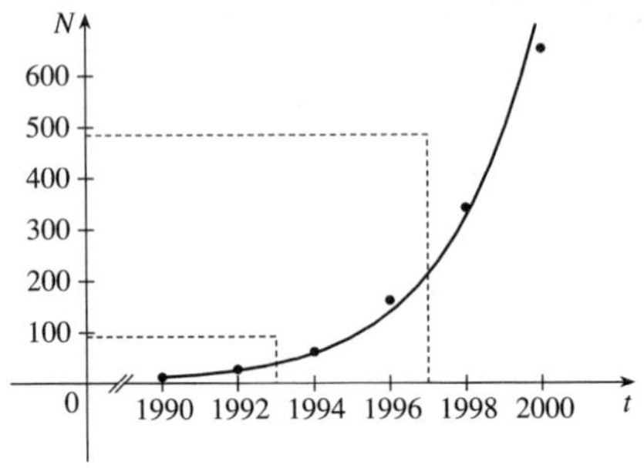

21. $f(x) = 3x^2 - x + 2$.

$f(2) = 3(2)^2 - 2 + 2 = 12 - 2 + 2 = 12$.

$f(-2) = 3(-2)^2 - (-2) + 2 = 12 + 2 + 2 = 16$.

$f(a) = 3a^2 - a + 2$.

$f(-a) = 3(-a)^2 - (-a) + 2 = 3a^2 + a + 2$.

$f(a+1) = 3(a+1)^2 - (a+1) + 2 = 3(a^2 + 2a + 1) - a - 1 + 2 = 3a^2 + 6a + 3 - a + 1 = 3a^2 + 5a + 4$.

$2f(a) = 2 \cdot f(a) = 2(3a^2 - a + 2) = 6a^2 - 2a + 4$.

$f(2a) = 3(2a)^2 - (2a) + 2 = 3(4a^2) - 2a + 2 = 12a^2 - 2a + 2$.

$f(a^2) = 3(a^2)^2 - (a^2) + 2 = 3(a^4) - a^2 + 2 = 3a^4 - a^2 + 2$.

$$[f(a)]^2 = \left[3a^2 - a + 2\right]^2 = \left(3a^2 - a + 2\right)\left(3a^2 - a + 2\right)$$
$$= 9a^4 - 3a^3 + 6a^2 - 3a^3 + a^2 - 2a + 6a^2 - 2a + 4 = 9a^4 - 6a^3 + 13a^2 - 4a + 4.$$

$f(a+h) = 3(a+h)^2 - (a+h) + 2 = 3(a^2 + 2ah + h^2) - a - h + 2 = 3a^2 + 6ah + 3h^2 - a - h + 2$.

23. $f(x) = 4 + 3x - x^2$, so $f(3+h) = 4 + 3(3+h) - (3+h)^2 = 4 + 9 + 3h - (9 + 6h + h^2) = 4 - 3h - h^2$,

and $\dfrac{f(3+h) - f(3)}{h} = \dfrac{(4 - 3h - h^2) - 4}{h} = \dfrac{h(-3-h)}{h} = -3 - h$.

25. $$\frac{f(x) - f(a)}{x - a} = \frac{\dfrac{1}{x} - \dfrac{1}{a}}{x - a} = \frac{\dfrac{a - x}{xa}}{x - a} = \frac{a - x}{xa(x - a)} = \frac{-1(x - a)}{xa(x - a)} = -\frac{1}{ax}$$

27. $f(x) = x/(3x - 1)$ is defined for all x except when $0 = 3x - 1 \Leftrightarrow x = \frac{1}{3}$, so the domain is $\left\{x \in \mathbb{R} \mid x \neq \frac{1}{3}\right\} = \left(-\infty, \frac{1}{3}\right) \cup \left(\frac{1}{3}, \infty\right)$.

29. $f(t) = \sqrt{t} + \sqrt[3]{t}$ is defined when $t \geq 0$. These values of t give real number results for $\sqrt{t}$, whereas any value of t gives a real number result for $\sqrt[3]{t}$. The domain is $[0, \infty)$.

31. $h(x) = 1 / \sqrt[4]{x^2 - 5x}$ is defined when $x^2 - 5x > 0 \quad \Leftrightarrow \quad x(x-5) > 0$. Note that $x^2 - 5x \neq 0$ since that would result in division by zero. The expression $x(x-5)$ is positive if $x < 0$ or $x > 5$. (See Appendix A for methods for solving inequalities.) Thus, the domain is $(-\infty, 0) \cup (5, \infty)$.

33. $f(x) = 5$ is defined for all real numbers, so the domain is $\mathbb{R}$, or $(-\infty, \infty)$. The graph of f is a horizontal line with y-intercept 5.

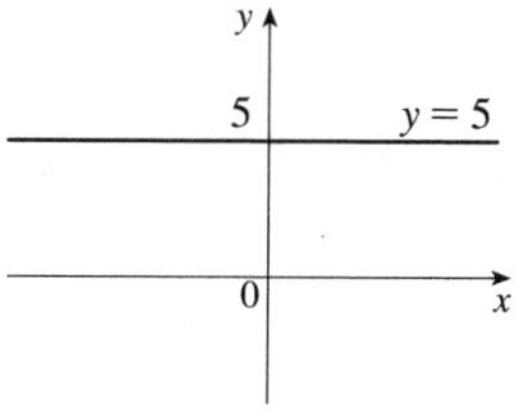

35. $f(t) = t^2 - 6t$ is defined for all real numbers, so the domain is $\mathbb{R}$, or $(-\infty, \infty)$. The graph of f is a parabola opening upward since the coefficient of t^2 is positive. To find the t-intercepts, let $y = 0$ and solve for t. $0 = t^2 - 6t = t(t-6) \quad \Rightarrow \quad t = 0$ and $t = 6$. The t-coordinate of the vertex is halfway between the t-intercepts, that is, at $t = 3$. Since $f(3) = 3^2 - 6 \cdot 3 = -9$, the vertex is $(3, -9)$.

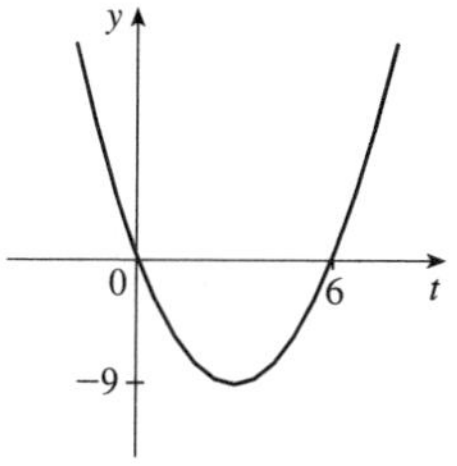

37. $g(x) = \sqrt{x - 5}$ is defined when $x - 5 \geq 0$ or $x \geq 5$, so the domain is $[5, \infty)$. Since $y = \sqrt{x-5} \quad \Rightarrow \quad y^2 = x - 5 \quad \Rightarrow \quad x = y^2 + 5$, we see that g is the top half of a parabola.

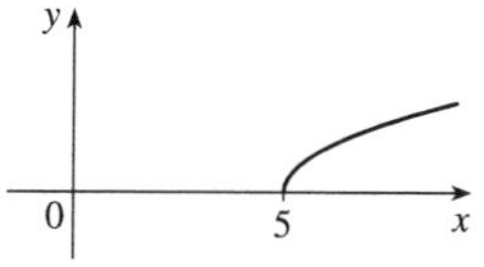

39. $G(x) = \dfrac{3x + |x|}{x}$. Since $|x| = \begin{cases} x & \text{if } x \geq 0 \\ -x & \text{if } x < 0 \end{cases}$, we have

$$G(x) = \begin{cases} \dfrac{3x + x}{x} & \text{if } x > 0 \\ \dfrac{3x - x}{x} & \text{if } x < 0 \end{cases} = \begin{cases} \dfrac{4x}{x} & \text{if } x > 0 \\ \dfrac{2x}{x} & \text{if } x < 0 \end{cases} = \begin{cases} 4 & \text{if } x > 0 \\ 2 & \text{if } x < 0 \end{cases}$$

Note that G is not defined for $x = 0$. The domain is $(-\infty, 0) \cup (0, \infty)$.

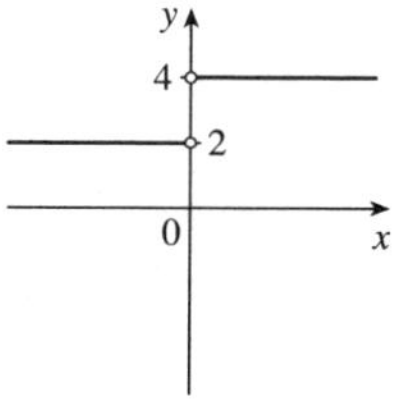

41. $f(x) = \begin{cases} x + 2 & \text{if } x < 0 \\ 1 - x & \text{if } x \geq 0 \end{cases}$

The domain is $\mathbb{R}$.

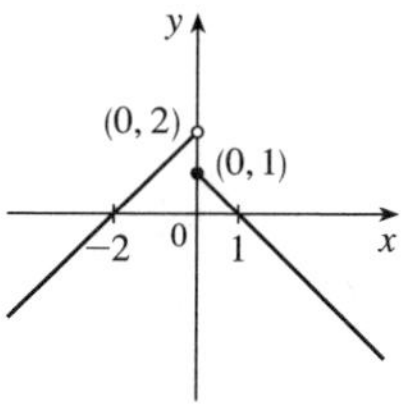

43. $f(x) = \begin{cases} x + 2 & \text{if } x \leq -1 \\ x^2 & \text{if } x > -1 \end{cases}$

Note that for $x = -1$, both $x + 2$ and x^2 are equal to 1. The domain is $\mathbb{R}$.

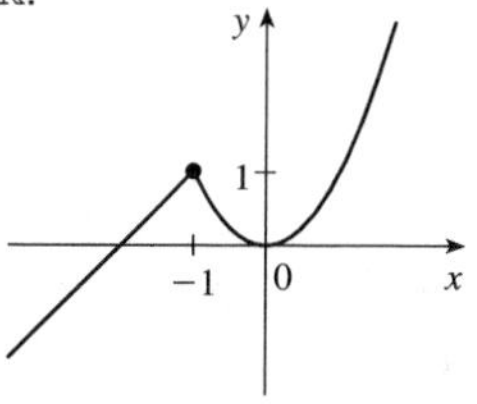

45. Recall that the slope m of a line between the two points (x_1, y_1) and (x_2, y_2) is $m = \dfrac{y_2 - y_1}{x_2 - x_1}$ and an equation of the line connecting those two points is $y - y_1 = m(x - x_1)$. The slope of this line segment is $\dfrac{-6 - 1}{4 - (-2)} = -\dfrac{7}{6}$, so an equation is $y - 1 = -\frac{7}{6}(x + 2)$. The function is $f(x) = -\frac{7}{6}x - \frac{4}{3}$, $-2 \leq x \leq 4$.

47. We need to solve the given equation for y. $x + (y-1)^2 = 0 \Leftrightarrow (y-1)^2 = -x \Leftrightarrow y - 1 = \pm\sqrt{-x} \Leftrightarrow$ $y = 1 \pm \sqrt{-x}$. The expression with the positive radical represents the top half of the parabola, and the one with the negative radical represents the bottom half. Hence, we want $f(x) = 1 - \sqrt{-x}$. Note that the domain is $x \le 0$.

49. For $-1 \le x \le 2$, the graph is the line with slope 1 and y-intercept 1, that is, the line $y = x + 1$. For $2 < x \le 4$, the graph is the line with slope $-\frac{3}{2}$ and x-intercept 4 [which corresponds to the point $(4, 0)$], so $y - 0 = -\frac{3}{2}(x - 4) = -\frac{3}{2}x + 6$. So the function is $f(x) = \begin{cases} x + 1 & \text{if } -1 \le x \le 2 \\ -\frac{3}{2}x + 6 & \text{if } 2 < x \le 4 \end{cases}$

51. Let the length and width of the rectangle be L and W. Then the perimeter is $2L + 2W = 20$ and the area is $A = LW$. Solving the first equation for W in terms of L gives $W = \dfrac{20 - 2L}{2} = 10 - L$. Thus, $A(L) = L(10 - L) = 10L - L^2$. Since lengths are positive, the domain of A is $0 < L < 10$. If we further restrict L to be larger than W, then $5 < L < 10$ would be the domain.

53. Let the length of a side of the equilateral triangle be x. Then by the Pythagorean Theorem, the height y of the triangle satisfies $y^2 + \left(\frac{1}{2}x\right)^2 = x^2$, so that $y^2 = x^2 - \frac{1}{4}x^2 = \frac{3}{4}x^2$ and $y = \frac{\sqrt{3}}{2}x$. Using the formula for the area A of a triangle, $A = \frac{1}{2}(\text{base})(\text{height})$, we obtain $A(x) = \frac{1}{2}(x)\left(\frac{\sqrt{3}}{2}x\right) = \frac{\sqrt{3}}{4}x^2$, with domain $x > 0$.

55. Let each side of the base of the box have length x, and let the height of the box be h. Since the volume is 2, we know that $2 = hx^2$, so that $h = 2/x^2$, and the surface area is $S = x^2 + 4xh$. Thus, $S(x) = x^2 + 4x(2/x^2) = x^2 + (8/x)$, with domain $x > 0$.

57. The height of the box is x and the length and width are $L = 20 - 2x$, $W = 12 - 2x$. Then $V = LWx$ and so
$V(x) = (20 - 2x)(12 - 2x)(x) = 4(10 - x)(6 - x)(x) = 4x(60 - 16x + x^2) = 4x^3 - 64x^2 + 240x$.
The sides L, W, and x must be positive. Thus, $L > 0 \Leftrightarrow 20 - 2x > 0 \Leftrightarrow x < 10$;
$W > 0 \Leftrightarrow 12 - 2x > 0 \Leftrightarrow x < 6$; and $x > 0$. Combining these restrictions gives us the domain $0 < x < 6$.

59. (a)

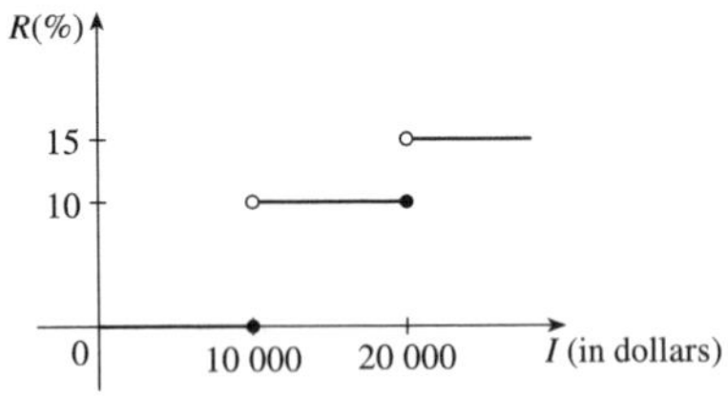

(b) On \$14 000, tax is assessed on \$4000, and $10\%(\$4000) = \$400$.
On \$26 000, tax is assessed on \$16 000, and
$10\%(\$10\,000) + 15\%(\$6000) = \$1000 + \$900 = \$1900$.

(c) As in part (b), there is \$1000 tax assessed on \$20 000 of income, so the graph of T is a line segment from $(10\,000, 0)$ to $(20\,000, 1000)$. The tax on \$30 000 is \$2500, so the graph of T for $x > 20\,000$ is the ray with initial point $(20\,000, 1000)$ that passes through $(30\,000, 2500)$.

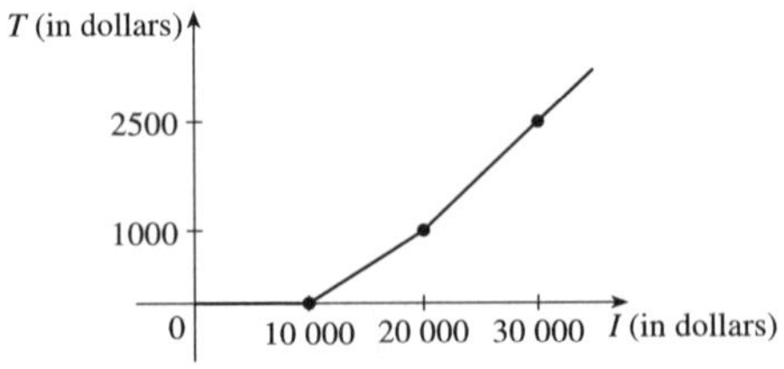

61. f is an odd function because its graph is symmetric about the origin. g is an even function because its graph is symmetric with respect to the y-axis.

63. (a) Because an even function is symmetric with respect to the y-axis, and the point $(5, 3)$ is on the graph of this even function, the point $(-5, 3)$ must also be on its graph.

(b) Because an odd function is symmetric with respect to the origin, and the point $(5, 3)$ is on the graph of this odd function, the point $(-5, -3)$ must also be on its graph.

65. $f(x) = \dfrac{x}{x^2+1}$.

$$f(-x) = \frac{-x}{(-x)^2+1} = \frac{-x}{x^2+1} = -\frac{x}{x^2+1} = -f(x).$$

So f is an odd function.

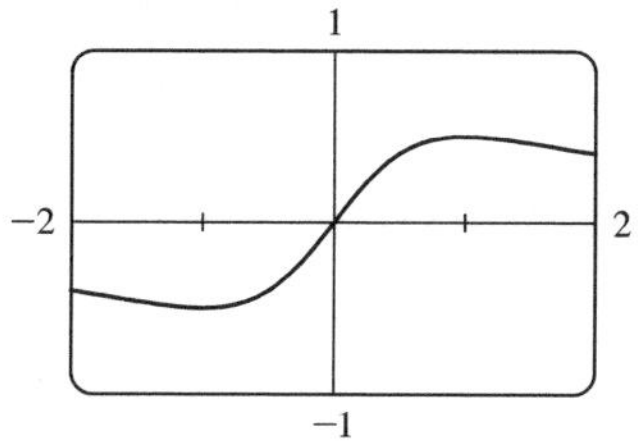

67. $f(x) = \dfrac{x}{x+1}$, so $f(-x) = \dfrac{-x}{-x+1} = \dfrac{x}{x-1}$.

Since this is neither $f(x)$ nor $-f(x)$, the function f is neither even nor odd.

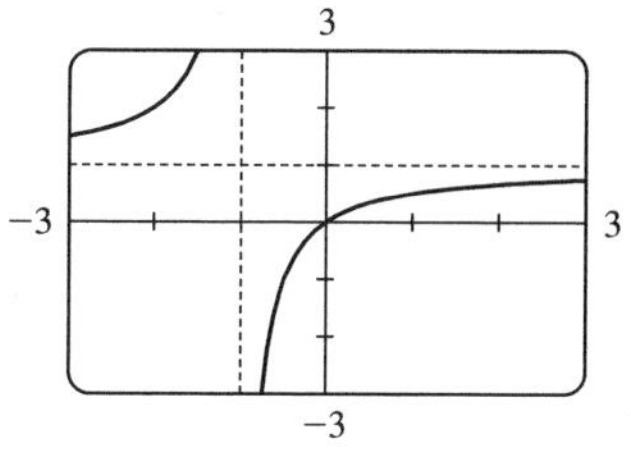

69. $f(x) = 1 + 3x^2 - x^4$. $f(-x) = 1 + 3(-x)^2 - (-x)^4 = 1 + 3x^2 - x^4 = f(x)$. So f is an even function.

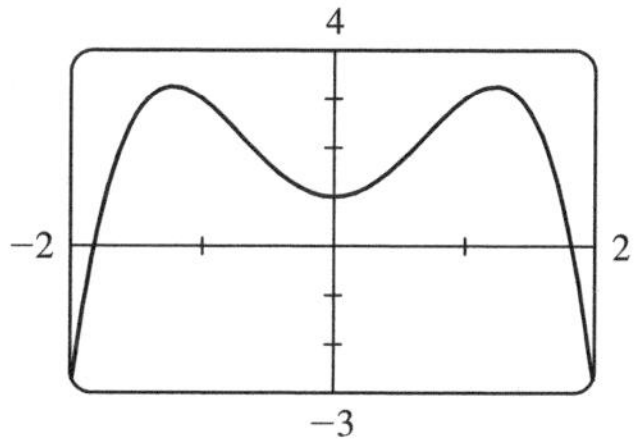

1.2 Mathematical Models: A Catalogue of Essential Functions

1. (a) $f(x) = \sqrt[5]{x}$ is a root function with $n = 5$.

(b) $g(x) = \sqrt{1-x^2}$ is an algebraic function because it is a root of a polynomial.

(c) $h(x) = x^9 + x^4$ is a polynomial of degree 9.

(d) $r(x) = \dfrac{x^2+1}{x^3+x}$ is a rational function because it is a ratio of polynomials.

(e) $s(x) = \tan 2x$ is a trigonometric function.

(f) $t(x) = \log_{10} x$ is a logarithmic function.

3. We notice from the figure that g and h are even functions (symmetric with respect to the y-axis) and that f is an odd function (symmetric with respect to the origin). So (b) $\left[y = x^5\right]$ must be f. Since g is flatter than h near the origin, we must have (c) $\left[y = x^8\right]$ matched with g and (a) $\left[y = x^2\right]$ matched with h.

5. (a) An equation for the family of linear functions with slope 2 is $y = f(x) = 2x + b$, where b is the y-intercept.

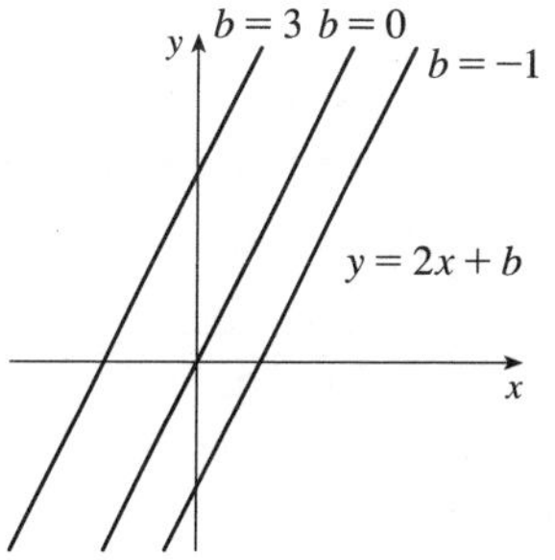

(b) $f(2) = 1$ means that the point $(2, 1)$ is on the graph of f. We can use the point-slope form of a line to obtain an equation for the family of linear functions through the point $(2, 1)$. $y - 1 = m(x - 2)$, which is equivalent to $y = mx + (1 - 2m)$ in slope-intercept form.

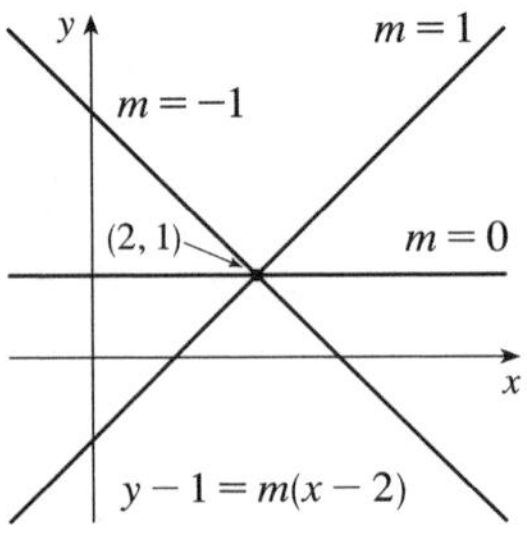

(c) To belong to both families, an equation must have slope $m = 2$, so the equation in part (b), $y = mx + (1 - 2m)$, becomes $y = 2x - 3$. It is the *only* function that belongs to both families.

7. All members of the family of linear functions $f(x) = c - x$ have graphs that are lines with slope -1. The y-intercept is c.

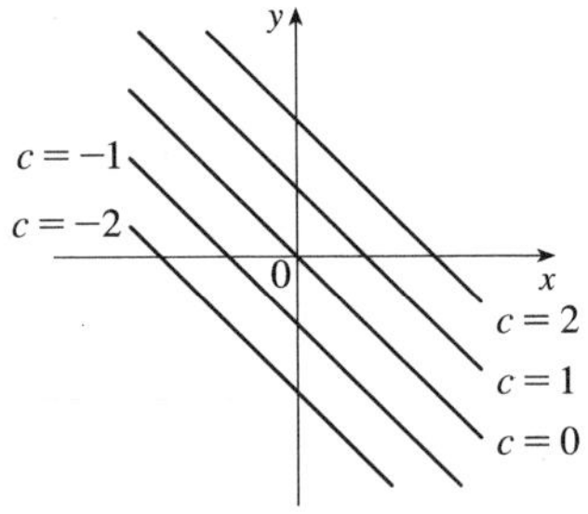

9. Since $f(-1) = f(0) = f(2) = 0$, f has zeros of -1, 0, and 2, so an equation for f is $f(x) = a[x - (-1)](x - 0)(x - 2)$, or $f(x) = ax(x + 1)(x - 2)$. Because $f(1) = 6$, we'll substitute 1 for x and 6 for $f(x)$.
$6 = a(1)(2)(-1) \quad \Rightarrow \quad -2a = 6 \quad \Rightarrow \quad a = -3$, so an equation for f is $f(x) = -3x(x + 1)(x - 2)$.

11. (a) $D = 200$, so $c = 0.0417D(a + 1) = 0.0417(200)(a + 1) = 8.34a + 8.34$. The slope is 8.34, which represents the change in mg of the dosage for a child for each change of 1 year in age.

(b) For a newborn, $a = 0$, so $c = 8.34$ mg.

13. (a)

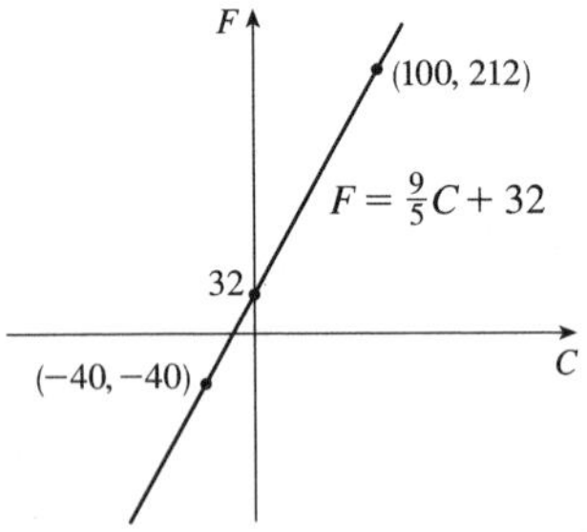

(b) The slope of $\frac{9}{5}$ means that F increases $\frac{9}{5}$ degrees for each increase of $1°C$. (Equivalently, F increases by 9 when C increases by 5 and F decreases by 9 when C decreases by 5.) The F-intercept of 32 is the Fahrenheit temperature corresponding to a Celsius temperature of 0.

15. (a) Using N in place of x and T in place of y, we find the slope to be $\dfrac{T_2 - T_1}{N_2 - N_1} = \dfrac{29 - 20}{180 - 112} = \dfrac{9}{68}$. So a linear equation is
$T - 29 = \frac{9}{68}(N - 180) \quad \Leftrightarrow \quad T - 29 = \frac{9}{68}N - \frac{405}{17} \quad \Leftrightarrow \quad T = \frac{9}{68}N + \frac{88}{17}$.

(b) The slope of $\frac{9}{68}$ means that the temperature in Celsius degrees increases nine sixty-eighths as rapidly as the number of cricket chirps per minute. Said differently, each increase of 68 cricket chirps per minute corresponds to an increase of $9°C$.

(c) When $N = 150$, the temperature is given approximately by $T = \frac{9}{68}(150) + \frac{88}{17} \approx 25°C$.

17. (a) We are given $\dfrac{\text{change in pressure}}{\text{1 metre change in depth}} = \dfrac{0.10}{1} = 0.10$. Using P for pressure and d for depth with the point $(d, P) = (0, 1.05)$, we have the slope-intercept form of the line, $P = 0.10d + 1.05$.

(b) When $P = 7$, then $7 = 0.10d + 1.05 \quad \Leftrightarrow \quad 0.10d = 5.95 \quad \Leftrightarrow \quad d = \frac{5.95}{0.10} = 59.5$ metres. Thus, the pressure is $7\ \text{kg/cm}^2$ at a depth of 59.5 metres.

19. (a) The data appear to be periodic and a sine or cosine function would make the best model. A model of the form $f(x) = a\cos(bx) + c$ seems appropriate.

(b) The data appear to be decreasing in a linear fashion. A model of the form $f(x) = mx + b$ seems appropriate.

Some values are given to many decimal places. These are the results given by several computer algebra systems — rounding is left to the reader.

21. (a)

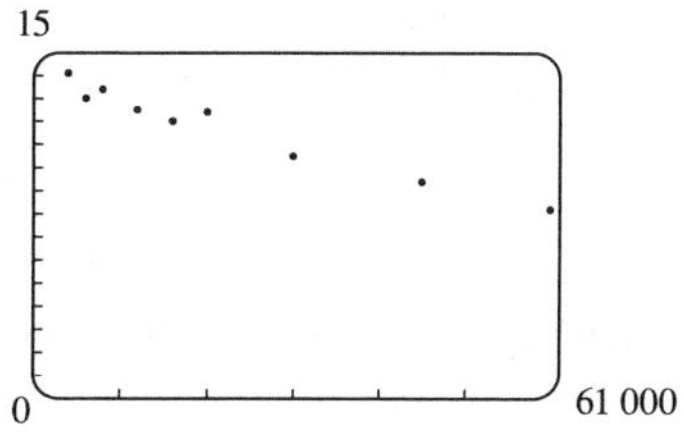

A linear model does seem appropriate.

(b) Using the points (4000, 14.1) and (60 000, 8.2), we obtain

$$y - 14.1 = \frac{8.2 - 14.1}{60\,000 - 4000}(x - 4000) \text{ or, equivalently,}$$

$$y \approx -0.000\,105\,357x + 14.521\,429.$$

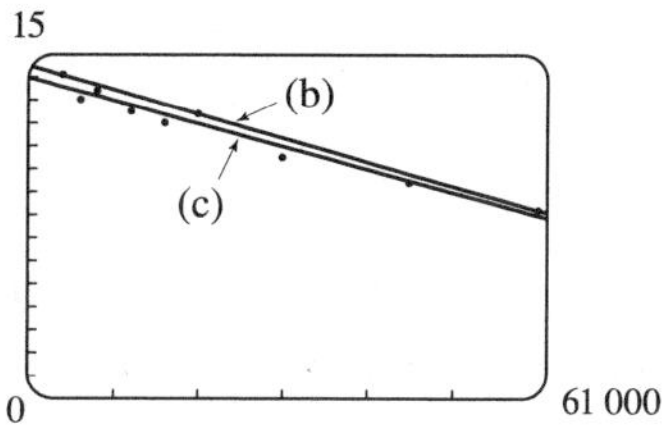

(c) Using a computing device, we obtain the least squares regression line $y = -0.000\,099\,785\,5x + 13.950\,764$.
The following commands and screens illustrate how to find the least squares regression line on a TI-83 Plus.
Enter the data into list one (L1) and list two (L2). Press STAT 1 to enter the editor.

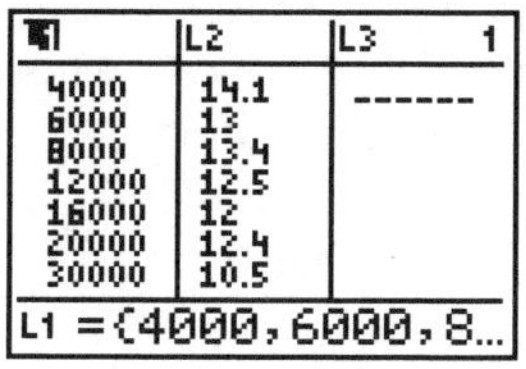

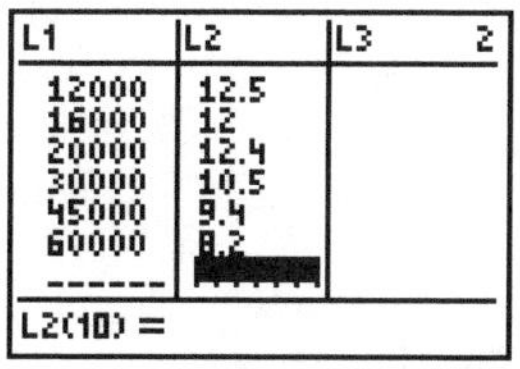

Find the regession line and store it in Y_1. Press 2nd QUIT STAT ▶ 4 VARS ▶ 1 1 ENTER .

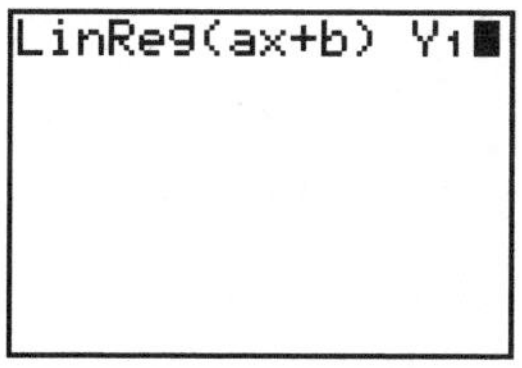

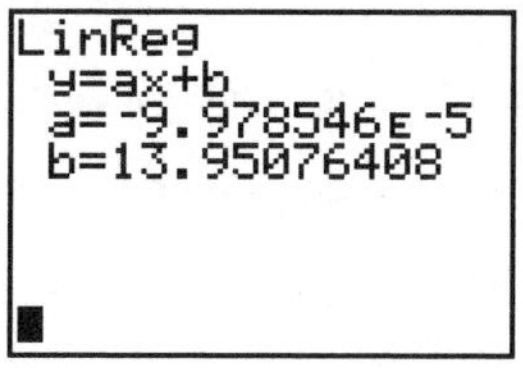

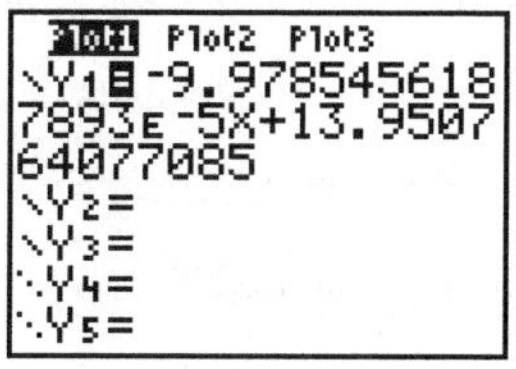

Note from the last figure that the regression line has been stored in Y_1 and that Plot1 has been turned on (Plot1 is highlighted). You can turn on Plot1 from the Y= menu by placing the cursor on Plot1 and pressing ENTER or by pressing 2nd STAT PLOT 1 ENTER .

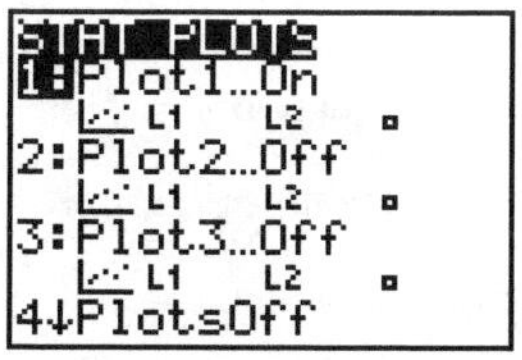

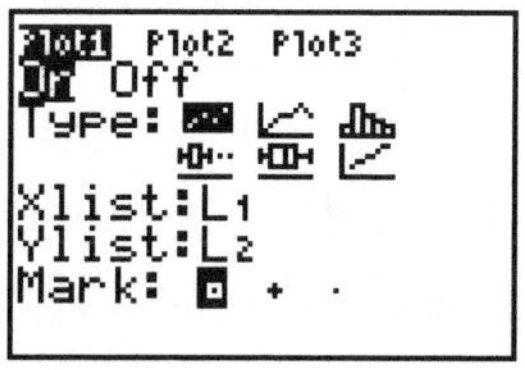

Now press ZOOM 9 to produce a graph of the data and the regression line. Note that choice 9 of the ZOOM menu automatically selects a window that displays all of the data.

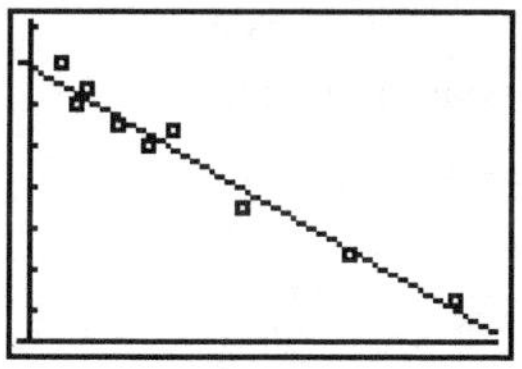

(d) When $x = 25\,000$, $y \approx 11.456$; or about 11.5 per 100 population.

(e) When $x = 80\,000$, $y \approx 5.968$; or about a 6% chance.

(f) When $x = 200\,000$, y is negative, so the model does not apply.

23. (a)

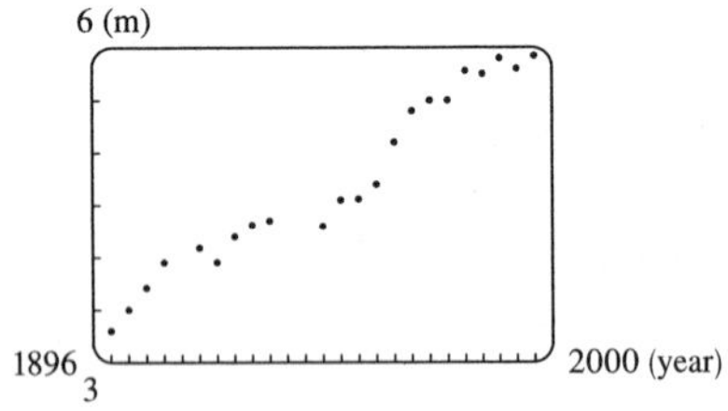

A linear model does seem appropriate.

(b)

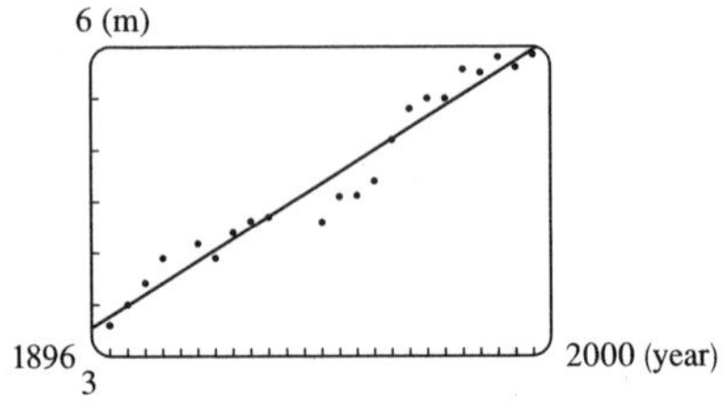

Using a computing device, we obtain the least squares regression line $y = 0.026\,942\,06x - 47.804\,298\,43$, where x is the year and y is the height in metres.

(c) When $x = 2000$, the model gives $y \approx 6.08$ m. Note that the actual winning height for the 2000 Olympics is *less than* the winning height for 1996—so much for that prediction.

(d) When $x = 2100$, $y \approx 8.77$ m. This would be an increase of 2.85 m from 1996 to 2100. Even though there was an increase of 2.62 m from 1900 to 1996, it is unlikely that a similar increase will occur over the next 100 years.

25.

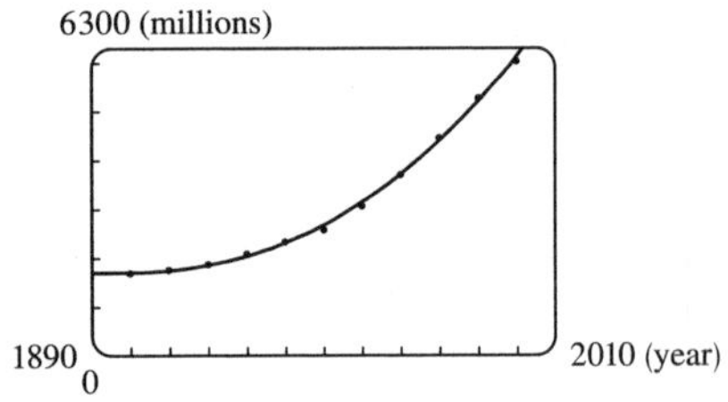

Using a computing device, we obtain the cubic function $y = ax^3 + bx^2 + cx + d$ with $a = 0.001\,293\,7$, $b = -7.061\,42$, $c = 12\,823$, and $d = -7\,743\,770$. When $x = 1925$, $y \approx 1914$ (million).

1.3 New Functions from Old Functions

1. (a) If the graph of f is shifted 3 units upward, its equation becomes $y = f(x) + 3$.

(b) If the graph of f is shifted 3 units downward, its equation becomes $y = f(x) - 3$.

(c) If the graph of f is shifted 3 units to the right, its equation becomes $y = f(x - 3)$.

(d) If the graph of f is shifted 3 units to the left, its equation becomes $y = f(x + 3)$.

(e) If the graph of f is reflected about the x-axis, its equation becomes $y = -f(x)$.

(f) If the graph of f is reflected about the y-axis, its equation becomes $y = f(-x)$.

(g) If the graph of f is stretched vertically by a factor of 3, its equation becomes $y = 3f(x)$.

(h) If the graph of f is shrunk vertically by a factor of 3, its equation becomes $y = \frac{1}{3}f(x)$.

3. (a) (graph 3) The graph of f is shifted 4 units to the right and has equation $y = f(x - 4)$.

(b) (graph 1) The graph of f is shifted 3 units upward and has equation $y = f(x) + 3$.

(c) (graph 4) The graph of f is shrunk vertically by a factor of 3 and has equation $y = \frac{1}{3}f(x)$.

(d) (graph 5) The graph of f is shifted 4 units to the left and reflected about the x-axis. Its equation is $y = -f(x + 4)$.

(e) (graph 2) The graph of f is shifted 6 units to the left and stretched vertically by a factor of 2. Its equation is $y = 2f(x + 6)$.

5. (a) To graph $y = f(2x)$ we shrink the graph of f horizontally by a factor of 2.

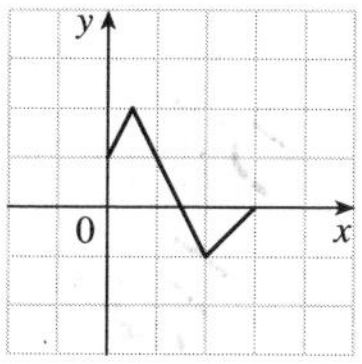

The point $(4, -1)$ on the graph of f corresponds to the point $\left(\frac{1}{2} \cdot 4, -1\right) = (2, -1)$.

(b) To graph $y = f\left(\frac{1}{2}x\right)$ we stretch the graph of f horizontally by a factor of 2.

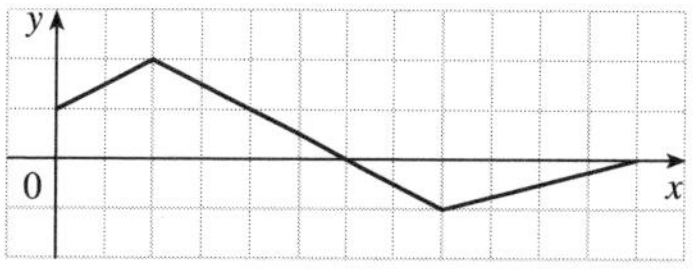

The point $(4, -1)$ on the graph of f corresponds to the point $(2 \cdot 4, -1) = (8, -1)$.

(c) To graph $y = f(-x)$ we reflect the graph of f about the y-axis.

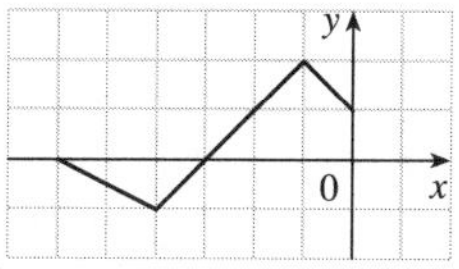

The point $(4, -1)$ on the graph of f corresponds to the point $(-1 \cdot 4, -1) = (-4, -1)$.

(d) To graph $y = -f(-x)$ we reflect the graph of f about the y-axis, then about the x-axis.

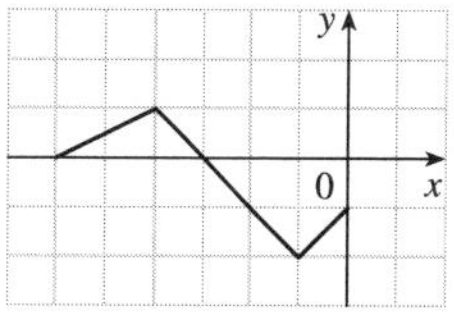

The point $(4, -1)$ on the graph of f corresponds to the point $(-1 \cdot 4, -1 \cdot -1) = (-4, 1)$.

7. The graph of $y = f(x) = \sqrt{3x - x^2}$ has been shifted 4 units to the left, reflected about the x-axis, and shifted downward 1 unit. Thus, a function describing the graph is

$$y = \underbrace{-1 \cdot}_{\substack{\text{reflect} \\ \text{about } x\text{-axis}}} \quad f \underbrace{(x+4)}_{\substack{\text{shift} \\ \text{4 units left}}} \quad \underbrace{-\ 1}_{\substack{\text{shift} \\ \text{1 unit left}}}$$

This function can be written as

$$y = -f(x+4) - 1 = -\sqrt{3(x+4) - (x+4)^2} - 1 = -\sqrt{3x + 12 - (x^2 + 8x + 16)} - 1 = -\sqrt{-x^2 - 5x - 4} - 1$$

9. $y = -x^3$: Start with the graph of $y = x^3$ and reflect about the x-axis. Note: Reflecting about the y-axis gives the same result since substituting $-x$ for x gives us $y = (-x)^3 = -x^3$.

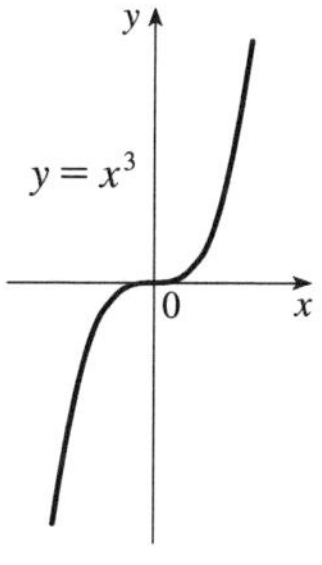

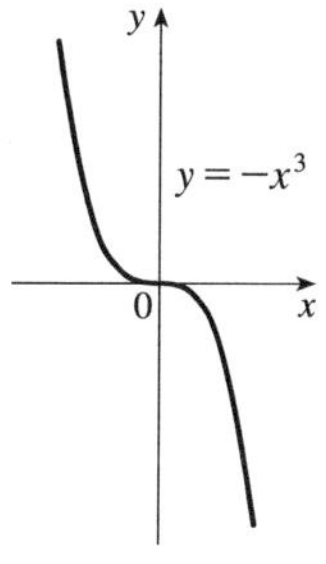

11. $y = (x+1)^2$: Start with the graph of $y = x^2$ and shift 1 unit to the left.

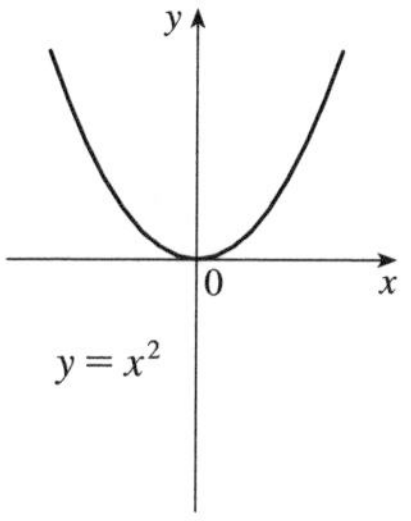

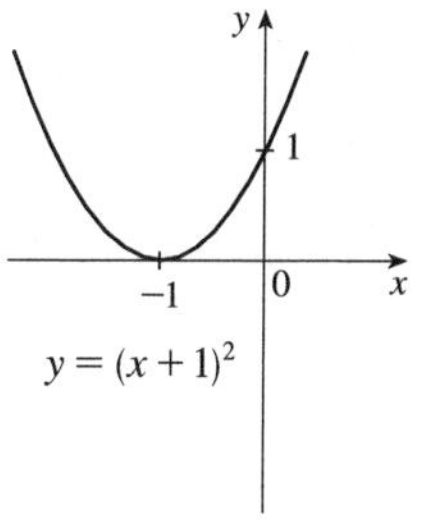

13. $y = 1 + 2\cos x$: Start with the graph of $y = \cos x$, stretch vertically by a factor of 2, and then shift 1 unit upward.

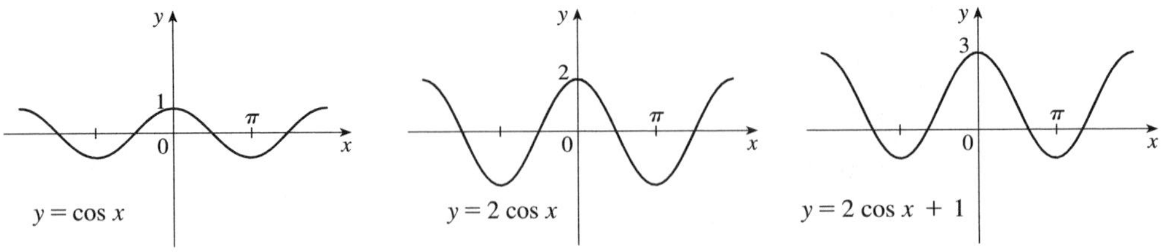

15. $y = \sin(x/2)$: Start with the graph of $y = \sin x$ and stretch horizontally by a factor of 2.

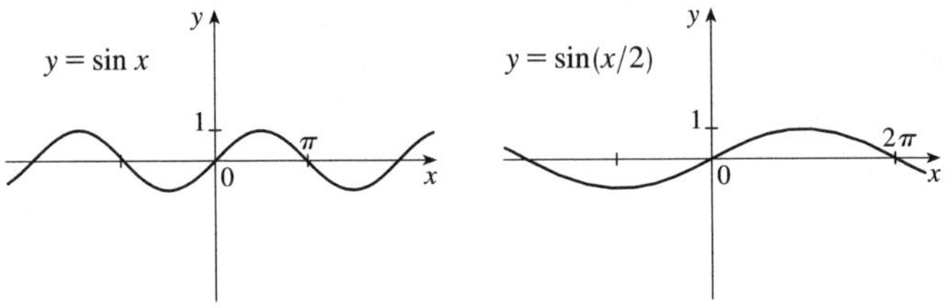

17. $y = \sqrt{x+3}$: Start with the graph of $y = \sqrt{x}$ and shift 3 units to the left.

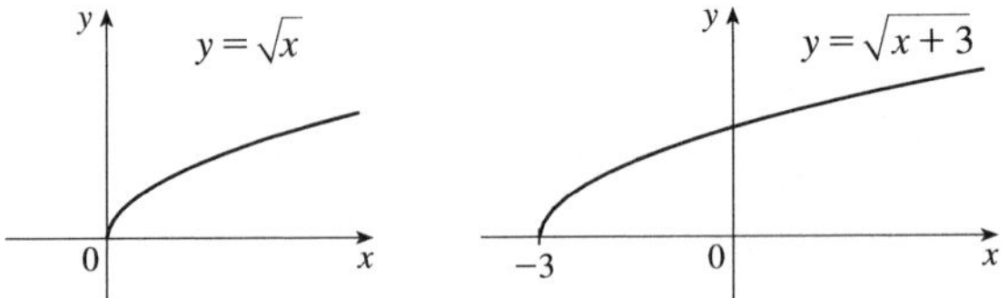

19. $y = \frac{1}{2}(x^2 + 8x) = \frac{1}{2}(x^2 + 8x + 16) - 8 = \frac{1}{2}(x+4)^2 - 8$: Start with the graph of $y = x^2$, compress vertically by a factor of 2, shift 4 units to the left, and then shift 8 units downward.

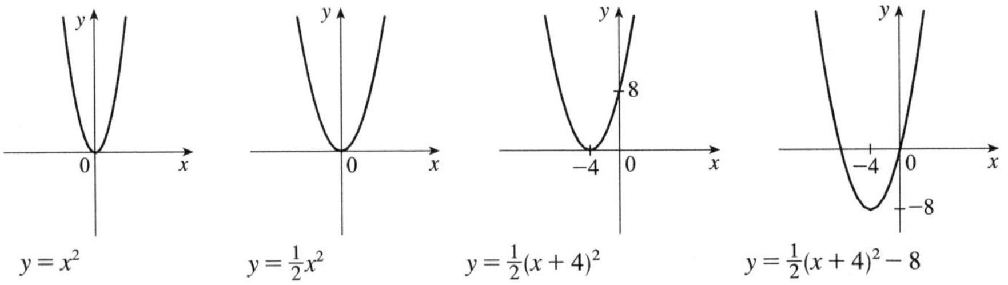

21. $y = 2/(x+1)$: Start with the graph of $y = 1/x$, shift 1 unit to the left, and then stretch vertically by a factor of 2.

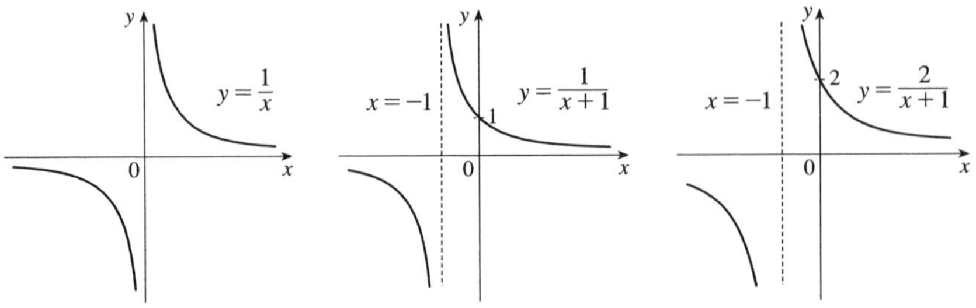

23. $y = |\sin x|$: Start with the graph of $y = \sin x$ and reflect all the parts of the graph below the x-axis about the x-axis.

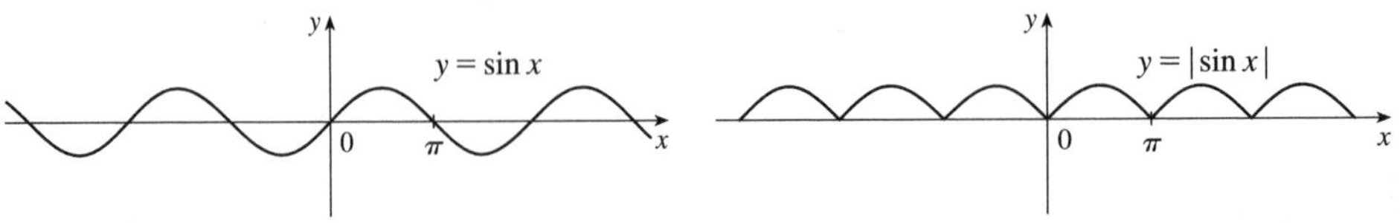

25. This is just like the solution to Example 4 except the amplitude of the curve (the 30°N curve in Figure 9 on June 21) is $14 - 12 = 2$. So the function is $L(t) = 12 + 2\sin\left[\frac{2\pi}{365}(t - 80)\right]$. March 31 is the 90th day of the year, so the model gives $L(90) \approx 12.34$ h. The daylight time (6:13 A.M. to 6:39 P.M.) is 12 hours and 26 minutes, or 12.43 h. The model value differs from the actual value by $\frac{12.43 - 12.34}{12.43} \approx 0.007$, less than 1%.

27. (a) To obtain $y = f(|x|)$, the portion of the graph of $y = f(x)$ to the right of the y-axis is reflected about the y-axis.

(b) $y = \sin|x|$

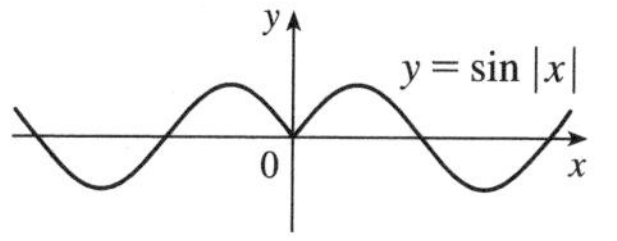

(c) $y = \sqrt{|x|}$

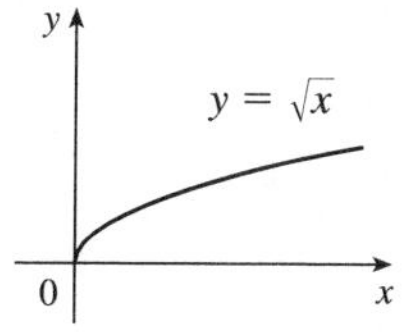

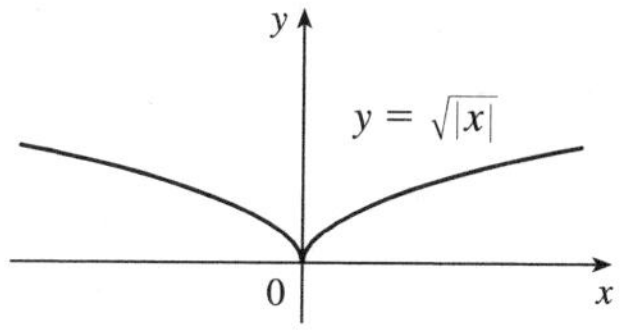

29. Assuming that successive horizontal and vertical gridlines are a unit apart, we can make a table of approximate values:

x	0	1	2	3	4	5	6
$f(x)$	2	1.7	1.3	1.0	0.7	0.3	0
$g(x)$	2	2.7	3	2.8	2.4	1.7	0
$f(x) + g(x)$	4	4.4	4.3	3.8	3.1	2.0	0

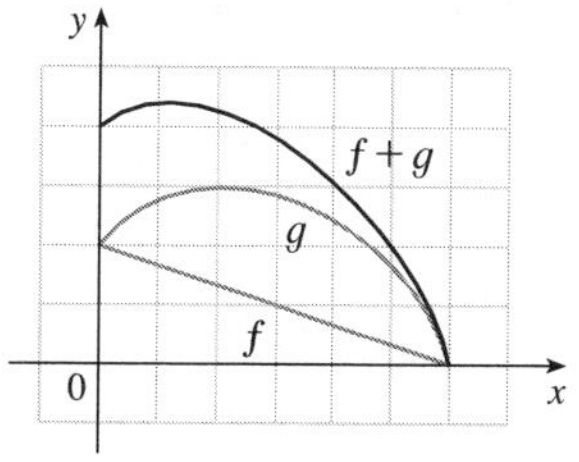

Connecting the points $(x, f(x) + g(x))$ with a smooth curve gives an approximation to the graph of $f + g$. Extra points can be plotted between those listed above if necessary.

31. $f(x) = x^3 + 2x^2$; $g(x) = 3x^2 - 1$. $D = \mathbb{R}$ for both f and g.

$(f + g)(x) = (x^3 + 2x^2) + (3x^2 - 1) = x^3 + 5x^2 - 1, \quad D = \mathbb{R}$.

$(f - g)(x) = (x^3 + 2x^2) - (3x^2 - 1) = x^3 - x^2 + 1, \quad D = \mathbb{R}$.

$(fg)(x) = (x^3 + 2x^2)(3x^2 - 1) = 3x^5 + 6x^4 - x^3 - 2x^2, \quad D = \mathbb{R}$.

$\left(\dfrac{f}{g}\right)(x) = \dfrac{x^3 + 2x^2}{3x^2 - 1}, \quad D = \left\{x \mid x \neq \pm\dfrac{1}{\sqrt{3}}\right\}$ since $3x^2 - 1 \neq 0$.

33. $f(x) = x, \quad g(x) = 1/x$

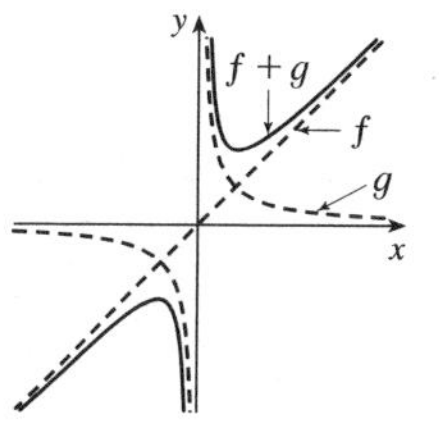

35. $f(x) = x^2 - 1$, $D = \mathbb{R}$; $g(x) = 2x + 1$, $D = \mathbb{R}$.

(a) $(f \circ g)(x) = f(g(x)) = f(2x + 1) = (2x + 1)^2 - 1 = (4x^2 + 4x + 1) - 1 = 4x^2 + 4x$, $D = \mathbb{R}$.

(b) $(g \circ f)(x) = g(f(x)) = g(x^2 - 1) = 2(x^2 - 1) + 1 = (2x^2 - 2) + 1 = 2x^2 - 1$, $D = \mathbb{R}$.

(c) $(f \circ f)(x) = f(f(x)) = f(x^2 - 1) = (x^2 - 1)^2 - 1 = (x^4 - 2x^2 + 1) - 1 = x^4 - 2x^2$, $D = \mathbb{R}$.

(d) $(g \circ g)(x) = g(g(x)) = g(2x + 1) = 2(2x + 1) + 1 = (4x + 2) + 1 = 4x + 3$, $D = \mathbb{R}$.

37. $f(x) = \sin x, \quad D = \mathbb{R}; \quad g(x) = 1 - \sqrt{x}, \quad D = [0, \infty)$.

(a) $(f \circ g)(x) = f(g(x)) = f(1 - \sqrt{x}) = \sin(1 - \sqrt{x})$, $D = [0, \infty)$.

(b) $(g \circ f)(x) = g(f(x)) = g(\sin x) = 1 - \sqrt{\sin x}$.

For $\sqrt{\sin x}$ to be defined, we must have

$$\sin x \geq 0 \quad \Leftrightarrow \quad x \in [0, \pi] \cup [2\pi, 3\pi] \cup [-2\pi, -\pi] \cup [4\pi, 5\pi] \cup [-4\pi, -3\pi] \cup \ \ldots,$$

so $D = \{x \mid x \in [2n\pi, \pi + 2n\pi]$, where n is an integer$\}$.

(c) $(f \circ f)(x) = f(f(x)) = f(\sin x) = \sin(\sin x)$, $D = \mathbb{R}$.

(d) $(g \circ g)(x) = g(g(x)) = g(1 - \sqrt{x}) = 1 - \sqrt{1 - \sqrt{x}}$,

$$D = \{x \geq 0 \mid 1 - \sqrt{x} \geq 0\} = \{x \geq 0 \mid 1 \geq \sqrt{x}\} = \{x \geq 0 \mid \sqrt{x} \leq 1\} = [0, 1].$$

39. $f(x) = x + \dfrac{1}{x}$, $D = \{x \mid x \neq 0\}$; $g(x) = \dfrac{x+1}{x+2}$, $D = \{x \mid x \neq -2\}$.

(a) $$(f \circ g)(x) = f(g(x)) = f\left(\frac{x+1}{x+2}\right) = \frac{x+1}{x+2} + \frac{1}{\dfrac{x+1}{x+2}} = \frac{x+1}{x+2} + \frac{x+2}{x+1}$$
$$= \frac{(x+1)(x+1)+(x+2)(x+2)}{(x+2)(x+1)} = \frac{(x^2+2x+1)+(x^2+4x+4)}{(x+2)(x+1)} = \frac{2x^2+6x+5}{(x+2)(x+1)}$$

Since $g(x)$ is not defined for $x = -2$ and $f(g(x))$ is not defined for $x = -2$ and $x = -1$, the domain of $(f \circ g)(x)$ is $D = \{x \mid x \neq -2, -1\}$.

(b) $$(g \circ f)(x) = g(f(x)) = g\left(x + \frac{1}{x}\right) = \frac{\left(x + \dfrac{1}{x}\right) + 1}{\left(x + \dfrac{1}{x}\right) + 2} = \frac{\dfrac{x^2+1+x}{x}}{\dfrac{x^2+1+2x}{x}} = \frac{x^2+x+1}{x^2+2x+1} = \frac{x^2+x+1}{(x+1)^2}$$

Since $f(x)$ is not defined for $x = 0$ and $g(f(x))$ is not defined for $x = -1$, the domain of $(g \circ f)(x)$ is $D = \{x \mid x \neq -1, 0\}$.

(c) $$(f \circ f)(x) = f(f(x)) = f\left(x + \frac{1}{x}\right) = \left(x + \frac{1}{x}\right) + \frac{1}{x + \frac{1}{x}} = x + \frac{1}{x} + \frac{1}{\frac{x^2+1}{x}} = x + \frac{1}{x} + \frac{x}{x^2+1}$$
$$= \frac{x(x)(x^2+1) + 1\,(x^2+1) + x(x)}{x(x^2+1)} = \frac{x^4 + x^2 + x^2 + 1 + x^2}{x(x^2+1)}$$
$$= \frac{x^4+3x^2+1}{x(x^2+1)}, \quad D = \{x \mid x \neq 0\}.$$

(d) $$(g \circ g)(x) = g(g(x)) = g\left(\frac{x+1}{x+2}\right) = \frac{\dfrac{x+1}{x+2} + 1}{\dfrac{x+1}{x+2} + 2} = \frac{\dfrac{x+1+1(x+2)}{x+2}}{\dfrac{x+1+2(x+2)}{x+2}} = \frac{x+1+x+2}{x+1+2x+4} = \frac{2x+3}{3x+5}$$

Since $g(x)$ is not defined for $x = -2$ and $g(g(x))$ is not defined for $x = -\frac{5}{3}$, the domain of $(g \circ g)(x)$ is $D = \left\{x \mid x \neq -2, -\frac{5}{3}\right\}$.

41. $$(f \circ g \circ h)(x) = f(g(h(x))) = f(g(x+3)) = f((x+3)^2 + 2)$$
$$= f(x^2 + 6x + 11) = \sqrt{(x^2+6x+11) - 1} = \sqrt{x^2+6x+10}$$

43. Let $g(x) = x^2 + 1$ and $f(x) = x^{10}$. Then $(f \circ g)(x) = f(g(x)) = f(x^2+1) = (x^2+1)^{10} = F(x)$.

45. Let $g(t) = \cos t$ and $f(t) = \sqrt{t}$. Then $(f \circ g)(t) = f(g(t)) = f(\cos t) = \sqrt{\cos t} = u(t)$.

47. Let $h(x) = x^2$, $g(x) = 3^x$, and $f(x) = 1 - x$. Then

$$(f \circ g \circ h)(x) = f(g(h(x))) = f(g(x^2)) = f\left(3^{x^2}\right) = 1 - 3^{x^2} = H(x).$$

49. Let $h(x) = \sqrt{x}$, $g(x) = \sec x$, and $f(x) = x^4$. Then

$$(f \circ g \circ h)(x) = f(g(h(x))) = f(g(\sqrt{x})) = f(\sec\sqrt{x}) = (\sec\sqrt{x})^4 = \sec^4(\sqrt{x}) = H(x).$$

51. (a) $g(2) = 5$, because the point $(2, 5)$ is on the graph of g. Thus, $f(g(2)) = f(5) = 4$, because the point $(5, 4)$ is on the graph of f.

(b) $g(f(0)) = g(0) = 3$

(c) $(f \circ g)(0) = f(g(0)) = f(3) = 0$

(d) $(g \circ f)(6) = g(f(6)) = g(6)$. This value is not defined, because there is no point on the graph of g that has x-coordinate 6.

(e) $(g \circ g)(-2) = g(g(-2)) = g(1) = 4$

(f) $(f \circ f)(4) = f(f(4)) = f(2) = -2$

53. (a) Using the relationship *distance* = *rate* · *time* with the radius r as the distance, we have $r(t) = 60t$.

(b) $A = \pi r^2 \quad \Rightarrow \quad (A \circ r)(t) = A(r(t)) = \pi(60t)^2 = 3600\pi t^2$. This formula gives us the extent of the rippled area (in cm^2) at any time t.

55. (a) From the figure, we have a right triangle with legs 6 and d, and hypotenuse s. By the Pythagorean Theorem, $d^2 + 6^2 = s^2 \quad \Rightarrow \quad s = f(d) = \sqrt{d^2 + 36}$.

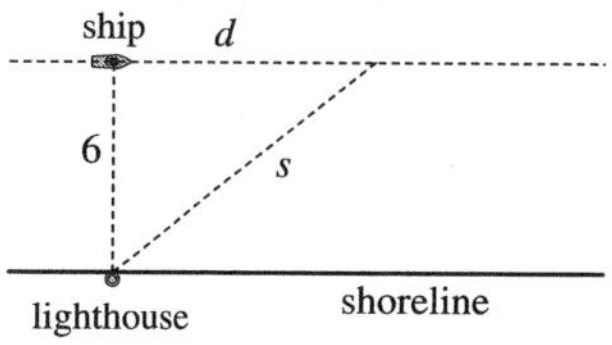

(b) Using $d = rt$, we get $d = (30 \text{ km/hr})(t \text{ hr}) = 30t$ (in km). Thus, $d = g(t) = 30t$.

(c) $(f \circ g)(t) = f(g(t)) = f(30t) = \sqrt{(30t)^2 + 36} = \sqrt{900t^2 + 36}$. This function represents the distance between the lighthouse and the ship as a function of the time elapsed since noon.

57. (a)

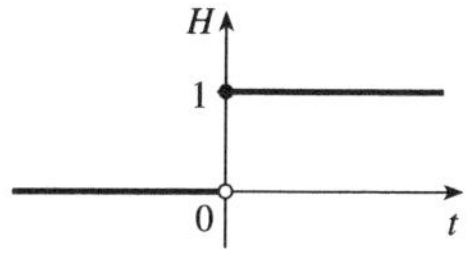

$$H(t) = \begin{cases} 0 & \text{if } t < 0 \\ 1 & \text{if } t \geq 0 \end{cases}$$

(b)

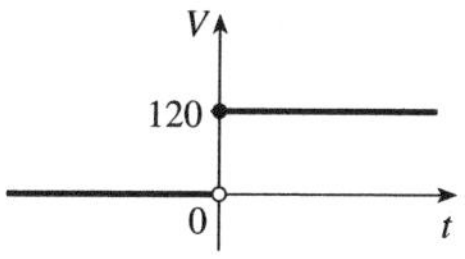

$$V(t) = \begin{cases} 0 & \text{if } t < 0 \\ 120 & \text{if } t \geq 0 \end{cases} \quad \text{so } V(t) = 120H(t).$$

(c)

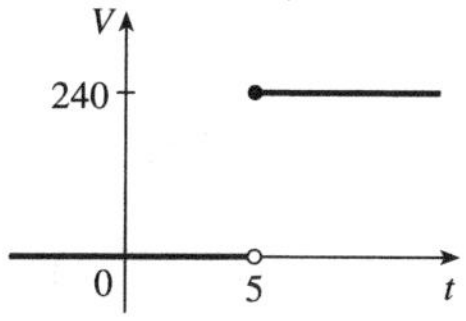

Starting with the formula in part (b), we replace 120 with 240 to reflect the different voltage. Also, because we are starting 5 units to the right of $t = 0$, we replace t with $t - 5$. Thus, the formula is $V(t) = 240H(t - 5)$.

59. If $f(x) = m_1 x + b_1$ and $g(x) = m_2 x + b_2$, then

$(f \circ g)(x) = f(g(x)) = f(m_2 x + b_2) = m_1(m_2 x + b_2) + b_1 = m_1 m_2 x + m_1 b_2 + b_1$.

So $f \circ g$ is a linear function with slope $m_1 m_2$.

61. (a) By examining the variable terms in g and h, we deduce that we must square g to get the terms $4x^2$ and $4x$ in h. If we let $f(x) = x^2 + c$, then $(f \circ g)(x) = f(g(x)) = f(2x + 1) = (2x + 1)^2 + c = 4x^2 + 4x + (1 + c)$. Since $h(x) = 4x^2 + 4x + 7$, we must have $1 + c = 7$. So $c = 6$ and $f(x) = x^2 + 6$.

(b) We need a function g so that $f(g(x)) = 3(g(x)) + 5 = h(x)$. But $h(x) = 3x^2 + 3x + 2 = 3(x^2 + x) + 2 = 3(x^2 + x - 1) + 5$, so we see that $g(x) = x^2 + x - 1$.

63. (a) If f and g are even functions, then $f(-x) = f(x)$ and $g(-x) = g(x)$.

(i) $(f + g)(-x) = f(-x) + g(-x) = f(x) + g(x) = (f + g)(x)$, so $f + g$ is an *even* function.

(ii) $(fg)(-x) = f(-x) \cdot g(-x) = f(x) \cdot g(x) = (fg)(x)$, so fg is an *even* function.

(b) If f and g are odd functions, then $f(-x) = -f(x)$ and $g(-x) = -g(x)$.

(i) $(f + g)(-x) = f(-x) + g(-x) = -f(x) + [-g(x)] = -[f(x) + g(x)] = -(f + g)(x)$, so $f + g$ is an *odd* function.

(ii) $(fg)(-x) = f(-x) \cdot g(-x) = -f(x) \cdot [-g(x)] = f(x) \cdot g(x) = (fg)(x)$, so fg is an *even* function.

65. We need to examine $h(-x)$.

$$h(-x) = (f \circ g)(-x) = f(g(-x)) = f(g(x)) \quad [\text{because } g \text{ is even}] \quad = h(x)$$

Because $h(-x) = h(x)$, h is an even function.

1.4 Graphing Calculators and Computers

1. $f(x) = \sqrt{x^3 - 5x^2}$

(a) $[-5, 5]$ by $[-5, 5]$

(There is no graph shown.)

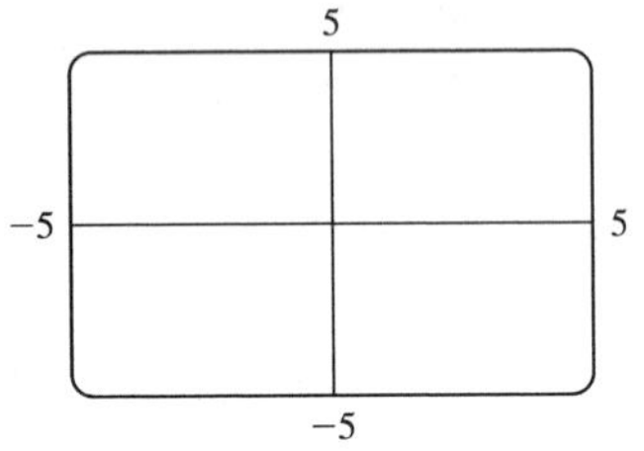

(b) $[0, 10]$ by $[0, 2]$

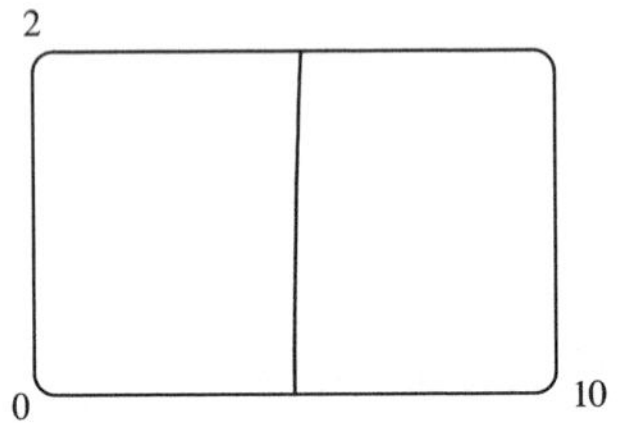

(c) $[0, 10]$ by $[0, 10]$

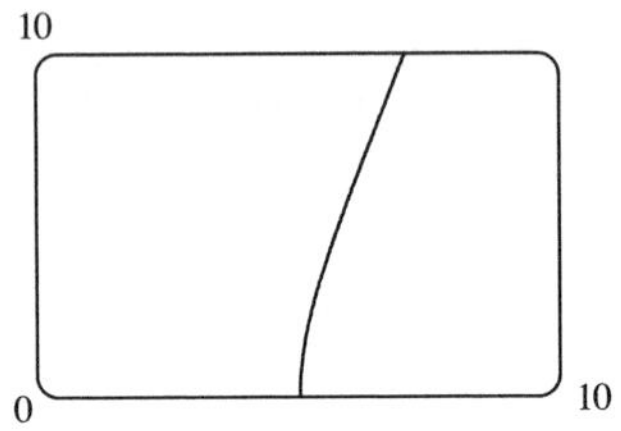

The most appropriate graph is produced in viewing rectangle (c).

3. Since the graph of $f(x) = 5 + 20x - x^2$ is a parabola opening downward, an appropriate viewing rectangle should include the maximum point.

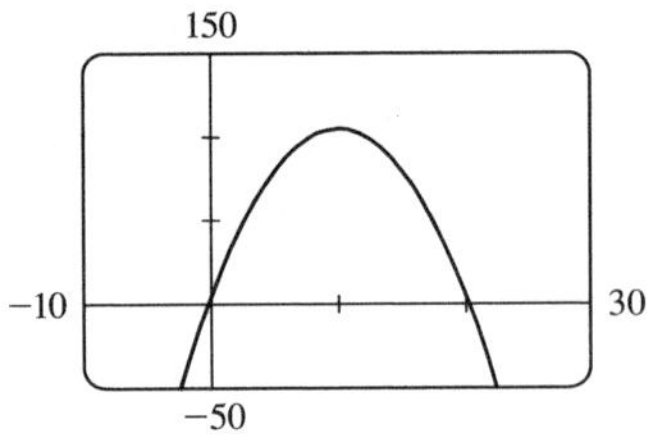

5. $f(x) = \sqrt[4]{81 - x^4}$ is defined when $81 - x^4 \geq 0 \Leftrightarrow x^4 \leq 81 \Leftrightarrow |x| \leq 3$, so the domain of f is $[-3, 3]$. Also $0 \leq \sqrt[4]{81 - x^4} \leq \sqrt[4]{81} = 3$, so the range is $[0, 3]$.

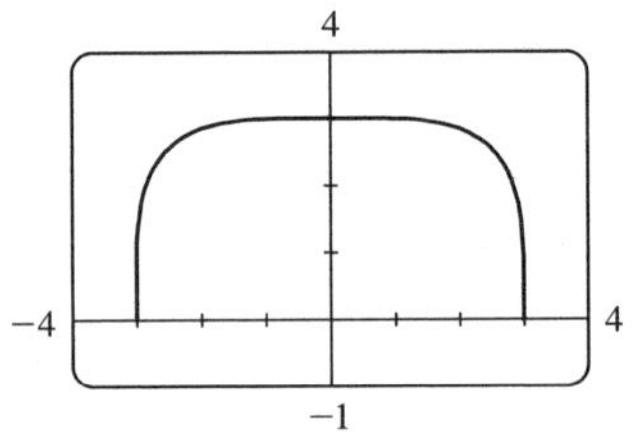

7. The graph of $f(x) = x^2 + (100/x)$ has a vertical asymptote of $x = 0$. As you zoom out, the graph of f looks more and more like that of $y = x^2$.

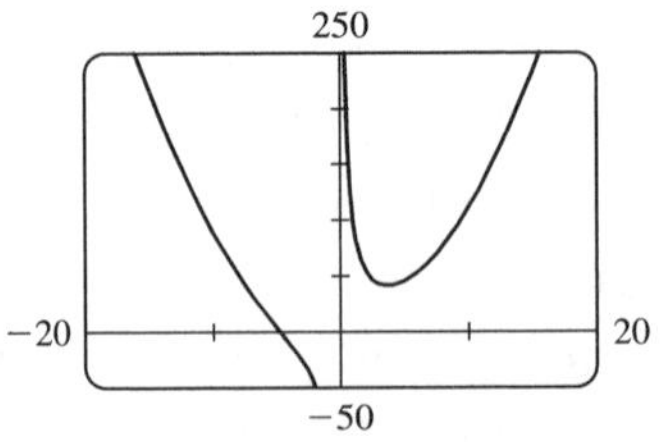

9. The period of $g(x) = \sin(1000x)$ is $\frac{2\pi}{1000} \approx 0.0063$ and its range is $[-1, 1]$. Since $f(x) = \sin^2(1000x)$ is the square of g, its range is $[0, 1]$ and a viewing rectangle of $[-0.01, 0.01]$ by $[0, 1.5]$ seems appropriate.

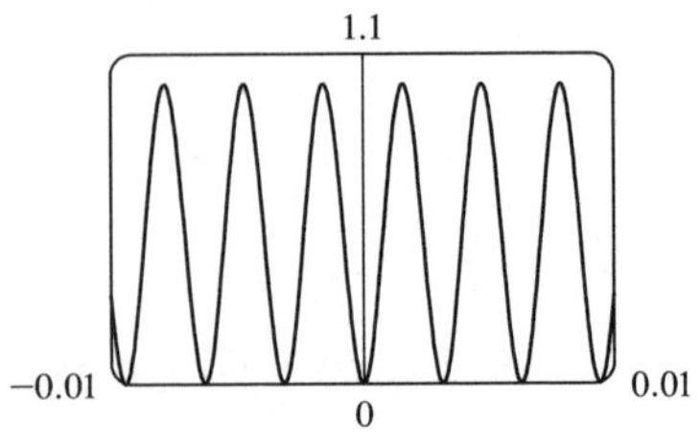

11. The domain of $y = \sqrt{x}$ is $x \geq 0$, so the domain of $f(x) = \sin\sqrt{x}$ is $[0, \infty)$ and the range is $[-1, 1]$. With a little trial-and-error experimentation, we find that an Xmax of 100 illustrates the general shape of f, so an appropriate viewing rectangle is $[0, 100]$ by $[-1.5, 1.5]$.

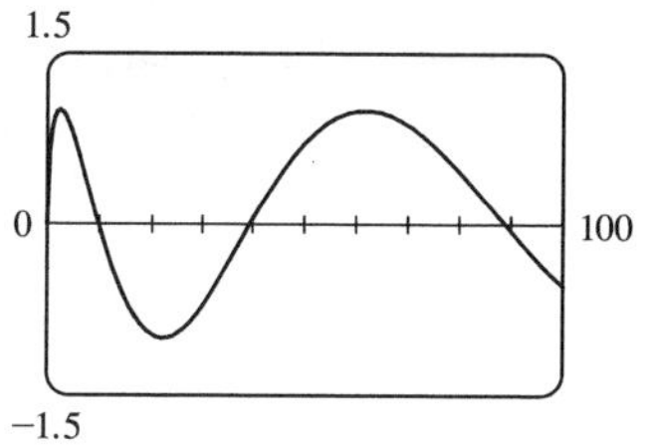

13. The first term, $10 \sin x$, has period 2π and range $[-10, 10]$. It will be the dominant term in any "large" graph of $y = 10 \sin x + \sin 100x$, as shown in the first figure. The second term, $\sin 100x$, has period $\frac{2\pi}{100} = \frac{\pi}{50}$ and range $[-1, 1]$. It causes the bumps in the first figure and will be the dominant term in any "small" graph, as shown in the view near the origin in the second figure.

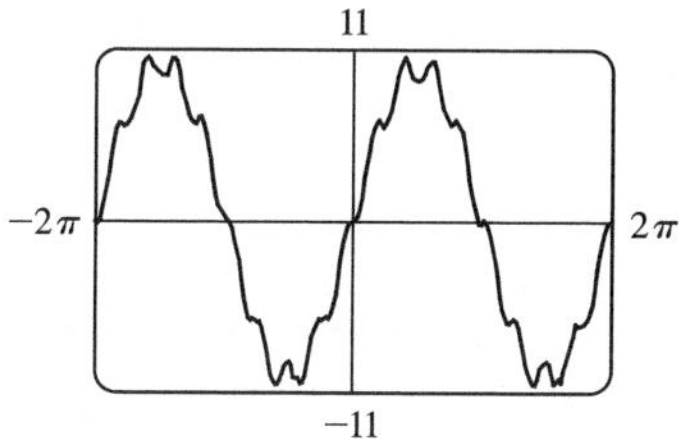

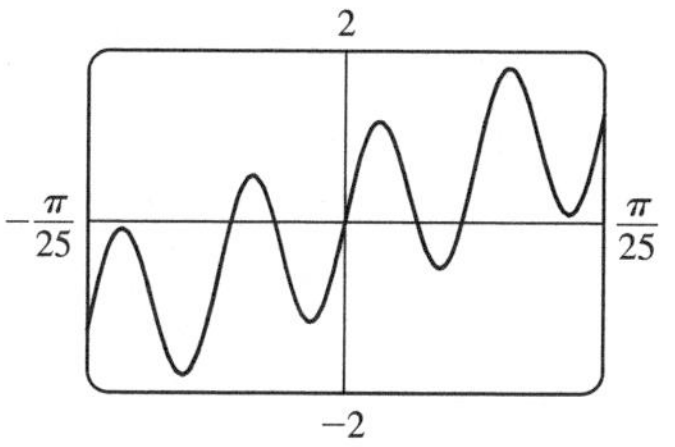

15. We must solve the given equation for y to obtain equations for the upper and lower halves of the ellipse.

$$4x^2 + 2y^2 = 1 \quad \Leftrightarrow \quad 2y^2 = 1 - 4x^2 \quad \Leftrightarrow \quad y^2 = \frac{1 - 4x^2}{2} \quad \Leftrightarrow$$

$$y = \pm\sqrt{\frac{1 - 4x^2}{2}}$$

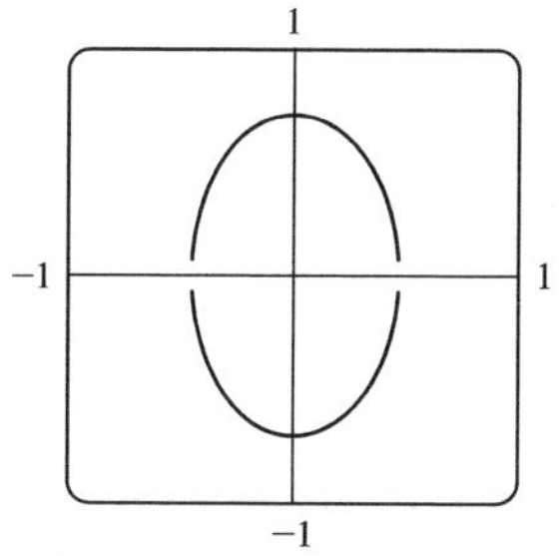

17. From the graph of $f(x) = x^3 - 9x^2 - 4$, we see that there is one solution of the equation $f(x) = 0$ and it is slightly larger than 9. By zooming in or using a root or zero feature, we obtain $x \approx 9.05$.

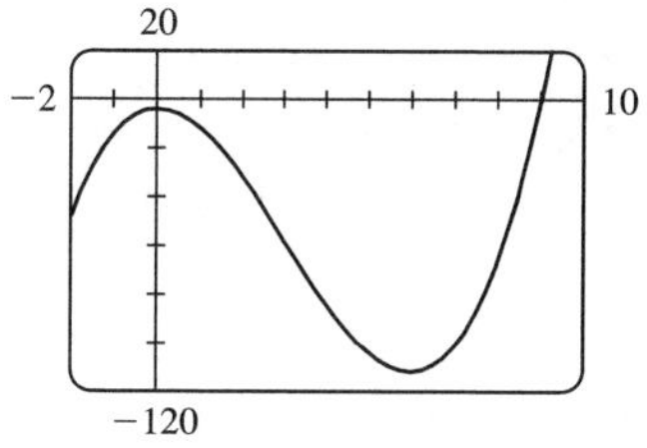

19. We see that the graphs of $f(x) = x^2$ and $g(x) = \sin x$ intersect twice. One solution is $x = 0$. The other solution of $f = g$ is the x-coordinate of the point of intersection in the first quadrant. Using an intersect feature or zooming in, we find this value to be approximately 0.88. Alternatively, we could find that value by finding the positive zero of $h(x) = x^2 - \sin x$.

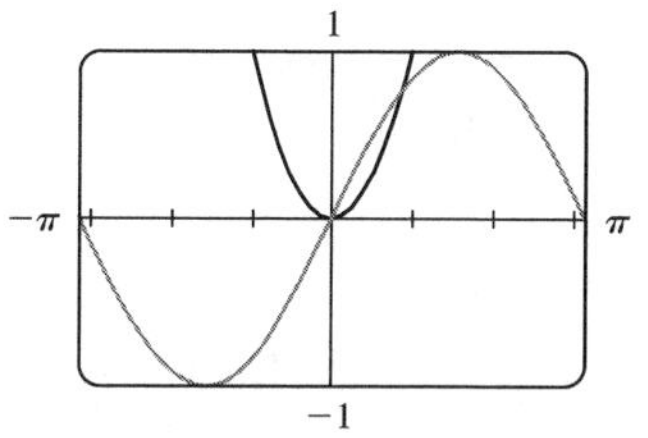

Note: After producing the graph on a TI-83 Plus, we can find the approximate value 0.88 by using the following keystrokes: [2nd] [CALC] [5] [ENTER] [ENTER] 1 [ENTER]. The "1" is just a guess for 0.88.

21. $g(x) = x^3/10$ is larger than $f(x) = 10x^2$ whenever $x > 100$.

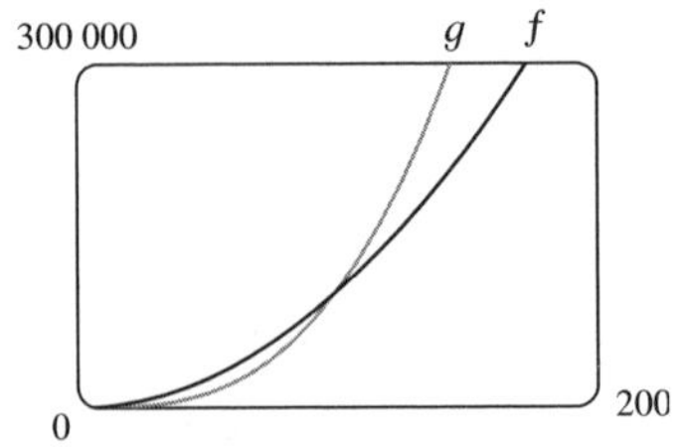

23.

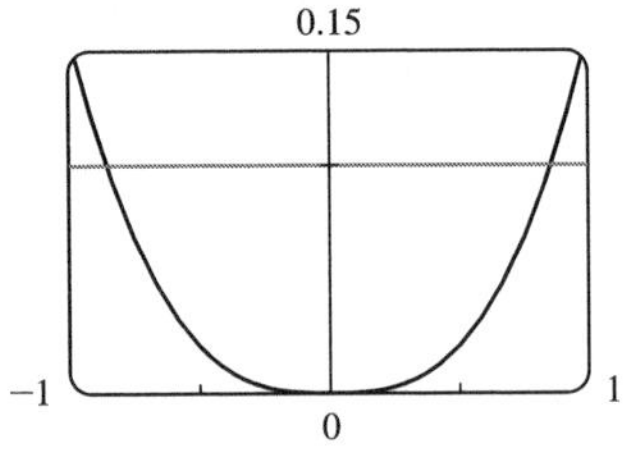

We see from the graphs of $y = |\sin x - x|$ and $y = 0.1$ that there are two solutions to the equation $|\sin x - x| = 0.1$: $x \approx -0.85$ and $x \approx 0.85$. The condition $|\sin x - x| < 0.1$ holds for any x lying between these two values.

25. (a) The root functions $y = \sqrt{x}$, $y = \sqrt[4]{x}$ and $y = \sqrt[6]{x}$

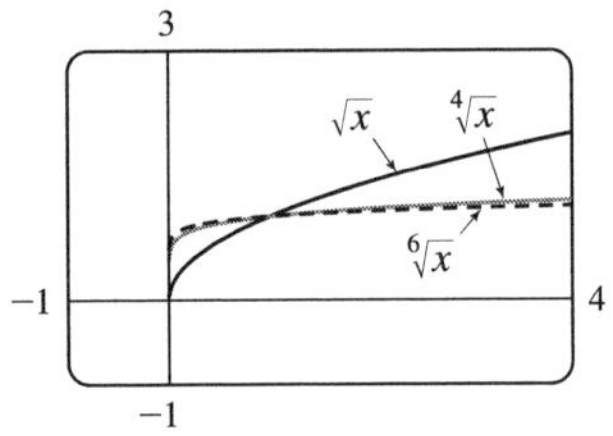

(b) The root functions $y = x$, $y = \sqrt[3]{x}$ and $y = \sqrt[5]{x}$

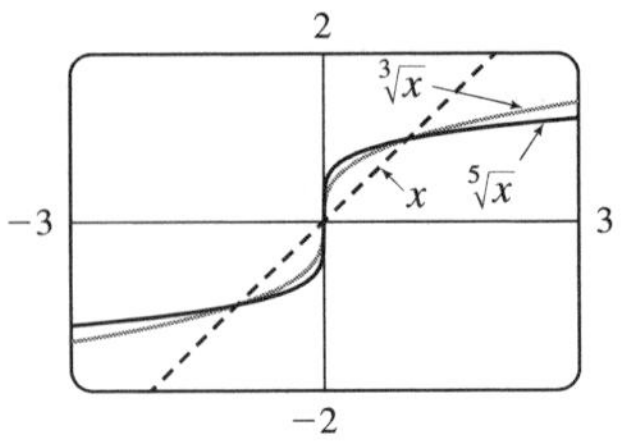

(c) The root functions $y = \sqrt{x}$, $y = \sqrt[3]{x}$, $y = \sqrt[4]{x}$ and $y = \sqrt[5]{x}$

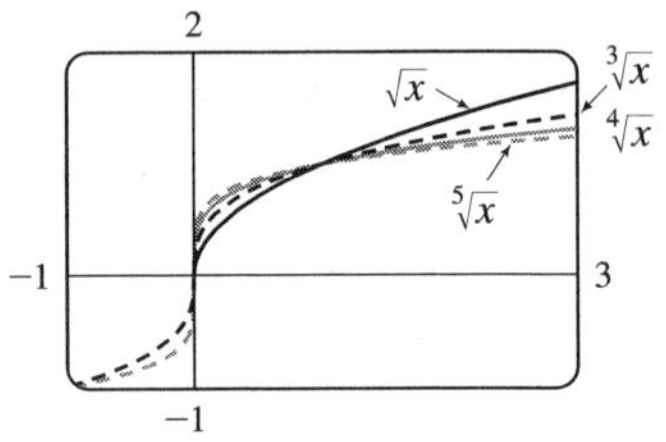

(d) • For any n, the nth root of 0 is 0 and the nth root of 1 is 1; that is, all nth root functions pass through the points $(0, 0)$ and $(1, 1)$.

• For odd n, the domain of the nth root function is $\mathbb{R}$, while for even n, it is $\{x \in \mathbb{R} \mid x \geq 0\}$.

• Graphs of even root functions look similar to that of $\sqrt{x}$, while those of odd root functions resemble that of $\sqrt[3]{x}$.

• As n increases, the graph of $\sqrt[n]{x}$ becomes steeper near 0 and flatter for $x > 1$.

27. $f(x) = x^4 + cx^2 + x$. If $c < -1.5$, there are three humps: two minimum points and a maximum point. These humps get flatter as c increases, until at $c = -1.5$ two of the humps disappear and there is only one minimum point. This single hump then moves to the right and approaches the origin as c increases.

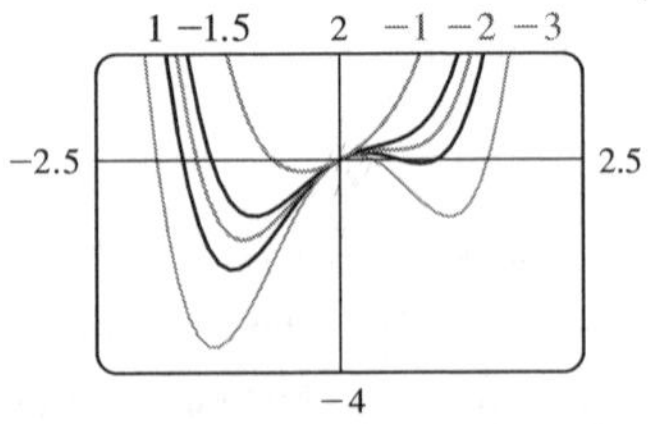

29. $y = x^n 2^{-x}$. As n increases, the maximum of the function moves further from the origin, and gets larger. Note, however, that regardless of n, the function approaches 0 as $x \to \infty$.

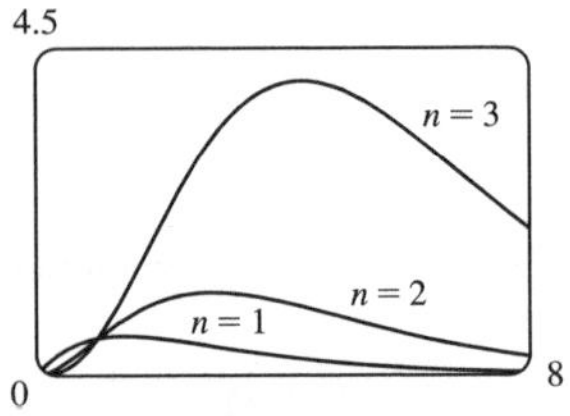

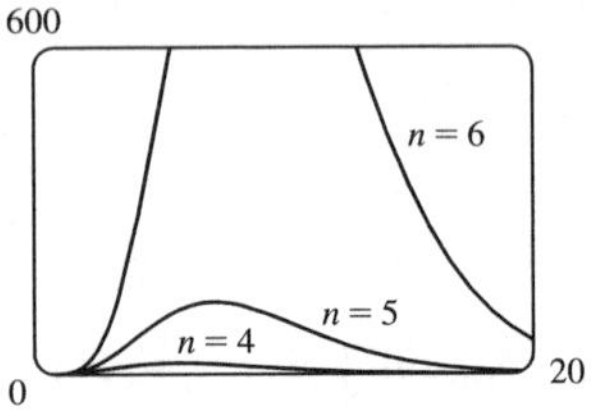

31. $y^2 = cx^3 + x^2$

If $c < 0$, the loop is to the right of the origin, and if c is positive, it is to the left. In both cases, the closer c is to 0, the larger the loop is. (In the limiting case, $c = 0$, the loop is "infinite," that is, it doesn't close.) Also, the larger $|c|$ is, the steeper the slope is on the loopless side of the origin.

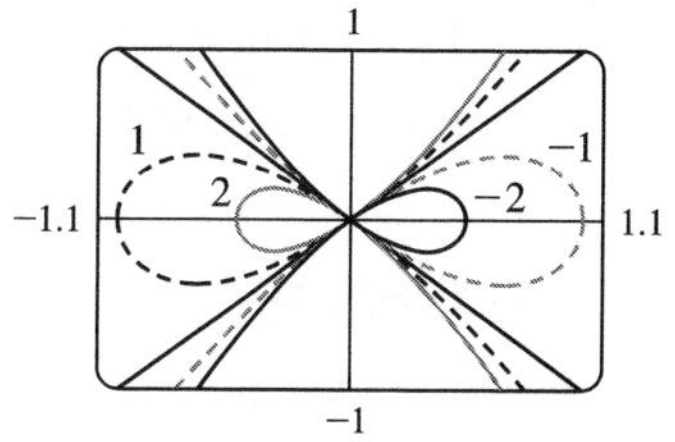

33. The graphing window is 95 pixels wide and we want to start with $x = 0$ and end with $x = 2\pi$. Since there are 94 "gaps" between pixels, the distance between pixels is $\frac{2\pi - 0}{94}$. Thus, the x-values that the calculator actually plots are $x = 0 + \frac{2\pi}{94} \cdot n$, where $n = 0, 1, 2, \ldots, 93, 94$. For $y = \sin 2x$, the actual points plotted by the calculator are $\left(\frac{2\pi}{94} \cdot n, \sin\left(2 \cdot \frac{2\pi}{94} \cdot n\right)\right)$ for $n = 0, 1, \ldots, 94$. For $y = \sin 96x$, the points plotted are $\left(\frac{2\pi}{94} \cdot n, \sin\left(96 \cdot \frac{2\pi}{94} \cdot n\right)\right)$ for $n = 0, 1, \ldots, 94$. But

$$\begin{aligned} \sin\left(96 \cdot \tfrac{2\pi}{94} \cdot n\right) &= \sin\left(94 \cdot \tfrac{2\pi}{94} \cdot n + 2 \cdot \tfrac{2\pi}{94} \cdot n\right) = \sin\left(2\pi n + 2 \cdot \tfrac{2\pi}{94} \cdot n\right) \\ &= \sin\left(2 \cdot \tfrac{2\pi}{94} \cdot n\right) \quad \text{[by periodicity of sine]}, \quad n = 0, 1, \ldots, 94 \end{aligned}$$

So the y-values, and hence the points, plotted for $y = \sin 96x$ are identical to those plotted for $y = \sin 2x$.

Note: Try graphing $y = \sin 94x$. Can you see why all the y-values are zero?

1.5 Exponential Functions

1. (a) $f(x) = a^x,\ a > 0$ (b) $\mathbb{R}$ (c) $(0, \infty)$ (d) See Figures 4(c), 4(b), and 4(a), respectively.

3. All of these graphs approach 0 as $x \to -\infty$, all of them pass through the point $(0, 1)$, and all of them are increasing and approach ∞ as $x \to \infty$. The larger the base, the faster the function increases for $x > 0$, and the faster it approaches 0 as $x \to -\infty$.

Note: The notation "$x \to \infty$" can be thought of as "x becomes large" at this point. More details on this notation are given in Chapter 2.

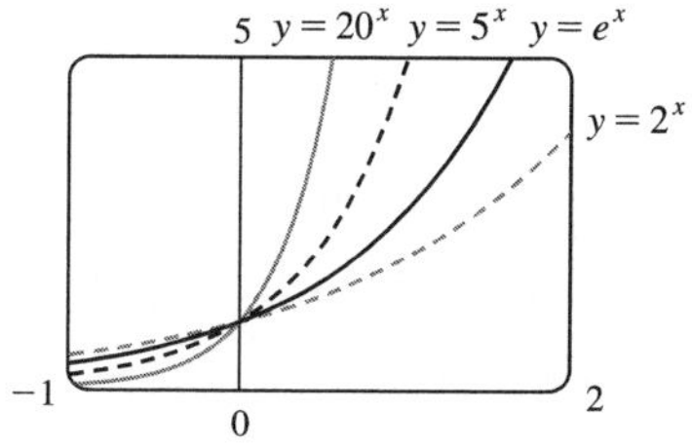

5. The functions with bases greater than 1 (3^x and 10^x) are increasing, while those with bases less than 1 $\left[\left(\frac{1}{3}\right)^x \text{ and } \left(\frac{1}{10}\right)^x\right]$ are decreasing. The graph of $\left(\frac{1}{3}\right)^x$ is the reflection of that of 3^x about the y-axis, and the graph of $\left(\frac{1}{10}\right)^x$ is the reflection of that of 10^x about the y-axis. The graph of 10^x increases more quickly than that of 3^x for $x > 0$, and approaches 0 faster as $x \to -\infty$.

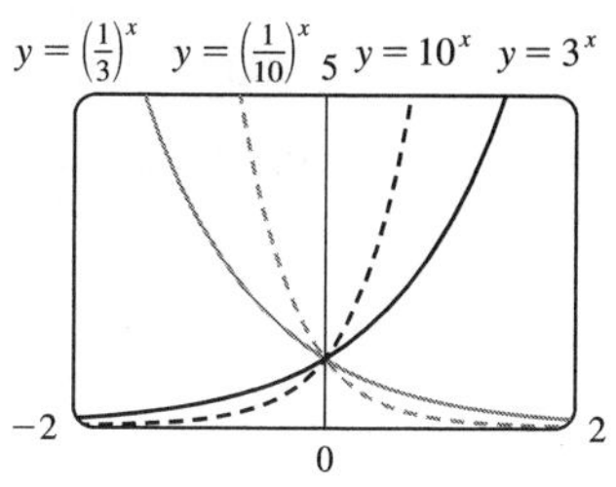

7. We start with the graph of $y = 4^x$ (Figure 3) and then shift 3 units downward. This shift doesn't affect the domain, but the range of $y = 4^x - 3$ is $(-3, \infty)$. There is a horizontal asymptote of $y = -3$.

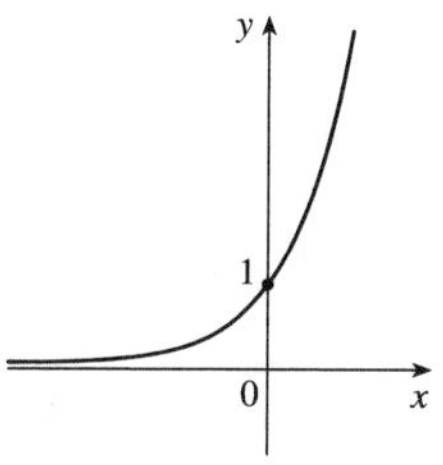

$y = 4^x$

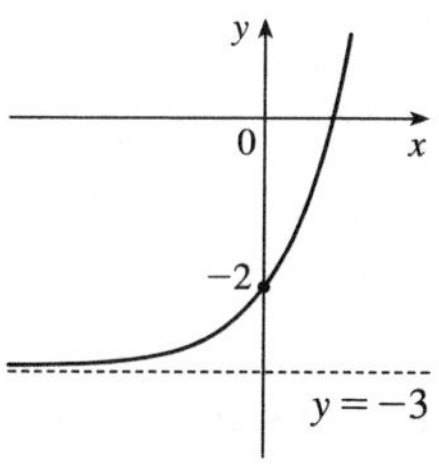

$y = 4^x - 3$

9. We start with the graph of $y = 2^x$ (Figure 3), reflect it about the y-axis, and then about the x-axis (or just rotate 180° to handle both reflections) to obtain the graph of $y = -2^{-x}$. In each graph, $y = 0$ is the horizontal asymptote.

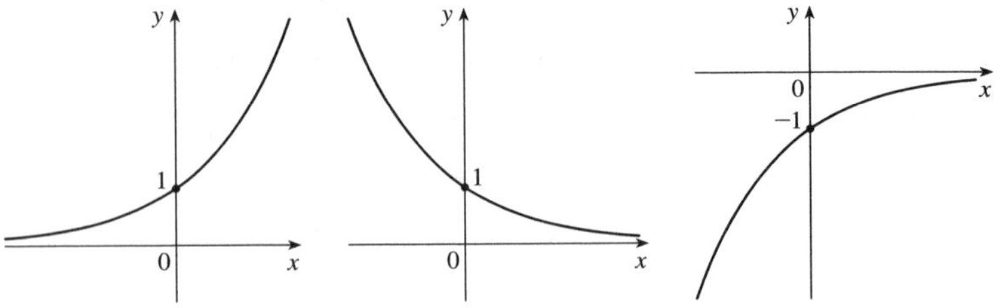

$y = 2^x$ $\quad$ $y = 2^{-x}$ $\quad$ $y = -2^{-x}$

11. We start with the graph of $y = e^x$ (Figure 13) and reflect about the y-axis to get the graph of $y = e^{-x}$. Then we compress the graph vertically by a factor of 2 to obtain the graph of $y = \frac{1}{2}e^{-x}$ and then reflect about the x-axis to get the graph of $y = -\frac{1}{2}e^{-x}$. Finally, we shift the graph upward one unit to get the graph of $y = 1 - \frac{1}{2}e^{-x}$.

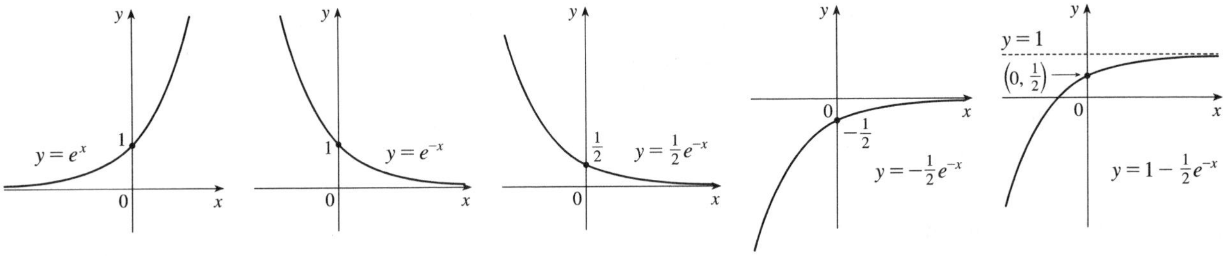

13. (a) To find the equation of the graph that results from shifting the graph of $y = e^x$ 2 units downward, we subtract 2 from the original function to get $y = e^x - 2$.

(b) To find the equation of the graph that results from shifting the graph of $y = e^x$ 2 units to the right, we replace x with $x - 2$ in the original function to get $y = e^{(x-2)}$.

(c) To find the equation of the graph that results from reflecting the graph of $y = e^x$ about the x-axis, we multiply the original function by -1 to get $y = -e^x$.

(d) To find the equation of the graph that results from reflecting the graph of $y = e^x$ about the y-axis, we replace x with $-x$ in the original function to get $y = e^{-x}$.

(e) To find the equation of the graph that results from reflecting the graph of $y = e^x$ about the x-axis and then about the y-axis, we first multiply the original function by -1 (to get $y = -e^x$) and then replace x with $-x$ in this equation to get $y = -e^{-x}$.

15. (a) The denominator $1 + e^x$ is never equal to zero because $e^x > 0$, so the domain of $f(x) = 1/(1 + e^x)$ is $\mathbb{R}$.

(b) $1 - e^x = 0 \quad\Leftrightarrow\quad e^x = 1 \quad\Leftrightarrow\quad x = 0$, so the domain of $f(x) = 1/(1 - e^x)$ is $(-\infty, 0) \cup (0, \infty)$.

17. Use $y = Ca^x$ with the points $(1, 6)$ and $(3, 24)$. $6 = Ca^1 \quad \left[C = \frac{6}{a}\right]$ and $24 = Ca^3 \quad\Rightarrow\quad 24 = \left(\dfrac{6}{a}\right)a^3 \quad\Rightarrow$

$4 = a^2 \quad\Rightarrow\quad a = 2$ [since $a > 0$] and $C = \frac{6}{2} = 3$. The function is $f(x) = 3 \cdot 2^x$.

19. If $f(x) = 5^x$, then $\dfrac{f(x+h) - f(x)}{h} = \dfrac{5^{x+h} - 5^x}{h} = \dfrac{5^x 5^h - 5^x}{h} = \dfrac{5^x\left(5^h - 1\right)}{h} = 5^x\left(\dfrac{5^h - 1}{h}\right)$.

21. $1\text{ m} = 100\text{ cm}$, $f(100) = 100^2\text{ cm} = 10\,000\text{ cm} = 100\text{ m}$.

$g(100) = 2^{100}\text{ cm} = 2^{100}/(100 \cdot 1000)\text{ km} \approx 1.27 \times 10^{25}\text{ km} > 10^{25}\text{ km}$.

23. The graph of g finally surpasses that of f at $x \approx 35.8$.

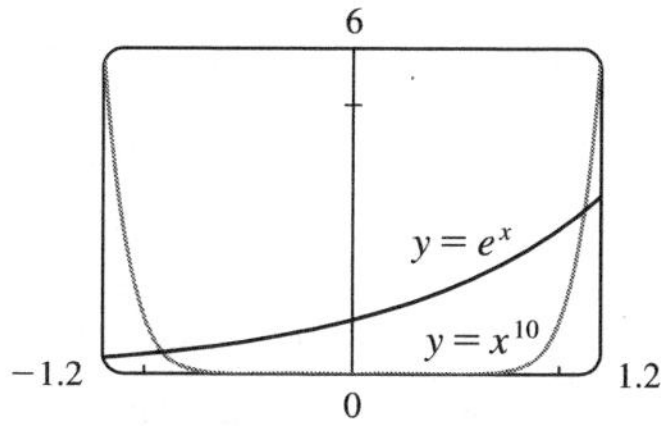

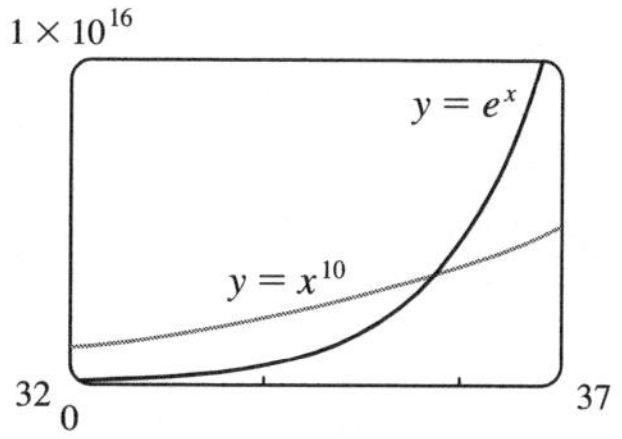

25. (a) Fifteen hours represents 5 doubling periods (one doubling period is three hours). $100 \cdot 2^5 = 3200$

(b) In t hours, there will be $t/3$ doubling periods. The initial population is 100, so the population y at time t is $y = 100 \cdot 2^{t/3}$.

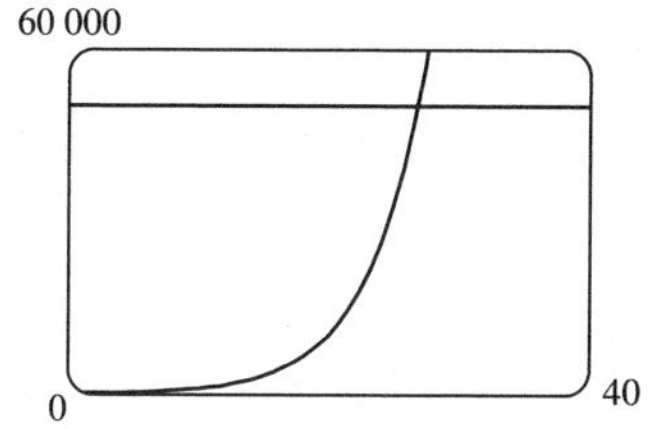

(c) $t = 20 \quad \Rightarrow \quad y = 100 \cdot 2^{20/3} \approx 10\,159$

(d) We graph $y_1 = 100 \cdot 2^{x/3}$ and $y_2 = 50\,000$. The two curves intersect at $x \approx 26.9$, so the population reaches 50 000 in about 26.9 hours.

27. (a) Fifteen days represents 3 half-life periods (one half-life period is 5 days). $200\left(\frac{1}{2}\right)^3 = 25$ mg

(b) In t hours, there will be $t/5$ half-life periods. The initial amount is 200 mg, so the amount remaining after t days is $y = 200\left(\frac{1}{2}\right)^{t/5}$, or equivalently, $y = 200 \cdot 2^{-t/5}$.

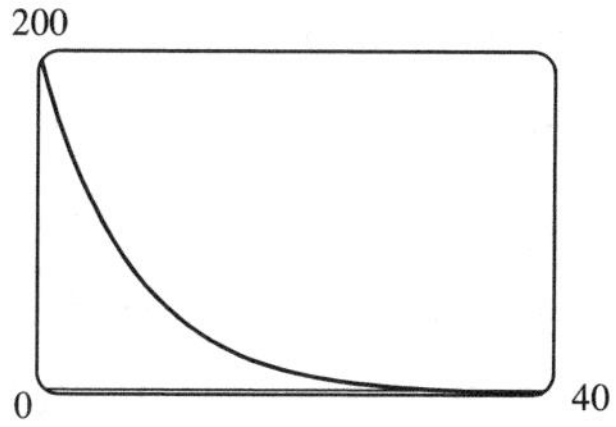

(c) $t = 3$ weeks $= 21$ days $\quad \Rightarrow \quad y = 200 \cdot 2^{-21/5} \approx 10.9$ mg

(d) We graph $y_1 = 200 \cdot 2^{-t/5}$ and $y_2 = 1$. The two curves intersect at $t \approx 38.2$, so the mass will be reduced to 1 mg in about 38.2 days.

29. An exponential model is $y = ab^t$, where $a = 3.154\,832\,569 \times 10^{-12}$ and $b = 1.017\,764\,706$. This model gives $y(1993) \approx 5498$ million and $y(2010) \approx 7417$ million.

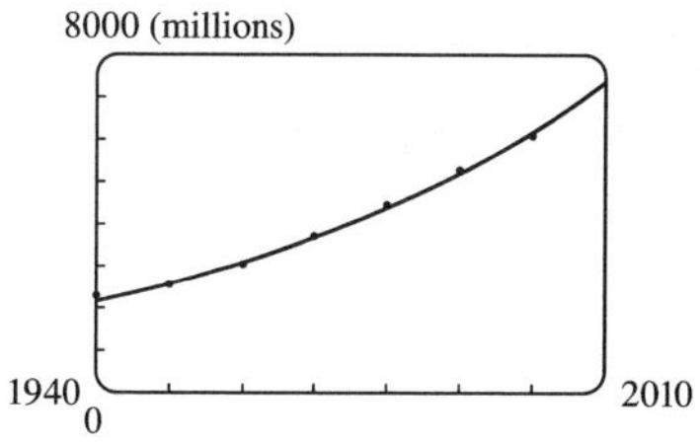

31.

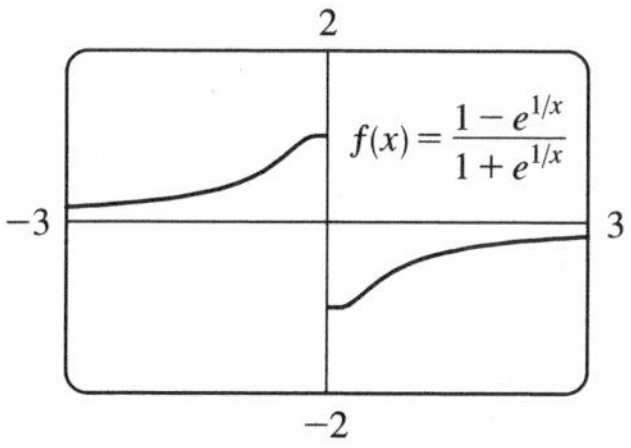

From the graph, it appears that f is an odd function (f is undefined for $x = 0$). To prove this, we must show that $f(-x) = -f(x)$.

$$f(-x) = \frac{1 - e^{1/(-x)}}{1 + e^{1/(-x)}} = \frac{1 - e^{(-1/x)}}{1 + e^{(-1/x)}} = \frac{1 - \dfrac{1}{e^{1/x}}}{1 + \dfrac{1}{e^{1/x}}} \cdot \frac{e^{1/x}}{e^{1/x}} = \frac{e^{1/x} - 1}{e^{1/x} + 1}$$

$$= -\frac{1 - e^{1/x}}{1 + e^{1/x}} = -f(x)$$

so f is an odd function.

1.6 Inverse Functions and Logarithms

1. (a) See Definition 1.

(b) It must pass the Horizontal Line Test.

3. f is not one-to-one because $2 \neq 6$, but $f(2) = 2.0 = f(6)$.

5. No horizontal line intersects the graph of f more than once. Thus, by the Horizontal Line Test, f is one-to-one.

7. The horizontal line $y = 0$ (the x-axis) intersects the graph of f in more than one point. Thus, by the Horizontal Line Test, f is not one-to-one.

9. The graph of $f(x) = \frac{1}{2}(x+5)$ is a line with slope $\frac{1}{2}$. It passes the Horizontal Line Test, so f is one-to-one.

Algebraic solution: If $x_1 \neq x_2$, then $x_1 + 5 \neq x_2 + 5 \Rightarrow \frac{1}{2}(x_1+5) \neq \frac{1}{2}(x_2+5) \Rightarrow f(x_1) \neq f(x_2)$, so f is one-to-one.

11. $g(x) = |x| \Rightarrow g(-1) = 1 = g(1)$, so g is not one-to-one.

13. A football will attain every height h up to its maximum height twice: once on the way up, and again on the way down. Thus, even if t_1 does not equal t_2, $f(t_1)$ may equal $f(t_2)$, so f is not 1-1.

15. Since $f(2) = 9$ and f is 1-1, we know that $f^{-1}(9) = 2$. Remember, if the point $(2,9)$ is on the graph of f, then the point $(9,2)$ is on the graph of f^{-1}.

17. First, we must determine x such that $g(x) = 4$. By inspection, we see that if $x = 0$, then $g(x) = 4$. Since g is 1-1 (g is an increasing function), it has an inverse, and $g^{-1}(4) = 0$.

19. We solve $C = \frac{5}{9}(F - 32)$ for F: $\frac{9}{5}C = F - 32 \Rightarrow F = \frac{9}{5}C + 32$. This gives us a formula for the inverse function, that is, the Fahrenheit temperature F as a function of the Celsius temperature C. $F \geq -459.67 \Rightarrow \frac{9}{5}C + 32 \geq -459.67 \Rightarrow \frac{9}{5}C \geq -491.67 \Rightarrow C \geq -273.15$, the domain of the inverse function.

21. $f(x) = \sqrt{10 - 3x} \Rightarrow y = \sqrt{10-3x}\ (y \geq 0) \Rightarrow y^2 = 10 - 3x \Rightarrow 3x = 10 - y^2 \Rightarrow x = -\frac{1}{3}y^2 + \frac{10}{3}$. Interchange x and y: $y = -\frac{1}{3}x^2 + \frac{10}{3}$. So $f^{-1}(x) = -\frac{1}{3}x^2 + \frac{10}{3}$. Note that the domain of f^{-1} is $x \geq 0$.

23. $y = f(x) = e^{x^3} \Rightarrow \ln y = x^3 \Rightarrow x = \sqrt[3]{\ln y}$. Interchange x and y: $y = \sqrt[3]{\ln x}$. So $f^{-1}(x) = \sqrt[3]{\ln x}$.

25. $y = f(x) = \ln(x+3) \Rightarrow x + 3 = e^y \Rightarrow x = e^y - 3$. Interchange x and y: $y = e^x - 3$. So $f^{-1}(x) = e^x - 3$.

27. $y = f(x) = x^4 + 1 \Rightarrow y - 1 = x^4 \Rightarrow x = \sqrt[4]{y-1}$ (not $\pm$ since $x \geq 0$). Interchange x and y: $y = \sqrt[4]{x-1}$. So $f^{-1}(x) = \sqrt[4]{x-1}$. The graph of $y = \sqrt[4]{x-1}$ is just the graph of $y = \sqrt[4]{x}$ shifted right one unit. From the graph, we see that f and f^{-1} are reflections about the line $y = x$.

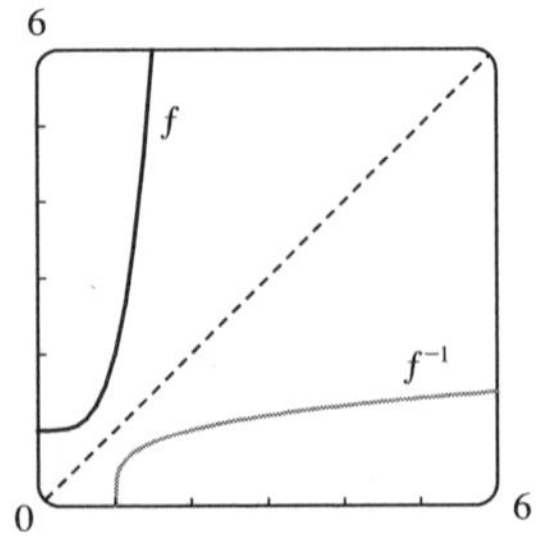

29. Reflect the graph of f about the line $y = x$. The points $(-1,-2)$, $(1,-1)$, $(2,2)$, and $(3,3)$ on f are reflected to $(-2,-1)$, $(-1,1)$, $(2,2)$, and $(3,3)$ on f^{-1}.

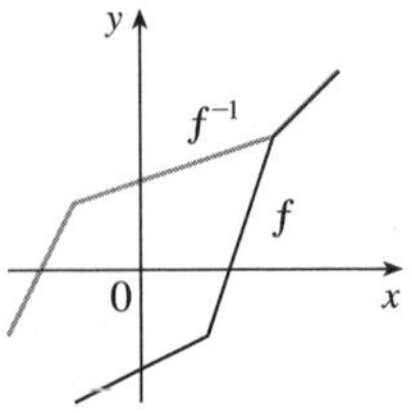

31. (a) It is defined as the inverse of the exponential function with base a, that is, $\log_a x = y \Leftrightarrow a^y = x$.

(b) $(0, \infty)$ (c) $\mathbb{R}$ (d) See Figure 11.

33. (a) $\log_2 64 = 6$ since $2^6 = 64$. (b) $\log_6 \frac{1}{36} = -2$ since $6^{-2} = \frac{1}{36}$.

35. (a) $\log_{10} 1.25 + \log_{10} 80 = \log_{10} (1.25 \cdot 80) = \log_{10} 100 = \log_{10} 10^2 = 2$

(b) $\log_5 10 + \log_5 20 - 3\log_5 2 = \log_5 (10 \cdot 20) - \log_5 2^3 = \log_5 \frac{200}{8} = \log_5 25 = \log_5 5^2 = 2$

37. $2\ln 4 - \ln 2 = \ln 4^2 - \ln 2 = \ln 16 - \ln 2 = \ln \frac{16}{2} = \ln 8$

39. $\ln(1+x^2) + \frac{1}{2}\ln x - \ln \sin x = \ln(1+x^2) + \ln x^{1/2} - \ln \sin x = \ln[(1+x^2)\sqrt{x}\,] - \ln \sin x = \ln \dfrac{(1+x^2)\sqrt{x}}{\sin x}$

41. To graph these functions, we use $\log_{1.5} x = \dfrac{\ln x}{\ln 1.5}$ and $\log_{50} x = \dfrac{\ln x}{\ln 50}$. These graphs all approach $-\infty$ as $x \to 0^+$, and they all pass through the point $(1, 0)$. Also, they are all increasing, and all approach ∞ as $x \to \infty$. The functions with larger bases increase extremely slowly, and the ones with smaller bases do so somewhat more quickly. The functions with large bases approach the y-axis more closely as $x \to 0^+$.

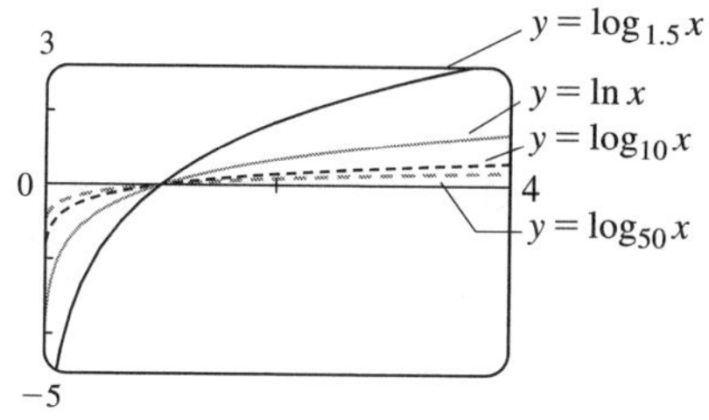

43. 1 m = 100 cm, so we need x such that $\log_2 x = 100 \;\Leftrightarrow\; x = 2^{100}$ cm. In kilometres, this is $2^{100}\text{ cm} \cdot \dfrac{1\text{ km}}{10^5\text{ cm}} \approx 1.27 \times 10^{25}$ km.

45. (a) Shift the graph of $y = \log_{10} x$ five units to the left to obtain the graph of $y = \log_{10}(x+5)$. Note the vertical asymptote of $x = -5$.

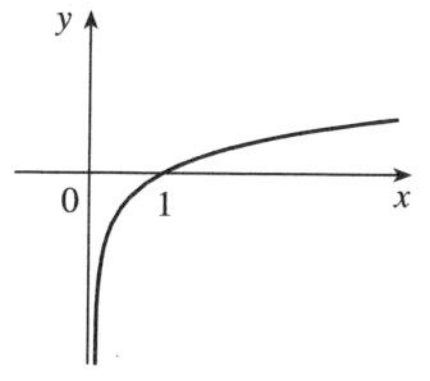

$y = \log_{10} x$

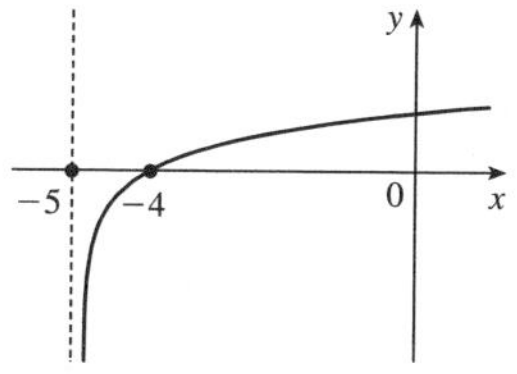

$y = \log_{10}(x+5)$

(b) Reflect the graph of $y = \ln x$ about the x-axis to obtain the graph of $y = -\ln x$.

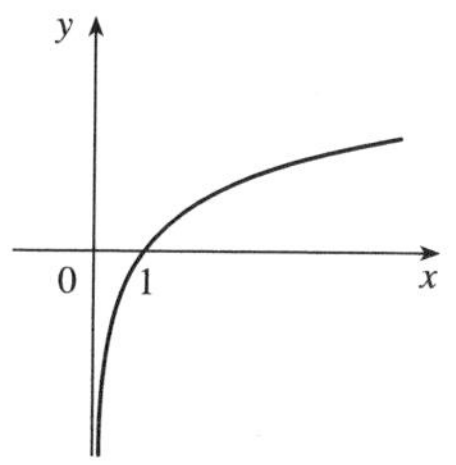

$y = \ln x$

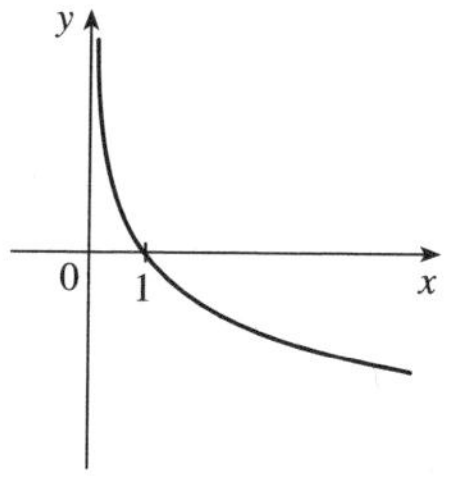

$y = -\ln x$

47. (a) $2\ln x = 1 \;\Rightarrow\; \ln x = \frac{1}{2} \;\Rightarrow\; x = e^{1/2} = \sqrt{e}$

(b) $e^{-x} = 5 \;\Rightarrow\; -x = \ln 5 \;\Rightarrow\; x = -\ln 5$

49. (a) $2^{x-5} = 3 \;\Leftrightarrow\; \log_2 3 = x - 5 \;\Leftrightarrow\; x = 5 + \log_2 3$.

Or: $2^{x-5} = 3 \;\Leftrightarrow\; \ln(2^{x-5}) = \ln 3 \;\Leftrightarrow\; (x-5)\ln 2 = \ln 3 \;\Leftrightarrow\; x - 5 = \dfrac{\ln 3}{\ln 2} \;\Leftrightarrow\; x = 5 + \dfrac{\ln 3}{\ln 2}$

(b) $\ln x + \ln(x-1) = \ln(x(x-1)) = 1 \;\Leftrightarrow\; x(x-1) = e^1 \;\Leftrightarrow\; x^2 - x - e = 0$. The quadratic formula (with $a = 1$, $b = -1$, and $c = -e$) gives $x = \frac{1}{2}\left(1 \pm \sqrt{1+4e}\right)$, but we reject the negative root since the natural logarithm is not defined for $x < 0$. So $x = \frac{1}{2}\left(1 + \sqrt{1+4e}\right)$.

51. (a) $e^x < 10 \;\Rightarrow\; \ln e^x < \ln 10 \;\Rightarrow\; x < \ln 10 \;\Rightarrow\; x \in (-\infty, \ln 10)$

(b) $\ln x > -1 \;\Rightarrow\; e^{\ln x} > e^{-1} \;\Rightarrow\; x > e^{-1} \;\Rightarrow\; x \in (1/e, \infty)$

53. (a) For $f(x) = \sqrt{3 - e^{2x}}$, we must have $3 - e^{2x} \geq 0 \;\Rightarrow\; e^{2x} \leq 3 \;\Rightarrow\; 2x \leq \ln 3 \;\Rightarrow\; x \leq \frac{1}{2}\ln 3$. Thus, the domain of f is $(-\infty, \frac{1}{2}\ln 3]$.

(b) $y = f(x) = \sqrt{3 - e^{2x}}$ [note that $y \geq 0$] $\;\Rightarrow\; y^2 = 3 - e^{2x} \;\Rightarrow\; e^{2x} = 3 - y^2 \;\Rightarrow\; 2x = \ln(3 - y^2) \;\Rightarrow\; x = \frac{1}{2}\ln(3 - y^2)$. Interchange x and y: $y = \frac{1}{2}\ln(3 - x^2)$. So $f^{-1}(x) = \frac{1}{2}\ln(3 - x^2)$. For the domain of f^{-1}, we must have $3 - x^2 > 0 \;\Rightarrow\; x^2 < 3 \;\Rightarrow\; |x| < \sqrt{3} \;\Rightarrow\; -\sqrt{3} < x < \sqrt{3} \;\Rightarrow\; 0 \leq x < \sqrt{3}$ since $x \geq 0$. Note that the domain of f^{-1}, $[0, \sqrt{3}\,)$, equals the range of f.

55. We see that the graph of $y = f(x) = \sqrt{x^3 + x^2 + x + 1}$ is increasing, so f is 1-1.

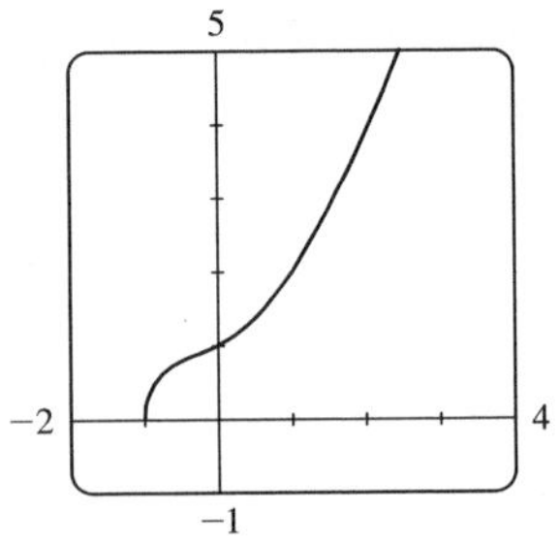

Enter $x = \sqrt{y^3 + y^2 + y + 1}$ and use your CAS to solve the equation for y.

Using Derive, we get two (irrelevant) solutions involving imaginary expressions, as well as one which can be simplified to the following:

$y = f^{-1}(x) = -\frac{\sqrt[3]{4}}{6}\left(\sqrt[3]{D - 27x^2 + 20} - \sqrt[3]{D + 27x^2 - 20} + \sqrt[3]{2}\right)$

where $D = 3\sqrt{3}\sqrt{27x^4 - 40x^2 + 16}$.

Maple and Mathematica each give two complex expressions and one real expression, and the real expression is equivalent to that given by Derive. For example, Maple's expression simplifies to $\frac{1}{6}\frac{M^{2/3} - 8 - 2M^{1/3}}{2M^{1/3}}$, where $M = 108x^2 + 12\sqrt{48 - 120x^2 + 81x^4} - 80$.

57. (a) $n = 100 \cdot 2^{t/3} \Rightarrow \frac{n}{100} = 2^{t/3} \Rightarrow \log_2\left(\frac{n}{100}\right) = \frac{t}{3} \Rightarrow t = 3\log_2\left(\frac{n}{100}\right)$. Using formula (10), we can write this as $t = f^{-1}(n) = 3 \cdot \frac{\ln(n/100)}{\ln 2}$. This function tells us how long it will take to obtain n bacteria (given the number n).

(b) $n = 50\,000 \Rightarrow t = f^{-1}(50\,000) = 3 \cdot \frac{\ln\left(\frac{50\,000}{100}\right)}{\ln 2} = 3\left(\frac{\ln 500}{\ln 2}\right) \approx 26.9$ hours

59. (a) To find the equation of the graph that results from shifting the graph of $y = \ln x$ 3 units upward, we add 3 to the original function to get $y = \ln x + 3$.

(b) To find the equation of the graph that results from shifting the graph of $y = \ln x$ 3 units to the left, we replace x with $x + 3$ in the original function to get $y = \ln(x + 3)$.

(c) To find the equation of the graph that results from reflecting the graph of $y = \ln x$ about the x-axis, we multiply the original equation by -1 to get $y = -\ln x$.

(d) To find the equation of the graph that results from reflecting the graph of $y = \ln x$ about the y-axis, we replace x with $-x$ in the original equation to get $y = \ln(-x)$.

(e) To find the equation of the graph that results from reflecting the graph of $y = \ln x$ about the line $y = x$, we interchange x and y in the original equation to get $x = \ln y \Leftrightarrow y = e^x$.

(f) To find the equation of the graph that results from reflecting the graph of $y = \ln x$ about the x-axis and then about the line $y = x$, we first multiply the original equation by -1 [to get $y = -\ln x$] and then interchange x and y in this equation to get $x = -\ln y \Leftrightarrow \ln y = -x \Leftrightarrow y = e^{-x}$.

(g) To find the equation of the graph that results from reflecting the graph of $y = \ln x$ about the y-axis and then about the line $y = x$, we first replace x with $-x$ in the original equation [to get $y = \ln(-x)$] and then interchange x and y to get $x = \ln(-y) \Leftrightarrow -y = e^x \Leftrightarrow y = -e^x$.

(h) To find the equation of the graph that results from shifting the graph of $y = \ln x$ 3 units to the left and then reflecting it about the line $y = x$, we first replace x with $x + 3$ in the original equation [to get $y = \ln(x + 3)$] and then interchange x and y in this equation to get $x = \ln(y + 3) \Leftrightarrow y + 3 = e^x \Leftrightarrow y = e^x - 3$.

1.7 Parametric Curves

1. $x = 1 + \sqrt{t}, \quad y = t^2 - 4t, \quad 0 \leq t \leq 5$

t	0	1	2	3	4	5
x	1	2	$1+\sqrt{2}$	$1+\sqrt{3}$	3	$1+\sqrt{5}$
			2.41	2.73		3.24
y	0	-3	-4	-3	0	5

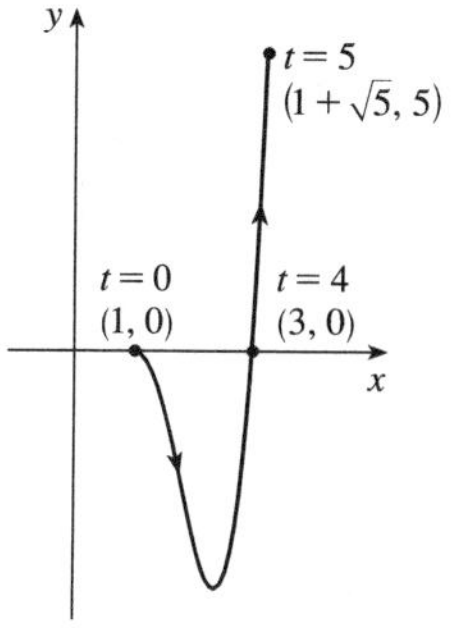

3. $x = 5\sin t, \quad y = t^2, \quad -\pi \leq t \leq \pi$

t	$-\pi$	$-\pi/2$	0	$\pi/2$	π
x	0	-5	0	5	0
y	π^2	$\pi^2/4$	0	$\pi^2/4$	π^2
	9.87	2.47		2.47	9.87

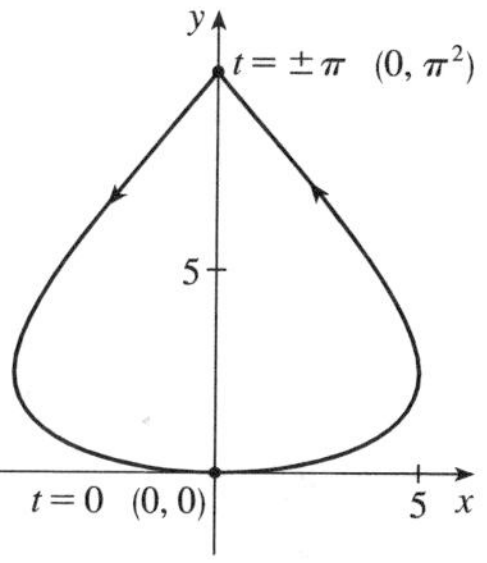

5. $x = 2t + 4, y = t - 1$

(a)

t	-3	-2	-1	0	1	2
x	-2	0	2	4	6	8
y	-4	-3	-2	-1	0	1

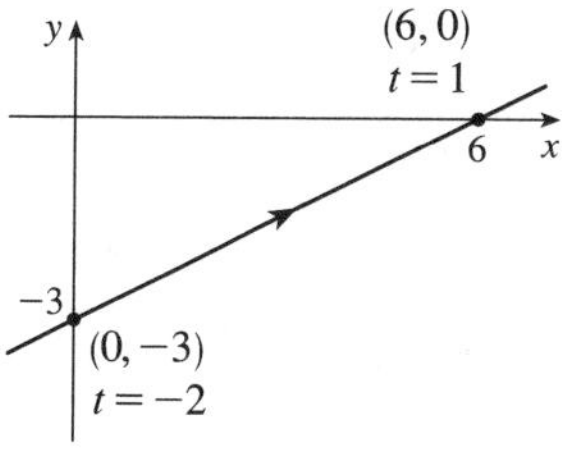

(b) $x = 2t + 4, y = t - 1 \quad \Rightarrow \quad x = 2(y + 1) + 4 = 2y + 6$, so $y = \frac{1}{2}x - 3$.

7. $x = \sqrt{t}, y = 1 - t$

(a)

t	0	1	2	3	4
x	0	1	1.414	1.732	2
y	1	0	-1	-2	-3

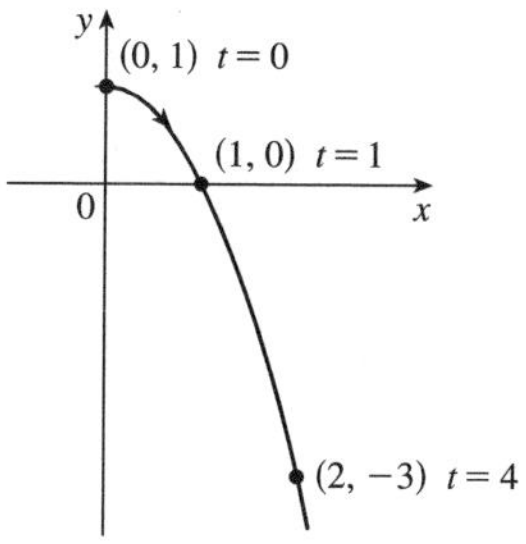

(b) $x = \sqrt{t} \quad \Rightarrow \quad t = x^2 \quad \Rightarrow \quad y = 1 - t = 1 - x^2$. Since $t \geq 0$, $x \geq 0$.

9. (a) $x = \sin\theta, y = \cos\theta, 0 \leq \theta \leq \pi$. $x^2 + y^2 = \sin^2\theta + \cos^2\theta = 1$. Since $0 \leq \theta \leq \pi$, we have $\sin\theta \geq 0$, so $x \geq 0$. Thus, the curve is the right half of the circle $x^2 + y^2 = 1$.

(b)

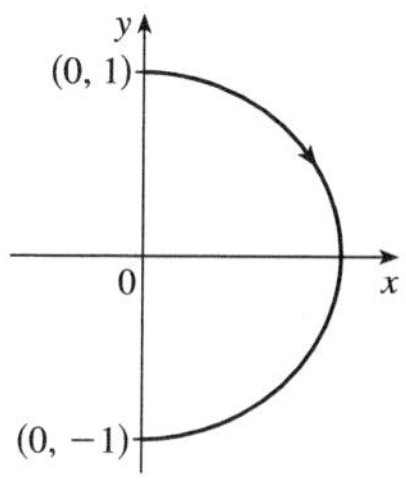

11. (a) $x = \sin t$, $y = \csc t$, $0 < t < \frac{\pi}{2}$. $y = \csc t = \dfrac{1}{\sin t} = \dfrac{1}{x}$.

For $0 < t < \frac{\pi}{2}$, we have $0 < x < 1$ and $y > 1$.

Thus, the curve is the portion of the hyperbola $y = 1/x$ with $y > 1$.

(b)

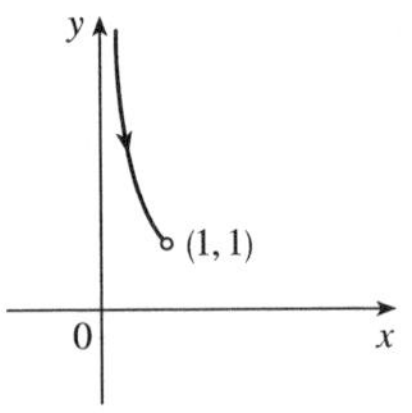

13. (a) $x = e^{2t} \;\Rightarrow\; 2t = \ln x \;\Rightarrow\; t = \frac{1}{2}\ln x$.

$y = t + 1 = \frac{1}{2}\ln x + 1$.

(b)

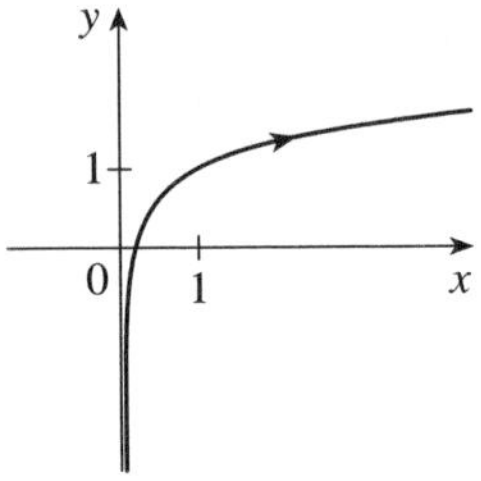

15. $x^2 + y^2 = \cos^2 \pi t + \sin^2 \pi t = 1$, $1 \le t \le 2$, so the particle moves counterclockwise along the circle $x^2 + y^2 = 1$ from $(-1, 0)$ to $(1, 0)$, along the lower half of the circle.

17. $x = 5\sin t$, $y = 2\cos t \;\Rightarrow\; \sin t = \dfrac{x}{5}$, $\cos t = \dfrac{y}{2}$. $\sin^2 t + \cos^2 t = 1 \;\Rightarrow\; \left(\frac{x}{5}\right)^2 + \left(\frac{y}{2}\right)^2 = 1$. The motion of the particle takes place on an ellipse centred at $(0, 0)$. As t goes from $-\pi$ to 5π, the particle starts at the point $(0, -2)$ and moves clockwise around the ellipse 3 times.

19. We must have $1 \le x \le 4$ and $2 \le y \le 3$. So the graph of the curve must be contained in the rectangle $[1, 4]$ by $[2, 3]$.

21. When $t = -1$, $(x, y) = (0, -1)$. As t increases to 0, x decreases to -1 and y increases to 0. As t increases from 0 to 1, x increases to 0 and y increases to 1. As t increases beyond 1, both x and y increase. For $t < -1$, x is positive and decreasing and y is negative and increasing. We could achieve greater accuracy by estimating x- and y-values for selected values of t from the given graphs and plotting the corresponding points.

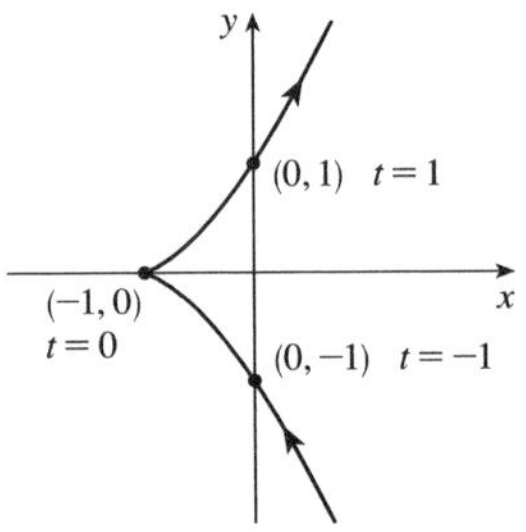

23. When $t = 0$ we see that $x = 0$ and $y = 0$, so the curve starts at the origin. As t increases from 0 to $\frac{1}{2}$, the graphs show that y increases from 0 to 1 while x increases from 0 to 1, decreases to 0 and to -1, then increases back to 0, so we arrive at the point $(0, 1)$. Similarly, as t increases from $\frac{1}{2}$ to 1, y decreases from 1 to 0 while x repeats its pattern, and we arrive back at the origin. We could achieve greater accuracy by estimating x- and y-values for selected values of t from the given graphs and plotting the corresponding points.

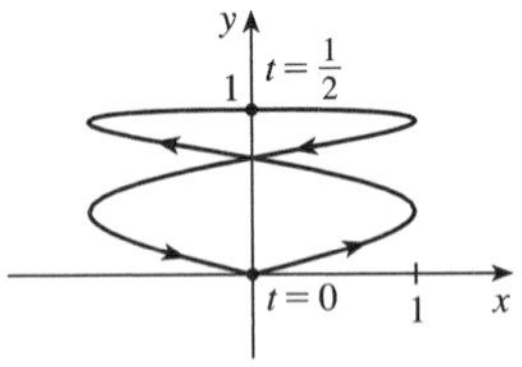

25. As in Example 6, we let $y = t$ and $x = t - 3t^3 + t^5$ and use a t-interval of $[-2\pi, 2\pi]$.

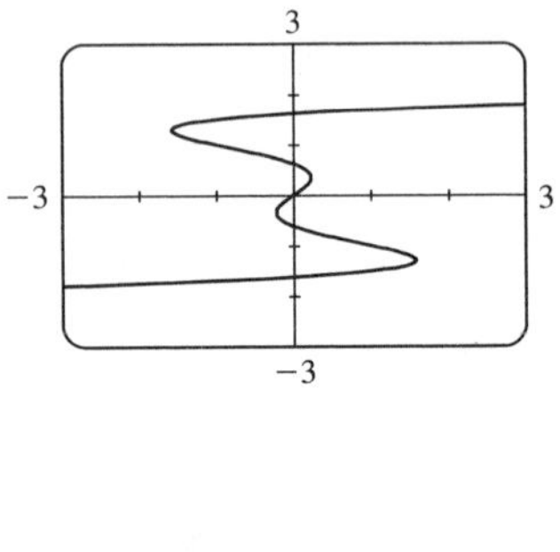

27. (a) $x = x_1 + (x_2 - x_1)t$, $y = y_1 + (y_2 - y_1)t$, $0 \le t \le 1$. Clearly the curve passes through $P_1(x_1, y_1)$ when $t = 0$ and through $P_2(x_2, y_2)$ when $t = 1$. For $0 < t < 1$, x is strictly between x_1 and x_2 and y is strictly between y_1 and y_2. For every value of t, x and y satisfy the relation $y - y_1 = \dfrac{y_2 - y_1}{x_2 - x_1}(x - x_1)$, which is the equation of the line through $P_1(x_1, y_1)$ and $P_2(x_2, y_2)$.

Finally, any point (x, y) on that line satisfies $\dfrac{y - y_1}{y_2 - y_1} = \dfrac{x - x_1}{x_2 - x_1}$; if we call that common value t, then the given parametric equations yield the point (x, y); and any (x, y) on the line between $P_1(x_1, y_1)$ and $P_2(x_2, y_2)$ yields a value of t in $[0, 1]$. So the given parametric equations exactly specify the line segment from $P_1(x_1, y_1)$ to $P_2(x_2, y_2)$.

(b) $x = -2 + [3 - (-2)]t = -2 + 5t$ and $y = 7 + (-1 - 7)t = 7 - 8t$ for $0 \le t \le 1$.

29. The circle $x^2 + y^2 = 4$ can be represented parametrically by $x = 2\cos t$, $y = 2\sin t$; $0 \le t \le 2\pi$. The circle $x^2 + (y - 1)^2 = 4$ can be represented by $x = 2\cos t$, $y = 1 + 2\sin t$; $0 \le t \le 2\pi$. This representation gives us the circle with a counterclockwise orientation starting at $(2, 1)$.

(a) To get a clockwise orientation, we could change the equations to $x = 2\cos t$, $y = 1 - 2\sin t$, $0 \le t \le 2\pi$.

(b) To get three times around in the counterclockwise direction, we use the original equations $x = 2\cos t$, $y = 1 + 2\sin t$ with the domain expanded to $0 \le t \le 6\pi$.

(c) To start at $(0, 3)$ using the original equations, we must have $x_1 = 0$; that is, $2\cos t = 0$. Hence, $t = \frac{\pi}{2}$. So we use $x = 2\cos t$, $y = 1 + 2\sin t$; $\frac{\pi}{2} \le t \le \frac{3\pi}{2}$.

Alternatively, if we want t to start at 0, we could change the equations of the curve. For example, we could use $x = -2\sin t$, $y = 1 + 2\cos t$, $0 \le t \le \pi$.

31. *Big circle:* It's centred at $(2, 2)$ with a radius of 2, so by Example 4, parametric equations are

$$x = 2 + 2\cos t, \qquad y = 2 + 2\sin t, \qquad 0 \le t \le 2\pi$$

Small circles: They are centred at $(1, 3)$ and $(3, 3)$ with a radius of 0.1. By Example 4, parametric equations are

(left) $\quad x = 1 + 0.1\cos t, \qquad y = 3 + 0.1\sin t, \qquad 0 \le t \le 2\pi$

and *(right)* $\quad x = 3 + 0.1\cos t, \qquad y = 3 + 0.1\sin t, \qquad 0 \le t \le 2\pi$

Semicircle: It's the lower half of a circle centred at $(2, 2)$ with radius 1. By Example 4, parametric equations are

$$x = 2 + 1\cos t, \qquad y = 2 + 1\sin t, \qquad \pi \le t \le 2\pi$$

To get all four graphs on the same screen with a typical graphing calculator, we need to change the last t-interval to $[0, 2\pi]$ in order to match the others. We can do this by changing t to $0.5t$. This change gives us the upper half. There are several ways to get the lower half—one is to change the "+" to a "−" in the y-assignment, giving us

$$x = 2 + 1\cos(0.5t), \qquad y = 2 - 1\sin(0.5t), \qquad 0 \le t \le 2\pi$$

33. (a) $x = t^3 \;\Rightarrow\; t = x^{1/3}$, so $y = t^2 = x^{2/3}$. We get the entire curve $y = x^{2/3}$ traversed in a left to right direction.

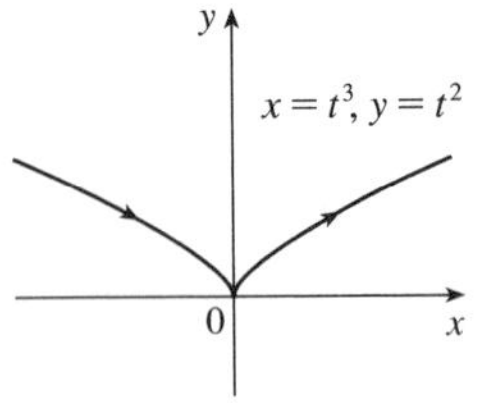

(b) $x = t^6 \;\Rightarrow\; t = x^{1/6}$, so $y = t^4 = x^{4/6} = x^{2/3}$. Since $x = t^6 \geq 0$, we only get the right half of the curve $y = x^{2/3}$.

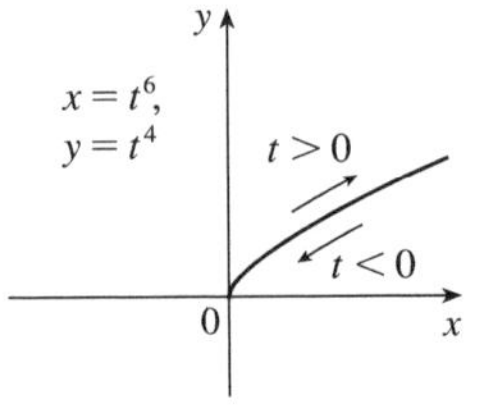

(c) $x = e^{-3t} = (e^{-t})^3$ [so $e^{-t} = x^{1/3}$],

$y = e^{-2t} = (e^{-t})^2 = (x^{1/3})^2 = x^{2/3}$.

If $t < 0$, then x and y are both larger than 1. If $t > 0$, then x and y are between 0 and 1. Since $x > 0$ and $y > 0$, the curve never quite reaches the origin.

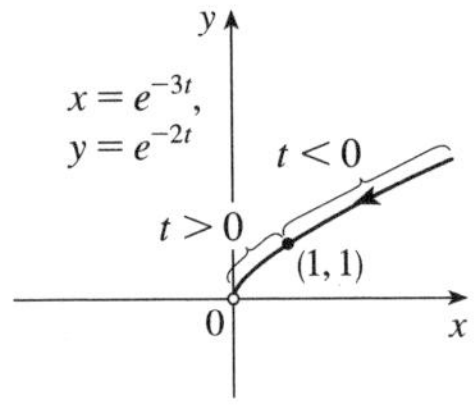

35. The case $\frac{\pi}{2} < \theta < \pi$ is illustrated. C has coordinates $(r\theta, r)$ as in Example 8, and Q has coordinates $(r\theta, r + r\cos(\pi - \theta)) = (r\theta, r(1 - \cos\theta))$ [since $\cos(\pi - \alpha) = \cos\pi\cos\alpha + \sin\pi\sin\alpha = -\cos\alpha$], so P has coordinates $(r\theta - r\sin(\pi - \theta), r(1 - \cos\theta)) = (r(\theta - \sin\theta), r(1 - \cos\theta))$ [since $\sin(\pi - \alpha) = \sin\pi\cos\alpha - \cos\pi\sin\alpha = \sin\alpha$]. Again we have the parametric equations $x = r(\theta - \sin\theta)$, $y = r(1 - \cos\theta)$.

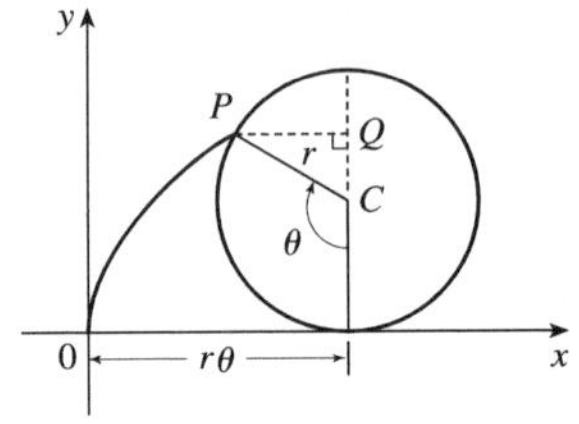

37. It is apparent that $x = |OQ|$ and $y = |QP| = |ST|$. From the diagram, $x = |OQ| = a\cos\theta$ and $y = |ST| = b\sin\theta$. Thus, the parametric equations are $x = a\cos\theta$ and $y = b\sin\theta$. To eliminate θ we rearrange: $\sin\theta = y/b \;\Rightarrow$ $\sin^2\theta = (y/b)^2$ and $\cos\theta = x/a \;\Rightarrow\; \cos^2\theta = (x/a)^2$. Adding the two equations: $\sin^2\theta + \cos^2\theta = 1 = x^2/a^2 + y^2/b^2$. Thus, we have an ellipse.

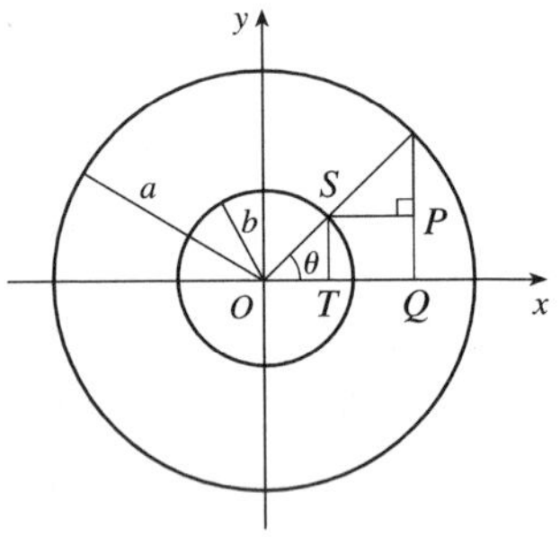

39. (a)

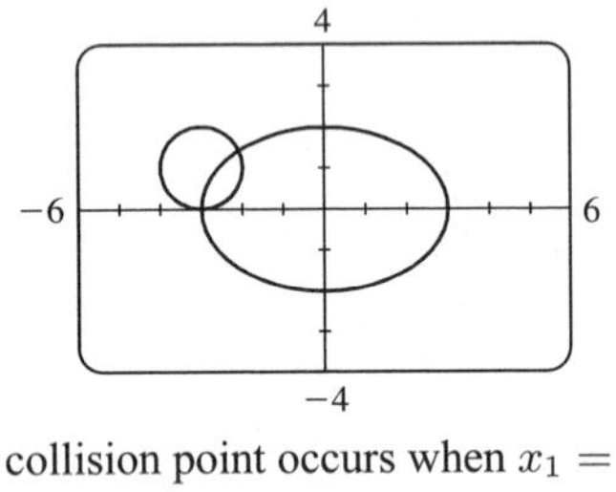

There are 2 points of intersection: $(-3, 0)$ and approximately $(-2.1, 1.4)$.

(b) A collision point occurs when $x_1 = x_2$ and $y_1 = y_2$ for the same t. So solve the equations:

$$3\sin t = -3 + \cos t \quad \textbf{(1)}$$
$$2\cos t = 1 + \sin t \quad \textbf{(2)}$$

From **(2)**, $\sin t = 2\cos t - 1$. Substituting into **(1)**, we get $3(2\cos t - 1) = -3 + \cos t \;\Rightarrow\; 5\cos t = 0 \;\; (*) \;\Rightarrow$ $\cos t = 0 \;\Rightarrow\; t = \frac{\pi}{2}$ or $\frac{3\pi}{2}$. We check that $t = \frac{3\pi}{2}$ satisfies **(1)** and **(2)** but $t = \frac{\pi}{2}$ does not. So the only collision point occurs when $t = \frac{3\pi}{2}$, and this gives the point $(-3, 0)$. [We could check our work by graphing x_1 and x_2 together as functions of t and, on another plot, y_1 and y_2 as functions of t. If we do so, we see that the only value of t for which *both* pairs of graphs intersect is $t = \frac{3\pi}{2}$.]

(c) The circle is centred at $(3, 1)$ instead of $(-3, 1)$. There are still 2 intersection points: $(3, 0)$ and $(2.1, 1.4)$, but there are no collision points, since $(*)$ in part (b) becomes $5\cos t = 6 \quad \Rightarrow \quad \cos t = \frac{6}{5} > 1$.

41. $x = t^2, y = t^3 - ct$. We use a graphing device to produce the graphs for various values of c with $-\pi \le t \le \pi$. Note that all the members of the family are symmetric about the x-axis. For $c < 0$, the graph does not cross itself, but for $c = 0$ it has a cusp at $(0, 0)$ and for $c > 0$ the graph crosses itself at $x = c$, só the loop grows larger as c increases.

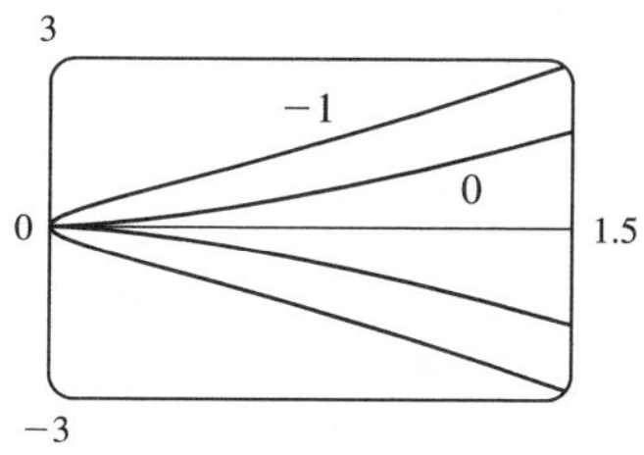

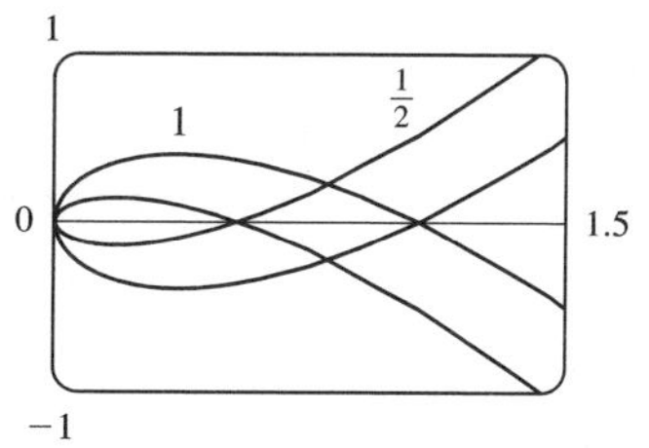

43. Note that all the Lissajous figures are symmetric about the x-axis. The parameters a and b simply stretch the graph in the x- and y-directions respectively. For $a = b = n = 1$ the graph is simply a circle with radius 1. For $n = 2$ the graph crosses itself at the origin and there are loops above and below the x-axis. In general, the figures have $n - 1$ points of intersection, all of which are on the y-axis, and a total of n closed loops.

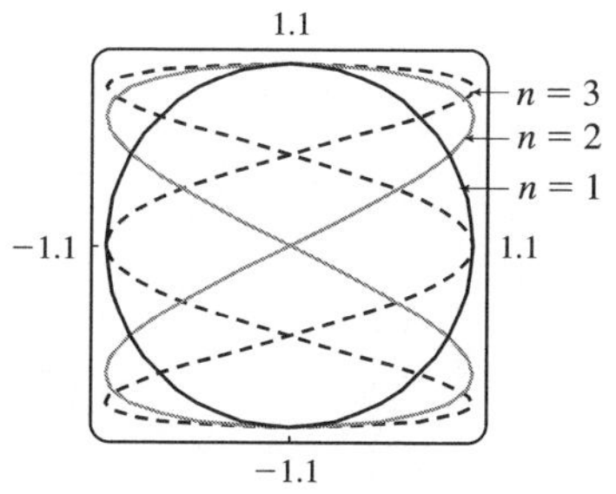

$a = b = 1$

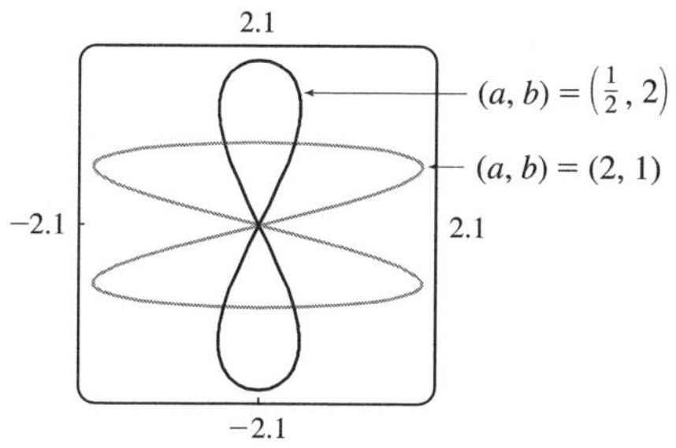

$n = 2$

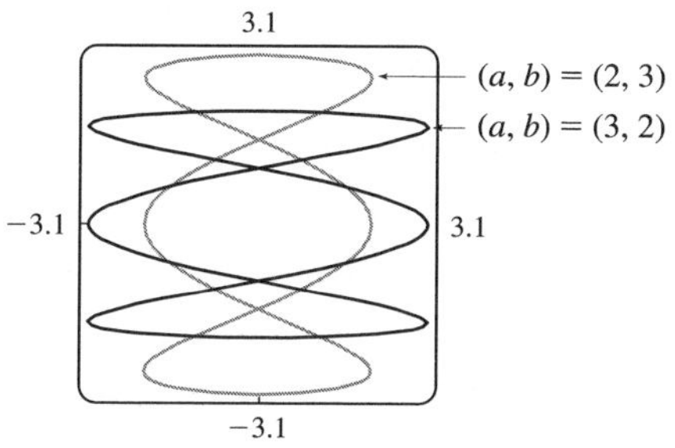

$n = 3$

1 Review

CONCEPT CHECK

1. (a) A **function** f is a rule that assigns to each element x in a set A exactly one element, called $f(x)$, in a set B. The set A is called the **domain** of the function. The **range** of f is the set of all possible values of $f(x)$ as x varies throughout the domain.

(b) If f is a function with domain A, then its **graph** is the set of ordered pairs $\{(x, f(x)) \mid x \in A\}$.

(c) Use the Vertical Line Test on page 17.

2. The four ways to represent a function are: verbally, numerically, visually, and algebraically. An example of each is given below.

Verbally: An assignment of students to chairs in a classroom (a description in words)

Numerically: A tax table that assigns an amount of tax to an income (a table of values)

Visually: A graphical history of a stock market average (a graph)

Algebraically: A relationship between distance, rate, and time: $d = rt$ (an explicit formula)

3. (a) An **even function** f satisfies $f(-x) = f(x)$ for every number x in its domain. It is symmetric with respect to the y-axis.

(b) An **odd function** g satisfies $g(-x) = -g(x)$ for every number x in its domain. It is symmetric with respect to the origin.

4. A function f is called **increasing** on an interval I if $f(x_1) < f(x_2)$ whenever $x_1 < x_2$ in I.

5. A **mathematical model** is a mathematical description (often by means of a function or an equation) of a real-world phenomenon.

6. (a) Linear function: $f(x) = 2x + 1$, $f(x) = ax + b$

(b) Power function: $f(x) = x^2$, $f(x) = x^a$

(c) Exponential function: $f(x) = 2^x$, $f(x) = a^x$

(d) Quadratic function: $f(x) = x^2 + x + 1$, $f(x) = ax^2 + bx + c$

(e) Polynomial of degree 5: $f(x) = x^5 + 2$

(f) Rational function: $f(x) = \dfrac{x}{x+2}$, $f(x) = \dfrac{P(x)}{Q(x)}$ where $P(x)$ and $Q(x)$ are polynomials

7.

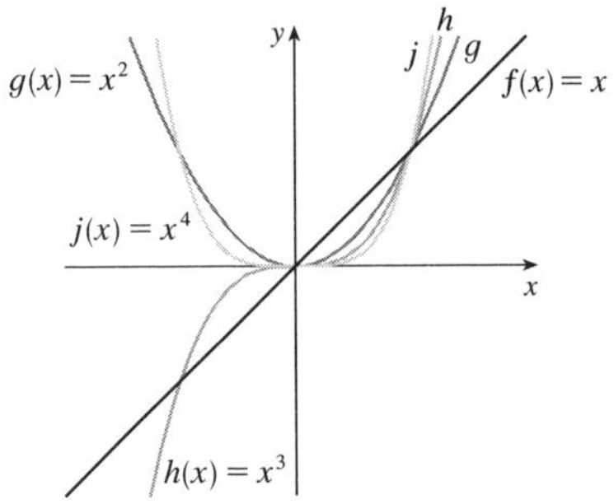

8. (a)

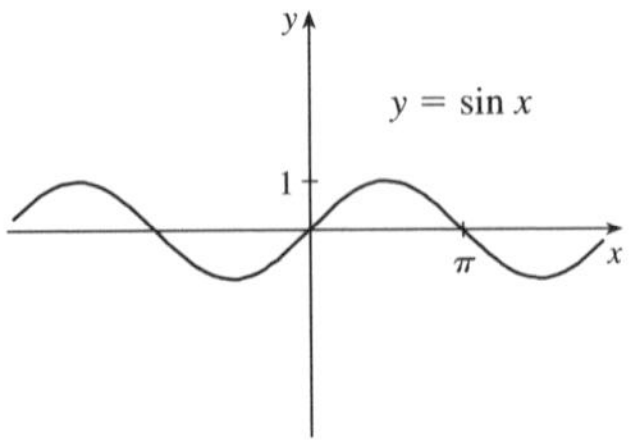

(b)

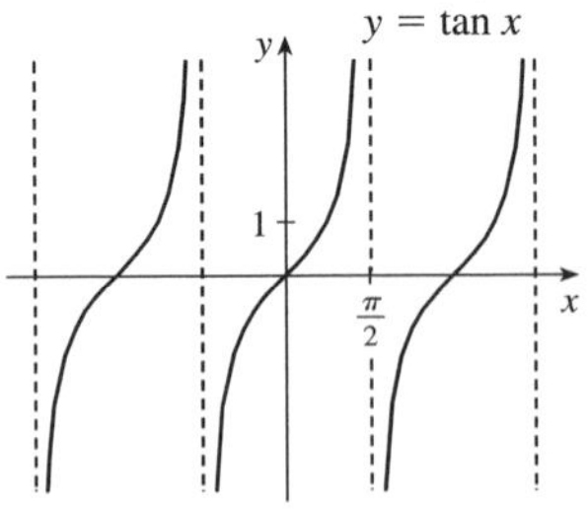

(c)

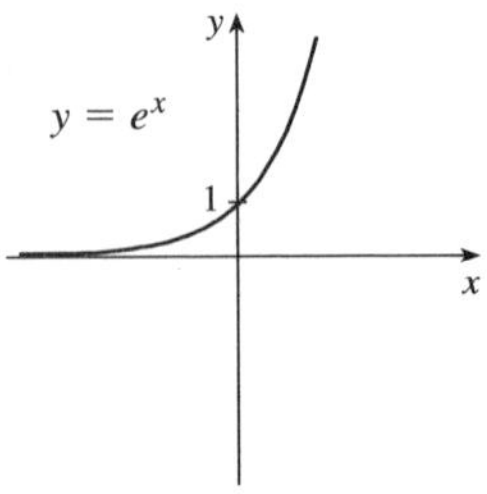

(d)

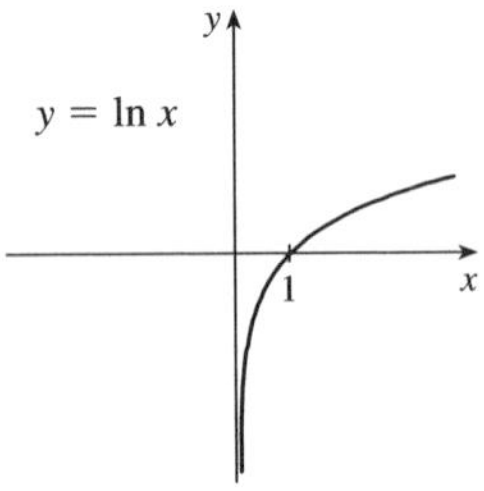

(e)

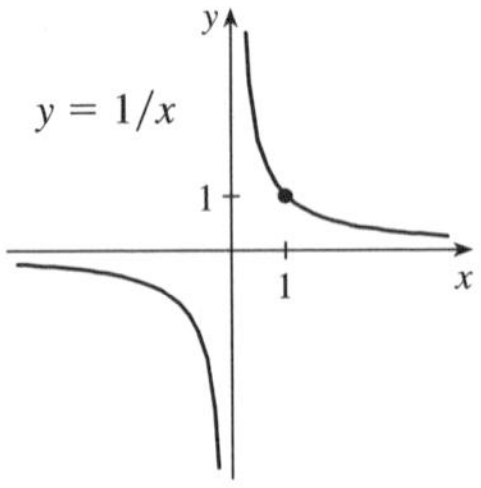

(f)

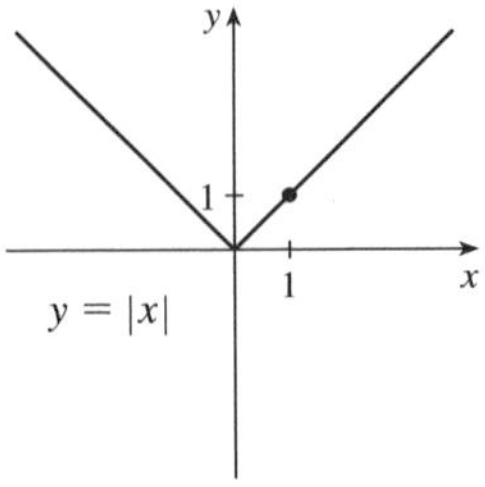

(g)

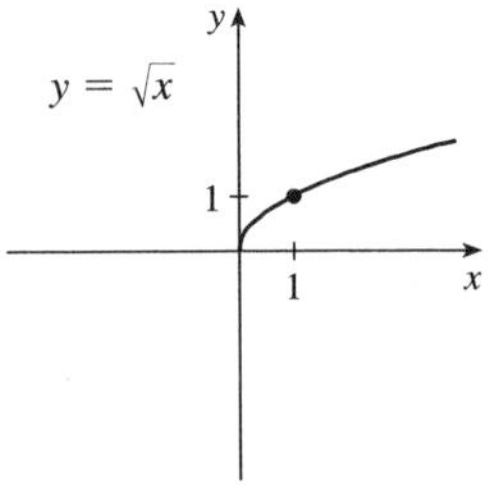

9. (a) The domain of $f + g$ is the intersection of the domain of f and the domain of g; that is, $A \cap B$.

(b) The domain of fg is also $A \cap B$.

(c) The domain of f/g must exclude values of x that make g equal to 0; that is, $\{x \in A \cap B \mid g(x) \neq 0\}$.

10. Given two functions f and g, the **composite** function $f \circ g$ is defined by $(f \circ g)(x) = f(g(x))$. The domain of $f \circ g$ is the set of all x in the domain of g such that $g(x)$ is in the domain of f.

11. (a) If the graph of f is shifted 2 units upward, its equation becomes $y = f(x) + 2$.

(b) If the graph of f is shifted 2 units downward, its equation becomes $y = f(x) - 2$.

(c) If the graph of f is shifted 2 units to the right, its equation becomes $y = f(x - 2)$.

(d) If the graph of f is shifted 2 units to the left, its equation becomes $y = f(x + 2)$.

(e) If the graph of f is reflected about the x-axis, its equation becomes $y = -f(x)$.

(f) If the graph of f is reflected about the y-axis, its equation becomes $y = f(-x)$.

(g) If the graph of f is stretched vertically by a factor of 2, its equation becomes $y = 2f(x)$.

(h) If the graph of f is shrunk vertically by a factor of 2, its equation becomes $y = \frac{1}{2}f(x)$.

(i) If the graph of f is stretched horizontally by a factor of 2, its equation becomes $y = f(\frac{1}{2})x$.

(j) If the graph of f is shrunk horizontally by a factor of 2, its equation becomes $y = f(2x)$.

12. (a) A function f is called a *one-to-one function* if it never takes on the same value twice; that is, if $f(x_1) \neq f(x_2)$ whenever $x_1 \neq x_2$. (Or, f is 1-1 if each output corresponds to only one input.)

Use the Horizontal Line Test: A function is one-to-one if and only if no horizontal line intersects its graph more than once.

(b) If f is a one-to-one function with domain A and range B, then its *inverse function* f^{-1} has domain B and range A and is defined by

$$f^{-1}(y) = x \quad \Leftrightarrow \quad f(x) = y$$

for any y in B. The graph of f^{-1} is obtained by reflecting the graph of f about the line $y = x$.

13. (a) A parametric curve is a set of points of the form $(x, y) = (f(t), g(t))$, where f and g are continuous functions of a variable t.

(b) Sketching a parametric curve, like sketching the graph of a function, is difficult to do in general. We can plot points on the curve by finding $f(t)$ and $g(t)$ for various values of t, either by hand or with a calculator or computer. Sometimes, when f and g are given by formulas, we can eliminate t from the equations $x = f(t)$ and $y = g(t)$ to get a Cartesian equation relating x and y. It may be easier to graph that equation than to work with the original formulas for x and y in terms of t.

(c) See the margin note on page 74.

TRUE-FALSE QUIZ

1. False. Let $f(x) = x^2$, $s = -1$, and $t = 1$. Then $f(s + t) = (-1 + 1)^2 = 0^2 = 0$, but $f(s) + f(t) = (-1)^2 + 1^2 = 2 \neq 0 = f(s + t)$.

3. False. Let $f(x) = x^2$. Then $f(3x) = (3x)^2 = 9x^2$ and $3f(x) = 3x^2$. So $f(3x) \neq 3f(x)$.

5. True. See the Vertical Line Test.

7. False. Let $f(x) = x^3$. Then f is one-to-one and $f^{-1}(x) = \sqrt[3]{x}$. But $1/f(x) = 1/x^3$, which is not equal to $f^{-1}(x)$.

9. True. The function $\ln x$ is an increasing function on $(0, \infty)$.

11. False. Let $x = e^2$ and $a = e$. Then $\dfrac{\ln x}{\ln a} = \dfrac{\ln e^2}{\ln e} = \dfrac{2 \ln e}{\ln e} = 2$ and $\ln \dfrac{x}{a} = \ln \dfrac{e^2}{e} = \ln e = 1$, so in general the statement is false. What *is* true, however, is that $\ln \dfrac{x}{a} = \ln x - \ln a$.

EXERCISES

1. (a) When $x = 2$, $y \approx 2.7$. Thus, $f(2) \approx 2.7$.

(b) $f(x) = 3 \quad \Rightarrow \quad x \approx 2.3, 5.6$

(c) The domain of f is $-6 \le x \le 6$, or $[-6, 6]$.

(d) The range of f is $-4 \le y \le 4$, or $[-4, 4]$.

(e) f is increasing on $[-4, 4]$, that is, on $-4 \le x \le 4$.

(f) f is not one-to-one since it fails the Horizontal Line Test.

(g) f is odd since its graph is symmetric about the origin.

3. (a)

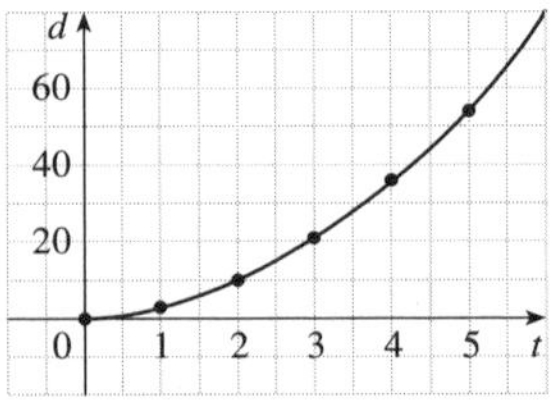

(b) From the graph, we see that the distance travelled after 4.5 seconds is slightly less than 45 metres.

5. $f(x) = \sqrt{4 - 3x^2}$. Domain: $4 - 3x^2 \ge 0 \quad \Rightarrow \quad 3x^2 \le 4 \quad \Rightarrow \quad x^2 \le \frac{4}{3} \quad \Rightarrow \quad |x| \le \frac{2}{\sqrt{3}}$.

Range: $y \ge 0$ and $y \le \sqrt{4} \quad \Rightarrow \quad 0 \le y \le 2$.

7. $y = 1 + \sin x$. Domain: $\mathbb{R}$. Range: $-1 \le \sin x \le 1 \quad \Rightarrow \quad 0 \le 1 + \sin x \le 2 \quad \Rightarrow \quad 0 \le y \le 2$.

9. (a) To obtain the graph of $y = f(x) + 8$, we shift the graph of $y = f(x)$ up 8 units.

(b) To obtain the graph of $y = f(x + 8)$, we shift the graph of $y = f(x)$ left 8 units.

(c) To obtain the graph of $y = 1 + 2f(x)$, we stretch the graph of $y = f(x)$ vertically by a factor of 2, and then shift the resulting graph 1 unit upward.

(d) To obtain the graph of $y = f(x - 2) - 2$, we shift the graph of $y = f(x)$ right 2 units (for the "-2" inside the parentheses), and then shift the resulting graph 2 units downward.

(e) To obtain the graph of $y = -f(x)$, we reflect the graph of $y = f(x)$ about the x-axis.

(f) To obtain the graph of $y = f^{-1}(x)$, we reflect the graph of $y = f(x)$ about the line $y = x$ (assuming f is one–to-one).

11. $y = -\sin 2x$: Start with the graph of $y = \sin x$, compress horizontally by a factor of 2, and reflect about the x-axis.

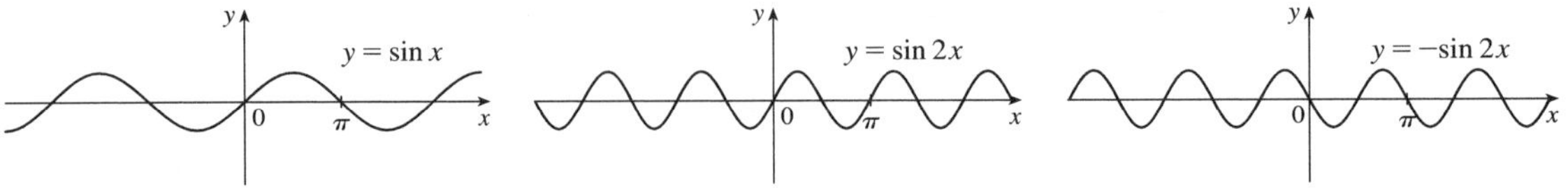

13. $y = (1 + e^x)/2$: Start with the graph of $y = e^x$, shift 1 unit upward, and compress vertically by a factor of 2.

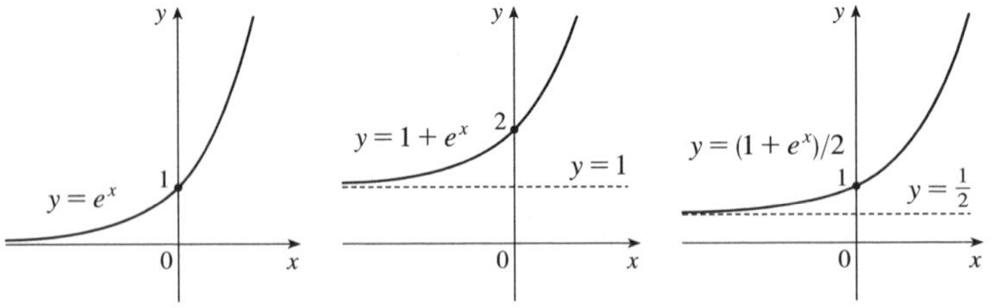

15. $f(x) = \dfrac{1}{x + 2}$:

Start with the graph of $f(x) = 1/x$ and shift 2 units to the left.

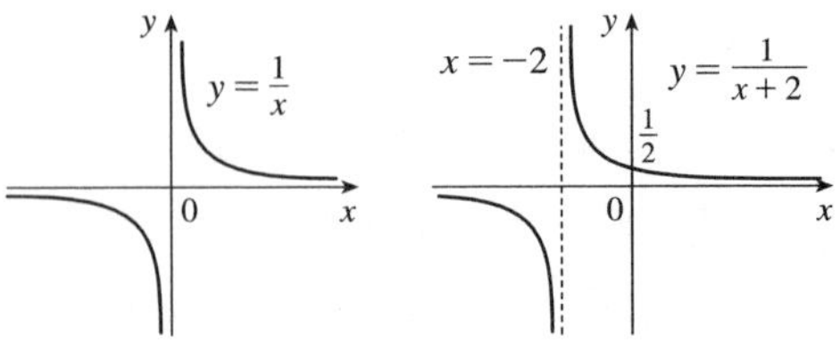

17. (a) The terms of f are a mixture of odd and even powers of x, so f is neither even nor odd.

(b) The terms of f are all odd powers of x, so f is odd.

(c) $f(-x) = e^{-(-x)^2} = e^{-x^2} = f(x)$, so f is even.

(d) $f(-x) = 1 + \sin(-x) = 1 - \sin x$. Now $f(-x) \ne f(x)$ and $f(-x) \ne -f(x)$, so f is neither even nor odd.

19. $f(x) = \ln x, \quad D = (0, \infty); \quad g(x) = x^2 - 9, \quad D = \mathbb{R}$.

(a) $(f \circ g)(x) = f(g(x)) = f(x^2 - 9) = \ln(x^2 - 9)$.

Domain: $x^2 - 9 > 0 \;\Rightarrow\; x^2 > 9 \;\Rightarrow\; |x| > 3 \;\Rightarrow\; x \in (-\infty, -3) \cup (3, \infty)$

(b) $(g \circ f)(x) = g(f(x)) = g(\ln x) = (\ln x)^2 - 9$. Domain: $x > 0$, or $(0, \infty)$

(c) $(f \circ f)(x) = f(f(x)) = f(\ln x) = \ln(\ln x)$. Domain: $\ln x > 0 \;\Rightarrow\; x > e^0 = 1$, or $(1, \infty)$

(d) $(g \circ g)(x) = g(g(x)) = g(x^2 - 9) = (x^2 - 9)^2 - 9$. Domain: $x \in \mathbb{R}$, or $(-\infty, \infty)$

21.

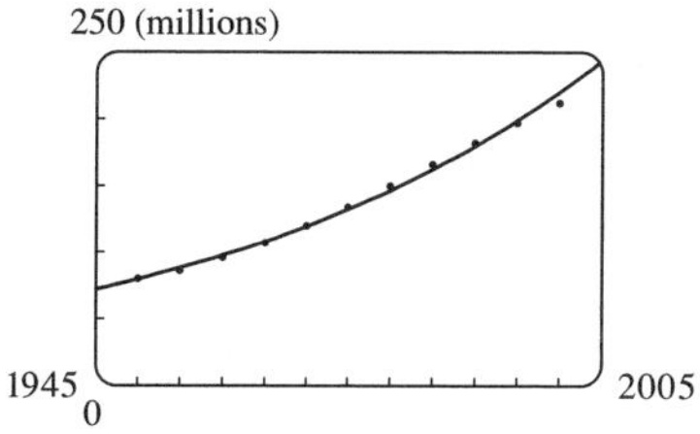

Many models appear to be plausible. Your choice depends on whether you think the population will keep increasing. Cubic and quartic models fit the data quite well, but are less reliable than an exponential model for predicting future population values. An exponential model, $y \approx (4.3763 \times 10^{-16})(1.0206)^x$, gives us an estimate of 270 million for the year 2010.

23. We need to know the value of x such that $f(x) = 2x + \ln x = 2$. Since $x = 1$ gives us $y = 2$, $f^{-1}(2) = 1$.

25. (a) $e^{2\ln 3} = \left(e^{\ln 3}\right)^2 = 3^2 = 9$

(b) $\log_{10} 25 + \log_{10} 4 = \log_{10}(25 \cdot 4) = \log_{10} 100 = \log_{10} 10^2 = 2$

27. (a) After 4 days, $\frac{1}{2}$ gram remains; after 8 days, $\frac{1}{4}$ g; after 12 days, $\frac{1}{8}$ g; after 16 days, $\frac{1}{16}$ g.

(b) $m(4) = \dfrac{1}{2}$, $m(8) = \dfrac{1}{2^2}$, $m(12) = \dfrac{1}{2^3}$, $m(16) = \dfrac{1}{2^4}$. From the pattern, we see that $m(t) = \dfrac{1}{2^{t/4}}$, or $2^{-t/4}$.

(c) $m = 2^{-t/4} \;\Rightarrow\; \log_2 m = -t/4 \;\Rightarrow\; t = -4\log_2 m$; this is the time elapsed when there are m grams of ^{100}Pd.

(d) $m = 0.01 \;\Rightarrow\; t = -4\log_2 0.01 = -4\left(\dfrac{\ln 0.01}{\ln 2}\right) \approx 26.6$ days

29. $f(x) = \ln(x^2 - c)$. If $c < 0$, the domain of f is $\mathbb{R}$. If $c = 0$, the domain of f is $(-\infty, 0) \cup (0, \infty)$. If $c > 0$, the domain of f is $(-\infty, -\sqrt{c}) \cup (\sqrt{c}, \infty)$. As c increases, the dip at $x = 0$ becomes deeper. For $c \geq 0$, the graph has asymptotes at $x = \pm\sqrt{c}$.

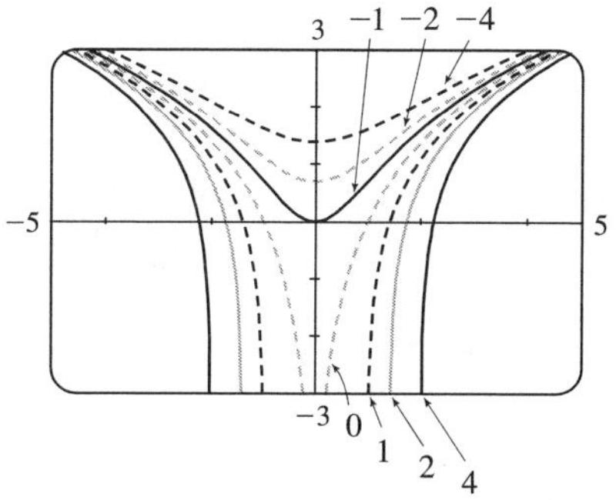

31. (a)

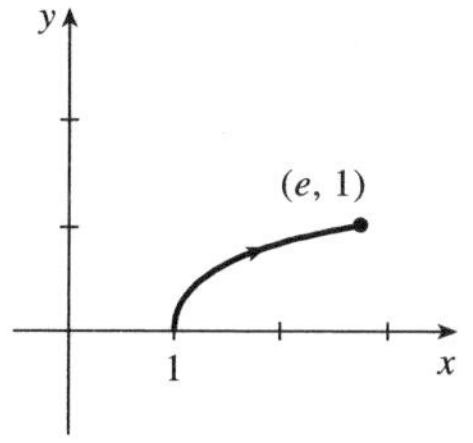

$0 \leq t \leq 1 \;\Rightarrow\; 0 \leq y \leq 1$ and $1 \leq x \leq e$.

(b) $x = e^t \;\Rightarrow\; t = \ln x$; $y = \sqrt{t}$ so $y = \sqrt{\ln x}$.

33. We sketch $x = t$, $y = 2t + \ln t$ (the function) and $x = 2t + \ln t$, $y = t$ (its inverse) for $t > 0$.

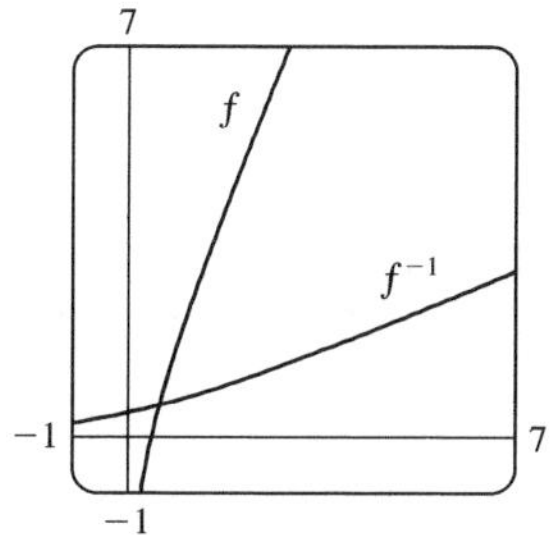

PRINCIPLES OF PROBLEM SOLVING

1. 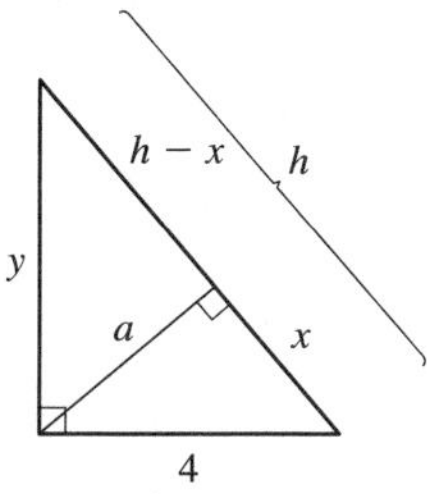

By using the area formula for a triangle, $\frac{1}{2}$ (base) (height), in two ways, we see that $\frac{1}{2}(4)(y) = \frac{1}{2}(h)(a)$, so $a = \dfrac{4y}{h}$. Since $4^2 + y^2 = h^2$, $y = \sqrt{h^2 - 16}$, and

$$a = \frac{4\sqrt{h^2 - 16}}{h}.$$

3. $|2x-1| = \begin{cases} 2x-1 & \text{if } x \ge \frac{1}{2} \\ 1-2x & \text{if } x < \frac{1}{2} \end{cases}$ and $|x+5| = \begin{cases} x+5 & \text{if } x \ge -5 \\ -x-5 & \text{if } x < -5 \end{cases}$

Therefore, we consider the three cases $x < -5$, $-5 \le x < \frac{1}{2}$, and $x \ge \frac{1}{2}$.

If $x < -5$, we must have $1 - 2x - (-x-5) = 3 \quad \Leftrightarrow \quad x = 3$, which is false, since we are considering $x < -5$.

If $-5 \le x < \frac{1}{2}$, we must have $1 - 2x - (x+5) = 3 \quad \Leftrightarrow \quad x = -\frac{7}{3}$.

If $x \ge \frac{1}{2}$, we must have $2x - 1 - (x+5) = 3 \quad \Leftrightarrow \quad x = 9$.

So the two solutions of the equation are $x = -\frac{7}{3}$ and $x = 9$.

5. $f(x) = \left|x^2 - 4|x| + 3\right|$. If $x \ge 0$, then $f(x) = \left|x^2 - 4x + 3\right| = |(x-1)(x-3)|$.

Case (i): If $0 < x \le 1$, then $f(x) = x^2 - 4x + 3$.

Case (ii): If $1 < x \le 3$, then $f(x) = -(x^2 - 4x + 3) = -x^2 + 4x - 3$.

Case (iii): If $x > 3$, then $f(x) = x^2 - 4x + 3$.

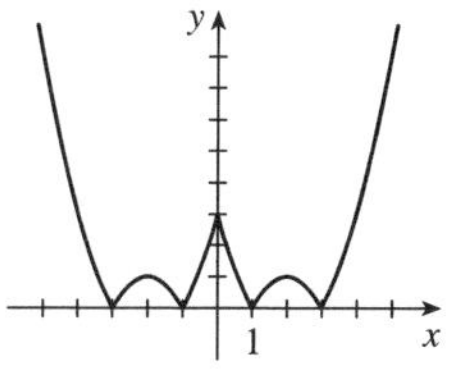

This enables us to sketch the graph for $x \ge 0$. Then we use the fact that f is an even function to reflect this part of the graph about the y-axis to obtain the entire graph. Or, we could consider also the cases $x < -3$, $-3 \le x < -1$, and $-1 \le x < 0$.

7. Remember that $|a| = a$ if $a \ge 0$ and that $|a| = -a$ if $a < 0$. Thus,

$$x + |x| = \begin{cases} 2x & \text{if } x \ge 0 \\ 0 & \text{if } x < 0 \end{cases} \quad \text{and} \quad y + |y| = \begin{cases} 2y & \text{if } y \ge 0 \\ 0 & \text{if } y < 0 \end{cases}$$

We will consider the equation $x + |x| = y + |y|$ in four cases.

(1) $x \ge 0, y \ge 0$	(2) $x \ge 0, y < 0$	(3) $x < 0, y \ge 0$	(4) $x < 0, y < 0$
$2x = 2y$	$2x = 0$	$0 = 2y$	$0 = 0$
$x = y$	$x = 0$	$0 = y$	

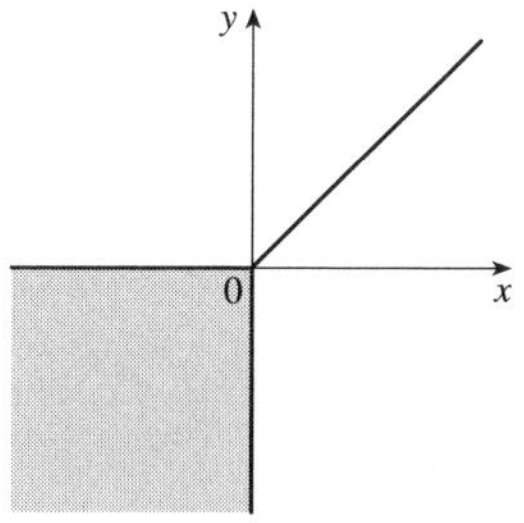

Case 1 gives us the line $y = x$ with nonnegative x and y.

Case 2 gives us the portion of the y-axis with y negative.

Case 3 gives us the portion of the x-axis with x negative.

Case 4 gives us the entire third quadrant.

9. $|x| + |y| \leq 1$. The boundary of the region has equation $|x| + |y| = 1$. In quadrants I, II, III, and IV, this becomes the lines $x + y = 1$, $-x + y = 1$, $-x - y = 1$, and $x - y = 1$ respectively.

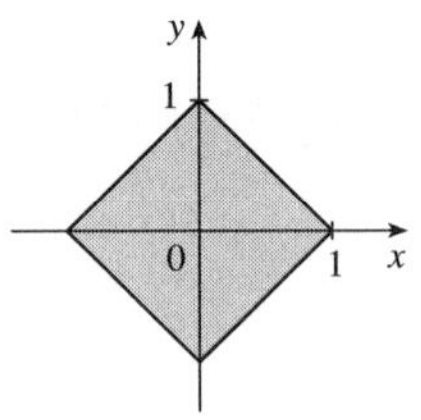

11. $(\log_2 3)(\log_3 4)(\log_4 5)\cdots(\log_{31} 32) = \left(\dfrac{\ln 3}{\ln 2}\right)\left(\dfrac{\ln 4}{\ln 3}\right)\left(\dfrac{\ln 5}{\ln 4}\right)\cdots\left(\dfrac{\ln 32}{\ln 31}\right) = \dfrac{\ln 32}{\ln 2} = \dfrac{\ln 2^5}{\ln 2} = \dfrac{5\ln 2}{\ln 2} = 5$

13. $\ln(x^2 - 2x - 2) \leq 0 \;\Rightarrow\; x^2 - 2x - 2 \leq e^0 = 1 \;\Rightarrow\; x^2 - 2x - 3 \leq 0 \;\Rightarrow\; (x-3)(x+1) \leq 0 \;\Rightarrow\; x \in [-1, 3]$.
Since the argument must be positive, $x^2 - 2x - 2 > 0 \;\Rightarrow\; \left[x - \left(1 - \sqrt{3}\right)\right]\left[x - \left(1 + \sqrt{3}\right)\right] > 0 \;\Rightarrow$
$x \in \left(-\infty, 1 - \sqrt{3}\right) \cup \left(1 + \sqrt{3}, \infty\right)$. The intersection of these intervals is $\left[-1, 1 - \sqrt{3}\right) \cup \left(1 + \sqrt{3}, 3\right]$.

15. Let d be the distance travelled on each half of the trip. Let t_1 and t_2 be the times taken for the first and second halves of the trip. For the first half of the trip we have $t_1 = d/60$ and for the second half we have $t_2 = d/120$. Thus, the average speed for the entire trip is $\dfrac{\text{total distance}}{\text{total time}} = \dfrac{2d}{t_1 + t_2} = \dfrac{2d}{\dfrac{d}{60} + \dfrac{d}{120}} \cdot \dfrac{120}{120} = \dfrac{240d}{2d + d} = \dfrac{240d}{3d} = 80$. The average speed for the entire trip is 80 km/h.

17. Let S_n be the statement that $7^n - 1$ is divisible by 6.

- S_1 is true because $7^1 - 1 = 6$ is divisible by 6.
- Assume S_k is true, that is, $7^k - 1$ is divisible by 6. In other words, $7^k - 1 = 6m$ for some positive integer m. Then $7^{k+1} - 1 = 7^k \cdot 7 - 1 = (6m + 1) \cdot 7 - 1 = 42m + 6 = 6(7m + 1)$, which is divisible by 6, so S_{k+1} is true.
- Therefore, by mathematical induction, $7^n - 1$ is divisible by 6 for every positive integer n.

19. $f_0(x) = x^2$ and $f_{n+1}(x) = f_0(f_n(x))$ for $n = 0, 1, 2, \ldots$.
$f_1(x) = f_0(f_0(x)) = f_0(x^2) = (x^2)^2 = x^4$, $f_2(x) = f_0(f_1(x)) = f_0(x^4) = (x^4)^2 = x^8$,
$f_3(x) = f_0(f_2(x)) = f_0(x^8) = (x^8)^2 = x^{16}, \ldots$. Thus, a general formula is $f_n(x) = x^{2^{n+1}}$.

2 ☐ LIMITS AND DERIVATIVES

2.1 The Tangent and Velocity Problems

1. (a) Using $P(15, 250)$, we construct the following table:

t	Q	slope $= m_{PQ}$
5	$(5, 694)$	$\frac{694-250}{5-15} = -\frac{444}{10} = -44.4$
10	$(10, 444)$	$\frac{444-250}{10-15} = -\frac{194}{5} = -38.8$
20	$(20, 111)$	$\frac{111-250}{20-15} = -\frac{139}{5} = -27.8$
25	$(25, 28)$	$\frac{28-250}{25-15} = -\frac{222}{10} = -22.2$
30	$(30, 0)$	$\frac{0-250}{30-15} = -\frac{250}{15} = -16.\overline{6}$

(b) Using the values of t that correspond to the points closest to P ($t = 10$ and $t = 20$), we have

$$\frac{-38.8 + (-27.8)}{2} = -33.3$$

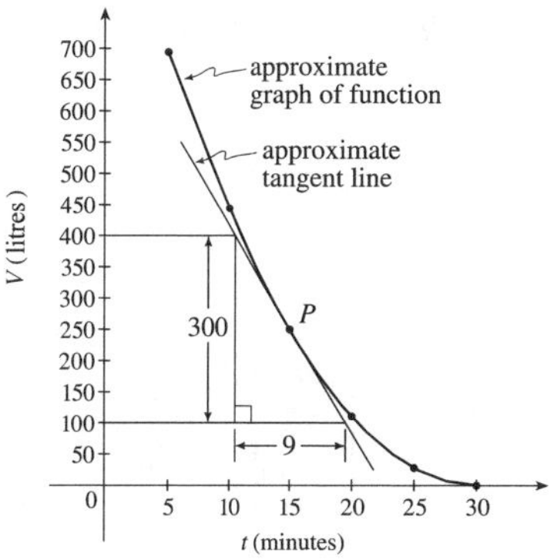

(c) From the graph, we can estimate the slope of the tangent line at P to be $\frac{-300}{9} = -33.\overline{3}$.

3. (a)

	x	Q	m_{PQ}
(i)	0.5	$(0.5, 0.333\,333)$	0.333 333
(ii)	0.9	$(0.9, 0.473\,684)$	0.263 158
(iii)	0.99	$(0.99, 0.497\,487)$	0.251 256
(iv)	0.999	$(0.999, 0.499\,750)$	0.250 125
(v)	1.5	$(1.5, 0.6)$	0.2
(vi)	1.1	$(1.1, 0.523\,810)$	0.238 095
(vii)	1.01	$(1.01, 0.502\,488)$	0.248 756
(viii)	1.001	$(1.001, 0.500\,250)$	0.249 875

(b) The slope appears to be $\frac{1}{4}$.

(c) $y - \frac{1}{2} = \frac{1}{4}(x - 1)$ or $y = \frac{1}{4}x + \frac{1}{4}$.

5. (a) $y = y(t) = 10t - 4.9t^2$. At $t = 1.5$, $y = 10(1.5) - 4.9(1.5)^2 = 3.975$. The average velocity between times 1.5 and $1.5 + h$ is $v_{\text{ave}} = \dfrac{y(1.5+h) - y(1.5)}{(1.5+h) - 1.5} = \dfrac{\left[10(1.5+h) - 4.9(1.5+h)^2\right] - 3.975}{h}$

$$= \frac{15 + 10h - 11.025 - 14.7h - 4.9h^2 - 3.975}{h} = \frac{-4.7h - 4.9h^2}{h} = -4.7 - 4.9h, \text{ if } h \neq 0.$$

(i) $[1.5, 2]$: $h = 0.5$, $v_{\text{ave}} = -7.15$ m/s

(ii) $[1.5, 1.6]$: $h = 0.1$, $v_{\text{ave}} = -5.19$ m/s

(iii) $[1.5, 1.55]$: $h = 0.05$, $v_{\text{ave}} = -4.945$ m/s

(iv) $[1.5, 1.51]$: $h = 0.01$, $v_{\text{ave}} = -4.749$ m/s

(b) The instantaneous velocity when $t = 1.5$ (h approaches 0) is -4.7 m/s.

7. (a) (i) On the interval $[1,3]$, $v_{\text{ave}} = \dfrac{s(3)-s(1)}{3-1} = \dfrac{10.7-1.4}{2} = \dfrac{9.3}{2} = 4.65 \text{ m/s}$.

(ii) On the interval $[2,3]$, $v_{\text{ave}} = \dfrac{s(3)-s(2)}{3-2} = \dfrac{10.7-5.1}{1} = 5.6 \text{ m/s}$.

(iii) On the interval $[3,5]$, $v_{\text{ave}} = \dfrac{s(5)-s(3)}{5-3} = \dfrac{25.8-10.7}{2} = \dfrac{15.1}{2} = 7.55 \text{ m/s}$.

(iv) On the interval $[3,4]$, $v_{\text{ave}} = \dfrac{s(4)-s(3)}{4-3} = \dfrac{17.7-10.7}{1} = 7 \text{ m/s}$.

(b)

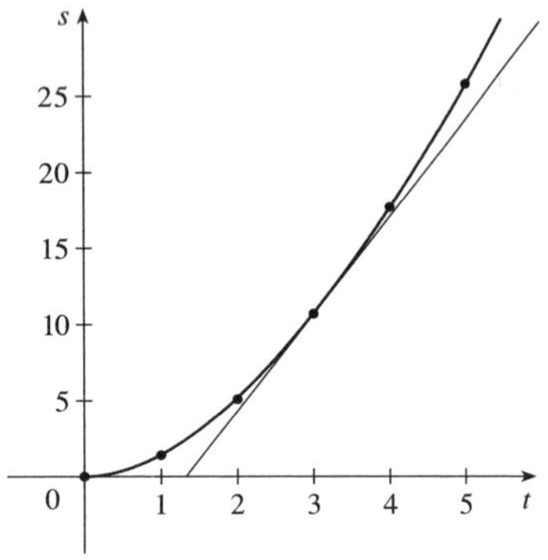

Using the points $(2,4)$ and $(5,23)$ from the approximate tangent line, the instantaneous velocity at $t = 3$ is about $\dfrac{23-4}{5-2} \approx 6.3 \text{ m/s}$.

9. (a) For the curve $y = \sin(10\pi/x)$ and the point $P(1,0)$:

x	Q	m_{PQ}
2	$(2,0)$	0
1.5	$(1.5, 0.8660)$	1.7321
1.4	$(1.4, -0.4339)$	-1.0847
1.3	$(1.3, -0.8230)$	-2.7433
1.2	$(1.2, 0.8660)$	4.3301
1.1	$(1.1, -0.2817)$	-2.8173

x	Q	m_{PQ}
0.5	$(0.5, 0)$	0
0.6	$(0.6, 0.8660)$	-2.1651
0.7	$(0.7, 0.7818)$	-2.6061
0.8	$(0.8, 1)$	-5
0.9	$(0.9, -0.3420)$	3.4202

As x approaches 1, the slopes do not appear to be approaching any particular value.

(b)

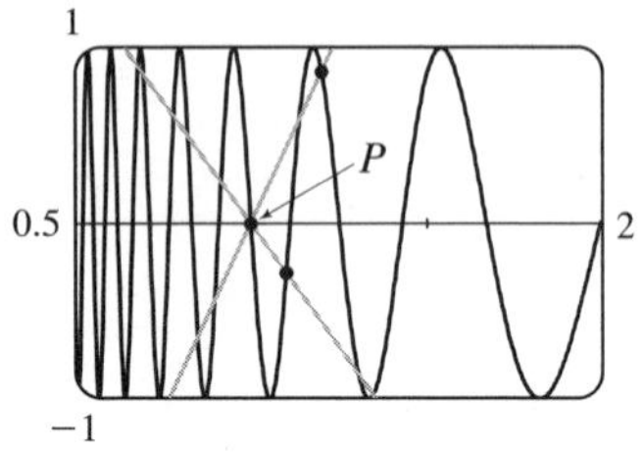

We see that problems with estimation are caused by the frequent oscillations of the graph. The tangent is so steep at P that we need to take x-values much closer to 1 in order to get accurate estimates of its slope.

(c) If we choose $x = 1.001$, then the point Q is $(1.001, -0.0314)$ and $m_{PQ} \approx -31.3794$. If $x = 0.999$, then Q is $(0.999, 0.0314)$ and $m_{PQ} = -31.4422$. The average of these slopes is -31.4108. So we estimate that the slope of the tangent line at P is about -31.4.

2.2 The Limit of a Function

1. As x approaches 2, $f(x)$ approaches 5. [Or, the values of $f(x)$ can be made as close to 5 as we like by taking x sufficiently close to 2 (but $x \neq 2$).] Yes, the graph could have a hole at $(2, 5)$ and be defined such that $f(2) = 3$.

3. (a) $f(x)$ approaches 2 as x approaches 1 from the left, so $\lim_{x \to 1^-} f(x) = 2$.

(b) $f(x)$ approaches 3 as x approaches 1 from the right, so $\lim_{x \to 1^+} f(x) = 3$.

(c) $\lim_{x \to 1} f(x)$ does not exist because the limits in part (a) and part (b) are not equal.

(d) $f(x)$ approaches 4 as x approaches 5 from the left and from the right, so $\lim_{x \to 5} f(x) = 4$.

(e) $f(5)$ is not defined, so it doesn't exist.

5. (a) $\lim_{t \to 0^-} g(t) = -1$ (b) $\lim_{t \to 0^+} g(t) = -2$

(c) $\lim_{t \to 0} g(t)$ does not exist because the limits in part (a) and part (b) are not equal.

(d) $\lim_{t \to 2^-} g(t) = 2$ (e) $\lim_{t \to 2^+} g(t) = 0$

(f) $\lim_{t \to 2} g(t)$ does not exist because the limits in part (d) and part (e) are not equal.

(g) $g(2) = 1$ (h) $\lim_{t \to 4} g(t) = 3$

7.

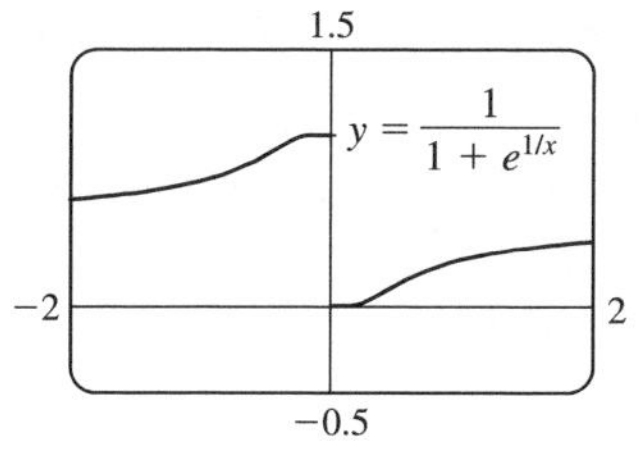

(a) $\lim_{x \to 0^-} f(x) = 1$

(b) $\lim_{x \to 0^+} f(x) = 0$

(c) $\lim_{x \to 0} f(x)$ does not exist because the limits in part (a) and part (b) are not equal.

9. $\lim_{x \to 1^-} f(x) = 2, \quad \lim_{x \to 1^+} f(x) = -2, \quad f(1) = 2$

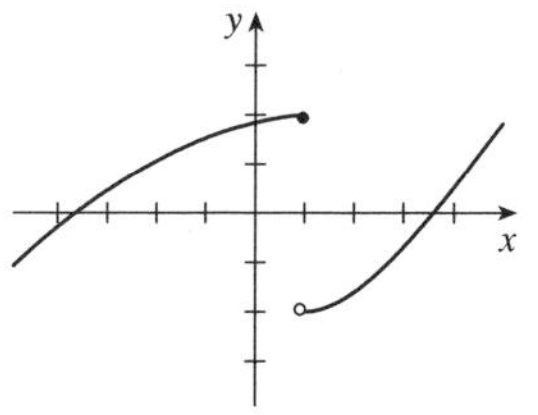

11. $\lim_{x \to 3^+} f(x) = 4, \quad \lim_{x \to 3^-} f(x) = 2, \lim_{x \to -2} f(x) = 2, \quad f(3) = 3, \quad f(-2) = 1$

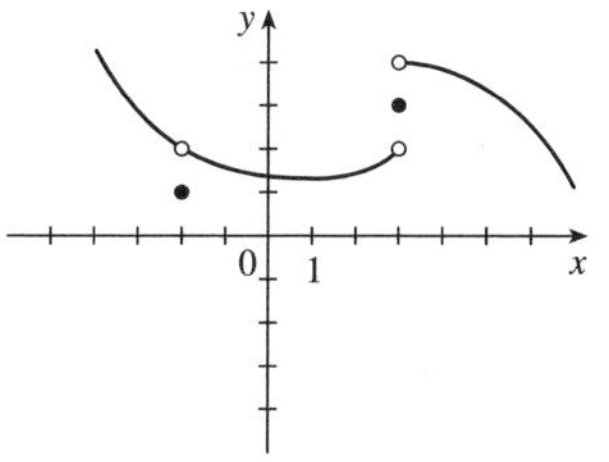

13. For $f(x) = \dfrac{x^2 - 2x}{x^2 - x - 2}$:

x	$f(x)$
2.5	0.714 286
2.1	0.677 419
2.05	0.672 131
2.01	0.667 774
2.005	0.667 221
2.001	0.666 778

x	$f(x)$
1.9	0.655 172
1.95	0.661 017
1.99	0.665 552
1.995	0.666 110
1.999	0.666 556

It appears that $\displaystyle\lim_{x\to 2} \frac{x^2 - 2x}{x^2 - x - 2} = 0.\bar{6} = \frac{2}{3}$.

15. For $f(x) = \dfrac{e^x - 1 - x}{x^2}$:

x	$f(x)$
1	0.718 282
0.5	0.594 885
0.1	0.517 092
0.05	0.508 439
0.01	0.501 671

x	$f(x)$
-1	0.367 879
-0.5	0.426 123
-0.1	0.483 742
-0.05	0.491 770
-0.01	0.498 337

It appears that $\displaystyle\lim_{x\to 0} \frac{e^x - 1 - x}{x^2} = 0.5 = \frac{1}{2}$.

17. For $f(x) = \dfrac{\sqrt{x+4} - 2}{x}$:

x	$f(x)$
1	0.236 068
0.5	0.242 641
0.1	0.248 457
0.05	0.249 224
0.01	0.249 844

x	$f(x)$
-1	0.267 949
-0.5	0.258 343
-0.1	0.251 582
-0.05	0.250 786
-0.01	0.250 156

It appears that $\displaystyle\lim_{x\to 0} \frac{\sqrt{x+4} - 2}{x} = 0.25 = \frac{1}{4}$.

19. For $f(x) = \dfrac{x^6 - 1}{x^{10} - 1}$:

x	$f(x)$
0.5	0.985 337
0.9	0.719 397
0.95	0.660 186
0.99	0.612 018
0.999	0.601 200

x	$f(x)$
1.5	0.183 369
1.1	0.484 119
1.05	0.540 783
1.01	0.588 022
1.001	0.598 800

It appears that $\displaystyle\lim_{x\to 1} \frac{x^6 - 1}{x^{10} - 1} = 0.6 = \frac{3}{5}$.

21. (a) From the graphs, it seems that $\displaystyle\lim_{x\to 0} \frac{\tan 4x}{x} = 4$.

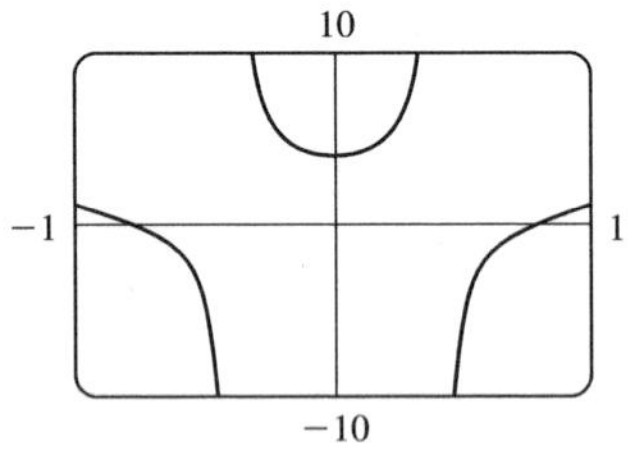

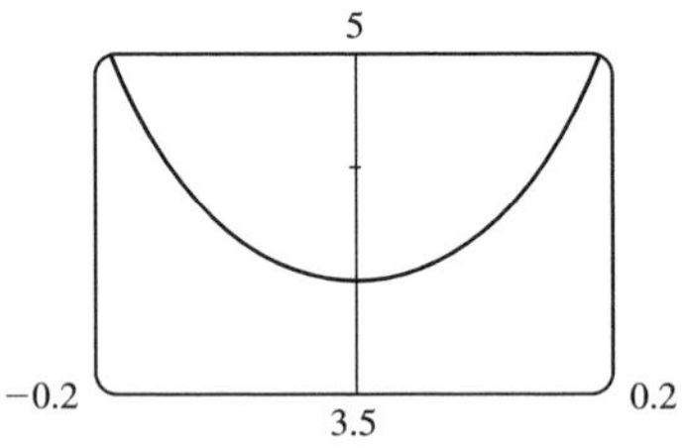

(b)

x	$f(x)$
± 0.1	4.227 932
± 0.01	4.002 135
± 0.001	4.000 021
± 0.0001	4.000 000

23. (a) Let $h(x) = (1 + x)^{1/x}$.

x	$h(x)$
-0.001	2.71964
$-0.000\,1$	2.71842
$-0.000\,01$	2.71830
$-0.000\,001$	2.71828
$0.000\,001$	2.71828
$0.000\,01$	2.71827
$0.000\,1$	2.71815
0.001	2.71692

(b)

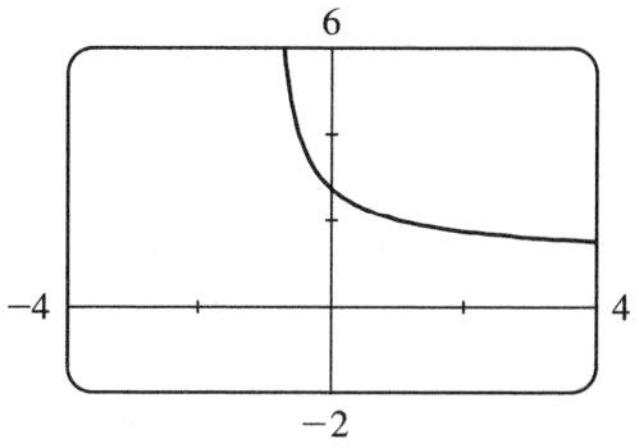

It appears that $\lim_{x \to 0} (1 + x)^{1/x} \approx 2.71828$, which is approximately e.

In Section 3.7 we will see that the value of the limit is exactly e.

25. For $f(x) = x^2 - (2^x/1000)$:

(a)

x	$f(x)$
1	0.998 000
0.8	0.638 259
0.6	0.358 484
0.4	0.158 680
0.2	0.038 851
0.1	0.008 928
0.05	0.001 465

It appears that $\lim_{x \to 0} f(x) = 0$.

(b)

x	$f(x)$
0.04	0.000 572
0.02	−0.000 614
0.01	−0.000 907
0.005	−0.000 978
0.003	−0.000 993
0.001	−0.001 000

It appears that $\lim_{x \to 0} f(x) = -0.001$.

27.

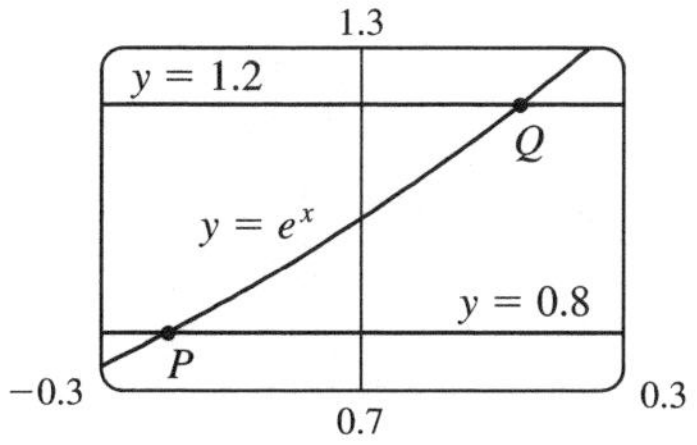

We need to have $0.8 < e^x < 1.2$. From the graph we obtain the approximate points of intersection $P(-0.223\,143\,6, 0.8)$ and $Q(0.182\,321\,56, 1.2)$. So if x is within 0.182 of 0, then y will be within 0.2 of 1. If we must have e^x within 0.1 of 1, we get $P(-0.105\,360\,5, 0.9)$ and $Q(0.095\,310\,18, 1.1)$. We would then need x to be within 0.095 of 0.

2.3 Calculating Limits Using the Limit Laws

1. (a) $\lim_{x \to a} [f(x) + h(x)] = \lim_{x \to a} f(x) + \lim_{x \to a} h(x) = -3 + 8 = 5$

(b) $\lim_{x \to a} [f(x)]^2 = \left[\lim_{x \to a} f(x)\right]^2 = (-3)^2 = 9$

(c) $\lim_{x \to a} \sqrt[3]{h(x)} = \sqrt[3]{\lim_{x \to a} h(x)} = \sqrt[3]{8} = 2$

(d) $\lim_{x \to a} \dfrac{1}{f(x)} = \dfrac{1}{\lim_{x \to a} f(x)} = \dfrac{1}{-3} = -\dfrac{1}{3}$

(e) $\lim_{x \to a} \dfrac{f(x)}{h(x)} = \dfrac{\lim_{x \to a} f(x)}{\lim_{x \to a} h(x)} = \dfrac{-3}{8} = -\dfrac{3}{8}$

(f) $\lim_{x \to a} \dfrac{g(x)}{f(x)} = \dfrac{\lim_{x \to a} g(x)}{\lim_{x \to a} f(x)} = \dfrac{0}{-3} = 0$

(g) The limit does not exist, since $\lim_{x \to a} g(x) = 0$ but $\lim_{x \to a} f(x) \neq 0$.

(h) $\lim_{x \to a} \dfrac{2f(x)}{h(x) - f(x)} = \dfrac{2 \lim_{x \to a} f(x)}{\lim_{x \to a} h(x) - \lim_{x \to a} f(x)} = \dfrac{2(-3)}{8 - (-3)} = -\dfrac{6}{11}$

3. $\lim\limits_{x\to 4}\left(5x^2-2x+3\right)=\lim\limits_{x\to 4}5x^2-\lim\limits_{x\to 4}2x+\lim\limits_{x\to 4}3$ [Limit Laws 2 and 1]

$=5\lim\limits_{x\to 4}x^2-2\lim\limits_{x\to 4}x+3$ [3 and 7]

$=5(4)^2-2(4)+3=75$ [9 and 8]

5. $\lim\limits_{t\to -2}(t+1)^9\left(t^2-1\right)=\lim\limits_{t\to -2}(t+1)^9\lim\limits_{t\to -2}\left(t^2-1\right)$ [4]

$=\left[\lim\limits_{t\to -2}(t+1)\right]^9\lim\limits_{t\to -2}\left(t^2-1\right)$ [6]

$=\left[\lim\limits_{t\to -2}t+\lim\limits_{t\to -2}1\right]^9\left[\lim\limits_{t\to -2}t^2-\lim\limits_{t\to -2}1\right]$ [1 and 2]

$=\left[(-2)+1\right]^9\left[(-2)^2-1\right]=-3$ [8, 7 and 9]

7. $\lim\limits_{x\to 1}\left(\dfrac{1+3x}{1+4x^2+3x^4}\right)^3=\left(\lim\limits_{x\to 1}\dfrac{1+3x}{1+4x^2+3x^4}\right)^3$ [6]

$=\left[\dfrac{\lim\limits_{x\to 1}(1+3x)}{\lim\limits_{x\to 1}(1+4x^2+3x^4)}\right]^3$ [5]

$=\left[\dfrac{\lim\limits_{x\to 1}1+3\lim\limits_{x\to 1}x}{\lim\limits_{x\to 1}1+4\lim\limits_{x\to 1}x^2+3\lim\limits_{x\to 1}x^4}\right]^3$ [2, 1, and 3]

$=\left[\dfrac{1+3(1)}{1+4(1)^2+3(1)^4}\right]^3=\left[\dfrac{4}{8}\right]^3=\left(\dfrac{1}{2}\right)^3=\dfrac{1}{8}$ [7, 8, and 9]

9. $\lim\limits_{x\to 2}\dfrac{x^2+x-6}{x-2}=\lim\limits_{x\to 2}\dfrac{(x+3)(x-2)}{x-2}=\lim\limits_{x\to 2}(x+3)=2+3=5$

11. $\lim\limits_{x\to 2}\dfrac{x^2-x+6}{x-2}$ does not exist since $x-2\to 0$ but $x^2-x+6\to 8$ as $x\to 2$.

13. $\lim\limits_{t\to -3}\dfrac{t^2-9}{2t^2+7t+3}=\lim\limits_{t\to -3}\dfrac{(t+3)(t-3)}{(2t+1)(t+3)}=\lim\limits_{t\to -3}\dfrac{t-3}{2t+1}=\dfrac{-3-3}{2(-3)+1}=\dfrac{-6}{-5}=\dfrac{6}{5}$

15. $\lim\limits_{h\to 0}\dfrac{(2+h)^3-8}{h}=\lim\limits_{h\to 0}\dfrac{(8+12h+6h^2+h^3)-8}{h}=\lim\limits_{h\to 0}\dfrac{12h+6h^2+h^3}{h}=\lim\limits_{h\to 0}\left(12+6h+h^2\right)=12+0+0=12$

17. By the formula for the sum of cubes, we have

$$\lim_{x\to -2}\frac{x+2}{x^3+8}=\lim_{x\to -2}\frac{x+2}{(x+2)(x^2-2x+4)}=\lim_{x\to -2}\frac{1}{x^2-2x+4}=\frac{1}{4+4+4}=\frac{1}{12}.$$

19. $\lim\limits_{x\to 7}\dfrac{\sqrt{x+2}-3}{x-7}=\lim\limits_{x\to 7}\dfrac{\sqrt{x+2}-3}{x-7}\cdot\dfrac{\sqrt{x+2}+3}{\sqrt{x+2}+3}=\lim\limits_{x\to 7}\dfrac{(x+2)-9}{(x-7)\left(\sqrt{x+2}+3\right)}$

$=\lim\limits_{x\to 7}\dfrac{x-7}{(x-7)\left(\sqrt{x+2}+3\right)}=\lim\limits_{x\to 7}\dfrac{1}{\sqrt{x+2}+3}=\dfrac{1}{\sqrt{9}+3}=\dfrac{1}{6}$

21. $\lim\limits_{x\to -4}\dfrac{\dfrac{1}{4}+\dfrac{1}{x}}{4+x}=\lim\limits_{x\to -4}\dfrac{\dfrac{x+4}{4x}}{4+x}=\lim\limits_{x\to -4}\dfrac{x+4}{4x(4+x)}=\lim\limits_{x\to -4}\dfrac{1}{4x}=\dfrac{1}{4(-4)}=-\dfrac{1}{16}$

23. (a)

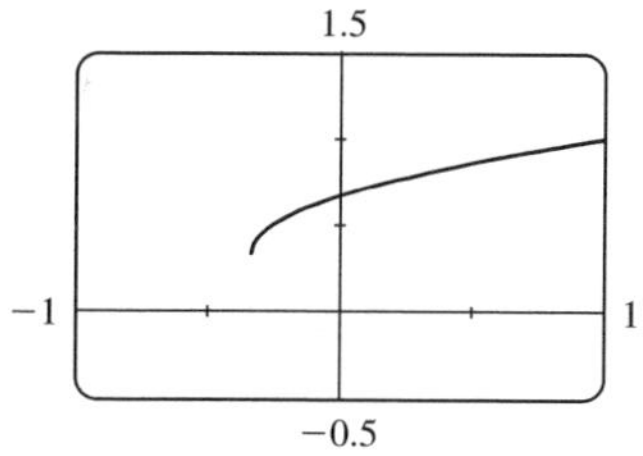

$\lim_{x\to 0} \dfrac{x}{\sqrt{1+3x}-1} \approx \dfrac{2}{3}$

(b)

x	$f(x)$
−0.001	0.666 166 3
−0.000 1	0.666 616 7
−0.000 01	0.666 661 7
−0.000 001	0.666 666 2
0.000 001	0.666 667 2
0.000 01	0.666 671 7
0.000 1	0.666 716 7
0.001	0.667 166 3

The limit appears to be $\dfrac{2}{3}$.

(c) $$\lim_{x\to 0}\left(\frac{x}{\sqrt{1+3x}-1}\cdot\frac{\sqrt{1+3x}+1}{\sqrt{1+3x}+1}\right) = \lim_{x\to 0}\frac{x\left(\sqrt{1+3x}+1\right)}{(1+3x)-1} = \lim_{x\to 0}\frac{x\left(\sqrt{1+3x}+1\right)}{3x}$$

$$= \frac{1}{3}\lim_{x\to 0}\left(\sqrt{1+3x}+1\right) \qquad \text{[Limit Law 3]}$$

$$= \frac{1}{3}\left[\sqrt{\lim_{x\to 0}(1+3x)} + \lim_{x\to 0} 1\right] \qquad \text{[1 and 11]}$$

$$= \frac{1}{3}\left(\sqrt{\lim_{x\to 0} 1 + 3\lim_{x\to 0} x} + 1\right) \qquad \text{[1, 3, and 7]}$$

$$= \frac{1}{3}\left(\sqrt{1+3\cdot 0}+1\right) \qquad \text{[7 and 8]}$$

$$= \frac{1}{3}(1+1) = \frac{2}{3}$$

25. Let $f(x) = -x^2$, $g(x) = x^2\cos 20\pi x$ and $h(x) = x^2$.

Then $-1 \le \cos 20\pi x \le 1 \;\Rightarrow\; -x^2 \le x^2\cos 20\pi x \le x^2 \;\Rightarrow$ $f(x) \le g(x) \le h(x)$. So since $\lim_{x\to 0} f(x) = \lim_{x\to 0} h(x) = 0$, by the Squeeze Theorem we have $\lim_{x\to 0} g(x) = 0$.

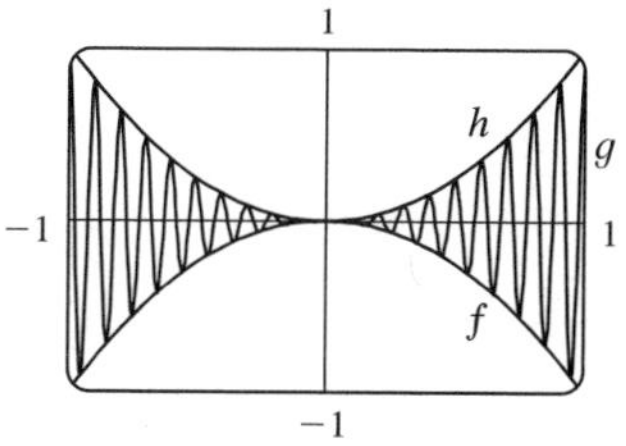

27. We have $\lim_{x\to 4}(4x-9) = 4(4)-9 = 7$ and $\lim_{x\to 4}\left(x^2-4x+7\right) = 4^2-4(4)+7 = 7$. Since $4x-9 \le f(x) \le x^2-4x+7$ for $x \ge 0$, $\lim_{x\to 4} f(x) = 7$ by the Squeeze Theorem.

29. $-1 \le \cos(2/x) \le 1 \;\Rightarrow\; -x^4 \le x^4\cos(2/x) \le x^4$. Since $\lim_{x\to 0}\left(-x^4\right) = 0$ and $\lim_{x\to 0} x^4 = 0$, we have $\lim_{x\to 0}\left[x^4\cos(2/x)\right] = 0$ by the Squeeze Theorem.

31. $$|x-3| = \begin{cases} x-3 & \text{if } x-3 \ge 0 \\ -(x-3) & \text{if } x-3 < 0 \end{cases} = \begin{cases} x-3 & \text{if } x \ge 3 \\ 3-x & \text{if } x < 3 \end{cases}$$

Thus, $\lim_{x\to 3^+}(2x+|x-3|) = \lim_{x\to 3^+}(2x+x-3) = \lim_{x\to 3^+}(3x-3) = 3(3)-3 = 6$ and

$\lim_{x\to 3^-}(2x+|x-3|) = \lim_{x\to 3^-}(2x+3-x) = \lim_{x\to 3^-}(x+3) = 3+3 = 6$. Since the left and right limits are equal,

$\lim_{x\to 3}(2x+|x-3|) = 6$.

33. Since $|x| = -x$ for $x < 0$, we have $\lim_{x\to 0^-}\left(\dfrac{1}{x}-\dfrac{1}{|x|}\right) = \lim_{x\to 0^-}\left(\dfrac{1}{x}-\dfrac{1}{-x}\right) = \lim_{x\to 0^-}\dfrac{2}{x}$, which does not exist since the denominator approaches 0 and the numerator does not.

35. (a) (i) If $x \to 1^+$, then $x > 1$ and $g(x) = x - 1$. Thus, $\lim\limits_{x\to 1^+} g(x) = \lim\limits_{x\to 1^+} (x-1) = 1 - 1 = 0$.

(ii) If $x \to 1^-$, then $x < 1$ and $g(x) = 1 - x^2$. Thus, $\lim\limits_{x\to 1^-} g(x) = \lim\limits_{x\to 1^-} (1 - x^2) = 1 - 1^2 = 0$.

Since the left- and right-hand limits of g at 1 are equal, $\lim\limits_{x\to 1} g(x) = 0$.

(iii) If $x \to 0$, then $-1 < x < 1$ and $g(x) = 1 - x^2$. Thus, $\lim\limits_{x\to 0} g(x) = \lim\limits_{x\to 0}(1 - x^2) = 1 - 0^2 = 1$.

(iv) If $x \to -1^-$, then $x < -1$ and $g(x) = -x$. Thus, $\lim\limits_{x\to -1^-} g(x) = \lim\limits_{x\to -1^-} (-x) = -(-1) = 1$.

(v) If $x \to -1^+$, then $-1 < x < 1$ and $g(x) = 1 - x^2$. Thus,

$$\lim_{x\to -1^+} g(x) = \lim_{x\to -1^+} (1 - x^2) = 1 - (-1)^2 = 1 - 1 = 0$$

(vi) $\lim\limits_{x\to -1} g(x)$ does not exist because the limits in part (iv) and part (v) are not equal.

(b)

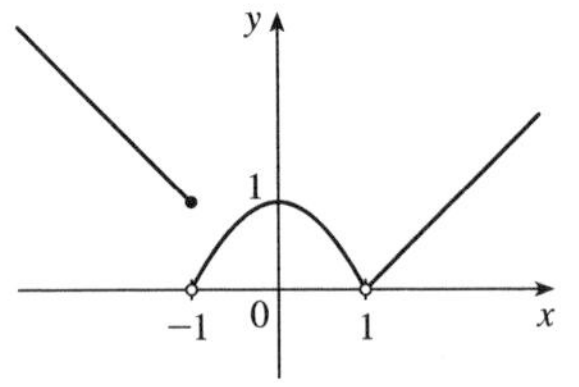

37. (a) (i) $[\![x]\!] = -2$ for $-2 \le x < -1$, so $\lim\limits_{x\to -2^+} [\![x]\!] = \lim\limits_{x\to -2^+} (-2) = -2$

(ii) $[\![x]\!] = -3$ for $-3 \le x < -2$, so $\lim\limits_{x\to -2^-} [\![x]\!] = \lim\limits_{x\to -2^-} (-3) = -3$.

The right and left limits are different, so $\lim\limits_{x\to -2} [\![x]\!]$ does not exist.

(iii) $[\![x]\!] = -3$ for $-3 \le x < -2$, so $\lim\limits_{x\to -2.4} [\![x]\!] = \lim\limits_{x\to -2.4} (-3) = -3$.

(b) (i) $[\![x]\!] = n - 1$ for $n - 1 \le x < n$, so $\lim\limits_{x\to n^-} [\![x]\!] = \lim\limits_{x\to n^-} (n-1) = n - 1$.

(ii) $[\![x]\!] = n$ for $n \le x < n + 1$, so $\lim\limits_{x\to n^+} [\![x]\!] = \lim\limits_{x\to n^+} n = n$.

(c) $\lim\limits_{x\to a} [\![x]\!]$ exists $\Leftrightarrow$ a is not an integer.

39. The graph of $f(x) = [\![x]\!] + [\![-x]\!]$ is the same as the graph of $g(x) = -1$ with holes at each integer, since $f(a) = 0$ for any integer a. Thus, $\lim\limits_{x\to 2^-} f(x) = -1$ and $\lim\limits_{x\to 2^+} f(x) = -1$, so $\lim\limits_{x\to 2} f(x) = -1$. However, $f(2) = [\![2]\!] + [\![-2]\!] = 2 + (-2) = 0$, so $\lim\limits_{x\to 2} f(x) \ne f(2)$.

41. Since $p(x)$ is a polynomial, $p(x) = a_0 + a_1x + a_2x^2 + \cdots + a_nx^n$. Thus, by the Limit Laws,

$$\lim_{x\to a} p(x) = \lim_{x\to a} \left(a_0 + a_1x + a_2x^2 + \cdots + a_nx^n\right) = a_0 + a_1 \lim_{x\to a} x + a_2 \lim_{x\to a} x^2 + \cdots + a_n \lim_{x\to a} x^n$$

$$= a_0 + a_1a + a_2a^2 + \cdots + a_na^n = p(a)$$

Thus, for any polynomial p, $\lim\limits_{x\to a} p(x) = p(a)$.

43. Let $f(x) = [\![x]\!]$ and $g(x) = -[\![x]\!]$. Then $\lim\limits_{x\to 3} f(x)$ and $\lim\limits_{x\to 3} g(x)$ do not exist [Example 9] but $\lim\limits_{x\to 3} [f(x) + g(x)] = \lim\limits_{x\to 3} ([\![x]\!] - [\![x]\!]) = \lim\limits_{x\to 3} 0 = 0$.

45. Since the denominator approaches 0 as $x \to -2$, the limit will exist only if the numerator also approaches 0 as $x \to -2$. In order for this to happen, we need $\lim_{x \to -2} \left(3x^2 + ax + a + 3\right) = 0 \iff$ $3(-2)^2 + a(-2) + a + 3 = 0 \iff 12 - 2a + a + 3 = 0 \iff a = 15$. With $a = 15$, the limit becomes

$$\lim_{x \to -2} \frac{3x^2 + 15x + 18}{x^2 + x - 2} = \lim_{x \to -2} \frac{3(x+2)(x+3)}{(x-1)(x+2)} = \lim_{x \to -2} \frac{3(x+3)}{x-1} = \frac{3(-2+3)}{-2-1} = \frac{3}{-3} = -1.$$

2.4 Continuity

1. From Definition 1, $\lim_{x \to 4} f(x) = f(4)$.

3. (a) The following are the numbers at which f is discontinuous and the type of discontinuity at that number: -4 (removable), -2 (jump), 2 (jump), 4 (infinite).

(b) f is continuous from the left at -2 since $\lim_{x \to -2^-} f(x) = f(-2)$. f is continuous from the right at 2 and 4 since $\lim_{x \to 2^+} f(x) = f(2)$ and $\lim_{x \to 4^+} f(x) = f(4)$. It is continuous from neither side at -4 since $f(-4)$ is undefined.

5. The graph of $y = f(x)$ must have a discontinuity at $x = 3$ and must show that $\lim_{x \to 3^-} f(x) = f(3)$.

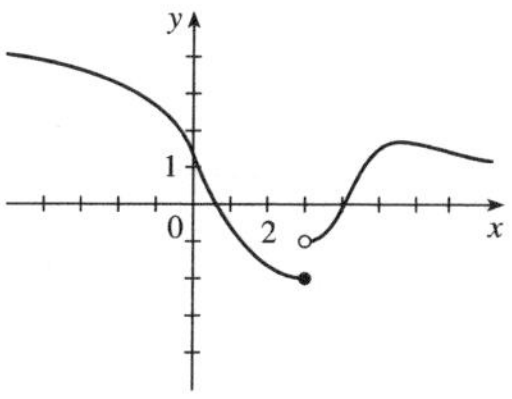

7. (a)

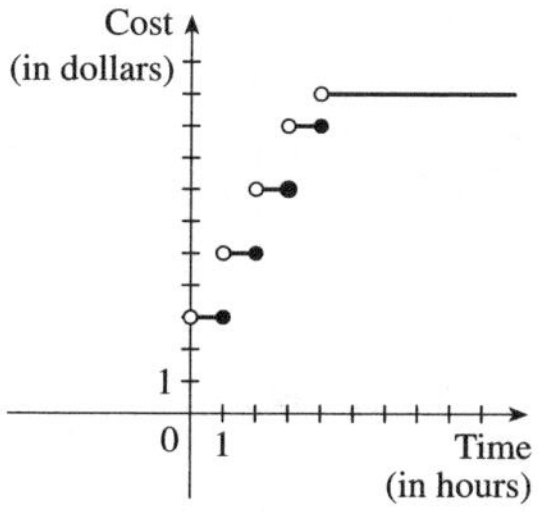

(b) There are discontinuities at times $t = 1$, 2, 3, and 4. A person parking in the lot would want to keep in mind that the charge will jump at the beginning of each hour.

9. Since f and g are continuous functions,

$$\begin{aligned} \lim_{x \to 3} [2f(x) - g(x)] &= 2 \lim_{x \to 3} f(x) - \lim_{x \to 3} g(x) && \text{[by Limit Laws 2 and 3]} \\ &= 2f(3) - g(3) && \text{[by continuity of } f \text{ and } g \text{ at } x = 3] \\ &= 2 \cdot 5 - g(3) = 10 - g(3) \end{aligned}$$

Since it is given that $\lim_{x \to 3} [2f(x) - g(x)] = 4$, we have $10 - g(3) = 4$, so $g(3) = 6$.

11. $\lim_{x \to -1} f(x) = \lim_{x \to -1} \left(x + 2x^3\right)^4 = \left(\lim_{x \to -1} x + 2 \lim_{x \to -1} x^3\right)^4 = \left[-1 + 2(-1)^3\right]^4 = (-3)^4 = 81 = f(-1)$.

By the definition of continuity, f is continuous at $a = -1$.

13. $f(x) = \ln|x - 2|$ is discontinuous at 2 since $f(2) = \ln 0$ is not defined.

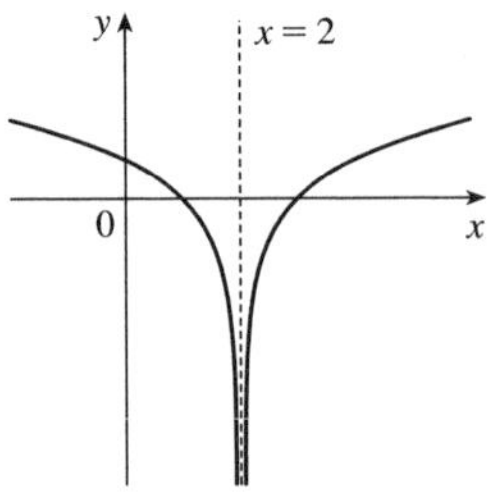

15. $f(x) = \begin{cases} e^x & \text{if } x < 0 \\ x^2 & \text{if } x \geq 0 \end{cases}$

The left-hand limit of f at $a = 0$ is $\lim_{x \to 0^-} f(x) = \lim_{x \to 0^-} e^x = 1$. The right-hand limit of f at $a = 0$ is $\lim_{x \to 0^+} f(x) = \lim_{x \to 0^+} x^2 = 0$. Since these limits are not equal, $\lim_{x \to 0} f(x)$ does not exist and f is discontinuous at 0.

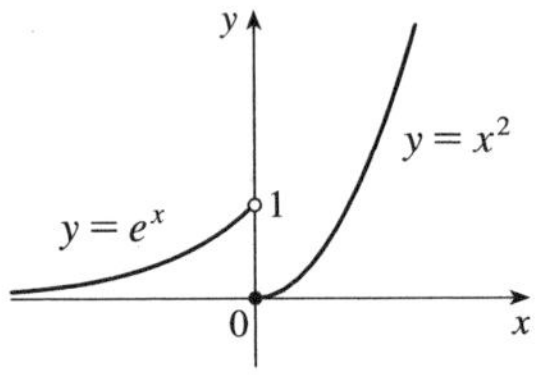

17. $F(x) = \dfrac{x}{x^2 + 5x + 6}$ is a rational function. So by Theorem 5 (or Theorem 7), F is continuous at every number in its domain, $\{x \mid x^2 + 5x + 6 \neq 0\} = \{x \mid (x+3)(x+2) \neq 0\} = \{x \mid x \neq -3, -2\}$ or $(-\infty, -3) \cup (-3, -2) \cup (-2, \infty)$.

19. By Theorem 5, the polynomial $5x$ is continuous on $(-\infty, \infty)$. By Theorems 9 and 7, $\sin 5x$ is continuous on $(-\infty, \infty)$. By Theorem 7, e^x is continuous on $(-\infty, \infty)$. By part 4 of Theorem 4, the product of e^x and $\sin 5x$ is continuous at all numbers which are in both of their domains, that is, on $(-\infty, \infty)$.

21. By Theorem 5, the polynomial $t^4 - 1$ is continuous on $(-\infty, \infty)$. By Theorem 7, $\ln x$ is continuous on its domain, $(0, \infty)$. By Theorem 9, $\ln(t^4 - 1)$ is continuous on its domain, which is $\{t \mid t^4 - 1 > 0\} = \{t \mid t^4 > 1\} = \{t \mid |t| > 1\} = (-\infty, -1) \cup (1, \infty)$.

23. The function $y = \dfrac{1}{1 + e^{1/x}}$ is discontinuous at $x = 0$ because the left- and right-hand limits at $x = 0$ are different.

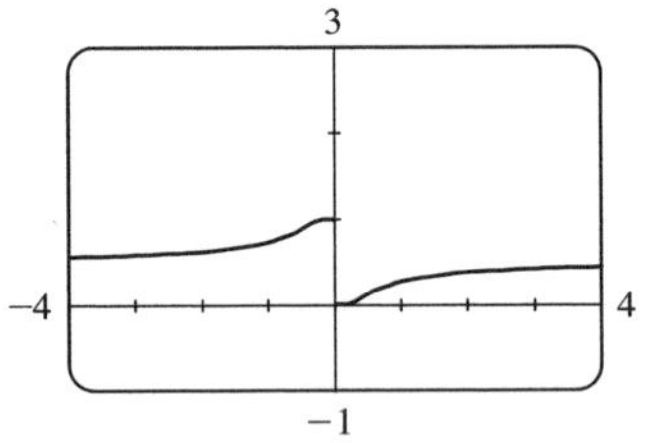

25. Because we are dealing with root functions, $5 + \sqrt{x}$ is continuous on $[0, \infty)$, $\sqrt{x + 5}$ is continuous on $[-5, \infty)$, so the quotient $f(x) = \dfrac{5 + \sqrt{x}}{\sqrt{5 + x}}$ is continuous on $[0, \infty)$. Since f is continuous at $x = 4$, $\lim_{x \to 4} f(x) = f(4) = \frac{7}{3}$.

27. Because $x^2 - x$ is continuous on $\mathbb{R}$, the composite function $f(x) = e^{x^2 - x}$ is continuous on $\mathbb{R}$, so $\lim_{x \to 1} f(x) = f(1) = e^{1 - 1} = e^0 = 1$.

29. $f(x) = \begin{cases} x^2 & \text{if } x < 1 \\ \sqrt{x} & \text{if } x \geq 1 \end{cases}$

By Theorem 5, since $f(x)$ equals the polynomial x^2 on $(-\infty, 1)$, f is continuous on $(-\infty, 1)$. By Theorem 7, since $f(x)$ equals the root function $\sqrt{x}$ on $(1, \infty)$, f is continuous on $(1, \infty)$. At $x = 1$, $\lim_{x \to 1^-} f(x) = \lim_{x \to 1^-} x^2 = 1$ and $\lim_{x \to 1^+} f(x) = \lim_{x \to 1^+} \sqrt{x} = 1$. Thus, $\lim_{x \to 1} f(x)$ exists and equals 1. Also, $f(1) = \sqrt{1} = 1$. Thus, f is continuous at $x = 1$. We conclude that f is continuous on $(-\infty, \infty)$.

31. $f(x) = \begin{cases} x+2 & \text{if } x<0 \\ e^x & \text{if } 0 \le x \le 1 \\ 2-x & \text{if } x>1 \end{cases}$

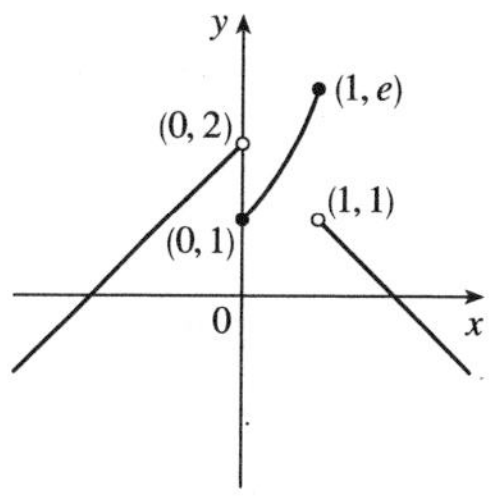

f is continuous on $(-\infty, 0)$ and $(1, \infty)$ since on each of these intervals it is a polynomial; it is continuous on $(0, 1)$ since it is an exponential.

Now $\lim\limits_{x\to 0^-} f(x) = \lim\limits_{x\to 0^-} (x+2) = 2$ and $\lim\limits_{x\to 0^+} f(x) = \lim\limits_{x\to 0^+} e^x = 1$, so f is discontinuous at 0. Since $f(0) = 1$, f is continuous from the right at 0. Also $\lim\limits_{x\to 1^-} f(x) = \lim\limits_{x\to 1^-} e^x = e$ and $\lim\limits_{x\to 1^+} f(x) = \lim\limits_{x\to 1^+} (2-x) = 1$, so f is discontinuous at 1. Since $f(1) = e$, f is continuous from the left at 1.

33. $f(x) = \begin{cases} cx^2+2x & \text{if } x<2 \\ x^3-cx & \text{if } x \ge 2 \end{cases}$

f is continuous on $(-\infty, 2)$ and $(2, \infty)$. Now $\lim\limits_{x\to 2^-} f(x) = \lim\limits_{x\to 2^-} \left(cx^2+2x\right) = 4c+4$ and $\lim\limits_{x\to 2^+} f(x) = \lim\limits_{x\to 2^+} \left(x^3-cx\right) = 8-2c$. So f is continuous $\Leftrightarrow$ $4c+4 = 8-2c$ $\Leftrightarrow$ $6c = 4$ $\Leftrightarrow$ $c = \frac{2}{3}$. Thus, for f to be continuous on $(-\infty, \infty)$, $c = \frac{2}{3}$.

35. $f(x) = x^3 - x^2 + x$ is continuous on the interval $[2, 3]$, $f(2) = 6$, and $f(3) = 21$. Since $6 < 10 < 21$, there is a number c in $(2, 3)$ such that $f(c) = 10$ by the Intermediate Value Theorem.

37. $f(x) = x^4 + x - 3$ is continuous on the interval $[1, 2]$, $f(1) = -1$, and $f(2) = 15$. Since $-1 < 0 < 15$, there is a number c in $(1, 2)$ such that $f(c) = 0$ by the Intermediate Value Theorem. Thus, there is a root of the equation $x^4 + x - 3 = 0$ in the interval $(1, 2)$.

39. $f(x) = \cos x - x$ is continuous on the interval $[0, 1]$, $f(0) = 1$, and $f(1) = \cos 1 - 1 \approx -0.46$. Since $-0.46 < 0 < 1$, there is a number c in $(0, 1)$ such that $f(c) = 0$ by the Intermediate Value Theorem. Thus, there is a root of the equation $\cos x - x = 0$, or $\cos x = x$, in the interval $(0, 1)$.

41. (a) $f(x) = \cos x - x^3$ is continuous on the interval $[0, 1]$, $f(0) = 1 > 0$, and $f(1) = \cos 1 - 1 \approx -0.46 < 0$. Since $1 > 0 > -0.46$, there is a number c in $(0, 1)$ such that $f(c) = 0$ by the Intermediate Value Theorem. Thus, there is a root of the equation $\cos x - x^3 = 0$, or $\cos x = x^3$, in the interval $(0, 1)$.

(b) $f(0.86) \approx 0.016 > 0$ and $f(0.87) \approx -0.014 < 0$, so there is a root between 0.86 and 0.87, that is, in the interval $(0.86, 0.87)$.

43. (a) Let $f(x) = 100e^{-x/100} - 0.01x^2$. Then $f(0) = 100 > 0$ and $f(100) = 100e^{-1} - 100 \approx -63.2 < 0$. So by the Intermediate Value Theorem, there is a number c in $(0, 100)$ such that $f(c) = 0$. This implies that $100e^{-c/100} = 0.01c^2$.

(b) Using the intersect feature of the graphing device, we find that the root of the equation is $x = 70.347$, correct to three decimal places.

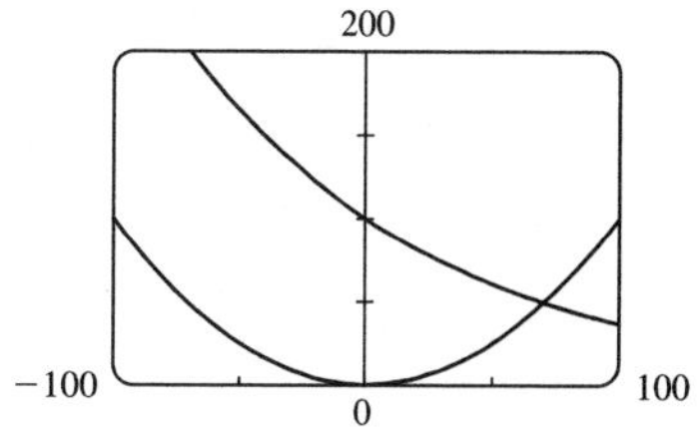

45. $\lim\limits_{h\to 0}\sin(a+h) = \lim\limits_{h\to 0}(\sin a\cos h + \cos a\sin h) = \lim\limits_{h\to 0}(\sin a\cos h) + \lim\limits_{h\to 0}(\cos a\sin h)$

$$= \left(\lim_{h\to 0}\sin a\right)\left(\lim_{h\to 0}\cos h\right) + \left(\lim_{h\to 0}\cos a\right)\left(\lim_{h\to 0}\sin h\right) = (\sin a)(1) + (\cos a)(0) = \sin a$$

47. If there is such a number, it satisfies the equation $x^3+1=x \;\Leftrightarrow\; x^3-x+1=0$. Let the left-hand side of this equation be called $f(x)$. Now $f(-2)=-5<0$, and $f(-1)=1>0$. Note also that $f(x)$ is a polynomial, and thus continuous. So by the Intermediate Value Theorem, there is a number c between -2 and -1 such that $f(c)=0$, so that $c=c^3+1$.

49. Define $u(t)$ to be the monk's distance from the monastery, as a function of time, on the first day, and define $d(t)$ to be his distance from the monastery, as a function of time, on the second day. Let D be the distance from the monastery to the top of the mountain. From the given information we know that $u(0)=0$, $u(12)=D$, $d(0)=D$ and $d(12)=0$. Now consider the function $u-d$, which is clearly continuous. We calculate that $(u-d)(0)=-D$ and $(u-d)(12)=D$. So by the Intermediate Value Theorem, there must be some time t_0 between 0 and 12 such that $(u-d)(t_0)=0 \;\Leftrightarrow\; u(t_0)=d(t_0)$. So at time t_0 after 7:00 A.M., the monk will be at the same place on both days.

2.5 Limits Involving Infinity

1. (a) As x approaches 2 (from the right or the left), the values of $f(x)$ become large.
(b) As x approaches 1 from the right, the values of $f(x)$ become large negative.
(c) As x becomes large, the values of $f(x)$ approach 5.
(d) As x becomes large negative, the values of $f(x)$ approach 3.

3. (a) $\lim\limits_{x\to 2} f(x)=\infty$ (b) $\lim\limits_{x\to -1^-} f(x)=\infty$ (c) $\lim\limits_{x\to -1^+} f(x)=-\infty$

(d) $\lim\limits_{x\to\infty} f(x)=1$ (e) $\lim\limits_{x\to -\infty} f(x)=2$ (f) Vertical: $x=-1$, $x=2$; Horizontal: $y=1$, $y=2$

5. $f(0)=0,\quad f(1)=1,\quad \lim\limits_{x\to\infty} f(x)=0,$

f is odd

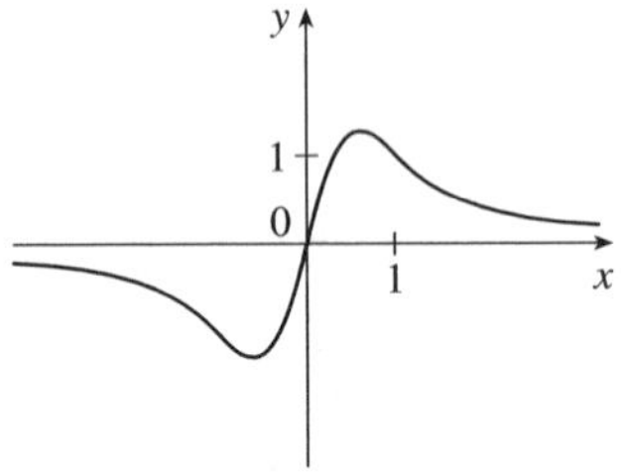

7. $\lim\limits_{x\to 2} f(x)=-\infty,\quad \lim\limits_{x\to\infty} f(x)=\infty,\quad \lim\limits_{x\to -\infty} f(x)=0,$

$\lim\limits_{x\to 0^+} f(x)=\infty,\quad \lim\limits_{x\to 0^-} f(x)=-\infty$

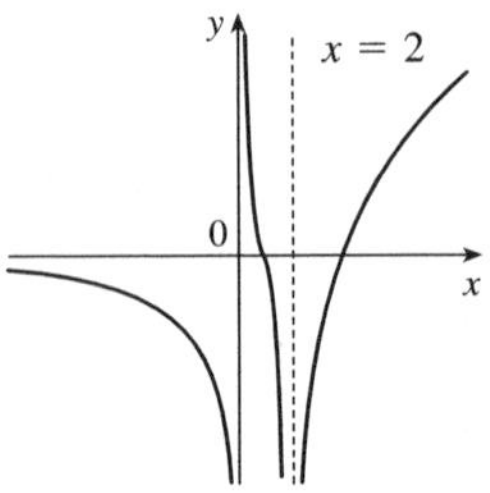

9. $f(0)=3,\quad \lim\limits_{x\to 0^-} f(x)=4,\quad \lim\limits_{x\to 0^+} f(x)=2,\quad \lim\limits_{x\to -\infty} f(x)=-\infty,$

$\lim\limits_{x\to 4^-} f(x)=-\infty,\quad \lim\limits_{x\to 4^+} f(x)=\infty,\quad \lim\limits_{x\to\infty} f(x)=3$

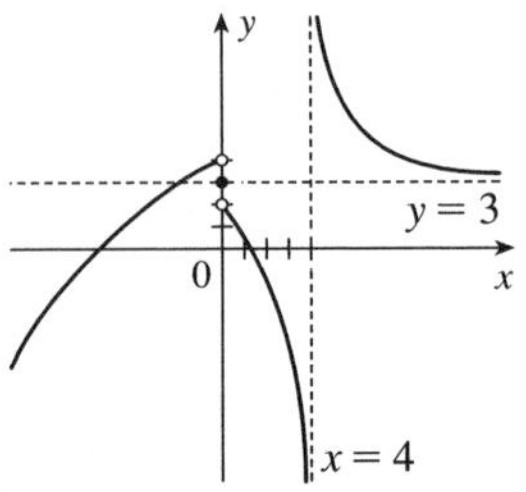

11. If $f(x) = x^2/2^x$, then a calculator gives $f(0) = 0$, $f(1) = 0.5$, $f(2) = 1$, $f(3) = 1.125$, $f(4) = 1$, $f(5) = 0.78125$, $f(6) = 0.5625$, $f(7) = 0.382\,812\,5$, $f(8) = 0.25$, $f(9) = 0.158\,203\,125$, $f(10) = 0.097\,656\,25$, $f(20) \approx 0.000\,381\,47$, $f(50) \approx 2.2204 \times 10^{-12}$, $f(100) \approx 7.8886 \times 10^{-27}$.

It appears that $\lim\limits_{x\to\infty} \left(x^2/2^x\right) = 0$.

13. Vertical: $x \approx -1.62$, $x \approx 0.62$, $x = 1$;

Horizontal: $y = 1$

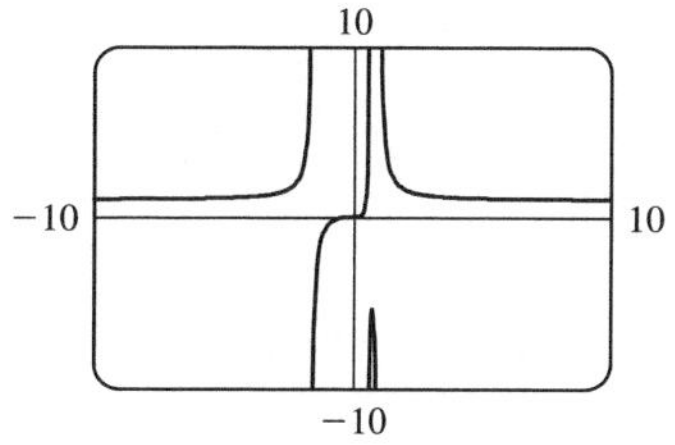

15. $\lim\limits_{x\to-3^+} \dfrac{x+2}{x+3} = -\infty$ since the numerator is negative and the denominator approaches 0 from the positive side as $x \to -3^+$.

17. $\lim\limits_{x\to1} \dfrac{2-x}{(x-1)^2} = \infty$ since the numerator is positive and the denominator approaches 0 through positive values as $x \to 1$.

19. $\lim\limits_{x\to(-\pi/2)^-} \sec x = \lim\limits_{x\to(-\pi/2)^-} (1/\cos x) = -\infty$ since $\cos x \to 0^-$ as $x \to (-\pi/2)^-$.

21. Divide both the numerator and denominator by x^3 (the highest power of x that occurs in the denominator).

$$\lim_{x\to\infty} \frac{x^3+5x}{2x^3-x^2+4} = \lim_{x\to\infty} \frac{\dfrac{x^3+5x}{x^3}}{\dfrac{2x^3-x^2+4}{x^3}} = \lim_{x\to\infty} \frac{1+\dfrac{5}{x^2}}{2-\dfrac{1}{x}+\dfrac{4}{x^3}} = \frac{\lim\limits_{x\to\infty}\left(1+\dfrac{5}{x^2}\right)}{\lim\limits_{x\to\infty}\left(2-\dfrac{1}{x}+\dfrac{4}{x^3}\right)}$$

$$= \frac{\lim\limits_{x\to\infty} 1 + 5\lim\limits_{x\to\infty} \dfrac{1}{x^2}}{\lim\limits_{x\to\infty} 2 - \lim\limits_{x\to\infty} \dfrac{1}{x} + 4\lim\limits_{x\to\infty} \dfrac{1}{x^3}} = \frac{1+5(0)}{2-0+4(0)} = \frac{1}{2}$$

23. First, multiply the factors in the denominator. Then divide both the numerator and denominator by u^4.

$$\lim_{u\to\infty} \frac{4u^4+5}{(u^2-2)(2u^2-1)} = \lim_{u\to\infty} \frac{4u^4+5}{2u^4-5u^2+2} = \lim_{u\to\infty} \frac{\dfrac{4u^4+5}{u^4}}{\dfrac{2u^4-5u^2+2}{u^4}} = \lim_{u\to\infty} \frac{4+\dfrac{5}{u^4}}{2-\dfrac{5}{u^2}+\dfrac{2}{u^4}}$$

$$= \frac{\lim\limits_{u\to\infty}\left(4+\dfrac{5}{u^4}\right)}{\lim\limits_{u\to\infty}\left(2-\dfrac{5}{u^2}+\dfrac{2}{u^4}\right)} = \frac{\lim\limits_{u\to\infty} 4 + 5\lim\limits_{u\to\infty} \dfrac{1}{u^4}}{\lim\limits_{u\to\infty} 2 - 5\lim\limits_{u\to\infty} \dfrac{1}{u^2} + 2\lim\limits_{u\to\infty} \dfrac{1}{u^4}} = \frac{4+5(0)}{2-5(0)+2(0)} = \frac{4}{2} = 2$$

25. $$\lim_{x\to\infty}\left(\sqrt{9x^2+x}-3x\right) = \lim_{x\to\infty} \frac{\left(\sqrt{9x^2+x}-3x\right)\left(\sqrt{9x^2+x}+3x\right)}{\sqrt{9x^2+x}+3x} = \lim_{x\to\infty} \frac{\left(\sqrt{9x^2+x}\right)^2-(3x)^2}{\sqrt{9x^2+x}+3x}$$

$$= \lim_{x\to\infty} \frac{(9x^2+x)-9x^2}{\sqrt{9x^2+x}+3x} = \lim_{x\to\infty} \frac{x}{\sqrt{9x^2+x}+3x} \cdot \frac{1/x}{1/x}$$

$$= \lim_{x\to\infty} \frac{x/x}{\sqrt{9x^2/x^2+x/x^2}+3x/x} = \lim_{x\to\infty} \frac{1}{\sqrt{9+1/x}+3} = \frac{1}{\sqrt{9}+3} = \frac{1}{3+3} = \frac{1}{6}$$

27. $\lim_{x\to\infty} \cos x$ does not exist because as x increases $\cos x$ does not approach any one value, but oscillates between 1 and -1.

29. Since $-1 \le \cos x \le 1$ and $e^{-2x} > 0$, we have $-e^{-2x} \le e^{-2x}\cos x \le e^{-2x}$. We know that $\lim_{x\to\infty}(-e^{-2x}) = 0$ and $\lim_{x\to\infty}(e^{-2x}) = 0$, so by the Squeeze Theorem, $\lim_{x\to\infty}(e^{-2x}\cos x) = 0$.

31. $\lim_{x\to\infty} \dfrac{x^7-1}{x^6+1} = \lim_{x\to\infty} \dfrac{1-1/x^7}{(1/x)+(1/x^7)} = \infty$ since $1 - \dfrac{1}{x^7} \to 1$ while $\dfrac{1}{x} + \dfrac{1}{x^7} \to 0^+$ as $x \to \infty$.

Or: Divide numerator and denominator by x^6 instead of x^7.

33. $\lim_{x\to-\infty}(x^3-5x^2) = -\infty$ since $x^3 \to -\infty$ and $-5x^2 \to -\infty$ as $x \to -\infty$.

Or: $\lim_{x\to-\infty}(x^3-5x^2) = \lim_{x\to-\infty} x^2(x-5) = -\infty$ since $x^2 \to \infty$ and $x-5 \to -\infty$.

35. $$\lim_{x\to\infty} \frac{2x^2+x-1}{x^2+x-2} = \lim_{x\to\infty} \frac{\dfrac{2x^2+x-1}{x^2}}{\dfrac{x^2+x-2}{x^2}} = \lim_{x\to\infty} \frac{2+\dfrac{1}{x}-\dfrac{1}{x^2}}{1+\dfrac{1}{x}-\dfrac{2}{x^2}} = \frac{\lim_{x\to\infty}\left(2+\dfrac{1}{x}-\dfrac{1}{x^2}\right)}{\lim_{x\to\infty}\left(1+\dfrac{1}{x}-\dfrac{2}{x^2}\right)}$$

$$= \frac{\lim_{x\to\infty} 2 + \lim_{x\to\infty}\dfrac{1}{x} - \lim_{x\to\infty}\dfrac{1}{x^2}}{\lim_{x\to\infty} 1 + \lim_{x\to\infty}\dfrac{1}{x} - 2\lim_{x\to\infty}\dfrac{1}{x^2}} = \frac{2+0-0}{1+0-2(0)} = 2,$$ so $y = 2$ is a horizontal asymptote.

$y = f(x) = \dfrac{2x^2+x-1}{x^2+x-2} = \dfrac{(2x-1)(x+1)}{(x+2)(x-1)}$, so

$\lim_{x\to-2^-} f(x) = \infty$, $\lim_{x\to-2^+} f(x) = -\infty$, $\lim_{x\to1^-} f(x) = -\infty$, and $\lim_{x\to1^+} f(x) = \infty$. Thus, $x = -2$ and $x = 1$ are vertical asymptotes.

The graph confirms our work.

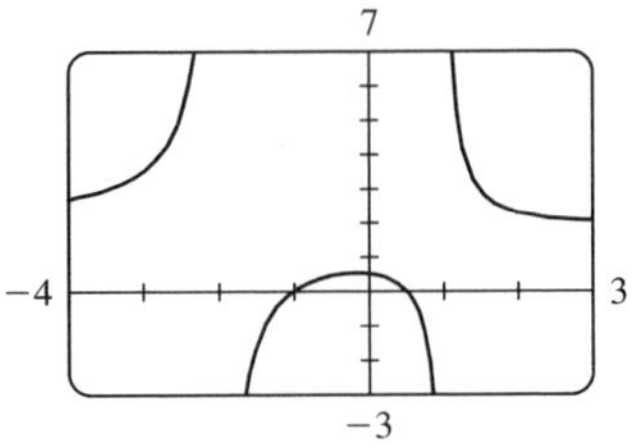

37. (a)

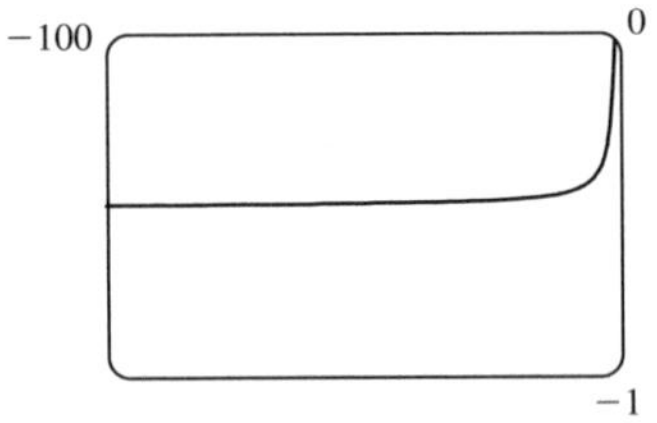

From the graph of $f(x) = \sqrt{x^2+x+1}+x$, we estimate the value of $\lim_{x\to-\infty} f(x)$ to be -0.5.

(b)

x	$f(x)$
$-10\,000$	$-0.499\,962\,5$
$-100\,000$	$-0.499\,996\,2$
$-1\,000\,000$	$-0.499\,999\,6$

From the table, we estimate the limit to be -0.5.

(c) $$\lim_{x\to-\infty}\left(\sqrt{x^2+x+1}+x\right) = \lim_{x\to-\infty}\left(\sqrt{x^2+x+1}+x\right)\left[\frac{\sqrt{x^2+x+1}-x}{\sqrt{x^2+x+1}-x}\right] = \lim_{x\to-\infty}\frac{(x^2+x+1)-x^2}{\sqrt{x^2+x+1}-x}$$

$$= \lim_{x\to-\infty}\frac{(x+1)(1/x)}{\left(\sqrt{x^2+x+1}-x\right)(1/x)} = \lim_{x\to-\infty}\frac{1+(1/x)}{-\sqrt{1+(1/x)+(1/x^2)}-1}$$

$$= \frac{1+0}{-\sqrt{1+0+0}-1} = -\frac{1}{2}$$

Note that for $x < 0$, we have $\sqrt{x^2} = |x| = -x$, so when we divide the radical by x, with $x < 0$, we get $\dfrac{1}{x}\sqrt{x^2+x+1} = -\dfrac{1}{\sqrt{x^2}}\sqrt{x^2+x+1} = -\sqrt{1+(1/x)+(1/x^2)}$.

39. From the graph, it appears $y = 1$ is a horizontal asymptote.

$$\lim_{x\to\infty} \frac{3x^3 + 500x^2}{x^3 + 500x^2 + 100x + 2000} = \lim_{x\to\infty} \frac{\dfrac{3x^3 + 500x^2}{x^3}}{\dfrac{x^3 + 500x^2 + 100x + 2000}{x^3}}$$

$$= \lim_{x\to\infty} \frac{3 + (500/x)}{1 + (500/x) + (100/x^2) + (2000/x^3)} = \frac{3 + 0}{1 + 0 + 0 + 0} = 3,$$

so $y = 3$ is a horizontal asymptote. The discrepancy can be explained by the choice in the viewing window. Try $[-100\,000, 100\,000]$ by $[-1, 4]$ to get a graph that lends credibility to our calculation that $y = 3$ is a horizontal asymptote.

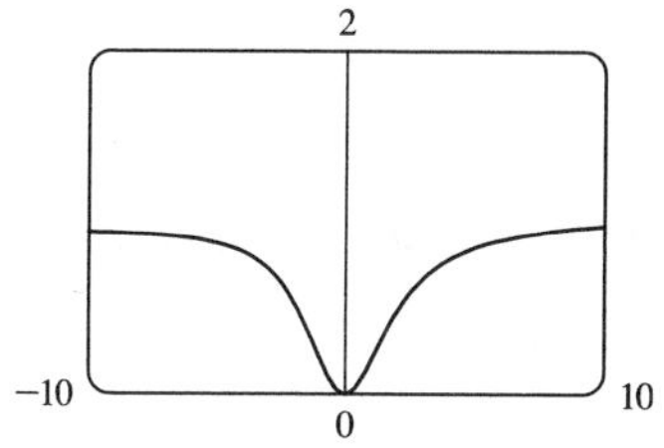

41. Let's look for a rational function.

(1) $\lim_{x\to\pm\infty} f(x) = 0 \;\Rightarrow\;$ degree of numerator < degree of denominator

(2) $\lim_{x\to 0} f(x) = -\infty \;\Rightarrow\;$ there is a factor of x^2 in the denominator (not just x, since that would produce a sign change at $x = 0$), and the function is negative near $x = 0$.

(3) $\lim_{x\to 3^-} f(x) = \infty$ and $\lim_{x\to 3^+} f(x) = -\infty \;\Rightarrow\;$ vertical asymptote at $x = 3$; there is a factor of $(x - 3)$ in the denominator.

(4) $f(2) = 0 \;\Rightarrow\;$ 2 is an x-intercept; there is at least one factor of $(x - 2)$ in the numerator.

Combining all of this information and putting in a negative sign to give us the desired left- and right-hand limits gives us $f(x) = \dfrac{2 - x}{x^2(x - 3)}$ as one possibility.

43. Divide the numerator and the denominator by the highest power of x in $Q(x)$.

(a) If $\deg P < \deg Q$, then the numerator $\to 0$ but the denominator doesn't. So $\lim_{x\to\infty} [P(x)/Q(x)] = 0$.

(b) If $\deg P > \deg Q$, then the numerator $\to \pm\infty$ but the denominator doesn't, so $\lim_{x\to\infty} [P(x)/Q(x)] = \pm\infty$ (depending on the ratio of the leading coefficients of P and Q).

45. $\displaystyle\lim_{x\to\infty} \frac{5\sqrt{x}}{\sqrt{x - 1}} \cdot \frac{1/\sqrt{x}}{1/\sqrt{x}} = \lim_{x\to\infty} \frac{5}{\sqrt{1 - (1/x)}} = \frac{5}{\sqrt{1 - 0}} = 5$ and

$\displaystyle\lim_{x\to\infty} \frac{10e^x - 21}{2e^x} \cdot \frac{1/e^x}{1/e^x} = \lim_{x\to\infty} \frac{10 - (21/e^x)}{2} = \frac{10 - 0}{2} = 5$. Since $\dfrac{5\sqrt{x}}{\sqrt{x - 1}} < f(x) < \dfrac{10e^x - 21}{2e^x}$, we have $\lim_{x\to\infty} f(x) = 5$ by the Squeeze Theorem.

47. (a) After t minutes, $25t$ litres of brine with 30 g of salt per litre has been pumped into the tank, so it contains $(5000 + 25t)$ litres of water and $25t \cdot 30 = 750t$ grams of salt. Therefore, the salt concentration at time t will be $C(t) = \dfrac{750t}{5000 + 25t} = \dfrac{30t}{200 + t}\,\dfrac{\text{g}}{\text{L}}$.

(b) $\displaystyle\lim_{t\to\infty} C(t) = \lim_{t\to\infty} \frac{30t}{200 + t} = \lim_{t\to\infty} \frac{30t/t}{200/t + t/t} = \frac{30}{0 + 1} = 30$. So the salt concentration approaches that of the brine being pumped into the tank.

49. (a) If $t = -x/10$, then $x = -10t$ and as $x \to \infty$, $t \to -\infty$.

Thus, $\lim\limits_{x\to\infty} e^{-x/10} = \lim\limits_{t\to-\infty} e^t = 0$ by Equation 8.

(b) $y = e^{-x/10}$ and $y = 0.1$ intersect at $x_1 \approx 23.03$.

If $x > x_1$, then $e^{-x/10} < 0.1$.

(c) $e^{-x/10} < 0.1 \;\;\Rightarrow\;\; -x/10 < \ln 0.1 \;\;\Rightarrow$

$x > -10 \ln \frac{1}{10} = -10 \ln 10^{-1} = 10 \ln 10$

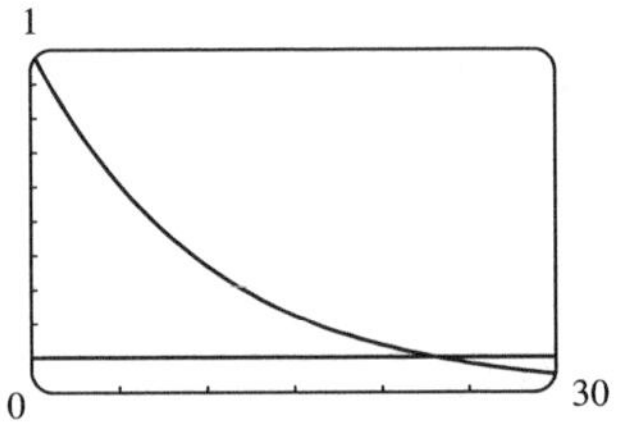

2.6 Tangents, Velocities, and Other Rates of Change

1. (a) This is just the slope of the line through two points: $m_{PQ} = \dfrac{\Delta y}{\Delta x} = \dfrac{f(x) - f(3)}{x - 3}$.

(b) This is the limit of the slope of the secant line PQ as Q approaches P: $m = \lim\limits_{x\to3} \dfrac{f(x) - f(3)}{x - 3}$.

3. The slope at D is the largest positive slope, followed by the positive slope at E. The slope at C is zero. The slope at B is steeper than at A (both are negative). In decreasing order, we have the slopes at: D, E, C, A, and B.

5. (a) (i) Using Definition 1,

$$m = \lim_{x\to a} \frac{f(x) - f(a)}{x - a} \lim_{x\to-3} \frac{f(x) - f(-3)}{x - (-3)} = \lim_{x\to-3} \frac{(x^2 + 2x) - (3)}{x - (-3)} = \lim_{x\to-3} \frac{(x + 3)(x - 1)}{x + 3}$$

$$= \lim_{x\to-3} (x - 1) = -4$$

(ii) Using Equation 2,

$$m = \lim_{h\to0} \frac{f(a + h) - f(a)}{h} = \lim_{h\to0} \frac{f(-3 + h) - f(-3)}{h} = \lim_{h\to0} \frac{[(-3 + h)^2 + 2(-3 + h)] - (3)}{h}$$

$$= \lim_{h\to0} \frac{9 - 6h + h^2 - 6 + 2h - 3}{h} = \lim_{h\to0} \frac{h(h - 4)}{h} = \lim_{h\to0} (h - 4) = -4$$

(b) Using the point-slope form of the equation of a line, an equation of the tangent line is $y - 3 = -4(x + 3)$. Solving for y gives us $y = -4x - 9$, which is the slope-intercept form of the equation of the tangent line.

(c)

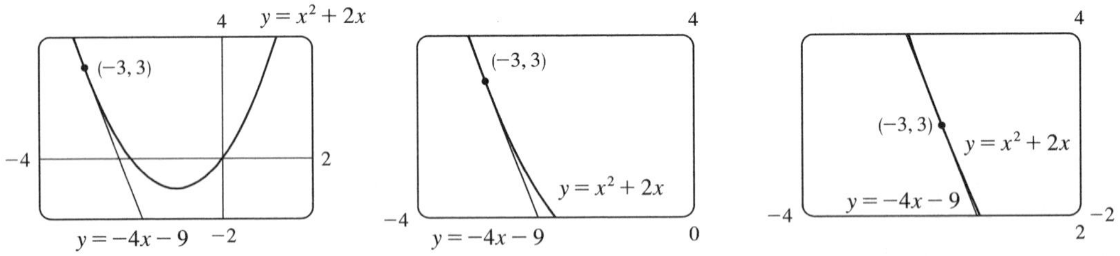

7. Using (1) with $f(x) = \dfrac{x - 1}{x - 2}$ and $P(3, 2)$,

$$m = \lim_{x\to a} \frac{f(x) - f(a)}{x - a} = \lim_{x\to3} \frac{\frac{x - 1}{x - 2} - 2}{x - 3} = \lim_{x\to3} \frac{\frac{x - 1 - 2(x - 2)}{x - 2}}{x - 3} = \lim_{x\to3} \frac{3 - x}{(x - 2)(x - 3)} = \lim_{x\to3} \frac{-1}{x - 2} = \frac{-1}{1} = -1.$$

Tangent line: $y - 2 = -1(x - 3) \;\;\Leftrightarrow\;\; y - 2 = -x + 3 \;\;\Leftrightarrow\;\; y = -x + 5$

9. Using (1), $m = \lim\limits_{x\to1} \dfrac{\sqrt{x} - \sqrt{1}}{x - 1} = \lim\limits_{x\to1} \dfrac{(\sqrt{x} - 1)(\sqrt{x} + 1)}{(x - 1)(\sqrt{x} + 1)} = \lim\limits_{x\to1} \dfrac{x - 1}{(x - 1)(\sqrt{x} + 1)} = \lim\limits_{x\to1} \dfrac{1}{\sqrt{x} + 1} = \dfrac{1}{2}$.

Tangent line: $y - 1 = \frac{1}{2}(x - 1) \;\;\Leftrightarrow\;\; y = \frac{1}{2}x + \frac{1}{2}$.

11. (a) Using (2) with $y = f(x) = 3 + 4x^2 - 2x^3$,

$$m = \lim_{h\to 0} \frac{f(a+h) - f(a)}{h} = \lim_{h\to 0} \frac{3 + 4(a+h)^2 - 2(a+h)^3 - (3 + 4a^2 - 2a^3)}{h}$$

$$= \lim_{h\to 0} \frac{3 + 4(a^2 + 2ah + h^2) - 2(a^3 + 3a^2h + 3ah^2 + h^3) - 3 - 4a^2 + 2a^3}{h}$$

$$= \lim_{h\to 0} \frac{3 + 4a^2 + 8ah + 4h^2 - 2a^3 - 6a^2h - 6ah^2 - 2h^3 - 3 - 4a^2 + 2a^3}{h}$$

$$= \lim_{h\to 0} \frac{8ah + 4h^2 - 6a^2h - 6ah^2 - 2h^3}{h} = \lim_{h\to 0} \frac{h(8a + 4h - 6a^2 - 6ah - 2h^2)}{h}$$

$$= \lim_{h\to 0}(8a + 4h - 6a^2 - 6ah - 2h^2) = 8a - 6a^2$$

(b) At $(1, 5)$: $m = 8(1) - 6(1)^2 = 2$, so an equation of the tangent line is $y - 5 = 2(x - 1) \Leftrightarrow y = 2x + 3$.

At $(2, 3)$: $m = 8(2) - 6(2)^2 = -8$, so an equation of the tangent line is $y - 3 = -8(x - 2) \Leftrightarrow y = -8x + 19$.

(c)

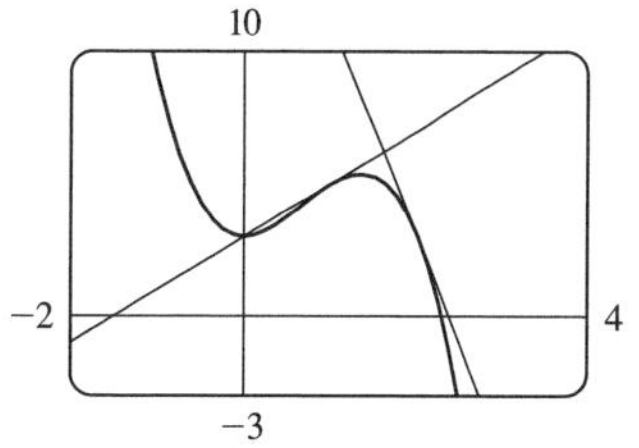

13. (a) The particle is moving to the right when s is increasing; that is, on the intervals $(0, 1)$ and $(4, 6)$. The particle is moving to the left when s is decreasing; that is, on the interval $(2, 3)$. The particle is standing still when s is constant; that is, on the intervals $(1, 2)$ and $(3, 4)$.

(b) The velocity of the particle is equal to the slope of the tangent line of the graph. Note that there is no slope at the corner points on the graph. On the interval $(0, 1)$, the slope is $\dfrac{3-0}{1-0} = 3$. On the interval $(2, 3)$, the slope is $\dfrac{1-3}{3-2} = -2$. On the interval $(4, 6)$, the slope is $\dfrac{3-1}{6-4} = 1$.

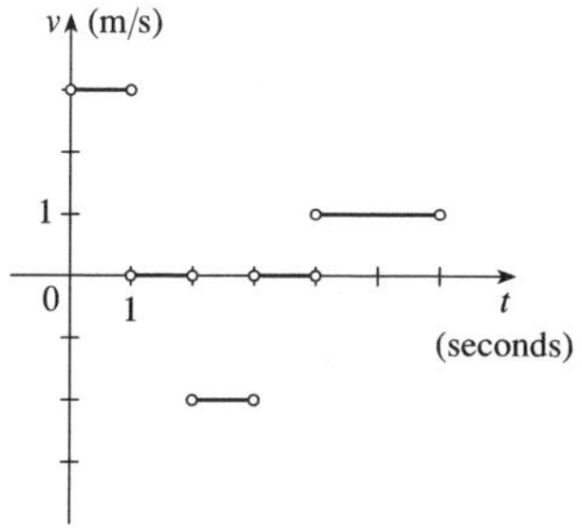

15. (a) Since the slope of the tangent at $t = 0$ is 0, the car's initial velocity was 0.

(b) The slope of the tangent is greater at C than at B, so the car was going faster at C.

(c) Near A, the tangent lines are becoming steeper as x increases, so the velocity was increasing, so the car was speeding up. Near B, the tangent lines are becoming less steep, so the car was slowing down. The steepest tangent near C is the one at C, so at C the car had just finished speeding up, and was about to start slowing down.

(d) Between D and E, the slope of the tangent is 0, so the car did not move during that time.

17. Let $s(t) = 10t - 4.9t^2$.

$$v(2) = \lim_{h\to 0} \frac{s(2+h) - s(2)}{h} = \lim_{h\to 0} \frac{[10(2+h) - 4.9(2+h)^2] - 0.4}{h} = \lim_{h\to 0} \frac{20 + 10h - 4.9(4 + 4h + h^2) - 0.4}{h}$$

$$= \lim_{h\to 0} \frac{-4.9h^2 - 9.6h}{h} = \lim_{h\to 0}(-4.9h - 9.6) = -9.6$$

Thus, the instantaneous velocity when $t = 2$ is -9.6 m/s.

19. $v(a) = \lim_{h\to 0} \dfrac{s(a+h)-s(a)}{h} = \lim_{h\to 0} \dfrac{\dfrac{1}{(a+h)^2} - \dfrac{1}{a^2}}{h} = \lim_{h\to 0} \dfrac{\dfrac{a^2-(a+h)^2}{a^2(a+h)^2}}{h} = \lim_{h\to 0} \dfrac{a^2-(a^2+2ah+h^2)}{ha^2(a+h)^2}$

$= \lim_{h\to 0} \dfrac{-(2ah+h^2)}{ha^2(a+h)^2} = \lim_{h\to 0} \dfrac{-h(2a+h)}{ha^2(a+h)^2} = \lim_{h\to 0} \dfrac{-(2a+h)}{a^2(a+h)^2} = \dfrac{-2a}{a^2\cdot a^2} = \dfrac{-2}{a^3}$ m/s

So $v(1) = \dfrac{-2}{1^3} = -2$ m/s, $v(2) = \dfrac{-2}{2^3} = -\dfrac{1}{4}$ m/s, and $v(3) = \dfrac{-2}{3^3} = -\dfrac{2}{27}$ m/s.

21. The sketch shows the graph for a room temperature of 22° and a refrigerator temperature of 3°. The initial rate of change is greater in magnitude than the rate of change after an hour.

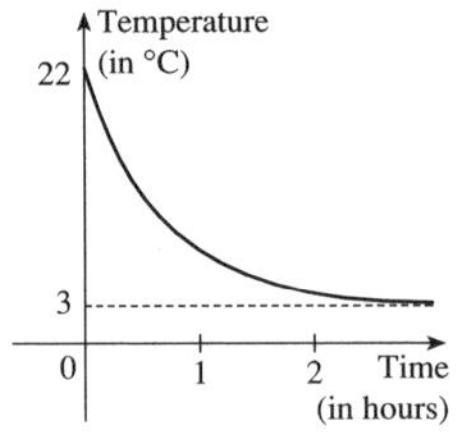

23. (a) (i) $[20, 23]$: $\dfrac{7.9-11.5}{23-20} = -1.2\,°\text{C/h}$

(ii) $[20, 22]$: $\dfrac{9.0-11.5}{22-20} = -1.25\,°\text{C/h}$

(iii) $[20, 21]$: $\dfrac{10.2-11.5}{21-20} = -1.3\,°\text{C/h}$

(b) In the figure, we estimate A to be $(18, 15.5)$ and B as $(23, 6)$. So the slope is $\dfrac{6-15.5}{23-18} = -1.9\,°\text{C/h}$ at 8:00 P.M.

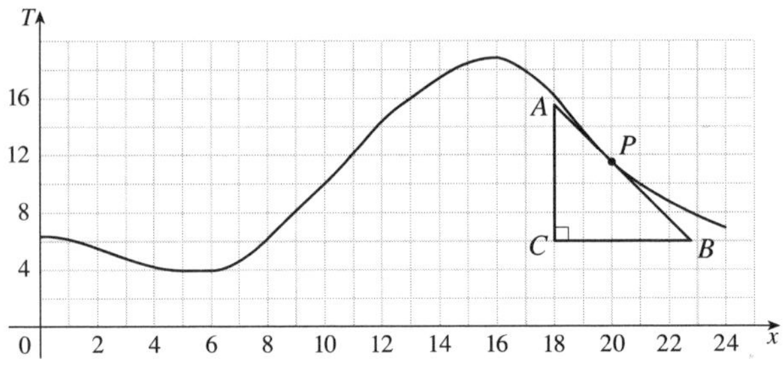

25. (a) (i) $[2000, 2002]$: $\dfrac{P(2002)-P(2000)}{2002-2000} = \dfrac{77-55}{2} = \dfrac{22}{2} = 11$ percent/year

(ii) $[2000, 2001]$: $\dfrac{P(2001)-P(2000)}{2001-2000} = \dfrac{68-55}{1} = 13$ percent/year

(iii) $[1999, 2000]$: $\dfrac{P(2000)-P(1999)}{2000-1999} = \dfrac{55-39}{1} = 16$ percent/year

(b) Using the values from (ii) and (iii), we have $\dfrac{13+16}{2} = 14.5$ percent/year.

(c) Estimating A as $(1999, 40)$ and B as $(2001, 70)$, the slope at 2000 is $\dfrac{70-40}{2001-1999} = \dfrac{30}{2} = 15$ percent/year.

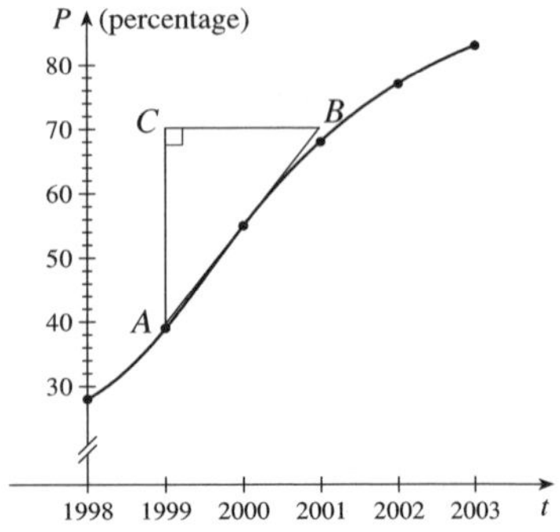

27. (a) (i) $\dfrac{\Delta C}{\Delta x} = \dfrac{C(105)-C(100)}{105-100} = \dfrac{6601.25-6500}{5} = \$20.25/\text{unit}.$

(ii) $\dfrac{\Delta C}{\Delta x} = \dfrac{C(101)-C(100)}{101-100} = \dfrac{6520.05-6500}{1} = \$20.05/\text{unit}.$

(b) $\dfrac{C(100+h)-C(100)}{h} = \dfrac{\left[5000+10(100+h)+0.05(100+h)^2\right]-6500}{h} = \dfrac{20h+0.05h^2}{h}$

$= 20 + 0.05h,\ h \neq 0$

So the instantaneous rate of change is $\lim\limits_{h\to 0} \dfrac{C(100+h)-C(100)}{h} = \lim\limits_{h\to 0}(20+0.05h) = \$20/\text{unit}$.

2.7 Derivatives

1. 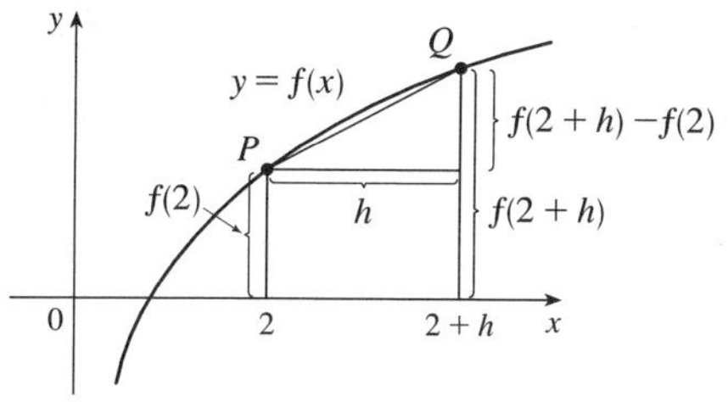

The line from $P(2, f(2))$ to $Q(2+h, f(2+h))$ is the line that has slope $\dfrac{f(2+h)-f(2)}{h}$.

3. $g'(0)$ is the only negative value. The slope at $x=4$ is smaller than the slope at $x=2$ and both are smaller than the slope at $x=-2$. Thus, $g'(0) < 0 < g'(4) < g'(2) < g'(-2)$.

5. We begin by drawing a curve through the origin with a slope of 3 to satisfy $f(0)=0$ and $f'(0)=3$. Since $f'(1)=0$, we will round off our figure so that there is a horizontal tangent directly over $x=1$. Lastly, we make sure that the curve has a slope of -1 as we pass over $x=2$. Two of the many possibilities are shown.

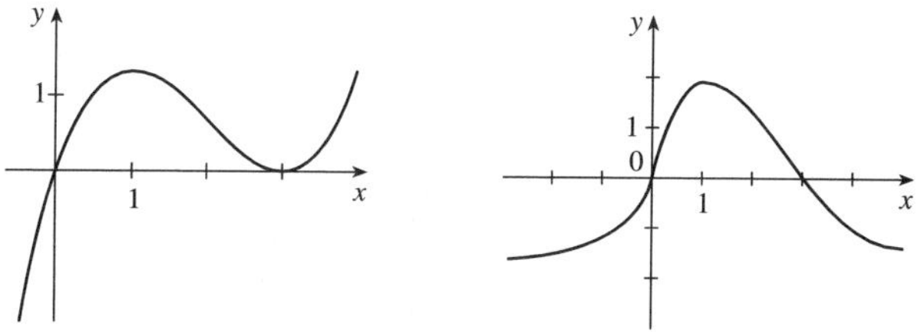

7. Using Definition 2 with $f(x) = 3x^2 - 5x$ and the point $(2, 2)$, we have

$$\begin{aligned} f'(2) &= \lim_{h\to 0} \frac{f(2+h)-f(2)}{h} = \lim_{h\to 0} \frac{\left[3(2+h)^2 - 5(2+h)\right]-2}{h} \\ &= \lim_{h\to 0} \frac{(12+12h+3h^2-10-5h)-2}{h} = \lim_{h\to 0} \frac{3h^2+7h}{h} = \lim_{h\to 0}(3h+7) = 7 \end{aligned}$$

So an equation of the tangent line at $(2,2)$ is $y-2=7(x-2)$ or $y=7x-12$.

9. (a) Using Definition 2 with $F(x) = 5x/(1+x^2)$ and the point $(2, 2)$, we have

$$\begin{aligned} F'(2) &= \lim_{h\to 0} \frac{F(2+h)-F(2)}{h} = \lim_{h\to 0} \frac{\dfrac{5(2+h)}{1+(2+h)^2} - 2}{h} \\ &= \lim_{h\to 0} \frac{\dfrac{5h+10}{h^2+4h+5} - 2}{h} = \lim_{h\to 0} \frac{\dfrac{5h+10-2(h^2+4h+5)}{h^2+4h+5}}{h} \\ &= \lim_{h\to 0} \frac{-2h^2-3h}{h(h^2+4h+5)} = \lim_{h\to 0} \frac{h(-2h-3)}{h(h^2+4h+5)} = \lim_{h\to 0} \frac{-2h-3}{h^2+4h+5} = \frac{-3}{5} \end{aligned}$$

So an equation of the tangent line at $(2,2)$ is $y-2=-\frac{3}{5}(x-2)$ or $y=-\frac{3}{5}x+\frac{16}{5}$.

(b) 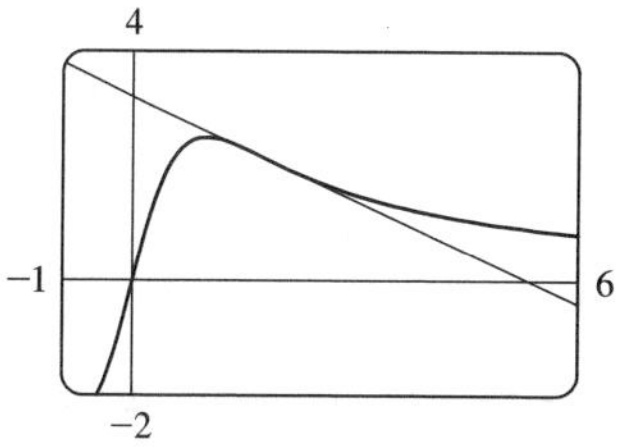

11. (a) $f'(1) = \lim_{h \to 0} \dfrac{f(1+h) - f(1)}{h} = \lim_{h \to 0} \dfrac{3^{1+h} - 3^1}{h}$.

So let $F(h) = \dfrac{3^{1+h} - 3}{h}$. We calculate:

h	$F(h)$	h	$F(h)$
0.1	3.484	-0.1	3.121
0.01	3.314	-0.01	3.278
0.001	3.298	-0.001	3.294
0.0001	3.296	-0.0001	3.296

We estimate that $f'(1) \approx 3.296$.

(b)

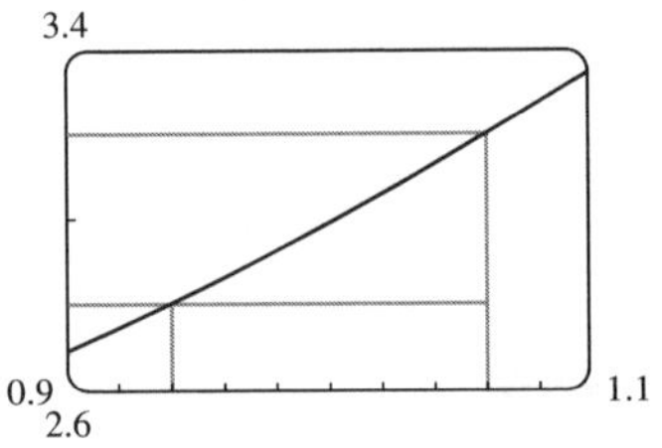

From the graph, we estimate that the slope of the tangent is about $\dfrac{3.2 - 2.8}{1.06 - 0.94} = \dfrac{0.4}{0.12} \approx 3.3$.

13. Use Definition 2 with $f(x) = 3 - 2x + 4x^2$.

$$f'(a) = \lim_{h \to 0} \frac{f(a+h) - f(a)}{h} = \lim_{h \to 0} \frac{[3 - 2(a+h) + 4(a+h)^2] - (3 - 2a + 4a^2)}{h}$$

$$= \lim_{h \to 0} \frac{(3 - 2a - 2h + 4a^2 + 8ah + 4h^2) - (3 - 2a + 4a^2)}{h}$$

$$= \lim_{h \to 0} \frac{-2h + 8ah + 4h^2}{h} = \lim_{h \to 0} \frac{h(-2 + 8a + 4h)}{h} = \lim_{h \to 0} (-2 + 8a + 4h) = -2 + 8a$$

15. Use Definition 2 with $f(t) = (2t + 1)/(t + 3)$.

$$f'(a) = \lim_{h \to 0} \frac{f(a+h) - f(a)}{h} = \lim_{h \to 0} \frac{\dfrac{2(a+h)+1}{(a+h)+3} - \dfrac{2a+1}{a+3}}{h} = \lim_{h \to 0} \frac{(2a + 2h + 1)(a+3) - (2a+1)(a+h+3)}{h(a+h+3)(a+3)}$$

$$= \lim_{h \to 0} \frac{(2a^2 + 6a + 2ah + 6h + a + 3) - (2a^2 + 2ah + 6a + a + h + 3)}{h(a+h+3)(a+3)}$$

$$= \lim_{h \to 0} \frac{5h}{h(a+h+3)(a+3)} = \lim_{h \to 0} \frac{5}{(a+h+3)(a+3)} = \frac{5}{(a+3)^2}$$

17. Use Definition 2 with $f(x) = 1/\sqrt{x+2}$.

$$f'(a) = \lim_{h \to 0} \frac{f(a+h) - f(a)}{h} = \lim_{h \to 0} \frac{\dfrac{1}{\sqrt{(a+h)+2}} - \dfrac{1}{\sqrt{a+2}}}{h} = \lim_{h \to 0} \frac{\dfrac{\sqrt{a+2} - \sqrt{a+h+2}}{\sqrt{a+h+2}\,\sqrt{a+2}}}{h}$$

$$= \lim_{h \to 0} \left[\frac{\sqrt{a+2} - \sqrt{a+h+2}}{h\sqrt{a+h+2}\,\sqrt{a+2}} \cdot \frac{\sqrt{a+2} + \sqrt{a+h+2}}{\sqrt{a+2} + \sqrt{a+h+2}} \right] = \lim_{h \to 0} \frac{(a+2) - (a+h+2)}{h\sqrt{a+h+2}\,\sqrt{a+2}\,\left(\sqrt{a+2} + \sqrt{a+h+2}\right)}$$

$$= \lim_{h \to 0} \frac{-h}{h\sqrt{a+h+2}\,\sqrt{a+2}\,\left(\sqrt{a+2} + \sqrt{a+h+2}\right)} = \lim_{h \to 0} \frac{-1}{\sqrt{a+h+2}\,\sqrt{a+2}\,\left(\sqrt{a+2} + \sqrt{a+h+2}\right)}$$

$$= \frac{-1}{\left(\sqrt{a+2}\right)^2 \left(2\sqrt{a+2}\right)} = -\frac{1}{2(a+2)^{3/2}}$$

Note that the answers to Exercises 19 – 24 are not unique.

19. By Definition 2, $\lim_{h \to 0} \dfrac{(1+h)^{10} - 1}{h} = f'(1)$, where $f(x) = x^{10}$ and $a = 1$.

Or: By Definition 2, $\lim_{h \to 0} \dfrac{(1+h)^{10} - 1}{h} = f'(0)$, where $f(x) = (1+x)^{10}$ and $a = 0$.

21. By Equation 3, $\lim_{x \to 5} \dfrac{2^x - 32}{x - 5} = f'(5)$, where $f(x) = 2^x$ and $a = 5$.

23. By Definition 2, $\lim\limits_{h\to 0} \dfrac{\cos(\pi+h)+1}{h} = f'(\pi)$, where $f(x) = \cos x$ and $a = \pi$.

Or: By Definition 2, $\lim\limits_{h\to 0} \dfrac{\cos(\pi+h)+1}{h} = f'(0)$, where $f(x) = \cos(\pi + x)$ and $a = 0$.

25. $v(5) = f'(5) = \lim\limits_{h\to 0} \dfrac{f(5+h)-f(5)}{h} = \lim\limits_{h\to 0} \dfrac{[100+50(5+h)-4.9(5+h)^2]-[100+50(5)-4.9(5)^2]}{h}$

$= \lim\limits_{h\to 0} \dfrac{(100+250+50h-4.9h^2-49h-122.5)-(100+250-122.5)}{h} = \lim\limits_{h\to 0} \dfrac{-4.9h^2+h}{h}$

$= \lim\limits_{h\to 0} \dfrac{h(-4.9h+1)}{h} = \lim\limits_{h\to 0}(-4.9h+1) = 1 \text{ m/s}$

The speed when $t = 5$ is $|1| = 1$ m/s.

27. (a) $f'(x)$ is the rate of change of the production cost with respect to the number of kilograms of gold produced. Its units are dollars per kilogram.

(b) After 50 kilograms of gold have been produced, the rate at which the production cost is increasing is \$36/kilogram. So the cost of producing the 50th (or 51st) kilogram is about \$36.

(c) In the short term, the values of $f'(x)$ will decrease because more efficient use is made of start-up costs as x increases. But eventually $f'(x)$ might increase due to large-scale operations.

29. (a) $f'(v)$ is the rate at which the fuel consumption is changing with respect to the speed. Its units are (L/h)/(km/h).

(b) The fuel consumption is decreasing by 0.05 (L/h)/(km/h) as the car's speed reaches 30 km/h. So if you increase your speed to 31 km/h, you could expect to decrease your fuel consumption by about 0.05 (L/h)/(km/h).

31. $T'(6)$ is the rate at which the temperature is changing at 6:00 P.M. To estimate the value of $T'(6)$, we will average the difference quotients obtained using the times $t = 4$ and $t = 8$. Let $A = \dfrac{T(4)-T(6)}{4-6} = \dfrac{38.3-32.8}{-2} = -2.75$ and $B = \dfrac{T(8)-T(6)}{8-6} = \dfrac{26.1-32.8}{2} = -3.35$. Then $T'(6) = \lim\limits_{t\to 6} \dfrac{T(t)-T(6)}{t-6} \approx \dfrac{A+B}{2} = \dfrac{-2.75-3.35}{2} = -3.05°\text{C/h}$.

33. (a) $S'(T)$ is the rate at which the oxygen solubility changes with respect to the water temperature. Its units are (mg/L)/°C.

(b) For $T = 16°$C, it appears that the tangent line to the curve goes through the points $(0, 14)$ and $(32, 6)$. So $S'(16) \approx \dfrac{6-14}{32-0} = -\dfrac{8}{32} = -0.25$ (mg/L)/°C. This means that as the temperature increases past 16°C, the oxygen solubility is decreasing at a rate of 0.25 (mg/L)/°C.

35. Since $f(x) = x\sin(1/x)$ when $x \neq 0$ and $f(0) = 0$, we have

$f'(0) = \lim\limits_{h\to 0} \dfrac{f(0+h)-f(0)}{h} = \lim\limits_{h\to 0} \dfrac{h\sin(1/h)-0}{h} = \lim\limits_{h\to 0} \sin(1/h)$. This limit does not exist since $\sin(1/h)$ takes the values -1 and 1 on any interval containing 0. (Compare with Example 4 in Section 2.2.)

2.8 The Derivative as a Function

1. It appears that f is an odd function, so f' will be an even function—that is, $f'(-a) = f'(a)$.

(a) $f'(-3) \approx 1.5$ (b) $f'(-2) \approx 1$

(c) $f'(-1) \approx 0$ (d) $f'(0) \approx -4$

(e) $f'(1) \approx 0$ (f) $f'(2) \approx 1$

(g) $f'(3) \approx 1.5$

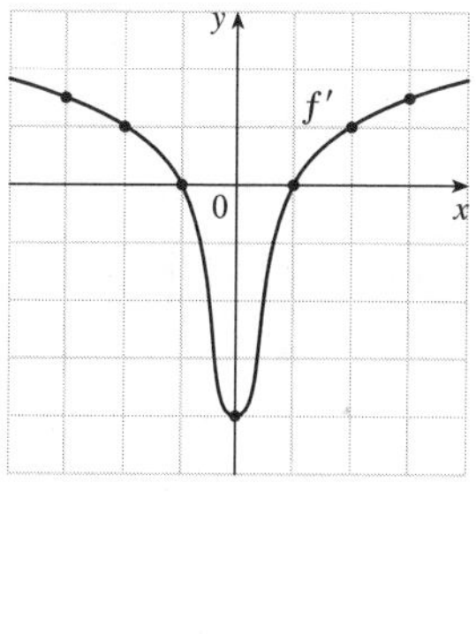

3. (a)$'$ = II, since from left to right, the slopes of the tangents to graph (a) start out negative, become 0, then positive, then 0, then negative again. The actual function values in graph II follow the same pattern.

(b)$'$ = IV, since from left to right, the slopes of the tangents to graph (b) start out at a fixed positive quantity, then suddenly become negative, then positive again. The discontinuities in graph IV indicate sudden changes in the slopes of the tangents.

(c)$'$ = I, since the slopes of the tangents to graph (c) are negative for $x < 0$ and positive for $x > 0$, as are the function values of graph I.

(d)$'$ = III, since from left to right, the slopes of the tangents to graph (d) are positive, then 0, then negative, then 0, then positive, then 0, then negative again, and the function values in graph III follow the same pattern.

Hints for Exercises 4–11: First plot x-intercepts on the graph of f' for any horizontal tangents on the graph of f. Look for any corners on the graph of f—there will be a discontinuity on the graph of f'. On any interval where f has a tangent with positive (or negative) slope, the graph of f' will be positive (or negative). If the graph of the function is linear, the graph of f' will be a horizontal line.

5.

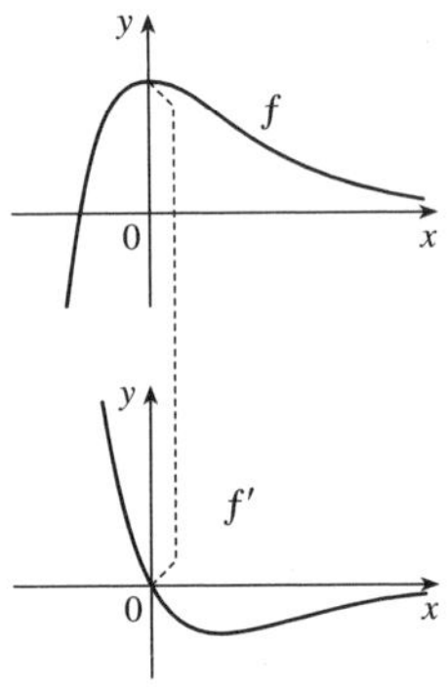

7.

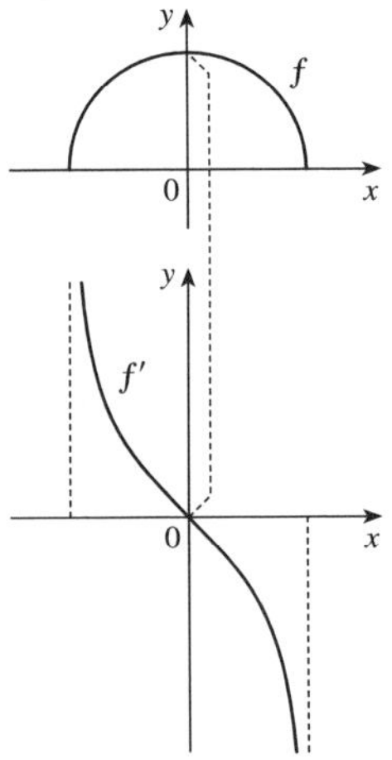

9.

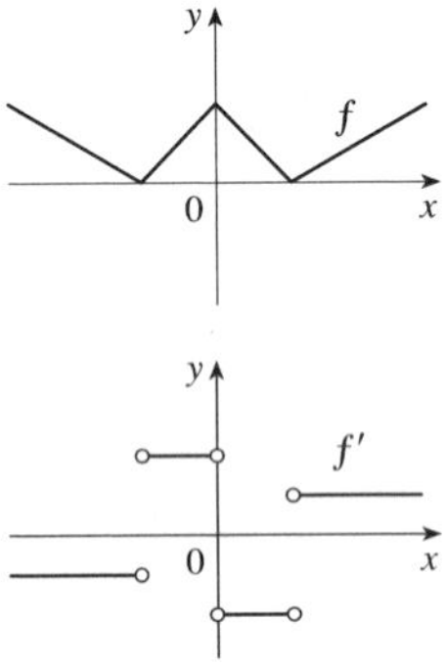

11.

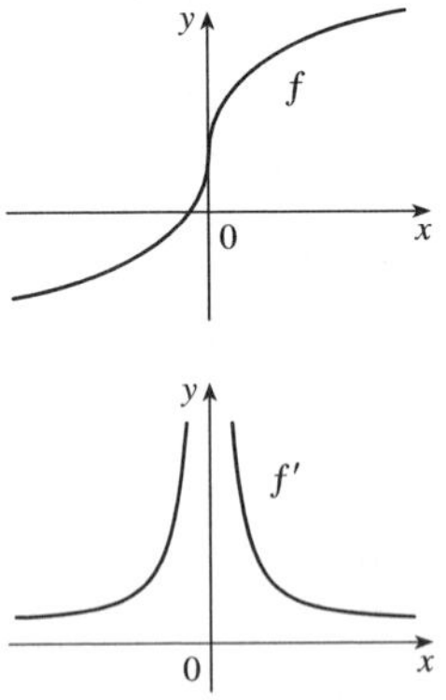

13. It appears that there are horizontal tangents on the graph of M for $t = 1963$ and $t = 1971$. Thus, there are zeros for those values of t on the graph of M'. The derivative is negative for the years 1963 to 1971.

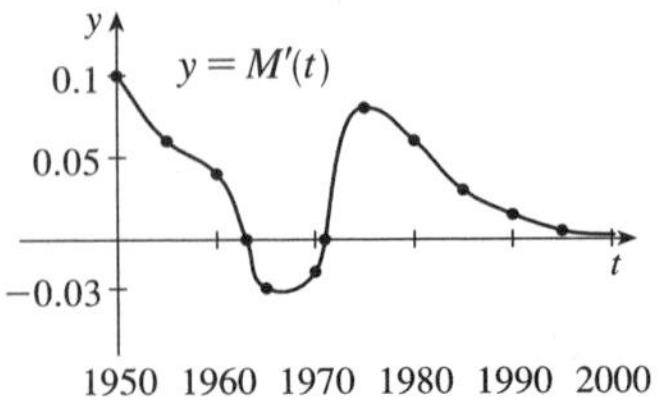

15.

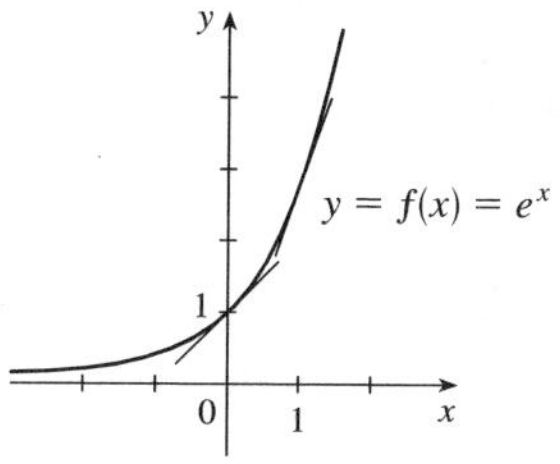

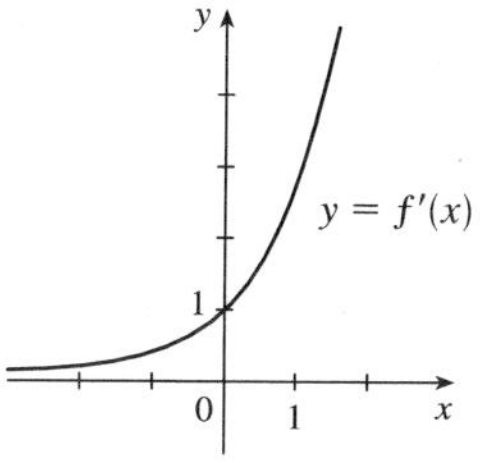

The slope at 0 appears to be 1 and the slope at 1 appears to be 2.7. As x decreases, the slope gets closer to 0. Since the graphs are so similar, we might guess that $f'(x) = e^x$.

17. (a) By zooming in, we estimate that $f'(0) = 0$, $f'\left(\frac{1}{2}\right) = 1$, $f'(1) = 2$, and $f'(2) = 4$.

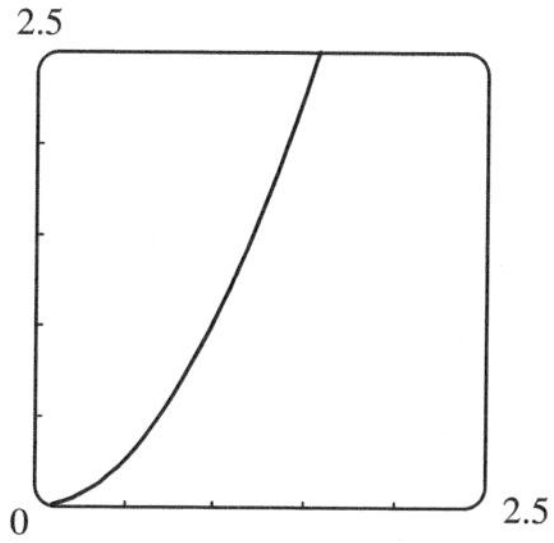

(b) By symmetry, $f'(-x) = -f'(x)$. So $f'\left(-\frac{1}{2}\right) = -1$, $f'(-1) = -2$, and $f'(-2) = -4$.

(c) It appears that $f'(x)$ is twice the value of x, so we guess that $f'(x) = 2x$.

(d) $$f'(x) = \lim_{h\to 0} \frac{f(x+h) - f(x)}{h} = \lim_{h\to 0} \frac{(x+h)^2 - x^2}{h}$$
$$= \lim_{h\to 0} \frac{(x^2 + 2hx + h^2) - x^2}{h} = \lim_{h\to 0} \frac{2hx + h^2}{h} = \lim_{h\to 0} \frac{h(2x+h)}{h} = \lim_{h\to 0}(2x+h) = 2x$$

19. $$f'(x) = \lim_{h\to 0} \frac{f(x+h) - f(x)}{h} = \lim_{h\to 0} \frac{[4 - 7(x+h)] - (4 - 7x)}{h} = \lim_{h\to 0} \frac{(4 - 7x - 7h) - (4 - 7x)}{h}$$
$$= \lim_{h\to 0} \frac{-7h}{h} = \lim_{h\to 0}(-7) = -7$$

Domain of f = domain of $f' = \mathbb{R}$.

21. $$f'(x) = \lim_{h\to 0} \frac{f(x+h) - f(x)}{h} = \lim_{h\to 0} \frac{\left[(x+h)^3 - 3(x+h) + 5\right] - (x^3 - 3x + 5)}{h}$$
$$= \lim_{h\to 0} \frac{(x^3 + 3x^2h + 3xh^2 + h^3 - 3x - 3h + 5) - (x^3 - 3x + 5)}{h} = \lim_{h\to 0} \frac{3x^2h + 3xh^2 + h^3 - 3h}{h}$$
$$= \lim_{h\to 0} \frac{h\left(3x^2 + 3xh + h^2 - 3\right)}{h} = \lim_{h\to 0}\left(3x^2 + 3xh + h^2 - 3\right) = 3x^2 - 3$$

Domain of f = domain of $f' = \mathbb{R}$.

23. $$g'(x) = \lim_{h\to 0} \frac{g(x+h) - g(x)}{h} = \lim_{h\to 0} \frac{\sqrt{1 + 2(x+h)} - \sqrt{1+2x}}{h}\left[\frac{\sqrt{1+2(x+h)} + \sqrt{1+2x}}{\sqrt{1+2(x+h)} + \sqrt{1+2x}}\right]$$
$$= \lim_{h\to 0} \frac{(1 + 2x + 2h) - (1 + 2x)}{h\left[\sqrt{1+2(x+h)} + \sqrt{1+2x}\right]} = \lim_{h\to 0} \frac{2}{\sqrt{1+2x+2h} + \sqrt{1+2x}} = \frac{2}{2\sqrt{1+2x}} = \frac{1}{\sqrt{1+2x}}$$

Domain of $g = \left[-\frac{1}{2}, \infty\right)$, domain of $g' = \left(-\frac{1}{2}, \infty\right)$.

25. $$G'(t) = \lim_{h\to 0} \frac{G(t+h)-G(t)}{h} = \lim_{h\to 0} \frac{\dfrac{4(t+h)}{(t+h)+1} - \dfrac{4t}{t+1}}{h} = \lim_{h\to 0} \frac{\dfrac{4(t+h)(t+1) - 4t(t+h+1)}{(t+h+1)(t+1)}}{h}$$

$$= \lim_{h\to 0} \frac{(4t^2+4ht+4t+4h)-(4t^2+4ht+4t)}{h(t+h+1)(t+1)} = \lim_{h\to 0} \frac{4h}{h(t+h+1)(t+1)}$$

$$= \lim_{h\to 0} \frac{4}{(t+h+1)(t+1)} = \frac{4}{(t+1)^2}$$

Domain of G = domain of $G' = (-\infty,-1)\cup(-1,\infty)$.

27. (a) $$f'(x) = \lim_{h\to 0} \frac{f(x+h)-f(x)}{h} = \lim_{h\to 0} \frac{[(x+h)^4+2(x+h)]-(x^4+2x)}{h}$$

$$= \lim_{h\to 0} \frac{x^4+4x^3h+6x^2h^2+4xh^3+h^4+2x+2h-x^4-2x}{h}$$

$$= \lim_{h\to 0} \frac{4x^3h+6x^2h^2+4xh^3+h^4+2h}{h} = \lim_{h\to 0} \frac{h(4x^3+6x^2h+4xh^2+h^3+2)}{h}$$

$$= \lim_{h\to 0}(4x^3+6x^2h+4xh^2+h^3+2) = 4x^3+2$$

(b) Notice that $f'(x)=0$ when f has a horizontal tangent, $f'(x)$ is positive when the tangents have positive slope, and $f'(x)$ is negative when the tangents have negative slope

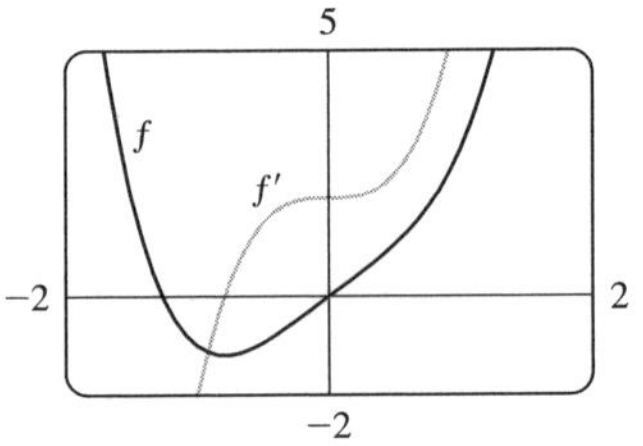

29. (a) $U'(t)$ is the rate at which the unemployment rate is changing with respect to time. Its units are percent per year.

(b) To find $U'(t)$, we use $\lim_{h\to 0} \frac{U(t+h)-U(t)}{h} \approx \frac{U(t+h)-U(t)}{h}$ for small values of h.

For 1995: $U'(1995) \approx \frac{U(1996)-U(1995)}{1996-1995} = \frac{8.0-8.1}{1} = -0.10$

For 1996: We estimate $U'(1996)$ by using $h=-1$ and $h=1$, and then average the two results to obtain a final estimate.

$h=-1 \;\Rightarrow\; U'(1996) \approx \frac{U(1995)-U(1996)}{1995-1996} = \frac{8.1-8.0}{-1} = -0.10;$

$h=1 \;\Rightarrow\; U'(1996) \approx \frac{U(1997)-U(1996)}{1997-1996} = \frac{8.2-8.0}{1} = 0.20.$

So we estimate that $U'(1996) \approx \frac{1}{2}[-0.10+0.20] = 0.05$.

t	1995	1996	1997	1998	1999	2000	2001	2002	2003	2004
$U'(t)$	−0.10	0.05	−0.05	−0.75	−0.85	0.10	0.15	−0.35	−0.45	− 0.60

31. f is not differentiable at $x=-4$, because the graph has a corner there, and at $x=0$, because there is a discontinuity there.

33. f is not differentiable at $x=-1$, because the graph has a vertical tangent there, and at $x=4$, because the graph has a corner there.

35. As we zoom in toward $(-1,0)$, the curve appears more and more like a straight line, so $f(x)=x+\sqrt{|x|}$ is differentiable at $x=-1$. But no matter how much we zoom in toward the origin, the curve doesn't straighten out—we can't eliminate the sharp point (a cusp). So f is not differentiable at $x=0$.

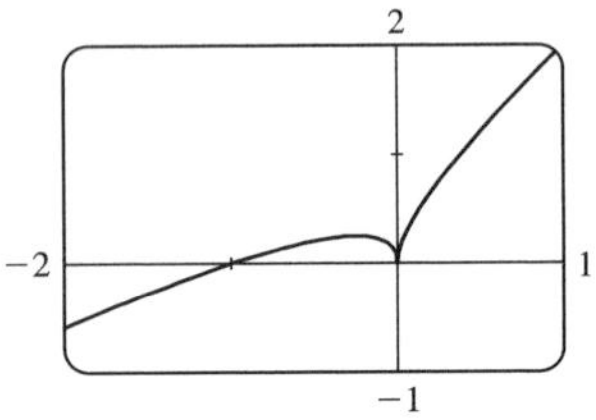

37. $a = f, b = f', c = f''$. We can see this because where a has a horizontal tangent, $b = 0$, and where b has a horizontal tangent, $c = 0$. We can immediately see that c can be neither f nor f', since at the points where c has a horizontal tangent, neither a nor b is equal to 0.

39. We can immediately see that a is the graph of the acceleration function, since at the points where a has a horizontal tangent, neither c nor b is equal to 0. Next, we note that $a = 0$ at the point where b has a horizontal tangent, so b must be the graph of the velocity function, and hence, $b' = a$. We conclude that c is the graph of the position function.

41. $$f'(x) = \lim_{h\to 0} \frac{f(x+h) - f(x)}{h} = \lim_{h\to 0} \frac{\left[1 + 4(x+h) - (x+h)^2\right] - (1 + 4x - x^2)}{h}$$
$$= \lim_{h\to 0} \frac{(1 + 4x + 4h - x^2 - 2xh - h^2) - (1 + 4x - x^2)}{h} = \lim_{h\to 0} \frac{4h - 2xh - h^2}{h} = \lim_{h\to 0} (4 - 2x - h) = 4 - 2x$$

$$f''(x) = \lim_{h\to 0} \frac{f'(x+h) - f'(x)}{h} = \lim_{h\to 0} \frac{[4 - 2(x+h)] - (4 - 2x)}{h} = \lim_{h\to 0} \frac{-2h}{h} = \lim_{h\to 0} (-2) = -2$$

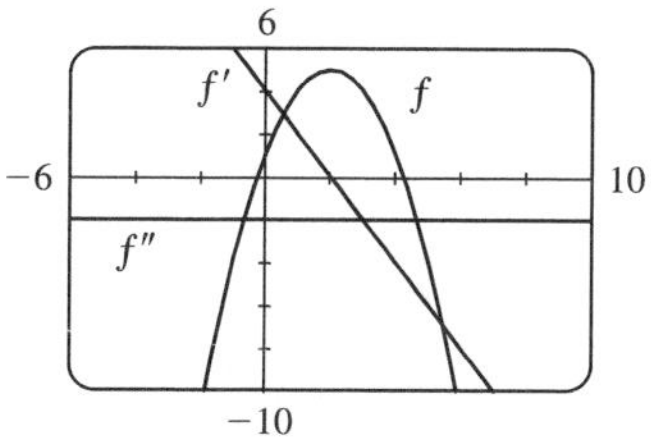

We see from the graph that our answers are reasonable because the graph of f' is that of a linear function and the graph of f'' is that of a constant function.

43. $$f'(x) = \lim_{h\to 0} \frac{f(x+h) - f(x)}{h} = \lim_{h\to 0} \frac{\left[2(x+h)^2 - (x+h)^3\right] - (2x^2 - x^3)}{h}$$
$$= \lim_{h\to 0} \frac{h(4x + 2h - 3x^2 - 3xh - h^2)}{h} = \lim_{h\to 0} (4x + 2h - 3x^2 - 3xh - h^2) = 4x - 3x^2$$

$$f''(x) = \lim_{h\to 0} \frac{f'(x+h) - f'(x)}{h} = \lim_{h\to 0} \frac{\left[4(x+h) - 3(x+h)^2\right] - (4x - 3x^2)}{h} = \lim_{h\to 0} \frac{h(4 - 6x - 3h)}{h}$$
$$= \lim_{h\to 0} (4 - 6x - 3h) = 4 - 6x$$

$$f'''(x) = \lim_{h\to 0} \frac{f''(x+h) - f''(x)}{h} = \lim_{h\to 0} \frac{[4 - 6(x+h)] - (4 - 6x)}{h} = \lim_{h\to 0} \frac{-6h}{h} = \lim_{h\to 0} (-6) = -6$$

$$f^{(4)}(x) = \lim_{h\to 0} \frac{f'''(x+h) - f'''(x)}{h} = \lim_{h\to 0} \frac{-6 - (-6)}{h} = \lim_{h\to 0} \frac{0}{h} = \lim_{h\to 0} (0) = 0$$

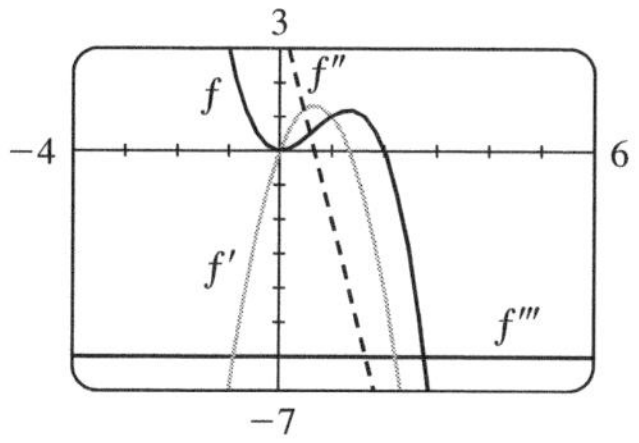

The graphs are consistent with the geometric interpretations of the derivatives because f' has zeros where f has a local minimum and a local maximum, f'' has a zero where f' has a local maximum, and f''' is a constant function equal to the slope of f''.

45. (a) Note that we have factored $x - a$ as the difference of two cubes in the third step.

$$f'(a) = \lim_{x\to a} \frac{f(x) - f(a)}{x - a} = \lim_{x\to a} \frac{x^{1/3} - a^{1/3}}{x - a} = \lim_{x\to a} \frac{x^{1/3} - a^{1/3}}{(x^{1/3} - a^{1/3})(x^{2/3} + x^{1/3}a^{1/3} + a^{2/3})}$$
$$= \lim_{x\to a} \frac{1}{x^{2/3} + x^{1/3}a^{1/3} + a^{2/3}} = \frac{1}{3a^{2/3}} \text{ or } \tfrac{1}{3}a^{-2/3}$$

(b) $f'(0) = \lim_{h\to 0} \dfrac{f(0+h) - f(0)}{h} = \lim_{h\to 0} \dfrac{\sqrt[3]{h} - 0}{h} = \lim_{h\to 0} \dfrac{1}{h^{2/3}}$. This function increases without bound, so the limit does not exist, and therefore $f'(0)$ does not exist.

(c) $\lim_{x\to 0} |f'(x)| = \lim_{x\to 0} \dfrac{1}{3x^{2/3}} = \infty$ and f is continuous at $x = 0$ (root function), so f has a vertical tangent at $x = 0$.

47. $f(x) = |x-6| = \begin{cases} x-6 & \text{if } x-6 \geq 6 \\ -(x-6) & \text{if } x-6 < 0 \end{cases} = \begin{cases} x-6 & \text{if } x \geq 6 \\ 6-x & \text{if } x < 6 \end{cases}$

So the right-hand limit is $\lim_{x\to 6^+} \dfrac{f(x)-f(6)}{x-6} = \lim_{x\to 6^+} \dfrac{|x-6|-0}{x-6} = \lim_{x\to 6^+} \dfrac{x-6}{x-6} = \lim_{x\to 6^+} 1 = 1$, and the left-hand limit is $\lim_{x\to 6^-} \dfrac{f(x)-f(6)}{x-6} = \lim_{x\to 6^-} \dfrac{|x-6|-0}{x-6} = \lim_{x\to 6^-} \dfrac{6-x}{x-6} = \lim_{x\to 6^-} (-1) = -1$. Since these limits are not equal, $f'(6) = \lim_{x\to 6} \dfrac{f(x)-f(6)}{x-6}$ does not exist and f is not differentiable at 6.

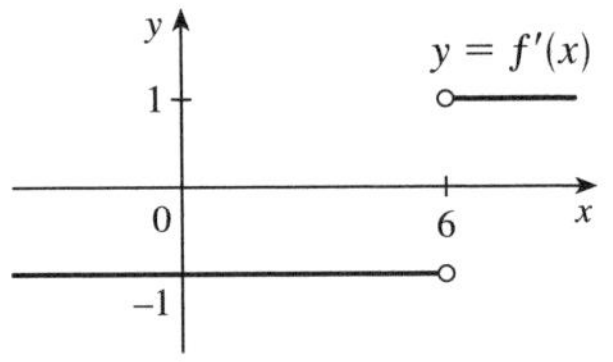

However, a formula for f' is $f'(x) = \begin{cases} 1 & \text{if } x > 6 \\ -1 & \text{if } x < 6 \end{cases}$

Another way of writing the formula is $f'(x) = \dfrac{x-6}{|x-6|}$.

49. (a) If f is even, then

$$\begin{aligned} f'(-x) &= \lim_{h\to 0} \frac{f(-x+h) - f(-x)}{h} = \lim_{h\to 0} \frac{f[-(x-h)] - f(-x)}{h} \\ &= \lim_{h\to 0} \frac{f(x-h) - f(x)}{h} = -\lim_{h\to 0} \frac{f(x-h) - f(x)}{-h} \qquad [\text{let } \Delta x = -h] \\ &= -\lim_{\Delta x\to 0} \frac{f(x+\Delta x) - f(x)}{\Delta x} = -f'(x) \end{aligned}$$

Therefore, f' is odd.

(b) If f is odd, then

$$\begin{aligned} f'(-x) &= \lim_{h\to 0} \frac{f(-x+h) - f(-x)}{h} = \lim_{h\to 0} \frac{f[-(x-h)] - f(-x)}{h} \\ &= \lim_{h\to 0} \frac{-f(x-h) + f(x)}{h} = \lim_{h\to 0} \frac{f(x-h) - f(x)}{-h} \qquad [\text{let } \Delta x = -h] \\ &= \lim_{\Delta x\to 0} \frac{f(x+\Delta x) - f(x)}{\Delta x} = f'(x) \end{aligned}$$

Therefore, f' is even.

51.

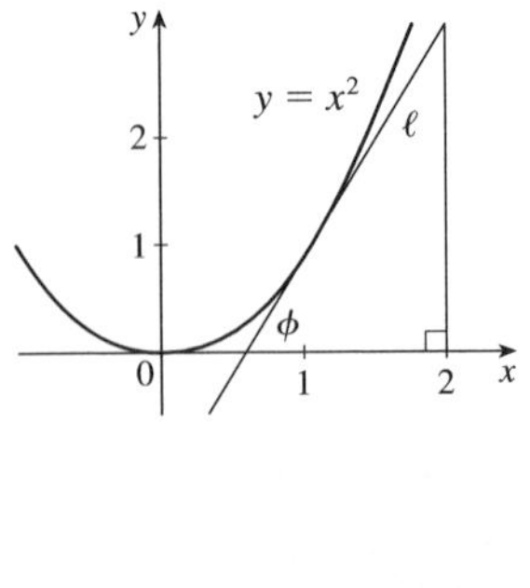

In the right triangle in the diagram, let Δy be the side opposite angle ϕ and Δx the side adjacent angle ϕ. Then the slope of the tangent line ℓ is $m = \Delta y/\Delta x = \tan\phi$. Note that $0 < \phi < \frac{\pi}{2}$. We know (see Exercise 17) that the derivative of $f(x) = x^2$ is $f'(x) = 2x$. So the slope of the tangent to the curve at the point $(1, 1)$ is 2. Thus, ϕ is the angle between 0 and $\frac{\pi}{2}$ whose tangent is 2; that is, $\phi = \tan^{-1} 2 \approx 63°$.

2.9 What Does f' Say about f?

1. (a) Since $f'(x) > 0$ on $(1, 5)$, f is increasing on this interval. Since $f'(x) < 0$ on $(0, 1)$ and $(5, 6)$, f is decreasing on these intervals.

(b) Since $f'(x) = 0$ at $x = 1$ and f' changes from negative to positive there, f changes from decreasing to increasing and has a local minimum at $x = 1$. Since $f'(x) = 0$ at $x = 5$ and f' changes from positive to negative there, f changes from increasing to decreasing and has a local maximum at $x = 5$.

(c) Since $f(0) = 0$, start at the origin. Draw a decreasing function on $(0, 1)$ with a local minimum at $x = 1$. Now draw an increasing function on $(1, 5)$ and the steepest slope should occur at $x = 3$ since that's where the largest value of f' occurs. Lastly, draw a decreasing function on $(5, 6)$ making sure you have a local maximum at $x = 5$.

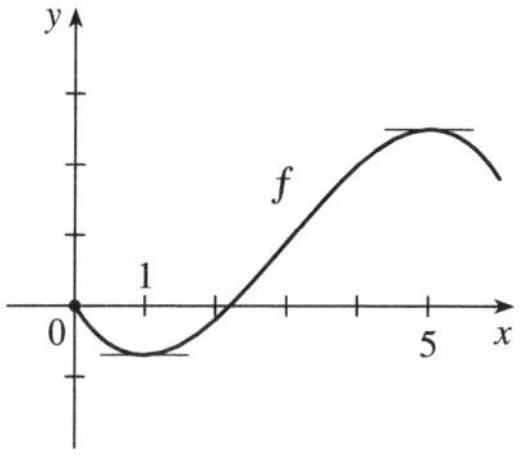

3. The derivative f' is increasing when the slopes of the tangent lines of f are becoming larger as x increases. This seems to be the case on the interval $(2, 5)$. The derivative is decreasing when the slopes of the tangent lines of f are becoming smaller as x increases, and this seems to be the case on $(-\infty, 2)$ and $(5, \infty)$. So f' is increasing on $(2, 5)$ and decreasing on $(-\infty, 2)$ and $(5, \infty)$.

5. If $D(t)$ is the size of the deficit as a function of time, then at the time of the speech $D'(t) > 0$, but $D''(t) < 0$ because $D''(t) = (D')'(t)$ is the rate of change of $D'(t)$.

7. (a) The rate of increase of the population is initially very small, then increases rapidly until about 1932 when it starts decreasing. The rate becomes negative by 1936, peaks in magnitude in 1937, and approaches 0 in 1940.

(b) Inflection points (IP) appear to be at $(1932, 2.5)$ and $(1937, 4.3)$. The rate of change of population density starts to decrease in 1932 and starts to increase in 1937. The rates of population increase and decrease have their maximum values at those points.

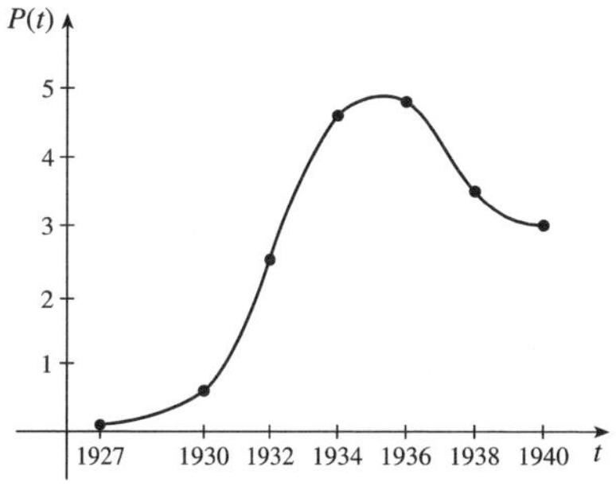

9. Most students learn more in the third hour of studying than in the eighth hour, so $K(3) - K(2)$ is larger than $K(8) - K(7)$. In other words, as you begin studying for a test, the rate of knowledge gain is large and then starts to taper off, so $K'(t)$ decreases and the graph of K is concave downward.

11. (a) f is increasing where f' is positive, that is, on $(0, 2)$, $(4, 6)$, and $(8, \infty)$; and decreasing where f' is negative, that is, on $(2, 4)$ and $(6, 8)$.

(b) f has local maxima where f' changes from positive to negative, at $x = 2$ and at $x = 6$, and local minima where f' changes from negative to positive, at $x = 4$ and at $x = 8$.

(c) f is concave upward (CU) where f' is increasing, that is, on $(3, 6)$ and $(6, \infty)$, and concave downward (CD) where f' is decreasing, that is, on $(0, 3)$.

(d) There is a point of inflection where f changes from being CD to being CU, that is, at $x = 3$.

(e)

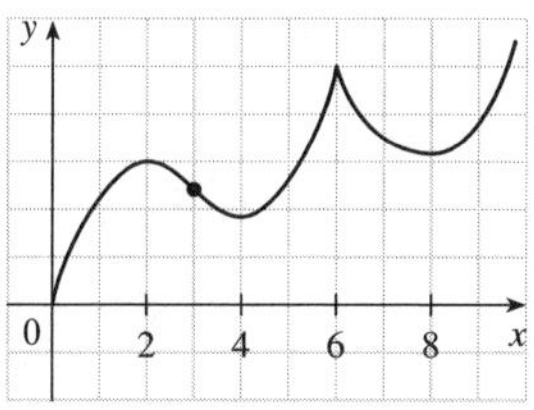

13. The function must be always decreasing (since the first derivative is always negative) and concave downward (since the second derivative is always negative).

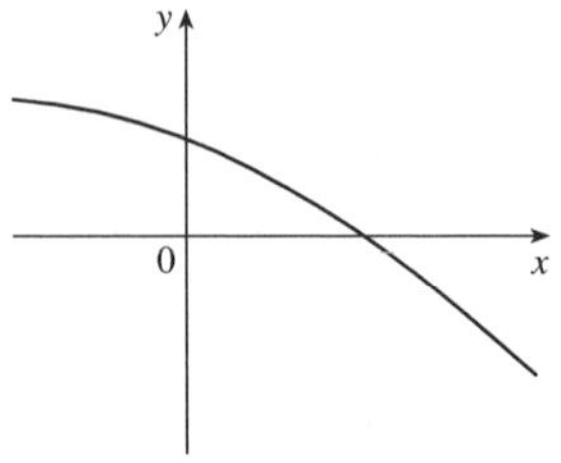

15. $f'(0) = f'(4) = 0 \;\Rightarrow\;$ horizontal tangents at $x = 0, 4$.

$f'(x) > 0$ if $x < 0 \;\Rightarrow\; f$ is increasing on $(-\infty, 0)$.

$f'(x) < 0$ if $0 < x < 4$ or if $x > 4 \;\Rightarrow\; f$ is decreasing on $(0, 4)$ and $(4, \infty)$.

$f''(x) > 0$ if $2 < x < 4 \;\Rightarrow\; f$ is concave upward on $(2, 4)$.

$f''(x) < 0$ if $x < 2$ or $x > 4 \;\Rightarrow\; f$ is concave downward on $(-\infty, 2)$ and $(4, \infty)$. There are inflection points when $x = 2$ and 4.

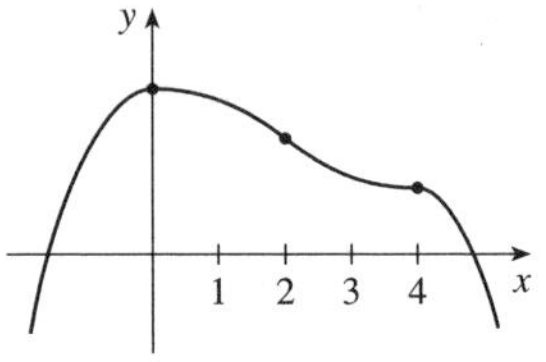

17. $f'(0) = f'(2) = f'(4) = 0 \;\Rightarrow\;$ horizontal tangents at $x = 0, 2, 4$.

$f'(x) > 0$ if $x < 0$ or $2 < x < 4 \;\Rightarrow\; f$ is increasing on $(-\infty, 0)$ and $(2, 4)$.

$f'(x) < 0$ if $0 < x < 2$ or $x > 4 \;\Rightarrow\; f$ is decreasing on $(0, 2)$ and $(4, \infty)$.

$f''(x) > 0$ if $1 < x < 3 \;\Rightarrow\; f$ is concave upward on $(1, 3)$.

$f''(x) < 0$ if $x < 1$ or $x > 3 \;\Rightarrow\; f$ is concave downward on $(-\infty, 1)$ and $(3, \infty)$. There are inflection points when $x = 1$ and 3.

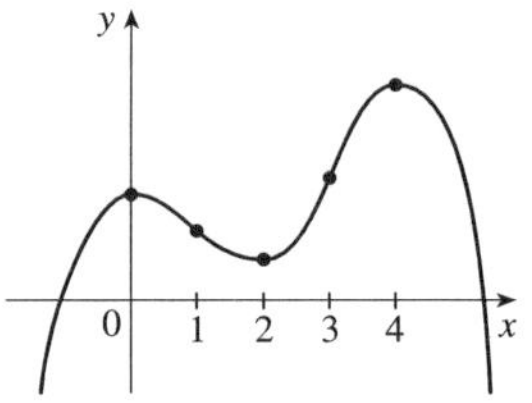

19. $f'(x) > 0$ if $|x| < 2 \;\Rightarrow\; f$ is increasing on $(-2, 2)$.

$f'(x) < 0$ if $|x| > 2 \;\Rightarrow\; f$ is decreasing on $(-\infty, -2)$ and $(2, \infty)$. $f'(-2) = 0 \;\Rightarrow\;$ horizontal tangent at $x = -2$. $\lim_{x \to 2} |f'(x)| = \infty \;\Rightarrow\;$ there is a vertical asymptote or vertical tangent (cusp) at $x = 2$. $f''(x) > 0$ if $x \neq 2 \;\Rightarrow\; f$ is concave upward on $(-\infty, 2)$ and $(2, \infty)$.

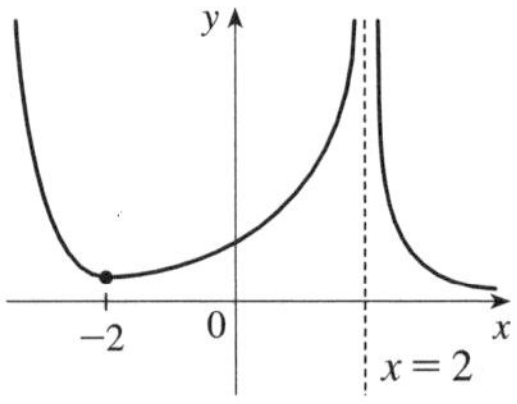

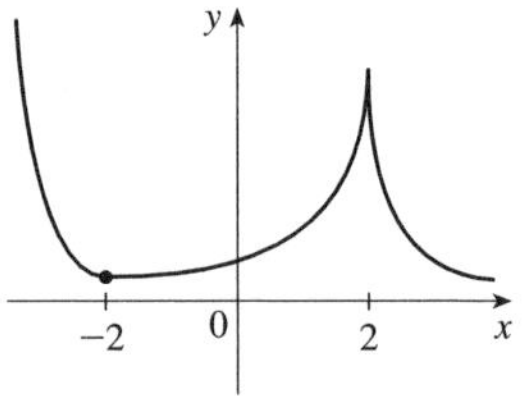

21. (a) Since e^{-x^2} is positive for all x, $f'(x) = xe^{-x^2}$ is positive where $x > 0$ and negative where $x < 0$. Thus, f is increasing on $(0, \infty)$ and decreasing on $(-\infty, 0)$.

(b) Since f changes from decreasing to increasing at $x = 0$, f has a minimum value there.

23. (a) To find the intervals on which f is increasing, we need to find the intervals on which $f'(x) = 3x^2 - 1$ is positive. $3x^2 - 1 > 0 \;\Leftrightarrow\; 3x^2 > 1 \;\Leftrightarrow\; x^2 > \frac{1}{3} \;\Leftrightarrow\; |x| > \sqrt{\frac{1}{3}}$, so $x \in \left(-\infty, -\sqrt{\frac{1}{3}}\right) \cup \left(\sqrt{\frac{1}{3}}, \infty\right)$. Thus, f is increasing on $\left(-\infty, -\sqrt{\frac{1}{3}}\right)$ and on $\left(\sqrt{\frac{1}{3}}, \infty\right)$. In a similar fashion, f is decreasing on $\left(-\sqrt{\frac{1}{3}}, \sqrt{\frac{1}{3}}\right)$.

(b) To find the intervals on which f is concave upward, we need to find the intervals on which $f''(x) = 6x$ is positive. $6x > 0 \;\Leftrightarrow\; x > 0$. So f is concave upward on $(0, \infty)$ and f is concave downward on $(-\infty, 0)$.

(c) There is an inflection point at $(0, 0)$ since f changes its direction of concavity at $x = 0$.

25. b is the antiderivative of f. For small x, f is negative, so the graph of its antiderivative must be decreasing. But both a and c are increasing for small x, so only b can be f's antiderivative. Also, f is positive where b is increasing, which supports our conclusion.

27. The graph of F will have a minimum at 0 and a maximum at 2, since $f = F'$ goes from negative to positive at $x = 0$, and from positive to negative at $x = 2$.

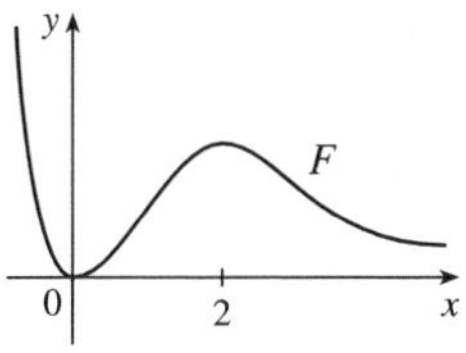

29.

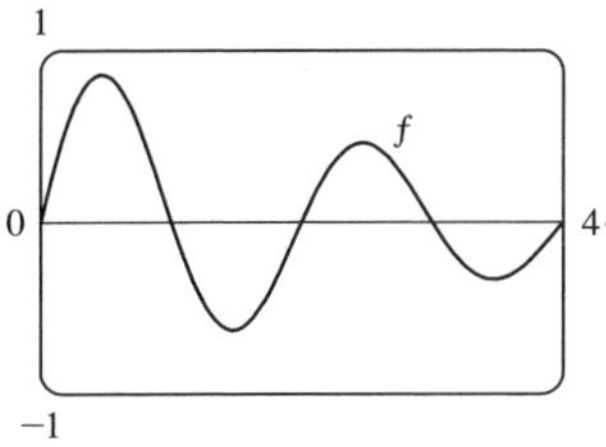

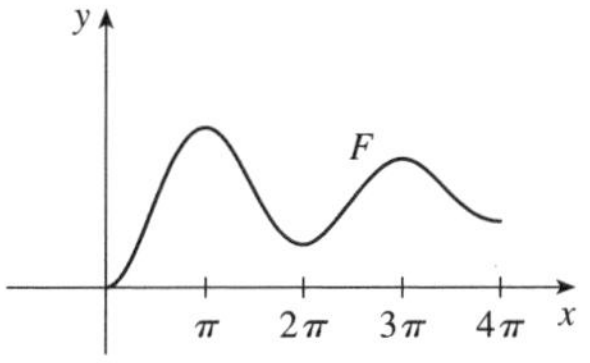

f is positive on $(0, \pi)$ and $(2\pi, 3\pi)$, so F is increasing on those intervals. f is negative on $(\pi, 2\pi)$ and $(3\pi, 4\pi)$, so F is decreasing on those intervals. f is zero at $x = 0, \pi, 2\pi, 3\pi$, and 4π, so F has horizontal tangents at those points.

2 Review

CONCEPT CHECK

1. (a) $\lim\limits_{x\to a} f(x) = L$: See Definition 2.2.1 and Figures 1 and 2 in Section 2.2.

(b) $\lim\limits_{x\to a^+} f(x) = L$: See the paragraph after Definition 2.2.2 and Figure 9(b) in Section 2.2.

(c) $\lim\limits_{x\to a^-} f(x) = L$: See Definition 2.2.2 and Figure 9(a) in Section 2.2.

(d) $\lim\limits_{x\to a} f(x) = \infty$: See Definition 2.5.1 and Figure 2 in Section 2.5.

(e) $\lim\limits_{x\to \infty} f(x) = L$: See Definition 2.5.4 and Figure 9 in Section 2.5.

2. In general, the limit of a function fails to exist when the function does not approach a fixed number. For each of the following functions, the limit fails to exist at $x = 2$.

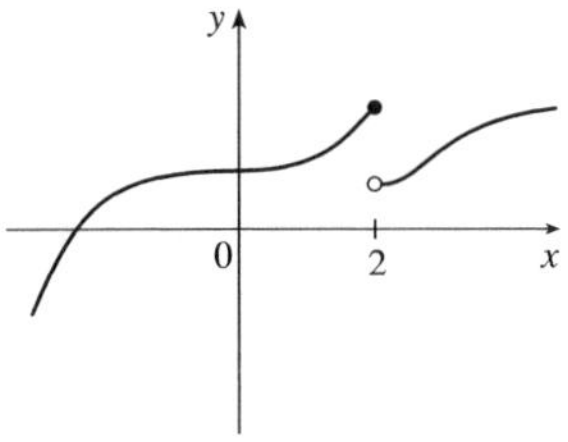

The left- and right-hand limits are not equal.

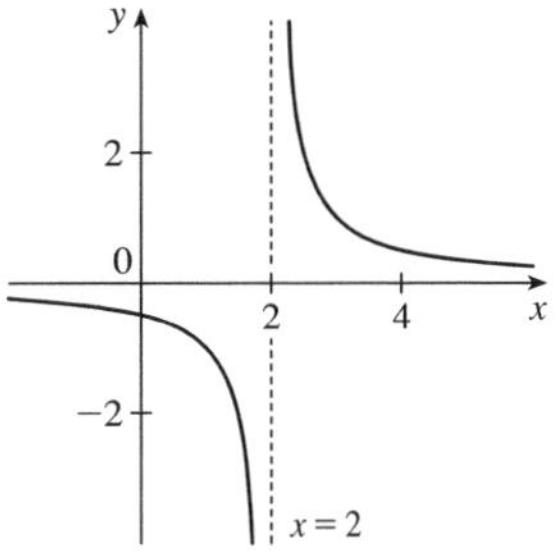

There is an infinite discontinuity.

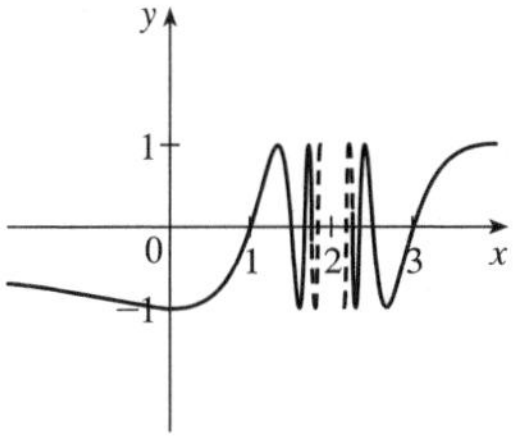

There are an infinite number of oscillations.

3. (a)–(g) See the statements of Limit Laws 1–6 and 11 in Section 2.3.

4. See Theorem 3 in Section 2.3.

5. (a) See Definition 2.5.2 and Figures 2–4 in Section 2.5.

(b) See Definition 2.5.5 and Figures 9 and 10 in Section 2.5.

6. (a) $y = x^4$: No asymptote

(b) $y = \sin x$: No asymptote

(c) $y = \tan x$: Vertical asymptotes $x = \frac{\pi}{2} + \pi n$, n an integer

(d) $y = \tan^{-1} x$: Horizontal asymptotes $y = \pm\frac{\pi}{2}$

(e) $y = e^x$: Horizontal asymptote $y = 0$ $\left(\lim\limits_{x\to-\infty} e^x = 0\right)$

(f) $y = \ln x$: Vertical asymptote $x = 0$ $\left(\lim\limits_{x\to0^+} \ln x = -\infty\right)$

(g) $y = 1/x$: Vertical asymptote $x = 0$, horizontal asymptote $y = 0$

(h) $y = \sqrt{x}$: No asymptote

7. (a) A function f is continuous at a number a if $f(x)$ approaches $f(a)$ as x approaches a; that is, $\lim\limits_{x\to a} f(x) = f(a)$.

(b) A function f is continuous on the interval $(-\infty, \infty)$ if f is continuous at every real number a. The graph of such a function has no breaks and every vertical line crosses it.

8. See Theorem 2.4.10.

9. See Definition 2.6.1.

10. See the paragraph containing Formula 3 in Section 2.6.

11. (a) The average rate of change of y with respect to x over the interval $[x_1, x_2]$ is $\dfrac{f(x_2) - f(x_1)}{x_2 - x_1}$.

(b) The instantaneous rate of change of y with respect to x at $x = x_1$ is $\lim\limits_{x_2\to x_1} \dfrac{f(x_2) - f(x_1)}{x_2 - x_1}$.

12. See Definition 2.7.2. The pages following the definition discuss interpretations of $f'(a)$ as the slope of a tangent line to the graph of f at $x = a$ and as an instantaneous rate of change of $f(x)$ with respect to x when $x = a$.

13. See the paragraphs before and after Example 7 in Section 2.8.

14. (a) A function f is differentiable at a number a if its derivative f' exists at $x = a$; that is, if $f'(a)$ exists.

(b) See Theorem 2.8.4. This theorem also tells us that if f is *not* continuous at a, then f is *not* differentiable at a.

(c)

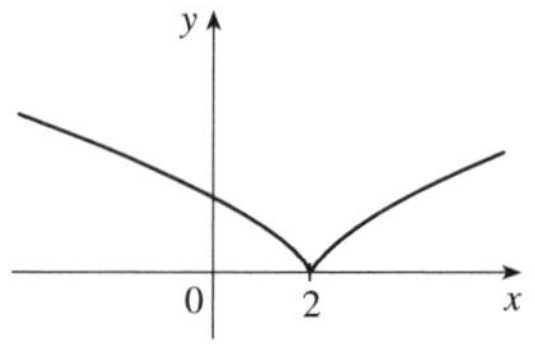

15. See the discussion and Figure 8 on page 162.

16. (a) See the first box in Section 2.9.

(b) See the second box in Section 2.9.

17. (a) An antiderivative of a function f is a function F such that $F' = f$.

(b) The antiderivative of a velocity function is a position function (the derivative of a position function is a velocity function). The antiderivative of an acceleration function is a velocity function (the derivative of a velocity function is an acceleration function).

TRUE-FALSE QUIZ

1. False. Limit Law 2 applies only if the individual limits exist (these don't).

3. True. Limit Law 5 applies.

5. False. Consider $\lim\limits_{x\to5} \dfrac{x(x-5)}{x-5}$ or $\lim\limits_{x\to5} \dfrac{\sin(x-5)}{x-5}$. The first limit exists and is equal to 5. By Example 3 in Section 2.2, we know that the latter limit exists (and it is equal to 1).

7. True. A polynomial is continuous everywhere, so $\lim\limits_{x\to b} p(x)$ exists and is equal to $p(b)$.

9. True. See Figure 11 in Section 2.5.

11. False. Consider $f(x) = \begin{cases} 1/(x-1) & \text{if } x \neq 1 \\ 2 & \text{if } x = 1 \end{cases}$

13. True. Use Theorem 2.4.8 with $a = 2$, $b = 5$, and $g(x) = 4x^2 - 11$. Note that $f(4) = 3$ is not needed.

15. False. See the note after Theorem 4 in Section 2.8.

17. False. $\dfrac{d^2y}{dx^2}$ is the second derivative while $\left(\dfrac{dy}{dx}\right)^2$ is the first derivative squared. For example, if $y = x$, then $\dfrac{d^2y}{dx^2} = 0$, but $\left(\dfrac{dy}{dx}\right)^2 = 1$.

EXERCISES

1. (a) (i) $\lim\limits_{x\to 2^+} f(x) = 3$ (ii) $\lim\limits_{x\to -3^+} f(x) = 0$

(iii) $\lim\limits_{x\to -3} f(x)$ does not exist since the left and right limits are not equal. (The left limit is -2.)

(iv) $\lim\limits_{x\to 4} f(x) = 2$

(v) $\lim\limits_{x\to 0} f(x) = \infty$ (vi) $\lim\limits_{x\to 2^-} f(x) = -\infty$

(vii) $\lim\limits_{x\to \infty} f(x) = 4$ (viii) $\lim\limits_{x\to -\infty} f(x) = -1$

(b) The equations of the horizontal asymptotes are $y = -1$ and $y = 4$.

(c) The equations of the vertical asymptotes are $x = 0$ and $x = 2$.

(d) f is discontinuous at $x = -3$, 0, 2, and 4. The discontinuities are jump, infinite, infinite, and removable, respectively.

3. Since the exponential function is continuous, $\lim\limits_{x\to 1} e^{x^3-x} = e^{1-1} = e^0 = 1$.

5. $$\lim_{x\to -3} \frac{x^2-9}{x^2+2x-3} = \lim_{x\to -3} \frac{(x+3)(x-3)}{(x+3)(x-1)} = \lim_{x\to -3} \frac{x-3}{x-1} = \frac{-3-3}{-3-1} = \frac{-6}{-4} = \frac{3}{2}$$

7. $$\lim_{h\to 0} \frac{(h-1)^3+1}{h} = \lim_{h\to 0} \frac{\left(h^3-3h^2+3h-1\right)+1}{h} = \lim_{h\to 0} \frac{h^3-3h^2+3h}{h} = \lim_{h\to 0} \left(h^2-3h+3\right) = 3$$

Another solution: Factor the numerator as a sum of two cubes and then simplify.

$$\lim_{h\to 0} \frac{(h-1)^3+1}{h} = \lim_{h\to 0} \frac{(h-1)^3+1^3}{h} = \lim_{h\to 0} \frac{[(h-1)+1]\left[(h-1)^2-1(h-1)+1^2\right]}{h}$$
$$= \lim_{h\to 0} \left[(h-1)^2-h+2\right] = 1-0+2 = 3$$

9. $\lim\limits_{r\to 9} \dfrac{\sqrt{r}}{(r-9)^4} = \infty$ since $(r-9)^4 \to 0$ as $r \to 9$ and $\dfrac{\sqrt{r}}{(r-9)^4} > 0$ for $r \neq 9$.

11. Let $t = \sin x$. Then as $x \to \pi^-$, $\sin x \to 0^+$, so $t \to 0^+$. Thus, $\lim\limits_{x\to \pi^-} \ln(\sin x) = \lim\limits_{t\to 0^+} \ln t = -\infty$.

13. $$\lim_{x\to\infty} \left(\sqrt{x^2+4x+1}-x\right) = \lim_{x\to\infty} \left[\frac{\sqrt{x^2+4x+1}-x}{1} \cdot \frac{\sqrt{x^2+4x+1}+x}{\sqrt{x^2+4x+1}+x}\right]$$
$$= \lim_{x\to\infty} \frac{\left(x^2+4x+1\right)-x^2}{\sqrt{x^2+4x+1}+x}$$
$$= \lim_{x\to\infty} \frac{(4x+1)/x}{\left(\sqrt{x^2+4x+1}+x\right)/x} \qquad \left[\text{divide by } x = \sqrt{x^2} \text{ for } x > 0\right]$$
$$= \lim_{x\to\infty} \frac{4+1/x}{\sqrt{1+4/x+1/x^2}+1} = \frac{4+0}{\sqrt{1+0+0}+1} = \frac{4}{2} = 2$$

15. Let $t = 1/x$. Then as $x \to 0^+$, $t \to \infty$, and $\lim_{x\to 0^+} \tan^{-1}(1/x) = \lim_{t\to\infty} \tan^{-1} t = \dfrac{\pi}{2}$.

17. From the graph of $y = \left(\cos^2 x\right)/x^2$, it appears that $y = 0$ is the horizontal asymptote and $x = 0$ is the vertical asymptote. Now $0 \le (\cos x)^2 \le 1 \;\Rightarrow$

$\dfrac{0}{x^2} \le \dfrac{\cos^2 x}{x^2} \le \dfrac{1}{x^2} \;\Rightarrow\; 0 \le \dfrac{\cos^2 x}{x^2} \le \dfrac{1}{x^2}$. But $\lim_{x\to\pm\infty} 0 = 0$ and

$\lim_{x\to\pm\infty} \dfrac{1}{x^2} = 0$, so by the Squeeze Theorem, $\lim_{x\to\pm\infty} \dfrac{\cos^2 x}{x^2} = 0$.

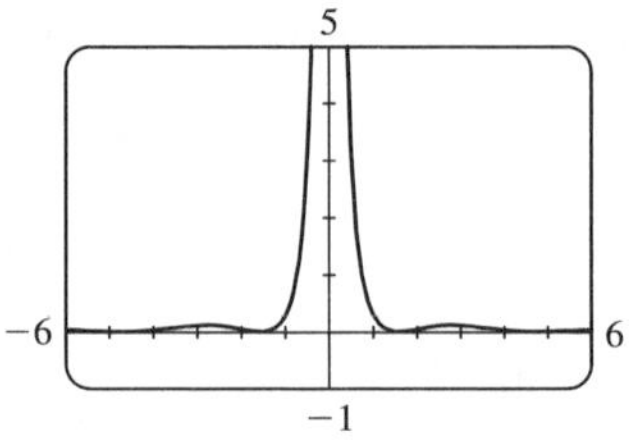

Thus, $y = 0$ is the horizontal asymptote. $\lim_{x\to 0} \dfrac{\cos^2 x}{x^2} = \infty$ because $\cos^2 x \to 1$ and $x^2 \to 0$ as $x \to 0$, so $x = 0$ is the vertical asymptote.

19. Since $2x - 1 \le f(x) \le x^2$ for $0 < x < 3$ and $\lim_{x\to 1}(2x-1) = 1 = \lim_{x\to 1} x^2$, we have $\lim_{x\to 1} f(x) = 1$ by the Squeeze Theorem.

21. (a) $f(x) = \sqrt{-x}$ if $x < 0$, $f(x) = 3 - x$ if $0 \le x < 3$, $f(x) = (x-3)^2$ if $x > 3$.

(i) $\lim_{x\to 0^+} f(x) = \lim_{x\to 0^+} (3 - x) = 3$

(ii) $\lim_{x\to 0^-} f(x) = \lim_{x\to 0^-} \sqrt{-x} = 0$

(iii) Because of (i) and (ii), $\lim_{x\to 0} f(x)$ does not exist.

(iv) $\lim_{x\to 3^-} f(x) = \lim_{x\to 3^-} (3 - x) = 0$

(v) $\lim_{x\to 3^+} f(x) = \lim_{x\to 3^+} (x-3)^2 = 0$

(vi) Because of (iv) and (v), $\lim_{x\to 3} f(x) = 0$.

(b) f is discontinuous at 0 since $\lim_{x\to 0} f(x)$ does not exist.

f is discontinuous at 3 since $f(3)$ does not exist.

(c)

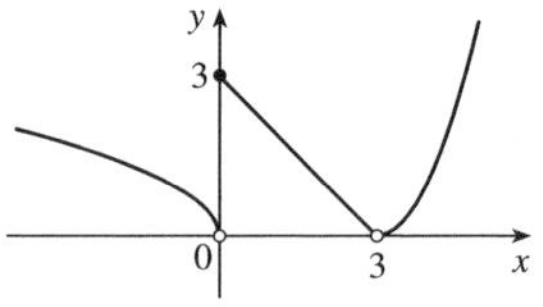

23. $f(x) = 2x^3 + x^2 + 2$ is a polynomial, so it is continuous on $[-2, -1]$ and $f(-2) = -10 < 0 < 1 = f(-1)$. So by the Intermediate Value Theorem there is a number c in $(-2, -1)$ such that $f(c) = 0$, that is, the equation $2x^3 + x^2 + 2 = 0$ has a root in $(-2, -1)$.

25. (a) $s = s(t) = 1 + 2t + t^2/4$. The average velocity over the time interval $[1, 1 + h]$ is

$$v_{\text{ave}} = \frac{s(1+h) - s(1)}{(1+h) - 1} = \frac{1 + 2(1+h) + (1+h)^2/4 - 13/4}{h} = \frac{10h + h^2}{4h} = \frac{10 + h}{4}.$$

So for the following intervals the average velocities are:

(i) $[1, 3]$: $h = 2$, $v_{\text{ave}} = (10 + 2)/4 = 3$ m/s

(ii) $[1, 2]$: $h = 1$, $v_{\text{ave}} = (10 + 1)/4 = 2.75$ m/s

(iii) $[1, 1.5]$: $h = 0.5$, $v_{\text{ave}} = (10 + 0.5)/4 = 2.625$ m/s

(iv) $[1, 1.1]$: $h = 0.1$, $v_{\text{ave}} = (10 + 0.1)/4 = 2.525$ m/s

(b) When $t = 1$, the instantaneous velocity is $\lim_{h\to 0} \dfrac{s(1+h) - s(1)}{h} = \lim_{h\to 0} \dfrac{10 + h}{4} = \dfrac{10}{4} = 2.5$ m/s.

27. Estimating the slopes of the tangent lines at $x = 2$, 3, and 5, we obtain approximate values 0.4, 2, and 0.1. Since the graph is concave downward at $x = 5$, $f''(5)$ is negative. Arranging the numbers in increasing order, we have:

$f''(5) < 0 < f'(5) < f'(2) < 1 < f'(3)$.

29. (a) Estimating $f'(1)$ from the triangle in the graph,

we get $\dfrac{\Delta y}{\Delta x} \approx \dfrac{-0.37}{0.50} = -0.74$.

To estimate $f'(1)$ numerically, we have

$$f'(1) = \lim_{h\to 0} \frac{f(1+h) - f(1)}{h} = \lim_{h\to 0} \frac{e^{-(1+h)^2} - e^{-1}}{h} = y$$

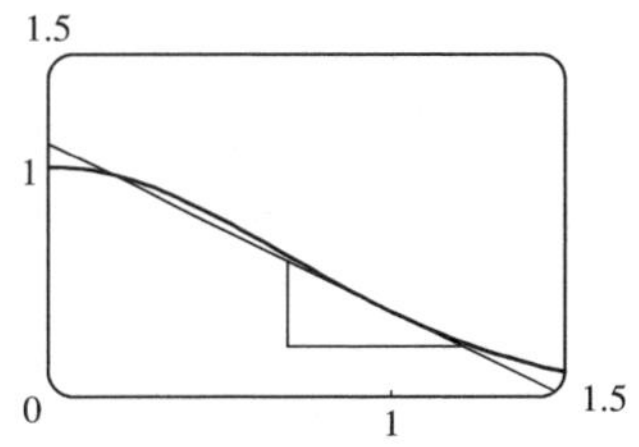

From the table, we have $f'(1) \approx -0.736$.

h	y
0.01	−0.732
0.001	−0.735
0.0001	−0.736
−0.01	−0.739
−0.001	−0.736
−0.0001	−0.736

(b) $y - e^{-1} \approx -0.736(x-1)$ or $y \approx -0.736x + 1.104$

(c) See the graph in part (a).

31. (a) $f'(r)$ is the rate at which the total cost changes with respect to the interest rate. Its units are dollars/(percent per year).

(b) The total cost of paying off the loan is increasing by \$1200/(percent per year) as the interest rate reaches 10%. So if the interest rate goes up from 10% to 11%, the cost goes up approximately \$1200.

(c) As r increases, C increases. So $f'(r)$ will always be positive.

33.

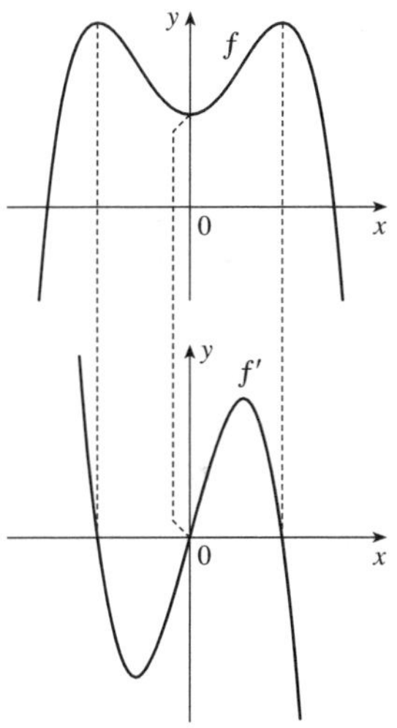

35. (a)
$$f'(x) = \lim_{h\to 0} \frac{f(x+h) - f(x)}{h} = \lim_{h\to 0} \frac{\sqrt{3-5(x+h)} - \sqrt{3-5x}}{h} \frac{\sqrt{3-5(x+h)} + \sqrt{3-5x}}{\sqrt{3-5(x+h)} + \sqrt{3-5x}}$$
$$= \lim_{h\to 0} \frac{[3-5(x+h)] - (3-5x)}{h\left(\sqrt{3-5(x+h)} + \sqrt{3-5x}\right)} = \lim_{h\to 0} \frac{-5}{\sqrt{3-5(x+h)} + \sqrt{3-5x}} = \frac{-5}{2\sqrt{3-5x}}$$

(b) Domain of f: (the radicand must be nonnegative) $3 - 5x \geq 0 \Rightarrow$

$5x \leq 3 \Rightarrow x \in \left(-\infty, \frac{3}{5}\right]$

Domain of f': exclude $\frac{3}{5}$ because it makes the denominator zero;

$x \in \left(-\infty, \frac{3}{5}\right)$

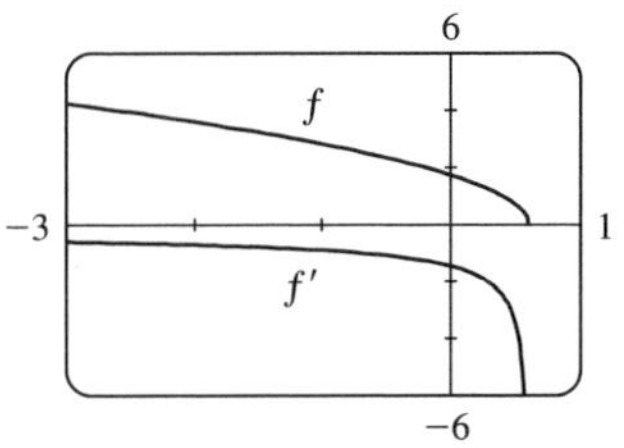

(c) Our answer to part (a) is reasonable because $f'(x)$ is always negative and f is always decreasing.

37. f is not differentiable: at $x = -4$ because f is not continuous, at $x = -1$ because f has a corner, at $x = 2$ because f is not continuous, and at $x = 5$ because f has a vertical tangent.

39. $E'(2002)$ is the rate at which the value of the euro in terms of the U.S. dollar is changing at midyear 2002 in dollars per year. To estimate the value of $E'(2002)$, we will average the difference quotients obtained using the times $t = 2001$ and $t = 2003$.

Let $A = \dfrac{E(2001) - E(2002)}{2001 - 2002} = \dfrac{0.847 - 0.986}{-1} = \dfrac{-0.139}{-1} = 0.139$ and

$B = \dfrac{E(2003) - E(2002)}{2003 - 2002} = \dfrac{1.149 - 0.986}{1} = \dfrac{0.163}{1} = 0.163$. Then

$E'(2002) = \displaystyle\lim_{t \to 2002} \frac{E(t) - E(2002)}{t - 2002} \approx \frac{A + B}{2} = \frac{0.139 + 0.163}{2} = \frac{0.302}{2} = 0.151$ dollars/year.

41. (a) $f'(x) > 0$ on $(-2, 0)$ and $(2, \infty)$ $\Rightarrow$ f is increasing on those intervals. $f'(x) < 0$ on $(-\infty, -2)$ and $(0, 2)$ $\Rightarrow$ f is decreasing on those intervals.

(b) $f'(x) = 0$ at $x = -2$, 0, and 2, so these are where local maxima or minima will occur. At $x = \pm 2$, f' changes from negative to positive, so f has local minima at those values. At $x = 0$, f' changes from positive to negative, so f has a local maximum there.

(c) f' is increasing on $(-\infty, -1)$ and $(1, \infty)$ $\Rightarrow$ $f'' > 0$ and f is concave upward on those intervals. f' is decreasing on $(-1, 1)$ $\Rightarrow$ $f'' < 0$ and f is concave downward on this interval.

(d)

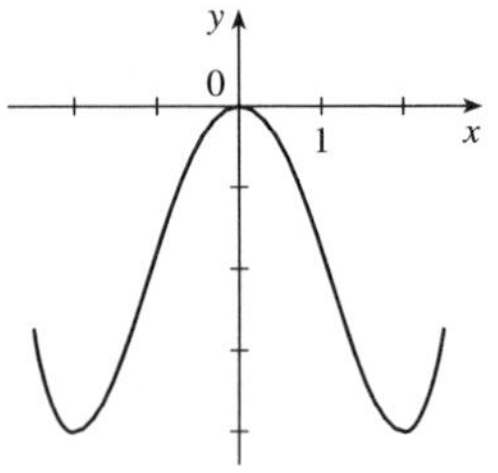

43. $f(0) = 0$, $f'(-2) = f'(1) = f'(9) = 0$, $\displaystyle\lim_{x \to \infty} f(x) = 0$, $\displaystyle\lim_{x \to 6} f(x) = -\infty$, $f'(x) < 0$ on $(-\infty, -2)$, $(1, 6)$, and $(9, \infty)$, $f'(x) > 0$ on $(-2, 1)$ and $(6, 9)$, $f''(x) > 0$ on $(-\infty, 0)$ and $(12, \infty)$, $f''(x) < 0$ on $(0, 6)$ and $(6, 12)$

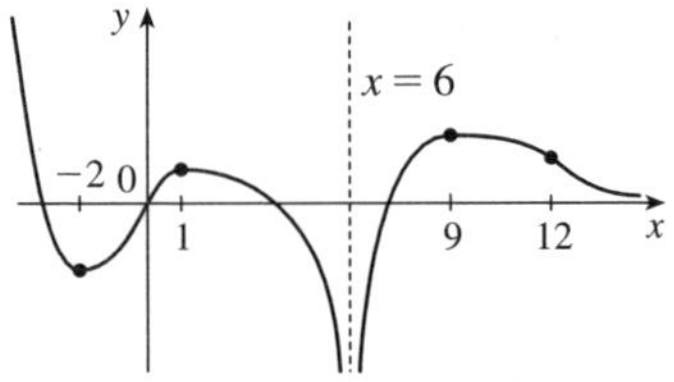

45. (a) Using the data closest to $t = 6$, we have $\dfrac{s(8) - s(6)}{8 - 6} = \dfrac{55 - 29}{2} = 13$ and $\dfrac{s(4) - s(6)}{4 - 6} = \dfrac{12 - 29}{-2} = 8.5$. Averaging these two values gives us $\dfrac{13 + 8.5}{2} = 10.75$ m/s as an estimate for the speed of the car after 6 seconds.

(b) From the graph, it appears that the inflection point is at $(8, 55)$.

(c) The velocity of the car is at a maximum at the inflection point.

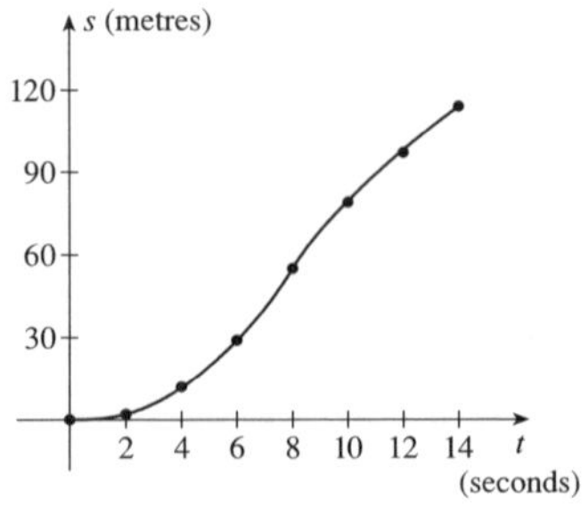

FOCUS ON PROBLEM SOLVING

1. Let $t = \sqrt[6]{x}$, so $x = t^6$. Then $t \to 1$ as $x \to 1$, so

$$\lim_{x\to 1}\frac{\sqrt[3]{x}-1}{\sqrt{x}-1} = \lim_{t\to 1}\frac{t^2-1}{t^3-1} = \lim_{t\to 1}\frac{(t-1)(t+1)}{(t-1)\left(t^2+t+1\right)} = \lim_{t\to 1}\frac{t+1}{t^2+t+1} = \frac{1+1}{1^2+1+1} = \frac{2}{3}.$$

Another method: Multiply both the numerator and the denominator by $(\sqrt{x}+1)\left(\sqrt[3]{x^2}+\sqrt[3]{x}+1\right)$.

3. For $-\frac{1}{2} < x < \frac{1}{2}$, we have $2x-1<0$ and $2x+1>0$, so $|2x-1| = -(2x-1)$ and $|2x+1| = 2x+1$.

Therefore, $\displaystyle\lim_{x\to 0}\frac{|2x-1|-|2x+1|}{x} = \lim_{x\to 0}\frac{-(2x-1)-(2x+1)}{x} = \lim_{x\to 0}\frac{-4x}{x} = \lim_{x\to 0}(-4) = -4.$

5. Since $[\![x]\!] \le x < [\![x]\!]+1$, we have $\dfrac{[\![x]\!]}{[\![x]\!]} \le \dfrac{x}{[\![x]\!]} < \dfrac{[\![x]\!]+1}{[\![x]\!]} \quad\Rightarrow\quad 1 \le \dfrac{x}{[\![x]\!]} < 1 + \dfrac{1}{[\![x]\!]}$ for $x \ge 1$. As $x\to\infty$, $[\![x]\!]\to\infty$,

so $\dfrac{1}{[\![x]\!]} \to 0$ and $1 + \dfrac{1}{[\![x]\!]} \to 1$. Thus, $\displaystyle\lim_{x\to\infty}\frac{x}{[\![x]\!]} = 1$ by the Squeeze Theorem.

7. f is continuous on $(-\infty, a)$ and (a,∞). To make f continuous on $\mathbb{R}$, we must have continuity at a. Thus,

$$\lim_{x\to a^+} f(x) = \lim_{x\to a^-} f(x) \quad\Rightarrow\quad \lim_{x\to a^+} x^2 = \lim_{x\to a^-}(x+1) \quad\Rightarrow\quad a^2 = a+1 \quad\Rightarrow\quad a^2-a-1=0 \quad\Rightarrow$$

[by the quadratic formula] $a = \left(1\pm\sqrt{5}\right)/2 \approx 1.618$ or -0.618.

9. (a) Consider $G(x) = T(x+180^\circ) - T(x)$. Fix any number a. If $G(a) = 0$, we are done: Temperature at $a =$ Temperature at $a+180^\circ$. If $G(a)>0$, then $G(a+180^\circ) = T(a+360^\circ) - T(a+180^\circ) = T(a) - T(a+180^\circ) = -G(a) < 0$. Also, G is continuous since temperature varies continuously. So, by the Intermediate Value Theorem, G has a zero on the interval $[a, a+180^\circ]$. If $G(a)<0$, then a similar argument applies.

(b) Yes. The same argument applies.

(c) The same argument applies for quantities that vary continuously, such as barometric pressure. But one could argue that altitude above sea level is sometimes discontinuous, so the result might not always hold for that quantity.

11. Let a be the x-coordinate of Q. Since the derivative of $y = 1-x^2$ is $y' = -2x$, the slope at Q is $-2a$. But since the triangle is equilateral, $\overline{AO}/\overline{OC} = \sqrt{3}/1$, so the slope at Q is $-\sqrt{3}$. Therefore, we must have that $-2a = -\sqrt{3} \quad\Rightarrow\quad a = \frac{\sqrt{3}}{2}$.

Thus, the point Q has coordinates $\left(\frac{\sqrt{3}}{2}, 1-\left(\frac{\sqrt{3}}{2}\right)^2\right) = \left(\frac{\sqrt{3}}{2}, \frac{1}{4}\right)$ and by symmetry, P has coordinates $\left(-\frac{\sqrt{3}}{2}, \frac{1}{4}\right)$.

13. (a) Put $x=0$ and $y=0$ in the equation: $f(0+0) = f(0)+f(0)+0^2\cdot 0 + 0\cdot 0^2 \quad\Rightarrow\quad f(0) = 2f(0)$. Subtracting $f(0)$ from each side of this equation gives $f(0)=0$.

(b) $\displaystyle f'(0) = \lim_{h\to 0}\frac{f(0+h)-f(0)}{h} = \lim_{h\to 0}\frac{\left[f(0)+f(h)+0^2h+0h^2\right]-f(0)}{h} = \lim_{h\to 0}\frac{f(h)}{h} = \lim_{x\to 0}\frac{f(x)}{x} = 1$

(c) $\displaystyle f'(x) = \lim_{h\to 0}\frac{f(x+h)-f(x)}{h} = \lim_{h\to 0}\frac{\left[f(x)+f(h)+x^2h+xh^2\right]-f(x)}{h} = \lim_{h\to 0}\frac{f(h)+x^2h+xh^2}{h}$

$\displaystyle \qquad = \lim_{h\to 0}\left[\frac{f(h)}{h}+x^2+xh\right] = 1+x^2$

15. $\lim_{x\to a} f(x) = \lim_{x\to a} \left(\frac{1}{2}[f(x)+g(x)] + \frac{1}{2}[f(x)-g(x)]\right) = \frac{1}{2}\lim_{x\to a}[f(x)+g(x)] + \frac{1}{2}\lim_{x\to a}[f(x)-g(x)]$

$= \frac{1}{2}\cdot 2 + \frac{1}{2}\cdot 1 = \frac{3}{2}$,

and $\lim_{x\to a} g(x) = \lim_{x\to a}\Big([f(x)+g(x)] - f(x)\Big) = \lim_{x\to a}[f(x)+g(x)] - \lim_{x\to a} f(x) = 2 - \frac{3}{2} = \frac{1}{2}$.

So $\lim_{x\to a}[f(x)g(x)] = \left[\lim_{x\to a} f(x)\right]\left[\lim_{x\to a} g(x)\right] = \frac{3}{2}\cdot\frac{1}{2} = \frac{3}{4}$.

Another solution: Since $\lim_{x\to a}[f(x)+g(x)]$ and $\lim_{x\to a}[f(x)-g(x)]$ exist, we must have

$\lim_{x\to a}[f(x)+g(x)]^2 = \left(\lim_{x\to a}[f(x)+g(x)]\right)^2$ and $\lim_{x\to a}[f(x)-g(x)]^2 = \left(\lim_{x\to a}[f(x)-g(x)]\right)^2$, so

$\lim_{x\to a}[f(x)\,g(x)] = \lim_{x\to a}\frac{1}{4}\left([f(x)+g(x)]^2 - [f(x)-g(x)]^2\right)$ [because all of the f^2 and g^2 cancel]

$= \frac{1}{4}\left(\lim_{x\to a}[f(x)+g(x)]^2 - \lim_{x\to a}[f(x)-g(x)]^2\right) = \frac{1}{4}(2^2 - 1^2) = \frac{3}{4}$.

17. We are given that $|f(x)| \le x^2$ for all x. In particular, $|f(0)| \le 0$, but $|a| \ge 0$ for all a. The only conclusion is

that $f(0) = 0$. Now $\left|\dfrac{f(x)-f(0)}{x-0}\right| = \left|\dfrac{f(x)}{x}\right| = \dfrac{|f(x)|}{|x|} \le \dfrac{x^2}{|x|} = \dfrac{|x^2|}{|x|} = |x| \;\Rightarrow\; -|x| \le \dfrac{f(x)-f(0)}{x-0} \le |x|$.

But $\lim_{x\to 0}(-|x|) = 0 = \lim_{x\to 0}|x|$, so by the Squeeze Theorem, $\lim_{x\to 0}\dfrac{f(x)-f(0)}{x-0} = 0$. So by the definition of a derivative,

f is differentiable at 0 and, furthermore, $f'(0) = 0$.

3 □ DIFFERENTIATION RULES

3.1 Derivatives of Polynomials and Exponential Functions

1. (a) e is the number such that $\lim_{h \to 0} \dfrac{e^h - 1}{h} = 1$.

(b)

x	$\dfrac{2.7^x - 1}{x}$
-0.001	0.9928
-0.0001	0.9932
0.001	0.9937
0.0001	0.9933

x	$\dfrac{2.8^x - 1}{x}$
-0.001	1.0291
-0.0001	1.0296
0.001	1.0301
0.0001	1.0297

From the tables (to two decimal places),

$$\lim_{h \to 0} \frac{2.7^h - 1}{h} = 0.99 \text{ and } \lim_{h \to 0} \frac{2.8^h - 1}{h} = 1.03.$$

Since $0.99 < 1 < 1.03$, $2.7 < e < 2.8$.

3. $f(x) = 186.5$ is a constant function, so its derivative is 0, that is, $f'(x) = 0$.

5. $f(x) = 9x^4 - 3x^2 + 8 \quad \Rightarrow \quad f'(x) = 9(4x^{4-1}) - 3(2x^{2-1}) + 0 = 36x^3 - 6x$

7. $f(t) = \frac{1}{4}(t^4 + 8) \quad \Rightarrow \quad f'(t) = \frac{1}{4}(t^4 + 8)' = \frac{1}{4}(4t^{4-1} + 0) = t^3$

9. $y = x^{-2/5} \quad \Rightarrow \quad y' = -\frac{2}{5}x^{(-2/5)-1} = -\frac{2}{5}x^{-7/5} = -\dfrac{2}{5x^{7/5}}$

11. $G(x) = \sqrt{x} - 2e^x = x^{1/2} - 2e^x \quad \Rightarrow \quad G'(x) = \frac{1}{2}x^{-1/2} - 2e^x = \dfrac{1}{2\sqrt{x}} - 2e^x$

13. $V(r) = \frac{4}{3}\pi r^3 \quad \Rightarrow \quad V'(r) = \frac{4}{3}\pi(3r^2) = 4\pi r^2$

15. $F(x) = (16x)^3 = 4096x^3 \quad \Rightarrow \quad F'(x) = 4096(3x^2) = 12\,288x^2$

17. $y = 4\pi^2 \quad \Rightarrow \quad y' = 0$ since $4\pi^2$ is a constant.

19. $y = \dfrac{x^2 + 4x + 3}{\sqrt{x}} = x^{3/2} + 4x^{1/2} + 3x^{-1/2} \quad \Rightarrow$

$y' = \frac{3}{2}x^{1/2} + 4\left(\frac{1}{2}\right)x^{-1/2} + 3\left(-\frac{1}{2}\right)x^{-3/2} = \frac{3}{2}\sqrt{x} + \dfrac{2}{\sqrt{x}} - \dfrac{3}{2x\sqrt{x}}$

$\left[\text{note that } x^{3/2} = x^{2/2} \cdot x^{1/2} = x\sqrt{x}\right]$

21. $v = t^2 - \dfrac{1}{\sqrt[4]{t^3}} = t^2 - t^{-3/4} \quad \Rightarrow \quad v' = 2t - \left(-\frac{3}{4}\right)t^{-7/4} = 2t + \dfrac{3}{4t^{7/4}} = 2t + \dfrac{3}{4t\sqrt[4]{t^3}}$

23. $z = \dfrac{A}{y^{10}} + Be^y = Ay^{-10} + Be^y \quad \Rightarrow \quad z' = -10Ay^{-11} + Be^y = -\dfrac{10A}{y^{11}} + Be^y$

25. $y = x^4 + 2e^x \quad \Rightarrow \quad y' = 4x^3 + 2e^x$. At $(0, 2)$, $y' = 2$ and an equation of the tangent line is $y - 2 = 2(x - 0)$ or $y = 2x + 2$. The slope of the normal line is $-\frac{1}{2}$ (the negative reciprocal of 2) and an equation of the normal line is $y - 2 = -\frac{1}{2}(x - 0)$ or $y = -\frac{1}{2}x + 2$.

27. $y = f(x) = x + \sqrt{x} \quad \Rightarrow \quad f'(x) = 1 + \frac{1}{2}x^{-1/2}$.

So the slope of the tangent line at $(1, 2)$ is $f'(1) = 1 + \frac{1}{2}(1) = \frac{3}{2}$ and its equation is $y - 2 = \frac{3}{2}(x - 1)$ or $y = \frac{3}{2}x + \frac{1}{2}$.

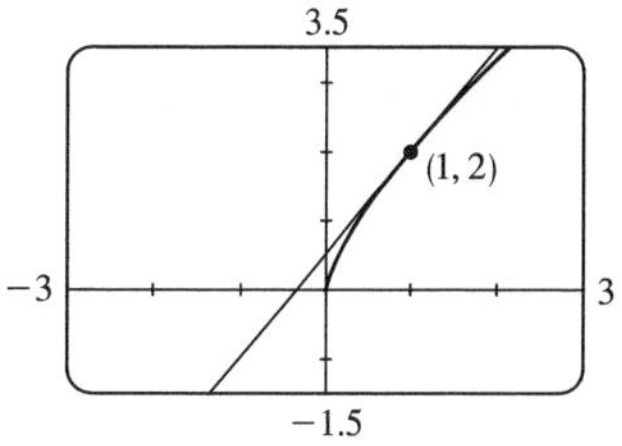

29. $f(x) = e^x - 5x \;\Rightarrow\; f'(x) = e^x - 5.$

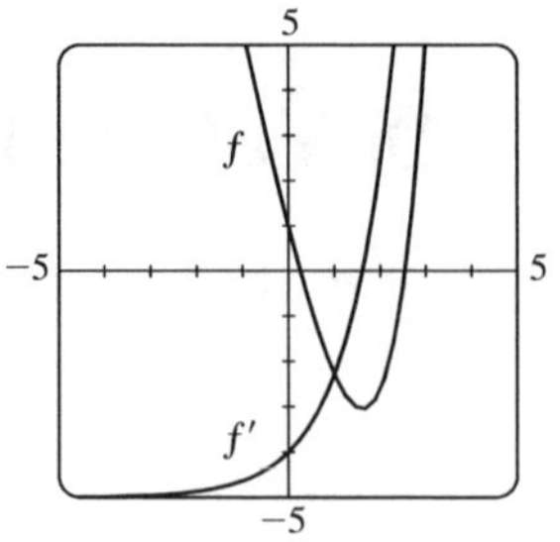

Notice that $f'(x) = 0$ when f has a horizontal tangent, f' is positive when f is increasing, and f' is negative when f is decreasing.

31. $f(x) = x - 3x^{1/3} \;\Rightarrow\; f'(x) = 1 - x^{-2/3} = 1 - 1/x^{2/3}.$

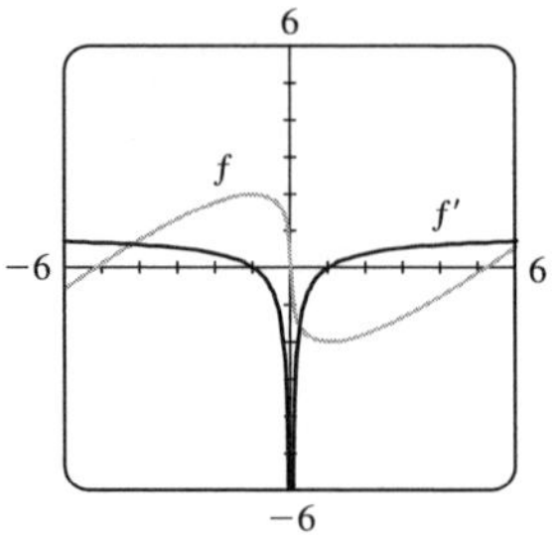

Note that $f'(x) = 0$ when f has a horizontal tangent, f' is positive when f is increasing, and f' is negative when f is decreasing.

33. To graphically estimate the value of $f'(1)$ for $f(x) = 3x^2 - x^3$, we'll graph f in the viewing rectangle $[1 - 0.1, 1 + 0.1]$ by $[f(0.9), f(1.1)]$, as shown in the figure. [When assigning values to the window variables, it is convenient to use $\mathrm{Y}_1(0.9)$ for $\mathrm{Y}_{\min}$ and $\mathrm{Y}_1(1.1)$ for $\mathrm{Y}_{\max}$.] If we have sufficiently zoomed in on the graph of f, we should obtain a graph that looks like a diagonal line; if not, graph again with $1 - 0.01$ and $1 + 0.01$, etc.

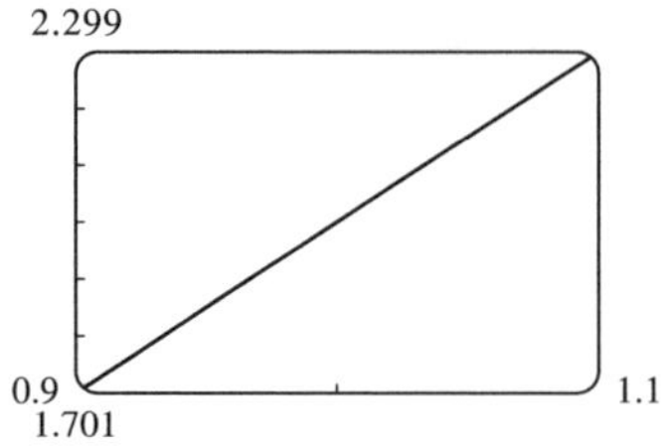

Estimated value:

$$f'(1) \approx \frac{2.299 - 1.701}{1.1 - 0.9} = \frac{0.589}{0.2} = 2.99.$$

Exact value: $f(x) = 3x^2 - x^3 \;\Rightarrow\; f'(x) = 6x - 3x^2$, so $f'(1) = 6 - 3 = 3$.

35. (a)

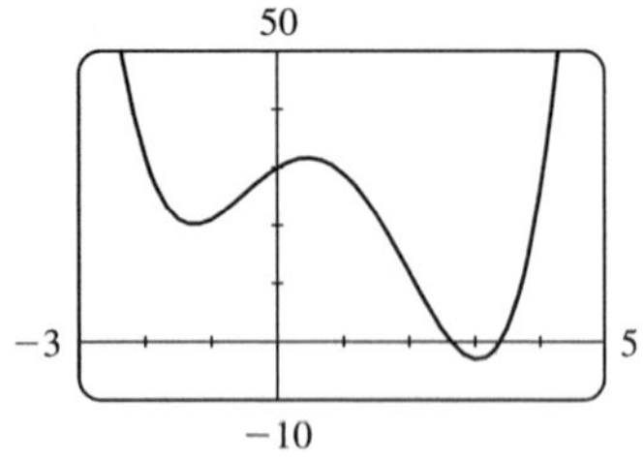

(b) From the graph in part (a), it appears that f' is zero at $x_1 \approx -1.25$, $x_2 \approx 0.5$, and $x_3 \approx 3$. The slopes are negative (so f' is negative) on $(-\infty, x_1)$ and (x_2, x_3). The slopes are positive (so f' is positive) on (x_1, x_2) and (x_3, ∞).

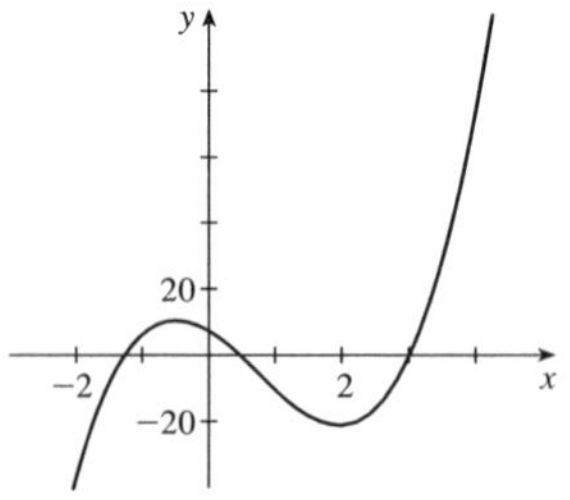

(c) $f(x) = x^4 - 3x^3 - 6x^2 + 7x + 30 \;\Rightarrow$

$f'(x) = 4x^3 - 9x^2 - 12x + 7$

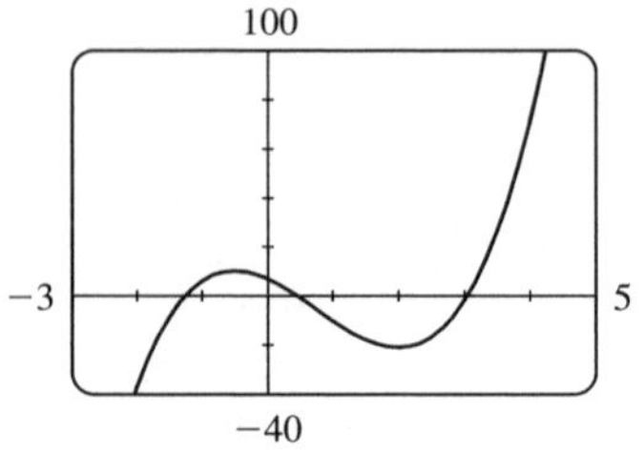

37. $f(x) = x^4 - 3x^3 + 16x \quad \Rightarrow \quad f'(x) = 4x^3 - 9x^2 + 16 \quad \Rightarrow \quad f''(x) = 12x^2 - 18x$

39. $f(x) = 2x - 5x^{3/4} \quad \Rightarrow \quad f'(x) = 2 - \frac{15}{4}x^{-1/4} \quad \Rightarrow \quad f''(x) = \frac{15}{16}x^{-5/4}$

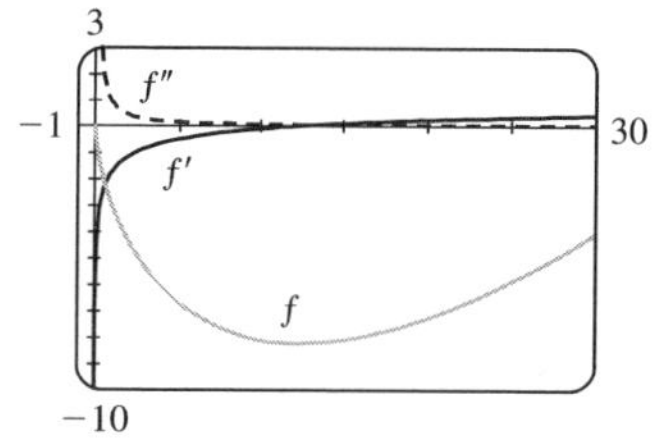

Note that f' is negative when f is decreasing and positive when f is increasing. f'' is always positive since f' is always increasing.

41. (a) $s = t^3 - 3t \quad \Rightarrow \quad v(t) = s'(t) = 3t^2 - 3 \quad \Rightarrow \quad a(t) = v'(t) = 6t$

(b) $a(2) = 6(2) = 12 \text{ m/s}^2$

(c) $v(t) = 3t^2 - 3 = 0$ when $t^2 = 1$, that is, $t = 1$ and $a(1) = 6 \text{ m/s}^2$.

43. $f(x) = 1 + 2e^x - 3x \quad \Rightarrow \quad f'(x) = 2e^x - 3$. $f'(x) > 0 \quad \Rightarrow \quad 2e^x - 3 > 0 \quad \Rightarrow \quad 2e^x > 3 \quad \Rightarrow$ $e^x > 1.5 \quad \Rightarrow \quad x > \ln 1.5 \approx 0.41$. f is increasing when f' is positive; that is, on $(\ln 1.5, \infty)$.

45. The curve $y = 2x^3 + 3x^2 - 12x + 1$ has a horizontal tangent when $y' = 6x^2 + 6x - 12 = 0 \quad \Leftrightarrow \quad 6(x^2 + x - 2) = 0 \quad \Leftrightarrow$ $6(x + 2)(x - 1) = 0 \quad \Leftrightarrow \quad x = -2$ or $x = 1$. The points on the curve are $(-2, 21)$ and $(1, -6)$.

47. $y = 6x^3 + 5x - 3 \quad \Rightarrow \quad m = y' = 18x^2 + 5$, but $x^2 \geq 0$ for all x, so $m \geq 5$ for all x.

49. The slope of the line $12x - y = 1$ (or $y = 12x - 1$) is 12, so the slope of both lines tangent to the curve is 12. $y = 1 + x^3 \quad \Rightarrow \quad y' = 3x^2$. Thus, $3x^2 = 12 \quad \Rightarrow \quad x^2 = 4 \quad \Rightarrow \quad x = \pm 2$, which are the x-coordinates at which the tangent lines have slope 12. The points on the curve are $(2, 9)$ and $(-2, -7)$, so the tangent line equations are $y - 9 = 12(x - 2)$ or $y = 12x - 15$ and $y + 7 = 12(x + 2)$ or $y = 12x + 17$.

51. The slope of $y = x^2 - 5x + 4$ is given by $m = y' = 2x - 5$. The slope of $x - 3y = 5 \quad \Leftrightarrow \quad y = \frac{1}{3}x - \frac{5}{3}$ is $\frac{1}{3}$, so the desired normal line must have slope $\frac{1}{3}$, and hence, the tangent line to the parabola must have slope -3. This occurs if $2x - 5 = -3 \quad \Rightarrow \quad 2x = 2 \quad \Rightarrow \quad x = 1$. When $x = 1$, $y = 1^2 - 5(1) + 4 = 0$, and an equation of the normal line is $y - 0 = \frac{1}{3}(x - 1)$ or $y = \frac{1}{3}x - \frac{1}{3}$.

53.

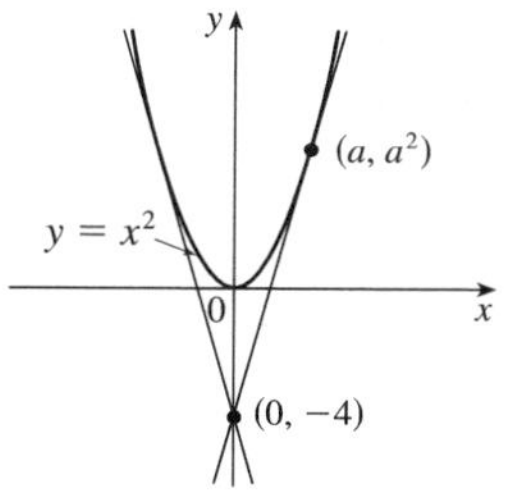

Let (a, a^2) be a point on the parabola at which the tangent line passes through the point $(0, -4)$. The tangent line has slope $2a$ and equation $y - (-4) = 2a(x - 0)$ $\Leftrightarrow \quad y = 2ax - 4$. Since (a, a^2) also lies on the line, $a^2 = 2a(a) - 4$, or $a^2 = 4$. So $a = \pm 2$ and the points are $(2, 4)$ and $(-2, 4)$.

55. $$f'(x) = \lim_{h \to 0} \frac{f(x + h) - f(x)}{h} = \lim_{h \to 0} \frac{\dfrac{1}{x + h} - \dfrac{1}{x}}{h} = \lim_{h \to 0} \frac{x - (x + h)}{hx(x + h)} = \lim_{h \to 0} \frac{-h}{hx(x + h)} = \lim_{h \to 0} \frac{-1}{x(x + h)} = -\frac{1}{x^2}$$

57. Let $P(x) = ax^2 + bx + c$. Then $P'(x) = 2ax + b$ and $P''(x) = 2a$. $P''(2) = 2 \quad \Rightarrow \quad 2a = 2 \quad \Rightarrow \quad a = 1$.

$P'(2) = 3 \quad \Rightarrow \quad 2(1)(2) + b = 3 \quad \Rightarrow \quad 4 + b = 3 \quad \Rightarrow \quad b = -1$.

$P(2) = 5 \quad \Rightarrow \quad 1(2)^2 + (-1)(2) + c = 5 \quad \Rightarrow \quad 2 + c = 5 \quad \Rightarrow \quad c = 3$. So $P(x) = x^2 - x + 3$.

59. (a) At this stage, we would guess that an antiderivative of x^2 must have x^3 in it. Differentiating x^3 gives us $3x^2$, so we know that we must divide x^3 by 3. That gives us $F(x) = \frac{1}{3}x^3$. Checking, we have $F'(x) = \frac{1}{3}(3x^2) = x^2 = f(x)$. Because we can add an arbitrary constant C to F without changing its derivative, we have an infinite number of antiderivatives of the form $F(x) = \frac{1}{3}x^3 + C$.

(b) As in part (a), antiderivatives of $f(x) = x^3$ and $f(x) = x^4$ are $F(x) = \frac{1}{4}x^4 + C$ and $F(x) = \frac{1}{5}x^5 + C$.

(c) Similarly, an antiderivative for $f(x) = x^n$ is $F(x) = \dfrac{1}{n+1}x^{n+1} + C$, since then

$F'(x) = \dfrac{1}{n+1}[(n+1)x^n] = x^n = f(x)$ for $n \neq -1$.

61. $y = f(x) = ax^2 \;\Rightarrow\; f'(x) = 2ax$. So the slope of the tangent to the parabola at $x = 2$ is $m = 2a(2) = 4a$. The slope of the given line, $2x + y = b \;\Leftrightarrow\; y = -2x + b$, is seen to be -2, so we must have $4a = -2 \;\Leftrightarrow\; a = -\frac{1}{2}$. So when $x = 2$, the point in question has y-coordinate $-\frac{1}{2}\cdot 2^2 = -2$. Now we simply require that the given line, whose equation is $2x + y = b$, pass through the point $(2, -2)$: $2(2) + (-2) = b \;\Leftrightarrow\; b = 2$. So we must have $a = -\frac{1}{2}$ and $b = 2$.

63. $y = f(x) = ax^3 + bx^2 + cx + d \;\Rightarrow\; f'(x) = 3ax^2 + 2bx + c$. The point $(-2, 6)$ is on f, so $f(-2) = 6 \;\Rightarrow$ $-8a + 4b - 2c + d = 6$ **(1)**. The point $(2, 0)$ is on f, so $f(2) = 0 \;\Rightarrow\; 8a + 4b + 2c + d = 0$ **(2)**. Since there are horizontal tangents at $(-2, 6)$ and $(2, 0)$, $f'(\pm 2) = 0$. $f'(-2) = 0 \;\Rightarrow\; 12a - 4b + c = 0$ **(3)** and $f'(2) = 0 \;\Rightarrow$ $12a + 4b + c = 0$ **(4)**. Subtracting equation **(3)** from **(4)** gives $8b = 0 \;\Rightarrow\; b = 0$. Adding **(1)** and **(2)** gives $8b + 2d = 6$, so $d = 3$ since $b = 0$. From **(3)** we have $c = -12a$, so **(2)** becomes $8a + 4(0) + 2(-12a) + 3 = 0 \;\Rightarrow\; 3 = 16a \;\Rightarrow$ $a = \frac{3}{16}$. Now $c = -12a = -12\left(\frac{3}{16}\right) = -\frac{9}{4}$ and the desired cubic function is $y = \frac{3}{16}x^3 - \frac{9}{4}x + 3$.

65. *Solution 1:* Let $f(x) = x^{1000}$. Then, by the definition of a derivative, $f'(1) = \displaystyle\lim_{x\to 1}\frac{f(x) - f(1)}{x - 1} = \lim_{x\to 1}\frac{x^{1000} - 1}{x - 1}$. But this is just the limit we want to find, and we know (from the Power Rule) that $f'(x) = 1000x^{999}$, so $f'(1) = 1000(1)^{999} = 1000$. So $\displaystyle\lim_{x\to 1}\frac{x^{1000} - 1}{x - 1} = 1000$.

Solution 2: Note that $(x^{1000} - 1) = (x - 1)(x^{999} + x^{998} + x^{997} + \cdots + x^2 + x + 1)$. So

$$\lim_{x\to 1}\frac{x^{1000} - 1}{x - 1} = \lim_{x\to 1}\frac{(x - 1)(x^{999} + x^{998} + x^{997} + \cdots + x^2 + x + 1)}{x - 1} = \lim_{x\to 1}(x^{999} + x^{998} + x^{997} + \cdots + x^2 + x + 1)$$
$$= \underbrace{1 + 1 + 1 + \cdots + 1 + 1 + 1}_{1000 \text{ ones}} = 1000, \text{ as above.}$$

67. $y = x^2 \;\Rightarrow\; y' = 2x$, so the slope of a tangent line at the point (a, a^2) is $y' = 2a$ and the slope of a normal line is $-1/(2a)$, for $a \neq 0$. The slope of the normal line through the points (a, a^2) and $(0, c)$ is $\dfrac{a^2 - c}{a - 0}$, so $\dfrac{a^2 - c}{a} = -\dfrac{1}{2a} \;\Rightarrow$ $a^2 - c = -\frac{1}{2} \;\Rightarrow\; a^2 = c - \frac{1}{2}$. The last equation has two solutions if $c > \frac{1}{2}$, one solution if $c = \frac{1}{2}$, and no solution if $c < \frac{1}{2}$. Since the y-axis is normal to $y = x^2$ regardless of the value of c (this is the case for $a = 0$), we have three normal lines if $c > \frac{1}{2}$ and one normal line if $c \leq \frac{1}{2}$.

3.2 The Product and Quotient Rules

1. Product Rule: $y = (x^2 + 1)(x^3 + 1) \;\Rightarrow$

$$y' = (x^2 + 1)(3x^2) + (x^3 + 1)(2x) = 3x^4 + 3x^2 + 2x^4 + 2x = 5x^4 + 3x^2 + 2x.$$

Multiplying first: $y = (x^2 + 1)(x^3 + 1) = x^5 + x^3 + x^2 + 1 \;\Rightarrow\; y' = 5x^4 + 3x^2 + 2x$ (equivalent).

3. By the Product Rule, $f(x) = x^2e^x \;\Rightarrow\; f'(x) = x^2\dfrac{d}{dx}(e^x) + e^x\dfrac{d}{dx}(x^2) = x^2e^x + e^x(2x) = xe^x(x + 2)$.

5. By the Quotient Rule, $y = \dfrac{e^x}{x^2} \;\Rightarrow$

$$y' = \frac{x^2 \dfrac{d}{dx}(e^x) - e^x \dfrac{d}{dx}(x^2)}{(x^2)^2} = \frac{x^2(e^x) - e^x(2x)}{x^4} = \frac{xe^x(x-2)}{x^4} = \frac{e^x(x-2)}{x^3}.$$

The notations $\overset{\text{PR}}{\Rightarrow}$ and $\overset{\text{QR}}{\Rightarrow}$ indicate the use of the Product and Quotient Rules, respectively.

7. $h(x) = \dfrac{x+2}{x-1} \;\overset{\text{QR}}{\Rightarrow}\; h'(x) = \dfrac{(x-1)(1) - (x+2)(1)}{(x-1)^2} = \dfrac{x-1-x-2}{(x-1)^2} = -\dfrac{3}{(x-1)^2}$

9. $F(y) = \left(\dfrac{1}{y^2} - \dfrac{3}{y^4}\right)(y + 5y^3) = (y^{-2} - 3y^{-4})(y + 5y^3) \;\overset{\text{PR}}{\Rightarrow}$

$$\begin{aligned} F'(y) &= (y^{-2} - 3y^{-4})(1 + 15y^2) + (y + 5y^3)(-2y^{-3} + 12y^{-5}) \\ &= (y^{-2} + 15 - 3y^{-4} - 45y^{-2}) + (-2y^{-2} + 12y^{-4} - 10 + 60y^{-2}) \\ &= 5 + 14y^{-2} + 9y^{-4} \text{ or } 5 + 14/y^2 + 9/y^4 \end{aligned}$$

11. $y = \dfrac{t^2}{3t^2 - 2t + 1} \;\overset{\text{QR}}{\Rightarrow}$

$$y' = \frac{(3t^2-2t+1)(2t) - t^2(6t-2)}{(3t^2-2t+1)^2} = \frac{2t[3t^2-2t+1-t(3t-1)]}{(3t^2-2t+1)^2} = \frac{2t(3t^2-2t+1-3t^2+t)}{(3t^2-2t+1)^2} = \frac{2t(1-t)}{(3t^2-2t+1)^2}$$

13. $y = (r^2 - 2r)e^r \;\overset{\text{PR}}{\Rightarrow}\; y' = (r^2 - 2r)(e^r) + e^r(2r - 2) = e^r(r^2 - 2r + 2r - 2) = e^r(r^2 - 2)$

15. $y = \dfrac{v^3 - 2v\sqrt{v}}{v} = v^2 - 2\sqrt{v} = v^2 - 2v^{1/2} \;\Rightarrow\; y' = 2v - 2\left(\tfrac{1}{2}\right)v^{-1/2} = 2v - v^{-1/2}.$

We can change the form of the answer as follows: $2v - v^{-1/2} = 2v - \dfrac{1}{\sqrt{v}} = \dfrac{2v\sqrt{v} - 1}{\sqrt{v}} = \dfrac{2v^{3/2} - 1}{\sqrt{v}}$

17. $f(x) = \dfrac{A}{B + Ce^x} \;\overset{\text{QR}}{\Rightarrow}\; f'(x) = \dfrac{(B + Ce^x)\cdot 0 - A(Ce^x)}{(B + Ce^x)^2} = -\dfrac{ACe^x}{(B + Ce^x)^2}$

19. $f(x) = \dfrac{x}{x + c/x} \;\Rightarrow\; f'(x) = \dfrac{(x + c/x)(1) - x(1 - c/x^2)}{\left(x + \dfrac{c}{x}\right)^2} = \dfrac{x + c/x - x + c/x}{\left(\dfrac{x^2 + c}{x}\right)^2} = \dfrac{2c/x}{\dfrac{(x^2+c)^2}{x^2}} \cdot \dfrac{x^2}{x^2} = \dfrac{2cx}{(x^2+c)^2}$

21. $y = 2xe^x \;\Rightarrow\; y' = 2(x \cdot e^x + e^x \cdot 1) = 2e^x((x+1)$. At $(0,0)$, $y' = 2e^0(0+1) = 2 \cdot 1 \cdot 1 = 2$, and an equation of the tangent line is $y - 0 = 2(x - 0)$, or $y = 2x$. The slope of the normal line is $-\frac{1}{2}$, so an equation of the normal line is $y - 0 = -\frac{1}{2}(x - 0)$, or $y = -\frac{1}{2}x$.

23. (a) $y = f(x) = \dfrac{1}{1 + x^2} \;\Rightarrow$

$f'(x) = \dfrac{(1+x^2)(0) - 1(2x)}{(1+x^2)^2} = \dfrac{-2x}{(1+x^2)^2}$. So the slope of the tangent line at the point $\left(-1, \frac{1}{2}\right)$ is $f'(-1) = \dfrac{2}{2^2} = \frac{1}{2}$ and its equation is $y - \frac{1}{2} = \frac{1}{2}(x+1)$ or $y = \frac{1}{2}x + 1$.

(b)

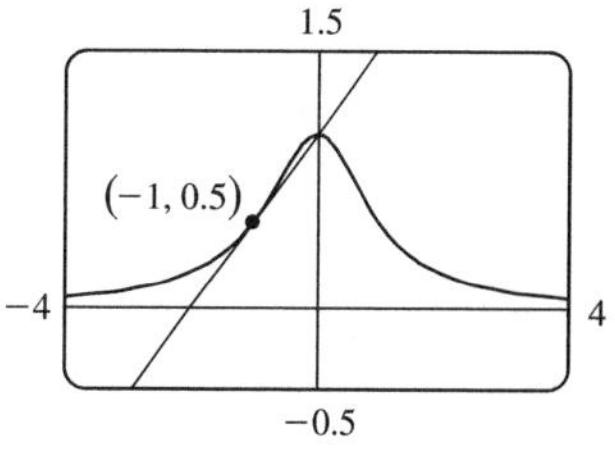

25. (a) $f(x) = \dfrac{e^x}{x^3} \;\Rightarrow\; f'(x) = \dfrac{x^3(e^x) - e^x(3x^2)}{(x^3)^2} = \dfrac{x^2e^x(x-3)}{x^6} = \dfrac{e^x(x-3)}{x^4}$

(b)

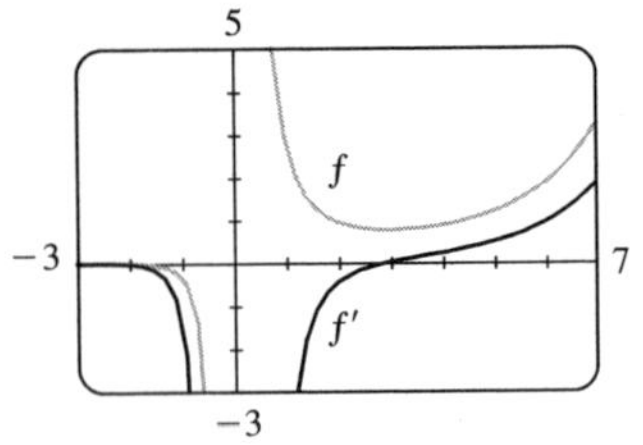

$f' = 0$ when f has a horizontal tangent line, f' is negative when f is decreasing, and f' is positive when f is increasing.

27. (a) $f(x) = (x-1)e^x \Rightarrow f'(x) = (x-1)e^x + e^x(1) = e^x(x-1+1) = xe^x$.
$f''(x) = x(e^x) + e^x(1) = e^x(x+1)$

(b)

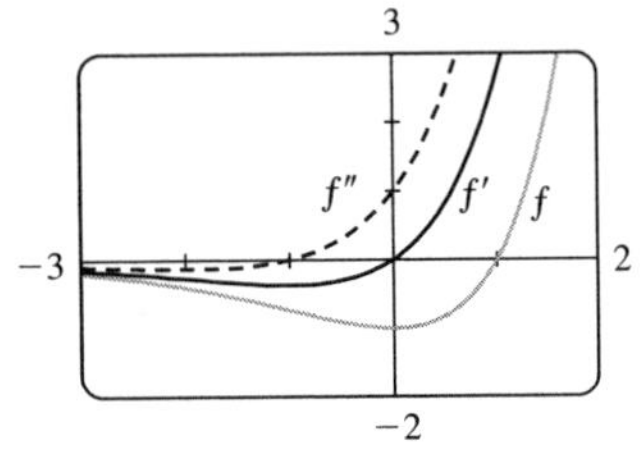

$f' = 0$ when f has a horizontal tangent and $f'' = 0$ when f' has a horizontal tangent. f' is negative when f is decreasing and positive when f is increasing. f'' is negative when f' is decreasing and positive when f' is increasing. f'' is negative when f is concave down and positive when f is concave up.

29. $f(x) = \dfrac{x^2}{1+x} \quad\Rightarrow\quad f'(x) = \dfrac{(1+x)(2x) - x^2(1)}{(1+x)^2} = \dfrac{2x + 2x^2 - x^2}{(1+x)^2} = \dfrac{x^2+2x}{x^2+2x+1} \quad\Rightarrow$

$$f''(x) = \frac{(x^2+2x+1)(2x+2) - (x^2+2x)(2x+2)}{(x^2+2x+1)^2} = \frac{(2x+2)(x^2+2x+1-x^2-2x)}{[(x+1)^2]^2}$$

$$= \frac{2(x+1)(1)}{(x+1)^4} = \frac{2}{(x+1)^3},$$

so $f''(1) = \dfrac{2}{(1+1)^3} = \dfrac{2}{8} = \dfrac{1}{4}$.

31. We are given that $f(5) = 1$, $f'(5) = 6$, $g(5) = -3$, and $g'(5) = 2$.

(a) $(fg)'(5) = f(5)g'(5) + g(5)f'(5) = (1)(2) + (-3)(6) = 2 - 18 = -16$

(b) $\left(\dfrac{f}{g}\right)'(5) = \dfrac{g(5)f'(5) - f(5)g'(5)}{[g(5)]^2} = \dfrac{(-3)(6) - (1)(2)}{(-3)^2} = -\dfrac{20}{9}$

(c) $\left(\dfrac{g}{f}\right)'(5) = \dfrac{f(5)g'(5) - g(5)f'(5)}{[f(5)]^2} = \dfrac{(1)(2) - (-3)(6)}{(1)^2} = 20$

33. $f(x) = e^x g(x) \quad\Rightarrow\quad f'(x) = e^x g'(x) + g(x)e^x = e^x[g'(x) + g(x)]$. $f'(0) = e^0[g'(0) + g(0)] = 1(5+2) = 7$

35. (a) From the graphs of f and g, we obtain the following values: $f(1) = 2$ since the point $(1,2)$ is on the graph of f; $g(1) = 1$ since the point $(1,1)$ is on the graph of g; $f'(1) = 2$ since the slope of the line segment between $(0,0)$ and $(2,4)$ is $\dfrac{4-0}{2-0} = 2$; $g'(1) = -1$ since the slope of the line segment between $(-2,4)$ and $(2,0)$ is $\dfrac{0-4}{2-(-2)} = -1$.

Now $u(x) = f(x)g(x)$, so $u'(1) = f(1)g'(1) + g(1)\,f'(1) = 2 \cdot (-1) + 1 \cdot 2 = 0$.

(b) $v(x) = f(x)/g(x)$, so $v'(5) = \dfrac{g(5)f'(5) - f(5)g'(5)}{[g(5)]^2} = \dfrac{2\left(-\frac{1}{3}\right) - 3 \cdot \frac{2}{3}}{2^2} = \dfrac{-\frac{8}{3}}{4} = -\dfrac{2}{3}$

37. (a) $y = xg(x) \quad\Rightarrow\quad y' = xg'(x) + g(x) \cdot 1 = xg'(x) + g(x)$

(b) $y = \dfrac{x}{g(x)} \quad\Rightarrow\quad y' = \dfrac{g(x) \cdot 1 - xg'(x)}{[g(x)]^2} = \dfrac{g(x) - xg'(x)}{[g(x)]^2}$

(c) $y = \dfrac{g(x)}{x} \quad\Rightarrow\quad y' = \dfrac{xg'(x) - g(x) \cdot 1}{(x)^2} = \dfrac{xg'(x) - g(x)}{x^2}$

39. If $P(t)$ denotes the population at time t and $A(t)$ the average annual income, then $T(t) = P(t)A(t)$ is the total personal income. The rate at which $T(t)$ is rising is given by $T'(t) = P(t)A'(t) + A(t)P'(t) \;\Rightarrow$

$$\begin{aligned} T'(1999) &= P(1999)A'(1999) + A(1999)P'(1999) = (961\,400)(\$1400/\text{yr}) + (\$30\,593)(9200/\text{yr}) \\ &= \$1\,345\,960\,000/\text{yr} + \$281\,455\,600/\text{yr} = \$1\,627\,415\,600/\text{yr} \end{aligned}$$

So the total personal income was rising by about \$1.627 billion per year in 1999.

The term $P(t)A'(t) \approx \$1.346$ billion represents the portion of the rate of change of total income due to the existing population's increasing income. The term $A(t)P'(t) \approx \$281$ million represents the portion of the rate of change of total income due to increasing population.

41. f is increasing when f' is positive. $f(x) = x^3e^x \;\Rightarrow\; f'(x) = x^3e^x + e^x(3x^2) = x^2e^x(x+3)$. Now $x^2 \geq 0$ and $e^x > 0$ for all x, so $f'(x) > 0$ when $x + 3 > 0$ and $x \neq 0$; that is, when $x \in (-3, 0) \cup (0, \infty)$. So f is increasing on $(-3, \infty)$.

43. If $y = f(x) = \dfrac{x}{x+1}$, then $f'(x) = \dfrac{(x+1)(1) - x(1)}{(x+1)^2} = \dfrac{1}{(x+1)^2}$. When $x = a$, the equation of the tangent line is $y - \dfrac{a}{a+1} = \dfrac{1}{(a+1)^2}(x - a)$. This line passes through $(1, 2)$ when $2 - \dfrac{a}{a+1} = \dfrac{1}{(a+1)^2}(1 - a) \;\Leftrightarrow$ $2(a+1)^2 - a(a+1) = 1 - a \;\Leftrightarrow\; 2a^2 + 4a + 2 - a^2 - a - 1 + a = 0 \;\Leftrightarrow\; a^2 + 4a + 1 = 0.$

The quadratic formula gives the roots of this equation as $a = \dfrac{-4 \pm \sqrt{4^2 - 4(1)(1)}}{2(1)} = \dfrac{-4 \pm \sqrt{12}}{2} = -2 \pm \sqrt{3}$,

so there are two such tangent lines. Since

$$\begin{aligned} f\left(-2 \pm \sqrt{3}\right) &= \frac{-2 \pm \sqrt{3}}{-2 \pm \sqrt{3} + 1} = \frac{-2 \pm \sqrt{3}}{-1 \pm \sqrt{3}} \cdot \frac{-1 \mp \sqrt{3}}{-1 \mp \sqrt{3}} \\ &= \frac{2 \pm 2\sqrt{3} \mp \sqrt{3} - 3}{1 - 3} = \frac{-1 \pm \sqrt{3}}{-2} = \frac{1 \mp \sqrt{3}}{2}, \end{aligned}$$

the lines touch the curve at $A\left(-2 + \sqrt{3}, \frac{1 - \sqrt{3}}{2}\right) \approx (-0.27, -0.37)$ and $B\left(-2 - \sqrt{3}, \frac{1 + \sqrt{3}}{2}\right) \approx (-3.73, 1.37)$.

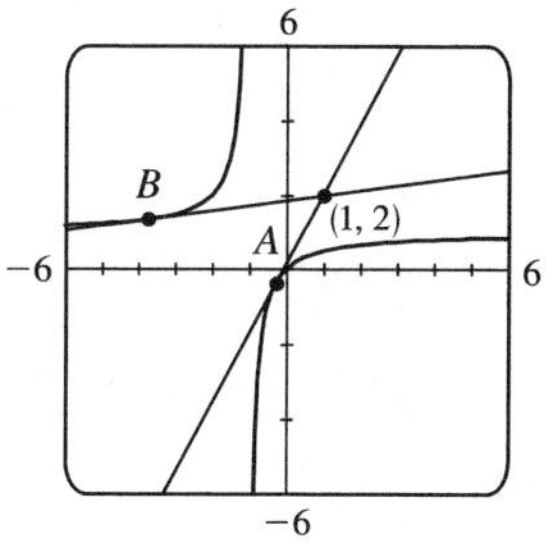

We will sometimes use the form $f'g + fg'$ rather than the form $fg' + gf'$ for the Product Rule.

45. (a) $(fgh)' = [(fg)h]' = (fg)'h + (fg)h' = (f'g + fg')h + (fg)h' = f'gh + fg'h + fgh'$

(b) Putting $f = g = h$ in part (a), we have $\dfrac{d}{dx}[f(x)]^3 = (fff)' = f'ff + ff'f + fff' = 3fff' = 3[f(x)]^2 f'(x)$.

(c) $\dfrac{d}{dx}\left(e^{3x}\right) = \dfrac{d}{dx}\left(e^x\right)^3 = 3(e^x)^2e^x = 3e^{2x}e^x = 3e^{3x}$

47. For $f(x) = x^2e^x$, $f'(x) = x^2e^x + e^x(2x) = e^x(x^2 + 2x)$. Similarly, we have

$$\begin{aligned} f''(x) &= e^x(x^2 + 4x + 2) \\ f'''(x) &= e^x(x^2 + 6x + 6) \\ f^{(4)}(x) &= e^x(x^2 + 8x + 12) \\ f^{(5)}(x) &= e^x(x^2 + 10x + 20) \end{aligned}$$

It appears that the coefficient of x in the quadratic term increases by 2 with each differentiation. The pattern for the constant terms seems to be $0 = 1 \cdot 0$, $2 = 2 \cdot 1$, $6 = 3 \cdot 2$, $12 = 4 \cdot 3$, $20 = 5 \cdot 4$. So a reasonable guess is that $f^{(n)}(x) = e^x[x^2 + 2nx + n(n-1)]$.

Proof: Let S_n be the statement that $f^{(n)}(x) = e^x[x^2 + 2nx + n(n-1)]$.

1. S_1 is true because $f'(x) = e^x(x^2 + 2x)$.
2. Assume that S_k is true; that is, $f^{(k)}(x) = e^x[x^2 + 2kx + k(k-1)]$. Then

$$\begin{aligned} f^{(k+1)}(x) &= \frac{d}{dx}\left[f^{(k)}(x)\right] = e^x(2x + 2k) + [x^2 + 2kx + k(k-1)]e^x \\ &= e^x[x^2 + (2k+2)x + (k^2 + k)] = e^x[x^2 + 2(k+1)x + (k+1)k] \end{aligned}$$

   This shows that S_{k+1} is true.
3. Therefore, by mathematical induction, S_n is true for all n; that is, $f^{(n)}(x) = e^x[x^2 + 2nx + n(n-1)]$ for every positive integer n.

3.3 Rates of Change in the Natural and Social Sciences

1. (a) $s = f(t) = t^3 - 12t^2 + 36t \quad\Rightarrow\quad v(t) = f'(t) = 3t^2 - 24t + 36$

(b) $v(3) = 27 - 72 + 36 = -9$ m/s

(c) The particle is at rest when $v(t) = 0$. $3t^2 - 24t + 36 = 0 \quad\Rightarrow\quad 3(t-2)(t-6) = 0 \quad\Rightarrow\quad t = 2, 6$.

(d) The particle is moving in the positive direction when $v(t) > 0$. $3(t-2)(t-6) > 0 \quad\Leftrightarrow\quad 0 \le t < 2$ or $t > 6$.

(e) Since the particle is moving forward and backward, we need to calculate the distance travelled in the intervals $[0, 2]$, $[2, 6]$, and $[6, 8]$ separately.

$|f(2) - f(0)| = |32 - 0| = 32.$

$|f(6) - f(2)| = |0 - 32| = 32.$

$|f(8) - f(6)| = |32 - 0| = 32.$

The total distance is $32 + 32 + 32 = 96$ m.

(f)

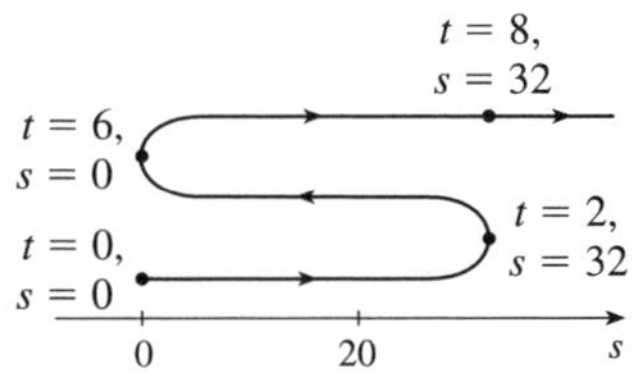

(g) $s = f(t) = t^3 - 12t^2 + 36t,\ t \ge 0 \quad\Rightarrow\quad v(t) = f'(t) = 3t^2 - 24t + 36$. $a(t) = v'(t) = 6t - 24$. $a(3) = 6(3) - 24 = -6$ (m/s)/s or m/s^2.

(h)

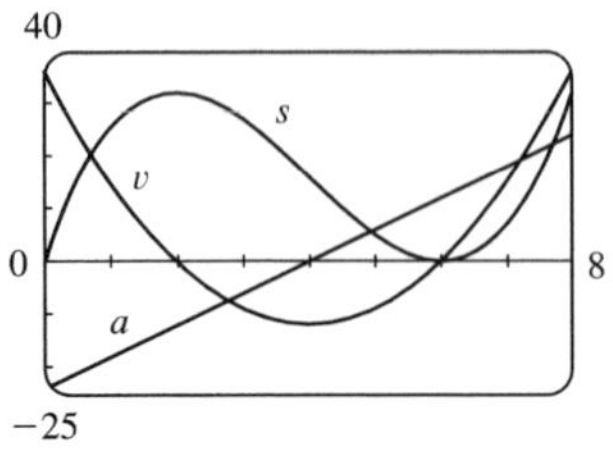

(i) The particle is speeding up when v and a have the same sign. This occurs when $2 < t < 4$ and when $t > 6$. It is slowing down when v and a have opposite signs; that is, when $0 \le t < 2$ and when $4 < t < 6$.

3. (a) From the figure, the velocity v is positive on the interval $(0, 2)$ and negative on the interval $(2, 3)$. The acceleration a is positive (negative) when the slope of the tangent line is positive (negative), so the acceleration is positive on the interval $(0, 1)$, and negative on the interval $(1, 3)$. The particle is speeding up when v and a have the same sign, that is, on the interval $(0, 1)$ when $v > 0$ and $a > 0$, and on the interval $(2, 3)$ when $v < 0$ and $a < 0$. The particle is slowing down when v and a have opposite signs, that is, on the interval $(1, 2)$ when $v > 0$ and $a < 0$.

(b) $v > 0$ on $(0, 3)$ and $v < 0$ on $(3, 4)$. $a > 0$ on $(1, 2)$ and $a < 0$ on $(0, 1)$ and $(2, 4)$. The particle is speeding up on $(1, 2)$ $[v > 0, a > 0]$ and on $(3, 4)$ $[v < 0, a < 0]$. The particle is slowing down on $(0, 1)$ and $(2, 3)$ $[v > 0, a < 0]$.

5. (a) $s(t) = t^3 - 4.5t^2 - 7t \quad \Rightarrow \quad v(t) = s'(t) = 3t^2 - 9t - 7 = 5 \quad \Leftrightarrow \quad 3t^2 - 9t - 12 = 0 \quad \Leftrightarrow$
$3(t-4)(t+1) = 0 \quad \Leftrightarrow \quad t = 4$ or -1. Since $t \geq 0$, the particle reaches a velocity of 5 m/s at $t = 4$ s.

(b) $a(t) = v'(t) = 6t - 9 = 0 \quad \Leftrightarrow \quad t = 1.5$. The acceleration changes from negative to positive, so the velocity changes from decreasing to increasing. Thus, at $t = 1.5$ s, the velocity has its minimum value.

7. (a) $h = 10t - 0.83t^2 \quad \Rightarrow \quad v(t) = \dfrac{dh}{dt} = 10 - 1.66t$, so $v(3) = 10 - 1.66(3) = 5.02$ m/s.

(b) $h = 25 \quad \Rightarrow \quad 10t - 0.83t^2 = 25 \quad \Rightarrow \quad 0.83t^2 - 10t + 25 = 0 \quad \Rightarrow \quad t = \dfrac{10 \pm \sqrt{17}}{1.66} \approx 3.54$ or 8.51.

The value $t_1 = \left(10 - \sqrt{17}\right)/1.66$ corresponds to the time it takes for the stone to rise 25 m and $t_2 = \left(10 + \sqrt{17}\right)/1.66$ corresponds to the time when the stone is 25 m high on the way down. Thus, $v(t_1) = 10 - 1.66\left[\left(10 - \sqrt{17}\right)/1.66\right] = \sqrt{17} \approx 4.12$ m/s.

9. (a) $A(x) = x^2 \quad \Rightarrow \quad A'(x) = 2x$. $A'(15) = 30\ \text{mm}^2/\text{mm}$ is the rate at which the area is increasing with respect to the side length as x reaches 15 mm.

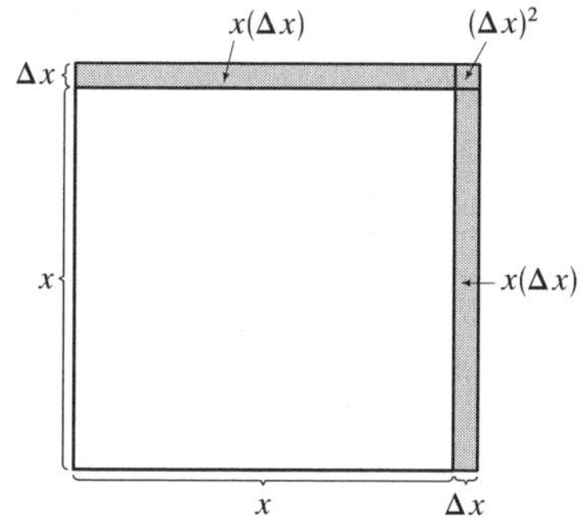

(b) The perimeter is $P(x) = 4x$, so $A'(x) = 2x = \frac{1}{2}(4x) = \frac{1}{2}P(x)$. The figure suggests that if Δx is small, then the change in the area of the square is approximately half of its perimeter (2 of the 4 sides) times Δx. From the figure, $\Delta A = 2x\,(\Delta x) + (\Delta x)^2$. If Δx is small, then $\Delta A \approx 2x\,(\Delta x)$ and so $\Delta A/\Delta x \approx 2x$.

11. (a) Using $A(r) = \pi r^2$, we find that the average rate of change is:

(i) $\dfrac{A(3) - A(2)}{3 - 2} = \dfrac{9\pi - 4\pi}{1} = 5\pi$ (ii) $\dfrac{A(2.5) - A(2)}{2.5 - 2} = \dfrac{6.25\pi - 4\pi}{0.5} = 4.5\pi$

(iii) $\dfrac{A(2.1) - A(2)}{2.1 - 2} = \dfrac{4.41\pi - 4\pi}{0.1} = 4.1\pi$

(b) $A(r) = \pi r^2 \quad \Rightarrow \quad A'(r) = 2\pi r$, so $A'(2) = 4\pi$.

(c) The circumference is $C(r) = 2\pi r = A'(r)$. The figure suggests that if Δr is small, then the change in the area of the circle (a ring around the outside) is approximately equal to its circumference times Δr. Straightening out this ring gives us a shape that is approximately rectangular with length $2\pi r$ and width Δr, so $\Delta A \approx 2\pi r(\Delta r)$.
Algebraically, $\Delta A = A(r + \Delta r) - A(r) = \pi(r + \Delta r)^2 - \pi r^2 = 2\pi r(\Delta r) + \pi(\Delta r)^2$.
So we see that if Δr is small, then $\Delta A \approx 2\pi r(\Delta r)$ and therefore, $\Delta A/\Delta r \approx 2\pi r$.

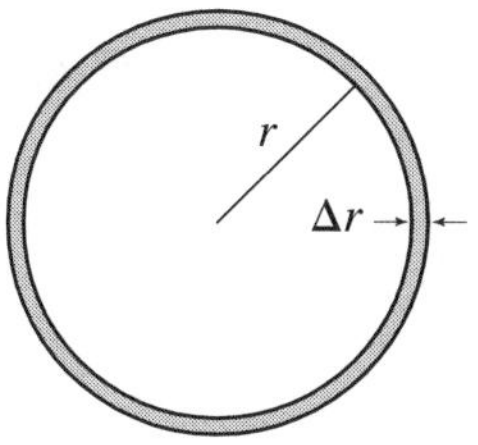

13. $S(r) = 4\pi r^2 \quad \Rightarrow \quad S'(r) = 8\pi r \quad \Rightarrow$

(a) $S'(20) = 160\pi\ \text{cm}^2/\text{cm}$ (b) $S'(40) = 320\pi\ \text{cm}^2/\text{cm}$ (c) $S'(60) = 480\pi\ \text{cm}^2/\text{cm}$

As the radius increases, the surface area grows at an increasing rate. In fact, the rate of change is linear with respect to the radius.

15. The mass is $f(x) = 3x^2$, so the linear density at x is $\rho(x) = f'(x) = 6x$.

(a) $\rho(1) = 6$ kg/m (b) $\rho(2) = 12$ kg/m (c) $\rho(3) = 18$ kg/m

Since ρ is an increasing function, the density will be the highest at the right end of the rod and lowest at the left end.

17. The quantity of charge is $Q(t) = t^3 - 2t^2 + 6t + 2$, so the current is $Q'(t) = 3t^2 - 4t + 6$.

(a) $Q'(0.5) = 3(0.5)^2 - 4(0.5) + 6 = 4.75$ A

(b) $Q'(1) = 3(1)^2 - 4(1) + 6 = 5$ A

The current is lowest when Q' has a minimum. $Q''(t) = 6t - 4 < 0$ when $t < \frac{2}{3}$. So the current decreases when $t < \frac{2}{3}$ and increases when $t > \frac{2}{3}$. Thus, the current is lowest at $t = \frac{2}{3}$ s.

19. (a) To find the rate of change of volume with respect to pressure, we first solve for V in terms of P.

$$PV = C \quad \Rightarrow \quad V = \frac{C}{P} \quad \Rightarrow \quad \frac{dV}{dP} = -\frac{C}{P^2}.$$

(b) From the formula for dV/dP in part (a), we see that as P increases, the absolute value of dV/dP decreases. Thus, the volume is decreasing more rapidly at the beginning.

(c) $\beta = -\dfrac{1}{V}\dfrac{dV}{dP} = -\dfrac{1}{V}\left(-\dfrac{C}{P^2}\right) = \dfrac{C}{(PV)P} = \dfrac{C}{CP} = \dfrac{1}{P}$

21. (a) **1920:** $m_1 = \dfrac{1860 - 1750}{1920 - 1910} = \dfrac{110}{10} = 11$, $m_2 = \dfrac{2070 - 1860}{1930 - 1920} = \dfrac{210}{10} = 21$,

$(m_1 + m_2)/2 = (11 + 21)/2 = 16$ million/year

1980: $m_1 = \dfrac{4450 - 3710}{1980 - 1970} = \dfrac{740}{10} = 74$, $m_2 = \dfrac{5280 - 4450}{1990 - 1980} = \dfrac{830}{10} = 83$,

$(m_1 + m_2)/2 = (74 + 83)/2 = 78.5$ million/year

(b) $P(t) = at^3 + bt^2 + ct + d$ (in millions of people), where $a \approx 0.001\,293\,706\,3$, $b \approx -7.061\,421\,911$, $c \approx 12\,822.97902$, and $d \approx -7\,743\,770.396$.

(c) $P(t) = at^3 + bt^2 + ct + d \quad \Rightarrow \quad P'(t) = 3at^2 + 2bt + c$ (in millions of people per year)

(d) $P'(1920) = 3(0.001\,293\,706\,3)(1920)^2 + 2(-7.061\,421\,911)(1920) + 12\,822.97902$

≈ 14.48 million/year [smaller than the answer in part (a), but close to it]

$P'(1980) \approx 75.29$ million/year (smaller, but close)

(e) $P'(1985) \approx 81.62$ million/year, so the rate of growth in 1985 was about 81.62 million/year.

23. (a) $[\mathrm{C}] = \dfrac{a^2kt}{akt+1} \quad \Rightarrow \quad$ rate of reaction $= \dfrac{d[\mathrm{C}]}{dt} = \dfrac{(akt+1)(a^2k) - (a^2kt)(ak)}{(akt+1)^2} = \dfrac{a^2k(akt+1-akt)}{(akt+1)^2} = \dfrac{a^2k}{(akt+1)^2}$

(b) If $x = [\mathrm{C}]$, then $a - x = a - \dfrac{a^2kt}{akt+1} = \dfrac{a^2kt + a - a^2kt}{akt+1} = \dfrac{a}{akt+1}$.

So $k(a-x)^2 = k\left(\dfrac{a}{akt+1}\right)^2 = \dfrac{a^2k}{(akt+1)^2} = \dfrac{d[\mathrm{C}]}{dt}$ [from part (a)] $= \dfrac{dx}{dt}$.

(c) As $t \to \infty$, $[\mathrm{C}] = \dfrac{a^2kt}{akt+1} = \dfrac{(a^2kt)/t}{(akt+1)/t} = \dfrac{a^2k}{ak + (1/t)} \to \dfrac{a^2k}{ak} = a$ moles/L.

(d) As $t \to \infty$, $\dfrac{d[\mathrm{C}]}{dt} = \dfrac{a^2k}{(akt+1)^2} \to 0$.

(e) As t increases, nearly all of the reactants A and B are converted into product C. In practical terms, the reaction virtually stops.

25. (a) Using $v = \dfrac{P}{4\eta l}(R^2 - r^2)$ with $R = 0.01$, $l = 3$, $P = 3000$, and $\eta = 0.027$, we have v as a function of r:

$v(r) = \dfrac{3000}{4(0.027)3}(0.01^2 - r^2)$. $v(0) = 0.\overline{925}$ cm/s, $v(0.005) = 0.69\overline{4}$ cm/s, $v(0.01) = 0$.

(b) $v(r) = \dfrac{P}{4\eta l}(R^2 - r^2) \;\Rightarrow\; v'(r) = \dfrac{P}{4\eta l}(-2r) = -\dfrac{Pr}{2\eta l}$. When $l = 3$, $P = 3000$, and $\eta = 0.027$, we have

$v'(r) = -\dfrac{3000r}{2(0.027)3}$. $v'(0) = 0$, $v'(0.005) = -92.\overline{592}$ (cm/s)/cm, and $v'(0.01) = -185.\overline{185}$ (cm/s)/cm.

(c) The velocity is greatest where $r = 0$ (at the center) and the velocity is changing most where $r = R = 0.01$ cm (at the edge).

27. (a) $C(x) = 2000 + 3x + 0.01x^2 + 0.0002x^3 \;\Rightarrow\; C'(x) = 3 + 0.02x + 0.0006x^2$

(b) $C'(100) = 3 + 0.02(100) + 0.0006(10\,000) = 3 + 2 + 6 = \$11/\text{pair}$. $C'(100)$ is the rate at which the cost is increasing as the 100th pair of jeans is produced. It predicts the cost of the 101st pair.

(c) The cost of manufacturing the 101st pair of jeans is

$C(101) - C(100) = (2000 + 303 + 102.01 + 206.0602) - (2000 + 300 + 100 + 200) = 11.0702 \approx \$11.07$.

29. (a) $A(x) = \dfrac{p(x)}{x} \;\Rightarrow\; A'(x) = \dfrac{xp'(x) - p(x)\cdot 1}{x^2} = \dfrac{xp'(x) - p(x)}{x^2}$. $A'(x) > 0 \;\Rightarrow\; A(x)$ is increasing; that is, the average productivity increases as the size of the workforce increases.

(b) $p'(x)$ is greater than the average productivity $\;\Rightarrow\; p'(x) > A(x) \;\Rightarrow\; p'(x) > \dfrac{p(x)}{x} \;\Rightarrow\; xp'(x) > p(x) \;\Rightarrow$

$xp'(x) - p(x) > 0 \;\Rightarrow\; \dfrac{xp'(x) - p(x)}{x^2} > 0 \;\Rightarrow\; A'(x) > 0.$

31. $PV = nRT \;\Rightarrow\; T = \dfrac{PV}{nR} = \dfrac{PV}{(10)(0.0821)} = \dfrac{1}{0.821}(PV)$. Using the Product Rule, we have

$$\frac{dT}{dt} = \frac{1}{0.821}\,[P(t)V'(t) + V(t)P'(t)] = \frac{1}{0.821}\,[(8)(-0.15) + (10)(0.10)] \approx -0.2436 \text{ K/min.}$$

33. (a) If the populations are stable, then the growth rates are neither positive nor negative; that is, $\dfrac{dC}{dt} = 0$ and $\dfrac{dW}{dt} = 0$.

(b) "The caribou go extinct" means that the population is zero, or mathematically, $C = 0$.

(c) We have the equations $\dfrac{dC}{dt} = aC - bCW$ and $\dfrac{dW}{dt} = -cW + dCW$. Let $dC/dt = dW/dt = 0$, $a = 0.05$, $b = 0.001$, $c = 0.05$, and $d = 0.0001$ to obtain $0.05C - 0.001CW = 0$ **(1)** and $-0.05W + 0.0001CW = 0$ **(2)**. Adding 10 times **(2)** to **(1)** eliminates the CW-terms and gives us $0.05C - 0.5W = 0 \;\Rightarrow\; C = 10W$. Substituting $C = 10W$ into **(1)** results in $0.05(10W) - 0.001(10W)W = 0 \;\Leftrightarrow\; 0.5W - 0.01W^2 = 0 \;\Leftrightarrow\; 50W - W^2 = 0 \;\Leftrightarrow$ $W(50 - W) = 0 \;\Leftrightarrow\; W = 0$ or 50. Since $C = 10W$, $C = 0$ or 500. Thus, the population pairs (C, W) that lead to stable populations are $(0, 0)$ and $(500, 50)$. So it is possible for the two species to live in harmony.

3.4 Derivatives of Trigonometric Functions

1. $f(x) = x - 3\sin x \;\Rightarrow\; f'(x) = 1 - 3\cos x$

3. $g(t) = t^3\cos t \;\Rightarrow\; g'(t) = t^3(-\sin t) + (\cos t)\cdot 3t^2 = 3t^2\cos t - t^3\sin t$ or $t^2(3\cos t - t\sin t)$

5. $y = \sec\theta\,\tan\theta \;\Rightarrow\; y' = \sec\theta\,(\sec^2\theta) + \tan\theta\,(\sec\theta\tan\theta) = \sec\theta\,(\sec^2\theta + \tan^2\theta)$. Using the identity $1 + \tan^2\theta = \sec^2\theta$, we can write alternative forms of the answer as $\sec\theta\,(1 + 2\tan^2\theta)$ or $\sec\theta\,(2\sec^2\theta - 1)$.

7. $y = \dfrac{x}{\cos x} \;\Rightarrow\; y' = \dfrac{(\cos x)(1) - (x)(-\sin x)}{(\cos x)^2} = \dfrac{\cos x + x\sin x}{\cos^2 x}$

9. $f(\theta) = \dfrac{\sec\theta}{1+\sec\theta} \quad\Rightarrow$

$$f'(\theta) = \frac{(1+\sec\theta)(\sec\theta\tan\theta) - (\sec\theta)(\sec\theta\tan\theta)}{(1+\sec\theta)^2} = \frac{(\sec\theta\tan\theta)[(1+\sec\theta) - \sec\theta]}{(1+\sec\theta)^2} = \frac{\sec\theta\tan\theta}{(1+\sec\theta)^2}$$

11. Using Exercise 3.2.45(a), $f(x) = xe^x \csc x \quad\Rightarrow$

$$f'(x) = (x)'e^x\csc x + x(e^x)'\csc x + xe^x(\csc x)' = 1e^x\csc x + xe^x\csc x + xe^x(-\cot x\csc x)$$
$$= e^x\csc x(1 + x - x\cot x)$$

13. $\dfrac{d}{dx}(\csc x) = \dfrac{d}{dx}\left(\dfrac{1}{\sin x}\right) = \dfrac{(\sin x)(0) - 1(\cos x)}{\sin^2 x} = \dfrac{-\cos x}{\sin^2 x} = -\dfrac{1}{\sin x}\cdot\dfrac{\cos x}{\sin x} = -\csc x\cot x$

15. $\dfrac{d}{dx}(\cot x) = \dfrac{d}{dx}\left(\dfrac{\cos x}{\sin x}\right) = \dfrac{(\sin x)(-\sin x) - (\cos x)(\cos x)}{\sin^2 x} = -\dfrac{\sin^2 x + \cos^2 x}{\sin^2 x} = -\dfrac{1}{\sin^2 x} = -\csc^2 x$

17. $y = \tan x \quad\Rightarrow\quad y' = \sec^2 x \quad\Rightarrow$ the slope of the tangent line at $\left(\frac{\pi}{4}, 1\right)$ is $\sec^2\left(\frac{\pi}{4}\right) = \left(\sqrt{2}\right)^2 = 2$ and an equation of the tangent line is $y - 1 = 2\left(x - \frac{\pi}{4}\right)$ or $y = 2x + 1 - \frac{\pi}{2}$.

19. (a) $y = x\cos x \quad\Rightarrow\quad y' = x(-\sin x) + \cos x(1) = \cos x - x\sin x$.

So the slope of the tangent at the point $(\pi, -\pi)$ is

$\cos\pi - \pi\sin\pi = -1 - \pi(0) = -1$, and an equation is

$y + \pi = -(x - \pi)$ or $y = -x$.

(b)

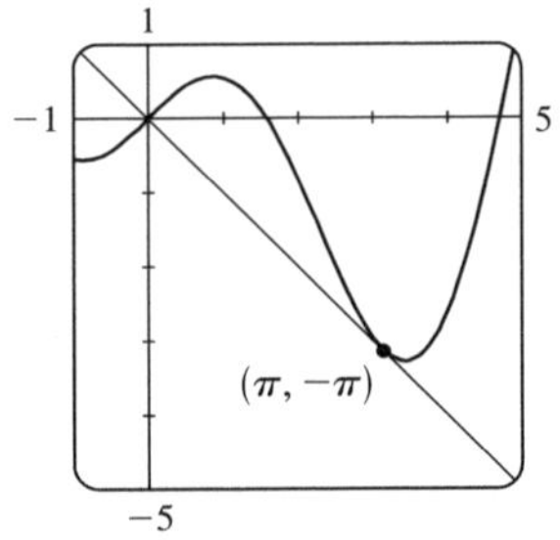

21. (a) $f(x) = 2x + \cot x \quad\Rightarrow\quad f'(x) = 2 - \csc^2 x$

(b)

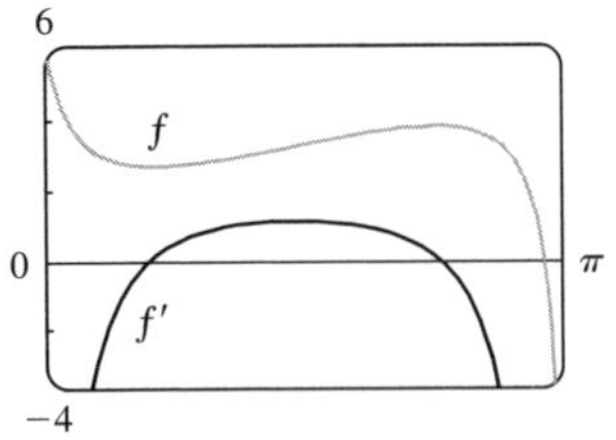

Notice that $f'(x) = 0$ when f has a horizontal tangent.

f' is positive when f is increasing and f' is negative when f is decreasing.

Also, $f'(x)$ is large negative when the graph of f is steep.

23. $H(\theta) = \theta\sin\theta \quad\Rightarrow\quad H'(\theta) = \theta(\cos\theta) + (\sin\theta)\cdot 1 = \theta\cos\theta + \sin\theta \quad\Rightarrow$

$H''(\theta) = \theta(-\sin\theta) + (\cos\theta)\cdot 1 + \cos\theta = -\theta\sin\theta + 2\cos\theta$

25. (a) $f(x) = \dfrac{\tan x - 1}{\sec x} \quad\Rightarrow$

$$f'(x) = \frac{\sec x(\sec^2 x) - (\tan x - 1)(\sec x\tan x)}{(\sec x)^2} = \frac{\sec x(\sec^2 x - \tan^2 x + \tan x)}{\sec^2 x} = \frac{1+\tan x}{\sec x}$$

(b) $f(x) = \dfrac{\tan x - 1}{\sec x} = \dfrac{\dfrac{\sin x}{\cos x} - 1}{\dfrac{1}{\cos x}} = \dfrac{\dfrac{\sin x - \cos x}{\cos x}}{\dfrac{1}{\cos x}} = \sin x - \cos x \quad\Rightarrow\quad f'(x) = \cos x - (-\sin x) = \cos x + \sin x$

(c) From part (a), $f'(x) = \dfrac{1+\tan x}{\sec x} = \dfrac{1}{\sec x} + \dfrac{\tan x}{\sec x} = \cos x + \sin x$, which is the expression for $f'(x)$ in part (b).

27. $f(x) = x + 2\sin x$ has a horizontal tangent when $f'(x) = 0 \quad\Leftrightarrow\quad 1 + 2\cos x = 0 \quad\Leftrightarrow\quad \cos x = -\frac{1}{2} \quad\Leftrightarrow$ $x = \frac{2\pi}{3} + 2\pi n$ or $\frac{4\pi}{3} + 2\pi n$, where n is an integer. Note that $\frac{4\pi}{3}$ and $\frac{2\pi}{3}$ are $\pm\frac{\pi}{3}$ units from π. This allows us to write the solutions in the more compact equivalent form $(2n+1)\pi \pm \frac{\pi}{3}$, n an integer.

29. $f(x) = x - 2\sin x$, $0 \le x \le 2\pi$. $f'(x) = 1 - 2\cos x$. So $f'(x) > 0 \quad\Leftrightarrow\quad 1 - 2\cos x > 0 \quad\Leftrightarrow\quad -2\cos x > -1 \quad\Leftrightarrow$ $\cos x < \frac{1}{2} \quad\Leftrightarrow\quad \frac{\pi}{3} < x < \frac{5\pi}{3} \quad\Rightarrow\quad f$ is increasing on $\left(\frac{\pi}{3}, \frac{5\pi}{3}\right)$.

31. (a) $x(t) = 8\sin t \quad\Rightarrow\quad v(t) = x'(t) = 8\cos t \quad\Rightarrow\quad a(t) = x''(t) = -8\sin t$

(b) The mass at time $t = \frac{2\pi}{3}$ has position $x\left(\frac{2\pi}{3}\right) = 8\sin\frac{2\pi}{3} = 8\left(\frac{\sqrt{3}}{2}\right) = 4\sqrt{3}$, velocity $v\left(\frac{2\pi}{3}\right) = 8\cos\frac{2\pi}{3} = 8\left(-\frac{1}{2}\right) = -4$, and acceleration $a\left(\frac{2\pi}{3}\right) = -8\sin\frac{2\pi}{3} = -8\left(\frac{\sqrt{3}}{2}\right) = -4\sqrt{3}$. Since $v\left(\frac{2\pi}{3}\right) < 0$, the particle is moving to the left. Because v and a have the same sign, the particle is speeding up.

33.

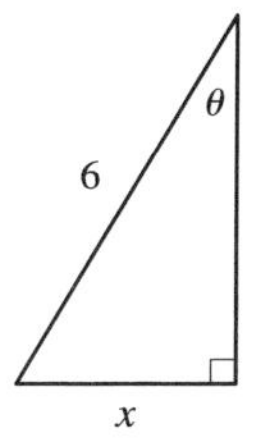

From the diagram we can see that $\sin\theta = x/6 \quad\Leftrightarrow\quad x = 6\sin\theta$. We want to find the rate of change of x with respect to θ; that is, $dx/d\theta$. Taking the derivative of the above expression, $dx/d\theta = 6(\cos\theta)$. So when $\theta = \frac{\pi}{3}$, $\frac{dx}{d\theta} = 6\cos\frac{\pi}{3} = 6\left(\frac{1}{2}\right) = 3$ m/rad.

35. $\dfrac{d}{dx}(\sin x) = \cos x \quad\Rightarrow\quad \dfrac{d^2}{dx^2}(\sin x) = -\sin x \quad\Rightarrow\quad \dfrac{d^3}{dx^3}(\sin x) = -\cos x \quad\Rightarrow\quad \dfrac{d^4}{dx^4}(\sin x) = \sin x$.

The derivatives of $\sin x$ occur in a cycle of four. Since $99 = 4(24) + 3$, we have $\dfrac{d^{99}}{dx^{99}}(\sin x) = \dfrac{d^3}{dx^3}(\sin x) = -\cos x$.

37. $y = A\sin x + B\cos x \quad\Rightarrow\quad y' = A\cos x - B\sin x \quad\Rightarrow\quad y'' = -A\sin x - B\cos x$. Substituting these expressions for y, y', and y'' into the given differential equation $y'' + y' - 2y = \sin x$ gives us

$(-A\sin x - B\cos x) + (A\cos x - B\sin x) - 2(A\sin x + B\cos x) = \sin x \quad\Leftrightarrow$

$-3A\sin x - B\sin x + A\cos x - 3B\cos x = \sin x \quad\Leftrightarrow\quad (-3A - B)\sin x + (A - 3B)\cos x = 1\sin x$, so we must have $-3A - B = 1$ and $A - 3B = 0$ (since 0 is the coefficient of $\cos x$ on the right side). Solving for A and B, we add the first equation to three times the second to get $B = -\frac{1}{10}$ and $A = -\frac{3}{10}$.

39.
$$\lim_{t\to 0}\frac{\tan 6t}{\sin 2t} = \lim_{t\to 0}\left(\frac{\sin 6t}{t}\cdot\frac{1}{\cos 6t}\cdot\frac{t}{\sin 2t}\right) = \lim_{t\to 0}\frac{6\sin 6t}{6t}\cdot\lim_{t\to 0}\frac{1}{\cos 6t}\cdot\lim_{t\to 0}\frac{2t}{2\sin 2t}$$
$$= 6\lim_{t\to 0}\frac{\sin 6t}{6t}\cdot\lim_{t\to 0}\frac{1}{\cos 6t}\cdot\frac{1}{2}\lim_{t\to 0}\frac{2t}{\sin 2t} = 6(1)\cdot\frac{1}{1}\cdot\frac{1}{2}(1) = 3$$

41.
$$\lim_{\theta\to 0}\frac{\sin\theta}{\theta + \tan\theta} = \frac{\lim\limits_{\theta\to 0}\dfrac{\sin\theta}{\theta}}{\lim\limits_{\theta\to 0}\dfrac{\theta + \tan\theta}{\theta}} = \frac{1}{\lim\limits_{\theta\to 0}\left(1 + \dfrac{\sin\theta}{\theta}\cdot\dfrac{1}{\cos\theta}\right)} = \frac{1}{1 + 1\cdot 1} = \frac{1}{2}$$

43. By the definition of radian measure, $s = r\theta$, where r is the radius of the circle.By drawing the bisector of the angle θ, we can see that $\sin\dfrac{\theta}{2} = \dfrac{d/2}{r} \quad\Rightarrow\quad d = 2r\sin\dfrac{\theta}{2}$. So $\displaystyle\lim_{\theta\to 0^+}\frac{s}{d} = \lim_{\theta\to 0^+}\frac{r\theta}{2r\sin(\theta/2)} = \lim_{\theta\to 0^+}\frac{2\cdot(\theta/2)}{2\sin(\theta/2)} = \lim_{\theta\to 0}\frac{\theta/2}{\sin(\theta/2)} = 1$.

[This is just the reciprocal of the limit $\lim\limits_{x\to 0}\frac{\sin x}{x} = 1$ combined with the fact that as $\theta \to 0$, $\frac{\theta}{2} \to 0$ also.]

3.5 The Chain Rule

1. Let $u = g(x) = 4x$ and $y = f(u) = \sin u$. Then $\dfrac{dy}{dx} = \dfrac{dy}{du}\dfrac{du}{dx} = (\cos u)(4) = 4\cos 4x$.

3. Let $u = g(x) = 1 - x^2$ and $y = f(u) = u^{10}$. Then $\dfrac{dy}{dx} = \dfrac{dy}{du}\dfrac{du}{dx} = (10u^9)(-2x) = -20x(1 - x^2)^9$.

5. Let $u = g(x) = \sqrt{x}$ and $y = f(u) = e^u$. Then $\dfrac{dy}{dx} = \dfrac{dy}{du}\dfrac{du}{dx} = (e^u)\left(\frac{1}{2}x^{-1/2}\right) = e^{\sqrt{x}} \cdot \dfrac{1}{2\sqrt{x}} = \dfrac{e^{\sqrt{x}}}{2\sqrt{x}}$.

7. $F(x) = \sqrt[4]{1 + 2x + x^3} = (1 + 2x + x^3)^{1/4} \quad \Rightarrow$

$$F'(x) = \tfrac{1}{4}(1 + 2x + x^3)^{-3/4} \cdot \frac{d}{dx}(1 + 2x + x^3) = \frac{1}{4(1 + 2x + x^3)^{3/4}} \cdot (2 + 3x^2)$$
$$= \frac{2 + 3x^2}{4(1 + 2x + x^3)^{3/4}} = \frac{2 + 3x^2}{4\sqrt[4]{(1 + 2x + x^3)^3}}$$

9. $g(t) = \dfrac{1}{(t^4 + 1)^3} = (t^4 + 1)^{-3} \quad \Rightarrow \quad g'(t) = -3(t^4 + 1)^{-4}(4t^3) = -12t^3(t^4 + 1)^{-4} = \dfrac{-12t^3}{(t^4 + 1)^4}$

11. $y = \cos(a^3 + x^3) \quad \Rightarrow \quad y' = -\sin(a^3 + x^3) \cdot 3x^2$ [a^3 is just a constant] $= -3x^2 \sin(a^3 + x^3)$

13. $h(t) = t^3 - 3^t \quad \Rightarrow \quad h'(t) = 3t^2 - 3^t \ln 3$ [by Formula 5]

15. $y = xe^{-x^2} \quad \Rightarrow \quad y' = xe^{-x^2}(-2x) + e^{-x^2} \cdot 1 = e^{-x^2}(-2x^2 + 1) = e^{-x^2}(1 - 2x^2)$

17. $G(x) = (3x - 2)^{10}(5x^2 - x + 1)^{12} \quad \Rightarrow$

$$\begin{aligned} G'(x) &= (3x-2)^{10}(12)(5x^2 - x + 1)^{11}(10x - 1) + (5x^2 - x + 1)^{12}(10)(3x - 2)^9(3) \\ &= 6(3x - 2)^9(5x^2 - x + 1)^{11}[2(3x - 2)(10x - 1) + 5(5x^2 - x + 1)] \\ &= 6(3x - 2)^9(5x^2 - x + 1)^{11}[(60x^2 - 46x + 4) + (25x^2 - 5x + 5)] \\ &= 6(3x - 2)^9(5x^2 - x + 1)^{11}(85x^2 - 51x + 9) \end{aligned}$$

19. $y = e^{x\cos x} \quad \Rightarrow \quad y' = e^{x\cos x} \cdot \dfrac{d}{dx}(x\cos x) = e^{x\cos x}[x(-\sin x) + (\cos x) \cdot 1] = e^{x\cos x}(\cos x - x\sin x)$

21. $F(z) = \sqrt{\dfrac{z-1}{z+1}} = \left(\dfrac{z-1}{z+1}\right)^{1/2} \quad \Rightarrow$

$$F'(z) = \frac{1}{2}\left(\frac{z-1}{z+1}\right)^{-1/2} \cdot \frac{d}{dz}\left(\frac{z-1}{z+1}\right) = \frac{1}{2}\left(\frac{z+1}{z-1}\right)^{1/2} \cdot \frac{(z+1)(1) - (z-1)(1)}{(z+1)^2}$$
$$= \frac{1}{2}\frac{(z+1)^{1/2}}{(z-1)^{1/2}} \cdot \frac{z+1-z+1}{(z+1)^2} = \frac{1}{2}\frac{(z+1)^{1/2}}{(z-1)^{1/2}} \cdot \frac{2}{(z+1)^2} = \frac{1}{(z-1)^{1/2}(z+1)^{3/2}}$$

23. $y = \sec^2 x + \tan^2 x = (\sec x)^2 + (\tan x)^2 \quad \Rightarrow$

$y' = 2(\sec x)(\sec x \tan x) + 2(\tan x)(\sec^2 x) = 2\sec^2 x \tan x + 2\sec^2 x \tan x = 4\sec^2 x \tan x$

25. $y = \dfrac{r}{\sqrt{r^2 + 1}} \quad \Rightarrow$

$$y' = \frac{\sqrt{r^2+1}\,(1) - r \cdot \frac{1}{2}(r^2+1)^{-1/2}(2r)}{\left(\sqrt{r^2+1}\right)^2} = \frac{\sqrt{r^2+1} - \dfrac{r^2}{\sqrt{r^2+1}}}{\left(\sqrt{r^2+1}\right)^2} = \frac{\dfrac{\sqrt{r^2+1}\,\sqrt{r^2+1} - r^2}{\sqrt{r^2+1}}}{\left(\sqrt{r^2+1}\right)^2}$$
$$= \frac{(r^2+1) - r^2}{\left(\sqrt{r^2+1}\right)^3} = \frac{1}{(r^2+1)^{3/2}} \quad \text{or} \quad (r^2+1)^{-3/2}$$

Another solution: Write y as a product and make use of the Product Rule. $y = r(r^2 + 1)^{-1/2} \quad \Rightarrow$

$y' = r \cdot -\frac{1}{2}(r^2+1)^{-3/2}(2r) + (r^2+1)^{-1/2} \cdot 1 = (r^2+1)^{-3/2}[-r^2 + (r^2+1)^1] = (r^2+1)^{-3/2}(1) = (r^2+1)^{-3/2}$.

The step that students usually have trouble with is factoring out $(r^2+1)^{-3/2}$. But this is no different than factoring out x^2 from $x^2 + x^5$; that is, we are just factoring out a factor with the *smallest* exponent that appears on it. In this case, $-\frac{3}{2}$ is smaller than $-\frac{1}{2}$.

27. Using Formula 5 and the Chain Rule, $y = 2^{\sin \pi x} \Rightarrow$

$$y' = 2^{\sin \pi x}(\ln 2) \cdot \frac{d}{dx}(\sin \pi x) = 2^{\sin \pi x}(\ln 2) \cdot \cos \pi x \cdot \pi = 2^{\sin \pi x}(\pi \ln 2) \cos \pi x$$

29. $y = \cot^2(\sin\theta) = [\cot(\sin\theta)]^2 \Rightarrow$

$$y' = 2[\cot(\sin\theta)] \cdot \frac{d}{d\theta}[\cot(\sin\theta)] = 2\cot(\sin\theta) \cdot [-\csc^2(\sin\theta) \cdot \cos\theta] = -2\cos\theta \cot(\sin\theta) \csc^2(\sin\theta)$$

31. $y = \sin\left(\tan\sqrt{\sin x}\right) \Rightarrow$

$$\begin{aligned} y' &= \cos\left(\tan\sqrt{\sin x}\right) \cdot \frac{d}{dx}\left(\tan\sqrt{\sin x}\right) = \cos\left(\tan\sqrt{\sin x}\right)\sec^2\sqrt{\sin x} \cdot \frac{d}{dx}(\sin x)^{1/2} \\ &= \cos\left(\tan\sqrt{\sin x}\right)\sec^2\sqrt{\sin x} \cdot \tfrac{1}{2}(\sin x)^{-1/2} \cdot \cos x = \cos\left(\tan\sqrt{\sin x}\right)\left(\sec^2\sqrt{\sin x}\right)\left(\frac{1}{2\sqrt{\sin x}}\right)(\cos x) \end{aligned}$$

33. $y = e^{\alpha x}\sin\beta x \Rightarrow y' = e^{\alpha x} \cdot \beta\cos\beta x + \sin\beta x \cdot \alpha e^{\alpha x} = e^{\alpha x}(\beta\cos\beta x + \alpha\sin\beta x) \Rightarrow$

$$\begin{aligned} y'' &= e^{\alpha x}(-\beta^2\sin\beta x + \alpha\beta\cos\beta x) + (\beta\cos\beta x + \alpha\sin\beta x) \cdot \alpha e^{\alpha x} \\ &= e^{\alpha x}(-\beta^2\sin\beta x + \alpha\beta\cos\beta x + \alpha\beta\cos\beta x + \alpha^2\sin\beta x) = e^{\alpha x}(\alpha^2\sin\beta x - \beta^2\sin\beta x + 2\alpha\beta\cos\beta x) \\ &= e^{\alpha x}\left[(\alpha^2 - \beta^2)\sin\beta x + 2\alpha\beta\cos\beta x\right] \end{aligned}$$

35. $y = \sin(\sin x) \Rightarrow y' = \cos(\sin x) \cdot \cos x$. At $(\pi, 0)$, $y' = \cos(\sin\pi) \cdot \cos\pi = \cos(0) \cdot (-1) = 1(-1) = -1$, and an equation of the tangent line is $y - 0 = -1(x - \pi)$, or $y = -x + \pi$.

37. (a) $y = \dfrac{2}{1 + e^{-x}} \Rightarrow y' = \dfrac{(1 + e^{-x})(0) - 2(-e^{-x})}{(1 + e^{-x})^2} = \dfrac{2e^{-x}}{(1 + e^{-x})^2}$.

At $(0, 1)$, $y' = \dfrac{2e^0}{(1 + e^0)^2} = \dfrac{2(1)}{(1 + 1)^2} = \dfrac{2}{2^2} = \dfrac{1}{2}$. So an equation of the tangent line is $y - 1 = \frac{1}{2}(x - 0)$ or $y = \frac{1}{2}x + 1$.

(b)

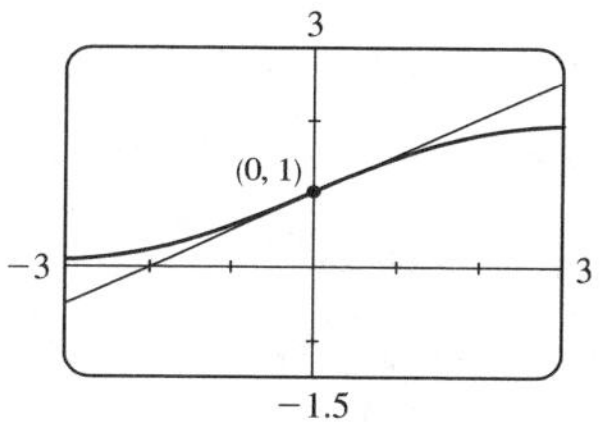

39. (a) $f(x) = x\sqrt{2 - x^2} = x(2 - x^2)^{1/2} \Rightarrow$

$$f'(x) = x \cdot \tfrac{1}{2}(2 - x^2)^{-1/2}(-2x) + (2 - x^2)^{1/2} \cdot 1 = (2 - x^2)^{-1/2}[-x^2 + (2 - x^2)] = \frac{2 - 2x^2}{\sqrt{2 - x^2}}$$

(b)

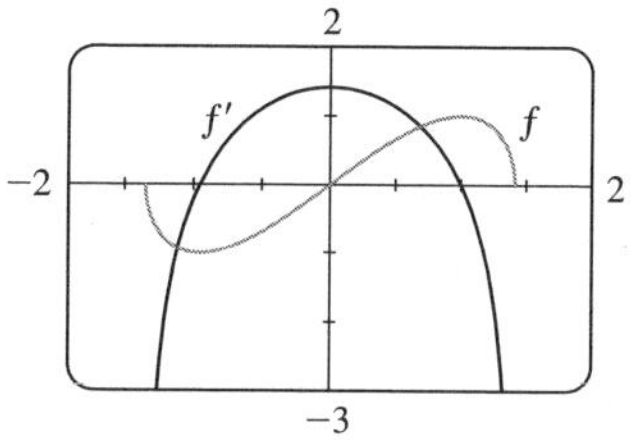

$f' = 0$ when f has a horizontal tangent line, f' is negative when f is decreasing, and f' is positive when f is increasing.

41. $F(x) = f(g(x)) \Rightarrow F'(x) = f'(g(x)) \cdot g'(x)$, so $F'(5) = f'(g(5)) \cdot g'(5) = f'(-2) \cdot 6 = 4 \cdot 6 = 24$

43. (a) $h(x) = f(g(x)) \Rightarrow h'(x) = f'(g(x)) \cdot g'(x)$, so $h'(1) = f'(g(1)) \cdot g'(1) = f'(2) \cdot 6 = 5 \cdot 6 = 30$.

(b) $H(x) = g(f(x)) \Rightarrow H'(x) = g'(f(x)) \cdot f'(x)$, so $H'(1) = g'(f(1)) \cdot f'(1) = g'(3) \cdot 4 = 9 \cdot 4 = 36$.

45. (a) $u(x) = f(g(x)) \Rightarrow u'(x) = f'(g(x))g'(x)$. So $u'(1) = f'(g(1))g'(1) = f'(3)g'(1)$. To find $f'(3)$, note that f is linear from $(2, 4)$ to $(6, 3)$, so its slope is $\dfrac{3 - 4}{6 - 2} = -\dfrac{1}{4}$. To find $g'(1)$, note that g is linear from $(0, 6)$ to $(2, 0)$, so its slope is $\dfrac{0 - 6}{2 - 0} = -3$. Thus, $f'(3)g'(1) = \left(-\frac{1}{4}\right)(-3) = \frac{3}{4}$.

(b) $v(x) = g(f(x)) \Rightarrow v'(x) = g'(f(x))f'(x)$. So $v'(1) = g'(f(1))f'(1) = g'(2)f'(1)$, which does not exist since $g'(2)$ does not exist.

(c) $w(x) = g(g(x)) \Rightarrow w'(x) = g'(g(x))g'(x)$. So $w'(1) = g'(g(1))g'(1) = g'(3)g'(1)$. To find $g'(3)$, note that g is linear from $(2, 0)$ to $(5, 2)$, so its slope is $\dfrac{2-0}{5-2} = \dfrac{2}{3}$. Thus, $g'(3)g'(1) = \left(\frac{2}{3}\right)(-3) = -2$.

47. $h(x) = f(g(x)) \Rightarrow h'(x) = f'(g(x))g'(x)$. So $h'(0.5) = f'(g(0.5))g'(0.5) = f'(0.1)g'(0.5)$.

We can estimate the derivatives by taking the average of two secant slopes.

For $f'(0.1)$: $m_1 = \dfrac{14.8 - 12.6}{0.1 - 0} = 22$, $m_2 = \dfrac{18.4 - 14.8}{0.2 - 0.1} = 36$. So $f'(0.1) \approx \dfrac{m_1 + m_2}{2} = \dfrac{22 + 36}{2} = 29$.

For $g'(0.5)$: $m_1 = \dfrac{0.10 - 0.17}{0.5 - 0.4} = -0.7$, $m_2 = \dfrac{0.05 - 0.10}{0.6 - 0.5} = -0.5$. So $g'(0.5) \approx \dfrac{m_1 + m_2}{2} = -0.6$.

Hence, $h'(0.5) = f'(0.1)g'(0.5) \approx (29)(-0.6) = -17.4$.

49. (a) $F(x) = f(e^x) \Rightarrow F'(x) = f'(e^x)\dfrac{d}{dx}(e^x) = f'(e^x)\,e^x$

(b) $G(x) = e^{f(x)} \Rightarrow G'(x) = e^{f(x)}\dfrac{d}{dx}f(x) = e^{f(x)}f'(x)$

51. $r(x) = f(g(h(x))) \Rightarrow r'(x) = f'(g(h(x))) \cdot g'(h(x)) \cdot h'(x)$, so
$r'(1) = f'(g(h(1))) \cdot g'(h(1)) \cdot h'(1) = f'(g(2)) \cdot g'(2) \cdot 4 = f'(3) \cdot 5 \cdot 4 = 6 \cdot 5 \cdot 4 = 120$

53. For the tangent line to be horizontal, $f'(x) = 0$. $f(x) = 2\sin x + \sin^2 x \Rightarrow f'(x) = 2\cos x + 2\sin x\cos x = 0 \Leftrightarrow$ $2\cos x\,(1 + \sin x) = 0 \Leftrightarrow \cos x = 0$ or $\sin x = -1$, so $x = \frac{\pi}{2} + 2n\pi$ or $\frac{3\pi}{2} + 2n\pi$, where n is any integer. Now $f\left(\frac{\pi}{2}\right) = 3$ and $f\left(\frac{3\pi}{2}\right) = -1$, so the points on the curve with a horizontal tangent are $\left(\frac{\pi}{2} + 2n\pi, 3\right)$ and $\left(\frac{3\pi}{2} + 2n\pi, -1\right)$, where n is any integer.

55.
$$\begin{aligned} y = Ae^{-x} + Bxe^{-x} \Rightarrow y' &= A(-e^{-x}) + B[x(-e^{-x}) + e^{-x} \cdot 1] \\ &= -Ae^{-x} + Be^{-x} - Bxe^{-x} = (B - A)e^{-x} - Bxe^{-x} \\ \Rightarrow y'' &= (B - A)(-e^{-x}) - B[x(-e^{-x}) + e^{-x} \cdot 1] \\ &= (A - B)e^{-x} - Be^{-x} + Bxe^{-x} = (A - 2B)e^{-x} + Bxe^{-x}, \end{aligned}$$

so
$$\begin{aligned} y'' + 2y' + y &= (A - 2B)e^{-x} + Bxe^{-x} + 2[(B - A)e^{-x} - Bxe^{-x}] + Ae^{-x} + Bxe^{-x} \\ &= [(A - 2B) + 2(B - A) + A]e^{-x} + [B - 2B + B]xe^{-x} = 0. \end{aligned}$$

57. The use of $D, D^2, \ldots, D^n$ is just a derivative notation (see text page 159). In general, $Df(2x) = 2f'(2x)$, $D^2 f(2x) = 4f''(2x), \ldots, D^n f(2x) = 2^n f^{(n)}(2x)$. Since $f(x) = \cos x$ and $50 = 4(12) + 2$, we have $f^{(50)}(x) = f^{(2)}(x) = -\cos x$, so $D^{50}\cos 2x = -2^{50}\cos 2x$.

59. $s(t) = 10 + \frac{1}{4}\sin(10\pi t) \Rightarrow$ the velocity after t seconds is $v(t) = s'(t) = \frac{1}{4}\cos(10\pi t)(10\pi) = \frac{5\pi}{2}\cos(10\pi t)$ cm/s.

61. (a) $B(t) = 4.0 + 0.35\sin\dfrac{2\pi t}{5.4} \Rightarrow \dfrac{dB}{dt} = \left(0.35\cos\dfrac{2\pi t}{5.4}\right)\left(\dfrac{2\pi}{5.4}\right) = \dfrac{0.7\pi}{5.4}\cos\dfrac{2\pi t}{5.4} = \dfrac{7\pi}{54}\cos\dfrac{2\pi t}{5.4}$

(b) At $t = 1$, $\dfrac{dB}{dt} = \dfrac{7\pi}{54}\cos\dfrac{2\pi}{5.4} \approx 0.16$.

63. $s(t) = 2e^{-1.5t}\sin 2\pi t \quad \Rightarrow$

$v(t) = s'(t) = 2[e^{-1.5t}(\cos 2\pi t)(2\pi) + (\sin 2\pi t)e^{-1.5t}(-1.5)] = 2e^{-1.5t}(2\pi\cos 2\pi t - 1.5\sin 2\pi t)$

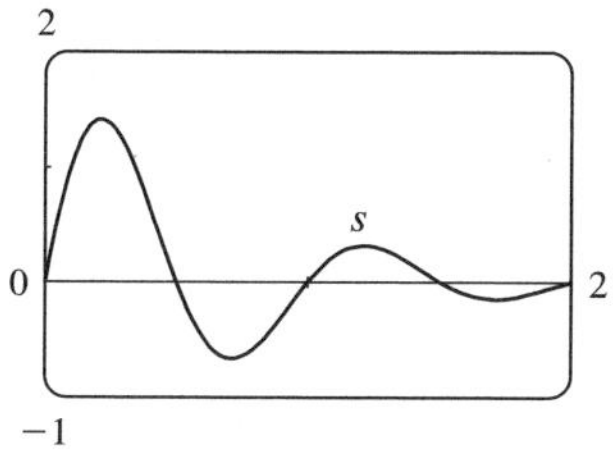

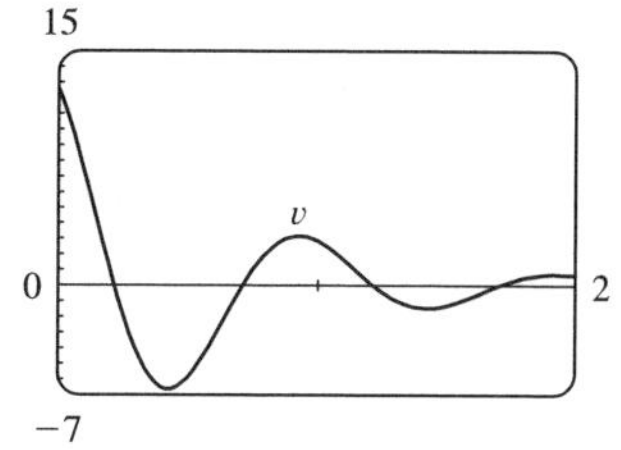

65. By the Chain Rule, $a(t) = \dfrac{dv}{dt} = \dfrac{dv}{ds}\dfrac{ds}{dt} = \dfrac{dv}{ds}v(t) = v(t)\dfrac{dv}{ds}$. The derivative dv/dt is the rate of change of the velocity with respect to time (in other words, the acceleration) whereas the derivative dv/ds is the rate of change of the velocity with respect to the displacement.

67. (a) Using a calculator or CAS, we obtain the model $Q = ab^t$ with $a \approx 100.012\,436\,9$ and $b \approx 0.000\,045\,145\,933$. We can change this model to one with base e and exponent $\ln b$ $[b^t = e^{t\ln b}$ from precalculus mathematics]: $Q = ae^{t\ln b} \approx 100.012\,437e^{-10.005\,531t}$.

(b) Use $Q'(t) = ab^t\ln b$ or the calculator command `nDeriv(Y₁,X,.04)` with Y_1=ab^x to get $Q'(0.04) \approx -670.63\ \mu$A. The result of Example 2 in Section 2.1 was $-670\ \mu$A.

69. $x = t^4 + 1,\ \ y = t^3 + t;\ t = -1.\quad \dfrac{dy}{dt} = 3t^2 + 1,\ \dfrac{dx}{dt} = 4t^3$, and $\dfrac{dy}{dx} = \dfrac{dy/dt}{dx/dt} = \dfrac{3t^2+1}{4t^3}$. When $t = -1$, $(x, y) = (2, -2)$ and $dy/dx = \frac{4}{-4} = -1$, so an equation of the tangent to the curve at the point corresponding to $t = -1$ is $y - (-2) = (-1)(x - 2)$, or $y = -x$.

71. $x = e^{\sqrt{t}}, y = t - t^2; t = 1.\quad \dfrac{dy}{dt} = 1 - 2t,\ \dfrac{dx}{dt} = e^{\sqrt{t}}\cdot\dfrac{1}{2}t^{-1/2}$, and $\dfrac{dy}{dx} = \dfrac{dy/dt}{dx/dt} = \dfrac{1-2t}{e^{\sqrt{t}}/(2\sqrt{t})}$.

When $t = 1$, $(x, y) = (e, 0)$ and $\dfrac{dy}{dx} = \dfrac{-1}{e/2} = -\dfrac{2}{e}$, so an equation of the tangent to the curve at the point corresponding to $t = 1$ is $y - 0 = -\dfrac{2}{e}(x - e)$, or $y = -\dfrac{2}{e}x + 2$.

73. (a) $x = t^2, y = t^3 - 3t \quad\Rightarrow\quad \dfrac{dy}{dx} = \dfrac{dy/dt}{dx/dt} = \dfrac{3t^2-3}{2t}$. At the point $(3, 0)$, $x = 3 \ \Rightarrow\ t^2 = 3 \ \Rightarrow\ t = \pm\sqrt{3} \ \Rightarrow$

$\dfrac{dy}{dx} = \dfrac{3(\pm\sqrt{3})^2 - 3}{2(\pm\sqrt{3})} = \dfrac{6}{2(\pm\sqrt{3})} = \pm\dfrac{3}{\sqrt{3}} = \pm\sqrt{3}$. When $t = \sqrt{3}$, an equation of the tangent line is $y - 0 = \sqrt{3}\,(x - 3)$ or $y = \sqrt{3}\,x - 3\sqrt{3}$. When $t = -\sqrt{3}$, an equation of the tangent line is $y - 0 = -\sqrt{3}\,(x - 3)$ or $y = -\sqrt{3}\,x + 3\sqrt{3}$.

(b) Horizontal tangent: $dy/dx = 0 \ \Leftrightarrow\ 3t^2 - 3 = 0 \ \Leftrightarrow$ $3(t^2 - 1) = 0 \ \Leftrightarrow\ t^2 = 1 \ \Leftrightarrow\ t = \pm 1$. $t = 1$ corresponds to the point $(x, y) = (t^2, t^3 - 3t) = (1^2, 1^3 - 3\cdot 1) = (1, -2)$ and $t = -1$ to $(1, 2)$.

Vertical tangent: dy/dx is undefined $\ \Leftrightarrow\ 2t = 0 \ \Leftrightarrow\ t = 0$. The value $t = 0$ corresponds to the origin; that is, $(0, 0)$.

(c)

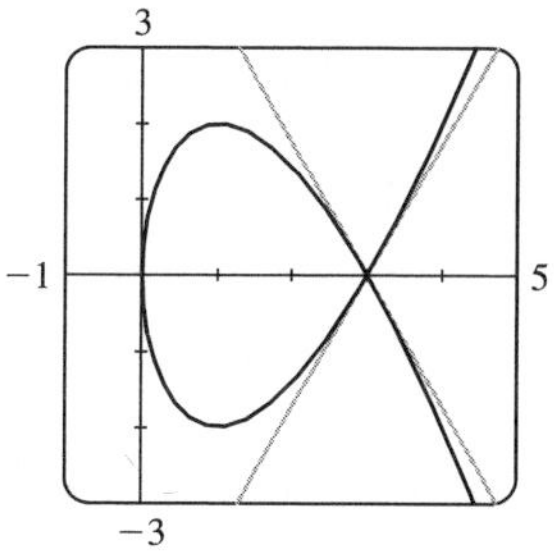

75. (a) Derive gives $g'(t) = \dfrac{45(t-2)^8}{(2t+1)^{10}}$ without simplifying. With either Maple or Mathematica, we first get

$g'(t) = 9\dfrac{(t-2)^8}{(2t+1)^9} - 18\dfrac{(t-2)^9}{(2t+1)^{10}}$, and the simplification command results in the above expression.

(b) Derive gives $y' = 2(x^3 - x + 1)^3(2x+1)^4(17x^3 + 6x^2 - 9x + 3)$ without simplifying. With either Maple or Mathematica, we first get $y' = 10(2x+1)^4(x^3 - x + 1)^4 + 4(2x+1)^5(x^3 - x + 1)^3(3x^2 - 1)$. If we use Mathematica's `Factor` or `Simplify`, or Maple's `factor`, we get the above expression, but Maple's `simplify` gives the polynomial expansion instead. For locating horizontal tangents, the factored form is the most helpful.

77. (a)
$$\begin{aligned} \frac{d}{dx}(\sin^n x \cos nx) &= n\sin^{n-1} x \cos x \cos nx + \sin^n x\,(-n\sin nx) && \text{[Product Rule]} \\ &= n\sin^{n-1} x\,(\cos nx \cos x - \sin nx \sin x) && \text{[factor out } n\sin^{n-1} x\text{]} \\ &= n\sin^{n-1} x \cos(nx + x) && \text{[Addition Formula for cosine]} \\ &= n\sin^{n-1} x \cos[(n+1)x] && \text{[factor out } x\text{]} \end{aligned}$$

(b)
$$\begin{aligned} \frac{d}{dx}(\cos^n x \cos nx) &= n\cos^{n-1} x\,(-\sin x)\cos nx + \cos^n x\,(-n\sin nx) && \text{[Product Rule]} \\ &= -n\cos^{n-1} x\,(\cos nx \sin x + \sin nx \cos x) && \text{[factor out } -n\cos^{n-1} x\text{]} \\ &= -n\cos^{n-1} x \sin(nx + x) && \text{[Addition Formula for sine]} \\ &= -n\cos^{n-1} x \sin[(n+1)x] && \text{[factor out } x\text{]} \end{aligned}$$

79. Since $\theta^\circ = \left(\frac{\pi}{180}\right)\theta$ rad, we have $\dfrac{d}{d\theta}(\sin\theta^\circ) = \dfrac{d}{d\theta}\left(\sin\frac{\pi}{180}\theta\right) = \frac{\pi}{180}\cos\frac{\pi}{180}\theta = \frac{\pi}{180}\cos\theta^\circ$.

81.
$$\begin{aligned} \frac{d^2y}{dx^2} &= \frac{d}{dx}\left(\frac{dy}{dx}\right) && \text{[Leibniz notation for the second derivative]} \\ &= \frac{d}{dx}\left(\frac{dy}{du}\frac{du}{dx}\right) && \text{[Chain Rule]} \\ &= \frac{dy}{du}\cdot\frac{d}{dx}\left(\frac{du}{dx}\right) + \frac{du}{dx}\cdot\frac{d}{dx}\left(\frac{dy}{du}\right) && \text{[Product Rule]} \\ &= \frac{dy}{du}\cdot\frac{d^2u}{dx^2} + \frac{du}{dx}\cdot\frac{d}{du}\left(\frac{dy}{du}\right)\cdot\frac{du}{dx} && [dy/du \text{ is a function of } u] \\ &= \frac{dy}{du}\frac{d^2u}{dx^2} + \frac{d^2y}{du^2}\left(\frac{du}{dx}\right)^2 \end{aligned}$$

Or: Using function notation for $y = f(u)$ and $u = g(x)$, we have $y = f(g(x))$, so

$y' = f'(g(x))\cdot g'(x)$ [by the Chain Rule] $\Rightarrow$

$(y')' = [f'(g(x))\cdot g'(x)]' = f'(g(x))\cdot g''(x) + g'(x)\cdot f''(g(x))\cdot g'(x) = f'(g(x))\cdot g''(x) + f''(g(x))\cdot[g'(x)]^2$.

3.6 Implicit Differentiation

1. (a) $\dfrac{d}{dx}(xy + 2x + 3x^2) = \dfrac{d}{dx}(4) \;\Rightarrow\; (x\cdot y' + y\cdot 1) + 2 + 6x = 0 \;\Rightarrow\; xy' = -y - 2 - 6x \;\Rightarrow$

$y' = \dfrac{-y - 2 - 6x}{x}$ or $y' = -6 - \dfrac{y+2}{x}$.

(b) $xy + 2x + 3x^2 = 4 \;\Rightarrow\; xy = 4 - 2x - 3x^2 \;\Rightarrow\; y = \dfrac{4 - 2x - 3x^2}{x} = \dfrac{4}{x} - 2 - 3x$, so $y' = -\dfrac{4}{x^2} - 3$.

(c) From part (a), $y' = \dfrac{-y - 2 - 6x}{x} = \dfrac{-(4/x - 2 - 3x) - 2 - 6x}{x} = \dfrac{-4/x - 3x}{x} = -\dfrac{4}{x^2} - 3$.

3. $\dfrac{d}{dx}(x^3 + x^2y + 4y^2) = \dfrac{d}{dx}(6) \;\Rightarrow\; 3x^2 + (x^2y' + y \cdot 2x) + 8yy' = 0 \;\Rightarrow\; x^2y' + 8yy' = -3x^2 - 2xy \;\Rightarrow$

$(x^2 + 8y)y' = -3x^2 - 2xy \;\Rightarrow\; y' = -\dfrac{3x^2 + 2xy}{x^2 + 8y} = -\dfrac{x(3x + 2y)}{x^2 + 8y}$

5. $\dfrac{d}{dx}(x^2y + xy^2) = \dfrac{d}{dx}(3x) \;\Rightarrow\; (x^2y' + y \cdot 2x) + (x \cdot 2yy' + y^2 \cdot 1) = 3 \;\Rightarrow\; x^2y' + 2xyy' = 3 - 2xy - y^2 \;\Rightarrow$

$y'(x^2 + 2xy) = 3 - 2xy - y^2 \;\Rightarrow\; y' = \dfrac{3 - 2xy - y^2}{x^2 + 2xy}$

7. $\sqrt{xy} = 1 + x^2y \;\Rightarrow\; \frac{1}{2}(xy)^{-1/2}(xy' + y \cdot 1) = 0 + x^2y' + y \cdot 2x \;\Rightarrow\; \dfrac{x}{2\sqrt{xy}}y' + \dfrac{y}{2\sqrt{xy}} = x^2y' + 2xy \;\Rightarrow$

$y'\left(\dfrac{x}{2\sqrt{xy}} - x^2\right) = 2xy - \dfrac{y}{2\sqrt{xy}} \;\Rightarrow\; y'\left(\dfrac{x - 2x^2\sqrt{xy}}{2\sqrt{xy}}\right) = \dfrac{4xy\sqrt{xy} - y}{2\sqrt{xy}} \;\Rightarrow\; y' = \dfrac{4xy\sqrt{xy} - y}{x - 2x^2\sqrt{xy}}$

9. $\dfrac{d}{dx}(4\cos x \sin y) = \dfrac{d}{dx}(1) \;\Rightarrow\; 4\left[\cos x \cdot \cos y \cdot y' + \sin y \cdot (-\sin x)\right] = 0 \;\Rightarrow$

$y'(4\cos x \cos y) = 4\sin x \sin y \;\Rightarrow\; y' = \dfrac{4\sin x \sin y}{4\cos x \cos y} = \tan x \tan y$

11. $\dfrac{d}{dx}(e^{x/y}) = \dfrac{d}{dx}(x - y) \;\Rightarrow\; e^{x/y} \cdot \dfrac{d}{dx}\left(\dfrac{x}{y}\right) = 1 - y' \;\Rightarrow$

$e^{x/y} \cdot \dfrac{y \cdot 1 - x \cdot y'}{y^2} = 1 - y' \;\Rightarrow\; e^{x/y} \cdot \dfrac{1}{y} - \dfrac{xe^{x/y}}{y^2} \cdot y' = 1 - y' \;\Rightarrow\; y' - \dfrac{xe^{x/y}}{y^2} \cdot y' = 1 - \dfrac{e^{x/y}}{y} \;\Rightarrow$

$y'\left(1 - \dfrac{xe^{x/y}}{y^2}\right) = \dfrac{y - e^{x/y}}{y} \;\Rightarrow\; y' = \dfrac{\dfrac{y - e^{x/y}}{y}}{\dfrac{y^2 - xe^{x/y}}{y^2}} = \dfrac{y(y - e^{x/y})}{y^2 - xe^{x/y}}$

13. $\dfrac{d}{dx}\{f(x) + x^2[f(x)]^3\} = \dfrac{d}{dx}(10) \;\Rightarrow\; f'(x) + x^2 \cdot 3[f(x)]^2 \cdot f'(x) + [f(x)]^3 \cdot 2x = 0$. If $x = 1$, we have

$f'(1) + 1^2 \cdot 3[f(1)]^2 \cdot f'(1) + [f(1)]^3 \cdot 2(1) = 0 \;\Rightarrow\; f'(1) + 1 \cdot 3 \cdot 2^2 \cdot f'(1) + 2^3 \cdot 2 = 0 \;\Rightarrow$

$f'(1) + 12f'(1) = -16 \;\Rightarrow\; 13f'(1) = -16 \;\Rightarrow\; f'(1) = -\frac{16}{13}$.

15. $x^2 + xy + y^2 = 3 \;\Rightarrow\; 2x + xy' + y \cdot 1 + 2yy' = 0 \;\Rightarrow\; xy' + 2yy' = -2x - y \;\Rightarrow\; y'(x + 2y) = -2x - y \;\Rightarrow$

$y' = \dfrac{-2x - y}{x + 2y}$. When $x = 1$ and $y = 1$, we have $y' = \dfrac{-2 - 1}{1 + 2 \cdot 1} = \dfrac{-3}{3} = -1$, so an equation of the tangent line is

$y - 1 = -1(x - 1)$ or $y = -x + 2$.

17. $x^2 + y^2 = (2x^2 + 2y^2 - x)^2 \;\Rightarrow\; 2x + 2yy' = 2(2x^2 + 2y^2 - x)(4x + 4yy' - 1)$. When $x = 0$ and $y = \frac{1}{2}$, we have

$0 + y' = 2(\frac{1}{2})(2y' - 1) \;\Rightarrow\; y' = 2y' - 1 \;\Rightarrow\; y' = 1$, so an equation of the tangent line is $y - \frac{1}{2} = 1(x - 0)$

or $y = x + \frac{1}{2}$.

19. $2(x^2 + y^2)^2 = 25(x^2 - y^2) \;\Rightarrow\; 4(x^2 + y^2)(2x + 2yy') = 25(2x - 2yy') \;\Rightarrow$

$4(x + yy')(x^2 + y^2) = 25(x - yy') \;\Rightarrow\; 4yy'(x^2 + y^2) + 25yy' = 25x - 4x(x^2 + y^2) \;\Rightarrow$

$y' = \dfrac{25x - 4x(x^2 + y^2)}{25y + 4y(x^2 + y^2)}$. When $x = 3$ and $y = 1$, we have $y' = \frac{75 - 120}{25 + 40} = -\frac{45}{65} = -\frac{9}{13}$, so an equation of the tangent line

is $y - 1 = -\frac{9}{13}(x - 3)$ or $y = -\frac{9}{13}x + \frac{40}{13}$.

21. (a) $y^2 = 5x^4 - x^2 \;\Rightarrow\; 2yy' = 5(4x^3) - 2x \;\Rightarrow\; y' = \dfrac{10x^3 - x}{y}$.

So at the point $(1, 2)$ we have $y' = \dfrac{10(1)^3 - 1}{2} = \dfrac{9}{2}$, and an equation

of the tangent line is $y - 2 = \frac{9}{2}(x - 1)$ or $y = \frac{9}{2}x - \frac{5}{2}$.

(b)

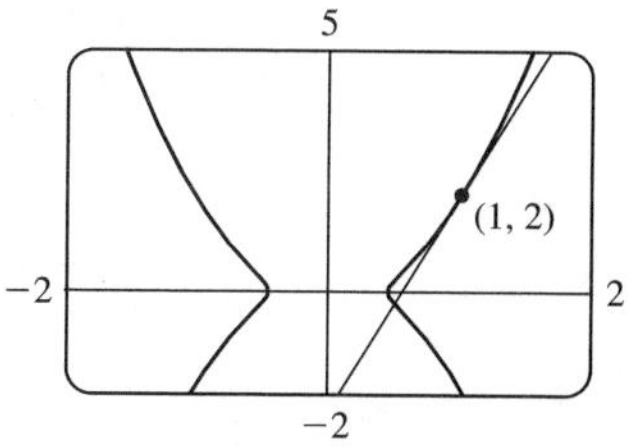

23. (a) There are eight points with horizontal tangents: four at $x \approx 1.57735$ and four at $x \approx 0.42265$.

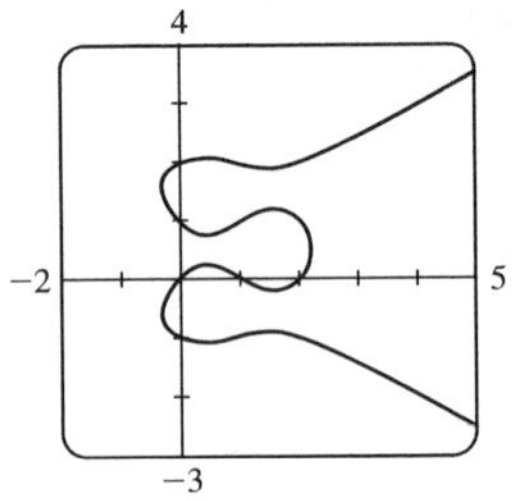

(b) $y' = \dfrac{3x^2 - 6x + 2}{2(2y^3 - 3y^2 - y + 1)} \quad \Rightarrow \quad y' = -1$ at $(0, 1)$ and $y' = \frac{1}{3}$ at $(0, 2)$.

Equations of the tangent lines are $y = -x + 1$ and $y = \frac{1}{3}x + 2$.

(c) $y' = 0 \quad \Rightarrow \quad 3x^2 - 6x + 2 = 0 \quad \Rightarrow \quad x = 1 \pm \frac{1}{3}\sqrt{3}$

(d) By multiplying the right side of the equation by $x - 3$, we obtain the first graph.By modifying the equation in other ways, we can generate the other graphs.

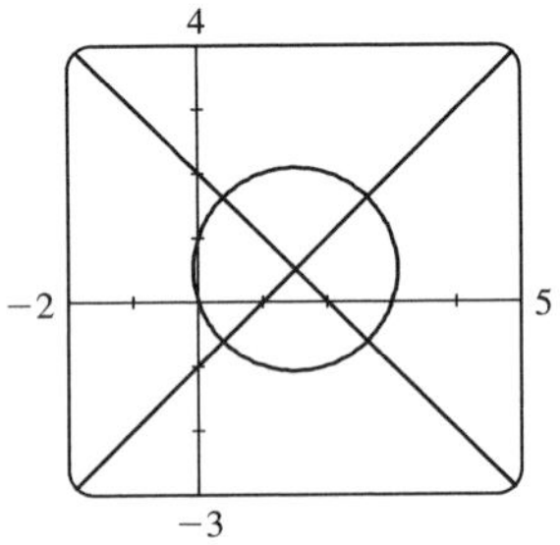

$y(y^2 - 1)(y - 2)$
$= x(x - 1)(x - 2)(x - 3)$

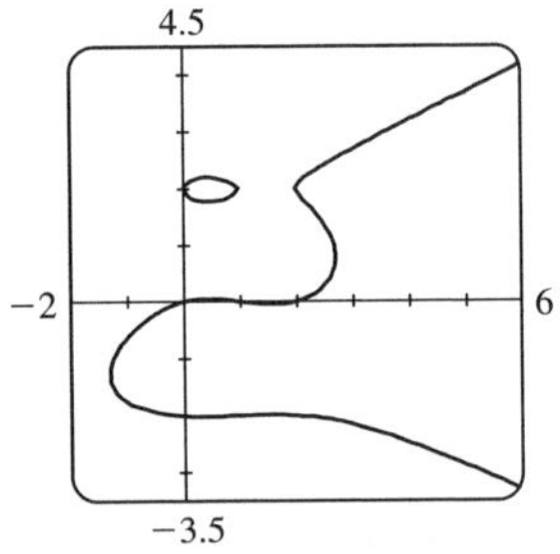

$y(y^2 - 4)(y - 2)$
$= x(x - 1)(x - 2)$

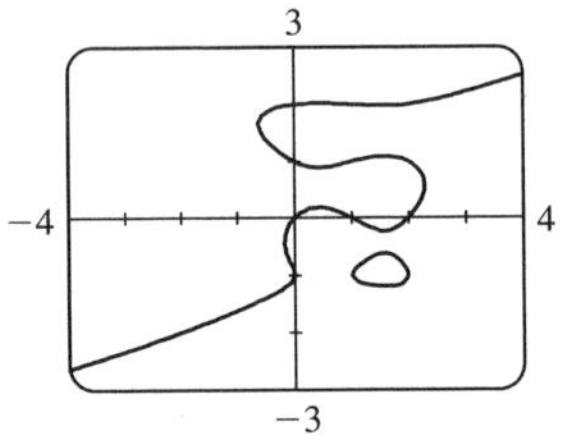

$y(y + 1)(y^2 - 1)(y - 2)$
$= x(x - 1)(x - 2)$

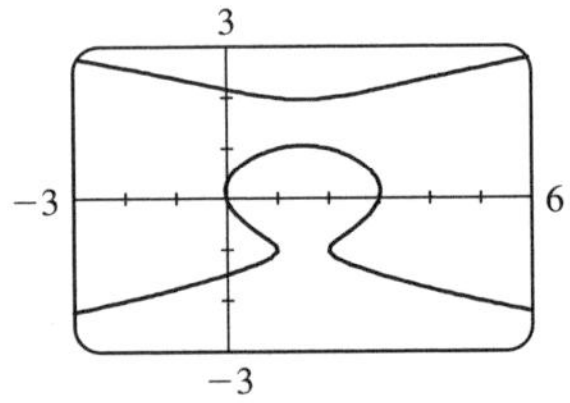

$(y + 1)(y^2 - 1)(y - 2)$
$= (x - 1)(x - 2)$

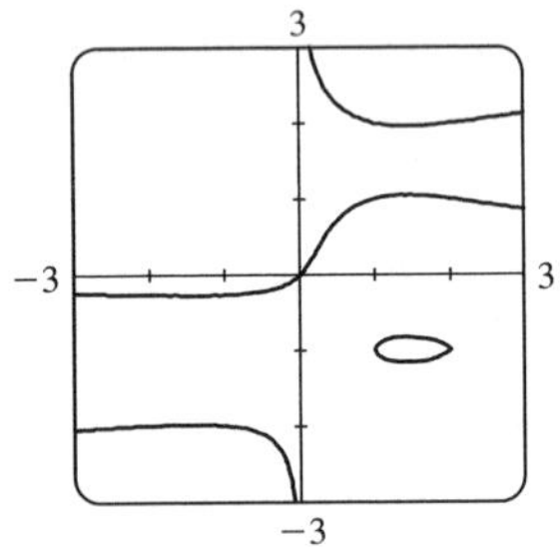

$x(y + 1)(y^2 - 1)(y - 2)$
$= y(x - 1)(x - 2)$

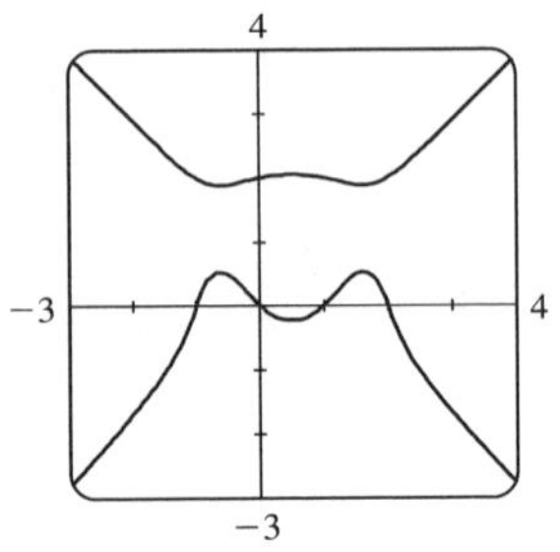

$y(y^2 + 1)(y - 2)$
$= x(x^2 - 1)(x - 2)$

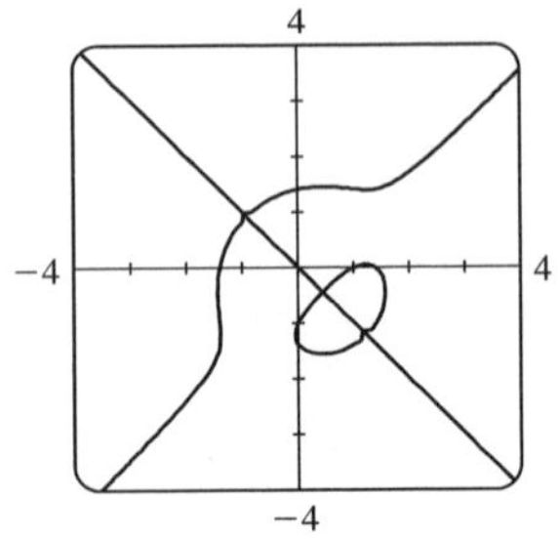

$y(y + 1)(y^2 - 2)$
$= x(x - 1)(x^2 - 2)$

25. From Exercise 19, a tangent to the lemniscate will be horizontal if $y' = 0 \Rightarrow 25x - 4x(x^2 + y^2) = 0 \Rightarrow$ $x[25 - 4(x^2 + y^2)] = 0 \Rightarrow x^2 + y^2 = \frac{25}{4}$ (1). (Note that when x is 0, y is also 0, and there is no horizontal tangent at the origin.) Substituting $\frac{25}{4}$ for $x^2 + y^2$ in the equation of the lemniscate, $2(x^2 + y^2)^2 = 25(x^2 - y^2)$, we get $x^2 - y^2 = \frac{25}{8}$ (2). Solving (1) and (2), we have $x^2 = \frac{75}{16}$ and $y^2 = \frac{25}{16}$, so the four points are $\left(\pm\frac{5\sqrt{3}}{4}, \pm\frac{5}{4}\right)$.

27. (a) $x^4 + y^4 = 16 \Rightarrow 4x^3 + 4y^3y' = 0 \Rightarrow y^3y' = -x^3 \Rightarrow y' = -x^3/y^3$

(b) $y'' = -\dfrac{y^3(3x^2) - (x^3)(3y^2y')}{(y^3)^2} = -\dfrac{3x^2y^3 - 3x^3y^2(-x^3/y^3)}{y^6} \cdot \dfrac{y}{y} = -\dfrac{3x^2y^4 + 3x^6}{y^7}$

(c) $y'' = -\dfrac{3x^2(y^4 + x^4)}{y^7} = -\dfrac{3x^2(16)}{y^7} = -48\dfrac{x^2}{y^7}$

29. $y = \tan^{-1}\sqrt{x} \Rightarrow y' = \dfrac{1}{1 + (\sqrt{x})^2} \cdot \dfrac{d}{dx}(\sqrt{x}) = \dfrac{1}{1+x}\left(\frac{1}{2}x^{-1/2}\right) = \dfrac{1}{2\sqrt{x}\,(1+x)}$

31. $y = \sin^{-1}(2x+1) \Rightarrow$

$$y' = \frac{1}{\sqrt{1-(2x+1)^2}} \cdot \frac{d}{dx}(2x+1) = \frac{1}{\sqrt{1-(4x^2+4x+1)}} \cdot 2 = \frac{2}{\sqrt{-4x^2-4x}} = \frac{1}{\sqrt{-x^2-x}}$$

33. $H(x) = (1+x^2)\arctan x \Rightarrow H'(x) = (1+x^2)\dfrac{1}{1+x^2} + (\arctan x)(2x) = 1 + 2x\arctan x$

35. $y = \arcsin(\tan\theta) \Rightarrow y' = \dfrac{1}{\sqrt{1-(\tan\theta)^2}} \cdot \dfrac{d}{d\theta}(\tan\theta) = \dfrac{\sec^2\theta}{\sqrt{1-\tan^2\theta}}$

37. $f(x) = e^x - x^2\arctan x \Rightarrow$

$$f'(x) = e^x - \left[x^2\left(\frac{1}{1+x^2}\right) + (\arctan x)(2x)\right]$$
$$= e^x - \frac{x^2}{1+x^2} - 2x\arctan x$$

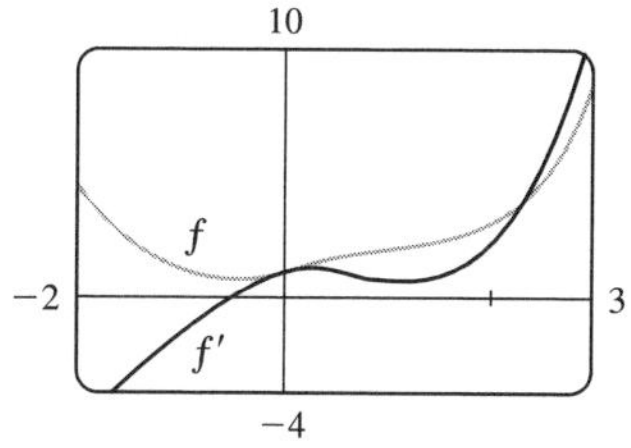

This is reasonable because the graphs show that f is increasing when f' is positive, and f' is zero when f has a minimum.

39. $2x^2 + y^2 = 3$ and $x = y^2$ intersect when $2x^2 + x - 3 = 0 \Leftrightarrow (2x+3)(x-1) = 0 \Leftrightarrow x = -\frac{3}{2}$ or 1, but $-\frac{3}{2}$ is extraneous since $x = y^2$ is nonnegative. When $x = 1$, $1 = y^2 \Rightarrow y = \pm 1$, so there are two points of intersection: $(1, \pm 1)$. $2x^2 + y^2 = 3 \Rightarrow 4x + 2yy' = 0 \Rightarrow y' = -2x/y$, and $x = y^2 \Rightarrow 1 = 2yy' \Rightarrow y' = 1/(2y)$. At $(1,1)$, the slopes are $m_1 = -2(1)/1 = -2$ and $m_2 = 1/(2 \cdot 1) = \frac{1}{2}$, so the curves are orthogonal (since m_1 and m_2 are negative reciprocals of each other). By symmetry, the curves are also orthogonal at $(1, -1)$.

41.

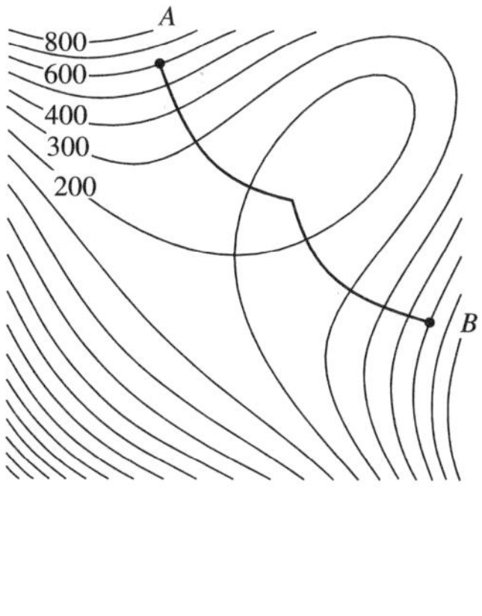

43. $x^2 + y^2 = r^2$ is a circle with centre O and $ax + by = 0$ is a line through O.

$x^2 + y^2 = r^2 \Rightarrow 2x + 2yy' = 0 \Rightarrow y' = -x/y$, so the slope of the tangent line at $P_0\,(x_0, y_0)$ is $-x_0/y_0$. The slope of the line OP_0 is y_0/x_0, which is the negative reciprocal of $-x_0/y_0$. Hence, the curves are orthogonal, and the families of curves are orthogonal trajectories of each other.

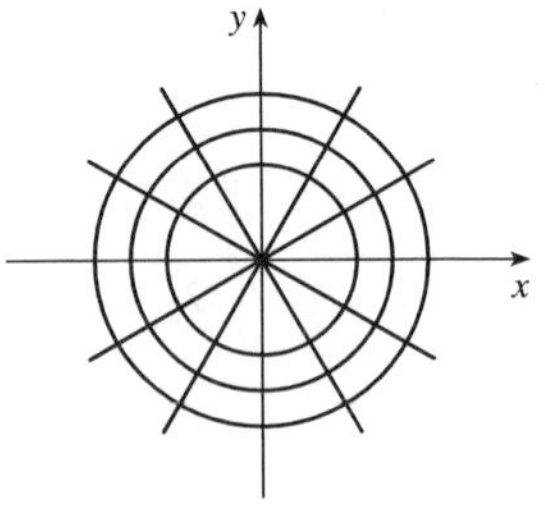

45. $y = cx^2 \Rightarrow y' = 2cx$ and $x^2 + 2y^2 = k \Rightarrow 2x + 4yy' = 0 \Rightarrow$

$2yy' = -x \Rightarrow y' = -\dfrac{x}{2(y)} = -\dfrac{x}{2(cx^2)} = -\dfrac{1}{2cx}$, so the curves are orthogonal.

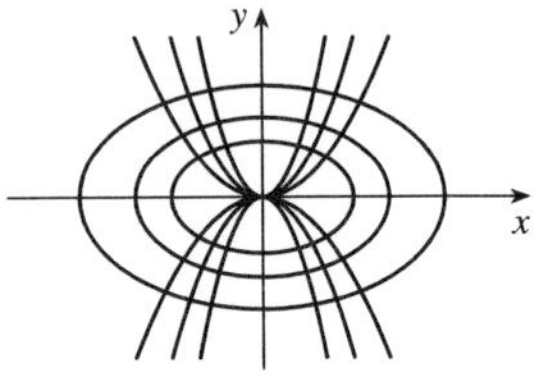

47. If the circle has radius r, its equation is $x^2 + y^2 = r^2 \Rightarrow 2x + 2yy' = 0 \Rightarrow y' = -\dfrac{x}{y}$, so the slope of the tangent line at $P(x_0, y_0)$ is $-\dfrac{x_0}{y_0}$. The negative reciprocal of that slope is $\dfrac{-1}{-x_0/y_0} = \dfrac{y_0}{x_0}$, which is the slope of OP, so the tangent line at P is perpendicular to the radius OP.

49. To find the points at which the ellipse $x^2 - xy + y^2 = 3$ crosses the x-axis, let $y = 0$ and solve for x.

$y = 0 \Rightarrow x^2 - x(0) + 0^2 = 3 \Leftrightarrow x = \pm\sqrt{3}$. So the graph of the ellipse crosses the x-axis at the points $(\pm\sqrt{3}, 0)$.

Using implicit differentiation to find y', we get $2x - xy' - y + 2yy' = 0 \Rightarrow y'(2y - x) = y - 2x \Leftrightarrow y' = \dfrac{y - 2x}{2y - x}$.

So y' at $(\sqrt{3}, 0)$ is $\dfrac{0 - 2\sqrt{3}}{2(0) - \sqrt{3}} = 2$ and y' at $(-\sqrt{3}, 0)$ is $\dfrac{0 + 2\sqrt{3}}{2(0) + \sqrt{3}} = 2$. Thus, the tangent lines at these points are parallel.

51. $x^2y^2 + xy = 2 \Rightarrow x^2 \cdot 2yy' + y^2 \cdot 2x + x \cdot y' + y \cdot 1 = 0 \Leftrightarrow y'(2x^2y + x) = -2xy^2 - y \Leftrightarrow$

$y' = -\dfrac{2xy^2 + y}{2x^2y + x}$. So $-\dfrac{2xy^2 + y}{2x^2y + x} = -1 \Leftrightarrow 2xy^2 + y = 2x^2y + x \Leftrightarrow y(2xy + 1) = x(2xy + 1) \Leftrightarrow$

$y(2xy + 1) - x(2xy + 1) = 0 \Leftrightarrow (2xy + 1)(y - x) = 0 \Leftrightarrow xy = -\frac{1}{2}$ or $y = x$. But $xy = -\frac{1}{2} \Rightarrow$

$x^2y^2 + xy = \frac{1}{4} - \frac{1}{2} \neq 2$, so we must have $x = y$. Then $x^2y^2 + xy = 2 \Rightarrow x^4 + x^2 = 2 \Leftrightarrow x^4 + x^2 - 2 = 0 \Leftrightarrow$

$(x^2 + 2)(x^2 - 1) = 0$. So $x^2 = -2$, which is impossible, or $x^2 = 1 \Leftrightarrow x = \pm 1$. Since $x = y$, the points on the curve where the tangent line has a slope of -1 are $(-1, -1)$ and $(1, 1)$.

53. (a) If $y = f^{-1}(x)$, then $f(y) = x$. Differentiating implicitly with respect to x and remembering that y is a function of x, we get $f'(y)\dfrac{dy}{dx} = 1$, so $\dfrac{dy}{dx} = \dfrac{1}{f'(y)} \Rightarrow (f^{-1})'(x) = \dfrac{1}{f'(f^{-1}(x))}$.

(b) $f(4) = 5 \Rightarrow f^{-1}(5) = 4$. By part (a), $(f^{-1})'(5) = 1/f'(f^{-1}(5)) = 1/f'(4) = 1/\left(\frac{2}{3}\right) = \frac{3}{2}$.

55. (a) $y = J(x)$ and $xy'' + y' + xy = 0 \Rightarrow xJ''(x) + J'(x) + xJ(x) = 0$. If $x = 0$, we have $0 + J'(0) + 0 = 0$, so $J'(0) = 0$.

(b) Differentiating $xy'' + y' + xy = 0$ implicitly, we get $xy''' + y'' \cdot 1 + y'' + xy' + y \cdot 1 = 0 \Rightarrow$ $xy''' + 2y'' + xy' + y = 0$, so $xJ'''(x) + 2J''(x) + xJ'(x) + J(x) = 0$. If $x = 0$, we have $0 + 2J''(0) + 0 + 1$ $[J(0) = 1$ is given$]$ $= 0 \Rightarrow 2J''(0) = -1 \Rightarrow J''(0) = -\frac{1}{2}$.

3.7 Derivatives of Logarithmic Functions

1. The differentiation formula for logarithmic functions, $\frac{d}{dx}(\log_a x) = \frac{1}{x \ln a}$, is simplest when $a = e$ because $\ln e = 1$.

3. $f(\theta) = \ln(\cos\theta) \quad \Rightarrow \quad f'(\theta) = \frac{1}{\cos\theta}\frac{d}{d\theta}(\cos\theta) = \frac{-\sin\theta}{\cos\theta} = -\tan\theta$

5. $f(x) = \sqrt[5]{\ln x} = (\ln x)^{1/5} \quad \Rightarrow \quad f'(x) = \frac{1}{5}(\ln x)^{-4/5}\frac{d}{dx}(\ln x) = \frac{1}{5(\ln x)^{4/5}} \cdot \frac{1}{x} = \frac{1}{5x\sqrt[5]{(\ln x)^4}}$

7. $f(x) = \log_2(1-3x) \quad \Rightarrow \quad f'(x) = \frac{1}{(1-3x)\ln 2}\frac{d}{dx}(1-3x) = \frac{-3}{(1-3x)\ln 2}$ or $\frac{3}{(3x-1)\ln 2}$

9. $f(x) = \sqrt{x}\ln x \quad \Rightarrow \quad f'(x) = \sqrt{x}\left(\frac{1}{x}\right) + (\ln x) \cdot \frac{1}{2\sqrt{x}} = \frac{1}{\sqrt{x}} + \frac{\ln x}{2\sqrt{x}} = \frac{2 + \ln x}{2\sqrt{x}}$

11. $F(t) = \ln\frac{(2t+1)^3}{(3t-1)^4} = \ln(2t+1)^3 - \ln(3t-1)^4 = 3\ln(2t+1) - 4\ln(3t-1) \quad \Rightarrow$

$F'(t) = 3 \cdot \frac{1}{2t+1} \cdot 2 - 4 \cdot \frac{1}{3t-1} \cdot 3 = \frac{6}{2t+1} - \frac{12}{3t-1}$, or combined, $\frac{-6(t+3)}{(2t+1)(3t-1)}$.

13. $y = \frac{\ln x}{1+x} \quad \Rightarrow \quad y' = \frac{(1+x)(1/x) - (\ln x)(1)}{(1+x)^2} = \frac{\frac{1+x}{x} - \frac{x\ln x}{x}}{(1+x)^2} = \frac{1 + x - x\ln x}{x(1+x)^2}$

15. $y = \ln\left|2 - x - 5x^2\right| \quad \Rightarrow \quad y' = \frac{1}{2 - x - 5x^2} \cdot (-1 - 10x) = \frac{-10x - 1}{2 - x - 5x^2}$ or $\frac{10x + 1}{5x^2 + x - 2}$

17. $y = \ln(e^{-x} + xe^{-x}) = \ln(e^{-x}(1+x)) = \ln(e^{-x}) + \ln(1+x) = -x + \ln(1+x) \quad \Rightarrow$

$y' = -1 + \frac{1}{1+x} = \frac{-1 - x + 1}{1+x} = -\frac{x}{1+x}$

19. $y = e^x \ln x \quad \Rightarrow \quad y' = e^x \cdot \frac{1}{x} + (\ln x) \cdot e^x = e^x\left(\frac{1}{x} + \ln x\right) \quad \Rightarrow$

$y'' = e^x\left(-\frac{1}{x^2} + \frac{1}{x}\right) + \left(\frac{1}{x} + \ln x\right)e^x = e^x\left(-\frac{1}{x^2} + \frac{1}{x} + \frac{1}{x} + \ln x\right) = e^x\left(\ln x + \frac{2}{x} - \frac{1}{x^2}\right)$

21. $f(x) = \frac{x}{1 - \ln(x-1)} \quad \Rightarrow$

$$f'(x) = \frac{[1 - \ln(x-1)] \cdot 1 - x \cdot \frac{-1}{x-1}}{[1 - \ln(x-1)]^2} = \frac{\frac{(x-1)[1 - \ln(x-1)] + x}{x-1}}{[1 - \ln(x-1)]^2} = \frac{x - 1 - (x-1)\ln(x-1) + x}{(x-1)[1 - \ln(x-1)]^2}$$

$$= \frac{2x - 1 - (x-1)\ln(x-1)}{(x-1)[1 - \ln(x-1)]^2}$$

$\text{Dom}(f) = \{x \mid x - 1 > 0 \text{ and } 1 - \ln(x-1) \neq 0\} = \{x \mid x > 1 \text{ and } \ln(x-1) \neq 1\}$

$= \{x \mid x > 1 \text{ and } x - 1 \neq e^1\} = \{x \mid x > 1 \text{ and } x \neq 1 + e\} = (1, 1+e) \cup (1+e, \infty)$

23. $y = \ln(x^2 - 3) \Rightarrow y' = \dfrac{1}{x^2-3} \cdot 2x = \dfrac{2x}{x^2-3}$.

$y'(2) = \dfrac{2(2)}{2^2-3} = 4$, so an equation of the tangent line at $(2, 0)$ is $y - 0 = 4(x - 2)$ or $y = 4x - 8$.

25. (a) The domain of $f(x) = x \ln x$ is $(0, \infty)$. $f'(x) = x\,(1/x) + (\ln x) \cdot 1 = 1 + \ln x$. So $f'(x) < 0$ when $1 + \ln x < 0 \Leftrightarrow \ln x < -1 \Leftrightarrow x < e^{-1}$. Therefore, f is decreasing on $(0, 1/e)$.

(b) $f'(x) = 1 + \ln x \Rightarrow f''(x) = 1/x > 0$ for $x > 0$. So the curve is concave upward on $(0, \infty)$.

27. $y = (2x+1)^5(x^4-3)^6 \Rightarrow \ln y = \ln\left((2x+1)^5(x^4-3)^6\right) \Rightarrow \ln y = 5\ln(2x+1) + 6\ln(x^4-3) \Rightarrow$

$\dfrac{1}{y}y' = 5 \cdot \dfrac{1}{2x+1} \cdot 2 + 6 \cdot \dfrac{1}{x^4-3} \cdot 4x^3 \Rightarrow$

$y' = y\left(\dfrac{10}{2x+1} + \dfrac{24x^3}{x^4-3}\right) = (2x+1)^5(x^4-3)^6\left(\dfrac{10}{2x+1} + \dfrac{24x^3}{x^4-3}\right)$.

[The answer could be simplified to $y' = 2(2x+1)^4(x^4-3)^5(29x^4 + 12x^3 - 15)$, but this is unnecessary.]

29. $y = \dfrac{\sin^2 x \tan^4 x}{(x^2+1)^2} \Rightarrow \ln y = \ln(\sin^2 x \tan^4 x) - \ln(x^2+1)^2 \Rightarrow$

$\ln y = \ln(\sin x)^2 + \ln(\tan x)^4 - \ln(x^2+1)^2 \Rightarrow \ln y = 2\ln|\sin x| + 4\ln|\tan x| - 2\ln(x^2+1) \Rightarrow$

$\dfrac{1}{y}y' = 2 \cdot \dfrac{1}{\sin x} \cdot \cos x + 4 \cdot \dfrac{1}{\tan x} \cdot \sec^2 x - 2 \cdot \dfrac{1}{x^2+1} \cdot 2x \Rightarrow y' = \dfrac{\sin^2 x \tan^4 x}{(x^2+1)^2}\left(2\cot x + \dfrac{4\sec^2 x}{\tan x} - \dfrac{4x}{x^2+1}\right)$

31. $y = x^x \Rightarrow \ln y = \ln x^x \Rightarrow \ln y = x \ln x \Rightarrow y'/y = x(1/x) + (\ln x) \cdot 1 \Rightarrow y' = y(1 + \ln x) \Rightarrow$

$y' = x^x(1 + \ln x)$

33. $y = (\cos x)^x \Rightarrow \ln y = \ln(\cos x)^x \Rightarrow \ln y = x \ln \cos x \Rightarrow \dfrac{1}{y}y' = x \cdot \dfrac{1}{\cos x} \cdot (-\sin x) + \ln \cos x \cdot 1 \Rightarrow$

$y' = y\left(\ln \cos x - \dfrac{x \sin x}{\cos x}\right) \Rightarrow y' = (\cos x)^x(\ln \cos x - x \tan x)$

35. $y = (\tan x)^{1/x} \Rightarrow \ln y = \ln(\tan x)^{1/x} \Rightarrow \ln y = \dfrac{1}{x}\ln \tan x \Rightarrow$

$\dfrac{1}{y}y' = \dfrac{1}{x} \cdot \dfrac{1}{\tan x} \cdot \sec^2 x + \ln \tan x \cdot \left(-\dfrac{1}{x^2}\right) \Rightarrow y' = y\left(\dfrac{\sec^2 x}{x \tan x} - \dfrac{\ln \tan x}{x^2}\right) \Rightarrow$

$y' = (\tan x)^{1/x}\left(\dfrac{\sec^2 x}{x \tan x} - \dfrac{\ln \tan x}{x^2}\right)$ or $y' = (\tan x)^{1/x} \cdot \dfrac{1}{x}\left(\csc x \sec x - \dfrac{\ln \tan x}{x}\right)$

37. $y = \ln(x^2 + y^2) \Rightarrow y' = \dfrac{1}{x^2+y^2}\dfrac{d}{dx}(x^2+y^2) \Rightarrow y' = \dfrac{2x + 2yy'}{x^2+y^2} \Rightarrow x^2y' + y^2y' = 2x + 2yy' \Rightarrow$

$x^2y' + y^2y' - 2yy' = 2x \Rightarrow (x^2 + y^2 - 2y)y' = 2x \Rightarrow y' = \dfrac{2x}{x^2+y^2-2y}$

39. $f(x) = \ln(x-1) \Rightarrow f'(x) = 1/(x-1) = (x-1)^{-1} \Rightarrow f''(x) = -(x-1)^{-2} \Rightarrow f'''(x) = 2(x-1)^{-3} \Rightarrow$

$f^{(4)}(x) = -2 \cdot 3(x-1)^{-4} \Rightarrow \cdots \Rightarrow f^{(n)}(x) = (-1)^{n-1} \cdot 2 \cdot 3 \cdot 4 \cdot \cdots \cdot (n-1)(x-1)^{-n} = (-1)^{n-1}\dfrac{(n-1)!}{(x-1)^n}$

41. If $f(x) = \ln(1+x)$, then $f'(x) = \dfrac{1}{1+x}$, so $f'(0) = 1$.

Thus, $\displaystyle\lim_{x\to 0}\frac{\ln(1+x)}{x} = \lim_{x\to 0}\frac{f(x)}{x} = \lim_{x\to 0}\frac{f(x) - f(0)}{x - 0} = f'(0) = 1$.

3.8 Linear Approximations and Differentials

1. As in Example 1, $T(0) = 85$, $T(10) = 78$, $T(20) = 72$, and

$$T'(20) \approx \frac{T(10) - T(20)}{10 - 20} = \frac{78 - 72}{-10} = -0.6°\text{C/min}.$$

$T(30) \approx T(20) + T'(20)(30 - 20) \approx 72 - 0.6(10) = 66°\text{C}.$

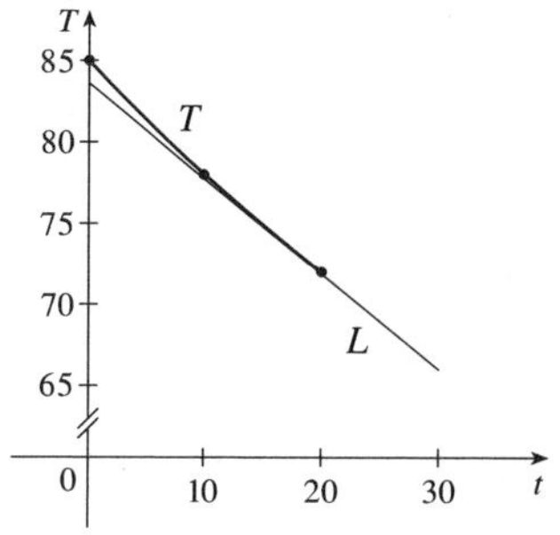

We would expect the temperature of the turkey to get closer to 24°C as time increases. Since the temperature decreased 7°C in the first 10 minutes and 6°C in the second 10 minutes, we can assume that the slopes of the tangent line are increasing through negative values: $-0.7, -0.6, \ldots$. Hence, the tangent lines are under the curve and 66°C is an underestimate. From the figure, we estimate the slope of the tangent line at $t = 20$ to be $\frac{83.5 - 66.25}{0 - 30} = -0.575$. Then the linear approximation becomes $T(30) \approx T(20) + T'(20) \cdot 10 \approx 72 - 0.575(10) = 66.25$.

3. Extend the tangent line at the point $(2030, 21)$ to the t-axis. Answers will vary based on this approximation—we'll use $t = 1900$ as our t-intercept. The linearization is then

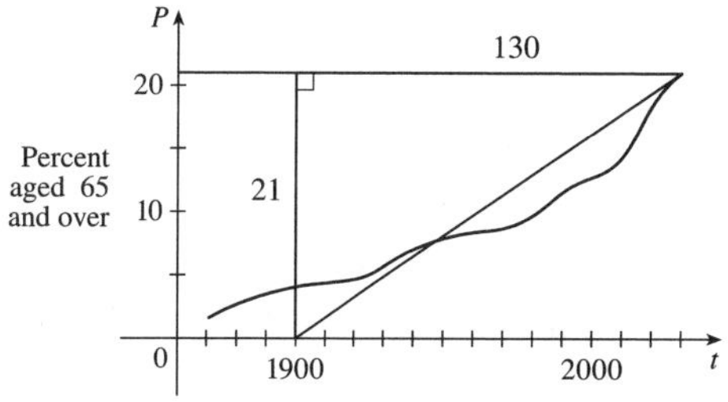

$$P(t) \approx P(2030) + P'(2030)(t - 2030) \approx 21 + \tfrac{21}{130}(t - 2030)$$

$P(2040) = 21 + \frac{21}{130}(2040 - 2030) \approx 22.6\%$

$P(2050) = 21 + \frac{21}{130}(2050 - 2030) \approx 24.2\%$

These predictions are probably too high since the tangent line lies above the graph at $t = 2030$.

5. $f(x) = x^4 + 3x^2 \quad \Rightarrow \quad f'(x) = 4x^3 + 6x$, so $f(-1) = 4$ and $f'(-1) = -10$.

Thus, $L(x) = f(-1) + f'(-1)(x - (-1)) = 4 + (-10)(x + 1) = -10x - 6$.

7. $f(x) = \cos x \quad \Rightarrow \quad f'(x) = -\sin x$, so $f\left(\frac{\pi}{2}\right) = 0$ and $f'\left(\frac{\pi}{2}\right) = -1$.

Thus, $L(x) = f\left(\frac{\pi}{2}\right) + f'\left(\frac{\pi}{2}\right)\left(x - \frac{\pi}{2}\right) = 0 - 1\left(x - \frac{\pi}{2}\right) = -x + \frac{\pi}{2}$.

9. $f(x) = \sqrt{1 - x} \quad \Rightarrow \quad f'(x) = \dfrac{-1}{2\sqrt{1 - x}}$, so $f(0) = 1$ and $f'(0) = -\frac{1}{2}$.

Therefore,

$\sqrt{1 - x} = f(x) \approx f(0) + f'(0)(x - 0) = 1 + \left(-\frac{1}{2}\right)(x - 0) = 1 - \frac{1}{2}x.$

So $\sqrt{0.9} = \sqrt{1 - 0.1} \approx 1 - \frac{1}{2}(0.1) = 0.95$

and $\sqrt{0.99} = \sqrt{1 - 0.01} \approx 1 - \frac{1}{2}(0.01) = 0.995$.

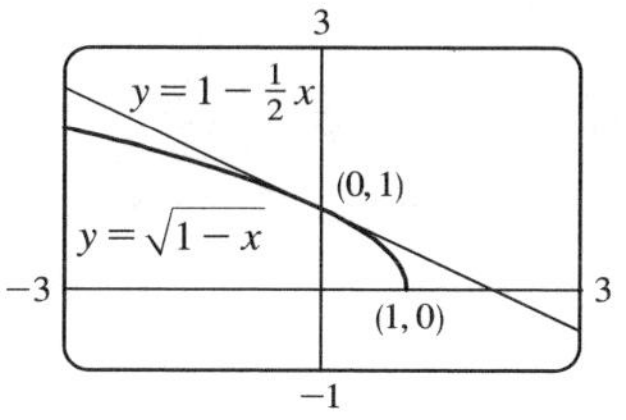

11. $f(x) = \sqrt[3]{1 - x} = (1 - x)^{1/3} \quad \Rightarrow \quad f'(x) = -\frac{1}{3}(1 - x)^{-2/3}$, so $f(0) = 1$ and $f'(0) = -\frac{1}{3}$. Thus, $f(x) \approx f(0) + f'(0)(x - 0) = 1 - \frac{1}{3}x$. We need $\sqrt[3]{1 - x} - 0.1 < 1 - \frac{1}{3}x < \sqrt[3]{1 - x} + 0.1$, which is true when $-1.204 < x < 0.706$.

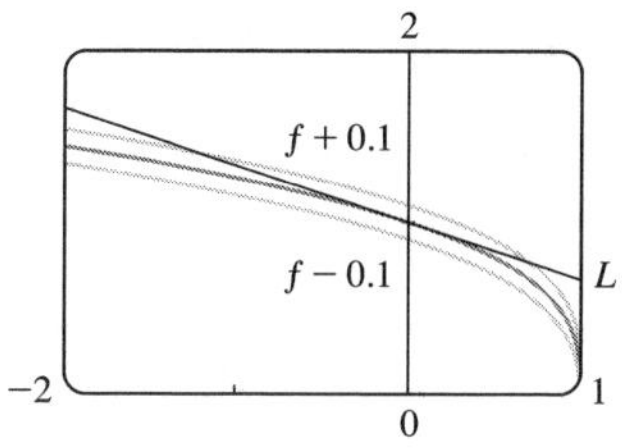

13. $f(x) = \dfrac{1}{(1+2x)^4} = (1+2x)^{-4} \;\Rightarrow$

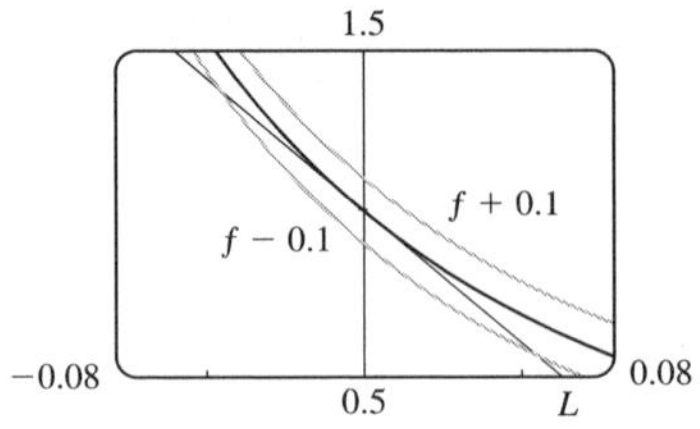

$f'(x) = -4(1+2x)^{-5}(2) = \dfrac{-8}{(1+2x)^5}$, so $f(0) = 1$ and $f'(0) = -8$.

Thus, $f(x) \approx f(0) + f'(0)(x-0) = 1 + (-8)(x-0) = 1 - 8x$.

We need $1/(1+2x)^4 - 0.1 < 1 - 8x < 1/(1+2x)^4 + 0.1$, which is true when $-0.045 < x < 0.055$.

15. To estimate $(2.001)^5$, we'll find the linearization of $f(x) = x^5$ at $a = 2$. Since $f'(x) = 5x^4$, $f(2) = 32$, and $f'(2) = 80$, we have $L(x) = 32 + 80(x-2) = 80x - 128$. Thus, $x^5 \approx 80x - 128$ when x is near 2 , so $(2.001)^5 \approx 80(2.001) - 128 = 160.08 - 128 = 32.08$.

17. To estimate $(8.06)^{2/3}$, we'll find the linearization of $f(x) = x^{2/3}$ at $a = 8$. Since $f'(x) = \frac{2}{3}x^{-1/3} = 2/\left(3\sqrt[3]{x}\right)$, $f(8) = 4$, and $f'(8) = \frac{1}{3}$, we have $L(x) = 4 + \frac{1}{3}(x-8) = \frac{1}{3}x + \frac{4}{3}$. Thus, $x^{2/3} \approx \frac{1}{3}x + \frac{4}{3}$ when x is near 8, so $(8.06)^{2/3} \approx \frac{1}{3}(8.06) + \frac{4}{3} = \frac{12.06}{3} = 4.02$.

19. $y = f(x) = \sec x \;\Rightarrow\; f'(x) = \sec x\,\tan x$, so $f(0) = 1$ and $f'(0) = 1 \cdot 0 = 0$. The linear approximation of f at 0 is $f(0) + f'(0)(x-0) = 1 + 0(x) = 1$. Since 0.08 is close to 0, approximating $\sec 0.08$ with 1 is reasonable.

21. $y = f(x) = \ln x \;\Rightarrow\; f'(x) = 1/x$, so $f(1) = 0$ and $f'(1) = 1$. The linear approximation of f at 1 is $f(1) + f'(1)(x-1) = 0 + 1(x-1) = x - 1$. Now $f(1.05) = \ln 1.05 \approx 1.05 - 1 = 0.05$, so the approximation is reasonable.

23. (a) The differential dy is defined in terms of dx by the equation $dy = f'(x)\,dx$. For $y = f(x) = x^2 \sin 2x$, $f'(x) = x^2 \cos 2x \cdot 2 + \sin 2x \cdot 2x = 2x(x\cos 2x + \sin 2x)$, so $dy = 2x(x\cos 2x + \sin 2x)\,dx$.

(b) For $y = f(t) = \ln\sqrt{1+t^2} = \frac{1}{2}\ln(1+t^2)$, $f'(t) = \dfrac{1}{2} \cdot \dfrac{1}{1+t^2} \cdot 2t = \dfrac{t}{1+t^2}$, so $dy = \dfrac{t}{1+t^2}\,dt$.

25. (a) $y = e^{x/10} \;\Rightarrow\; dy = e^{x/10} \cdot \frac{1}{10}\,dx = \frac{1}{10}e^{x/10}dx$

(b) $x = 0$ and $dx = 0.1 \;\Rightarrow\; dy = \frac{1}{10}e^{0/10}(0.1) = 0.01$.

$\Delta y = f(x + \Delta x) - f(x) = e^{(x+\Delta x)/10} - e^{x} = e^{(0+0.1)/10} - e^0 = e^{1/100} - 1 \approx 0.0101$

27. (a) If x is the edge length, then $V = x^3 \;\Rightarrow\; dV = 3x^2\,dx$. When $x = 30$ and $dx = 0.1$, $dV = 3(30)^2(0.1) = 270$, so the maximum possible error in computing the volume of the cube is about 270 cm^3. The relative error is calculated by dividing the change in V, ΔV, by V. We approximate ΔV with dV.

$$\text{Relative error} = \frac{\Delta V}{V} \approx \frac{dV}{V} = \frac{3x^2\,dx}{x^3} = 3\,\frac{dx}{x} = 3\left(\frac{0.1}{30}\right) = 0.01.$$

Percentage error = relative error × 100% = 0.01 × 100% = 1%.

(b) $S = 6x^2 \;\Rightarrow\; dS = 12x\,dx$. When $x = 30$ and $dx = 0.1$, $dS = 12(30)(0.1) = 36$, so the maximum possible error in computing the surface area of the cube is about 36 cm^2.

$$\text{Relative error} = \frac{\Delta S}{S} \approx \frac{dS}{S} = \frac{12x\,dx}{6x^2} = 2\,\frac{dx}{x} = 2\left(\frac{0.1}{30}\right) = 0.00\overline{6}.$$

Percentage error = relative error ×100% = $0.00\overline{6} \times 100\% = 0.\overline{6}\%$.

29. (a) For a sphere of radius r, the circumference is $C = 2\pi r$ and the surface area is $S = 4\pi r^2$, so

$$r = \frac{C}{2\pi} \quad \Rightarrow \quad S = 4\pi\left(\frac{C}{2\pi}\right)^2 = \frac{C^2}{\pi} \quad \Rightarrow \quad dS = \frac{2}{\pi}C\,dC.$$ When $C = 84$ and $dC = 0.5$, $dS = \frac{2}{\pi}(84)(0.5) = \frac{84}{\pi}$,

so the maximum error is about $\frac{84}{\pi} \approx 27\text{ cm}^2$. Relative error $\approx \frac{dS}{S} = \frac{84/\pi}{84^2/\pi} = \frac{1}{84} \approx 0.012$

(b) $V = \frac{4}{3}\pi r^3 = \frac{4}{3}\pi\left(\frac{C}{2\pi}\right)^3 = \frac{C^3}{6\pi^2} \quad \Rightarrow \quad dV = \frac{1}{2\pi^2}C^2\,dC$. When $C = 84$ and $dC = 0.5$,

$dV = \frac{1}{2\pi^2}(84)^2(0.5) = \frac{1764}{\pi^2}$, so the maximum error is about $\frac{1764}{\pi^2} \approx 179\text{ cm}^3$.

The relative error is approximately $\frac{dV}{V} = \frac{1764/\pi^2}{(84)^3/(6\pi^2)} = \frac{1}{56} \approx 0.018$.

31. $F = kR^4 \quad \Rightarrow \quad dF = 4kR^3\,dR \quad \Rightarrow \quad \frac{dF}{F} = \frac{4kR^3\,dR}{kR^4} = 4\left(\frac{dR}{R}\right)$. Thus, the relative change in F is about 4 times the relative change in R. So a 5% increase in the radius corresponds to a 20% increase in blood flow.

33. (a) The graph shows that $f'(1) = 2$, so $L(x) = f(1) + f'(1)(x - 1) = 5 + 2(x - 1) = 2x + 3$.
$f(0.9) \approx L(0.9) = 4.8$ and $f(1.1) \approx L(1.1) = 5.2$.

(b) From the graph, we see that $f'(x)$ is positive and decreasing. This means that the slopes of the tangent lines are positive, but the tangents are becoming less steep. So the tangent lines lie *above* the curve. Thus, the estimates in part (a) are too large.

3 Review

CONCEPT CHECK

1. (a) The Power Rule: If n is any real number, then $\frac{d}{dx}(x^n) = nx^{n-1}$. The derivative of a variable base raised to a constant power is the power times the base raised to the power minus one.

(b) The Constant Multiple Rule: If c is a constant and f is a differentiable function, then $\frac{d}{dx}[cf(x)] = c\frac{d}{dx}f(x)$. The derivative of a constant times a function is the constant times the derivative of the function.

(c) The Sum Rule: If f and g are both differentiable, then $\frac{d}{dx}[f(x) + g(x)] = \frac{d}{dx}f(x) + \frac{d}{dx}g(x)$. The derivative of a sum of functions is the sum of the derivatives.

(d) The Difference Rule: If f and g are both differentiable, then $\frac{d}{dx}[f(x) - g(x)] = \frac{d}{dx}f(x) - \frac{d}{dx}g(x)$. The derivative of a difference of functions is the difference of the derivatives.

(e) The Product Rule: If f and g are both differentiable, then $\frac{d}{dx}[f(x)g(x)] = f(x)\frac{d}{dx}g(x) + g(x)\frac{d}{dx}f(x)$. The derivative of a product of two functions is the first function times the derivative of the second function plus the second function times the derivative of the first function.

(f) The Quotient Rule: If f and g are both differentiable, then $\frac{d}{dx}\left[\frac{f(x)}{g(x)}\right] = \frac{g(x)\frac{d}{dx}f(x) - f(x)\frac{d}{dx}g(x)}{[g(x)]^2}$.
The derivative of a quotient of functions is the denominator times the derivative of the numerator minus the numerator times the derivative of the denominator, all divided by the square of the denominator.

(g) The Chain Rule: If f and g are both differentiable and $F = f \circ g$ is the composite function defined by $F(x) = f(g(x))$, then F is differentiable and F' is given by the product $F'(x) = f'(g(x))g'(x)$. The derivative of a composite function is the derivative of the outer function evaluated at the inner function times the derivative of the inner function.

2. (a) $y = x^n \Rightarrow y' = nx^{n-1}$

(b) $y = e^x \Rightarrow y' = e^x$

(c) $y = a^x \Rightarrow y' = a^x \ln a$

(d) $y = \ln x \Rightarrow y' = 1/x$

(e) $y = \log_a x \Rightarrow y' = 1/(x \ln a)$

(f) $y = \sin x \Rightarrow y' = \cos x$

(g) $y = \cos x \Rightarrow y' = -\sin x$

(h) $y = \tan x \Rightarrow y' = \sec^2 x$

(i) $y = \csc x \Rightarrow y' = -\csc x \cot x$

(j) $y = \sec x \Rightarrow y' = \sec x \tan x$

(k) $y = \cot x \Rightarrow y' = -\csc^2 x$

(l) $y = \sin^{-1} x \Rightarrow y' = 1/\sqrt{1 - x^2}$

(m) $y = \tan^{-1} x \Rightarrow y' = 1/(1 + x^2)$

3. (a) e is the number such that $\lim\limits_{h \to 0} \dfrac{e^h - 1}{h} = 1$.

(b) $e = \lim\limits_{x \to 0} (1 + x)^{1/x}$

(c) The differentiation formula for $y = a^x$ $[y' = a^x \ln a]$ is simplest when $a = e$ because $\ln e = 1$.

(d) The differentiation formula for $y = \log_a x$ $[y' = 1/(x \ln a)]$ is simplest when $a = e$ because $\ln e = 1$.

4. (a) Implicit differentiation consists of differentiating both sides of an equation involving x and y with respect to x, and then solving the resulting equation for y'.

(b) Logarithmic differentiation consists of taking natural logarithms of both sides of an equation $y = f(x)$, simplifying, differentiating implicitly with respect to x, and then solving the resulting equation for y'.

5. The linearization L of f at $x = a$ is $L(x) = f(a) + f'(a)(x - a)$.

TRUE-FALSE QUIZ

1. True. This is the Sum Rule.

3. True. This is the Chain Rule.

5. False. $\dfrac{d}{dx} f(\sqrt{x}\,) = \dfrac{f'(\sqrt{x}\,)}{2\sqrt{x}}$ by the Chain Rule.

7. False. $\dfrac{d}{dx} 10^x = 10^x \ln 10$

9. True. $\dfrac{d}{dx}(\tan^2 x) = 2 \tan x \sec^2 x$, and $\dfrac{d}{dx}(\sec^2 x) = 2 \sec x (\sec x \tan x) = 2 \tan x \sec^2 x$.

11. True. $g(x) = x^5 \Rightarrow g'(x) = 5x^4 \Rightarrow g'(2) = 5(2)^4 = 80$, and by the definition of the derivative,

$$\lim_{x \to 2} \frac{g(x) - g(2)}{x - 2} = g'(2) = 80.$$

EXERCISES

1. $y = (x^4 - 3x^2 + 5)^3 \Rightarrow$

$$y' = 3(x^4 - 3x^2 + 5)^2 \frac{d}{dx}(x^4 - 3x^2 + 5) = 3(x^4 - 3x^2 + 5)^2(4x^3 - 6x) = 6x(x^4 - 3x^2 + 5)^2(2x^2 - 3)$$

3. $y = \sqrt{x} + \dfrac{1}{\sqrt[3]{x^4}} = x^{1/2} + x^{-4/3} \Rightarrow y' = \frac{1}{2}x^{-1/2} - \frac{4}{3}x^{-7/3} = \dfrac{1}{2\sqrt{x}} - \dfrac{4}{3\sqrt[3]{x^7}}$

5. $y = 2x\sqrt{x^2+1} \quad\Rightarrow$

$$y' = 2x \cdot \tfrac{1}{2}(x^2+1)^{-1/2}(2x) + \sqrt{x^2+1}\,(2) = \frac{2x^2}{\sqrt{x^2+1}} + 2\sqrt{x^2+1} = \frac{2x^2+2(x^2+1)}{\sqrt{x^2+1}} = \frac{2(2x^2+1)}{\sqrt{x^2+1}}$$

7. $y = e^{\sin 2\theta} \quad\Rightarrow\quad y' = e^{\sin 2\theta}\dfrac{d}{d\theta}(\sin 2\theta) = e^{\sin 2\theta}(\cos 2\theta)(2) = 2\cos 2\theta\, e^{\sin 2\theta}$

9. $y = \dfrac{t}{1-t^2} \quad\Rightarrow\quad y' = \dfrac{(1-t^2)(1)-t(-2t)}{(1-t^2)^2} = \dfrac{1-t^2+2t^2}{(1-t^2)^2} = \dfrac{t^2+1}{(1-t^2)^2}$

11. $y = xe^{-1/x} \quad\Rightarrow\quad y' = xe^{-1/x}(1/x^2) + e^{-1/x}\cdot 1 = e^{-1/x}(1/x+1)$

13. $\dfrac{d}{dx}(xy^4 + x^2y) = \dfrac{d}{dx}(x+3y) \quad\Rightarrow\quad x\cdot 4y^3y' + y^4\cdot 1 + x^2\cdot y' + y\cdot 2x = 1 + 3y' \quad\Rightarrow$

$$y'(4xy^3 + x^2 - 3) = 1 - y^4 - 2xy \quad\Rightarrow\quad y' = \frac{1-y^4-2xy}{4xy^3+x^2-3}$$

15. $y = \dfrac{\sec 2\theta}{1+\tan 2\theta} \quad\Rightarrow$

$$y' = \frac{(1+\tan 2\theta)(\sec 2\theta\,\tan 2\theta\cdot 2) - (\sec 2\theta)(\sec^2 2\theta\cdot 2)}{(1+\tan 2\theta)^2} = \frac{2\sec 2\theta\,[(1+\tan 2\theta)\tan 2\theta - \sec^2 2\theta]}{(1+\tan 2\theta)^2}$$

$$= \frac{2\sec 2\theta\,(\tan 2\theta + \tan^2 2\theta - \sec^2 2\theta)}{(1+\tan 2\theta)^2} = \frac{2\sec 2\theta\,(\tan 2\theta - 1)}{(1+\tan 2\theta)^2} \qquad \left[1+\tan^2 x = \sec^2 x\right]$$

17. $y = e^{cx}(c\sin x - \cos x) \quad\Rightarrow$

$$y' = e^{cx}(c\cos x + \sin x) + ce^{cx}(c\sin x - \cos x)$$
$$= e^{cx}(c^2\sin x - c\cos x + c\cos x + \sin x) = e^{cx}(c^2\sin x + \sin x) = e^{cx}\sin x\,(c^2+1)$$

19. $y = \log_5(1+2x) \quad\Rightarrow\quad y' = \dfrac{1}{(1+2x)\ln 5}\dfrac{d}{dx}(1+2x) = \dfrac{2}{(1+2x)\ln 5}$

21. $\sin(xy) = x^2 - y \quad\Rightarrow\quad \cos(xy)(xy' + y\cdot 1) = 2x - y' \quad\Rightarrow\quad x\cos(xy)y' + y' = 2x - y\cos(xy) \quad\Rightarrow$

$$y'[x\cos(xy)+1] = 2x - y\cos(xy) \quad\Rightarrow\quad y' = \frac{2x - y\cos(xy)}{x\cos(xy)+1}$$

23. $y = 3^{x\ln x} \quad\Rightarrow\quad y' = 3^{x\ln x}\cdot\ln 3\cdot\dfrac{d}{dx}(x\ln x) = 3^{x\ln x}\cdot\ln 3\left(x\cdot\dfrac{1}{x} + \ln x\cdot 1\right) = 3^{x\ln x}\cdot\ln 3(1+\ln x)$

25. $y = \ln\sin x - \tfrac{1}{2}\sin^2 x \quad\Rightarrow\quad y' = \dfrac{1}{\sin x}\cdot\cos x - \tfrac{1}{2}\cdot 2\sin x\cdot\cos x = \cot x - \sin x\cos x$

27. $y = x\tan^{-1}(4x) \quad\Rightarrow\quad y' = x\cdot\dfrac{1}{1+(4x)^2}\cdot 4 + \tan^{-1}(4x)\cdot 1 = \dfrac{4x}{1+16x^2} + \tan^{-1}(4x)$

29. $y = \ln|\sec 5x + \tan 5x| \quad\Rightarrow$

$$y' = \frac{1}{\sec 5x + \tan 5x}(\sec 5x\,\tan 5x\cdot 5 + \sec^2 5x\cdot 5) = \frac{5\sec 5x\,(\tan 5x + \sec 5x)}{\sec 5x + \tan 5x} = 5\sec 5x$$

31. $y = \tan^2(\sin\theta) = [\tan(\sin\theta)]^2 \quad\Rightarrow\quad y' = 2[\tan(\sin\theta)]\cdot\sec^2(\sin\theta)\cdot\cos\theta$

33. $y = \sin\left(\tan\sqrt{1+x^3}\right) \quad\Rightarrow\quad y' = \cos\left(\tan\sqrt{1+x^3}\right)\left(\sec^2\sqrt{1+x^3}\right)\left[3x^2/\left(2\sqrt{1+x^3}\right)\right]$

35. $f(t) = \sqrt{4t+1} \;\Rightarrow\; f'(t) = \frac{1}{2}(4t+1)^{-1/2} \cdot 4 = 2(4t+1)^{-1/2} \;\Rightarrow$

$f''(t) = 2(-\frac{1}{2})(4t+1)^{-3/2} \cdot 4 = -4/(4t+1)^{3/2}$, so $f''(2) = -4/9^{3/2} = -\frac{4}{27}$.

37. $f(x) = 2^x \;\Rightarrow\; f'(x) = 2^x \ln 2 \;\Rightarrow\; f''(x) = (2^x \ln 2) \ln 2 = 2^x (\ln 2)^2 \;\Rightarrow$

$f'''(x) = (2^x \ln 2)(\ln 2)^2 = 2^x (\ln 2)^3 \;\Rightarrow\; \cdots \;\Rightarrow\; f^{(n)}(x) = (2^x \ln 2)(\ln 2)^{n-1} = 2^x (\ln 2)^n$

39. $y = 4\sin^2 x \;\Rightarrow\; y' = 4 \cdot 2\sin x \cos x$. At $\left(\frac{\pi}{6}, 1\right)$, $y' = 8 \cdot \frac{1}{2} \cdot \frac{\sqrt{3}}{2} = 2\sqrt{3}$, so an equation of the tangent line is $y - 1 = 2\sqrt{3}\left(x - \frac{\pi}{6}\right)$, or $y = 2\sqrt{3}\,x + 1 - \pi\sqrt{3}/3$.

41. $y = (2+x)e^{-x} \;\Rightarrow\; y' = (2+x)(-e^{-x}) + e^{-x} \cdot 1 = e^{-x}[-(2+x)+1] = e^{-x}(-x-1)$.

At $(0, 2)$, $y' = 1(-1) = -1$, so an equation of the tangent line is $y - 2 = -1(x - 0)$, or $y = -x + 2$.

The slope of the normal line is 1, so an equation of the normal line is $y - 2 = 1(x - 0)$, or $y = x + 2$.

43. (a) $f(x) = x\sqrt{5-x} \;\Rightarrow$

$$f'(x) = x\left[\tfrac{1}{2}(5-x)^{-1/2}(-1)\right] + \sqrt{5-x} = \frac{-x}{2\sqrt{5-x}} + \sqrt{5-x} \cdot \frac{2\sqrt{5-x}}{2\sqrt{5-x}}$$

$$= \frac{-x}{2\sqrt{5-x}} + \frac{2(5-x)}{2\sqrt{5-x}} = \frac{-x+10-2x}{2\sqrt{5-x}} = \frac{10-3x}{2\sqrt{5-x}}$$

(b) At $(1, 2)$: $f'(1) = \frac{7}{4}$. So an equation of the tangent line is $y - 2 = \frac{7}{4}(x - 1)$ or $y = \frac{7}{4}x + \frac{1}{4}$.

At $(4, 4)$: $f'(4) = -\frac{2}{2} = -1$. So an equation of the tangent line is $y - 4 = -1(x - 4)$ or $y = -x + 8$.

(c)

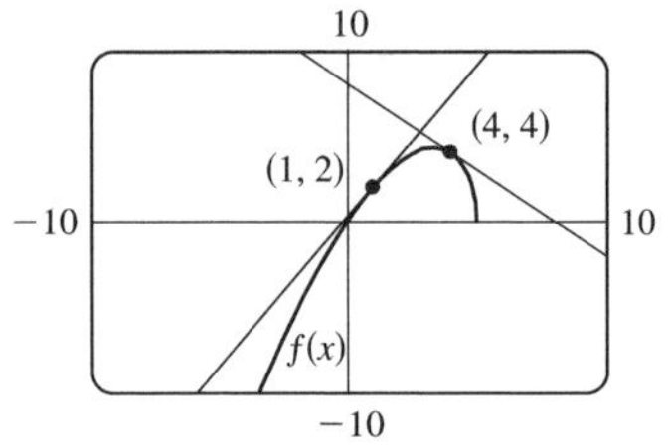

(d)

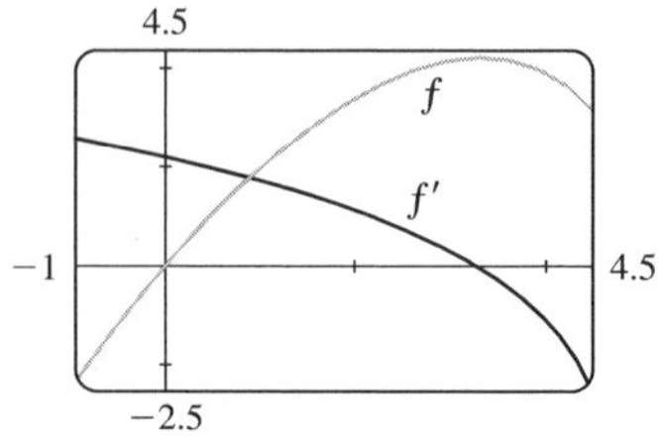

The graphs look reasonable, since f' is positive where f has tangents with positive slope, and f' is negative where f has tangents with negative slope.

45. $f(x) = xe^{\sin x} \;\Rightarrow\; f'(x) = x[e^{\sin x}(\cos x)] + e^{\sin x}(1) = e^{\sin x}(x\cos x + 1)$. As a check on our work, we notice from the graphs that $f'(x) > 0$ when f is increasing. Also, we see in the larger viewing rectangle a certain similarity in the graphs of f and f': the sizes of the oscillations of f and f' are linked.

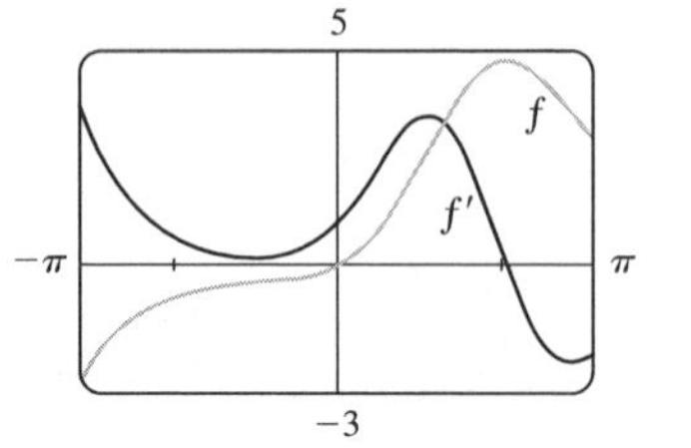

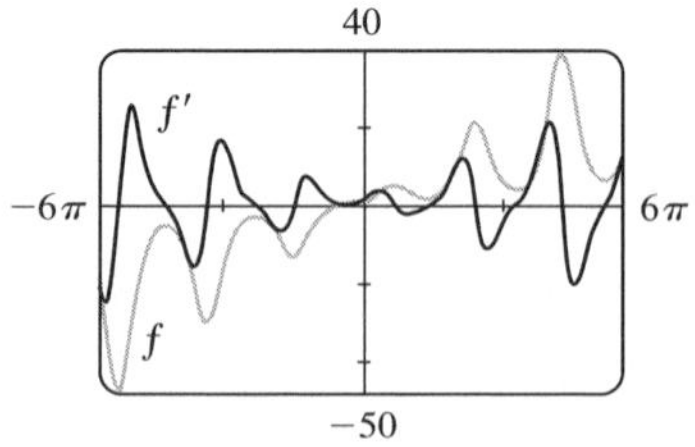

47. (a) $h(x) = f(x)g(x) \;\Rightarrow\; h'(x) = f(x)g'(x) + g(x)f'(x) \;\Rightarrow$
$h'(2) = f(2)g'(2) + g(2)f'(2) = (3)(4) + (5)(-2) = 12 - 10 = 2$

(b) $F(x) = f(g(x)) \;\Rightarrow\; F'(x) = f'(g(x))g'(x) \;\Rightarrow\; F'(2) = f'(g(2))g'(2) = f'(5)(4) = 11 \cdot 4 = 44$

49. $f(x) = x^2 g(x) \;\Rightarrow\; f'(x) = x^2 g'(x) + g(x)(2x) = x[xg'(x) + 2g(x)]$

51. $f(x) = [g(x)]^2 \;\Rightarrow\; f'(x) = 2[g(x)]^1 \cdot g'(x) = 2g(x)g'(x)$

53. $f(x) = g(e^x) \;\Rightarrow\; f'(x) = g'(e^x)e^x$

55. $f(x) = \ln|g(x)| \;\Rightarrow\; f'(x) = \dfrac{1}{g(x)}g'(x) = \dfrac{g'(x)}{g(x)}$

57. $h(x) = \dfrac{f(x)g(x)}{f(x)+g(x)} \;\Rightarrow$

$$\begin{aligned} h'(x) &= \frac{[f(x)+g(x)]\,[f(x)g'(x)+g(x)f'(x)] - f(x)g(x)\,[f'(x)+g'(x)]}{[f(x)+g(x)]^2} \\ &= \frac{[f(x)]^2\,g'(x) + f(x)g(x)f'(x) + f(x)g(x)g'(x) + [g(x)]^2\,f'(x) - f(x)g(x)f'(x) - f(x)g(x)g'(x)}{[f(x)+g(x)]^2} \\ &= \frac{f'(x)[g(x)]^2 + g'(x)[f(x)]^2}{[f(x)+g(x)]^2} \end{aligned}$$

59. $y = [\ln(x+4)]^2 \;\Rightarrow\; y' = 2[\ln(x+4)]^1 \cdot \dfrac{1}{x+4} \cdot 1 = 2\,\dfrac{\ln(x+4)}{x+4}$ and $y' = 0 \;\Leftrightarrow\; \ln(x+4) = 0 \;\Leftrightarrow$ $x + 4 = e^0 \;\Rightarrow\; x+4 = 1 \;\Leftrightarrow\; x = -3$, so the tangent is horizontal at the point $(-3, 0)$.

61. $x^2 + 2y^2 = 1 \;\Rightarrow\; 2x + 4yy' = 0 \;\Rightarrow\; y' = -x/(2y) = 1 \;\Leftrightarrow\; x = -2y$. Since the points lie on the ellipse, we have $(-2y)^2 + 2y^2 = 1 \;\Rightarrow\; 6y^2 = 1 \;\Rightarrow\; y = \pm\frac{1}{\sqrt{6}}$. The points are $\left(-\frac{2}{\sqrt{6}}, \frac{1}{\sqrt{6}}\right)$ and $\left(\frac{2}{\sqrt{6}}, -\frac{1}{\sqrt{6}}\right)$.

63. $y = f(x) = ax^2 + bx + c \;\Rightarrow\; f'(x) = 2ax + b$. We know that $f'(-1) = 6$ and $f'(5) = -2$, so $-2a + b = 6$ and $10a + b = -2$. Subtracting the first equation from the second gives $12a = -8 \;\Rightarrow\; a = -\frac{2}{3}$. Substituting $-\frac{2}{3}$ for a in the first equation gives $b = \frac{14}{3}$. Now $f(1) = 4 \;\Rightarrow\; 4 = a + b + c$, so $c = 4 + \frac{2}{3} - \frac{14}{3} = 0$ and hence, $f(x) = -\frac{2}{3}x^2 + \frac{14}{3}x$.

65. $s(t) = Ae^{-ct}\cos(\omega t + \delta) \;\Rightarrow$

$$\begin{aligned} v(t) = s'(t) &= A\{e^{-ct}\,[-\omega\sin(\omega t+\delta)] + \cos(\omega t+\delta)(-ce^{-ct})\} \\ &= -Ae^{-ct}\,[\omega\sin(\omega t+\delta) + c\cos(\omega t+\delta)] \;\Rightarrow \end{aligned}$$

$$\begin{aligned} a(t) = v'(t) &= -A\{e^{-ct}[\omega^2\cos(\omega t+\delta) - c\omega\sin(\omega t+\delta)] + [\omega\sin(\omega t+\delta) + c\cos(\omega t+\delta)](-ce^{-ct})\} \\ &= -Ae^{-ct}[\omega^2\cos(\omega t+\delta) - c\omega\sin(\omega t+\delta) - c\omega\sin(\omega t+\delta) - c^2\cos(\omega t+\delta)] \\ &= -Ae^{-ct}[(\omega^2 - c^2)\cos(\omega t+\delta) - 2c\omega\sin(\omega t+\delta)] \\ &= Ae^{-ct}[(c^2 - \omega^2)\cos(\omega t+\delta) + 2c\omega\sin(\omega t+\delta)] \end{aligned}$$

67. The linear density ρ is the rate of change of mass m with respect to length x. $m = x(1+\sqrt{x}\,) = x + x^{3/2} \;\Rightarrow$ $\rho = dm/dx = 1 + \frac{3}{2}\sqrt{x}$, so the linear density when $x = 4$ is $1 + \frac{3}{2}\sqrt{4} = 4$ kg/m.

69. (a) $C(x) = 920 + 2x - 0.02x^2 + 0.00007x^3 \;\Rightarrow\; C'(x) = 2 - 0.04x + 0.00021x^2$

(b) $C'(100) = 2 - 4 + 2.1 = \$0.10/\text{unit}$. This value represents the rate at which costs are increasing as the hundredth unit is produced, and is the approximate cost of producing the 101st unit.

(c) The cost of producing the 101st item is $C(101) - C(100) = 990.10107 - 990 = \$0.10107$, slightly larger than $C'(100)$.

(d) $C''(x) = -0.04 + 0.00042x = 0 \;\Rightarrow\; x = \frac{0.04}{0.00042} \approx 95.24$ and C'' changes from negative to positive at this value of x. This is the value of x at which the marginal cost is minimized.

71. (a) $f(x) = \sqrt[3]{1+3x} = (1+3x)^{1/3} \quad \Rightarrow \quad f'(x) = (1+3x)^{-2/3}$, so the linearization of f at $a = 0$ is $L(x) = f(0) + f'(0)(x-0) = 1^{1/3} + 1^{-2/3}x = 1 + x$. Thus, $\sqrt[3]{1+3x} \approx 1 + x \quad \Rightarrow$ $\sqrt[3]{1.03} = \sqrt[3]{1+3(0.01)} \approx 1 + (0.01) = 1.01$.

(b) The linear approximation is $\sqrt[3]{1+3x} \approx 1 + x$, so for the required accuracy we want $\sqrt[3]{1+3x} - 0.1 < 1 + x < \sqrt[3]{1+3x} + 0.1$. From the graph, it appears that this is true when $-0.23 < x < 0.40$.

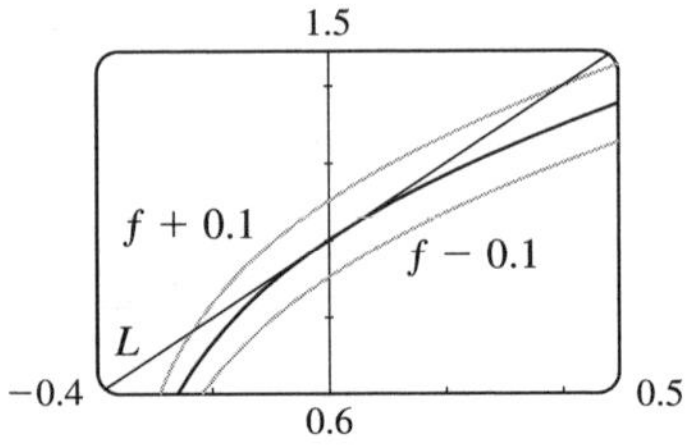

73. $\displaystyle \lim_{\theta \to \pi/3} \frac{\cos\theta - 0.5}{\theta - \pi/3} = \left[\frac{d}{d\theta}\cos\theta\right]_{\theta = \pi/3} = -\sin\frac{\pi}{3} = -\frac{\sqrt{3}}{2}$

75.
$$\begin{aligned}
\lim_{x\to 0} \frac{\sqrt{1+\tan x} - \sqrt{1+\sin x}}{x^3} &= \lim_{x\to 0} \frac{\left(\sqrt{1+\tan x} - \sqrt{1+\sin x}\right)\left(\sqrt{1+\tan x} + \sqrt{1+\sin x}\right)}{x^3\left(\sqrt{1+\tan x} + \sqrt{1+\sin x}\right)} \\
&= \lim_{x\to 0} \frac{(1+\tan x) - (1+\sin x)}{x^3\left(\sqrt{1+\tan x} + \sqrt{1+\sin x}\right)} = \lim_{x\to 0} \frac{\sin x\,(1/\cos x - 1)}{x^3\left(\sqrt{1+\tan x} + \sqrt{1+\sin x}\right)} \cdot \frac{\cos x}{\cos x} \\
&= \lim_{x\to 0} \frac{\sin x\,(1-\cos x)}{x^3\left(\sqrt{1+\tan x} + \sqrt{1+\sin x}\right)\cos x} \cdot \frac{1+\cos x}{1+\cos x} \\
&= \lim_{x\to 0} \frac{\sin x \cdot \sin^2 x}{x^3\left(\sqrt{1+\tan x} + \sqrt{1+\sin x}\right)\cos x\,(1+\cos x)} \\
&= \left(\lim_{x\to 0}\frac{\sin x}{x}\right)^3 \lim_{x\to 0} \frac{1}{\left(\sqrt{1+\tan x} + \sqrt{1+\sin x}\right)\cos x\,(1+\cos x)} \\
&= 1^3 \cdot \frac{1}{\left(\sqrt{1}+\sqrt{1}\right)\cdot 1 \cdot (1+1)} = \frac{1}{4}
\end{aligned}$$

FOCUS ON PROBLEM SOLVING

1. We must find a value x_0 such that the normal lines to the parabola $y = x^2$ at $x = \pm x_0$ intersect at a point one unit from the points $(\pm x_0, x_0^2)$. The normals to $y = x^2$ at $x = \pm x_0$ have slopes $-\dfrac{1}{\pm 2x_0}$ and pass through $(\pm x_0, x_0^2)$ respectively, so the normals have the equations $y - x_0^2 = -\dfrac{1}{2x_0}(x - x_0)$ and $y - x_0^2 = \dfrac{1}{2x_0}(x + x_0)$. The common y-intercept is $x_0^2 + \frac{1}{2}$. We want to find the value of x_0 for which the distance from $(0, x_0^2 + \frac{1}{2})$ to (x_0, x_0^2) equals 1. The square of the distance is $(x_0 - 0)^2 + \left[x_0^2 - (x_0^2 + \frac{1}{2})\right]^2 = x_0^2 + \frac{1}{4} = 1 \quad \Leftrightarrow \quad x_0 = \pm\frac{\sqrt{3}}{2}$. For these values of x_0, the y-intercept is $x_0^2 + \frac{1}{2} = \frac{5}{4}$, so the center of the circle is at $(0, \frac{5}{4})$.

Another solution: Let the center of the circle be $(0, a)$. Then the equation of the circle is $x^2 + (y - a)^2 = 1$. Solving with the equation of the parabola, $y = x^2$, we get $x^2 + (x^2 - a)^2 = 1 \quad \Leftrightarrow \quad x^2 + x^4 - 2ax^2 + a^2 = 1 \quad \Leftrightarrow$ $x^4 + (1 - 2a)x^2 + a^2 - 1 = 0$. The parabola and the circle will be tangent to each other when this quadratic equation in x^2 has equal roots; that is, when the discriminant is 0. Thus, $(1 - 2a)^2 - 4(a^2 - 1) = 0 \quad \Leftrightarrow$ $1 - 4a + 4a^2 - 4a^2 + 4 = 0 \quad \Leftrightarrow \quad 4a = 5$, so $a = \frac{5}{4}$. The center of the circle is $(0, \frac{5}{4})$.

3.

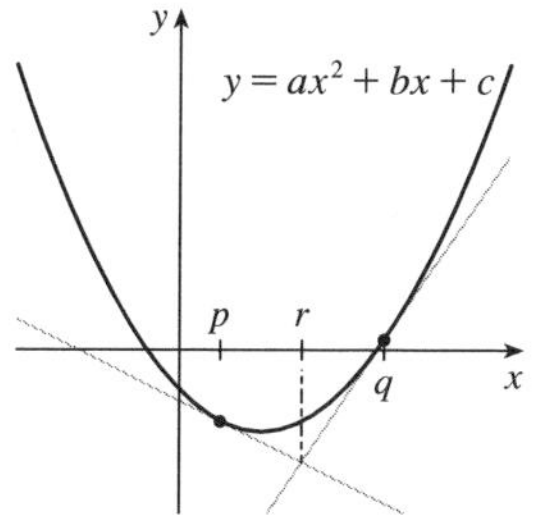

We must show that r (in the figure) is halfway between p and q, that is, $r = (p + q)/2$. For the parabola $y = ax^2 + bx + c$, the slope of the tangent line is given by $y' = 2ax + b$. An equation of the tangent line at $x = p$ is $y - (ap^2 + bp + c) = (2ap + b)(x - p)$. Solving for y gives us

$$y = (2ap + b)x - 2ap^2 - bp + (ap^2 + bp + c)$$

$$\text{or} \quad y = (2ap + b)x + c - ap^2 \qquad \textbf{(1)}$$

Similarly, an equation of the tangent line at $x = q$ is

$$y = (2aq + b)x + c - aq^2 \qquad \textbf{(2)}$$

We can eliminate y and solve for x by subtracting equation **(1)** from equation **(2)**.

$$\begin{aligned} [(2aq + b) - (2ap + b)]x - aq^2 + ap^2 &= 0 \\ (2aq - 2ap)x &= aq^2 - ap^2 \\ 2a(q - p)x &= a(q^2 - p^2) \\ x &= \frac{a(q + p)(q - p)}{2a(q - p)} = \frac{p + q}{2} \end{aligned}$$

Thus, the x-coordinate of the point of intersection of the two tangent lines, namely r, is $(p + q)/2$.

5. We can assume without loss of generality that $\theta = 0$ at time $t = 0$, so that $\theta = 12\pi t$ rad. [The angular velocity of the wheel is 360 rpm $= 360 \cdot (2\pi \text{ rad})/(60 \text{ s}) = 12\pi$ rad/s.] Then the position of A as a function of time is

$A = (40\cos\theta, 40\sin\theta) = (40\cos 12\pi t, 40\sin 12\pi t)$, so $\sin\alpha = \dfrac{y}{1.2 \text{ m}} = \dfrac{40\sin\theta}{120} = \dfrac{\sin\theta}{3} = \frac{1}{3}\sin 12\pi t$.

(a) Differentiating the expression for $\sin\alpha$, we get $\cos\alpha \cdot \dfrac{d\alpha}{dt} = \frac{1}{3} \cdot 12\pi \cdot \cos 12\pi t = 4\pi\cos\theta$. When $\theta = \frac{\pi}{3}$, we have

$$\sin\alpha = \tfrac{1}{3}\sin\theta = \tfrac{\sqrt{3}}{6}, \text{ so } \cos\alpha = \sqrt{1 - \left(\tfrac{\sqrt{3}}{6}\right)^2} = \sqrt{\tfrac{11}{12}} \text{ and } \frac{d\alpha}{dt} = \frac{4\pi\cos\frac{\pi}{3}}{\cos\alpha} = \frac{2\pi}{\sqrt{11/12}} = \frac{4\pi\sqrt{3}}{\sqrt{11}} \approx 6.56 \text{ rad/s.}$$

(b) By the Law of Cosines, $|AP|^2 = |OA|^2 + |OP|^2 - 2\,|OA|\,|OP|\cos\theta \;\Rightarrow$

$120^2 = 40^2 + |OP|^2 - 2\cdot 40\,|OP|\cos\theta \;\Rightarrow\; |OP|^2 - (80\cos\theta)\,|OP| - 12\,800 = 0 \;\Rightarrow$

$|OP| = \frac{1}{2}\left(80\cos\theta \pm \sqrt{6400\cos^2\theta + 51\,200}\,\right) = 40\cos\theta \pm 40\sqrt{\cos^2\theta + 8} = 40\left(\cos\theta + \sqrt{8+\cos^2\theta}\,\right)$ cm

[since $|OP| > 0$]. As a check, note that $|OP| = 160$ cm when $\theta = 0$ and $|OP| = 80\sqrt{2}$ cm when $\theta = \frac{\pi}{2}$.

(c) By part (b), the x-coordinate of P is given by $x = 40\left(\cos\theta + \sqrt{8+\cos^2\theta}\,\right)$, so

$$\frac{dx}{dt} = \frac{dx}{d\theta}\frac{d\theta}{dt} = 40\left(-\sin\theta - \frac{2\cos\theta\sin\theta}{2\sqrt{8+\cos^2\theta}}\right)\cdot 12\pi = -480\pi\sin\theta\left(1 + \frac{\cos\theta}{\sqrt{8+\cos^2\theta}}\right)\text{ cm/s.}$$

In particular, $dx/dt = 0$ cm/s when $\theta = 0$ and $dx/dt = -480\pi$ cm/s when $\theta = \frac{\pi}{2}$.

7. Consider the statement that $\dfrac{d^n}{dx^n}\left(e^{ax}\sin bx\right) = r^n e^{ax}\sin(bx + n\theta)$. For $n = 1$,

$\dfrac{d}{dx}\left(e^{ax}\sin bx\right) = ae^{ax}\sin bx + be^{ax}\cos bx$, and

$$\begin{aligned} re^{ax}\sin(bx+\theta) &= re^{ax}[\sin bx\cos\theta + \cos bx\sin\theta] = re^{ax}\left(\frac{a}{r}\sin bx + \frac{b}{r}\cos bx\right) \\ &= ae^{ax}\sin bx + be^{ax}\cos bx \end{aligned}$$

since $\tan\theta = \dfrac{b}{a} \;\Rightarrow\; \sin\theta = \dfrac{b}{r}$ and $\cos\theta = \dfrac{a}{r}$. So the statement is true for $n = 1$.

Assume it is true for $n = k$. Then

$$\begin{aligned} \frac{d^{k+1}}{dx^{k+1}}\left(e^{ax}\sin bx\right) &= \frac{d}{dx}\left[r^k e^{ax}\sin(bx+k\theta)\right] = r^k ae^{ax}\sin(bx+k\theta) + r^k e^{ax} b\cos(bx+k\theta) \\ &= r^k e^{ax}[a\sin(bx+k\theta) + b\cos(bx+k\theta)] \end{aligned}$$

But

$$\begin{aligned} \sin[bx + (k+1)\theta] &= \sin[(bx+k\theta)+\theta] = \sin(bx+k\theta)\cos\theta + \sin\theta\cos(bx+k\theta) \\ &= \frac{a}{r}\sin(bx+k\theta) + \frac{b}{r}\cos(bx+k\theta) \end{aligned}$$

Hence, $a\sin(bx+k\theta) + b\cos(bx+k\theta) = r\sin[bx+(k+1)\theta]$. So

$$\begin{aligned} \frac{d^{k+1}}{dx^{k+1}}\left(e^{ax}\sin bx\right) &= r^k e^{ax}[a\sin(bx+k\theta) + b\cos(bx+k\theta)] = r^k e^{ax}[r\sin(bx+(k+1)\theta)] \\ &= r^{k+1}e^{ax}[\sin(bx+(k+1)\theta)] \end{aligned}$$

Therefore, the statement is true for all n by mathematical induction.

9. It seems from the figure that as P approaches the point $(0, 2)$ from the right, $x_T \to \infty$ and $y_T \to 2^+$. As P approaches the point $(3, 0)$ from the left, it appears that $x_T \to 3^+$ and $y_T \to \infty$. So we guess that $x_T \in (3, \infty)$ and $y_T \in (2, \infty)$. It is more difficult to estimate the range of values for x_N and y_N. We might perhaps guess that $x_N \in (0, 3)$, and $y_N \in (-\infty, 0)$ or $(-2, 0)$.

In order to actually solve the problem, we implicitly differentiate the equation of the ellipse to find the equation of the tangent line: $\dfrac{x^2}{9} + \dfrac{y^2}{4} = 1 \;\Rightarrow\; \dfrac{2x}{9} + \dfrac{2y}{4}y' = 0$, so $y' = -\dfrac{4}{9}\dfrac{x}{y}$. So at the point (x_0, y_0) on the ellipse, an equation of the tangent line is $y - y_0 = -\dfrac{4}{9}\dfrac{x_0}{y_0}(x - x_0)$ or $4x_0x + 9y_0y = 4x_0^2 + 9y_0^2$. This can be written as $\dfrac{x_0x}{9} + \dfrac{y_0y}{4} = \dfrac{x_0^2}{9} + \dfrac{y_0^2}{4} = 1$, because (x_0, y_0) lies on the ellipse. So an equation of the tangent line is $\dfrac{x_0x}{9} + \dfrac{y_0y}{4} = 1$.

Therefore, the x-intercept x_T for the tangent line is given by $\dfrac{x_0x_T}{9} = 1 \Leftrightarrow x_T = \dfrac{9}{x_0}$, and the y-intercept y_T is given by $\dfrac{y_0y_T}{4} = 1 \Leftrightarrow y_T = \dfrac{4}{y_0}$.

So as x_0 takes on all values in $(0, 3)$, x_T takes on all values in $(3, \infty)$, and as y_0 takes on all values in $(0, 2)$, y_T takes on all values in $(2, \infty)$. At the point (x_0, y_0) on the ellipse, the slope of the normal line is $-\dfrac{1}{y'(x_0, y_0)} = \dfrac{9}{4}\dfrac{y_0}{x_0}$, and its equation is $y - y_0 = \dfrac{9}{4}\dfrac{y_0}{x_0}(x - x_0)$. So the x-intercept x_N for the normal line is given by $0 - y_0 = \dfrac{9}{4}\dfrac{y_0}{x_0}(x_N - x_0) \Rightarrow$ $x_N = -\dfrac{4x_0}{9} + x_0 = \dfrac{5x_0}{9}$, and the y-intercept y_N is given by $y_N - y_0 = \dfrac{9}{4}\dfrac{y_0}{x_0}(0 - x_0) \Rightarrow y_N = -\dfrac{9y_0}{4} + y_0 = -\dfrac{5y_0}{4}$.

So as x_0 takes on all values in $(0, 3)$, x_N takes on all values in $\left(0, \frac{5}{3}\right)$, and as y_0 takes on all values in $(0, 2)$, y_N takes on all values in $\left(-\frac{5}{2}, 0\right)$.

11. $y = \dfrac{x}{\sqrt{a^2 - 1}} - \dfrac{2}{\sqrt{a^2 - 1}} \arctan \dfrac{\sin x}{a + \sqrt{a^2 - 1} + \cos x}$. Let $k = a + \sqrt{a^2 - 1}$. Then

$$\begin{aligned} y' &= \frac{1}{\sqrt{a^2 - 1}} - \frac{2}{\sqrt{a^2 - 1}} \cdot \frac{1}{1 + \sin^2 x/(k + \cos x)^2} \cdot \frac{\cos x(k + \cos x) + \sin^2 x}{(k + \cos x)^2} \\ &= \frac{1}{\sqrt{a^2 - 1}} - \frac{2}{\sqrt{a^2 - 1}} \cdot \frac{k \cos x + \cos^2 x + \sin^2 x}{(k + \cos x)^2 + \sin^2 x} = \frac{1}{\sqrt{a^2 - 1}} - \frac{2}{\sqrt{a^2 - 1}} \cdot \frac{k \cos x + 1}{k^2 + 2k \cos x + 1} \\ &= \frac{k^2 + 2k \cos x + 1 - 2k \cos x - 2}{\sqrt{a^2 - 1}\,(k^2 + 2k \cos x + 1)} = \frac{k^2 - 1}{\sqrt{a^2 - 1}\left(k^2 + 2k \cos x + 1\right)} \end{aligned}$$

But $k^2 = 2a^2 + 2a\sqrt{a^2 - 1} - 1 = 2a\left(a + \sqrt{a^2 - 1}\right) - 1 = 2ak - 1$, so $k^2 + 1 = 2ak$, and $k^2 - 1 = 2(ak - 1)$.

So $y' = \dfrac{2(ak - 1)}{\sqrt{a^2 - 1}\,(2ak + 2k \cos x)} = \dfrac{ak - 1}{\sqrt{a^2 - 1}k\,(a + \cos x)}$. But $ak - 1 = a^2 + a\sqrt{a^2 - 1} - 1 = k\sqrt{a^2 - 1}$, so $y' = 1/(a + \cos x)$.

13. 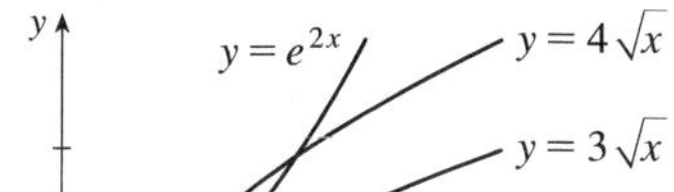

Let $f(x) = e^{2x}$ and $g(x) = k\sqrt{x}$ $(k > 0)$. From the graphs of f and g, we see that f will intersect g exactly once when f and g share a tangent line. Thus, we must have $f = g$ and $f' = g'$ at $x = a$. $f(a) = g(a) \Rightarrow$ $e^{2a} = k\sqrt{a}$ $(\star)$ and $f'(a) = g'(a) \Rightarrow 2e^{2a} = \dfrac{k}{2\sqrt{a}} \Rightarrow$ $e^{2a} = \dfrac{k}{4\sqrt{a}}$. So we must have $k\sqrt{a} = \dfrac{k}{4\sqrt{a}} \Rightarrow \left(\sqrt{a}\right)^2 = \dfrac{k}{4k} \Rightarrow$ $a = \frac{1}{4}$. From $(\star)$, $e^{2(1/4)} = k\sqrt{1/4} \Rightarrow k = 2e^{1/2} = 2\sqrt{e} \approx 3.297$.

15. (a) If the two lines L_1 and L_2 have slopes m_1 and m_2 and angles of inclination ϕ_1 and ϕ_2, then $m_1 = \tan\,\phi_1$ and $m_2 = \tan\,\phi_2$. The triangle in the figure shows that $\phi_1 + \alpha + (180° - \phi_2) = 180°$ and so $\alpha = \phi_2 - \phi_1$. Therefore, using the identity for $\tan(x - y)$, we have

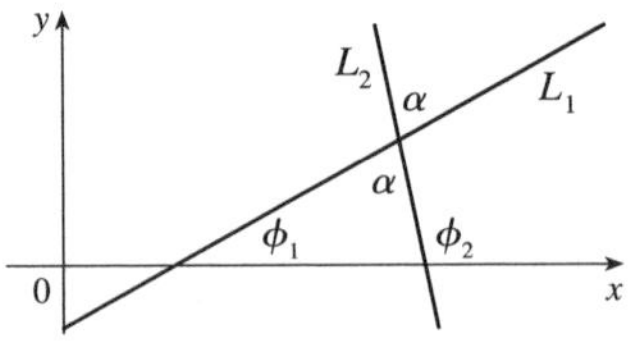

$$\tan\alpha = \tan(\phi_2 - \phi_1) = \frac{\tan\phi_2 - \tan\phi_1}{1 + \tan\phi_2\tan\phi_1} \text{ and so } \tan\,\alpha = \frac{m_2 - m_1}{1 + m_1 m_2}.$$

(b) (i) The parabolas intersect when $x^2 = (x-2)^2 \;\Rightarrow\; x = 1$. If $y = x^2$, then $y' = 2x$, so the slope of the tangent to $y = x^2$ at $(1,1)$ is $m_1 = 2(1) = 2$. If $y = (x-2)^2$, then $y' = 2(x-2)$, so the slope of the tangent to $y = (x-2)^2$ at $(1,1)$ is $m_2 = 2(1-2) = -2$. Therefore, $\tan\alpha = \dfrac{m_2 - m_1}{1 + m_1 m_2} = \dfrac{-2-2}{1 + 2(-2)} = \frac{4}{3}$ and so $\alpha = \tan^{-1}\left(\frac{4}{3}\right) \approx 53°$ [or $127°$].

(ii) $x^2 - y^2 = 3$ and $x^2 - 4x + y^2 + 3 = 0$ intersect when $x^2 - 4x + \left(x^2 - 3\right) + 3 = 0 \;\Leftrightarrow\; 2x(x-2) = 0 \;\Rightarrow\; x = 0$ or 2, but 0 is extraneous. If $x = 2$, then $y = \pm 1$. If $x^2 - y^2 = 3$ then $2x - 2yy' = 0 \;\Rightarrow\; y' = x/y$ and $x^2 - 4x + y^2 + 3 = 0 \;\Rightarrow\; 2x - 4 + 2yy' = 0 \;\Rightarrow\; y' = \dfrac{2-x}{y}$. At $(2,1)$ the slopes are $m_1 = 2$ and $m_2 = 0$, so $\tan\alpha = \frac{0-2}{1+2\cdot 0} = -2 \;\Rightarrow\; \alpha \approx 117°$. At $(2,-1)$ the slopes are $m_1 = -2$ and $m_2 = 0$, so $\tan\alpha = \dfrac{0-(-2)}{1+(-2)\,(0)} = 2 \;\Rightarrow\; \alpha \approx 63°$ [or $117°$].

17. Since $\angle ROQ = \angle OQP = \theta$, the triangle QOR is isosceles, so $|QR| = |RO| = x$. By the Law of Cosines, $x^2 = x^2 + r^2 - 2rx\cos\theta$. Hence, $2rx\cos\theta = r^2$, so $x = \dfrac{r^2}{2r\cos\theta} = \dfrac{r}{2\cos\theta}$. Note that as $y \to 0^+$, $\theta \to 0^+$ (since $\sin\theta = y/r$), and hence $x \to \dfrac{r}{2\cos 0} = \dfrac{r}{2}$. Thus, as P is taken closer and closer to the x-axis, the point R approaches the midpoint of the radius AO.

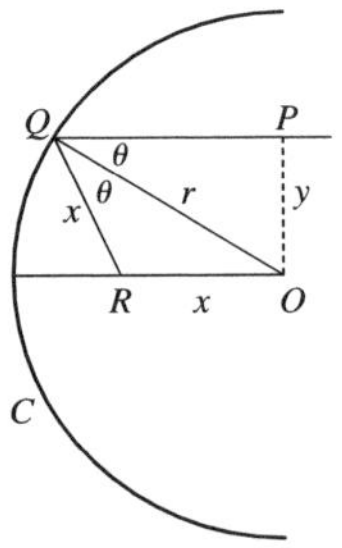

19. $y = x^4 - 2x^2 - x \;\Rightarrow\; y' = 4x^3 - 4x - 1$. The equation of the tangent line at $x = a$ is $y - (a^4 - 2a^2 - a) = (4a^3 - 4a - 1)(x - a)$ or $y = (4a^3 - 4a - 1)x + \left(-3a^4 + 2a^2\right)$ and similarly for $x = b$. So if at $x = a$ and $x = b$ we have the same tangent line, then $4a^3 - 4a - 1 = 4b^3 - 4b - 1$ and $-3a^4 + 2a^2 = -3b^4 + 2b^2$. The first equation gives $a^3 - b^3 = a - b \;\Rightarrow\; (a-b)(a^2 + ab + b^2) = (a - b)$. Assuming $a \neq b$, we have $1 = a^2 + ab + b^2$. The second equation gives $3(a^4 - b^4) = 2(a^2 - b^2) \;\Rightarrow\; 3(a^2 - b^2)(a^2 + b^2) = 2(a^2 - b^2)$ which is true if $a = -b$. Substituting into $1 = a^2 + ab + b^2$ gives $1 = a^2 - a^2 + a^2 \;\Rightarrow\; a = \pm 1$ so that $a = 1$ and $b = -1$ or vice versa. Thus, the points $(1, -2)$ and $(-1, 0)$ have a common tangent line.

As long as there are only two such points, we are done. So we show that these are in fact the only two such points. Suppose that $a^2 - b^2 \neq 0$. Then $3(a^2 - b^2)(a^2 + b^2) = 2(a^2 - b^2)$ gives $3(a^2 + b^2) = 2$ or $a^2 + b^2 = \frac{2}{3}$.

Thus, $ab = (a^2 + ab + b^2) - (a^2 + b^2) = 1 - \frac{2}{3} = \frac{1}{3}$, so $b = \dfrac{1}{3a}$. Hence, $a^2 + \dfrac{1}{9a^2} = \dfrac{2}{3}$, so $9a^4 + 1 = 6a^2 \Rightarrow$ $0 = 9a^4 - 6a^2 + 1 = (3a^2 - 1)^2$. So $3a^2 - 1 = 0 \Rightarrow a^2 = \frac{1}{3} \Rightarrow b^2 = \dfrac{1}{9a^2} = \frac{1}{3} = a^2$, contradicting our assumption that $a^2 \neq b^2$.

21. Because of the periodic nature of the lattice points, it suffices to consider the points in the 5×2 grid shown. We can see that the minimum value of r occurs when there is a line with slope $\frac{2}{5}$ which touches the circle centred at $(3, 1)$ and the circles centred at $(0, 0)$ and $(5, 2)$.

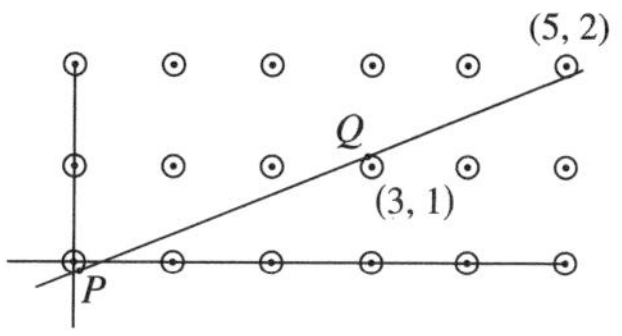

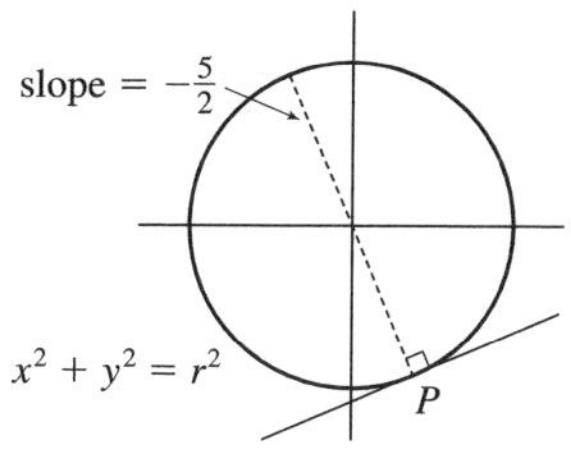

To find P, the point at which the line is tangent to the circle at $(0, 0)$, we simultaneously solve $x^2 + y^2 = r^2$ and $y = -\frac{5}{2}x \Rightarrow x^2 + \frac{25}{4}x^2 = r^2 \Rightarrow x^2 = \frac{4}{29}r^2 \Rightarrow x = \frac{2}{\sqrt{29}}r$, $y = -\frac{5}{\sqrt{29}}r$. To find Q, we either use symmetry or solve $(x-3)^2 + (y-1)^2 = r^2$ and $y - 1 = -\frac{5}{2}(x - 3)$. As above, we get $x = 3 - \frac{2}{\sqrt{29}}r$, $y = 1 + \frac{5}{\sqrt{29}}r$. Now the slope of the line PQ is $\frac{2}{5}$, so $m_{PQ} = \dfrac{1 + \frac{5}{\sqrt{29}}r - \left(-\frac{5}{\sqrt{29}}r\right)}{3 - \frac{2}{\sqrt{29}}r - \frac{2}{\sqrt{29}}r} = \dfrac{1 + \frac{10}{\sqrt{29}}r}{3 - \frac{4}{\sqrt{29}}r} = \dfrac{\sqrt{29} + 10r}{3\sqrt{29} - 4r} = \dfrac{2}{5} \Rightarrow$ $5\sqrt{29} + 50r = 6\sqrt{29} - 8r \Leftrightarrow 58r = \sqrt{29} \Leftrightarrow r = \frac{\sqrt{29}}{58}$. So the minimum value of r for which any line with slope $\frac{2}{5}$ intersects circles with radius r centred at the lattice points on the plane is $r = \frac{\sqrt{29}}{58} \approx 0.093$.

4 ☐ APPLICATIONS OF DIFFERENTIATION

4.1 Related Rates

1. $V = x^3 \quad\Rightarrow\quad \dfrac{dV}{dt} = \dfrac{dV}{dx}\,\dfrac{dx}{dt} = 3x^2\,\dfrac{dx}{dt}$

3. Let s denote the side of a square. The square's area A is given by $A = s^2$. Differentiating with respect to t gives us $\dfrac{dA}{dt} = 2s\,\dfrac{ds}{dt}$. When $A = 16$, $s = 4$. Substitution 4 for s and 6 for $\dfrac{ds}{dt}$ gives us $\dfrac{dA}{dt} = 2(4)(6) = 48\text{ cm}^2/\text{s}$.

5. $y = x^3 + 2x \quad\Rightarrow\quad \dfrac{dy}{dt} = \dfrac{dy}{dx}\,\dfrac{dx}{dt} = (3x^2+2)(5) = 5(3x^2+2)$. When $x = 2$, $\dfrac{dy}{dt} = 5(14) = 70$.

7. $z^2 = x^2 + y^2 \quad\Rightarrow\quad 2z\,\dfrac{dz}{dt} = 2x\,\dfrac{dx}{dt} + 2y\,\dfrac{dy}{dt} \quad\Rightarrow\quad \dfrac{dz}{dt} = \dfrac{1}{z}\left(x\,\dfrac{dx}{dt} + y\,\dfrac{dy}{dt}\right)$. When $x = 5$ and $y = 12$,

$z^2 = 5^2 + 12^2 \quad\Rightarrow\quad z^2 = 169 \quad\Rightarrow\quad z = \pm 13$. For $\dfrac{dx}{dt} = 2$ and $\dfrac{dy}{dt} = 3$, $\dfrac{dz}{dt} = \dfrac{1}{\pm 13}(5 \cdot 2 + 12 \cdot 3) = \pm\dfrac{46}{13}$.

9. (a) Given: the rate of decrease of the surface area is $1\text{ cm}^2/\text{min}$. If we let t be time (in minutes) and S be the surface area (in cm^2), then we are given that $dS/dt = -1\text{ cm}^2/\text{s}$.

(b) Unknown: the rate of decrease of the diameter when the diameter is 10 cm. If we let x be the diameter, then we want to find dx/dt when $x = 10$ cm.

(c)

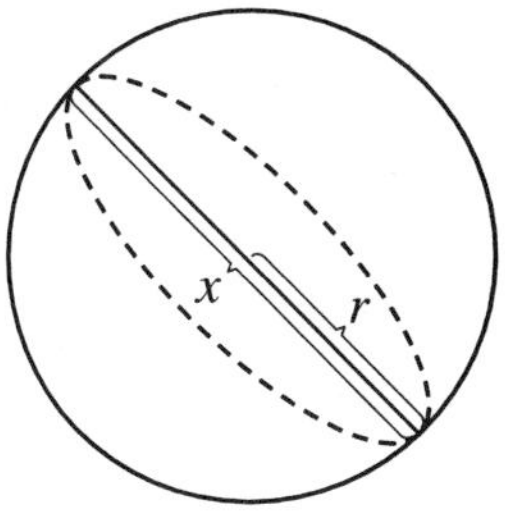

(d) If the radius is r and the diameter $x = 2r$, then $r = \frac{1}{2}x$ and

$S = 4\pi r^2 = 4\pi\left(\frac{1}{2}x\right)^2 = \pi x^2 \quad\Rightarrow\quad \dfrac{dS}{dt} = \dfrac{dS}{dx}\,\dfrac{dx}{dt} = 2\pi x\,\dfrac{dx}{dt}$.

(e) $-1 = \dfrac{dS}{dt} = 2\pi x\,\dfrac{dx}{dt} \quad\Rightarrow\quad \dfrac{dx}{dt} = -\dfrac{1}{2\pi x}$. When $x = 10$, $\dfrac{dx}{dt} = -\dfrac{1}{20\pi}$. So the rate of decrease is $\frac{1}{20\pi}$ cm/min.

11. (a) Given: a plane flying horizontally at an altitude of 2 km and a speed of 800 km/h passes directly over a radar station. If we let t be time (in hours) and x be the horizontal distance travelled by the plane (in km), then we are given that $dx/dt = 800$ km/h.

(b) Unknown: the rate at which the distance from the plane to the station is increasing when it is 3 km from the station. If we let y be the distance from the plane to the station, then we want to find dy/dt when $y = 3$ km.

(c)

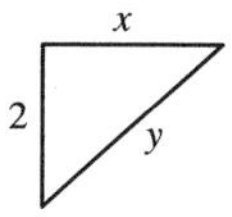

(d) By the Pythagorean Theorem, $y^2 = x^2 + 2^2 \quad\Rightarrow\quad 2y\,(dy/dt) = 2x\,(dx/dt)$.

(e) $\dfrac{dy}{dt} = \dfrac{x}{y}\,\dfrac{dx}{dt} = \dfrac{x}{y}(800)$. Since $y^2 = x^2 + 4$, when $y = 3$, $x = \sqrt{5}$, so $\dfrac{dy}{dt} = \frac{\sqrt{5}}{3}(800) \approx 596$ km/h.

13.

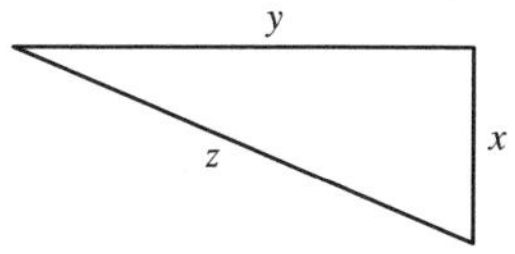

We are given that $\dfrac{dx}{dt} = 30$ km/h and $\dfrac{dy}{dt} = 72$ km/h. $z^2 = x^2 + y^2 \quad\Rightarrow$

$$2z\,\frac{dz}{dt} = 2x\,\frac{dx}{dt} + 2y\,\frac{dy}{dt} \quad\Rightarrow\quad z\,\frac{dz}{dt} = x\,\frac{dx}{dt} + y\,\frac{dy}{dt} \quad\Rightarrow\quad \frac{dz}{dt} = \frac{1}{z}\left(x\,\frac{dx}{dt} + y\,\frac{dy}{dt}\right).$$

After 2 hours, $x = 2\,(30) = 60$ and $y = 2\,(72) = 144 \quad\Rightarrow\quad z = \sqrt{60^2 + 144^2} = 156$, so

$$\frac{dz}{dt} = \frac{1}{z}\left(x\,\frac{dx}{dt} + y\,\frac{dy}{dt}\right) = \frac{60(30) + 144(72)}{156} = 78 \text{ km/h}.$$

15.

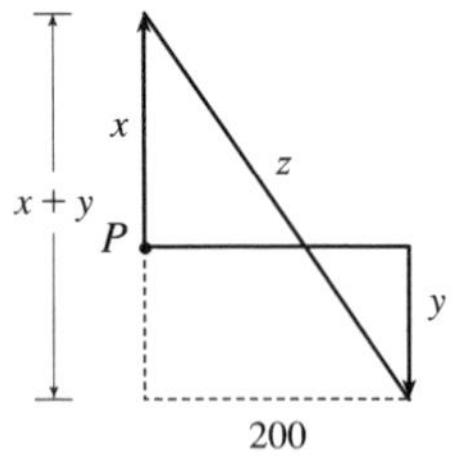

We are given that $\frac{dx}{dt} = 1.2$ m/s and $\frac{dy}{dt} = 1.6$ m/s. $z^2 = (x+y)^2 + 200^2 \Rightarrow$

$2z\,\frac{dz}{dt} = 2(x+y)\left(\frac{dx}{dt} + \frac{dy}{dt}\right)$. 15 minutes after the woman starts, we have

$x = (1.2 \text{ m/s})(20 \text{ min})(60 \text{ s/min}) = 1440$ m and $y = 1.6 \cdot 15 \cdot 60 = 1440 \Rightarrow$

$z = \sqrt{(1440 + 1440)^2 + 200^2} = \sqrt{8\,334\,400}$, so

$$\frac{dz}{dt} = \frac{x+y}{z}\left(\frac{dx}{dt} + \frac{dy}{dt}\right) = \frac{1440 + 1440}{\sqrt{8\,334\,400}}(1.2 + 1.6) = \frac{8064}{\sqrt{8\,334\,400}} \approx 2.79 \text{ m/s}.$$

17. $A = \frac{1}{2}bh$, where b is the base and h is the altitude. We are given that $\frac{dh}{dt} = 1$ cm/min and $\frac{dA}{dt} = 2$ cm^2/min. Using the Product Rule, we have $\frac{dA}{dt} = \frac{1}{2}\left(b\,\frac{dh}{dt} + h\,\frac{db}{dt}\right)$. When $h = 10$ and $A = 100$, we have $100 = \frac{1}{2}b(10) \Rightarrow \frac{1}{2}b = 10 \Rightarrow$ $b = 20$, so $2 = \frac{1}{2}\left(20 \cdot 1 + 10\,\frac{db}{dt}\right) \Rightarrow 4 = 20 + 10\,\frac{db}{dt} \Rightarrow \frac{db}{dt} = \frac{4 - 20}{10} = -1.6$ cm/min.

19.

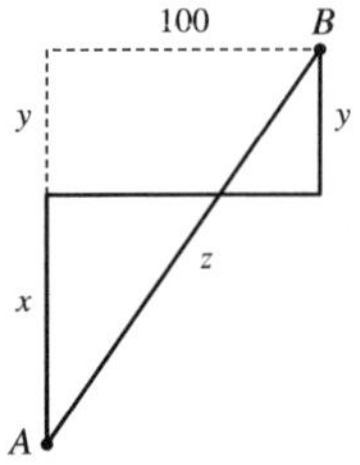

We are given that $\frac{dx}{dt} = 35$ km/h and $\frac{dy}{dt} = 25$ km/h. $z^2 = (x+y)^2 + 100^2 \Rightarrow$

$2z\,\frac{dz}{dt} = 2(x+y)\left(\frac{dx}{dt} + \frac{dy}{dt}\right)$. At 4:00 P.M., $x = 4(35) = 140$ and

$y = 4(25) = 100 \Rightarrow z = \sqrt{(140 + 100)^2 + 100^2} = \sqrt{67\,600} = 260$, so

$$\frac{dz}{dt} = \frac{x+y}{z}\left(\frac{dx}{dt} + \frac{dy}{dt}\right) = \frac{140 + 100}{260}(35 + 25) = \frac{720}{13} \approx 55.4 \text{ km/h}.$$

21. Using Q for the origin, we are given $\frac{dx}{dt} = -0.5$ m/s and need to find $\frac{dy}{dt}$ when $x = -3$. Using the Pythagorean Theorem twice, we have $\sqrt{x^2 + 4^2} + \sqrt{y^2 + 4^2} = 12$, the total length of the rope. Differentiating with respect to t, we get $\frac{x}{\sqrt{x^2 + 4^2}}\,\frac{dx}{dt} + \frac{y}{\sqrt{y^2 + 4^2}}\,\frac{dy}{dt} = 0$, so $\frac{dy}{dt} = -\frac{x\sqrt{y^2 + 4^2}}{y\sqrt{x^2 + 4^2}}\,\frac{dx}{dt}$.

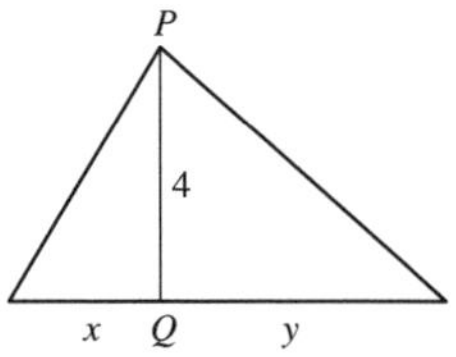

Now when $x = -3$, $12 = \sqrt{(-3)^2 + 4^2} + \sqrt{y^2 + 4^2} = 5 + \sqrt{y^2 + 4^2} \Leftrightarrow \sqrt{y^2 + 4^2} = 7$, and $y = \sqrt{7^2 - 4^2} = \sqrt{33}$.

So when $x = -3$, $\frac{dy}{dt} = -\frac{(-3)(7)}{\sqrt{33}\,(5)}(-0.5) = -\frac{2.1}{\sqrt{33}} \approx -0.37$ m/s. So cart B is moving towards Q at about 0.37 m/s.

23. By similar triangles, $\frac{100}{50} = \frac{b}{h}$, so $b = 2h$. The trough has volume

$V = \frac{1}{2}bh(6) = 3(2h)h = 6h^2 \Rightarrow 1.2 = \frac{dV}{dt} = 12h\,\frac{dh}{dt} \Rightarrow \frac{dh}{dt} = \frac{1}{10h}$.

When $h = 0.3$, $\frac{dh}{dt} = \frac{1}{10 \cdot 0.3} = \frac{1}{3}$ m/min.

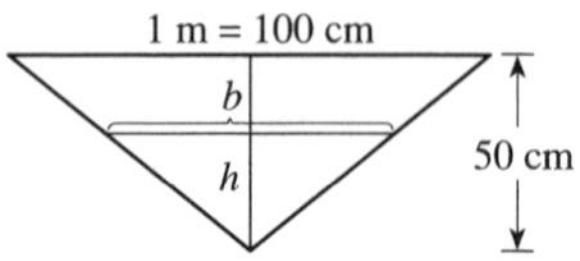

25. We are given that $\frac{dV}{dt} = 3 \text{ m}^3/\text{min}$. $V = \frac{1}{3}\pi r^2 h = \frac{1}{3}\pi\left(\frac{h}{2}\right)^2 h = \frac{\pi h^3}{12}$ $\Rightarrow$

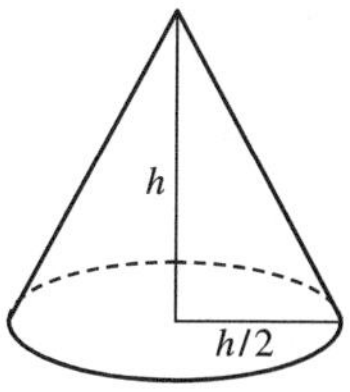

$\frac{dV}{dt} = \frac{dV}{dh}\frac{dh}{dt}$ $\Rightarrow$ $3 = \frac{\pi h^2}{4}\frac{dh}{dt}$ $\Rightarrow$ $\frac{dh}{dt} = \frac{12}{\pi h^2}$. When $h = 3$ m,

$\frac{dh}{dt} = \frac{12}{3^2\pi} = \frac{4}{3\pi} \approx 0.42 \text{ m/min}$.

27. $A = \frac{1}{2}bh$, but $b = 5$ m and $\sin\theta = \frac{h}{4}$ $\Rightarrow$ $h = 4\sin\theta$, so $A = \frac{1}{2}(5)(4\sin\theta) = 10\sin\theta$.

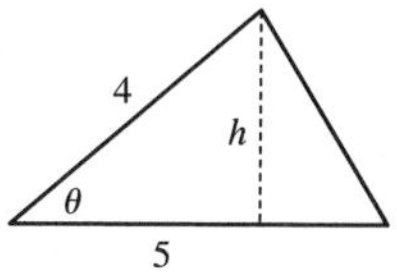

We are given $\frac{d\theta}{dt} = 0.06$ rad/s, so $\frac{dA}{dt} = \frac{dA}{d\theta}\frac{d\theta}{dt} = (10\cos\theta)(0.06) = 0.6\cos\theta$.

When $\theta = \frac{\pi}{3}$, $\frac{dA}{dt} = 0.6\left(\cos\frac{\pi}{3}\right) = (0.6)\left(\frac{1}{2}\right) = 0.3 \text{ m}^2/\text{s}$.

29. Differentiating both sides of $PV = C$ with respect to t and using the Product Rule gives us $P\frac{dV}{dt} + V\frac{dP}{dt} = 0$ $\Rightarrow$ $\frac{dV}{dt} = -\frac{V}{P}\frac{dP}{dt}$. When $V = 600$, $P = 150$ and $\frac{dP}{dt} = 20$, so we have $\frac{dV}{dt} = -\frac{600}{150}(20) = -80$. Thus, the volume is decreasing at a rate of 80 cm^3/min.

31. With $R_1 = 80$ and $R_2 = 100$, $\frac{1}{R} = \frac{1}{R_1} + \frac{1}{R_2} = \frac{1}{80} + \frac{1}{100} = \frac{180}{8000} = \frac{9}{400}$, so $R = \frac{400}{9}$. Differentiating $\frac{1}{R} = \frac{1}{R_1} + \frac{1}{R_2}$ with respect to t, we have $-\frac{1}{R^2}\frac{dR}{dt} = -\frac{1}{R_1^2}\frac{dR_1}{dt} - \frac{1}{R_2^2}\frac{dR_2}{dt}$ $\Rightarrow$ $\frac{dR}{dt} = R^2\left(\frac{1}{R_1^2}\frac{dR_1}{dt} + \frac{1}{R_2^2}\frac{dR_2}{dt}\right)$. When $R_1 = 80$ and $R_2 = 100$, $\frac{dR}{dt} = \frac{400^2}{9^2}\left[\frac{1}{80^2}(0.3) + \frac{1}{100^2}(0.2)\right] = \frac{107}{810} \approx 0.132 \ \Omega/\text{s}$.

33. (a) By the Pythagorean Theorem, $1200^2 + y^2 = \ell^2$. Differentiating with respect to t, we obtain

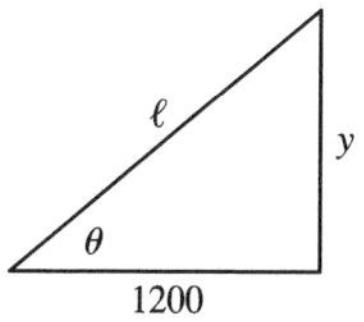

$2y\frac{dy}{dt} = 2\ell\frac{d\ell}{dt}$. We know that $\frac{dy}{dt} = 200$ m/s, so when $y = 900$ m,

$\ell = \sqrt{1200^2 + 900^2} = \sqrt{2\,250\,000} = 1500$ m and $\frac{d\ell}{dt} = \frac{y}{\ell}\frac{dy}{dt} = \frac{900}{1500}(200) = 120$ m/s.

(b) Here $\tan\theta = \frac{y}{1200}$ $\Rightarrow$ $\frac{d}{dt}(\tan\theta) = \frac{d}{dt}\left(\frac{y}{1200}\right)$ $\Rightarrow$ $\sec^2\theta\frac{d\theta}{dt} = \frac{1}{1200}\frac{dy}{dt}$ $\Rightarrow$ $\frac{d\theta}{dt} = \frac{\cos^2\theta}{1200}\frac{dy}{dt}$. When $y = 900$ m, $\frac{dy}{dt} = 200$ m/s, $\ell = 1500$ and $\cos\theta = \frac{1200}{\ell} = \frac{1200}{1500} = \frac{4}{5}$, so $\frac{d\theta}{dt} = \frac{(4/5)^2}{1200}(200) \approx 0.107$ rad/s.

35. We are given that $\frac{dx}{dt} = 300$ km/h. By the Law of Cosines,

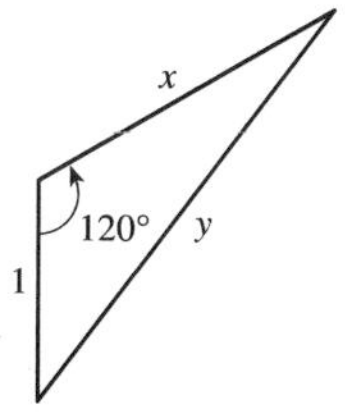

$y^2 = x^2 + 1^2 - 2(1)(x)\cos 120° = x^2 + 1 - 2x\left(-\frac{1}{2}\right) = x^2 + x + 1$, so

$2y\frac{dy}{dt} = 2x\frac{dx}{dt} + \frac{dx}{dt}$ $\Rightarrow$ $\frac{dy}{dt} = \frac{2x+1}{2y}\frac{dx}{dt}$. After 1 minute, $x = \frac{300}{60} = 5$ km $\Rightarrow$

$y = \sqrt{5^2 + 5 + 1} = \sqrt{31}$ km $\Rightarrow$ $\frac{dy}{dt} = \frac{2(5)+1}{2\sqrt{31}}(300) = \frac{1650}{\sqrt{31}} \approx 296$ km/h.

37. Let the distance between the runner and the friend be ℓ. Then by the Law of Cosines, $\ell^2 = 200^2 + 100^2 - 2 \cdot 200 \cdot 100 \cdot \cos\theta = 50\,000 - 40\,000\cos\theta$ $(\star)$. Differentiating implicitly with respect to t, we obtain $2\ell\,\dfrac{d\ell}{dt} = -40\,000(-\sin\theta)\,\dfrac{d\theta}{dt}$. Now if D is the distance run when the angle is θ radians, then by the formula for the length of an arc on a circle, $s = r\theta$, we have $D = 100\theta$, so $\theta = \dfrac{1}{100}D \;\Rightarrow\; \dfrac{d\theta}{dt} = \dfrac{1}{100}\dfrac{dD}{dt} = \dfrac{7}{100}$. To substitute into the expression for $\dfrac{d\ell}{dt}$, we must know $\sin\theta$ at the time when $\ell = 200$, which we find from $(\star)$: $200^2 = 50\,000 - 40\,000\cos\theta \;\Leftrightarrow$

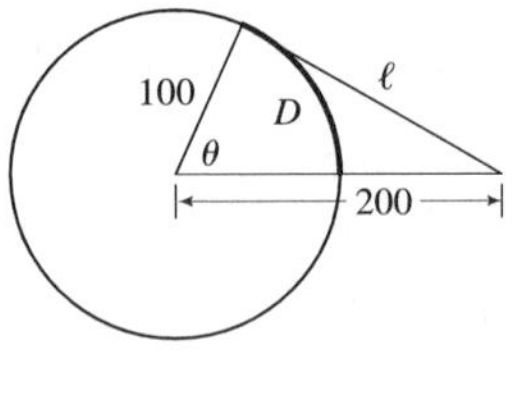

$$\cos\theta = \frac{1}{4} \;\Rightarrow\; \sin\theta = \sqrt{1 - \left(\tfrac{1}{4}\right)^2} = \frac{\sqrt{15}}{4}.$$ Substituting, we get $2(200)\,\dfrac{d\ell}{dt} = 40\,000\,\dfrac{\sqrt{15}}{4}\left(\dfrac{7}{100}\right) \;\Rightarrow$

$d\ell/dt = \frac{7\sqrt{15}}{4} \approx 6.78$ m/s. Whether the distance between them is increasing or decreasing depends on the direction in which the runner is running.

4.2 Maximum and Minimum Values

1. A function f has an **absolute minimum** at $x = c$ if $f(c)$ is the smallest function value on the entire domain of f, whereas f has a **local minimum** at c if $f(c)$ is the smallest function value when x is near c.

3. Absolute maximum at b; absolute minimum at d; local maxima at b and e; local minima at d and s; neither a maximum nor a minimum at a, c, r, and t.

5. Absolute maximum value is $f(4) = 4$; absolute minimum value is $f(7) = 0$; local maximum values are $f(4) = 4$ and $f(6) = 3$; local minimum values are $f(2) = 1$ and $f(5) = 2$.

7. Absolute minimum at 2, absolute maximum at 3, local minimum at 4

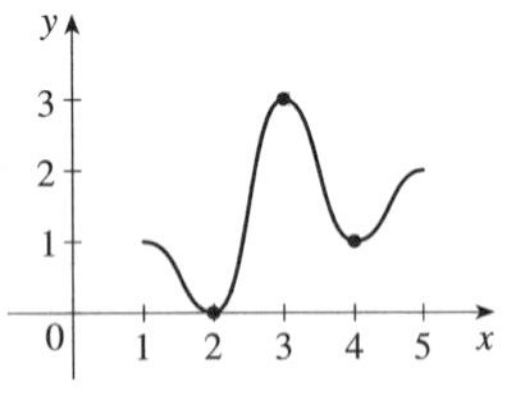

9. Absolute maximum at 5, absolute minimum at 2, local maximum at 3, local minima at 2 and 4

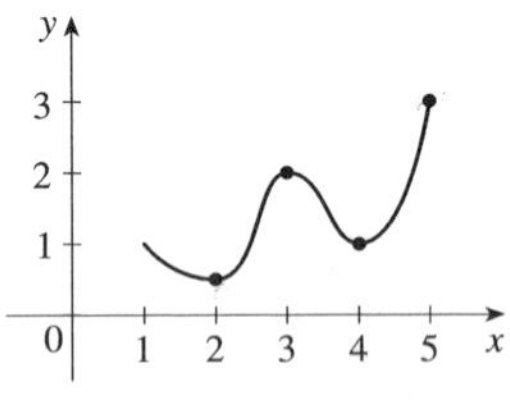

11. (a)

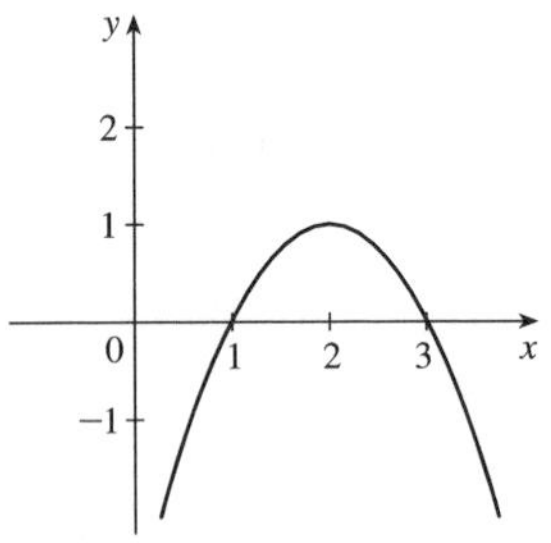

(b)

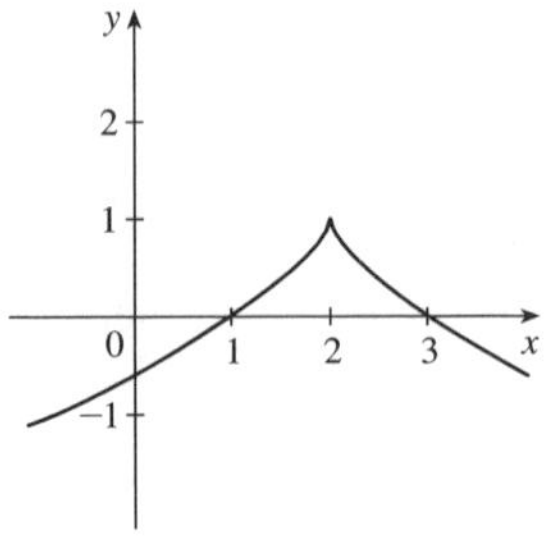

(c)

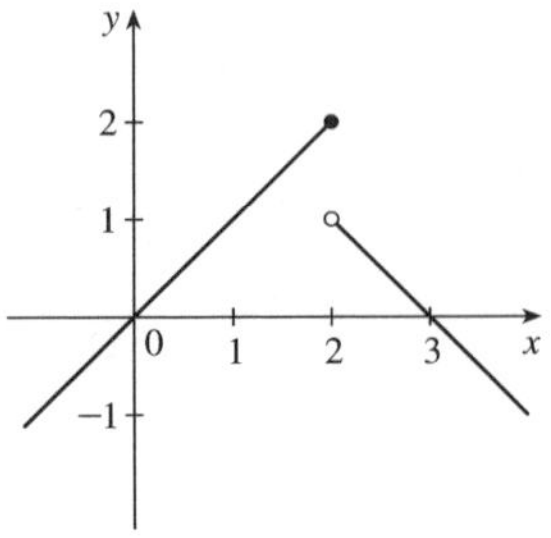

13. (a) *Note:* By the Extreme Value Theorem, f must *not* be continuous; because if it were, it would attain an absolute minimum.

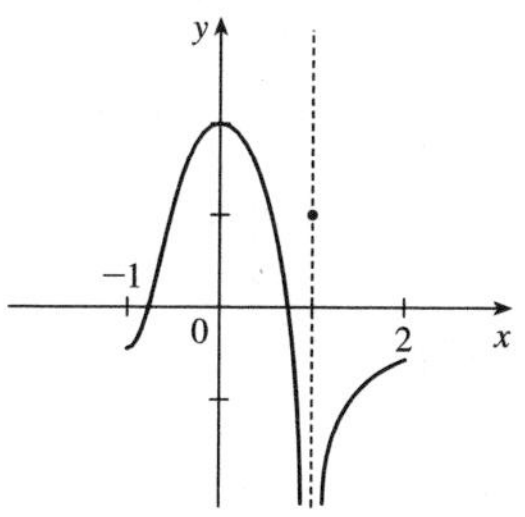

(b)

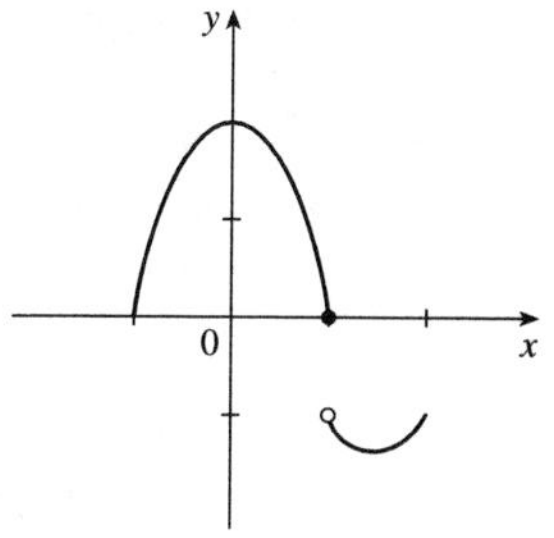

15. $f(x) = 8 - 3x$, $x \geq 1$. Absolute maximum $f(1) = 5$; no local maximum. No absolute or local minimum.

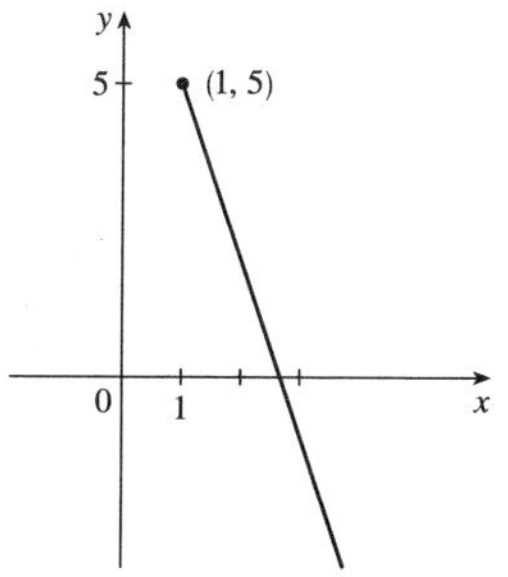

17. $f(x) = x^2$, $0 < x < 2$. No absolute or local maximum or minimum value.

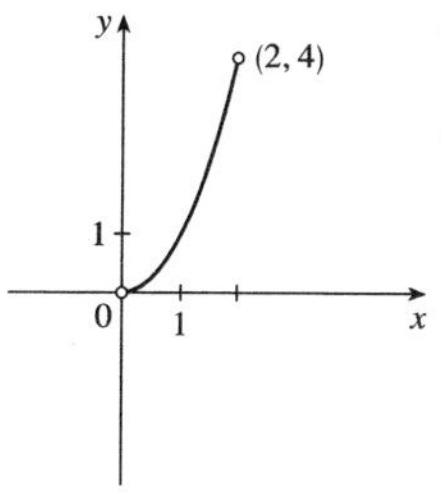

19. $f(\theta) = \sin\theta$, $-2\pi \leq \theta \leq 2\pi$. Absolute and local maxima $f\left(-\frac{3\pi}{2}\right) = f\left(\frac{\pi}{2}\right) = 1$. Absolute and local minima $f\left(-\frac{\pi}{2}\right) = f\left(\frac{3\pi}{2}\right) = -1$.

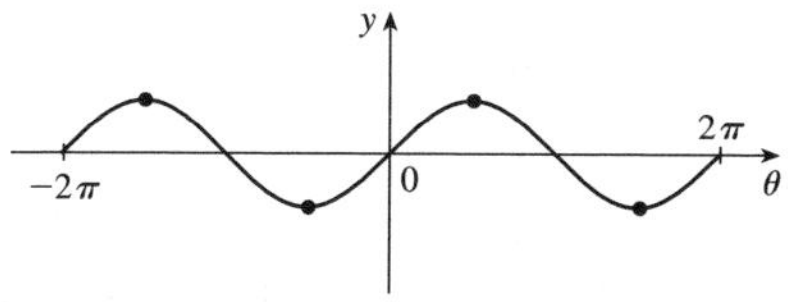

19. $f(x) = 1 - \sqrt{x}$. Absolute maximum $f(0) = 1$; no local maximum. No absolute or local minimum.

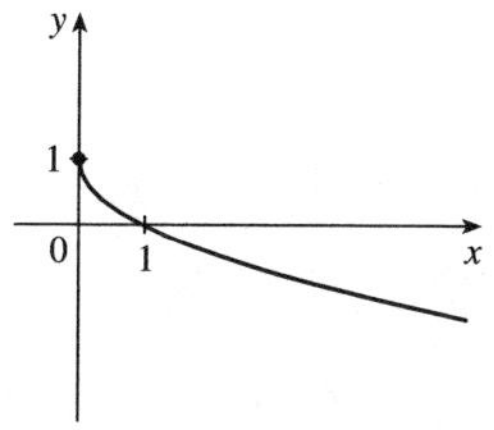

23. $f(x) = 5x^2 + 4x \;\Rightarrow\; f'(x) = 10x + 4$. $f'(x) = 0 \;\Rightarrow\; x = -\frac{2}{5}$, so $-\frac{2}{5}$ is the only critical number.

25. $f(x) = x^3 + 3x^2 - 24x \;\Rightarrow\; f'(x) = 3x^2 + 6x - 24 = 3(x^2 + 2x - 8)$.
$f'(x) = 0 \;\Rightarrow\; 3(x+4)(x-2) = 0 \;\Rightarrow\; x = -4, 2$. These are the only critical numbers.

27. $s(t) = 3t^4 + 4t^3 - 6t^2 \;\Rightarrow\; s'(t) = 12t^3 + 12t^2 - 12t$. $s'(t) = 0 \;\Rightarrow\; 12t(t^2 + t - 1) \;\Rightarrow$
$t = 0$ or $t^2 + t - 1 = 0$. Using the quadratic formula to solve the latter equation gives us
$t = \dfrac{-1 \pm \sqrt{1^2 - 4(1)(-1)}}{2(1)} = \dfrac{-1 \pm \sqrt{5}}{2} \approx 0.618, -1.618$. The three critical numbers are 0, $\dfrac{-1 \pm \sqrt{5}}{2}$.

29. $f(r) = \dfrac{r}{r^2 + 1} \;\Rightarrow\; f'(r) = \dfrac{(r^2+1)1 - r(2r)}{(r^2+1)^2} = \dfrac{-r^2 + 1}{(r^2+1)^2} = 0 \;\Leftrightarrow\; r^2 = 1 \;\Leftrightarrow\; r = \pm 1$, so these are the critical numbers. Note that $f'(r)$ always exists since $r^2 + 1 \neq 0$.

31. $F(x) = x^{4/5}(x-4)^2 \quad \Rightarrow$

$$F'(x) = x^{4/5} \cdot 2(x-4) + (x-4)^2 \cdot \tfrac{4}{5}x^{-1/5} = \tfrac{1}{5}x^{-1/5}(x-4)\,[5 \cdot x \cdot 2 + (x-4) \cdot 4]$$
$$= \frac{(x-4)(14x-16)}{5x^{1/5}} = \frac{2(x-4)(7x-8)}{5x^{1/5}}$$

$F'(x) = 0 \;\Rightarrow\; x = 4, \frac{8}{7}$. $F'(0)$ does not exist. Thus, the three critical numbers are 0, $\frac{8}{7}$, and 4.

33. $f(\theta) = 2\cos\theta + \sin^2\theta \;\Rightarrow\; f'(\theta) = -2\sin\theta + 2\sin\theta\cos\theta$. $f'(\theta) = 0 \;\Rightarrow\; 2\sin\theta(\cos\theta - 1) = 0 \;\Rightarrow\; \sin\theta = 0$ or $\cos\theta = 1 \;\Rightarrow\; \theta = n\pi$ (n an integer) or $\theta = 2n\pi$. The solutions $\theta = n\pi$ include the solutions $\theta = 2n\pi$, so the critical numbers are $\theta = n\pi$.

35. $f(x) = x\ln x \;\Rightarrow\; f'(x) = x(1/x) + (\ln x)\cdot 1 = \ln x + 1$. $f'(x) = 0 \;\Leftrightarrow\; \ln x = -1 \;\Leftrightarrow\; x = e^{-1} = 1/e$. Therefore, the only critical number is $x = 1/e$.

37. $f(x) = 3x^2 - 12x + 5$, $[0,3]$. $f'(x) = 6x - 12 = 0 \;\Leftrightarrow\; x = 2$. Applying the Closed Interval Method, we find that $f(0) = 5$, $f(2) = -7$, and $f(3) = -4$. So $f(0) = 5$ is the absolute maximum value and $f(2) = -7$ is the absolute minimum value.

39. $f(x) = 2x^3 - 3x^2 - 12x + 1$, $[-2,3]$. $f'(x) = 6x^2 - 6x - 12 = 6(x^2 - x - 2) = 6(x-2)(x+1) = 0 \;\Leftrightarrow\; x = 2, -1$. $f(-2) = -3$, $f(-1) = 8$, $f(2) = -19$, and $f(3) = -8$. So $f(-1) = 8$ is the absolute maximum value and $f(2) = -19$ is the absolute minimum value.

41. $f(x) = x^4 - 2x^2 + 3$, $[-2,3]$. $f'(x) = 4x^3 - 4x = 4x(x^2-1) = 4x(x+1)(x-1) = 0 \;\Leftrightarrow\; x = -1, 0, 1$. $f(-2) = 11$, $f(-1) = 2$, $f(0) = 3$, $f(1) = 2$, $f(3) = 66$. So $f(3) = 66$ is the absolute maximum value and $f(\pm 1) = 2$ is the absolute minimum value.

43. $f(x) = x^2 + \dfrac{2}{x}$, $\left[\frac{1}{2}, 2\right]$. $f'(x) = 2x - \dfrac{2}{x^2} = 2\dfrac{x^3 - 1}{x^2} = 0 \;\Leftrightarrow\; x^3 - 1 = 0 \;\Leftrightarrow\; (x-1)(x^2 + x + 1) = 0$, but $x^2 + x + 1 \neq 0$, so $x = 1$. The denominator is 0 at $x = 0$, but not in the desired interval. $f\left(\frac{1}{2}\right) = \frac{17}{4} = 4.25$, $f(1) = 3$, $f(2) = 5$. So $f(2) = 5$ is the absolute maximum and $f(1) = 3$ is the absolute minimum.

45. $f(x) = \sin x + \cos x$, $\left[0, \frac{\pi}{3}\right]$. $f'(x) = \cos x - \sin x = 0 \;\Leftrightarrow\; \sin x = \cos x \;\Rightarrow\; \dfrac{\sin x}{\cos x} = 1 \;\Rightarrow\; \tan x = 1 \;\Rightarrow\; x = \frac{\pi}{4}$. $f(0) = 1$, $f\left(\frac{\pi}{4}\right) = \sqrt{2} \approx 1.41$, $f\left(\frac{\pi}{3}\right) = \frac{\sqrt{3}+1}{2} \approx 1.37$. So $f\left(\frac{\pi}{4}\right) = \sqrt{2}$ is the absolute maximum value and $f(0) = 1$ is the absolute minimum value.

47. $f(x) = xe^{-x^2/8}$, $[-1,4]$. $f'(x) = x \cdot e^{-x^2/8} \cdot \left(-\frac{x}{4}\right) + e^{-x^2/8} \cdot 1 = e^{-x^2/8}\left(-\frac{x^2}{4} + 1\right)$. Since $e^{-x^2/8}$ is never 0, $f'(x) = 0 \;\Rightarrow\; -x^2/4 + 1 = 0 \;\Rightarrow\; 1 = x^2/4 \;\Rightarrow\; x^2 = 4 \;\Rightarrow\; x = \pm 2$, but -2 is not in the given interval, $[-1,4]$. $f(-1) = -e^{-1/8} \approx -0.88$, $f(2) = 2e^{-1/2} \approx 1.21$, and $f(4) = 4e^{-2} \approx 0.54$. So $f(2) = 2e^{-1/2}$ is the absolute maximum value and $f(-1) = -e^{-1/8}$ is the absolute minimum value.

49. $f(x) = x^a(1-x)^b, \;\; 0 \le x \le 1, a > 0, b > 0.$

$$f'(x) = x^a \cdot b(1-x)^{b-1}(-1) + (1-x)^b \cdot ax^{a-1} = x^{a-1}(1-x)^{b-1}\left[x \cdot b(-1) + (1-x)\cdot a\right]$$
$$= x^{a-1}(1-x)^{b-1}(a - ax - bx)$$

At the endpoints, we have $f(0) = f(1) = 0$ [the minimum value of f]. In the interval $(0,1)$, $f'(x) = 0 \;\Leftrightarrow\; x = \dfrac{a}{a+b}$.

$$f\left(\frac{a}{a+b}\right) = \left(\frac{a}{a+b}\right)^a\left(1 - \frac{a}{a+b}\right)^b = \frac{a^a}{(a+b)^a}\left(\frac{a+b-a}{a+b}\right)^b = \frac{a^a}{(a+b)^a}\cdot\frac{b^b}{(a+b)^b} = \frac{a^a b^b}{(a+b)^{a+b}}.$$

So $f\left(\dfrac{a}{a+b}\right) = \dfrac{a^a b^b}{(a+b)^{a+b}}$ is the absolute maximum value.

51. (a)

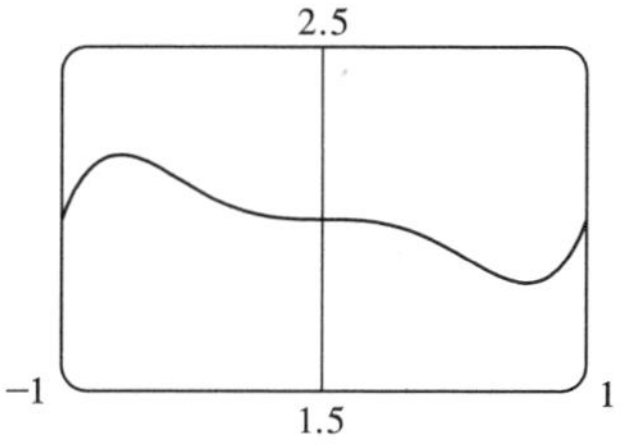

From the graph, it appears that the absolute maximum value is about $f(-0.77) = 2.19$, and the absolute minimum value is about $f(0.77) = 1.81$.

(b) $f(x) = x^5 - x^3 + 2 \;\Rightarrow\; f'(x) = 5x^4 - 3x^2 = x^2(5x^2 - 3)$. So $f'(x) = 0 \;\Rightarrow\; x = 0, \pm\sqrt{\frac{3}{5}}$.

$$f\left(-\sqrt{\tfrac{3}{5}}\right) = \left(-\sqrt{\tfrac{3}{5}}\right)^5 - \left(-\sqrt{\tfrac{3}{5}}\right)^3 + 2 = -\left(\tfrac{3}{5}\right)^2\sqrt{\tfrac{3}{5}} + \tfrac{3}{5}\sqrt{\tfrac{3}{5}} + 2 = \left(\tfrac{3}{5} - \tfrac{9}{25}\right)\sqrt{\tfrac{3}{5}} + 2 = \tfrac{6}{25}\sqrt{\tfrac{3}{5}} + 2 \text{ (maximum)}$$

and similarly, $f\left(\sqrt{\frac{3}{5}}\right) = -\frac{6}{25}\sqrt{\frac{3}{5}} + 2$ (minimum).

53. (a)

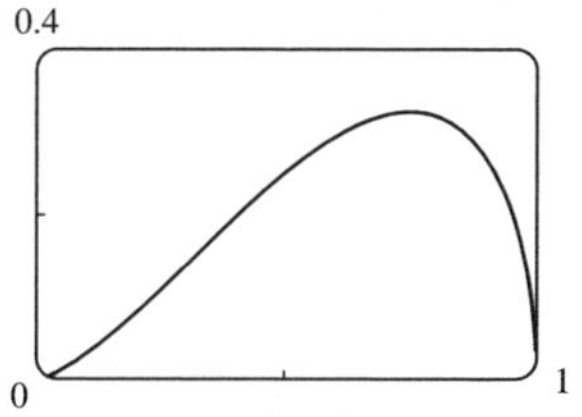

From the graph, it appears that the absolute maximum value is about $f(0.75) = 0.32$, and the absolute minimum value is $f(0) = f(1) = 0$; that is, at both endpoints.

(b) $f(x) = x\sqrt{x - x^2} \;\Rightarrow\; f'(x) = x \cdot \dfrac{1-2x}{2\sqrt{x-x^2}} + \sqrt{x - x^2} = \dfrac{(x - 2x^2) + (2x - 2x^2)}{2\sqrt{x-x^2}} = \dfrac{3x - 4x^2}{2\sqrt{x-x^2}}.$

So $f'(x) = 0 \;\Rightarrow\; 3x - 4x^2 = 0 \;\Rightarrow\; x(3 - 4x) = 0 \;\Rightarrow\; x = 0$ or $\frac{3}{4}$.

$f(0) = f(1) = 0$ (minimum), and $f\left(\frac{3}{4}\right) = \frac{3}{4}\sqrt{\frac{3}{4} - \left(\frac{3}{4}\right)^2} = \frac{3}{4}\sqrt{\frac{3}{16}} = \frac{3\sqrt{3}}{16}$ (maximum).

55. The density is defined as $\rho = \dfrac{\text{mass}}{\text{volume}} = \dfrac{1000}{V(T)}$ (in g/cm^3). But a critical point of ρ will also be a critical point of V $\left[\text{since } \dfrac{d\rho}{dT} = -1000V^{-2}\dfrac{dV}{dT} \text{ and } V \text{ is never } 0\right]$, and V is easier to differentiate than ρ.

$V(T) = 999.87 - 0.064\,26T + 0.008\,504\,3T^2 - 0.000\,067\,9T^3 \;\Rightarrow$

$V'(T) = -0.06426 + 0.017\,008\,6T - 0.000\,203\,7T^2$. Setting this equal to 0 and using the quadratic formula to find T, we get $T = \dfrac{-0.017\,008\,6 \pm \sqrt{0.017\,008\,6^2 - 4 \cdot 0.000\,203\,7 \cdot 0.06426}}{2(-0.000\,203\,7)} \approx 3.9665^\circ\text{C}$ or 79.5318°C. Since we are only interested in the region $0^\circ\text{C} \le T \le 30^\circ\text{C}$, we check the density ρ at the endpoints and at 3.9665°C:

$\rho(0) \approx \dfrac{1000}{999.87} \approx 1.00013$; $\rho(30) \approx \dfrac{1000}{1003.7628} \approx 0.99625$; $\rho(3.9665) \approx \dfrac{1000}{999.7447} \approx 1.000\,255$. So water has its maximum density at about 3.9665°C.

57. We apply the Closed Interval Method to the continuous function

$I(t) = 0.000\,090\,45t^5 + 0.001\,438t^4 - 0.06561t^3 + 0.4598t^2 - 0.6270t + 99.33$ on $[0, 10]$. Its derivative is $I'(t) = 0.000\,452\,25t^4 + 0.005\,752t^3 - 0.19683t^2 + 0.9196t - 0.6270$. Since I' exists for all t, the only critical numbers of I occur when $I'(t) = 0$. We use a rootfinder on a computer algebra system (or a graphing device) to find that $I'(t) = 0$ when $t \approx -29.7186, 0.8231, 5.1309$, or 11.0459, but only the second and third roots lie in the interval $[0, 10]$. The values of I at these critical numbers are $I(0.8231) \approx 99.09$ and $I(5.1309) \approx 100.67$. The values of I at the endpoints of the interval are $I(0) = 99.33$ and $I(10) \approx 96.86$. Comparing these four numbers, we see that food was most expensive at $t \approx 5.1309$ (corresponding roughly to August, 1989) and cheapest at $t = 10$ (midyear 1994).

59. (a) $v(r) = k(r_0 - r)r^2 = kr_0r^2 - kr^3 \;\Rightarrow\; v'(r) = 2kr_0r - 3kr^2$. $v'(r) = 0 \;\Rightarrow\; kr(2r_0 - 3r) = 0 \;\Rightarrow$ $r = 0$ or $\frac{2}{3}r_0$ (but 0 is not in the interval). Evaluating v at $\frac{1}{2}r_0$, $\frac{2}{3}r_0$, and r_0, we get $v\left(\frac{1}{2}r_0\right) = \frac{1}{8}kr_0^3$, $v\left(\frac{2}{3}r_0\right) = \frac{4}{27}kr_0^3$, and $v(r_0) = 0$. Since $\frac{4}{27} > \frac{1}{8}$, v attains its maximum value at $r = \frac{2}{3}r_0$. This supports the statement in the text.

(b) From part (a), the maximum value of v is $\frac{4}{27}kr_0^3$.

(c)

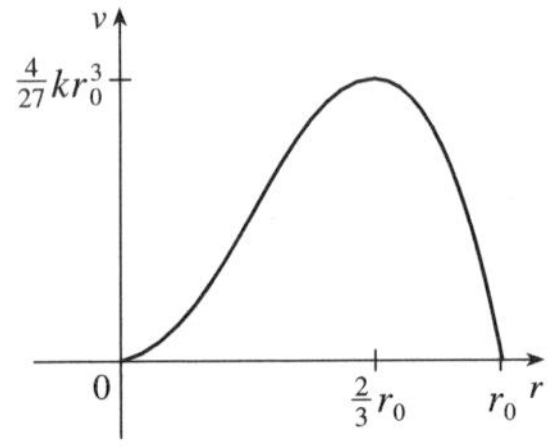

4.3 Derivatives and the Shapes of Curves

1. $\dfrac{f(8) - f(0)}{8 - 0} = \dfrac{6 - 4}{8} = \dfrac{1}{4}$. The values of c which satisfy $f'(c) = \frac{1}{4}$ seem to be about $c = 0.8, 3.2, 4.4$, and 6.1.

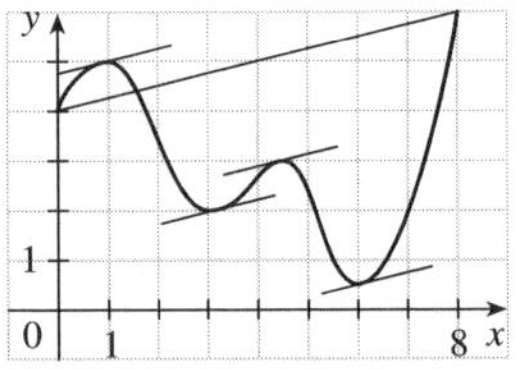

3. (a) Use the Increasing/Decreasing (I/D) Test.

(b) Use the Concavity Test.

(c) At any value of x where the concavity changes, we have an inflection point at $(x, f(x))$.

5. There is an inflection point at $x = 1$ because $f''(x)$ changes from negative to positive there, and so the graph of f changes from concave downward to concave upward. There is an inflection point at $x = 7$ because $f''(x)$ changes from positive to negative there, and so the graph of f changes from concave upward to concave downward.

7. (a) $f(x) = x^3 - 12x + 1 \;\Rightarrow\; f'(x) = 3x^2 - 12 = 3(x + 2)(x - 2)$.

We don't need to include "3" in the chart to determine the sign of $f'(x)$.

Interval	$x + 2$	$x - 2$	$f'(x)$	f
$x < -2$	$-$	$-$	$+$	increasing on $(-\infty, -2)$
$-2 < x < 2$	$+$	$-$	$-$	decreasing on $(-2, 2)$
$x > 2$	$+$	$+$	$+$	increasing on $(2, \infty)$

So f is increasing on $(-\infty, -2)$ and $(2, \infty)$ and f is decreasing on $(-2, 2)$.

(b) f changes from increasing to decreasing at $x = -2$ and from decreasing to increasing at $x = 2$. Thus, $f(-2) = 17$ is a local maximum value and $f(2) = -15$ is a local minimum value.

(c) $f''(x) = 6x$. $f''(x) > 0 \iff x > 0$ and $f''(x) < 0 \iff x < 0$. Thus, f is concave upward on $(0, \infty)$ and concave downward on $(-\infty, 0)$. There is an inflection point where the concavity changes, at $(0, f(0)) = (0, 1)$.

9. (a) $f(x) = x - 2\sin x$ on $(0, 3\pi)$ $\Rightarrow$ $f'(x) = 1 - 2\cos x$. $f'(x) > 0 \iff 1 - 2\cos x > 0 \iff \cos x < \frac{1}{2} \iff$ $\frac{\pi}{3} < x < \frac{5\pi}{3}$ or $\frac{7\pi}{3} < x < 3\pi$. $f'(x) < 0 \iff \cos x > \frac{1}{2} \iff 0 < x < \frac{\pi}{3}$ or $\frac{5\pi}{3} < x < \frac{7\pi}{3}$. So f is increasing on $\left(\frac{\pi}{3}, \frac{5\pi}{3}\right)$ and $\left(\frac{7\pi}{3}, 3\pi\right)$, and f is decreasing on $\left(0, \frac{\pi}{3}\right)$ and $\left(\frac{5\pi}{3}, \frac{7\pi}{3}\right)$.

(b) f changes from increasing to decreasing at $x = \frac{5\pi}{3}$, and from decreasing to increasing at $x = \frac{\pi}{3}$ and at $x = \frac{7\pi}{3}$. Thus, $f\left(\frac{5\pi}{3}\right) = \frac{5\pi}{3} + \sqrt{3} \approx 6.97$ is a local maximum value and $f\left(\frac{\pi}{3}\right) = \frac{\pi}{3} - \sqrt{3} \approx -0.68$ and $f\left(\frac{7\pi}{3}\right) = \frac{7\pi}{3} - \sqrt{3} \approx 5.60$ are local minimum values.

(c) $f''(x) = 2\sin x > 0 \iff 0 < x < \pi$ and $2\pi < x < 3\pi$, $f''(x) < 0 \iff \pi < x < 2\pi$. Thus, f is concave upward on $(0, \pi)$ and $(2\pi, 3\pi)$, and f is concave downward on $(\pi, 2\pi)$. There are inflection points at (π, π) and $(2\pi, 2\pi)$.

11. (a) $y = f(x) = xe^x \Rightarrow f'(x) = xe^x + e^x = e^x(x + 1)$. So $f'(x) > 0 \iff x + 1 > 0 \iff x > -1$. Thus, f is increasing on $(-1, \infty)$ and decreasing on $(-\infty, -1)$.

(b) f changes from decreasing to increasing at its only critical number, $x = -1$. Thus, $f(-1) = -e^{-1}$ is a local minimum value.

(c) $f'(x) = e^x(x + 1) \Rightarrow f''(x) = e^x(1) + (x + 1)e^x = e^x(x + 2)$. So $f''(x) > 0 \iff x + 2 > 0 \iff x > -2$. Thus, f is concave upward on $(-2, \infty)$ and concave downward on $(-\infty, -2)$. Since the concavity changes direction at $x = -2$, the point $\left(-2, -2e^{-2}\right)$ is an inflection point.

13. (a) $y = f(x) = \dfrac{\ln x}{\sqrt{x}}$. (Note that f is only defined for $x > 0$.)

$$f'(x) = \frac{\sqrt{x}\,(1/x) - \ln x\left(\frac{1}{2}x^{-1/2}\right)}{x} = \frac{\dfrac{1}{\sqrt{x}} - \dfrac{\ln x}{2\sqrt{x}}}{x} \cdot \frac{2\sqrt{x}}{2\sqrt{x}} = \frac{2 - \ln x}{2x^{3/2}} > 0 \iff 2 - \ln x > 0 \iff$$

$\ln x < 2 \iff x < e^2$. Therefore f is increasing on $\left(0, e^2\right)$ and decreasing on $\left(e^2, \infty\right)$.

(b) f changes from increasing to decreasing at $x = e^2$, so $f\left(e^2\right) = \dfrac{\ln e^2}{\sqrt{e^2}} = \dfrac{2}{e}$ is a local maximum value.

(c)
$$f''(x) = \frac{2x^{3/2}(-1/x) - (2 - \ln x)(3x^{1/2})}{(2x^{3/2})^2} = \frac{-2x^{1/2} + 3x^{1/2}(\ln x - 2)}{4x^3}$$
$$= \frac{x^{1/2}(-2 + 3\ln x - 6)}{4x^3} = \frac{3\ln x - 8}{4x^{5/2}}$$

$f''(x) = 0 \iff \ln x = \frac{8}{3} \iff x = e^{8/3}$. $f''(x) > 0 \iff x > e^{8/3}$, so f is concave upward on $(e^{8/3}, \infty)$ and concave downward on $(0, e^{8/3})$. There is an inflection point at $\left(e^{8/3}, \frac{8}{3}e^{-4/3}\right) \approx (14.39, 0.70)$.

15. $f(x) = x + \sqrt{1 - x} \Rightarrow f'(x) = 1 + \frac{1}{2}(1 - x)^{-1/2}(-1) = 1 - \dfrac{1}{2\sqrt{1 - x}}$. Note that f is defined for $1 - x \geq 0$; that is, for $x \leq 1$. $f'(x) = 0 \Rightarrow 2\sqrt{1 - x} = 1 \Rightarrow \sqrt{1 - x} = \frac{1}{2} \Rightarrow 1 - x = \frac{1}{4} \Rightarrow x = \frac{3}{4}$. f' does not exist at $x = 1$, but we can't have a local maximum or minimum at an endpoint.

First Derivative Test: $f'(x) > 0 \Rightarrow x < \frac{3}{4}$ and $f'(x) < 0 \Rightarrow \frac{3}{4} < x < 1$. Since f' changes from positive to negative at $x = \frac{3}{4}$, $f\left(\frac{3}{4}\right) = \frac{5}{4}$ is a local maximum value.

Second Derivative Test: $f''(x) = -\frac{1}{2}\left(-\frac{1}{2}\right)(1 - x)^{-3/2}(-1) = -\dfrac{1}{4\left(\sqrt{1 - x}\right)^3}$. $f''\left(\frac{3}{4}\right) = -2 < 0 \Rightarrow$ $f\left(\frac{3}{4}\right) = \frac{5}{4}$ is a local maximum value.

Preference: The First Derivative Test may be slightly easier to apply in this case.

17. (a) By the Second Derivative Test, if $f'(2) = 0$ and $f''(2) = -5 < 0$, f has a local maximum at $x = 2$.

(b) If $f'(6) = 0$, we know that f has a horizontal tangent at $x = 6$. Knowing that $f''(6) = 0$ does not provide any additional information since the Second Derivative Test fails. For example, the first and second derivatives of $y = (x-6)^4$, $y = -(x-6)^4$, and $y = (x-6)^3$ all equal zero for $x = 6$, but the first has a local minimum at $x = 6$, the second has a local maximum at $x = 6$, and the third has an inflection point at $x = 6$.

19. (a) $f(x) = 2x^3 - 3x^2 - 12x \Rightarrow f'(x) = 6x^2 - 6x - 12 = 6(x^2 - x - 2) = 6(x-2)(x+1)$.
$f'(x) > 0 \Leftrightarrow x < -1$ or $x > 2$ and $f'(x) < 0 \Leftrightarrow -1 < x < 2$. So f is increasing on $(-\infty, -1)$ and $(2, \infty)$, and f is decreasing on $(-1, 2)$.

(b) Since f changes from increasing to decreasing at $x = -1$, $f(-1) = 7$ is a local maximum value. Since f changes from decreasing to increasing at $x = 2$, $f(2) = -20$ is a local minimum value.

(c) $f''(x) = 6(2x-1) \Rightarrow f''(x) > 0$ on $(\frac{1}{2}, \infty)$ and $f''(x) < 0$ on $(-\infty, \frac{1}{2})$. So f is concave upward on $(\frac{1}{2}, \infty)$ and concave downward on $(-\infty, \frac{1}{2})$. There is a change in concavity at $x = \frac{1}{2}$, and we have an inflection point at $(\frac{1}{2}, -\frac{13}{2})$.

(d)

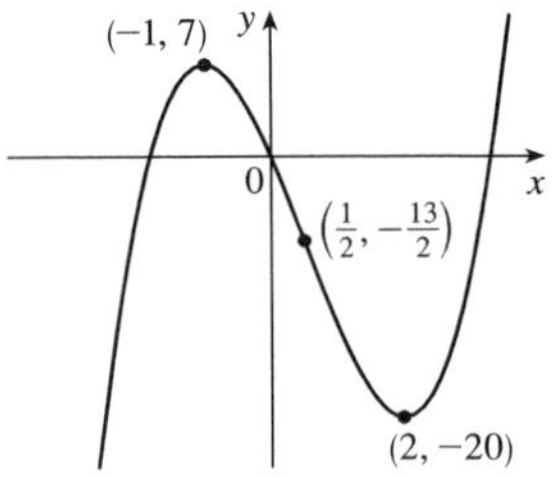

21. (a) $h(x) = 3x^5 - 5x^3 + 3 \Rightarrow h'(x) = 15x^4 - 15x^2 = 15x^2(x^2 - 1) = 0$ when $x = 0, \pm 1$. Since $15x^2$ is nonnegative, $h'(x) > 0 \Leftrightarrow x^2 > 1 \Leftrightarrow |x| > 1 \Leftrightarrow x > 1$ or $x < -1$, so h is increasing on $(-\infty, -1)$ and $(1, \infty)$ and decreasing on $(-1, 1)$, with a horizontal tangent at $x = 0$.

(b) Local maximum value $h(-1) = 5$, local minimum value $h(1) = 1$

(c) $h''(x) = 60x^3 - 30x = 30x(2x^2 - 1)$
$$= 60x\left(x + \tfrac{1}{\sqrt{2}}\right)\left(x - \tfrac{1}{\sqrt{2}}\right) \Rightarrow$$
$h''(x) > 0$ when $x > \frac{1}{\sqrt{2}}$ or $-\frac{1}{\sqrt{2}} < x < 0$, so h is CU on $\left(-\frac{1}{\sqrt{2}}, 0\right)$ and $\left(\frac{1}{\sqrt{2}}, \infty\right)$ and CD on $\left(-\infty, -\frac{1}{\sqrt{2}}\right)$ and $\left(0, \frac{1}{\sqrt{2}}\right)$. Inflection points at $(0, 3)$ and $\left(\pm\frac{1}{\sqrt{2}}, 3 \mp \frac{7}{8}\sqrt{2}\right)$ [about $(-0.71, 4.24)$ and $(0.71, 1.76)$].

(d)

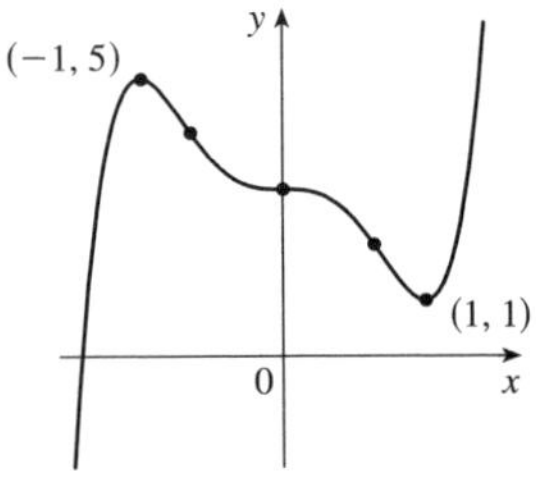

23. (a) $A(x) = x\sqrt{x+3} \Rightarrow A'(x) = x \cdot \frac{1}{2}(x+3)^{-1/2} + \sqrt{x+3} \cdot 1 = \dfrac{x}{2\sqrt{x+3}} + \sqrt{x+3} = \dfrac{x + 2(x+3)}{2\sqrt{x+3}} = \dfrac{3x+6}{2\sqrt{x+3}}$.
The domain of A is $[-3, \infty)$. $A'(x) > 0$ for $x > -2$ and $A'(x) < 0$ for $-3 < x < -2$, so A is increasing on $(-2, \infty)$ and decreasing on $(-3, -2)$.

(b) $A(-2) = -2$ is a local minimum value.

(c)
$$A''(x) = \frac{2\sqrt{x+3} \cdot 3 - (3x+6) \cdot \dfrac{1}{\sqrt{x+3}}}{(2\sqrt{x+3})^2}$$
$$= \frac{6(x+3) - (3x+6)}{4(x+3)^{3/2}} = \frac{3x+12}{4(x+3)^{3/2}} = \frac{3(x+4)}{4(x+3)^{3/2}}$$

$A''(x) > 0$ for all $x > -3$, so A is concave upward on $(-3, \infty)$. There is no inflection point.

(d)

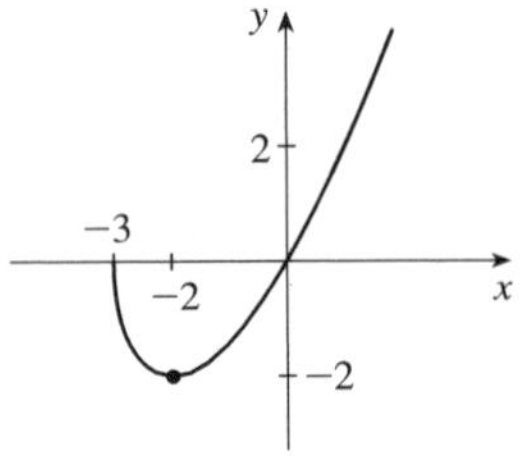

25. (a) $C(x) = x^{1/3}(x+4) = x^{4/3} + 4x^{1/3} \Rightarrow C'(x) = \frac{4}{3}x^{1/3} + \frac{4}{3}x^{-2/3} = \frac{4}{3}x^{-2/3}(x+1) = \dfrac{4(x+1)}{3\sqrt[3]{x^2}}$. $C'(x) > 0$ if $-1 < x < 0$ or $x > 0$ and $C'(x) < 0$ for $x < -1$, so C is increasing on $(-1, \infty)$ and C is decreasing on $(-\infty, -1)$.

(b) $C(-1) = -3$ is a local minimum value.

(c) $C''(x) = \frac{4}{9}x^{-2/3} - \frac{8}{9}x^{-5/3} = \frac{4}{9}x^{-5/3}(x-2) = \dfrac{4(x-2)}{9\sqrt[3]{x^5}}$. $C''(x) < 0$ for $0 < x < 2$ and $C''(x) > 0$ for $x < 0$ and $x > 2$, so C is concave downward on $(0, 2)$ and concave upward on $(-\infty, 0)$ and $(2, \infty)$. There are inflection points at $(0, 0)$ and $(2, 6\sqrt[3]{2}) \approx (2, 7.56)$.

(d)

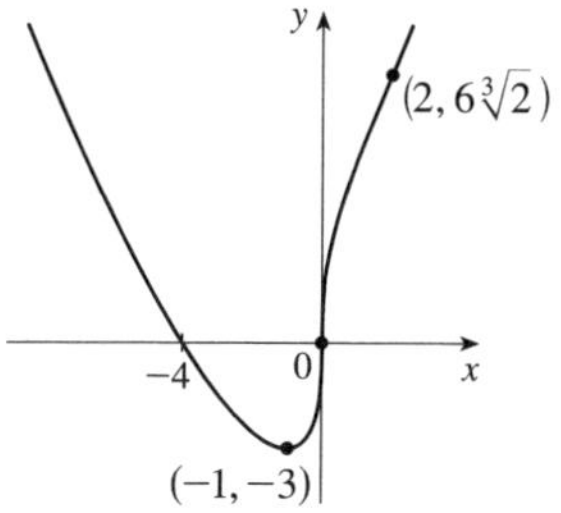

27. (a) $f(x) = 2\cos x + \sin^2 x$, $-\pi \le x \le \pi$. $f'(x) = -2\sin x + 2\sin x\, \cos x = 2\sin x\,(\cos x - 1)$. Since $\cos x \le 1$, $\cos x - 1 \le 0$, so the sign of $f'(x)$ is the opposite of the sign of $\sin x$. Thus, $f'(x) > 0 \Leftrightarrow \sin x < 0 \Leftrightarrow -\pi < x < 0$, so f is increasing on $(-\pi, 0)$ and decreasing on $(0, \pi)$.

(b) f changes from increasing to decreasing at $x = 0$, so $f(0) = 2$ is a local maximum. The absolute minimum value of -2 occurs at $x = \pm\pi$ (the endpoints), but there is no local minimum.

(c) $f'(x) = -2\sin x + 2\sin x\, \cos x = -2\sin x + \sin 2x \Rightarrow$

$$f''(x) = -2\cos x + 2\cos 2x = 2(2\cos^2 x - \cos x - 1)$$
$$= 2(2\cos x + 1)(\cos x - 1) > 0 \Leftrightarrow$$

$\cos x < -\frac{1}{2}$ $[\cos x - 1 \le 0]$ $\Leftrightarrow$ $x \in \left(-\pi, -\frac{2\pi}{3}\right)$ and $\left(\frac{2\pi}{3}, \pi\right)$, so f is CU on these intervals and CD on $\left(-\frac{2\pi}{3}, \frac{2\pi}{3}\right)$. Note that $f'' < 0$ on $\left(-\frac{2\pi}{3}, 0\right)$ and $\left(0, \frac{2\pi}{3}\right)$, so f is CD on these intervals by the Concavity Test. In fact, since f' is decreasing on $\left(-\frac{2\pi}{3}, \frac{2\pi}{3}\right)$, f is CD on $\left(-\frac{2\pi}{3}, \frac{2\pi}{3}\right)$. There are IPs at $\left(\pm\frac{2\pi}{3}, -\frac{1}{4}\right)$.

(d)

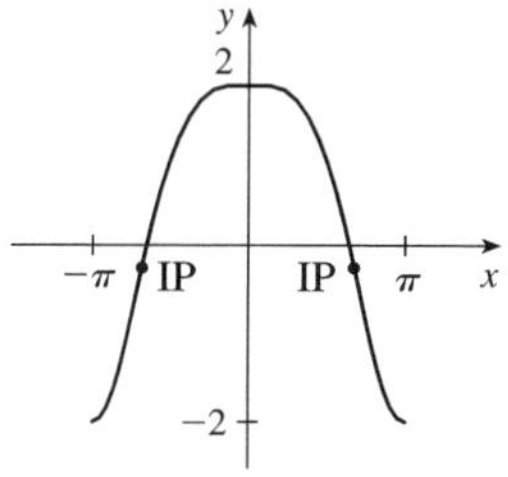

29. $f(x) = \dfrac{x^2}{x^2 - 1} = \dfrac{x^2}{(x+1)(x-1)}$ has domain $(-\infty, -1) \cup (-1, 1) \cup (1, \infty)$.

(a) $\displaystyle\lim_{x\to\pm\infty} f(x) = \lim_{x\to\pm\infty} \frac{x^2/x^2}{(x^2-1)/x^2} = \lim_{x\to\pm\infty} \frac{1}{1 - 1/x^2} = \frac{1}{1-0} = 1$, so $y = 1$ is a HA.

$\displaystyle\lim_{x\to -1^-} \frac{x^2}{x^2-1} = \infty$ since $x^2 \to 1$ and $(x^2 - 1) \to 0^+$ as $x \to -1^-$, so $x = -1$ is a VA.

$\displaystyle\lim_{x\to 1^+} \frac{x^2}{x^2-1} = \infty$ since $x^2 \to 1$ and $(x^2 - 1) \to 0^+$ as $x \to 1^+$, so $x = 1$ is a VA.

(b) $f(x) = \dfrac{x^2}{x^2-1} \Rightarrow f'(x) = \dfrac{(x^2-1)(2x) - x^2(2x)}{(x^2-1)^2} = \dfrac{2x[(x^2-1) - x^2]}{(x^2-1)^2} = \dfrac{-2x}{(x^2-1)^2}$. Since $(x^2-1)^2$ is positive for all x in the domain of f, the sign of the derivative is determined by the sign of $-2x$. Thus, $f'(x) > 0$ if $x < 0$ $(x \ne -1)$ and $f'(x) < 0$ if $x > 0$ $(x \ne 1)$. So f is increasing on $(-\infty, -1)$ and $(-1, 0)$, and f is decreasing on $(0, 1)$ and $(1, \infty)$.

(c) $f'(x) = 0 \Rightarrow x = 0$ and $f(0) = 0$ is a local maximum value.

(d) $f''(x) = \dfrac{(x^2-1)^2(-2) - (-2x)\cdot 2(x^2-1)(2x)}{[(x^2-1)^2]^2}$

$= \dfrac{2(x^2-1)[-(x^2-1)+4x^2]}{(x^2-1)^4} = \dfrac{2(3x^2+1)}{(x^2-1)^3}.$

The sign of $f''(x)$ is determined by the denominator; that is, $f''(x) > 0$ if $|x| > 1$ and $f''(x) < 0$ if $|x| < 1$. Thus, f is CU on $(-\infty, -1)$ and $(1, \infty)$, and f is CD on $(-1, 1)$. There are no inflection points.

(e)

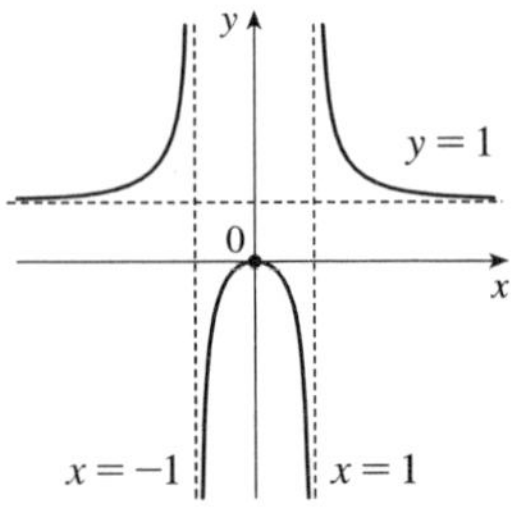

31. (a) $\lim_{x\to-\infty}\left(\sqrt{x^2+1} - x\right) = \infty$ and

$\lim_{x\to\infty}\left(\sqrt{x^2+1} - x\right) = \lim_{x\to\infty}\left(\sqrt{x^2+1} - x\right)\dfrac{\sqrt{x^2+1}+x}{\sqrt{x^2+1}+x} = \lim_{x\to\infty}\dfrac{1}{\sqrt{x^2+1}+x} = 0$, so $y = 0$ is a HA.

(b) $f(x) = \sqrt{x^2+1} - x \Rightarrow f'(x) = \dfrac{x}{\sqrt{x^2+1}} - 1$. Since $\dfrac{x}{\sqrt{x^2+1}} < 1$ for all x, $f'(x) < 0$, so f is decreasing on $\mathbb{R}$.

(c) No minimum or maximum

(d) $f''(x) = \dfrac{(x^2+1)^{1/2}(1) - x\cdot\frac{1}{2}(x^2+1)^{-1/2}(2x)}{\left(\sqrt{x^2+1}\right)^2}$

$= \dfrac{(x^2+1)^{1/2} - \dfrac{x^2}{(x^2+1)^{1/2}}}{x^2+1} = \dfrac{(x^2+1)-x^2}{(x^2+1)^{3/2}}$

$= \dfrac{1}{(x^2+1)^{3/2}} > 0$, so f is CU on $\mathbb{R}$. No IP

(e)

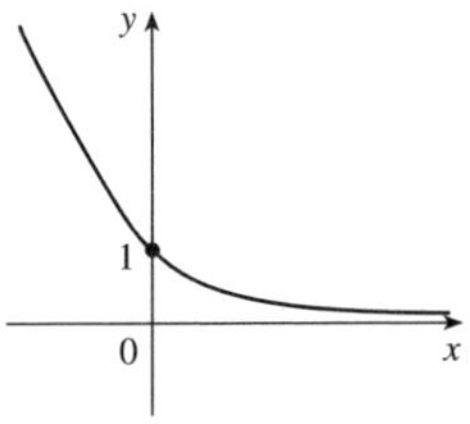

33. $f(x) = \ln(1 - \ln x)$ is defined when $x > 0$ (so that $\ln x$ is defined) and $1 - \ln x > 0$ [so that $\ln(1-\ln x)$ is defined]. The second condition is equivalent to $1 > \ln x \Leftrightarrow x < e$, so f has domain $(0, e)$.

(a) As $x \to 0^+$, $\ln x \to -\infty$, so $1 - \ln x \to \infty$ and $f(x) \to \infty$. As $x \to e^-$, $\ln x \to 1^-$, so $1 - \ln x \to 0^+$ and $f(x) \to -\infty$. Thus, $x = 0$ and $x = e$ are vertical asymptotes. There is no horizontal asymptote.

(b) $f'(x) = \dfrac{1}{1-\ln x}\left(-\dfrac{1}{x}\right) = -\dfrac{1}{x(1-\ln x)} < 0$ on $(0, e)$. Thus, f is decreasing on its domain, $(0, e)$.

(c) $f'(x) \neq 0$ on $(0, e)$, so f has no local maximum or minimum value.

(d) $f''(x) = -\dfrac{-[x(1-\ln x)]'}{[x(1-\ln x)]^2} = \dfrac{x(-1/x)+(1-\ln x)}{x^2(1-\ln x)^2}$

$= -\dfrac{\ln x}{x^2(1-\ln x)^2}$

so $f''(x) > 0 \Leftrightarrow \ln x < 0 \Leftrightarrow 0 < x < 1$. Thus, f is CU on $(0, 1)$ and CD on $(1, e)$. There is an inflection point at $(1, 0)$.

(e)

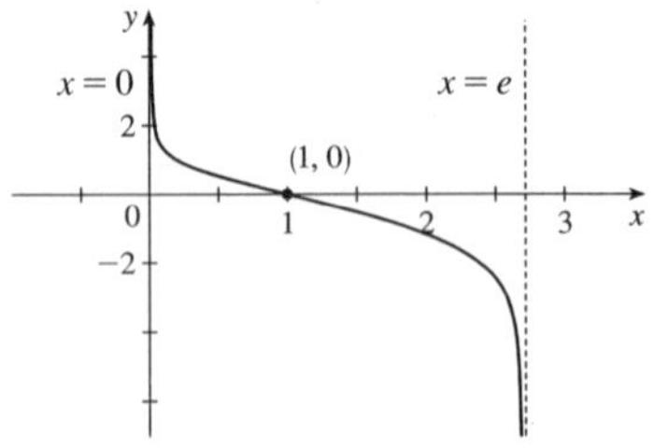

35. (a) $\lim_{x\to\pm\infty} e^{-1/(x+1)} = 1$ since $-1/(x+1) \to 0$, so $y = 1$ is a HA. $\lim_{x\to-1^+} e^{-1/(x+1)} = 0$ since $-1/(x+1) \to -\infty$,

$\lim_{x\to-1^-} e^{-1/(x+1)} = \infty$ since $-1/(x+1) \to \infty$, so $x = -1$ is a VA.

(b) $f(x) = e^{-1/(x+1)} \Rightarrow f'(x) = e^{-1/(x+1)}\left[-(-1)\dfrac{1}{(x+1)^2}\right]$ [Reciprocal Rule] $= e^{-1/(x+1)}/(x+1)^2 \Rightarrow$

$f'(x) > 0$ for all x except -1, so f is increasing on $(-\infty, -1)$ and $(-1, \infty)$.

(c) No local maximum or minimum

(d) $f''(x) = \dfrac{(x+1)^2 e^{-1/(x+1)}\left[1/(x+1)^2\right] - e^{-1/(x+1)}\left[2(x+1)\right]}{\left[(x+1)^2\right]^2}$

$= \dfrac{e^{-1/(x+1)}\left[1-(2x+2)\right]}{(x+1)^4} = -\dfrac{e^{-1/(x+1)}(2x+1)}{(x+1)^4} \quad \Rightarrow$

$f''(x) > 0 \quad \Leftrightarrow \quad 2x+1 < 0 \quad \Leftrightarrow \quad x < -\frac{1}{2}$, so f is CU on $(-\infty, -1)$ and $\left(-1, -\frac{1}{2}\right)$, and CD on $\left(-\frac{1}{2}, \infty\right)$. f has an IP at $\left(-\frac{1}{2}, e^{-2}\right)$.

(e)

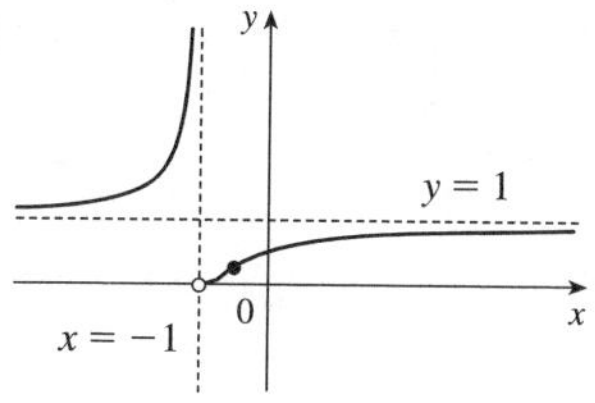

37. The nonnegative factors $(x+1)^2$ and $(x-6)^4$ do not affect the sign of $f'(x) = (x+1)^2(x-3)^5(x-6)^4$.

So $f'(x) > 0 \quad \Rightarrow \quad (x-3)^5 > 0 \quad \Rightarrow \quad x - 3 > 0 \quad \Rightarrow \quad x > 3$. Thus, f is increasing on the interval $(3, \infty)$.

39. (a)

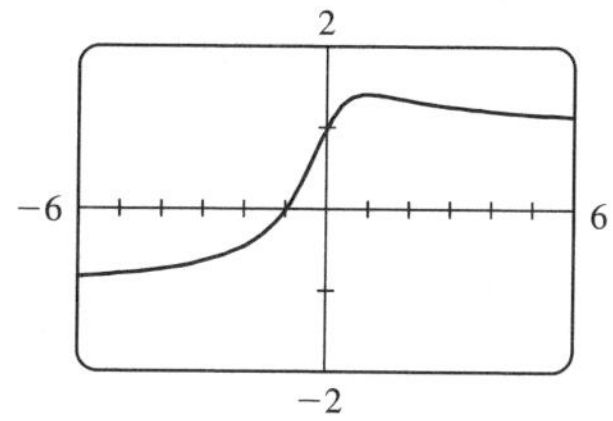

From the graph, we get an estimate of $f(1) \approx 1.41$ as a local maximum value, and no local minimum value.

$f(x) = \dfrac{x+1}{\sqrt{x^2+1}} \quad \Rightarrow \quad f'(x) = \dfrac{1-x}{(x^2+1)^{3/2}}$.

$f'(x) = 0 \quad \Leftrightarrow \quad x = 1$. $f(1) = \frac{2}{\sqrt{2}} = \sqrt{2}$ is the exact value.

(b) From the graph in part (a), f increases most rapidly somewhere between $x = -\frac{1}{2}$ and $x = -\frac{1}{4}$. To find the exact value, we need to find the maximum value of f', which we can do by finding the critical numbers of f'.

$f''(x) = \dfrac{2x^2 - 3x - 1}{(x^2+1)^{5/2}} = 0 \quad \Leftrightarrow \quad x = \dfrac{3 \pm \sqrt{17}}{4}$. $x = \dfrac{3+\sqrt{17}}{4}$ corresponds to the *minimum* value of f'. The maximum value of f' is at $\left(\frac{3-\sqrt{17}}{4}, \sqrt{\frac{7}{6} - \frac{\sqrt{17}}{6}}\right) \approx (-0.28, 0.69)$.

41. $f(x) = \cos x + \frac{1}{2}\cos 2x \quad \Rightarrow \quad f'(x) = -\sin x - \sin 2x \quad \Rightarrow \quad f''(x) = -\cos x - 2\cos 2x$

(a)

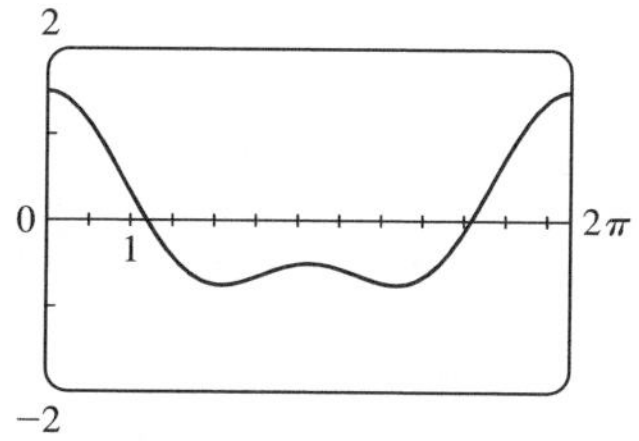

From the graph of f, it seems that f is CD on $(0, 1)$, CU on $(1, 2.5)$, CD on $(2.5, 3.7)$, CU on $(3.7, 5.3)$, and CD on $(5.3, 2\pi)$. The points of inflection appear to be at $(1, 0.4)$, $(2.5, -0.6)$, $(3.7, -0.6)$, and $(5.3, 0.4)$.

(b)

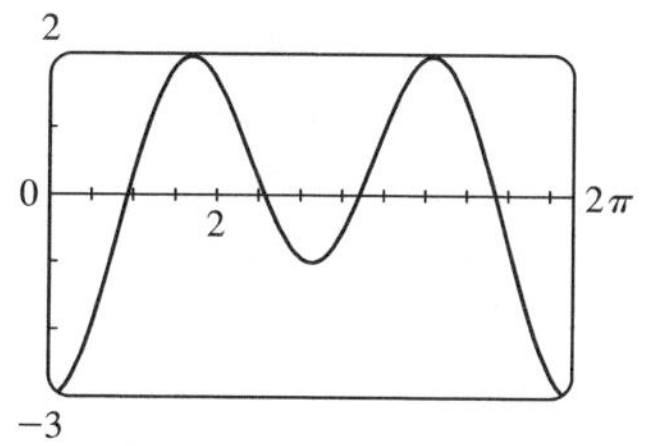

From the graph of f'' (and zooming in near the zeros), it seems that f is CD on $(0, 0.94)$, CU on $(0.94, 2.57)$, CD on $(2.57, 3.71)$, CU on $(3.71, 5.35)$, and CD on $(5.35, 2\pi)$. Refined estimates of the inflection points are $(0.94, 0.44)$, $(2.57, -0.63)$, $(3.71, -0.63)$, and $(5.35, 0.44)$.

43. In Maple, we define f and then use the command `plot(diff(diff(f,x),x),x=-3..3);`. In Mathematica, we define f and then use `Plot[Dt[Dt[f,x],x],{x,-3,3}]`.We see that $f'' > 0$ for $x > 0.1$ and $f'' < 0$ for $x < 0.1$. So f is concave up on $(0.1, \infty)$ and concave down on $(-\infty, 0.1)$.

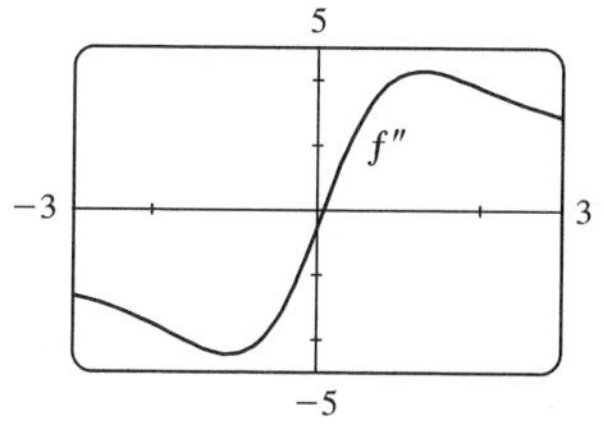

45. $y = -\dfrac{W}{24EI}x^4 + \dfrac{WL}{12EI}x^3 - \dfrac{WL^2}{24EI}x^2 = -\dfrac{W}{24EI}x^2\left(x^2 - 2Lx + L^2\right)$

$= \dfrac{-W}{24EI}x^2(x-L)^2 = cx^2(x-L)^2$

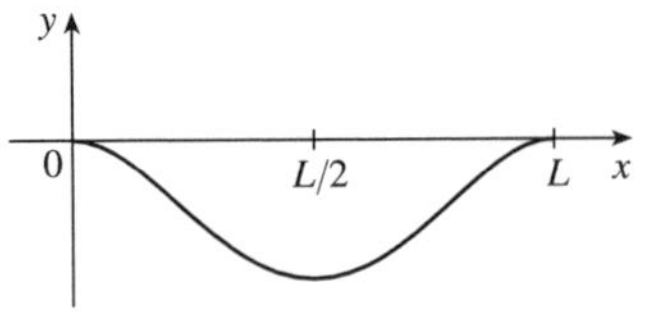

where $c = -\dfrac{W}{24EI}$ is a negative constant and $0 \le x \le L$. We sketch

$f(x) = cx^2(x-L)^2$ for $c = -1$. $f(0) = f(L) = 0$.

$f'(x) = cx^2[2(x-L)] + (x-L)^2(2cx) = 2cx(x-L)\,[x + (x-L)] = 2cx(x-L)(2x-L)$. So for $0 < x < L$,

$f'(x) > 0 \;\Leftrightarrow\; x(x-L)(2x-L) < 0$ [since $c < 0$] $\;\Leftrightarrow\; L/2 < x < L$ and $f'(x) < 0 \;\Leftrightarrow\; 0 < x < L/2$.

Thus, f is increasing on $(L/2, L)$ and decreasing on $(0, L/2)$, and there is a local and absolute minimum at the point $(L/2, f(L/2)) = \left(L/2, cL^4/16\right)$.

$f'(x) = 2c\,[x(x-L)(2x-L)] \;\Rightarrow$

$f''(x) = 2c\,[1(x-L)(2x-L) + x(1)(2x-L) + x(x-L)(2)] = 2c\left(6x^2 - 6Lx + L^2\right) = 0 \;\Leftrightarrow$

$x = \dfrac{6L \pm \sqrt{12L^2}}{12} = \frac{1}{2}L \pm \frac{\sqrt{3}}{6}L$, and these are the x-coordinates of the two inflection points.

47.

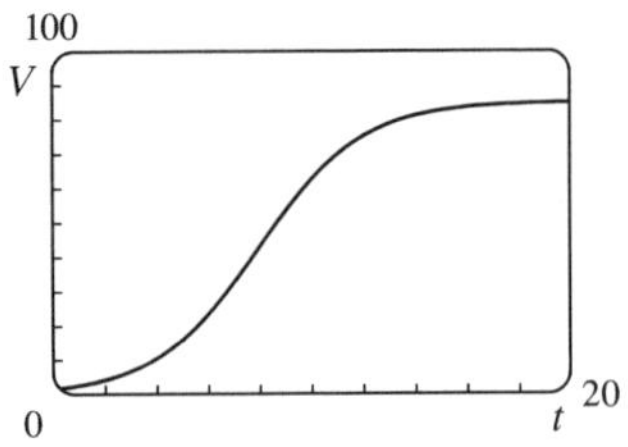

From the graph, we estimate that the most rapid increase in the percentage of households in the United States with at least one VCR occurs at about $t = 8$. To maximize the first derivative, we need to determine the values for which the second derivative is 0. We'll use $V(t) = \dfrac{a}{1 + be^{ct}}$, and substitute $a = 85$, $b = 53$, and $c = -0.5$ later.

$V'(t) = -\dfrac{a(bce^{ct})}{(1 + be^{ct})^2}$ [by the Reciprocal Rule] and

$$V''(t) = -abc \cdot \frac{\left(1 + be^{ct}\right)^2 \cdot ce^{ct} - e^{ct} \cdot 2\left(1 + be^{ct}\right) \cdot bce^{ct}}{\left[(1 + be^{ct})^2\right]^2}$$

$$= \frac{-abc \cdot ce^{ct}(1 + be^{ct})[(1 + be^{ct}) - 2be^{ct}]}{(1 + be^{ct})^4} = \frac{-abc^2e^{ct}(1 - be^{ct})}{(1 + be^{ct})^3}$$

So $V''(t) = 0 \;\Leftrightarrow\; 1 = be^{ct} \;\Leftrightarrow\; e^{ct} = 1/b \;\Leftrightarrow\; ct = \ln(1/b) \;\Leftrightarrow\; t = (1/c)\ln(1/b) = -2\ln\frac{1}{53} \approx 7.94$ years, which corresponds to roughly midyear 1988.

49. $f(x) = ax^3 + bx^2 + cx + d \;\Rightarrow\; f'(x) = 3ax^2 + 2bx + c$. We are given that $f(1) = 0$ and $f(-2) = 3$, so $f(1) = a + b + c + d = 0$ and $f(-2) = -8a + 4b - 2c + d = 3$. Also $f'(1) = 3a + 2b + c = 0$ and $f'(-2) = 12a - 4b + c = 0$ by Fermat's Theorem. Solving these four equations, we get $a = \frac{2}{9}$, $b = \frac{1}{3}$, $c = -\frac{4}{3}$, $d = \frac{7}{9}$, so the function is $f(x) = \frac{1}{9}\left(2x^3 + 3x^2 - 12x + 7\right)$.

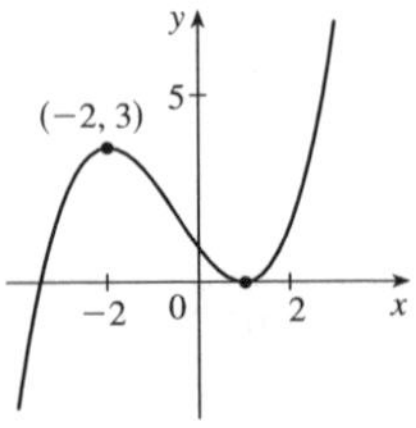

51. $f(x) = \tan x - x \;\Rightarrow\; f'(x) = \sec^2 x - 1 > 0$ for $0 < x < \frac{\pi}{2}$ since $\sec^2 x > 1$ for $0 < x < \frac{\pi}{2}$. So f is increasing on $\left(0, \frac{\pi}{2}\right)$. Thus, $f(x) > f(0) = 0$ for $0 < x < \frac{\pi}{2} \;\Rightarrow\; \tan x - x > 0 \;\Rightarrow\; \tan x > x$ for $0 < x < \frac{\pi}{2}$.

53. We are given that f is differentiable (and therefore continuous) everywhere. In particular, we can apply the Mean Value Theorem on the interval $[0, 4]$. There exists a number c in $(0, 4)$ such that $f(4) - f(0) = f'(c)(4 - 0)$, so $f(4) = f(0) + 4f'(c) = -3 + 4f'(c)$. We are given that $f'(x) \le 5$ for all x, so in particular we know that $f'(c) \le 5$. Multiplying both sides of this inequality by 4, we have $4f'(c) \le 20$, so $f(4) = -3 + 4f'(c) \le -3 + 20 = 17$. The largest possible value for $f(4)$ is 17.

55. Let $g(t)$ and $h(t)$ be the position functions of the two runners and let $f(t) = g(t) - h(t)$. By hypothesis, $f(0) = g(0) - h(0) = 0$ and $f(b) = g(b) - h(b) = 0$, where b is the finishing time. Then by the Mean Value Theorem, there is a time c, with $0 < c < b$, such that $f'(c) = \dfrac{f(b) - f(0)}{b - 0}$. But $f(b) = f(0) = 0$, so $f'(c) = 0$. Since $f'(c) = g'(c) - h'(c) = 0$, we have $g'(c) = h'(c)$. So at time c, both runners have the same speed $g'(c) = h'(c)$.

57. Let the cubic function be $f(x) = ax^3 + bx^2 + cx + d \;\Rightarrow\; f'(x) = 3ax^2 + 2bx + c \;\Rightarrow\; f''(x) = 6ax + 2b$. So f is CU when $6ax + 2b > 0 \;\Leftrightarrow\; x > -b/(3a)$, CD when $x < -b/(3a)$, and so the only point of inflection occurs when $x = -b/(3a)$. If the graph has three x-intercepts x_1, x_2 and x_3, then the expression for $f(x)$ must factor as $f(x) = a(x - x_1)(x - x_2)(x - x_3)$. Multiplying these factors together gives us $f(x) = a\left[x^3 - (x_1 + x_2 + x_3)x^2 + (x_1x_2 + x_1x_3 + x_2x_3)x - x_1x_2x_3\right]$. Equating the coefficients of the x^2-terms for the two forms of f gives us $b = -a(x_1 + x_2 + x_3)$. Hence, the x-coordinate of the point of inflection is

$$-\frac{b}{3a} = -\frac{-a(x_1 + x_2 + x_3)}{3a} = \frac{x_1 + x_2 + x_3}{3}.$$

4.4 Graphing with Calculus and Calculators

1. $f(x) = 4x^4 - 7x^2 + 4x + 6 \;\Rightarrow\; f'(x) = 16x^3 - 14x + 4 \;\Rightarrow\; f''(x) = 48x^2 - 14$

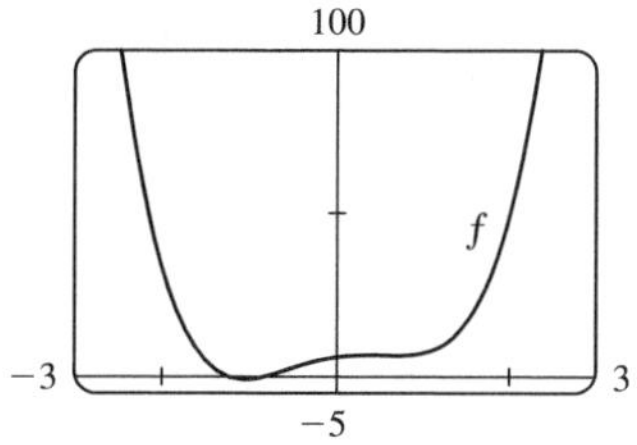

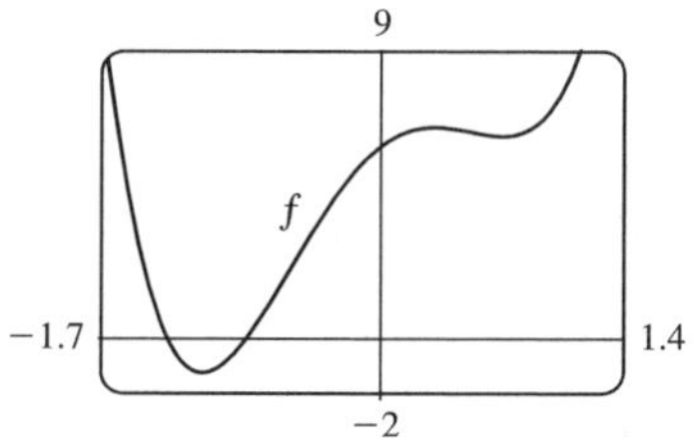

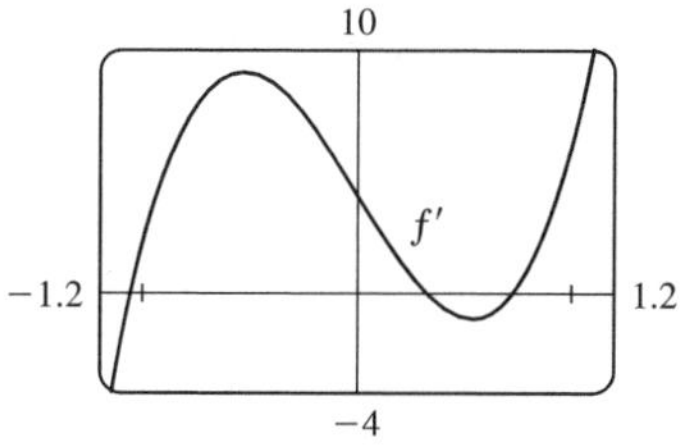

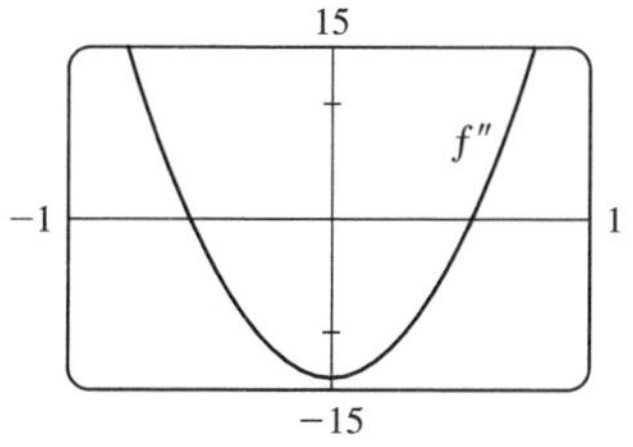

After finding suitable viewing rectangles (by ensuring that we have located all of the x-values where either $f' = 0$ or $f'' = 0$) we estimate from the graph of f' that f is increasing on $(-1.1, 0.3)$ and $(0.7, \infty)$ and decreasing on $(-\infty, -1.1)$ and $(0.3, 0.7)$, with a local maximum of $f(0.3) \approx 6.6$ and minima of $f(-1.1) \approx -1.1$ and $f(0.7) \approx 6.3$. We estimate from the graph of f'' that f is CU on $(-\infty, -0.5)$ and $(0.5, \infty)$ and CD on $(-0.5, 0.5)$, and that f has inflection points at about $(-0.5, 2.1)$ and $(0.5, 6.5)$.

3. $f(x) = \sqrt[3]{x^2 - 3x - 5} \Rightarrow f'(x) = \dfrac{1}{3}\dfrac{2x-3}{(x^2-3x-5)^{2/3}} \Rightarrow f''(x) = -\dfrac{2}{9}\dfrac{x^2-3x+24}{(x^2-3x-5)^{5/3}}$

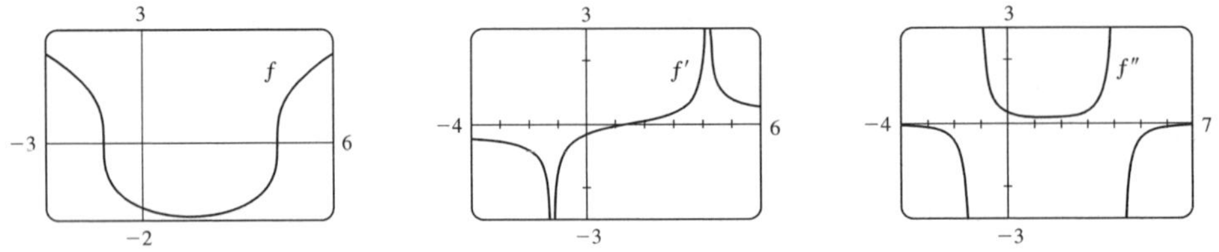

Note: With some CAS's, including Maple, it is necessary to define $f(x) = \dfrac{x^2-3x-5}{|x^2-3x-5|}\left|x^2-3x-5\right|^{1/3}$, since the CAS does not compute real cube roots of negative numbers. We estimate from the graph of f' that f is increasing on $(1.5, \infty)$, and decreasing on $(-\infty, 1.5)$. f has no maximum. Minimum value: $f(1.5) \approx -1.9$.
From the graph of f'', we estimate that f is CU on $(-1.2, 4.2)$ and CD on $(-\infty, -1.2)$ and $(4.2, \infty)$. IP at $(-1.2, 0)$ and $(4.2, 0)$.

5. $f(x) = \dfrac{x}{x^3 - x^2 - 4x + 1} \Rightarrow f'(x) = \dfrac{-2x^3 + x^2 + 1}{(x^3 - x^2 - 4x + 1)^2} \Rightarrow f''(x) = \dfrac{2(3x^5 - 3x^4 + 5x^3 - 6x^2 + 3x + 4)}{(x^3 - x^2 - 4x + 1)^3}$

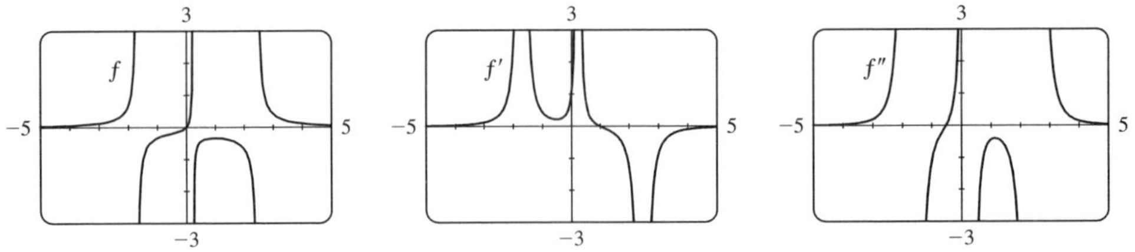

We estimate from the graph of f that $y = 0$ is a horizontal asymptote, and that there are vertical asymptotes at $x = -1.7$, $x = 0.24$, and $x = 2.46$. From the graph of f', we estimate that f is increasing on $(-\infty, -1.7)$, $(-1.7, 0.24)$, and $(0.24, 1)$, and that f is decreasing on $(1, 2.46)$ and $(2.46, \infty)$. There is a local maximum value at $f(1) = -\frac{1}{3}$. From the graph of f'', we estimate that f is CU on $(-\infty, -1.7)$, $(-0.506, 0.24)$, and $(2.46, \infty)$, and that f is CD on $(-1.7, -0.506)$ and $(0.24, 2.46)$. There is an inflection point at $(-0.506, -0.192)$.

7. $f(x) = x^2 - 4x + 7\cos x$, $-4 \le x \le 4$. $f'(x) = 2x - 4 - 7\sin x \Rightarrow f''(x) = 2 - 7\cos x$.
$f(x) = 0 \Leftrightarrow x \approx 1.10$; $f'(x) = 0 \Leftrightarrow x \approx -1.49, -1.07$, or 2.89; $f''(x) = 0 \Leftrightarrow x = \pm\cos^{-1}\left(\frac{2}{7}\right) \approx \pm 1.28$.

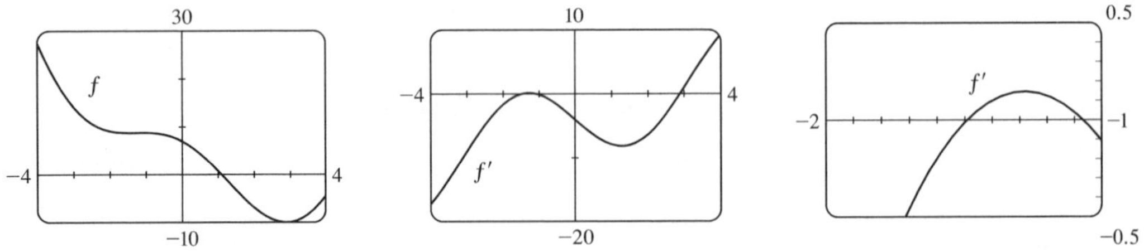

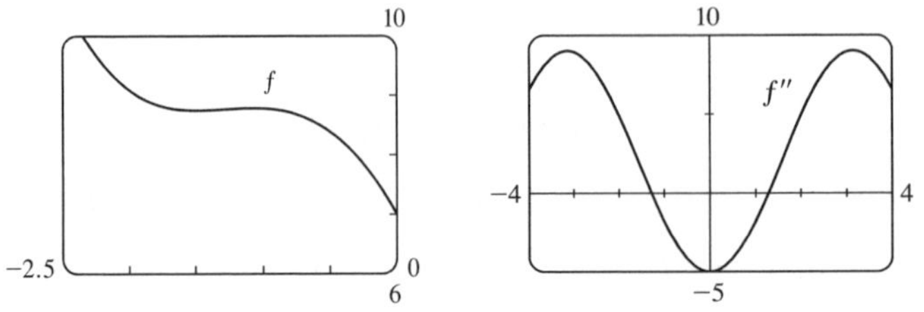

From the graphs of f', we estimate that f is decreasing ($f' < 0$) on $(-4, -1.49)$, increasing on $(-1.49, -1.07)$, decreasing on $(-1.07, 2.89)$, and increasing on $(2.89, 4)$, with local minimum values $f(-1.49) \approx 8.75$ and $f(2.89) \approx -9.99$ and local maximum value $f(-1.07) \approx 8.79$ (notice the second graph of f). From the graph of f'', we estimate that f is CU ($f'' > 0$) on $(-4, -1.28)$, CD on $(-1.28, 1.28)$, and CU on $(1.28, 4)$. There are inflection points at about $(-1.28, 8.77)$ and $(1.28, -1.48)$.

9. $f(x) = 1 + \dfrac{1}{x} + \dfrac{8}{x^2} + \dfrac{1}{x^3} \Rightarrow f'(x) = -\dfrac{1}{x^2} - \dfrac{16}{x^3} - \dfrac{3}{x^4} = -\dfrac{1}{x^4}(x^2 + 16x + 3) \Rightarrow$

$f''(x) = \dfrac{2}{x^3} + \dfrac{48}{x^4} + \dfrac{12}{x^5} = \dfrac{2}{x^5}(x^2 + 24x + 6)$.

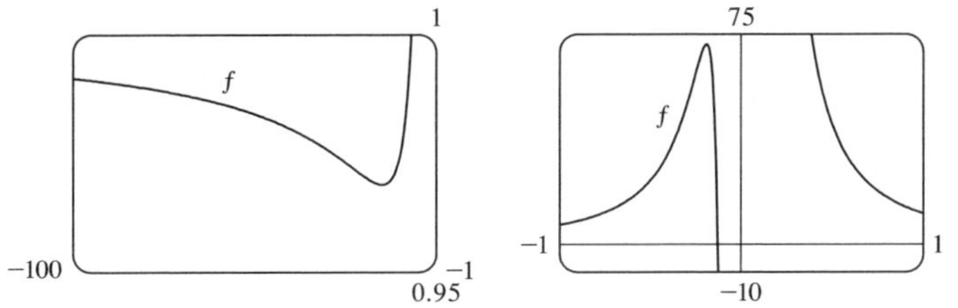

From the graphs, it appears that f increases on $(-15.8, -0.2)$ and decreases on $(-\infty, -15.8)$, $(-0.2, 0)$, and $(0, \infty)$; that f has a local minimum value of $f(-15.8) \approx 0.97$ and a local maximum value of $f(-0.2) \approx 72$; that f is CD on $(-\infty, -24)$ and $(-0.25, 0)$ and is CU on $(-24, -0.25)$ and $(0, \infty)$; and that f has IPs at $(-24, 0.97)$ and $(-0.25, 60)$.

To find the exact values, note that $f' = 0 \Rightarrow x = \dfrac{-16 \pm \sqrt{256 - 12}}{2} = -8 \pm \sqrt{61}$ [≈ -0.19 and -15.81]. f' is positive (f is increasing) on $\left(-8 - \sqrt{61}, -8 + \sqrt{61}\right)$ and f' is negative (f is decreasing) on $\left(-\infty, -8 - \sqrt{61}\right)$, $\left(-8 + \sqrt{61}, 0\right)$, and $(0, \infty)$. $f'' = 0 \Rightarrow x = \dfrac{-24 \pm \sqrt{576 - 24}}{2} = -12 \pm \sqrt{138}$ [≈ -0.25 and -23.75]. f'' is positive (f is CU) on $\left(-12 - \sqrt{138}, -12 + \sqrt{138}\right)$ and $(0, \infty)$ and f'' is negative (f is CD) on $\left(-\infty, -12 - \sqrt{138}\right)$ and $\left(-12 + \sqrt{138}, 0\right)$.

11.

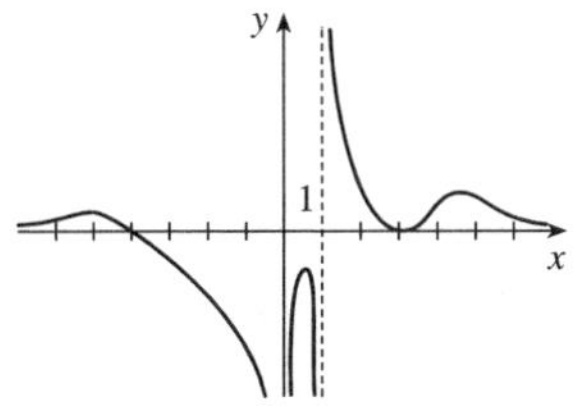

$f(x) = \dfrac{(x+4)(x-3)^2}{x^4(x-1)}$ has VA at $x = 0$ and at $x = 1$ since $\lim\limits_{x \to 0} f(x) = -\infty$, $\lim\limits_{x \to 1^-} f(x) = -\infty$ and $\lim\limits_{x \to 1^+} f(x) = \infty$.

$$f(x) = \frac{\dfrac{x+4}{x} \cdot \dfrac{(x-3)^2}{x^2}}{\dfrac{x^4}{x^3} \cdot (x-1)} \quad \text{[dividing numerator and denominator by } x^3\text{]}$$

$$= \frac{(1 + 4/x)(1 - 3/x)^2}{x(x-1)} \to 0 \text{ as } x \to \pm\infty, \text{ so } f \text{ is asymptotic to the } x\text{-axis.}$$

Since f is undefined at $x = 0$, it has no y-intercept. $f(x) = 0 \Rightarrow (x+4)(x-3)^2 = 0 \Rightarrow x = -4$ or $x = 3$, so f has x-intercepts -4 and 3. Note, however, that the graph of f is only tangent to the x-axis and does not cross it at $x = 3$, since f is positive as $x \to 3^-$ and as $x \to 3^+$.

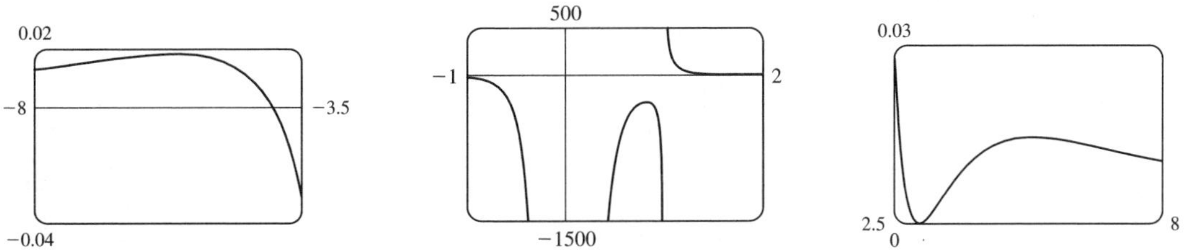

From these graphs, it appears that f has three maximum values and one minimum value. The maximum values are approximately $f(-5.6) = 0.0182$, $f(0.82) = -281.5$ and $f(5.2) = 0.0145$ and we know (since the graph is tangent to the x-axis at $x = 3$) that the minimum value is $f(3) = 0$.

13. $f(x) = \dfrac{x^2(x+1)^3}{(x-2)^2(x-4)^4} \quad \Rightarrow \quad f'(x) = -\dfrac{x(x+1)^2(x^3+18x^2-44x-16)}{(x-2)^3(x-4)^5}$ (from CAS).

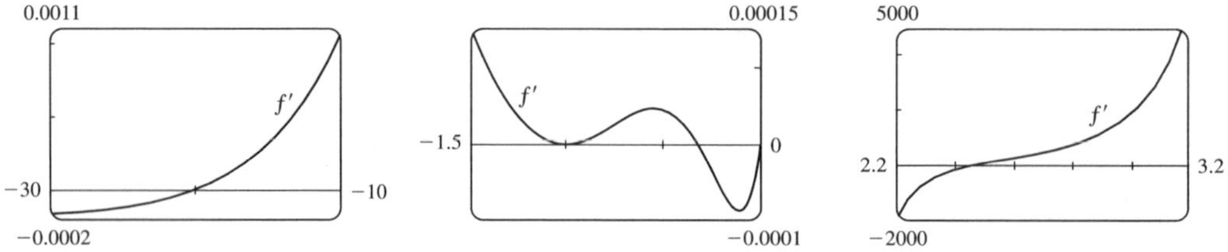

From the graphs of f', it seems that the critical points which indicate extrema occur at $x \approx -20$, -0.3, and 2.5, as estimated in Example 3. (There is another critical point at $x = -1$, but the sign of f' does not change there.) We differentiate again, obtaining $f''(x) = 2\dfrac{(x+1)(x^6+36x^5+6x^4-628x^3+684x^2+672x+64)}{(x-2)^4(x-4)^6}$.

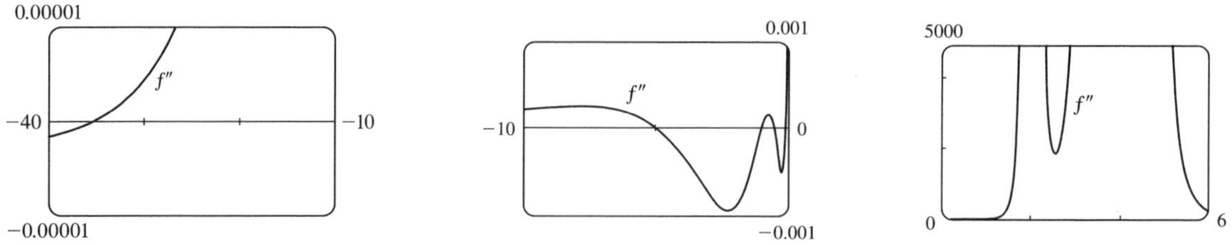

From the graphs of f'', it appears that f is CU on $(-35.3, -5.0)$, $(-1, -0.5)$, $(-0.1, 2)$, $(2, 4)$ and $(4, \infty)$ and CD on $(-\infty, -35.3)$, $(-5.0, -1)$ and $(-0.5, -0.1)$. We check back on the graphs of f to find the y-coordinates of the inflection points, and find that these points are approximately $(-35.3, -0.015)$, $(-5.0, -0.005)$, $(-1, 0)$, $(-0.5, 0.00001)$, and $(-0.1, 0.000\,006\,6)$.

15. $y = f(x) = \dfrac{\sin^2 x}{\sqrt{x^2+1}}$ with $0 \le x \le 3\pi$. From a CAS, $y' = \dfrac{\sin x\left[2(x^2+1)\cos x - x\sin x\right]}{(x^2+1)^{3/2}}$ and

$$y'' = \frac{(4x^4+6x^2+5)\cos^2 x - 4x(x^2+1)\sin x\cos x - 2x^4 - 2x^2 - 3}{(x^2+1)^{5/2}}.$$

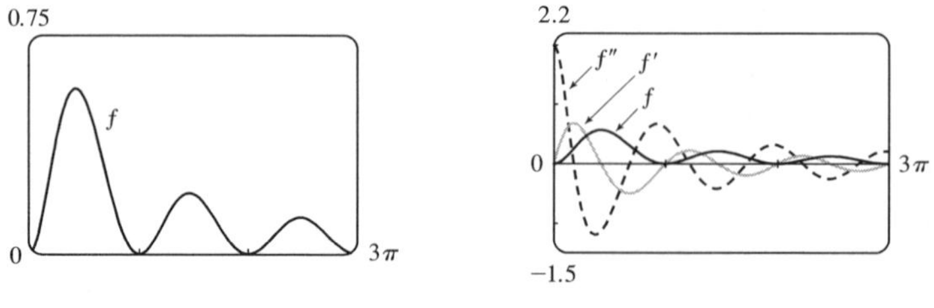

From the graph of f' and the formula for y', we determine that $y' = 0$ when $x = \pi$, 2π, 3π, or $x \approx 1.3$, 4.6, or 7.8. So f is increasing on $(0, 1.3)$, $(\pi, 4.6)$, and $(2\pi, 7.8)$. f is decreasing on $(1.3, \pi)$, $(4.6, 2\pi)$, and $(7.8, 3\pi)$. Local maximum values: $f(1.3) \approx 0.6$, $f(4.6) \approx 0.21$, and $f(7.8) \approx 0.13$. Local minimum values: $f(\pi) = f(2\pi) = 0$. From the graph of f'', we see that $y'' = 0 \quad \Leftrightarrow \quad x \approx 0.6, 2.1, 3.8, 5.4, 7.0$, or 8.6. So f is CU on $(0, 0.6)$, $(2.1, 3.8)$, $(5.4, 7.0)$, and $(8.6, 3\pi)$. f is CD on $(0.6, 2.1)$, $(3.8, 5.4)$, and $(7.0, 8.6)$. There are IP at $(0.6, 0.25)$, $(2.1, 0.31)$, $(3.8, 0.10)$, $(5.4, 0.11)$, $(7.0, 0.061)$, and $(8.6, 0.065)$.

17. $y = f(x) = \dfrac{1 - e^{1/x}}{1 + e^{1/x}}$. From a CAS, $y' = \dfrac{2e^{1/x}}{x^2(1 + e^{1/x})^2}$ and $y'' = \dfrac{-2e^{1/x}(1 - e^{1/x} + 2x + 2xe^{1/x})}{x^4(1 + e^{1/x})^3}$.

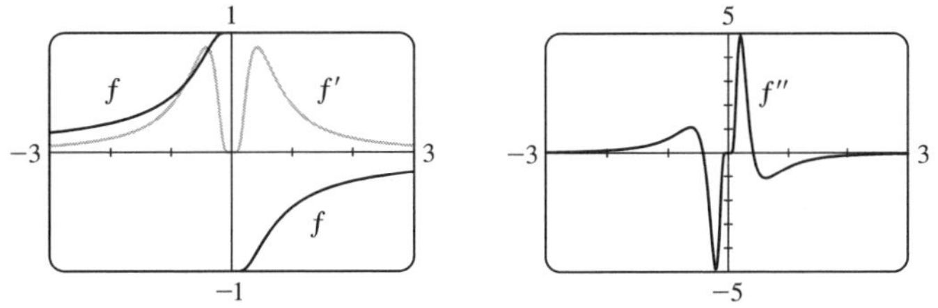

f is an odd function defined on $(-\infty, 0) \cup (0, \infty)$. Its graph has no x- or y-intercepts. Since $\lim\limits_{x\to\pm\infty} f(x) = 0$, the x-axis is a HA. $f'(x) > 0$ for $x \neq 0$, so f is increasing on $(-\infty, 0)$ and $(0, \infty)$. It has no local extreme values. $f''(x) = 0$ for $x \approx \pm 0.417$, so f is CU on $(-\infty, -0.417)$, CD on $(-0.417, 0)$, CU on $(0, 0.417)$, and CD on $(0.417, \infty)$. f has IPs at $(-0.417, 0.834)$ and $(0.417, -0.834)$.

19.

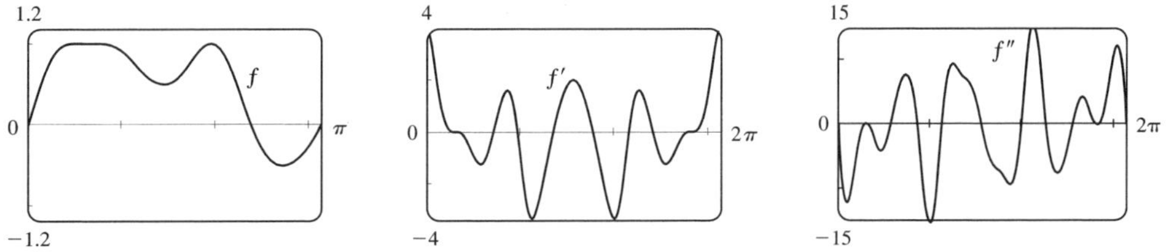

From the graph of $f(x) = \sin(x + \sin 3x)$ in the viewing rectangle $[0, \pi]$ by $[-1.2, 1.2]$, it looks like f has two maxima and two minima. If we calculate and graph $f'(x) = [\cos(x + \sin 3x)](1 + 3\cos 3x)$ on $[0, 2\pi]$, we see that the graph of f' appears to be almost tangent to the x-axis at about $x = 0.7$. The graph of $f'' = -[\sin(x + \sin 3x)](1 + 3\cos 3x)^2 + \cos(x + \sin 3x)(-9\sin 3x)$ is even more interesting near this x-value: it seems to just touch the x-axis.

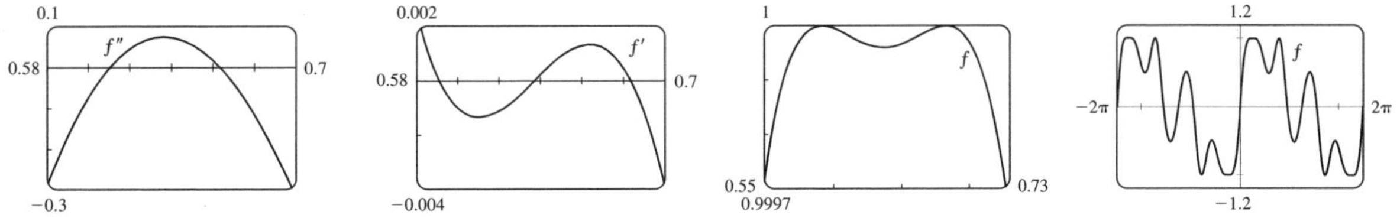

If we zoom in on this place on the graph of f'', we see that f'' actually does cross the axis twice near $x = 0.65$, indicating a change in concavity for a very short interval. If we look at the graph of f' on the same interval, we see that it changes sign three times near $x = 0.65$, indicating that what we had thought was a broad extremum at about $x = 0.7$ actually consists of three extrema (two maxima and a minimum). These maximum values are roughly $f(0.59) = 1$ and $f(0.68) = 1$, and the minimum value is roughly $f(0.64) = 0.99996$. There are also a maximum value of about $f(1.96) = 1$ and minimum values of about $f(1.46) = 0.49$ and $f(2.73) = -0.51$. The points of inflection on $(0, \pi)$ are about $(0.61, 0.99998)$, $(0.66, 0.99998)$, $(1.17, 0.72)$, $(1.75, 0.77)$, and $(2.28, 0.34)$. On $(\pi, 2\pi)$, they are about $(4.01, -0.34)$, $(4.54, -0.77)$, $(5.11, -0.72)$, $(5.62, -0.99998)$, and $(5.67, -0.99998)$. There are also IP at $(0, 0)$ and $(\pi, 0)$. Note that the function is odd and periodic with period 2π, and it is also rotationally symmetric about all points of the form $((2n + 1)\pi, 0)$, n an integer.

21.

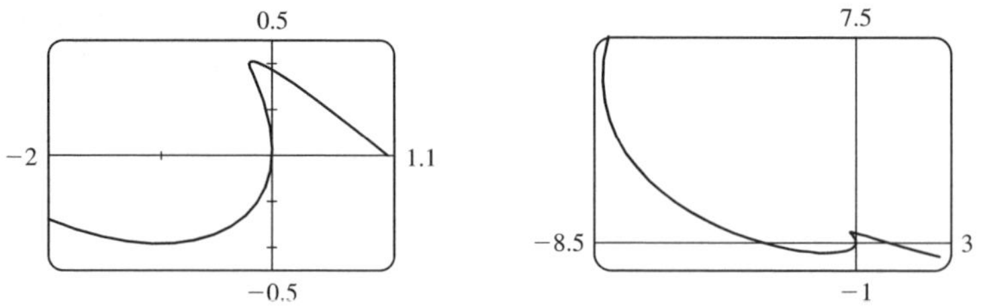

We graph the curve $x = t^4 - 2t^3 - 2t^2$, $y = t^3 - t$ in the viewing rectangle $[-2, 1.1]$ by $[-0.5, 0.5]$. This rectangle corresponds approximately to $t \in [-1, 0.8]$.We estimate that the curve has horizontal tangents at about $(-1, -0.4)$ and $(-0.17, 0.39)$ and vertical tangents at about $(0, 0)$ and $(-0.19, 0.37)$. We calculate $\dfrac{dy}{dx} = \dfrac{dy/dt}{dx/dt} = \dfrac{3t^2 - 1}{4t^3 - 6t^2 - 4t}$. The horizontal tangents occur when $dy/dt = 3t^2 - 1 = 0 \quad \Leftrightarrow \quad t = \pm\frac{1}{\sqrt{3}}$, so both horizontal tangents are shown in our graph. $t = \frac{1}{\sqrt{3}}$ corresponds to the point $\left(\frac{-2\sqrt{3}-5}{9}, \frac{-2\sqrt{3}}{9}\right) \approx (-0.94, -0.38)$ and $t = -\frac{1}{\sqrt{3}}$ corresponds to $\left(\frac{2\sqrt{3}-5}{9}, \frac{2\sqrt{3}}{9}\right) \approx (-0.17, 0.38)$. The vertical tangents occur when $dx/dt = 2t(2t^2 - 3t - 2) = 0 \quad \Leftrightarrow$ $2t(2t + 1)(t - 2) = 0 \quad \Leftrightarrow \quad t = 0, -\frac{1}{2}$ or 2. It seems that we have missed one vertical tangent, and indeed if we plot the curve on the t-interval $[-1.2, 2.2]$ we see that there is another vertical tangent at $(-8, 6)$. The t-values and points at which there are vertical tangents are $t = 0$, $(0, 0)$; $t = -\frac{1}{2}$, $\left(-\frac{3}{16}, \frac{3}{8}\right)$; and $t = 2$, $(-8, 6)$.

23. $x = t^3 - ct$, $y = t^2$. For $c = 0$, there is a cusp at $(0, 0)$. For $c < 0$, there is a local minimum at $(0, 0)$. For $c > 0$, there is a loop whose size increases as c increases ($c = \frac{1}{2}$ and $c = 1$ are shown in the figure). The curve intersects itself on the y-axis; that is, when $x = 0 \quad \Leftrightarrow \quad t^3 - ct = 0 \quad \Leftrightarrow \quad t(t^2 - c) = 0 \quad \Leftrightarrow \quad t = 0, \pm\sqrt{c}$. Substituting $\pm\sqrt{c}$ for t gives us $y = c$, so the point of intersection is $(0, c)$.

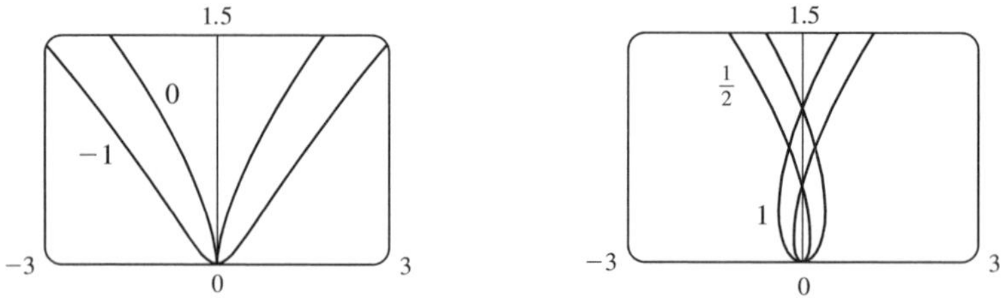

From the second figure, we see that the left- and rightmost points of the loop occur when there are vertical tangent lines. $dx/dt = 0 \quad \Rightarrow \quad 3t^2 - c = 0 \quad \Rightarrow \quad t = \pm\sqrt{c/3}$. The rightmost point occurs when $t = -\sqrt{c/3}$ and has coordinates $\left(\dfrac{2c\sqrt{3c}}{9}, \dfrac{c}{3}\right)$. The leftmost point occurs when $t = \sqrt{c/3}$ and has coordinates $\left(-\dfrac{2c\sqrt{3c}}{9}, \dfrac{c}{3}\right)$.

25. $f(x) = x^4 + cx^2 = x^2(x^2 + c)$. Note that f is an even function. For $c \geq 0$, the only x-intercept is the point $(0, 0)$. We calculate $f'(x) = 4x^3 + 2cx = 4x(x^2 + \frac{1}{2}c) \quad \Rightarrow \quad f''(x) = 12x^2 + 2c$. If $c \geq 0$, $x = 0$ is the only critical point and there is no inflection point. As we can see from the examples, there is no change in the basic shape of the graph for $c \geq 0$; it merely becomes steeper as c increases. For $c = 0$, the graph is the simple curve $y = x^4$. For $c < 0$, there are x-intercepts at 0 and at $\pm\sqrt{-c}$. Also, there is a maximum at $(0, 0)$, and there are minima at $\left(\pm\sqrt{-\frac{1}{2}c}, -\frac{1}{4}c^2\right)$. As $c \to -\infty$, the x-coordinates of these minima get larger in absolute value, and the minimum points move downward. There are inflection points at $\left(\pm\sqrt{-\frac{1}{6}c}, -\frac{5}{36}c^2\right)$, which also move away from the origin as $c \to -\infty$.

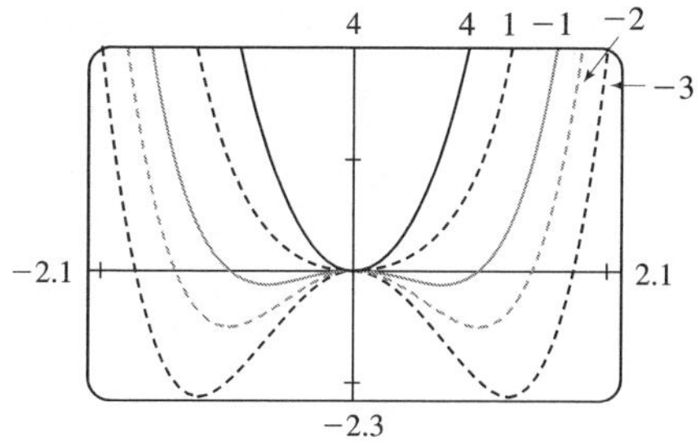

27.

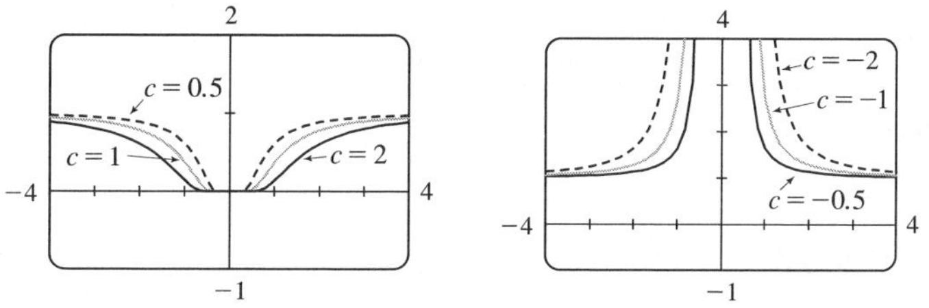

$c = 0$ is a transitional value—we get the graph of $y = 1$. For $c > 0$, we see that there is a HA at $y = 1$, and that the graph spreads out as c increases. At first glance there appears to be a minimum at $(0, 0)$, but $f(0)$ is undefined, so there is no minimum or maximum. For $c < 0$, we still have the HA at $y = 1$, but the range is $(1, \infty)$ rather than $(0, 1)$. We also have a VA at $x = 0$. $f(x) = e^{-c/x^2} \;\Rightarrow\; f'(x) = e^{-c/x^2}\left(\dfrac{2c}{x^3}\right) \;\Rightarrow\; f''(x) = \dfrac{2c(2c - 3x^2)}{x^6 e^{c/x^2}}$. $f'(x) \neq 0$ and $f'(x)$ exists for all $x \neq 0$ (and 0 is not in the domain of f), so there are no maxima or minima. $f''(x) = 0 \;\Rightarrow\; x = \pm\sqrt{2c/3}$, so if $c > 0$, the inflection points spread out as c increases, and if $c < 0$, there are no IP. For $c > 0$, there are IP at $\left(\pm\sqrt{2c/3}, e^{-3/2}\right)$. Note that the y-coordinate of the IP is constant.

29. Note that $c = 0$ is a transitional value at which the graph consists of the x-axis. Also, we can see that if we substitute $-c$ for c, the function $f(x) = \dfrac{cx}{1 + c^2x^2}$ will be reflected in the x-axis, so we investigate only positive values of c (except $c = -1$, as a demonstration of this reflective property). Also, f is an odd function. $\lim\limits_{x \to \pm\infty} f(x) = 0$, so $y = 0$ is a horizontal asymptote for all c. We calculate $f'(x) = \dfrac{(1 + c^2x^2)c - cx(2c^2x)}{(1 + c^2x^2)^2} = -\dfrac{c(c^2x^2 - 1)}{(1 + c^2x^2)^2}$. $f'(x) = 0 \;\Leftrightarrow\; c^2x^2 - 1 = 0 \;\Leftrightarrow\; x = \pm 1/c$.

So there is an absolute maximum value of $f(1/c) = \frac{1}{2}$ and an absolute minimum value of $f(-1/c) = -\frac{1}{2}$. These extrema have the same value regardless of c, but the maximum points move closer to the y-axis as c increases.

$$\begin{aligned} f''(x) &= \frac{(-2c^3x)(1 + c^2x^2)^2 - (-c^3x^2 + c)[2(1 + c^2x^2)(2c^2x)]}{(1 + c^2x^2)^4} \\ &= \frac{(-2c^3x)(1 + c^2x^2) + (c^3x^2 - c)(4c^2x)}{(1 + c^2x^2)^3} = \frac{2c^3x(c^2x^2 - 3)}{(1 + c^2x^2)^3} \end{aligned}$$

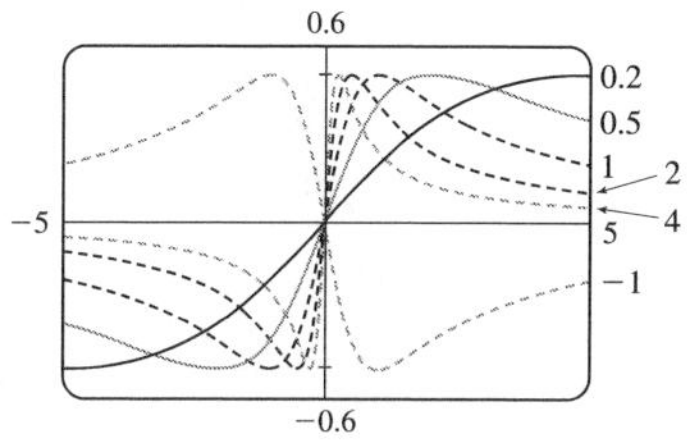

$f''(x) = 0 \;\Leftrightarrow\; x = 0$ or $\pm\sqrt{3}/c$, so there are inflection points at $(0, 0)$ and at $\left(\pm\sqrt{3}/c, \pm\sqrt{3}/4\right)$.

Again, the y-coordinate of the inflection points does not depend on c, but as c increases, both inflection points approach the y-axis.

31. $f(x) = cx + \sin x \quad \Rightarrow \quad f'(x) = c + \cos x \quad \Rightarrow \quad f''(x) = -\sin x$

$f(-x) = -f(x)$, so f is an odd function and its graph is symmetric with respect to the origin.

$f(x) = 0 \quad \Leftrightarrow \quad \sin x = -cx$, so 0 is always an x-intercept.

$f'(x) = 0 \quad \Leftrightarrow \quad \cos x = -c$, so there is no critical number when $|c| > 1$. If $|c| \le 1$, then there are infinitely many critical numbers. If x_1 is the unique solution of $\cos x = -c$ in the interval $[0, \pi]$, then the critical numbers are $2n\pi \pm x_1$, where n ranges over the integers. (Special cases: When $c = 1$, $x_1 = 0$; when $c = 0$, $x = \frac{\pi}{2}$; and when $c = -1$, $x_1 = \pi$.)

$f''(x) < 0 \quad \Leftrightarrow \quad \sin x > 0$, so f is CD on intervals of the form $(2n\pi, (2n+1)\pi)$. f is CU on intervals of the form $((2n-1)\pi, 2n\pi)$. The inflection points of f are the points $(2n\pi, 2n\pi c)$, where n is an integer.

If $c \ge 1$, then $f'(x) \ge 0$ for all x, so f is increasing and has no extremum. If $c \le -1$, then $f'(x) \le 0$ for all x, so f is decreasing and has no extremum. If $|c| < 1$, then $f'(x) > 0 \quad \Leftrightarrow \quad \cos x > -c \quad \Leftrightarrow \quad x$ is in an interval of the form $(2n\pi - x_1, 2n\pi + x_1)$ for some integer n. These are the intervals on which f is increasing. Similarly, we find that f is decreasing on the intervals of the form $(2n\pi + x_1, 2(n+1)\pi - x_1)$. Thus, f has local maxima at the points $2n\pi + x_1$, where f has the values $c(2n\pi + x_1) + \sin x_1 = c(2n\pi + x_1) + \sqrt{1 - c^2}$, and f has local minima at the points $2n\pi - x_1$, where we have $f(2n\pi - x_1) = c(2n\pi - x_1) - \sin x_1 = c(2n\pi - x_1) - \sqrt{1 - c^2}$.

The transitional values of c are -1 and 1. The inflection points move vertically, but not horizontally, when c changes. When $|c| \ge 1$, there is no extremum. For $|c| < 1$, the maxima are spaced 2π apart horizontally, as are the minima. The horizontal spacing between maxima and adjacent minima is regular (and equals π) when $c = 0$, but the horizontal space between a local maximum and the nearest local minimum shrinks as $|c|$ approaches 1.

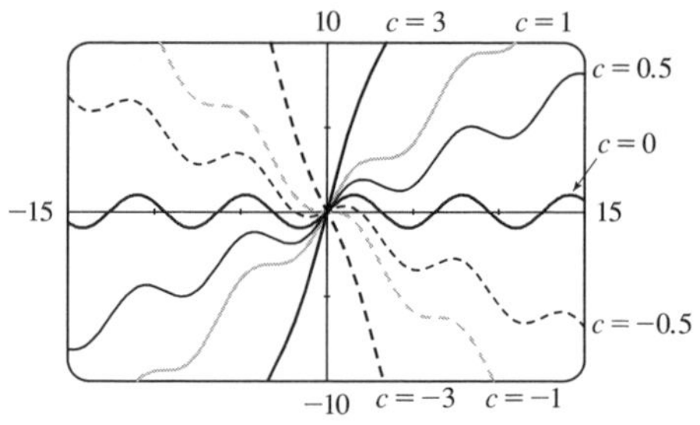

33. (a) $f(x) = cx^4 - 2x^2 + 1$. For $c = 0$, $f(x) = -2x^2 + 1$, a parabola whose vertex, $(0, 1)$, is the absolute maximum. For $c > 0$, $f(x) = cx^4 - 2x^2 + 1$ opens upward with two minimum points. As $c \to 0$, the minimum points spread apart and move downward; they are below the x-axis for $0 < c < 1$ and above for $c > 1$. For $c < 0$, the graph opens downward, and has an absolute maximum at $x = 0$ and no local minimum.

(b) $f'(x) = 4cx^3 - 4x = 4cx(x^2 - 1/c)$ $(c \ne 0)$. If $c \le 0$, 0 is the only critical number. $f''(x) = 12cx^2 - 4$, so $f''(0) = -4$ and there is a local maximum at $(0, f(0)) = (0, 1)$, which lies on $y = 1 - x^2$. If $c > 0$, the critical numbers are 0 and $\pm 1/\sqrt{c}$. As before, there is a local maximum at $(0, f(0)) = (0, 1)$, which lies on $y = 1 - x^2$. $f''(\pm 1/\sqrt{c}) = 12 - 4 = 8 > 0$, so there is a local minimum at $x = \pm 1/\sqrt{c}$. Here $f(\pm 1/\sqrt{c}) = c(1/c^2) - 2/c + 1 = -1/c + 1$. But $(\pm 1/\sqrt{c}, -1/c + 1)$ lies on $y = 1 - x^2$ since $1 - (\pm 1/\sqrt{c})^2 = 1 - 1/c$.

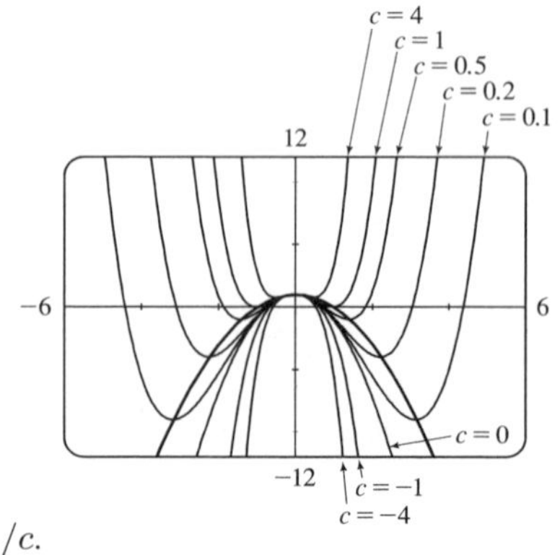

4.5 Indeterminate Forms and l'Hospital's Rule

Note: The use of l'Hospital's Rule is indicated by an H above the equal sign: $\overset{\text{H}}{=}$

1. (a) $\lim\limits_{x \to a} \dfrac{f(x)}{g(x)}$ is an indeterminate form of type $\dfrac{0}{0}$.

(b) $\lim\limits_{x \to a} \dfrac{f(x)}{p(x)} = 0$ because the numerator approaches 0 while the denominator becomes large.

(c) $\lim\limits_{x \to a} \dfrac{h(x)}{p(x)} = 0$ because the numerator approaches a finite number while the denominator becomes large.

(d) If $\lim\limits_{x \to a} p(x) = \infty$ and $f(x) \to 0$ through positive values, then $\lim\limits_{x \to a} \dfrac{p(x)}{f(x)} = \infty$. [For example, take $a = 0$, $p(x) = 1/x^2$, and $f(x) = x^2$.] If $f(x) \to 0$ through negative values, then $\lim\limits_{x \to a} \dfrac{p(x)}{f(x)} = -\infty$. [For example, take $a = 0$, $p(x) = 1/x^2$, and $f(x) = -x^2$.] If $f(x) \to 0$ through both positive and negative values, then the limit might not exist. [For example, take $a = 0$, $p(x) = 1/x^2$, and $f(x) = x$.]

(e) $\lim\limits_{x \to a} \dfrac{p(x)}{q(x)}$ is an indeterminate form of type $\dfrac{\infty}{\infty}$.

3. (a) When x is near a, $f(x)$ is near 0 and $p(x)$ is large, so $f(x) - p(x)$ is large negative. Thus, $\lim\limits_{x \to a} [f(x) - p(x)] = -\infty$.

(b) $\lim\limits_{x \to a} [p(x) - q(x)]$ is an indeterminate form of type $\infty - \infty$.

(c) When x is near a, $p(x)$ and $q(x)$ are both large, so $p(x) + q(x)$ is large. Thus, $\lim\limits_{x \to a} [p(x) + q(x)] = \infty$.

5. This limit has the form $\frac{0}{0}$. We can simply factor the numerator to evaluate this limit.

$$\lim_{x \to -1} \frac{x^2 - 1}{x + 1} = \lim_{x \to -1} \frac{(x + 1)(x - 1)}{x + 1} = \lim_{x \to -1} (x - 1) = -2$$

7. This limit has the form $\frac{0}{0}$. $\displaystyle \lim_{x \to (\pi/2)^+} \frac{\cos x}{1 - \sin x} \overset{\text{H}}{=} \lim_{x \to (\pi/2)^+} \frac{-\sin x}{-\cos x} = \lim_{x \to (\pi/2)^+} \tan x = -\infty$.

9. $\displaystyle \lim_{x \to 0} \frac{e^x - 1}{\sin x}$ is an indeterminate form of type $\frac{0}{0}$, so we'll apply l'Hospital's Rule. $\displaystyle \lim_{x \to 0} \frac{e^x - 1}{\sin x} \overset{\text{H}}{=} \lim_{x \to 0} \frac{e^x}{\cos x} = \frac{1}{1} = 1$

11. This limit has the form $\frac{0}{0}$. $\displaystyle \lim_{x \to 0} \frac{\tan px}{\tan qx} \overset{\text{H}}{=} \lim_{x \to 0} \frac{p \sec^2 px}{q \sec^2 qx} = \frac{p(1)^2}{q(1)^2} = \frac{p}{q}$

13. $\lim\limits_{x \to 0^+} [(\ln x)/x] = -\infty$ since $\ln x \to -\infty$ as $x \to 0^+$ and dividing by small values of x just increases the magnitude of the quotient $(\ln x)/x$. L'Hospital's Rule does not apply.

15. This limit has the form $\frac{0}{0}$. $\displaystyle \lim_{t \to 0} \frac{5^t - 3^t}{t} \overset{\text{H}}{=} \lim_{t \to 0} \frac{5^t \ln 5 - 3^t \ln 3}{1} = \ln 5 - \ln 3 = \ln \tfrac{5}{3}$

17. This limit has the form $\frac{0}{0}$. $\displaystyle \lim_{x \to 0} \frac{e^x - 1 - x}{x^2} \overset{\text{H}}{=} \lim_{x \to 0} \frac{e^x - 1}{2x} \overset{\text{H}}{=} \lim_{x \to 0} \frac{e^x}{2} = \frac{1}{2}$

19. This limit has the form $\frac{\infty}{\infty}$. $\displaystyle \lim_{x \to \infty} \frac{x}{\ln(1 + 2e^x)} \overset{\text{H}}{=} \lim_{x \to \infty} \frac{1}{\dfrac{1}{1 + 2e^x} \cdot 2e^x} = \lim_{x \to \infty} \frac{1 + 2e^x}{2e^x} \overset{\text{H}}{=} \lim_{x \to \infty} \frac{2e^x}{2e^x} = 1$

21. This limit has the form $\frac{0}{0}$. $\displaystyle \lim_{x \to 1} \frac{1 - x + \ln x}{1 + \cos \pi x} \overset{\text{H}}{=} \lim_{x \to 1} \frac{-1 + 1/x}{-\pi \sin \pi x} \overset{\text{H}}{=} \lim_{x \to 1} \frac{-1/x^2}{-\pi^2 \cos \pi x} = \frac{-1}{-\pi^2 (-1)} = -\frac{1}{\pi^2}$

23. This limit has the form $\frac{0}{0}$. $\displaystyle\lim_{x\to1}\frac{x^a-ax+a-1}{(x-1)^2}\overset{\text{H}}{=}\lim_{x\to1}\frac{ax^{a-1}-a}{2(x-1)}\overset{\text{H}}{=}\lim_{x\to1}\frac{a(a-1)x^{a-2}}{2}=\frac{a(a-1)}{2}$

25. This limit has the form $0\cdot(-\infty)$. We need to write this product as a quotient, but keep in mind that we will have to differentiate both the numerator and the denominator. If we differentiate $\dfrac{1}{\ln x}$, we get a complicated expression that results in a more difficult limit. Instead we write the quotient as $\dfrac{\ln x}{x^{-1/2}}$.

$$\lim_{x\to0^+}\sqrt{x}\ln x=\lim_{x\to0^+}\frac{\ln x}{x^{-1/2}}\overset{\text{H}}{=}\lim_{x\to0^+}\frac{1/x}{-\frac{1}{2}x^{-3/2}}\cdot\frac{-2x^{3/2}}{-2x^{3/2}}=\lim_{x\to0^+}(-2\sqrt{x})=0$$

27. $\displaystyle\lim_{x\to\infty}e^{-x}\ln x=\lim_{x\to\infty}\frac{\ln x}{e^x}\overset{\text{H}}{=}\lim_{x\to\infty}\frac{1/x}{e^x}=\lim_{x\to\infty}\frac{1}{xe^x}=0$

29. This limit has the form $\infty\cdot0$. $\displaystyle\lim_{x\to\infty}x^3e^{-x^2}=\lim_{x\to\infty}\frac{x^3}{e^{x^2}}\overset{\text{H}}{=}\lim_{x\to\infty}\frac{3x^2}{2xe^{x^2}}=\lim_{x\to\infty}\frac{3x}{2e^{x^2}}\overset{\text{H}}{=}\lim_{x\to\infty}\frac{3}{4xe^{x^2}}=0$

31. As $x\to\infty$, $1/x\to0$, and $e^{1/x}\to1$. So the limit has the form $\infty-\infty$ and we will change the form to a product by factoring out x.

$$\lim_{x\to\infty}\left(xe^{1/x}-x\right)=\lim_{x\to\infty}x\left(e^{1/x}-1\right)=\lim_{x\to\infty}\frac{e^{1/x}-1}{1/x}\overset{\text{H}}{=}\lim_{x\to\infty}\frac{e^{1/x}(-1/x^2)}{-1/x^2}=\lim_{x\to\infty}e^{1/x}=e^0=1$$

33. The limit has the form $\infty-\infty$ and we will change the form to a product by factoring out x.

$$\lim_{x\to\infty}(x-\ln x)=\lim_{x\to\infty}x\left(1-\frac{\ln x}{x}\right)=\infty\text{ since }\lim_{x\to\infty}\frac{\ln x}{x}\overset{\text{H}}{=}\lim_{x\to\infty}\frac{1/x}{1}=0.$$

35. $y=x^{x^2}\quad\Rightarrow\quad\ln y=x^2\ln x$, so $\displaystyle\lim_{x\to0^+}\ln y=\lim_{x\to0^+}x^2\ln x=\lim_{x\to0^+}\frac{\ln x}{1/x^2}\overset{\text{H}}{=}\lim_{x\to0^+}\frac{1/x}{-2/x^3}=\lim_{x\to0^+}\left(-\frac{1}{2}x^2\right)=0\quad\Rightarrow$

$\displaystyle\lim_{x\to0^+}x^{x^2}=\lim_{x\to0^+}e^{\ln y}=e^0=1.$

37. $y=(1-2x)^{1/x}\quad\Rightarrow\quad\ln y=\dfrac{1}{x}\ln(1-2x)$, so $\displaystyle\lim_{x\to0}\ln y=\lim_{x\to0}\frac{\ln(1-2x)}{x}\overset{\text{H}}{=}\lim_{x\to0}\frac{-2/(1-2x)}{1}=-2\quad\Rightarrow$

$\displaystyle\lim_{x\to0}(1-2x)^{1/x}=\lim_{x\to0}e^{\ln y}=e^{-2}.$

39. $y=(\cos x)^{1/x^2}\quad\Rightarrow\quad\ln y=\dfrac{1}{x^2}\ln\cos x\quad\Rightarrow$

$$\lim_{x\to0^+}\ln y=\lim_{x\to0^+}\frac{\ln\cos x}{x^2}\overset{\text{H}}{=}\lim_{x\to0^+}\frac{-\tan x}{2x}\overset{\text{H}}{=}\lim_{x\to0^+}\frac{-\sec^2x}{2}=-\frac{1}{2}\quad\Rightarrow$$

$$\lim_{x\to0^+}(\cos x)^{1/x^2}=\lim_{x\to0^+}e^{\ln y}=e^{-1/2}=1/\sqrt{e}$$

41. From the graph, it appears that $\displaystyle\lim_{x\to\infty}x\,[\ln(x+5)-\ln x]=5$. To prove this, we first note that

$\ln(x+5)-\ln x=\ln\dfrac{x+5}{x}=\ln\left(1+\dfrac{5}{x}\right)\to\ln1=0$ as $x\to\infty$. Thus,

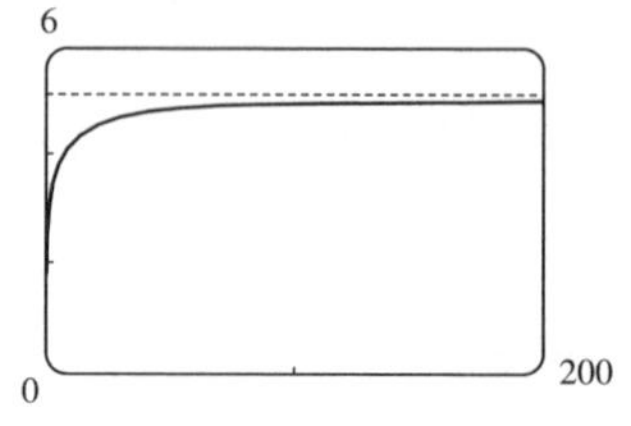

$$\lim_{x\to\infty}x\,[\ln(x+5)-\ln x]=\lim_{x\to\infty}\frac{\ln(x+5)-\ln x}{1/x}\overset{\text{H}}{=}\lim_{x\to\infty}\frac{\dfrac{1}{x+5}-\dfrac{1}{x}}{-1/x^2}$$

$$=\lim_{x\to\infty}\left[\frac{x-(x+5)}{x(x+5)}\cdot\frac{-x^2}{1}\right]=\lim_{x\to\infty}\frac{5x^2}{x^2+5x}=5$$

43. 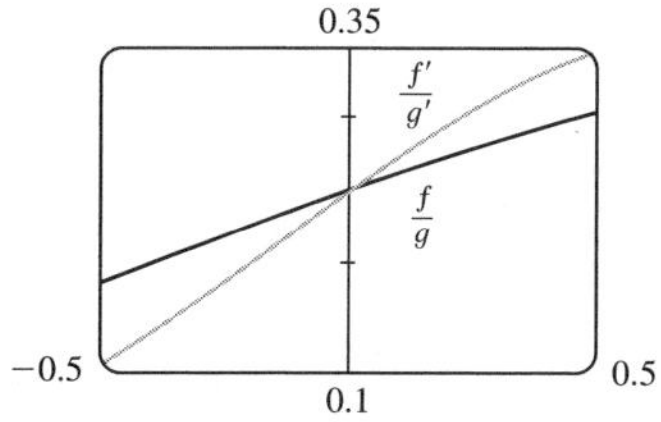

From the graph, it appears that $\lim\limits_{x\to 0}\frac{f(x)}{g(x)} = \lim\limits_{x\to 0}\frac{f'(x)}{g'(x)} = 0.25$. We calculate $\lim\limits_{x\to 0}\frac{f(x)}{g(x)} = \lim\limits_{x\to 0}\frac{e^x - 1}{x^3 + 4x} \overset{\text{H}}{=} \lim\limits_{x\to 0}\frac{e^x}{3x^2+4} = \frac{1}{4}$.

45. $\lim\limits_{x\to\infty} xe^{-x} = \lim\limits_{x\to\infty}(x/e^x) \overset{\text{H}}{=} \lim\limits_{x\to\infty}(1/e^x) = 0$, so $y = 0$ is a HA. $\lim\limits_{x\to-\infty} xe^{-x} = -\infty$.

$f(x) = xe^{-x} \quad\Rightarrow\quad f'(x) = x(-e^{-x}) + e^{-x}\cdot 1 = e^{-x}(1-x) > 0 \quad\Leftrightarrow\quad 1 - x > 0 \quad\Leftrightarrow\quad x < 1$,

so f is increasing on $(-\infty, 1)$ and decreasing on $(1, \infty)$. By the FDT, $f(1) = 1/e$ is a local maximum.

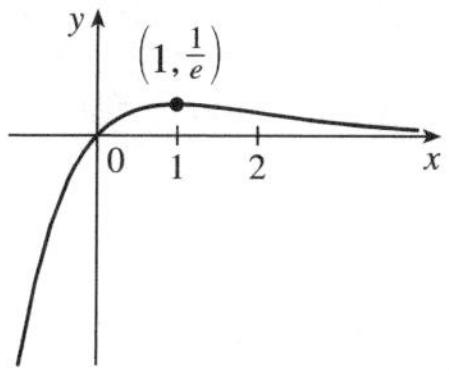

$f''(x) = e^{-x}(-1) + (1-x)(-e^{-x}) = e^{-x}(-1-1+x) = e^{-x}(x-2) > 0 \quad\Leftrightarrow$

$x - 2 > 0 \quad\Leftrightarrow\quad x > 2$, so f is CU on $(2, \infty)$ and CD on $(-\infty, 2)$. IP is $(2, 2/e^2)$.

47. 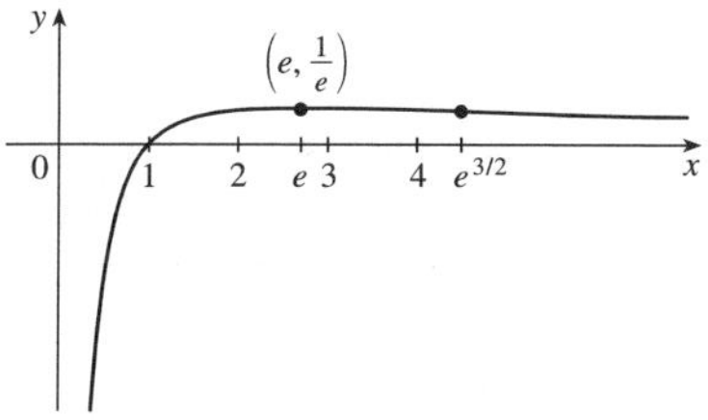

$\lim\limits_{x\to\infty}\frac{\ln x}{x} \overset{\text{H}}{=} \lim\limits_{x\to\infty}\frac{1/x}{1} = 0$, so $y = 0$ is a HA. Also $\lim\limits_{x\to 0^+}\frac{\ln x}{x} = -\infty$ since $\ln x \to -\infty$ and $x \to 0^+$, so $x = 0$ is a VA.

$f(x) = \frac{\ln x}{x} \quad\Rightarrow\quad f'(x) = \frac{x(1/x) - (\ln x)(1)}{x^2} = \frac{1 - \ln x}{x^2} = 0$ when $\ln x = 1 \quad\Leftrightarrow\quad x = e$.

$f'(x) > 0 \quad\Leftrightarrow\quad 1 - \ln x > 0 \quad\Leftrightarrow\quad \ln x < 1 \quad\Leftrightarrow\quad 0 < x < e$. $f'(x) < 0 \quad\Leftrightarrow\quad x > e$.

So f is increasing on $(0, e)$ and decreasing on (e, ∞). By the FDT, $f(e) = 1/e$ is a local maximum.

$f''(x) = \frac{x^2(-1/x) - (1 - \ln x)(2x)}{(x^2)^2} = \frac{x(-1-2+2\ln x)}{x^4} = \frac{2\ln x - 3}{x^3}$, so $f''(x) > 0 \quad\Leftrightarrow\quad 2\ln x - 3 > 0 \quad\Leftrightarrow$

$\ln x > \frac{3}{2} \quad\Leftrightarrow\quad x > e^{3/2}$. $f''(x) < 0 \quad\Leftrightarrow\quad 0 < x < e^{3/2}$. So f is CU on $\left(e^{3/2}, \infty\right)$ and CD on $\left(0, e^{3/2}\right)$. There is an inflection point at $\left(e^{3/2}, \frac{3}{2}e^{-3/2}\right)$.

49. (a) $f(x) = x^2 \ln x$. The domain of f is $(0, \infty)$.

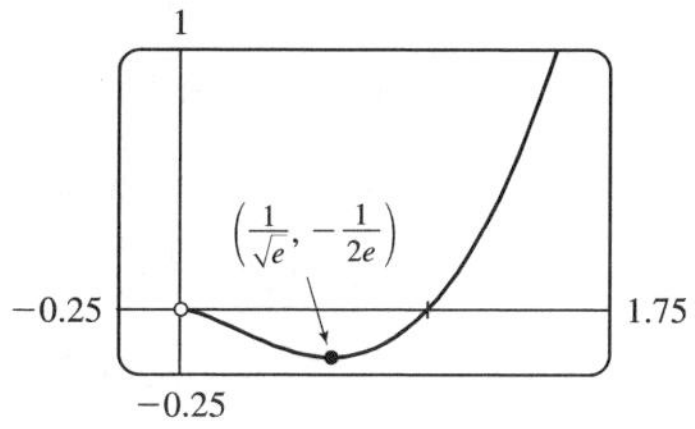

(b) $\lim\limits_{x\to 0^+} x^2\ln x = \lim\limits_{x\to 0^+}\frac{\ln x}{1/x^2} \overset{\text{H}}{=} \lim\limits_{x\to 0^+}\frac{1/x}{-2/x^3} = \lim\limits_{x\to 0^+}\left(-\frac{x^2}{2}\right) = 0$. There is a hole at $(0, 0)$.

(c) It appears that there is an IP at about $(0.2, -0.06)$ and a local minimum at $(0.6, -0.18)$. $f(x) = x^2\ln x \quad\Rightarrow$

$f'(x) = x^2(1/x) + (\ln x)(2x) = x(2\ln x + 1) > 0 \quad\Leftrightarrow\quad \ln x > -\frac{1}{2} \quad\Leftrightarrow\quad x > e^{-1/2}$, so f is increasing on $(1/\sqrt{e}, \infty)$, decreasing on $(0, 1/\sqrt{e})$. By the FDT, $f(1/\sqrt{e}) = -1/(2e)$ is a local minimum value. This point is approximately $(0.6065, -0.1839)$, which agrees with our estimate.

$f''(x) = x(2/x) + (2\ln x + 1) = 2\ln x + 3 > 0 \quad\Leftrightarrow\quad \ln x > -\frac{3}{2} \quad\Leftrightarrow\quad x > e^{-3/2}$, so f is CU on $\left(e^{-3/2}, \infty\right)$ and CD on $\left(0, e^{-3/2}\right)$. IP is $\left(e^{-3/2}, -3/(2e^3)\right) \approx (0.2231, -0.0747)$.

51. (a) $f(x) = x^{1/x}$

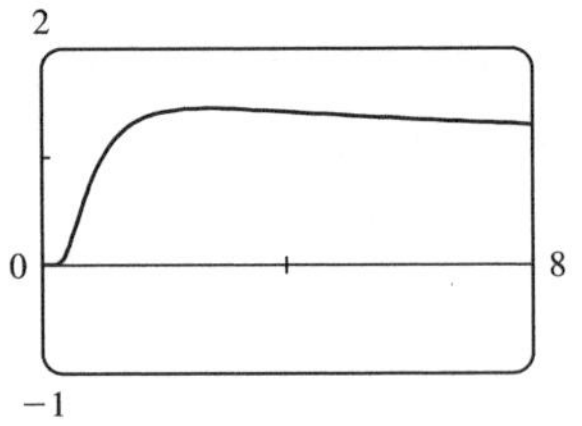

(b) Recall that $a^b = e^{b \ln a}$. $\lim\limits_{x \to 0^+} x^{1/x} = \lim\limits_{x \to 0^+} e^{(1/x)\ln x}$. As $x \to 0^+$, $\dfrac{\ln x}{x} \to -\infty$, so $x^{1/x} = e^{(1/x)\ln x} \to 0$. This indicates that there is a hole at $(0, 0)$. As $x \to \infty$, we have the indeterminate form ∞^0. $\lim\limits_{x \to \infty} x^{1/x} = \lim\limits_{x \to \infty} e^{(1/x)\ln x}$, but $\lim\limits_{x \to \infty} \dfrac{\ln x}{x} \overset{\text{H}}{=} \lim\limits_{x \to \infty} \dfrac{1/x}{1} = 0$, so $\lim\limits_{x \to \infty} x^{1/x} = e^0 = 1$. This indicates that $y = 1$ is a HA.

(c) Estimated maximum: $(2.72, 1.45)$. No estimated minimum. We use logarithmic differentiation to find any critical numbers. $y = x^{1/x} \quad \Rightarrow \quad \ln y = \dfrac{1}{x} \ln x \quad \Rightarrow \quad \dfrac{y'}{y} = \dfrac{1}{x} \cdot \dfrac{1}{x} + (\ln x)\left(-\dfrac{1}{x^2}\right) \quad \Rightarrow$

$y' = x^{1/x}\left(\dfrac{1 - \ln x}{x^2}\right) = 0 \quad \Rightarrow \quad \ln x = 1 \quad \Rightarrow \quad x = e$. For $0 < x < e$, $y' > 0$ and for $x > e$, $y' < 0$, so $f(e) = e^{1/e}$ is a local maximum value. This point is approximately $(2.7183, 1.4447)$, which agrees with our estimate.

(d)

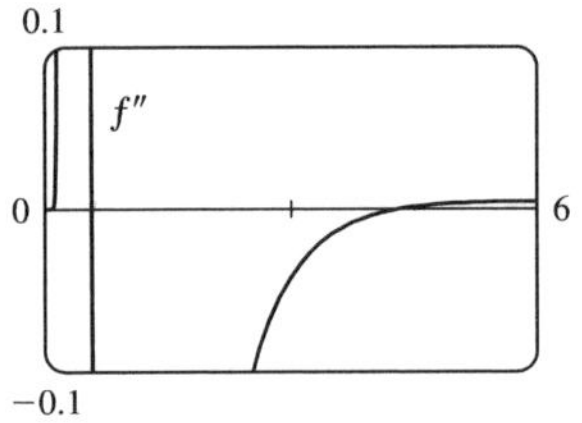

From the graph, we see that $f''(x) = 0$ at $x \approx 0.58$ and $x \approx 4.37$. Since f'' changes sign at these values, they are x-coordinates of inflection points.

53. If $c < 0$, then $\lim\limits_{x \to -\infty} f(x) = \lim\limits_{x \to -\infty} xe^{-cx} = \lim\limits_{x \to -\infty} \dfrac{x}{e^{cx}} \overset{\text{H}}{=} \lim\limits_{x \to -\infty} \dfrac{1}{ce^{cx}} = 0$, and $\lim\limits_{x \to \infty} f(x) = \infty$.

If $c > 0$, then $\lim\limits_{x \to -\infty} f(x) = -\infty$, and $\lim\limits_{x \to \infty} f(x) \overset{\text{H}}{=} \lim\limits_{x \to \infty} \dfrac{1}{ce^{cx}} = 0$.

If $c = 0$, then $f(x) = x$, so $\lim\limits_{x \to \pm\infty} f(x) = \pm\infty$, respectively.

So we see that $c = 0$ is a transitional value. We now exclude the case $c = 0$, since we know how the function behaves in that case. To find the maxima and minima of f, we differentiate: $f(x) = xe^{-cx} \quad \Rightarrow$

$f'(x) = x(-ce^{-cx}) + e^{-cx} = (1 - cx)e^{-cx}$. This is 0 when $1 - cx = 0 \quad \Leftrightarrow \quad x = 1/c$. If $c < 0$ then this represents a minimum value of $f(1/c) = 1/(ce)$, since $f'(x)$ changes from negative to positive at $x = 1/c$; and if $c > 0$, it represents a maximum value. As $|c|$ increases, the maximum or minimum point gets closer to the origin. To find the inflection points, we differentiate again: $f'(x) = e^{-cx}(1 - cx) \quad \Rightarrow$

$f''(x) = e^{-cx}(-c) + (1 - cx)(-ce^{-cx}) = (cx - 2)ce^{-cx}$. This changes sign when $cx - 2 = 0 \quad \Leftrightarrow \quad x = 2/c$. So as $|c|$ increases, the points of inflection get closer to the origin.

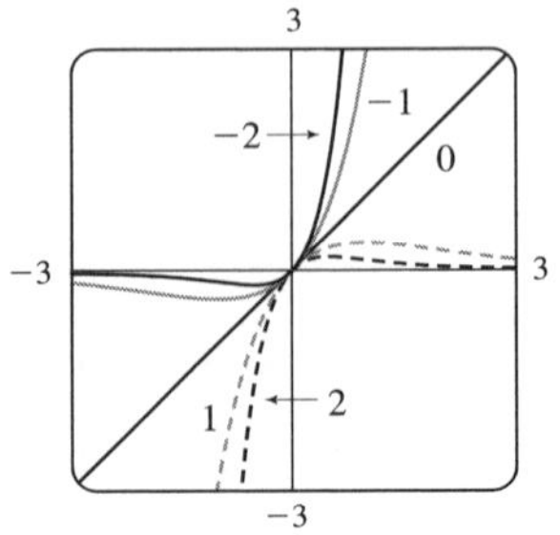

55. $\lim\limits_{x \to \infty} \dfrac{e^x}{x^n} \overset{\text{H}}{=} \lim\limits_{x \to \infty} \dfrac{e^x}{nx^{n-1}} \overset{\text{H}}{=} \lim\limits_{x \to \infty} \dfrac{e^x}{n(n-1)x^{n-2}} \overset{\text{H}}{=} \cdots \overset{\text{H}}{=} \lim\limits_{x \to \infty} \dfrac{e^x}{n!} = \infty$

57. First we will find $\lim\limits_{n\to\infty}\left(1+\dfrac{r}{n}\right)^{nt}$, which is of the form 1^∞. $y=\left(1+\dfrac{r}{n}\right)^{nt} \Rightarrow \ln y = nt\ln\left(1+\dfrac{r}{n}\right)$, so

$$\lim_{n\to\infty}\ln y=\lim_{n\to\infty} nt\ln\left(1+\frac{r}{n}\right)=t\lim_{n\to\infty}\frac{\ln(1+r/n)}{1/n}\overset{\text{H}}{=}t\lim_{n\to\infty}\frac{(-r/n^2)}{(1+r/n)(-1/n^2)}=t\lim_{n\to\infty}\frac{r}{1+i/n}=tr \Rightarrow$$

$\lim\limits_{n\to\infty} y=e^{rt}$. Thus, as $n\to\infty$, $A=A_0\left(1+\dfrac{r}{n}\right)^{nt}\to A_0e^{rt}$.

59.
$$\lim_{E\to0^+}P(E)=\lim_{E\to0^+}\left(\frac{e^E+e^{-E}}{e^E-e^{-E}}-\frac{1}{E}\right)$$

$$=\lim_{E\to0^+}\frac{E(e^E+e^{-E})-1(e^E-e^{-E})}{(e^E-e^{-E})\,E}=\lim_{E\to0^+}\frac{Ee^E+Ee^{-E}-e^E+e^{-E}}{Ee^E-Ee^{-E}}\qquad\left[\text{form is }\tfrac{0}{0}\right]$$

$$\overset{\text{H}}{=}\lim_{E\to0^+}\frac{Ee^E+e^E\cdot1+E(-e^{-E})+e^{-E}\cdot1-e^E+(-e^{-E})}{Ee^E+e^E\cdot1-[E(-e^{-E})+e^{-E}\cdot1]}$$

$$=\lim_{E\to0^+}\frac{Ee^E-Ee^{-E}}{Ee^E+e^E+Ee^{-E}-e^{-E}}=\lim_{E\to0^+}\frac{e^E-e^{-E}}{e^E+\dfrac{e^E}{E}+e^{-E}-\dfrac{e^{-E}}{E}}\qquad[\text{divide by }E]$$

$$=\frac{0}{2+L},\text{ where }L=\lim_{E\to0^+}\frac{e^E-e^{-E}}{E}\qquad\left[\text{form is }\tfrac{0}{0}\right]$$

$$\overset{\text{H}}{=}\lim_{E\to0^+}\frac{e^E+e^{-E}}{1}=\frac{1+1}{1}=2$$

Thus, $\lim\limits_{E\to0^+}P(E)=\dfrac{0}{2+2}=0$.

61. We see that both numerator and denominator approach 0, so we can use l'Hospital's Rule:

$$\lim_{x\to a}\frac{\sqrt{2a^3x-x^4}-a\sqrt[3]{aax}}{a-\sqrt[4]{ax^3}}\overset{\text{H}}{=}\lim_{x\to a}\frac{\frac{1}{2}(2a^3x-x^4)^{-1/2}(2a^3-4x^3)-a\left(\frac{1}{3}\right)(aax)^{-2/3}a^2}{-\frac{1}{4}(ax^3)^{-3/4}(3ax^2)}$$

$$=\frac{\frac{1}{2}(2a^3a-a^4)^{-1/2}(2a^3-4a^3)-\frac{1}{3}a^3(a^2a)^{-2/3}}{-\frac{1}{4}(aa^3)^{-3/4}(3aa^2)}$$

$$=\frac{(a^4)^{-1/2}(-a^3)-\frac{1}{3}a^3(a^3)^{-2/3}}{-\frac{3}{4}a^3(a^4)^{-3/4}}=\frac{-a-\frac{1}{3}a}{-\frac{3}{4}}=\tfrac{4}{3}\left(\tfrac{4}{3}a\right)=\tfrac{16}{9}a$$

63. Since $f(2)=0$, the given limit has the form $\frac{0}{0}$.

$$\lim_{x\to0}\frac{f(2+3x)+f(2+5x)}{x}\overset{\text{H}}{=}\lim_{x\to0}\frac{f'(2+3x)\cdot3+f'(2+5x)\cdot5}{1}=f'(2)\cdot3+f'(2)\cdot5=8f'(2)=8\cdot7=56$$

65. Since $\lim\limits_{h\to0}[f(x+h)-f(x-h)]=f(x)-f(x)=0$ (f is differentiable and hence continuous) and $\lim\limits_{h\to0}2h=0$, we use l'Hospital's Rule:

$$\lim_{h\to0}\frac{f(x+h)-f(x-h)}{2h}\overset{\text{H}}{=}\lim_{h\to0}\frac{f'(x+h)(1)-f'(x-h)(-1)}{2}=\frac{f'(x)+f'(x)}{2}=\frac{2f'(x)}{2}=f'(x)$$

$\dfrac{f(x+h)-f(x-h)}{2h}$ is the slope of the secant line between $(x-h,f(x-h))$ and $(x+h,f(x+h))$. As $h\to0$, this line gets closer to the tangent line and its slope approaches $f'(x)$.

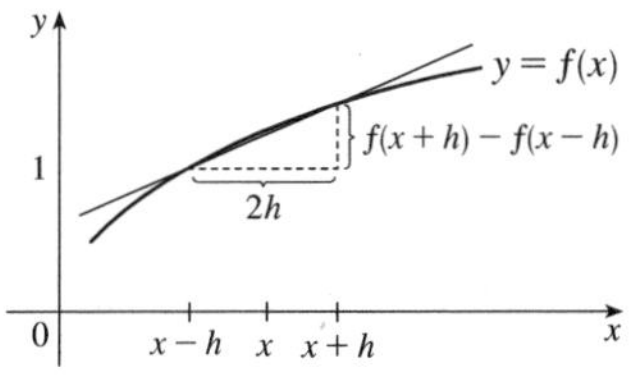

4.6 Optimization Problems

1. (a)

First Number	Second Number	Product
1	22	22
2	21	42
3	20	60
4	19	76
5	18	90
6	17	102
7	16	112
8	15	120
9	14	126
10	13	130
11	12	132

We needn't consider pairs where the first number is larger than the second, since we can just interchange the numbers in such cases. The answer appears to be 11 and 12, but we have considered only integers in the table.

(b) Call the two numbers x and y. Then $x + y = 23$, so $y = 23 - x$. Call the product P. Then $P = xy = x(23 - x) = 23x - x^2$, so we wish to maximize the function $P(x) = 23x - x^2$. Since $P'(x) = 23 - 2x$, we see that $P'(x) = 0 \quad \Leftrightarrow \quad x = \frac{23}{2} = 11.5$. Thus, the maximum value of P is $P(11.5) = (11.5)^2 = 132.25$ and it occurs when $x = y = 11.5$.

Or: Note that $P''(x) = -2 < 0$ for all x, so P is everywhere concave downward and the local maximum at $x = 11.5$ must be an absolute maximum.

3. The two numbers are x and $\dfrac{100}{x}$, where $x > 0$. Minimize $f(x) = x + \dfrac{100}{x}$. $f'(x) = 1 - \dfrac{100}{x^2} = \dfrac{x^2 - 100}{x^2}$. The critical number is $x = 10$. Since $f'(x) < 0$ for $0 < x < 10$ and $f'(x) > 0$ for $x > 10$, there is an absolute minimum at $x = 10$. The numbers are 10 and 10.

5. If the rectangle has dimensions x and y, then its perimeter is $2x + 2y = 100$ m, so $y = 50 - x$. Thus, the area is $A = xy = x(50 - x)$. We wish to maximize the function $A(x) = x(50 - x) = 50x - x^2$, where $0 < x < 50$. Since $A'(x) = 50 - 2x = -2(x - 25)$, $A'(x) > 0$ for $0 < x < 25$ and $A'(x) < 0$ for $25 < x < 50$. Thus, A has an absolute maximum at $x = 25$, and $A(25) = 25^2 = 625\text{ m}^2$. The dimensions of the rectangle that maximize its area are $x = y = 25$ m. (The rectangle is a square.)

7. (a)

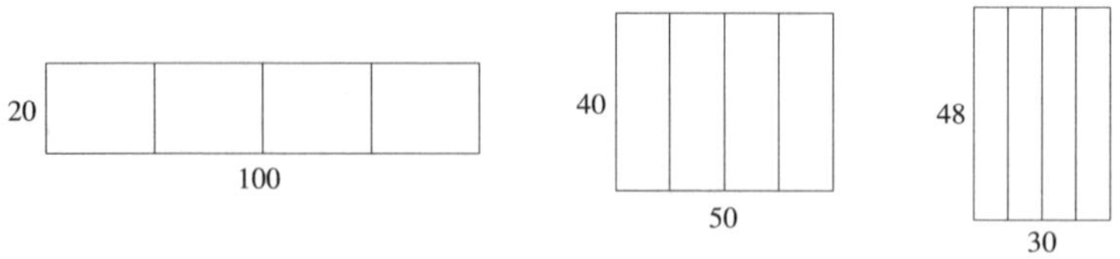

The areas of the three figures are 2000, 2000, and 1440 m^2. There appears to be a maximum area of at least 2000 m^2.

(b) Let x denote the length of each of two sides and three dividers.

Let y denote the length of the other two sides.

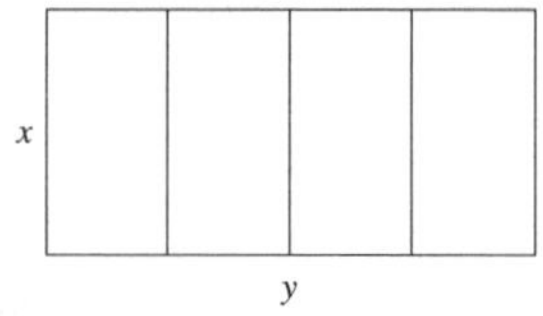

(c) Area $A = \text{length} \times \text{width} = y \cdot x$

(d) Length of fencing $= 300 \quad \Rightarrow \quad 5x + 2y = 300$

(e) $5x + 2y = 300 \quad \Rightarrow \quad y = 150 - \frac{5}{2}x \quad \Rightarrow \quad A(x) = \left(150 - \frac{5}{2}x\right)x = 150x - \frac{5}{2}x^2$

(f) $A'(x) = 150 - 5x = 0 \quad \Rightarrow \quad x = 30$. Since $A''(x) = -5 < 0$, there is an absolute maximum when $x = 30$. Then $y = 150 - \frac{5}{2}(30) = 75$. The largest area is $30(75) = 2250\text{ m}^2$. These values of x and y are between the values in the first and second figures in part (a). Our original estimate was low.

9. Let b be the length of the base of the box and h the height. The surface area is $1200 = b^2 + 4hb \Rightarrow$ $h = (1200 - b^2)/(4b)$. The volume is $V = b^2 h = b^2(1200 - b^2)/4b = 300b - b^3/4 \Rightarrow V'(b) = 300 - \frac{3}{4}b^2$. $V'(b) = 0 \Rightarrow 300 = \frac{3}{4}b^2 \Rightarrow b^2 = 400 \Rightarrow b = \sqrt{400} = 20$. Since $V'(b) > 0$ for $0 < b < 20$ and $V'(b) < 0$ for $b > 20$, there is an absolute maximum when $b = 20$ by the First Derivative Test for Absolute Extreme Values (see page 308). If $b = 20$, then $h = (1200 - 20^2)/(4 \cdot 20) = 10$, so the largest possible volume is $b^2 h = (20)^2(10) = 4000\ \text{cm}^3$.

11. (a) Let the rectangle have sides x and y and area A, so $A = xy$ or $y = A/x$. The problem is to minimize the perimeter $= 2x + 2y = 2x + 2A/x = P(x)$. Now $P'(x) = 2 - 2A/x^2 = 2(x^2 - A)/x^2$. So the critical number is $x = \sqrt{A}$. Since $P'(x) < 0$ for $0 < x < \sqrt{A}$ and $P'(x) > 0$ for $x > \sqrt{A}$, there is an absolute minimum at $x = \sqrt{A}$. The sides of the rectangle are $\sqrt{A}$ and $A/\sqrt{A} = \sqrt{A}$, so the rectangle is a square.

(b) Let p be the perimeter and x and y the lengths of the sides, so $p = 2x + 2y \Rightarrow 2y = p - 2x \Rightarrow y = \frac{1}{2}p - x$. The area is $A(x) = x\left(\frac{1}{2}p - x\right) = \frac{1}{2}px - x^2$. Now $A'(x) = 0 \Rightarrow \frac{1}{2}p - 2x = 0 \Rightarrow$ $2x = \frac{1}{2}p \Rightarrow x = \frac{1}{4}p$. Since $A''(x) = -2 < 0$, there is an absolute maximum for A when $x = \frac{1}{4}p$ by the Second Derivative Test. The sides of the rectangle are $\frac{1}{4}p$ and $\frac{1}{2}p - \frac{1}{4}p = \frac{1}{4}p$, so the rectangle is a square.

13.

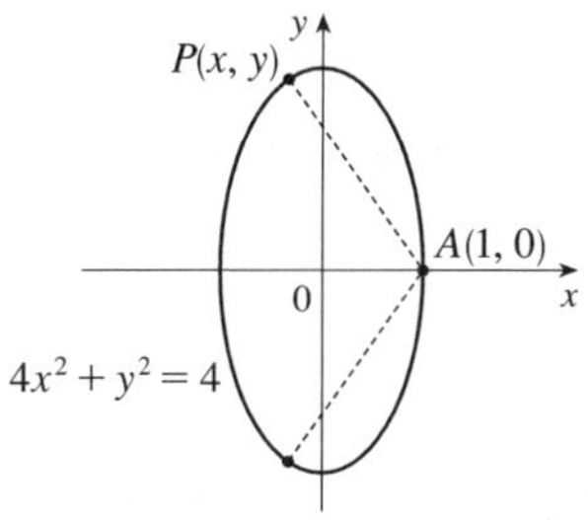

From the figure, we see that there are two points that are furthest away from $A(1, 0)$. The distance d from A to an arbitrary point $P(x, y)$ on the ellipse is $d = \sqrt{(x-1)^2 + (y-0)^2}$ and the square of the distance is $S = d^2 = x^2 - 2x + 1 + y^2 = x^2 - 2x + 1 + (4 - 4x^2) = -3x^2 - 2x + 5$. $S' = -6x - 2$ and $S' = 0 \Rightarrow x = -\frac{1}{3}$. Now $S'' = -6 < 0$, so we know that S has a maximum at $x = -\frac{1}{3}$. Since $-1 \le x \le 1$, $S(-1) = 4$, $S\left(-\frac{1}{3}\right) = \frac{16}{3}$, and $S(1) = 0$, we see that the maximum distance is $\sqrt{\frac{16}{3}}$. The corresponding y-values are $y = \pm\sqrt{4 - 4\left(-\frac{1}{3}\right)^2} = \pm\sqrt{\frac{32}{9}} = \pm\frac{4}{3}\sqrt{2} \approx \pm 1.89$. The points are $\left(-\frac{1}{3}, \pm\frac{4}{3}\sqrt{2}\right)$.

15.

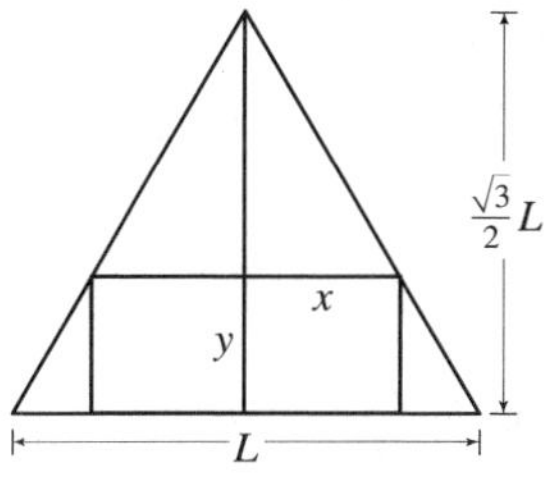

The height h of the equilateral triangle with sides of length L is $\frac{\sqrt{3}}{2}L$, since $h^2 + (L/2)^2 = L^2 \Rightarrow h^2 = L^2 - \frac{1}{4}L^2 = \frac{3}{4}L^2 \Rightarrow$ $h = \frac{\sqrt{3}}{2}L$. Using similar triangles, $\dfrac{\frac{\sqrt{3}}{2}L - y}{x} = \dfrac{\frac{\sqrt{3}}{2}L}{L/2} = \sqrt{3} \Rightarrow$

$$\sqrt{3}\,x = \frac{\sqrt{3}}{2}L - y \Rightarrow y = \frac{\sqrt{3}}{2}L - \sqrt{3}\,x \Rightarrow y = \frac{\sqrt{3}}{2}(L - 2x).$$

The area of the inscribed rectangle is $A(x) = (2x)y = \sqrt{3}\,x(L - 2x) = \sqrt{3}\,Lx - 2\sqrt{3}\,x^2$, where $0 \le x \le L/2$. Now $0 = A'(x) = \sqrt{3}\,L - 4\sqrt{3}\,x \Rightarrow x = \sqrt{3}\,L/(4\sqrt{3}) = L/4$. Since $A(0) = A(L/2) = 0$, the maximum occurs when $x = L/4$, and $y = \frac{\sqrt{3}}{2}L - \frac{\sqrt{3}}{4}L = \frac{\sqrt{3}}{4}L$, so the dimensions are $L/2$ and $\frac{\sqrt{3}}{4}L$.

17.

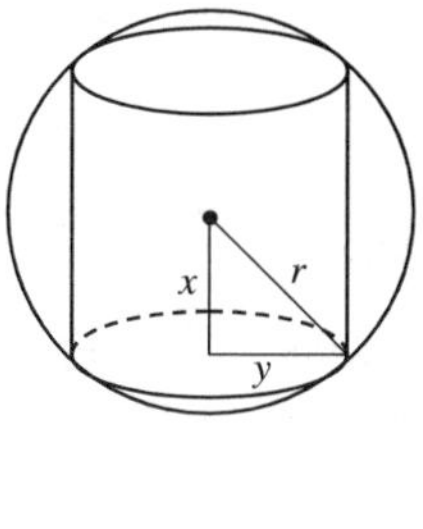

The cylinder has volume $V = \pi y^2(2x)$. Also $x^2 + y^2 = r^2 \Rightarrow$ $y^2 = r^2 - x^2$, so $V(x) = \pi(r^2 - x^2)(2x) = 2\pi(r^2 x - x^3)$, where $0 \le x \le r$. $V'(x) = 2\pi(r^2 - 3x^2) = 0 \Rightarrow x = r/\sqrt{3}$. Now $V(0) = V(r) = 0$, so there is a maximum when $x = r/\sqrt{3}$ and $V(r/\sqrt{3}) = \pi(r^2 - r^2/3)(2r/\sqrt{3}) = 4\pi r^3/(3\sqrt{3})$.

19.

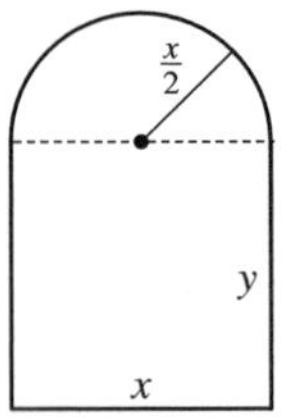

Perimeter $= 10 \;\Rightarrow\; 2y + x + \pi\left(\frac{x}{2}\right) = 10 \;\Rightarrow$

$y = \frac{1}{2}\left(10 - x - \frac{\pi x}{2}\right) = 5 - \frac{x}{2} - \frac{\pi x}{4}$. The area is the area of the rectangle plus the area of the semicircle, or $xy + \frac{1}{2}\pi\left(\frac{x}{2}\right)^2$, so

$$A(x) = x\left(5 - \frac{x}{2} - \frac{\pi x}{4}\right) + \tfrac{1}{8}\pi x^2 = 5x - \tfrac{1}{2}x^2 - \tfrac{\pi}{8}x^2.$$

$A'(x) = 5 - \left(1 + \frac{\pi}{4}\right)x = 0 \;\Rightarrow\; x = \dfrac{5}{1 + \pi/4} = \dfrac{20}{4 + \pi}$. $A''(x) = -\left(1 + \dfrac{\pi}{4}\right) < 0$, so this gives a maximum. The dimensions are $x = \dfrac{20}{4+\pi}$ m and $y = 5 - \dfrac{10}{4+\pi} - \dfrac{5\pi}{4+\pi} = \dfrac{20 + 5\pi - 10 - 5\pi}{4+\pi} = \dfrac{10}{4+\pi}$ m, so the height of the rectangle is half the base.

21.

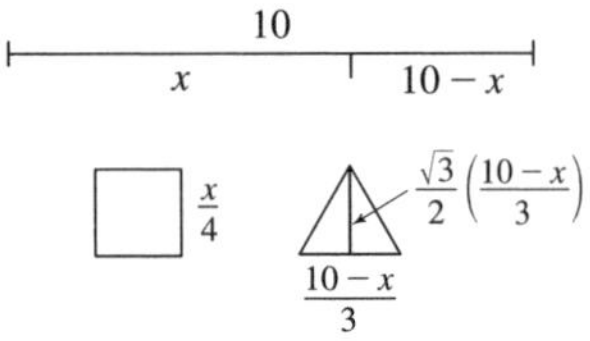

Let x be the length of the wire used for the square. The total area is

$$A(x) = \left(\frac{x}{4}\right)^2 + \frac{1}{2}\left(\frac{10-x}{3}\right)\frac{\sqrt{3}}{2}\left(\frac{10-x}{3}\right)$$

$$= \tfrac{1}{16}x^2 + \tfrac{\sqrt{3}}{36}(10 - x)^2, \quad 0 \le x \le 10$$

$A'(x) = \frac{1}{8}x - \frac{\sqrt{3}}{18}(10 - x) = 0 \;\Leftrightarrow\; \frac{9}{72}x + \frac{4\sqrt{3}}{72}x - \frac{40\sqrt{3}}{72} = 0 \;\Leftrightarrow\; x = \frac{40\sqrt{3}}{9 + 4\sqrt{3}}$. Now $A(0) = \left(\frac{\sqrt{3}}{36}\right)100 \approx 4.81$, $A(10) = \frac{100}{16} = 6.25$ and $A\left(\frac{40\sqrt{3}}{9+4\sqrt{3}}\right) \approx 2.72$, so

(a) The maximum area occurs when $x = 10$ m, and all the wire is used for the square.

(b) The minimum area occurs when $x = \frac{40\sqrt{3}}{9+4\sqrt{3}} \approx 4.35$ m.

23.

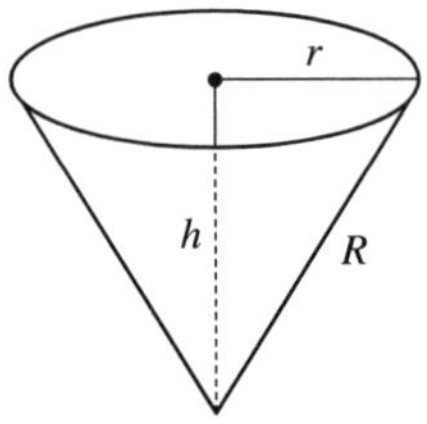

$h^2 + r^2 = R^2 \;\Rightarrow\; V = \frac{\pi}{3}r^2 h = \frac{\pi}{3}(R^2 - h^2)h = \frac{\pi}{3}(R^2 h - h^3)$.

$V'(h) = \frac{\pi}{3}(R^2 - 3h^2) = 0$ when $h = \frac{1}{\sqrt{3}}R$. This gives an absolute maximum, since $V'(h) > 0$ for $0 < h < \frac{1}{\sqrt{3}}R$ and $V'(h) < 0$ for $h > \frac{1}{\sqrt{3}}R$. The maximum volume is

$$V\left(\tfrac{1}{\sqrt{3}}R\right) = \tfrac{\pi}{3}\left(\tfrac{1}{\sqrt{3}}R^3 - \tfrac{1}{3\sqrt{3}}R^3\right) = \tfrac{2}{9\sqrt{3}}\pi R^3.$$

25.

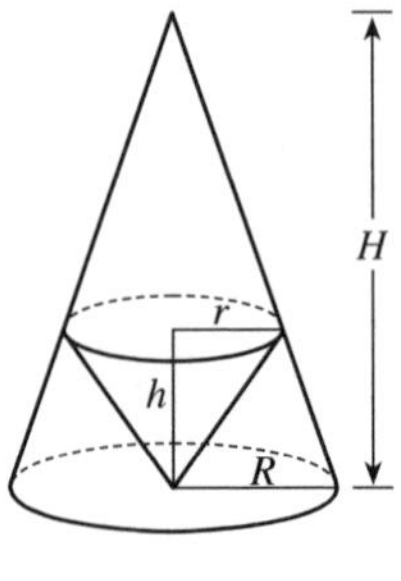

By similar triangles, $\dfrac{H}{R} = \dfrac{H - h}{r}$ **(1)**. The volume of the inner cone is $V = \frac{1}{3}\pi r^2 h$, so we'll solve **(1)** for h. $\dfrac{Hr}{R} = H - h \;\Rightarrow$

$$h = H - \frac{Hr}{R} = \frac{HR - Hr}{R} = \frac{H}{R}(R - r) \quad \textbf{(2)}.$$

Thus, $V(r) = \dfrac{\pi}{3}r^2 \cdot \dfrac{H}{R}(R - r) = \dfrac{\pi H}{3R}(Rr^2 - r^3) \;\Rightarrow$

$$V'(r) = \frac{\pi H}{3R}(2Rr - 3r^2) = \frac{\pi H}{3R}r(2R - 3r).$$

$V'(r) = 0 \;\Rightarrow\; r = 0$ or $2R = 3r \;\Rightarrow\; r = \frac{2}{3}R$ and from **(2)**, $h = \dfrac{H}{R}\left(R - \frac{2}{3}R\right) = \dfrac{H}{R}\left(\frac{1}{3}R\right) = \frac{1}{3}H$.

$V'(r)$ changes from positive to negative at $r = \frac{2}{3}R$, so the inner cone has a maximum volume of $V = \frac{1}{3}\pi r^2 h = \frac{1}{3}\pi\left(\frac{2}{3}R\right)^2\left(\frac{1}{3}H\right) = \frac{4}{27} \cdot \frac{1}{3}\pi R^2 H$, which is approximately 15% of the volume of the larger cone.

27. $P(R) = \dfrac{E^2 R}{(R+r)^2} \quad \Rightarrow$

$$P'(R) = \frac{(R+r)^2 \cdot E^2 - E^2 R \cdot 2(R+r)}{[(R+r)^2]^2} = \frac{\left(R^2 + 2Rr + r^2\right) E^2 - 2E^2 R^2 - 2E^2 Rr}{(R+r)^4}$$
$$= \frac{E^2 r^2 - E^2 R^2}{(R+r)^4} = \frac{E^2(r^2 - R^2)}{(R+r)^4} = \frac{E^2(r+R)(r-R)}{(R+r)^4} = \frac{E^2(r-R)}{(R+r)^3}$$

$$P'(R) = 0 \quad \Rightarrow \quad R = r \quad \Rightarrow \quad P(r) = \frac{E^2 r}{(r+r)^2} = \frac{E^2 r}{4r^2} = \frac{E^2}{4r}.$$

The expression for $P'(R)$ shows that $P'(R) > 0$ for $R < r$ and $P'(R) < 0$ for $R > r$. Thus, the maximum value of the power is $E^2/(4r)$, and this occurs when $R = r$.

29. $S = 6sh - \frac{3}{2}s^2 \cot\theta + 3s^2 \frac{\sqrt{3}}{2} \csc\theta$

(a) $\dfrac{dS}{d\theta} = \frac{3}{2}s^2 \csc^2\theta - 3s^2 \frac{\sqrt{3}}{2} \csc\theta \cot\theta$ or $\frac{3}{2}s^2 \csc\theta \left(\csc\theta - \sqrt{3}\cot\theta\right)$.

(b) $\dfrac{dS}{d\theta} = 0$ when $\csc\theta - \sqrt{3}\cot\theta = 0 \quad \Rightarrow \quad \dfrac{1}{\sin\theta} - \sqrt{3}\,\dfrac{\cos\theta}{\sin\theta} = 0 \quad \Rightarrow \quad \cos\theta = \frac{1}{\sqrt{3}}$. The First Derivative Test shows that the minimum surface area occurs when $\theta = \cos^{-1}\left(\frac{1}{\sqrt{3}}\right) \approx 55°$.

(c)

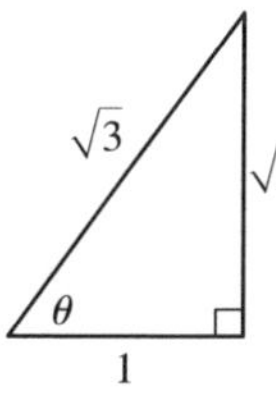

If $\cos\theta = \frac{1}{\sqrt{3}}$, then $\cot\theta = \frac{1}{\sqrt{2}}$ and $\csc\theta = \frac{\sqrt{3}}{\sqrt{2}}$, so the surface area is

$$S = 6sh - \tfrac{3}{2}s^2 \tfrac{1}{\sqrt{2}} + 3s^2 \tfrac{\sqrt{3}}{2} \tfrac{\sqrt{3}}{\sqrt{2}} = 6sh - \tfrac{3}{2\sqrt{2}}s^2 + \tfrac{9}{2\sqrt{2}}s^2$$
$$= 6sh + \tfrac{6}{2\sqrt{2}}s^2 = 6s\left(h + \tfrac{1}{2\sqrt{2}}s\right)$$

31.

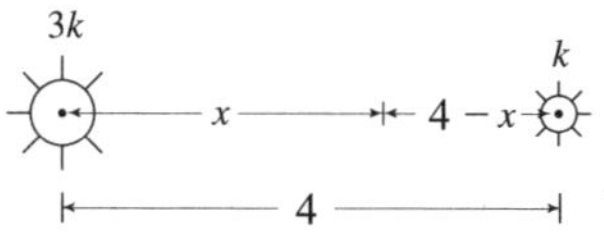

The total illumination is $I(x) = \dfrac{3k}{x^2} + \dfrac{k}{(4-x)^2}$, $0 < x < 4$. Then

$$I'(x) = \frac{-6k}{x^3} + \frac{2k}{(4-x)^3} = 0 \quad \Rightarrow \quad 6k(4-x)^3 = 2kx^3 \quad \Rightarrow$$

$$3(4-x)^3 = x^3 \quad \Rightarrow \quad \sqrt[3]{3}\,(4-x) = x \quad \Rightarrow \quad 4\sqrt[3]{3} - \sqrt[3]{3}\,x = x \quad \Rightarrow \quad 4\sqrt[3]{3} = x + \sqrt[3]{3}\,x \quad \Rightarrow \quad 4\sqrt[3]{3} = \left(1 + \sqrt[3]{3}\right)x \quad \Rightarrow$$

$x = \dfrac{4\sqrt[3]{3}}{1 + \sqrt[3]{3}} \approx 2.4$ m. This gives a minimum since $I''(x) > 0$ for $0 < x < 4$.

33.

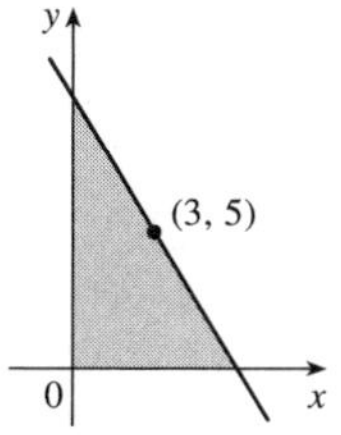

The line with slope m (where $m < 0$) through $(3, 5)$ has equation $y - 5 = m(x - 3)$ or $y = mx + (5 - 3m)$. The y-intercept is $5 - 3m$ and the x-intercept is $-5/m + 3$. So the triangle has area $A(m) = \frac{1}{2}(5 - 3m)(-5/m + 3) = 15 - 25/(2m) - \frac{9}{2}m$.

Now $A'(m) = \dfrac{25}{2m^2} - \dfrac{9}{2} = 0 \quad \Leftrightarrow \quad m^2 = \frac{25}{9} \quad \Rightarrow \quad m = -\frac{5}{3}$ (since $m < 0$).

$A''(m) = -\dfrac{25}{m^3} > 0$, so there is an absolute minimum when $m = -\frac{5}{3}$. Thus, an equation of the line is $y - 5 = -\frac{5}{3}(x - 3)$ or $y = -\frac{5}{3}x + 10$.

35.

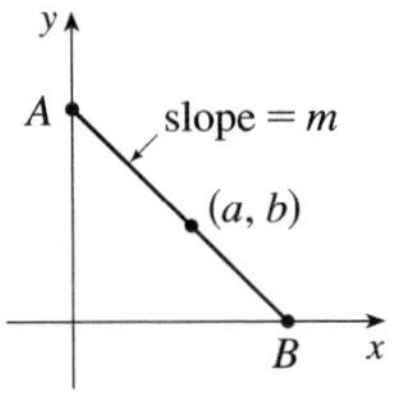

Every line segment in the first quadrant passing through (a,b) with endpoints on the x- and y-axes satisfies an equation of the form $y-b=m(x-a)$, where $m<0$. By setting $x=0$ and then $y=0$, we find its endpoints, $A(0,b-am)$ and $B\left(a-\frac{b}{m},0\right)$. The distance d from A to B is given by $d=\sqrt{\left[\left(a-\frac{b}{m}\right)-0\right]^2+[0-(b-am)]^2}$.

It follows that the square of the length of the line segment, as a function of m, is given by

$S(m)=\left(a-\dfrac{b}{m}\right)^2+(am-b)^2=a^2-\dfrac{2ab}{m}+\dfrac{b^2}{m^2}+a^2m^2-2abm+b^2$. Thus,

$$\begin{aligned} S'(m)&=\frac{2ab}{m^2}-\frac{2b^2}{m^3}+2a^2m-2ab=\frac{2}{m^3}(abm-b^2+a^2m^4-abm^3)\\ &=\frac{2}{m^3}[b(am-b)+am^3(am-b)]=\frac{2}{m^3}(am-b)(b+am^3)\end{aligned}$$

Thus, $S'(m)=0 \quad\Leftrightarrow\quad m=b/a$ or $m=-\sqrt[3]{\frac{b}{a}}$. Since $b/a>0$ and $m<0$, m must equal $-\sqrt[3]{\frac{b}{a}}$. Since $\dfrac{2}{m^3}<0$, we see that $S'(m)<0$ for $m<-\sqrt[3]{\frac{b}{a}}$ and $S'(m)>0$ for $m>-\sqrt[3]{\frac{b}{a}}$. Thus, S has its absolute minimum value when $m=-\sqrt[3]{\frac{b}{a}}$. That value is

$$\begin{aligned} S\left(-\sqrt[3]{\tfrac{b}{a}}\right)&=\left(a+b\sqrt[3]{\tfrac{a}{b}}\right)^2+\left(-a\sqrt[3]{\tfrac{b}{a}}-b\right)^2=\left(a+\sqrt[3]{ab^2}\right)^2+\left(\sqrt[3]{a^2b}+b\right)^2\\ &=a^2+2a^{4/3}b^{2/3}+a^{2/3}b^{4/3}+a^{4/3}b^{2/3}+2a^{2/3}b^{4/3}+b^2=a^2+3a^{4/3}b^{2/3}+3a^{2/3}b^{4/3}+b^2\end{aligned}$$

The last expression is of the form $x^3+3x^2y+3xy^2+y^3 \quad \left[=(x+y)^3\right]$ with $x=a^{2/3}$ and $y=b^{2/3}$, so we can write it as $(a^{2/3}+b^{2/3})^3$ and the shortest such line segment has length $\sqrt{S}=(a^{2/3}+b^{2/3})^{3/2}$.

37.

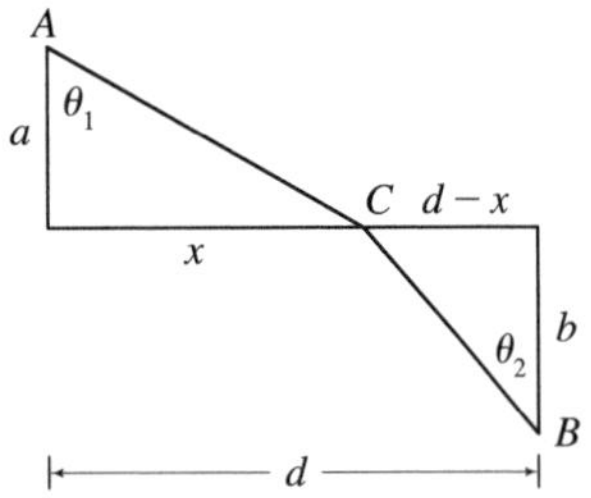

The total time is

$$\begin{aligned} T(x)&=(\text{time from } A \text{ to } C)+(\text{time from } C \text{ to } B)\\ &=\frac{\sqrt{a^2+x^2}}{v_1}+\frac{\sqrt{b^2+(d-x)^2}}{v_2},\quad 0<x<d\end{aligned}$$

$$T'(x)=\frac{x}{v_1\sqrt{a^2+x^2}}-\frac{d-x}{v_2\sqrt{b^2+(d-x)^2}}=\frac{\sin\theta_1}{v_1}-\frac{\sin\theta_2}{v_2}$$

The minimum occurs when $T'(x)=0 \;\Rightarrow\; \dfrac{\sin\theta_1}{v_1}=\dfrac{\sin\theta_2}{v_2}$.

[*Note:* $T''(x)>0$]

39.

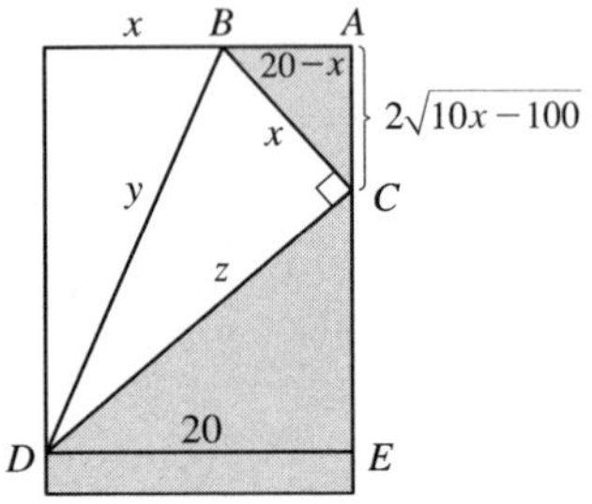

$y^2=x^2+z^2$, but triangles CDE and BCA are similar, so $z/20=x/\left(2\sqrt{10x-100}\right) \;\Rightarrow\; z=10x/\sqrt{10x-100}$. Thus, we minimize $f(x)=y^2=x^2+100x^2/(10x-100)=x^3/(x-10)$, $10<x\le 20$.

$$f'(x)=\frac{(x-10)(3x^2)-x^3}{(x-10)^2}=\frac{x^2[3(x-10)-x]}{(x-10)^2}=\frac{2x^2(x-15)}{(x-10)^2}=0 \text{ when}$$

$x=15$. $f'(x)<0$ when $x<15$, $f'(x)>0$ when $x>15$, so the minimum occurs when $x=15$ cm.

41. 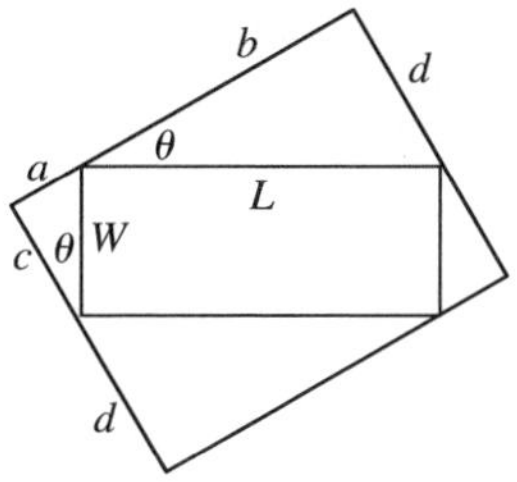

In the small triangle with sides a and c and hypotenuse W, $\sin\theta = \dfrac{a}{W}$ and $\cos\theta = \dfrac{c}{W}$. In the triangle with sides b and d and hypotenuse L, $\sin\theta = \dfrac{d}{L}$ and $\cos\theta = \dfrac{b}{L}$. Thus, $a = W\sin\theta$, $c = W\cos\theta$, $d = L\sin\theta$, and $b = L\cos\theta$, so the area of the circumscribed rectangle is

$$\begin{aligned} A(\theta) &= (a+b)(c+d) = (W\sin\theta + L\cos\theta)(W\cos\theta + L\sin\theta) \\ &= W^2\sin\theta\,\cos\theta + WL\sin^2\theta + LW\cos^2\theta + L^2\sin\theta\,\cos\theta \\ &= LW\sin^2\theta + LW\cos^2\theta + \left(L^2+W^2\right)\sin\theta\,\cos\theta \\ &= LW\left(\sin^2\theta + \cos^2\theta\right) + \left(L^2+W^2\right)\cdot\tfrac{1}{2}\cdot 2\sin\theta\,\cos\theta \\ &= LW + \tfrac{1}{2}\left(L^2+W^2\right)\sin 2\theta, \quad 0 \le \theta \le \tfrac{\pi}{2} \end{aligned}$$

This expression shows, without calculus, that the maximum value of $A(\theta)$ occurs when $\sin 2\theta = 1 \iff 2\theta = \frac{\pi}{2} \Rightarrow \theta = \frac{\pi}{4}$. So the maximum area is $A\left(\frac{\pi}{4}\right) = LW + \frac{1}{2}\left(L^2+W^2\right) = \frac{1}{2}\left(L^2+2LW+W^2\right) = \frac{1}{2}(L+W)^2$.

43. 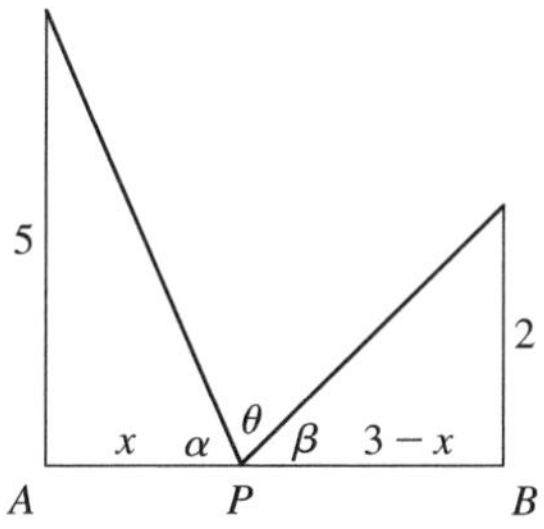

From the figure, $\tan\alpha = \dfrac{5}{x}$ and $\tan\beta = \dfrac{2}{3-x}$. Since

$\alpha + \beta + \theta = 180^\circ = \pi$, $\theta = \pi - \tan^{-1}\left(\dfrac{5}{x}\right) - \tan^{-1}\left(\dfrac{2}{3-x}\right) \Rightarrow$

$$\begin{aligned} \frac{d\theta}{dx} &= -\frac{1}{1+\left(\frac{5}{x}\right)^2}\left(-\frac{5}{x^2}\right) - \frac{1}{1+\left(\frac{2}{3-x}\right)^2}\left[\frac{2}{(3-x)^2}\right] \\ &= \frac{x^2}{x^2+25}\cdot\frac{5}{x^2} - \frac{(3-x)^2}{(3-x)^2+4}\cdot\frac{2}{(3-x)^2}. \end{aligned}$$ Now

$\dfrac{d\theta}{dx} = 0 \Rightarrow \dfrac{5}{x^2+25} = \dfrac{2}{x^2-6x+13} \Rightarrow 2x^2+50 = 5x^2-30x+65 \Rightarrow$

$3x^2 - 30x + 15 = 0 \Rightarrow x^2 - 10x + 5 = 0 \Rightarrow x = 5 \pm 2\sqrt{5}$. We reject the root with the + sign, since it is larger than 3. $d\theta/dx > 0$ for $x < 5 - 2\sqrt{5}$ and $d\theta/dx < 0$ for $x > 5 - 2\sqrt{5}$, so θ is maximized when $|AP| = x = 5 - 2\sqrt{5} \approx 0.53$.

45. (a) 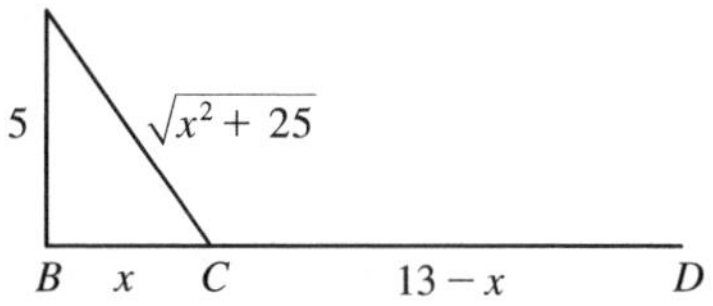

If k = energy/km over land, then energy/km over water = $1.4k$.

So the total energy is $E = 1.4k\sqrt{25+x^2} + k(13-x)$, $0 \le x \le 13$,

and so $\dfrac{dE}{dx} = \dfrac{1.4kx}{(25+x^2)^{1/2}} - k$.

Set $\dfrac{dE}{dx} = 0$: $1.4kx = k\left(25+x^2\right)^{1/2} \Rightarrow 1.96x^2 = x^2+25 \Rightarrow 0.96x^2 = 25 \Rightarrow x = \frac{5}{\sqrt{0.96}} \approx 5.1$. Testing against the value of E at the endpoints: $E(0) = 1.4k(5) + 13k = 20k$, $E(5.1) \approx 17.9k$, $E(13) \approx 19.5k$. Thus, to minimize energy, the bird should fly to a point about 5.1 km from B.

(b) If W/L is large, the bird would fly to a point C that is closer to B than to D to minimize the energy used flying over water. If W/L is small, the bird would fly to a point C that is closer to D than to B to minimize the distance of the flight.

$E = W\sqrt{25+x^2} + L(13-x) \Rightarrow \dfrac{dE}{dx} = \dfrac{Wx}{\sqrt{25+x^2}} - L = 0$ when $\dfrac{W}{L} = \dfrac{\sqrt{25+x^2}}{x}$. By the same sort of argument as in part (a), this ratio will give the minimal expenditure of energy if the bird heads for the point x km from B.

(c) For flight direct to D, $x = 13$, so from part (b), $W/L = \frac{\sqrt{25+13^2}}{13} \approx 1.07$. There is no value of W/L for which the bird should fly directly to B. But note that $\lim_{x\to 0^+} (W/L) = \infty$, so if the point at which E is a minimum is close to B, then W/L is large.

(d) Assuming that the birds instinctively choose the path that minimizes the energy expenditure, we can use the equation for $dE/dx = 0$ from part (a) with $1.4k = c$, $x = 4$, and $k = 1$: $c(4) = 1 \cdot (25 + 4^2)^{1/2} \Rightarrow c = \sqrt{41}/4 \approx 1.6$.

47. (a) Distance = rate × time, so time = distance/rate. $T_1 = \frac{D}{c_1}$, $T_2 = \frac{2\,|PR|}{c_1} + \frac{|RS|}{c_2} = \frac{2h\sec\theta}{c_1} + \frac{D - 2h\tan\theta}{c_2}$, $T_3 = \frac{2\sqrt{h^2 + D^2/4}}{c_1} = \frac{\sqrt{4h^2+D^2}}{c_1}$.

(b) $\frac{dT_2}{d\theta} = \frac{2h}{c_1}\cdot\sec\theta\tan\theta - \frac{2h}{c_2}\sec^2\theta = 0$ when $2h\sec\theta\left(\frac{1}{c_1}\tan\theta - \frac{1}{c_2}\sec\theta\right) = 0 \Rightarrow \frac{1}{c_1}\frac{\sin\theta}{\cos\theta} - \frac{1}{c_2}\frac{1}{\cos\theta} = 0$ $\Rightarrow \frac{\sin\theta}{c_1\cos\theta} = \frac{1}{c_2\cos\theta} \Rightarrow \sin\theta = \frac{c_1}{c_2}$. The First Derivative Test shows that this gives a minimum.

(c) Using part (a) with $D = 1$ and $T_1 = 0.26$, we have $T_1 = \frac{D}{c_1} \Rightarrow c_1 = \frac{1}{0.26} \approx 3.85$ km/s. $T_3 = \frac{\sqrt{4h^2+D^2}}{c_1} \Rightarrow$ $4h^2 + D^2 = T_3^2c_1^2 \Rightarrow h = \frac{1}{2}\sqrt{T_3^2c_1^2 - D^2} = \frac{1}{2}\sqrt{(0.34)^2(1/0.26)^2 - 1^2} \approx 0.42$ km. To find c_2, we use $\sin\theta = \frac{c_1}{c_2}$ from part (b) and $T_2 = \frac{2h\sec\theta}{c_1} + \frac{D - 2h\tan\theta}{c_2}$ from part (a). From the figure,

$\sin\theta = \frac{c_1}{c_2} \Rightarrow \sec\theta = \frac{c_2}{\sqrt{c_2^2 - c_1^2}}$ and $\tan\theta = \frac{c_1}{\sqrt{c_2^2-c_1^2}}$, so

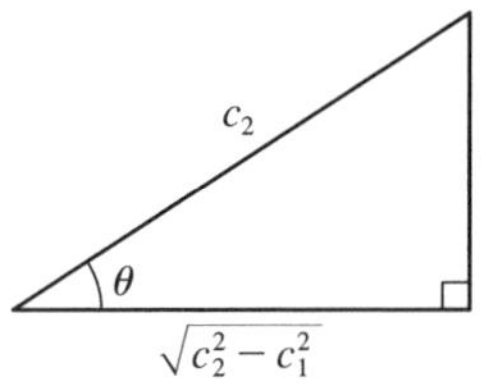

$T_2 = \frac{2hc_2}{c_1\sqrt{c_2^2-c_1^2}} + \frac{D\sqrt{c_2^2-c_1^2} - 2hc_1}{c_2\sqrt{c_2^2-c_1^2}}$. Using the values for T_2 [given as 0.32], h, c_1, and D, we can graph $Y_1 = T_2$ and $Y_2 = \frac{2hc_2}{c_1\sqrt{c_2^2-c_1^2}} + \frac{D\sqrt{c_2^2-c_1^2} - 2hc_1}{c_2\sqrt{c_2^2-c_1^2}}$ and find their intersection points. Doing so gives us $c_2 \approx 4.10$ and 7.66, but if $c_2 = 4.10$, then $\theta = \arcsin(c_1/c_2) \approx 69.6°$, which implies that point S is to the left of point R in the diagram. So $c_2 = 7.66$ km/s.

4.7 Applications to Business and Economics

1. (a) $C(0)$ represents the fixed costs of production, such as rent, utilities, machinery etc., which are incurred even when nothing is produced.

(b) The inflection point is the point at which $C''(x)$ changes from negative to positive; that is, the marginal cost $C'(x)$ changes from decreasing to increasing. Thus, the marginal cost is minimized at the inflection point.

(c) The marginal cost function is $C'(x)$. We graph it as in Example 1 in Section 2.8.

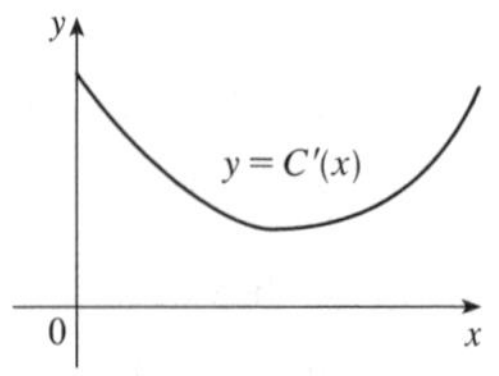

3. $c(x) = 21.4 - 0.002x$ and $c(x) = C(x)/x \Rightarrow C(x) = 21.4x - 0.002x^2$. $C'(x) = 21.4 - 0.004x$ and $C'(1000) = 17.4$. This means that the cost of producing the 1001st unit is about \$17.40.

5. (a) $C(x) = 16\,000 + 200x + 4x^{3/2}$, $C(1000) = 16\,000 + 200\,000 + 40\,000\sqrt{10} \approx 216\,000 + 126\,491$, so $C(1000) \approx \$342\,491$. $c(x) = C(x)/x = \dfrac{16\,000}{x} + 200 + 4x^{1/2}$, $c(1000) \approx \$342.49/\text{unit}$. $C'(x) = 200 + 6x^{1/2}$, $C'(1000) = 200 + 60\sqrt{10} \approx \$389.74/\text{unit}$.

(b) We must have $C'(x) = c(x) \quad \Leftrightarrow \quad 200 + 6x^{1/2} = \dfrac{16\,000}{x} + 200 + 4x^{1/2} \quad \Leftrightarrow \quad 2x^{3/2} = 16\,000 \quad \Leftrightarrow$ $x = (8000)^{2/3} = 400$ units. To check that this is a minimum, we calculate $c'(x) = \dfrac{-16\,000}{x^2} + \dfrac{2}{\sqrt{x}} = \dfrac{2}{x^2}(x^{3/2} - 8000)$. This is negative for $x < (8000)^{2/3} = 400$, zero at $x = 400$, and positive for $x > 400$, so c is decreasing on $(0, 400)$ and increasing on $(400, \infty)$. Thus, c has an absolute minimum at $x = 400$. [*Note:* $c''(x)$ is *not* positive for all $x > 0$.]

(c) The minimum average cost is $c(400) = 40 + 200 + 80 = \$320/\text{unit}$.

7. (a) $C(x) = 3700 + 5x - 0.04x^2 + 0.0003x^3 \quad \Rightarrow \quad C'(x) = 5 - 0.08x + 0.0009x^2$ (marginal cost).
$c(x) = \dfrac{C(x)}{x} = \dfrac{3700}{x} + 5 - 0.04x + 0.0003x^2$ (average cost).

(b)

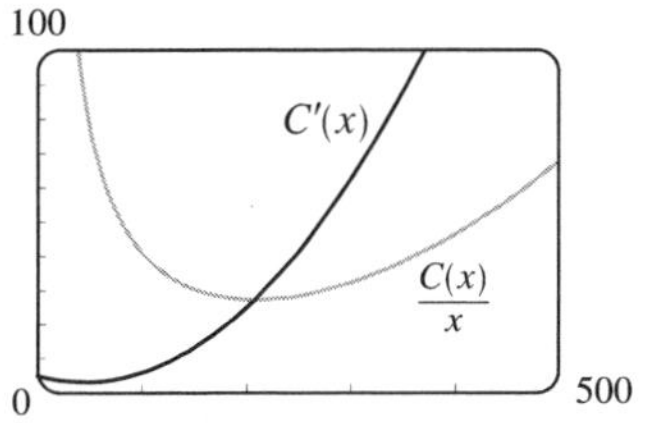

The graphs intersect at $(208.51, 27.45)$, so the production level that minimizes average cost is about 209 units.

(c) $c'(x) = -\dfrac{3700}{x^2} - 0.04 + 0.0006x = 0 \quad \Rightarrow \quad 3700 + 0.04x^2 - 0.0006x^3 = 0 \quad \Rightarrow \quad x_1 \approx 208.51$.
$c(x_1) \approx \$27.45/\text{unit}$.

(d) The marginal cost is given by $C'(x)$, so to find its minimum value we'll find the derivative of C'; that is, C''.
$C''(x) = -0.08 + 0.0018x = 0 \quad \Rightarrow \quad x_1 = \frac{800}{18} = 44.\overline{4}$. $C'(x_1) = \$3.22/\text{unit}$.
$C'''(x) = 0.0018 > 0$ for all x, so this is the minimum marginal cost. C''' is the second derivative of C'.

9. $C(x) = 680 + 4x + 0.01x^2$, $p(x) = 12 - x/500$. Then $R(x) = xp(x) = 12x - x^2/500$. If the profit is maximum, then $R'(x) = C'(x)$ [See the box preceding Example 2.] $\Leftrightarrow \quad 12 - x/250 = 4 + 0.02x \quad \Leftrightarrow \quad 8 = 0.024x \quad \Leftrightarrow$ $x = 8/0.024 = \frac{1000}{3}$. The profit is maximized if $P''(x) < 0$, but since $P''(x) = R''(x) - C''(x)$, we can just check the condition $R''(x) < C''(x)$. Now $R''(x) = -\frac{1}{250} < 0.02 = C''(x)$, so $x = \frac{1000}{3}$ gives a maximum.

11. The marginal cost function, $C'(x)$, starts to increase when $C''(x)$ becomes positive.
$C(x) = 0.0008x^3 - 0.72x^2 + 325.3x + 78\,000 \quad \Rightarrow \quad C'(x) = 0.0024x^2 - 1.44x + 325.3 \quad \Rightarrow$
$C''(x) = 0.0048x - 1.44$. $\quad C'' > 0 \quad \Rightarrow \quad 0.0048x > 1.44 \quad \Rightarrow \quad x > \dfrac{1.44}{0.0048} = 300$. Thus, the marginal cost function starts to increase when the production level is 300.

13. (a) $C(x) = 1200 + 12x - 0.1x^2 + 0.0005x^3$.
$R(x) = xp(x) = 29x - 0.00021x^2$.
Since the profit is maximized when $R'(x) = C'(x)$, we examine the curves R and C in the figure, looking for x-values at which the slopes of the tangent lines are equal. It appears that $x = 200$ is a good estimate.

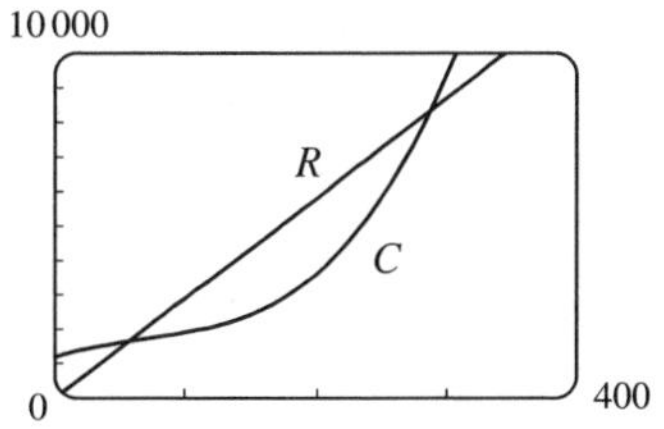

(b) $R'(x) = C'(x) \Rightarrow 29 - 0.00042x = 12 - 0.2x + 0.0015x^2 \Rightarrow 0.0015x^2 - 0.19958x - 17 = 0 \Rightarrow$ $x \approx 192.06$ (for $x > 0$). As in Exercise 9, $R''(x) < C''(x) \Rightarrow -0.00042 < -0.2 + 0.003x \Leftrightarrow$ $0.003x > 0.19958 \Leftrightarrow x > 66.5$. Our value of 192 is in this range, so we have a maximum profit when we produce 192 metres of fabric.

15. (a) We are given that the demand function p is linear and $p(27\,000) = 10$, $p(33\,000) = 8$, so the slope is $\frac{10-8}{27\,000 - 33\,000} = -\frac{1}{3000}$ and an equation of the line is $y - 10 = \left(-\frac{1}{3000}\right)(x - 27\,000) \Rightarrow$ $y = p(x) = -\frac{1}{3000}x + 19 = 19 - (x/3000)$.

(b) The revenue is $R(x) = xp(x) = 19x - (x^2/3000) \Rightarrow R'(x) = 19 - (x/1500) = 0$ when $x = 28\,500$. Since $R''(x) = -1/1500 < 0$, the maximum revenue occurs when $x = 28\,500 \Rightarrow$ the price is $p(28\,500) = \$9.50$.

17. (a) As in Example 3, we see that the demand function p is linear. We are given that $p(1000) = 450$ and deduce that $p(1100) = 440$, since a \$10 reduction in price increases sales by 100 per week. The slope for p is $\frac{440 - 450}{1100 - 1000} = -\frac{1}{10}$, so an equation is $p - 450 = -\frac{1}{10}(x - 1000)$ or $p(x) = -\frac{1}{10}x + 550$.

(b) $R(x) = xp(x) = -\frac{1}{10}x^2 + 550x$. $R'(x) = -\frac{1}{5}x + 550 = 0$ when $x = 5(550) = 2750$. $p(2750) = 275$, so the rebate should be $450 - 275 = \$175$.

(c) $C(x) = 68\,000 + 150x \Rightarrow P(x) = R(x) - C(x) = -\frac{1}{10}x^2 + 550x - 68\,000 - 150x = -\frac{1}{10}x^2 + 400x - 68\,000$, $P'(x) = -\frac{1}{5}x + 400 = 0$ when $x = 2000$. $p(2000) = 350$. Therefore, the rebate to maximize profits should be $450 - 350 = \$100$.

19. If the reorder quantity is x, then the manager places $\dfrac{800}{x}$ orders per year. Storage costs for the year are $\frac{1}{2}x \cdot 4 = 2x$ dollars. Handling costs are \$100 per delivery, for a total of $\dfrac{800}{x} \cdot 100 = \dfrac{80\,000}{x}$ dollars. The total costs for the year are $C(x) = 2x + \dfrac{80\,000}{x}$. To minimize $C(x)$, we calculate $C'(x) = 2 - \dfrac{80\,000}{x^2} = \dfrac{2}{x^2}(x^2 - 40\,000)$. This is negative when $x < 200$, zero when $x = 200$, and positive when $x > 200$, so C is decreasing on $(0, 200)$ and increasing on $(200, \infty)$, reaching its absolute minimum when $x = 200$. Thus, the optimal reorder quantity is 200 cases. The manager will place 4 orders per year for a total cost of $C(200) = \$800$.

4.8 Newton's Method

1. (a)

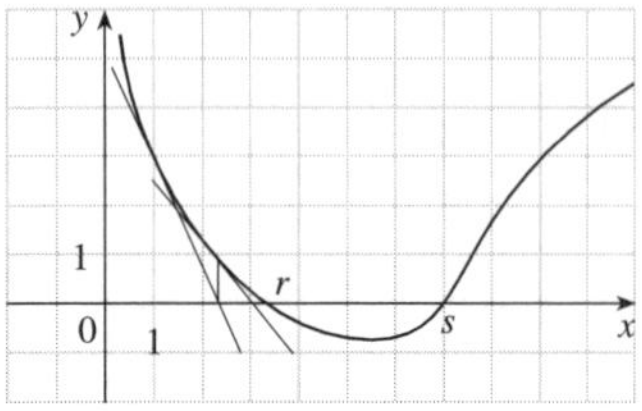

The tangent line at $x = 1$ intersects the x-axis at $x \approx 2.3$, so $x_2 \approx 2.3$. The tangent line at $x = 2.3$ intersects the x-axis at $x \approx 3$, so $x_3 \approx 3.0$.

(b) $x_1 = 5$ would *not* be a better first approximation than $x_1 = 1$ since the tangent line is nearly horizontal. In fact, the second approximation for $x_1 = 5$ appears to be to the left of $x = 1$.

3. Since $x_1 = 3$ and $y = 5x - 4$ is tangent to $y = f(x)$ at $x = 3$, we simply need to find where the tangent line intersects the x-axis. $y = 0 \Rightarrow 5x_2 - 4 = 0 \Rightarrow x_2 = \frac{4}{5}$.

5. $f(x) = x^4 - 20 \quad \Rightarrow \quad f'(x) = 4x^3$, so $x_{n+1} = x_n - \dfrac{f(x_n)}{f'(x_n)} = x_n - \dfrac{x_n^4 - 20}{4x_n^3}$.

Now $x_1 = 2 \quad \Rightarrow \quad x_2 = 2 - \dfrac{2^4 - 20}{4(2)^3} = 2.125 \quad \Rightarrow \quad x_3 = 2.125 - \dfrac{(2.125)^4 - 20}{4(2.125)^3} \approx 2.1148$.

7. $f(x) = x^3 + x + 3 \quad \Rightarrow \quad f'(x) = 3x^2 + 1$, so $x_{n+1} = x_n - \dfrac{x_n^3 + x_n + 3}{3x_n^2 + 1}$.

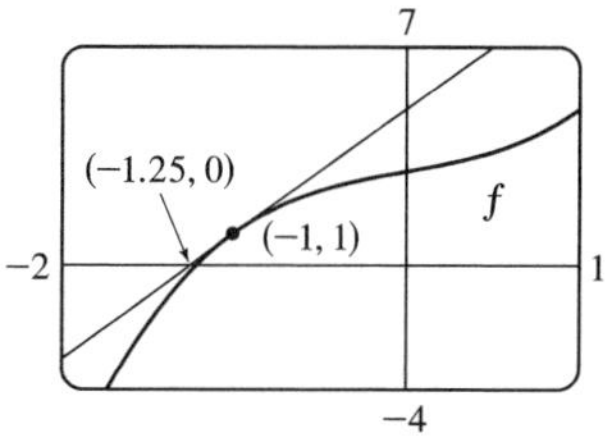

Now $x_1 = -1 \quad \Rightarrow$

$x_2 = -1 - \dfrac{(-1)^3 + (-1) + 3}{3(-1)^2 + 1} = -1 - \dfrac{-1 - 1 + 3}{3 + 1} = -1 - \dfrac{1}{4} = -1.25$.

Newton's method follows the tangent line at $(-1, 1)$ down to its intersection with the x-axis at $(-1.25, 0)$, giving the second approximation $x_2 = -1.25$.

9. To approximate $x = \sqrt[3]{30}$ (so that $x^3 = 30$), we can take $f(x) = x^3 - 30$. So $f'(x) = 3x^2$, and thus,

$x_{n+1} = x_n - \dfrac{x_n^3 - 30}{3x_n^2}$. Since $\sqrt[3]{27} = 3$ and 27 is close to 30, we'll use $x_1 = 3$. We need to find approximations until they agree to eight decimal places. $x_1 = 3 \quad \Rightarrow \quad x_2 \approx 3.111\,111\,11$, $x_3 \approx 3.107\,237\,34$, $x_4 \approx 3.107\,232\,51 \approx x_5$. So $\sqrt[3]{30} \approx 3.107\,232\,51$, to eight decimal places.

Here is a quick and easy method for finding the iterations for Newton's method on a programmable calculator. (The screens shown are from the TI-83 Plus, but the method is similar on other calculators.) Assign $f(x) = x^3 - 30$ to Y_1, and $f'(x) = 3x^2$ to Y_2. Now store $x_1 = 3$ in X and then enter $\text{X} - \text{Y}_1/\text{Y}_2 \to \text{X}$ to get $x_2 = 3.\overline{1}$. By successively pressing the ENTER key, you get the approximations $x_3, x_4, \ldots$.

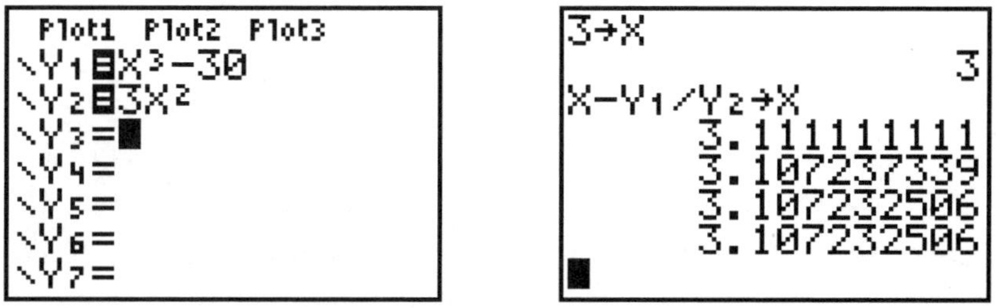

In Derive, load the utility file `SOLVE`. Enter `NEWTON(x^3-30,x,3)` and then `APPROXIMATE` to get [3, 3.111 111 11, 3.107 237 33, 3.107 232 50, 3.107 232 50]. You can request a specific iteration by adding a fourth argument. For example, `NEWTON(x^3-30,x,3,2)` gives [3, 3.111 111 11, 3.107 237 33].

In Maple, make the assignments $f := x \to x\hat{}3 - 30$;, $g := x \to x - f(x)/D(f)(x)$;, and $x := 3.$;. Repeatedly execute the command $x := g(x)$; to generate successive approximations.

In Mathematica, make the assignments $f[x\_] := x\hat{}3 - 30$, $g[x\_] := x - f[x]/f'[x]$, and $x = 3$. Repeatedly execute the command $x = g[x]$ to generate successive approximations.

11. $\sin x = x^2$, so $f(x) = \sin x - x^2 \quad \Rightarrow \quad f'(x) = \cos x - 2x \quad \Rightarrow$

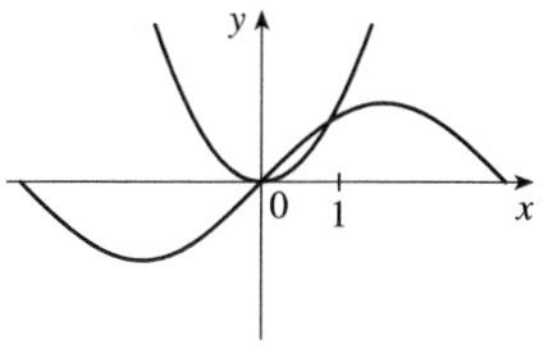

$x_{n+1} = x_n - \dfrac{\sin x_n - x_n^2}{\cos x_n - 2x_n}$. From the figure, the positive root of $\sin x = x^2$ is near 1. $x_1 = 1 \quad \Rightarrow \quad x_2 \approx 0.891\,396$, $x_3 \approx 0.876\,985$, $x_4 \approx 0.876\,726 \approx x_5$. So the positive root is $0.876\,726$, to six decimal places.

13. 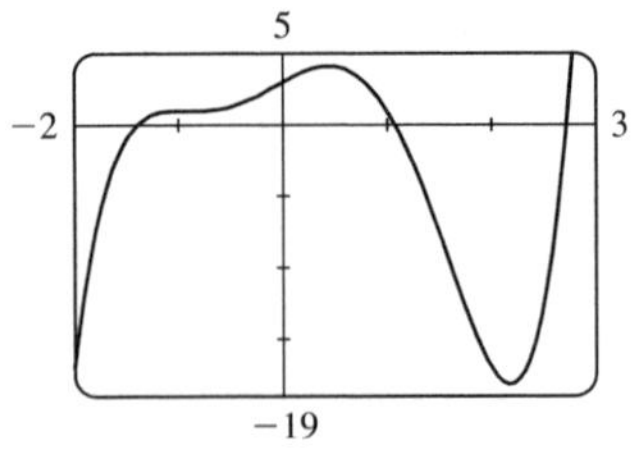

$f(x) = x^5 - x^4 - 5x^3 - x^2 + 4x + 3 \;\Rightarrow$

$f'(x) = 5x^4 - 4x^3 - 15x^2 - 2x + 4 \;\Rightarrow$

$x_{n+1} = x_n - \dfrac{x_n^5 - x_n^4 - 5x_n^3 - x_n^2 + 4x_n + 3}{5x_n^4 - 4x_n^3 - 15x_n^2 - 2x_n + 4}$. From the graph of f, there appear to be roots near -1.4, 1.1, and 2.7.

$x_1 = -1.4$	$x_1 = 1.1$	$x_1 = 2.7$
$x_2 \approx -1.392\,109\,70$	$x_2 \approx 1.077\,804\,02$	$x_2 \approx 2.720\,462\,50$
$x_3 \approx -1.391\,946\,98$	$x_3 \approx 1.077\,394\,42$	$x_3 \approx 2.719\,878\,70$
$x_4 \approx -1.391\,946\,91 \approx x_5$	$x_4 \approx 1.077\,394\,28 \approx x_5$	$x_4 \approx 2.719\,878\,22 \approx x_5$

To eight decimal places, the roots of the equation are $-1.391\,946\,91$, $1.077\,394\,28$, and $2.719\,878\,22$.

15. 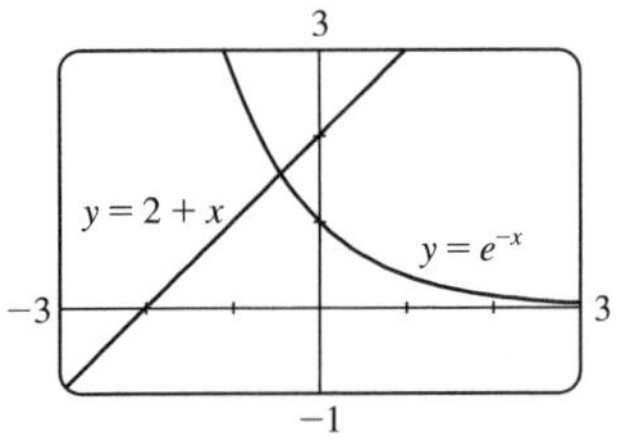

Solving $e^{-x} = 2 + x$ is the same as solving $f(x) = e^{-x} - x - 2 = 0$.

$f'(x) = -e^{-x} - 1 \;\Rightarrow\; x_{n+1} = x_n - \dfrac{e^{-x_n} - x_n - 2}{-e^{-x_n} - 1}$. From the graph of $y = e^{-x}$ and $y = 2 + x$, there appears to be a root near $x = -0.5$. Now $x_1 = -0.5 \;\Rightarrow\; x_2 \approx -0.443\,851\,67$, $x_3 \approx -0.442\,854\,70$, $x_4 \approx -0.442\,854\,40 \approx x_5$. To eight decimal places, the root of the equation is $-0.442\,854\,40$.

17. 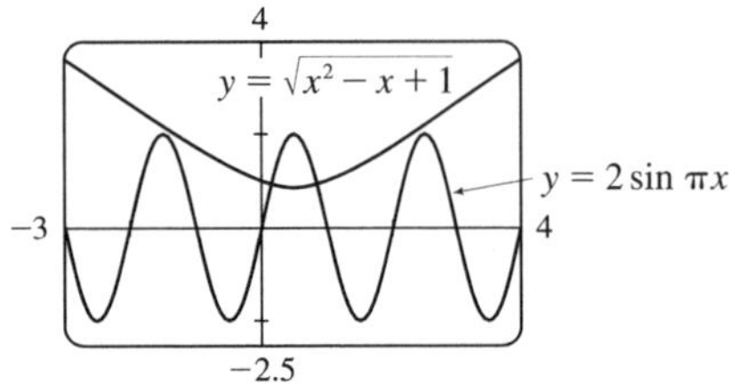

From the graph, we see that there are roots of this equation near 0.2 and 0.8. $f(x) = \sqrt{x^2 - x + 1} - 2\sin \pi x \;\Rightarrow$

$f'(x) = \dfrac{2x - 1}{2\sqrt{x^2 - x + 1}} - 2\pi\cos \pi x$, so

$$x_{n+1} = x_n - \frac{\sqrt{x_n^2 - x_n + 1} - 2\sin \pi x_n}{\dfrac{2x_n - 1}{2\sqrt{x_n^2 - x_n + 1}} - 2\pi\cos \pi x_n}.$$

Taking $x_1 = 0.2$, we get $x_2 \approx 0.152\,120\,15$, $x_3 \approx 0.154\,380\,67$, $x_4 \approx 0.154\,385\,00 \approx x_5$. Taking $x_1 = 0.8$, we get $x_2 \approx 0.847\,879\,85$, $x_3 \approx 0.845\,619\,33$, $x_4 \approx 0.845\,615\,00 \approx x_5$. To eight decimal places, the roots of the equation are $0.154\,385\,00$ and $0.845\,615\,00$.

19. 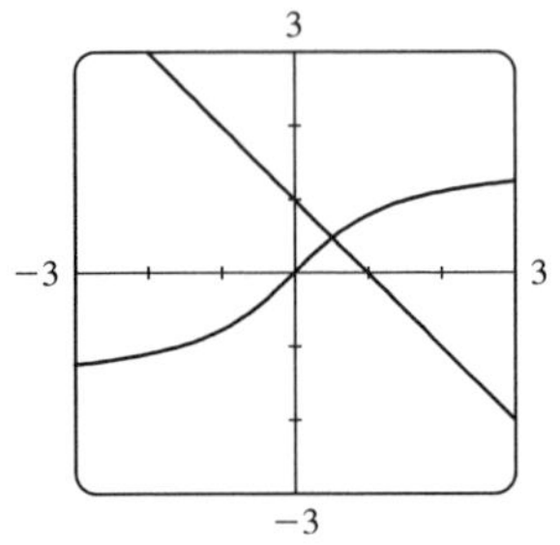

From the graph of $y = \tan^{-1} x$ and $y = 1 - x$, there appears to be a point of intersection near $x = 0.5$. Solving $\tan^{-1} x = 1 - x$ is the same as solving $f(x) = \tan^{-1} x + x - 1 = 0$. $f(x) = \tan^{-1} x + x - 1 \;\Rightarrow$

$f'(x) = \dfrac{1}{1 + x^2} + 1$, so $x_{n+1} = x_n - \dfrac{\tan^{-1} x_n + x_n - 1}{1/(1 + x_n^2) + 1}$.

Now $x_1 = 0.5 \;\Rightarrow\; x_2 \approx 0.520\,195\,77$, $x_3 \approx 0.520\,268\,99 \approx x_4$. To eight decimal places, the root of the equation is $0.520\,268\,99$.

21. (a) $f(x) = x^2 - a \;\Rightarrow\; f'(x) = 2x$, so Newton's method gives

$$x_{n+1} = x_n - \frac{x_n^2 - a}{2x_n} = x_n - \frac{1}{2}x_n + \frac{a}{2x_n} = \frac{1}{2}x_n + \frac{a}{2x_n} = \frac{1}{2}\left(x_n + \frac{a}{x_n}\right).$$

(b) Using (a) with $a = 1000$ and $x_1 = \sqrt{900} = 30$, we get $x_2 \approx 31.666\,667$, $x_3 \approx 31.622\,807$, and $x_4 \approx 31.622\,777 \approx x_5$. So $\sqrt{1000} \approx 31.622\,777$.

23. $f(x) = x^3 - 3x + 6 \Rightarrow f'(x) = 3x^2 - 3$. If $x_1 = 1$, then $f'(x_1) = 0$ and the tangent line used for approximating x_2 is horizontal. Attempting to find x_2 results in trying to divide by zero.

25. For $f(x) = x^{1/3}$, $f'(x) = \frac{1}{3}x^{-2/3}$ and

$$x_{n+1} = x_n - \frac{f(x_n)}{f'(x_n)} = x_n - \frac{x_n^{1/3}}{\frac{1}{3}x_n^{-2/3}} = x_n - 3x_n = -2x_n.$$

Therefore, each successive approximation becomes twice as large as the previous one in absolute value, so the sequence of approximations fails to converge to the root, which is 0. In the figure, we have $x_1 = 0.5$, $x_2 = -2(0.5) = -1$, and $x_3 = -2(-1) = 2$.

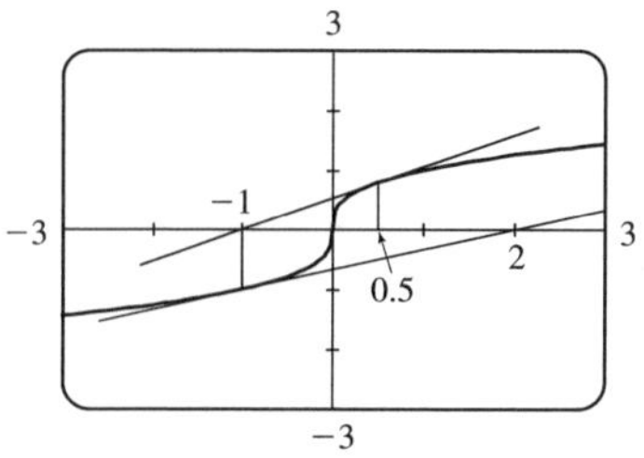

27.

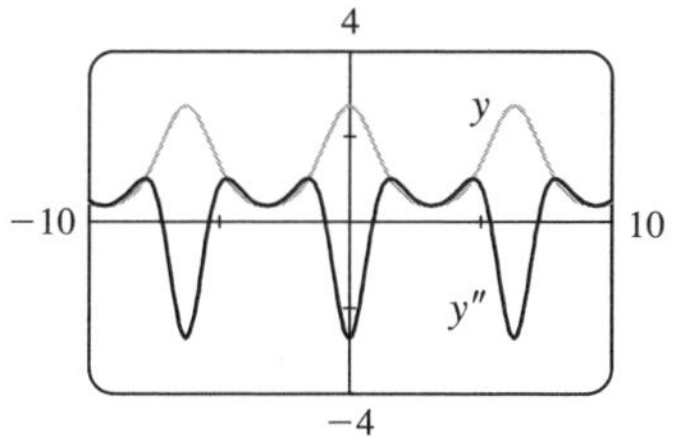

From the figure, we see that $y = f(x) = e^{\cos x}$ is periodic with period 2π. To find the x-coordinates of the IP, we only need to approximate the zeros of y'' on $[0, \pi]$. $f'(x) = -e^{\cos x}\sin x \Rightarrow f''(x) = e^{\cos x}\left(\sin^2 x - \cos x\right)$. Since $e^{\cos x} \neq 0$, we will use Newton's method with $g(x) = \sin^2 x - \cos x$, $g'(x) = 2\sin x\,\cos x + \sin x$, and $x_1 = 1$. $x_2 \approx 0.904\,173$, $x_3 \approx 0.904\,557 \approx x_4$. Thus, $(0.904\,557, 1.855\,277)$ is the IP.

29.

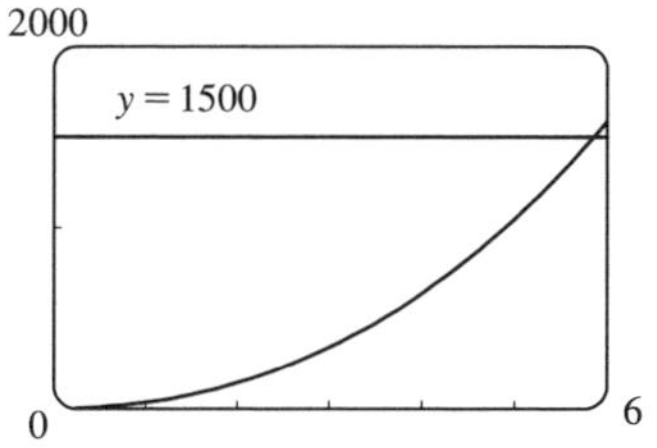

The volume of the silo, in terms of its radius, is

$V(r) = \pi r^2(10) + \frac{1}{2}\left(\frac{4}{3}\pi r^3\right) = 10\pi r^2 + \frac{2}{3}\pi r^3$.

From a graph of V, we see that $V(r) = 1500$ at $r \approx 5.9$ m. Now we use Newton's method to solve the equation $V(r) - 1500 = 0$.

$$\frac{dV}{dr} = 20\pi r + 2\pi r^2, \text{ so } r_{n+1} = r_n - \frac{10\pi r_n^2 + \frac{2}{3}\pi r_n^3 - 1500}{20\pi r_n + 2\pi r_n^2}.$$

Taking $r_1 = 5.9$, we get $r_2 \approx 5.8597$, $r_3 \approx 5.8595 \approx r_4$. So in order for the silo to hold 1500 m^3 of grain, its radius must be about 5.86 m.

31. In this case, $A = 18\,000$, $R = 375$, and $n = 5(12) = 60$. So the formula $A = \dfrac{R}{i}\left[1 - (1+i)^{-n}\right]$ becomes

$18\,000 = \dfrac{375}{x}\left[1 - (1+x)^{-60}\right] \Leftrightarrow 48x = 1 - (1+x)^{-60}$ [multiply each term by $(1+x)^{60}$] $\Leftrightarrow$

$48x(1+x)^{60} - (1+x)^{60} + 1 = 0$. Let the LHS be called $f(x)$, so that

$$\begin{aligned} f'(x) &= 48x(60)(1+x)^{59} + 48(1+x)^{60} - 60(1+x)^{59} \\ &= 12(1+x)^{59}\left[4x(60) + 4(1+x) - 5\right] = 12(1+x)^{59}(244x - 1) \end{aligned}$$

$x_{n+1} = x_n - \dfrac{48x_n(1+x_n)^{60} - (1+x_n)^{60} + 1}{12(1+x_n)^{59}(244x_n - 1)}$. An interest rate of 1% per month seems like a reasonable estimate for $x = i$. So let $x_1 = 1\% = 0.01$, and we get $x_2 \approx 0.008\,220\,2$, $x_3 \approx 0.007\,680\,2$, $x_4 \approx 0.007\,629\,1$, $x_5 \approx 0.007\,628\,6 \approx x_6$. Thus, the dealer is charging a monthly interest rate of 0.76286% (or 9.55% per year, compounded monthly).

4.9 Antiderivatives

1. $f(x) = 6x^2 - 8x + 3 \Rightarrow F(x) = 6\dfrac{x^{2+1}}{2+1} - 8\dfrac{x^{1+1}}{1+1} + 3x + C = 2x^3 - 4x^2 + 3x + C$

Check: $F'(x) = 2 \cdot 3x^2 - 4 \cdot 2x + 3 + 0 = 6x^2 - 8x + 3 = f(x)$

3. $f(x) = 5x^{1/4} - 7x^{3/4} \Rightarrow F(x) = 5\dfrac{x^{1/4+1}}{\frac{1}{4}+1} - 7\dfrac{x^{3/4+1}}{\frac{3}{4}+1} + C = 5\dfrac{x^{5/4}}{5/4} - 7\dfrac{x^{7/4}}{7/4} + C = 4x^{5/4} - 4x^{7/4} + C$

5. $f(x) = \sqrt[3]{x} + \dfrac{5}{x^6} = x^{1/3} + 5x^{-6}$ has domain $(-\infty, 0) \cup (0, \infty)$, so

$$F(x) = \begin{cases} \dfrac{x^{1/3+1}}{\frac{1}{3}+1} + 5\,\dfrac{x^{-6+1}}{-6+1} + C_1 = \frac{3}{4}x^{4/3} - x^{-5} + C_1 & \text{if } x < 0 \\ \frac{3}{4}x^{4/3} - x^{-5} + C_2 & \text{if } x > 0 \end{cases}$$

See Example 1(b) for a similar problem.

7. $f(u) = \dfrac{u^4 + 3\sqrt{u}}{u^2} = \dfrac{u^4}{u^2} + \dfrac{3u^{1/2}}{u^2} = u^2 + 3u^{-3/2} \Rightarrow$

$F(u) = \dfrac{u^3}{3} + 3\dfrac{u^{-3/2+1}}{-3/2+1} + C = \dfrac{1}{3}u^3 + 3\dfrac{u^{-1/2}}{-1/2} + C = \dfrac{1}{3}u^3 - \dfrac{6}{\sqrt{u}} + C$

9. $g(\theta) = \cos\theta - 5\sin\theta \Rightarrow G(\theta) = \sin\theta - 5(-\cos\theta) + C = \sin\theta + 5\cos\theta + C$

11. $f(x) = 2x + 5(1 - x^2)^{-1/2} = 2x + \dfrac{5}{\sqrt{1 - x^2}} \Rightarrow F(x) = x^2 + 5\sin^{-1}x + C$

13. $f(x) = 5x^4 - 2x^5 \Rightarrow F(x) = 5 \cdot \dfrac{x^5}{5} - 2 \cdot \dfrac{x^6}{6} + C = x^5 - \frac{1}{3}x^6 + C.$

$F(0) = 4 \Rightarrow 0^5 - \frac{1}{3} \cdot 0^6 + C = 4 \Rightarrow C = 4$, so $F(x) = x^5 - \frac{1}{3}x^6 + 4.$

The graph confirms our answer since $f(x) = 0$ when F has a local maximum, f is positive when F is increasing, and f is negative when F is decreasing.

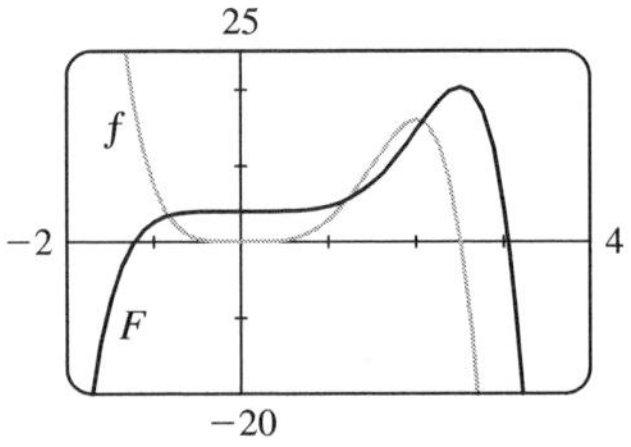

15. $f''(x) = 6x + 12x^2 \Rightarrow f'(x) = 6 \cdot \dfrac{x^2}{2} + 12 \cdot \dfrac{x^3}{3} + C = 3x^2 + 4x^3 + C \Rightarrow$

$f(x) = 3 \cdot \dfrac{x^3}{3} + 4 \cdot \dfrac{x^4}{4} + Cx + D = x^3 + x^4 + Cx + D$ [C and D are just arbitrary constants]

17. $f''(x) = 1 + x^{4/5} \Rightarrow f'(x) = x + \frac{5}{9}x^{9/5} + C \Rightarrow$

$f(x) = \frac{1}{2}x^2 + \frac{5}{9} \cdot \frac{5}{14}x^{14/5} + Cx + D = \frac{1}{2}x^2 + \frac{25}{126}x^{14/5} + Cx + D$

19. $f'(x) = \sqrt{x}(6 + 5x) = 6x^{1/2} + 5x^{3/2} \Rightarrow f(x) = 4x^{3/2} + 2x^{5/2} + C.$

$f(1) = 6 + C$ and $f(1) = 10 \Rightarrow C = 4$, so $f(x) = 4x^{3/2} + 2x^{5/2} + 4.$

21. $f'(t) = 2\cos t + \sec^2 t \Rightarrow f(t) = 2\sin t + \tan t + C$ because $-\pi/2 < t < \pi/2$.

$f\left(\frac{\pi}{3}\right) = 2\left(\sqrt{3}/2\right) + \sqrt{3} + C = 2\sqrt{3} + C$ and $f\left(\frac{\pi}{3}\right) = 4 \Rightarrow C = 4 - 2\sqrt{3}$, so $f(t) = 2\sin t + \tan t + 4 - 2\sqrt{3}.$

23. $f''(x) = 24x^2 + 2x + 10 \Rightarrow f'(x) = 8x^3 + x^2 + 10x + C.$ $f'(1) = 8 + 1 + 10 + C$ and $f'(1) = -3 \Rightarrow$ $19 + C = -3 \Rightarrow C = -22$, so $f'(x) = 8x^3 + x^2 + 10x - 22$ and hence, $f(x) = 2x^4 + \frac{1}{3}x^3 + 5x^2 - 22x + D.$

$f(1) = 2 + \frac{1}{3} + 5 - 22 + D$ and $f(1) = 5 \Rightarrow D = 22 - \frac{7}{3} = \frac{59}{3}$, so $f(x) = 2x^4 + \frac{1}{3}x^3 + 5x^2 - 22x + \frac{59}{3}.$

25. $f''(\theta) = \sin\theta + \cos\theta \Rightarrow f'(\theta) = -\cos\theta + \sin\theta + C.$ $f'(0) = -1 + C$ and $f'(0) = 4 \Rightarrow C = 5$, so $f'(\theta) = -\cos\theta + \sin\theta + 5$ and hence, $f(\theta) = -\sin\theta - \cos\theta + 5\theta + D.$ $f(0) = -1 + D$ and $f(0) = 3 \Rightarrow D = 4$, so $f(\theta) = -\sin\theta - \cos\theta + 5\theta + 4.$

27. $f''(x) = x^{-2}, x > 0 \Rightarrow f'(x) = -1/x + C \Rightarrow f(x) = -\ln|x| + Cx + D = -\ln x + Cx + D$ (since $x > 0$). $f(1) = 0 \Rightarrow C + D = 0$ and $f(2) = 0 \Rightarrow -\ln 2 + 2C + D = 0 \Rightarrow$ $-\ln 2 + 2C - C = 0$ [since $D = -C$] $\Rightarrow -\ln 2 + C = 0 \Rightarrow C = \ln 2$ and $D = -\ln 2$. So $f(x) = -\ln x + (\ln 2)x - \ln 2$.

29. Given $f'(x) = 2x + 1$, we have $f(x) = x^2 + x + C$. Since f passes through $(1, 6)$, $f(1) = 6 \Rightarrow 1^2 + 1 + C = 6 \Rightarrow$ $C = 4$. Therefore, $f(x) = x^2 + x + 4$ and $f(2) = 2^2 + 2 + 4 = 10$.

31.

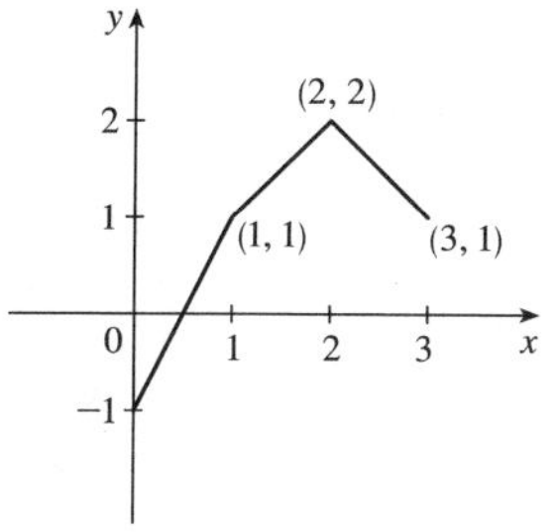

$$f'(x) = \begin{cases} 2 & \text{if } 0 \le x < 1 \\ 1 & \text{if } 1 < x < 2 \\ -1 & \text{if } 2 < x \le 3 \end{cases} \Rightarrow f(x) = \begin{cases} 2x + C & \text{if } 0 \le x < 1 \\ x + D & \text{if } 1 < x < 2 \\ -x + E & \text{if } 2 < x \le 3 \end{cases}$$

$f(0) = -1 \Rightarrow 2(0) + C = -1 \Rightarrow C = -1$. Starting at the point $(0, -1)$ and moving to the right on a line with slope 2 gets us to the point $(1, 1)$. The slope for $1 < x < 2$ is 1, so we get to the point $(2, 2)$. Here we have used the fact that f is continuous. We can include the point $x = 1$ on either the first or the second part of f. The line connecting $(1, 1)$ to $(2, 2)$ is $y = x$, so $D = 0$. The slope for $2 < x \le 3$ is -1, so we get to $(3, 1)$. $f(3) = 1 \Rightarrow -3 + E = 1 \Rightarrow E = 4$. Thus,

$$f(x) = \begin{cases} 2x - 1 & \text{if } 0 \le x \le 1 \\ x & \text{if } 1 < x < 2 \\ -x + 4 & \text{if } 2 \le x \le 3 \end{cases}$$

Note that $f'(x)$ does not exist at $x = 1$ or at $x = 2$.

33.

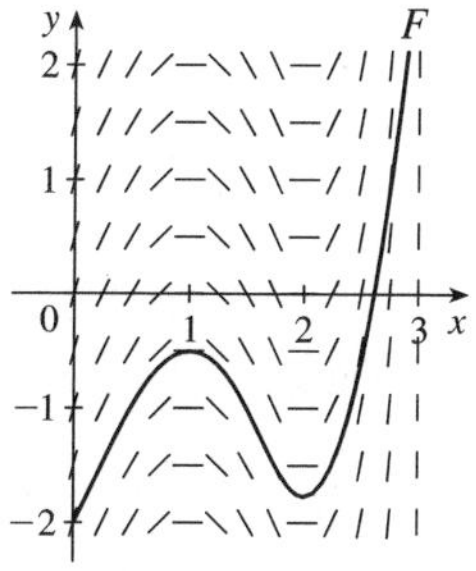

35.

x	$f(x)$
0	1
0.5	0.959
1.0	0.841
1.5	0.665
2.0	0.455
2.5	0.239
3.0	0.047

x	$f(x)$
3.5	−0.100
4.0	−0.189
4.5	−0.217
5.0	−0.192
5.5	−0.128
6.0	−0.047

We compute slopes [values of $f(x) = (\sin x)/x$ for $0 < x < 2\pi$] as in the table [$\lim_{x \to 0^+} f(x) = 1$] and draw a direction field as in Example 6. Then we use the direction field to graph F starting at $(0, 0)$.

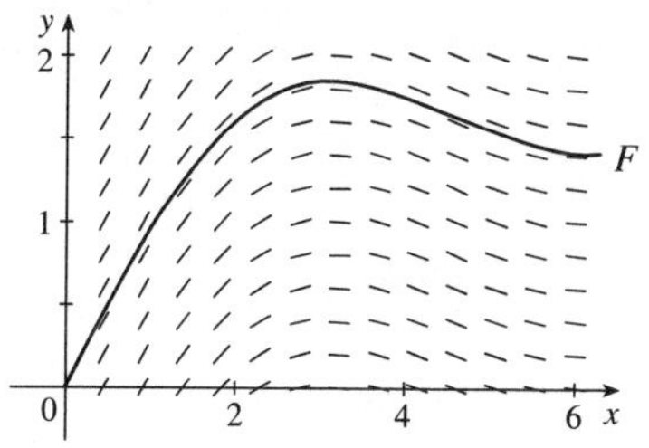

37. 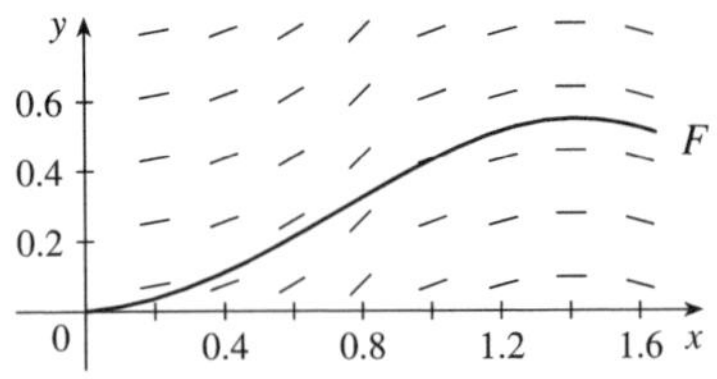

Remember that the given table values of f are the slopes of F at any x. For example, at $x = 1.4$, the slope of F is $f(1.4) = 0$.

39. $v(t) = s'(t) = \sin t - \cos t \;\Rightarrow\; s(t) = -\cos t - \sin t + C$. $s(0) = -1 + C$ and $s(0) = 0 \;\Rightarrow\; C = 1$, so $s(t) = -\cos t - \sin t + 1$.

41. (a) We first observe that since the stone is dropped 450 m above the ground, $v(0) = 0$ and $s(0) = 450$.
$v'(t) = a(t) = -9.8 \;\Rightarrow\; v(t) = -9.8t + C$. Now $v(0) = 0 \;\Rightarrow\; C = 0$, so $v(t) = -9.8t \;\Rightarrow$ $s(t) = -4.9t^2 + D$. Last, $s(0) = 450 \;\Rightarrow\; D = 450 \;\Rightarrow\; s(t) = 450 - 4.9t^2$.

(b) The stone reaches the ground when $s(t) = 0$. $450 - 4.9t^2 = 0 \;\Rightarrow\; t^2 = 450/4.9 \;\Rightarrow\; t_1 = \sqrt{450/4.9} \approx 9.58$ s.

(c) The velocity with which the stone strikes the ground is $v(t_1) = -9.8\sqrt{450/4.9} \approx -93.9$ m/s.

(d) This is just reworking parts (a) and (b) with $v(0) = -5$. Using $v(t) = -9.8t + C$, $v(0) = -5 \;\Rightarrow\; 0 + C = -5 \;\Rightarrow$ $v(t) = -9.8t - 5$. So $s(t) = -4.9t^2 - 5t + D$ and $s(0) = 450 \;\Rightarrow\; D = 450 \;\Rightarrow\; s(t) = -4.9t^2 - 5t + 450$. Solving $s(t) = 0$ by using the quadratic formula gives us $t = \left(5 \pm \sqrt{8845}\right)/(-9.8) \;\Rightarrow\; t_1 \approx 9.09$ s.

43. By Exercise 42 with $a = -9.8$, $s(t) = -4.9t^2 + v_0t + s_0$ and $v(t) = s'(t) = -9.8t + v_0$. So
$[v(t)]^2 = (-9.8t + v_0)^2 = (9.8)^2t^2 - 19.6v_0t + v_0^2 = v_0^2 + 96.04t^2 - 19.6v_0t = v_0^2 - 19.6\left(-4.9t^2 + v_0t\right)$.
But $-4.9t^2 + v_0t$ is just $s(t)$ without the s_0 term; that is, $s(t) - s_0$. Thus, $[v(t)]^2 = v_0^2 - 19.6\,[s(t) - s_0]$.

45. Marginal cost $= 1.92 - 0.002x = C'(x) \;\Rightarrow\; C(x) = 1.92x - 0.001x^2 + K$. But $C(1) = 1.92 - 0.001 + K = 562 \;\Rightarrow$ $K = 560.081$. Therefore, $C(x) = 1.92x - 0.001x^2 + 560.081 \;\Rightarrow\; C(100) = 742.081$, so the cost of producing 100 items is \$742.08.

47. Using Exercise 42 with $a = -9.8$, $v_0 = 0$, and $s_0 = h$ (the height of the cliff), we know that the height at time t is $s(t) = -4.9t^2 + h$. $v(t) = s'(t) = -9.8t$ and $v(t) = -40 \;\Rightarrow\; -9.8t = -40 \;\Rightarrow\; t_f = \frac{40}{9.8} \approx 4.08$ s, so $0 = s(t_f) = -4.9(t_f)^2 + h \;\Rightarrow\; h = 4.9(t_f)^2 \approx 81.6$ m.

49. $a(t) = k$, the initial velocity is 50 km/h $= 50 \cdot \frac{1000}{3600} = \frac{125}{9}$ m/s, and the final velocity (after 5 seconds) is 80 km/h $= 80 \cdot \frac{1000}{3600} = \frac{200}{9}$ m/s. So $v(t) = kt + C$ and $v(0) = \frac{125}{9} \;\Rightarrow\; C = \frac{125}{9}$. Thus, $v(t) = kt + \frac{125}{9} \;\Rightarrow$ $v(5) = 5k + \frac{125}{9}$. But $v(5) = \frac{200}{9}$, so $5k + \frac{125}{9} = \frac{200}{9} \;\Rightarrow\; 5k = \frac{75}{9} \;\Rightarrow\; k = \frac{5}{3} \approx 1.67\ \text{m/s}^2$.

51. Let the acceleration be $a(t) = k\ \text{km/h}^2$. We have $v(0) = 100$ km/h and we can take the initial position $s(0)$ to be 0. We want the time t_f for which $v(t) = 0$ to satisfy $s(t) < 0.08$ km. In general, $v'(t) = a(t) = k$, so $v(t) = kt + C$, where $C = v(0) = 100$. Now $s'(t) = v(t) = kt + 100$, so $s(t) = \frac{1}{2}kt^2 + 100t + D$, where $D = s(0) = 0$.
Thus, $s(t) = \frac{1}{2}kt^2 + 100t$. Since $v(t_f) = 0$, we have $kt_f + 100 = 0$ or $t_f = -100/k$, so

$$s(t_f) = \frac{1}{2}k\left(-\frac{100}{k}\right)^2 + 100\left(-\frac{100}{k}\right) = 10{,}000\left(\frac{1}{2k} - \frac{1}{k}\right) = -\frac{5000}{k}.$$ The condition $s(t_f)$ must satisfy is

$$-\frac{5000}{k} < 0.08 \;\Rightarrow\; -\frac{5000}{0.08} > k \quad [k \text{ is negative}] \;\Rightarrow\; k < -62\,500\ \text{km/h}^2,$$ or equivalently,

$k < -\frac{3125}{648} \approx -4.82\ \text{m/s}^2$.

53. (a) The Mean Value Theorem says that there exists a number c in the interval (x_1, x_2) such that $H'(c) = \dfrac{H(x_2) - H(x_1)}{x_2 - x_1}$.

Since $H = G - F$ and G and F are antiderivatives of f, $H'(c) = G'(c) - F'(c) = f(c) - f(c) = 0$. So now

$\dfrac{H(x_2) - H(x_1)}{x_2 - x_1} = 0 \quad \Rightarrow \quad H(x_2) - H(x_1) = 0 \; (x_2 \neq x_1) \quad \Rightarrow \quad H(x_2) = H(x_1)$. Since this is true for any

$x_1 < x_2$ in I, H must be a constant function.

(b) We have $H = G - F$ and $H(x) = C$, so $C = G - F \quad \Rightarrow \quad G(x) = F(x) + C$. Thus, any antiderivative G can be expressed as $F(x) + C$.

55. (a) First note that $145 \text{ km/h} = 145 \times \frac{1000}{3600} \text{ m/s} = \frac{725}{18} \text{ m/s}$. Then $a(t) = 1.2 \text{ m/s}^2 \quad \Rightarrow \quad v(t) = 1.2t + C$, but

$v(0) = 0 \quad \Rightarrow \quad C = 0$. Now $1.2t = \frac{725}{18}$ when $t_1 = \frac{3625}{108} \approx 33.6$ s, so it takes about 34 s to reach about 40.3 m/s.

Therefore, taking $s(0) = 0$, we have $s(t) = 0.6t^2$, $0 \leq t \leq t_1$. So $s(t_1) \approx 676$ m. 15 minutes $= 15(60) = 900$ s, so for

$t_1 < t \leq 900 + t_1$, we have $v(t) = \frac{725}{18} \text{ m/s} \quad \Rightarrow \quad s(900 + t_1) = \frac{725}{18}(900) + 676 \approx 36\,926$ m.

(b) As in part (a), the train accelerates for 33.6 s and travels 676 m while doing so. Similarly, it decelerates for 33.6 s and travels 676 m at the end of its trip. During the remaining $900 - 67.1 = 832.9$ s it travels at $\frac{725}{18}$ m/s, so the distance travelled is $\frac{725}{18} \cdot 832.9 \approx 33\,546$ m. Thus, the total distance is about $676 + 33\,546 + 676 = 34\,898$ m.

(c) 72 km $= 72\,000$ m. Subtract $2(676)$ to take care of the speeding up and slowing down, and we have about $70\,648$ m at $\frac{725}{18}$ m/s for a trip of $70\,648/\left(\frac{725}{18}\right) \approx 1754$ s at 145 km/h. The total time is about $1754 + 2t_1 \approx 1821 \text{ s} = 30$ min 21 s.

(d) $37.5(60) = 2250$ s. $\;2250 - 2t_1 \approx 2183$ s at maximum speed. $\;2183\left(\frac{725}{18}\right) + 2(676) \approx 89\,278$ total metres.

4 Review

CONCEPT CHECK

1. A function f has an **absolute maximum** at $x = c$ if $f(c)$ is the largest function value on the entire domain of f, whereas f has a **local maximum** at c if $f(c)$ is the largest function value when x is near c. See Figure 4 in Section 4.2.

2. (a) See Theorem 4.2.3.

(b) See the Closed Interval Method before Example 6 in Section 4.2.

3. (a) See Theorem 4.2.4.

(b) See Definition 4.2.5.

4. See the Mean Value Theorem in Section 4.3. Geometric interpretation—there is some point P on the graph of a function f [on the interval (a, b)] where the tangent line is parallel to the secant line that connects $(a, f(a))$ and $(b, f(b))$.

5. (a) See the I/D Test before Example 2 in Section 4.3.

(b) A function f is concave upward on an interval I if f' is an increasing function on I (or, equivalently, the graph of f lies above all of its tangent lines on I).

(c) See the Concavity Test before Example 4 in Section 4.3.

(d) An inflection point is a point where a curve changes its direction of concavity. They can be found by determining the points at which the second derivative changes sign.

6. (a) See the First Derivative Test after Example 2 in Section 4.3.

(b) See the Second Derivative Test before Example 4 in Section 4.3.

(c) See the note before Example 5 in Section 4.3.

7. (a) See l'Hospital's Rule and the three notes that follow it in Section 4.5.

(b) Write fg as $\dfrac{f}{1/g}$ or $\dfrac{g}{1/f}$.

(c) Convert the difference into a quotient using a common denominator, rationalizing, factoring, or some other method.

(d) Convert the power to a product by taking the natural logarithm of both sides of $y = f^g$ or by writing f^g as $e^{g \ln f}$.

8. Without calculus you could get misleading graphs that fail to show the most interesting features of a function. See the discussion following Figure 3 in Section 4.5 and the first paragraph in Section 4.4.

9. (a) See Figure 3 in Section 4.8.

(b) $x_2 = x_1 - \dfrac{f(x_1)}{f'(x_1)}$

(c) $x_{n+1} = x_n - \dfrac{f(x_n)}{f'(x_n)}$

(d) Newton's method is likely to fail or to work very slowly when $f'(x_1)$ is close to 0.

10. (a) See the definition at the beginning of Section 4.9.

(b) If F_1 and F_2 are both antiderivatives of f on an interval I, then they differ by a constant.

TRUE-FALSE QUIZ

1. False. For example, take $f(x) = x^3$, then $f'(x) = 3x^2$ and $f'(0) = 0$, but $f(0) = 0$ is not a maximum or minimum; $(0, 0)$ is an inflection point.

3. False. For example, $f(x) = x$ is continuous on $(0, 1)$ but attains neither a maximum nor a minimum value on $(0, 1)$. Don't confuse this with f being continuous on the *closed* interval $[a, b]$, which would make the statement true.

5. True. This is an example of part (b) of the I/D Test.

7. False. $f'(x) = g'(x) \;\Rightarrow\; f(x) = g(x) + C$. For example, if $f(x) = x + 2$ and $g(x) = x + 1$, then $f'(x) = g'(x) = 1$, but $f(x) \neq g(x)$.

9. True. The graph of one such function is sketched.

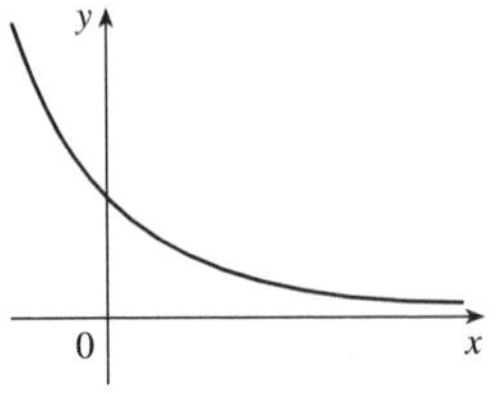

11. True. By the Mean Value Theorem, there exists a number c in $(0, 1)$ such that $f(1) - f(0) = f'(c)(1 - 0) = f'(c)$. Since $f'(c)$ is nonzero, $f(1) - f(0) \neq 0$, so $f(1) \neq f(0)$.

13. False. $\displaystyle\lim_{x \to 0} \frac{x}{e^x} = \frac{\lim_{x \to 0} x}{\lim_{x \to 0} e^x} = \frac{0}{1} = 0$, not 1.

EXERCISES

1. $f(x) = 10 + 27x - x^3, 0 \le x \le 4$. $f'(x) = 27 - 3x^2 = -3(x^2 - 9) = -3(x+3)(x-3) = 0$ only when $x = 3$ (since -3 is not in the domain). $f'(x) > 0$ for $x < 3$ and $f'(x) < 0$ for $x > 3$, so $f(3) = 64$ is a local maximum value. Checking the endpoints, we find $f(0) = 10$ and $f(4) = 54$. Thus, $f(0) = 10$ is the absolute minimum value and $f(3) = 64$ is the absolute maximum value.

3. $f(x) = \dfrac{x}{x^2 + x + 1}, -2 \le x \le 0$. $f'(x) = \dfrac{(x^2+x+1)(1) - x(2x+1)}{(x^2+x+1)^2} = \dfrac{1 - x^2}{(x^2+x+1)^2} = 0 \Leftrightarrow x = -1$ (since 1 is not in the domain). $f'(x) < 0$ for $-2 < x < -1$ and $f'(x) > 0$ for $-1 < x < 0$, so $f(-1) = -1$ is a local and absolute minimum value. $f(-2) = -\frac{2}{3}$ and $f(0) = 0$, so $f(0) = 0$ is an absolute maximum value.

5. (a) $f(x) = 2 - 2x - x^3$ is a polynomial, so there is no asymptote.

(b) $f'(x) = -2 - 3x^2 = -1(3x^2 + 2) < 0$, so f is decreasing on $\mathbb{R}$.

(c) No local extrema

(d) $f''(x) = -6x < 0$ on $(0, \infty)$ and $f''(x) > 0$ on $(-\infty, 0)$, so f is CD on $(0, \infty)$ and CU on $(-\infty, 0)$. There is an IP at $(0, 2)$.

(e)

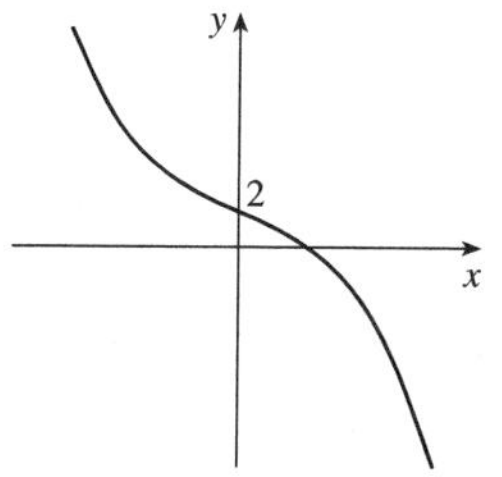

7. (a) $f(x) = x + \sqrt{1 - x}$ has no asymptote.

(b) $f'(x) = 1 - 1/(2\sqrt{1-x}) = 0 \Leftrightarrow 2\sqrt{1-x} = 1 \Leftrightarrow$ $1 - x = \frac{1}{4} \Leftrightarrow x = \frac{3}{4}$ and $f'(x) > 0 \Leftrightarrow x < \frac{3}{4}$, so f is increasing on $(-\infty, \frac{3}{4})$ and decreasing on $(\frac{3}{4}, 1)$.

(c) $f(\frac{3}{4}) = \frac{3}{4} + \sqrt{1 - \frac{3}{4}} = \frac{3}{4} + \sqrt{\frac{1}{4}} = \frac{3}{4} + \frac{1}{2} = \frac{5}{4}$ is a local maximum.

(d) $f''(x) = -\dfrac{1}{4(1-x)^{3/2}} < 0$ on the domain of f, so f is CD on $(-\infty, 1)$. No IP

(e)

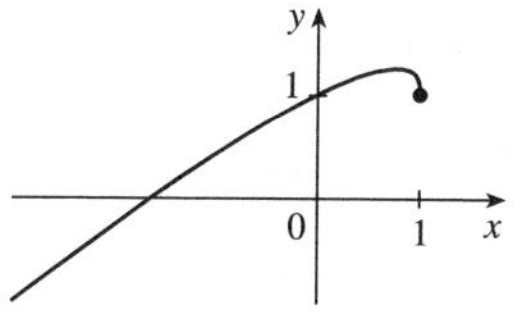

9. (a) $y = f(x) = \sin^2 x - 2\cos x$ has no asymptote.

(b) $y' = 2\sin x \cos x + 2\sin x = 2\sin x(\cos x + 1)$. $y' = 0 \Leftrightarrow \sin x = 0$ or $\cos x = -1 \Leftrightarrow x = n\pi$ or $x = (2n+1)\pi$. $y' > 0$ when $\sin x > 0$, since $\cos x + 1 \ge 0$ for all x. Therefore, $y' > 0$ (and so f is increasing) on $(2n\pi, (2n+1)\pi)$; $y' < 0$ (and so f is decreasing) on $((2n-1)\pi, 2n\pi)$ or equivalently, $((2n+1)\pi, (2n+2)\pi)$.

(c) Local maxima are $f((2n+1)\pi) = 2$; local minima are $f(2n\pi) = -2$.

(d) $y' = \sin 2x + 2\sin x \Rightarrow$

$$y'' = 2\cos 2x + 2\cos x = 2(2\cos^2 x - 1) + 2\cos x = 4\cos^2 x + 2\cos x - 2$$
$$= 2(2\cos^2 x + \cos x - 1) = 2(2\cos x - 1)(\cos x + 1)$$

$y'' = 0 \Leftrightarrow \cos x = \frac{1}{2}$ or $-1 \Leftrightarrow x = 2n\pi \pm \frac{\pi}{3}$ or $x = (2n+1)\pi$. $y'' > 0$ (and so f is CU) on $(2n\pi - \frac{\pi}{3}, 2n\pi + \frac{\pi}{3})$; $y'' \le 0$ (and so f is CD) on $(2n\pi + \frac{\pi}{3}, 2n\pi + \frac{5\pi}{3})$. There are inflection points at $(2n\pi \pm \frac{\pi}{3}, -\frac{1}{4})$.

(e)

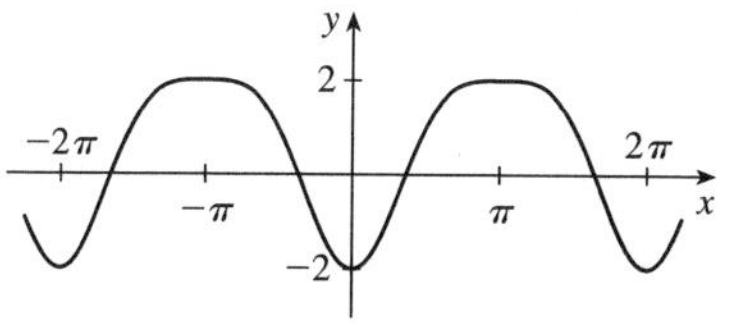

11. (a) $\lim_{x\to\pm\infty}\left(e^x + e^{-3x}\right) = \infty$, no asymptote.

(b) $y = f(x) = e^x + e^{-3x} \;\Rightarrow\; f'(x) = e^x - 3e^{-3x} = e^{-3x}\left(e^{4x} - 3\right) > 0 \;\Leftrightarrow$ $e^{4x} > 3 \;\Leftrightarrow\; 4x > \ln 3 \;\Leftrightarrow\; x > \frac{1}{4}\ln 3$, so f is increasing on $\left(\frac{1}{4}\ln 3, \infty\right)$ and decreasing on $\left(-\infty, \frac{1}{4}\ln 3\right)$.

(c) $f\left(\frac{1}{4}\ln 3\right) = 3^{1/4} + 3^{-3/4} \approx 1.75$ is a local and absolute minimum.

(d) $f''(x) = e^x + 9e^{-3x} > 0$, so f is CU on $(-\infty, \infty)$. No IP

(e)

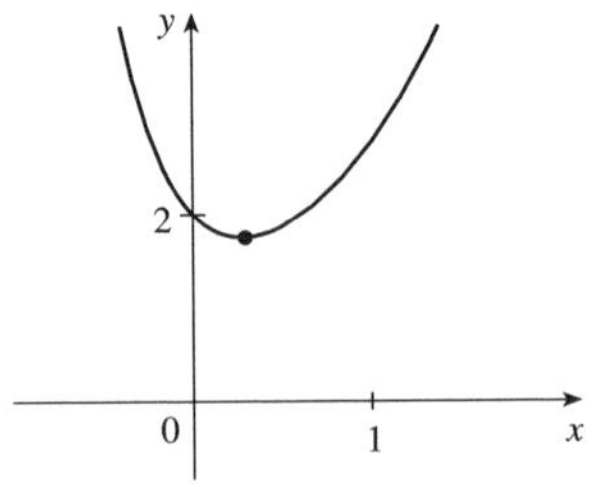

13. $f(x) = \dfrac{x^2 - 1}{x^3} \;\Rightarrow\; f'(x) = \dfrac{x^3(2x) - \left(x^2 - 1\right)3x^2}{x^6} = \dfrac{3 - x^2}{x^4} \;\Rightarrow$

$f''(x) = \dfrac{x^4(-2x) - \left(3 - x^2\right)4x^3}{x^8} = \dfrac{2x^2 - 12}{x^5}$

Estimates: From the graphs of f' and f'', it appears that f is increasing on $(-1.73, 0)$ and $(0, 1.73)$ and decreasing on $(-\infty, -1.73)$ and $(1.73, \infty)$; f has a local maximum of about $f(1.73) = 0.38$ and a local minimum of about $f(-1.7) = -0.38$; f is CU on $(-2.45, 0)$ and $(2.45, \infty)$, and CD on $(-\infty, -2.45)$ and $(0, 2.45)$; and f has inflection points at about $(-2.45, -0.34)$ and $(2.45, 0.34)$.

Exact: Now $f'(x) = \dfrac{3 - x^2}{x^4}$ is positive for $0 < x^2 < 3$, that is, f is increasing on $\left(-\sqrt{3}, 0\right)$ and $\left(0, \sqrt{3}\right)$; and $f'(x)$ is negative (and so f is decreasing) on $\left(-\infty, -\sqrt{3}\right)$ and $\left(\sqrt{3}, \infty\right)$. $f'(x) = 0$ when $x = \pm\sqrt{3}$.

f' goes from positive to negative at $x = \sqrt{3}$, so f has a local maximum of $f\left(\sqrt{3}\right) = \frac{\left(\sqrt{3}\right)^2 - 1}{\left(\sqrt{3}\right)^3} = \frac{2\sqrt{3}}{9}$; and since f is odd, we know that maxima on the interval $(0, \infty)$ correspond to minima on $(-\infty, 0)$, so f has a local minimum of $f\left(-\sqrt{3}\right) = -\frac{2\sqrt{3}}{9}$. Also, $f''(x) = \dfrac{2x^2 - 12}{x^5}$ is positive (so f is CU) on $\left(-\sqrt{6}, 0\right)$ and $\left(\sqrt{6}, \infty\right)$, and negative (so f is CD) on $\left(-\infty, -\sqrt{6}\right)$ and $\left(0, \sqrt{6}\right)$. There are IP at $\left(\sqrt{6}, \frac{5\sqrt{6}}{36}\right)$ and $\left(-\sqrt{6}, -\frac{5\sqrt{6}}{36}\right)$.

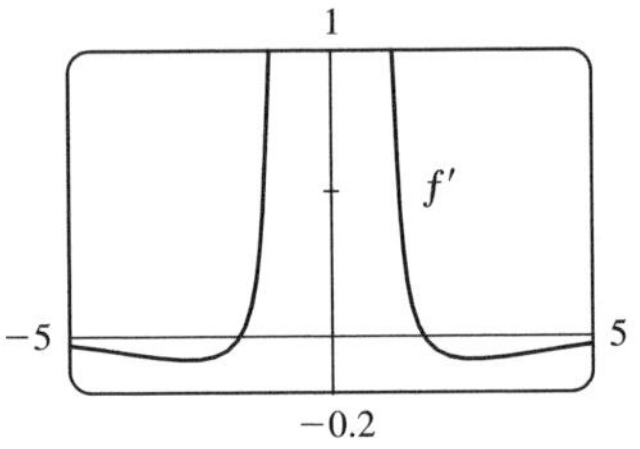

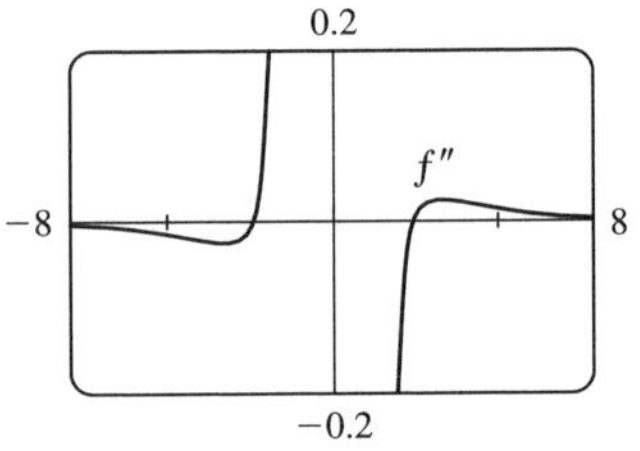

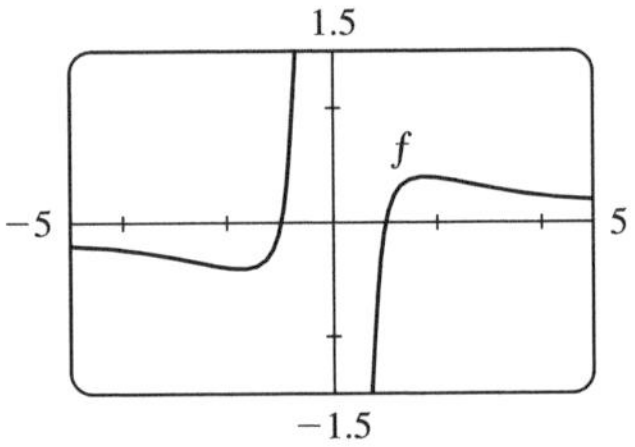

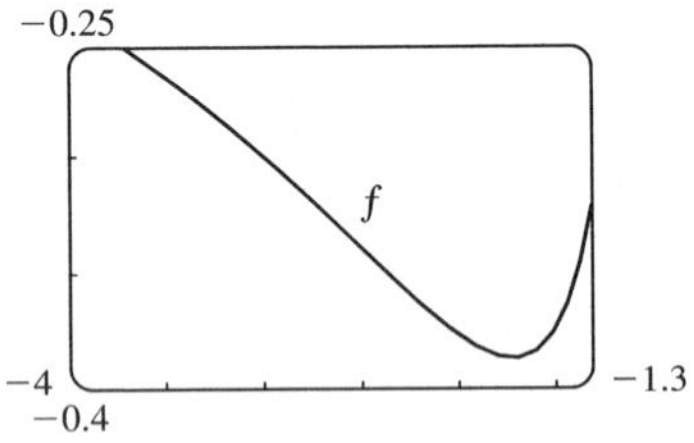

15. $f(x) = 3x^6 - 5x^5 + x^4 - 5x^3 - 2x^2 + 2 \;\Rightarrow\; f'(x) = 18x^5 - 25x^4 + 4x^3 - 15x^2 - 4x \;\Rightarrow$

$f''(x) = 90x^4 - 100x^3 + 12x^2 - 30x - 4$

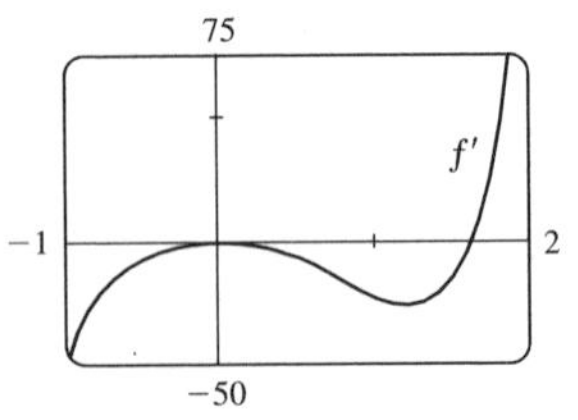

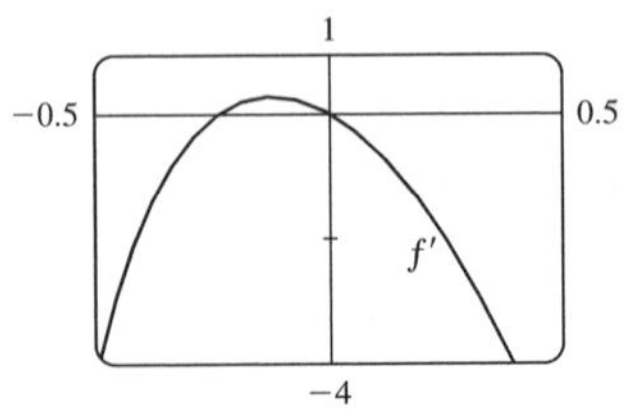

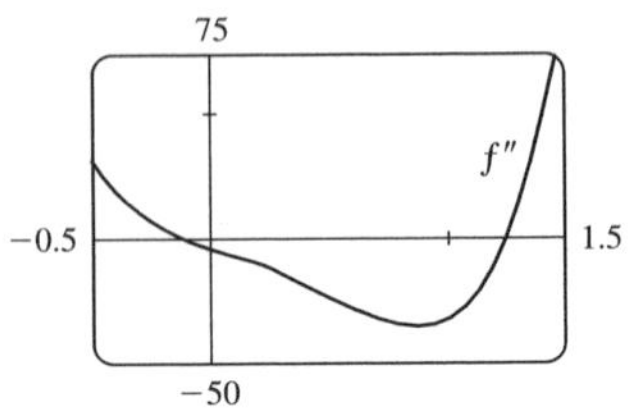

From the graphs of f' and f'', it appears that f is increasing on $(-0.23, 0)$ and $(1.62, \infty)$ and decreasing on $(-\infty, -0.23)$ and $(0, 1.62)$; f has a local maximum of about $f(0) = 2$ and local minima of about $f(-0.23) = 1.96$ and $f(1.62) = -19.2$; f is CU on $(-\infty, -0.12)$ and $(1.24, \infty)$ and CD on $(-0.12, 1.24)$; and f has inflection points at about $(-0.12, 1.98)$ and $(1.24, -12.1)$.

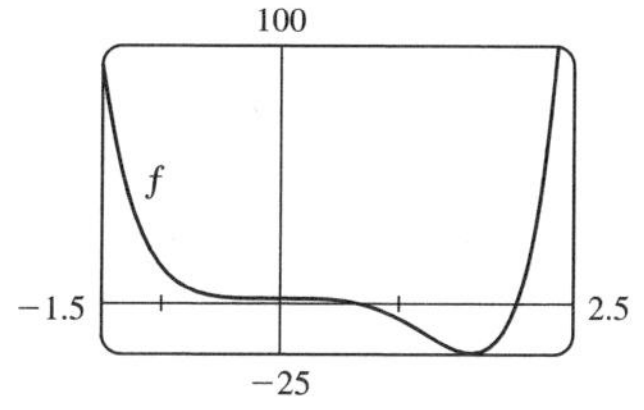

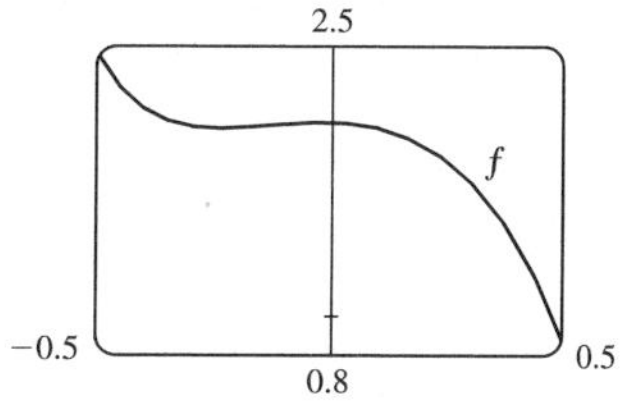

17.

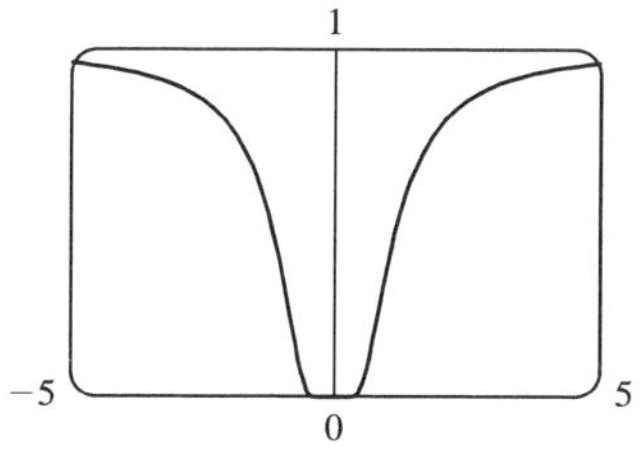

From the graph, we estimate the points of inflection to be about $(\pm 0.82, 0.22)$.

$f(x) = e^{-1/x^2} \quad\Rightarrow\quad f'(x) = 2x^{-3}e^{-1/x^2} \quad\Rightarrow$

$f''(x) = 2\left[x^{-3}(2x^{-3})e^{-1/x^2} + e^{-1/x^2}(-3x^{-4})\right] = 2x^{-6}e^{-1/x^2}\,(2 - 3x^2).$

This is 0 when $2 - 3x^2 = 0 \quad\Leftrightarrow\quad x = \pm\sqrt{\frac{2}{3}}$, so the inflection points are $\left(\pm\sqrt{\frac{2}{3}}, e^{-3/2}\right)$.

19. $f(x) = \arctan(\cos(3\arcsin x))$. We use a CAS to compute f' and f'', and to graph f, f', and f'':

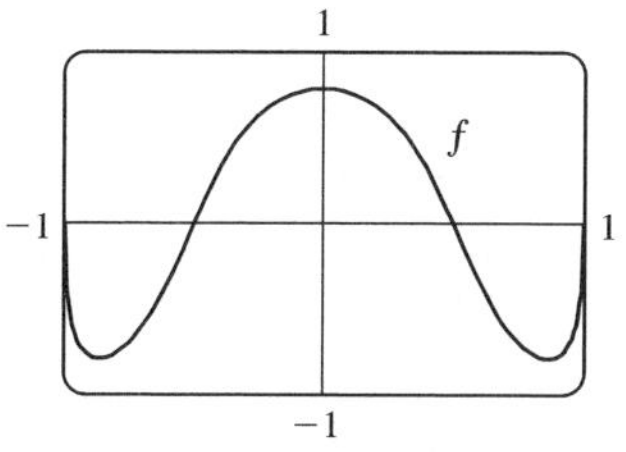

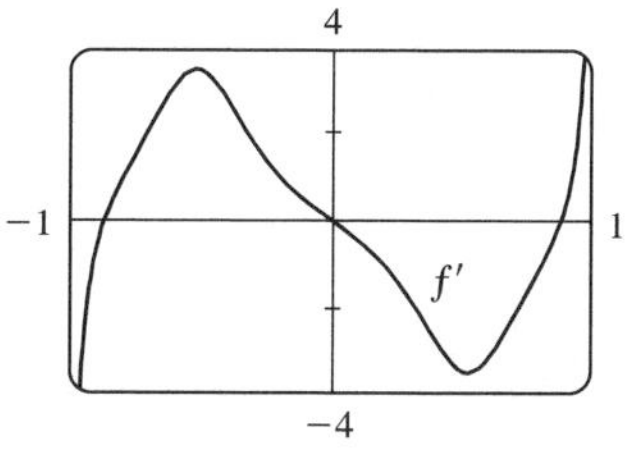

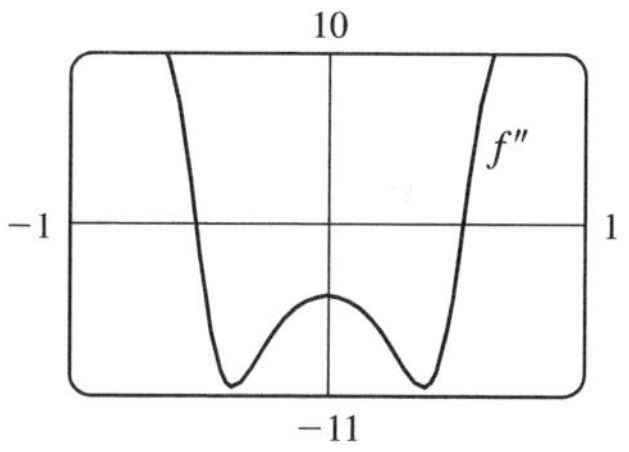

From the graph of f', it appears that the only maximum occurs at $x = 0$ and there are minima at $x = \pm 0.87$. From the graph of f'', it appears that there are inflection points at $x = \pm 0.52$.

21. The family of functions $f(x) = \ln(\sin x + C)$ all have the same period and all have maximum values at $x = \frac{\pi}{2} + 2\pi n$. Since the domain of ln is $(0, \infty)$, f has a graph only if $\sin x + C > 0$ somewhere. Since $-1 \le \sin x \le 1$, this happens if $C > -1$, that is, f has no graph if $C \le -1$. Similarly, if $C > 1$, then $\sin x + C > 0$ and f is continuous on $(-\infty, \infty)$. As C increases, the graph of f is shifted vertically upward and flattens out.

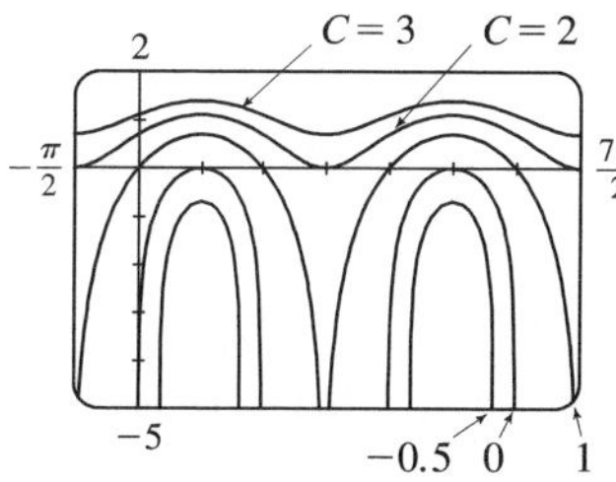

If $-1 < C \le 1$, f is defined where $\sin x + C > 0 \quad\Leftrightarrow\quad \sin x > -C \quad\Leftrightarrow\quad \sin^{-1}(-C) < x < \pi - \sin^{-1}(-C)$.

Since the period is 2π, the domain of f is $\left(2n\pi + \sin^{-1}(-C), (2n+1)\pi - \sin^{-1}(-C)\right)$, n an integer.

23. For $(1, 6)$ to be on the curve $y = x^3 + ax^2 + bx + 1$, we have that $6 = 1 + a + b + 1 \Rightarrow b = 4 - a$. Now $y' = 3x^2 + 2ax + b$ and $y'' = 6x + 2a$. Also, for $(1, 6)$ to be an inflection point it must be true that $y''(1) = 6(1) + 2a = 0 \Rightarrow a = -3 \Rightarrow b = 4 - (-3) = 7$. Note that with $a = -3$, we have $y'' = 6x - 6 = 6(x - 1)$, so y'' changes sign at $x = 1$, proving that $(1, 6)$ is a point of inflection. [This does not follow from the fact that $y''(1) = 0$.]

25. $$\lim_{x \to 0} \frac{\tan \pi x}{\ln(1 + x)} \overset{\text{H}}{=} \lim_{x \to 0} \frac{\pi \sec^2 \pi x}{1/(1 + x)} = \frac{\pi \cdot 1^2}{1/1} = \pi$$

27. $$\lim_{x \to 0} \frac{e^{4x} - 1 - 4x}{x^2} \overset{\text{H}}{=} \lim_{x \to 0} \frac{4e^{4x} - 4}{2x} \overset{\text{H}}{=} \lim_{x \to 0} \frac{16e^{4x}}{2} = \lim_{x \to 0} 8e^{4x} = 8 \cdot 1 = 8$$

29. $$\lim_{x \to \infty} x^3 e^{-x} = \lim_{x \to \infty} \frac{x^3}{e^x} \overset{\text{H}}{=} \lim_{x \to \infty} \frac{3x^2}{e^x} \overset{\text{H}}{=} \lim_{x \to \infty} \frac{6x}{e^x} \overset{\text{H}}{=} \lim_{x \to \infty} \frac{6}{e^x} = 0$$

31. $$\lim_{x \to 1^+} \left(\frac{x}{x - 1} - \frac{1}{\ln x} \right) = \lim_{x \to 1^+} \left(\frac{x \ln x - x + 1}{(x - 1) \ln x} \right) \overset{\text{H}}{=} \lim_{x \to 1^+} \frac{x \cdot (1/x) + \ln x - 1}{(x - 1) \cdot (1/x) + \ln x}$$
$$= \lim_{x \to 1^+} \frac{\ln x}{1 - 1/x + \ln x} \overset{\text{H}}{=} \lim_{x \to 1^+} \frac{1/x}{1/x^2 + 1/x} = \frac{1}{1 + 1} = \frac{1}{2}$$

33. We are given $d\theta/dt = -0.25$ rad/h. $\tan \theta = 400/x \Rightarrow$ $x = 400 \cot \theta \Rightarrow \dfrac{dx}{dt} = -400 \csc^2 \theta \, \dfrac{d\theta}{dt}$. When $\theta = \frac{\pi}{6}$,
$$\frac{dx}{dt} = -400(2)^2(-0.25) = 400 \text{ m/h}.$$

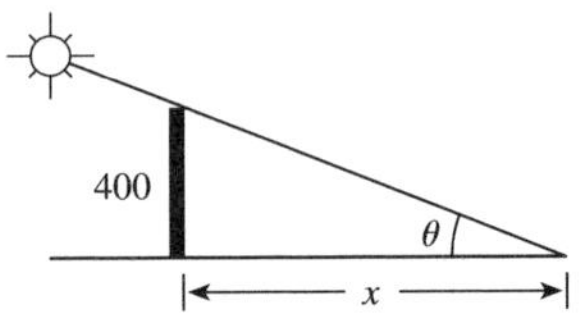

35. Given $dh/dt = 2$ and $dx/dt = 5$, find dz/dt. $z^2 = x^2 + h^2 \Rightarrow$
$$2z \frac{dz}{dt} = 2x \frac{dx}{dt} + 2h \frac{dh}{dt} \Rightarrow \frac{dz}{dt} = \frac{1}{z}(5x + 2h).$$ When $t = 3$,
$h = 15 + 3(2) = 21$ and $x = 5(3) = 15 \Rightarrow z = \sqrt{15^2 + 21^2} = \sqrt{666}$,
$$\text{so } \frac{dz}{dt} = \frac{1}{\sqrt{666}}[5(15) + 2(21)] = \frac{117}{\sqrt{666}} = \frac{39}{\sqrt{74}} \approx 4.53 \text{ m/s}.$$

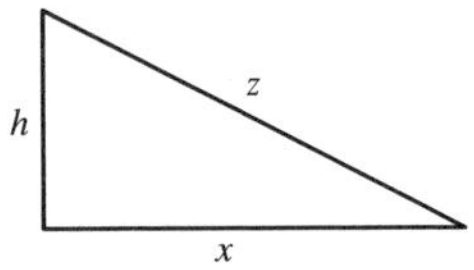

37. Call the two integers x and y. Then $x + 4y = 1000$, so $x = 1000 - 4y$. Their product is $P = xy = (1000 - 4y)y$, so our problem is to maximize the function $P(y) = 1000y - 4y^2$, where $0 < y < 250$ and y is an integer. $P'(y) = 1000 - 8y$, so $P'(y) = 0 \Leftrightarrow y = 125$. $P''(y) = -8 < 0$, so $P(125) = 62\,500$ is an absolute maximum. Since the optimal y turned out to be an integer, we have found the desired pair of numbers, namely $x = 1000 - 4(125) = 500$ and $y = 125$.

39.

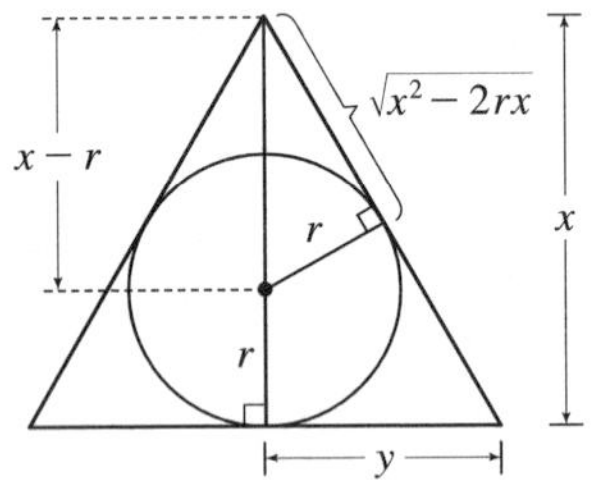

By similar triangles, $\dfrac{y}{x} = \dfrac{r}{\sqrt{x^2 - 2rx}}$, so the area of the triangle is
$$A(x) = \tfrac{1}{2}(2y)x = xy = \frac{rx^2}{\sqrt{x^2 - 2rx}} \Rightarrow$$
$$A'(x) = \frac{2rx\sqrt{x^2 - 2rx} - rx^2(x - r)/\sqrt{x^2 - 2rx}}{x^2 - 2rx}$$
$$= \frac{rx^2(x - 3r)}{(x^2 - 2rx)^{3/2}} = 0 \text{ when } x = 3r.$$

$A'(x) < 0$ when $2r < x < 3r$, $A'(x) > 0$ when $x > 3r$. So $x = 3r$ gives a minimum and $A(3r) = r(9r^2)/(\sqrt{3}\,r) = 3\sqrt{3}\,r^2$.

41. 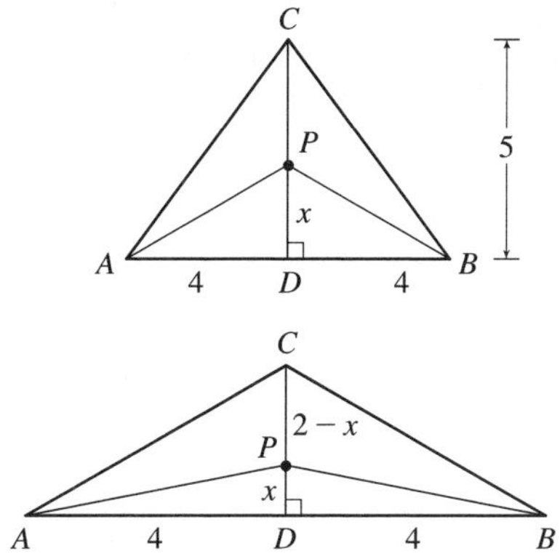

We minimize $L(x) = |PA| + |PB| + |PC| = 2\sqrt{x^2+16} + (5-x)$, $0 \le x \le 5$. $L'(x) = 2x/\sqrt{x^2+16} - 1 = 0 \quad \Leftrightarrow \quad 2x = \sqrt{x^2+16} \quad \Leftrightarrow$ $4x^2 = x^2 + 16 \quad \Leftrightarrow \quad x = \frac{4}{\sqrt{3}}$. $L(0) = 13$, $L\left(\frac{4}{\sqrt{3}}\right) \approx 11.9$, $L(5) \approx 12.8$, so the minimum occurs when $x = \frac{4}{\sqrt{3}} \approx 2.3$.

If $|CD| = 2$, $L(x)$ changes from $(5-x)$ to $(2-x)$ with $0 \le x \le 2$. But we still get $L'(x) = 0 \quad \Leftrightarrow \quad x = \frac{4}{\sqrt{3}}$, which isn't in the interval $[0, 2]$. Now $L(0) = 10$ and $L(2) = 2\sqrt{20} = 4\sqrt{5} \approx 8.9$. The minimum occurs when $P = C$.

43. $v = K\sqrt{\dfrac{L}{C} + \dfrac{C}{L}} \quad \Rightarrow \quad \dfrac{dv}{dL} = \dfrac{K}{2\sqrt{(L/C) + (C/L)}}\left(\dfrac{1}{C} - \dfrac{C}{L^2}\right) = 0 \quad \Leftrightarrow \quad \dfrac{1}{C} = \dfrac{C}{L^2} \quad \Leftrightarrow \quad L^2 = C^2 \quad \Leftrightarrow \quad L = C.$

This gives the minimum velocity since $v' < 0$ for $0 < L < C$ and $v' > 0$ for $L > C$.

45. Let x denote the number of \$1 decreases in ticket price. Then the ticket price is $\$12 - \$1(x)$, and the average attendance is $11\,000 + 1000(x)$. Now the revenue per game is

$$R(x) = (\text{price per person}) \times (\text{number of people per game})$$
$$= (12 - x)(11\,000 + 1000x) = -1000x^2 + 1000x + 132\,000$$

for $0 \le x \le 4$ (since the seating capacity is 15 000) $\Rightarrow \quad R'(x) = -2000x + 1000 = 0 \quad \Leftrightarrow \quad x = 0.5$. This is a maximum since $R''(x) = -2000 < 0$ for all x. Now we must check the value of $R(x) = (12 - x)(11\,000 + 1000x)$ at $x = 0.5$ and at the endpoints of the domain to see which value of x gives the maximum value of R. $R(0) = (12)(11\,000) = 132\,000$, $R(0.5) = (11.5)(11\,500) = 132\,250$, and $R(4) = (8)(15\,000) = 120\,000$. Thus, the maximum revenue of \$132 250 per game occurs when the average attendance is 11 500 and the ticket price is \$11.50.

47. $f(t) = \cos t + t - t^2 \quad \Rightarrow \quad f'(t) = -\sin t + 1 - 2t$. $f'(t)$ exists for all t, so to find the maximum of f, we can examine the zeros of f'. From the graph of f', we see that a good choice for t_1 is $t_1 = 0.3$. Use $g(t) = -\sin t + 1 - 2t$ and $g'(t) = -\cos t - 2$ to obtain $t_2 \approx 0.335\,352\,93$, $t_3 \approx 0.335\,418\,03 \approx t_4$. Since $f''(t) = -\cos t - 2 < 0$ for all t, $f(0.335\,418\,03) \approx 1.167\,185\,57$ is the absolute maximum.

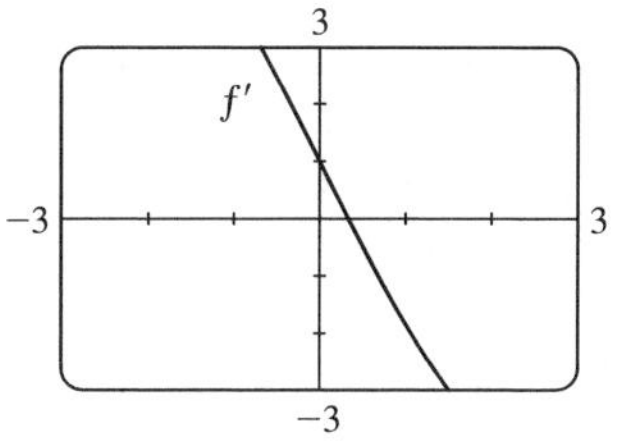

49. $f(x) = e^x - (2/\sqrt{x}) = e^x - 2x^{-1/2} \quad \Rightarrow \quad F(x) = e^x - 2\dfrac{x^{-1/2+1}}{-1/2+1} + C = e^x - 2\dfrac{x^{1/2}}{1/2} + C = e^x - 4\sqrt{x} + C$

51. $f'(x) = 2/(1 + x^2) \quad \Rightarrow \quad f(x) = 2\arctan x + C$. $f(0) = 2\arctan 0 + C = 0 + C = C$ and $f(0) = -1 \quad \Rightarrow \quad C = -1$. Therefore, $f(x) = 2\arctan x - 1$.

53. $f''(x) = 1 - 6x + 48x^2 \quad \Rightarrow \quad f'(x) = x - 3x^2 + 16x^3 + C$. $f'(0) = C$ and $f'(0) = 2 \quad \Rightarrow \quad C = 2$, so $f'(x) = x - 3x^2 + 16x^3 + 2$ and hence, $f(x) = \frac{1}{2}x^2 - x^3 + 4x^4 + 2x + D$.

$f(0) = D$ and $f(0) = 1 \quad \Rightarrow \quad D = 1$, so $f(x) = \frac{1}{2}x^2 - x^3 + 4x^4 + 2x + 1$.

55. (a) Since f is 0 just to the left of the y-axis, we must have a minimum of F at the same place since we are increasing through $(0, 0)$ on F. There must be a local maximum to the left of $x = -3$, since f changes from positive to negative there.

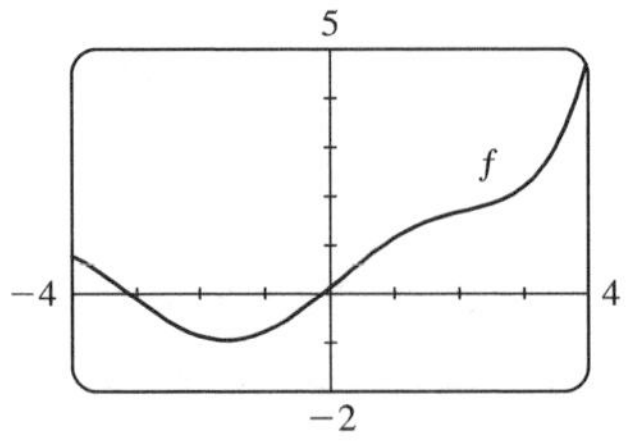

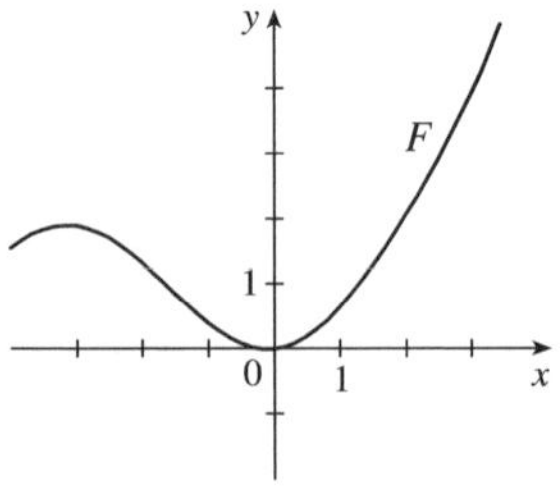

(b) $f(x) = 0.1e^x + \sin x \quad \Rightarrow$

$F(x) = 0.1e^x - \cos x + C$. $F(0) = 0 \quad \Rightarrow$

$0.1 - 1 + C = 0 \quad \Rightarrow \quad C = 0.9$, so

$F(x) = 0.1e^x - \cos x + 0.9$.

(c)

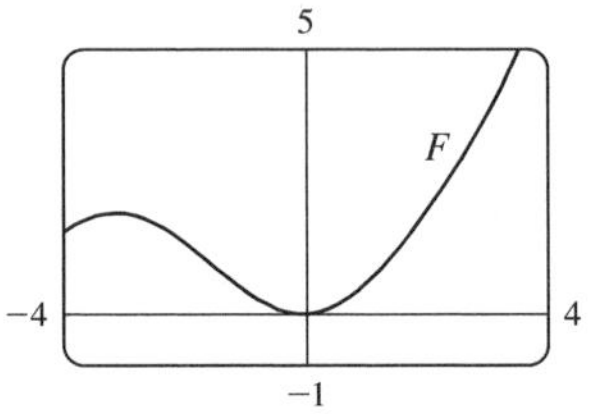

57. Choosing the positive direction to be upward, we have $a(t) = -9.8 \quad \Rightarrow \quad v(t) = -9.8t + v_0$, but $v(0) = 0 = v_0 \quad \Rightarrow$ $v(t) = -9.8t = s'(t) \quad \Rightarrow \quad s(t) = -4.9t^2 + s_0$, but $s(0) = s_0 = 500 \quad \Rightarrow \quad s(t) = -4.9t^2 + 500$. When $s = 0$, $-4.9t^2 + 500 = 0 \quad \Rightarrow \quad t_1 = \sqrt{\frac{500}{4.9}} \approx 10.1 \quad \Rightarrow \quad v(t_1) = -9.8\sqrt{\frac{500}{4.9}} \approx -98.995$ m/s. Since the canister has been designed to withstand an impact velocity of 100 m/s, the canister will *not burst.*

59. (a)

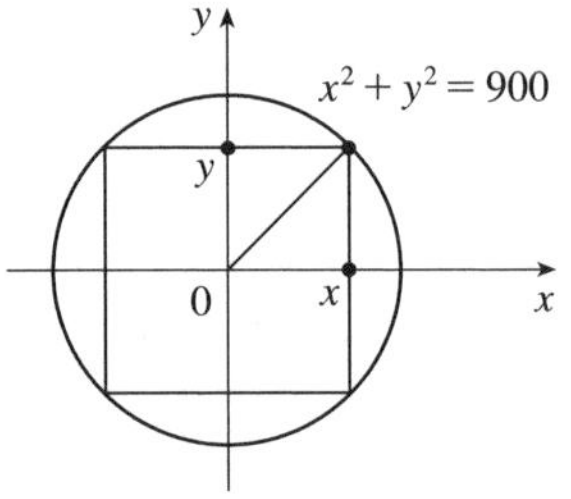

The cross-sectional area of the rectangular beam is

$A = 2x \cdot 2y = 4xy = 4x\sqrt{900 - x^2}$, $0 \leq x \leq 30$, so

$$\frac{dA}{dx} = 4x\left(\tfrac{1}{2}\right)\left(900 - x^2\right)^{-1/2}(-2x) + \left(900 - x^2\right)^{1/2} \cdot 4$$
$$= \frac{-4x^2}{(900 - x^2)^{1/2}} + 4(900 - x^2)^{1/2} = \frac{4[-x^2 + (900 - x^2)]}{(900 - x^2)^{1/2}}.$$

$\frac{dA}{dx} = 0$ when $-x^2 + \left(900 - x^2\right) = 0 \quad \Rightarrow \quad x^2 = 450 \quad \Rightarrow \quad x = \sqrt{450} \approx 21.21 \quad \Rightarrow$
$y = \sqrt{900 - \left(\sqrt{450}\right)^2} = \sqrt{450}$. Since $A(0) = A(30) = 0$, the rectangle of maximum area is a square.

(b)

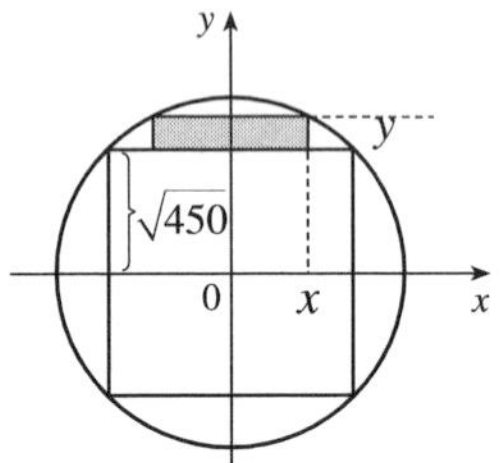

The cross-sectional area of each rectangular plank (shaded in the figure) is

$A = 2x\left(y - \sqrt{450}\right) = 2x\left[\sqrt{900 - x^2} - \sqrt{450}\right]$, $0 \leq x \leq \sqrt{450}$, so

$$\frac{dA}{dx} = 2\left(\sqrt{900 - x^2} - \sqrt{450}\right) + 2x\left(\tfrac{1}{2}\right)(900 - x^2)^{-1/2}(-2x)$$
$$= 2(900 - x^2)^{1/2} - 2\sqrt{450} - \frac{2x^2}{(900 - x^2)^{1/2}}$$

Set $\frac{dA}{dx} = 0$: $\quad (900 - x^2) - \sqrt{450}\,(900 - x^2)^{1/2} - x^2 = 0 \quad \Rightarrow \quad 900 - 2x^2 = \sqrt{450}\,(900 - x^2)^{1/2} \quad \Rightarrow$
$810\,000 - 3600x^2 + 4x^4 = 450(900 - x^2) \quad \Rightarrow \quad 4x^4 - 3150x^2 + 405\,000 = 0 \quad \Rightarrow$

$2x^4 - 1575x^2 + 202\,500 = 0 \quad \Rightarrow \quad x^2 = \dfrac{1575 \pm \sqrt{860\,625}}{4} \approx 625.67$ or $161.83 \quad \Rightarrow \quad x \approx 25.01$ or 12.72.

But $25.01 > \sqrt{450}$, so $x_1 \approx 12.72 \quad \Rightarrow \quad y - \sqrt{450} = \sqrt{900 - x_1^2} - \sqrt{450} \approx 5.96$. Each plank should have dimensions about 25.44 cm by 5.96 cm.

(c) From the figure in part (a), the width is $2x$ and the depth is $2y$, so the strength is $S = k(2x)(2y)^2 = 8kxy^2 = 8kx(900 - x^2) = 7200kx - 8kx^3$, $0 \le x \le 30$. $dS/dx = 7200k - 24kx^2 = 0$ when $24kx^2 = 7200k \;\Rightarrow\; x^2 = 300 \;\Rightarrow\; x = \sqrt{300} \;\Rightarrow\; y = \sqrt{900 - (\sqrt{300})^2} = \sqrt{600} = \sqrt{2}\,x$. Since $S(0) = S(30) = 0$, the maximum strength occurs when $x = \sqrt{300}$. The dimensions should be $2\sqrt{300} \approx 34.64$ cm by $2\sqrt{600} \approx 48.99$ cm.

61. (a) $I = \dfrac{k\cos\theta}{d^2} = \dfrac{k(h/d)}{d^2} = k\dfrac{h}{d^3} = k\dfrac{h}{\left(\sqrt{20^2 + h^2}\right)^3} = k\dfrac{h}{(400 + h^2)^{3/2}} \;\Rightarrow$

$$\frac{dI}{dh} = k\,\frac{(400 + h^2)^{3/2} - h\frac{3}{2}(400 + h^2)^{1/2}\cdot 2h}{[(400 + h^2)^{3/2}]^2} = \frac{k(400 + h^2)^{1/2}(400 + h^2 - 3h^2)}{(400 + h^2)^{3/2}}$$

$$= \frac{k(400 - 2h^2)}{(400 + h^2)^{5/2}} \qquad [k \text{ is the constant of proportionality}]$$

Set $dI/dh = 0$: $400 - 2h^2 = 0 \;\Rightarrow\; h^2 = 200 \;\Rightarrow\; h = \sqrt{200} = 10\sqrt{2}$. By the First Derivative Test, I has a local maximum at $h = 10\sqrt{2} \approx 14$ m.

(b)

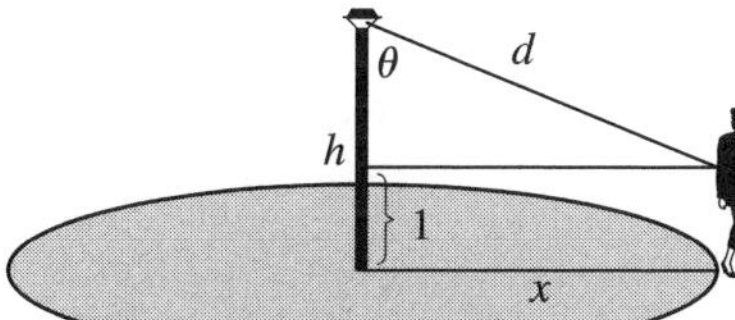

$$\frac{dx}{dt} = 1 \text{ m/s}$$

$$I = \frac{k\cos\theta}{d^2} = \frac{k[(h-1)/d]}{d^2} = \frac{k(h-1)}{d^3} = \frac{k(h-1)}{[(h-1)^2 + x^2]^{3/2}} = k(h-1)\left[(h-1)^2 + x^2\right]^{-3/2}$$

$$\frac{dI}{dt} = \frac{dI}{dx}\cdot\frac{dx}{dt} = k(h-1)\left(-\tfrac{3}{2}\right)\left[(h-1)^2 + x^2\right]^{-5/2}\cdot 2x\cdot\frac{dx}{dt}$$

$$= k(h-1)(-3x)\left[(h-1)^2 + x^2\right]^{-5/2}\cdot 1 = \frac{-3xk(h-1)}{[(h-1)^2 + x^2]^{5/2}}$$

$$\left.\frac{dI}{dt}\right|_{x=20} = -\frac{60k(h-1)}{[(h-1)^2 + 400]^{5/2}}$$

FOCUS ON PROBLEM SOLVING

1. Let $y = f(x) = e^{-x^2}$. The area of the rectangle under the curve from $-x$ to x is $A(x) = 2xe^{-x^2}$ where $x \geq 0$. We maximize $A(x)$: $A'(x) = 2e^{-x^2} - 4x^2e^{-x^2} = 2e^{-x^2}(1 - 2x^2) = 0 \quad \Rightarrow \quad x = \frac{1}{\sqrt{2}}$. This gives a maximum since $A'(x) > 0$ for $0 \leq x < \frac{1}{\sqrt{2}}$ and $A'(x) < 0$ for $x > \frac{1}{\sqrt{2}}$. We next determine the points of inflection of $f(x)$. Notice that $f'(x) = -2xe^{-x^2} = -A(x)$. So $f''(x) = -A'(x)$ and hence, $f''(x) < 0$ for $-\frac{1}{\sqrt{2}} < x < \frac{1}{\sqrt{2}}$ and $f''(x) > 0$ for $x < -\frac{1}{\sqrt{2}}$ and $x > \frac{1}{\sqrt{2}}$. So $f(x)$ changes concavity at $x = \pm\frac{1}{\sqrt{2}}$, and the two vertices of the rectangle of largest area are at the inflection points.

3. First, we recognize some symmetry in the inequality: $\dfrac{e^{x+y}}{xy} \geq e^2 \quad \Leftrightarrow \quad \dfrac{e^x}{x} \cdot \dfrac{e^y}{y} \geq e \cdot e$. This suggests that we need to show that $\dfrac{e^x}{x} \geq e$ for $x > 0$. If we can do this, then the inequality $\dfrac{e^y}{y} \geq e$ is true, and the given inequality follows. $f(x) = \dfrac{e^x}{x} \quad \Rightarrow$ $f'(x) = \dfrac{xe^x - e^x}{x^2} = \dfrac{e^x(x-1)}{x^2} = 0 \quad \Rightarrow \quad x = 1$. By the First Derivative Test, we have a minimum of $f(1) = e$, so $e^x/x \geq e$ for all x.

5. First we show that $x(1-x) \leq \frac{1}{4}$ for all x. Let $f(x) = x(1-x) = x - x^2$. Then $f'(x) = 1 - 2x$. This is 0 when $x = \frac{1}{2}$ and $f'(x) > 0$ for $x < \frac{1}{2}$, $f'(x) < 0$ for $x > \frac{1}{2}$, so the absolute maximum of f is $f\left(\frac{1}{2}\right) = \frac{1}{4}$. Thus, $x(1-x) \leq \frac{1}{4}$ for all x.

Now suppose that the given assertion is false, that is, $a(1-b) > \frac{1}{4}$ and $b(1-a) > \frac{1}{4}$. Multiply these inequalities: $a(1-b)b(1-a) > \frac{1}{16} \quad \Rightarrow \quad [a(1-a)]\,[b(1-b)] > \frac{1}{16}$. But we know that $a(1-a) \leq \frac{1}{4}$ and $b(1-b) \leq \frac{1}{4} \quad \Rightarrow$ $[a(1-a)]\,[b(1-b)] \leq \frac{1}{16}$. Thus, we have a contradiction, so the given assertion is proved.

7. Differentiating $x^2 + xy + y^2 = 12$ implicitly with respect to x gives $2x + y + x\dfrac{dy}{dx} + 2y\dfrac{dy}{dx} = 0$, so $\dfrac{dy}{dx} = -\dfrac{2x+y}{x+2y}$.

At a highest or lowest point, $\dfrac{dy}{dx} = 0 \quad \Leftrightarrow \quad y = -2x$. Substituting $-2x$ for y in the original equation gives $x^2 + x(-2x) + (-2x)^2 = 12$, so $3x^2 = 12$ and $x = \pm 2$. If $x = 2$, then $y = -2x = -4$, and if $x = -2$ then $y = 4$. Thus, the highest and lowest points are $(-2, 4)$ and $(2, -4)$.

9. Let $L = \displaystyle\lim_{x \to 0} \frac{ax^2 + \sin bx + \sin cx + \sin dx}{3x^2 + 5x^4 + 7x^6}$. Now L has the indeterminate form of type $\frac{0}{0}$, so we can apply l'Hospital's Rule. $L = \displaystyle\lim_{x \to 0} \frac{2ax + b\cos bx + c\cos cx + d\cos dx}{6x + 20x^3 + 42x^5}$. The denominator approaches 0 as $x \to 0$, so the numerator must also approach 0 (because the limit exists). But the numerator approaches $0 + b + c + d$, so $b + c + d = 0$. Apply l'Hospital's Rule again. $L = \displaystyle\lim_{x \to 0} \frac{2a - b^2\sin bx - c^2\sin cx - d^2\sin dx}{6 + 60x^2 + 210x^4} = \frac{2a - 0}{6 + 0} = \frac{2a}{6}$, which must equal 8.

$\dfrac{2a}{6} = 8 \quad \Rightarrow \quad a = 24$. Thus, $a + b + c + d = a + (b + c + d) = 24 + 0 = 24$.

11. We first show that $\dfrac{x}{1+x^2} < \tan^{-1} x$ for $x > 0$. Let $f(x) = \tan^{-1} x - \dfrac{x}{1+x^2}$. Then $f'(x) = \dfrac{1}{1+x^2} - \dfrac{1(1+x^2) - x(2x)}{(1+x^2)^2} = \dfrac{(1+x^2) - (1-x^2)}{(1+x^2)^2} = \dfrac{2x^2}{(1+x^2)^2} > 0$ for $x > 0$. So $f(x)$ is increasing on $(0, \infty)$. Hence, $0 < x \quad \Rightarrow \quad 0 = f(0) < f(x) = \tan^{-1} x - \dfrac{x}{1+x^2}$. So $\dfrac{x}{1+x^2} < \tan^{-1} x$ for $0 < x$. We next show that $\tan^{-1} x < x$ for $x > 0$. Let $h(x) = x - \tan^{-1} x$. Then $h'(x) = 1 - \dfrac{1}{1+x^2} = \dfrac{x^2}{1+x^2} > 0$. Hence, $h(x)$ is increasing on $(0, \infty)$. So for $0 < x$, $0 = h(0) < h(x) = x - \tan^{-1} x$. Hence, $\tan^{-1} x < x$ for $x > 0$, and we conclude that $\dfrac{x}{1+x^2} < \tan^{-1} x < x$ for $x > 0$.

13. $f(x) = (a^2 + a - 6)\cos 2x + (a-2)x + \cos 1 \quad\Rightarrow\quad f'(x) = -(a^2+a-6)\sin 2x\,(2) + (a-2)$. The derivative exists for all x, so the only possible critical points will occur where $f'(x) = 0 \quad\Leftrightarrow\quad 2(a-2)(a+3)\sin 2x = a-2 \quad\Leftrightarrow$ either $a = 2$ or $2(a+3)\sin 2x = 1$, with the latter implying that $\sin 2x = \dfrac{1}{2(a+3)}$. Since the range of $\sin 2x$ is $[-1, 1]$, this equation has no solution whenever either $\dfrac{1}{2(a+3)} < -1$ or $\dfrac{1}{2(a+3)} > 1$. Solving these inequalities, we get $-\frac{7}{2} < a < -\frac{5}{2}$.

15. $y = x^2 \quad\Rightarrow\quad y' = 2x$, so the slope of the tangent line at $P(a, a^2)$ is $2a$ and the slope of the normal line is $-\dfrac{1}{2a}$ for $a \neq 0$. An equation of the normal line is $y - a^2 = -\dfrac{1}{2a}(x - a)$. Substitute x^2 for y to find the x-coordinates of the two points of intersection of the parabola and the normal line. $x^2 - a^2 = -\dfrac{x}{2a} + \dfrac{1}{2} \quad\Rightarrow\quad 2ax^2 + x - 2a^3 - a = 0 \quad\Rightarrow$

$$x = \frac{-1 \pm \sqrt{1 - 4(2a)(-2a^3 - a)}}{2(2a)} = \frac{-1 \pm \sqrt{1 + 16a^4 + 8a^2}}{4a} = \frac{-1 \pm \sqrt{(4a^2+1)^2}}{4a} = \frac{-1 \pm (4a^2+1)}{4a}$$

$$= \frac{4a^2}{4a} \text{ or } \frac{-4a^2 - 2}{4a}, \text{ or equivalently, } a \text{ or } -a - \frac{1}{2a}.$$

So the point Q has coordinates $\left(-a - \dfrac{1}{2a}, \left(-a - \dfrac{1}{2a}\right)^2\right)$. The square S of the distance from P to Q is given by

$$S = \left(-a - \frac{1}{2a} - a\right)^2 + \left[\left(-a - \frac{1}{2a}\right)^2 - a^2\right]^2 = \left(-2a - \frac{1}{2a}\right)^2 + \left[\left(a^2 + 1 + \frac{1}{4a^2}\right) - a^2\right]^2$$

$$= \left(4a^2 + 2 + \frac{1}{4a^2}\right) + \left(1 + \frac{1}{4a^2}\right)^2 = \left(4a^2 + 2 + \frac{1}{4a^2}\right) + 1 + \frac{2}{4a^2} + \frac{1}{16a^4} = 4a^2 + 3 + \frac{3}{4a^2} + \frac{1}{16a^4}$$

$S' = 8a - \dfrac{6}{4a^3} - \dfrac{4}{16a^5} = 8a - \dfrac{3}{2a^3} - \dfrac{1}{4a^5} = \dfrac{32a^6 - 6a^2 - 1}{4a^5}$. The only real positive zero of the equation $S' = 0$ is $a = \dfrac{1}{\sqrt{2}}$. Since $S'' = 8 + \dfrac{9}{2a^4} + \dfrac{5}{4a^6} > 0$, $a = \dfrac{1}{\sqrt{2}}$ corresponds to the shortest possible length of the line segment PQ.

17. (a) Let $y = |AD|$, $x = |AB|$, and $1/x = |AC|$, so that $|AB| \cdot |AC| = 1$.

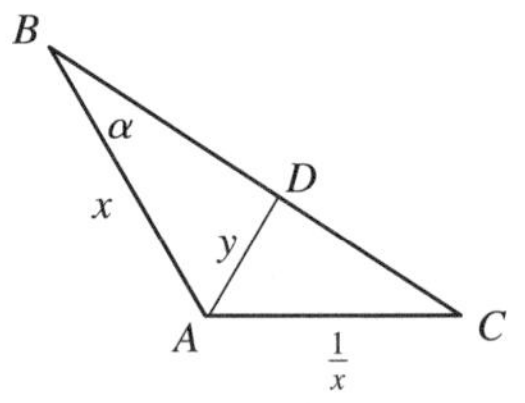

We compute the area $\mathcal{A}$ of $\triangle ABC$ in two ways.

First, $\mathcal{A} = \frac{1}{2}|AB|\,|AC| \sin\frac{2\pi}{3} = \frac{1}{2} \cdot 1 \cdot \frac{\sqrt{3}}{2} = \frac{\sqrt{3}}{4}$.

Second,

$$\begin{aligned} \mathcal{A} &= (\text{area of } \triangle ABD) + (\text{area of } \triangle ACD) \\ &= \tfrac{1}{2}|AB|\,|AD| \sin\tfrac{\pi}{3} + \tfrac{1}{2}|AD|\,|AC| \sin\tfrac{\pi}{3} \\ &= \tfrac{1}{2}xy\tfrac{\sqrt{3}}{2} + \tfrac{1}{2}y(1/x)\tfrac{\sqrt{3}}{2} = \tfrac{\sqrt{3}}{4}y(x + 1/x) \end{aligned}$$

Equating the two expressions for the area, we get $\frac{\sqrt{3}}{4}y\left(x + \dfrac{1}{x}\right) = \frac{\sqrt{3}}{4} \quad\Leftrightarrow\quad y = \dfrac{1}{x + 1/x} = \dfrac{x}{x^2+1}$, $x > 0$.

Another method: Use the Law of Sines on the triangles ABD and ABC. In $\triangle ABD$, we have $\angle A + \angle B + \angle D = 180° \quad\Leftrightarrow\quad 60° + \alpha + \angle D = 180° \quad\Leftrightarrow\quad \angle D = 120° - \alpha$. Thus,

$\dfrac{x}{y} = \dfrac{\sin(120° - \alpha)}{\sin\alpha} = \dfrac{\sin 120° \cos\alpha - \cos 120° \sin\alpha}{\sin\alpha} = \dfrac{\frac{\sqrt{3}}{2}\cos\alpha + \frac{1}{2}\sin\alpha}{\sin\alpha} \quad\Rightarrow\quad \dfrac{x}{y} = \frac{\sqrt{3}}{2}\cot\alpha + \frac{1}{2}$, and by a similar argument with $\triangle ABC$, $\frac{\sqrt{3}}{2}\cot\alpha = x^2 + \frac{1}{2}$. Eliminating $\cot\alpha$ gives $\dfrac{x}{y} = \left(x^2 + \frac{1}{2}\right) + \frac{1}{2} \quad\Rightarrow$

$y = \dfrac{x}{x^2+1}$, $x > 0$.

(b) We differentiate our expression for y with respect to x to find the maximum:

$\dfrac{dy}{dx} = \dfrac{(x^2+1) - x(2x)}{(x^2+1)^2} = \dfrac{1-x^2}{(x^2+1)^2} = 0$ when $x = 1$. This indicates a maximum by the First Derivative Test, since $y'(x) > 0$ for $0 < x < 1$ and $y'(x) < 0$ for $x > 1$, so the maximum value of y is $y(1) = \frac{1}{2}$.

19. Let $s_A(t)$ and $s_B(t)$ be the position functions for cars A and B and let $f(t) = s_A(t) - s_B(t)$. Since A passed B twice (B passed A once), there must be three values of t such that $f(t) = 0$. Let these times be denoted t_1, t_2, and t_3. By the Mean Value Theorem, we know that for some number c_1 in (t_1, t_2), $f'(c_1) = \dfrac{f(t_2) - f(t_1)}{t_2 - t_1}$, but $f(t_2) - f(t_1) = 0$, so $f'(c_1) = 0 \quad \Leftrightarrow \quad s'_A(c_1) - s'_B(c_1) = 0 \quad \Leftrightarrow \quad v_A(c_1) - v_B(c_1) = 0 \quad \Rightarrow \quad v_A(c_1) = v_B(c_1)$.

By a similar argument, there exists some number c_2 in (t_2, t_3) such that $v_A(c_2) = v_B(c_2)$. Now let $g(t) = v_A(t) - v_B(t)$ and apply the Mean Value Theorem on $[c_1, c_2]$ with $c_1 < c < c_2$. $g'(c) = \dfrac{g(c_2) - g(c_1)}{c_2 - c_1}$, but $g(c_2) = g(c_1) = 0$, so $g'(c) = 0 \quad \Rightarrow \quad v'_A(c) - v'_B(c) = 0 \quad \Rightarrow \quad a_A(c) - a_B(c) = 0 \quad \Rightarrow \quad a_A(c) = a_B(c)$; that is, A and B had equal accelerations at $t = c$.

21.

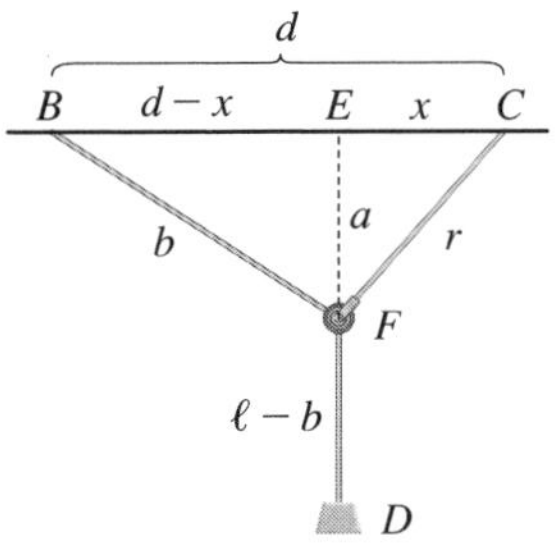

Let $a = |EF|$ and $b = |BF|$ as shown in the figure. Since $\ell = |BF| + |FD|$, $|FD| = \ell - b$. Now

$$\begin{aligned}
|ED| &= |EF| + |FD| = a + \ell - b \\
&= \sqrt{r^2 - x^2} + \ell - \sqrt{(d-x)^2 + a^2} \\
&= \sqrt{r^2 - x^2} + \ell - \sqrt{(d-x)^2 + \left(\sqrt{r^2 - x^2}\right)^2} \\
&= \sqrt{r^2 - x^2} + \ell - \sqrt{d^2 - 2dx + x^2 + r^2 - x^2}
\end{aligned}$$

Let $f(x) = \sqrt{r^2 - x^2} + \ell - \sqrt{d^2 + r^2 - 2dx}$.

$$f'(x) = \tfrac{1}{2}\left(r^2 - x^2\right)^{-1/2}(-2x) - \tfrac{1}{2}\left(d^2 + r^2 - 2dx\right)^{-1/2}(-2d) = \frac{-x}{\sqrt{r^2 - x^2}} + \frac{d}{\sqrt{d^2 + r^2 - 2dx}}.$$

$$f'(x) = 0 \quad \Rightarrow \quad \frac{x}{\sqrt{r^2 - x^2}} = \frac{d}{\sqrt{d^2 + r^2 - 2dx}} \quad \Rightarrow \quad \frac{x^2}{r^2 - x^2} = \frac{d^2}{d^2 + r^2 - 2dx} \quad \Rightarrow$$

$$d^2x^2 + r^2x^2 - 2dx^3 = d^2r^2 - d^2x^2 \quad \Rightarrow \quad 0 = 2dx^3 - 2d^2x^2 - r^2x^2 + d^2r^2 \quad \Rightarrow$$

$$0 = 2dx^2(x - d) - r^2\left(x^2 - d^2\right) \quad \Rightarrow \quad 0 = 2dx^2(x - d) - r^2(x + d)(x - d) \quad \Rightarrow \quad 0 = (x - d)\left[2dx^2 - r^2(x + d)\right]$$

But $d > r > x$, so $x \neq d$. Thus, we solve $2dx^2 - r^2x - dr^2 = 0$ for x:

$x = \dfrac{-(-r^2) \pm \sqrt{(-r^2)^2 - 4(2d)(-dr^2)}}{2(2d)} = \dfrac{r^2 \pm \sqrt{r^4 + 8d^2r^2}}{4d}$. Because $\sqrt{r^4 + 8d^2r^2} > r^2$, the "negative" can be discarded. Thus, $x = \dfrac{r^2 + \sqrt{r^2}\,\sqrt{r^2 + 8d^2}}{4d} = \dfrac{r^2 + r\sqrt{r^2 + 8d^2}}{4d} \quad [r > 0] \quad = \dfrac{r}{4d}\left(r + \sqrt{r^2 + 8d^2}\,\right)$. The maximum value of $|ED|$ occurs at this value of x.

23.

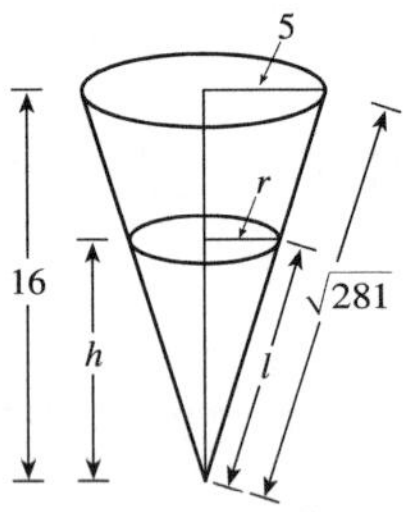

By similar triangles, $\frac{r}{5} = \frac{h}{16} \Rightarrow r = \frac{5h}{16}$. The volume of the cone is $V = \frac{1}{3}\pi r^2 h = \frac{1}{3}\pi\left(\frac{5h}{16}\right)^2 h = \frac{25\pi}{768}h^3$, so $\frac{dV}{dt} = \frac{25\pi}{256}h^2\frac{dh}{dt}$. Now the rate of change of the volume is also equal to the difference of what is being added ($2\ \text{cm}^3/\text{min}$) and what is oozing out ($k\pi rl$, where πrl is the area of the cone and k is a proportionality constant). Thus, $\frac{dV}{dt} = 2 - k\pi rl$.

Equating the two expressions for $\frac{dV}{dt}$ and substituting $h = 10$, $\frac{dh}{dt} = -0.3$, $r = \frac{5(10)}{16} = \frac{25}{8}$, and $\frac{l}{\sqrt{281}} = \frac{10}{16} \Leftrightarrow$ $l = \frac{5}{8}\sqrt{281}$, we get $\frac{25\pi}{256}(10)^2(-0.3) = 2 - k\pi\frac{25}{8}\cdot\frac{5}{8}\sqrt{281} \Leftrightarrow \frac{125k\pi\sqrt{281}}{64} = 2 + \frac{750\pi}{256}$. Solving for k gives us $k = \frac{256 + 375\pi}{250\pi\sqrt{281}}$. To maintain a certain height, the rate of oozing, $k\pi rl$, must equal the rate of the liquid being poured in; that is, $\frac{dV}{dt} = 0$. $k\pi rl = \frac{256 + 375\pi}{250\pi\sqrt{281}}\cdot\pi\cdot\frac{25}{8}\cdot\frac{5\sqrt{281}}{8} = \frac{256 + 375\pi}{128} \approx 11.204\ \text{cm}^3/\text{min}$.

5 □ INTEGRALS

5.1 Areas and Distances

1. (a) Since f is *increasing*, we can obtain a *lower* estimate by using *left* endpoints. We are instructed to use five rectangles, so $n = 5$.

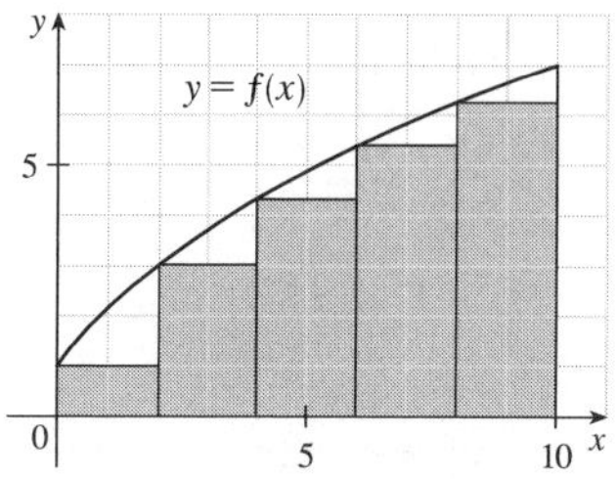

$$\begin{aligned} L_5 &= \sum_{i=1}^{5} f(x_{i-1})\,\Delta x \qquad \left[\Delta x = \tfrac{b-a}{n} = \tfrac{10-0}{5} = 2\right] \\ &= f(x_0)\cdot 2 + f(x_1)\cdot 2 + f(x_2)\cdot 2 + f(x_3)\cdot 2 + f(x_4)\cdot 2 \\ &= 2\,[f(0) + f(2) + f(4) + f(6) + f(8)] \\ &\approx 2(1 + 3 + 4.3 + 5.4 + 6.3) = 2(20) = 40 \end{aligned}$$

Since f is *increasing*, we can obtain an *upper* estimate by using *right* endpoints.

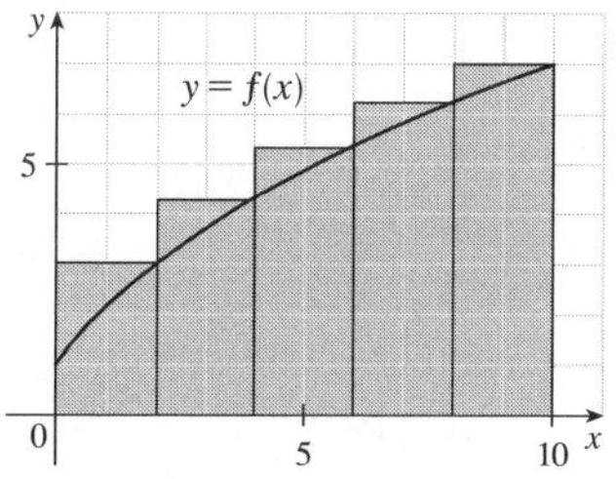

$$\begin{aligned} R_5 &= \sum_{i=1}^{5} f(x_i)\,\Delta x \\ &= 2\,[f(x_1) + f(x_2) + f(x_3) + f(x_4) + f(x_5)] \\ &= 2\,[f(2) + f(4) + f(6) + f(8) + f(10)] \\ &\approx 2(3 + 4.3 + 5.4 + 6.3 + 7) = 2(26) = 52 \end{aligned}$$

Comparing R_5 to L_5, we see that we have added the area of the rightmost upper rectangle, $f(10)\cdot 2$, to the sum and subtracted the area of the leftmost lower rectangle, $f(0)\cdot 2$, from the sum.

(b)

$$\begin{aligned} L_{10} &= \sum_{i=1}^{10} f(x_{i-1})\,\Delta x \qquad \left[\Delta x = \tfrac{10-0}{10} = 1\right] \\ &= 1\,[f(x_0) + f(x_1) + \cdots + f(x_9)] \\ &= f(0) + f(1) + \cdots + f(9) \\ &\approx 1 + 2.1 + 3 + 3.7 + 4.3 + 4.9 + 5.4 + 5.8 + 6.3 + 6.7 \\ &= 43.2 \end{aligned}$$

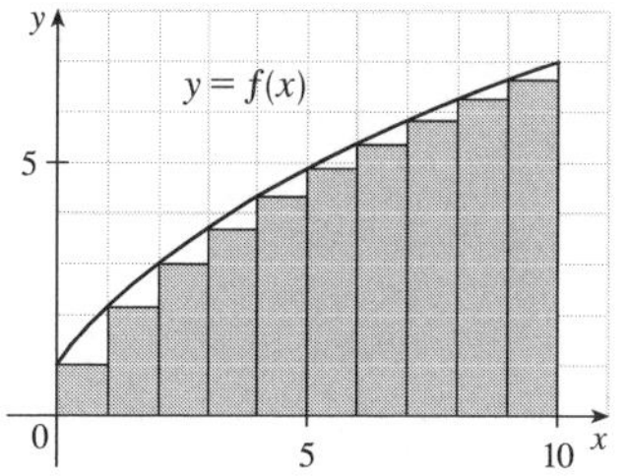

$$\begin{aligned} R_{10} &= \sum_{i=1}^{10} f(x_i)\,\Delta x = f(1) + f(2) + \cdots + f(10) \\ &= L_{10} + 1\cdot f(10) - 1\cdot f(0) \qquad \left[\begin{array}{c}\text{add rightmost upper rectangle,} \\ \text{subtract leftmost lower rectangle}\end{array}\right] \\ &= 43.2 + 7 - 1 = 49.2 \end{aligned}$$

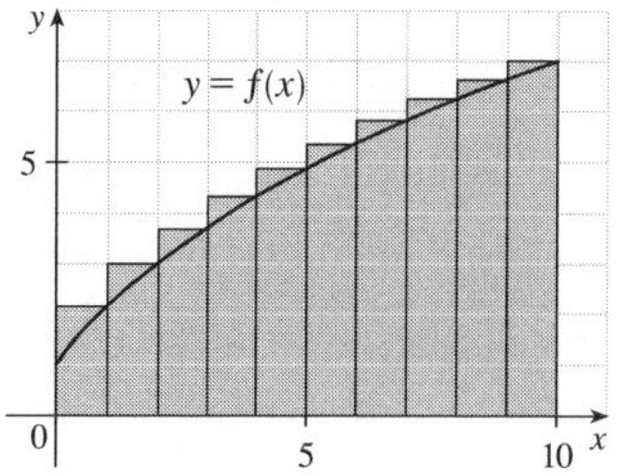

3. (a) $R_4 = \sum_{i=1}^{4} f(x_i)\,\Delta x \quad [\Delta x = \frac{5-1}{4} = 1]$

$= f(x_1)\cdot 1 + f(x_2)\cdot 1 + f(x_3)\cdot 1 + f(x_4)\cdot 1$

$= f(2) + f(3) + f(4) + f(5)$

$= \frac{1}{2} + \frac{1}{3} + \frac{1}{4} + \frac{1}{5} = \frac{77}{60} = 1.28\overline{3}$

Since f is *decreasing* on $[1, 5]$, an *underestimate* is obtained by using the *right* endpoint approximation, R_4.

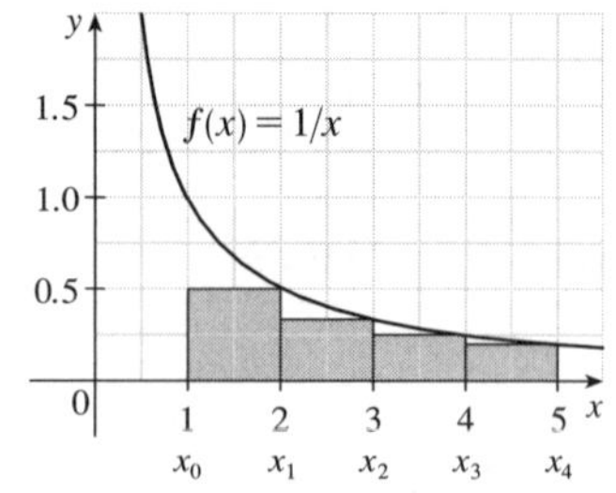

(b) $L_4 = \sum_{i=1}^{4} f(x_{i-1})\,\Delta x$

$= f(1) + f(2) + f(3) + f(4)$

$= 1 + \frac{1}{2} + \frac{1}{3} + \frac{1}{4} = \frac{25}{12} = 2.08\overline{3}$

L_4 is an overestimate. Alternatively, we could just add the area of the leftmost upper rectangle and subtract the area of the rightmost lower rectangle; that is, $L_4 = R_4 + f(1)\cdot 1 - f(5)\cdot 1$.

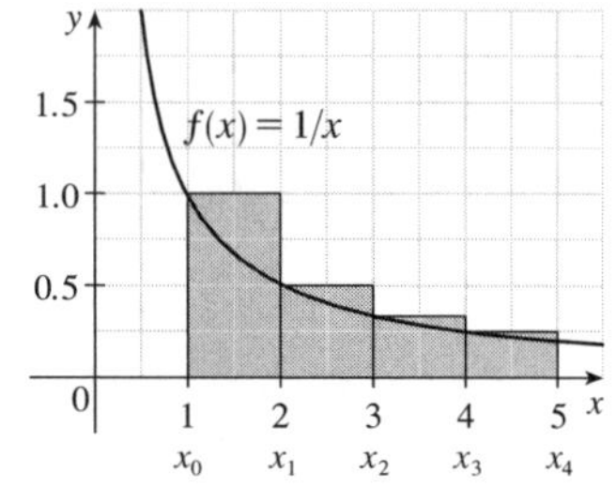

5. (a) $f(x) = 1 + x^2$ and $\Delta x = \dfrac{2-(-1)}{3} = 1 \;\Rightarrow$

$R_3 = 1\cdot f(0) + 1\cdot f(1) + 1\cdot f(2) = 1\cdot 1 + 1\cdot 2 + 1\cdot 5 = 8.$

$\Delta x = \dfrac{2-(-1)}{6} = 0.5 \;\Rightarrow$

$R_6 = 0.5[f(-0.5) + f(0) + f(0.5) + f(1) + f(1.5) + f(2)]$

$= 0.5(1.25 + 1 + 1.25 + 2 + 3.25 + 5)$

$= 0.5(13.75) = 6.875$

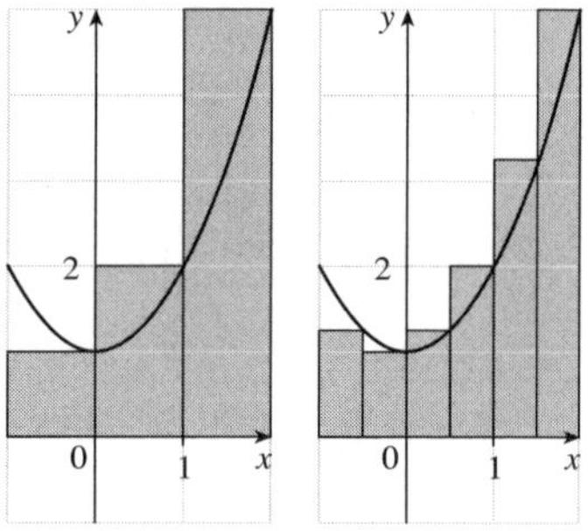

(b) $L_3 = 1\cdot f(-1) + 1\cdot f(0) + 1\cdot f(1) = 1\cdot 2 + 1\cdot 1 + 1\cdot 2 = 5$

$L_6 = 0.5[f(-1) + f(-0.5) + f(0) + f(0.5) + f(1) + f(1.5)]$

$= 0.5(2 + 1.25 + 1 + 1.25 + 2 + 3.25)$

$= 0.5(10.75) = 5.375$

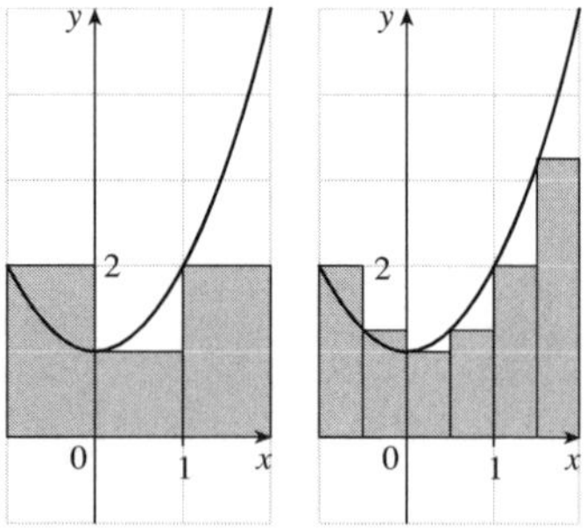

(c) $M_3 = 1\cdot f(-0.5) + 1\cdot f(0.5) + 1\cdot f(1.5)$

$= 1\cdot 1.25 + 1\cdot 1.25 + 1\cdot 3.25 = 5.75$

$M_6 = 0.5[f(-0.75) + f(-0.25) + f(0.25)$

$+ f(0.75) + f(1.25) + f(1.75)]$

$= 0.5(1.5625 + 1.0625 + 1.0625 + 1.5625 + 2.5625 + 4.0625)$

$= 0.5(11.875) = 5.9375$

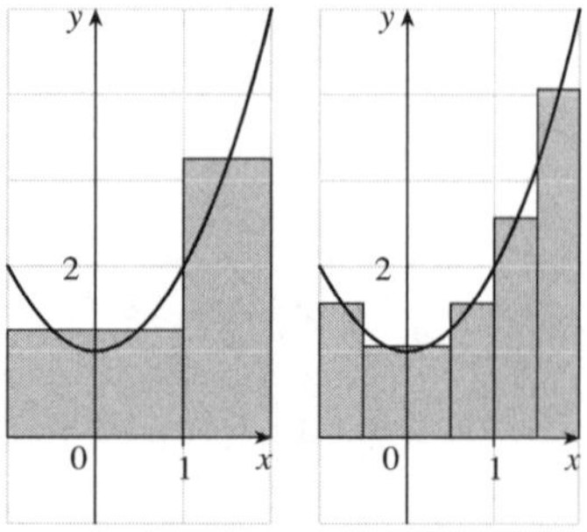

(d) M_6 appears to be the best estimate.

7. Here is one possible algorithm (ordered sequence of operations) for calculating the sums:

1 Let SUM = 0, X_MIN = 0, X_MAX = 1, N = 10 (depending on which sum we are calculating), DELTA_X = (X_MAX - X_MIN)/N, and RIGHT_ENDPOINT = X_MIN + DELTA_X.

2 Repeat steps 2a, 2b in sequence until RIGHT_ENDPOINT > X_MAX.

2a Add (RIGHT_ENDPOINT)^4 to SUM.

2b Add DELTA_X to RIGHT_ENDPOINT.

At the end of this procedure, (DELTA_X)·(SUM) is equal to the answer we are looking for. We find that

$R_{10} = \frac{1}{10}\sum_{i=1}^{10}\left(\frac{i}{10}\right)^4 \approx 0.2533$, $R_{30} = \frac{1}{30}\sum_{i=1}^{30}\left(\frac{i}{30}\right)^4 \approx 0.2170$, $R_{50} = \frac{1}{50}\sum_{i=1}^{50}\left(\frac{i}{50}\right)^4 \approx 0.2101$, and

$R_{100} = \frac{1}{100}\sum_{i=1}^{100}\left(\frac{i}{100}\right)^4 \approx 0.2050$. It appears that the exact area is 0.2.

Shown below is program SUMRIGHT and its output from a TI-83 Plus calculator. To generalize the program, we have input (rather than assign) values for Xmin, Xmax, and N. Also, the function, x^4, is assigned to Y_1, enabling us to evaluate any right sum merely by changing Y_1 and running the program.

```
PROGRAM:SUMRIGHT
:0→S
:Prompt Xmin
:Prompt Xmax
:Prompt N
:(Xmax-Xmin)/N→D
:Xmin+D→R
:For(I,1,N)
:S+Y1(R)→S
:R+D→R
:End
:D*S→Z
:Disp Z
```

```
prgmSUMRIGHT
Xmin=?0
Xmax=?1
N=?10
                .25333
                  Done
```

9. In Maple, we have to perform a number of steps before getting a numerical answer. After loading the student package [command: `with(student);`] we use the command `left_sum:=leftsum(1/(x^2+1),x=0..1,10` [or `30`, or `50`]`);` which gives us the expression in summation notation. To get a numerical approximation to the sum, we use `evalf(left_sum);`. Mathematica does not have a special command for these sums, so we must type them in manually. For example, the first left sum is given by `(1/10)*Sum[1/(((i-1)/10)^2+1)],{i,1,10}]`, and we use the `N` command on the resulting output to get a numerical approximation.

In Derive, we use the `LEFT_RIEMANN` command to get the left sums, but must define the right sums ourselves. (We can define a new function using `LEFT_RIEMANN` with k ranging from 1 to n instead of from 0 to $n-1$.)

(a) With $f(x) = \frac{1}{x^2+1}$, $0 \le x \le 1$, the left sums are of the form $L_n = \frac{1}{n}\sum_{i=1}^{n}\frac{1}{\left(\frac{i-1}{n}\right)^2+1}$. Specifically, $L_{10} \approx 0.8100$, $L_{30} \approx 0.7937$, and $L_{50} \approx 0.7904$. The right sums are of the form $R_n = \frac{1}{n}\sum_{i=1}^{n}\frac{1}{\left(\frac{i}{n}\right)^2+1}$. Specifically, $R_{10} \approx 0.7600$, $R_{30} \approx 0.7770$, and $R_{50} \approx 0.7804$.

(b) In Maple, we use the `leftbox` (with the same arguments as `left_sum`) and `rightbox` commands to generate the graphs.

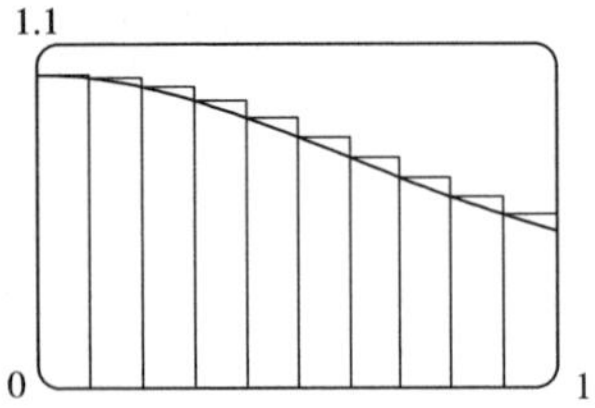

left endpoints, $n = 10$

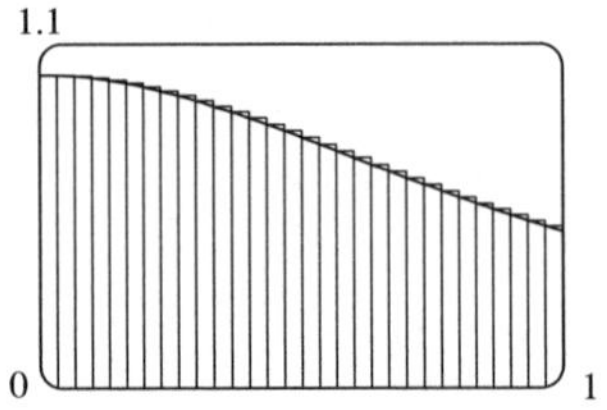

left endpoints, $n = 30$

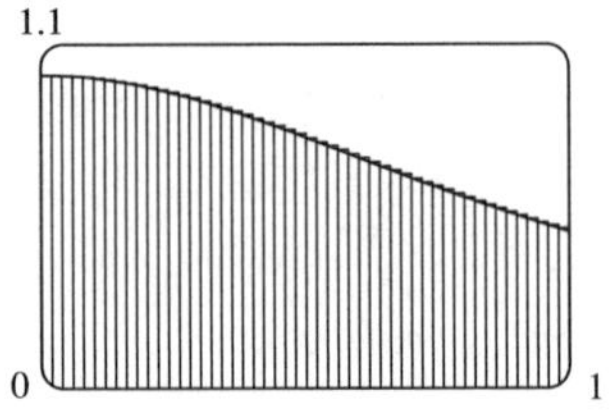

left endpoints, $n = 50$

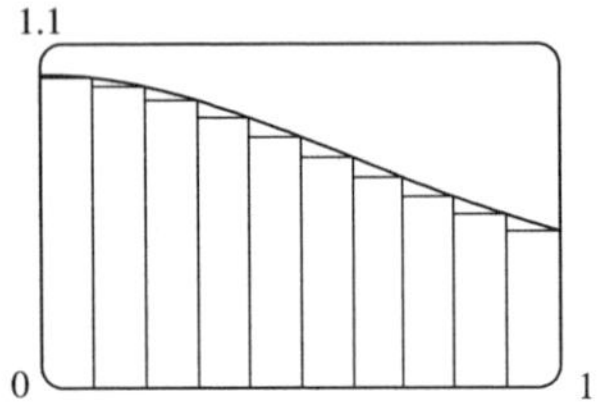

right endpoints, $n = 10$

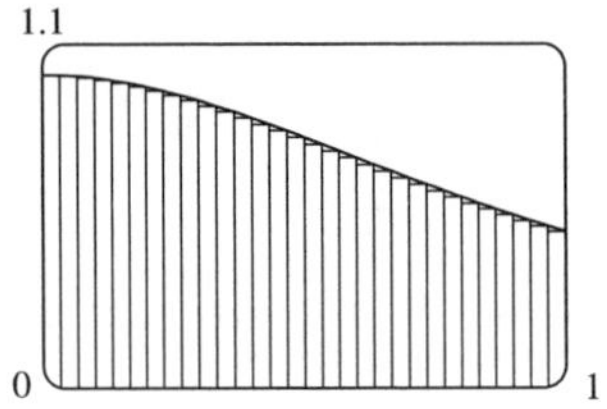

right endpoints, $n = 30$

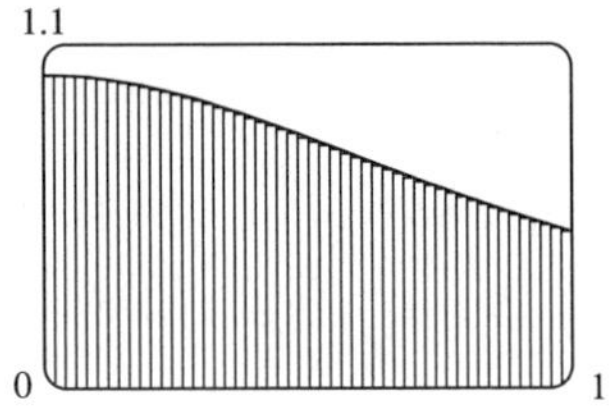

right endpoints, $n = 50$

(c) We know that since $y = 1/(x^2+1)$ is a decreasing function on $(0, 1)$, all of the left sums are larger than the actual area, and all of the right sums are smaller than the actual area. Since the left sum with $n = 50$ is about $0.7904 < 0.791$ and the right sum with $n = 50$ is about $0.7804 > 0.780$, we conclude that $0.780 < R_{50} < \text{exact area} < L_{50} < 0.791$, so the exact area is between 0.780 and 0.791.

11. Since v is an increasing function, L_6 will give us a lower estimate and R_6 will give us an upper estimate.

$$\begin{aligned} L_6 &= (0\text{ m/s})(0.5\text{ s}) + (1.9)(0.5) + (3.3)(0.5) + (4.5)(0.5) + (5.5)(0.5) + (5.9)(0.5) \\ &= 0.5(21.1) = 10.55\text{ m} \end{aligned}$$

$$R_6 = 0.5(1.9 + 3.3 + 4.5 + 5.5 + 5.9 + 6.2) = 0.5(27.3) = 13.65\text{ m}$$

13. Lower estimate for oil leakage: $R_5 = (7.6 + 6.8 + 6.2 + 5.7 + 5.3)(2) = (31.6)(2) = 63.2$ L.

Upper estimate for oil leakage: $L_5 = (8.7 + 7.6 + 6.8 + 6.2 + 5.7)(2) = (35)(2) = 70$ L.

15. For a decreasing function, using left endpoints gives us an overestimate and using right endpoints results in an underestimate. We will use M_6 to get an estimate. $\Delta t = 1$, so

$$\begin{aligned} M_6 &= 1[v(0.5) + v(1.5) + v(2.5) + v(3.5) + v(4.5) + v(5.5)] \\ &\approx 14 + 10 + 7 + 4.5 + 2.5 + 1 = 39\text{ m} \end{aligned}$$

For a very rough check on the above calculation, we can draw a line from $(0, 17)$ to $(6, 0)$ and calculate the area of the triangle: $\frac{1}{2}(17)(6) = 51$. This is clearly an overestimate, so our midpoint estimate of 39 is reasonable.

17. $f(x) = \sqrt[4]{x}$, $1 \le x \le 16$. $\Delta x = (16 - 1)/n = 15/n$ and $x_i = 1 + i\,\Delta x = 1 + 15i/n$.

$$A = \lim_{n\to\infty} R_n = \lim_{n\to\infty} \sum_{i=1}^{n} f(x_i)\,\Delta x = \lim_{n\to\infty} \sum_{i=1}^{n} \sqrt[4]{1 + \tfrac{15i}{n}} \cdot \tfrac{15}{n}.$$

19. $\lim_{n\to\infty}\sum_{i=1}^{n}\frac{\pi}{4n}\tan\frac{i\pi}{4n}$ can be interpreted as the area of the region lying under the graph of $y=\tan x$ on the interval $\left[0,\frac{\pi}{4}\right]$, since for $y=\tan x$ on $\left[0,\frac{\pi}{4}\right]$ with $\Delta x=\frac{\pi/4-0}{n}=\frac{\pi}{4n}$, $x_i=0+i\,\Delta x=\frac{i\pi}{4n}$, and $x_i^*=x_i$, the expression for the area is $A=\lim_{n\to\infty}\sum_{i=1}^{n}f\left(x_i^*\right)\Delta x=\lim_{n\to\infty}\sum_{i=1}^{n}\tan\left(\frac{i\pi}{4n}\right)\frac{\pi}{4n}$. Note that this answer is not unique, since the expression for the area is the same for the function $y=\tan(x-k\pi)$ on the interval $\left[k\pi,k\pi+\frac{\pi}{4}\right]$, where k is any integer.

21. (a) $y=f(x)=x^5$. $\Delta x=\frac{2-0}{n}=\frac{2}{n}$ and $x_i=0+i\,\Delta x=\frac{2i}{n}$.

$$A=\lim_{n\to\infty}R_n=\lim_{n\to\infty}\sum_{i=1}^{n}f(x_i)\,\Delta x=\lim_{n\to\infty}\sum_{i=1}^{n}\left(\frac{2i}{n}\right)^5\cdot\frac{2}{n}=\lim_{n\to\infty}\sum_{i=1}^{n}\frac{32i^5}{n^5}\cdot\frac{2}{n}=\lim_{n\to\infty}\frac{64}{n^6}\sum_{i=1}^{n}i^5.$$

(b) $\sum_{i=1}^{n}i^5\overset{\text{CAS}}{=}\frac{n^2(n+1)^2(2n^2+2n-1)}{12}$

(c) $$\lim_{n\to\infty}\frac{64}{n^6}\cdot\frac{n^2(n+1)^2(2n^2+2n-1)}{12}=\frac{64}{12}\lim_{n\to\infty}\frac{(n^2+2n+1)(2n^2+2n-1)}{n^2\cdot n^2}$$
$$=\frac{16}{3}\lim_{n\to\infty}\left(1+\frac{2}{n}+\frac{1}{n^2}\right)\left(2+\frac{2}{n}-\frac{1}{n^2}\right)=\tfrac{16}{3}\cdot1\cdot2=\tfrac{32}{3}$$

23. $y=f(x)=\cos x$. $\Delta x=\frac{b-0}{n}=\frac{b}{n}$ and $x_i=0+i\,\Delta x=\frac{bi}{n}$.

$$A=\lim_{n\to\infty}R_n=\lim_{n\to\infty}\sum_{i=1}^{n}f(x_i)\,\Delta x=\lim_{n\to\infty}\sum_{i=1}^{n}\cos\left(\frac{bi}{n}\right)\cdot\frac{b}{n}\overset{\text{CAS}}{=}\lim_{n\to\infty}\left[\frac{b\sin\left(b\left(\frac{1}{2n}+1\right)\right)}{2n\sin\left(\frac{b}{2n}\right)}-\frac{b}{2n}\right]\overset{\text{CAS}}{=}\sin b$$

If $b=\frac{\pi}{2}$, then $A=\sin\frac{\pi}{2}=1$.

5.2 The Definite Integral

1. $R_4=\sum_{i=1}^{4}f(x_i)\,\Delta x\quad[x_i^*=x_i$ is a right endpoint and $\Delta x=0.5]$

$=0.5\,[f(0.5)+f(1)+f(1.5)+f(2)]\quad[f(x)=2-x^2]$

$=0.5\,[1.75+1+(-0.25)+(-2)]$

$=0.5(0.5)=0.25$

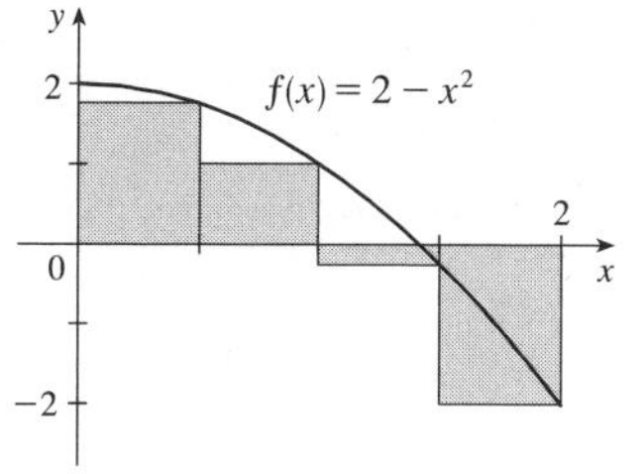

The Riemann sum represents the sum of the areas of the two rectangles above the x-axis minus the sum of the areas of the two rectangles below the x-axis; that is, the *net area* of the rectangles with respect to the x-axis.

3. $M_5=\sum_{i=1}^{5}f(\overline{x}_i)\,\Delta x\quad[x_i^*=\overline{x}_i=\frac{1}{2}(x_{i-1}+x_i)$ is a midpoint and $\Delta x=1]$

$=1\,[f(1.5)+f(2.5)+f(3.5)$
$+f(4.5)+f(5.5)]\quad[f(x)=\sqrt{x}-2]$

$\approx-0.856\,759$

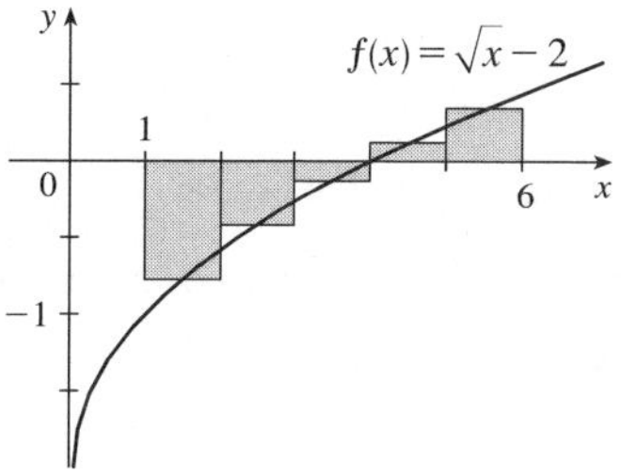

The Riemann sum represents the sum of the areas of the two rectangles above the x-axis minus the sum of the areas of the three rectangles below the x-axis.

5. $\Delta x = (b - a)/n = (8 - 0)/4 = 8/4 = 2$.

(a) Using the right endpoints to approximate $\int_0^8 f(x)\,dx$, we have

$$\sum_{i=1}^{4} f(x_i)\,\Delta x = 2[f(2) + f(4) + f(6) + f(8)] \approx 2[1 + 2 + (-2) + 1] = 4.$$

(b) Using the left endpoints to approximate $\int_0^8 f(x)\,dx$, we have

$$\sum_{i=1}^{4} f(x_{i-1})\,\Delta x = 2[f(0) + f(2) + f(4) + f(6)] \approx 2[2 + 1 + 2 + (-2)] = 6.$$

(c) Using the midpoint of each subinterval to approximate $\int_0^8 f(x)\,dx$, we have

$$\sum_{i=1}^{4} f(\overline{x}_i)\,\Delta x = 2[f(1) + f(3) + f(5) + f(7)] \approx 2[3 + 2 + 1 + (-1)] = 10.$$

7. Since f is increasing, $L_5 \le \int_0^{25} f(x)\,dx \le R_5$.

$$\begin{aligned}\text{Lower estimate} = L_5 &= \sum_{i=1}^{5} f(x_{i-1})\,\Delta x = 5[f(0) + f(5) + f(10) + f(15) + f(20)] \\ &= 5(-42 - 37 - 25 - 6 + 15) = 5(-95) = -475\end{aligned}$$

$$\begin{aligned}\text{Upper estimate} = R_5 &= \sum_{i=1}^{5} f(x_i)\,\Delta x = 5[f(5) + f(10) + f(15) + f(20) + f(25)] \\ &= 5(-37 - 25 - 6 + 15 + 36) = 5(-17) = -85\end{aligned}$$

9. $\Delta x = (10 - 2)/4 = 2$, so the endpoints are 2, 4, 6, 8, and 10, and the midpoints are 3, 5, 7, and 9. The Midpoint Rule gives

$$\int_2^{10} \sqrt{x^3 + 1}\,dx \approx \sum_{i=1}^{4} f(\overline{x}_i)\,\Delta x = 2\left(\sqrt{3^3 + 1} + \sqrt{5^3 + 1} + \sqrt{7^3 + 1} + \sqrt{9^3 + 1}\right) \approx 124.1644.$$

11. $\Delta x = (2 - 1)/10 = 0.1$, so the endpoints are $1.0, 1.1, \ldots, 2.0$, and the midpoints are $1.05, 1.15, \ldots, 1.95$.

The Midpoint Rule gives

$$\int_1^2 \sqrt{1 + x^2}\,dx \approx \sum_{i=1}^{10} f(\overline{x}_i)\,\Delta x = 0.1\left[\sqrt{1 + (1.05)^2} + \sqrt{1 + (1.15)^2} + \cdots + \sqrt{1 + (1.95)^2}\right] \approx 1.8100.$$

13. In Maple, we use the command `with(student);` to load the sum and box commands, then `m:=middlesum(sqrt(1+x^2),x=1..2,10);` which gives us the sum in summation notation, then `M:=evalf(m);` which gives $M_{10} \approx 1.810\,014\,14$, confirming the result of Exercise 11. The command `middlebox(sqrt(1+x^2),x=1..2,10);` generates the graph. Repeating for $n = 20$ and $n = 30$ gives $M_{20} \approx 1.810\,072\,63$ and $M_{30} \approx 1.810\,083\,47$.

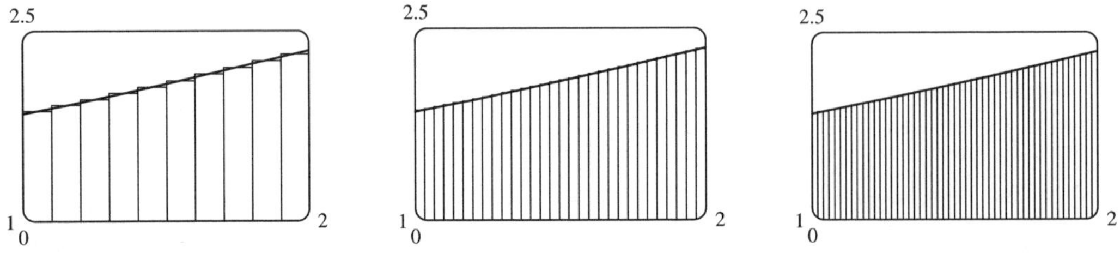

15. We'll create the table of values to approximate $\int_0^{\pi} \sin x\,dx$ by using the program in the solution to Exercise 5.1.7 with $Y_1 = \sin x$, Xmin $= 0$, Xmax $= \pi$, and $n = 5$, 10, 50, and 100.

n	R_n
5	1.933 766
10	1.983 524
50	1.999 342
100	1.999 836

The values of R_n appear to be approaching 2.

17. On $[0,\pi]$, $\lim\limits_{n\to\infty}\sum\limits_{i=1}^{n} x_i\sin x_i\,\Delta x=\int_0^{\pi} x\sin x\,dx$.

19. On $[1,8]$, $\lim\limits_{n\to\infty}\sum\limits_{i=1}^{n}\sqrt{2x_i^*+(x_i^*)^2}\,\Delta x=\int_1^{8}\sqrt{2x+x^2}\,dx$.

21. Note that $\Delta x=\dfrac{5-(-1)}{n}=\dfrac{6}{n}$ and $x_i=-1+i\,\Delta x=-1+\dfrac{6i}{n}$.

$$\begin{aligned}\int_{-1}^{5}(1+3x)\,dx&=\lim_{n\to\infty}\sum_{i=1}^{n}f(x_i)\,\Delta x=\lim_{n\to\infty}\sum_{i=1}^{n}\left[1+3\left(-1+\frac{6i}{n}\right)\right]\frac{6}{n}\\&=\lim_{n\to\infty}\frac{6}{n}\sum_{i=1}^{n}\left[-2+\frac{18i}{n}\right]=\lim_{n\to\infty}\frac{6}{n}\left[\sum_{i=1}^{n}(-2)+\sum_{i=1}^{n}\frac{18i}{n}\right]\\&=\lim_{n\to\infty}\frac{6}{n}\left[-2n+\frac{18}{n}\sum_{i=1}^{n}i\right]=\lim_{n\to\infty}\frac{6}{n}\left[-2n+\frac{18}{n}\cdot\frac{n(n+1)}{2}\right]\\&=\lim_{n\to\infty}\left[-12+\frac{108}{n^2}\cdot\frac{n(n+1)}{2}\right]=\lim_{n\to\infty}\left[-12+54\frac{n+1}{n}\right]\\&=\lim_{n\to\infty}\left[-12+54\left(1+\frac{1}{n}\right)\right]=-12+54\cdot 1=42\end{aligned}$$

23. Note that $\Delta x=\dfrac{2-0}{n}=\dfrac{2}{n}$ and $x_i=0+i\,\Delta x=\dfrac{2i}{n}$.

$$\begin{aligned}\int_{0}^{2}(2-x^2)\,dx&=\lim_{n\to\infty}\sum_{i=1}^{n}f(x_i)\,\Delta x=\lim_{n\to\infty}\sum_{i=1}^{n}\left(2-\frac{4i^2}{n^2}\right)\left(\frac{2}{n}\right)\\&=\lim_{n\to\infty}\frac{2}{n}\left[\sum_{i=1}^{n}2-\frac{4}{n^2}\sum_{i=1}^{n}i^2\right]=\lim_{n\to\infty}\frac{2}{n}\left(2n-\frac{4}{n^2}\sum_{i=1}^{n}i^2\right)\\&=\lim_{n\to\infty}\left[4-\frac{8}{n^3}\cdot\frac{n(n+1)(2n+1)}{6}\right]=\lim_{n\to\infty}\left(4-\frac{4}{3}\cdot\frac{n+1}{n}\cdot\frac{2n+1}{n}\right)\\&=\lim_{n\to\infty}\left[4-\frac{4}{3}\left(1+\frac{1}{n}\right)\left(2+\frac{1}{n}\right)\right]=4-\tfrac{4}{3}\cdot 1\cdot 2=\tfrac{4}{3}\end{aligned}$$

25. Note that $\Delta x=\dfrac{2-1}{n}=\dfrac{1}{n}$ and $x_i=1+i\,\Delta x=1+i(1/n)=1+i/n$.

$$\begin{aligned}\int_{1}^{2}x^3\,dx&=\lim_{n\to\infty}\sum_{i=1}^{n}f(x_i)\,\Delta x=\lim_{n\to\infty}\sum_{i=1}^{n}\left(1+\frac{i}{n}\right)^3\left(\frac{1}{n}\right)=\lim_{n\to\infty}\frac{1}{n}\sum_{i=1}^{n}\left(\frac{n+i}{n}\right)^3\\&=\lim_{n\to\infty}\frac{1}{n^4}\sum_{i=1}^{n}\left(n^3+3n^2i+3ni^2+i^3\right)=\lim_{n\to\infty}\frac{1}{n^4}\left[\sum_{i=1}^{n}n^3+\sum_{i=1}^{n}3n^2i+\sum_{i=1}^{n}3ni^2+\sum_{i=1}^{n}i^3\right]\\&=\lim_{n\to\infty}\frac{1}{n^4}\left[n\cdot n^3+3n^2\sum_{i=1}^{n}i+3n\sum_{i=1}^{n}i^2+\sum_{i=1}^{n}i^3\right]\\&=\lim_{n\to\infty}\left[1+\frac{3}{n^2}\cdot\frac{n(n+1)}{2}+\frac{3}{n^3}\cdot\frac{n(n+1)(2n+1)}{6}+\frac{1}{n^4}\cdot\frac{n^2(n+1)^2}{4}\right]\\&=\lim_{n\to\infty}\left[1+\frac{3}{2}\cdot\frac{n+1}{n}+\frac{1}{2}\cdot\frac{n+1}{n}\cdot\frac{2n+1}{n}+\frac{1}{4}\cdot\frac{(n+1)^2}{n^2}\right]\\&=\lim_{n\to\infty}\left[1+\frac{3}{2}\left(1+\frac{1}{n}\right)+\frac{1}{2}\left(1+\frac{1}{n}\right)\left(2+\frac{1}{n}\right)+\frac{1}{4}\left(1+\frac{1}{n}\right)^2\right]=1+\frac{3}{2}+\tfrac{1}{2}\cdot 2+\tfrac{1}{4}=3.75\end{aligned}$$

27. $f(x) = \dfrac{x}{1+x^5}$, $a = 2$, $b = 6$, and $\Delta x = \dfrac{6-2}{n} = \dfrac{4}{n}$. Using Equation 3, we get $x_i^* = x_i = 2 + i\,\Delta x = 2 + \dfrac{4i}{n}$, so

$$\int_2^6 \frac{x}{1+x^5}\,dx = \lim_{n\to\infty} R_n = \lim_{n\to\infty} \sum_{i=1}^{n} \frac{2+\dfrac{4i}{n}}{1+\left(2+\dfrac{4i}{n}\right)^5} \cdot \frac{4}{n}.$$

29. $\Delta x = (\pi - 0)/n = \pi/n$ and $x_i^* = x_i = \pi i/n$.

$$\int_0^{\pi} \sin 5x\,dx = \lim_{n\to\infty} \sum_{i=1}^{n} (\sin 5x_i)\left(\frac{\pi}{n}\right) = \lim_{n\to\infty} \sum_{i=1}^{n} \left(\sin \frac{5\pi i}{n}\right)\frac{\pi}{n} \overset{\text{CAS}}{=} \pi \lim_{n\to\infty} \frac{1}{n} \cot\left(\frac{5\pi}{2n}\right) \overset{\text{CAS}}{=} \pi\left(\frac{2}{5\pi}\right) = \frac{2}{5}$$

31. (a) Think of $\int_0^2 f(x)\,dx$ as the area of a trapezoid with bases 1 and 3 and height 2. The area of a trapezoid is $A = \frac{1}{2}(b+B)h$, so $\int_0^2 f(x)\,dx = \frac{1}{2}(1+3)2 = 4$.

(b) $\int_0^5 f(x)\,dx = \underset{\text{trapezoid}}{\int_0^2 f(x)\,dx} + \underset{\text{rectangle}}{\int_2^3 f(x)\,dx} + \underset{\text{triangle}}{\int_3^5 f(x)\,dx}$

$= \frac{1}{2}(1+3)2 + 3\cdot 1 + \frac{1}{2}\cdot 2\cdot 3 = 4+3+3 = 10$

(c) $\int_5^7 f(x)\,dx$ is the negative of the area of the triangle with base 2 and height 3. $\int_5^7 f(x)\,dx = -\frac{1}{2}\cdot 2\cdot 3 = -3$.

(d) $\int_7^9 f(x)\,dx$ is the negative of the area of a trapezoid with bases 3 and 2 and height 2, so it equals $-\frac{1}{2}(B+b)h = -\frac{1}{2}(3+2)2 = -5$. Thus,
$\int_0^9 f(x)\,dx = \int_0^5 f(x)\,dx + \int_5^7 f(x)\,dx + \int_7^9 f(x)\,dx = 10 + (-3) + (-5) = 2$.

33. $\int_0^3 \left(\frac{1}{2}x - 1\right)dx$ can be interpreted as the area of the triangle above the x-axis minus the area of the triangle below the x-axis; that is,

$\frac{1}{2}(1)\left(\frac{1}{2}\right) - \frac{1}{2}(2)(1) = \frac{1}{4} - 1 = -\frac{3}{4}$.

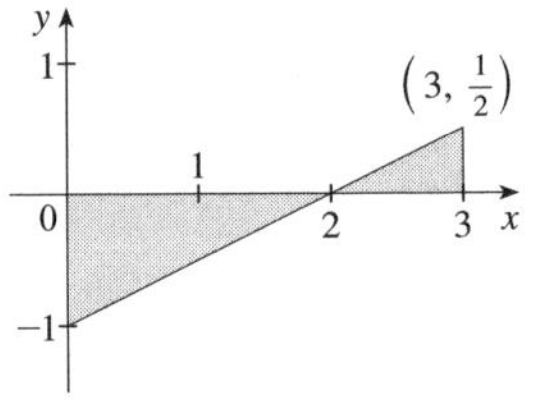

35. $\int_{-3}^{0}\left(1+\sqrt{9-x^2}\right)dx$ can be interpreted as the area under the graph of $f(x) = 1 + \sqrt{9-x^2}$ between $x = -3$ and $x = 0$. This is equal to one-quarter the area of the circle with radius 3, plus the area of the rectangle, so

$\int_{-3}^{0}\left(1+\sqrt{9-x^2}\right)dx = \frac{1}{4}\pi\cdot 3^2 + 1\cdot 3 = 3 + \frac{9}{4}\pi$.

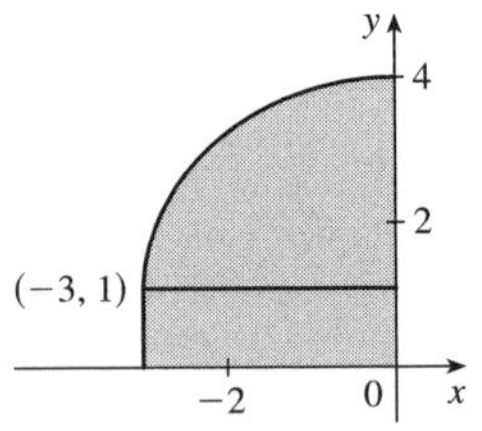

37. $\int_{-1}^{2} |x|\,dx$ can be interpreted as the sum of the areas of the two shaded triangles; that is, $\frac{1}{2}(1)(1) + \frac{1}{2}(2)(2) = \frac{1}{2} + \frac{4}{2} = \frac{5}{2}$.

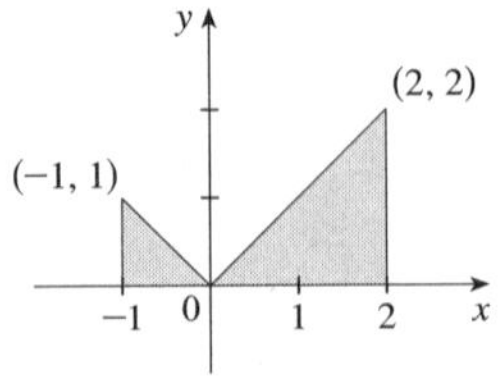

39. $\int_9^4 \sqrt{t}\,dt = -\int_4^9 \sqrt{t}\,dt$ [because we reversed the limits of integration]

$= -\int_4^9 \sqrt{x}\,dx$ [we can use any letter without changing the value of the integral]

$= -\frac{38}{3}$

41. $\int_{-2}^{2} f(x)\,dx + \int_2^5 f(x)\,dx - \int_{-2}^{-1} f(x)\,dx = \int_{-2}^{5} f(x)\,dx + \int_{-1}^{-2} f(x)\,dx$ [by Property 5 and reversing limits]

$= \int_{-1}^{5} f(x)\,dx$ [Property 5]

43. $\int_0^9 [2f(x) + 3g(x)]\,dx = 2\int_0^9 f(x)\,dx + 3\int_0^9 g(x)\,dx = 2(37) + 3(16) = 122$

45. $\int_0^1 (5 - 6x^2)\,dx = \int_0^1 5\,dx - 6\int_0^1 x^2\,dx = 5(1-0) - 6(\frac{1}{3}) = 5 - 2 = 3$

47. $\int_1^3 e^{x+2}\,dx = \int_1^3 e^x \cdot e^2\,dx = e^2\int_1^3 e^x\,dx = e^2(e^3 - e) = e^5 - e^3$

49. On $[1,3]$, $\ln 1 \le \ln x \le \ln 3 \quad\Rightarrow\quad 0(3-1) \le \int_1^3 \ln x\,dx \le (\ln 3)(3-1) \quad\Rightarrow\quad 0 \le \int_1^3 \ln x\,dx \le 2\ln 3$.

51. $\lim\limits_{n\to\infty}\sum\limits_{i=1}^{n}\dfrac{i^4}{n^5} = \lim\limits_{n\to\infty}\sum\limits_{i=1}^{n}\dfrac{i^4}{n^4}\cdot\dfrac{1}{n} = \lim\limits_{n\to\infty}\sum\limits_{i=1}^{n}\left(\dfrac{i}{n}\right)^4\dfrac{1}{n}$. At this point, we need to recognize the limit as being of the form $\lim\limits_{n\to\infty}\sum\limits_{i=1}^{n} f(x_i)\,\Delta x$, where $\Delta x = (1-0)/n = 1/n$, $x_i = 0 + i\,\Delta x = i/n$, and $f(x) = x^4$. Thus, the definite integral is $\int_0^1 x^4\,dx$.

5.3 Evaluating Definite Integrals

1. $\displaystyle\int_{-1}^{3} x^5\,dx = \left[\frac{x^6}{6}\right]_{-1}^{3} = \frac{3^6}{6} - \frac{(-1)^6}{6} = \frac{729-1}{6} = \frac{364}{3}$

3. $\int_0^2 (6x^2 - 4x + 5)\,dx = \left[6\cdot\frac{1}{3}x^3 - 4\cdot\frac{1}{2}x^2 + 5x\right]_0^2 = \left[2x^3 - 2x^2 + 5x\right]_0^2 = (16 - 8 + 10) - 0 = 18$

5. $\displaystyle\int_0^4 \sqrt{x}\,dx = \int_0^4 x^{1/2}\,dx = \left[\frac{x^{3/2}}{3/2}\right]_0^4 = \left[\frac{2x^{3/2}}{3}\right]_0^4 = \frac{2(4)^{3/2}}{3} - 0 = \frac{2\cdot 8}{3} = \frac{16}{3}$

7. $\int_{-1}^0 (2x - e^x)\,dx = \left[x^2 - e^x\right]_{-1}^0 = (0-1) - (1 - e^{-1}) = -2 + 1/e$

9. $\int_{-2}^2 (3u+1)^2\,du = \int_{-2}^2 (9u^2 + 6u + 1)\,du = \left[9\cdot\frac{1}{3}u^3 + 6\cdot\frac{1}{2}u^2 + u\right]_{-2}^2 = \left[3u^3 + 3u^2 + u\right]_{-2}^2$

$= (24 + 12 + 2) - (-24 + 12 - 2) = 38 - (-14) = 52$

11. $\displaystyle\int_{-2}^{-1}\left(4y^3 + \frac{2}{y^3}\right)dy = \left[4\cdot\tfrac{1}{4}y^4 + 2\cdot\frac{1}{-2}y^{-2}\right]_{-2}^{-1} = \left[y^4 - \frac{1}{y^2}\right]_{-2}^{-1} = (1-1) - \left(16 - \tfrac{1}{4}\right) = -\tfrac{63}{4}$

13. $\int_0^1 x(\sqrt[3]{x} + \sqrt[4]{x})\,dx = \int_0^1 (x^{4/3} + x^{5/4})\,dx = \left[\frac{3}{7}x^{7/3} + \frac{4}{9}x^{9/4}\right]_0^1 = \left(\frac{3}{7} + \frac{4}{9}\right) - 0 = \frac{55}{63}$

15. $\int_0^{\pi/4} \sec^2 t\,dt = \left[\tan t\right]_0^{\pi/4} = \tan\frac{\pi}{4} - \tan 0 = 1 - 0 = 1$

17. $\displaystyle\int_1^9 \frac{1}{2x}\,dx = \frac{1}{2}\int_1^9 \frac{1}{x}\,dx = \tfrac{1}{2}\Big[\ln|x|\Big]_1^9 = \tfrac{1}{2}(\ln 9 - \ln 1) = \tfrac{1}{2}\ln 9 - 0 = \ln 9^{1/2} = \ln 3$

19. $\displaystyle\int_{1/2}^{\sqrt{3}/2} \frac{6}{\sqrt{1-t^2}}\,dt = 6\int_{1/2}^{\sqrt{3}/2} \frac{1}{\sqrt{1-t^2}}\,dt = 6\left[\sin^{-1} t\right]_{1/2}^{\sqrt{3}/2} = 6\left[\sin^{-1}\left(\tfrac{\sqrt{3}}{2}\right) - \sin^{-1}\left(\tfrac{1}{2}\right)\right]$

$= 6\left(\frac{\pi}{3} - \frac{\pi}{6}\right) = 6\left(\frac{\pi}{6}\right) = \pi$

21. $\displaystyle\int_1^{64} \frac{1 + \sqrt[3]{x}}{\sqrt{x}}\,dx = \int_1^{64}\left(\frac{1}{x^{1/2}} + \frac{x^{1/3}}{x^{1/2}}\right)dx = \int_1^{64}\left(x^{-1/2} + x^{(1/3)-(1/2)}\right)dx = \int_1^{64}\left(x^{-1/2} + x^{-1/6}\right)dx$

$= \left[2x^{1/2} + \frac{6}{5}x^{5/6}\right]_1^{64} = \left(16 + \frac{192}{5}\right) - \left(2 + \frac{6}{5}\right) = 14 + \frac{186}{5} = \frac{256}{5}$

23. $\int_{-1}^1 e^{u+1}\,du = \left[e^{u+1}\right]_{-1}^1 = e^2 - e^0 = e^2 - 1$ [or start with $e^{u+1} = e^u e^1$]

25. $\displaystyle\int_0^{\pi/4} \frac{1 + \cos^2\theta}{\cos^2\theta}\,d\theta = \int_0^{\pi/4}\left(\frac{1}{\cos^2\theta} + \frac{\cos^2\theta}{\cos^2\theta}\right)d\theta = \int_0^{\pi/4}(\sec^2\theta + 1)\,d\theta$

$= \left[\tan\theta + \theta\right]_0^{\pi/4} = \left(\tan\frac{\pi}{4} + \frac{\pi}{4}\right) - (0 + 0) = 1 + \frac{\pi}{4}$

27. $\displaystyle\int_1^e \frac{x^2+x+1}{x}\,dx = \int_1^e \left(x+1+\frac{1}{x}\right)dx = \left[\tfrac{1}{2}x^2+x+\ln|x|\right]_1^e$

$= \left(\tfrac{1}{2}e^2+e+\ln e\right) - \left(\tfrac{1}{2}+1+\ln 1\right) = \tfrac{1}{2}e^2+e-\tfrac{1}{2}$

29. $f(x)=1/x^2$ is not continuous on the interval $[-1,3]$, so the Evaluation Theorem does not apply. In fact, f has an infinite discontinuity at $x=0$, so $\int_{-1}^3 (1/x^2)\,dx$ does not exist.

31. It appears that the area under the graph is about $\frac{2}{3}$ of the area of the viewing rectangle, or about $\frac{2}{3}\pi \approx 2.1$. The actual area is $\int_0^\pi \sin x\,dx = [-\cos x]_0^\pi = (-\cos\pi)-(-\cos 0) = -(-1)+1 = 2$.

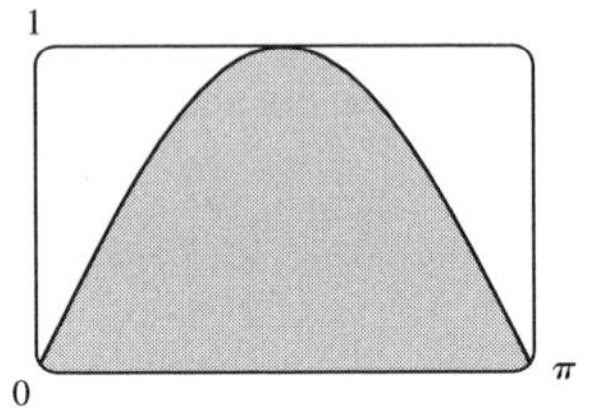

33. The graph shows that $y=x+x^2-x^4$ has x-intercepts at $x=0$ and at $x=a\approx 1.32$. So the area of the region that lies under the curve and above the x-axis is

$\int_0^a (x+x^2-x^4)\,dx = \left[\tfrac{1}{2}x^2+\tfrac{1}{3}x^3-\tfrac{1}{5}x^5\right]_0^a$

$= \left(\tfrac{1}{2}a^2+\tfrac{1}{3}a^3-\tfrac{1}{5}a^5\right)-0$

≈ 0.84

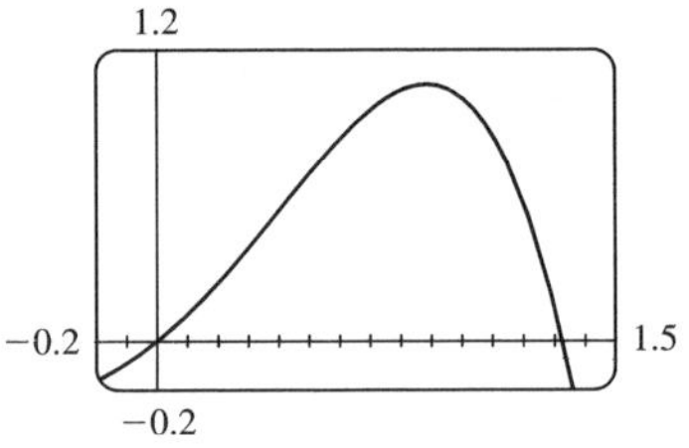

35. $\int_{-1}^2 x^3\,dx = \left[\tfrac{1}{4}x^4\right]_{-1}^2 = 4-\tfrac{1}{4} = \tfrac{15}{4} = 3.75$

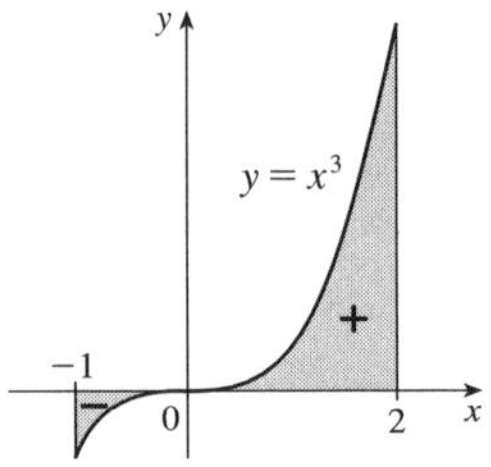

37. $\displaystyle\frac{d}{dx}\left[\sqrt{x^2+1}+C\right] = \frac{d}{dx}\left[(x^2+1)^{1/2}+C\right] = \tfrac{1}{2}(x^2+1)^{-1/2}\cdot 2x = \frac{x}{\sqrt{x^2+1}}$

39. $\int x\sqrt{x}\,dx = \int x^{3/2}\,dx = \frac{2}{5}x^{5/2}+C$.

The members of the family in the figure correspond to $C=5, 3, 0, -2$, and -4.

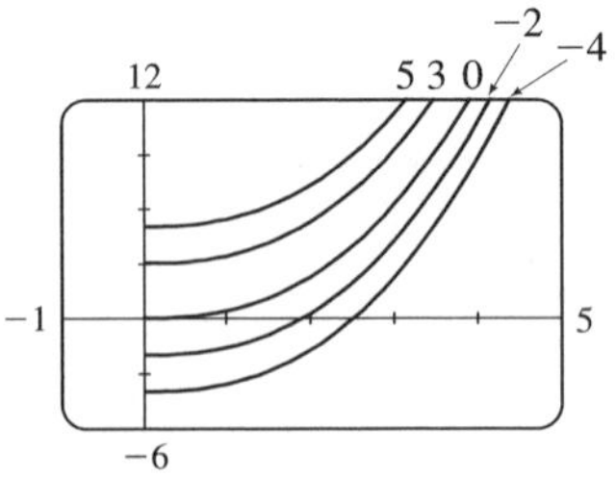

41. $\displaystyle\int (1-t)(2+t^2)\,dt = \int (2-2t+t^2-t^3)\,dt = 2t-2\frac{t^2}{2}+\frac{t^3}{3}-\frac{t^4}{4}+C = 2t-t^2+\tfrac{1}{3}t^3-\tfrac{1}{4}t^4+C$

43. $\displaystyle\int \frac{\sin x}{1-\sin^2 x}\,dx = \int \frac{\sin x}{\cos^2 x}\,dx = \int \frac{1}{\cos x}\cdot\frac{\sin x}{\cos x}\,dx = \int \sec x\,\tan x\,dx = \sec x + C$

45. $A = \int_0^2 (2y-y^2)\,dy = \left[y^2-\tfrac{1}{3}y^3\right]_0^2 = \left(4-\tfrac{8}{3}\right)-0 = \tfrac{4}{3}$

47. If $w'(t)$ is the rate of change of weight in kilograms per year, then $w(t)$ represents the weight in kilograms of the child at age t. We know from the Net Change Theorem that $\int_5^{10} w'(t)\,dt = w(10)-w(5)$, so the integral represents the increase in the child's weight (in kilograms) between the ages of 5 and 10.

49. Since $r(t)$ is the rate at which oil leaks, we can write $r(t) = -V'(t)$, where $V(t)$ is the volume of oil at time t. [Note that the minus sign is needed because V is decreasing, so $V'(t)$ is negative, but $r(t)$ is positive.] Thus, by the Net Change Theorem, $\int_0^{120} r(t)\,dt = -\int_0^{120} V'(t)\,dt = -[V(120) - V(0)] = V(0) - V(120)$, which is the number of litres of oil that leaked from the tank in the first two hours (120 minutes).

51. By the Net Change Theorem, $\int_{1000}^{5000} R'(x)\,dx = R(5000) - R(1000)$, so it represents the increase in revenue when production is increased from 1000 units to 5000 units.

53. In general, the unit of measurement for $\int_a^b f(x)\,dx$ is the product of the unit for $f(x)$ and the unit for x. Since $f(x)$ is measured in newtons and x is measured in metres, the units for $\int_0^{100} f(x)\,dx$ are newton-metres. (A newton-metre is abbreviated N·m and is called a joule.)

55. (a) Displacement $= \int_0^3 (3t-5)\,dt = \left[\frac{3}{2}t^2 - 5t\right]_0^3 = \frac{27}{2} - 15 = -\frac{3}{2}$ m

(b) Distance travelled $= \int_0^3 |3t-5|\,dt = \int_0^{5/3}(5-3t)\,dt + \int_{5/3}^3 (3t-5)\,dt$

$= \left[5t - \frac{3}{2}t^2\right]_0^{5/3} + \left[\frac{3}{2}t^2 - 5t\right]_{5/3}^3 = \frac{25}{3} - \frac{3}{2}\cdot\frac{25}{9} + \frac{27}{2} - 15 - \left(\frac{3}{2}\cdot\frac{25}{9} - \frac{25}{3}\right) = \frac{41}{6}$ m

57. (a) $v'(t) = a(t) = t + 4 \;\Rightarrow\; v(t) = \frac{1}{2}t^2 + 4t + C \;\Rightarrow\; v(0) = C = 5 \;\Rightarrow\; v(t) = \frac{1}{2}t^2 + 4t + 5$ m/s

(b) Distance travelled $= \int_0^{10} |v(t)|\,dt = \int_0^{10} \left|\frac{1}{2}t^2 + 4t + 5\right|dt = \int_0^{10}\left(\frac{1}{2}t^2 + 4t + 5\right)dt$

$= \left[\frac{1}{6}t^3 + 2t^2 + 5t\right]_0^{10} = \frac{500}{3} + 200 + 50 = 416\frac{2}{3}$ m

59. Since $m'(x) = \rho(x)$, $m = \int_0^4 \rho(x)\,dx = \int_0^4 (9 + 2\sqrt{x}\,)\,dx = \left[9x + \frac{4}{3}x^{3/2}\right]_0^4 = 36 + \frac{32}{3} - 0 = \frac{140}{3} = 46\frac{2}{3}$ kg.

61. Let s be the position of the car. We know from Equation 2 that $s(100) - s(0) = \int_0^{100} v(t)\,dt$. We use the Midpoint Rule for $0 \le t \le 100$ with $n = 5$. Note that the length of each of the five time intervals is 20 seconds $= \frac{20}{3600}$ hour $= \frac{1}{180}$ hour. So the distance travelled is

$$\begin{aligned}\int_0^{100} v(t)\,dt &\approx \tfrac{1}{180}[v(10) + v(30) + v(50) + v(70) + v(90)]\\ &= \tfrac{1}{180}(61 + 93 + 82 + 85 + 75)\\ &= \tfrac{396}{180} = 2.2 \text{ kilometres}\end{aligned}$$

63. From the Net Change Theorem, the increase in cost if the production level is raised from 2000 metres to 4000 metres is $C(4000) - C(2000) = \int_{2000}^{4000} C'(x)\,dx$.

$$\begin{aligned}\int_{2000}^{4000} C'(x)\,dx &= \int_{2000}^{4000} \left(3 - 0.01x + 0.000\,006x^2\right)dx\\ &= \left[3x - 0.005x^2 + 0.000\,002x^3\right]_{2000}^{4000} = 60\,000 - 2000 = \$58\,000\end{aligned}$$

65. (a) We can find the area between the Lorenz curve and the line $y = x$ by subtracting the area under $y = L(x)$ from the area under $y = x$. Thus,

$$\begin{aligned}\text{coefficient of inequality} &= \frac{\text{area between Lorenz curve and line } y = x}{\text{area under line } y = x} = \frac{\int_0^1 [x - L(x)]\,dx}{\int_0^1 x\,dx}\\ &= \frac{\int_0^1 [x - L(x)]\,dx}{\left[x^2/2\right]_0^1} = \frac{\int_0^1 [x - L(x)]\,dx}{1/2} = 2\int_0^1 [x - L(x)]\,dx\end{aligned}$$

(b) $L(x) = \frac{5}{12}x^2 + \frac{7}{12}x \;\Rightarrow\; L(50\%) = L\left(\frac{1}{2}\right) = \frac{5}{48} + \frac{7}{24} = \frac{19}{48} = 0.3958\overline{3}$, so the bottom 50% of the households receive at most about 40% of the income. Using the result in part (a),

$$\begin{aligned}\text{coefficient of inequality} &= 2\int_0^1 [x - L(x)]\,dx = 2\int_0^1 \left(x - \tfrac{5}{12}x^2 - \tfrac{7}{12}x\right)dx \\ &= 2\int_0^1 \left(\tfrac{5}{12}x - \tfrac{5}{12}x^2\right)dx = 2\int_0^1 \tfrac{5}{12}\left(x - x^2\right)dx \\ &= \tfrac{5}{6}\left[\tfrac{1}{2}x^2 - \tfrac{1}{3}x^3\right]_0^1 = \tfrac{5}{6}\left(\tfrac{1}{2} - \tfrac{1}{3}\right) = \tfrac{5}{6}\left(\tfrac{1}{6}\right) = \tfrac{5}{36}\end{aligned}$$

67. The second derivative is the derivative of the first derivative, so we'll apply the Net Change Theorem with $F = h'$. $\int_1^2 h''(u)\,du = \int_1^2 (h')'(u)\,du = h'(2) - h'(1) = 5 - 2 = 3$. The other information is unnecessary.

5.4 The Fundamental Theorem of Calculus

1. The precise version of this statement is given by the Fundamental Theorem of Calculus. See the statement of this theorem and the paragraph that follows it on page 381.

3. (a) $g(x) = \int_0^x f(t)\,dt$.

$g(0) = \int_0^0 f(t)\,dt = 0$

$g(1) = \int_0^1 f(t)\,dt = 1 \cdot 2 = 2$ [rectangle],

$g(2) = \int_0^2 f(t)\,dt = \int_0^1 f(t)\,dt + \int_1^2 f(t)\,dt = g(1) + \int_1^2 f(t)\,dt$

$\quad = 2 + 1 \cdot 2 + \frac{1}{2} \cdot 1 \cdot 2 = 5$ [rectangle plus triangle],

$g(3) = \int_0^3 f(t)\,dt = g(2) + \int_2^3 f(t)\,dt = 5 + \frac{1}{2} \cdot 1 \cdot 4 = 7$,

$g(6) = g(3) + \int_3^6 f(t)\,dt$ [the integral is negative since f lies under the x-axis]

$\quad = 7 + \left[-\left(\frac{1}{2} \cdot 2 \cdot 2 + 1 \cdot 2\right)\right] = 7 - 4 = 3$

(b) g is increasing on $(0, 3)$ because as x increases from 0 to 3, we keep adding more area.

(c) g has a maximum value when we start subtracting area; that is, at $x = 3$.

(d)

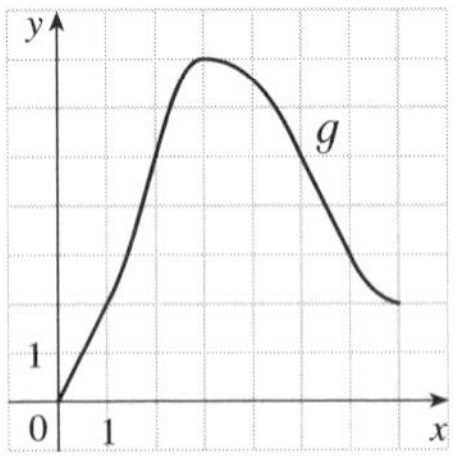

5.

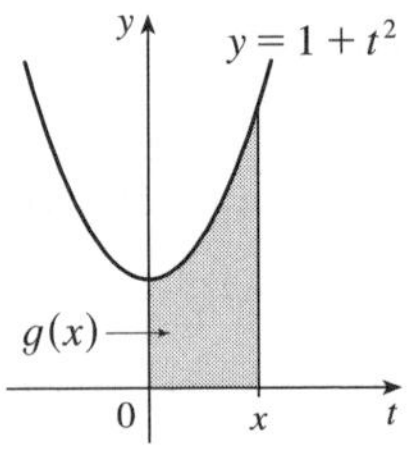

(a) By FTC1, $g(x) = \int_0^x (1 + t^2)\,dt \;\Rightarrow\; g'(x) = f(x) = 1 + x^2$.

(b) By FTC2, $g(x) = \int_0^x (1 + t^2)\,dt = \left[t + \frac{1}{3}t^3\right]_0^x = \left(x + \frac{1}{3}x^3\right) - 0 \;\Rightarrow\; g'(x) = 1 + x^2$.

7. $f(t) = \sqrt{1 + 2t}$ and $g(x) = \int_0^x \sqrt{1 + 2t}\,dt$, so by FTC1, $g'(x) = f(x) = \sqrt{1 + 2x}$.

9. $f(t) = t^2 \sin t$ and $g(y) = \int_2^y t^2 \sin t\,dt$, so by FTC1, $g'(y) = f(y) = y^2 \sin y$.

11. Let $u = \dfrac{1}{x}$. Then $\dfrac{du}{dx} = -\dfrac{1}{x^2}$. Also, $\dfrac{dh}{dx} = \dfrac{dh}{du}\dfrac{du}{dx}$, so

$$h'(x) = \frac{d}{dx}\int_2^{1/x} \arctan t\,dt = \frac{d}{du}\int_2^u \arctan t\,dt \cdot \frac{du}{dx} = \arctan u\,\frac{du}{dx} = -\frac{\arctan(1/x)}{x^2}.$$

13. Let $u = \sqrt{x}$. Then $\dfrac{du}{dx} = \dfrac{1}{2\sqrt{x}}$. Also, $\dfrac{dy}{dx} = \dfrac{dy}{du}\dfrac{du}{dx}$, so

$$y' = \frac{d}{dx}\int_3^{\sqrt{x}} \frac{\cos t}{t}\,dt = \frac{d}{du}\int_3^u \frac{\cos t}{t}\,dt \cdot \frac{du}{dx} = \frac{\cos u}{u}\cdot\frac{1}{2\sqrt{x}} = \frac{\cos\sqrt{x}}{\sqrt{x}}\cdot\frac{1}{2\sqrt{x}} = \frac{\cos\sqrt{x}}{2x}.$$

15. $g(x) = \displaystyle\int_{2x}^{3x} \frac{u^2-1}{u^2+1}\,du = \int_{2x}^{0} \frac{u^2-1}{u^2+1}\,du + \int_0^{3x} \frac{u^2-1}{u^2+1}\,du = -\int_0^{2x} \frac{u^2-1}{u^2+1}\,du + \int_0^{3x} \frac{u^2-1}{u^2+1}\,du \;\Rightarrow$

$$g'(x) = -\frac{(2x)^2-1}{(2x)^2+1}\cdot\frac{d}{dx}(2x) + \frac{(3x)^2-1}{(3x)^2+1}\cdot\frac{d}{dx}(3x) = -2\cdot\frac{4x^2-1}{4x^2+1} + 3\cdot\frac{9x^2-1}{9x^2+1}$$

17. $F(x) = \displaystyle\int_1^x f(t)\,dt \;\Rightarrow\; F'(x) = f(x) = \int_1^{x^2} \frac{\sqrt{1+u^4}}{u}\,du \quad \left[\text{since } f(t) = \int_1^{t^2} \frac{\sqrt{1+u^4}}{u}\,du\right] \;\Rightarrow$

$$F''(x) = f'(x) = \frac{\sqrt{1+(x^2)^4}}{x^2}\cdot\frac{d}{dx}(x^2) = \frac{\sqrt{1+x^8}}{x^2}\cdot 2x = \frac{2\sqrt{1+x^8}}{x}.$$ So $F''(2) = \sqrt{1+2^8} = \sqrt{257}$.

19. (a) By FTC1, $g'(x) = f(x)$. So $g'(x) = f(x) = 0$ at $x = 1, 3, 5, 7$, and 9. g has local maxima at $x = 1$ and 5 (since $f = g'$ changes from positive to negative there) and local minima at $x = 3$ and 7. There is no local maximum or minimum at $x = 9$, since f is not defined for $x > 9$.

(b) We can see from the graph that $\left|\int_0^1 f\,dt\right| < \left|\int_1^3 f\,dt\right| < \left|\int_3^5 f\,dt\right| < \left|\int_5^7 f\,dt\right| < \left|\int_7^9 f\,dt\right|$. So $g(1) = \left|\int_0^1 f\,dt\right|$, $g(5) = \int_0^5 f\,dt = g(1) - \left|\int_1^3 f\,dt\right| + \left|\int_3^5 f\,dt\right|$, and $g(9) = \int_0^9 f\,dt = g(5) - \left|\int_5^7 f\,dt\right| + \left|\int_7^9 f\,dt\right|$. Thus, $g(1) < g(5) < g(9)$, and so the absolute maximum of $g(x)$ occurs at $x = 9$.

(c) g is concave downward on those intervals where $g'' < 0$. But $g'(x) = f(x)$, so $g''(x) = f'(x)$, which is negative on (approximately) $\left(\frac{1}{2}, 2\right)$, $(4, 6)$ and $(8, 9)$. So g is concave downward on these intervals.

(d)

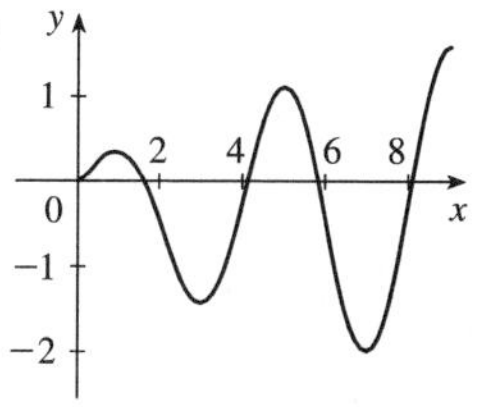

21. By FTC2, $\int_1^4 f'(x)\,dx = f(4) - f(1)$, so $17 = f(4) - 12 \;\Rightarrow\; f(4) = 17 + 12 = 29$.

23. (a) The Fresnel function $S(x) = \int_0^x \sin\left(\frac{\pi}{2}t^2\right)dt$ has local maximum values where $0 = S'(x) = \sin\left(\frac{\pi}{2}x^2\right)$ and S' changes from positive to negative. For $x > 0$, this happens when $\frac{\pi}{2}x^2 = (2n-1)\pi$ [odd multiples of π] $\Leftrightarrow$ $x^2 = 2(2n-1)$ $\Leftrightarrow$ $x = \sqrt{4n-2}$, n any positive integer. For $x < 0$, S' changes from positive to negative where $\frac{\pi}{2}x^2 = 2n\pi$ [even multiples of π] $\Leftrightarrow$ $x^2 = 4n$ $\Leftrightarrow$ $x = -2\sqrt{n}$. S' does not change sign at $x = 0$.

(b) S is concave upward on those intervals where $S''(x) > 0$. Differentiating our expression for $S'(x)$, we get $S''(x) = \cos\left(\frac{\pi}{2}x^2\right)\left(2\frac{\pi}{2}x\right) = \pi x\cos\left(\frac{\pi}{2}x^2\right)$. For $x > 0$, $S''(x) > 0$ where $\cos\left(\frac{\pi}{2}x^2\right) > 0$ $\Leftrightarrow$ $0 < \frac{\pi}{2}x^2 < \frac{\pi}{2}$ or $\left(2n - \frac{1}{2}\right)\pi < \frac{\pi}{2}x^2 < \left(2n + \frac{1}{2}\right)\pi$, n any integer $\Leftrightarrow$ $0 < x < 1$ or $\sqrt{4n-1} < x < \sqrt{4n+1}$, n any positive integer. For $x < 0$, $S''(x) > 0$ where $\cos\left(\frac{\pi}{2}x^2\right) < 0$ $\Leftrightarrow$ $\left(2n - \frac{3}{2}\right)\pi < \frac{\pi}{2}x^2 < \left(2n - \frac{1}{2}\right)\pi$, n any integer $\Leftrightarrow$ $4n - 3 < x^2 < 4n - 1$ $\Leftrightarrow$ $\sqrt{4n-3} < |x| < \sqrt{4n-1}$ $\Rightarrow$ $\sqrt{4n-3} < -x < \sqrt{4n-1}$ $\Rightarrow$ $-\sqrt{4n-3} > x > -\sqrt{4n-1}$, so the intervals of upward concavity for $x < 0$ are $\left(-\sqrt{4n-1}, -\sqrt{4n-3}\right)$, n any positive integer. To summarize: S is concave upward on the intervals $(0, 1)$, $\left(-\sqrt{3}, -1\right)$, $\left(\sqrt{3}, \sqrt{5}\right)$, $\left(-\sqrt{7}, -\sqrt{5}\right)$, $\left(\sqrt{7}, 3\right)$, ...

(c) In Maple, we use `plot({int(sin(Pi*t^2/2),t=0..x),0.2},x=0..2);`. Note that Maple recognizes the Fresnel function, calling it `FresnelS(x)`. In Mathematica, we use `Plot[{Integrate[Sin[Pi*t^2/2],{t,0,x}],0.2},{x,0,2}]`. In Derive, we load the utility file `FRESNEL` and plot `FRESNEL_SIN(x)`. From the graphs, we see that $\int_0^x \sin\left(\frac{\pi}{2}t^2\right) dt = 0.2$ at $x \approx 0.74$.

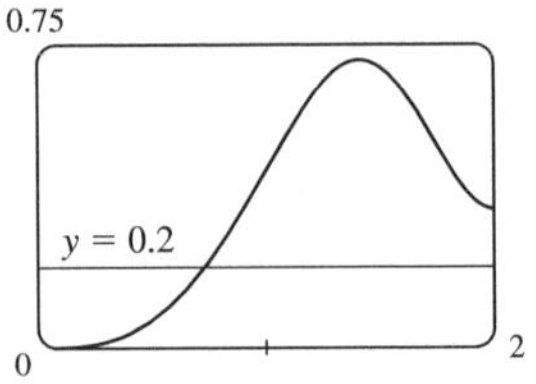

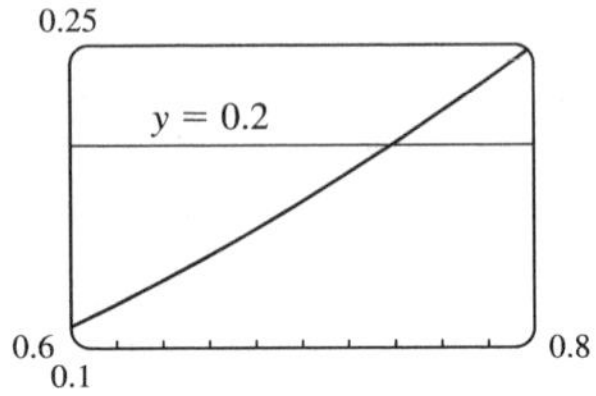

25. By FTC2, $\int_1^x f'(t)\,dt = f(x) - f(1) \;\Rightarrow\; f(x) = f(1) + \int_1^x f'(t)\,dt = f(1) + \int_1^x (2^t/t)\,dt$. This integral cannot be expressed in a simpler form. Since we want $f(1) = 0$, we have $f(x) = \int_1^x (2^t/t)\,dt$.

27. Using FTC1, we differentiate both sides of $6 + \displaystyle\int_a^x \frac{f(t)}{t^2}\,dt = 2\sqrt{x}$ to get $\dfrac{f(x)}{x^2} = 2\dfrac{1}{2\sqrt{x}} \;\Rightarrow\; f(x) = x^{3/2}$. To find a, we substitute $x = a$ in the original equation to obtain $6 + \displaystyle\int_a^a \frac{f(t)}{t^2}\,dt = 2\sqrt{a} \;\Rightarrow\; 6 + 0 = 2\sqrt{a} \;\Rightarrow$ $3 = \sqrt{a} \;\Rightarrow\; a = 9$.

29. (a) Let $F(t) = \int_0^t f(s)\,ds$. Then, by FTC1, $F'(t) = f(t) =$ rate of depreciation, so $F(t)$ represents the loss in value over the interval $[0, t]$.

(b) $C(t) = \dfrac{1}{t}\left[A + \displaystyle\int_0^t f(s)\,ds\right] = \dfrac{A + F(t)}{t}$ represents the average expenditure per unit of t during the interval $[0, t]$, assuming that there has been only one overhaul during that time period. The company wants to minimize average expenditure.

(c) $C(t) = \dfrac{1}{t}\left[A + \displaystyle\int_0^t f(s)\,ds\right]$. Using FTC1, we have $C'(t) = -\dfrac{1}{t^2}\left[A + \displaystyle\int_0^t f(s)\,ds\right] + \dfrac{1}{t}f(t)$.

$C'(t) = 0 \;\Rightarrow\; t\,f(t) = A + \displaystyle\int_0^t f(s)\,ds \;\Rightarrow\; f(t) = \frac{1}{t}\left[A + \int_0^t f(s)\,ds\right] = C(t)$.

5.5 The Substitution Rule

1. Let $u = 3x$. Then $du = 3\,dx$, so $dx = \frac{1}{3}du$. Thus, $\int \cos 3x\,dx = \int \cos u\left(\frac{1}{3}\,du\right) = \frac{1}{3}\int \cos u\,du = \frac{1}{3}\sin u + C = \frac{1}{3}\sin 3x + C$. Don't forget that it is often very easy to check an indefinite integration by differentiating your answer. In this case, $\dfrac{d}{dx}\left(\frac{1}{3}\sin 3x + C\right) = \frac{1}{3}(\cos 3x)\cdot 3 = \cos 3x$, the desired result.

3. Let $u = x^3 + 1$. Then $du = 3x^2\,dx$ and $x^2\,dx = \frac{1}{3}\,du$, so

$$\int x^2\sqrt{x^3+1}\,dx = \int \sqrt{u}\left(\frac{1}{3}\,du\right) = \frac{1}{3}\frac{u^{3/2}}{3/2} + C = \frac{1}{3}\cdot\frac{2}{3}u^{3/2} + C = \frac{2}{9}(x^3+1)^{3/2} + C.$$

5. Let $u = 1 + 2x$. Then $du = 2\,dx$ and $dx = \frac{1}{2}\,du$, so

$$\int \frac{4}{(1+2x)^3}\,dx = 4\int u^{-3}\left(\tfrac{1}{2}\,du\right) = 2\frac{u^{-2}}{-2} + C = -\frac{1}{u^2} + C = -\frac{1}{(1+2x)^2} + C.$$

7. Let $u = x^2 + 3$. Then $du = 2x\,dx$, so $\int 2x(x^2+3)^4\,dx = \int u^4\,du = \frac{1}{5}u^5 + C = \frac{1}{5}(x^2+3)^5 + C$.

9. Let $u = 3x - 2$. Then $du = 3\,dx$ and $dx = \frac{1}{3}\,du$, so $\int (3x-2)^{20}\,dx = \int u^{20}\left(\frac{1}{3}\,du\right) = \frac{1}{3}\cdot\frac{1}{21}u^{21} + C = \frac{1}{63}(3x-2)^{21} + C$.

11. Let $u = \ln x$. Then $du = \dfrac{dx}{x}$, so $\displaystyle\int \frac{(\ln x)^2}{x}\,dx = \int u^2\,du = \frac{1}{3}u^3 + C = \frac{1}{3}(\ln x)^3 + C$.

13. Let $u = 5 - 3x$. Then $du = -3\,dx$ and $dx = -\frac{1}{3}\,du$, so

$$\int \frac{dx}{5-3x} = \int \frac{1}{u}\left(-\tfrac{1}{3}\,du\right) = -\tfrac{1}{3}\ln|u| + C = -\tfrac{1}{3}\ln|5-3x| + C.$$

15. Let $u = 1 + x + 2x^2$. Then $du = (1+4x)\,dx$, so

$$\int \frac{1+4x}{\sqrt{1+x+2x^2}}\,dx = \int \frac{du}{\sqrt{u}} = \int u^{-1/2}\,du = \frac{u^{1/2}}{1/2} + C = 2\sqrt{1+x+2x^2} + C.$$

17. Let $u = 3\theta$. Then $du = 3\,d\theta$, so $\int \sin 3\theta\,d\theta = \int \sin u\left(\frac{1}{3}\,du\right) = \frac{1}{3}(-\cos u) + C = -\frac{1}{3}\cos 3\theta + C$.

19. Let $u = 1 + e^x$. Then $du = e^x\,dx$, so $\int e^x\sqrt{1+e^x}\,dx = \int \sqrt{u}\,du = \frac{2}{3}u^{3/2} + C = \frac{2}{3}(1+e^x)^{3/2} + C$.

Or: Let $u = \sqrt{1+e^x}$. Then $u^2 = 1 + e^x$ and $2u\,du = e^x\,dx$, so

$\int e^x\sqrt{1+e^x}\,dx = \int u \cdot 2u\,du = \frac{2}{3}u^3 + C = \frac{2}{3}(1+e^x)^{3/2} + C$.

21. Let $u = \cos x$. Then $du = -\sin x\,dx$, so $\int \cos^4 x \sin x\,dx = \int u^4(-du) = -\frac{1}{5}u^5 + C = -\frac{1}{5}\cos^5 x + C$.

23. Let $u = \cot x$. Then $du = -\csc^2 x\,dx$ and $\csc^2 x\,dx = -du$, so

$$\int \sqrt{\cot x}\csc^2 x\,dx = \int \sqrt{u}\,(-du) = -\frac{u^{3/2}}{3/2} + C = -\tfrac{2}{3}(\cot x)^{3/2} + C.$$

25. Let $u = \sin^{-1} x$. Then $du = \dfrac{1}{\sqrt{1-x^2}}\,dx$, so $\displaystyle\int \frac{dx}{\sqrt{1-x^2}\sin^{-1} x} = \int \frac{1}{u}\,du = \ln|u| + C = \ln\left|\sin^{-1} x\right| + C$.

27. Let $u = \sec x$. Then $du = \sec x\,\tan x\,dx$, so

$\int \sec^3 x\,\tan x\,dx = \int \sec^2 x\,(\sec x\,\tan x)\,dx = \int u^2\,du = \frac{1}{3}u^3 + C = \frac{1}{3}\sec^3 x + C$.

29. $\displaystyle\int \frac{e^x+1}{e^x}\,dx = \int \left(\frac{e^x}{e^x} + \frac{1}{e^x}\right)dx = \int (1 + e^{-x})\,dx = x - e^{-x} + C$ [Substitute $u = -x$.]

31. $\displaystyle\int \frac{\sin 2x}{1+\cos^2 x}\,dx = 2\int \frac{\sin x\cos x}{1+\cos^2 x}\,dx = 2I$. Let $u = \cos x$. Then $du = -\sin x\,dx$, so

$$2I = -2\int \frac{u\,du}{1+u^2} = -2\cdot\tfrac{1}{2}\ln(1+u^2) + C = -\ln(1+u^2) + C = -\ln(1+\cos^2 x) + C.$$

Or: Let $u = 1 + \cos^2 x$.

33. Let $u = 1 + x^2$. Then $du = 2x\,dx$, so

$$\int \frac{1+x}{1+x^2}\,dx = \int \frac{1}{1+x^2}\,dx + \int \frac{x}{1+x^2}\,dx = \tan^{-1} x + \int \frac{\frac{1}{2}du}{u} = \tan^{-1} x + \tfrac{1}{2}\ln|u| + C$$

$$= \tan^{-1} x + \tfrac{1}{2}\ln\left|1+x^2\right| + C = \tan^{-1} x + \tfrac{1}{2}\ln\left(1+x^2\right) + C \quad [\text{since } 1 + x^2 > 0].$$

35. $f(x) = \dfrac{3x-1}{(3x^2-2x+1)^4}$.

$u = 3x^2 - 2x + 1 \quad\Rightarrow\quad du = (6x-2)\,dx = 2(3x-1)\,dx$, so

$$\int \frac{3x-1}{(3x^2-2x+1)^4}\,dx = \int \frac{1}{u^4}\left(\frac{1}{2}\,du\right) = \frac{1}{2}\int u^{-4}\,du$$

$$= -\frac{1}{6}u^{-3} + C = -\frac{1}{6(3x^2-2x+1)^3} + C$$

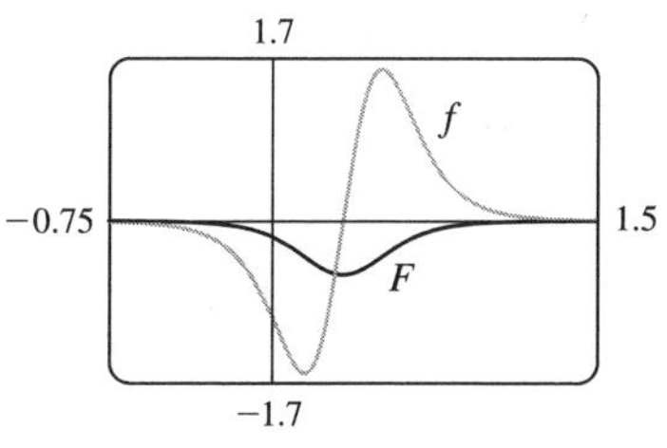

Notice that at $x = \frac{1}{3}$, f changes from negative to positive, and F has a local minimum.

37. $f(x) = \sin^3 x \cos x$. $u = \sin x \Rightarrow du = \cos x\,dx$, so

$$\int \sin^3 x \cos x\,dx = \int u^3\,du = \tfrac{1}{4}u^4 + C = \tfrac{1}{4}\sin^4 x + C$$

Note that at $x = \frac{\pi}{2}$, f changes from positive to negative and F has a local maximum. Also, both f and F are periodic with period π, so at $x = 0$ and at $x = \pi$, f changes from negative to positive and F has local minima.

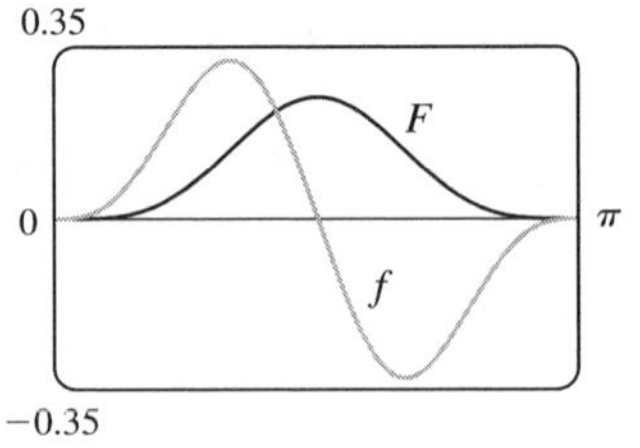

39. Let $u = x - 1$, so $du = dx$. When $x = 0$, $u = -1$; when $x = 2$, $u = 1$. Thus, $\int_0^2 (x-1)^{25}\,dx = \int_{-1}^1 u^{25}\,du = 0$ by Theorem 6(b), since $f(u) = u^{25}$ is an odd function.

41. Let $u = 1 + 2x^3$, so $du = 6x^2\,dx$. When $x = 0$, $u = 1$; when $x = 1$, $u = 3$. Thus,

$$\int_0^1 x^2 (1 + 2x^3)^5\,dx = \int_1^3 u^5 \left(\tfrac{1}{6}\,du\right) = \tfrac{1}{6}\left[\tfrac{1}{6}u^6\right]_1^3 = \tfrac{1}{36}(3^6 - 1^6) = \tfrac{1}{36}(729 - 1) = \tfrac{728}{36} = \tfrac{182}{9}$$

43. Let $u = t/4$, so $du = \frac{1}{4}\,dt$. When $t = 0$, $u = 0$; when $t = \pi$, $u = \pi/4$. Thus,

$\int_0^\pi \sec^2(t/4)\,dt = \int_0^{\pi/4} \sec^2 u\,(4\,du) = 4\big[\tan u\big]_0^{\pi/4} = 4\left(\tan\frac{\pi}{4} - \tan 0\right) = 4(1 - 0) = 4.$

45. Let $u = \sqrt{x}$, so $du = \dfrac{1}{2\sqrt{x}}\,dx$. When $x = 1$, $u = 1$; when $x = 4$, $u = 2$. Thus,

$$\int_1^4 \frac{e^{\sqrt{x}}}{\sqrt{x}}\,dx = \int_1^2 e^u (2\,du) = 2\left[e^u\right]_1^2 = 2(e^2 - e)$$

47. Let $u = x - 1$, so $u + 1 = x$ and $du = dx$. When $x = 1$, $u = 0$; when $x = 2$, $u = 1$. Thus,

$\int_1^2 x\sqrt{x-1}\,dx = \int_0^1 (u+1)\sqrt{u}\,du = \int_0^1 (u^{3/2} + u^{1/2})\,du = \left[\frac{2}{5}u^{5/2} + \frac{2}{3}u^{3/2}\right]_0^1 = \frac{2}{5} + \frac{2}{3} = \frac{16}{15}.$

49. Let $u = e^z + z$, so $du = (e^z + 1)\,dz$. When $z = 0$, $u - 1$; when $z = 1$, $u = e + 1$. Thus,

$$\int_0^1 \frac{e^z + 1}{e^z + z}\,dz = \int_1^{e+1} \frac{1}{u}\,du = \Big[\ln|u|\Big]_1^{e+1} = \ln|e+1| - \ln|1| = \ln(e+1).$$

51. $\int_{-\pi/6}^{\pi/6} \tan^3\theta\,d\theta = 0$ by Theorem 6(b), since $f(\theta) = \tan^3\theta$ is an odd function.

53. Let $u = \ln x$, so $du = \dfrac{dx}{x}$. When $x = e$, $u = 1$; when $x = e^4$; $u = 4$. Thus,

$$\int_e^{e^4} \frac{dx}{x\sqrt{\ln x}} = \int_1^4 u^{-1/2}\,du = 2\Big[u^{1/2}\Big]_1^4 = 2(2 - 1) = 2.$$

55. From the graph, it appears that the area under the curve is about $1 + \left(\text{a little more than } \frac{1}{2}\cdot 1 \cdot 0.7\right)$, or about 1.4. The exact area is given by $A = \int_0^1 \sqrt{2x+1}\,dx$. Let $u = 2x + 1$, so $du = 2\,dx$. The limits change to $2 \cdot 0 + 1 = 1$ and $2 \cdot 1 + 1 = 3$, and

$A = \int_1^3 \sqrt{u}\left(\frac{1}{2}\,du\right) = \frac{1}{2}\left[\frac{2}{3}u^{3/2}\right]_1^3 = \frac{1}{3}\left(3\sqrt{3} - 1\right) = \sqrt{3} - \frac{1}{3} \approx 1.399.$

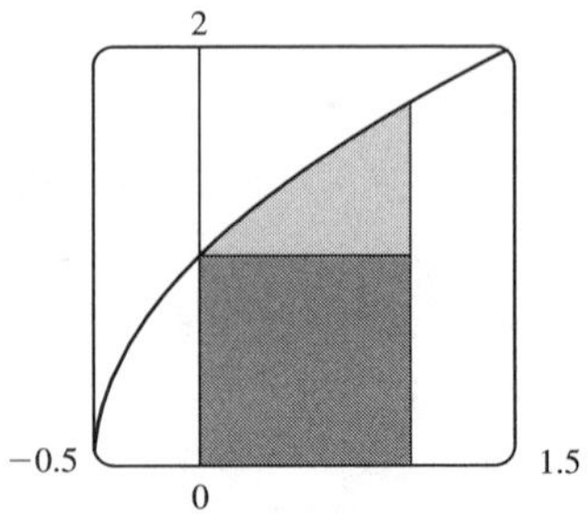

57. First write the integral as a sum of two integrals:

$I = \int_{-2}^2 (x+3)\sqrt{4 - x^2}\,dx = I_1 + I_2 = \int_{-2}^2 x\sqrt{4 - x^2}\,dx + \int_{-2}^2 3\sqrt{4 - x^2}\,dx$. $I_1 = 0$ by Theorem 6(b), since $f(x) = x\sqrt{4 - x^2}$ is an odd function and we are integrating from $x = -2$ to $x = 2$. We interpret I_2 as three times the area of a semicircle with radius 2, so $I = 0 + 3 \cdot \frac{1}{2}\left(\pi \cdot 2^2\right) = 6\pi$.

59. First Figure Let $u = \sqrt{x}$, so $x = u^2$ and $dx = 2u\,du$. When $x = 0$, $u = 0$; when $x = 1$, $u = 1$. Thus,
$A_1 = \int_0^1 e^{\sqrt{x}}\,dx = \int_0^1 e^u(2u\,du) = 2\int_0^1 ue^u\,du$.

Second Figure $A_2 = \int_0^1 2xe^x\,dx = 2\int_0^1 ue^u\,du$.

Third Figure Let $u = \sin x$, so $du = \cos x\,dx$. When $x = 0$, $u = 0$; when $x = \frac{\pi}{2}$, $u = 1$. Thus,

$$A_3 = \int_0^{\pi/2} e^{\sin x}\sin 2x\,dx = \int_0^{\pi/2} e^{\sin x}(2\sin x\,\cos x)\,dx = \int_0^1 e^u(2u\,du) = 2\int_0^1 ue^u\,du.$$

Since $A_1 = A_2 = A_3$, all three areas are equal.

61. The volume of inhaled air in the lungs at time t is

$$V(t) = \int_0^t f(u)\,du = \int_0^t \frac{1}{2}\sin\left(\tfrac{2\pi}{5}\,u\right)du = \int_0^{2\pi t/5} \tfrac{1}{2}\sin v\left(\tfrac{5}{2\pi}\,dv\right) \quad \left[\text{substitute } v = \tfrac{2\pi}{5}u,\ dv = \tfrac{2\pi}{5}\,du\right]$$

$$= \tfrac{5}{4\pi}\left[-\cos v\right]_0^{2\pi t/5} = \tfrac{5}{4\pi}\left[-\cos\left(\tfrac{2\pi}{5}t\right) + 1\right] = \tfrac{5}{4\pi}\left[1 - \cos\left(\tfrac{2\pi}{5}t\right)\right] \text{ litres}$$

63. Let $u = 2x$. Then $du = 2\,dx$, so $\int_0^2 f(2x)\,dx = \int_0^4 f(u)\left(\frac{1}{2}\,du\right) = \frac{1}{2}\int_0^4 f(u)\,du = \frac{1}{2}(10) = 5$.

65. Let $u = -x$. Then $du = -dx$, so

$\int_a^b f(-x)\,dx = \int_{-a}^{-b} f(u)(-du) = \int_{-b}^{-a} f(u)\,du = \int_{-b}^{-a} f(x)\,dx$

From the diagram, we see that the equality follows from the fact that we are reflecting the graph of f, and the limits of integration, about the y-axis.

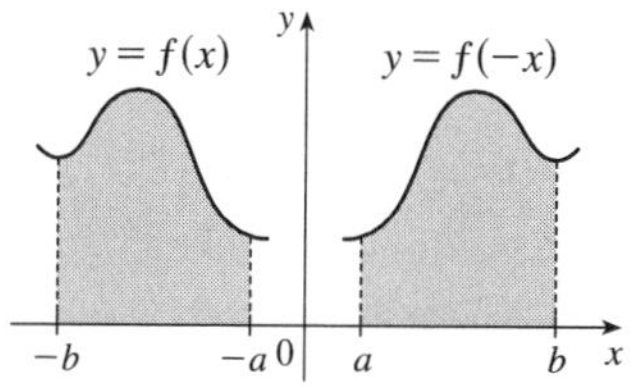

67. Let $u = 1 - x$. Then $x = 1 - u$ and $dx = -du$, so
$\int_0^1 x^a(1-x)^b\,dx = \int_1^0 (1-u)^a\,u^b(-du) = \int_0^1 u^b(1-u)^a\,du = \int_0^1 x^b(1-x)^a\,dx$.

5.6 Integration by Parts

1. Let $u = \ln x$, $dv = x\,dx$ $\Rightarrow$ $du = dx/x$, $v = \frac{1}{2}x^2$. Then by Equation 2, $\int u\,dv = uv - \int v\,du$,

$$\int x\ln x\,dx = \tfrac{1}{2}x^2\ln x - \int \tfrac{1}{2}x^2(dx/x) = \tfrac{1}{2}x^2\ln x - \tfrac{1}{2}\int x\,dx = \tfrac{1}{2}x^2\ln x - \tfrac{1}{2}\cdot\tfrac{1}{2}x^2 + C$$
$$= \tfrac{1}{2}x^2\ln x - \tfrac{1}{4}x^2 + C$$

Note: A mnemonic device which is helpful for selecting u when using integration by parts is the LIATE principle of precedence for u:

Logarithmic
Inverse trigonometric
Algebraic
Trigonometric
Exponential

If the integrand has several factors, then we try to choose among them a u which appears as high as possible on the list. For example, in $\int xe^{2x}\,dx$ the integrand is xe^{2x}, which is the product of an algebraic function (x) and an exponential function (e^{2x}). Since Algebraic appears before Exponential, we choose $u = x$. Sometimes the integration turns out to be similar regardless of the selection of u and dv, but it is advisable to refer to LIATE when in doubt.

3. Let $u = x$, $dv = \cos 5x\,dx$ $\Rightarrow$ $du = dx$, $v = \frac{1}{5}\sin 5x$. Then by Equation 2,
$\int x\cos 5x\,dx = \frac{1}{5}x\sin 5x - \int \frac{1}{5}\sin 5x\,dx = \frac{1}{5}x\sin 5x + \frac{1}{25}\cos 5x + C$.

5. Let $u = r$, $dv = e^{r/2}\,dr$ $\Rightarrow$ $du = dr$, $v = 2e^{r/2}$. Then $\int re^{r/2}\,dr = 2re^{r/2} - \int 2e^{r/2}\,dr = 2re^{r/2} - 4e^{r/2} + C$.

7. First let $u = x^2$, $dv = \cos 3x\,dx \Rightarrow du = 2x\,dx$, $v = \frac{1}{3}\sin 3x$. Then by Equation 2, the original integral $I = \int x^2 \cos 3x\,dx = \frac{1}{3}x^2 \sin 3x - \frac{2}{3}\int x \sin 3x\,dx$. To evaluate the last integral, we next let $U = x$, $dV = \sin 3x\,dx \Rightarrow dU = dx$, $V = -\frac{1}{3}\cos 3x$. So $\int x \sin 3x\,dx = -\frac{1}{3}x\cos 3x + \frac{1}{3}\int \cos 3x\,dx = -\frac{1}{3}x \cos 3x + \frac{1}{9}\sin 3x + C_1$. Substituting for $\int x \sin 3x\,dx$, we get $I = \frac{1}{3}x^2 \sin 3x - \frac{2}{3}\left(-\frac{1}{3}x\cos 3x + \frac{1}{9}\sin 3x + C_1\right) = \frac{1}{3}x^2 \sin 3x + \frac{2}{9}x\cos 3x - \frac{2}{27}\sin 3x + C$, where $C = -\frac{2}{3}C_1$.

9. Let $u = \ln(2x+1)$, $dv = dx \Rightarrow du = \dfrac{2}{2x+1}\,dx$, $v = x$. Then

$$\begin{aligned}\int \ln(2x+1)\,dx &= x\ln(2x+1) - \int \frac{2x}{2x+1}\,dx = x\ln(2x+1) - \int \frac{(2x+1)-1}{2x+1}\,dx \\ &= x\ln(2x+1) - \int\left(1 - \frac{1}{2x+1}\right)dx = x\ln(2x+1) - x + \tfrac{1}{2}\ln(2x+1) + C \\ &= \tfrac{1}{2}(2x+1)\ln(2x+1) - x + C\end{aligned}$$

11. Let $u = \arctan 4t$, $dv = dt \Rightarrow du = \dfrac{4}{1+(4t)^2}\,dt = \dfrac{4}{1+16t^2}\,dt$, $v = t$. Then

$$\int \arctan 4t\,dt = t\arctan 4t - \int \frac{4t}{1+16t^2}\,dt = t\arctan 4t - \frac{1}{8}\int \frac{32t}{1+16t^2}\,dt = t\arctan 4t - \tfrac{1}{8}\ln(1+16t^2) + C.$$

13. First let $u = \sin 3\theta$, $dv = e^{2\theta}\,d\theta \Rightarrow du = 3\cos 3\theta\,d\theta$, $v = \frac{1}{2}e^{2\theta}$. Then $I = \int e^{2\theta}\sin 3\theta\,d\theta = \frac{1}{2}e^{2\theta}\sin 3\theta - \frac{3}{2}\int e^{2\theta}\cos 3\theta\,d\theta$. Next let $U = \cos 3\theta$, $dV = e^{2\theta}\,d\theta \Rightarrow dU = -3\sin 3\theta\,d\theta$, $V = \frac{1}{2}e^{2\theta}$ to get $\int e^{2\theta}\cos 3\theta\,d\theta = \frac{1}{2}e^{2\theta}\cos 3\theta + \frac{3}{2}\int e^{2\theta}\sin 3\theta\,d\theta$. Substituting in the previous formula gives $I = \frac{1}{2}e^{2\theta}\sin 3\theta - \frac{3}{4}e^{2\theta}\cos 3\theta - \frac{9}{4}\int e^{2\theta}\sin 3\theta\,d\theta = \frac{1}{2}e^{2\theta}\sin 3\theta - \frac{3}{4}e^{2\theta}\cos 3\theta - \frac{9}{4}I \Rightarrow$ $\frac{13}{4}I = \frac{1}{2}e^{2\theta}\sin 3\theta - \frac{3}{4}e^{2\theta}\cos 3\theta + C_1$. Hence, $I = \frac{1}{13}e^{2\theta}(2\sin 3\theta - 3\cos 3\theta) + C$, where $C = \frac{4}{13}C_1$.

15. Let $u = t$, $dv = \sin 3t\,dt \Rightarrow du = dt$, $v = -\frac{1}{3}\cos 3t$. Then $\int_0^\pi t\sin 3t\,dt = \left[-\frac{1}{3}t\cos 3t\right]_0^\pi + \frac{1}{3}\int_0^\pi \cos 3t\,dt = \left(\frac{1}{3}\pi - 0\right) + \frac{1}{9}\left[\sin 3t\right]_0^\pi = \frac{\pi}{3}$.

17. Let $u = \ln x$, $dv = x^{-2}\,dx \Rightarrow du = \dfrac{1}{x}\,dx$, $v = -x^{-1}$. By (6),

$$\int_1^2 \frac{\ln x}{x^2}\,dx = \left[-\frac{\ln x}{x}\right]_1^2 + \int_1^2 x^{-2}\,dx = -\tfrac{1}{2}\ln 2 + \ln 1 + \left[-\frac{1}{x}\right]_1^2 = -\tfrac{1}{2}\ln 2 + 0 - \tfrac{1}{2} + 1 = \tfrac{1}{2} - \tfrac{1}{2}\ln 2.$$

19. Let $u = y$, $dv = \dfrac{dy}{e^{2y}} = e^{-2y}dy \Rightarrow du = dy$, $v = -\dfrac{1}{2}e^{-2y}$. Then

$$\int_0^1 \frac{y}{e^{2y}}\,dy = \left[-\tfrac{1}{2}ye^{-2y}\right]_0^1 + \tfrac{1}{2}\int_0^1 e^{-2y}dy = \left(-\tfrac{1}{2}e^{-2} + 0\right) - \tfrac{1}{4}\left[e^{-2y}\right]_0^1 = -\tfrac{1}{2}e^{-2} - \tfrac{1}{4}e^{-2} + \tfrac{1}{4} = \tfrac{1}{4} - \tfrac{3}{4}e^{-2}.$$

21. Let $u = \sin^{-1}x$, $dv = dx \Rightarrow du = \dfrac{dx}{\sqrt{1-x^2}}$, $v = x$. By Formula 6,

$$I = \int_0^{1/2}\sin^{-1}x\,dx = \left[x\sin^{-1}x\right]_0^{1/2} - \int_0^{1/2}\frac{x\,dx}{\sqrt{1-x^2}} = \frac{1}{2}\cdot\frac{\pi}{6} - \int_0^{1/2}\frac{x\,dx}{\sqrt{1-x^2}}.$$ To evaluate the last integral, let $t = 1 - x^2$, so $dt = -2x\,dx$ and $x\,dx = -\frac{1}{2}\,dt$. When $x = 0$, $t = 1$; when $x = \frac{1}{2}$, $t = \frac{3}{4}$. So

$$\int_0^{1/2}\frac{x\,dx}{\sqrt{1-x^2}} = \int_1^{3/4}\frac{1}{\sqrt{t}}\left(-\frac{1}{2}\,dt\right) = \frac{1}{2}\int_{3/4}^1 t^{-1/2}dt = \tfrac{1}{2}\left[2t^{1/2}\right]_{3/4}^1 = \sqrt{1} - \sqrt{\tfrac{3}{4}} = 1 - \tfrac{\sqrt{3}}{2}.$$

Thus, $I = \frac{\pi}{12} - \left(1 - \frac{\sqrt{3}}{2}\right) = \frac{\pi}{12} - 1 + \frac{\sqrt{3}}{2} = \frac{1}{12}\left(\pi - 12 + 6\sqrt{3}\right)$.

23. Let $u = (\ln x)^2$, $dv = dx \;\Rightarrow\; du = \dfrac{2}{x}\ln x\,dx$, $v = x$. By Formula 6, $I = \int_1^2 (\ln x)^2\,dx = \left[x(\ln x)^2\right]_1^2 - 2\int_1^2 \ln x\,dx$.

To evaluate the last integral, let $U = \ln x$, $dV = dx \;\Rightarrow\; dU = \dfrac{1}{x}\,dx$, $V = x$. Thus,

$$I = [x(\ln x)]_1^2 - 2\left([x\ln x]_1^2 - \int_1^2 dx\right) = [x(\ln x) - 2x\ln x + 2x]_1^2$$
$$= \left(2(\ln 2)^2 - 4\ln 2 + 4\right) - (0 - 0 + 2) = 2(\ln 2)^2 - 4\ln 2 + 2$$

25. Let $w = \sqrt{x}$, so that $x = w^2$ and $dx = 2w\,dw$. Thus, $\int \sin\sqrt{x}\,dx = \int 2w\sin w\,dw$. Now use parts with $u = 2w$, $dv = \sin w\,dw$, $du = 2\,dw$, $v = -\cos w$ to get

$$\int 2w\sin w\,dw = -2w\cos w + \int 2\cos w\,dw = -2w\cos w + 2\sin w + C$$
$$= -2\sqrt{x}\cos\sqrt{x} + 2\sin\sqrt{x} + C = 2\left(\sin\sqrt{x} - \sqrt{x}\cos\sqrt{x}\right) + C$$

27. Let $x = \theta^2$, so that $dx = 2\theta\,d\theta$. Thus, $\int_{\sqrt{\pi/2}}^{\sqrt{\pi}} \theta^3\cos\left(\theta^2\right)\,d\theta = \int_{\sqrt{\pi/2}}^{\sqrt{\pi}} \theta^2\cos\left(\theta^2\right)\cdot\frac{1}{2}(2\theta\,d\theta) = \frac{1}{2}\int_{\pi/2}^{\pi} x\cos x\,dx$. Now use parts with $u = x$, $dv = \cos x\,dx$, $du = dx$, $v = \sin x$ to get

$$\frac{1}{2}\int_{\pi/2}^{\pi} x\cos x\,dx = \frac{1}{2}\left([x\sin x]_{\pi/2}^{\pi} - \int_{\pi/2}^{\pi}\sin x\,dx\right) = \frac{1}{2}\,[x\sin x + \cos x]_{\pi/2}^{\pi}$$
$$= \frac{1}{2}(\pi\sin\pi + \cos\pi) - \frac{1}{2}\left(\frac{\pi}{2}\sin\frac{\pi}{2} + \cos\frac{\pi}{2}\right) = \frac{1}{2}(\pi\cdot 0 - 1) - \frac{1}{2}\left(\frac{\pi}{2}\cdot 1 + 0\right) = -\frac{1}{2} - \frac{\pi}{4}$$

29. Let $u = x$, $dv = \cos\pi x\,dx \;\Rightarrow\; du = dx$, $v = (\sin\pi x)/\pi$. Then

$$\int x\cos\pi x\,dx = x\cdot\frac{\sin\pi x}{\pi} - \int\frac{\sin\pi x}{\pi}\,dx = \frac{x\sin\pi x}{\pi} + \frac{\cos\pi x}{\pi^2} + C.$$

We see from the graph that this is reasonable, since F has extreme values where f is 0.

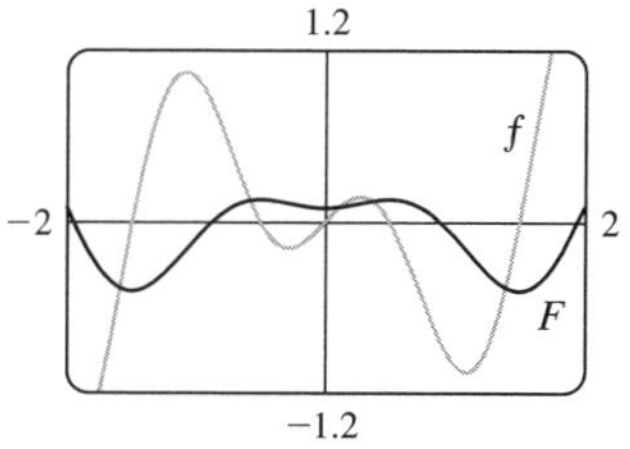

31. Let $u = 2x + 3$, $dv = e^x\,dx \;\Rightarrow\; du = 2\,dx$, $v = e^x$. Then $\int(2x+3)e^x\,dx = (2x+3)e^x - 2\int e^x\,dx = (2x+3)e^x - 2e^x + C = (2x+1)\,e^x + C$. We see from the graph that this is reasonable, since F has a minimum where f changes from negative to positive.

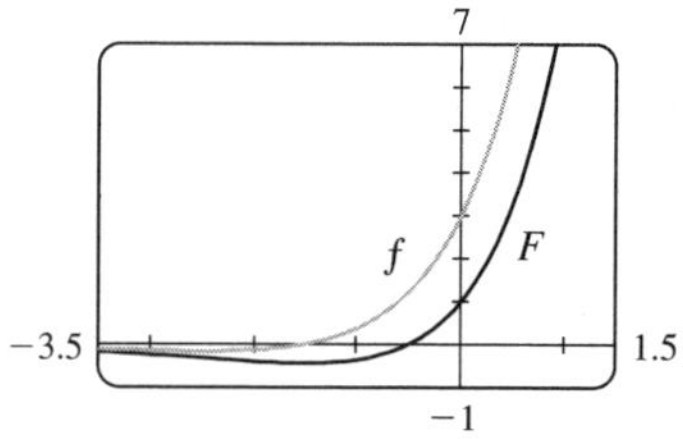

33. (a) Take $n = 2$ in Example 6 to get $\displaystyle\int\sin^2 x\,dx = -\frac{1}{2}\cos x\sin x + \frac{1}{2}\int 1\,dx = \frac{x}{2} - \frac{\sin 2x}{4} + C$.

(b) $\int\sin^4 x\,dx = -\frac{1}{4}\cos x\sin^3 x + \frac{3}{4}\int\sin^2 x\,dx = -\frac{1}{4}\cos x\sin^3 x + \frac{3}{8}x - \frac{3}{16}\sin 2x + C$.

35. (a) From Example 6, $\displaystyle\int\sin^n x\,dx = -\frac{1}{n}\cos x\sin^{n-1}x + \frac{n-1}{n}\int\sin^{n-2}x\,dx$. Using (6),

$$\int_0^{\pi/2}\sin^n x\,dx = \left[-\frac{\cos x\sin^{n-1}x}{n}\right]_0^{\pi/2} + \frac{n-1}{n}\int_0^{\pi/2}\sin^{n-2}x\,dx$$
$$= (0 - 0) + \frac{n-1}{n}\int_0^{\pi/2}\sin^{n-2}x\,dx = \frac{n-1}{n}\int_0^{\pi/2}\sin^{n-2}x\,dx$$

(b) Using $n = 3$ in part (a), we have $\int_0^{\pi/2}\sin^3 x\,dx = \frac{2}{3}\int_0^{\pi/2}\sin x\,dx = \left[-\frac{2}{3}\cos x\right]_0^{\pi/2} = \frac{2}{3}$.

Using $n = 5$ in part (a), we have $\int_0^{\pi/2}\sin^5 x\,dx = \frac{4}{5}\int_0^{\pi/2}\sin^3 x\,dx = \frac{4}{5}\cdot\frac{2}{3} = \frac{8}{15}$.

(c) The formula holds for $n = 1$ (that is, $2n + 1 = 3$) by (b). Assume it holds for some $k \geq 1$. Then
$\int_0^{\pi/2} \sin^{2k+1} x\,dx = \dfrac{2 \cdot 4 \cdot 6 \cdot \cdots \cdot (2k)}{3 \cdot 5 \cdot 7 \cdot \cdots \cdot (2k+1)}$. By Example 6,

$$\int_0^{\pi/2} \sin^{2k+3} x\,dx = \frac{2k+2}{2k+3}\int_0^{\pi/2} \sin^{2k+1} x\,dx = \frac{2k+2}{2k+3} \cdot \frac{2 \cdot 4 \cdot 6 \cdot \cdots \cdot (2k)}{3 \cdot 5 \cdot 7 \cdot \cdots \cdot (2k+1)}$$

$$= \frac{2 \cdot 4 \cdot 6 \cdot \cdots \cdot (2k)[2\,(k+1)]}{3 \cdot 5 \cdot 7 \cdot \cdots \cdot (2k+1)[2\,(k+1)+1]},$$

so the formula holds for $n = k + 1$. By induction, the formula holds for all $n \geq 1$.

37. Let $u = (\ln x)^n$, $dv = dx \;\Rightarrow\; du = n(\ln x)^{n-1}(dx/x)$, $v = x$. By Equation 2,
$\int (\ln x)^n\,dx = x(\ln x)^n - \int nx(\ln x)^{n-1}(dx/x) = x(\ln x)^n - n\int(\ln x)^{n-1}\,dx$.

39. Take $n = 3$ in Exercise 37 to get $\int (\ln x)^3\,dx = x\,(\ln x)^3 - 3\int(\ln x)^2\,dx = x(\ln x)^3 - 3x(\ln x)^2 + 6x\ln x - 6x + C$ [by Exercise 23].
Or: Instead of using Exercise 23, apply Exercise 37 again with $n = 2$.

41. Since $v(t) > 0$ for all t, the desired distance is $s(t) = \int_0^t v(w)dw = \int_0^t w^2 e^{-w}\,dw$.

First let $u = w^2$, $dv = e^{-w}\,dw \;\Rightarrow\; du = 2w\,dw$, $v = -e^{-w}$. Then $s(t) = \left[-w^2e^{-w}\right]_0^t + 2\int_0^t we^{-w}\,dw$.

Next let $U = w$, $dV = e^{-w}\,dw \;\Rightarrow\; dU = dw$, $V = -e^{-w}$. Then

$$\begin{aligned} s(t) &= -t^2e^{-t} + 2\left(\left[-we^{-w}\right]_0^t + \int_0^t e^{-w}\,dw\right) = -t^2e^{-t} + 2\left(-te^{-t} + 0 + \left[-e^{-w}\right]_0^t\right) \\ &= -t^2e^{-t} + 2\left(-te^{-t} - e^{-t} + 1\right) = -t^2e^{-t} - 2te^{-t} - 2e^{-t} + 2 \\ &= 2 - e^{-t}\left(t^2 + 2t + 2\right) \text{ metres} \end{aligned}$$

43. For $I = \int_1^4 xf''(x)\,dx$, let $u = x$, $dv = f''(x)\,dx \;\Rightarrow\; du = dx$, $v = f'(x)$. Then
$I = [xf'(x)]_1^4 - \int_1^4 f'(x)\,dx = 4f'(4) - 1 \cdot f'(1) - [f(4) - f(1)] = 4 \cdot 3 - 1 \cdot 5 - (7 - 2) = 12 - 5 - 5 = 2$.
We used the fact that f'' is continuous to guarantee that I exists.

45. Suppose $f(0) = g(0) = 0$ and let $u = f(x)$, $dv = g''(x)\,dx \;\Rightarrow\; du = f'(x)\,dx$, $v = g'(x)$.
Then $\int_0^a f(x)g''(x)\,dx = [f(x)g'(x)]_0^a - \int_0^a f'(x)g'(x)\,dx = f(a)g'(a) - \int_0^a f'(x)g'(x)\,dx$.
Now let $U = f'(x)$, $dV = g'(x)\,dx \;\Rightarrow\; dU = f''(x)\,dx$ and $V = g(x)$, so
$\int_0^a f'(x)g'(x)\,dx = [f'(x)g(x)]_0^a - \int_0^a f''(x)g(x)\,dx = f'(a)g(a) - \int_0^a f''(x)g(x)\,dx$.
Combining the two results, we get $\int_0^a f(x)g''(x)\,dx = f(a)g'(a) - f'(a)g(a) + \int_0^a f''(x)g(x)\,dx$.

5.7 Additional Techniques of Integration

The symbols $\overset{s}{=}$ and $\overset{c}{=}$ indicate the use of the substitutions $\{u = \sin x, du = \cos x\,dx\}$ and $\{u = \cos x, du = -\sin x\,dx\}$, respectively.

1. $$\begin{aligned} \int \sin^3 x\cos^2 x\,dx &= \int \sin^2 x\cos^2 x\sin x\,dx = \int (1 - \cos^2 x)\cos^2 x\sin x\,dx \overset{c}{=} \int (1 - u^2)u^2(-du) \\ &= \int (u^2 - 1)u^2\,du = \int (u^4 - u^2)\,du = \tfrac{1}{5}u^5 - \tfrac{1}{3}u^3 + C = \tfrac{1}{5}\cos^5 x - \tfrac{1}{3}\cos^3 x + C \end{aligned}$$

3. $$\begin{aligned} \int_{\pi/2}^{3\pi/4} \sin^5 x\cos^3 x\,dx &= \int_{\pi/2}^{3\pi/4} \sin^5 x\cos^2 x\cos x\,dx = \int_{\pi/2}^{3\pi/4} \sin^5 x\,(1 - \sin^2 x)\cos x\,dx \\ &\overset{s}{=} \int_1^{\sqrt{2}/2} u^5(1 - u^2)\,du = \int_1^{\sqrt{2}/2} (u^5 - u^7)\,du = \left[\tfrac{1}{6}u^6 - \tfrac{1}{8}u^8\right]_1^{\sqrt{2}/2} \\ &= \left(\tfrac{1/8}{6} - \tfrac{1/16}{8}\right) - \left(\tfrac{1}{6} - \tfrac{1}{8}\right) = -\tfrac{11}{384} \end{aligned}$$

5. $\int_0^{2\pi} \cos^2(6\theta)\,d\theta = \frac{1}{2}\int_0^{2\pi} [1 + \cos(12\theta)]\,d\theta = \frac{1}{2}\left[\theta + \frac{1}{12}\sin(12\theta)\right]_0^{2\pi} = \frac{1}{2}\left[(2\pi + 0) - (0 + 0)\right] = \pi$

7. Let $u = \sec x$. Then $du = \sec x \, \tan x \, dx$, so

$$\begin{aligned} \int \tan^3 x \, \sec x \, dx &= \int (\tan^2 x)(\tan x \, \sec x) \, dx = \int (\sec^2 x - 1)(\sec x \, \tan x \, dx) \\ &= \int (u^2 - 1) \, du = \tfrac{1}{3}u^3 - u + C = \tfrac{1}{3}\sec^3 x - \sec x + C \end{aligned}$$

9. $x = 3 \sin \theta$, where $-\pi/2 \le \theta \le \pi/2$. Then $dx = 3 \cos \theta \, d\theta$ and $\sqrt{9 - x^2} = \sqrt{9 - 9 \sin^2 \theta} = \sqrt{9 \cos^2 \theta} = 3 \, |\cos \theta| = 3 \cos \theta$. (Note that $\cos \theta \ge 0$ because $-\pi/2 \le \theta \le \pi/2$.) Thus, substitution gives

$$\int \frac{\sqrt{9 - x^2}}{x^2} \, dx = \int \frac{3 \cos \theta}{9 \sin^2 \theta} 3 \cos \theta \, d\theta = \int \frac{\cos^2 \theta}{\sin^2 \theta} \, d\theta = \int \cot^2 \theta \, d\theta = \int (\csc^2 \theta - 1) \, d\theta = -\cot \theta - \theta + C$$

Since this is an indefinite integral, we must return to the original variable x. This can be done either by using trigonometric identities to express $\cot \theta$ in terms of $\sin \theta = x/3$ or by drawing a diagram, as shown, where θ is interpreted as an angle of a right triangle.

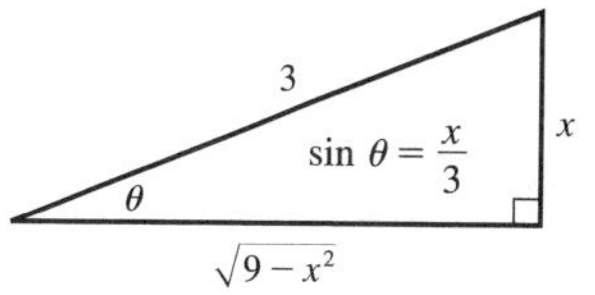

Since $\sin \theta = x/3$, we label the opposite side and the hypotenuse as having lengths x and 3. Then the Pythagorean Theorem gives the length of the adjacent side as $\sqrt{9 - x^2}$, so we can simply read the value of $\cot \theta$ from the figure: $\cot \theta = \dfrac{\sqrt{9 - x^2}}{x}$. (Although $\theta > 0$ in the diagram, this expression for $\cot \theta$ is valid even when $\theta < 0$.) Since $\sin \theta = x/3$, we have $\theta = \sin^{-1}(x/3)$ and so $\displaystyle\int \frac{\sqrt{9 - x^2}}{x^2} \, dx = -\frac{\sqrt{9 - x^2}}{x} - \sin^{-1}\!\left(\frac{x}{3}\right) + C.$

11. $x = 2 \tan \theta$, where $-\pi/2 < \theta < \pi/2$. Then $dx = 2 \sec^2 \theta \, d\theta$ and

$$\begin{aligned} \sqrt{x^2 + 4} &= \sqrt{(2 \tan \theta)^2 + 4} = \sqrt{4 \tan^2 \theta + 4} \\ &= \sqrt{4(\tan^2 \theta + 1)} = 2\sqrt{\sec^2 \theta} = 2\,|\sec \theta| \\ &= 2 \sec \theta \quad [\text{since } \sec \theta \ge 0 \text{ for } -\pi/2 < \theta < \pi/2]. \end{aligned}$$

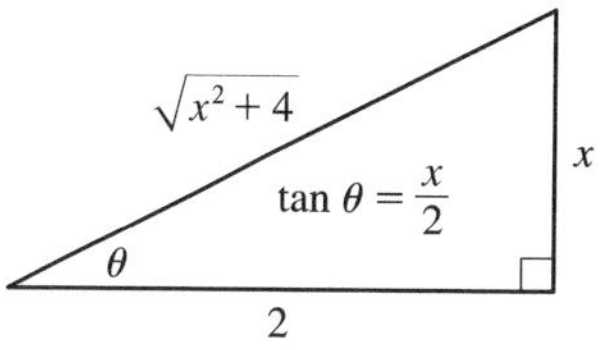

Thus, substitution gives

$$\begin{aligned} \int \frac{1}{x^2 \sqrt{x^2 + 4}} \, dx &= \int \frac{1}{4 \tan^2 \theta \, (2 \sec \theta)} \left(2 \sec^2 \theta \, d\theta\right) = \frac{1}{4} \int \frac{\sec \theta}{\tan^2 \theta} \, d\theta = \frac{1}{4} \int \frac{1}{\cos \theta} \cdot \frac{\cos^2 \theta}{\sin^2 \theta} \, d\theta \\ &= \frac{1}{4} \int \frac{\cos \theta}{\sin^2 \theta} \, d\theta = \frac{1}{4} \int \frac{1}{u^2} \, du \quad [u = \sin \theta, du = \cos \theta \, d\theta] \\ &= \frac{1}{4}\left(-\frac{1}{u}\right) + C = -\frac{1}{4} \frac{1}{\sin \theta} + C = -\frac{1}{4} \cdot \frac{\sqrt{x^2 + 4}}{x} + C = -\frac{\sqrt{x^2 + 4}}{4x} + C \end{aligned}$$

13. Let $t = \sec \theta$, so $dt = \sec \theta \tan \theta \, d\theta$, $t = \sqrt{2} \;\Rightarrow\; \theta = \frac{\pi}{4}$, and $t = 2 \;\Rightarrow\; \theta = \frac{\pi}{3}$. Then

$$\begin{aligned} \int_{\sqrt{2}}^{2} \frac{1}{t^3 \sqrt{t^2 - 1}} \, dt &= \int_{\pi/4}^{\pi/3} \frac{1}{\sec^3 \theta \tan \theta} \sec \theta \tan \theta \, d\theta = \int_{\pi/4}^{\pi/3} \frac{1}{\sec^2 \theta} \, d\theta = \int_{\pi/4}^{\pi/3} \cos^2 \theta \, d\theta \\ &= \int_{\pi/4}^{\pi/3} \tfrac{1}{2}(1 + \cos 2\theta) \, d\theta = \tfrac{1}{2}\left[\theta + \tfrac{1}{2} \sin 2\theta\right]_{\pi/4}^{\pi/3} \\ &= \tfrac{1}{2}\left[\left(\tfrac{\pi}{3} + \tfrac{1}{2}\tfrac{\sqrt{3}}{2}\right) - \left(\tfrac{\pi}{4} + \tfrac{1}{2} \cdot 1\right)\right] = \tfrac{1}{2}\left(\tfrac{\pi}{12} + \tfrac{\sqrt{3}}{4} - \tfrac{1}{2}\right) = \tfrac{\pi}{24} + \tfrac{\sqrt{3}}{8} - \tfrac{1}{4} \end{aligned}$$

15. (a) $\dfrac{2}{x^2 + 3x - 4} = \dfrac{2}{(x + 4)(x - 1)} = \dfrac{A}{x + 4} + \dfrac{B}{x - 1}$

(b) $x^2 + x + 1$ is irreducible, so $\dfrac{x^2}{(x - 1)(x^2 + x + 1)} = \dfrac{A}{x - 1} + \dfrac{Bx + C}{x^2 + x + 1}$.

17. $\dfrac{x-9}{(x+5)(x-2)} = \dfrac{A}{x+5} + \dfrac{B}{x-2}$. Multiply both sides by $(x+5)(x-2)$ to get $x - 9 = A(x-2) + B(x+5)$ $(*)$, or equivalently, $x - 9 = (A+B)x - 2A + 5B$. Equating coefficients of x on each side of the equation gives us $1 = A + B$ **(1)** and equating constants gives us $-9 = -2A + 5B$ **(2)**. Adding two times **(1)** to **(2)** gives us $-7 = 7B \quad \Leftrightarrow \quad B = -1$ and hence, $A = 2$. [Alternatively, to find the coefficients A and B, we may use substitution as follows: substitute 2 for x in $(*)$ to get $-7 = 7B \quad \Leftrightarrow \quad B = -1$, then substitute -5 for x in $(*)$ to get $-14 = -7A \quad \Leftrightarrow \quad A = 2$.] Thus,

$$\int \frac{x-9}{(x+5)(x-2)}\,dx = \int \left(\frac{2}{x+5} + \frac{-1}{x-2}\right)dx = 2\ln|x+5| - \ln|x-2| + C.$$

To find the constants in problems involving partial fractions, we may use the coefficient comparison method or the substitution method (as in the solution for Exercise 17) or a combination of both methods.

19. $\dfrac{1}{x^2-1} = \dfrac{1}{(x+1)(x-1)} = \dfrac{A}{x+1} + \dfrac{B}{x-1}$. Multiply both sides by $(x+1)(x-1)$ to get $1 = A(x-1) + B(x+1)$. Substituting 1 for x gives $1 = 2B \quad \Leftrightarrow \quad B = \frac{1}{2}$. Substituting -1 for x gives $1 = -2A \quad \Leftrightarrow \quad A = -\frac{1}{2}$. Thus,

$$\begin{aligned}\int_2^3 \frac{1}{x^2-1}\,dx &= \int_2^3 \left(\frac{-1/2}{x+1} + \frac{1/2}{x-1}\right)dx = \left[-\tfrac{1}{2}\ln|x+1| + \tfrac{1}{2}\ln|x-1|\right]_2^3 \\ &= \left(-\tfrac{1}{2}\ln 4 + \tfrac{1}{2}\ln 2\right) - \left(-\tfrac{1}{2}\ln 3 + \tfrac{1}{2}\ln 1\right) = \tfrac{1}{2}(\ln 2 + \ln 3 - \ln 4) \quad \left[\text{or } \tfrac{1}{2}\ln\tfrac{3}{2}\right]\end{aligned}$$

21. $\dfrac{10}{(x-1)(x^2+9)} = \dfrac{A}{x-1} + \dfrac{Bx+C}{x^2+9}$. Multiply both sides by $(x-1)(x^2+9)$ to get $10 = A(x^2+9) + (Bx+C)(x-1)$ $(*)$. Substituting 1 for x gives $10 = 10A \quad \Leftrightarrow \quad A = 1$. Substituting 0 for x gives $10 = 9A - C \quad \Rightarrow$ $C = 9(1) - 10 = -1$. The coefficients of the x^2-terms in $(*)$ must be equal, so $0 = A + B \quad \Rightarrow \quad B = -1$. Thus,

$$\begin{aligned}\int \frac{10}{(x-1)(x^2+9)}\,dx &= \int \left(\frac{1}{x-1} + \frac{-x-1}{x^2+9}\right)dx = \int \left(\frac{1}{x-1} - \frac{x}{x^2+9} - \frac{1}{x^2+9}\right)dx \\ &= \ln|x-1| - \tfrac{1}{2}\ln(x^2+9) \;\; [\text{let } u = x^2+9] \; - \tfrac{1}{3}\tan^{-1}\left(\tfrac{x}{3}\right) \;\; [\text{Formula 10}] + C\end{aligned}$$

23. $\dfrac{x^3+x^2+2x+1}{(x^2+1)(x^2+2)} = \dfrac{Ax+B}{x^2+1} + \dfrac{Cx+D}{x^2+2}$. Multiply both sides by $(x^2+1)(x^2+2)$ to get

$x^3 + x^2 + 2x + 1 = (Ax+B)(x^2+2) + (Cx+D)(x^2+1) \quad \Leftrightarrow$

$x^3 + x^2 + 2x + 1 = \left(Ax^3 + Bx^2 + 2Ax + 2B\right) + \left(Cx^3 + Dx^2 + Cx + D\right) \quad \Leftrightarrow$

$x^3 + x^2 + 2x + 1 = (A+C)x^3 + (B+D)x^2 + (2A+C)x + (2B+D)$. Comparing coefficients gives us the following system of equations:

$$\begin{aligned} A + C &= 1 \quad \textbf{(1)} \qquad & B + D &= 1 \quad \textbf{(2)} \\ 2A + C &= 2 \quad \textbf{(3)} \qquad & 2B + D &= 1 \quad \textbf{(4)} \end{aligned}$$

Subtracting equation **(1)** from equation **(3)** gives us $A = 1$, so $C = 0$. Subtracting equation **(2)** from equation **(4)** gives us $B = 0$, so $D = 1$. Thus, $I = \displaystyle\int \frac{x^3+x^2+2x+1}{(x^2+1)(x^2+2)}\,dx = \int \left(\frac{x}{x^2+1} + \frac{1}{x^2+2}\right)dx$. For $\displaystyle\int \frac{x}{x^2+1}\,dx$, let $u = x^2+1$ so $du = 2x\,dx$ and then $\displaystyle\int \frac{x}{x^2+1}\,dx = \frac{1}{2}\int \frac{1}{u}\,du = \frac{1}{2}\ln|u| + C = \frac{1}{2}\ln(x^2+1) + C$. For $\displaystyle\int \frac{1}{x^2+2}\,dx$, use Formula 10 with $a = \sqrt{2}$. So $\displaystyle\int \frac{1}{x^2+2}\,dx = \int \frac{1}{x^2+\left(\sqrt{2}\right)^2}\,dx = \frac{1}{\sqrt{2}}\tan^{-1}\frac{x}{\sqrt{2}} + C$.

Thus, $I = \dfrac{1}{2}\ln(x^2+1) + \dfrac{1}{\sqrt{2}}\tan^{-1}\dfrac{x}{\sqrt{2}} + C$.

25. $\displaystyle\int \frac{x}{x-6}\,dx = \int \frac{(x-6)+6}{x-6}\,dx = \int \left(1 + \frac{6}{x-6}\right)dx = x + 6\ln|x-6| + C$

27.

$$\begin{array}{r|l} & x \\ \hline x^2+4 & x^3 + 0x^2 + 0x + 4 \\ & x^3 \quad\quad + 4x \\ \hline & -4x + 4 \end{array}$$

By long division, $\dfrac{x^3+4}{x^2+4} = x + \dfrac{-4x+4}{x^2+4}$. Thus,

$$\int \frac{x^3+4}{x^2+4}\,dx = \int \left(x + \frac{-4x+4}{x^2+4}\right) dx = \int \left(x - \frac{4x}{x^2+4} + \frac{4}{x^2+2^2}\right) dx$$

$$= \tfrac{1}{2}x^2 - 4\cdot\tfrac{1}{2}\ln\left|x^2+4\right| + 4\cdot\tfrac{1}{2}\tan^{-1}\left(\tfrac{x}{2}\right) + C$$

$$= \tfrac{1}{2}x^2 - 2\ln(x^2+4) + 2\tan^{-1}\left(\tfrac{x}{2}\right) + C$$

29. Let $u = \sqrt{x}$, so $u^2 = x$ and $dx = 2u\,du$. Thus,

$$\int_9^{16} \frac{\sqrt{x}}{x-4}\,dx = \int_3^4 \frac{u}{u^2-4}2u\,du = 2\int_3^4 \frac{u^2}{u^2-4}\,du = 2\int_3^4 \left(1 + \frac{4}{u^2-4}\right) du \quad \text{[by long division]}$$

$$= 2 + 8\int_3^4 \frac{du}{(u+2)(u-2)}. \quad (*)$$

Multiply $\dfrac{1}{(u+2)(u-2)} = \dfrac{A}{u+2} + \dfrac{B}{u-2}$ by $(u+2)(u-2)$ to get $1 = A(u-2) + B(u+2)$. Equating coefficients we get $A + B = 0$ and $-2A + 2B = 1$. Solving gives us $B = \frac{1}{4}$ and $A = -\frac{1}{4}$, so $\dfrac{1}{(u+2)(u-2)} = \dfrac{-1/4}{u+2} + \dfrac{1/4}{u-2}$ and $(*)$ is

$$2 + 8\int_3^4 \left(\frac{-1/4}{u+2} + \frac{1/4}{u-2}\right) du = 2 + 8\left[-\tfrac{1}{4}\ln|u+2| + \tfrac{1}{4}\ln|u-2|\right]_3^4 = 2 + \Big[2\ln|u-2| - 2\ln|u+2|\Big]_3^4$$

$$= 2 + 2\left[\ln\left|\frac{u-2}{u+2}\right|\right]_3^4 = 2 + 2\left(\ln\frac{2}{6} - \ln\frac{1}{5}\right) = 2 + 2\ln\frac{2/6}{1/5}$$

$$= 2 + 2\ln\tfrac{5}{3} \text{ or } 2 + \ln\left(\tfrac{5}{3}\right)^2 = 2 + \ln\tfrac{25}{9}$$

31. $x^2 + x + 1 = x^2 + x + \frac{1}{4} + 1 - \frac{1}{4}$ [add and subtract the square of one-half the coefficient of x to complete the square]

$$= x^2 + x + \tfrac{1}{4} + \tfrac{3}{4} = \left(x + \tfrac{1}{2}\right)^2 + \left(\tfrac{\sqrt{3}}{2}\right)^2$$

So $I = \displaystyle\int \frac{dx}{x^2+x+1} = \int \frac{1}{\left(x+\frac{1}{2}\right)^2 + \left(\frac{\sqrt{3}}{2}\right)^2}\,dx$. Now let $u = x + \frac{1}{2} \Rightarrow du = dx$ and

$$I = \int \frac{1}{u^2 + \left(\frac{\sqrt{3}}{2}\right)^2}\,du = \frac{1}{\frac{\sqrt{3}}{2}}\tan^{-1}\frac{u}{\frac{\sqrt{3}}{2}} + C = \frac{2}{\sqrt{3}}\tan^{-1}\frac{2\left(x+\frac{1}{2}\right)}{\sqrt{3}} + C = \frac{2}{\sqrt{3}}\tan^{-1}\frac{2x+1}{\sqrt{3}} + C.$$

5.8 Integration Using Tables and Computer Algebra Systems

Keep in mind that there are several ways to approach many of these exercises, and different methods can lead to different forms of the answer.

1. Using long division, $\dfrac{x^3 - x^2 + x - 1}{x^2+9} = x - 1 - 8\,\dfrac{x-1}{x^2+9}$.

$$I = \int \left(x - 1 - 8\,\frac{x-1}{x^2+9}\right) dx = \int (x-1)dx - 8\left[\int \frac{x}{x^2+9}\,dx - \frac{1}{x^2+9}\,dx\right]$$

Using Formula 17 with $a = 3$, we have

$I \overset{17}{=} \frac{1}{2}x^2 - x - 8\cdot\frac{1}{2}\ln(x^2+9) + 8\cdot\frac{1}{3}\tan^{-1}(x/3) + C = \frac{1}{2}x^2 - x - 4\ln(x^2+9) + \frac{8}{3}\tan^{-1}(x/3) + C.$

3. Let $u = \pi x \Rightarrow du = \pi\,dx$, so

$$\textstyle\int \sec^3(\pi x)\,dx = \frac{1}{\pi}\int \sec^3 u\,du \overset{71}{=} \frac{1}{\pi}\left(\frac{1}{2}\sec u\tan u + \frac{1}{2}\ln|\sec u + \tan u|\right) + C$$

$$= \tfrac{1}{2\pi}\sec\pi x\tan\pi x + \tfrac{1}{2\pi}\ln|\sec\pi x + \tan\pi x| + C$$

5. Let $u = 2x$ and $a = 3$. Then $du = 2\,dx$ and

$$\int \frac{dx}{x^2\sqrt{4x^2+9}} = \int \frac{\frac{1}{2}\,du}{\frac{u^2}{4}\sqrt{u^2+a^2}} = 2\int \frac{du}{u^2\sqrt{a^2+u^2}} \overset{28}{=} -2\,\frac{\sqrt{a^2+u^2}}{a^2u} + C$$

$$= -2\,\frac{\sqrt{4x^2+9}}{9\cdot 2x} + C = -\frac{\sqrt{4x^2+9}}{9x} + C$$

7. $\int x^3\sin x\,dx \overset{84}{=} -x^3\cos x + 3\int x^2\cos x\,dx$, $\int x^2\cos x\,dx \overset{85}{=} x^2\sin x - 2\int x\sin x\,dx$, and

$\int x\sin x\,dx \overset{82}{=} \sin x - x\cos x + C$. Substituting, we get

$\int x^3\sin x\,dx = -x^3\cos x + 3\left[x^2\sin x - 2(\sin x - x\cos x)\right] + C = -x^3\cos x + 3x^2\sin x - 6\sin x + 6x\cos x + C.$

So $\int_0^\pi x^3\sin x\,dx = \left[-x^3\cos x + 3x^2\sin x - 6\sin x + 6x\cos x\right]_0^\pi = \left(-\pi^3\cdot -1 + 6\pi\cdot -1\right) - (0) = \pi^3 - 6\pi.$

9. Let $u = x^2$. Then $du = 2x\,dx$, so

$$\int x\sin^{-1}(x^2)\,dx = \tfrac{1}{2}\int \sin^{-1}u\,du \overset{87}{=} \tfrac{1}{2}\left(u\sin^{-1}u + \sqrt{1-u^2}\right) + C$$

$$= \tfrac{1}{2}\left[x^2\sin^{-1}(x^2) + \sqrt{1-x^4}\right] + C$$

11. Let $z = 6 + 4y - 4y^2 = 6 - (4y^2 - 4y + 1) + 1 = 7 - (2y-1)^2$, $u = 2y - 1$, and $a = \sqrt{7}$. Then $z = a^2 - u^2$, $du = 2\,dy$, and

$$\int y\sqrt{6+4y-4y^2}\,dy = \int y\sqrt{z}\,dy = \int \tfrac{1}{2}(u+1)\sqrt{a^2-u^2}\,\tfrac{1}{2}\,du$$

$$= \tfrac{1}{4}\int u\sqrt{a^2-u^2}\,du + \tfrac{1}{4}\int \sqrt{a^2-u^2}\,du$$

$$= \tfrac{1}{4}\int \sqrt{a^2-u^2}\,du - \tfrac{1}{8}\int(-2u)\sqrt{a^2-u^2}\,du$$

$$\overset{30}{=} \frac{u}{8}\sqrt{a^2-u^2} + \frac{a^2}{8}\sin^{-1}\left(\frac{u}{a}\right) - \frac{1}{8}\int\sqrt{w}\,dw \qquad \begin{bmatrix} w = a^2-u^2, \\ dw = -2u\,du \end{bmatrix}$$

$$= \frac{2y-1}{8}\sqrt{6+4y-4y^2} + \frac{7}{8}\sin^{-1}\frac{2y-1}{\sqrt{7}} - \frac{1}{8}\cdot\frac{2}{3}w^{3/2} + C$$

$$= \frac{2y-1}{8}\sqrt{6+4y-4y^2} + \frac{7}{8}\sin^{-1}\frac{2y-1}{\sqrt{7}} - \frac{1}{12}(6+4y-4y^2)^{3/2} + C$$

This can be rewritten as

$$\sqrt{6+4y-4y^2}\left[\frac{1}{8}(2y-1) - \frac{1}{12}(6+4y-4y^2)\right] + \frac{7}{8}\sin^{-1}\frac{2y-1}{\sqrt{7}} + C$$

$$= \left(\frac{1}{3}y^2 - \frac{1}{12}y - \frac{5}{8}\right)\sqrt{6+4y-4y^2} + \frac{7}{8}\sin^{-1}\left(\frac{2y-1}{\sqrt{7}}\right) + C$$

$$= \frac{1}{24}(8y^2 - 2y - 15)\sqrt{6+4y-4y^2} + \frac{7}{8}\sin^{-1}\left(\frac{2y-1}{\sqrt{7}}\right) + C$$

13. Let $u = \sin x$. Then $du = \cos x\,dx$, so

$$\int \sin^2 x\,\cos x\,\ln(\sin x)\,dx = \int u^2\ln u\,du \overset{101}{=} \frac{u^{2+1}}{(2+1)^2}\left[(2+1)\ln u - 1\right] + C = \tfrac{1}{9}u^3(3\ln u - 1) + C$$

$$= \tfrac{1}{9}\sin^3 x\left[3\ln(\sin x) - 1\right] + C$$

15. Let $u = e^x$ and $a = \sqrt{3}$. Then $du = e^x\,dx$ and

$$\int \frac{e^x}{3-e^{2x}}\,dx = \int \frac{du}{a^2-u^2} \overset{19}{=} \frac{1}{2a}\ln\left|\frac{u+a}{u-a}\right| + C = \frac{1}{2\sqrt{3}}\ln\left|\frac{e^x+\sqrt{3}}{e^x-\sqrt{3}}\right| + C.$$

17. $$\int \frac{x^4\,dx}{\sqrt{x^{10}-2}} = \int \frac{x^4\,dx}{\sqrt{(x^5)^2-2}} = \frac{1}{5}\int \frac{du}{\sqrt{u^2-2}} \qquad [u = x^5,\ du = 5x^4\,dx]$$

$$\overset{43}{=} \tfrac{1}{5}\ln\left|u + \sqrt{u^2-2}\right| + C = \tfrac{1}{5}\ln\left|x^5 + \sqrt{x^{10}-2}\right| + C$$

19. Let $u = \ln x$ and $a = 2$. Then $du = \dfrac{dx}{x}$ and

$$\int \frac{\sqrt{4 + (\ln x)^2}}{x}\,dx = \int \sqrt{a^2 + u^2}\,du \overset{21}{=} \frac{u}{2}\sqrt{a^2 + u^2} + \frac{a^2}{2}\ln\left(u + \sqrt{a^2 + u^2}\right) + C$$
$$= \tfrac{1}{2}(\ln x)\sqrt{4 + (\ln x)^2} + 2\ln\left[\ln x + \sqrt{4 + (\ln x)^2}\right] + C$$

21. Let $u = e^x$. Then $x = \ln u$, $dx = du/u$, so

$$\int \sqrt{e^{2x} - 1}\,dx = \int \frac{\sqrt{u^2 - 1}}{u}\,du \overset{41}{=} \sqrt{u^2 - 1} - \cos^{-1}(1/u) + C = \sqrt{e^{2x} - 1} - \cos^{-1}(e^{-x}) + C.$$

23. (a) $\dfrac{d}{du}\left[\dfrac{1}{b^3}\left(a + bu - \dfrac{a^2}{a + bu} - 2a\ln|a + bu|\right) + C\right] = \dfrac{1}{b^3}\left[b + \dfrac{ba^2}{(a + bu)^2} - \dfrac{2ab}{(a + bu)}\right]$

$$= \frac{1}{b^3}\left[\frac{b(a + bu)^2 + ba^2 - (a + bu)2ab}{(a + bu)^2}\right] = \frac{1}{b^3}\left[\frac{b^3u^2}{(a + bu)^2}\right] = \frac{u^2}{(a + bu)^2}$$

(b) Let $t = a + bu \;\Rightarrow\; dt = b\,du$. Note that $u = \dfrac{t - a}{b}$ and $du = \dfrac{1}{b}\,dt$.

$$\int \frac{u^2\,du}{(a + bu)^2} = \frac{1}{b^3}\int \frac{(t - a)^2}{t^2}\,dt = \frac{1}{b^3}\int \frac{t^2 - 2at + a^2}{t^2}\,dt$$
$$= \frac{1}{b^3}\int\left(1 - \frac{2a}{t} + \frac{a^2}{t^2}\right)dt = \frac{1}{b^3}\left(t - 2a\ln|t| - \frac{a^2}{t}\right) + C$$
$$= \frac{1}{b^3}\left(a + bu - \frac{a^2}{a + bu} - 2a\ln|a + bu|\right) + C$$

25. Maple, Mathematica and Derive all give $\int x^2\sqrt{5 - x^2}\,dx = -\frac{1}{4}x(5 - x^2)^{3/2} + \frac{5}{8}x\sqrt{5 - x^2} + \frac{25}{8}\sin^{-1}\left(\frac{1}{\sqrt{5}}x\right)$. Using Formula 31, we get $\int x^2\sqrt{5 - x^2}\,dx = \frac{1}{8}x(2x^2 - 5)\sqrt{5 - x^2} + \frac{1}{8}(5^2)\sin^{-1}\left(\frac{1}{\sqrt{5}}x\right) + C$. But $-\frac{1}{4}x(5 - x^2)^{3/2} + \frac{5}{8}x\sqrt{5 - x^2} = \frac{1}{8}x\sqrt{5 - x^2}\left[5 - 2(5 - x^2)\right] = \frac{1}{8}x(2x^2 - 5)\sqrt{5 - x^2}$, and the $\sin^{-1}$ terms are the same in each expression, so the answers are equivalent.

27. Maple and Derive both give $\int \sin^3 x\cos^2 x\,dx = -\frac{1}{5}\sin^2 x\cos^3 x - \frac{2}{15}\cos^3 x$ (although Derive factors the expression), and Mathematica gives $\int \sin^3 x\cos^2 x\,dx = -\frac{1}{8}\cos x - \frac{1}{48}\cos 3x + \frac{1}{80}\cos 5x$. We can use a CAS to show that both of these expressions are equal to $-\frac{1}{3}\cos^3 x + \frac{1}{5}\cos^5 x$. Using Formula 86, we write

$$\int \sin^3 x\cos^2 x\,dx = -\tfrac{1}{5}\sin^2 x\cos^3 x + \tfrac{2}{5}\int \sin x\cos^2 x\,dx = -\tfrac{1}{5}\sin^2 x\cos^3 x + \tfrac{2}{5}\left(-\tfrac{1}{3}\cos^3 x\right) + C$$
$$= -\tfrac{1}{5}\sin^2 x\cos^3 x - \tfrac{2}{15}\cos^3 x + C$$

29. Maple gives $\int x\sqrt{1 + 2x}\,dx = \frac{1}{10}(1 + 2x)^{5/2} - \frac{1}{6}(1 + 2x)^{3/2}$, Mathematica gives $\sqrt{1 + 2x}\left(\frac{2}{5}x^2 + \frac{1}{15}x - \frac{1}{15}\right)$, and Derive gives $\frac{1}{15}(1 + 2x)^{3/2}(3x - 1)$. The first two expressions can be simplified to Derive's result. If we use Formula 54, we get

$$\int x\sqrt{1 + 2x}\,dx = \tfrac{2}{15(2)^2}(3\cdot 2x - 2\cdot 1)(1 + 2x)^{3/2} + C = \tfrac{1}{30}(6x - 2)(1 + 2x)^{3/2} + C$$
$$= \tfrac{1}{15}(3x - 1)(1 + 2x)^{3/2}$$

31. Maple gives $\int \tan^5 x\,dx = \frac{1}{4}\tan^4 x - \frac{1}{2}\tan^2 x + \frac{1}{2}\ln(1 + \tan^2 x)$, Mathematica gives $\int \tan^5 x\,dx = \frac{1}{4}[-1 - 2\cos(2x)]\sec^4 x - \ln(\cos x)$, and Derive gives $\int \tan^5 x\,dx = \frac{1}{4}\tan^4 x - \frac{1}{2}\tan^2 x - \ln(\cos x)$. These expressions are equivalent, and none includes absolute value bars or a constant of integration. Note that Mathematica's and Derive's expressions suggest that the integral is undefined where $\cos x < 0$, which is not the case.

Using Formula 75, $\int \tan^5 x\,dx = \frac{1}{5 - 1}\tan^{5-1} x - \int \tan^{5-2} x\,dx = \frac{1}{4}\tan^4 x - \int \tan^3 x\,dx$. Using Formula 69, $\int \tan^3 x\,dx = \frac{1}{2}\tan^2 x + \ln|\cos x| + C$, so $\int \tan^5 x\,dx = \frac{1}{4}\tan^4 x - \frac{1}{2}\tan^2 x - \ln|\cos x| + C$.

33. Derive gives $I = \int 2^x \sqrt{4^x - 1}\, dx = \dfrac{2^{x-1}\sqrt{2^{2x}-1}}{\ln 2} - \dfrac{\ln\left(\sqrt{2^{2x}-1} + 2^x\right)}{2\ln 2}$ immediately. Neither Maple nor Mathematica is able to evaluate I in its given form. However, if we instead write I as $\int 2^x \sqrt{(2^x)^2 - 1}\, dx$, both systems give the same answer as Derive (after minor simplification). Our trick works because the CAS now recognizes 2^x as a promising substitution.

5.9 Approximate Integration

1. (a) $\Delta x = (b-a)/n = (4-0)/2 = 2$

$$L_2 = \sum_{i=1}^{2} f(x_{i-1})\,\Delta x = f(x_0)\cdot 2 + f(x_1)\cdot 2 = 2\,[f(0) + f(2)] = 2(0.5 + 2.5) = 6$$

$$R_2 = \sum_{i=1}^{2} f(x_i)\,\Delta x = f(x_1)\cdot 2 + f(x_2)\cdot 2 = 2\,[f(2) + f(4)] = 2(2.5 + 3.5) = 12$$

$$M_2 = \sum_{i=1}^{2} f(\overline{x}_i)\Delta x = f(\overline{x}_1)\cdot 2 + f(\overline{x}_2)\cdot 2 = 2\,[f(1) + f(3)] \approx 2(1.6 + 3.2) = 9.6$$

(b) 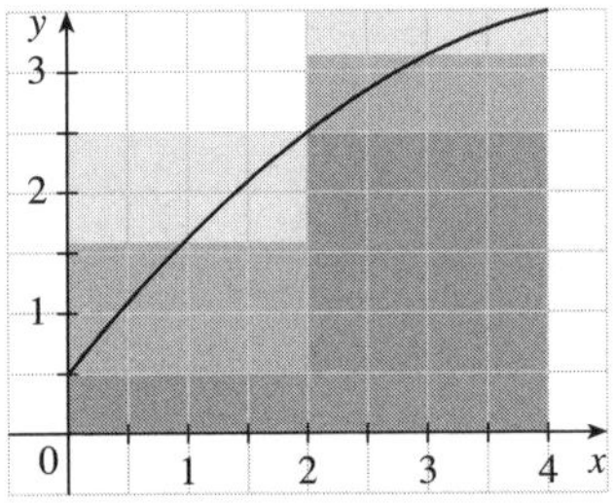

L_2 is an underestimate, since the area under the small rectangles is less than the area under the curve, and R_2 is an overestimate, since the area under the large rectangles is greater than the area under the curve. It appears that M_2 is an overestimate, though it is fairly close to I. See the solution to Exercise 37 for a proof of the fact that if f is concave down on $[a, b]$, then the Midpoint Rule is an overestimate of $\int_a^b f(x)\,dx$.

(c) $T_2 = \left(\frac{1}{2}\,\Delta x\right)[f(x_0) + 2f(x_1) + f(x_2)] = \frac{2}{2}[f(0) + 2f(2) + f(4)] = 0.5 + 2(2.5) + 3.5 = 9.$
This approximation is an underestimate, since the graph is concave down. Thus, $T_2 = 9 < I$. See the solution to Exercise 37 for a general proof of this conclusion.

(d) For any n, we will have $L_n < T_n < I < M_n < R_n$.

3. $f(x) = \cos(x^2)$, $\Delta x = \frac{1-0}{4} = \frac{1}{4}$

(a) $T_4 = \frac{1}{4\cdot 2}\left[f(0) + 2f\left(\frac{1}{4}\right) + 2f\left(\frac{2}{4}\right) + 2f\left(\frac{3}{4}\right) + f(1)\right] \approx 0.895\,759$

(b) $M_4 = \frac{1}{4}\left[f\left(\frac{1}{8}\right) + f\left(\frac{3}{8}\right) + f\left(\frac{5}{8}\right) + f\left(\frac{7}{8}\right)\right] \approx 0.908\,907$

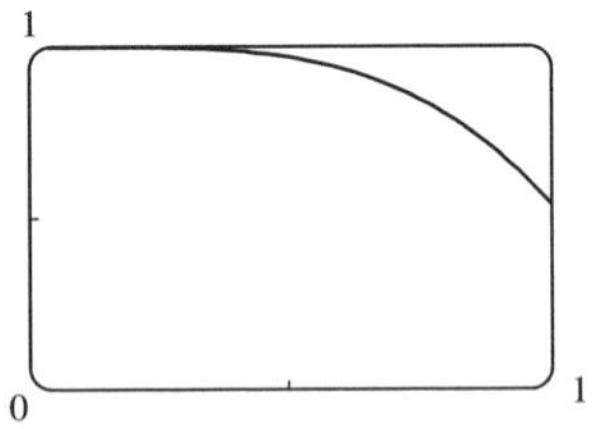

The graph shows that f is concave down on $[0, 1]$. So T_4 is an underestimate and M_4 is an overestimate. We can conclude that $0.895\,759 < \int_0^1 \cos(x^2)\,dx < 0.908\,907$.

5. $f(x) = x^2 \sin x$, $\Delta x = \dfrac{b-a}{n} = \dfrac{\pi - 0}{8} = \dfrac{\pi}{8}$

(a) $M_8 = \frac{\pi}{8}\left[f\left(\frac{\pi}{16}\right) + f\left(\frac{3\pi}{16}\right) + f\left(\frac{5\pi}{16}\right) + \cdots + f\left(\frac{15\pi}{16}\right)\right] \approx 5.932\,957$

(b) $S_8 = \frac{\pi}{8\cdot 3}\left[f(0) + 4f\left(\frac{\pi}{8}\right) + 2f\left(\frac{2\pi}{8}\right) + 4f\left(\frac{3\pi}{8}\right) + 2f\left(\frac{4\pi}{8}\right) + 4f\left(\frac{5\pi}{8}\right) + 2f\left(\frac{6\pi}{8}\right) + 4f\left(\frac{7\pi}{8}\right) + f(\pi)\right]$

$\approx 5.869\,247$

Actual: $\int_0^\pi x^2 \sin x\,dx \overset{84}{=} \left[-x^2\cos x\right]_0^\pi + 2\int_0^\pi x\cos x\,dx \overset{83}{=} \left[-\pi^2(-1) - 0\right] + 2[\cos x + x\sin x]_0^\pi$

$$= \pi^2 + 2[(-1+0) - (1+0)] = \pi^2 - 4 \approx 5.869\,604$$

Errors: $E_M = \text{actual} - M_8 = \int_0^\pi x^2 \sin x\,dx - M_8 \approx -0.063\,353$

$E_S = \text{actual} - S_8 = \int_0^\pi x^2 \sin x\,dx - S_8 \approx 0.000\,357$

7. $f(x) = e^{-x^2}$, $\Delta x = \dfrac{1-0}{10} = \dfrac{1}{10}$

(a) $T_{10} = \frac{1}{10 \cdot 2}[f(0) + 2f(0.1) + 2f(0.2) + \cdots + 2f(0.8) + 2f(0.9) + f(1)] \approx 0.746\,211$

(b) $M_{10} = \frac{1}{10}[f(0.05) + f(0.15) + f(0.25) + \cdots + f(0.75) + f(0.85) + f(0.95)] \approx 0.747\,131$

(c) $S_{10} = \frac{1}{10 \cdot 3}[f(0) + 4f(0.1) + 2f(0.2) + 4f(0.3) + 2f(0.4) + 4f(0.5)$
$+ 2f(0.6) + 4f(0.7) + 2f(0.8) + 4f(0.9) + f(1)] \approx 0.746\,825$

9. $f(x) = \dfrac{\ln x}{1+x}$, $\Delta x = \dfrac{2-1}{10} = \dfrac{1}{10}$

(a) $T_{10} = \frac{1}{10 \cdot 2}[f(1) + 2f(1.1) + 2f(1.2) + \cdots + 2f(1.8) + 2f(1.9) + f(2)] \approx 0.146\,879$

(b) $M_{10} = \frac{1}{10}[f(1.05) + f(1.15) + f(1.25) + \cdots + f(1.75) + f(1.85) + f(1.95)] \approx 0.147\,391$

(c) $S_{10} = \frac{1}{10 \cdot 3}[f(1) + 4f(1.1) + 2f(1.2) + 4f(1.3) + 2f(1.4) + 4f(1.5)$
$+ 2f(1.6) + 4f(1.7) + 2f(1.8) + 4f(1.9) + f(2)] \approx 0.147\,219$

11. $f(t) = e^{\sqrt{t}} \sin t$, $\Delta t = \dfrac{4-0}{8} = \dfrac{1}{2}$

(a) $T_8 = \frac{1}{2 \cdot 2}\left[f(0) + 2f\left(\frac{1}{2}\right) + 2f(1) + 2f\left(\frac{3}{2}\right) + 2f(2) + 2f\left(\frac{5}{2}\right) + 2f(3) + 2f\left(\frac{7}{2}\right) + f(4)\right] \approx 4.513\,618$

(b) $M_8 = \frac{1}{2}\left[f\left(\frac{1}{4}\right) + f\left(\frac{3}{4}\right) + f\left(\frac{5}{4}\right) + f\left(\frac{7}{4}\right) + f\left(\frac{9}{4}\right) + f\left(\frac{11}{4}\right) + f\left(\frac{13}{4}\right) + f\left(\frac{15}{4}\right)\right] \approx 4.748\,256$

(c) $S_8 = \frac{1}{2 \cdot 3}\left[f(0) + 4f\left(\frac{1}{2}\right) + 2f(1) + 4f\left(\frac{3}{2}\right) + 2f(2) + 4f\left(\frac{5}{2}\right) + 2f(3) + 4f\left(\frac{7}{2}\right) + f(4)\right] \approx 4.675\,111$

13. $f(x) = \dfrac{\cos x}{x}$, $\Delta x = \dfrac{5-1}{8} = \dfrac{1}{2}$

(a) $T_8 = \frac{1}{2 \cdot 2}\left[f(1) + 2f\left(\frac{3}{2}\right) + 2f(2) + 2f\left(\frac{5}{2}\right) + 2f(3) + 2f\left(\frac{7}{2}\right) + 2f(4) + 2f\left(\frac{9}{2}\right) + f(5)\right] \approx -0.495\,333$

(b) $M_8 = \frac{1}{2}\left[f\left(\frac{5}{4}\right) + f\left(\frac{7}{4}\right) + f\left(\frac{9}{4}\right) + f\left(\frac{11}{4}\right) + f\left(\frac{13}{4}\right) + f\left(\frac{15}{4}\right) + f\left(\frac{17}{4}\right) + f\left(\frac{19}{4}\right)\right] \approx -0.543\,321$

(c) $S_8 = \frac{1}{2 \cdot 3}\left[f(1) + 4f\left(\frac{3}{2}\right) + 2f(2) + 4f\left(\frac{5}{2}\right) + 2f(3) + 4f\left(\frac{7}{2}\right) + 2f(4) + 4f\left(\frac{9}{2}\right) + f(5)\right] \approx -0.526\,123$

15. $f(y) = \dfrac{1}{1+y^5}$, $\Delta y = \dfrac{3-0}{6} = \dfrac{1}{2}$

(a) $T_6 = \frac{1}{2 \cdot 2}\left[f(0) + 2f\left(\frac{1}{2}\right) + 2f\left(\frac{2}{2}\right) + 2f\left(\frac{3}{2}\right) + 2f\left(\frac{4}{2}\right) + 2f\left(\frac{5}{2}\right) + f(3)\right] \approx 1.064\,275$

(b) $M_6 = \frac{1}{2}\left[f\left(\frac{1}{4}\right) + f\left(\frac{3}{4}\right) + f\left(\frac{5}{4}\right) + f\left(\frac{7}{4}\right) + f\left(\frac{9}{4}\right) + f\left(\frac{11}{4}\right)\right] \approx 1.067\,416$

(c) $S_6 = \frac{1}{2 \cdot 3}\left[f(0) + 4f\left(\frac{1}{2}\right) + 2f\left(\frac{2}{2}\right) + 4f\left(\frac{3}{2}\right) + 2f\left(\frac{4}{2}\right) + 4f\left(\frac{5}{2}\right) + f(3)\right] \approx 1.074\,915$

17. (a) $f(x) = e^{-x^2}$, $\Delta x = \dfrac{2-0}{10} = \dfrac{1}{5}$

$T_{10} = \frac{1}{5 \cdot 2}\{f(0) + 2[f(0.2) + f(0.4) + \cdots + f(1.8)] + f(2)\} \approx 0.881\,839$

$M_{10} = \frac{1}{5}[f(0.1) + f(0.3) + f(0.5) + \cdots + f(1.7) + f(1.9)] \approx 0.882\,202$

(b) $f(x) = e^{-x^2}$, $f'(x) = -2xe^{-x^2}$, $f''(x) = (4x^2 - 2)e^{-x^2}$, $f'''(x) = 4x(3 - 2x^2)e^{-x^2}$.

$f'''(x) = 0 \quad \Leftrightarrow \quad x = 0$ or $x = \pm\sqrt{\frac{3}{2}}$. So to find the maximum value of $|f''(x)|$ on $[0, 2]$, we need only consider its values at $x = 0$, $x = 2$, and $x = \sqrt{\frac{3}{2}}$. $|f''(0)| = 2$, $|f''(2)| \approx 0.2564$ and $\left|f''\left(\sqrt{\frac{3}{2}}\right)\right| \approx 0.8925$. Thus, taking $K = 2$, $a = 0$, $b = 2$, and $n = 10$ in Theorem 3, we get $|E_T| \le 2 \cdot 2^3/(12 \cdot 10^2) = \frac{1}{75} = 0.01\overline{3}$, and $|E_M| \le |E_T|/2 \le 0.00\overline{6}$.

(c) Take $K = 2$ [as in part (b)] in Theorem 3. $|E_T| \le \dfrac{K(b-a)^3}{12n^2} \le 10^{-5} \quad \Leftrightarrow \quad \dfrac{2(2-0)^3}{12n^2} \le 10^{-5} \quad \Leftrightarrow \quad \frac{3}{4}n^2 \ge 10^5 \quad \Leftrightarrow$ $n \ge 365.1\ldots \quad \Leftrightarrow \quad n \ge 366$. Take $n = 366$ for T_n. For E_M, again take $K = 2$ in Theorem 3 to get $|E_M| \le 10^{-5} \quad \Leftrightarrow$ $\frac{3}{2}n^2 \ge 10^5 \quad \Leftrightarrow \quad n \ge 258.2 \quad \Rightarrow \quad n \ge 259$. Take $n = 259$ for M_n.

19. (a) $T_{10} = \frac{1}{10 \cdot 2}\{f(0) + 2[f(0.1) + f(0.2) + \cdots + f(0.9)] + f(1)\} \approx 1.719\,713\,49$

$S_{10} = \frac{1}{10 \cdot 3}[f(0) + 4f(0.1) + 2f(0.2) + 4f(0.3) + \cdots + 4f(0.9) + f(1)] \approx 1.718\,282\,78$

Since $I = \int_0^1 e^x\,dx = [e^x]_0^1 = e - 1 \approx 1.718\,281\,83$, $E_T = I - T_{10} \approx -0.001\,431\,66$ and $E_S = I - S_{10} \approx -0.000\,000\,95$.

(b) $f(x) = e^x \Rightarrow f''(x) = e^x \leq e$ for $0 \leq x \leq 1$. Taking $K = e$, $a = 0$, $b = 1$, and $n = 10$ in Theorem 3, we get $|E_T| \leq e(1)^3/(12 \cdot 10^2) \approx 0.002\,265 > 0.001\,431\,66$ [actual $|E_T|$ from (a)]. $f^{(4)}(x) = e^x < e$ for $0 \leq x \leq 1$. Using Theorem 4, we have $|E_S| \leq e(1)^5/(180 \cdot 10^4) \approx 0.000\,001\,5 > 0.000\,000\,95$ [actual $|E_S|$ from (a)]. We see that the actual errors are about two-thirds the size of the error estimates.

(c) From part (b), we take $K = e$ to get $|E_T| \leq \dfrac{K(b-a)^3}{12n^2} \leq 0.00001 \Rightarrow n^2 \geq \dfrac{e(1^3)}{12(0.00001)} \Rightarrow n \geq 150.5$. Take $n = 151$ for T_n. Now $|E_M| \leq \dfrac{K(b-a)^3}{24n^2} \leq 0.00001 \Rightarrow n \geq 106.4$. Take $n = 107$ for M_n. Finally, $|E_S| \leq \dfrac{K(b-a)^5}{180n^4} \leq 0.00001 \Rightarrow n^4 \geq \dfrac{e(1^5)}{180(0.00001)} \Rightarrow n \geq 6.23$. Take $n = 8$ for S_n (since n has to be even for Simpson's Rule).

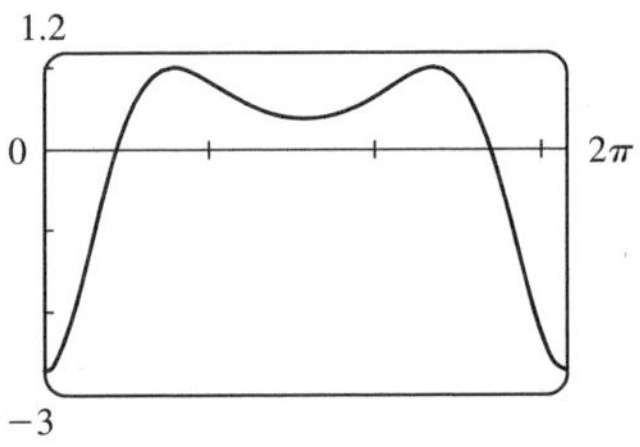

21. (a) Using a CAS, we differentiate $f(x) = e^{\cos x}$ twice, and find that $f''(x) = e^{\cos x}(\sin^2 x - \cos x)$. From the graph, we see that the maximum value of $|f''(x)|$ occurs at the endpoints of the interval $[0, 2\pi]$. Since $f''(0) = -e$, we can use $K = e$ or $K = 2.8$.

(b) A CAS gives $M_{10} \approx 7.954\,926\,518$. (In Maple, use `student[middlesum]`.)

(c) Using Theorem 3 for the Midpoint Rule, with $K = e$, we get $|E_M| \leq \dfrac{e(2\pi - 0)^3}{24 \cdot 10^2} \approx 0.280\,945\,995$. With $K = 2.8$, we get $|E_M| \leq \dfrac{2.8(2\pi - 0)^3}{24 \cdot 10^2} = 0.289\,391\,916$.

(d) A CAS gives $I \approx 7.954\,926\,521$.

(e) The actual error is only about 3×10^{-9}, much less than the estimate in part (c).

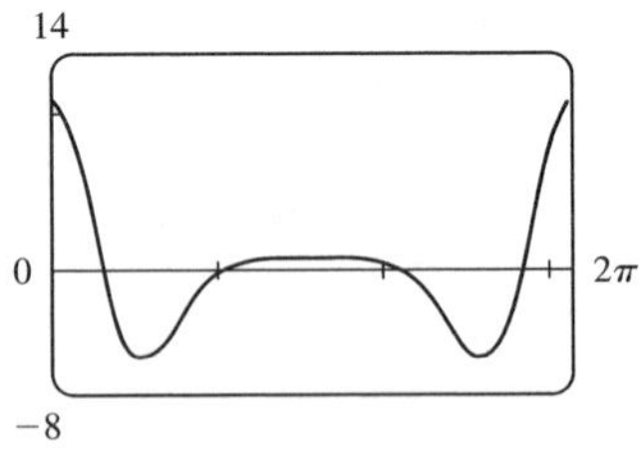

(f) We use the CAS to differentiate twice more, and then graph $f^{(4)}(x) = e^{\cos x}(\sin^4 x - 6\sin^2 x\,\cos x + 3 - 7\sin^2 x + \cos x)$. From the graph, we see that the maximum value of $\left|f^{(4)}(x)\right|$ occurs at the endpoints of the interval $[0, 2\pi]$. Since $f^{(4)}(0) = 4e$, we can use $K = 4e$ or $K = 10.9$.

(g) A CAS gives $S_{10} \approx 7.953\,789\,422$. (In Maple, use `student[simpson]`.)

(h) Using Theorem 4 with $K = 4e$, we get $|E_S| \leq \dfrac{4e(2\pi - 0)^5}{180 \cdot 10^4} \approx 0.059\,153\,618$. With $K = 10.9$, we get $|E_S| \leq \dfrac{10.9(2\pi - 0)^5}{180 \cdot 10^4} \approx 0.059\,299\,814$.

(i) The actual error is about $7.954\,926\,521 - 7.953\,789\,422 \approx 0.00114$. This is quite a bit smaller than the estimate in part (h), though the difference is not nearly as great as it was in the case of the Midpoint Rule.

(j) To ensure that $|E_S| \leq 0.0001$, we use Theorem 4: $|E_S| \leq \dfrac{4e(2\pi)^5}{180 \cdot n^4} \leq 0.0001 \Rightarrow \dfrac{4e(2\pi)^5}{180 \cdot 0.0001} \leq n^4 \Rightarrow n^4 \geq 5\,915\,362 \Leftrightarrow n \geq 49.3$. So we must take $n \geq 50$ to ensure that $|I - S_n| \leq 0.0001$. ($K = 10.9$ leads to the same value of n.)

23. $I = \int_0^1 x^3\,dx = \left[\frac{1}{4}x^4\right]_0^1 = 0.25.$ $f(x) = x^3.$

$n = 4$: $L_4 = \frac{1}{4}\left[0^3 + \left(\frac{1}{4}\right)^3 + \left(\frac{2}{4}\right)^3 + \left(\frac{3}{4}\right)^3\right] = 0.140\,625$

$R_4 = \frac{1}{4}\left[\left(\frac{1}{4}\right)^3 + \left(\frac{2}{4}\right)^3 + \left(\frac{3}{4}\right)^3 + 1^3\right] = 0.390\,625$

$T_4 = \frac{1}{4 \cdot 2}\left[0^3 + 2\left(\frac{1}{4}\right)^3 + 2\left(\frac{2}{4}\right)^3 + 2\left(\frac{3}{4}\right)^3 + 1^3\right] = 0.265\,625,$

$M_4 = \frac{1}{4}\left[\left(\frac{1}{8}\right)^3 + \left(\frac{3}{8}\right)^3 + \left(\frac{5}{8}\right)^3 + \left(\frac{7}{8}\right)^3\right] = 0.242\,187\,5,$

$E_L = I - L_4 = \frac{1}{4} - 0.140\,625 = 0.109\,375,\ E_R = \frac{1}{4} - 0.390\,625 = -0.140\,625,$

$E_T = \frac{1}{4} - 0.265\,625 = -0.015\,625,\ E_M = \frac{1}{4} - 0.242\,187\,5 = 0.007\,812\,5$

$n = 8$: $L_8 = \frac{1}{8}\left[f(0) + f\left(\frac{1}{8}\right) + f\left(\frac{2}{8}\right) + \cdots + f\left(\frac{7}{8}\right)\right] \approx 0.191\,406$

$R_8 = \frac{1}{8}\left[f\left(\frac{1}{8}\right) + f\left(\frac{2}{8}\right) + \cdots + f\left(\frac{7}{8}\right) + f(1)\right] \approx 0.316\,406$

$T_8 = \frac{1}{8 \cdot 2}\left\{f(0) + 2\left[f\left(\frac{1}{8}\right) + f\left(\frac{2}{8}\right) + \cdots + f\left(\frac{7}{8}\right)\right] + f(1)\right\} \approx 0.253\,906$

$M_8 = \frac{1}{8}\left[f\left(\frac{1}{16}\right) + f\left(\frac{3}{16}\right) + \cdots + f\left(\frac{13}{16}\right) + f\left(\frac{15}{16}\right)\right] = 0.248\,047$

$E_L \approx \frac{1}{4} - 0.191\,406 \approx 0.058\,594,\ E_R \approx \frac{1}{4} - 0.316\,406 \approx -0.066\,406,$

$E_T \approx \frac{1}{4} - 0.253\,906 \approx -0.003\,906,\ E_M \approx \frac{1}{4} - 0.248\,047 \approx 0.001\,953.$

$n = 16$: $L_{16} = \frac{1}{16}\left[f(0) + f\left(\frac{1}{16}\right) + f\left(\frac{2}{16}\right) + \cdots + f\left(\frac{15}{16}\right)\right] \approx 0.219\,727$

$R_{16} = \frac{1}{16}\left[f\left(\frac{1}{16}\right) + f\left(\frac{2}{16}\right) + \cdots + f\left(\frac{15}{16}\right) + f(1)\right] \approx 0.282\,227$

$T_{16} = \frac{1}{16 \cdot 2}\left\{f(0) + 2\left[f\left(\frac{1}{16}\right) + f\left(\frac{2}{16}\right) + \cdots + f\left(\frac{15}{16}\right)\right] + f(1)\right\} \approx 0.250\,977$

$M_{16} = \frac{1}{16}\left[f\left(\frac{1}{32}\right) + f\left(\frac{3}{32}\right) + \cdots + f\left(\frac{31}{32}\right)\right] \approx 0.249\,512$

$E_L \approx \frac{1}{4} - 0.219\,727 \approx 0.030\,273,\ E_R \approx \frac{1}{4} - 0.282\,227 \approx -0.032\,227,$

$E_T \approx \frac{1}{4} - 0.250\,977 \approx -0.000\,977,\ E_M \approx \frac{1}{4} - 0.249\,512 \approx 0.000\,488.$

n	L_n	R_n	T_n	M_n
4	0.140 625	0.390 625	0.265 625	0.242 188
8	0.191 406	0.316 406	0.253 906	0.248 047
16	0.219 727	0.282 227	0.250 977	0.249 512

n	E_L	E_R	E_T	E_M
4	0.109 375	−0.140 625	−0.015 625	0.007 813
8	0.058 594	−0.066 406	−0.003 906	0.001 953
16	0.030 273	−0.032 227	−0.000 977	0.000 488

Observations:

1. E_L and E_R are always opposite in sign, as are E_T and E_M.
2. As n is doubled, E_L and E_R are decreased by about a factor of 2, and E_T and E_M are decreased by a factor of about 4.
3. The Midpoint approximation is about twice as accurate as the Trapezoidal approximation.
4. All the approximations become more accurate as the value of n increases.
5. The Midpoint and Trapezoidal approximations are much more accurate than the endpoint approximations.

25. $\Delta x = (4 - 0)/4 = 1$

(a) $T_4 = \frac{1}{2}[f(0) + 2f(1) + 2f(2) + 2f(3) + f(4)] \approx \frac{1}{2}[0 + 2(3) + 2(5) + 2(3) + 1] = 11.5$

(b) $M_4 = 1 \cdot [f(0.5) + f(1.5) + f(2.5) + f(3.5)] \approx 1 + 4.5 + 4.5 + 2 = 12$

(c) $S_4 = \frac{1}{3}[f(0) + 4f(1) + 2f(2) + 4f(3) + f(4)] \approx \frac{1}{3}[0 + 4(3) + 2(5) + 4(3) + 1] = 11.\overline{6}$

27. By the Net Change Theorem, the increase in velocity is equal to $\int_0^6 a(t)\,dt$. We use Simpson's Rule with $n = 6$ and $\Delta t = (6-0)/6 = 1$ to estimate this integral:

$$\int_0^6 a(t)\,dt \approx S_6 = \tfrac{1}{3}[a(0) + 4a(1) + 2a(2) + 4a(3) + 2a(4) + 4a(5) + a(6)]$$
$$\approx \tfrac{1}{3}[0 + 4(0.2) + 2(2.0) + 4(5.0) + 2(6.4) + 4(4.7) + 0] = \tfrac{1}{3}(56.4) = 18.8 \text{ m/s}$$

29. By the Net Change Theorem, the energy used is equal to $\int_0^6 P(t)\,dt$. We use Simpson's Rule with $n = 12$ and $\Delta t = (6-0)/12 = \frac{1}{2}$ to estimate this integral:

$$\int_0^6 P(t)\,dt \approx S_{12} = \tfrac{1/2}{3}[P(0) + 4P(0.5) + 2P(1) + 4P(1.5) + 2P(2) + 4P(2.5)$$
$$+ 2P(3) + 4P(3.5) + 2P(4) + 4P(4.5) + 2P(5) + 4P(5.5) + P(6)]$$
$$= \tfrac{1}{6}[17\,888 + 4(17\,398) + 2(17\,110) + 4(16\,881) + 2(16\,832) + 4(16\,950) + 2(16\,833)$$
$$+ 4(16\,835) + 2(17\,065) + 4(17\,264) + 2(17\,577) + 4(17\,992) + 18\,216]$$
$$= \tfrac{1}{6}(620\,218) = 103\,369.\overline{6} \approx 1.0337 \times 10^5 \text{ megawatt-hours}$$

31. (a) We are given the function values at the endpoints of 8 intervals of length 0.4, so we'll use the Midpoint Rule with $n = 8/2 = 4$ and $\Delta x = (3.2 - 0)/4 = 0.8$.

$$\int_0^{3.2} f(x)\,dx \approx M_4 = 0.8[f(0.4) + f(1.2) + f(2.0) + f(2.8)]$$
$$= 0.8[6.5 + 6.4 + 7.6 + 8.8]$$
$$= 0.8(29.3) = 23.44$$

(b) $-4 \le f''(x) \le 1 \quad\Rightarrow\quad |f''(x)| \le 4$, so use $K = 4$, $a = 0$, $b = 3.2$, and $n = 4$ in Theorem 3. So

$$|E_M| \le \frac{4(3.2-0)^3}{24(4)^2} = \frac{128}{375} = 0.341\overline{3}.$$

33. $I(\theta) = \dfrac{N^2 \sin^2 k}{k^2}$, where $k = \dfrac{\pi N d \sin\theta}{\lambda}$, $N = 10\,000$, $d = 10^{-4}$, and $\lambda = 632.8 \times 10^{-9}$. So $I(\theta) = \dfrac{(10^4)^2 \sin^2 k}{k^2}$,

where $k = \dfrac{\pi(10^4)(10^{-4})\sin\theta}{632.8 \times 10^{-9}}$. Now $n = 10$ and $\Delta\theta = \dfrac{10^{-6} - (-10^{-6})}{10} = 2 \times 10^{-7}$, so

$$M_{10} = 2 \times 10^{-7}[I(-0.000\,000\,9) + I(-0.000\,000\,7) + \cdots + I(0.000\,000\,9)] \approx 59.4.$$

35. Consider the function f whose graph is shown. The area $\int_0^2 f(x)\,dx$ is close to 2. The Trapezoidal Rule gives

$$T_2 = \tfrac{2-0}{2\cdot 2}\,[f(0) + 2f(1) + f(2)] = \tfrac{1}{2}\,[1 + 2\cdot 1 + 1] = 2.$$

The Midpoint Rule gives

$$M_2 = \tfrac{2-0}{2}\,[f(0.5) + f(1.5)] = 1[0+0] = 0,$$

so the Trapezoidal Rule is more accurate.

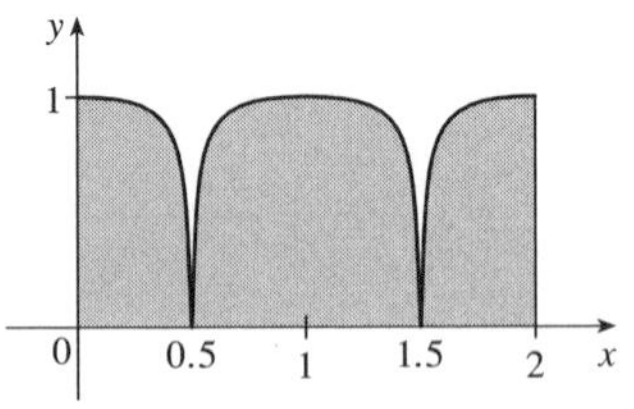

37. Since the Trapezoidal and Midpoint approximations on the interval $[a, b]$ are the sums of the Trapezoidal and Midpoint approximations on the subintervals $[x_{i-1}, x_i]$, $i = 1, 2, \ldots, n$, we can focus our attention on one such interval. The condition $f''(x) < 0$ for $a \le x \le b$ means that the graph of f is concave down as in Figure 5. In that figure, T_n is the area of the trapezoid $AQRD$, $\int_a^b f(x)\,dx$ is the area of the region $AQPRD$, and M_n is the area of the trapezoid $ABCD$, so $T_n < \int_a^b f(x)\,dx < M_n$. In general, the condition $f'' < 0$ implies that the graph of f on $[a, b]$ lies above the chord joining the points $(a, f(a))$ and $(b, f(b))$. Thus, $\int_a^b f(x)\,dx > T_n$. Since M_n is the area under a tangent to the graph, and since $f'' < 0$ implies that the tangent lies above the graph, we also have $M_n > \int_a^b f(x)\,dx$. Thus, $T_n < \int_a^b f(x)\,dx < M_n$.

39. $T_n = \frac{1}{2}\,\Delta x\,[f(x_0) + 2f(x_1) + \cdots + 2f(x_{n-1}) + f(x_n)]$ and

$M_n = \Delta x\,[f(\overline{x}_1) + f(\overline{x}_2) + \cdots + f(\overline{x}_{n-1}) + f(\overline{x}_n)]$, where $\overline{x}_i = \frac{1}{2}(x_{i-1} + x_i)$. Now

$$T_{2n} = \tfrac{1}{2}\left(\tfrac{1}{2}\Delta x\right)[f(x_0) + 2f(\overline{x}_1) + 2f(x_1) + 2f(\overline{x}_2) + 2f(x_2) + \cdots + 2f(\overline{x}_{n-1}) + 2f(x_{n-1}) + 2f(\overline{x}_n) + f(x_n)]$$

so
$$\begin{aligned}\tfrac{1}{2}(T_n + M_n) &= \tfrac{1}{2}T_n + \tfrac{1}{2}M_n \\ &= \tfrac{1}{4}\,\Delta x\,[f(x_0) + 2f(x_1) + \cdots + 2f(x_{n-1}) + f(x_n)] \\ &\quad + \tfrac{1}{4}\Delta x\,[2f(\overline{x}_1) + 2f(\overline{x}_2) + \cdots + 2f(\overline{x}_{n-1}) + 2f(\overline{x}_n)] \\ &= T_{2n}\end{aligned}$$

5.10 Improper Integrals

1. (a) Since $\int_1^\infty x^4 e^{-x^4}\,dx$ has an infinite interval of integration, it is an improper integral of Type I.

(b) Since $y = \sec x$ has an infinite discontinuity at $x = \frac{\pi}{2}$, $\int_0^{\pi/2} \sec x\,dx$ is a Type II improper integral.

(c) Since $y = \dfrac{x}{(x-2)(x-3)}$ has an infinite discontinuity at $x = 2$, $\displaystyle\int_0^2 \frac{x}{x^2 - 5x + 6}\,dx$ is a Type II improper integral.

(d) Since $\displaystyle\int_{-\infty}^0 \frac{1}{x^2+5}\,dx$ has an infinite interval of integration, it is an improper integral of Type I.

3. The area under the graph of $y = 1/x^3 = x^{-3}$ between $x = 1$ and $x = t$ is

$A(t) = \int_1^t x^{-3}\,dx = \left[-\frac{1}{2}x^{-2}\right]_1^t = -\frac{1}{2}t^{-2} - \left(-\frac{1}{2}\right) = \frac{1}{2} - 1/(2t^2)$. So the area for $1 \le x \le 10$ is $A(10) = 0.5 - 0.005 = 0.495$, the area for $1 \le x \le 100$ is $A(100) = 0.5 - 0.00005 = 0.49995$, and the area for $1 \le x \le 1000$ is $A(1000) = 0.5 - 0.000\,000\,5 = 0.499\,999\,5$. The total area under the curve for $x \ge 1$ is $\lim\limits_{t\to\infty} A(t) = \lim\limits_{t\to\infty}\left[\frac{1}{2} - 1/(2t^2)\right] = \frac{1}{2}$.

5. $I = \displaystyle\int_1^\infty \frac{1}{(3x+1)^2}\,dx = \lim_{t\to\infty}\int_1^t \frac{1}{(3x+1)^2}\,dx$. Now

$\displaystyle\int \frac{1}{(3x+1)^2}\,dx = \frac{1}{3}\int \frac{1}{u^2}\,du \quad [u = 3x+1,\ du = 3\,dx] \quad = -\frac{1}{3u} + C = -\frac{1}{3(3x+1)} + C,$

so $I = \displaystyle\lim_{t\to\infty}\left[-\frac{1}{3(3x+1)}\right]_1^t = \lim_{t\to\infty}\left[-\frac{1}{3(3t+1)} + \frac{1}{12}\right] = 0 + \frac{1}{12} = \frac{1}{12}.$ Convergent

7. $\displaystyle\int_{-\infty}^{-1} \frac{1}{\sqrt{2-w}}\,dw = \lim_{t\to-\infty}\int_t^{-1} \frac{1}{\sqrt{2-w}}\,dw = \lim_{t\to-\infty}\left[-2\sqrt{2-w}\right]_t^{-1} \quad [u = 2-w,\ du = -dw]$

$\displaystyle = \lim_{t\to-\infty}\left[-2\sqrt{3} + 2\sqrt{2-t}\right] = \infty.$ Divergent

9. $\int_4^\infty e^{-y/2}\,dy = \lim\limits_{t\to\infty}\int_4^t e^{-y/2}\,dy = \lim\limits_{t\to\infty}\left[-2e^{-y/2}\right]_4^t = \lim\limits_{t\to\infty}\left(-2e^{-t/2} + 2e^{-2}\right) = 0 + 2e^{-2} = 2e^{-2}.$

Convergent

11. $\int_0^\infty \cos x\,dx = \lim\limits_{t\to\infty}\,[\sin x]_0^t = \lim\limits_{t\to\infty}\sin t$, which does not exist. Divergent

13. $\int_{-\infty}^{\infty} xe^{-x^2}\,dx = \int_{-\infty}^{0} xe^{-x^2}\,dx + \int_{0}^{\infty} xe^{-x^2}\,dx.$

$\int_{-\infty}^{0} xe^{-x^2}\,dx = \lim\limits_{t\to-\infty}\left(-\frac{1}{2}\right)\left[e^{-x^2}\right]_t^0 = \lim\limits_{t\to-\infty}\left(-\frac{1}{2}\right)\left(1-e^{-t^2}\right) = -\frac{1}{2}\cdot 1 = -\frac{1}{2}$, and

$\int_{0}^{\infty} xe^{-x^2}\,dx = \lim\limits_{t\to\infty}\left(-\frac{1}{2}\right)\left[e^{-x^2}\right]_0^t = \lim\limits_{t\to\infty}\left(-\frac{1}{2}\right)\left(e^{-t^2}-1\right) = -\frac{1}{2}\cdot(-1) = \frac{1}{2}.$

Therefore, $\int_{-\infty}^{\infty} xe^{-x^2}\,dx = -\frac{1}{2}+\frac{1}{2} = 0.$ Convergent

15. $\int_{-\infty}^{1} xe^{2x}\,dx = \lim\limits_{t\to-\infty}\int_t^1 xe^{2x}\,dx = \lim\limits_{t\to-\infty}\left[\frac{1}{2}xe^{2x} - \frac{1}{4}e^{2x}\right]_t^1$ [by parts with $u = x$ and $dv = e^{2x}\,dx$]

$= \lim\limits_{t\to-\infty}\left[\frac{1}{2}e^2 - \frac{1}{4}e^2 - \frac{1}{2}te^{2t} + \frac{1}{4}e^{2t}\right] = \frac{1}{4}e^2 - 0 + 0 = \frac{1}{4}e^2$

since $\lim\limits_{t\to-\infty} te^{2t} = \lim\limits_{t\to-\infty}\dfrac{t}{e^{-2t}} \overset{\text{H}}{=} \lim\limits_{t\to-\infty}\dfrac{1}{-2e^{-2t}} = \lim\limits_{t\to-\infty} -\frac{1}{2}e^{2t} = 0.$ Convergent

17. $\displaystyle\int_1^{\infty}\frac{\ln x}{x}\,dx = \lim_{t\to\infty}\left[\frac{(\ln x)^2}{2}\right]_1^t$ (by substitution with $u = \ln x$, $du = dx/x$) $= \displaystyle\lim_{t\to\infty}\frac{(\ln t)^2}{2} = \infty.$ Divergent

19. Integrate by parts with $u = \ln x$, $dv = dx/x^2 \;\Rightarrow\; du = dx/x$, $v = -1/x$.

$$\int_1^{\infty}\frac{\ln x}{x^2}\,dx = \lim_{t\to\infty}\int_1^t \frac{\ln x}{x^2}\,dx = \lim_{t\to\infty}\left[-\frac{\ln x}{x} - \frac{1}{x}\right]_1^t = \lim_{t\to\infty}\left(-\frac{\ln t}{t} - \frac{1}{t} + 0 + 1\right)$$
$$= -0 - 0 + 0 + 1 = 1$$

since $\displaystyle\lim_{t\to\infty}\frac{\ln t}{t} \overset{\text{H}}{=} \lim_{t\to\infty}\frac{1/t}{1} = 0.$ Convergent

21. $\displaystyle\int_{-\infty}^{\infty}\frac{x^2}{9+x^6}\,dx = \int_{-\infty}^{0}\frac{x^2}{9+x^6}\,dx + \int_0^{\infty}\frac{x^2}{9+x^6}\,dx = 2\int_0^{\infty}\frac{x^2}{9+x^6}\,dx$ [since the integrand is even].

Now $\displaystyle\int\frac{x^2\,dx}{9+x^6}\;\begin{bmatrix} u = x^3 \\ du = 3x^2\,dx\end{bmatrix} = \int\frac{\frac{1}{3}\,du}{9+u^2}\;\begin{bmatrix} u = 3v \\ du = 3\,dv\end{bmatrix} = \int\frac{\frac{1}{3}(3\,dv)}{9+9v^2} = \frac{1}{9}\int\frac{dv}{1+v^2}$

$$= \frac{1}{9}\tan^{-1}v + C = \frac{1}{9}\tan^{-1}\left(\frac{u}{3}\right) + C = \frac{1}{9}\tan^{-1}\left(\frac{x^3}{3}\right) + C,$$

so $\displaystyle 2\int_0^{\infty}\frac{x^2}{9+x^6}\,dx = 2\lim_{t\to\infty}\int_0^t\frac{x^2}{9+x^6}\,dx = 2\lim_{t\to\infty}\left[\frac{1}{9}\tan^{-1}\left(\frac{x^3}{3}\right)\right]_0^t$

$\displaystyle = 2\lim_{t\to\infty}\frac{1}{9}\tan^{-1}\left(\frac{t^3}{3}\right) = \frac{2}{9}\cdot\frac{\pi}{2} = \frac{\pi}{9}.$ Convergent

23. $\displaystyle\int_0^1\frac{3}{x^5}\,dx = \lim_{t\to0^+}\int_t^1 3x^{-5}\,dx = \lim_{t\to0^+}\left[-\frac{3}{4x^4}\right]_t^1 = -\frac{3}{4}\lim_{t\to0^+}\left(1-\frac{1}{t^4}\right) = \infty.$ Divergent

25. $\displaystyle\int_{-2}^{14}\frac{dx}{\sqrt[4]{x+2}} = \lim_{t\to-2^+}\int_t^{14}(x+2)^{-1/4}\,dx = \lim_{t\to-2^+}\left[\frac{4}{3}(x+2)^{3/4}\right]_t^{14} = \frac{4}{3}\lim_{t\to-2^+}\left[16^{3/4} - (t+2)^{3/4}\right]$

$= \frac{4}{3}(8-0) = \frac{32}{3}.$ Convergent

27. There is an infinite discontinuity at $x = 1$. $\int_0^{33}(x-1)^{-1/5}\,dx = \int_0^1 (x-1)^{-1/5}\,dx + \int_1^{33}(x-1)^{-1/5}\,dx$. Here

$\int_0^1 (x-1)^{-1/5}\,dx = \lim\limits_{t\to 1^-} \int_0^t (x-1)^{-1/5}\,dx = \lim\limits_{t\to 1^-}\left[\frac{5}{4}(x-1)^{4/5}\right]_0^t = \lim\limits_{t\to 1^-}\left[\frac{5}{4}(t-1)^{4/5} - \frac{5}{4}\right] = -\frac{5}{4}$ and

$\int_1^{33}(x-1)^{-1/5}\,dx = \lim\limits_{t\to 1^+}\int_t^{33}(x-1)^{-1/5}\,dx = \lim\limits_{t\to 1^+}\left[\frac{5}{4}(x-1)^{4/5}\right]_t^{33} = \lim\limits_{t\to 1^+}\left[\frac{5}{4}\cdot 16 - \frac{5}{4}(t-1)^{4/5}\right] = 20$. Thus,

$\int_0^{33}(x-1)^{-1/5}\,dx = -\frac{5}{4} + 20 = \frac{75}{4}$. Convergent

29. There is an infinite discontinuity at $x = 0$. $\displaystyle\int_{-1}^{1}\frac{e^x}{e^x-1}\,dx = \int_{-1}^{0}\frac{e^x}{e^x-1}\,dx + \int_{0}^{1}\frac{e^x}{e^x-1}\,dx$.

$$\int_{-1}^{0}\frac{e^x}{e^x-1}\,dx = \lim_{t\to 0^-}\int_{-1}^{t}\frac{e^x}{e^x-1}\,dx = \lim_{t\to 0^-}\Big[\ln|e^x-1|\Big]_{-1}^{t} = \lim_{t\to 0^-}\Big[\ln|e^t-1| - \ln|e^{-1}-1|\Big] = -\infty,$$

so $\displaystyle\int_{-1}^{1}\frac{e^x}{e^x-1}\,dx$ is divergent. The integral $\displaystyle\int_{0}^{1}\frac{e^x}{e^x-1}\,dx$ also diverges since

$$\int_{0}^{1}\frac{e^x}{e^x-1}\,dx = \lim_{t\to 0^+}\int_{t}^{1}\frac{e^x}{e^x-1}\,dx = \lim_{t\to 0^+}\Big[\ln|e^x-1|\Big]_{t}^{1} = \lim_{t\to 0^+}\Big[\ln|e-1| - \ln|e^t-1|\Big] = \infty.$$

Divergent

31. $I = \int_0^2 z^2\ln z\,dz = \lim\limits_{t\to 0^+}\int_t^2 z^2 \ln z\,dz \overset{101}{=} \lim\limits_{t\to 0^+}\left[\frac{z^3}{3^2}(3\ln z - 1)\right]_t^2$

$= \lim\limits_{t\to 0^+}\left[\frac{8}{9}(3\ln 2 - 1) - \frac{1}{9}t^3(3\ln t - 1)\right] = \frac{8}{3}\ln 2 - \frac{8}{9} - \frac{1}{9}\lim\limits_{t\to 0^+}\left[t^3(3\ln t - 1)\right] = \frac{8}{3}\ln 2 - \frac{8}{9} - \frac{1}{9}L.$

Now $L = \lim\limits_{t\to 0^+}\left[t^3(3\ln t - 1)\right] = \lim\limits_{t\to 0^+}\dfrac{3\ln t - 1}{t^{-3}} \overset{\text{H}}{=} \lim\limits_{t\to 0^+}\dfrac{3/t}{-3/t^4} = \lim\limits_{t\to 0^+}(-t^3) = 0.$

Thus, $L = 0$ and $I = \frac{8}{3}\ln 2 - \frac{8}{9}$. Convergent

33.

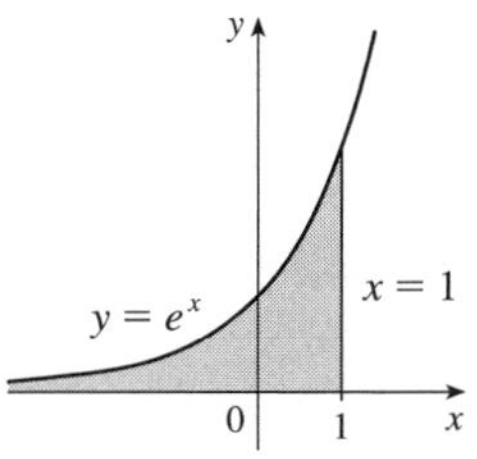

Area $= \int_{-\infty}^{1} e^x\,dx = \lim\limits_{t\to -\infty}\left[e^x\right]_t^1$

$= e - \lim\limits_{t\to -\infty} e^t = e$

35.

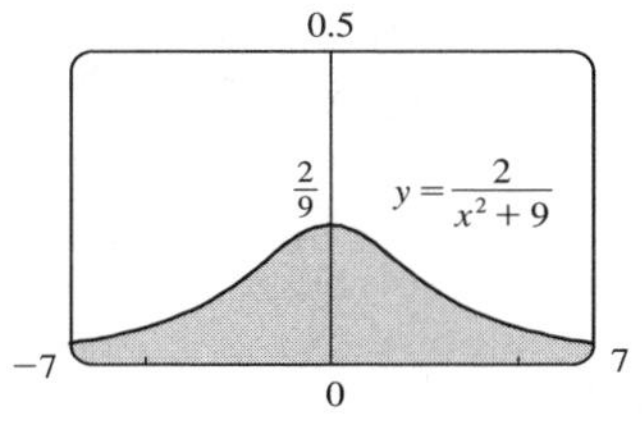

$$\text{Area} = \int_{-\infty}^{\infty}\frac{2}{x^2+9}\,dx = 2\cdot 2\int_0^{\infty}\frac{1}{x^2+9}\,dx$$

$$= 4\lim_{t\to\infty}\int_0^t \frac{1}{x^2+9}\,dx = 4\lim_{t\to\infty}\left[\frac{1}{3}\tan^{-1}\frac{x}{3}\right]_0^t$$

$$= \frac{4}{3}\lim_{t\to\infty}\left[\tan^{-1}\frac{t}{3} - 0\right] = \frac{4}{3}\cdot\frac{\pi}{2} = \frac{2\pi}{3}$$

37.

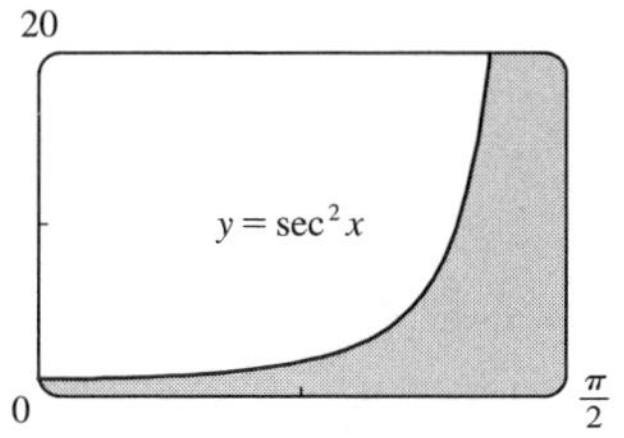

Area $= \int_0^{\pi/2}\sec^2 x\,dx = \lim\limits_{t\to(\pi/2)^-}\int_0^t \sec^2 x\,dx$

$= \lim\limits_{t\to(\pi/2)^-}\left[\tan x\right]_0^t = \lim\limits_{t\to(\pi/2)^-}(\tan t - 0)$

$= \infty$

Infinite area

39. (a)

t	$\int_1^t g(x)\,dx$
2	0.447 453
5	0.577 101
10	0.621 306
100	0.668 479
1 000	0.672 957
10 000	0.673 407

$g(x) = \dfrac{\sin^2 x}{x^2}$.

It appears that the integral is convergent.

(b) $-1 \le \sin x \le 1 \;\Rightarrow\; 0 \le \sin^2 x \le 1 \;\Rightarrow\; 0 \le \dfrac{\sin^2 x}{x^2} \le \dfrac{1}{x^2}$. Since $\displaystyle\int_1^\infty \frac{1}{x^2}\,dx$ is convergent (Equation 2 with $p = 2 > 1$), $\displaystyle\int_1^\infty \frac{\sin^2 x}{x^2}\,dx$ is convergent by the Comparison Theorem.

(c)

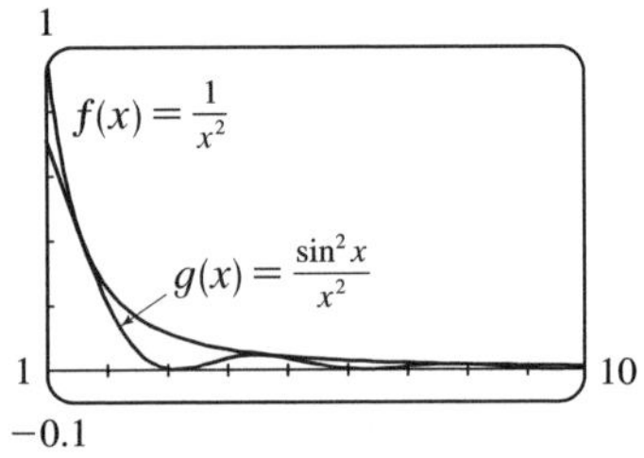

Since $\int_1^\infty f(x)\,dx$ is finite and the area under $g(x)$ is less than the area under $f(x)$ on any interval $[1, t]$, $\int_1^\infty g(x)\,dx$ must be finite; that is, the integral is convergent.

41. For $x \ge 1$, $\dfrac{\cos^2 x}{1 + x^2} \le \dfrac{1}{1 + x^2} < \dfrac{1}{x^2}$. $\displaystyle\int_1^\infty \frac{1}{x^2}\,dx$ is convergent by Equation 2 with $p = 2 > 1$, so $\displaystyle\int_1^\infty \frac{\cos^2 x}{1 + x^2}\,dx$ is convergent by the Comparison Theorem.

43. For $x \ge 1$, $x + e^{2x} > e^{2x} > 0 \;\Rightarrow\; \dfrac{1}{x + e^{2x}} \le \dfrac{1}{e^{2x}} = e^{-2x}$ on $[1, \infty)$.

$$\int_1^\infty e^{-2x}\,dx = \lim_{t\to\infty}\left[-\frac{1}{2}e^{-2x}\right]_1^t = \lim_{t\to\infty}\left[-\frac{1}{2}e^{-2t} + \frac{1}{2}e^{-2}\right] = \frac{1}{2}e^{-2}.$$

Therefore, $\displaystyle\int_1^\infty e^{-2x}\,dx$ is convergent, and by the Comparison Theorem, $\displaystyle\int_1^\infty \frac{dx}{x + e^{2x}}$ is also convergent.

45. $\dfrac{1}{x \sin x} \ge \dfrac{1}{x}$ on $\left(0, \frac{\pi}{2}\right]$ since $0 \le \sin x \le 1$. $\displaystyle\int_0^{\pi/2} \frac{dx}{x} = \lim_{t\to 0^+} \int_t^{\pi/2} \frac{dx}{x} = \lim_{t\to 0^+} \left[\ln x\right]_t^{\pi/2}$.

But $\ln t \to -\infty$ as $t \to 0^+$, so $\displaystyle\int_0^{\pi/2} \frac{dx}{x}$ is divergent, and by the Comparison Theorem, $\displaystyle\int_0^{\pi/2} \frac{dx}{x \sin x}$ is also divergent.

47. $\displaystyle\int_0^\infty \frac{dx}{\sqrt{x}\,(1+x)} = \int_0^1 \frac{dx}{\sqrt{x}\,(1+x)} + \int_1^\infty \frac{dx}{\sqrt{x}\,(1+x)} = \lim_{t\to 0^+} \int_t^1 \frac{dx}{\sqrt{x}\,(1+x)} + \lim_{t\to\infty} \int_1^t \frac{dx}{\sqrt{x}\,(1+x)}$. Now

$$\int \frac{dx}{\sqrt{x}\,(1+x)} = \int \frac{2u\,du}{u(1+u^2)} \quad [u = \sqrt{x}, x = u^2, dx = 2u\,du]$$

$$= 2\int \frac{du}{1+u^2} = 2\tan^{-1} u + C = 2\tan^{-1}\sqrt{x} + C,$$

so

$$\int_0^\infty \frac{dx}{\sqrt{x}\,(1+x)} = \lim_{t\to 0^+}\left[2\tan^{-1}\sqrt{x}\right]_t^1 + \lim_{t\to\infty}\left[2\tan^{-1}\sqrt{x}\right]_1^t$$

$$= \lim_{t\to 0^+}\left[2\left(\tfrac{\pi}{4}\right) - 2\tan^{-1}\sqrt{t}\right] + \lim_{t\to\infty}\left[2\tan^{-1}\sqrt{t} - 2\left(\tfrac{\pi}{4}\right)\right] = \tfrac{\pi}{2} - 0 + 2\left(\tfrac{\pi}{2}\right) - \tfrac{\pi}{2} = \pi.$$

49. If $p = 1$, then $\displaystyle\int_0^1 \frac{dx}{x^p} = \lim_{t\to 0^+} \int_t^1 \frac{dx}{x} = \lim_{t\to 0^+} [\ln x]_t^1 = \infty$. Divergent.

If $p \neq 1$, then $\displaystyle\int_0^1 \frac{dx}{x^p} = \lim_{t\to 0^+} \int_t^1 \frac{dx}{x^p}$ (note that the integral is not improper if $p < 0$)

$$= \lim_{t\to 0^+} \left[\frac{x^{-p+1}}{-p+1}\right]_t^1 = \lim_{t\to 0^+} \frac{1}{1-p}\left[1 - \frac{1}{t^{p-1}}\right]$$

If $p > 1$, then $p - 1 > 0$, so $\dfrac{1}{t^{p-1}} \to \infty$ as $t \to 0^+$, and the integral diverges.

If $p < 1$, then $p - 1 < 0$, so $\dfrac{1}{t^{p-1}} \to 0$ as $t \to 0^+$ and $\displaystyle\int_0^1 \frac{dx}{x^p} = \frac{1}{1-p}\left[\lim_{t\to 0^+}\left(1 - t^{1-p}\right)\right] = \frac{1}{1-p}$.

Thus, the integral converges if and only if $p < 1$, and in that case its value is $\dfrac{1}{1-p}$.

51. (a) $I = \int_{-\infty}^{\infty} x\,dx = \int_{-\infty}^{0} x\,dx + \int_0^{\infty} x\,dx$, and $\int_0^{\infty} x\,dx = \lim\limits_{t\to\infty} \int_0^t x\,dx = \lim\limits_{t\to\infty} \left[\frac{1}{2}x^2\right]_0^t = \lim\limits_{t\to\infty} \left[\frac{1}{2}t^2 - 0\right] = \infty$, so I is divergent.

(b) $\int_{-t}^{t} x\,dx = \left[\frac{1}{2}x^2\right]_{-t}^{t} = \frac{1}{2}t^2 - \frac{1}{2}t^2 = 0$, so $\lim\limits_{t\to\infty} \int_{-t}^{t} x\,dx = 0$. Therefore, $\int_{-\infty}^{\infty} x\,dx \neq \lim\limits_{t\to\infty} \int_{-t}^{t} x\,dx$.

53. We would expect a small percentage of bulbs to burn out in the first few hundred hours, most of the bulbs to burn out after close to 700 hours, and a few overachievers to burn on and on.

(a)

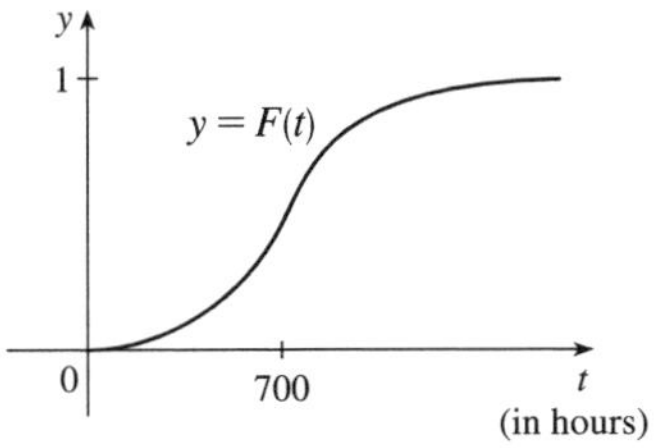

(b) $r(t) = F'(t)$ is the rate at which the fraction $F(t)$ of burnt-out bulbs increases as t increases. This could be interpreted as a fractional burnout rate.

(c) $\int_0^{\infty} r(t)\,dt = \lim\limits_{x\to\infty} F(x) = 1$, since all of the bulbs will eventually burn out.

55. $\displaystyle I = \int_0^{\infty} te^{kt}\,dt = \lim_{s\to\infty}\left[\frac{1}{k^2}(kt-1)\,e^{kt}\right]_0^s$ [Formula 96, or parts]

$$= \lim_{s\to\infty}\left[\left(\frac{1}{k}se^{ks} - \frac{1}{k^2}e^{ks}\right) - \left(-\frac{1}{k^2}\right)\right].$$

Since $k < 0$ the first two terms approach 0 (you can verify that the first term does so with l'Hospital's Rule), so the limit is equal to $1/k^2$. Thus, $M = -kI = -k\left(1/k^2\right) = -1/k = -1/(-0.000\,121) \approx 8264.5$ years.

57. $\displaystyle I = \int_a^{\infty} \frac{1}{x^2+1}\,dx = \lim_{t\to\infty}\int_a^t \frac{1}{x^2+1}\,dx = \lim_{t\to\infty}\left[\tan^{-1}x\right]_a^t = \lim_{t\to\infty}\left(\tan^{-1}t - \tan^{-1}a\right) = \frac{\pi}{2} - \tan^{-1}a$.

$I < 0.001 \;\Rightarrow\; \frac{\pi}{2} - \tan^{-1}a < 0.001 \;\Rightarrow\; \tan^{-1}a > \frac{\pi}{2} - 0.001 \;\Rightarrow\; a > \tan\left(\frac{\pi}{2} - 0.001\right) \approx 1000.$

59. We use integration by parts: let $u = x$, $dv = xe^{-x^2}\,dx \quad \Rightarrow \quad du = dx$, $v = -\frac{1}{2}e^{-x^2}$. So

$$\int_0^\infty x^2 e^{-x^2}\,dx = \lim_{t\to\infty}\left[-\tfrac{1}{2}xe^{-x^2}\right]_0^t + \tfrac{1}{2}\int_0^\infty e^{-x^2}\,dx$$
$$= \lim_{t\to\infty}\left[-t\big/\left(2e^{t^2}\right)\right] + \tfrac{1}{2}\int_0^\infty e^{-x^2}\,dx = \tfrac{1}{2}\int_0^\infty e^{-x^2}\,dx$$

(The limit is 0 by l'Hospital's Rule.)

61. For the first part of the integral, let $x = 2\tan\theta \quad \Rightarrow \quad dx = 2\sec^2\theta\,d\theta$.

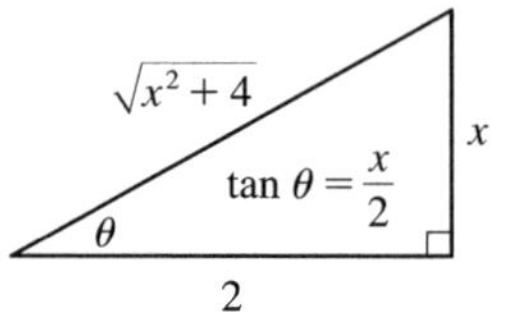

$$\int \frac{1}{\sqrt{x^2+4}}\,dx = \int \frac{2\sec^2\theta}{2\sec\theta}\,d\theta = \int \sec\theta\,d\theta = \ln|\sec\theta + \tan\theta|.$$

From the figure, $\tan\theta = \dfrac{x}{2}$, and $\sec\theta = \dfrac{\sqrt{x^2+4}}{2}$. So

$$I = \int_0^\infty \left(\frac{1}{\sqrt{x^2+4}} - \frac{C}{x+2}\right)dx = \lim_{t\to\infty}\left[\ln\left|\frac{\sqrt{x^2+4}}{2} + \frac{x}{2}\right| - C\ln|x+2|\right]_0^t$$
$$= \lim_{t\to\infty}\left[\ln\frac{\sqrt{t^2+4}+t}{2} - C\ln(t+2) - (\ln 1 - C\ln 2)\right]$$
$$= \lim_{t\to\infty}\left[\ln\left(\frac{\sqrt{t^2+4}+t}{2\,(t+2)^C}\right) + \ln 2^C\right] = \ln\left(\lim_{t\to\infty}\frac{t+\sqrt{t^2+4}}{(t+2)^C}\right) + \ln 2^{C-1}$$

Now $L = \displaystyle\lim_{t\to\infty}\frac{t+\sqrt{t^2+4}}{(t+2)^C} \overset{\text{H}}{=} \lim_{t\to\infty}\frac{1+t/\sqrt{t^2+4}}{C\,(t+2)^{C-1}} = \frac{2}{C\displaystyle\lim_{t\to\infty}(t+2)^{C-1}}$.

If $C < 1$, $L = \infty$ and I diverges.

If $C = 1$, $L = 2$ and I converges to $\ln 2 + \ln 2^0 = \ln 2$.

If $C > 1$, $L = 0$ and I diverges to $-\infty$.

5 Review

CONCEPT CHECK

1. (a) $\sum_{i=1}^n f(x_i^*)\,\Delta x$ is an expression for a Riemann sum of a function f.
x_i^* is a point in the ith subinterval $[x_{i-1}, x_i]$ and Δx is the length of the subintervals.

(b) See Figure 1 in Section 5.2.

(c) In Section 5.2, see Figure 3 and the paragraph beside it.

2. (a) See Definition 5.2.2.

(b) See Figure 2 in Section 5.2.

(c) In Section 5.2, see Figure 4 and the paragraph above it.

3. (a) See the Evaluation Theorem at the beginning of Section 5.3.

(b) See the Net Change Theorem after Example 6 in Section 5.3.

4. $\int_{t_1}^{t_2} r(t)\,dt$ represents the change in the amount of water in the reservoir between time t_1 and time t_2.

5. (a) $\int_{60}^{120} v(t)\,dt$ represents the change in position of the particle from $t = 60$ to $t = 120$ seconds.

(b) $\int_{60}^{120} |v(t)|\,dt$ represents the total distance travelled by the particle from $t = 60$ to 120 seconds.

(c) $\int_{60}^{120} a(t)\,dt$ represents the change in the velocity of the particle from $t = 60$ to $t = 120$ seconds.

6. (a) $\int f(x)\,dx$ is the family of functions $\{F \mid F' = f\}$. Any two such functions differ by a constant.

(b) The connection is given by the Evaluation Theorem: $\int_a^b f(x)\,dx = \left[\int f(x)\,dx\right]_a^b$ if f is continuous.

7. See the Fundamental Theorem of Calculus after Example 5 in Section 5.4.

8. (a) See the Substitution Rule (5.5.4). This says that it is permissible to operate with the dx after an integral sign as if it were a differential.

(b) See Formula 5.6.1 or 5.6.2. We try to choose $u = f(x)$ to be a function that becomes simpler when differentiated (or at least not more complicated) as long as $dv = g'(x)\,dx$ can be readily integrated to give v.

9. See the Midpoint Rule, the Trapezoidal Rule, and Simpson's Rule, as well as their associated error bounds, all in Section 5.9. We would expect the best estimate to be given by Simpson's Rule.

10. See Definitions 1(a), (b), and (c) in Section 5.10.

11. See Definitions 3(b), (a), and (c) in Section 5.10.

12. See the Comparison Theorem after Example 8 in Section 5.10.

13. The precise version of this statement is given by the Fundamental Theorem of Calculus. See the statement of this theorem and the paragraph that follows it in Section 5.4.

TRUE-FALSE QUIZ

1. True by Property 2 of the Integral in Section 5.2.

3. True by Property 3 of the Integral in Section 5.2.

5. False. For example, let $f(x) = x^2$. Then $\int_0^1 \sqrt{x^2}\,dx = \int_0^1 x\,dx = \frac{1}{2}$, but $\sqrt{\int_0^1 x^2\,dx} = \sqrt{\frac{1}{3}} = \frac{1}{\sqrt{3}}$.

7. True by Comparison Property 7 of the Integral in Section 5.2.

9. True. The integrand is an odd function that is continuous on $[-1, 1]$, so the result follows from Theorem 5.5.6(b).

11. False. This is an improper integral, since the denominator vanishes at $x = 1$.

$$\int_0^4 \frac{x}{x^2 - 1}\,dx = \int_0^1 \frac{x}{x^2 - 1}\,dx + \int_1^4 \frac{x}{x^2 - 1}\,dx \text{ and}$$

$$\int_0^1 \frac{x}{x^2 - 1}\,dx = \lim_{t \to 1^-} \int_0^t \frac{x}{x^2 - 1}\,dx = \lim_{t \to 1^-} \left[\tfrac{1}{2}\ln\left|x^2 - 1\right|\right]_0^t = \lim_{t \to 1^-} \tfrac{1}{2}\ln\left|t^2 - 1\right| = \infty$$

So the integral diverges.

13. False. See the remarks and Figure 4 before Example 1 in Section 5.2, and notice that $y = x - x^3 < 0$ for $1 < x \le 2$.

15. False. For example, the function $y = |x|$ is continuous on $\mathbb{R}$, but has no derivative at $x = 0$.

17. False. See Exercise 51 in Section 5.10.

19. False. If $f(x) = 1/x$, then f is continuous and decreasing on $[1, \infty)$ with $\lim_{x \to \infty} f(x) = 0$, but $\int_1^\infty f(x)\,dx$ is divergent.

21. False. Take $f(x) = 1$ for all x and $g(x) = -1$ for all x. Then $\int_a^\infty f(x)\,dx = \infty$ [divergent] and $\int_a^\infty g(x)\,dx = -\infty$ [divergent], but $\int_a^\infty [f(x) + g(x)]\,dx = 0$ [convergent].

EXERCISES

(a) 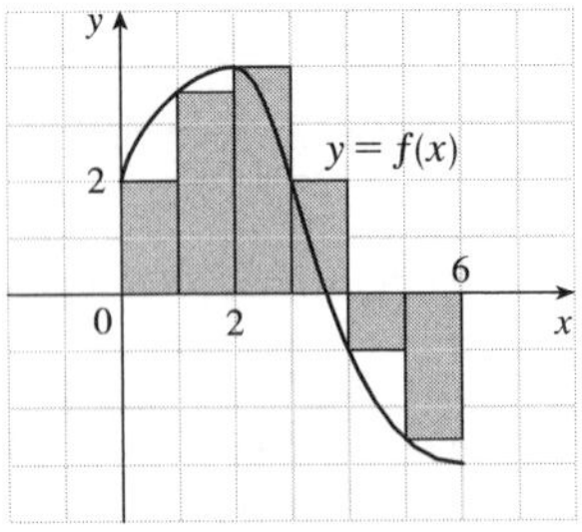

$$L_6 = \sum_{i=1}^{6} f(x_{i-1})\,\Delta x \quad [\Delta x = \tfrac{6-0}{6} = 1]$$
$$= f(x_0)\cdot 1 + f(x_1)\cdot 1 + f(x_2)\cdot 1 + f(x_3)\cdot 1 + f(x_4)\cdot 1 + f(x_5)\cdot 1$$
$$\approx 2 + 3.5 + 4 + 2 + (-1) + (-2.5) = 8$$

The Riemann sum represents the sum of the areas of the four rectangles above the x-axis minus the sum of the areas of the two rectangles below the x-axis.

(b) 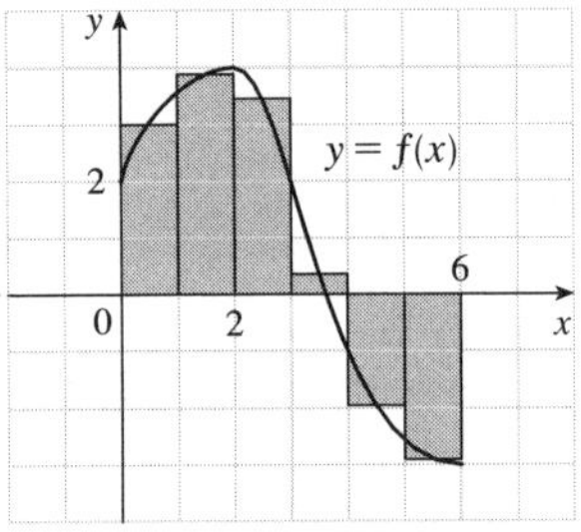

$$M_6 = \sum_{i=1}^{6} f(\overline{x}_i)\,\Delta x \quad [\Delta x = \tfrac{6-0}{6} = 1]$$
$$= f(\overline{x}_1)\cdot 1 + f(\overline{x}_2)\cdot 1 + f(\overline{x}_3)\cdot 1 + f(\overline{x}_4)\cdot 1 + f(\overline{x}_5)\cdot 1 + f(\overline{x}_6)\cdot 1$$
$$= f(0.5) + f(1.5) + f(2.5) + f(3.5) + f(4.5) + f(5.5)$$
$$\approx 3 + 3.9 + 3.4 + 0.3 + (-2) + (-2.9) = 5.7$$

The Riemann sum represents the sum of the areas of the four rectangles above the x-axis minus the sum of the areas of the two rectangles below the x-axis.

3. $\int_0^1 \left(x + \sqrt{1-x^2}\right) dx = \int_0^1 x\,dx + \int_0^1 \sqrt{1-x^2}\,dx = I_1 + I_2$.

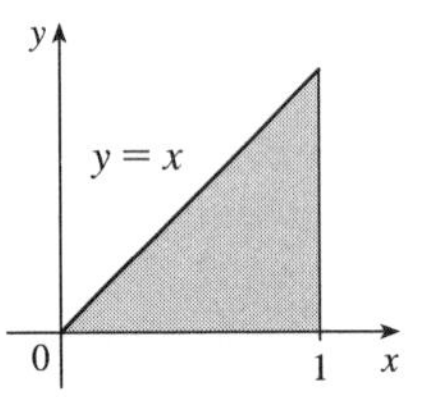

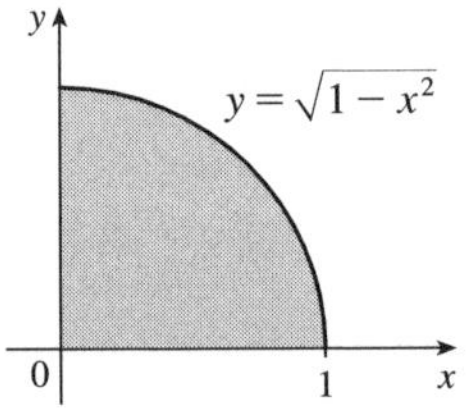

I_1 can be interpreted as the area of the triangle shown in the figure and I_2 can be interpreted as the area of the quarter-circle. Area $= \frac{1}{2}(1)(1) + \frac{1}{4}(\pi)(1)^2 = \frac{1}{2} + \frac{\pi}{4}$.

5. $\int_0^6 f(x)\,dx = \int_0^4 f(x)\,dx + \int_4^6 f(x)\,dx \;\Rightarrow\; 10 = 7 + \int_4^6 f(x)\,dx \;\Rightarrow\; \int_4^6 f(x)\,dx = 10 - 7 = 3$

7. First note that either a or b must be the graph of $\int_0^x f(t)\,dt$, since $\int_0^0 f(t)\,dt = 0$, and $c(0) \neq 0$. Now notice that $b > 0$ when c is increasing, and that $c > 0$ when a is increasing. It follows that c is the graph of $f(x)$, b is the graph of $f'(x)$, and a is the graph of $\int_0^x f(t)\,dt$.

9. $\int_1^2 (8x^3 + 3x^2)\,dx = \left[8\cdot\frac{1}{4}x^4 + 3\cdot\frac{1}{3}x^3\right]_1^2 = \left[2x^4 + x^3\right]_1^2 = (2\cdot 2^4 + 2^3) - (2+1) = 40 - 3 = 37$

11. $\int_0^1 (1 - x^9)\,dx = \left[x - \frac{1}{10}x^{10}\right]_0^1 = \left(1 - \frac{1}{10}\right) - 0 = \frac{9}{10}$

13. $\displaystyle\int_1^9 \frac{\sqrt{u} - 2u^2}{u}\,du = \int_1^9 (u^{-1/2} - 2u)\,du = \left[2u^{1/2} - u^2\right]_1^9 = (6 - 81) - (2 - 1) = -76$

15. $u = x^2 + 1$, $du = 2x\,dx$, so $\displaystyle\int_0^1 \frac{x}{x^2+1}\,dx = \int_1^2 \frac{1}{u}\left(\frac{1}{2}\,du\right) = \frac{1}{2}\left[\ln u\right]_1^2 = \frac{1}{2}\ln 2.$

17. Let $u = v^3$, so $du = 3v^2\,dv$. When $v = 0$, $u = 0$; when $v = 1$, $u = 1$. Thus,
$\int_0^1 v^2\cos(v^3)\,dv = \int_0^1 \cos u\left(\frac{1}{3}\,du\right) = \frac{1}{3}\left[\sin u\right]_0^1 = \frac{1}{3}(\sin 1 - 0) = \frac{1}{3}\sin 1.$

19. $\int_0^1 e^{\pi t}\,dt = \left[\frac{1}{\pi}e^{\pi t}\right]_0^1 = \frac{1}{\pi}(e^\pi - 1)$

21. Let $u = x^2 + 4x$. Then $du = (2x+4)\,dx = 2(x+2)\,dx$, so

$$\int \frac{x+2}{\sqrt{x^2+4x}}\,dx = \int u^{-1/2}\left(\tfrac{1}{2}\,du\right) = \tfrac{1}{2}\cdot 2u^{1/2} + C = \sqrt{u} + C = \sqrt{x^2+4x} + C.$$

23. $$\int_0^5 \frac{x}{x+10}\,dx = \int_0^5 \left(1 - \frac{10}{x+10}\right)dx = \Big[x - 10\ln(x+10)\Big]_0^5$$
$$= 5 - 10\ln 15 + 10\ln 10 = 5 + 10\ln\tfrac{10}{15} = 5 + 10\ln\tfrac{2}{3}$$

25. $$\int_0^{\pi/2} \frac{\cos\theta}{1+\sin\theta}\,d\theta = \Big[\ln(1+\sin\theta)\Big]_0^{\pi/2} = \ln 2 - \ln 1 = \ln 2$$

27. $$\int_1^4 x^{3/2}\ln x\,dx \quad \begin{bmatrix} u = \ln x, & dv = x^{3/2}\,dx, \\ du = dx/x & v = \frac{2}{5}x^{5/2} \end{bmatrix} = \frac{2}{5}\Big[x^{5/2}\ln x\Big]_1^4 - \frac{2}{5}\int_1^4 x^{3/2}\,dx$$
$$= \tfrac{2}{5}(32\ln 4 - \ln 1) - \tfrac{2}{5}\left[\tfrac{2}{5}x^{5/2}\right]_1^4$$
$$= \tfrac{2}{5}(64\ln 2) - \tfrac{4}{25}(32 - 1)$$
$$= \tfrac{128}{5}\ln 2 - \tfrac{124}{25} \quad \left(\text{or } \tfrac{64}{5}\ln 4 - \tfrac{124}{25}\right)$$

29. $\dfrac{1}{t^2+6t+8} = \dfrac{1}{(t+2)(t+4)} = \dfrac{A}{t+2} + \dfrac{B}{t+4}$. Multiply both sides by $(t+2)(t+4)$ to get $1 = A(t+4) + B(t+2)$. Substituting -4 for t gives $1 = -2B \quad\Leftrightarrow\quad B = -\frac{1}{2}$. Substituting -2 for t gives $1 = 2A \quad\Leftrightarrow\quad A = \frac{1}{2}$. Thus,

$$\int \frac{dt}{t^2+6t+8} = \int\left(\frac{1/2}{t+2} - \frac{1/2}{t+4}\right)dt = \tfrac{1}{2}\ln|t+2| - \tfrac{1}{2}\ln|t+4| + C = \tfrac{1}{2}\ln\left|\frac{t+2}{t+4}\right| + C.$$

31. Let $w = \sqrt[3]{x}$. Then $w^3 = x$ and $3w^2\,dw = dx$, so $\int e^{\sqrt[3]{x}}\,dx = \int e^w \cdot 3w^2\,dw = 3I$. To evaluate I, let $u = w^2$, $dv = e^w\,dw \quad\Rightarrow\quad du = 2w\,dw$, $v = e^w$, so $I = \int w^2 e^w\,dw = w^2e^w - \int 2we^w\,dw$. Now let $U = w$, $dV = e^w\,dw \quad\Rightarrow$ $dU = dw$, $V = e^w$. Thus, $I = w^2e^w - 2\left[we^w - \int e^w\,dw\right] = w^2e^w - 2we^w + 2e^w + C_1$, and hence $3I = 3e^w(w^2 - 2w + 2) + C = 3e^{\sqrt[3]{x}}(x^{2/3} - 2x^{1/3} + 2) + C$.

33. Let $u = 1 + \sec\theta$. Then $du = \sec\theta\,\tan\theta\,d\theta$, so

$$\int \frac{\sec\theta\,\tan\theta}{1+\sec\theta}\,d\theta = \int \frac{1}{1+\sec\theta}(\sec\theta\,\tan\theta\,d\theta) = \int \frac{1}{u}\,du = \ln|u| + C = \ln|1+\sec\theta| + C.$$

35. Let $u = 1 + \sin x$. Then $du = \cos x\,dx$, so

$$\int \frac{\cos x\,dx}{\sqrt{1+\sin x}} = \int u^{-1/2}\,du = 2u^{1/2} + C = 2\sqrt{1+\sin x} + C.$$

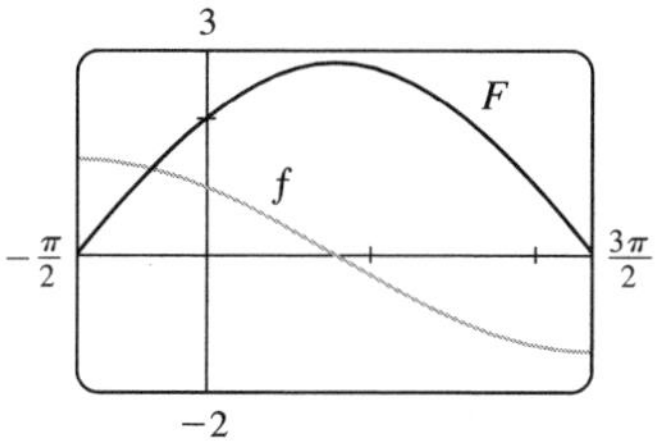

37. From the graph, it appears that the area under the curve $y = x\sqrt{x}$ between $x = 0$ and $x = 4$ is somewhat less than half the area of an 8×4 rectangle, so perhaps about 13 or 14. To find the exact value, we evaluate

$$\int_0^4 x\sqrt{x}\,dx = \int_0^4 x^{3/2}\,dx = \left[\tfrac{2}{5}x^{5/2}\right]_0^4 = \tfrac{2}{5}(4)^{5/2} = \tfrac{64}{5} = 12.8.$$

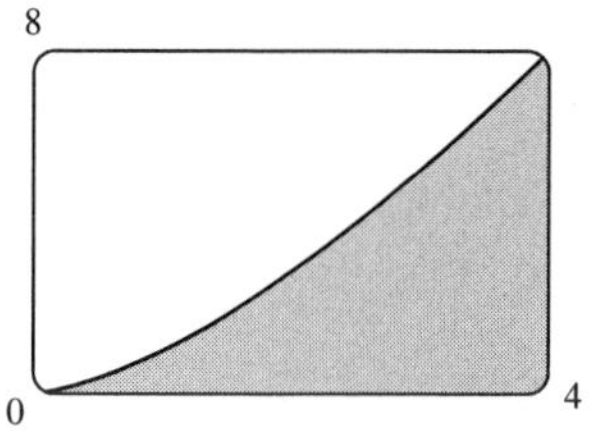

39. By FTC1, $F(x) = \int_1^x \sqrt{1+t^4}\,dt \quad\Rightarrow\quad F'(x) = \sqrt{1+x^4}$.

41. $y = \int_{\sqrt{x}}^{x} \frac{e^t}{t}\,dt = \int_{\sqrt{x}}^{1} \frac{e^t}{t}\,dt + \int_1^x \frac{e^t}{t}\,dt = -\int_1^{\sqrt{x}} \frac{e^t}{t}\,dt + \int_1^x \frac{e^t}{t}\,dt \;\Rightarrow$

$\frac{dy}{dx} = -\frac{d}{dx}\left(\int_1^{\sqrt{x}} \frac{e^t}{t}\,dt\right) + \frac{d}{dx}\left(\int_1^x \frac{e^t}{t}\,dt\right)$. Let $u = \sqrt{x}$. Then

$$\frac{d}{dx}\int_1^{\sqrt{x}} \frac{e^t}{t}\,dt = \frac{d}{dx}\int_1^u \frac{e^t}{t}\,dt = \frac{d}{du}\left(\int_1^u \frac{e^t}{t}\,dt\right)\frac{du}{dx} = \frac{e^u}{u}\cdot\frac{1}{2\sqrt{x}} = \frac{e^{\sqrt{x}}}{\sqrt{x}}\cdot\frac{1}{2\sqrt{x}} = \frac{e^{\sqrt{x}}}{2x},$$

so $\frac{dy}{dx} = -\frac{e^{\sqrt{x}}}{2x} + \frac{e^x}{x}$.

43. $u = e^x \;\Rightarrow\; du = e^x\,dx$, so

$$\int e^x\sqrt{1-e^{2x}}\,dx = \int \sqrt{1-u^2}\,du \overset{30}{=} \tfrac{1}{2}u\sqrt{1-u^2} + \tfrac{1}{2}\sin^{-1}u + C = \tfrac{1}{2}\left[e^x\sqrt{1-e^{2x}} + \sin^{-1}(e^x)\right] + C$$

45. $\int \sqrt{x^2+x+1}\,dx = \int \sqrt{x^2+x+\frac{1}{4}+\frac{3}{4}}\,dx = \int \sqrt{\left(x+\frac{1}{2}\right)^2+\frac{3}{4}}\,dx$

$$= \int \sqrt{u^2+\left(\tfrac{\sqrt{3}}{2}\right)^2}\,du \quad \left[u = x+\tfrac{1}{2},\; du = dx\right]$$

$$\overset{21}{=} \tfrac{1}{2}u\sqrt{u^2+\tfrac{3}{4}} + \tfrac{3}{8}\ln\left(u+\sqrt{u^2+\tfrac{3}{4}}\right) + C$$

$$= \frac{2x+1}{4}\sqrt{x^2+x+1} + \tfrac{3}{8}\ln\left(x+\tfrac{1}{2}+\sqrt{x^2+x+1}\right) + C$$

47. $f(x) = \sqrt{1+x^4}$, $\Delta x = \frac{b-a}{n} = \frac{1-0}{10} = \frac{1}{10}$.

(a) $T_{10} = \frac{1}{10\cdot 2}\{f(0) + 2[f(0.1)+f(0.2)+\cdots+f(0.9)] + f(1)\} \approx 1.090\,608$

(b) $M_{10} = \frac{1}{10}\left[f\left(\frac{1}{20}\right)+f\left(\frac{3}{20}\right)+f\left(\frac{5}{20}\right)+\cdots+f\left(\frac{19}{20}\right)\right] \approx 1.088\,840$

(c) $S_{10} = \frac{1}{10\cdot 3}[f(0)+4f(0.1)+2f(0.2)+\cdots+4f(0.9)+f(1)] \approx 1.089\,429$

f is concave upward, so the Trapezoidal Rule gives us an overestimate, the Midpoint Rule gives an underestimate, and we cannot tell whether Simpson's Rule gives us an overestimate or an underestimate.

49. $f(x) = \left(1+x^4\right)^{1/2}$, $f'(x) = \frac{1}{2}\left(1+x^4\right)^{-1/2}\left(4x^3\right) = 2x^3\left(1+x^4\right)^{-1/2}$, $f''(x) = \left(2x^6+6x^2\right)\left(1+x^4\right)^{-3/2}$.

A graph of f'' on $[0,1]$ shows that it has its maximum at $x=1$, so $|f''(x)| \le f''(1) = \sqrt{8}$ on $[0,1]$. By taking $K = \sqrt{8}$, we find that the error in Exercise 47(a) is bounded by $\frac{K(b-a)^3}{12n^2} = \frac{\sqrt{8}}{1200} \approx 0.0024$, and in (b) by about $\frac{1}{2}(0.0024) = 0.0012$.

Note: Another way to estimate K is to let $x=1$ in the factor $2x^6+6x^2$ (maximizing the numerator) and let $x=0$ in the factor $\left(1+x^4\right)^{-3/2}$ (minimizing the denominator). Doing so gives us $K=8$ and errors of $0.00\overline{6}$ and $0.00\overline{3}$.

Using $K=8$ for the Trapezoidal Rule, we have $|E_T| \le \frac{K(b-a)^3}{12n^2} \le 0.00001 \;\Leftrightarrow\; \frac{8(1-0)^3}{12n^2} \le \frac{1}{100\,000} \;\Leftrightarrow$

$n^2 \ge \frac{800\,000}{12} \;\Leftrightarrow\; n \gtrsim 258.2$, so we should take $n = 259$.

For the Midpoint Rule, $|E_M| \le \frac{K(b-a)^3}{24n^2} \le 0.00001 \;\Leftrightarrow\; n^2 \ge \frac{800\,000}{24} \;\Leftrightarrow\; n \gtrsim 182.6$, so we should take $n = 183$.

51. (a) $f(x) = \sin(\sin x)$. A CAS gives

$$f^{(4)}(x) = \sin(\sin x)\left[\cos^4 x + 7\cos^2 x - 3\right] + \cos(\sin x)\left[6\cos^2 x \sin x + \sin x\right]$$

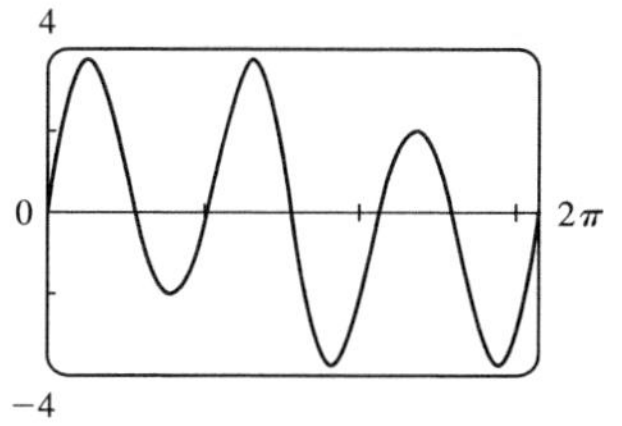

From the graph, we see that $\left|f^{(4)}(x)\right| < 3.8$ for $x \in [0, \pi]$.

(b) We use Simpson's Rule with $f(x) = \sin(\sin x)$ and $\Delta x = \frac{\pi}{10}$:

$$\int_0^\pi f(x)\,dx \approx \tfrac{\pi}{10\cdot 3}\left[f(0) + 4f\left(\tfrac{\pi}{10}\right) + 2f\left(\tfrac{2\pi}{10}\right) + \cdots + 4f\left(\tfrac{9\pi}{10}\right) + f(\pi)\right] \approx 1.786\,721$$

From part (a), we know that $\left|f^{(4)}(x)\right| < 3.8$ on $[0, \pi]$, so we use Theorem 5.9.4 with $K = 3.8$, and estimate the error as $|E_S| \le \dfrac{3.8(\pi - 0)^5}{180(10)^4} \approx 0.000\,646$.

(c) If we want the error to be less than 0.00001, we must have $|E_S| \le \frac{3.8\pi^5}{180n^4} \le 0.00001$, so $n^4 \ge \frac{3.8\pi^5}{180(0.00001)} \approx 646\,041.6 \quad\Rightarrow\quad n \ge 28.35$. Since n must be even for Simpson's Rule, we must have $n \ge 30$ to ensure the desired accuracy.

53. If $1 \le x \le 3$, then $\sqrt{1^2+3} \le \sqrt{x^2+3} \le \sqrt{3^2+3} \quad\Rightarrow\quad 2 \le \sqrt{x^2+3} \le 2\sqrt{3}$, so $2(3-1) \le \int_1^3 \sqrt{x^2+3}\,dx \le 2\sqrt{3}(3-1)$; that is, $4 \le \int_1^3 \sqrt{x^2+3}\,dx \le 4\sqrt{3}$.

55.
$$\int_1^\infty \frac{1}{(2x+1)^3}\,dx = \lim_{t\to\infty}\int_1^t \frac{1}{(2x+1)^3}\,dx = \lim_{t\to\infty}\int_1^t \tfrac{1}{2}(2x+1)^{-3}\,2\,dx$$
$$= \lim_{t\to\infty}\left[-\frac{1}{4(2x+1)^2}\right]_1^t = -\frac{1}{4}\lim_{t\to\infty}\left[\frac{1}{(2t+1)^2} - \frac{1}{9}\right] = -\frac{1}{4}\left(0 - \frac{1}{9}\right) = \frac{1}{36}$$

57. $\int_{-\infty}^0 e^{-2x}\,dx = \lim\limits_{t\to-\infty}\int_t^0 e^{-2x}\,dx = \lim\limits_{t\to-\infty}\left[-\frac{1}{2}e^{-2x}\right]_t^0 = \lim\limits_{t\to-\infty}\left(-\frac{1}{2} + \frac{1}{2}e^{-2t}\right) = \infty$. Divergent

59. Let $u = \ln x$. Then $du = dx/x$, so $\displaystyle\int \frac{dx}{x\sqrt{\ln x}} = \int \frac{du}{\sqrt{u}} = 2\sqrt{u} + C = 2\sqrt{\ln x} + C$.

Thus, $\displaystyle\int_1^e \frac{dx}{x\sqrt{\ln x}} = \lim_{t\to 1^+}\int_t^e \frac{dx}{x\sqrt{\ln x}} = \lim_{t\to 1^+}\left[2\sqrt{\ln x}\right]_t^e = \lim_{t\to 1^+}\left(2\sqrt{\ln e} - 2\sqrt{\ln t}\right) = 2\cdot 1 - 2\cdot 0 = 2.$

61. $\dfrac{x^3}{x^5+2} \le \dfrac{x^3}{x^5} = \dfrac{1}{x^2}$ for x in $[1, \infty)$. $\displaystyle\int_1^\infty \frac{1}{x^2}\,dx$ is convergent by (5.10.2) with $p = 2 > 1$. Therefore, $\displaystyle\int_1^\infty \frac{x^3}{x^5+2}\,dx$ is convergent by the Comparison Theorem.

63. (a) Displacement $= \int_0^5 (t^2 - t)\,dt = \left[\frac{1}{3}t^3 - \frac{1}{2}t^2\right]_0^5 = \frac{125}{3} - \frac{25}{2} = \frac{175}{6} = 29.1\overline{6}$ metres

(b) Distance travelled $= \int_0^5 \left|t^2 - t\right|dt = \int_0^5 |t(t-1)|\,dt = \int_0^1 (t - t^2)\,dt + \int_1^5 (t^2 - t)\,dt$

$$= \left[\tfrac{1}{2}t^2 - \tfrac{1}{3}t^3\right]_0^1 + \left[\tfrac{1}{3}t^3 - \tfrac{1}{2}t^2\right]_1^5$$
$$= \tfrac{1}{2} - \tfrac{1}{3} - 0 + \left(\tfrac{125}{3} - \tfrac{25}{2}\right) - \left(\tfrac{1}{3} - \tfrac{1}{2}\right) = \tfrac{177}{6} = 29.5 \text{ metres}$$

65. Note that $r(t) = b'(t)$, where $b(t) =$ the number of barrels of oil consumed up to time t. So, by the Net Change Theorem, $\int_0^3 r(t)\,dt = b(3) - b(0)$ represents the number of barrels of oil consumed from Jan. 1, 2000, through Jan. 1, 2003.

67. Both numerator and denominator approach 0 as $a \to 0$, so we use l'Hospital's Rule. (Note that we are differentiating *with respect to* a, since that is the quantity which is changing.) We also use FTC1:

$$\lim_{a\to 0} T(x,t) = \lim_{a\to 0} \frac{C\int_0^a e^{-(x-u)^2/(4kt)}\,du}{a\sqrt{4\pi kt}} \overset{\text{H}}{=} \lim_{a\to 0} \frac{Ce^{-(x-a)^2/(4kt)}}{\sqrt{4\pi kt}} = \frac{Ce^{-x^2/(4kt)}}{\sqrt{4\pi kt}}$$

69. Using FTC1, we differentiate both sides of the given equation, $\int_0^x f(t)\,dt = xe^{2x} + \int_0^x e^{-t} f(t)\,dt$, and get

$$f(x) = e^{2x} + 2xe^{2x} + e^{-x} f(x) \quad\Rightarrow\quad f(x)\left(1 - e^{-x}\right) = e^{2x} + 2xe^{2x} \quad\Rightarrow\quad f(x) = \frac{e^{2x}(1+2x)}{1-e^{-x}}.$$

71. Let $u = f(x)$ and $du = f'(x)\,dx$. So $2\int_a^b f(x)f'(x)\,dx = 2\int_{f(a)}^{f(b)} u\,du = \left[u^2\right]_{f(a)}^{f(b)} = [f(b)]^2 - [f(a)]^2$.

73. By the Fundamental Theorem of Calculus,

$$\int_0^\infty f'(x)\,dx = \lim_{t\to\infty} \int_0^t f'(x)\,dx = \lim_{t\to\infty} [f(t) - f(0)] = \lim_{t\to\infty} f(t) - f(0) = 0 - f(0) = -f(0).$$

FOCUS ON PROBLEM SOLVING

1.

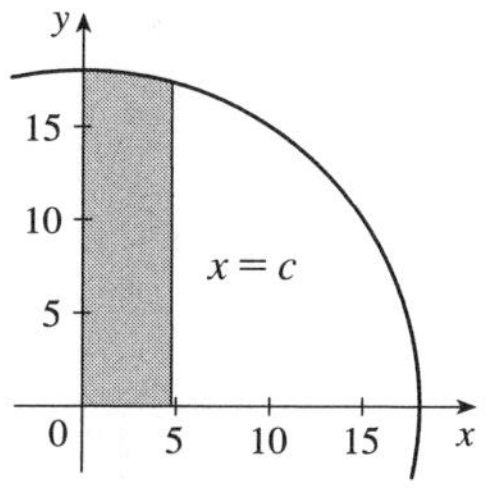

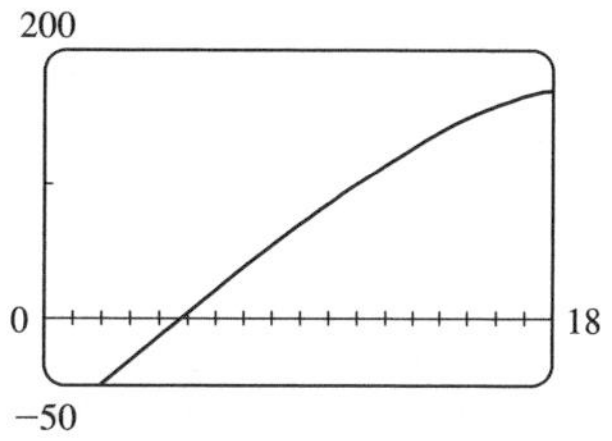

By symmetry, the problem can be reduced to finding the line $x = c$ such that the shaded area is one-third of the area of the quarter-circle. An equation of the semicircle is $y = \sqrt{18^2 - x^2}$, so we require that $\int_0^c \sqrt{324 - x^2}\,dx = \frac{1}{3} \cdot \frac{1}{4}\pi(18)^2 \quad \Leftrightarrow$ $\left[\frac{1}{2}x\sqrt{324 - x^2} + 162\sin^{-1}(x/18)\right]_0^c = 27\pi$ [by Formula 30] $\Leftrightarrow \quad \frac{1}{2}c\sqrt{324 - c^2} + 162\sin^{-1}(c/18) = 27\pi$.

This equation would be difficult to solve exactly, so we plot the left-hand side as a function of c, and find that the equation holds for $c \approx 4.77$. So the cuts should be made at distances of about 4.77 cm from the centre of the pizza.

3. Differentiating both sides of the equation $x\sin\pi x = \int_0^{x^2} f(t)\,dt$ (using FTC1 and the Chain Rule for the right side) gives $\sin\pi x + \pi x\cos\pi x = 2xf(x^2)$. Letting $x = 2$ so that $f(x^2) = f(4)$, we obtain $\sin 2\pi + 2\pi\cos 2\pi = 4f(4)$, so $f(4) = \frac{1}{4}(0 + 2\pi \cdot 1) = \frac{\pi}{2}$.

5. By FTC2, $\int_0^1 f'(x)\,dx = f(1) - f(0) = 1 - 0 = 1$.

7. Differentiating the given equation, $\int_0^x f(t)\,dt = [f(x)]^2$, using FTC1 gives $f(x) = 2f(x)f'(x) \quad \Rightarrow$ $f(x)\,[2f'(x) - 1] = 0$, so $f(x) = 0$ or $f'(x) = \frac{1}{2}$. Since $f(x)$ is never 0, we must have $f'(x) = \frac{1}{2}$ and $f'(x) = \frac{1}{2} \quad \Rightarrow$ $f(x) = \frac{1}{2}x + C$. To find C, we substitute into the given equation to get $\int_0^x (\frac{1}{2}t + C)\,dt = (\frac{1}{2}x + C)^2 \quad \Leftrightarrow$ $\frac{1}{4}x^2 + Cx = \frac{1}{4}x^2 + Cx + C^2$. It follows that $C^2 = 0$, so $C = 0$, and $f(x) = \frac{1}{2}x$.

9. By l'Hospital's Rule and the Fundamental Theorem, using the notation $\exp(y) = e^y$,

$$\lim_{x \to 0} \frac{\int_0^x (1 - \tan 2t)^{1/t}\,dt}{x} \overset{\text{H}}{=} \lim_{x \to 0} \frac{(1 - \tan 2x)^{1/x}}{1} = \exp\left(\lim_{x \to 0} \frac{\ln(1 - \tan 2x)}{x}\right)$$

$$\overset{\text{H}}{=} \exp\left(\lim_{x \to 0} \frac{-2\sec^2 2x}{1 - \tan 2x}\right) = \exp\left(\frac{-2 \cdot 1^2}{1 - 0}\right) = e^{-2}.$$

11. Such a function cannot exist. $f'(x) > 3$ for all x means that f is differentiable (and hence continuous) for all x. So by FTC2, $\int_1^4 f'(x)\,dx = f(4) - f(1) = 7 - (-1) = 8$. However, if $f'(x) > 3$ for all x, then $\int_1^4 f'(x)\,dx \ge 3 \cdot (4 - 1) = 9$ by Comparison Property 8 in Section 5.2.

Another solution: By the Mean Value Theorem, there exists a number $c \in (1, 4)$ such that

$f'(c) = \dfrac{f(4) - f(1)}{4 - 1} = \dfrac{7 - (-1)}{3} = \dfrac{8}{3} \quad \Rightarrow \quad 8 = 3f'(c)$. But $f'(x) > 3 \quad \Rightarrow \quad 3f'(c) > 9$, so such a function cannot exist.

13. $f(x) = 2 + x - x^2 = (-x + 2)(x + 1) = 0 \quad \Leftrightarrow \quad x = 2$ or $x = -1$. $f(x) \geq 0$ for $x \in [-1, 2]$ and $f(x) < 0$ everywhere else. The integral $\int_a^b (2 + x - x^2)\, dx$ has a maximum on the interval where the integrand is positive, which is $[-1, 2]$. So $a = -1$, $b = 2$. (Any larger interval gives a smaller integral since $f(x) < 0$ outside $[-1, 2]$. Any smaller interval also gives a smaller integral since $f(x) \geq 0$ in $[-1, 2]$.)

15. By FTC1, $\dfrac{d}{dx}\displaystyle\int_0^x \left(\int_1^{\sin t} \sqrt{1 + u^4}\, du\right) dt = \int_1^{\sin x} \sqrt{1 + u^4}\, du$. Again using FTC1,

$$\frac{d^2}{dx^2}\int_0^x \left(\int_1^{\sin t} \sqrt{1 + u^4}\, du\right) dt = \frac{d}{dx}\int_1^{\sin x} \sqrt{1 + u^4}\, du = \sqrt{1 + \sin^4 x}\,\cos x$$

17. The given integral represents the difference of the shaded areas, which appears to be 0. It can be calculated by integrating with respect to either x or y, so we find x in terms of y for each curve: $y = \sqrt[3]{1 - x^7} \quad \Rightarrow \quad x = \sqrt[7]{1 - y^3}$ and $y = \sqrt[7]{1 - x^3} \quad \Rightarrow \quad x = \sqrt[3]{1 - y^7}$, so

$\int_0^1 \left(\sqrt[3]{1 - y^7} - \sqrt[7]{1 - y^3}\right) dy = \int_0^1 \left(\sqrt[7]{1 - x^3} - \sqrt[3]{1 - x^7}\right) dx$. But this equation is of the form $z = -z$. So $\int_0^1 \left(\sqrt[3]{1 - x^7} - \sqrt[7]{1 - x^3}\right) dx = 0$.

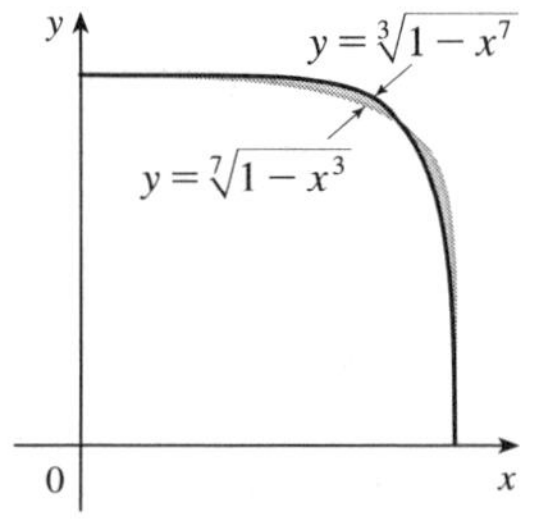

19. In accordance with the hint, we let $I_k = \int_0^1 (1 - x^2)^k\, dx$, and we find an expression for I_{k+1} in terms of I_k. We integrate I_{k+1} by parts with $u = (1 - x^2)^{k+1}$, $dv = dx \quad \Rightarrow \quad du = (k + 1)(1 - x^2)^k(-2x)$, $v = x$, and then split the remaining integral into identifiable quantities:

$$\begin{aligned} I_{k+1} &= \left[x(1 - x^2)^{k+1}\right]_0^1 + 2(k + 1)\int_0^1 x^2(1 - x^2)^k\, dx = (2k + 2)\int_0^1 (1 - x^2)^k \left[1 - (1 - x^2)\right] d \\ &= (2k + 2)(I_k - I_{k+1}) \end{aligned}$$

So $I_{k+1}\left[1 + (2k + 2)\right] = (2k + 2)I_k \Rightarrow I_{k+1} = \dfrac{2k + 2}{2k + 3} I_k$.

Now to complete the proof, we use induction: $I_0 = 1 = \dfrac{2^0 (0!)^2}{1!}$, so the formula holds for $n = 0$. Now suppose it holds for $n = k$. Then

$$\begin{aligned} I_{k+1} &= \frac{2k + 2}{2k + 3} I_k = \frac{2k + 2}{2k + 3}\left[\frac{2^{2k}(k!)^2}{(2k + 1)!}\right] = \frac{2(k + 1)2^{2k}(k!)^2}{(2k + 3)(2k + 1)!} = \frac{2(k + 1)2^{2k}(k!)^2}{(2k + 3)(2k + 1)!} \cdot \frac{2(k + 1)}{2k + 2} \\ &= \frac{[2(k + 1)]^2\, 2^{2k}(k!)^2}{(2k + 3)(2k + 2)(2k + 1)!} = \frac{2^{2(k+1)}\,[(k + 1)!]^2}{[2(k + 1) + 1]!} \end{aligned}$$

So by induction, the formula holds for all integers $n \geq 0$.

21. (a) The tangent to the curve $y = f(x)$ at $x = x_0$ has the equation $y - f(x_0) = f'(x_0)(x - x_0)$. The y-intercept of this tangent line is $f(x_0) - f'(x_0)x_0$. Thus, L is the distance from the point $(0, f(x_0) - f'(x_0)x_0)$ to the point $(x_0, f(x_0))$; that is, $L^2 = x_0^2 + [f'(x_0)]^2 x_0^2$, so $[f'(x_0)]^2 = \dfrac{L^2 - x_0^2}{x_0^2}$ and $f'(x_0) = -\dfrac{\sqrt{L^2 - x_0^2}}{x_0}$ for $0 < x_0 < L$.

(b) $\dfrac{dy}{dx} = -\dfrac{\sqrt{L^2 - x^2}}{x} \quad \Rightarrow \quad y = \displaystyle\int \left(-\frac{\sqrt{L^2 - x^2}}{x} \right) dx.$

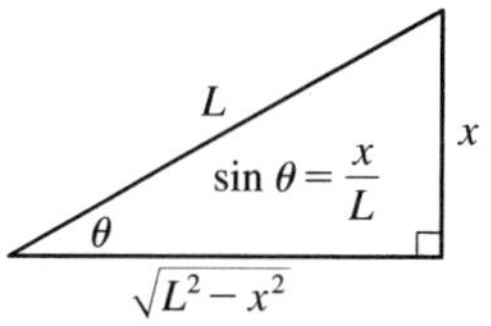

Let $x = L\sin\theta$. Then $dx = L\cos\theta\, d\theta$ and

$$y = \int \frac{-L\cos\theta\, L\cos\theta\, d\theta}{L\sin\theta} = L\int \frac{\sin^2\theta - 1}{\sin\theta}\, d\theta = L\int (\sin\theta - \csc\theta)\, d\theta$$

$$= -L\cos\theta - L\ln|\csc\theta - \cot\theta| + C = -\sqrt{L^2 - x^2} - L\ln\left(\frac{L}{x} - \frac{\sqrt{L^2 - x^2}}{x} \right) + C$$

When $x = L$, $y = 0$, and $0 = -0 - L\ln(1 - 0) + C$, so $C = 0$. Therefore, $y = -\sqrt{L^2 - x^2} - L\ln\left(\dfrac{L - \sqrt{L^2 - x^2}}{x} \right)$.

6 □ APPLICATIONS OF INTEGRATION

6.1 More about Areas

1. $A = \int_{x=0}^{x=4} (y_T - y_B)\,dx = \int_0^4 \left[(5x - x^2) - x\right] dx = \int_0^4 (4x - x^2)\,dx$

$= \left[2x^2 - \frac{1}{3}x^3\right]_0^4 = \left(32 - \frac{64}{3}\right) - (0) = \frac{32}{3}$

3. $A = \int_{y=-1}^{y=1} (x_R - x_L)\,dy = \int_{-1}^{1} \left[e^y - (y^2 - 2)\right] dy$

$= \int_{-1}^{1} (e^y - y^2 + 2)\,dy = \left[e^y - \frac{1}{3}y^3 + 2y\right]_{-1}^{1} = \left(e^1 - \frac{1}{3} + 2\right) - \left(e^{-1} + \frac{1}{3} - 2\right) = e - \dfrac{1}{e} + \dfrac{10}{3}$

5. $A = \int_{-1}^{2} \left[(9 - x^2) - (x + 1)\right] dx$

$= \int_{-1}^{2} (8 - x - x^2)\,dx$

$= \left[8x - \dfrac{x^2}{2} - \dfrac{x^3}{3}\right]_{-1}^{2}$

$= \left(16 - 2 - \frac{8}{3}\right) - \left(-8 - \frac{1}{2} + \frac{1}{3}\right)$

$= 22 - 3 + \frac{1}{2} = \frac{39}{2}$

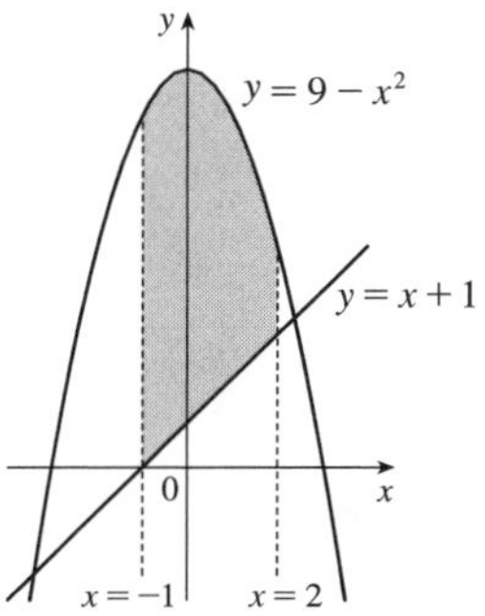

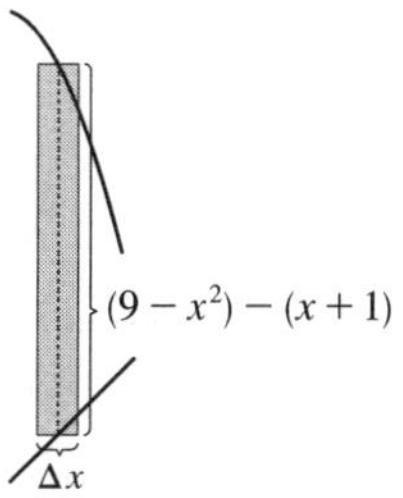

7. The curves intersect when $x = x^2 \;\Rightarrow\; x^2 - x = 0 \;\Leftrightarrow\; x(x-1) = 0 \;\Leftrightarrow\; x = 0, 1$.

$A = \int_0^1 (x - x^2)\,dx = \left[\frac{1}{2}x^2 - \frac{1}{3}x^3\right]_0^1$

$= \frac{1}{2} - \frac{1}{3} = \frac{1}{6}$

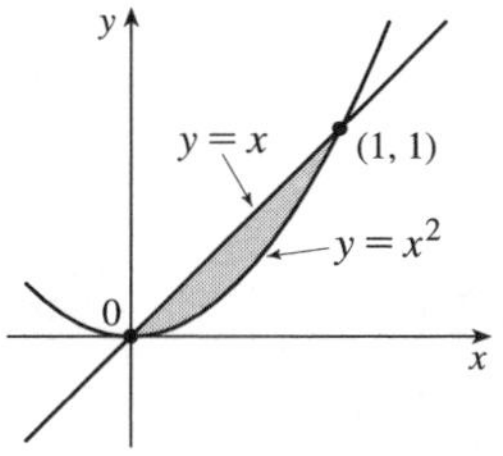

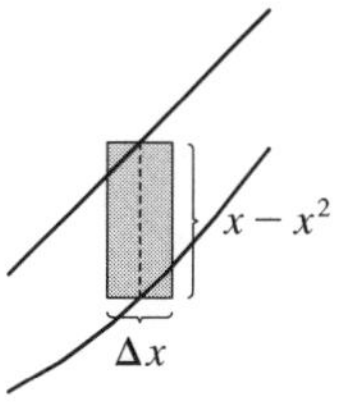

9. $12 - x^2 = x^2 - 6 \;\Leftrightarrow\; 2x^2 = 18 \;\Leftrightarrow\; x^2 = 9 \;\Leftrightarrow\; x = \pm 3$, so

$$A = \int_{-3}^{3} \left[(12 - x^2) - (x^2 - 6)\right] dx = 2\int_0^3 (18 - 2x^2)\,dx \qquad \text{[by symmetry]}$$

$$= 2\left[18x - \tfrac{2}{3}x^3\right]_0^3 = 2\left[(54 - 18) - 0\right] = 2(36) = 72$$

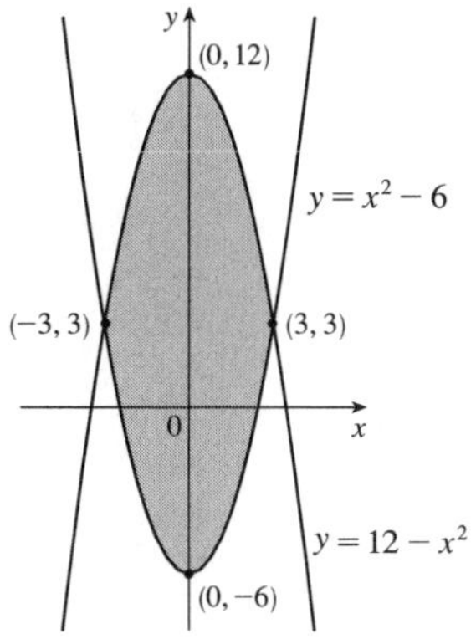

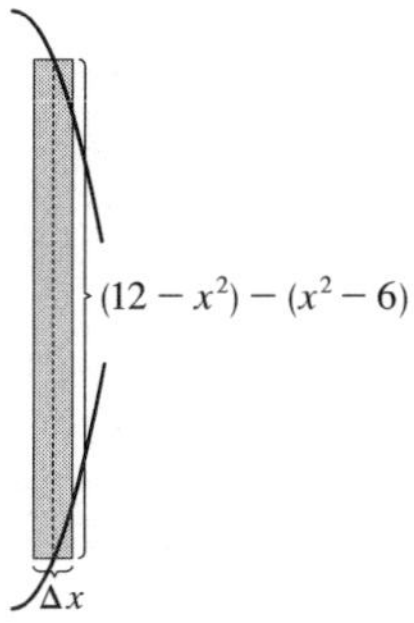

11. $2y^2 = 1 - y \quad \Leftrightarrow \quad 2y^2 + y - 1 = 0 \quad \Leftrightarrow \quad (2y-1)(y+1) = 0 \quad \Leftrightarrow \quad y = \frac{1}{2}$ or -1, so $x = \frac{1}{2}$ or 2 and

$$A = \int_{-1}^{1/2} \left[(1-y) - 2y^2\right] dy = \int_{-1}^{1/2} (1 - y - 2y^2)\, dy = \left[y - \tfrac{1}{2}y^2 - \tfrac{2}{3}y^3\right]_{-1}^{1/2}$$

$$= \left(\tfrac{1}{2} - \tfrac{1}{8} - \tfrac{1}{12}\right) - \left(-1 - \tfrac{1}{2} + \tfrac{2}{3}\right) = \tfrac{7}{24} - \left(-\tfrac{5}{6}\right) = \tfrac{7}{24} + \tfrac{20}{24} = \tfrac{27}{24} = \tfrac{9}{8}$$

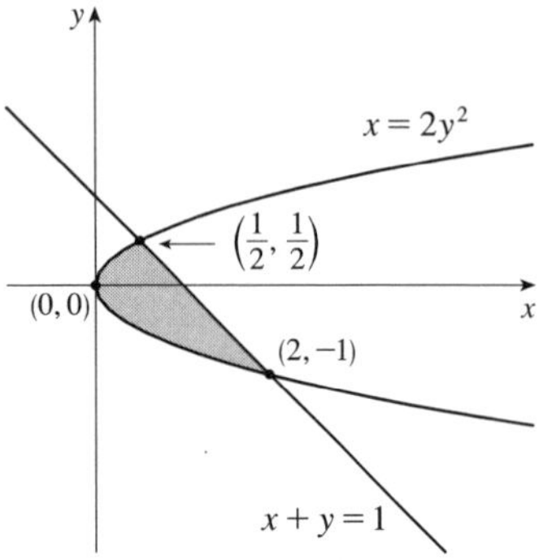

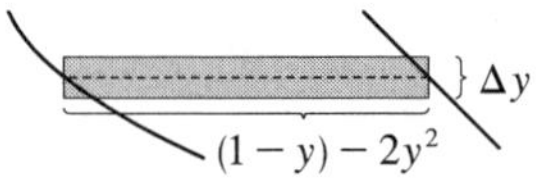

13. The curves intersect when $1 - y^2 = y^2 - 1 \quad \Leftrightarrow \quad 2 = 2y^2 \quad \Leftrightarrow \quad y^2 = 1 \quad \Leftrightarrow \quad y = \pm 1$.

$$\begin{aligned} A &= \int_{-1}^{1} \left[(1-y^2) - (y^2-1)\right] dy \\ &= \int_{-1}^{1} 2(1-y^2)\, dy = 2 \cdot 2 \int_{0}^{1} (1-y^2)\, dy \\ &= 4\left[y - \tfrac{1}{3}y^3\right]_0^1 = 4\left(1 - \tfrac{1}{3}\right) = \tfrac{8}{3} \end{aligned}$$

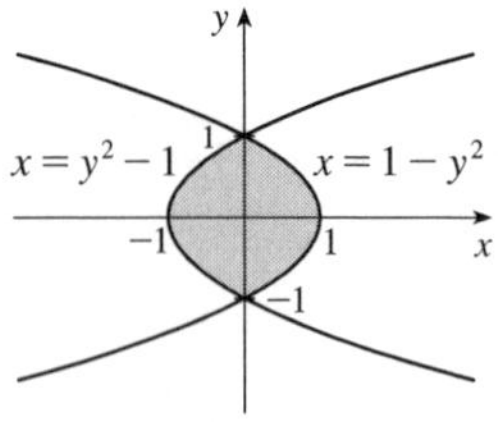

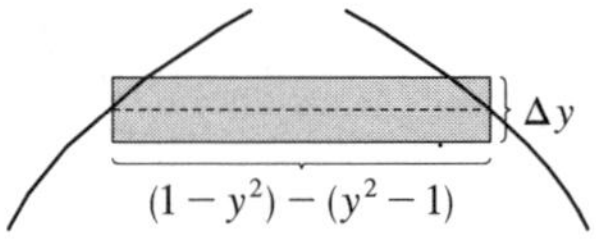

15. $1/x = x \quad \Leftrightarrow \quad 1 = x^2 \quad \Leftrightarrow \quad x = \pm 1$ and $1/x = \frac{1}{4}x \quad \Leftrightarrow \quad 4 = x^2 \quad \Leftrightarrow \quad x = \pm 2$, so for $x > 0$,

$$A = \int_0^1 \left(x - \frac{1}{4}x\right) dx + \int_1^2 \left(\frac{1}{x} - \frac{1}{4}x\right) dx = \int_0^1 \left(\frac{3}{4}x\right) dx + \int_1^2 \left(\frac{1}{x} - \frac{1}{4}x\right) dx = \left[\frac{3}{8}x^2\right]_0^1 + \left[\ln|x| - \frac{1}{8}x^2\right]_1^2$$

$$= \tfrac{3}{8} + \left(\ln 2 - \tfrac{1}{2}\right) - \left(0 - \tfrac{1}{8}\right) = \ln 2$$

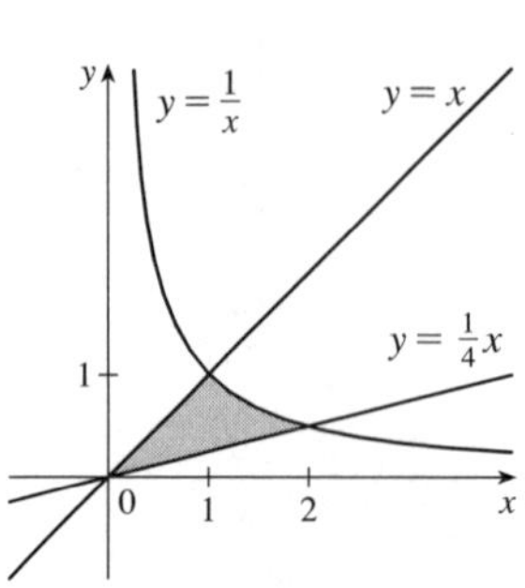

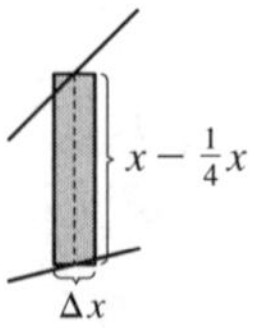

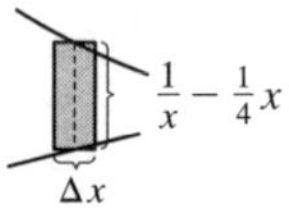

17.

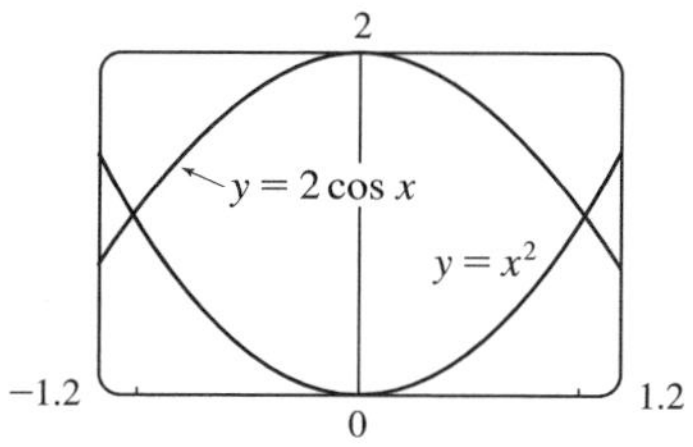

From the graph, we see that the curves intersect at $x = \pm a \approx \pm 1.02$, with $2\cos x > x^2$ on $(-a, a)$. So the area of the region bounded by the curves is

$$A = \int_{-a}^{a} (2\cos x - x^2)\,dx = 2\int_0^a (2\cos x - x^2)\,dx$$

$$= 2\left[2\sin x - \tfrac{1}{3}x^3\right]_0^a \approx 2.70$$

19.

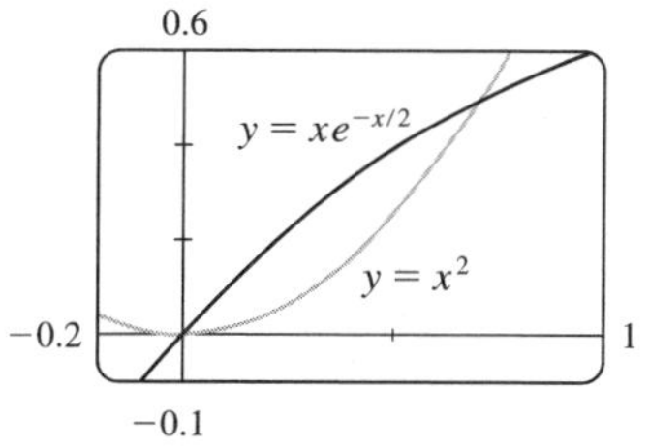

From the graph, we see that the curves intersect at $x = 0$ and $x = a \approx 0.70$, with $xe^{-x/2} > x^2$ on $(0, a)$. So the area A of the region bounded by the curves is

$$A = \int_0^a (xe^{-x/2} - x^2)\,dx$$

$$= \left[4\left(-\tfrac{1}{2}x - 1\right)e^{-x/2} - \tfrac{1}{3}x^3\right]_0^a \quad \text{[Formula 96 with } a = -\tfrac{1}{2}\text{]}$$

$$\approx 0.08$$

21. As in Example 4, we approximate the distance between the two cars after ten seconds using Simpson's Rule with $\Delta t = 1\text{ s} = \frac{1}{3600}\text{ h}$.

$$\text{distance}_{\text{Kelly}} - \text{distance}_{\text{Chris}} = \int_0^{10} v_K\,dt - \int_0^{10} v_C\,dt = \int_0^{10} (v_K - v_C)\,dt \approx S_{10}$$

$$= \tfrac{1}{3 \cdot 3600}[(0-0) + 4(35-32) + 2(59-51) + 4(83-74) + 2(98-86) + 4(114-99)$$
$$+ 2(128-110) + 4(138-120) + 2(150-130) + 4(157-138) + (163-144)]$$

$$= \tfrac{1}{10\,800}(391) \approx 0.0362\text{ km}$$

So after 10 seconds, Kelly's car is about 36 metres ahead of Chris's.

23. If x = distance from left end of pool and $w = w(x)$ = width at x, then Simpson's Rule with $n = 8$ and $\Delta x = 2$ gives

$\text{Area} = \int_0^{16} w\,dx \approx \frac{2}{3}[0 + 4(6.2) + 2(7.2) + 4(6.8) + 2(5.6) + 4(5.0) + 2(4.8) + 4(4.8) + 0] = \frac{2}{3}(126.4) \approx 84\text{ m}^2$.

25. For $0 \le t \le 10$, $b(t) > d(t)$, so the area between the curves is given by

$$\int_0^{10} [b(t) - d(t)]\,dt = \int_0^{10} (2200e^{0.024t} - 1460e^{0.018t})\,dt = \left[\frac{2200}{0.024}e^{0.024t} - \frac{1460}{0.018}e^{0.018t}\right]_0^{10}$$

$$= \left(\frac{275\,000}{3}e^{0.24} - \frac{730\,000}{9}e^{0.18}\right) - \left(\frac{275\,000}{3} - \frac{730\,000}{9}\right) \approx 8868\text{ people}$$

This area represents the increase in population over a 10-year period.

27. $\cos x = \sin 2x = 2\sin x\cos x \;\Leftrightarrow\; 2\sin x\cos x - \cos x = 0 \;\Leftrightarrow\; \cos x\,(2\sin x - 1) = 0 \;\Leftrightarrow$
$2\sin x = 1$ or $\cos x = 0 \;\Leftrightarrow\; x = \frac{\pi}{6}$ or $\frac{\pi}{2}$.

$$A = \int_0^{\pi/6} (\cos x - \sin 2x)\,dx + \int_{\pi/6}^{\pi/2} (\sin 2x - \cos x)\,dx$$

$$= \left[\sin x + \tfrac{1}{2}\cos 2x\right]_0^{\pi/6} + \left[-\tfrac{1}{2}\cos 2x - \sin x\right]_{\pi/6}^{\pi/2}$$

$$= \left(\tfrac{1}{2} + \tfrac{1}{2}\cdot\tfrac{1}{2}\right) - \left(0 + \tfrac{1}{2}\cdot 1\right) + \left[-\tfrac{1}{2}\cdot(-1) - 1\right] - \left(-\tfrac{1}{2}\cdot\tfrac{1}{2} - \tfrac{1}{2}\right)$$

$$= \tfrac{3}{4} - \tfrac{1}{2} - \tfrac{1}{2} + \tfrac{3}{4} = \tfrac{1}{2}$$

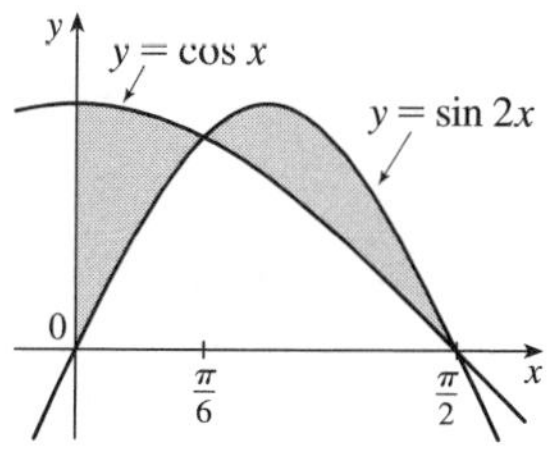

29. Let the equation of the large circle be $x^2 + y^2 = R^2$. Then the equation of the small circle is $x^2 + (y - b)^2 = r^2$, where $b = \sqrt{R^2 - r^2}$ is the distance between the centres of the circles. The desired area is

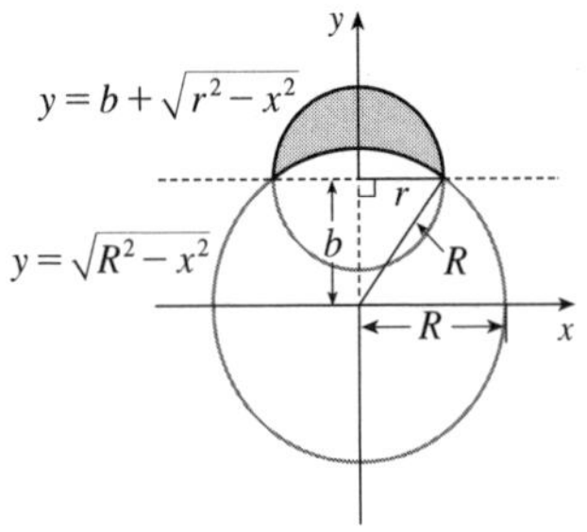

$$A = \int_{-r}^{r} \left[\left(b + \sqrt{r^2 - x^2}\,\right) - \sqrt{R^2 - x^2}\,\right] dx$$

$$= 2\int_0^r \left(b + \sqrt{r^2 - x^2} - \sqrt{R^2 - x^2}\,\right) dx$$

$$= 2\int_0^r b\,dx + 2\int_0^r \sqrt{r^2 - x^2}\,dx - 2\int_0^r \sqrt{R^2 - x^2}\,dx$$

The first integral is just $2br = 2r\sqrt{R^2 - r^2}$. The second integral represents the area of a quarter-circle of radius r, so its value is $\frac{1}{4}\pi r^2$. To evaluate the other integral, note that

$$\int \sqrt{a^2 - x^2}\,dx = \int a^2 \cos^2\theta\,d\theta \;\; [x = a\sin\theta,\, dx = a\cos\theta\,d\theta] \;\; = \left(\tfrac{1}{2}a^2\right)\int (1 + \cos 2\theta)\,d\theta$$

$$= \tfrac{1}{2}a^2\left(\theta + \tfrac{1}{2}\sin 2\theta\right) + C = \tfrac{1}{2}a^2(\theta + \sin\theta\,\cos\theta) + C$$

$$= \tfrac{a^2}{2}\arcsin\left(\tfrac{x}{a}\right) + \tfrac{a^2}{2}\left(\tfrac{x}{a}\right)\tfrac{\sqrt{a^2 - x^2}}{a} + C = \tfrac{a^2}{2}\arcsin\left(\tfrac{x}{a}\right) + \tfrac{x}{2}\sqrt{a^2 - x^2} + C$$

Thus, the desired area is

$$A = 2r\sqrt{R^2 - r^2} + 2\left(\tfrac{1}{4}\pi r^2\right) - \left[R^2\arcsin(x/R) + x\sqrt{R^2 - x^2}\,\right]_0^r$$

$$= 2r\sqrt{R^2 - r^2} + \tfrac{1}{2}\pi r^2 - \left[R^2\arcsin(r/R) + r\sqrt{R^2 - r^2}\,\right] = r\sqrt{R^2 - r^2} + \tfrac{\pi}{2}r^2 - R^2\arcsin(r/R)$$

31. By symmetry of the ellipse about the x- and y-axes,

$$A = 4\int_0^a y\,dx = 4\int_{\pi/2}^0 b\sin\theta\,(-a\sin\theta)\,d\theta \qquad \begin{bmatrix} x = a\cos\theta = 0 & \Rightarrow & \theta = \frac{\pi}{2}, \\ x = a\cos\theta = a & \Rightarrow & \theta = 0 \end{bmatrix}$$

$$= 4ab\int_0^{\pi/2} \sin^2\theta\,d\theta = 4ab\int_0^{\pi/2} \tfrac{1}{2}(1 - \cos 2\theta)\,d\theta$$

$$= 2ab\left[\theta - \tfrac{1}{2}\sin 2\theta\right]_0^{\pi/2} = 2ab\left(\tfrac{\pi}{2}\right) = \pi ab$$

Note that the formula for the area of a circle, $A = \pi r^2$, is just a special case of this formula with $a = b = r$.

33. 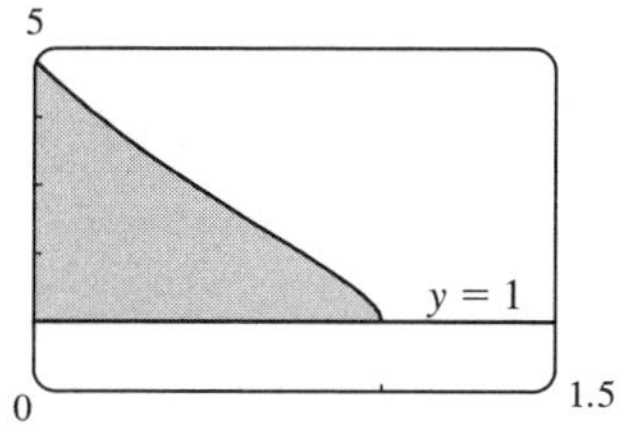

$x = \cos t$, $y = e^t$, $0 \le t \le \frac{\pi}{2}$. When $x = 0$, $t = \pi/2$; when $x = 1$, $t = 0$.

$$A = \int_0^1 (y-1)\,dx = \int_{\pi/2}^0 (e^t - 1)(-\sin t)\,dt$$
$$= \int_0^{\pi/2} (e^t \sin t - \sin t)\,dt$$

By Example 4 in Section 5.6 or Integration Formula 98, $\int e^x \sin x\,dx = \frac{1}{2}e^x(\sin x - \cos x) + C$, so

$$\int_0^{\pi/2} (e^t \sin t - \sin t)\,dt = \left[\tfrac{1}{2}e^t(\sin t - \cos t) + \cos t\right]_0^{\pi/2} = \left[\tfrac{1}{2}e^{\pi/2}(1-0) + 0\right] - \left[\tfrac{1}{2}e^0(0-1) + 1\right]$$
$$= \tfrac{1}{2}e^{\pi/2} - \tfrac{1}{2} = \tfrac{1}{2}(e^{\pi/2} - 1)$$

35. By symmetry, the area of the region enclosed by the loop is twice the area above the x-axis inside the loop. $y = 0 \iff t^3 - 3t = 0 \iff t(t^2 - 3) = 0 \iff t = 0, \pm\sqrt{3}$. The top half of the loop is described by $x = t^2$, $y = t^3 - 3t$, $-\sqrt{3} \le t \le 0$, so, using the Substitution Rule with $y = t^3 - 3t$ and $dx = 2t\,dt$, we find that

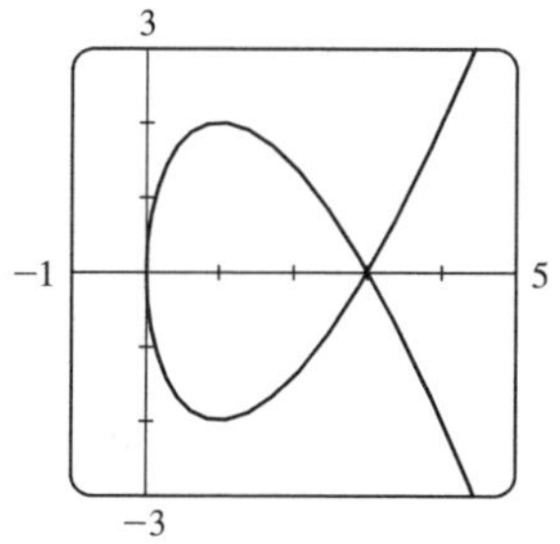

$$\text{area} = 2\int_0^3 y\,dx = 2\int_0^{-\sqrt{3}} (t^3 - 3t)2t\,dt = 4\int_0^{-\sqrt{3}} (t^4 - 3t^2)\,dt$$
$$= 4\left[\tfrac{1}{5}t^5 - t^3\right]_0^{-\sqrt{3}} = 4\left[\tfrac{1}{5}\left(-3^{1/2}\right)^5 - \left(-3^{1/2}\right)^3\right]$$
$$= 4\left[\tfrac{1}{5}(-9\sqrt{3}) - (-3\sqrt{3})\right] = \tfrac{24}{5}\sqrt{3} \approx 8.31.$$

37. We first assume that $c > 0$, since c can be replaced by $-c$ in both equations without changing the graphs, and if $c = 0$ the curves do not enclose a region. We see from the graph that the enclosed area A lies between $x = -c$ and $x = c$, and by symmetry, it is equal to four times the area in the first quadrant. The enclosed area is

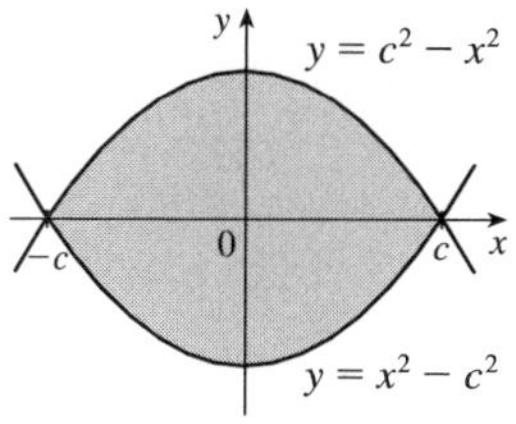

$$A = 4\int_0^c (c^2 - x^2)\,dx = 4\left[c^2 x - \tfrac{1}{3}x^3\right]_0^c$$
$$= 4\left(c^3 - \tfrac{1}{3}c^3\right) = 4\left(\tfrac{2}{3}c^3\right) = \tfrac{8}{3}c^3$$

So $A = 576 \iff \frac{8}{3}c^3 = 576 \iff c^3 = 216 \iff c = \sqrt[3]{216} = 6$.

Note that $c = -6$ is another solution, since the graphs are the same.

39. 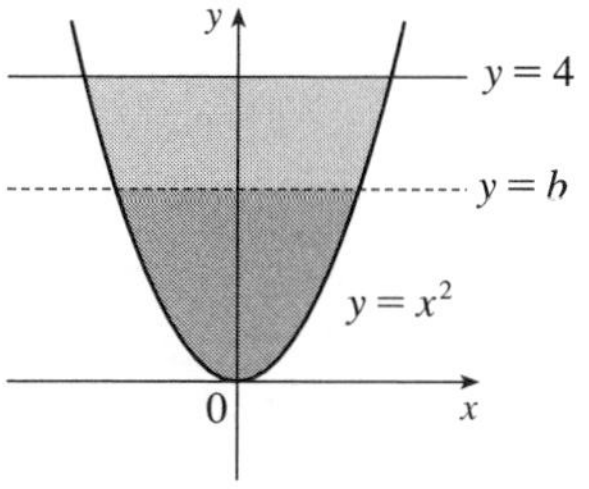

By the symmetry of the problem, we consider only the first quadrant, where $y = x^2 \Rightarrow x = \sqrt{y}$. We are looking for a number b such that

$$\int_0^b \sqrt{y}\,dy = \int_b^4 \sqrt{y}\,dy \Rightarrow \tfrac{2}{3}\left[y^{3/2}\right]_0^b = \tfrac{2}{3}\left[y^{3/2}\right]_b^4 \Rightarrow$$

$$b^{3/2} = 4^{3/2} - b^{3/2} \Rightarrow 2b^{3/2} = 8 \Rightarrow b^{3/2} = 4 \Rightarrow b = 4^{2/3} \approx 2.52.$$

41. The area under the graph of f from 0 to t is equal to $\int_0^t f(x)\,dx$, so the requirement is that $\int_0^t f(x)\,dx = t^3$ for all t. We differentiate both sides of this equation with respect to t (with the help of FTC1) to get $f(t) = 3t^2$. This function is positive and continuous, as required.

43. The curve and the line will determine a region when they intersect at two or more points. So we solve the equation $x/(x^2+1) = mx \;\Rightarrow\; x = x(mx^2+m) \;\Rightarrow$ $x(mx^2+m) - x = 0 \;\Rightarrow\; x(mx^2+m-1) = 0 \;\Rightarrow$ $x = 0$ or $mx^2+m-1 = 0 \;\Rightarrow$ $x = 0$ or $x^2 = \dfrac{1-m}{m} \;\Rightarrow\; x = 0$ or $x = \pm\sqrt{\dfrac{1}{m} - 1}$.

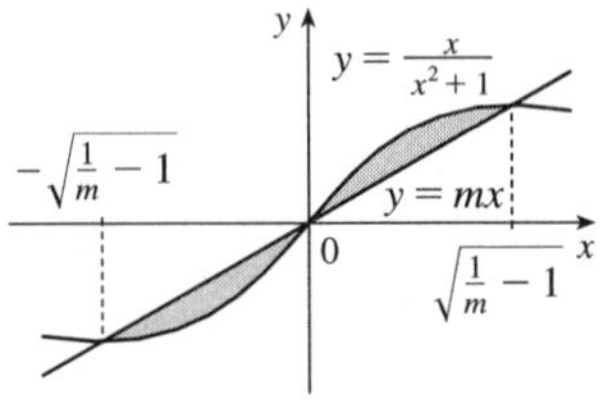

Note that if $m = 1$, this has only the solution $x = 0$, and no region is determined. But if $1/m - 1 > 0 \;\Leftrightarrow\; 1/m > 1 \;\Leftrightarrow$ $0 < m < 1$, then there are two solutions. [Another way of seeing this is to observe that the slope of the tangent to $y = x/(x^2+1)$ at the origin is $y' = 1$ and therefore we must have $0 < m < 1$.] Note that we cannot just integrate between the positive and negative roots, since the curve and the line cross at the origin. Since mx and $x/(x^2+1)$ are both odd functions, the total area is twice the area between the curves on the interval $\left[0, \sqrt{1/m - 1}\right]$. So the total area enclosed is

$$\begin{aligned} 2\int_0^{\sqrt{1/m-1}} \left[\frac{x}{x^2+1} - mx\right] dx &= 2\left[\frac{1}{2}\ln(x^2+1) - \frac{1}{2}mx^2\right]_0^{\sqrt{1/m-1}} \\ &= [\ln(1/m - 1 + 1) - m(1/m - 1)] - (\ln 1 - 0) \\ &= \ln(1/m) - 1 + m = m - \ln m - 1 \end{aligned}$$

6.2 Volumes

1. A cross-section is a disc with radius $1/x$, so its area is $A(x) = \pi(1/x)^2$.

$$\begin{aligned} V &= \int_1^2 A(x)\,dx = \int_1^2 \pi\left(\frac{1}{x}\right)^2 dx \\ &= \pi\int_1^2 \frac{1}{x^2}\,dx = \pi\left[-\frac{1}{x}\right]_1^2 \\ &= \pi\left[-\tfrac{1}{2} - (-1)\right] = \frac{\pi}{2} \end{aligned}$$

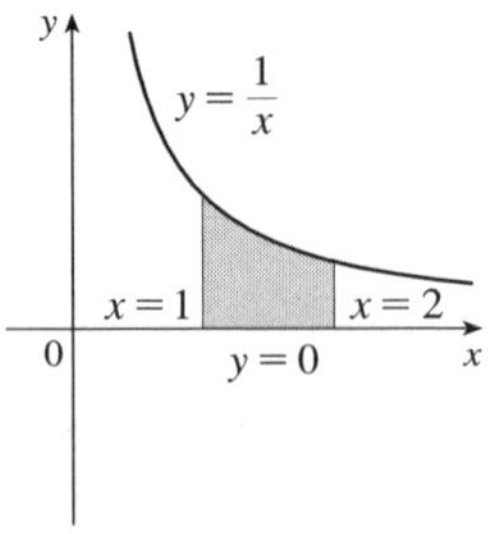

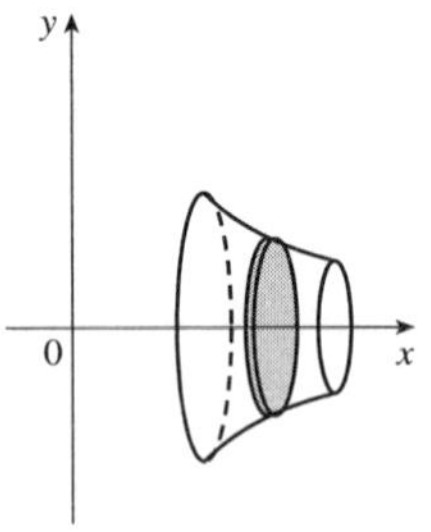

3. A cross-section is a disc with radius $\sqrt{y}$, so its area is $A(y) = \pi\left(\sqrt{y}\right)^2$.

$$\begin{aligned} V &= \int_0^4 A(y)\,dy = \int_0^4 \pi\left(\sqrt{y}\right)^2 dy \\ &= \pi\int_0^4 y\,dy = \pi\left[\tfrac{1}{2}y^2\right]_0^4 = 8\pi \end{aligned}$$

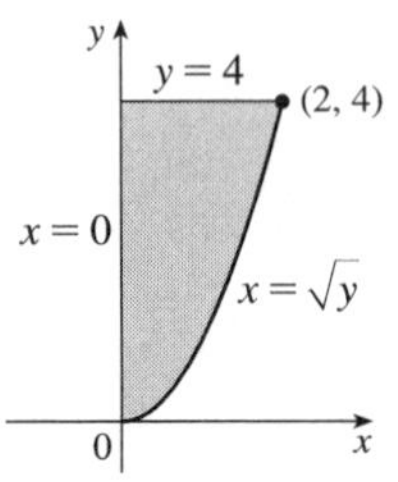

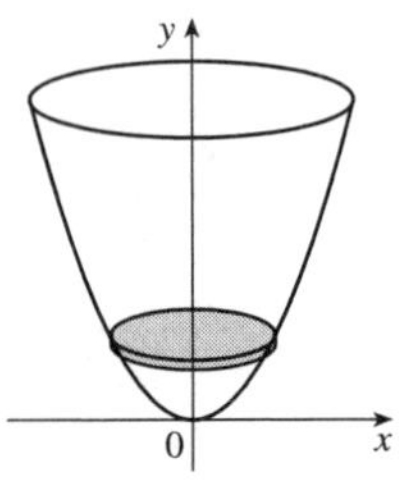

5. A cross-section is a washer (annulus) with inner radius x^3 and outer radius x, so its area is $A(x) = \pi(x)^2 - \pi(x^3)^2 = \pi(x^2 - x^6)$.

$$V = \int_0^1 A(x)\,dx = \int_0^1 \pi(x^2 - x^6)\,dx$$
$$= \pi\left[\tfrac{1}{3}x^3 - \tfrac{1}{7}x^7\right]_0^1 = \pi\left(\tfrac{1}{3} - \tfrac{1}{7}\right) = \tfrac{4\pi}{21}$$

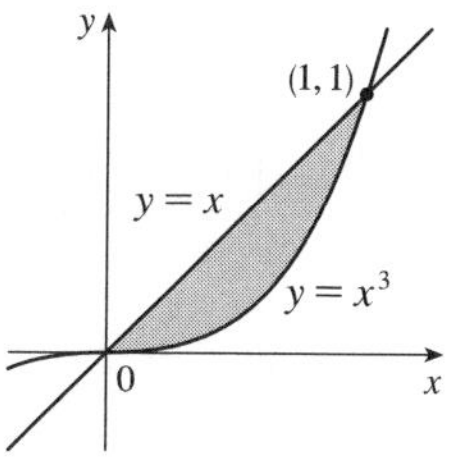

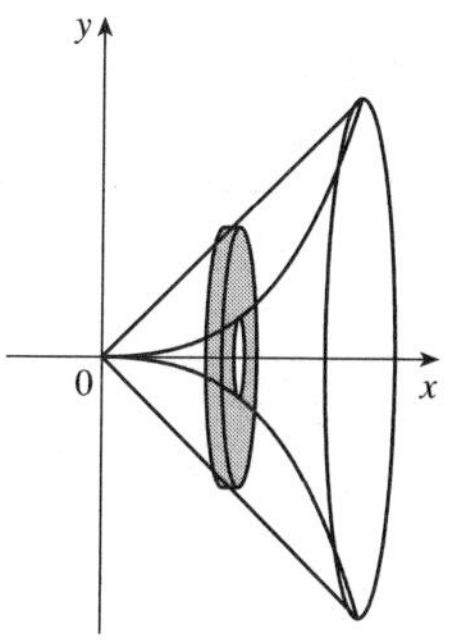

7. A cross-section is a washer with inner radius y^2 and outer radius $2y$, so its area is $A(y) = \pi(2y)^2 - \pi(y^2)^2 = \pi(4y^2 - y^4)$.

$$V = \int_0^2 A(y)\,dy = \pi\int_0^2 (4y^2 - y^4)\,dy$$
$$= \pi\left[\tfrac{4}{3}y^3 - \tfrac{1}{5}y^5\right]_0^2 = \pi\left(\tfrac{32}{3} - \tfrac{32}{5}\right) = \tfrac{64\pi}{15}$$

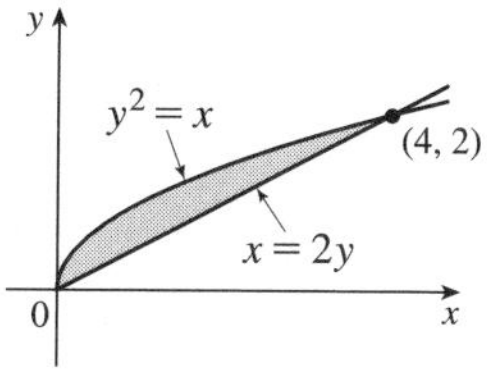

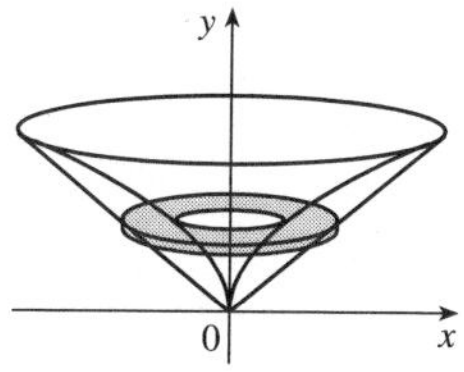

9. A cross-section is a washer with inner radius $1 - \sqrt{x}$ and outer radius $1 - x$, so its area is

$$A(x) = \pi(1-x)^2 - \pi(1-\sqrt{x})^2$$
$$= \pi\left[(1 - 2x + x^2) - (1 - 2\sqrt{x} + x)\right]$$
$$= \pi\left(-3x + x^2 + 2\sqrt{x}\right).$$

$$V = \int_0^1 A(x)\,dx = \pi\int_0^1 \left(-3x + x^2 + 2\sqrt{x}\right)dx$$
$$= \pi\left[-\tfrac{3}{2}x^2 + \tfrac{1}{3}x^3 + \tfrac{4}{3}x^{3/2}\right]_0^1 = \pi\left(-\tfrac{3}{2} + \tfrac{5}{3}\right) = \tfrac{\pi}{6}$$

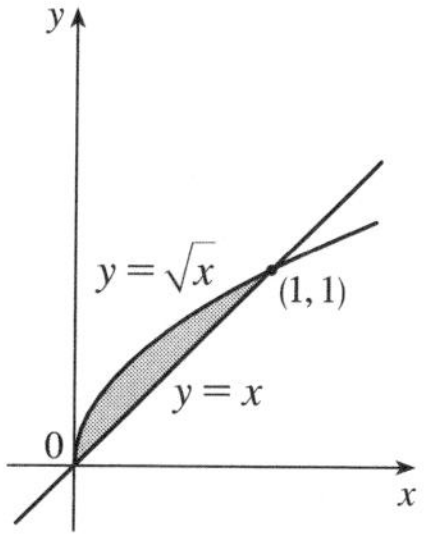

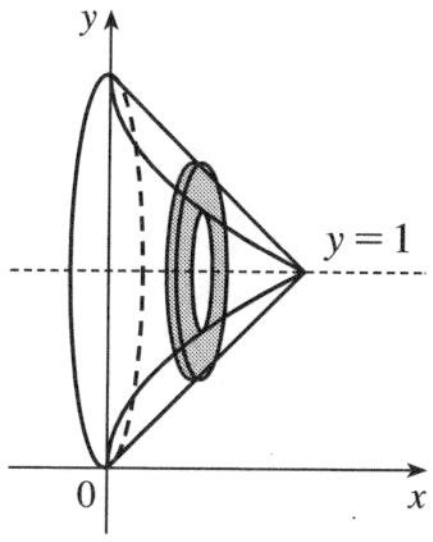

11. $y = x^2 \;\Rightarrow\; x = \sqrt{y}$ for $x \geq 0$. The outer radius is the distance from $x = -1$ to $x = \sqrt{y}$ and the inner radius is the distance from $x = -1$ to $x = y^2$.

$$V = \int_0^1 \pi\left\{\left[\sqrt{y} - (-1)\right]^2 - \left[y^2 - (-1)\right]^2\right\}dy = \pi\int_0^1 \left[(\sqrt{y}+1)^2 - (y^2+1)^2\right]dy$$
$$= \pi\int_0^1 \left(y + 2\sqrt{y} + 1 - y^4 - 2y^2 - 1\right)dy = \pi\int_0^1 \left(y + 2\sqrt{y} - y^4 - 2y^2\right)dy$$
$$= \pi\left[\tfrac{1}{2}y^2 + \tfrac{4}{3}y^{3/2} - \tfrac{1}{5}y^5 - \tfrac{2}{3}y^3\right]_0^1 = \pi\left(\tfrac{1}{2} + \tfrac{4}{3} - \tfrac{1}{5} - \tfrac{2}{3}\right) = \tfrac{29}{30}\pi$$

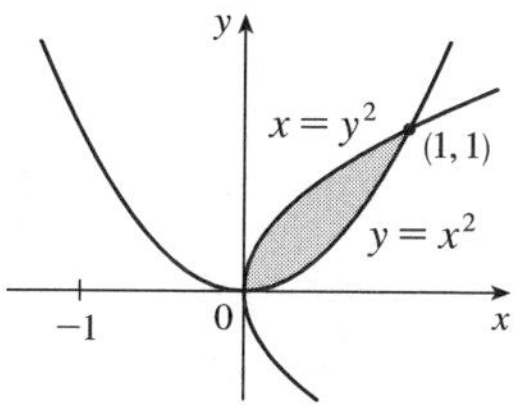

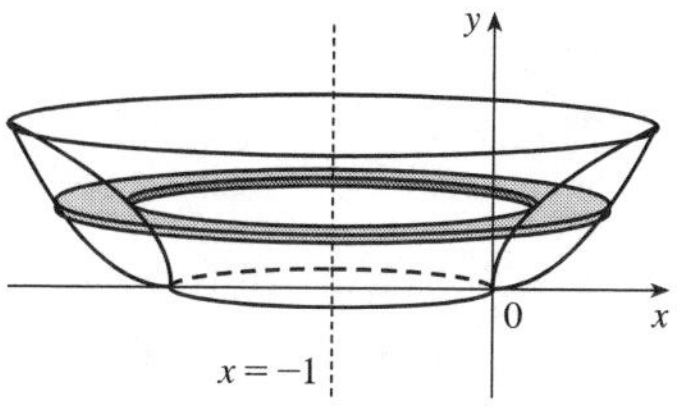

13. $y = \sqrt{x} \;\Rightarrow\; x = y^2$ and $y = x^3 \;\Rightarrow\; x = \sqrt[3]{y}$. A cross-section is a washer with inner radius $1 - \sqrt[3]{y}$ and outer radius $1 - y^2$, so its area is

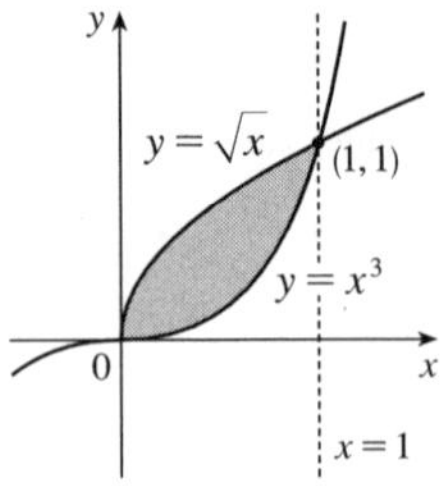

$A(y) = \pi(1 - y^2)^2 - \pi\left(1 - \sqrt[3]{y}\right)^2$.

$$V = \int_0^1 A(y)\,dy = \int_0^1 \left[\pi(1-y^2)^2 - \pi(1-\sqrt[3]{y})^2\right] dy$$

$$= \pi \int_0^1 [(1 - 2y^2 + y^4) - (1 - 2y^{1/3} + y^{2/3})]\,dy$$

$$= \pi \int_0^1 (-2y^2 + y^4 + 2y^{1/3} - y^{2/3})\,dy = \pi\left[-\tfrac{2}{3}y^3 + \tfrac{1}{5}y^5 + \tfrac{3}{2}y^{4/3} - \tfrac{3}{5}y^{5/3}\right]_0^1$$

$$= \pi\left(-\tfrac{2}{3} + \tfrac{1}{5} + \tfrac{3}{2} - \tfrac{3}{5}\right) = \tfrac{13\pi}{30}$$

15. $V = \pi \displaystyle\int_0^{\pi/4} (1 - \tan^3 x)^2\,dx$

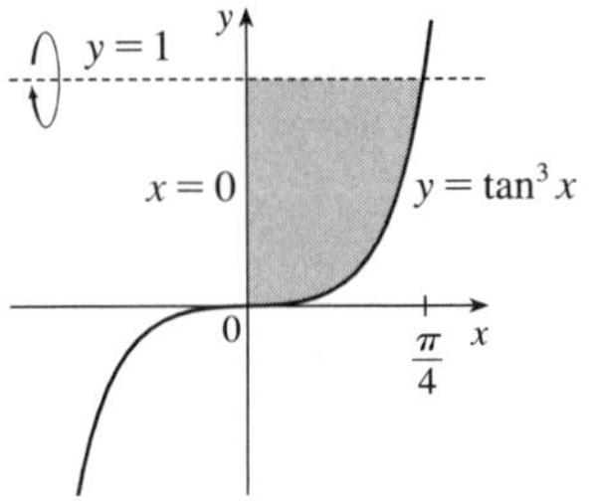

17.

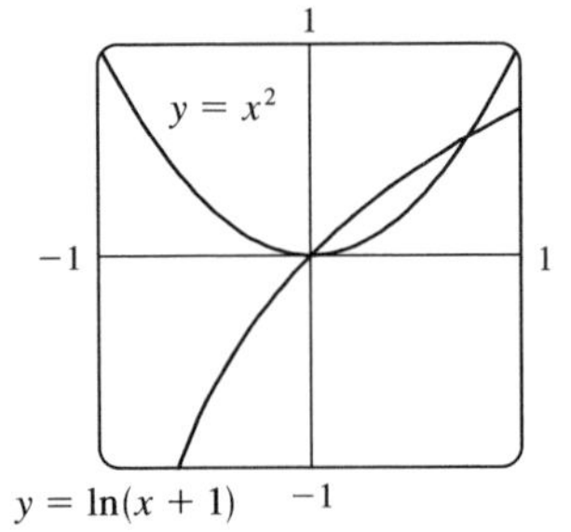

$y = x^2$ and $y = \ln(x+1)$ intersect at $x = 0$ and at $x = a \approx 0.747$.

$$V = \pi \int_0^a \left\{[\ln(x+1)]^2 - \left(x^2\right)^2\right\} dx \approx 0.132$$

19. $V = \pi \displaystyle\int_0^{\pi} \left\{\left[\sin^2 x - (-1)\right]^2 - [0 - (-1)]^2\right\} dx \overset{\text{CAS}}{=} \frac{11}{8}\pi^2$

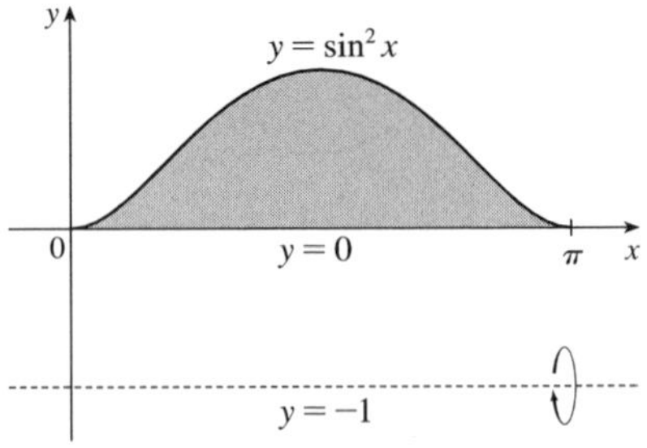

21. (a) $\pi \int_0^{\pi/2} \cos^2 x\,dx$ describes the volume of the solid obtained by rotating the region $\mathcal{R} = \{(x,y) \mid 0 \le x \le \frac{\pi}{2}, 0 \le y \le \cos x\}$ of the xy-plane about the x-axis.

(b) $\pi \displaystyle\int_0^1 (y^4 - y^8)\,dy = \pi \int_0^1 \left[\left(y^2\right)^2 - \left(y^4\right)^2\right] dy$ describes the volume of the solid obtained by rotating the region $\mathcal{R} = \{(x,y) \mid 0 \le y \le 1, y^4 \le x \le y^2\}$ of the xy-plane about the y-axis.

23. There are 10 subintervals over the 15-cm length, so we'll use $n = 10/2 = 5$ for the Midpoint Rule.

$$V = \int_0^{15} A(x)\,dx \approx M_5 = \tfrac{15-0}{5}[A(1.5) + A(4.5) + A(7.5) + A(10.5) + A(13.5)]$$

$$= 3(18 + 79 + 106 + 128 + 39) = 3 \cdot 370 = 1110 \text{ cm}^3$$

25. We'll form a right circular cone with height h and base radius r by revolving the line $y = \frac{r}{h}x$ about the x-axis.

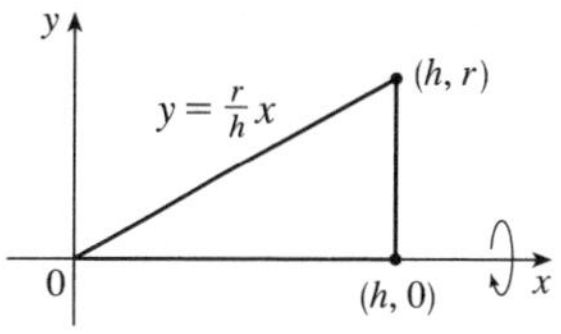

$$V = \pi \int_0^h \left(\frac{r}{h}x\right)^2 dx = \pi \int_0^h \frac{r^2}{h^2}x^2\,dx = \pi \frac{r^2}{h^2}\left[\frac{1}{3}x^3\right]_0^h$$

$$= \pi \frac{r^2}{h^2}\left(\frac{1}{3}h^3\right) = \frac{1}{3}\pi r^2 h$$

Another solution: Revolve $x = -\dfrac{r}{h}y + r$ about the y-axis.

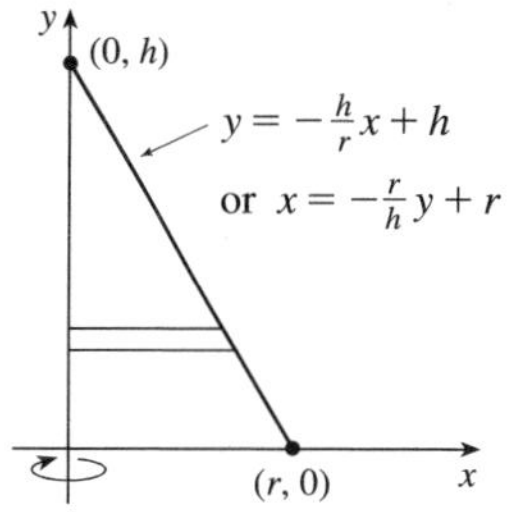

$$V = \pi \int_0^h \left(-\frac{r}{h}y + r\right)^2 dy \overset{*}{=} \pi \int_0^h \left[\frac{r^2}{h^2}y^2 - \frac{2r^2}{h}y + r^2\right] dy$$

$$= \pi\left[\frac{r^2}{3h^2}y^3 - \frac{r^2}{h}y^2 + r^2 y\right]_0^h = \pi\left(\tfrac{1}{3}r^2h - r^2h + r^2h\right) = \tfrac{1}{3}\pi r^2 h$$

* Or use substitution with $u = r - \dfrac{r}{h}y$ and $du = -\dfrac{r}{h}\,dy$ to get

$$\pi \int_r^0 u^2\left(-\frac{h}{r}\,du\right) = -\pi\frac{h}{r}\left[\frac{1}{3}u^3\right]_r^0 = -\pi\frac{h}{r}\left(-\frac{1}{3}r^3\right) = \frac{1}{3}\pi r^2 h.$$

27. $x^2 + y^2 = r^2 \quad \Leftrightarrow \quad x^2 = r^2 - y^2$

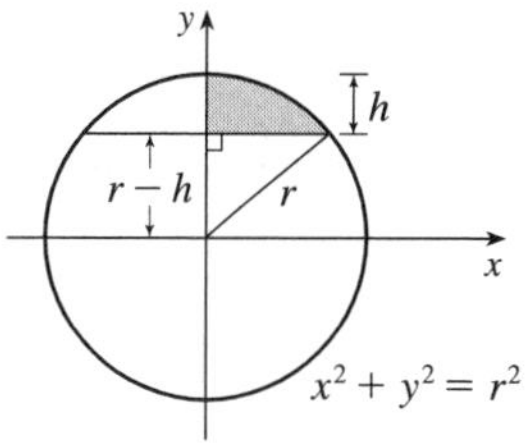

$$V = \pi \int_{r-h}^{r} (r^2 - y^2)\,dy = \pi\left[r^2y - \frac{y^3}{3}\right]_{r-h}^{r}$$

$$= \pi\left\{\left[r^3 - \frac{r^3}{3}\right] - \left[r^2(r-h) - \frac{(r-h)^3}{3}\right]\right\}$$

$$= \pi\left\{\tfrac{2}{3}r^3 - \tfrac{1}{3}(r-h)\left[3r^2 - (r-h)^2\right]\right\}$$

$$= \tfrac{1}{3}\pi\left\{2r^3 - (r-h)\left[3r^2 - (r^2 - 2rh + h^2)\right]\right\}$$

$$= \tfrac{1}{3}\pi\left\{2r^3 - (r-h)\left[2r^2 + 2rh - h^2\right]\right\}$$

$$= \tfrac{1}{3}\pi\left(2r^3 - 2r^3 - 2r^2h + rh^2 + 2r^2h + 2rh^2 - h^3\right)$$

$$= \tfrac{1}{3}\pi\left(3rh^2 - h^3\right) = \tfrac{1}{3}\pi h^2(3r - h), \text{ or, equivalently, } \pi h^2\left(r - \frac{h}{3}\right)$$

29. For a cross-section at height y, we see from similar triangles that $\dfrac{\alpha/2}{b/2} = \dfrac{h-y}{h}$, so $\alpha = b\left(1 - \frac{y}{h}\right)$.

Similarly, for cross-sections having $2b$ as their base and β replacing α, $\beta = 2b\left(1 - \frac{y}{h}\right)$. So

$$V = \int_0^h A(y)\,dy = \int_0^h \left[b\left(1 - \frac{y}{h}\right)\right]\left[2b\left(1 - \frac{y}{h}\right)\right] dy$$

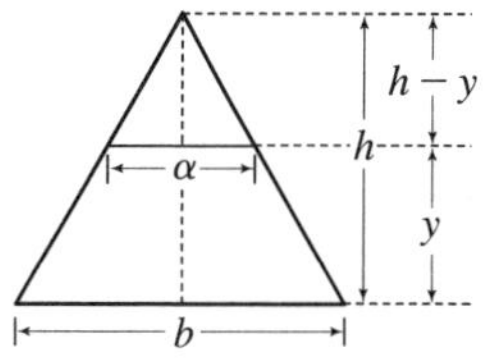

$$= \int_0^h 2b^2\left(1 - \frac{y}{h}\right)^2 dy = 2b^2 \int_0^h \left(1 - \frac{2y}{h} + \frac{y^2}{h^2}\right) dy$$

$$= 2b^2\left[y - \frac{y^2}{h} + \frac{y^3}{3h^2}\right]_0^h = 2b^2\left[h - h + \tfrac{1}{3}h\right]$$

$$= \tfrac{2}{3}b^2h \quad \left[= \tfrac{1}{3}Bh \text{ where } B \text{ is the area of the base, as with any pyramid.}\right]$$

31. A cross-section at height z is a triangle similar to the base, so we'll multiply the legs of the base triangle, 3 and 4, by a proportionality factor of $(5-z)/5$. Thus, the triangle at height z has area

$$A(z) = \frac{1}{2}\cdot 3\left(\frac{5-z}{5}\right)\cdot 4\left(\frac{5-z}{5}\right) = 6\left(1-\frac{z}{5}\right)^2, \text{ so}$$

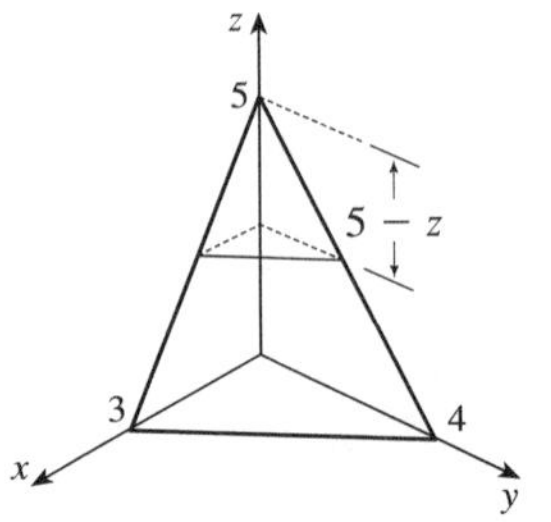

$$V = \int_0^5 A(z)\,dz = 6\int_0^5 (1-z/5)^2\,dz$$

$$= 6\int_1^0 u^2(-5\,du) \quad \left[u = 1 - z/5,\ du = -\tfrac{1}{5}dz\right]$$

$$= -30\left[\tfrac{1}{3}u^3\right]_1^0 = -30\left(-\tfrac{1}{3}\right) = 10\text{ cm}^3$$

33. If l is a leg of the isosceles right triangle and $2y$ is the hypotenuse, then $l^2 + l^2 = (2y)^2 \Rightarrow 2l^2 = 4y^2 \Rightarrow l^2 = 2y^2$.

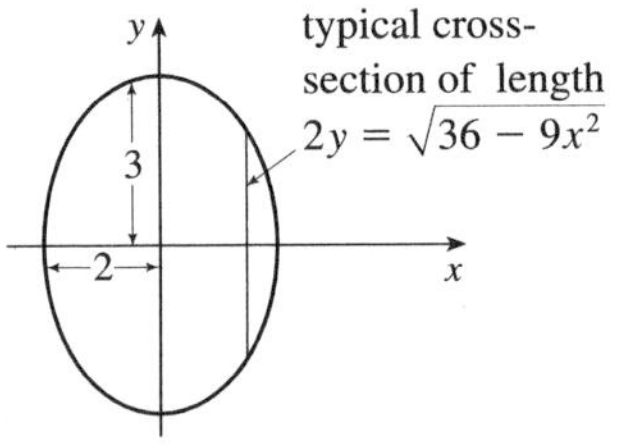

$$V = \int_{-2}^2 A(x)\,dx = 2\int_0^2 A(x)\,dx = 2\int_0^2 \tfrac{1}{2}(l)(l)\,dx = 2\int_0^2 y^2\,dx$$

$$= 2\int_0^2 \tfrac{1}{4}(36-9x^2)\,dx = \tfrac{9}{2}\int_0^2 (4-x^2)\,dx$$

$$= \tfrac{9}{2}\left[4x - \tfrac{1}{3}x^3\right]_0^2 = \tfrac{9}{2}\left(8-\tfrac{8}{3}\right) = 24$$

35. The cross-section of the base corresponding to the coordinate y has length $2x = 2\sqrt{y}$. The square has area $A(y) = \left(2\sqrt{y}\right)^2 = 4y$, so $V = \int_0^1 A(y)\,dy = \int_0^1 4y\,dy = \left[2y^2\right]_0^1 = 2$.

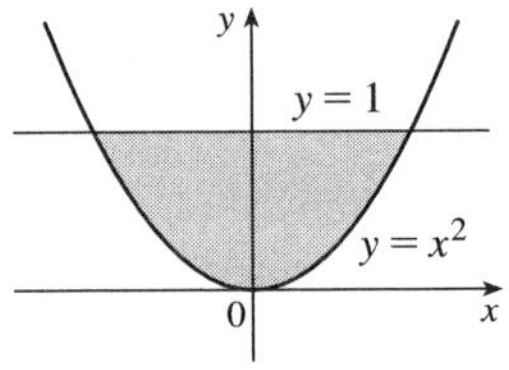

37. A typical cross-section perpendicular to the y-axis in the base has length $\ell(y) = 3 - \frac{3}{2}y$. This length is the leg of an isosceles right triangle, so

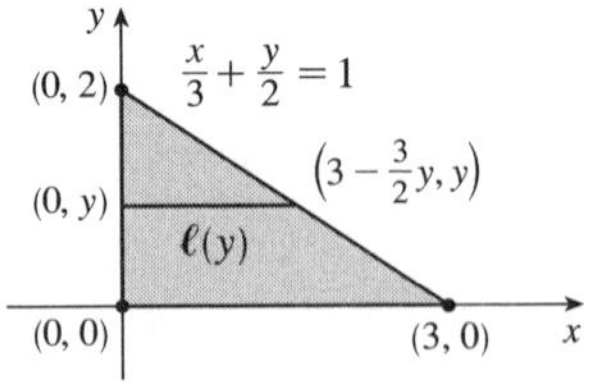

$$A(y) = \tfrac{1}{2}\left[\ell(y)\right]^2 \quad \left[\tfrac{1}{2}bh \text{ with base} = \text{height}\right]$$

$$= \tfrac{1}{2}\left[3\left(1-\tfrac{1}{2}y\right)\right]^2 = \tfrac{9}{2}\left(1-\tfrac{1}{2}y\right)^2$$

Thus,

$$V = \int_0^2 A(y)\,dy = \tfrac{9}{2}\int_1^0 u^2(-2\,du) \qquad \left[u = 1 - \tfrac{1}{2}y,\ du = -\tfrac{1}{2}\,dy\right]$$

$$= -9\left[\tfrac{1}{3}u^3\right]_1^0 = -9\left(-\tfrac{1}{3}\right) = 3$$

39. (a) The torus is obtained by rotating the circle $(x - R)^2 + y^2 = r^2$ about the y-axis. Solving for x, we see that the right half of the circle is given by $x = R + \sqrt{r^2 - y^2} = f(y)$ and the left half by $x = R - \sqrt{r^2 - y^2} = g(y)$. So

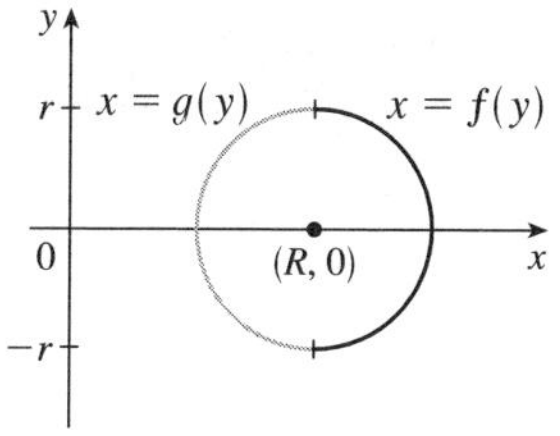

$$V = \pi \int_{-r}^{r} \{[f(y)]^2 - [g(y)]^2\}\, dy$$

$$= 2\pi \int_0^r \left[\left(R^2 + 2R\sqrt{r^2 - y^2} + r^2 - y^2\right) - \left(R^2 - 2R\sqrt{r^2 - y^2} + r^2 - y^2\right)\right] dy$$

$$= 2\pi \int_0^r 4R\sqrt{r^2 - y^2}\, dy = 8\pi R \int_0^r \sqrt{r^2 - y^2}\, dy$$

(b) Observe that the integral represents a quarter of the area of a circle with radius r, so $8\pi R \int_0^r \sqrt{r^2 - y^2}\, dy = 8\pi R \cdot \frac{1}{4}\pi r^2 = 2\pi^2 r^2 R$.

41. (a) Volume$(S_1) = \int_0^h A(z)\, dz =$ Volume(S_2) since the cross-sectional area $A(z)$ at height z is the same for both solids.

(b) By Cavalieri's Principle, the volume of the cylinder in the figure is the same as that of a right circular cylinder with radius r and height h, that is, $\pi r^2 h$.

43. The volume is obtained by rotating the area common to two circles of radius r, as shown. The volume of the right half is

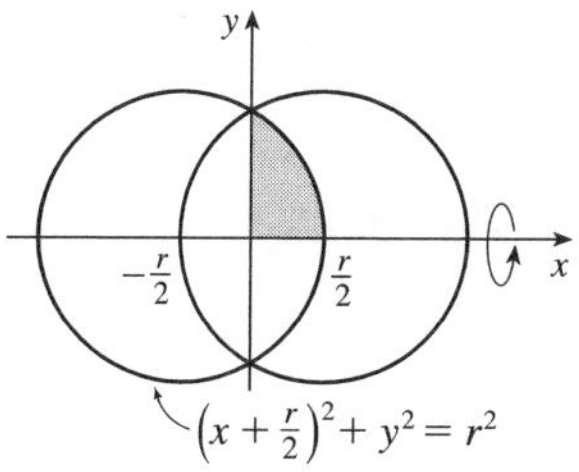

$$V_{\text{right}} = \pi \int_0^{r/2} y^2\, dx = \pi \int_0^{r/2} \left[r^2 - \left(\tfrac{1}{2}r + x\right)^2\right] dx$$

$$= \pi\left[r^2 x - \tfrac{1}{3}\left(\tfrac{1}{2}r + x\right)^3\right]_0^{r/2} = \pi\left[\left(\tfrac{1}{2}r^3 - \tfrac{1}{3}r^3\right) - \left(0 - \tfrac{1}{24}r^3\right)\right] = \tfrac{5}{24}\pi r^3$$

So by symmetry, the total volume is twice this, or $\frac{5}{12}\pi r^3$.

Another solution: We observe that the volume is the twice the volume of a cap of a sphere, so we can use the formula from Exercise 27 with $h = \frac{1}{2}r$: $V = 2 \cdot \frac{1}{3}\pi h^2(3r - h) = \frac{2}{3}\pi\left(\frac{1}{2}r\right)^2\left(3r - \frac{1}{2}r\right) = \frac{5}{12}\pi r^3$.

45. Take the x-axis to be the axis of the cylindrical hole of radius r. A quarter of the cross-section through y, perpendicular to the y-axis, is the rectangle shown. Using the Pythagorean Theorem twice, we see that the dimensions of this rectangle are $x = \sqrt{R^2 - y^2}$ and $z = \sqrt{r^2 - y^2}$, so $\frac{1}{4}A(y) = xz = \sqrt{r^2 - y^2}\sqrt{R^2 - y^2}$, and

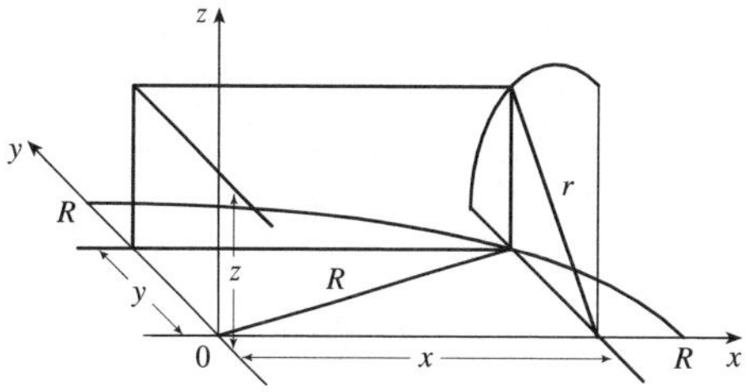

$$V = \int_{-r}^{r} A(y)\, dy = \int_{-r}^{r} 4\sqrt{r^2 - y^2}\sqrt{R^2 - y^2}\, dy = 8\int_0^r \sqrt{r^2 - y^2}\sqrt{R^2 - y^2}\, dy$$

47. (a) The radius of the barrel is the same at each end by symmetry, since the function $y = R - cx^2$ is even. Since the barrel is obtained by rotating the graph of the function y about the x-axis, this radius is equal to the value of y at $x = \frac{1}{2}h$, which is $R - c\left(\frac{1}{2}h\right)^2 = R - d = r$.

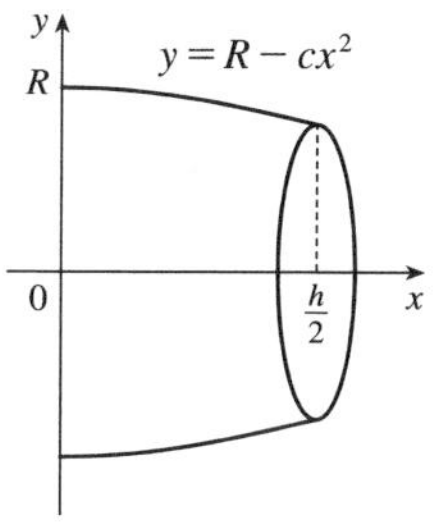

(b) The barrel is symmetric about the y-axis, so its volume is twice the volume of that part of the barrel for $x > 0$. Also, the barrel is a volume of rotation, so

$$V = 2\int_0^{h/2} \pi y^2\,dx = 2\pi\int_0^{h/2} \left(R - cx^2\right)^2 dx = 2\pi\left[R^2x - \tfrac{2}{3}Rcx^3 + \tfrac{1}{5}c^2x^5\right]_0^{h/2}$$
$$= 2\pi\left(\tfrac{1}{2}R^2h - \tfrac{1}{12}Rch^3 + \tfrac{1}{160}c^2h^5\right)$$

Trying to make this look more like the expression we want, we rewrite it as $V = \frac{1}{3}\pi h\left[2R^2 + \left(R^2 - \frac{1}{2}Rch^2 + \frac{3}{80}c^2h^4\right)\right]$.
But $R^2 - \frac{1}{2}Rch^2 + \frac{3}{80}c^2h^4 = \left(R - \frac{1}{4}ch^2\right)^2 - \frac{1}{40}c^2h^4 = (R-d)^2 - \frac{2}{5}\left(\frac{1}{4}ch^2\right)^2 = r^2 - \frac{2}{5}d^2$.
Substituting this back into V, we see that $V = \frac{1}{3}\pi h\left(2R^2 + r^2 - \frac{2}{5}d^2\right)$, as required.

49. 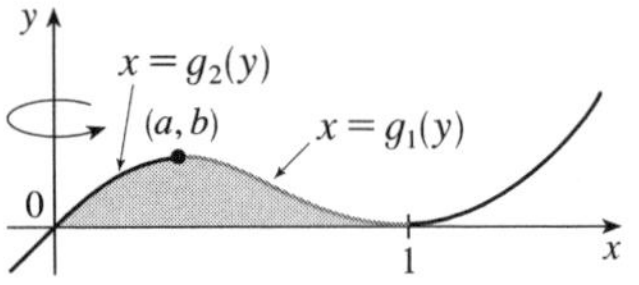

If we were to use the "washer method," we would first have to locate the local maximum point (a, b) of $y = x(x-1)^2$ using the methods of Chapter 4. Then we would have to solve the equation $y = x(x-1)^2$ for x in terms of y to obtain the functions $x = g_1(y)$ and $x = g_2(y)$ shown in the figure above. This step would be difficult because it involves the cubic formula. Finally we would find the volume using $V = \pi\int_0^b \left\{[g_1(y)]^2 - [g_2(y)]^2\right\} dy$.

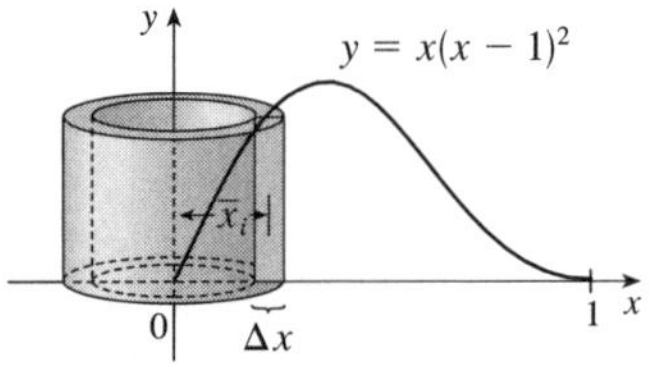

Instead, we use cylindrical shells. As in Example 9, we rotate an approximating rectangle with width Δx about the y-axis, to get a cylindrical shell whose average radius is $\overline{x}_i$ and whose volume is $2\pi\overline{x}_i\left[\overline{x}_i(\overline{x}_i - 1)^2\right]\Delta x$. So the total volume is

$$V = \lim_{n\to\infty}\sum_{i=1}^{n} 2\pi\,\overline{x}_i\left[\overline{x}_i(\overline{x}_i-1)^2\right]\Delta x = \int_0^1 2\pi x\left[x(x-1)^2\right]dx$$
$$= 2\pi\int_0^1 \left(x^4 - 2x^3 + x^2\right)dx = 2\pi\left[\frac{x^5}{5} - 2\frac{x^4}{4} + \frac{x^3}{3}\right]_0^1$$
$$= 2\pi\left(\frac{1}{5} - \frac{1}{2} + \frac{1}{3}\right) = 2\pi\left(\frac{1}{30}\right) = \frac{\pi}{15}$$

51. (a) Let $y = f(x)$ denote the curve. Using cylindrical shells, $V = \int_2^{10} 2\pi x f(x)\,dx = 2\pi\int_2^{10} x f(x)\,dx = 2\pi I_1$.
Now use Simpson's Rule to approximate I_1:

$$I_1 \approx S_8 = \tfrac{10-2}{3(8)}\left[2f(2) + 4\cdot 3f(3) + 2\cdot 4f(4) + 4\cdot 5f(5) + 2\cdot 6f(6)\right.$$
$$\left. + 4\cdot 7f(7) + 2\cdot 8f(8) + 4\cdot 9f(9) + 10f(10)\right]$$
$$\approx \tfrac{1}{3}[2(0) + 12(1.5) + 8(1.9) + 20(2.2) + 12(3.0) + 28(3.8) + 16(4.0) + 36(3.1) + 10(0)]$$
$$= \tfrac{1}{3}(395.2)$$

Thus, $V \approx 2\pi\cdot\frac{1}{3}(395.2) \approx 827.7$ or 828 cubic units.

(b) Using discs, $V = \int_2^{10} \pi[f(x)]^2\,dx = \pi\int_2^{10}[f(x)]^2\,dx = \pi I_2$. Now use Simpson's Rule to approximate I_2:

$$I_2 \approx S_8 = \tfrac{10-2}{3(8)}\left\{[f(2)]^2 + 4[f(3)]^2 + 2[f(4)]^2 + 4[f(5)]^2 + 2[f(6)]^2\right.$$
$$\left. + 4[f(7)]^2 + 2[f(8)]^2 + 4[f(9)]^2 + [f(10)]^2\right\}$$
$$\approx \tfrac{1}{3}\left[(0)^2 + 4(1.5)^2 + 2(1.9)^2 + 4(2.2)^2 + 2(3.0)^2 + 4(3.8)^2 + 2(4.0)^2 + 4(3.1)^2 + (0)^2\right]$$
$$= \tfrac{1}{3}(181.78)$$

Thus, $V \approx \pi\cdot\frac{1}{3}(181.78) \approx 190.4$ or 190 cubic units.

53.

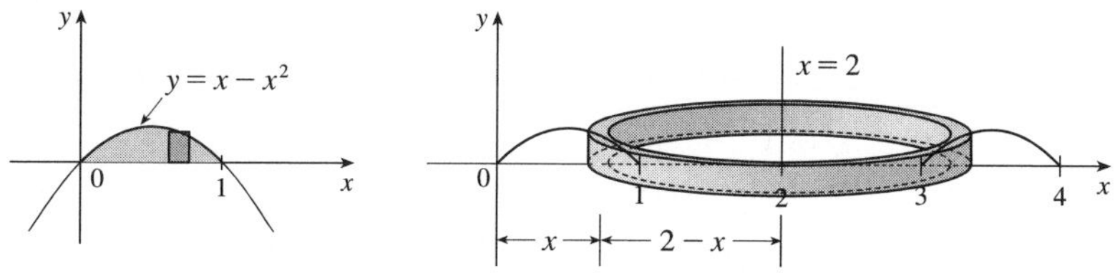

$$V = \int_0^1 (\text{circumference})\,(\text{height})\,(\text{thickness}) = \int_0^1 [2\pi(2-x)]\left(x - x^2\right)dx$$

$$= 2\pi \int_0^1 \left(x^3 - 3x^2 + 2x\right)dx = 2\pi\left[\tfrac{1}{4}x^4 - x^3 + x^2\right]_0^1 = 2\pi\left(\tfrac{1}{4}\right) = \tfrac{\pi}{2}$$

See the solution for Exercise 49 as to why the method of cylindrical shells is preferable to slicing.

6.3 Arc Length

1. $y = 2 - 3x \;\Rightarrow\; L = \int_{-2}^{1} \sqrt{1 + (dy/dx)^2}\,dx = \int_{-2}^{1} \sqrt{1 + (-3)^2}\,dx = \sqrt{10}\,[1 - (-2)] = 3\sqrt{10}.$

The arc length can be calculated using the distance formula, since the curve is a line segment, so

$$L = [\text{distance from } (-2, 8) \text{ to } (1, -1)] = \sqrt{[1 - (-2)]^2 + [(-1) - 8]^2} = \sqrt{90} = 3\sqrt{10}$$

3. $y = \cos x \;\Rightarrow\; dy/dx = -\sin x \;\Rightarrow\; 1 + (dy/dx)^2 = 1 + \sin^2 x$. So $L = \int_0^{2\pi} \sqrt{1 + \sin^2 x}\,dx.$

5.

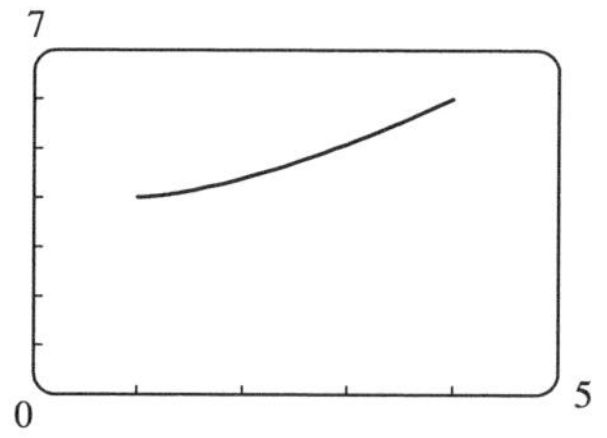

$x = 1 + 3t^2, \quad y = 4 + 2t^3, \quad 0 \le t \le 1.$

$dx/dt = 6t$ and $dy/dt = 6t^2$, so $(dx/dt)^2 + (dy/dt)^2 = 36t^2 + 36t^4$.

Thus, $L = \int_0^1 \sqrt{36t^2 + 36t^4}\,dt$

$$= \int_0^1 6t\sqrt{1 + t^2}\,dt = 6\int_1^2 \sqrt{u}\left(\tfrac{1}{2}du\right) \quad [u = 1 + t^2,\, du = 2t\,dt]$$

$$= 3\left[\tfrac{2}{3}u^{3/2}\right]_1^2 = 2(2^{3/2} - 1) = 2\left(2\sqrt{2} - 1\right)$$

7.

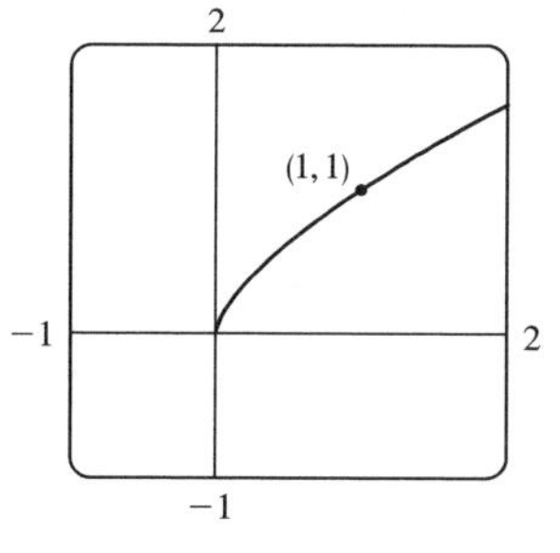

$$x = y^{3/2} \;\Rightarrow\; 1 + (dx/dy)^2 = 1 + \left(\tfrac{3}{2}y^{1/2}\right)^2 = 1 + \tfrac{9}{4}y.$$

$$L = \int_0^1 \sqrt{1 + \tfrac{9}{4}y}\,dy = \int_1^{13/4} \sqrt{u}\left(\tfrac{4}{9}\,du\right) \quad \left[u = 1 + \tfrac{9}{4}\,y,\, du = \tfrac{9}{4}\,dy\right]$$

$$= \tfrac{4}{9}\cdot\tfrac{2}{3}\left[u^{3/2}\right]_1^{13/4} = \tfrac{8}{27}\left(\tfrac{13\sqrt{13}}{8} - 1\right) = \tfrac{13\sqrt{13} - 8}{27}.$$

9.

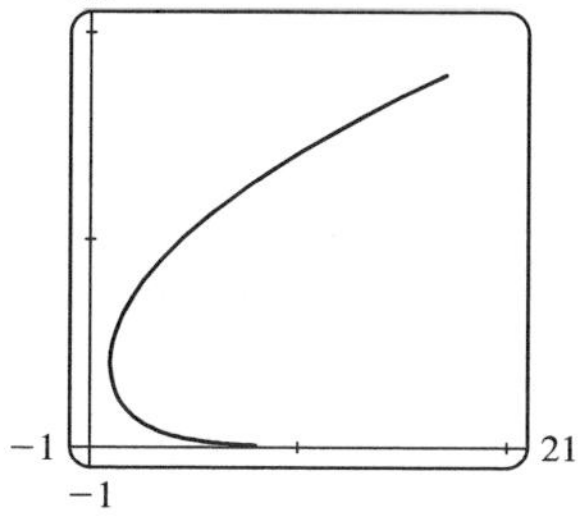

$x = e^t - t,\; y = 4e^{t/2},\; -8 \le t \le 3$

$$(dx/dt)^2 + (dy/dt)^2 = (e^t - 1)^2 + (2e^{t/2})^2 = e^{2t} - 2e^t + 1 + 4e^t$$

$$= e^{2t} + 2e^t + 1 = (e^t + 1)^2$$

$$L = \int_{-8}^{3} \sqrt{(e^t + 1)^2}\,dt = \int_{-8}^{3} (e^t + 1)\,dt = \left[e^t + t\right]_{-8}^{3t}$$

$$= (e^3 + 3) - (e^{-8} - 8) = e^3 - e^{-8} + 11$$

11. $x = \ln t$ and $y = e^{-t} \;\Rightarrow\; \dfrac{dx}{dt} = \dfrac{1}{t}$ and $\dfrac{dy}{dt} = -e^{-t} \;\Rightarrow\; L = \int_1^2 \sqrt{t^{-2} + e^{-2t}}\,dt$.

Using Simpson's Rule with $n = 10$, $\Delta t = (2-1)/10 = 0.1$ and $f(t) = \sqrt{t^{-2} + e^{-2t}}$, we get

$$L \approx S_{10} = \tfrac{0.1}{3}[f(1.0) + 4f(1.1) + 2f(1.2) + 4f(1.3) + 2f(1.4) + 4f(1.5) + 2f(1.6) + 4f(1.7) + 2f(1.8) + 4f(1.9) + f(2.0)] \approx 0.731\,371$$

The value of the integral produced by a calculator is 0.731 368 (to six decimal places).

13. $x = \sin t,\ y = t^2 \;\Rightarrow\; (dx/dt)^2 + (dy/dt)^2 = (\cos t)^2 + (2t)^2 = \cos^2 t + 4t^2 \;\Rightarrow\; L = \int_0^{2\pi} \sqrt{\cos^2 t + 4t^2}\,dt$.

Using Simpson's Rule with $n = 10$, $\Delta t = \dfrac{2\pi - 0}{10} = \dfrac{\pi}{5}$, and $f(t) = \sqrt{\cos^2 t + 4t^2}\,dt$, we get

$$L \approx S_{10} = \tfrac{2\pi - 0}{3(10)}\left[f(0) + 4f(\tfrac{\pi}{5}) + 2f(\tfrac{2\pi}{5}) + 4f(\tfrac{3\pi}{5}) + 2f(\tfrac{4\pi}{5}) + 4f(\pi) + 2f(\tfrac{6\pi}{5}) + 4f(\tfrac{7\pi}{5}) + 2f(\tfrac{8\pi}{5}) + 4f(\tfrac{9\pi}{5}) + f(2\pi)\right] \approx 40.056\,222$$

The value of the integral produced by a calculator is 40.051 156 (to six decimal places).

15. (a)

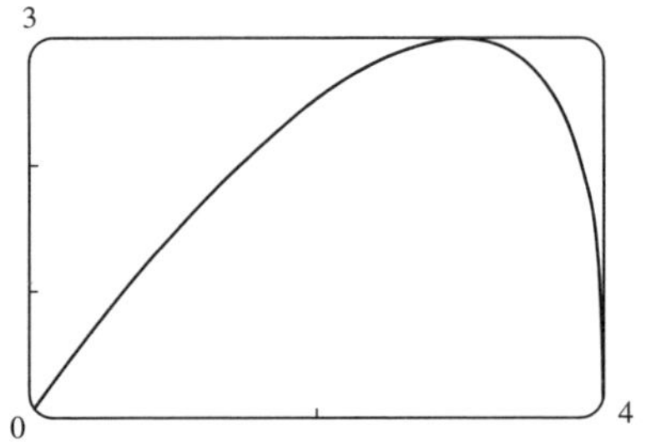

(b)

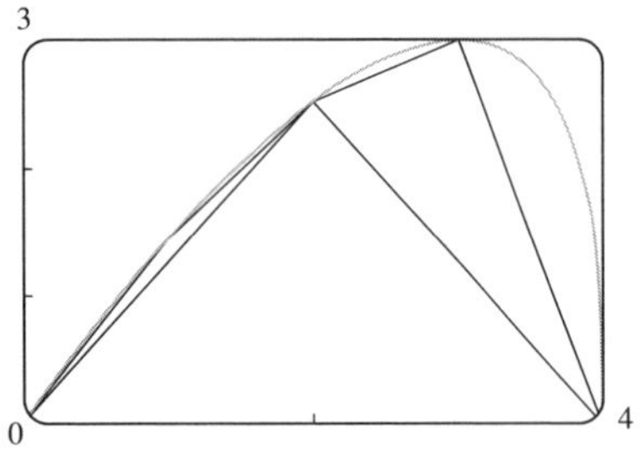

Let $f(x) = y = x\sqrt[3]{4 - x}$. The polygon with one side is just the line segment joining the points $(0, f(0)) = (0, 0)$ and $(4, f(4)) = (4, 0)$, and its length is 4. The polygon with two sides joins the points $(0, 0)$, $(2, f(2)) = \left(2, 2\sqrt[3]{2}\right)$ and $(4, 0)$.

Its length is

$$\sqrt{(2-0)^2 + \left(2\sqrt[3]{2} - 0\right)^2} + \sqrt{(4-2)^2 + \left(0 - 2\sqrt[3]{2}\right)^2} = 2\sqrt{4 + 2^{8/3}} \approx 6.43$$

Similarly, the inscribed polygon with four sides joins the points $(0, 0)$, $\left(1, \sqrt[3]{3}\right)$, $\left(2, 2\sqrt[3]{2}\right)$, $(3, 3)$, and $(4, 0)$, so its length is

$$\sqrt{1 + \left(\sqrt[3]{3}\right)^2} + \sqrt{1 + \left(2\sqrt[3]{2} - \sqrt[3]{3}\right)^2} + \sqrt{1 + \left(3 - 2\sqrt[3]{2}\right)^2} + \sqrt{1 + 9} \approx 7.50$$

(c) Using the arc length formula with $\dfrac{dy}{dx} = x\left[\tfrac{1}{3}(4 - x)^{-2/3}(-1)\right] + \sqrt[3]{4 - x} = \dfrac{12 - 4x}{3(4 - x)^{2/3}}$, the length of the curve is

$$L = \int_0^4 \sqrt{1 + \left(\frac{dy}{dx}\right)^2}\,dx = \int_0^4 \sqrt{1 + \left[\frac{12 - 4x}{3(4 - x)^{2/3}}\right]^2}\,dx.$$

(d) According to a CAS, the length of the curve is $L \approx 7.7988$. The actual value is larger than any of the approximations in part (b). This is always true, since any approximating straight line between two points on the curve is shorter than the length of the curve between the two points.

17. $x = t^3 \quad \Rightarrow \quad dx/dt = 3t^2$ and $y = t^4 \quad \Rightarrow \quad dy/dt = 4t^3$. So
$L = \int_0^1 \sqrt{9t^4 + 16t^6}\, dt = \int_0^1 \sqrt{t^4(9 + 16t^2)}\, dt = \int_0^1 t^2\sqrt{9 + 16t^2}\, dt.$
Now use Formula 22 from the table of integrals to evaluate L.

$$\begin{aligned} L &= \int_0^4 \left(\tfrac{1}{4}u\right)^2 \sqrt{a^2 + u^2}\left(\tfrac{1}{4}u\right) \qquad [a = 3,\, u = 4t,\, du = 4\,dt] \\ &= \tfrac{1}{64}\int_0^4 u^2\sqrt{a^2 + u^2}\, du = \tfrac{1}{64}\left[\tfrac{u}{8}\left(9 + 2u^2\right)\sqrt{9 + u^2} - \tfrac{81}{8}\ln\left(u + \sqrt{9 + u^2}\,\right)\right]_0^4 \\ &= \tfrac{1}{64}\left\{\left[\tfrac{1}{2}\cdot 41\cdot 5 - \tfrac{81}{8}\ln(4 + 5)\right] - \left[0 - \tfrac{81}{8}\ln 3\right]\right\} \\ &= \tfrac{1}{64}\left[\tfrac{205}{2} - \tfrac{81}{8}(2\ln 3) + \tfrac{81}{8}\ln 3\right] \qquad \left[\ln 9 = \ln 3^2 = 2\ln 3\right] \\ &= \tfrac{1}{64}\left(\tfrac{205}{2} - \tfrac{81}{8}\ln 3\right) = \tfrac{205}{128} - \tfrac{81}{512}\ln 3 \approx 1.428. \end{aligned}$$

19. $y = \ln(\cos x) \quad \Rightarrow \quad y' = \dfrac{1}{\cos x}(-\sin x) = -\tan x \quad \Rightarrow \quad 1 + (y')^2 = 1 + \tan^2 x = \sec^2 x.$

So $L = \int_0^{\pi/4} \sec x\, dx \overset{14}{=} \Big[\ln|\sec x + \tan x|\Big]_0^{\pi/4} = \ln\left(\sqrt{2} + 1\right) - \ln(1 + 0) = \ln\left(\sqrt{2} + 1\right) \approx 0.881.$

21. The prey hits the ground when $y = 0 \quad \Leftrightarrow \quad 180 - \frac{1}{45}x^2 = 0 \quad \Leftrightarrow \quad x^2 = 45 \cdot 180 \quad \Rightarrow \quad x = \sqrt{8100} = 90$, since x must be positive. $y' = -\frac{2}{45}x \quad \Rightarrow \quad 1 + (y')^2 = 1 + \frac{4}{45^2}x^2$, so the distance travelled by the prey is

$L = \int_0^{90} \sqrt{1 + \frac{4}{45^2}x^2}\, dx = \int_0^4 \sqrt{1 + u^2}\left(\frac{45}{2}\, du\right) \qquad \left[u = \frac{2}{45}x,\, du = \frac{2}{45}\, dx\right]$

$\overset{21}{=} \frac{45}{2}\left[\frac{1}{2}u\sqrt{1 + u^2} + \frac{1}{2}\ln\left(u + \sqrt{1 + u^2}\,\right)\right]_0^4 = \frac{45}{2}\left[2\sqrt{17} + \frac{1}{2}\ln\left(4 + \sqrt{17}\right)\right] = 45\sqrt{17} + \frac{45}{4}\ln\left(4 + \sqrt{17}\,\right) \approx 209.1$ m

23. The sine wave has amplitude 2 and period 30, since it goes through two periods in a distance of 60 cm, so its equation is $y = 2\sin\left(\frac{2\pi}{30}x\right) = 2\sin\left(\frac{\pi}{15}x\right)$. The width w of the flat metal sheet needed to make the panel is the arc length of the sine curve from $x = 0$ to $x = 60$. We set up the integral to evaluate w using the arc length formula with $\frac{dy}{dx} = \frac{2\pi}{15}\cos\left(\frac{\pi}{15}x\right)$:
$L = \int_0^{60} \sqrt{1 + \left[\frac{2\pi}{15}\cos\left(\frac{\pi}{15}x\right)\right]^2}\, dx$. This integral would be very difficult to evaluate exactly, so we use a CAS, and find that $L \approx 62.55$ cm.

25. $x = a\sin\theta,\, y = b\cos\theta,\, 0 \le \theta \le 2\pi$.

$$\begin{aligned} \left(\frac{dx}{d\theta}\right)^2 + \left(\frac{dy}{d\theta}\right)^2 &= (a\cos\theta)^2 + (-b\sin\theta)^2 = a^2\cos^2\theta + b^2\sin^2\theta = a^2\left(1 - \sin^2\theta\right) + b^2\sin^2\theta \\ &= a^2 - \left(a^2 - b^2\right)\sin^2\theta = a^2 - c^2\sin^2\theta = a^2\left(1 - \frac{c^2}{a^2}\sin^2\theta\right) = a^2\left(1 - e^2\sin^2\theta\right) \end{aligned}$$

So $L = 4\int_0^{\pi/2} \sqrt{a^2\left(1 - e^2\sin^2\theta\right)}\, d\theta$ [by symmetry] $= 4a\int_0^{\pi/2} \sqrt{1 - e^2\sin^2\theta}\, d\theta$.

27. (a) Notice that $0 \le t \le 2\pi$ does not give the complete curve because $x(0) \ne x(2\pi)$. In fact, we must take $t \in [0, 4\pi]$ in order to obtain the complete curve, since the first term in each of the parametric equations has period 2π and the second has period $\frac{2\pi}{11/2} = \frac{4\pi}{11}$, and the least common integer multiple of these two numbers is 4π.

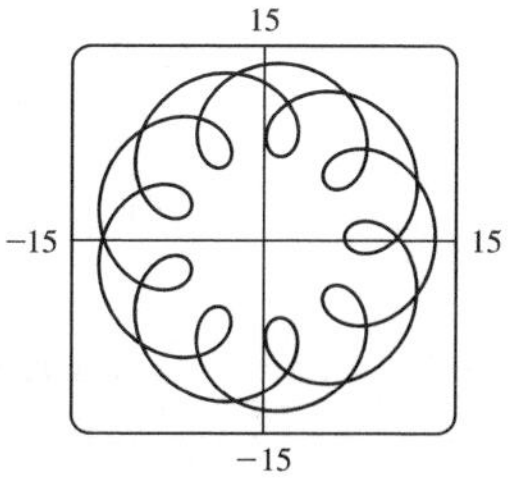

(b) We use the CAS to find the derivatives dx/dt and dy/dt, and then use Formula 1 to find the arc length. Recent versions of Maple express the integral $\int_0^{4\pi} \sqrt{(dx/dt)^2 + (dy/dt)^2}\, dt$ as $88E\left(2\sqrt{2}\, i\right)$, where $E(x)$ is the elliptic integral $\displaystyle\int_0^1 \frac{\sqrt{1 - x^2t^2}}{\sqrt{1 - t^2}}\, dt$ and i is the imaginary number $\sqrt{-1}$. Some earlier versions of Maple (as well as Mathematica) cannot do the integral exactly, so we use the command `evalf(Int(sqrt(diff(x,t)^2+diff(y,t)^2),t=0..4*Pi));` to estimate the length, and find that the arc length is approximately 294.03. Derive's `Para_arc_length` function in the utility file `Int_apps` simplifies the integral to $11\int_0^{4\pi}\sqrt{-4\cos t\,\cos\left(\frac{11t}{2}\right)-4\sin t\,\sin\left(\frac{11t}{2}\right)+5}\,dt$.

6.4 Average Value of a Function

1. $g_{\text{ave}}=\dfrac{1}{\frac{\pi}{2}-0}\int_0^{\pi/2}\cos x\,dx=\frac{2}{\pi}\left[\sin x\right]_0^{\pi/2}=\frac{2}{\pi}(1-0)=\frac{2}{\pi}$

3. $f_{\text{ave}}=\frac{1}{5-0}\int_0^5 te^{-t^2}\,dt=\frac{1}{5}\int_0^{-25}e^u\left(-\frac{1}{2}\,du\right)\qquad\left[u=-t^2,\,du=-2t\,dt,\,t\,dt=-\frac{1}{2}\,du\right]$

$=-\frac{1}{10}\left[e^u\right]_0^{-25}=-\frac{1}{10}\left(e^{-25}-1\right)=\frac{1}{10}\left(1-e^{-25}\right)$

5. (a) $f_{\text{ave}}=\dfrac{1}{5-2}\displaystyle\int_2^5(x-3)^2\,dx=\frac{1}{3}\left[\frac{1}{3}(x-3)^3\right]_2^5$

$=\frac{1}{9}\left[2^3-(-1)^3\right]=\frac{1}{9}(8+1)=1$

(b) $f(c)=f_{\text{ave}}\iff(c-3)^2=1\iff$

$c-3=\pm1\iff c=2\text{ or }4$

(c)

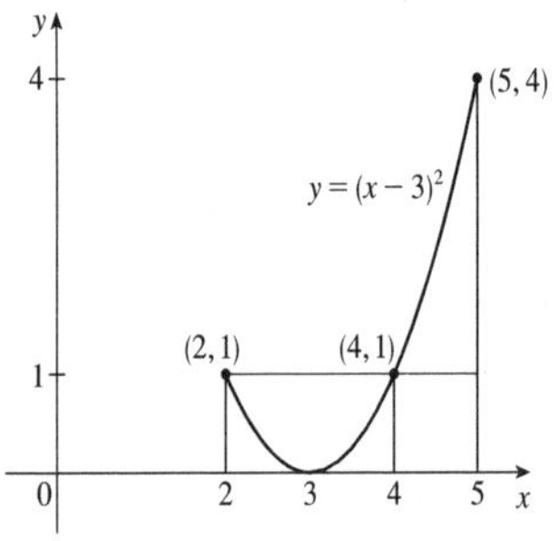

7. (a) $f_{\text{ave}}=\dfrac{1}{\pi-0}\displaystyle\int_0^\pi(2\sin x-\sin 2x)\,dx$

$=\frac{1}{\pi}\left[-2\cos x+\frac{1}{2}\cos 2x\right]_0^\pi$

$=\frac{1}{\pi}\left[\left(2+\frac{1}{2}\right)-\left(-2+\frac{1}{2}\right)\right]=\frac{4}{\pi}$

(b) $f(c)=f_{\text{ave}}\iff 2\sin c-\sin 2c=\frac{4}{\pi}\iff$

$c_1\approx1.238$ or $c_2\approx2.808$

(c)

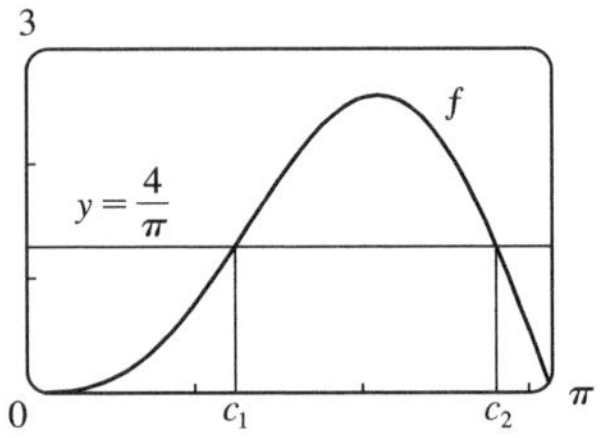

9. f is continuous on $[1,3]$, so by the Mean Value Theorem for Integrals there exists a number c in $[1,3]$ such that $\int_1^3 f(x)\,dx=f(c)(3-1)\Rightarrow 8=2f(c)$; that is, there is a number c such that $f(c)=\frac{8}{2}=4$.

11. $f_{\text{ave}}=\dfrac{1}{50-20}\displaystyle\int_{20}^{50}f(x)\,dx\approx\frac{1}{30}M_3=\frac{1}{30}\cdot\frac{50-20}{3}[f(25)+f(35)+f(45)]=\frac{1}{3}(38+29+48)=\frac{115}{3}=38\frac{1}{3}$

13. Let $t=0$ and $t=12$ correspond to 9 A.M. and 9 P.M., respectively.

$$T_{\text{ave}}=\frac{1}{12-0}\int_0^{12}\left[20+6\sin\tfrac{1}{12}\pi t\right]dt=\frac{1}{12}\left[20t-6\cdot\tfrac{12}{\pi}\cos\tfrac{1}{12}\pi t\right]_0^{12}$$
$$=\frac{1}{12}\left[20\cdot12+6\cdot\tfrac{12}{\pi}+6\cdot\tfrac{12}{\pi}\right]=\left(20+\tfrac{12}{\pi}\right)^\circ\text{C}\approx24^\circ\text{C}$$

15. (a) We want to calculate the square root of the average value of $[E(t)]^2=[155\sin(120\pi t)]^2=155^2\sin^2(120\pi t)$. First, we calculate the average value itself, by integrating $[E(t)]^2$ over one cycle (between $t=0$ and $t=\frac{1}{60}$, since there are

60 cycles per second) and dividing by $\left(\frac{1}{60}-0\right)$:

$$[E(t)]^2_{\text{ave}} = \frac{1}{1/60}\int_0^{1/60}\left[155^2\sin^2(120\pi t)\right]dt = 60\cdot 155^2\int_0^{1/60}\tfrac{1}{2}[1-\cos(240\pi t)]\,dt$$

$$= 60\cdot 155^2\left(\tfrac{1}{2}\right)\left[t-\tfrac{1}{240\pi}\sin(240\pi t)\right]_0^{1/60} = 60\cdot 155^2\left(\tfrac{1}{2}\right)\left[\left(\tfrac{1}{60}-0\right)-(0-0)\right] = \tfrac{155^2}{2}$$

The RMS value is just the square root of this quantity, which is $\frac{155}{\sqrt{2}} \approx 110$ V.

(b) $220 = \sqrt{[E(t)]^2_{\text{ave}}} \;\Rightarrow$

$$220^2 = [E(t)]^2_{\text{ave}} = \frac{1}{1/60}\int_0^{1/60} A^2\sin^2(120\pi t)\,dt = 60A^2\int_0^{1/60}\tfrac{1}{2}[1-\cos(240\pi t)]\,dt$$

$$= 30A^2\left[t-\tfrac{1}{240\pi}\sin(240\pi t)\right]_0^{1/60} = 30A^2\left[\left(\tfrac{1}{60}-0\right)-(0-0)\right] = \tfrac{1}{2}A^2$$

Thus, $220^2 = \frac{1}{2}A^2 \;\Rightarrow\; A = 220\sqrt{2} \approx 311$ V.

17. $V_{\text{ave}} = \frac{1}{5}\int_0^5 V(t)\,dt = \frac{1}{5}\int_0^5 \frac{5}{4\pi}\left[1-\cos\left(\frac{2}{5}\pi t\right)\right]dt = \frac{1}{4\pi}\int_0^5\left[1-\cos\left(\frac{2}{5}\pi t\right)\right]dt$

$= \frac{1}{4\pi}\left[t-\frac{5}{2\pi}\sin\left(\frac{2}{5}\pi t\right)\right]_0^5 = \frac{1}{4\pi}\left[(5-0)-0\right] = \frac{5}{4\pi} \approx 0.4$ L

19. Let $F(x) = \int_a^x f(t)\,dt$ for x in $[a,b]$. Then F is continuous on $[a,b]$ and differentiable on (a,b), so by the Mean Value Theorem there is a number c in (a,b) such that $F(b)-F(a) = F'(c)(b-a)$. But $F'(x) = f(x)$ by the Fundamental Theorem of Calculus. Therefore, $\int_a^b f(t)\,dt - 0 = f(c)(b-a)$.

6.5 Applications to Physics and Engineering

1. $W = \displaystyle\int_a^b f(x)\,dx = \int_0^9 \frac{10}{(1+x)^2}\,dx = 10\int_1^{10}\frac{1}{u^2}\,du \quad [u=1+x,\ du=dx] \;= 10\left[-\frac{1}{u}\right]_1^{10} = 10\left(-\tfrac{1}{10}+1\right) = 9$ J

3. The force function is given by $F(x)$ (in newtons) and the work (in joules) is the area under the curve, given by $\int_0^8 F(x)\,dx = \int_0^4 F(x)\,dx + \int_4^8 F(x)\,dx = \frac{1}{2}(4)(30) + (4)(30) = 180$ J.

5. $10 = f(x) = kx = \frac{1}{3}k$ [4 inches $= \frac{1}{3}$ foot], so $k = 30$ lb/ft and $f(x) = 30x$. Now 6 inches $= \frac{1}{2}$ foot, so $W = \int_0^{1/2} 30x\,dx = \left[15x^2\right]_0^{1/2} = \frac{15}{4}$ ft-lb.

7. (a) If $\int_0^{0.12} kx\,dx = 2$ J, then $2 = \left[\frac{1}{2}kx^2\right]_0^{0.12} = \frac{1}{2}k(0.0144) = 0.0072k$ and $k = \frac{2}{0.0072} = \frac{2500}{9} \approx 277.78$ N/m.
Thus, the work needed to stretch the spring from 35 cm to 40 cm is
$\int_{0.05}^{0.10}\frac{2500}{9}x\,dx = \left[\frac{1250}{9}x^2\right]_{1/20}^{1/10} = \frac{1250}{9}\left(\frac{1}{100}-\frac{1}{400}\right) = \frac{25}{24} \approx 1.04$ J.

(b) $f(x) = kx$, so $30 = \frac{2500}{9}x$ and $x = \frac{270}{2500}$ m $= 10.8$ cm

In Exercises 9–16, n is the number of subintervals of length Δx, and x_i^* is a sample point in the ith subinterval $[x_{i-1}, x_i]$.

9. (a) The portion of the rope from x ft to $(x+\Delta x)$ ft below the top of the building weighs $\frac{1}{2}\Delta x$ lb and must be lifted x_i^* ft, so its contribution to the total work is $\frac{1}{2}x_i^*\,\Delta x$ ft-lb. The total work is

$$W = \lim_{n\to\infty}\sum_{i=1}^{n}\tfrac{1}{2}x_i^*\,\Delta x = \int_0^{50}\tfrac{1}{2}x\,dx = \left[\tfrac{1}{4}x^2\right]_0^{50} = \tfrac{2500}{4} = 625 \text{ ft-lb}$$

Notice that the exact height of the building does not matter (as long as it is more than 50 ft).

(b) When half the rope is pulled to the top of the building, the work to lift the top half of the rope is

$W_1 = \int_0^{25} \frac{1}{2}x\,dx = \left[\frac{1}{4}x^2\right]_0^{25} = \frac{625}{4}$ ft-lb. The bottom half of the rope is lifted 25 ft and the work needed to accomplish that is $W_2 = \int_{25}^{50} \frac{1}{2} \cdot 25\,dx = \frac{25}{2}[x]_{25}^{50} = \frac{625}{2}$ ft-lb. The total work done in pulling half the rope to the top of the building is $W = W_1 + W_2 = \frac{625}{2} + \frac{625}{4} = \frac{3}{4} \cdot 625 = \frac{1875}{4}$ ft-lb.

11. The work needed to lift the cable is $\lim_{n\to\infty} \sum_{i=1}^{n} 2x_i^*\,\Delta x = \int_0^{500} 2x\,dx = \left[x^2\right]_0^{500} = 250\,000$ ft-lb. The work needed to lift the coal is 800 lb · 500 ft $= 400\,000$ ft-lb. Thus, the total work required is $250\,000 + 400\,000 = 650\,000$ ft-lb.

13. At a height of x metres $(0 \le x \le 12)$, the mass of the rope is $(0.8 \text{ kg/m})(12 - x \text{ m}) = (9.6 - 0.8x)$ kg and the mass of the water is $\left(\frac{36}{12} \text{ kg/m}\right)(12 - x \text{ m}) = (36 - 3x)$ kg. The mass of the bucket is 10 kg, so the total mass is $(9.6 - 0.8x) + (36 - 3x) + 10 = (55.6 - 3.8x)$ kg, and hence, the total force is $9.8(55.6 - 3.8x)$ N. The work needed to lift the bucket Δx m through the ith subinterval of $[0, 12]$ is $9.8(55.6 - 3.8x_i^*)\Delta x$, so the total work is

$$W = \lim_{n\to\infty} \sum_{i=1}^{n} 9.8(55.6 - 3.8x_i^*)\,\Delta x = \int_0^{12} (9.8)(55.6 - 3.8x)\,dx = 9.8\Big[55.6x - 1.9x^2\Big]_0^{12} = 9.8(393.6) \approx 3857 \text{ J}$$

15. A "slice" of water Δx m thick and lying at a depth of x_i^* m $\left(\text{where } 0 \le x_i^* \le \frac{1}{2}\right)$ has volume $(2 \times 1 \times \Delta x)$ m^3, a mass of $2000\,\Delta x$ kg, weighs about $(9.8)(2000\,\Delta x) = 19\,600\,\Delta x$ N, and thus requires about $19\,600 x_i^*\,\Delta x$ J of work for its removal.

So $W = \lim_{n\to\infty} \sum_{i=1}^{n} 19\,600 x_i^*\,\Delta x = \int_0^{1/2} 19\,600x\,dx = \left[9800x^2\right]_0^{1/2} = 2450$ J.

17. (a) A rectangular "slice" of water Δx m thick and lying x ft above the bottom has width x ft and volume $8x\,\Delta x$ m^3. It weighs about $(9.8 \times 1000)(8x\,\Delta x)$ N, and must be lifted $(5 - x)$ m by the pump, so the work needed is about $(9.8 \times 10^3)(5 - x)(8x\,\Delta x)$ J. The total work required is

$$W \approx \int_0^3 (9.8 \times 10^3)(5 - x)8x\,dx = (9.8 \times 10^3)\int_0^3 (40x - 8x^2)\,dx = (9.8 \times 10^3)\left[20x^2 - \tfrac{8}{3}x^3\right]_0^3$$
$$= (9.8 \times 10^3)(180 - 72) = (9.8 \times 10^3)(108) = 1058.4 \times 10^3 \approx 1.06 \times 10^6 \text{ J}$$

(b) If only 4.7×10^5 J of work is done, then only the water above a certain level (call it h) will be pumped out. So we use the same formula as in part (a), except that the work is fixed, and we are trying to find the lower limit of integration:

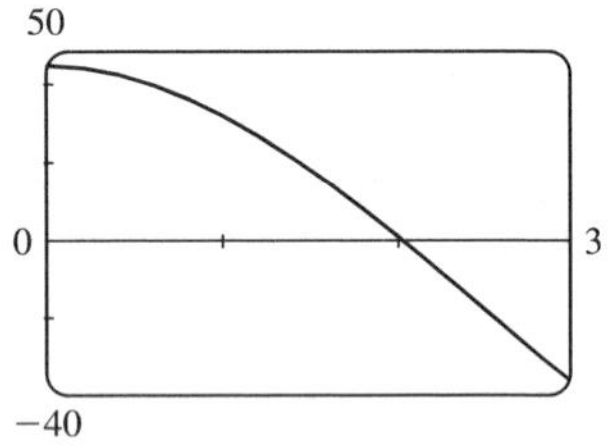

$4.7 \times 10^5 \approx \int_h^3 (9.8 \times 10^3)(5 - x)8x\,dx = (9.8 \times 10^3)\left[20x^2 - \frac{8}{3}x^3\right]_h^3 \quad \Leftrightarrow$

$\frac{4.7}{9.8} \times 10^2 \approx 48 = \left(20 \cdot 3^2 - \frac{8}{3} \cdot 3^3\right) - \left(20h^2 - \frac{8}{3}h^3\right) \quad \Leftrightarrow$

$2h^3 - 15h^2 + 45 = 0$. To find the solution of this equation, we plot $2h^3 - 15h^2 + 45$ between $h = 0$ and $h = 3$. We see that the equation is satisfied for $h \approx 2.0$. So the depth of water remaining in the tank is about 2.0 m.

19. $V = \pi r^2 x$, so V is a function of x and P can also be regarded as a function of x. If $V_1 = \pi r^2 x_1$ and $V_2 = \pi r^2 x_2$, then

$$W = \int_{x_1}^{x_2} F(x)\,dx = \int_{x_1}^{x_2} \pi r^2 P(V(x))\,dx = \int_{x_1}^{x_2} P(V(x))\,dV(x) \qquad [\text{Let } V(x) = \pi r^2 x, \text{ so } dV(x) = \pi r^2\,dx.]$$
$$= \int_{V_1}^{V_2} P(V)\,dV \quad \text{by the Substitution Rule.}$$

21. (a) $W = \displaystyle\int_a^b F(r)\,dr = \int_a^b G\frac{m_1 m_2}{r^2}\,dr = G m_1 m_2\left[\frac{-1}{r}\right]_a^b = G m_1 m_2\left(\frac{1}{a} - \frac{1}{b}\right)$

(b) By part (a), $W = GMm\left(\frac{1}{R} - \frac{1}{R + 1\,000\,000}\right)$ where M = mass of Earth in kg, R = radius of Earth in m, and m = mass of satellite in kg. (Note that 1000 km = 1 000 000 m.) Thus,

$$W = (6.67 \times 10^{-11})(5.98 \times 10^{24})(1000) \times \left(\frac{1}{6.37 \times 10^6} - \frac{1}{7.37 \times 10^6}\right) \approx 8.50 \times 10^9 \text{ J}$$

23. Since an equation for the shape is $x^2 + y^2 = 10^2$ ($x \geq 0$), we have $y = \sqrt{100 - x^2}$. Thus, the area of the ith strip is $2\sqrt{100 - (x_i^*)^2}\,\Delta x$ and the pressure on the strip is $\rho g x_i^*$, so the hydrostatic force on the strip is $\rho g x_i^* \cdot 2\sqrt{100 - (x_i^*)^2}\,\Delta x$ and the total force on the plate $\approx \sum_{i=1}^{n} \rho g x_i^* \cdot 2\sqrt{100 - (x_i^*)^2}\,\Delta x$. The total force

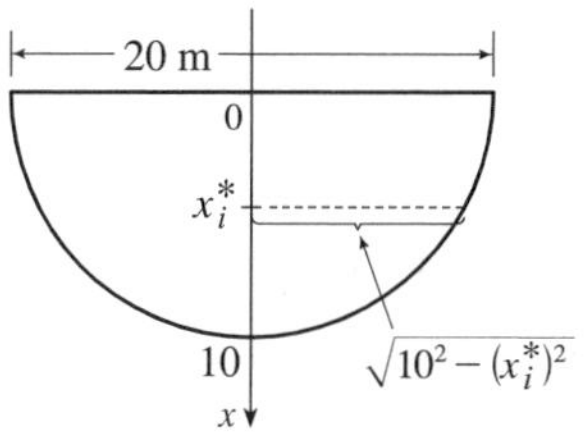

$$\begin{aligned} F &= \lim_{n\to\infty} \sum_{i=1}^{n} \rho g x_i^* \cdot 2\sqrt{100 - (x_i^*)^2}\,\Delta x = \int_0^{10} 2\rho g x\sqrt{100 - x^2}\,dx \\ &= -\rho g \textstyle\int_0^{10} (100 - x^2)^{1/2}(-2x)\,dx = -\rho g\left[\frac{2}{3}(100 - x^2)^{3/2}\right]_0^{10} = -\frac{2}{3}\rho g(0 - 1000) \\ &= \tfrac{2000}{3}\rho g \approx \tfrac{2000}{3} \cdot 1000 \cdot 9.8 \approx 6.5 \times 10^6 \text{ N} \qquad \left[\rho \approx 1000 \text{ kg/m}^3 \text{ and } g \approx 9.8 \text{ m/s}^2\right] \end{aligned}$$

25. Using similar triangles, $\frac{4 \text{ ft wide}}{8 \text{ ft high}} = \frac{a \text{ ft wide}}{x_i^* \text{ ft high}}$, so $a = \frac{1}{2}x_i^*$ and the width of the ith rectangular strip is $12 + 2a = 12 + x_i^*$. The area of the strip is $(12 + x_i^*)\,\Delta x$. The pressure on the strip is δx_i^*.

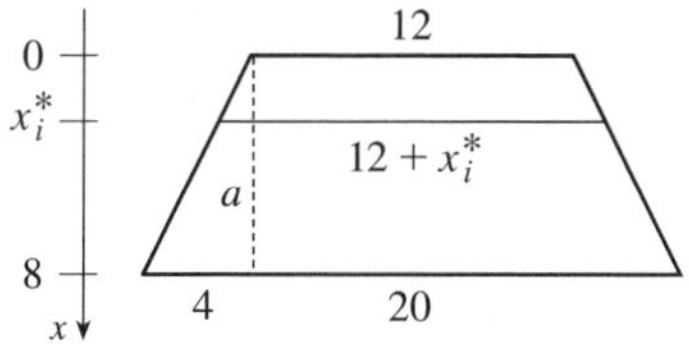

$$\begin{aligned} F &= \lim_{n\to\infty} \sum_{i=1}^{n} \delta x_i^*(12 + x_i^*)\,\Delta x = \textstyle\int_0^8 \delta x \cdot (12 + x)\,dx = \delta \int_0^8 (12x + x^2)\,dx \\ &= \delta\left[6x^2 + \tfrac{x^3}{3}\right]_0^8 = \delta\left(384 + \tfrac{512}{3}\right) = (62.5)\tfrac{1664}{3} \approx 3.47 \times 10^4 \text{ lb} \end{aligned}$$

27. By similar triangles, $\frac{8}{4\sqrt{3}} = \frac{w_i}{x_i^*} \Rightarrow w_i = \frac{2x_i^*}{\sqrt{3}}$. The area of the ith rectangular strip is $\frac{2x_i^*}{\sqrt{3}}\,\Delta x$ and the pressure on it is $\rho g\left(4\sqrt{3} - x_i^*\right)$.

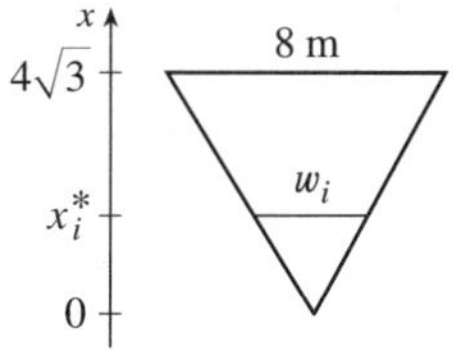

$$\begin{aligned} F &= \int_0^{4\sqrt{3}} \rho g\left(4\sqrt{3} - x\right)\frac{2x}{\sqrt{3}}\,dx = 8\rho g\int_0^{4\sqrt{3}} x\,dx - \frac{2\rho g}{\sqrt{3}}\int_0^{4\sqrt{3}} x^2\,dx \\ &= 4\rho g\left[x^2\right]_0^{4\sqrt{3}} - \frac{2\rho g}{3\sqrt{3}}\left[x^3\right]_0^{4\sqrt{3}} = 192\rho g - \frac{2\rho g}{3\sqrt{3}}\,64 \cdot 3\sqrt{3} = 192\rho g - 128\rho g = 64\rho g \\ &\approx 64(840)(9.8) \approx 5.27 \times 10^5 \text{ N} \end{aligned}$$

29. (a) The area of a strip is $10\,\Delta x$ and the pressure on it is $\rho g x_i$.

$$\begin{aligned} F &= \textstyle\int_0^1 \rho g x 10\,dx = 10\rho g\left[\frac{1}{2}x^2\right]_0^1 = 10\rho g \cdot \frac{1}{2} = 5\rho g \\ &= 5(1000)(9.8) = 49\,000 \text{ N} = 4.9 \times 10^4 \text{ N} \end{aligned}$$

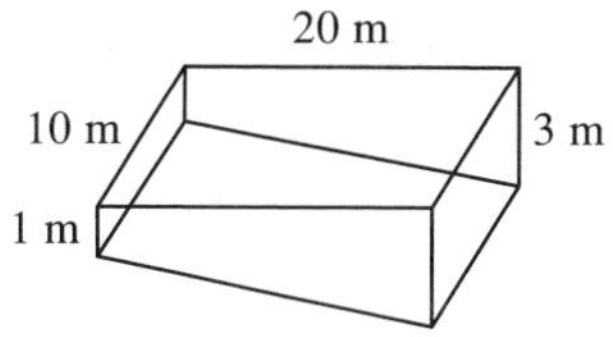

(b) $F = \int_0^3 \rho g x 10\,dx = 10\rho g\left[\frac{1}{2}x^2\right]_0^3 = 10\rho g \cdot \frac{9}{2} = 45\rho g = 45(1000)(9.8) = 441\,000 \text{ N} \approx 4.4 \times 10^5$ N.

(c) For the first 1 m, the length of the side is constant at 20 m. For $1 < x \le 3$, we can use similar triangles to find the length a:

$\frac{a}{20} = \frac{3-x}{2} \quad\Rightarrow\quad a = 20 \cdot \frac{3-x}{2} = 10(3-x)$.

$$F = \int_0^1 \rho g x\, 20\, dx + \int_1^3 \rho g x\, 10(3-x)\, dx = 20\rho g\left[\tfrac{1}{2}x^2\right]_0^1 + 10\rho g \int_1^3 (3x - x^2)\, dx$$

$$= 20\rho g\left(\tfrac{1}{2}\right) + 10\rho g\left[\tfrac{3}{2}x^2 - \tfrac{1}{3}x^3\right]_1^3 = 10\rho g + 10\rho g\left[\left(\tfrac{27}{2} - 9\right) - \left(\tfrac{3}{2} - \tfrac{1}{3}\right)\right]$$

$$= 10\rho g + 10\rho g\left(\tfrac{10}{3}\right) = \tfrac{130}{3}\rho g = \tfrac{130}{3}(1000)(9.8)\text{ N} \approx 424\,667\text{ N} \approx 4.2 \times 10^5\text{ N}$$

(d) For any right triangle with hypotenuse on the bottom,

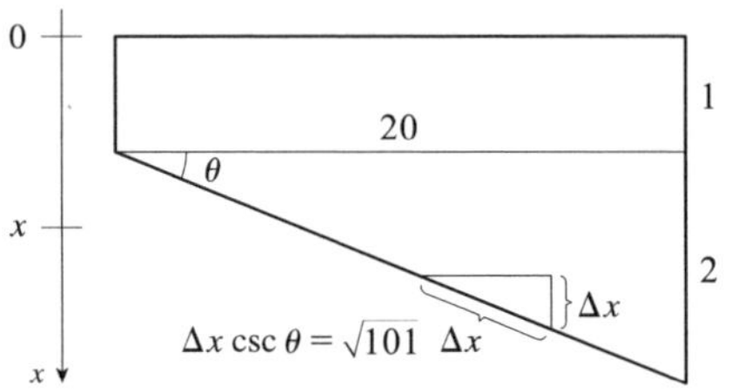

$\sin\theta = \frac{\Delta x}{\text{hypotenuse}} \quad\Rightarrow$

$\text{hypotenuse} = \Delta x \csc\theta = \Delta x\, \frac{\sqrt{20^2 + 2^2}}{2} = \sqrt{101}\,\Delta x$.

$$F = \int_1^3 \rho g x 10\sqrt{101}\, dx = 10\sqrt{101}\,\rho g\left[\tfrac{1}{2}x^2\right]_1^3$$

$$= 5\sqrt{101}\,\rho g(9-1) = 40\sqrt{101}(1000)(9.8) \approx 3\,939\,551\text{ N} \approx 3.9 \times 10^6\text{ N}$$

31. $F = \int_2^5 \rho g x \cdot w(x)\, dx$, where $w(x)$ is the width of the plate at depth x. Since $n = 6$, $\Delta x = \frac{5-2}{6} = \frac{1}{2}$, and

$$F \approx S_6 = \rho g \cdot \tfrac{1/2}{3}[2 \cdot w(2) + 4 \cdot 2.5 \cdot w(2.5) + 2 \cdot 3 \cdot w(3) + 4 \cdot 3.5 \cdot w(3.5) + 2 \cdot 4 \cdot w(4) + 4 \cdot 4.5 \cdot w(4.5) + 5 \cdot w(5)]$$

$$= \tfrac{1}{6}\rho g(2 \cdot 0 + 10 \cdot 0.8 + 6 \cdot 1.7 + 14 \cdot 2.4 + 8 \cdot 2.9 + 18 \cdot 3.3 + 5 \cdot 3.6)$$

$$= \tfrac{1}{6}(1000)(9.8)(152.4) \approx 2.5 \times 10^5\text{ N}$$

33. $m = \sum_{i=1}^{3} m_i = 6 + 5 + 10 = 21$.

$M_x = \sum_{i=1}^{3} m_i y_i = 6(5) + 5(-2) + 10(-1) = 10$; $M_y = \sum_{i=1}^{3} m_i x_i = 6(1) + 5(3) + 10(-2) = 1$.

$\overline{x} = \frac{M_y}{m} = \frac{1}{21}$ and $\overline{y} = \frac{M_x}{m} = \frac{10}{21}$, so the centre of mass of the system is $\left(\frac{1}{21}, \frac{10}{21}\right)$.

35. Since the region in the figure is symmetric about the y-axis, we know that $\overline{x} = 0$. The region is "bottom-heavy," so we know that $\overline{y} < 2$, and we might guess that $\overline{y} = 1.5$.

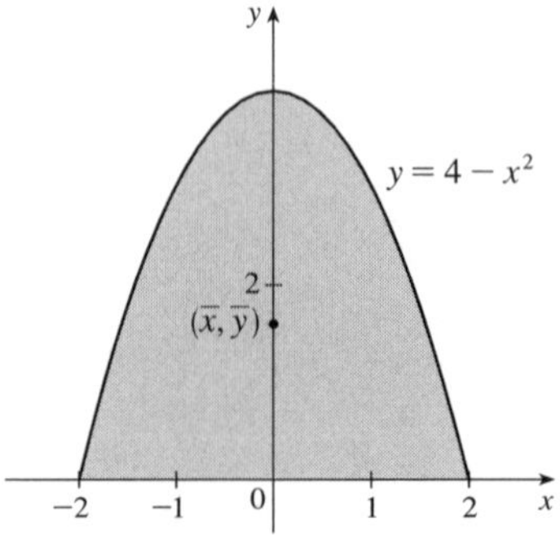

$A = \int_{-2}^{2}(4 - x^2)\, dx = 2\int_0^2 (4 - x^2)\, dx = 2\left[4x - \tfrac{1}{3}x^3\right]_0^2$
$= 2\left(8 - \tfrac{8}{3}\right) = \tfrac{32}{3}$

$\overline{x} = \frac{1}{A}\int_{-2}^{2} x(4 - x^2)\, dx = 0$ since $f(x) = x(4 - x^2)$ is an odd function (or since the region is symmetric about the y-axis).

$$\overline{y} = \frac{1}{A}\int_{-2}^{2}\frac{1}{2}(4-x^2)^2\,dx = \frac{3}{32}\cdot\frac{1}{2}\cdot 2\int_0^2(16-8x^2+x^4)\,dx = \frac{3}{32}\left[16x - \frac{8}{3}x^3 + \frac{1}{5}x^5\right]_0^2$$

$$= \frac{3}{32}\left(32 - \frac{64}{3} + \frac{32}{5}\right) = 3\left(1 - \frac{2}{3} + \frac{1}{5}\right) = 3\left(\frac{8}{15}\right) = \frac{8}{5}$$

Thus, the centroid is $(\overline{x}, \overline{y}) = \left(0, \frac{8}{5}\right)$.

37. The region in the figure is "right-heavy" and "bottom-heavy," so we know $\overline{x} > 0.5$ and $\overline{y} < 1$, and we might guess that $\overline{x} = 0.6$ and $\overline{y} = 0.9$.

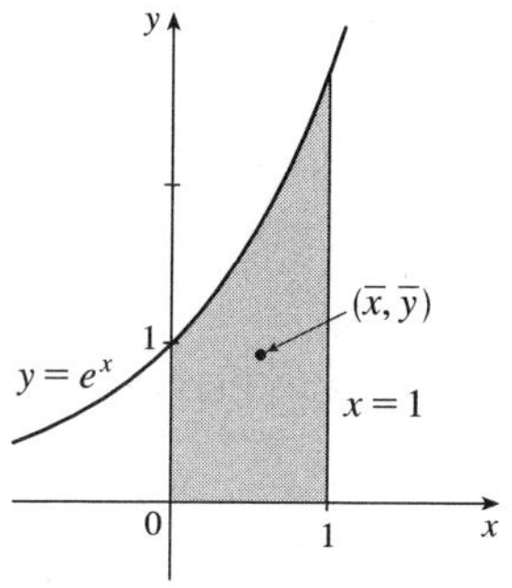

$A = \int_0^1 e^x\,dx = [e^x]_0^1 = e - 1$,

$\overline{x} = \frac{1}{A}\int_0^1 xe^x\,dx = \frac{1}{e-1}[xe^x - e^x]_0^1$ [by parts]

$= \frac{1}{e-1}[0-(-1)] = \frac{1}{e-1}$,

$\overline{y} = \frac{1}{A}\int_0^1 \frac{1}{2}(e^x)^2\,dx = \frac{1}{e-1}\cdot\frac{1}{4}\left[e^{2x}\right]_0^1 = \frac{1}{4(e-1)}(e^2-1) = \frac{e+1}{4}$.

Thus, the centroid is $(\overline{x}, \overline{y}) = \left(\frac{1}{e-1}, \frac{e+1}{4}\right) \approx (0.58, 0.93)$.

39. By symmetry, $M_y = 0$ and $\overline{x} = 0$. $A = \frac{1}{2}bh = \frac{1}{2}\cdot 2\cdot 2 = 2$.

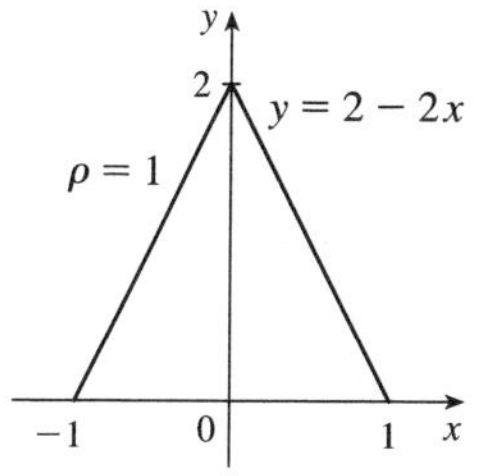

$M_x = \rho\int_{-1}^{1}\frac{1}{2}(2-2x)^2\,dx = 2\rho\int_0^1\frac{1}{2}(2-2x)^2\,dx$

$= \left(2\cdot 1\cdot\frac{1}{2}\cdot 2^2\right)\int_0^1(1-x)^2\,dx = 4\int_1^0 u^2(-du)$ $[u = 1-x,\ du = -dx]$

$= -4\left[\frac{1}{3}u^3\right]_1^0 = -4\left(-\frac{1}{3}\right) = \frac{4}{3}$

$\overline{y} = \frac{1}{m}M_x = \frac{1}{\rho A}M_x = \frac{1}{1\cdot 2}\cdot\frac{4}{3} = \frac{2}{3}$. Thus, the centroid is $(\overline{x}, \overline{y}) = \left(0, \frac{2}{3}\right)$.

41. (a)

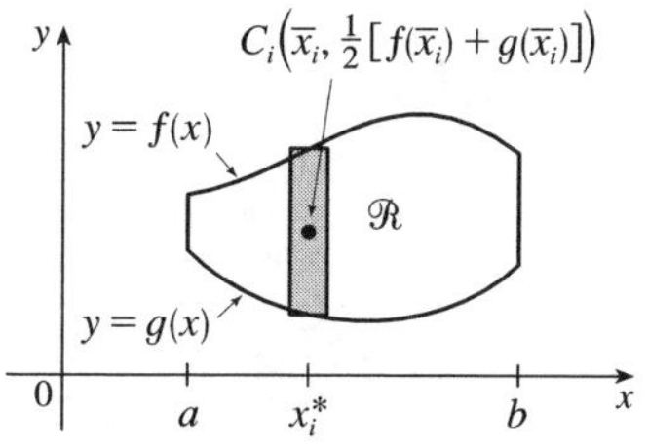

Suppose the region lies between two curves $y = f(x)$ and $y = g(x)$ where $f(x) \geq g(x)$, as illustrated in the figure. Use n subintervals determined by points x_i with $a = x_0 < x_1 < \cdots < x_n = b$ and choose $x_i^* = \overline{x}_i$ to be the midpoint of the ith subinterval; that is, $\overline{x}_i = \frac{1}{2}(x_{i-1} + x_i)$. Then the centroid of the ith approximating rectangle R_i is its centre $C_i = \left(\overline{x}_i, \frac{1}{2}[f(\overline{x}_i) + g(\overline{x}_i)]\right)$.

Its area is $[f(\overline{x}_i) - g(\overline{x}_i)]\,\Delta x$, so its mass is $\rho[f(\overline{x}_i) - g(\overline{x}_i)]\,\Delta x$.
Thus, $M_y(R_i) = \rho[f(\overline{x}_i) - g(\overline{x}_i)]\,\Delta x\cdot\overline{x}_i = \rho\overline{x}_i[f(\overline{x}_i) - g(\overline{x}_i)]\,\Delta x$ and
$M_x(R_i) = \rho[f(\overline{x}_i) - g(\overline{x}_i)]\,\Delta x\cdot\frac{1}{2}[f(\overline{x}_i) + g(\overline{x}_i)] = \rho\cdot\frac{1}{2}\left\{[f(\overline{x}_i)]^2 - [g(\overline{x}_i)]^2\right\}\Delta x$. Summing over i and taking the limit as $n\to\infty$, we get $M_y = \lim\limits_{n\to\infty}\sum\limits_{i=1}^{n}\rho\overline{x}_i[f(\overline{x}_i) - g(\overline{x}_i)]\,\Delta x = \rho\int_a^b x[f(x) - g(x)]\,dx$ and

$M_x = \lim\limits_{n\to\infty}\sum\limits_{i=1}^{n}\rho\cdot\frac{1}{2}\left[f(\overline{x}_i)^2 - g(\overline{x}_i)^2\right]\Delta x = \rho\int_a^b\frac{1}{2}\left\{[f(x)]^2 - [g(x)]^2\right\}dx$. Thus,

$$\overline{x} = \frac{M_y}{m} = \frac{M_y}{\rho A} = \frac{1}{A}\int_a^b x[f(x) - g(x)]\,dx \text{ and } \overline{y} = \frac{M_x}{m} = \frac{M_x}{\rho A} = \frac{1}{A}\int_a^b \frac{1}{2}\left\{[f(x)]^2 - [g(x)]^2\right\}dx.$$

(b)

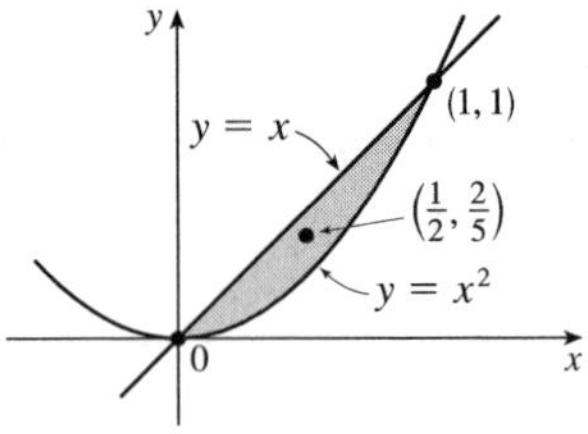

The region is sketched in the figure. We take $f(x) = x$, $g(x) = x^2$, $a = 0$, and $b = 1$ in the formulas in part (a). First we note that the area of the region is $A = \int_0^1 (x - x^2)\,dx = \left[\frac{1}{2}x^2 - \frac{1}{3}x^3\right]_0^1 = \frac{1}{6}$.

Therefore, $\overline{x} = \frac{1}{A}\int_0^1 x[f(x) - g(x)]\,dx = \frac{1}{1/6}\int_0^1 x(x - x^2)\,dx = 6\int_0^1 (x^2 - x^3)\,dx = 6\left[\frac{1}{3}x^3 - \frac{1}{4}x^4\right]_0^1 = \frac{1}{2}$

and $\overline{y} = \frac{1}{A}\int_0^1 \frac{1}{2}\left\{[f(x)]^2 - [g(x)]^2\right\} dx = \frac{1}{1/6}\int_0^1 \frac{1}{2}(x^2 - x^4)dx = 3\left[\frac{1}{3}x^3 - \frac{1}{5}x^5\right]_0^1 = \frac{2}{5}$.

The centroid is $\left(\frac{1}{2}, \frac{2}{5}\right)$.

6.6 Applications to Economics and Biology

1. By the Net Change Theorem, $C(2000) - C(0) = \int_0^{2000} C'(x)\,dx \quad\Rightarrow$

$$\begin{aligned} C(2000) &= 20\,000 + \int_0^{2000} (5 - 0.008x + 0.000\,009x^2)\,dx = 20\,000 + \left[5x - 0.004x^2 + 0.000\,003x^3\right]_0^{2000} \\ &= 20\,000 + 10\,000 - 0.004(4\,000\,000) + 0.000\,003(8\,000\,000\,000) = 30\,000 - 16\,000 + 24\,000 \\ &= \$38\,000 \end{aligned}$$

3. If the production level is raised from 1200 units to 1600 units, then the increase in cost is

$$\begin{aligned} C(1600) - C(1200) &= \int_{1200}^{1600} C'(x)\,dx = \int_{1200}^{1600} (74 + 1.1x - 0.002x^2 + 0.00004x^3)\,dx \\ &= \left[74x + 0.55x^2 - \tfrac{0.002}{3}x^3 + 0.00001x^4\right]_{1200}^{1600} \\ &= 64\,331\,733.33 - 20\,464\,800 = \$43\,866\,933.33 \end{aligned}$$

5. $p(x) = 10 \quad\Rightarrow\quad \dfrac{450}{x+8} = 10 \quad\Rightarrow\quad x + 8 = 45 \quad\Rightarrow\quad x = 37.$

$$\begin{aligned} \text{Consumer surplus} &= \int_0^{37} [p(x) - 10]\,dx = \int_0^{37} \left(\frac{450}{x+8} - 10\right) dx \\ &= \left[450\ln(x+8) - 10x\right]_0^{37} = (450\ln 45 - 370) - 450\ln 8 \\ &= 450\ln\left(\tfrac{45}{8}\right) - 370 \approx \$407.25 \end{aligned}$$

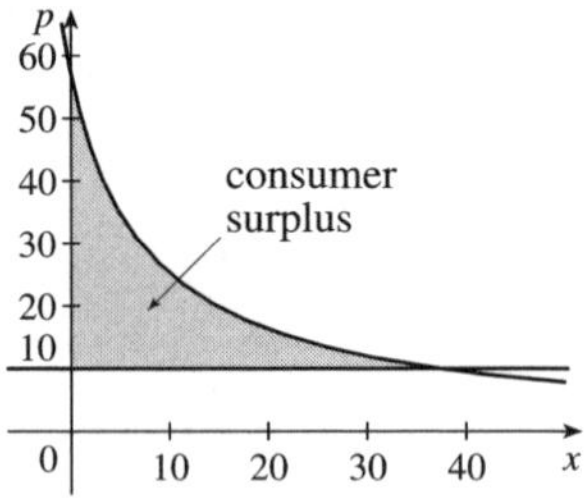

7. $P = p_S(x) \quad\Rightarrow\quad 400 = 200 + 0.2x^{3/2} \quad\Rightarrow\quad 200 = 0.2x^{3/2} \quad\Rightarrow\quad 1000 = x^{3/2} \quad\Rightarrow\quad x = 1000^{2/3} = 100.$

$$\begin{aligned} \text{Producer surplus} &= \int_0^{100} [P - p_S(x)]\,dx = \int_0^{100} \left[400 - (200 + 0.2x^{3/2})\right] dx = \int_0^{100} \left(200 - \tfrac{1}{5}x^{3/2}\right) dx \\ &= \left[200x - \tfrac{2}{25}x^{5/2}\right]_0^{100} = 20\,000 - 8000 = \$12\,000 \end{aligned}$$

9. $p(x) = \dfrac{800\,000e^{-x/5000}}{x + 20\,000} = 16 \quad\Rightarrow\quad x = x_1 \approx 3727.04.$

Consumer surplus $= \int_0^{x_1} [p(x) - 16]\,dx \approx \$37\,753$

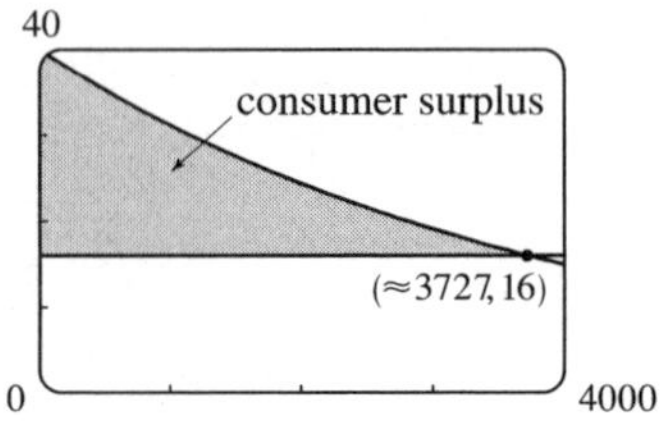

11. $f(8) - f(4) = \int_4^8 f'(t)\,dt = \int_4^8 \sqrt{t}\,dt = \left[\frac{2}{3}t^{3/2}\right]_4^8 = \frac{2}{3}\left(16\sqrt{2} - 8\right) \approx \$9.75$ million

13. $F = \dfrac{\pi P R^4}{8\eta l} = \dfrac{\pi(4000)(0.008)^4}{8(0.027)(2)} \approx 1.19 \times 10^{-4}\ \text{cm}^3/\text{s}$

15. $\int_0^{12} c(t)\,dt = \int_0^{12} \frac{1}{4}t(12-t)\,dt = \int_0^{12}\left(3t - \frac{1}{4}t^2\right)dt = \left[\frac{3}{2}t^2 - \frac{1}{12}t^3\right]_0^{12} = (216 - 144) = 72\ \text{mg}\cdot\text{s/L}.$

Thus, the cardiac output is $F = \dfrac{A}{\int_0^{12} c(t)\,dt} = \dfrac{8\ \text{mg}}{72\ \text{mg}\cdot\text{s/L}} = \dfrac{1}{9}\ \text{L/s} = \dfrac{60}{9}\ \text{L/min}.$

6.7 Probability

1. (a) $\int_{30\,000}^{40\,000} f(x)\,dx$ is the probability that a randomly chosen tire will have a lifetime between 30 000 and 40 000 km.

(b) $\int_{25\,000}^{\infty} f(x)\,dx$ is the probability that a randomly chosen tire will have a lifetime of at least 25 000 km.

3. (a) In general, we must satisfy the two conditions that are mentioned before Example 1—namely, (1) $f(x) \geq 0$ for all x, and (2) $\int_{-\infty}^{\infty} f(x)\,dx = 1$. For $0 \leq x \leq 4$, we have $f(x) = \frac{3}{64}x\sqrt{16 - x^2} \geq 0$, so $f(x) \geq 0$ for all x. Also,

$$\begin{aligned}\int_{-\infty}^{\infty} f(x)\,dx &= \int_0^4 \tfrac{3}{64}x\sqrt{16-x^2}\,dx = -\tfrac{3}{128}\int_0^4 (16-x^2)^{1/2}(-2x)\,dx = -\tfrac{3}{128}\left[\tfrac{2}{3}(16-x^2)^{3/2}\right]_0^4 \\ &= -\tfrac{1}{64}\left[(16-x^2)^{3/2}\right]_0^4 = -\tfrac{1}{64}(0-64) = 1.\end{aligned}$$

Therefore, f is a probability density function.

(b)
$$\begin{aligned}P(X < 2) &= \int_{-\infty}^{2} f(x)\,dx = \int_0^2 \tfrac{3}{64}x\sqrt{16-x^2}\,dx = -\tfrac{3}{128}\int_0^2 (16-x^2)^{1/2}(-2x)\,dx \\ &= -\tfrac{3}{128}\left[\tfrac{2}{3}(16-x^2)^{3/2}\right]_0^2 = -\tfrac{1}{64}\left[(16-x^2)^{3/2}\right]_0^2 = -\tfrac{1}{64}\left(12^{3/2} - 16^{3/2}\right) \\ &= \tfrac{1}{64}\left(64 - 12\sqrt{12}\right) = \tfrac{1}{64}\left(64 - 24\sqrt{3}\right) = 1 - \tfrac{3}{8}\sqrt{3} \approx 0.350\,481\end{aligned}$$

5. (a) In general, we must satisfy the two conditions that are mentioned before Example 1—namely, (1) $f(x) \geq 0$ for all x, and (2) $\int_{-\infty}^{\infty} f(x)\,dx = 1$. Since $f(x) = 0$ or $f(x) = 0.1$, condition (1) is satisfied. For condition (2), we see that $\int_{-\infty}^{\infty} f(x)\,dx = \int_0^{10} 0.1\,dx = \left[\frac{1}{10}x\right]_0^{10} = 1$. Thus, $f(x)$ is a probability density function for the spinner's values.

(b) Since all the numbers between 0 and 10 are equally likely to be selected, we expect the mean to be halfway between the endpoints of the interval; that is, $x = 5$.

$$\mu = \int_{-\infty}^{\infty} x f(x)\,dx = \int_0^{10} x(0.1)\,dx = \left[\tfrac{1}{20}x^2\right]_0^{10} = \tfrac{100}{20} = 5, \quad \text{as expected.}$$

7. We need to find m so that $\int_m^{\infty} f(t)\,dt = \frac{1}{2} \;\Rightarrow\; \lim\limits_{x\to\infty}\int_m^x \frac{1}{5}e^{-t/5}\,dt = \frac{1}{2} \;\Rightarrow\; \lim\limits_{x\to\infty}\left[\frac{1}{5}(-5)e^{-t/5}\right]_m^x = \frac{1}{2} \;\Rightarrow$ $(-1)(0 - e^{-m/5}) = \frac{1}{2} \;\Rightarrow\; e^{-m/5} = \frac{1}{2} \;\Rightarrow\; -m/5 = \ln\frac{1}{2} \;\Rightarrow\; m = -5\ln\frac{1}{2} = 5\ln 2 \approx 3.47$ min.

9. We use an exponential density function with $\mu = 2.5$ min.

(a) $P(X > 4) = \int_4^{\infty} f(t)\,dt = \lim\limits_{x\to\infty}\int_4^x \frac{1}{2.5}e^{-t/2.5}\,dt = \lim\limits_{x\to\infty}\left[-e^{-t/2.5}\right]_4^x = 0 + e^{-4/2.5} \approx 0.202$

(b) $P(0 \le X \le 2) = \int_0^2 f(t)\,dt = \left[-e^{-t/2.5}\right]_0^2 = -e^{-2/2.5} + 1 \approx 0.551$

(c) We need to find a value a so that $P(X \ge a) = 0.02$, or, equivalently, $P(0 \le X \le a) = 0.98 \Leftrightarrow$

$\int_0^a f(t)\,dt = 0.98 \Leftrightarrow \left[-e^{-t/2.5}\right]_0^a = 0.98 \Leftrightarrow -e^{-a/2.5} + 1 = 0.98 \Leftrightarrow e^{-a/2.5} = 0.02 \Leftrightarrow$

$-a/2.5 = \ln 0.02 \Leftrightarrow a = -2.5 \ln \frac{1}{50} = 2.5 \ln 50 \approx 9.78$ min ≈ 10 min. The ad should say that if you aren't served within 10 minutes, you get a free hamburger.

11. $P(X \ge 5) = \displaystyle\int_5^\infty \frac{1}{1.9\sqrt{2\pi}} \exp\left(-\frac{(x-4.3)^2}{2 \cdot 1.9^2}\right) dx$. To avoid the improper integral we approximate it by the integral from 5 to 50. Thus, $P(X \ge 5) \approx \displaystyle\int_5^{50} \frac{1}{1.9\sqrt{2\pi}} \exp\left(-\frac{(x-4.3)^2}{2 \cdot 1.9^2}\right) dx \approx 0.356$ (using a calculator or computer to estimate the integral), so about 36 percent of the households throw out at least 5 kg of paper a week.

Note: We can't evaluate $1 - P(0 \le X \le 5)$ for this problem since a significant amount of area lies to the left of $X = 0$.

13. $P(\mu - 2\sigma \le X \le \mu + 2\sigma) = \displaystyle\int_{\mu-2\sigma}^{\mu+2\sigma} \frac{1}{\sigma\sqrt{2\pi}} \exp\left(-\frac{(x-\mu)^2}{2\sigma^2}\right) dx$. Substituting $t = \dfrac{x-\mu}{\sigma}$ and $dt = \dfrac{1}{\sigma}\,dx$ gives us

$\displaystyle\int_{-2}^{2} \frac{1}{\sigma\sqrt{2\pi}} e^{-t^2/2}(\sigma\,dt) = \frac{1}{\sqrt{2\pi}} \int_{-2}^{2} e^{-t^2/2}\,dt \approx 0.9545.$

6 Review

CONCEPT CHECK

1. (a) See Section 6.1, Figure 2 and Equations 6.1.1 and 6.1.2.

(b) Instead of using "top minus bottom" and integrating from left to right, we use "right minus left" and integrate from bottom to top. See Figures 9 and 10 in Section 6.1.

2. The numerical value of the area represents the number of metres by which Sue is ahead of Kathy after 1 minute.

3. (a) See the discussion in Section 6.2, near Figures 2 and 3, ending in the Definition of Volume.

(b) See the discussion between Examples 5 and 6 in Section 6.2. If the cross-section is a disc, find the radius in terms of x or y and use $A = \pi(\text{radius})^2$. If the cross-section is a washer, find the inner radius r_{in} and outer radius r_{out} and use $A = \pi\left(r_{\text{out}}^2\right) - \pi\left(r_{\text{in}}^2\right)$.

4. (a) The length of a curve is defined to be the limit of the lengths of the inscribed polygons, as described near Figure 3 in Section 6.3.

(b) See Equation 6.3.1.

(c) See Equations 6.3.2 and 6.3.3.

5. (a) The average value of a function f on an interval $[a, b]$ is $f_{\text{ave}} = \dfrac{1}{b-a}\displaystyle\int_a^b f(x)\,dx$.

(b) The Mean Value Theorem for Integrals says that there is a number c at which the value of f is exactly equal to the average value of the function, that is, $f(c) = f_{\text{ave}}$. For a geometric interpretation of the Mean Value Theorem for Integrals, see Figure 2 in Section 6.4 and the discussion that accompanies it.

6. $\int_0^6 f(x)\,dx$ represents the amount of work done. Its units are newton-metres, or joules.

7. Let $c(x)$ be the cross-sectional length of the wall (measured parallel to the surface of the fluid) at depth x. Then the hydrostatic force against the wall is given by $F = \int_a^b \delta x c(x)\,dx$, where a and b are the lower and upper limits for x at points of the wall and δ is the weight density of the fluid.

8. (a) The centre of mass is the point at which the plate balances horizontally.
(b) See Equations 6.5.12.

9. See Figure 3 in Section 6.6, and the discussion which precedes it.

10. (a) See the definition in the first paragraph of the subsection *Cardiac Output* in Section 6.6.
(b) See the discussion in the second paragraph of the subsection *Cardiac Output* in Section 6.6.

11. A probability density function f is a function on the domain of a continuous random variable X such that $\int_a^b f(x)\,dx$ measures the probability that X lies between a and b. Such a function f has nonnegative values and satisfies the relation $\int_D f(x)\,dx = 1$, where D is the domain of the corresponding random variable X. If $D = \mathbb{R}$, or if we define $f(x) = 0$ for real numbers $x \notin D$, then $\int_{-\infty}^{\infty} f(x)\,dx = 1$. (Of course, to work with f in this way, we must assume that the integrals of f exist.)

12. (a) $\int_0^{60} f(x)\,dx$ represents the probability that the weight of a randomly chosen female college student is less than 60 kg.
(b) $\mu = \int_{-\infty}^{\infty} xf(x)\,dx = \int_0^{\infty} xf(x)\,dx$
(c) The median of f is the number m such that $\int_m^{\infty} f(x)\,dx = \frac{1}{2}$.

13. See the discussion near Equation 3 in Section 6.7.

EXERCISES

1. $A = \int_0^1 \left[(e^x - 1) - (x^2 - x)\right] dx$

$= \int_0^1 (e^x - 1 - x^2 + x)\,dx = \left[e^x - x - \frac{1}{3}x^3 + \frac{1}{2}x^2\right]_0^1$

$= \left(e - 1 - \frac{1}{3} + \frac{1}{2}\right) - (1 - 0 - 0 + 0) = e - \frac{11}{6}$

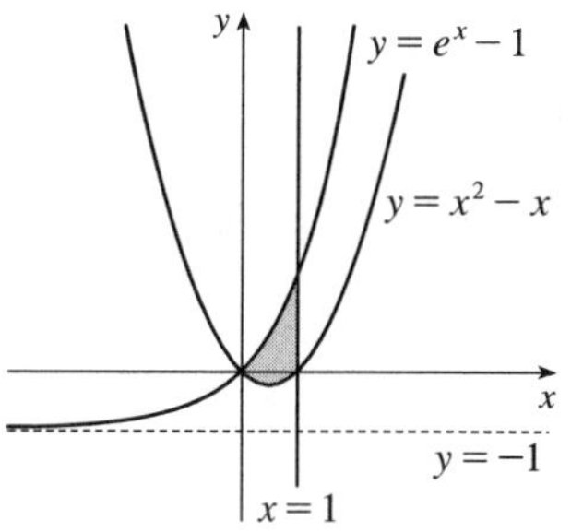

3. $x = 2\theta - \sin\theta \;\Rightarrow\; dx = (2 - \cos\theta)\,d\theta$

$A = \int_0^{2\pi} y\,dx = \int_0^{2\pi} \left[(2 - \cos\theta)(2 - \cos\theta)\right] d\theta$

$= \int_0^{2\pi} (4 - 4\cos\theta + \cos^2\theta)\,d\theta = \int_0^{2\pi} \left(4 - 4\cos\theta + \frac{1}{2} + \frac{1}{2}\cos 2\theta\right) d\theta$

$= \left[4\theta - 4\sin\theta + \frac{1}{2}\theta + \frac{1}{4}\sin 2\theta\right]_0^{2\pi} = (8\pi - 0 + \pi + 0) - (0) = 9\pi$

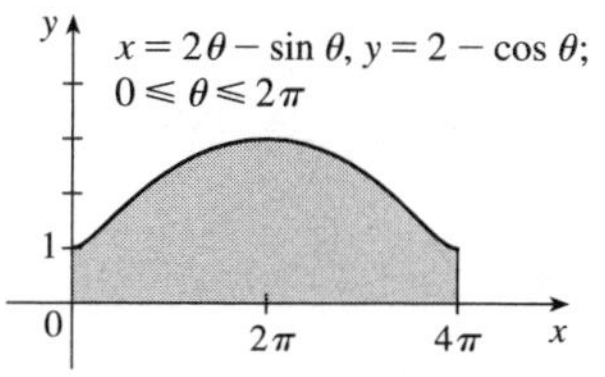

5. (a) Using the Midpoint Rule on $[0, 1]$ with $f(x) = \tan(x^2)$ and $n = 4$, we estimate

$$A = \int_0^1 \tan(x^2)\,dx \approx \tfrac{1}{4}\left[\tan\left(\left(\tfrac{1}{8}\right)^2\right) + \tan\left(\left(\tfrac{3}{8}\right)^2\right) + \tan\left(\left(\tfrac{5}{8}\right)^2\right) + \tan\left(\left(\tfrac{7}{8}\right)^2\right)\right] \approx \tfrac{1}{4}(1.53) \approx 0.38$$

(b) Using the Midpoint Rule on $[0, 1]$ with $f(x) = \pi \tan^2(x^2)$ (for discs) and $n = 4$, we estimate

$$V = \int_0^1 f(x)\,dx \approx \tfrac{1}{4}\pi\left[\tan^2\left(\left(\tfrac{1}{8}\right)^2\right) + \tan^2\left(\left(\tfrac{3}{8}\right)^2\right) + \tan^2\left(\left(\tfrac{5}{8}\right)^2\right) + \tan^2\left(\left(\tfrac{7}{8}\right)^2\right)\right] \approx \tfrac{\pi}{4}(1.114) \approx 0.87$$

7. (a) A cross-section is a washer with inner radius x^2 and outer radius x.

$$V = \int_0^1 \pi\left[(x)^2 - (x^2)^2\right] dx = \int_0^1 \pi(x^2 - x^4)\,dx = \pi\left[\tfrac{1}{3}x^3 - \tfrac{1}{5}x^5\right]_0^1 = \pi\left[\tfrac{1}{3} - \tfrac{1}{5}\right] = \tfrac{2\pi}{15}$$

(b) A cross-section is a washer with inner radius y and outer radius $\sqrt{y}$.

$$V = \int_0^1 \pi\left[\left(\sqrt{y}\right)^2 - y^2\right] dy = \int_0^1 \pi(y - y^2)\,dy = \pi\left[\tfrac{1}{2}y^2 - \tfrac{1}{3}y^3\right]_0^1 = \pi\left[\tfrac{1}{2} - \tfrac{1}{3}\right] = \tfrac{\pi}{6}$$

(c) A cross-section is a washer with inner radius $2 - x$ and outer radius $2 - x^2$.

$$V = \int_0^1 \pi\left[(2 - x^2)^2 - (2 - x)^2\right] dx = \int_0^1 \pi(x^4 - 5x^2 + 4x)\,dx = \pi\left[\tfrac{1}{5}x^5 - \tfrac{5}{3}x^3 + 2x^2\right]_0^1 = \pi\left[\tfrac{1}{5} - \tfrac{5}{3} + 2\right] = \tfrac{8\pi}{15}$$

9. (a) The solid is obtained by rotating the region $\mathcal{R} = \left\{(x, y) \mid 0 \le x \le \frac{\pi}{2}, 0 \le y \le \sqrt{2}\cos x\right\}$ about the x-axis.

(b) The solid is obtained by rotating the region $\mathcal{R} = \left\{(x, y) \mid 0 \le x \le 1, 2 - \sqrt{x} \le y \le 2 - x^2\right\}$ about the x-axis.

Or: The solid is obtained by rotating the region $\mathcal{R} = \left\{(x, y) \mid 0 \le x \le 1, x^2 \le y \le \sqrt{x}\right\}$ about the line $y = 2$.

11. Take the base to be the disc $x^2 + y^2 \le 9$. Then $V = \int_{-3}^{3} A(x)\,dx$, where $A(x_0)$ is the area of the isosceles right triangle whose hypotenuse lies along the line $x = x_0$ in the xy-plane. The length of the hypotenuse is $2\sqrt{9 - x^2}$ and the length of each leg is $\sqrt{2}\sqrt{9 - x^2}$. $A(x) = \frac{1}{2}\left(\sqrt{2}\sqrt{9 - x^2}\right)^2 = 9 - x^2$, so

$$V = 2\int_0^3 A(x)\,dx = 2\int_0^3 (9 - x^2)\,dx = 2\left[9x - \tfrac{1}{3}x^3\right]_0^3 = 2(27 - 9) = 36$$

13. Equilateral triangles with sides measuring $\frac{1}{4}x$ metres have height $\frac{1}{4}x\sin 60^\circ = \frac{\sqrt{3}}{8}x$. Therefore,

$$A(x) = \tfrac{1}{2} \cdot \tfrac{1}{4}x \cdot \tfrac{\sqrt{3}}{8}x = \tfrac{\sqrt{3}}{64}x^2. \quad V = \int_0^{20} A(x)\,dx = \tfrac{\sqrt{3}}{64}\int_0^{20} x^2\,dx = \tfrac{\sqrt{3}}{64}\left[\tfrac{1}{3}x^3\right]_0^{20} = \tfrac{8000\sqrt{3}}{64 \cdot 3} = \tfrac{125\sqrt{3}}{3}\ \text{m}^3.$$

15. $x = 3t^2$, $y = 2t^3$, $0 \le t \le 2$.

$$L = \int_0^2 \sqrt{(dx/dt)^2 + (dy/dt)^2}\,dt = \int_0^2 \sqrt{(6t)^2 + (6t^2)^2}\,dt = \int_0^2 \sqrt{36t^2 + 36t^4}\,dt = 6\int_0^2 t\sqrt{1 + t^2}\,dt$$
$$= 6\int_1^5 \sqrt{u}\left(\tfrac{1}{2}du\right)\ \left[u = 1 + t^2, du = 2t\,dt\right] = 3\left[\tfrac{2}{3}u^{3/2}\right]_1^5 = 2\left(5\sqrt{5} - 1\right)$$

17. $y = \frac{1}{6}(x^2 + 4)^{3/2} \Rightarrow dy/dx = \frac{1}{4}(x^2 + 4)^{1/2}(2x) \Rightarrow$

$$1 + (dy/dx)^2 = 1 + \left[\tfrac{1}{2}x(x^2 + 4)^{1/2}\right]^2 = 1 + \tfrac{1}{4}x^2(x^2 + 4) = \tfrac{1}{4}x^4 + x^2 + 1 = \left(\tfrac{1}{2}x^2 + 1\right)^2.$$

Thus, $L = \int_0^3 \sqrt{\left(\frac{1}{2}x^2 + 1\right)^2}\,dx = \int_0^3 \left(\frac{1}{2}x^2 + 1\right)dx = \left[\frac{1}{6}x^3 + x\right]_0^3 = \frac{15}{2}$.

19. $f(x) = kx \Rightarrow 30\ \text{N} = k(15 - 12)\ \text{cm} \Rightarrow k = 10\ \text{N/cm} = 1000\ \text{N/m}$. $20\ \text{cm} - 12\ \text{cm} = 0.08\ \text{m} \Rightarrow$

$$W = \int_0^{0.08} kx\,dx = 1000\int_0^{0.08} x\,dx = 500\left[x^2\right]_0^{0.08} = 500(0.08)^2 = 3.2\ \text{N·m} = 3.2\ \text{J}.$$

21. (a) The parabola has equation $y = ax^2$ with vertex at the origin and passing through $(4, 4)$. $4 = a \cdot 4^2 \;\Rightarrow\; a = \frac{1}{4} \;\Rightarrow\; y = \frac{1}{4}x^2 \;\Rightarrow\; x^2 = 4y \;\Rightarrow\; x = 2\sqrt{y}$.

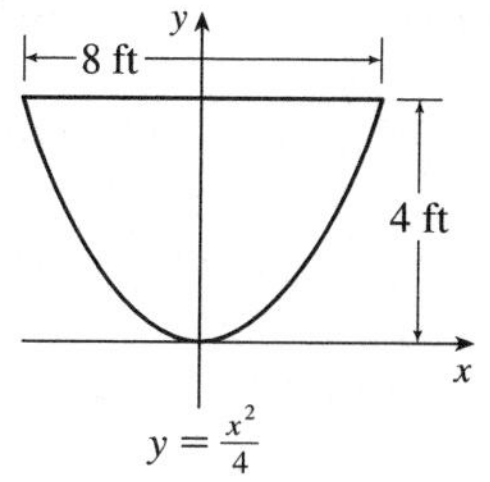

Each circular disc has radius $2\sqrt{y}$ and is moved $4 - y$ ft.

$$W = \int_0^4 \pi\left(2\sqrt{y}\right)^2 62.5\,(4 - y)\,dy = 250\pi \int_0^4 y(4 - y)\,dy$$

$$= 250\pi\left[2y^2 - \tfrac{1}{3}y^3\right]_0^4 = 250\pi\left(32 - \tfrac{64}{3}\right) = \tfrac{8000\pi}{3} \approx 8378 \text{ ft-lb}$$

(b) In part (a) we knew the final water level (0) but not the amount of work done. Here we use the same equation, except with the work fixed, and the lower limit of integration (that is, the final water level — call it h) unknown: $W = 4000 \;\Leftrightarrow$ $250\pi\left[2y^2 - \frac{1}{3}y^3\right]_h^4 = 4000 \;\Leftrightarrow\; \frac{16}{\pi} = \left[\left(32 - \frac{64}{3}\right) - \left(2h^2 - \frac{1}{3}h^3\right)\right] \;\Leftrightarrow$ $h^3 - 6h^2 + 32 - \frac{48}{\pi} = 0$. We graph the function $f(h) = h^3 - 6h^2 + 32 - \frac{48}{\pi}$ on the interval $[0, 4]$ to see where it is 0. From the graph, $f(h) = 0$ for $h \approx 2.1$. So the depth of water remaining is about 2.1 ft.

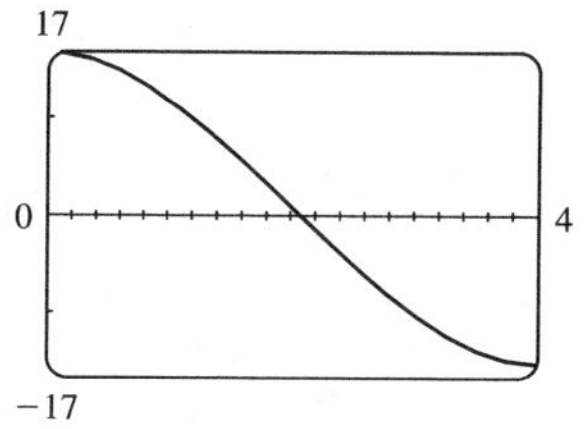

23. As in Example 4 of Section 6.5, $\dfrac{a}{1 - x} = \dfrac{\frac{1}{2}}{1} \;\Rightarrow\; 2a = 1 - x$ and $w = 2(0.5 + a) = 1 + 2a = 1 + 1 - x = 2 - x$.

Thus, $F = \int_0^1 \rho g x(2 - x)\,dx = \rho g\left[x^2 - \frac{1}{3}x^3\right]_0^1 = \rho g\left(1 - \frac{1}{3}\right) = \frac{2}{3}\rho g = \frac{2}{3}(1000)(9.8) \approx 6533$ N.

25. $x = 100 \;\Rightarrow\; P = 2000 - 0.1(100) - 0.01(100)^2 = 1890$

$$\text{Consumer surplus} = \int_0^{100} [p(x) - P]\,dx = \int_0^{100} \left(2000 - 0.1x - 0.01x^2 - 1890\right) dx$$

$$= \left[110x - 0.05x^2 - \tfrac{0.01}{3}x^3\right]_0^{100} = 11\,000 - 500 - \tfrac{10\,000}{3} \approx \$7166.67$$

27. $\displaystyle\lim_{h\to 0} f_{\text{ave}} = \lim_{h\to 0} \frac{1}{(x + h) - x} \int_x^{x+h} f(t)\,dt = \lim_{h\to 0} \frac{F(x + h) - F(x)}{h}$, where $F(x) = \int_a^x f(t)\,dt$. But we recognize this limit as being $F'(x)$ by the definition of a derivative. Therefore, $\lim_{h\to 0} f_{\text{ave}} = F'(x) = f(x)$ by FTC1.

29. $f(x) = \begin{cases} \frac{\pi}{20}\sin\left(\frac{\pi}{10}x\right) & \text{if } 0 \le x \le 10 \\ 0 & \text{if } x < 0 \text{ or } x > 10 \end{cases}$

(a) $f(x) \ge 0$ for all real numbers x and

$$\int_{-\infty}^{\infty} f(x)\,dx = \int_0^{10} \tfrac{\pi}{20}\sin\left(\tfrac{\pi}{10}x\right) dx = \tfrac{\pi}{20} \cdot \tfrac{10}{\pi}\left[-\cos\left(\tfrac{\pi}{10}x\right)\right]_0^{10} = \tfrac{1}{2}(-\cos\pi + \cos 0) = \tfrac{1}{2}(1 + 1) = 1$$

Therefore, f is a probability density function.

(b) $P(X < 4) = \int_{-\infty}^4 f(x)\,dx = \int_0^4 \frac{\pi}{20}\sin\left(\frac{\pi}{10}x\right) dx = \frac{1}{2}\left[-\cos\left(\frac{\pi}{10}x\right)\right]_0^4 = \frac{1}{2}\left(-\cos\frac{2\pi}{5} + \cos 0\right)$

$$\approx \tfrac{1}{2}(-0.309\,017 + 1) \approx 0.3455$$

(c) $\mu = \int_{-\infty}^{\infty} x f(x)\,dx = \int_0^{10} \frac{\pi}{20} x \sin\left(\frac{\pi}{10}x\right) dx$

$= \int_0^{\pi} \frac{\pi}{20} \cdot \frac{10}{\pi} u(\sin u)\left(\frac{10}{\pi}\right) du \quad \left[u = \frac{\pi}{10}x,\ du = \frac{\pi}{10}\,dx\right]$

$= \frac{5}{\pi} \int_0^{\pi} u \sin u\,du \overset{82}{=} \frac{5}{\pi}\left[\sin u - u\cos u\right]_0^{\pi} = \frac{5}{\pi}[0 - \pi(-1)] = 5$

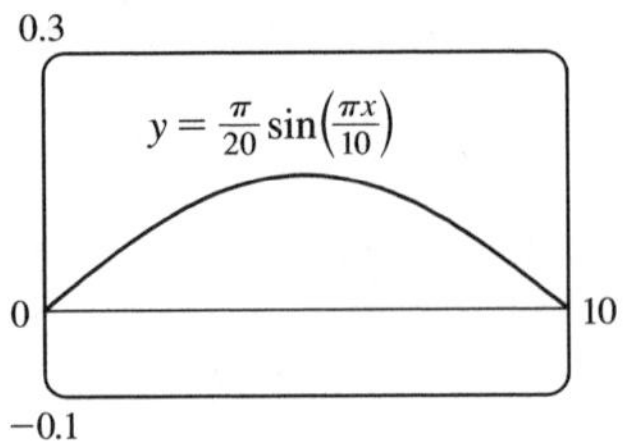

This answer is expected because the graph of f is symmetric about the line $x = 5$.

31. (a) The probability density function is $f(t) = \begin{cases} 0 & \text{if } t < 0 \\ \frac{1}{8}e^{-t/8} & \text{if } t \geq 0 \end{cases}$

$P(0 \leq X \leq 3) = \int_0^3 \frac{1}{8}e^{-t/8}\,dt = \left[-e^{-t/8}\right]_0^3 = -e^{-3/8} + 1 \approx 0.3127$

(b) $P(X > 10) = \int_{10}^{\infty} \frac{1}{8}e^{-t/8}\,dt = \lim_{x\to\infty}\left[-e^{-t/8}\right]_{10}^{x} = \lim_{x\to\infty}\left(-e^{-x/8} + e^{-10/8}\right) = 0 + e^{-5/4} \approx 0.2865$

(c) We need to find m such that $P(X \geq m) = \frac{1}{2} \Rightarrow \int_m^{\infty} \frac{1}{8}e^{-t/8}\,dt = \frac{1}{2} \Rightarrow \lim_{x\to\infty}\left[-e^{-t/8}\right]_m^x = \frac{1}{2} \Rightarrow$ $\lim_{x\to\infty}\left(-e^{-x/8} + e^{-m/8}\right) = \frac{1}{2} \Rightarrow e^{-m/8} = \frac{1}{2} \Rightarrow -m/8 = \ln\frac{1}{2} \Rightarrow m = -8\ln\frac{1}{2} = 8\ln 2 \approx 5.55$ minutes.

FOCUS ON PROBLEM SOLVING

1. The volume generated from $x = 0$ to $x = b$ is $\int_0^b \pi[f(x)]^2\,dx$. Hence, we are given that $b^2 = \int_0^b \pi[f(x)]^2\,dx$ for all $b > 0$. Differentiating both sides of this equation with respect to b using the Fundamental Theorem of Calculus gives $2b = \pi[f(b)]^2 \Rightarrow f(b) = \sqrt{2b/\pi}$, since f is positive. Therefore, $f(x) = \sqrt{2x/\pi}$.

3. (a) $V = \pi h^2(r - h/3) = \frac{1}{3}\pi h^2(3r - h)$. See the solution to Exercise 6.2.27.

(b) The smaller segment has height $h = 1 - x$ and so by part (a) its volume is

$V = \frac{1}{3}\pi(1-x)^2\,[3(1) - (1-x)] = \frac{1}{3}\pi(x-1)^2(x+2)$. This volume must be $\frac{1}{3}$ of the total volume of the sphere, which is $\frac{4}{3}\pi(1)^3$. So $\frac{1}{3}\pi(x-1)^2(x+2) = \frac{1}{3}\left(\frac{4}{3}\pi\right) \Rightarrow (x^2 - 2x + 1)(x+2) = \frac{4}{3} \Rightarrow x^3 - 3x + 2 = \frac{4}{3} \Rightarrow$ $3x^3 - 9x + 2 = 0$. Using Newton's method with $f(x) = 3x^3 - 9x + 2$, $f'(x) = 9x^2 - 9$, we get

$x_{n+1} = x_n - \dfrac{3x_n^3 - 9x_n + 2}{9x_n^2 - 9}$. Taking $x_1 = 0$, we get $x_2 \approx 0.2222$, and $x_3 \approx 0.2261 \approx x_4$, so, correct to four decimal places, $x \approx 0.2261$.

(c) With $r = 0.5$ and $s = 0.75$, the equation $x^3 - 3rx^2 + 4r^3 s = 0$ becomes $x^3 - 3(0.5)x^2 + 4(0.5)^3(0.75) = 0 \Rightarrow$ $x^3 - \frac{3}{2}x^2 + 4\left(\frac{1}{8}\right)\frac{3}{4} = 0 \Rightarrow 8x^3 - 12x^2 + 3 = 0$. We use Newton's method with $f(x) = 8x^3 - 12x^2 + 3$, $f'(x) = 24x^2 - 24x$, so $x_{n+1} = x_n - \dfrac{8x_n^3 - 12x_n^2 + 3}{24x_n^2 - 24x_n}$. Take $x_1 = 0.5$. Then $x_2 \approx 0.6667$, and $x_3 \approx 0.6736 \approx x_4$.

So to four decimal places the depth is 0.6736 m.

(d) (i) From part (a) with $r = 12$ cm, the volume of water in the bowl is

$V = \frac{1}{3}\pi h^2(3r - h) = \frac{1}{3}\pi h^2(36 - h) = 12\pi h^2 - \frac{1}{3}\pi h^3$. We are given that $\dfrac{dV}{dt} = 3\ \text{cm}^3/\text{s}$ and we want to find $\dfrac{dh}{dt}$ when $h = 7$. Now $\dfrac{dV}{dt} = 24\pi h\,\dfrac{dh}{dt} - \pi h^2\dfrac{dh}{dt}$, so $\dfrac{dh}{dt} = \dfrac{3}{\pi(24h - h^2)}$. When $h = 7$, we have

$$\frac{dh}{dt} = \frac{3}{\pi(24\cdot 7 - 7^2)} = \frac{3}{119\pi} \approx 0.008\ \text{cm/s}.$$

(ii) From part (a), the volume of water required to fill the hemispherical bowl from the instant that the water is 8 cm deep is $V = \frac{1}{2}\cdot\frac{4}{3}\pi(12)^3 - \frac{1}{3}\pi(8)^2(36 - 8) = 1152\pi - \frac{1792}{3}\pi = \frac{1664}{3}\pi$. To find the time required to fill the bowl we divide this volume by the rate: Time $= \frac{1664\pi/3}{3} = \frac{1664}{9}\pi \approx 581\ \text{s} \approx 9.7$ min.

5. We are given that the rate of change of the volume of water is $\dfrac{dV}{dt} = -kA(x)$, where k is some positive constant and $A(x)$ is the area of the surface when the water has depth x. Now we are concerned with the rate of change of the depth of the water with respect to time, that is, $\dfrac{dx}{dt}$. But by the Chain Rule, $\dfrac{dV}{dt} = \dfrac{dV}{dx}\dfrac{dx}{dt}$, so the first equation can be written

$\dfrac{dV}{dx}\dfrac{dx}{dt} = -kA(x)$ $(\star)$. Also, we know that the total volume of water up to a depth x is $V(x) = \int_0^x A(s)\,ds$, where $A(s)$ is the area of a cross-section of the water at a depth s. Differentiating this equation with respect to x, we get $dV/dx = A(x)$. Substituting this into equation $\star$, we get $A(x)(dx/dt) = -kA(x) \;\Rightarrow\; dx/dt = -k$, a constant.

7. To find the height of the pyramid, we use similar triangles. The first figure shows a cross-section of the pyramid passing through the top and through two opposite corners of the square base. Now $|BD| = b$, since it is a radius of the sphere, which has diameter $2b$ since it is tangent to the opposite sides of the square base. Also, $|AD| = b$ since $\triangle ADB$ is isosceles. So the height is $|AB| = \sqrt{b^2 + b^2} = \sqrt{2}b$.

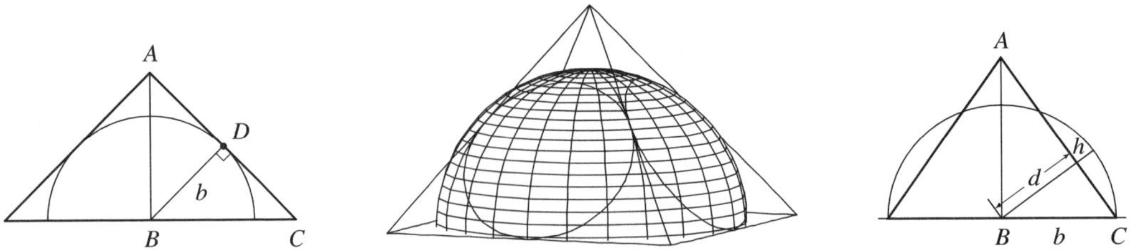

We first observe that the shared volume is equal to half the volume of the sphere, minus the sum of the four equal volumes (caps of the sphere) cut off by the triangular faces of the pyramid. See Exercise 6.2.27 for a derivation of the formula for the volume of a cap of a sphere. To use the formula, we need to find the perpendicular distance h of each triangular face from the surface of the sphere. We first find the distance d from the centre of the sphere to one of the triangular faces. The third figure shows a cross-section of the pyramid through the top and through the midpoints of opposite sides of the square base. From similar triangles we find that

$$\frac{d}{b} = \frac{|AB|}{|AC|} = \frac{\sqrt{2}\,b}{\sqrt{b^2 + \left(\sqrt{2}\,b\right)^2}} \quad\Rightarrow\quad d = \frac{\sqrt{2}\,b^2}{\sqrt{3b^2}} = \frac{\sqrt{6}}{3}\,b$$

So $h = b - d = b - \frac{\sqrt{6}}{3}b = \frac{3-\sqrt{6}}{3}b$. So, using the formula $V = \pi h^2(r - h/3)$ from Exercise 6.2.27 with $r = b$, we find that the volume of each of the caps is $\pi\left(\frac{3-\sqrt{6}}{3}b\right)^2\left(b - \frac{3-\sqrt{6}}{3\cdot 3}b\right) = \frac{15-6\sqrt{6}}{9}\cdot\frac{6+\sqrt{6}}{9}\pi b^3 = \left(\frac{2}{3} - \frac{7}{27}\sqrt{6}\right)\pi b^3$. So, using our first observation, the shared volume is $V = \frac{1}{2}\left(\frac{4}{3}\pi b^3\right) - 4\left(\frac{2}{3} - \frac{7}{27}\sqrt{6}\right)\pi b^3 = \left(\frac{28}{27}\sqrt{6} - 2\right)\pi b^3$.

9. $x = \displaystyle\int_1^t \frac{\cos u}{u}\,du$, $y = \displaystyle\int_1^t \frac{\sin u}{u}\,du$, so by FTC1, we have $\dfrac{dx}{dt} = \dfrac{\cos t}{t}$ and $\dfrac{dy}{dt} = \dfrac{\sin t}{t}$. Vertical tangent lines occur when $dx/dt = 0 \;\Leftrightarrow\; \cos t = 0 \;\Leftrightarrow\; t = \frac{\pi}{2} + n\pi$. The parameter value corresponding to the origin, $(x, y) = (0, 0)$, is $t = 1$, so the nearest vertical tangent occurs when $t = \frac{\pi}{2}$. Therefore, the arc length between these points is

$$L = \int_1^{\pi/2}\sqrt{\left(\frac{dx}{dt}\right)^2 + \left(\frac{dy}{dt}\right)^2}\,dt = \int_1^{\pi/2}\sqrt{\frac{\cos^2 t}{t^2} + \frac{\sin^2 t}{t^2}}\,dt = \int_1^{\pi/2}\frac{dt}{t} = \left[\ln t\right]_1^{\pi/2} = \ln\frac{\pi}{2}$$

11.

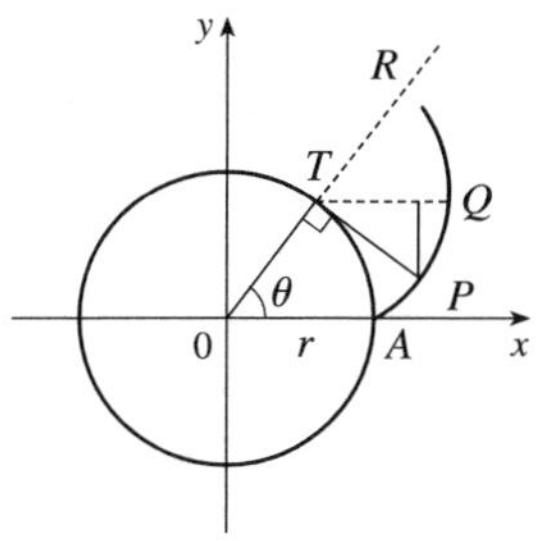

The coordinates of T are $(x_1, y_1) = (r\cos\theta, r\sin\theta)$. Since TP was unwound from arc TA, TP has length $r\theta$. Also $\angle PTQ = \angle PTR - \angle QTR = \frac{1}{2}\pi - \theta$, so P has coordinates

$$x = x_1 + |TP|\cos\angle PTQ = r\cos\theta + r\theta\cos\left(\tfrac{1}{2}\pi - \theta\right) = r(\cos\theta + \theta\sin\theta),$$

$$y = y_1 + |TP|\sin\angle PTQ = r\sin\theta - r\theta\sin\left(\tfrac{1}{2}\pi - \theta\right) = r(\sin\theta - \theta\cos\theta).$$

13. We can assume that the cut is made along a vertical line $x = b > 0$, that the disc's boundary is the circle $x^2 + y^2 = 1$, and that the centre of mass of the smaller piece (to the right of $x = b$) is $\left(\frac{1}{2}, 0\right)$. We wish to find b to two decimal places. We have $\dfrac{1}{2} = \overline{x} = \dfrac{\int_b^1 x \cdot 2\sqrt{1 - x^2}\,dx}{\int_b^1 2\sqrt{1 - x^2}\,dx}$. Evaluating the numerator gives us

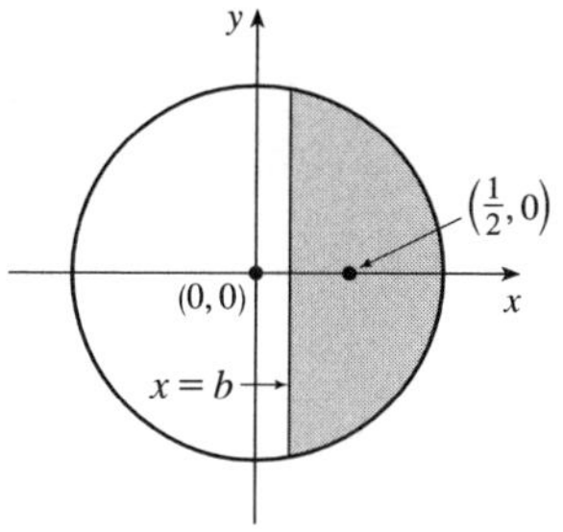

$$-\int_b^1 (1 - x^2)^{1/2}(-2x)\,dx = -\tfrac{2}{3}\left[\left(1 - x^2\right)^{3/2}\right]_b^1 = -\tfrac{2}{3}\left[0 - \left(1 - b^2\right)^{3/2}\right] = \tfrac{2}{3}(1 - b^2)^{3/2}$$

Using Formula 30 in the Table of Integrals, we find that the denominator is

$\left[x\sqrt{1 - x^2} + \sin^{-1}x\right]_b^1 = \left(0 + \frac{\pi}{2}\right) - \left(b\sqrt{1 - b^2} + \sin^{-1}b\right)$. Thus, we have $\dfrac{1}{2} = \overline{x} = \dfrac{\frac{2}{3}(1 - b^2)^{3/2}}{\frac{\pi}{2} - b\sqrt{1 - b^2} - \sin^{-1}b}$, or, equivalently, $\frac{2}{3}(1 - b^2)^{3/2} = \frac{\pi}{4} - \frac{1}{2}b\sqrt{1 - b^2} - \frac{1}{2}\sin^{-1}b$. Solving this equation numerically with a calculator or CAS, we obtain $b \approx 0.138173$, or $b = 0.14$ m to two decimal places.

7 □ DIFFERENTIAL EQUATIONS

7.1 Modelling with Differential Equations

1. $y = x - x^{-1} \Rightarrow y' = 1 + x^{-2}$. To show that y is a solution of the differential equation, we will substitute the expressions for y and y' in the left-hand side of the equation and show that the left-hand side is equal to the right-hand side.

$$\text{LHS} = xy' + y = x(1 + x^{-2}) + (x - x^{-1}) = x + x^{-1} + x - x^{-1} = 2x = \text{RHS}$$

3. (a) $y = e^{rx} \Rightarrow y' = re^{rx} \Rightarrow y'' = r^2e^{rx}$. Substituting these expressions into the differential equation $2y'' + y' - y = 0$, we get $2r^2e^{rx} + re^{rx} - e^{rx} = 0 \Rightarrow (2r^2 + r - 1)e^{rx} = 0 \Rightarrow$ $(2r - 1)(r + 1) = 0$ [since e^{rx} is never zero] $\Rightarrow r = \frac{1}{2}$ or -1.

(b) Let $r_1 = \frac{1}{2}$ and $r_2 = -1$, so we need to show that every member of the family of functions $y = ae^{x/2} + be^{-x}$ is a solution of the differential equation $2y'' + y' - y = 0$.

$y = ae^{x/2} + be^{-x} \Rightarrow y' = \frac{1}{2}ae^{x/2} - be^{-x} \Rightarrow y'' = \frac{1}{4}ae^{x/2} + be^{-x}$.

$$\begin{aligned}\text{LHS} = 2y'' + y' - y &= 2(\tfrac{1}{4}ae^{x/2} + be^{-x}) + (\tfrac{1}{2}ae^{x/2} - be^{-x}) - (ae^{x/2} + be^{-x}) \\ &= \tfrac{1}{2}ae^{x/2} + 2be^{-x} + \tfrac{1}{2}ae^{x/2} - be^{-x} - ae^{x/2} - be^{-x} \\ &= (\tfrac{1}{2}a + \tfrac{1}{2}a - a)e^{x/2} + (2b - b - b)e^{-x} \\ &= 0 = \text{RHS}\end{aligned}$$

5. (a) $y = \sin x \Rightarrow y' = \cos x \Rightarrow y'' = -\sin x$.

$\text{LHS} = y'' + y = -\sin x + \sin x = 0 \neq \sin x$, so $y = \sin x$ **is not** a solution of the differential equation.

(b) $y = \cos x \Rightarrow y' = -\sin x \Rightarrow y'' = -\cos x$.

$\text{LHS} = y'' + y = -\cos x + \cos x = 0 \neq \sin x$, so $y = \cos x$ **is not** a solution of the differential equation.

(c) $y = \frac{1}{2}x\sin x \Rightarrow y' = \frac{1}{2}(x\cos x + \sin x) \Rightarrow y'' = \frac{1}{2}(-x\sin x + \cos x + \cos x)$.

$\text{LHS} = y'' + y = \frac{1}{2}(-x\sin x + 2\cos x) + \frac{1}{2}x\sin x = \cos x \neq \sin x$, so $y = \frac{1}{2}x\sin x$ **is not** a solution of the differential equation.

(d) $y = -\frac{1}{2}x\cos x \Rightarrow y' = -\frac{1}{2}(-x\sin x + \cos x) \Rightarrow y'' = -\frac{1}{2}(-x\cos x - \sin x - \sin x)$.

$\text{LHS} = y'' + y = -\frac{1}{2}(-x\cos x - 2\sin x) + \left(-\frac{1}{2}x\cos x\right) = \sin x = \text{RHS}$, so $y = -\frac{1}{2}x\cos x$ **is** a solution of the differential equation.

7. (a) Since the derivative $y' = -y^2$ is always negative (or 0 if $y = 0$), the function y must be decreasing (or equal to 0) on any interval on which it is defined.

(b) $y = \dfrac{1}{x + C} \Rightarrow y' = -\dfrac{1}{(x + C)^2}$. $\text{LHS} = y' = -\dfrac{1}{(x + C)^2} = -\left(\dfrac{1}{x + C}\right)^2 = -y^2 = \text{RHS}$

(c) $y = 0$ is a solution of $y' = -y^2$ that is not a member of the family in part (b).

(d) If $y(x) = \dfrac{1}{x + C}$, then $y(0) = \dfrac{1}{0 + C} = \dfrac{1}{C}$. Since $y(0) = 0.5$, $\dfrac{1}{C} = \dfrac{1}{2} \Rightarrow C = 2$, so $y = \dfrac{1}{x + 2}$.

9. (a) $\dfrac{dP}{dt} = 1.2P\left(1 - \dfrac{P}{4200}\right)$. Now $\dfrac{dP}{dt} > 0 \Rightarrow 1 - \dfrac{P}{4200} > 0$ [assuming that $P > 0$] $\Rightarrow \dfrac{P}{4200} < 1 \Rightarrow$ $P < 4200 \Rightarrow$ the population is increasing for $0 < P < 4200$.

(b) $\dfrac{dP}{dt} < 0 \Rightarrow P > 4200$

(c) $\dfrac{dP}{dt} = 0 \Rightarrow P = 4200$ or $P = 0$

11. (a) This function is increasing *and* also decreasing. But $dy/dt = e^t(y-1)^2 \geq 0$ for all t, implying that the graph of the solution of the differential equation cannot be decreasing on any interval.

(b) When $y = 1$, $dy/dt = 0$, but the graph does not have a horizontal tangent line.

13. (a) P increases most rapidly at the beginning, since there are usually many simple, easily-learned sub-skills associated with learning a skill. As t increases, we would expect dP/dt to remain positive, but decrease. This is because as time progresses, the only points left to learn are the more difficult ones.

(b) $\dfrac{dP}{dt} = k(M-P)$ is always positive, so the level of performance P is increasing. As P gets close to M, dP/dt gets close to 0; that is, the performance levels off, as explained in part (a).

(c)

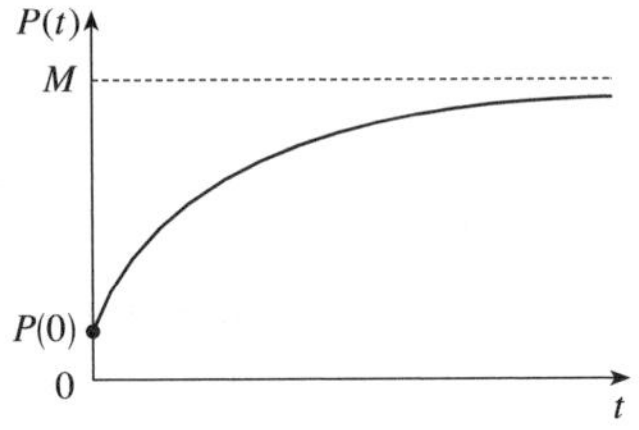

7.2 Direction Fields and Euler's Method

1. (a)

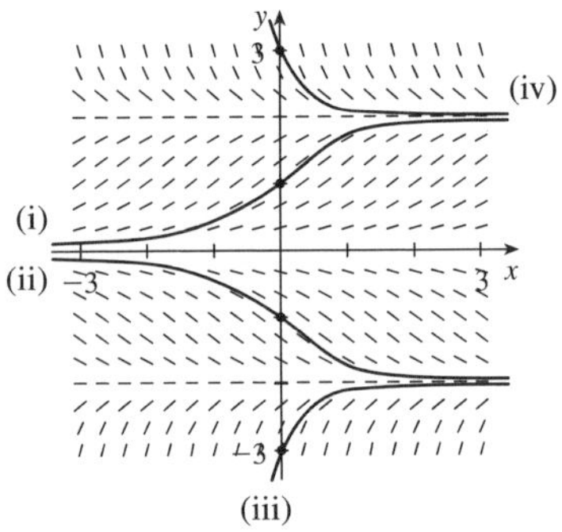

(b) It appears that the constant functions $y = 0$, $y = -2$, and $y = 2$ are equilibrium solutions. Note that these three values of y satisfy the given differential equation $y' = y\left(1 - \frac{1}{4}y^2\right)$.

3. $y' = y - 2$. The slopes at each point are independent of x, so the slopes are the same along each line parallel to the x-axis. Thus, III is the direction field for this equation. Note that for $y = 2$, $y' = 0$.

5. $y' = x + y - 1 = 0$ on the line $y = -x + 1$. Direction field IV satisfies this condition. Notice also that on the line $y = -x$ we have $y' = -1$, which is true in IV.

7. (a) $y(0) = 1$

(b) $y(0) = 2$

(c) $y(0) = -1$

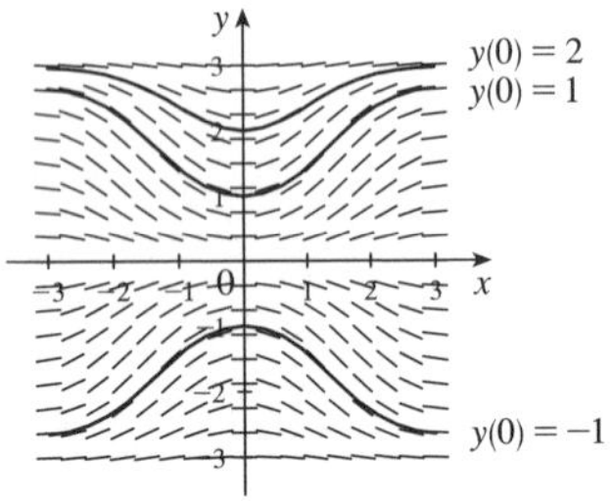

9.

x	y	$y' = 1 + y$
0	0	1
0	1	2
0	2	3
0	−3	−2
0	−2	−1

Note that for $y = -1$, $y' = 0$. The three solution curves sketched go through $(0, 0)$, $(0, -1)$, and $(0, -2)$.

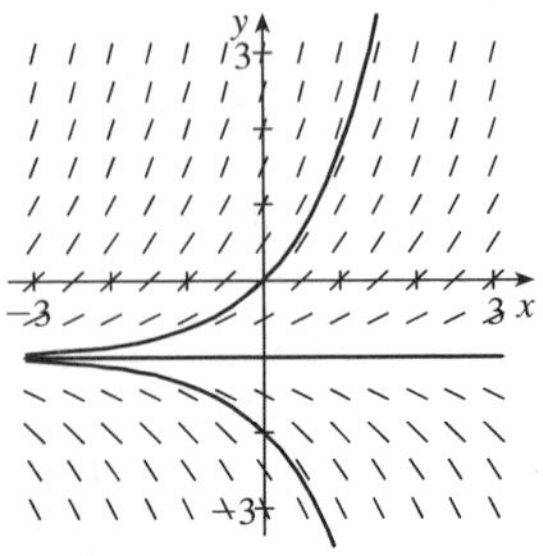

11.

x	y	$y' = y - 2x$
-2	-2	2
-2	2	6
2	2	-2
2	-2	-6

Note that $y' = 0$ for any point on the line $y = 2x$. The slopes are positive to the left of the line and negative to the right of the line. The solution curve in the graph passes through $(1, 0)$.

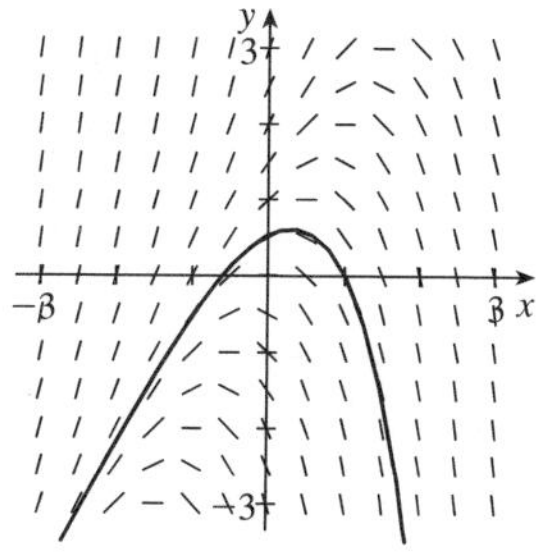

13.

x	y	$y' = y + xy$
0	± 2	± 2
1	± 2	± 4
-3	± 2	∓ 4

Note that $y' = y(x + 1) = 0$ for any point on $y = 0$ or on $x = -1$. The slopes are positive when the factors y and $x + 1$ have the same sign and negative when they have opposite signs. The solution curve in the graph passes through $(0, 1)$.

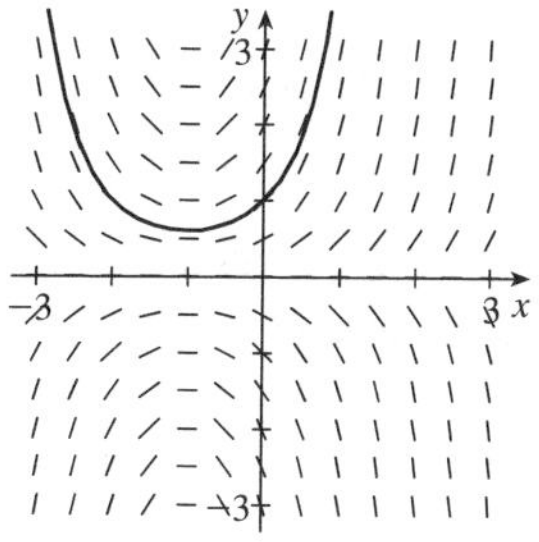

15. In Maple, we can use either `directionfield` (in Maple's share library) or `DEtools[DEplot]` to plot the direction field. To plot the solution, we can either use the initial-value option in `directionfield`, or actually solve the equation.

In Mathematica, we use `PlotVectorField` for the direction field, and the `Plot[Evaluate[...]]` construction to plot the solution, which is

$y = 2\arctan\left(e^{x^3/3} \cdot \tan\frac{1}{2}\right)$.

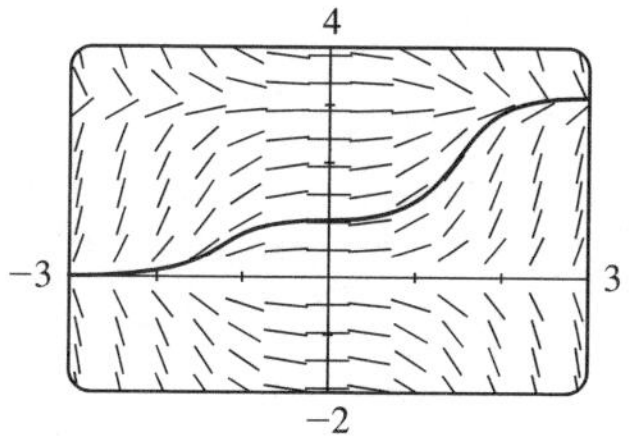

In Derive, use `Direction_Field` (in utility file `ODE_APPR`) to plot the direction field. Then use `DSOLVE1(-x^2*SIN(y),1,x,y,0,1)` (in utility file `ODE1`) to solve the equation. Simplify each result.

17.

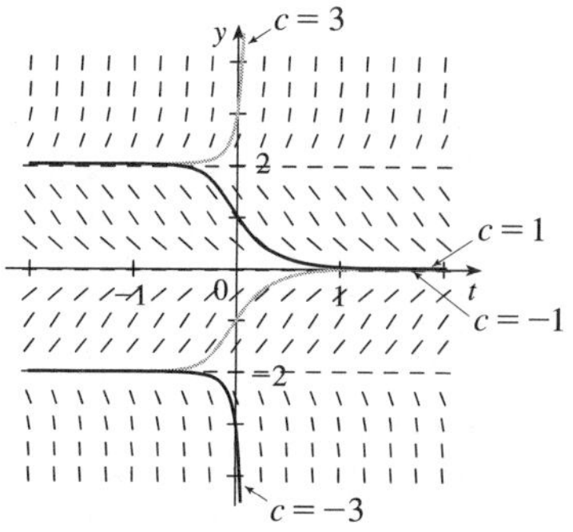

The direction field is for the differential equation $y' = y^3 - 4y$.

$L = \lim\limits_{t \to \infty} y(t)$ exists for $-2 \le c \le 2$;

$L = \pm 2$ for $c = \pm 2$ and $L = 0$ for $-2 < c < 2$.

For other values of c, L does not exist.

19. (a) $y' = F(x, y) = y$ and $y(0) = 1 \;\;\Rightarrow\;\; x_0 = 0,\ y_0 = 1$.

(i) $h = 0.4$ and $y_1 = y_0 + hF(x_0, y_0) \;\;\Rightarrow\;\; y_1 = 1 + 0.4 \cdot 1 = 1.4$. $x_1 = x_0 + h = 0 + 0.4 = 0.4$, so $y_1 = y\,(0.4) = 1.4$.

(ii) $h = 0.2 \;\;\Rightarrow\;\; x_1 = 0.2$ and $x_2 = 0.4$, so we need to find y_2.

$y_1 = y_0 + hF(x_0, y_0) = 1 + 0.2y_0 = 1 + 0.2 \cdot 1 = 1.2$,

$y_2 = y_1 + hF(x_1, y_1) = 1.2 + 0.2y_1 = 1.2 + 0.2 \cdot 1.2 = 1.44$.

(iii) $h = 0.1 \;\;\Rightarrow\;\; x_4 = 0.4$, so we need to find y_4. $y_1 = y_0 + hF(x_0, y_0) = 1 + 0.1y_0 = 1 + 0.1 \cdot 1 = 1.1$,

$y_2 = y_1 + hF(x_1, y_1) = 1.1 + 0.1y_1 = 1.1 + 0.1 \cdot 1.1 = 1.21$,

$y_3 = y_2 + hF(x_2, y_2) = 1.21 + 0.1y_2 = 1.21 + 0.1 \cdot 1.21 = 1.331$,

$y_4 = y_3 + hF(x_3, y_3) = 1.331 + 0.1y_3 = 1.331 + 0.1 \cdot 1.331 = 1.4641$.

(b)

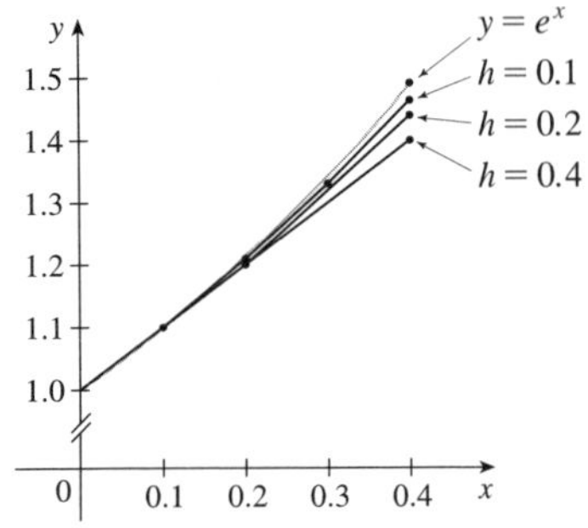

We see that the estimates are underestimates since they are all below the graph of $y = e^x$.

(c) (i) For $h = 0.4$: (exact value) − (approximate value) $= e^{0.4} - 1.4 \approx 0.0918$

(ii) For $h = 0.2$: (exact value) − (approximate value) $= e^{0.4} - 1.44 \approx 0.0518$

(iii) For $h = 0.1$: (exact value) − (approximate value) $= e^{0.4} - 1.4641 \approx 0.0277$

Each time the step size is halved, the error estimate also appears to be halved (approximately).

21. $h = 0.5$, $x_0 = 1$, $y_0 = 0$, and $F(x, y) = y - 2x$.

Note that $x_1 = x_0 + h = 1 + 0.5 = 1.5$, $x_2 = 2$, and $x_3 = 2.5$.

$y_1 = y_0 + hF(x_0, y_0) = 0 + 0.5F(1, 0) = 0.5[0 - 2(1)] = -1$.

$y_2 = y_1 + hF(x_1, y_1) = -1 + 0.5F(1.5, -1) = -1 + 0.5[-1 - 2(1.5)] = -3$.

$y_3 = y_2 + hF(x_2, y_2) = -3 + 0.5F(2, -3) = -3 + 0.5[-3 - 2(2)] = -6.5$.

$y_4 = y_3 + hF(x_3, y_3) = -6.5 + 0.5F(2.5, -6.5) = -6.5 + 0.5[-6.5 - 2(2.5)] = -12.25$.

23. $h = 0.1$, $x_0 = 0$, $y_0 = 1$, and $F(x, y) = y + xy$.

Note that $x_1 = x_0 + h = 0 + 0.1 = 0.1$, $x_2 = 0.2$, $x_3 = 0.3$, and $x_4 = 0.4$.

$y_1 = y_0 + hF(x_0, y_0) = 1 + 0.1F(0, 1) = 1 + 0.1[1 + (0)(1)] = 1.1$.

$y_2 = y_1 + hF(x_1, y_1) = 1.1 + 0.1F(0.1, 1.1) = 1.1 + 0.1[1.1 + (0.1)(1.1)] = 1.221$.

$y_3 = y_2 + hF(x_2, y_2) = 1.221 + 0.1F(0.2, 1.221) = 1.221 + 0.1[1.221 + (0.2)(1.221)] = 1.36752$.

$$\begin{aligned} y_4 = y_3 + hF(x_3, y_3) &= 1.36752 + 0.1F(0.3, 1.36752) \\ &= 1.36752 + 0.1[1.36752 + (0.3)(1.36752)] = 1.545\,297\,6. \end{aligned}$$

$$\begin{aligned} y_5 = y_4 + hF(x_4, y_4) &= 1.545\,297\,6 + 0.1F(0.4, 1.545\,297\,6) \\ &= 1.545\,297\,6 + 0.1[1.545\,297\,6 + (0.4)(1.545\,297\,6)] = 1.761\,639\,264. \end{aligned}$$

Thus, $y(0.5) \approx 1.7616$.

25. (a) $dy/dx + 3x^2y = 6x^2 \Rightarrow y' = 6x^2 - 3x^2y$. Store this expression in Y_1 and use the following simple program to evaluate $y(1)$ for each part, using $\text{H} = h = 1$ and $\text{N} = 1$ for part (i), $\text{H} = 0.1$ and $\text{N} = 10$ for part (ii), and so forth.

$h \to$ H: $0 \to$ X: $3 \to$ Y:

For(I, 1, N): $\text{Y} + \text{H} \times \text{Y}_1 \to$ Y: $\text{X} + \text{H} \to$ X:

End(loop):

Display Y. [To see all iterations, include this statement in the loop.]

(i) $\text{H} = 1, \text{N} = 1 \Rightarrow y(1) = 3$

(ii) $\text{H} = 0.1, \text{N} = 10 \Rightarrow y(1) \approx 2.3928$

(iii) $\text{H} = 0.01, \text{N} = 100 \Rightarrow y(1) \approx 2.3701$

(iv) $\text{H} = 0.001, \text{N} = 1000 \Rightarrow y(1) \approx 2.3681$

(b) $y = 2 + e^{-x^3} \Rightarrow y' = -3x^2e^{-x^3}$

$$\text{LHS} = y' + 3x^2y = -3x^2e^{-x^3} + 3x^2\left(2 + e^{-x^3}\right) = -3x^2e^{-x^3} + 6x^2 + 3x^2e^{-x^3} = 6x^2 = \text{RHS}$$

$y(0) = 2 + e^{-0} = 2 + 1 = 3$

(c) The exact value of $y(1)$ is $2 + e^{-1^3} = 2 + e^{-1}$.

(i) For $h = 1$: (exact value) − (approximate value) $= 2 + e^{-1} - 3 \approx -0.6321$

(ii) For $h = 0.1$: (exact value) − (approximate value) $= 2 + e^{-1} - 2.3928 \approx -0.0249$

(iii) For $h = 0.01$: (exact value) − (approximate value) $= 2 + e^{-1} - 2.3701 \approx -0.0022$

(iv) For $h = 0.001$: (exact value) − (approximate value) $= 2 + e^{-1} - 2.3681 \approx -0.0002$

In (ii)–(iv), it seems that when the step size is divided by 10, the error estimate is also divided by 10 (approximately).

27. (a) $R\dfrac{dQ}{dt} + \dfrac{1}{C}Q = E(t)$ becomes $5Q' + \dfrac{1}{0.05}Q = 60$ or $Q' + 4Q = 12$.

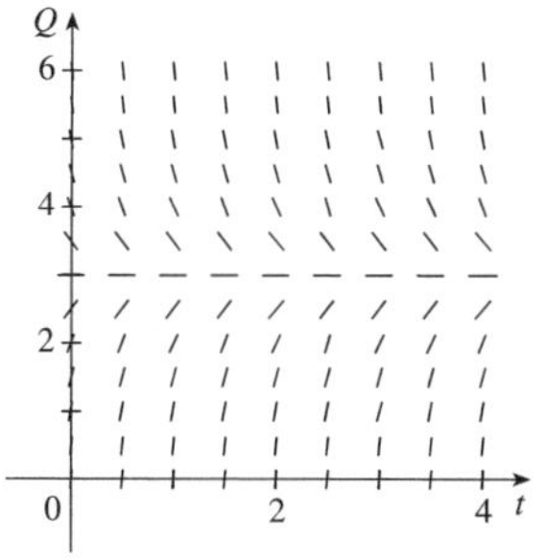

(b) From the graph, it appears that the limiting value of the charge Q is about 3.

(c) If $Q' = 0$, then $4Q = 12 \Rightarrow Q = 3$ is an equilibrium solution.

(d)

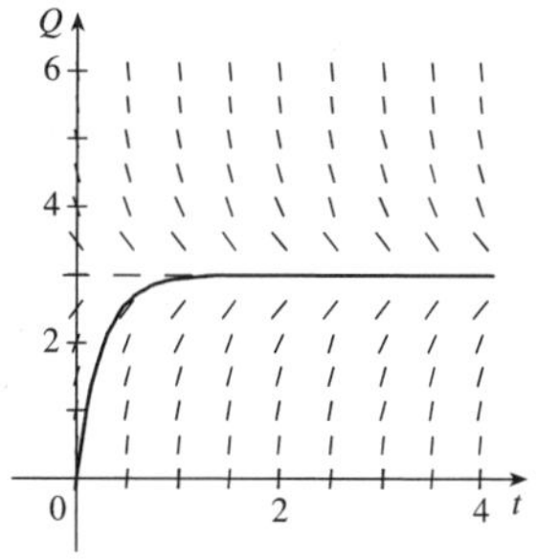

(e) $Q' + 4Q = 12 \Rightarrow Q' = 12 - 4Q$. Now $Q(0) = 0$, so $t_0 = 0$ and $Q_0 = 0$.

$$Q_1 = Q_0 + hF(t_0, Q_0) = 0 + 0.1(12 - 4 \cdot 0) = 1.2$$

$$Q_2 = Q_1 + hF(t_1, Q_1) = 1.2 + 0.1(12 - 4 \cdot 1.2) = 1.92$$

$$Q_3 = Q_2 + hF(t_2, Q_2) = 1.92 + 0.1(12 - 4 \cdot 1.92) = 2.352$$

$$Q_4 = Q_3 + hF(t_3, Q_3) = 2.352 + 0.1(12 - 4 \cdot 2.352) = 2.6112$$

$$Q_5 = Q_4 + hF(t_4, Q_4) = 2.6112 + 0.1(12 - 4 \cdot 2.6112) = 2.76672$$

Thus, $Q_5 = Q(0.5) \approx 2.77$ C.

7.3 Separable Equations

1. $\dfrac{dy}{dx} = y^2 \;\Rightarrow\; \dfrac{dy}{y^2} = dx \;\; [y \neq 0] \;\Rightarrow\; \displaystyle\int \frac{dy}{y^2} = \int dx \;\Rightarrow\; -\frac{1}{y} = x + C \;\Rightarrow\; -y = \frac{1}{x+C} \;\Rightarrow\; y = \frac{-1}{x+C}$, and

$y = 0$ is also a solution.

3. $(x^2+1)y' = xy \;\Rightarrow\; \dfrac{dy}{dx} = \dfrac{xy}{x^2+1} \;\Rightarrow\; \dfrac{dy}{y} = \dfrac{x\,dx}{x^2+1} \;\; [y \neq 0] \;\Rightarrow\; \displaystyle\int \frac{dy}{y} = \int \frac{x\,dx}{x^2+1} \;\Rightarrow$

$\ln|y| = \frac{1}{2}\ln(x^2+1) + C \;\; [u = x^2+1,\, du = 2x\,dx] \;\; = \ln(x^2+1)^{1/2} + \ln e^C = \ln\left(e^C\sqrt{x^2+1}\right) \;\Rightarrow$

$|y| = e^C\sqrt{x^2+1} \;\Rightarrow\; y = K\sqrt{x^2+1}$, where $K = \pm e^C$ is a constant. (In our derivation, K was nonzero, but we can restore the excluded case $y = 0$ by allowing K to be zero.)

5. $(1+\tan y)\,y' = x^2+1 \;\Rightarrow\; (1+\tan y)\dfrac{dy}{dx} = x^2+1 \;\Rightarrow\; \left(1 + \dfrac{\sin y}{\cos y}\right) dy = (x^2+1)\,dx \;\Rightarrow$

$\displaystyle\int \left(1 - \frac{-\sin y}{\cos y}\right) dy = \int (x^2+1)\,dx \;\Rightarrow\; y - \ln|\cos y| = \tfrac{1}{3}x^3 + x + C$. Note: The left side is equivalent to $y + \ln|\sec y|$.

7. $\dfrac{du}{dt} = 2 + 2u + t + tu \;\Rightarrow\; \dfrac{du}{dt} = (1+u)(2+t) \;\Rightarrow\; \displaystyle\int \frac{du}{1+u} = \int (2+t)dt \;\; [u \neq -1] \;\Rightarrow$

$\ln|1+u| = \frac{1}{2}t^2 + 2t + C \;\Rightarrow\; |1+u| = e^{t^2/2 + 2t + C} = Ke^{t^2/2 + 2t}$, where $K = e^C \;\Rightarrow\; 1 + u = \pm Ke^{t^2/2+2t} \;\Rightarrow$

$u = -1 \pm Ke^{t^2/2+2t}$ where $K > 0$. $u = -1$ is also a solution, so $u = -1 + Ae^{t^2/2+2t}$, where A is an arbitrary constant.

9. $\dfrac{du}{dt} = \dfrac{2t + \sec^2 t}{2u}$, $u(0) = -5$. $\int 2u\,du = \int (2t + \sec^2 t)\,dt \;\Rightarrow\; u^2 = t^2 + \tan t + C$,

where $[u(0)]^2 = 0^2 + \tan 0 + C \;\Rightarrow\; C = (-5)^2 = 25$. Therefore, $u^2 = t^2 + \tan t + 25$, so $u = \pm\sqrt{t^2 + \tan t + 25}$.

Since $u(0) = -5$, we must have $u = -\sqrt{t^2 + \tan t + 25}$.

11. $x\cos x = (2y + e^{3y})\,y' \;\Rightarrow\; x\cos x\,dx = (2y + e^{3y})\,dy \;\Rightarrow\; \int (2y + e^{3y})\,dy = \int x\cos x\,dx \;\Rightarrow$

$y^2 + \frac{1}{3}e^{3y} = x\sin x + \cos x + C$ [where the second integral is evaluated using integration by parts]. Now $y(0) = 0 \;\Rightarrow$

$0 + \frac{1}{3} = 0 + 1 + C \;\Rightarrow\; C = -\frac{2}{3}$. Thus, a solution is $y^2 + \frac{1}{3}e^{3y} = x\sin x + \cos x - \frac{2}{3}$.

We cannot solve explicitly for y.

13. $y'\tan x = a + y,\; 0 < x < \pi/2 \;\Rightarrow\; \dfrac{dy}{dx} = \dfrac{a+y}{\tan x} \;\Rightarrow\; \dfrac{dy}{a+y} = \cot x\,dx \;\; [a + y \neq 0] \;\Rightarrow$

$\displaystyle\int \frac{dy}{a+y} = \int \frac{\cos x}{\sin x}\,dx \;\Rightarrow\; \ln|a+y| = \ln|\sin x| + C \;\Rightarrow\; |a+y| = e^{\ln|\sin x| + C} = e^{\ln|\sin x|}\cdot e^C = e^C|\sin x| \;\Rightarrow$

$a + y = K\sin x$, where $K = \pm e^C$. (In our derivation, K was nonzero, but we can restore the excluded case $y = -a$ by allowing K to be zero.) $y(\pi/3) = a \;\Rightarrow$

$a + a = K\sin\left(\dfrac{\pi}{3}\right) \;\Rightarrow\; 2a = K\dfrac{\sqrt{3}}{2} \;\Rightarrow\; K = \dfrac{4a}{\sqrt{3}}$. Thus, $a + y = \dfrac{4a}{\sqrt{3}}\sin x$ and so $y = \dfrac{4a}{\sqrt{3}}\sin x - a$.

15. $\dfrac{dy}{dx} = 4x^3 y$, $y(0) = 7$. $\dfrac{dy}{y} = 4x^3\,dx$ [if $y \neq 0$] $\Rightarrow$ $\displaystyle\int \frac{dy}{y} = \int 4x^3\,dx$ $\Rightarrow$ $\ln|y| = x^4 + C$ $\Rightarrow$

$e^{\ln|y|} = e^{x^4+C}$ $\Rightarrow$ $|y| = e^{x^4}e^C$ $\Rightarrow$ $y = Ae^{x^4}$; $y(0) = 7$ $\Rightarrow$ $A = 7$ $\Rightarrow$ $y = 7e^{x^4}$.

17. (a) $y' = 2x\sqrt{1-y^2}$ $\Rightarrow$ $\dfrac{dy}{dx} = 2x\sqrt{1-y^2}$ $\Rightarrow$ $\dfrac{dy}{\sqrt{1-y^2}} = 2x\,dx$ $\Rightarrow$ $\displaystyle\int \frac{dy}{\sqrt{1-y^2}} = \int 2x\,dx$ $\Rightarrow$

$\sin^{-1} y = x^2 + C$ for $-\frac{\pi}{2} \le x^2 + C \le \frac{\pi}{2}$.

(b) $y(0) = 0$ $\Rightarrow$ $\sin^{-1} 0 = 0^2 + C$ $\Rightarrow$

$C = 0$, so $\sin^{-1} y = x^2$ and $y = \sin(x^2)$ for

$-\sqrt{\pi/2} \le x \le \sqrt{\pi/2}$.

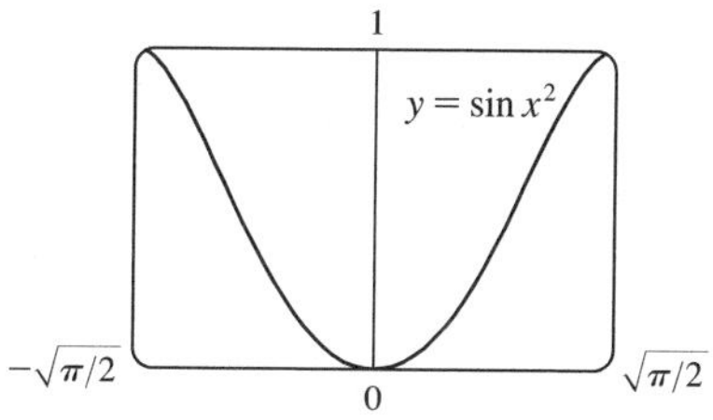

(c) For $\sqrt{1-y^2}$ to be a real number, we must have $-1 \le y \le 1$; that is, $-1 \le y(0) \le 1$. Thus, the initial-value problem $y' = 2x\sqrt{1-y^2}$, $y(0) = 2$ does *not* have a solution.

19. $\dfrac{dy}{dx} = \dfrac{\sin x}{\sin y}$, $y(0) = \frac{\pi}{2}$. So $\int \sin y\,dy = \int \sin x\,dx$ $\Leftrightarrow$ $-\cos y = -\cos x + C$ $\Leftrightarrow$ $\cos y = \cos x - C$. From the initial condition, we need $\cos\frac{\pi}{2} = \cos 0 - C$ $\Rightarrow$ $0 = 1 - C$ $\Rightarrow$ $C = 1$, so the solution is $\cos y = \cos x - 1$. Note that we cannot take $\cos^{-1}$ of both sides, since that would unnecessarily restrict the solution to the case where $-1 \le \cos x - 1$ $\Leftrightarrow$ $0 \le \cos x$, as $\cos^{-1}$ is defined only on $[-1, 1]$. Instead we plot the graph using Maple's `plots[implicitplot]` or Mathematica's `Plot[Evaluate[···]]`.

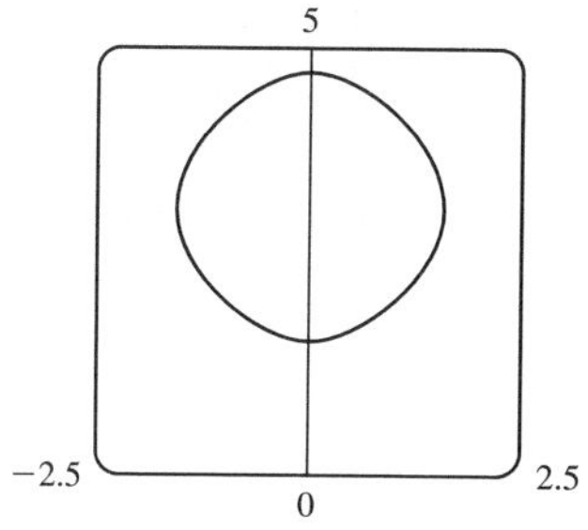

21. (a)

x	y	$y' = 1/y$
0	0.5	2
0	−0.5	−2
0	1	1
0	−1	−1
0	2	0.5

x	y	$y' = 1/y$
0	−2	−0.5
0	4	0.25
0	3	$0.\overline{3}$
0	0.25	4
0	$0.\overline{3}$	3

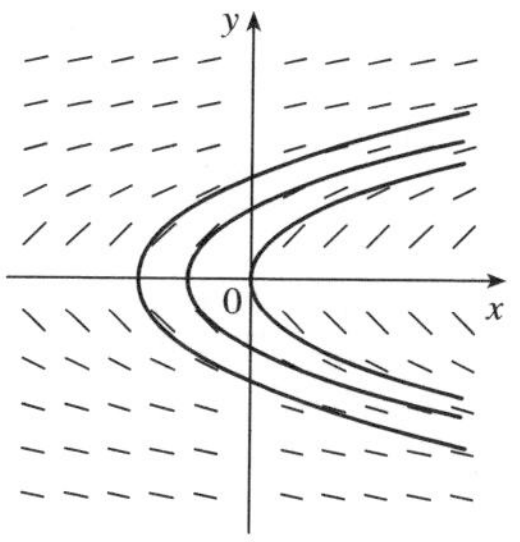

(b) $y' = 1/y$ $\Rightarrow$ $dy/dx = 1/y$ $\Rightarrow$ $y\,dy = dx$ $\Rightarrow$ $\int y\,dy = \int dx$ $\Rightarrow$ $\frac{1}{2}y^2 = x + C$ $\Rightarrow$ $y^2 = 2(x + C)$ or $y = \pm\sqrt{2(x + C)}$.

(c)

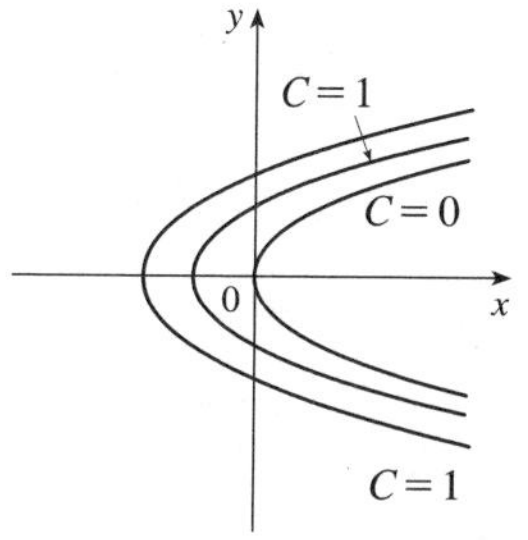

23. The curves $x^2 + 2y^2 = k^2$ form a family of ellipses with major axis on the x-axis. Differentiating gives

$\frac{d}{dx}(x^2 + 2y^2) = \frac{d}{dx}(k^2) \;\Rightarrow\; 2x + 4yy' = 0 \;\Rightarrow\; 4yy' = -2x \;\Rightarrow\; y' = \frac{-x}{2y}$. Thus, the slope of the tangent line at any point (x, y) on one of the ellipses is $y' = \frac{-x}{2y}$, so the orthogonal trajectories must satisfy $y' = \frac{2y}{x} \;\Leftrightarrow$

$\frac{dy}{dx} = \frac{2y}{x} \;\Leftrightarrow\; \frac{dy}{y} = 2 = \frac{dx}{x} \;\Leftrightarrow\; \int \frac{dy}{y} = 2\int \frac{dx}{x} \;\Leftrightarrow\; \ln|y| = 2\ln|x| + C_1 \;\Leftrightarrow\; \ln|y| = \ln|x|^2 + C_1 \;\Leftrightarrow$

$|y| = e^{\ln x^2 + C_1} \;\Leftrightarrow\; y = \pm x^2 \cdot e^{C_1} = Cx^2$. This is a family of parabolas.

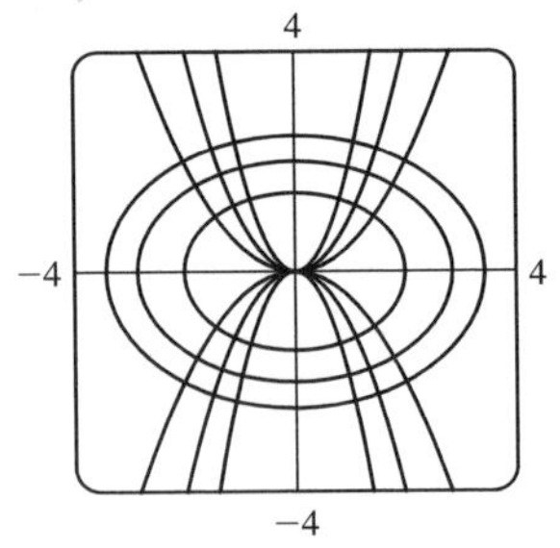

25. The curves $y = k/x$ form a family of hyperbolas with asymptotes $x = 0$ and $y = 0$. Differentiating gives

$\frac{d}{dx}(y) = \frac{d}{dx}\left(\frac{k}{x}\right) \;\Rightarrow\; y' = -\frac{k}{x^2} \;\Rightarrow\; y' = -\frac{xy}{x^2}$ [since $y = k/x \;\Rightarrow\; xy = k$] $\Rightarrow\; y' = -\frac{y}{x}$. Thus, the slope of the tangent line at any point (x, y) on one of the hyperbolas is $y' = -y/x$, so the orthogonal trajectories must satisfy

$y' = x/y \;\Leftrightarrow\; \frac{dy}{dx} = \frac{x}{y} \;\Leftrightarrow\; y\,dy = x\,dx \;\Leftrightarrow\; \int y\,dy = \int x\,dx \;\Leftrightarrow\; \frac{1}{2}y^2 = \frac{1}{2}x^2 + C_1 \;\Leftrightarrow\; y^2 = x^2 + C_2 \;\Leftrightarrow$

$x^2 - y^2 = C$. This is a family of hyperbolas with asymptotes $y = \pm x$.

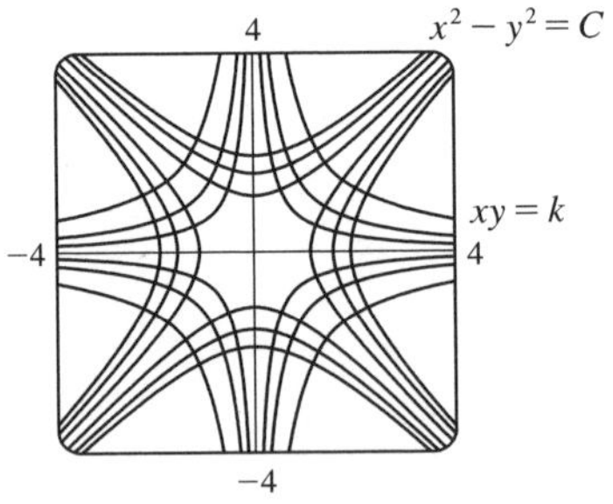

27. From Exercise 7.2.27, $\frac{dQ}{dt} = 12 - 4Q \;\Leftrightarrow\; \int \frac{dQ}{12 - 4Q} = \int dt \;\Leftrightarrow\; -\frac{1}{4}\ln|12 - 4Q| = t + C \;\Leftrightarrow$

$\ln|12 - 4Q| = -4t - 4C \;\Leftrightarrow\; |12 - 4Q| = e^{-4t-4C} \;\Leftrightarrow\; 12 - 4Q = Ke^{-4t}$ $[K = \pm e^{-4C}]$ $\Leftrightarrow$

$4Q = 12 - Ke^{-4t} \;\Leftrightarrow\; Q = 3 - Ae^{-4t}$ $[A = K/4]$. $Q(0) = 0 \;\Leftrightarrow\; 0 = 3 - A \;\Leftrightarrow\; A = 3 \;\Leftrightarrow$

$Q(t) = 3 - 3e^{-4t}$. As $t \to \infty$, $Q(t) \to 3 - 0 = 3$ (the limiting value).

29. $\frac{dP}{dt} = k(M - P) \;\Leftrightarrow\; \int \frac{dP}{P - M} = \int (-k)\,dt \;\Leftrightarrow\; \ln|P - M| = -kt + C \;\Leftrightarrow\; |P - M| = e^{-kt+C} \;\Leftrightarrow$

$P - M = Ae^{-kt}$ $[A = \pm e^C]$ $\Leftrightarrow\; P = M + Ae^{-kt}$. If we assume that performance is at level 0 when $t = 0$, then $P(0) = 0 \;\Leftrightarrow\; 0 = M + A \;\Leftrightarrow\; A = -M \;\Leftrightarrow\; P(t) = M - Me^{-kt}$. $\lim_{t\to\infty} P(t) = M - M \cdot 0 = M$.

31. (a) If $a = b$, then $\dfrac{dx}{dt} = k(a-x)(b-x)^{1/2}$ becomes $\dfrac{dx}{dt} = k(a-x)^{3/2} \Rightarrow (a-x)^{-3/2}dx = k\,dt \Rightarrow$

$\int (a-x)^{-3/2}dx = \int k\,dt \Rightarrow 2(a-x)^{-1/2} = kt + C$ [by substitution] $\Rightarrow \dfrac{2}{kt+C} = \sqrt{a-x} \Rightarrow$

$\left(\dfrac{2}{kt+C}\right)^2 = a - x \Rightarrow x(t) = a - \dfrac{4}{(kt+C)^2}$. The initial concentration of HBr is 0, so $x(0) = 0 \Rightarrow$

$0 = a - \dfrac{4}{C^2} \Rightarrow \dfrac{4}{C^2} = a \Rightarrow C^2 = \dfrac{4}{a} \Rightarrow C = 2/\sqrt{a}$ (C is positive since $kt + C = 2(a-x)^{-1/2} > 0$). Thus,

$x(t) = a - \dfrac{4}{(kt + 2/\sqrt{a})^2}$.

(b) $\dfrac{dx}{dt} = k(a-x)(b-x)^{1/2} \Rightarrow \dfrac{dx}{(a-x)\sqrt{b-x}} = k\,dt \Rightarrow \displaystyle\int \frac{dx}{(a-x)\sqrt{b-x}} = \int k\,dt$ $(\star)$. From the hint,

$u = \sqrt{b-x} \Rightarrow u^2 = b - x \Rightarrow 2u\,du = -dx$, so

$$\int \frac{dx}{(a-x)\sqrt{b-x}} = \int \frac{-2u\,du}{[a-(b-u^2)]u} = -2\int \frac{du}{a-b+u^2} = -2\int \frac{du}{\left(\sqrt{a-b}\right)^2 + u^2}$$

$$\overset{17}{=} -2\left(\frac{1}{\sqrt{a-b}}\tan^{-1}\frac{u}{\sqrt{a-b}}\right)$$

So $(\star)$ becomes $\dfrac{-2}{\sqrt{a-b}}\tan^{-1}\dfrac{\sqrt{b-x}}{\sqrt{a-b}} = kt + C$. Now $x(0) = 0 \Rightarrow C = \dfrac{-2}{\sqrt{a-b}}\tan^{-1}\dfrac{\sqrt{b}}{\sqrt{a-b}}$ and we have

$\dfrac{-2}{\sqrt{a-b}}\tan^{-1}\dfrac{\sqrt{b-x}}{\sqrt{a-b}} = kt - \dfrac{2}{\sqrt{a-b}}\tan^{-1}\dfrac{\sqrt{b}}{\sqrt{a-b}} \Rightarrow \dfrac{2}{\sqrt{a-b}}\left(\tan^{-1}\sqrt{\dfrac{b}{a-b}} - \tan^{-1}\sqrt{\dfrac{b-x}{a-b}}\right) = kt \Rightarrow$

$t(x) = \dfrac{2}{k\sqrt{a-b}}\left(\tan^{-1}\sqrt{\dfrac{b}{a-b}} - \tan^{-1}\sqrt{\dfrac{b-x}{a-b}}\right)$.

33. (a) $\dfrac{dC}{dt} = r - kC \Rightarrow \dfrac{dC}{dt} = -(kC - r) \Rightarrow \displaystyle\int \frac{dC}{kC-r} = \int -dt \Rightarrow (1/k)\ln|kC - r| = -t + M_1 \Rightarrow$

$\ln|kC - r| = -kt + M_2 \Rightarrow |kC - r| = e^{-kt+M_2} \Rightarrow kC - r = M_3e^{-kt} \Rightarrow kC = M_3e^{-kt} + r \Rightarrow$

$C(t) = M_4e^{-kt} + r/k$. $C(0) = C_0 \Rightarrow C_0 = M_4 + r/k \Rightarrow M_4 = C_0 - r/k \Rightarrow$

$C(t) = (C_0 - r/k)e^{-kt} + r/k$.

(b) If $C_0 < r/k$, then $C_0 - r/k < 0$ and the formula for $C(t)$ shows that $C(t)$ increases and $\lim\limits_{t\to\infty} C(t) = r/k$.

As t increases, the formula for $C(t)$ shows how the role of C_0 steadily diminishes as that of r/k increases.

35. (a) Let $y(t)$ be the amount of salt (in kg) after t minutes. Then $y(0) = 15$. The amount of liquid in the tank is 1000 L at all times, so the concentration at time t (in minutes) is $y(t)/1000$ kg/L and $\dfrac{dy}{dt} = -\left[\dfrac{y(t)}{1000}\dfrac{\text{kg}}{\text{L}}\right]\left(10\dfrac{\text{L}}{\text{min}}\right) = -\dfrac{y(t)}{100}\dfrac{\text{kg}}{\text{min}}$.

$\displaystyle\int \frac{dy}{y} = -\frac{1}{100}\int dt \Rightarrow \ln y = -\frac{t}{100} + C$, and $y(0) = 15 \Rightarrow \ln 15 = C$, so $\ln y = \ln 15 - \dfrac{t}{100}$.

It follows that $\ln\left(\dfrac{y}{15}\right) = -\dfrac{t}{100}$ and $\dfrac{y}{15} = e^{-t/100}$, so $y = 15e^{-t/100}$ kg.

(b) After 20 minutes, $y = 15e^{-20/100} = 15e^{-0.2} \approx 12.3$ kg.

37. Let $y(t)$ be the amount of alcohol in the vat after t minutes. Then $y(0) = 0.04(2000) = 80$ L. The amount of beer in the vat is 2000 Llons at all times, so the percentage at time t (in minutes) is $y(t)/2000 \times 100$, and the change in the amount of alcohol with respect to time t is

$$\frac{dy}{dt} = \text{rate in} - \text{rate out} = 0.06\left(20\,\frac{\text{L}}{\text{min}}\right) - \frac{y(t)}{2000}\left(20\,\frac{\text{L}}{\text{min}}\right) = 1.2 - \frac{y}{100} = \frac{120-y}{100}\,\frac{\text{L}}{\text{min}}$$

Hence, $\displaystyle\int \frac{dy}{120-y} = \int \frac{dt}{100}$ and $-\ln|120-y| = \frac{1}{100}t + C$. Because $y(0) = 80$, we have $-\ln 40 = C$, so $-\ln|120-y| = \frac{1}{100}t - \ln 40 \;\Rightarrow\; \ln|120-y| = -t/100 + \ln 40 \;\Rightarrow\; \ln|120-y| = \ln e^{-t/100} + \ln 40 \;\Rightarrow\; \ln|120-y| = \ln(40e^{-t/100}) \;\Rightarrow\; |120-y| = 40e^{-t/100}$. Since y is continuous, $y(0) = 80$, and the right-hand side is never zero, we deduce that $120 - y$ is always positive. Thus, $120 - y = 40e^{-t/100} \;\Rightarrow\; y = 120 - 40e^{-t/100}$. The percentage of alcohol is $p(t) = y(t)/2000 \times 100 = y(t)/20 = 6 - 2e^{-t/100}$. The percentage of alcohol after one hour is $p(60) = 6 - 2e^{-60/100} \approx 4.9$.

39. Assume that the raindrop begins at rest, so that $v(0) = 0$. $dm/dt = km$ and $(mv)' = gm \;\Rightarrow\; mv' + vm' = gm \;\Rightarrow\; mv' + v(km) = gm \;\Rightarrow\; v' + vk = g \;\Rightarrow\; \dfrac{dv}{dt} = g - kv \;\Rightarrow\; \displaystyle\int \frac{dv}{g-kv} = \int dt \;\Rightarrow\; -(1/k)\ln|g-kv| = t + C \;\Rightarrow\; \ln|g-kv| = -kt - kC \;\Rightarrow\; g - kv = Ae^{-kt}$. $v(0) = 0 \;\Rightarrow\; A = g$. So $kv = g - ge^{-kt} \;\Rightarrow\; v = (g/k)\left(1 - e^{-kt}\right)$. Since $k > 0$, as $t \to \infty$, $e^{-kt} \to 0$ and therefore, $\displaystyle\lim_{t\to\infty} v(t) = g/k$.

41. (a) The rate of growth of the area is jointly proportional to $\sqrt{A(t)}$ and $M - A(t)$; that is, the rate is proportional to the product of those two quantities. So for some constant k, $dA/dt = k\sqrt{A}\,(M - A)$. We are interested in the maximum of the function dA/dt (when the tissue grows the fastest), so we differentiate, using the Chain Rule and then substituting for dA/dt from the differential equation:

$$\frac{d}{dt}\left(\frac{dA}{dt}\right) = k\left[\sqrt{A}\,(-1)\frac{dA}{dt} + (M-A)\cdot\tfrac{1}{2}A^{-1/2}\frac{dA}{dt}\right] = \tfrac{1}{2}kA^{-1/2}\frac{dA}{dt}\,[-2A + (M-A)]$$
$$= \tfrac{1}{2}kA^{-1/2}\left[k\sqrt{A}(M-A)\right][M-3A] = \tfrac{1}{2}k^2(M-A)(M-3A)$$

This is 0 when $M - A = 0$ [this situation never actually occurs, since the graph of $A(t)$ is asymptotic to the line $y = M$, as in the logistic model] and when $M - 3A = 0 \;\Leftrightarrow\; A(t) = M/3$. This represents a maximum by the First Derivative Test, since $\dfrac{d}{dt}\left(\dfrac{dA}{dt}\right)$ goes from positive to negative when $A(t) = M/3$.

(b) From the CAS, we get $A(t) = M\left(\dfrac{Ce^{\sqrt{M}kt} - 1}{Ce^{\sqrt{M}kt} + 1}\right)^2$. To get C in terms of the initial area A_0 and the maximum area M, we substitute $t = 0$ and $A = A_0 = A(0)$: $A_0 = M\left(\dfrac{C-1}{C+1}\right)^2 \;\Leftrightarrow\; (C+1)\sqrt{A_0} = (C-1)\sqrt{M} \;\Leftrightarrow\; C\sqrt{A_0} + \sqrt{A_0} = C\sqrt{M} - \sqrt{M} \;\Leftrightarrow\; \sqrt{M} + \sqrt{A_0} = C\sqrt{M} - C\sqrt{A_0} \;\Leftrightarrow\; \sqrt{M} + \sqrt{A_0} = C\left(\sqrt{M} - \sqrt{A_0}\right) \;\Leftrightarrow\; C = \dfrac{\sqrt{M} + \sqrt{A_0}}{\sqrt{M} - \sqrt{A_0}}$. (Notice that if $A_0 = 0$, then $C = 1$.)

7.4 Exponential Growth and Decay

1. The relative growth rate is $\dfrac{1}{P}\dfrac{dP}{dt} = 0.7944$, so $\dfrac{dP}{dt} = 0.7944P$ and, by Theorem 2, $P(t) = P(0)e^{0.7944t} = 2e^{0.7944t}$. Thus,

$P(6) = 2e^{0.7944(6)} \approx 234.99$ or about 235 members.

3. (a) By Theorem 2, $P(t) = P(0)e^{kt} = 100e^{kt}$. Now $P(1) = 100e^{k(1)} = 420 \;\Rightarrow\; e^k = \frac{420}{100} \;\Rightarrow\; k = \ln 4.2$.

So $P(t) = 100e^{(\ln 4.2)t} = 100(4.2)^t$.

(b) $P(3) = 100(4.2)^3 = 7408.8 \approx 7409$ bacteria

(c) $dP/dt = kP \;\Rightarrow\; P'(3) = k \cdot P(3) = (\ln 4.2)\left(100(4.2)^3\right)$ [from part (a)] $\approx 10\,632$ bacteria/hour

(d) $P(t) = 100(4.2)^t = 10{,}000 \;\Rightarrow\; (4.2)^t = 100 \;\Rightarrow\; t = (\ln 100)/(\ln 4.2) \approx 3.2$ hours

5. (a) Let the population (in millions) in the year t be $P(t)$. Since the initial time is the year 1750, we substitute $t - 1750$ for t in Theorem 2, so the exponential model gives $P(t) = P(1750)e^{k(t-1750)}$. Then $P(1800) = 980 = 790e^{k(1800-1750)} \;\Rightarrow$ $\frac{980}{790} = e^{k(50)} \;\Rightarrow\; \ln\frac{980}{790} = 50k \;\Rightarrow\; k = \frac{1}{50}\ln\frac{980}{790} \approx 0.004\,3104$. So with this model, we have $P(1900) = 790e^{k(1900-1750)} \approx 1508$ million, and $P(1950) = 790e^{k(1950-1750)} \approx 1871$ million. Both of these estimates are much too low.

(b) In this case, the exponential model gives $P(t) = P(1850)e^{k(t-1850)} \;\Rightarrow\; P(1900) = 1650 = 1260e^{k(1900-1850)} \;\Rightarrow$ $\ln\frac{1650}{1260} = k(50) \;\Rightarrow\; k = \frac{1}{50}\ln\frac{1650}{1260} \approx 0.005\,393$. So with this model, we estimate $P(1950) = 1260e^{k(1950-1850)} \approx 2161$ million. This is still too low, but closer than the estimate of $P(1950)$ in part (a).

(c) The exponential model gives $P(t) = P(1900)e^{k(t-1900)} \;\Rightarrow\; P(1950) = 2560 = 1650e^{k(1950-1900)} \;\Rightarrow$ $\ln\frac{2560}{1650} = k(50) \;\Rightarrow\; k = \frac{1}{50}\ln\frac{2560}{1650} \approx 0.008\,785$. With this model, we estimate $P(2000) = 1650e^{k(2000-1900)} \approx 3972$ million. This is much too low. The discrepancy is explained by the fact that the world birth rate (average yearly number of births per person) is about the same as always, whereas the mortality rate (especially the infant mortality rate) is much lower, owing mostly to advances in medical science and to the wars in the first part of the twentieth century. The exponential model assumes, among other things, that the birth and mortality rates will remain constant.

7. (a) If $y = [N_2O_5]$ then by Theorem 2, $\dfrac{dy}{dt} = -0.0005y \;\Rightarrow\; y(t) = y(0)e^{-0.0005t} = Ce^{-0.0005t}$.

(b) $y(t) = Ce^{-0.0005t} = 0.9C \;\Rightarrow\; e^{-0.0005t} = 0.9 \;\Rightarrow\; -0.0005t = \ln 0.9 \;\Rightarrow\; t = -2000\ln 0.9 \approx 211$ s

9. (a) If $y(t)$ is the mass (in mg) remaining after t years, then $y(t) = y(0)e^{kt} = 100e^{kt}$. $y(30) = 100e^{30k} = \frac{1}{2}(100) \;\Rightarrow$ $e^{30k} = \frac{1}{2} \;\Rightarrow\; k = -(\ln 2)/30 \;\Rightarrow\; y(t) = 100e^{-(\ln 2)t/30} = 100 \cdot 2^{-t/30}$

(b) $y(100) = 100 \cdot 2^{-100/30} \approx 9.92$ mg

(c) $100e^{-(\ln 2)t/30} = 1 \;\Rightarrow\; -(\ln 2)t/30 = \ln\frac{1}{100} \;\Rightarrow\; t = -30\,\frac{\ln 0.01}{\ln 2} \approx 199.3$ years

11. Let $y(t)$ be the level of radioactivity. Thus, $y(t) = y(0)e^{-kt}$ and k is determined by using the half-life:

$y(5730) = \frac{1}{2}y(0) \;\Rightarrow\; y(0)e^{-k(5730)} = \frac{1}{2}y(0) \;\Rightarrow\; e^{-5730k} = \frac{1}{2} \;\Rightarrow\; -5730k = \ln\frac{1}{2} \;\Rightarrow$

$k = -\dfrac{\ln\frac{1}{2}}{5730} = \dfrac{\ln 2}{5730}$. If 74% of the ^{14}C remains, then we know that $y(t) = 0.74y(0) \;\Rightarrow\; 0.74 = e^{-t(\ln 2)/5730} \;\Rightarrow$

$\ln 0.74 = -\dfrac{t\ln 2}{5730} \;\Rightarrow\; t = -\dfrac{5730(\ln 0.74)}{\ln 2} \approx 2489 \approx 2500$ years.

13. (a) Using Newton's Law of Cooling, $\dfrac{dT}{dt} = k(T - T_s)$, we have $\dfrac{dT}{dt} = k(T - 22)$. Now let $y = T - 22$, so $y(0) = T(0) - 22 = 85 - 22 = 63$, so y is a solution of the initial-value problem $dy/dt = ky$ with $y(0) = 63$ and by Theorem 2 we have $y(t) = y(0)e^{kt} = 63e^{kt}$.

$y(30) = 63e^{30k} = 65 - 22 \;\Rightarrow\; e^{30k} = \frac{43}{63} \;\Rightarrow\; k = \frac{1}{30}\ln\frac{43}{63}$, so $y(t) = 63e^{\frac{1}{30}t\ln\left(\frac{43}{63}\right)}$ and

$y(45) = 63e^{\frac{45}{30}\ln\left(\frac{43}{63}\right)} \approx 36°\text{C}$. Thus, $T(45) \approx 36 + 22 = 58°\text{C}$.

(b) $T(t) = 40 \;\Rightarrow\; y(t) = 18$. $y(t) = 63e^{\frac{1}{30}t\ln\left(\frac{43}{63}\right)} = 18 \;\Rightarrow\; e^{\frac{1}{30}t\ln\left(\frac{43}{63}\right)} = \frac{18}{63} = \frac{2}{7} \;\Rightarrow\; \frac{1}{30}t\ln\frac{43}{63} = \ln\frac{2}{7} \;\Rightarrow$

$t = \dfrac{30\ln\frac{2}{7}}{\ln\frac{43}{63}} \approx 98$ min.

15. $\dfrac{dT}{dt} = k(T - 20)$. Letting $y = T - 20$, we get $\dfrac{dy}{dt} = ky$, so $y(t) = y(0)e^{kt}$. $y(0) = T(0) - 20 = 5 - 20 = -15$, so $y(25) = y(0)e^{25k} = -15e^{25k}$, and $y(25) = T(25) - 20 = 10 - 20 = -10$, so $-15e^{25k} = -10 \;\Rightarrow\; e^{25k} = \frac{2}{3}$. Thus, $25k = \ln\left(\frac{2}{3}\right)$ and $k = \frac{1}{25}\ln\left(\frac{2}{3}\right)$, so $y(t) = y(0)e^{kt} = -15e^{(1/25)\ln(2/3)t}$. More simply, $e^{25k} = \frac{2}{3} \;\Rightarrow\; e^{k} = \left(\frac{2}{3}\right)^{1/25} \;\Rightarrow$ $e^{kt} = \left(\frac{2}{3}\right)^{t/25} \;\Rightarrow\; y(t) = -15\cdot\left(\frac{2}{3}\right)^{t/25}$.

(a) $T(50) = 20 + y(50) = 20 - 15\cdot\left(\frac{2}{3}\right)^{50/25} = 20 - 15\cdot\left(\frac{2}{3}\right)^2 = 20 - \frac{20}{3} = 13.\bar{3}\,°\text{C}$

(b) $15 = T(t) = 20 + y(t) = 20 - 15\cdot\left(\frac{2}{3}\right)^{t/25} \;\Rightarrow\; 15\cdot\left(\frac{2}{3}\right)^{t/25} = 5 \;\Rightarrow\; \left(\frac{2}{3}\right)^{t/25} = \frac{1}{3} \;\Rightarrow$ $(t/25)\ln\left(\frac{2}{3}\right) = \ln\left(\frac{1}{3}\right) \;\Rightarrow\; t = 25\ln\left(\frac{1}{3}\right)/\ln\left(\frac{2}{3}\right) \approx 67.74$ min.

17. (a) Let $P(h)$ be the pressure at altitude h. Then $dP/dh = kP \;\Rightarrow\; P(h) = P(0)e^{kh} = 101.3e^{kh}$.

$P(1000) = 101.3e^{1000k} = 87.14 \;\Rightarrow\; 1000k = \ln\left(\frac{87.14}{101.3}\right) \;\Rightarrow\; k = \frac{1}{1000}\ln\left(\frac{87.14}{101.3}\right) \;\Rightarrow$

$P(h) = 101.3\,e^{\frac{1}{1000}h\ln\left(\frac{87.14}{101.3}\right)}$, so $P(3000) = 101.3e^{3\ln\left(\frac{87.14}{101.3}\right)} \approx 64.5$ kPa.

(b) $P(6187) = 101.3\,e^{\frac{6187}{1000}\ln\left(\frac{87.14}{101.3}\right)} \approx 39.9$ kPa

19. (a) Using $A = A_0\left(1 + \dfrac{r}{n}\right)^{nt}$ with $A_0 = 3000$, $r = 0.05$, and $t = 5$, we have:

(i) Annually: $n = 1$;	$A = 3000\left(1 + \frac{0.05}{1}\right)^{1\cdot5} = \$3828.84$
(ii) Semiannually: $n = 2$;	$A = 3000\left(1 + \frac{0.05}{2}\right)^{2\cdot5} = \$3840.25$
(iii) Monthly: $n = 12$;	$A = 3000\left(1 + \frac{0.05}{12}\right)^{12\cdot5} = \$3850.08$
(iv) Weekly: $n = 52$;	$A = 3000\left(1 + \frac{0.05}{52}\right)^{52\cdot5} = \$3851.61$
(v) Daily: $n = 365$;	$A = 3000\left(1 + \frac{0.05}{365}\right)^{365\cdot5} = \$3852.01$
(vi) Continuously:	$A = 3000e^{(0.05)5} = \$3852.08$

(b) $dA/dt = 0.05A$ and $A(0) = 3000$.

21. (a) $\dfrac{dP}{dt} = kP - m = k\left(P - \dfrac{m}{k}\right)$. Let $y = P - \dfrac{m}{k}$, so $\dfrac{dy}{dt} = \dfrac{dP}{dt}$ and the differential equation becomes $\dfrac{dy}{dt} = ky$.

The solution is $y = y_0e^{kt} \;\Rightarrow\; P - \dfrac{m}{k} = \left(P_0 - \dfrac{m}{k}\right)e^{kt} \;\Rightarrow\; P(t) = \dfrac{m}{k} + \left(P_0 - \dfrac{m}{k}\right)e^{kt}$.

(b) Since $k > 0$, there will be an exponential expansion $\Leftrightarrow\; P_0 - \dfrac{m}{k} > 0 \;\Leftrightarrow\; m < kP_0$.

(c) The population will be constant if $P_0 - \dfrac{m}{k} = 0 \;\Leftrightarrow\; m = kP_0$. It will decline if $P_0 - \dfrac{m}{k} < 0 \;\Leftrightarrow\; m > kP_0$.

(d) $P_0 = 8\,000\,000$, $k = \alpha - \beta = 0.016$, $m = 210\,000 \;\Rightarrow\; m > kP_0\ (= 128\,000)$, so by part (c), the population was declining.

7.5 The Logistic Equation

1. (a) $dP/dt = 0.05P - 0.0005P^2 = 0.05P(1 - 0.01P) = 0.05P(1 - P/100)$. Comparing to Equation 1, $dP/dt = kP(1 - P/K)$, we see that the carrying capacity is $K = 100$ and the value of k is 0.05.

(b) The slopes close to 0 occur where P is near 0 or 100. The largest slopes appear to be on the line $P = 50$. The solutions are increasing for $0 < P_0 < 100$ and decreasing for $P_0 > 100$.

(c)

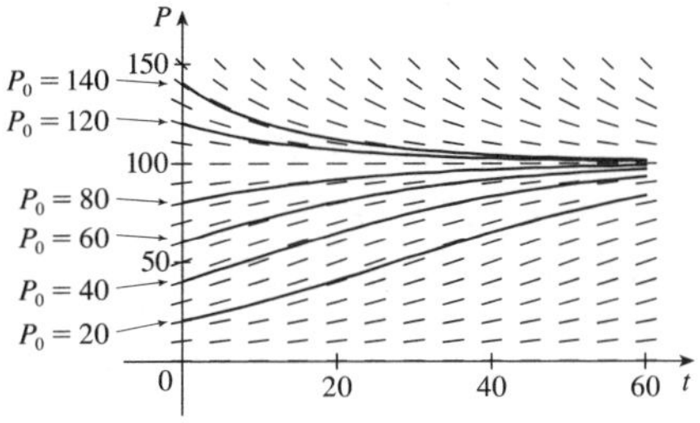

All of the solutions approach $P = 100$ as t increases. As in part (b), the solutions differ since for $0 < P_0 < 100$ they are increasing, and for $P_0 > 100$ they are decreasing. Also, some have an IP and some don't. It appears that the solutions which have $P_0 = 20$ and $P_0 = 40$ have inflection points at $P = 50$.

(d) The equilibrium solutions are $P = 0$ (trivial solution) and $P = 100$. The increasing solutions move away from $P = 0$ and all nonzero solutions approach $P = 100$ as $t \to \infty$.

3. (a) $\dfrac{dy}{dt} = ky\left(1 - \dfrac{y}{K}\right) \Rightarrow y(t) = \dfrac{K}{1 + Ae^{-kt}}$ with $A = \dfrac{K - y(0)}{y(0)}$. With $K = 8 \times 10^7$, $k = 0.71$, and $y(0) = 2 \times 10^7$, we get the model $y(t) = \dfrac{8 \times 10^7}{1 + 3e^{-0.71t}}$, so $y(1) = \dfrac{8 \times 10^7}{1 + 3e^{-0.71}} \approx 3.23 \times 10^7$ kg.

(b) $y(t) = 4 \times 10^7 \Rightarrow \dfrac{8 \times 10^7}{1 + 3e^{-0.71t}} = 4 \times 10^7 \Rightarrow 2 = 1 + 3e^{-0.71t} \Rightarrow e^{-0.71t} = \frac{1}{3} \Rightarrow -0.71t = \ln\frac{1}{3} \Rightarrow$ $t = \dfrac{\ln 3}{0.71} \approx 1.55$ years

5. (a) We will assume that the difference in the birth and death rates is 20 million/year. Let $t = 0$ correspond to the year 1990 and use a unit of 1 billion for all calculations. $k \approx \dfrac{1}{P}\dfrac{dP}{dt} = \dfrac{1}{5.3}(0.02) = \dfrac{1}{265}$, so

$$\frac{dP}{dt} = kP\left(1 - \frac{P}{K}\right) = \frac{1}{265}P\left(1 - \frac{P}{100}\right), \qquad P \text{ in billions}$$

(b) $A = \dfrac{K - P_0}{P_0} = \dfrac{100 - 5.3}{5.3} = \dfrac{947}{53} \approx 17.8679$. $P(t) = \dfrac{K}{1 + Ae^{-kt}} = \dfrac{100}{1 + \frac{947}{53}e^{-(1/265)t}}$, so $P(10) \approx 5.49$ billion.

(c) $P(110) \approx 7.81$, and $P(510) \approx 27.72$. The predictions are 7.81 billion in the year 2100 and 27.72 billion in 2500.

(d) If $K = 50$, then $P(t) = \dfrac{50}{1 + \frac{447}{53}e^{-(1/265)t}}$. So $P(10) \approx 5.48$, $P(110) \approx 7.61$, and $P(510) \approx 22.41$. The predictions become 5.48 billion in the year 2000, 7.61 billion in 2100, and 22.41 billion in the year 2500.

7. (a) Our assumption is that $\dfrac{dy}{dt} = ky(1 - y)$, where y is the fraction of the population that has heard the rumour.

(b) Using the logistic equation (1), $\dfrac{dP}{dt} = kP\left(1 - \dfrac{P}{K}\right)$, we substitute $y = \dfrac{P}{K}$, $P = Ky$, and $\dfrac{dP}{dt} = K\dfrac{dy}{dt}$, to obtain $K\dfrac{dy}{dt} = k(Ky)(1 - y) \Leftrightarrow \dfrac{dy}{dt} = ky(1 - y)$, our equation in part (a).

Now the solution to (1) is $P(t) = \dfrac{K}{1 + Ae^{-kt}}$, where $A = \dfrac{K - P_0}{P_0}$.

We use the same substitution to obtain $Ky = \dfrac{K}{1 + \dfrac{K - Ky_0}{Ky_0}e^{-kt}} \Rightarrow y = \dfrac{y_0}{y_0 + (1 - y_0)e^{-kt}}$.

Alternatively, we could use the same steps as outlined in "The Analytic Solution," following Example 2.

(c) Let t be the number of hours since 8 A.M. Then $y_0 = y(0) = \frac{80}{1000} = 0.08$ and $y(4) = \frac{1}{2}$, so

$\frac{1}{2} = y(4) = \frac{0.08}{0.08 + 0.92e^{-4k}}$. Thus, $0.08 + 0.92e^{-4k} = 0.16$, $e^{-4k} = \frac{0.08}{0.92} = \frac{2}{23}$, and $e^{-k} = \left(\frac{2}{23}\right)^{1/4}$, so

$y = \frac{0.08}{0.08 + 0.92(2/23)^{t/4}} = \frac{2}{2 + 23(2/23)^{t/4}}$. Solving this equation for t, we get

$$2y + 23y\left(\frac{2}{23}\right)^{t/4} = 2 \;\Rightarrow\; \left(\frac{2}{23}\right)^{t/4} = \frac{2 - 2y}{23y} \;\Rightarrow\; \left(\frac{2}{23}\right)^{t/4} = \frac{2}{23} \cdot \frac{1-y}{y} \;\Rightarrow\; \left(\frac{2}{23}\right)^{t/4-1} = \frac{1-y}{y}.$$

It follows that $\frac{t}{4} - 1 = \frac{\ln[(1-y)/y]}{\ln\frac{2}{23}}$, so $t = 4\left[1 + \frac{\ln((1-y)/y)}{\ln\frac{2}{23}}\right]$.

When $y = 0.9$, $\frac{1-y}{y} = \frac{1}{9}$, so $t = 4\left(1 - \frac{\ln 9}{\ln\frac{2}{23}}\right) \approx 7.6$ h or 7 h 36 min. Thus, 90% of the population will have heard the rumour by 3:36 P.M.

9. (a) $\frac{dP}{dt} = kP\left(1 - \frac{P}{K}\right) \;\Rightarrow\; \frac{d^2P}{dt^2} = k\left[P\left(-\frac{1}{K}\frac{dP}{dt}\right) + \left(1 - \frac{P}{K}\right)\frac{dP}{dt}\right] = k\,\frac{dP}{dt}\left(-\frac{P}{K} + 1 - \frac{P}{K}\right)$

$$= k\left[kP\left(1 - \frac{P}{K}\right)\right]\left(1 - \frac{2P}{K}\right) = k^2P\left(1 - \frac{P}{K}\right)\left(1 - \frac{2P}{K}\right)$$

(b) P grows fastest when P' has a maximum, that is, when $P'' = 0$. From part (a), $P'' = 0 \;\Leftrightarrow\; P = 0$, $P = K$, or $P = K/2$. Since $0 < P < K$, we see that $P'' = 0 \;\Leftrightarrow\; P = K/2$.

11. (a) The term -15 represents a harvesting of fish at a constant rate—in this case, 15 fish/week. This is the rate at which fish are caught.

(b)

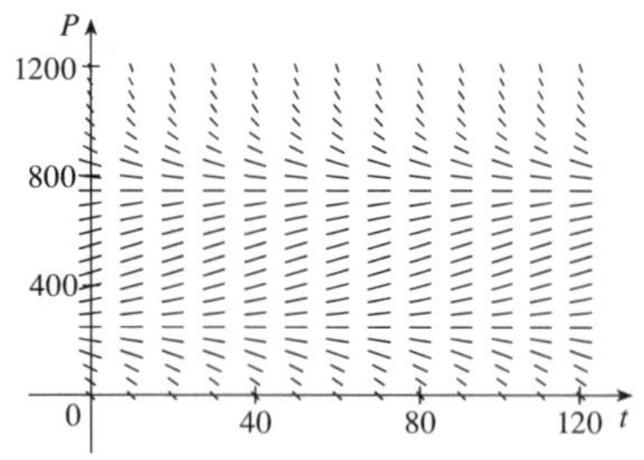

(c) From the graph in part (b), it appears that $P(t) = 250$ and $P(t) = 750$ are the equilibrium solutions. We confirm this analytically by solving the equation $dP/dt = 0$ as follows: $0.08P(1 - P/1000) - 15 = 0 \;\Rightarrow$ $0.08P - 0.00008P^2 - 15 = 0 \;\Rightarrow\; -0.00008(P^2 - 1000P + 187\,500) = 0 \;\Rightarrow\; (P - 250)(P - 750) = 0 \;\Rightarrow$ $P = 250$ or 750.

(d)

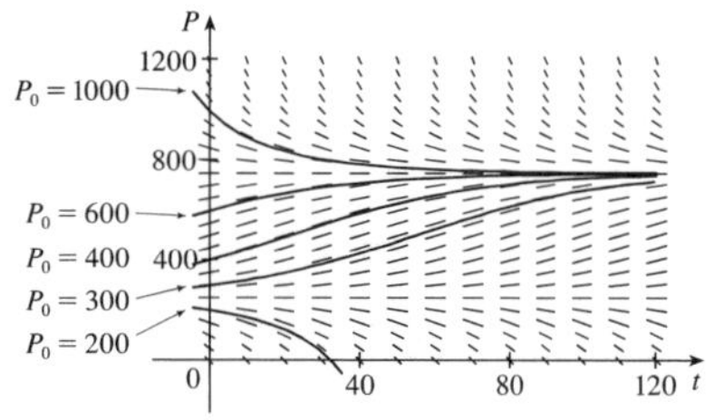

For $0 < P_0 < 250$, $P(t)$ decreases to 0. For $P_0 = 250$, $P(t)$ remains constant. For $250 < P_0 < 750$, $P(t)$ increases and approaches 750. For $P_0 = 750$, $P(t)$ remains constant. For $P_0 > 750$, $P(t)$ decreases and approaches 750.

(e) $\dfrac{dP}{dt} = 0.08P\left(1 - \dfrac{P}{1000}\right) - 15 \quad\Leftrightarrow\quad -\dfrac{100\,000}{8} \cdot \dfrac{dP}{dt} = \left(0.08P - 0.00008P^2 - 15\right) \cdot \left(-\dfrac{100\,000}{8}\right) \quad\Leftrightarrow$

$-12\,500\,\dfrac{dP}{dt} = P^2 - 1000P + 187\,500 \quad\Leftrightarrow\quad \dfrac{dP}{(P-250)(P-750)} = -\dfrac{1}{12\,500}\,dt \quad\Leftrightarrow$

$\displaystyle\int\left(\frac{-1/500}{P-250} + \frac{1/500}{P-750}\right)dP = -\frac{1}{12\,500}\,dt \quad\Leftrightarrow\quad \int\left(\frac{1}{P-250} - \frac{1}{P-750}\right)dP = \tfrac{1}{25}\,dt \quad\Leftrightarrow$

$\ln|P-250| - \ln|P-750| = \frac{1}{25}t + C \quad\Leftrightarrow\quad \ln\left|\dfrac{P-250}{P-750}\right| = \frac{1}{25}t + C \quad\Leftrightarrow\quad \left|\dfrac{P-250}{P-750}\right| = e^{t/25+C} = ke^{t/25} \quad\Leftrightarrow$

$\dfrac{P-250}{P-750} = ke^{t/25} \quad\Leftrightarrow\quad P - 250 = Pke^{t/25} - 750ke^{t/25} \quad\Leftrightarrow\quad P - Pke^{t/25} = 250 - 750ke^{t/25} \quad\Leftrightarrow$

$P(t) = \dfrac{250 - 750ke^{t/25}}{1 - ke^{t/25}}$. If $t = 0$ and $P = 200$, then $200 = \dfrac{250 - 750k}{1-k} \quad\Leftrightarrow\quad 200 - 200k = 250 - 750k \quad\Leftrightarrow$

$550k = 50 \quad\Leftrightarrow\quad k = \frac{1}{11}$. Similarly, if $t = 0$ and $P = 300$, then $k = -\frac{1}{9}$. Simplifying P with these two values of k gives us

$$P(t) = \frac{250\left(3e^{t/25} - 11\right)}{e^{t/25} - 11} \text{ and } P(t) = \frac{750\left(e^{t/25} + 3\right)}{e^{t/25} + 9}.$$

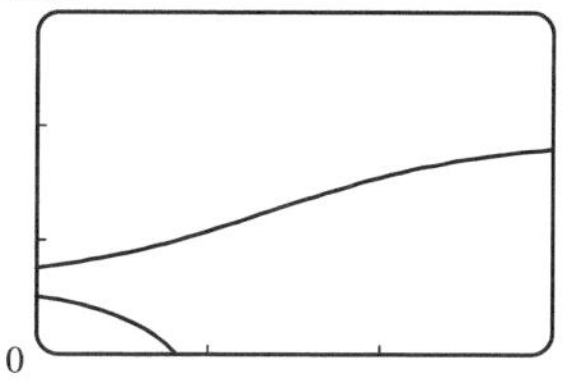

13. (a) $\dfrac{dP}{dt} = (kP)\left(1 - \dfrac{P}{K}\right)\left(1 - \dfrac{m}{P}\right)$. If $m < P < K$, then $dP/dt = (+)(+)(+) = + \quad\Rightarrow\quad P$ is increasing.

If $0 < P < m$, then $dP/dt = (+)(+)(-) = - \quad\Rightarrow\quad P$ is decreasing.

(b)

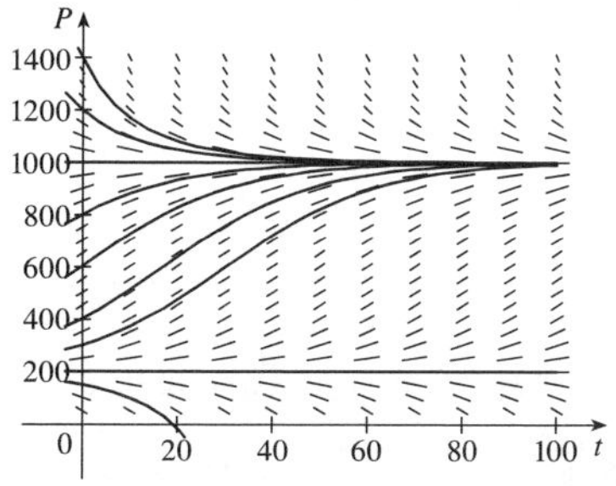

$k = 0.08$, $K = 1000$, and $m = 200 \quad\Rightarrow$

$$\frac{dP}{dt} = 0.08P\left(1 - \frac{P}{1000}\right)\left(1 - \frac{200}{P}\right)$$

For $0 < P_0 < 200$, the population dies out. For $P_0 = 200$, the population is steady. For $200 < P_0 < 1000$, the population increases and approaches 1000. For $P_0 > 1000$, the population decreases and approaches 1000.

The equilibrium solutions are $P(t) = 200$ and $P(t) = 1000$.

(c) $\dfrac{dP}{dt} = kP\left(1 - \dfrac{P}{K}\right)\left(1 - \dfrac{m}{P}\right) = kP\left(\dfrac{K-P}{K}\right)\left(\dfrac{P-m}{P}\right) = \dfrac{k}{K}(K-P)(P-m) \quad\Leftrightarrow$

$\displaystyle\int \frac{dP}{(K-P)(P-m)} = \int \frac{k}{K}\,dt.$

By partial fractions, $\dfrac{1}{(K-P)(P-m)} = \dfrac{A}{K-P} + \dfrac{B}{P-m}$, so $A(P-m) + B(K-P) = 1$.

If $P = m$, $B = \dfrac{1}{K-m}$; if $P = K$, $A = \dfrac{1}{K-m}$, so $\displaystyle\frac{1}{K-m}\int\left(\frac{1}{K-P} + \frac{1}{P-m}\right)dP = \int\frac{k}{K}\,dt \quad\Rightarrow$

$\dfrac{1}{K-m}\left(-\ln|K-P| + \ln|P-m|\right) = \dfrac{k}{K}t + M \quad\Rightarrow\quad \dfrac{1}{K-m}\ln\left|\dfrac{P-m}{K-P}\right| = \dfrac{k}{K}t + M \quad\Rightarrow$

$\ln\left|\dfrac{P-m}{K-P}\right| = (K-m)\dfrac{k}{K}t + M_1 \quad\Leftrightarrow\quad \dfrac{P-m}{K-P} = De^{(K-m)(k/K)t} \quad [D = \pm e^{M_1}].$

Let $t = 0$: $\dfrac{P_0 - m}{K - P_0} = D$. So $\dfrac{P-m}{K-P} = \dfrac{P_0 - m}{K - P_0}e^{(K-m)(k/K)t}$. Solving for P, we get

$$P(t) = \frac{m(K - P_0) + K(P_0 - m)e^{(K-m)(k/K)t}}{K - P_0 + (P_0 - m)e^{(K-m)(k/K)t}}.$$

(d) If $P_0 < m$, then $P_0 - m < 0$. Let $N(t)$ be the numerator of the expression for $P(t)$ in part (c). Then $N(0) = P_0\,(K - m) > 0$, and $P_0 - m < 0 \;\Leftrightarrow\; \lim\limits_{t\to\infty} K(P_0 - m)\,e^{(K-m)(k/K)t} = -\infty \;\Rightarrow\; \lim\limits_{t\to\infty} N(t) = -\infty$. Since N is continuous, there is a number t such that $N(t) = 0$ and thus $P(t) = 0$. So the species will become extinct.

15. (a) $dP/dt = kP\cos(rt - \phi) \;\Rightarrow\; (dP)/P = k\cos(rt - \phi)\,dt \;\Rightarrow\; \int (dP)/P = k\int \cos(rt - \phi)\,dt \;\Rightarrow$ $\ln P = (k/r)\sin(rt - \phi) + C$. (Since this is a growth model, $P > 0$ and we can write $\ln P$ instead of $\ln|P|$.) Since $P(0) = P_0$, we obtain $\ln P_0 = (k/r)\sin(-\phi) + C = -(k/r)\sin\phi + C \;\Rightarrow\; C = \ln P_0 + (k/r)\sin\phi$. Thus, $\ln P = (k/r)\sin(rt - \phi) + \ln P_0 + (k/r)\sin\phi$, which we can rewrite as $\ln(P/P_0) = (k/r)[\sin(rt - \phi) + \sin\phi]$ or, after exponentiation, $P(t) = P_0 e^{(k/r)[\sin(rt-\phi)+\sin\phi]}$.

(b) As k increases, the amplitude increases, but the minimum value stays the same.

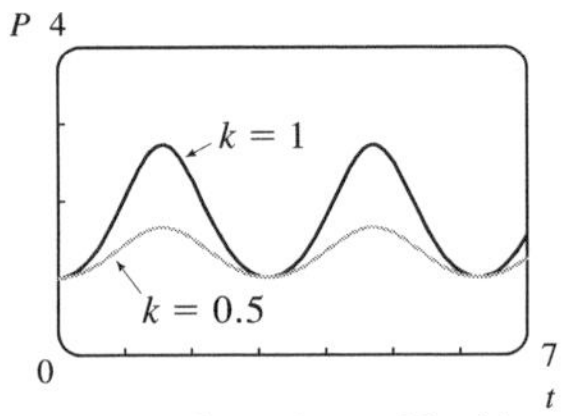

Comparing values of k with $P_0 = 1$, $r = 2$, and $\phi = \pi/2$

As r increases, the amplitude and the period decrease.

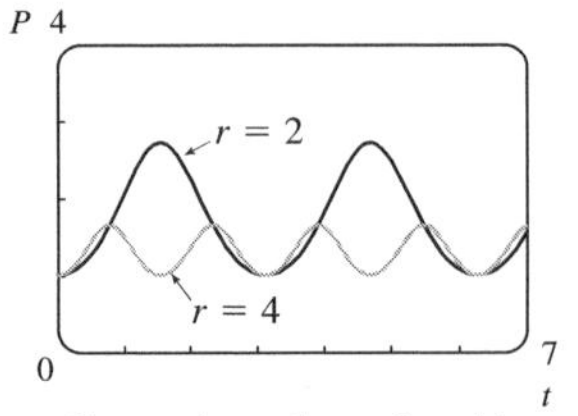

Comparing values of r with $P_0 = 1$, $k = 1$, and $\phi = \pi/2$

A change in ϕ produces slight adjustments in the phase shift and amplitude.

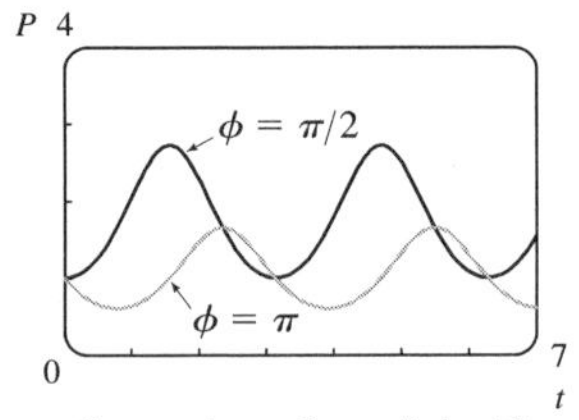

Comparing values of ϕ with $P_0 = 1$, $k = 1$, and $r = 2$

$P(t)$ oscillates between $P_0 e^{(k/r)(1+\sin\phi)}$ and $P_0 e^{(k/r)(-1+\sin\phi)}$ (the extreme values are attained when $rt - \phi$ is an odd multiple of $\frac{\pi}{2}$), so $\lim\limits_{t\to\infty} P(t)$ does not exist.

7.6 Predator-Prey Systems

1. (a) $dx/dt = -0.05x + 0.0001xy$. If $y = 0$, we have $dx/dt = -0.05x$, which indicates that in the absence of y, x declines at a rate proportional to itself. So x represents the predator population and y represents the prey population. The growth of the prey population, $0.1y$ (from $dy/dt = 0.1y - 0.005xy$), is restricted only by encounters with predators (the term $-0.005xy$). The predator population increases only through the term $0.0001xy$; that is, by encounters with the prey and not through additional food sources.

(b) $dy/dt = -0.015y + 0.00008xy$. If $x = 0$, we have $dy/dt = -0.015y$, which indicates that in the absence of x, y would decline at a rate proportional to itself. So y represents the predator population and x represents the prey population. The growth of the prey population, $0.2x$ (from $dx/dt = 0.2x - 0.0002x^2 - 0.006xy = 0.2x(1 - 0.001x) - 0.006xy$), is restricted by a carrying capacity of 1000 [from the term $1 - 0.001x = 1 - x/1000$] and by encounters with predators (the term $-0.006xy$). The predator population increases only through the term $0.00008xy$; that is, by encounters with the prey and not through additional food sources.

3. (a) At $t = 0$, there are about 300 rabbits and 100 foxes. At $t = t_1$, the number of foxes reaches a minimum of about 20 while the number of rabbits is about 1000. At $t = t_2$, the number of rabbits reaches a maximum of about 2400, while the number of foxes rebounds to 100. At $t = t_3$, the number of rabbits decreases to about 1000 and the number of foxes reaches a maximum of about 315. As t increases, the number of foxes decreases greatly to 100, and the number of rabbits decreases to 300 (the initial populations), and the cycle starts again.

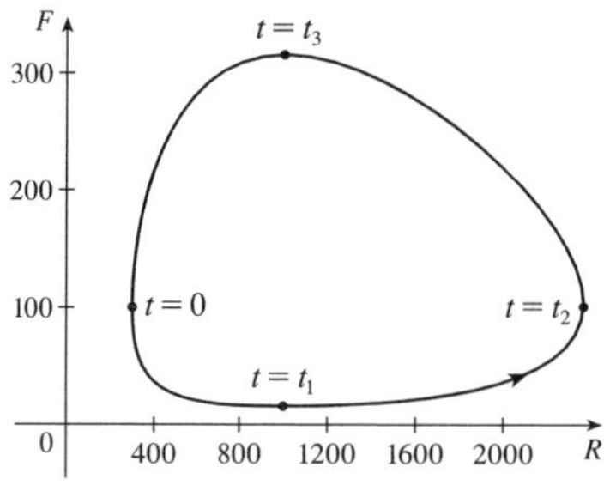

(b)

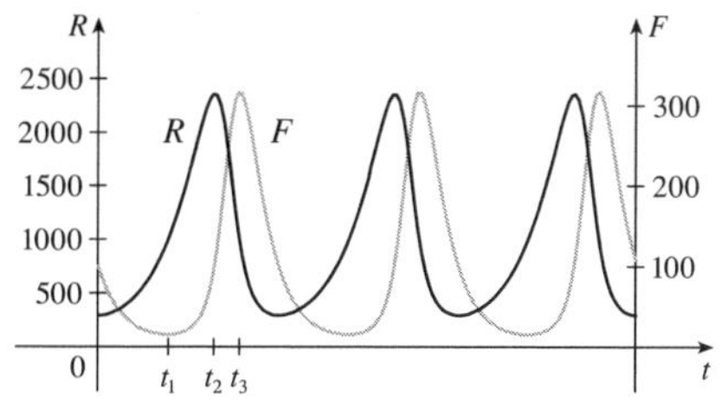

5.

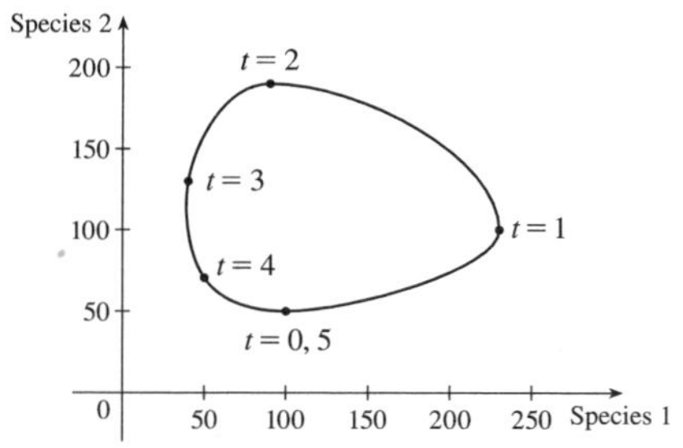

7. $\dfrac{dW}{dR} = \dfrac{-0.02W + 0.00002RW}{0.08R - 0.001RW} \quad\Leftrightarrow\quad (0.08 - 0.001W)R\,dW = (-0.02 + 0.00002R)W\,dR \quad\Leftrightarrow$

$\dfrac{0.08 - 0.001W}{W}\,dW = \dfrac{-0.02 + 0.00002R}{R}\,dR \quad\Leftrightarrow\quad \displaystyle\int\left(\frac{0.08}{W} - 0.001\right)dW = \int\left(-\frac{0.02}{R} + 0.00002\right)dR \quad\Leftrightarrow$

$0.08\,\ln|W| - 0.001W = -0.02\,\ln|R| + 0.00002R + K \quad\Leftrightarrow\quad 0.08\,\ln W + 0.02\,\ln R = 0.001W + 0.00002R + K \quad\Leftrightarrow$

$\ln\left(W^{0.08}R^{0.02}\right) = 0.00002R + 0.001W + K \quad\Leftrightarrow\quad W^{0.08}R^{0.02} = e^{0.00002R + 0.001W + K} \quad\Leftrightarrow$

$R^{0.02}W^{0.08} = Ce^{0.00002R}e^{0.001W} \quad\Leftrightarrow\quad \dfrac{R^{0.02}W^{0.08}}{e^{0.00002R}e^{0.001W}} = C$. In general, if $\dfrac{dy}{dx} = \dfrac{-ry + bxy}{kx - axy}$, then $C = \dfrac{x^r y^k}{e^{bx}e^{ay}}$.

9. (a) Letting $W = 0$ gives us $dR/dt = 0.08R(1 - 0.0002R)$. $dR/dt = 0 \quad\Leftrightarrow\quad R = 0$ or 5000. Since $dR/dt > 0$ for $0 < R < 5000$, we would expect the rabbit population to *increase* to 5000 for these values of R. Since $dR/dt < 0$ for $R > 5000$, we would expect the rabbit population to *decrease* to 5000 for these values of R. Hence, in the absence of wolves, we would expect the rabbit population to stabilize at 5000.

(b) R and W are constant $\Rightarrow$ $R' = 0$ and $W' = 0$ $\Rightarrow$

$$\left\{\begin{array}{l} 0 = 0.08R(1 - 0.0002R) - 0.001RW \\ 0 = -0.02W + 0.00002RW \end{array}\right\} \Rightarrow \left\{\begin{array}{l} 0 = R[0.08(1 - 0.0002R) - 0.001W] \\ 0 = W(-0.02 + 0.00002R) \end{array}\right.$$

The second equation is true if $W = 0$ or $R = \frac{0.02}{0.00002} = 1000$. If $W = 0$ in the first equation, then either $R = 0$ or $R = \frac{1}{0.0002} = 5000$ [as in part (a)]. If $R = 1000$, then $0 = 1000[0.08(1 - 0.0002 \cdot 1000) - 0.001W]$ $\Leftrightarrow$ $0 = 80(1 - 0.2) - W$ $\Leftrightarrow$ $W = 64$.

Case (i): $W = 0$, $R = 0$: both populations are zero

Case (ii): $W = 0$, $R = 5000$: see part (a)

Case (iii): $R = 1000$, $W = 64$: the predator/prey interaction balances and the populations are stable.

(c) The populations of wolves and rabbits fluctuate around 64 and 1000, respectively, and eventually stabilize at those values.

(d)

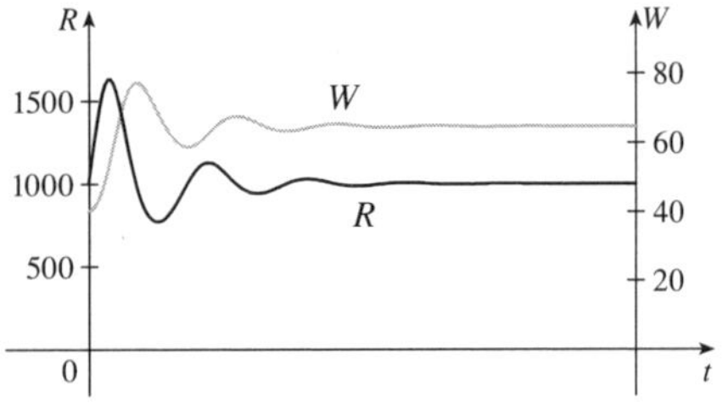

7 Review

CONCEPT CHECK

1. (a) A differential equation is an equation that contains an unknown function and one or more of its derivatives.

(b) The order of a differential equation is the order of the highest derivative that occurs in the equation.

(c) An initial condition is a condition of the form $y(t_0) = y_0$.

2. $y' = x^2 + y^2 \geq 0$ for all x and y. $y' = 0$ only at the origin, so there is a horizontal tangent at $(0, 0)$, but nowhere else. The graph of the solution is increasing on every interval.

3. See the paragraph preceding Example 1 in Section 7.2.

4. See the paragraph after Figure 14 in Section 7.2.

5. A separable equation is a first-order differential equation in which the expression for dy/dx can be factored as a function of x times a function of y, that is, $dy/dx = g(x)f(y)$. We can solve the equation by integrating both sides of the equation $dy/f(y) = g(x)dx$ and solving for y.

6. (a) $\frac{dy}{dt} = ky$; the relative growth rate, $\frac{1}{y}\frac{dy}{dt}$, is constant.

(b) The equation in part (a) is an appropriate model for population growth, assuming that there is enough room and nutrition to support the growth.

(c) If $y(0) = y_0$, then the solution is $y(t) = y_0 e^{kt}$.

7. (a) $dP/dt = kP(1 - P/K)$, where K is the carrying capacity.

(b) The equation in part (a) is an appropriate model for population growth, assuming that the population grows at a rate proportional to the size of the population in the beginning, but eventually levels off and approaches its carrying capacity because of limited resources.

8. (a) $dF/dt = kF - aFS$ and $dS/dt = -rS + bFS$.

(b) In the absence of sharks, an ample food supply would support exponential growth of the fish population, that is, $dF/dt = kF$, where k is a positive constant. In the absence of fish, we assume that the shark population would decline at a rate proportional to itself, that is, $dS/dt = -rS$, where r is a positive constant.

TRUE-FALSE QUIZ

1. True. Since $y^4 \geq 0$, $y' = -1 - y^4 < 0$ and the solutions are decreasing functions.

3. False. $x + y$ cannot be written in the form $g(x)f(y)$.

5. True. By comparing $\frac{dy}{dt} = 2y\left(1 - \frac{y}{5}\right)$ with the logistic differential equation (7.5.1), we see that the carrying capacity is 5; that is, $\lim\limits_{t\to\infty} y = 5$.

EXERCISES

1. (a)

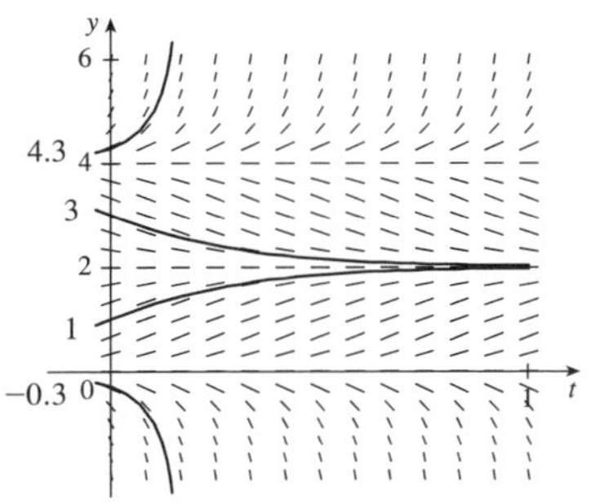

(b) $\lim\limits_{t\to\infty} y(t)$ appears to be finite for $0 \leq c \leq 4$. In fact $\lim\limits_{t\to\infty} y(t) = 4$ for $c = 4$, $\lim\limits_{t\to\infty} y(t) = 2$ for $0 < c < 4$, and $\lim\limits_{t\to\infty} y(t) = 0$ for $c = 0$. The equilibrium solutions are $y(t) = 0$, $y(t) = 2$, and $y(t) = 4$.

3. (a)

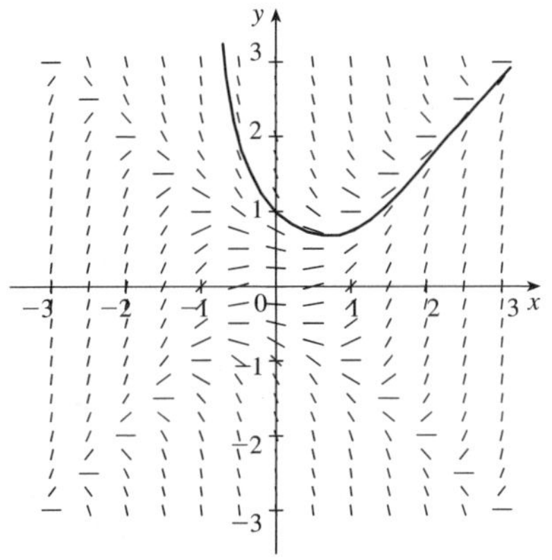

(b) $h = 0.1$, $x_0 = 0$, $y_0 = 1$ and $F(x, y) = x^2 - y^2$. So $y_n = y_{n-1} + 0.1\left(x_{n-1}^2 - y_{n-1}^2\right)$. Thus,

$y_1 = 1 + 0.1\left(0^2 - 1^2\right) = 0.9$,

$y_2 = 0.9 + 0.1\left(0.1^2 - 0.9^2\right) = 0.82$,

$y_3 = 0.82 + 0.1\left(0.2^2 - 0.82^2\right) = 0.75676$. This is close to our graphical estimate of $y(0.3) \approx 0.8$.

We estimate that when $x = 0.3$, $y = 0.8$, so $y(0.3) \approx 0.8$.

(c) The centres of the horizontal line segments of the direction field are located on the lines $y = x$ and $y = -x$. When a solution curve crosses one of these lines, it has a local maximum or minimum.

5. $(3y^2 + 2y)\, y' = x \cos x \quad \Rightarrow \quad (3y^2 + 2y)\, dy = (x \cos x)\, dx \quad \Rightarrow \quad \int (3y^2 + 2y)\, dy = \int (x \cos x)\, dx \quad \Rightarrow$ $y^3 + y^2 = \cos x + x \sin x + C$. For the last step, use integration by parts or Formula 83 in the Table of Integrals.

7. $\dfrac{dr}{dt} + 2tr = r \quad\Rightarrow\quad \dfrac{dr}{dt} = r - 2tr = r(1-2t) \quad\Rightarrow\quad \displaystyle\int \frac{dr}{r} = \int (1-2t)\,dt \quad\Rightarrow\quad \ln|r| = t - t^2 + C \quad\Rightarrow$

$|r| = e^{t-t^2+C} = ke^{t-t^2}$. Since $r(0) = 5$, $5 = ke^0 = k$. Thus, $r(t) = 5e^{t-t^2}$.

9. $\dfrac{d}{dx}(y) = \dfrac{d}{dx}(ke^x) \quad\Rightarrow\quad y' = ke^x = y$, so the orthogonal trajectories must have $y' = -\dfrac{1}{y} \quad\Rightarrow\quad \dfrac{dy}{dx} = -\dfrac{1}{y} \quad\Rightarrow$

$y\,dy = -dx \quad\Rightarrow\quad \int y\,dy = -\int dx \quad\Rightarrow\quad \frac{1}{2}y^2 = -x + C \quad\Rightarrow\quad x = C - \frac{1}{2}y^2$, which are parabolas with a horizontal axis.

11. (a) $y(t) = y(0)e^{kt} = 200e^{kt} \quad\Rightarrow\quad y(0.5) = 200e^{0.5k} = 360 \quad\Rightarrow\quad e^{0.5k} = 1.8 \quad\Rightarrow\quad 0.5k = \ln 1.8 \quad\Rightarrow$

$k = 2\ln 1.8 = \ln(1.8)^2 = \ln 3.24 \quad\Rightarrow\quad y(t) = 200e^{(\ln 3.24)t} = 200(3.24)^t$

(b) $y(4) = 200(3.24)^4 \approx 22\,040$ bacteria

(c) $y'(t) = 200(3.24)^t \cdot \ln 3.24$, so $y'(4) = 200(3.24)^4 \cdot \ln 3.24 \approx 25\,910$ bacteria per hour

(d) $200(3.24)^t = 10\,000 \quad\Rightarrow\quad (3.24)^t = 50 \quad\Rightarrow\quad t\ln 3.24 = \ln 50 \quad\Rightarrow\quad t = \ln 50/\ln 3.24 \approx 3.33$ hours

13. (a) $C'(t) = -kC(t) \quad\Rightarrow\quad C(t) = C(0)e^{-kt}$ by Theorem 10.4.2. But $C(0) = C_0$, so $C(t) = C_0e^{-kt}$.

(b) $C(30) = \frac{1}{2}C_0$ since the concentration is reduced by half. Thus, $\frac{1}{2}C_0 = C_0e^{-30k} \quad\Rightarrow \ln\frac{1}{2} = -30k \quad\Rightarrow$

$k = -\frac{1}{30}\ln\frac{1}{2} = \frac{1}{30}\ln 2$. Since 10% of the original concentration remains if 90% is eliminated, we want the value of t such that $C(t) = \frac{1}{10}C_0$. Therefore, $\frac{1}{10}C_0 = C_0e^{-t(\ln 2)/30} \quad\Rightarrow\quad \ln 0.1 = -t(\ln 2)/30 \quad\Rightarrow\quad t = -\frac{30}{\ln 2}\ln 0.1 \approx 100$ h.

15. (a) $\dfrac{dL}{dt} \propto L_\infty - L \quad\Rightarrow\quad \dfrac{dL}{dt} = k(L_\infty - L) \quad\Rightarrow\quad \displaystyle\int \frac{dL}{L_\infty - L} = \int k\,dt \quad\Rightarrow\quad -\ln|L_\infty - L| = kt + C \quad\Rightarrow$

$\ln|L_\infty - L| = -kt - C \quad\Rightarrow\quad |L_\infty - L| = e^{-kt-C} \quad\Rightarrow\quad L_\infty - L = Ae^{-kt} \quad\Rightarrow\quad L = L_\infty - Ae^{-kt}$. At $t = 0$,

$L = L(0) = L_\infty - A \quad\Rightarrow\quad A = L_\infty - L(0) \quad\Rightarrow\quad L(t) = L_\infty - [L_\infty - L(0)]\,e^{-kt}$.

(b) $L_\infty = 53$ cm, $L(0) = 10$ cm, and $k = 0.2 \quad\Rightarrow\quad L(t) = 53 - (53 - 10)e^{-0.2t} = 53 - 43e^{-0.2t}$.

17. Let P represent the population and I the number of infected people.

The rate of spread dI/dt is jointly proportional to I and to $P - I$, so for some constant k, $dI/dt = kI(P - I) \quad\Rightarrow$

$I = \dfrac{I_0P}{I_0 + (P - I_0)e^{-kPt}}$ [from the discussion of logistic growth in Section 7.5].

Now, measuring t in days, we substitute $t = 7$, $P = 5000$, $I_0 = 160$ and $I(7) = 1200$ to find k:

$1200 = \dfrac{160 \cdot 5000}{160 + (5000 - 160)e^{-5000 \cdot 7 \cdot k}} \quad\Leftrightarrow\quad 3 = \dfrac{2000}{160 + 4840e^{-35\,000k}} \quad\Leftrightarrow\quad 480 + 14\,520e^{-35\,000k} = 2000 \quad\Leftrightarrow$

$e^{-35\,000k} = \dfrac{2000 - 480}{14\,520} \quad\Leftrightarrow\quad -35\,000k = \ln\dfrac{38}{363} \quad\Leftrightarrow\quad k = \dfrac{-1}{35\,000}\ln\dfrac{38}{363} \approx 0.000\,064\,48$. Next, let

$I = 5000 \times 80\% = 4000$, and solve for t: $4000 = \dfrac{160 \cdot 5000}{160 + (5000 - 160)e^{-k \cdot 5000 \cdot t}} \quad\Leftrightarrow\quad 1 = \dfrac{200}{160 + 4840e^{-5000kt}} \quad\Leftrightarrow$

$160 + 4840e^{-5000kt} = 200 \quad\Leftrightarrow\quad e^{-5000kt} = \dfrac{200 - 160}{4840} \quad\Leftrightarrow\quad -5000kt = \ln\dfrac{1}{121} \quad\Leftrightarrow$

$t = \dfrac{-1}{5000k}\ln\dfrac{1}{121} = \dfrac{1}{\frac{1}{7}\ln\frac{38}{363}} \cdot \ln\dfrac{1}{121} = 7 \cdot \dfrac{\ln 121}{\ln\frac{363}{38}} \approx 14.875$. So it takes about 15 days for 80% of the population to be infected.

19. $\dfrac{dh}{dt} = -\dfrac{R}{V}\left(\dfrac{h}{k+h}\right) \Rightarrow \displaystyle\int \frac{k+h}{h}\,dh = \int\left(-\frac{R}{V}\right)dt \Rightarrow \int\left(1+\frac{k}{h}\right)dh = -\frac{R}{V}\int 1\,dt \Rightarrow$

$h + k\ln h = -\dfrac{R}{V}t + C$. This equation gives a relationship between h and t, but it is not possible to isolate h and express it in terms of t.

21. (a) $dx/dt = 0.4x(1 - 0.000\,005x) - 0.002xy$, $dy/dt = -0.2y + 0.000\,008xy$. If $y = 0$, then $dx/dt = 0.4x(1 - 0.000\,005x)$, so $dx/dt = 0 \Leftrightarrow x = 0$ or $x = 200\,000$, which shows that the insect population increases logistically with a carrying capacity of 200 000. Since $dx/dt > 0$ for $0 < x < 200\,000$ and $dx/dt < 0$ for $x > 200\,000$, we expect the insect population to stabilize at 200 000.

(b) x and y are constant $\Rightarrow$ $x' = 0$ and $y' = 0$ $\Rightarrow$

$$\left\{\begin{array}{l} 0 = 0.4x(1 - 0.000\,005x) - 0.002xy \\ 0 = -0.2y + 0.000\,008xy \end{array}\right\} \Rightarrow \left\{\begin{array}{l} 0 = 0.4x[(1 - 0.000\,005x) - 0.005y] \\ 0 = y(-0.2 + 0.000\,008x) \end{array}\right.$$

The second equation is true if $y = 0$ or $x = \frac{0.2}{0.000\,008} = 25\,000$. If $y = 0$ in the first equation, then either $x = 0$ or $x = \frac{1}{0.000\,005} = 200\,000$. If $x = 25\,000$, then $0 = 0.4(25\,000)[(1 - 0.000\,005 \cdot 25\,000) - 0.005y] \Rightarrow$ $0 = 10\,000[(1 - 0.125) - 0.005y] \Rightarrow 0 = 8750 - 50y \Rightarrow y = 175$.

Case (i): $y = 0$, $x = 0$: Zero populations

Case (ii): $y = 0$, $x = 200\,000$: In the absence of birds, the insect population is always 200 000.

Case (iii): $x = 25\,000$, $y = 175$: The predator/prey interaction balances and the populations are stable.

(c) The populations of the birds and insects fluctuate around 175 and 25 000, respectively, and eventually stabilize at those values.

(d)

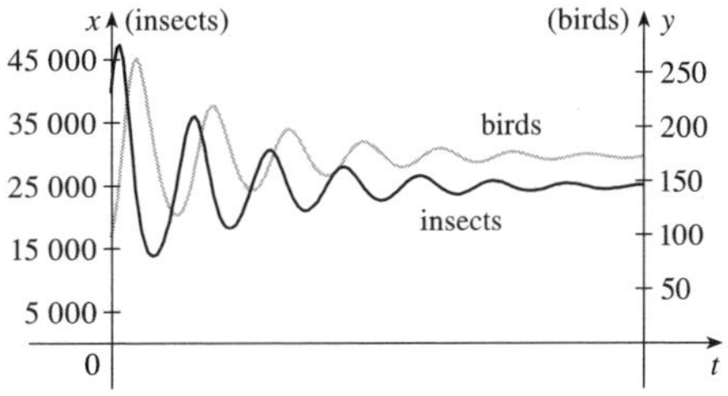

FOCUS ON PROBLEM SOLVING

1. We use the Fundamental Theorem of Calculus to differentiate the given equation:

$[f(x)]^2 = 100 + \int_0^x \left\{[f(t)]^2 + [f'(t)]^2\right\} dt \;\Rightarrow\; 2f(x)f'(x) = [f(x)]^2 + [f'(x)]^2 \;\Rightarrow$

$[f(x)]^2 + [f'(x)]^2 - 2f(x)f'(x) = 0 \;\Rightarrow\; [f(x) - f'(x)]^2 = 0 \;\Leftrightarrow\; f(x) = f'(x)$. We can solve this as a separable equation, or else use Theorem 7.4.2 with $k = 1$, which says that the solutions are $f(x) = Ce^x$. Now $[f(0)]^2 = 100$, so $f(0) = C = \pm 10$, and hence $f(x) = \pm 10e^x$ are the only functions satisfying the given equation.

3. $f'(x) = \lim\limits_{h\to 0} \dfrac{f(x+h) - f(x)}{h} = \lim\limits_{h\to 0} \dfrac{f(x)\,[f(h) - 1]}{h}$ $\quad$ [since $f(x+h) = f(x)f(h)$]

$= f(x) \lim\limits_{h\to 0} \dfrac{f(h) - 1}{h} = f(x) \lim\limits_{h\to 0} \dfrac{f(h) - f(0)}{h - 0} = f(x)f'(0) = f(x)$

Therefore, $f'(x) = f(x)$ for all x and from Theorem 7.4.2 we get $f(x) = Ae^x$. Now $f(0) = 1 \;\Rightarrow\; A = 1 \;\Rightarrow$ $f(x) = e^x$.

5. Let $y(t)$ denote the temperature of the peach pie t minutes after 5:00 P.M. and R the temperature of the room. Newton's Law of Cooling gives us $dy/dt = k(y - R)$. Solving for y we get $\dfrac{dy}{y - R} = k\,dt \;\Rightarrow\; \ln|y - R| = kt + C \;\Rightarrow$ $|y - R| = e^{kt+C} \;\Rightarrow\; y - R = \pm e^{kt} \cdot e^C \;\Rightarrow\; y = Me^{kt} + R$, where M is a nonzero constant. We are given temperatures at three times.

$$y(0) = 100 \;\Rightarrow\; 100 = M + R \;\Rightarrow\; R = 100 - M$$

$$y(10) = 80 \;\Rightarrow\; 80 = Me^{10k} + R \quad \textbf{(1)}$$

$$y(20) = 65 \;\Rightarrow\; 65 = Me^{20k} + R \quad \textbf{(2)}$$

Substituting $100 - M$ for R in **(1)** and **(2)** gives us

$$-20 = Me^{10k} - M \;\textbf{(3)} \quad \text{and} \quad -35 = Me^{20k} - M \;\textbf{(4)}$$

Dividing **(3)** by **(4)** gives us $\dfrac{-20}{-35} = \dfrac{M(e^{10k} - 1)}{M(e^{20k} - 1)} \;\Rightarrow\; \dfrac{4}{7} = \dfrac{e^{10k} - 1}{e^{20k} - 1} \;\Rightarrow\; 4e^{20k} - 4 = 7e^{10k} - 7 \;\Rightarrow$ $4e^{20k} - 7e^{10k} + 3 = 0$. This is a quadratic equation in e^{10k}. $(4e^{10k} - 3)(e^{10k} - 1) = 0 \;\Rightarrow\; e^{10k} = \frac{3}{4}$ or $1 \;\Rightarrow$ $10k = \ln\frac{3}{4}$ or $\ln 1 \;\Rightarrow\; k = \frac{1}{10}\ln\frac{3}{4}$ since k is a nonzero constant of proportionality. Substituting $\frac{3}{4}$ for e^{10k} in **(3)** gives us $-20 = M \cdot \frac{3}{4} - M \;\Rightarrow\; -20 = -\frac{1}{4}M \;\Rightarrow\; M = 80$. Now $R = 100 - M$ so $R = 20°\text{C}$.

7. (a) While running from $(L, 0)$ to (x, y), the dog travels a distance

$s = \displaystyle\int_x^L \sqrt{1 + (dy/dx)^2}\, dx = -\int_L^x \sqrt{1 + (dy/dx)^2}\, dx$, so

$\dfrac{ds}{dx} = -\sqrt{1 + (dy/dx)^2}$. The dog and rabbit run at the same speed, so the rabbit's position when the dog has travelled a distance s is $(0, s)$.

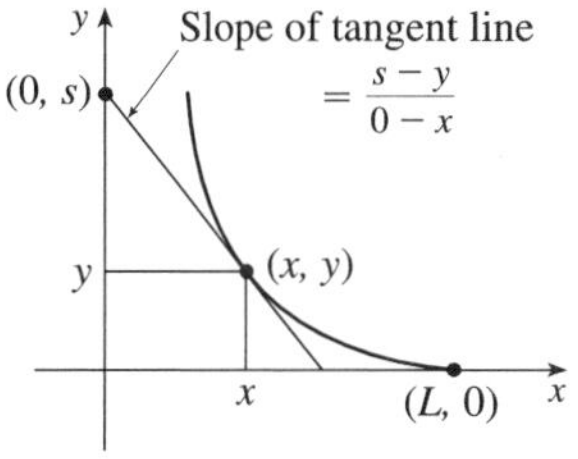

Since the dog runs straight for the rabbit, $\dfrac{dy}{dx} = \dfrac{s - y}{0 - x}$ (see the figure).

Thus, $s = y - x\dfrac{dy}{dx} \;\Rightarrow\; \dfrac{ds}{dx} = \dfrac{dy}{dx} - \left(x\dfrac{d^2y}{dx^2} + 1\dfrac{dy}{dx}\right) = -x\dfrac{d^2y}{dx^2}$. Equating the two expressions for $\dfrac{ds}{dx}$ gives us

$x\dfrac{d^2y}{dx^2} = \sqrt{1 + \left(\dfrac{dy}{dx}\right)^2}$, as claimed.

(b) Letting $z = \dfrac{dy}{dx}$, we obtain the differential equation $x\dfrac{dz}{dx} = \sqrt{1+z^2}$, or $\dfrac{dz}{\sqrt{1+z^2}} = \dfrac{dx}{x}$. Integrating:

$\ln x = \displaystyle\int \frac{dz}{\sqrt{1+z^2}} \overset{25}{=} \ln\!\left(z+\sqrt{1+z^2}\,\right) + C$. When $x = L$, $z = dy/dx = 0$, so $\ln L = \ln 1 + C$. Therefore, $C = \ln L$,

so $\ln x = \ln\!\left(\sqrt{1+z^2}+z\right) + \ln L = \ln\!\left[L\left(\sqrt{1+z^2}+z\right)\right] \;\Rightarrow\; x = L\left(\sqrt{1+z^2}+z\right) \;\Rightarrow\; \sqrt{1+z^2} = \dfrac{x}{L} - z \;\Rightarrow$

$1 + z^2 = \left(\dfrac{x}{L}\right)^2 - \dfrac{2xz}{L} + z^2 \;\Rightarrow\; \left(\dfrac{x}{L}\right)^2 - 2z\left(\dfrac{x}{L}\right) - 1 = 0 \;\Rightarrow\; z = \dfrac{(x/L)^2 - 1}{2(x/L)} = \dfrac{x^2 - L^2}{2Lx} = \dfrac{x}{2L} - \dfrac{L}{2}\dfrac{1}{x}$

[for $x > 0$]. Since $z = \dfrac{dy}{dx}$, $y = \dfrac{x^2}{4L} - \dfrac{L}{2}\ln x + C_1$. Since $y = 0$ when $x = L$, $0 = \dfrac{L}{4} - \dfrac{L}{2}\ln L + C_1 \;\Rightarrow$

$C_1 = \dfrac{L}{2}\ln L - \dfrac{L}{4}$. Thus, $y = \dfrac{x^2}{4L} - \dfrac{L}{2}\ln x + \dfrac{L}{2}\ln L - \dfrac{L}{4} = \dfrac{x^2 - L^2}{4L} - \dfrac{L}{2}\ln\!\left(\dfrac{x}{L}\right)$.

(c) As $x \to 0^+$, $y \to \infty$, so the dog never catches the rabbit.

9. (a) We are given that $V = \frac{1}{3}\pi r^2 h$, $dV/dt = 1500\pi\ \text{m}^3/\text{h}$, and $r = 1.5h = \frac{3}{2}h$. So $V = \frac{1}{3}\pi\left(\frac{3}{2}h\right)^2 h = \frac{3}{4}\pi h^3 \;\Rightarrow$

$\dfrac{dV}{dt} = \frac{3}{4}\pi\cdot 3h^2\dfrac{dh}{dt} = \frac{9}{4}\pi h^2\dfrac{dh}{dt}$. Therefore, $\dfrac{dh}{dt} = \dfrac{4(dV/dt)}{9\pi h^2} = \dfrac{6000\pi}{9\pi h^2} = \dfrac{2000}{3h^2}$ $(\star)$ $\Rightarrow\; \int 3h^2\,dh = \int 2000\,dt \;\Rightarrow$

$h^3 = 2000t + C$. When $t = 0$, $h = 20$. Thus, $C = 20^3 = 8000$, so $h^3 = 2000t + 8000$. Let $h = 30$. Then

$30^3 = 27\,000 = 2000t + 8000 \;\Rightarrow\; 2000t = 19\,000 \;\Rightarrow\; t = 9.5$, so the time required is 9.5 hours.

(b) The floor area of the silo is $F = \pi\cdot 60^2 = 3600\pi\ \text{m}^2$, and the area of the base of the pile is $A = \pi r^2 = \pi\left(\frac{3}{2}h\right)^2 = \frac{9\pi}{4}h^2$.

So the area of the floor which is not covered when $h = 20$ is $F - A = 3600\pi - 900\pi = 2700\pi \approx 8482\ \text{m}^2$. Now

$A = \frac{9\pi}{4}h^2 \;\Rightarrow\; dA/dt = \frac{9\pi}{4}\cdot 2h\,(dh/dt)$, and from $(\star)$ in part (a) we know that when $h = 20$,

$dh/dt = \frac{2000}{3(20)^2} = \frac{5}{3}$ m/h. Therefore, $dA/dt = \frac{9\pi}{4}(2)(20)\left(\frac{5}{3}\right) = 150\pi \approx 471\ \text{m}^2/\text{h}$.

(c) At $h = 27$ m, $dV/dt = 1500\pi - 500\pi = 1000\pi\ \text{m}^3/\text{h}$. From $(\star)$ in part (a),

$\dfrac{dh}{dt} = \dfrac{4(dV/dt)}{9\pi h^2} = \dfrac{4(1000\pi)}{9\pi h^2} = \dfrac{4000}{9h^2} \;\Rightarrow\; \int 9h^2\,dh = \int 4000\,dt \;\Rightarrow\; 3h^3 = 4000t + C$. When $t = 0$, $h = 27$;

therefore, $C = 3\cdot 27^3 = 59\,049$. So $3h^3 = 4000t + 59\,049$. At the top, $h = 30 \;\Rightarrow\; 3(30)^3 = 4000t + 59\,049 \;\Rightarrow$

$t = \frac{21\,951}{4000} \approx 5.5$. The pile reaches the top after about 5.5 h.

11. Let $P(a, b)$ be any point on the curve. If m is the slope of the tangent line at P, then $m = y'(a)$, and an equation of the normal line at P is $y - b = -\dfrac{1}{m}(x - a)$, or equivalently, $y = -\dfrac{1}{m}x + b + \dfrac{a}{m}$. The y-intercept is always 6, so

$b + \dfrac{a}{m} = 6 \;\Rightarrow\; \dfrac{a}{m} = 6 - b \;\Rightarrow\; m = \dfrac{a}{6-b}$. We will solve the equivalent differential equation $\dfrac{dy}{dx} = \dfrac{x}{6-y} \;\Rightarrow$

$(6 - y)\,dy = x\,dx \;\Rightarrow\; \displaystyle\int (6-y)\,dy = \int x\,dx \;\Rightarrow\; 6y - \tfrac{1}{2}y^2 = \tfrac{1}{2}x^2 + C \;\Rightarrow\; 12y - y^2 = x^2 + K$.

Since $(3, 2)$ is on the curve, $12(2) - 2^2 = 3^2 + K \;\Rightarrow\; K = 11$. So the curve is given by $12y - y^2 = x^2 + 11 \;\Rightarrow$

$x^2 + y^2 - 12y + 36 = -11 + 36 \;\Rightarrow\; x^2 + (y-6)^2 = 25$, a circle with centre $(0, 6)$ and radius 5.

8 □ INFINITE SEQUENCES AND SERIES

8.1 Sequences

1. (a) A sequence is an ordered list of numbers. It can also be defined as a function whose domain is the set of positive integers.

(b) The terms a_n approach 8 as n becomes large. In fact, we can make a_n as close to 8 as we like by taking n sufficiently large.

(c) The terms a_n become large as n becomes large. In fact, we can make a_n as large as we like by taking n sufficiently large.

3. The first six terms of $a_n = \dfrac{n}{2n+1}$ are $\dfrac{1}{3}, \dfrac{2}{5}, \dfrac{3}{7}, \dfrac{4}{9}, \dfrac{5}{11}, \dfrac{6}{13}$. It appears that the sequence is approaching $\dfrac{1}{2}$.

$\lim\limits_{n\to\infty} \dfrac{n}{2n+1} = \lim\limits_{n\to\infty} \dfrac{1}{2+1/n} = \dfrac{1}{2}$

5. $\left\{1, -\frac{2}{3}, \frac{4}{9}, -\frac{8}{27}, \ldots\right\}$. Each term is $-\frac{2}{3}$ times the preceding one, so $a_n = \left(-\frac{2}{3}\right)^{n-1}$.

7. $\{2, 7, 12, 17, \ldots\}$. Each term is larger than the preceding one by 5, so $a_n = a_1 + d(n-1) = 2 + 5(n-1) = 5n - 3$.

9. $a_n = \dfrac{3+5n^2}{n+n^2} = \dfrac{(3+5n^2)/n^2}{(n+n^2)/n^2} = \dfrac{5+3/n^2}{1+1/n}$, so $a_n \to \dfrac{5+0}{1+0} = 5$ as $n \to \infty$. Converges

11. $a_n = \dfrac{2^n}{3^{n+1}} = \dfrac{1}{3}\left(\dfrac{2}{3}\right)^n$, so $\lim\limits_{n\to\infty} a_n = \frac{1}{3} \lim\limits_{n\to\infty} \left(\frac{2}{3}\right)^n = \frac{1}{3} \cdot 0 = 0$ by (8) with $r = \frac{2}{3}$. Converges

13. $a_n = \dfrac{(n+2)!}{n!} = \dfrac{(n+2)(n+1)n!}{n!} = (n+2)(n+1)$, so $a_n \to \infty$ as $n \to \infty$ and the sequence diverges.

15. $a_n = \dfrac{(-1)^{n-1} n}{n^2+1} = \dfrac{(-1)^{n-1}}{n+1/n}$, so $0 \le |a_n| = \dfrac{1}{n+1/n} \le \dfrac{1}{n} \to 0$ as $n \to \infty$, so $a_n \to 0$ by the Squeeze Theorem and Theorem 4. Converges

17. $a_n = \dfrac{e^n + e^{-n}}{e^{2n} - 1} \cdot \dfrac{e^{-n}}{e^{-n}} = \dfrac{1 + e^{-2n}}{e^n - e^{-n}} \to \dfrac{1+0}{e^n - 0} \to 0$ as $n \to \infty$. Converges

19. $\lim\limits_{x\to\infty} \dfrac{x}{2^x} \overset{\text{H}}{=} \lim\limits_{x\to\infty} \dfrac{1}{(\ln 2)2^x} = 0$, so by Theorem 2, $\{n2^{-n}\}$ converges to 0.

21. $0 \le \dfrac{\cos^2 n}{2^n} \le \dfrac{1}{2^n}$ [since $0 \le \cos^2 n \le 1$], so since $\lim\limits_{n\to\infty} \dfrac{1}{2^n} = 0$, $\left\{\dfrac{\cos^2 n}{2^n}\right\}$ converges to 0 by the Squeeze Theorem.

23. $y = \left(1 + \dfrac{2}{x}\right)^x \Rightarrow \ln y = x \ln\left(1 + \dfrac{2}{x}\right)$, so

$$\lim_{x\to\infty} \ln y = \lim_{x\to\infty} \frac{\ln(1+2/x)}{1/x} \overset{\text{H}}{=} \lim_{x\to\infty} \frac{\left(\dfrac{1}{1+2/x}\right)\left(-\dfrac{2}{x^2}\right)}{-1/x^2} = \lim_{x\to\infty} \frac{2}{1+2/x} = 2 \Rightarrow$$

$\lim\limits_{x\to\infty} \left(1 + \dfrac{2}{x}\right)^x = \lim\limits_{x\to\infty} e^{\ln y} = e^2$, so by Theorem 2, $\lim\limits_{n\to\infty} \left(1 + \dfrac{2}{n}\right)^n = e^2$. Convergent

25. $\{0, 1, 0, 0, 1, 0, 0, 0, 1, \ldots\}$ diverges since the sequence takes on only two values, 0 and 1, and never stays arbitrarily close to either one (or any other value) for n sufficiently large.

27. $a_n = \ln(2n^2+1) - \ln(n^2+1) = \ln\left(\dfrac{2n^2+1}{n^2+1}\right) = \ln\left(\dfrac{2+1/n^2}{1+1/n^2}\right) \to \ln 2$ as $n \to \infty$. Convergent

29.

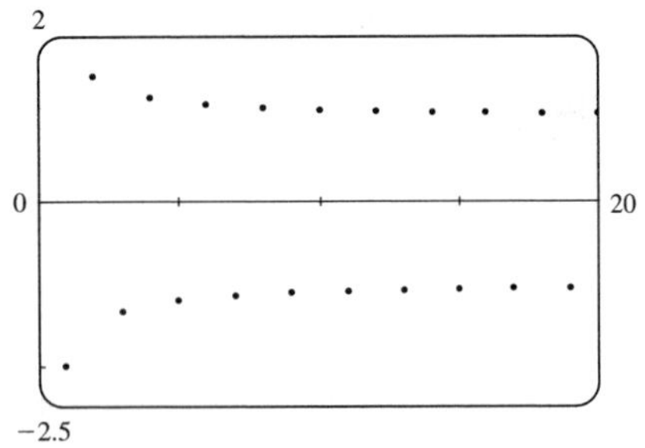

From the graph, we see that the sequence $\left\{(-1)^n \dfrac{n+1}{n}\right\}$ is divergent, since it oscillates between 1 and -1 (approximately).

31.

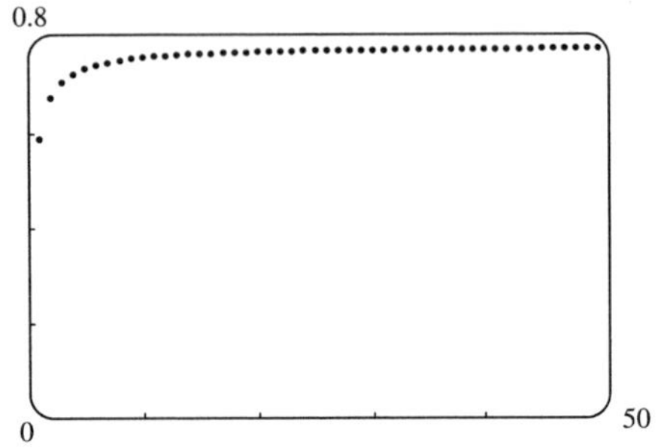

From the graph, it appears that the sequence converges to about 0.78.

$$\lim_{n\to\infty} \frac{2n}{2n+1} = \lim_{n\to\infty} \frac{2}{2+1/n} = 1, \text{ so}$$

$$\lim_{n\to\infty} \arctan\left(\frac{2n}{2n+1}\right) = \arctan 1 = \frac{\pi}{4}.$$

33.

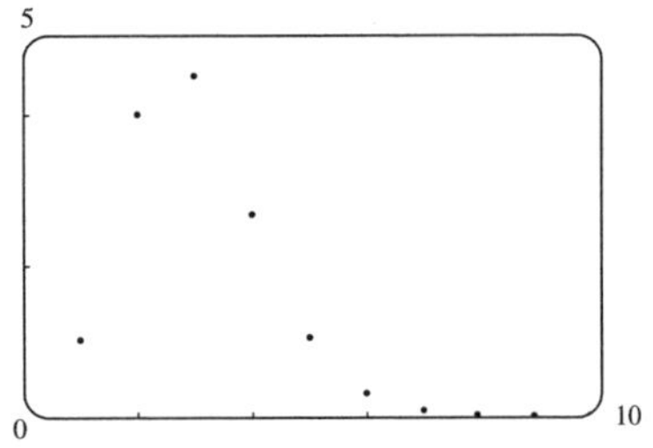

From the graph, it appears that the sequence converges to 0.

$$0 < a_n = \frac{n^3}{n!} = \frac{n}{n} \cdot \frac{n}{(n-1)} \cdot \frac{n}{(n-2)} \cdot \frac{1}{(n-3)} \cdot \cdots \cdot \frac{1}{3} \cdot \frac{1}{2} \cdot \frac{1}{1}$$

$$\leq \frac{n^2}{(n-1)(n-2)(n-3)} \quad [\text{for } n \geq 4]$$

$$= \frac{1/n}{(1-1/n)(1-2/n)(1-3/n)} \to 0 \text{ as } n \to \infty$$

So by the Squeeze Theorem, $\{n^3/n!\}$ converges to 0.

35. (a) $a_n = 1000(1.06)^n \Rightarrow a_1 = 1060$, $a_2 = 1123.60$, $a_3 = 1191.02$, $a_4 = 1262.48$, and $a_5 = 1338.23$.

(b) $\lim\limits_{n\to\infty} a_n = 1000 \lim\limits_{n\to\infty} (1.06)^n$, so the sequence diverges by (6) with $r = 1.06 > 1$.

37. (a) $a_1 = 1$, $a_2 = 4 - a_1 = 4 - 1 = 3$, $a_3 = 4 - a_2 = 4 - 3 = 1$, $a_4 = 4 - a_3 = 4 - 1 = 3$, $a_5 = 4 - a_4 = 4 - 3 = 1$. Since the terms of the sequence alternate between 1 and 3, the sequence is divergent.

(b) $a_1 = 2$, $a_2 = 4 - a_1 = 4 - 2 = 2$, $a_3 = 4 - a_2 = 4 - 2 = 2$. Since all of the terms are 2, $\lim\limits_{n\to\infty} a_n = 2$ and hence, the sequence is convergent.

39. (a) Let a_n be the number of rabbit pairs in the nth month. Clearly $a_1 = 1 = a_2$. In the nth month, each pair that is 2 or more months old (that is, a_{n-2} pairs) will produce a new pair to add to the a_{n-1} pairs already present. Thus, $a_n = a_{n-1} + a_{n-2}$, so that $\{a_n\} = \{f_n\}$, the Fibonacci sequence.

(b) $a_n = \dfrac{f_{n+1}}{f_n} \Rightarrow a_{n-1} = \dfrac{f_n}{f_{n-1}} = \dfrac{f_{n-1} + f_{n-2}}{f_{n-1}} = 1 + \dfrac{f_{n-2}}{f_{n-1}} = 1 + \dfrac{1}{f_{n-1}/f_{n-2}} = 1 + \dfrac{1}{a_{n-2}}$. If $L = \lim\limits_{n\to\infty} a_n$, then $L = \lim\limits_{n\to\infty} a_{n-1}$ and $L = \lim\limits_{n\to\infty} a_{n-2}$, so L must satisfy $L = 1 + \dfrac{1}{L} \Rightarrow L^2 - L - 1 = 0 \Rightarrow L = \frac{1+\sqrt{5}}{2}$ (since L must be positive).

41. $a_n = \dfrac{1}{2n+3}$ is decreasing since $a_{n+1} = \dfrac{1}{2(n+1)+3} = \dfrac{1}{2n+5} < \dfrac{1}{2n+3} = a_n$ for each $n \geq 1$. The sequence is bounded since $0 < a_n \leq \frac{1}{5}$ for all $n \geq 1$. Note that $a_1 = \frac{1}{5}$.

43. $a_n = \cos(n\pi/2)$ is not monotonic. The first few terms are $0, -1, 0, 1, 0, -1, 0, 1, \ldots$. In fact, the sequence consists of the terms $0, -1, 0, 1$ repeated over and over again in that order. The sequence is bounded since $|a_n| \le 1$ for all $n \ge 1$.

45. Since $\{a_n\}$ is a decreasing sequence, $a_n > a_{n+1}$ for all $n \ge 1$. Because all of its terms lie between 5 and 8, $\{a_n\}$ is a bounded sequence. By the Monotonic Sequence Theorem, $\{a_n\}$ is convergent; that is, $\{a_n\}$ has a limit L. L must be less than 8 since $\{a_n\}$ is decreasing, so $5 \le L < 8$.

47. We show by induction that $\{a_n\}$ is increasing and bounded above by 3. Let P_n be the proposition that $a_{n+1} > a_n$ and $0 < a_n < 3$. Clearly P_1 is true. Assume that P_n is true.

Then $a_{n+1} > a_n \;\Rightarrow\; \dfrac{1}{a_{n+1}} < \dfrac{1}{a_n} \;\Rightarrow\; -\dfrac{1}{a_{n+1}} > -\dfrac{1}{a_n}$. Now $a_{n+2} = 3 - \dfrac{1}{a_{n+1}} > 3 - \dfrac{1}{a_n} = a_{n+1} \;\Leftrightarrow\; P_{n+1}$.

This proves that $\{a_n\}$ is increasing and bounded above by 3, so $1 = a_1 < a_n < 3$, that is, $\{a_n\}$ is bounded, and hence convergent by the Monotonic Sequence Theorem. If $L = \lim\limits_{n\to\infty} a_n$, then $\lim\limits_{n\to\infty} a_{n+1} = L$ also, so L must satisfy

$L = 3 - 1/L \;\Rightarrow\; L^2 - 3L + 1 = 0 \;\Rightarrow\; L = \frac{3 \pm \sqrt{5}}{2}$. But $L > 1$, so $L = \frac{3+\sqrt{5}}{2}$.

49. $(0.8)^n < 0.000\,001 \;\Rightarrow\; \ln(0.8)^n < \ln(0.000\,001) \;\Rightarrow\; n\ln(0.8) < \ln(0.000\,001) \;\Rightarrow$

$n > \dfrac{\ln(0.000\,001)}{\ln(0.8)} \;\Rightarrow\; n > 61.9$, so n must be at least 62 to satisfy the given inequality.

51. (a) Suppose $\{p_n\}$ converges to p. Then $p_{n+1} = \dfrac{bp_n}{a+p_n} \;\Rightarrow\; \lim\limits_{n\to\infty} p_{n+1} = \dfrac{b\lim\limits_{n\to\infty} p_n}{a + \lim\limits_{n\to\infty} p_n} \;\Rightarrow\; p = \dfrac{bp}{a+p} \;\Rightarrow$

$p^2 + ap = bp \;\Rightarrow\; p(p+a-b) = 0 \;\Rightarrow\; p = 0$ or $p = b - a$.

(b) $p_{n+1} = \dfrac{bp_n}{a+p_n} = \dfrac{\left(\frac{b}{a}\right)p_n}{1 + \frac{p_n}{a}} < \left(\dfrac{b}{a}\right)p_n$ since $1 + \dfrac{p_n}{a} > 1$.

(c) By part (b), $p_1 < \left(\dfrac{b}{a}\right)p_0$, $p_2 < \left(\dfrac{b}{a}\right)p_1 < \left(\dfrac{b}{a}\right)^2 p_0$, $p_3 < \left(\dfrac{b}{a}\right)p_2 < \left(\dfrac{b}{a}\right)^3 p_0$, etc. In general, $p_n < \left(\dfrac{b}{a}\right)^n p_0$,

so $\lim\limits_{n\to\infty} p_n \le \lim\limits_{n\to\infty} \left(\dfrac{b}{a}\right)^n \cdot p_0 = 0$ since $b < a$. $\left[\text{By result 8, } \lim\limits_{n\to\infty} r^n = 0 \text{ if } -1 < r < 1. \text{ Here } r = \dfrac{b}{a} \in (0,1).\right]$

(d) Let $a < b$. We first show, by induction, that if $p_0 < b - a$, then $p_n < b - a$ and $p_{n+1} > p_n$.

For $n = 0$, we have $p_1 - p_0 = \dfrac{bp_0}{a+p_0} - p_0 = \dfrac{p_0(b-a-p_0)}{a+p_0} > 0$ since $p_0 < b - a$. So $p_1 > p_0$.

Now we suppose the assertion is true for $n = k$, that is, $p_k < b - a$ and $p_{k+1} > p_k$. Then

$b - a - p_{k+1} = b - a - \dfrac{bp_k}{a+p_k} = \dfrac{a(b-a) + bp_k - ap_k - bp_k}{a+p_k} = \dfrac{a(b-a-p_k)}{a+p_k} > 0$ because $p_k < b - a$. So

$p_{k+1} < b - a$. And $p_{k+2} - p_{k+1} = \dfrac{bp_{k+1}}{a+p_{k+1}} - p_{k+1} = \dfrac{p_{k+1}(b-a-p_{k+1})}{a+p_{k+1}} > 0$ since $p_{k+1} < b - a$. Therefore,

$p_{k+2} > p_{k+1}$. Thus, the assertion is true for $n = k+1$. It is therefore true for all n by mathematical induction. A similar proof by induction shows that if $p_0 > b - a$, then $p_n > b - a$ and $\{p_n\}$ is decreasing.

In either case the sequence $\{p_n\}$ is bounded and monotonic, so it is convergent by the Monotonic Sequence Theorem. It then follows from part (a) that $\lim\limits_{n\to\infty} p_n = b - a$.

8.2 Series

1. (a) A sequence is an ordered list of numbers whereas a series is the *sum* of a list of numbers.

(b) A series is convergent if the sequence of partial sums is a convergent sequence. A series is divergent if it is not convergent.

3.

n	s_n
1	−2.40000
2	−1.92000
3	−2.01600
4	−1.99680
5	−2.00064
6	−1.99987
7	−2.00003
8	−1.99999
9	−2.00000
10	−2.00000

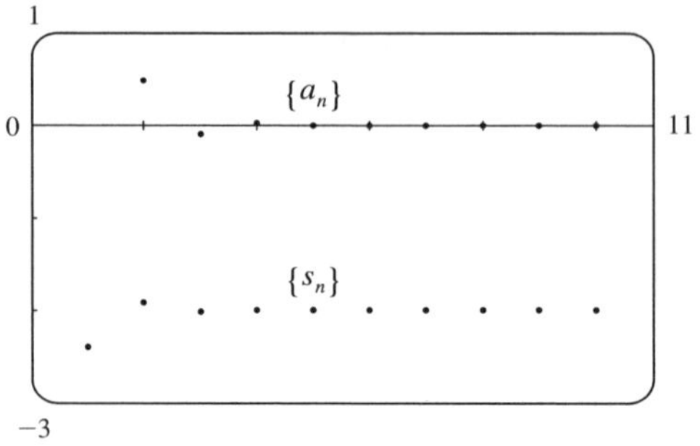

From the graph and the table, it seems that the series converges to -2. In fact, it is a geometric series with $a = -2.4$ and $r = -\frac{1}{5}$, so its sum is $\sum\limits_{n=1}^{\infty} \dfrac{12}{(-5)^n} = \dfrac{-2.4}{1-\left(-\frac{1}{5}\right)} = \dfrac{-2.4}{1.2} = -2.$

Note that the dot corresponding to $n = 1$ is part of both $\{a_n\}$ and $\{s_n\}$.

TI-86 Note: To graph $\{a_n\}$ and $\{s_n\}$, set your calculator to Param mode and DrawDot mode. (DrawDot is under GRAPH, MORE, FORMT (F3).) Now under `E(t)=` make the assignments: `xt1=t, yt1=12/(-5)^t, xt2=t, yt2=sum seq(yt1,t,1,t,1)`. (sum and seq are under LIST, OPS (F5), MORE.) Under WIND use `1,10,1,0,10,1,-3,1,1` to obtain a graph similar to the one above. Then use TRACE (F4) to see the values.

5.

n	s_n
1	1.55741
2	−0.62763
3	−0.77018
4	0.38764
5	−2.99287
6	−3.28388
7	−2.41243
8	−9.21214
9	−9.66446
10	−9.01610

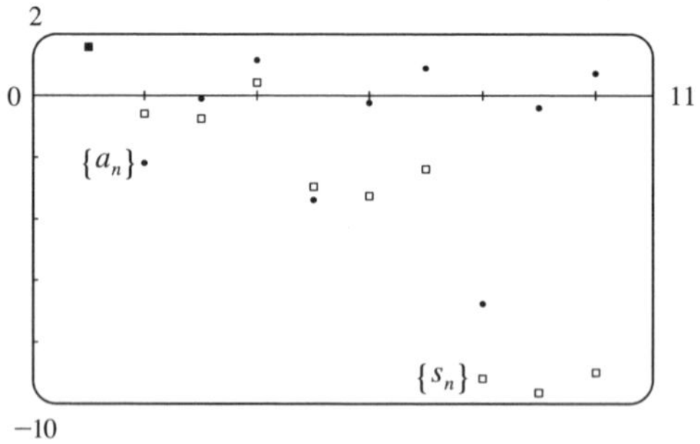

The series $\sum\limits_{n=1}^{\infty} \tan n$ diverges, since its terms do not approach 0.

7.

n	s_n
1	0.64645
2	0.80755
3	0.87500
4	0.91056
5	0.93196
6	0.94601
7	0.95581
8	0.96296
9	0.96838
10	0.97259

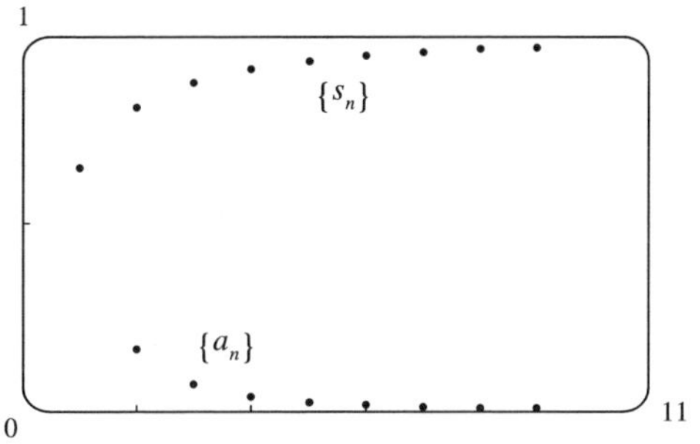

From the graph, it seems that the series converges to 1. To find the sum, we write

$$\begin{aligned} s_n &= \sum_{i=1}^{n}\left(\frac{1}{i^{1.5}} - \frac{1}{(i+1)^{1.5}}\right) \\ &= \left(1 - \frac{1}{2^{1.5}}\right) + \left(\frac{1}{2^{1.5}} - \frac{1}{3^{1.5}}\right) + \left(\frac{1}{3^{1.5}} - \frac{1}{4^{1.5}}\right) + \cdots + \left(\frac{1}{n^{1.5}} - \frac{1}{(n+1)^{1.5}}\right) \\ &= 1 - \frac{1}{(n+1)^{1.5}} \end{aligned}$$

So the sum is $\lim\limits_{n\to\infty} s_n = 1 - 0 = 1$.

9. (a) $\lim\limits_{n\to\infty} a_n = \lim\limits_{n\to\infty} \dfrac{2n}{3n+1} = \dfrac{2}{3}$, so the *sequence* $\{a_n\}$ is convergent by (8.1.1).

(b) Since $\lim\limits_{n\to\infty} a_n = \frac{2}{3} \neq 0$, the *series* $\sum\limits_{n=1}^{\infty} a_n$ is divergent by the Test for Divergence (7).

11. $5 - \frac{10}{3} + \frac{20}{9} - \frac{40}{27} + \cdots$ is a geometric series with $a = 5$ and $r = -\frac{2}{3}$. Since $|r| = \frac{2}{3} < 1$, the series converges to

$$\frac{a}{1-r} = \frac{5}{1-(-2/3)} = \frac{5}{5/3} = 3.$$

13. $\sum\limits_{n=1}^{\infty} 5\left(\frac{2}{3}\right)^{n-1}$ is a geometric series with $a = 5$ and $r = \frac{2}{3}$. Since $|r| = \frac{2}{3} < 1$, the series converges to

$$\frac{a}{1-r} = \frac{5}{1-2/3} = \frac{5}{1/3} = 15.$$

15. $\sum\limits_{n=0}^{\infty} \dfrac{\pi^n}{3^{n+1}} = \dfrac{1}{3}\sum\limits_{n=0}^{\infty}\left(\dfrac{\pi}{3}\right)^n$ is a geometric series with ratio $r = \dfrac{\pi}{3}$. Since $|r| > 1$, the series diverges.

17. $\sum\limits_{n=1}^{\infty} \dfrac{1}{2n} = \dfrac{1}{2}\sum\limits_{n=1}^{\infty}\dfrac{1}{n}$ diverges since each of its partial sums is $\dfrac{1}{2}$ times the corresponding partial sum of the harmonic series $\sum\limits_{n=1}^{\infty}\dfrac{1}{n}$, which diverges. $\left[\text{If } \sum\limits_{n=1}^{\infty}\dfrac{1}{2n} \text{ were to converge, then } \sum\limits_{n=1}^{\infty}\dfrac{1}{n} \text{ would also have to converge by Theorem 8(i).}\right]$ In general, constant multiples of divergent series are divergent.

19. $\sum\limits_{n=1}^{\infty} \dfrac{n}{n+5}$ diverges since $\lim\limits_{n\to\infty} a_n = \lim\limits_{n\to\infty} \dfrac{n}{n+5} = 1 \neq 0$. [Use (7), the Test for Divergence.]

21. Converges.

$$\sum_{n=1}^{\infty} \frac{1+2^n}{3^n} = \sum_{n=1}^{\infty} \left(\frac{1}{3^n} + \frac{2^n}{3^n}\right) = \sum_{n=1}^{\infty} \left[\left(\frac{1}{3}\right)^n + \left(\frac{2}{3}\right)^n\right] \qquad \text{[sum of two convergent geometric series]}$$

$$= \frac{1/3}{1-1/3} + \frac{2/3}{1-2/3} = \frac{1}{2} + 2 = \frac{5}{2}$$

23. $\sum_{n=1}^{\infty} \sqrt[n]{2} = 2 + \sqrt{2} + \sqrt[3]{2} + \sqrt[4]{2} + \cdots$ diverges by the Test for Divergence since

$\lim_{n\to\infty} a_n = \lim_{n\to\infty} \sqrt[n]{2} = \lim_{n\to\infty} 2^{1/n} = 2^0 = 1 \neq 0.$

25. $\lim_{n\to\infty} a_n = \lim_{n\to\infty} \arctan n = \frac{\pi}{2} \neq 0$, so the series diverges by the Test for Divergence.

27. Using partial fractions, the partial sums of the series $\sum_{n=2}^{\infty} \frac{2}{n^2-1}$ are

$$s_n = \sum_{i=2}^{n} \frac{2}{(i-1)(i+1)} = \sum_{i=2}^{n} \left(\frac{1}{i-1} - \frac{1}{i+1}\right)$$

$$= \left(1 - \frac{1}{3}\right) + \left(\frac{1}{2} - \frac{1}{4}\right) + \left(\frac{1}{3} - \frac{1}{5}\right) + \cdots + \left(\frac{1}{n-3} - \frac{1}{n-1}\right) + \left(\frac{1}{n-2} - \frac{1}{n}\right)$$

This sum is a telescoping series and $s_n = 1 + \frac{1}{2} - \frac{1}{n-1} - \frac{1}{n}$.

Thus, $\sum_{n=2}^{\infty} \frac{2}{n^2-1} = \lim_{n\to\infty} s_n = \lim_{n\to\infty} \left(1 + \frac{1}{2} - \frac{1}{n-1} - \frac{1}{n}\right) = \frac{3}{2}$.

29. For the series $\sum_{n=1}^{\infty} \frac{3}{n(n+3)}$, $s_n = \sum_{i=1}^{n} \frac{3}{i(i+3)} = \sum_{i=1}^{n} \left(\frac{1}{i} - \frac{1}{i+3}\right)$ [using partial fractions]. The latter sum is

$$\left(1 - \tfrac{1}{4}\right) + \left(\tfrac{1}{2} - \tfrac{1}{5}\right) + \left(\tfrac{1}{3} - \tfrac{1}{6}\right) + \left(\tfrac{1}{4} - \tfrac{1}{7}\right) + \cdots + \left(\tfrac{1}{n-3} - \tfrac{1}{n}\right) + \left(\tfrac{1}{n-2} - \tfrac{1}{n+1}\right) + \left(\tfrac{1}{n-1} - \tfrac{1}{n+2}\right) + \left(\tfrac{1}{n} - \tfrac{1}{n+3}\right)$$

$$= 1 + \tfrac{1}{2} + \tfrac{1}{3} - \tfrac{1}{n+1} - \tfrac{1}{n+2} - \tfrac{1}{n+3} \qquad \text{[telescoping series]}$$

Thus, $\sum_{n=1}^{\infty} \frac{3}{n(n+3)} = \lim_{n\to\infty} s_n = \lim_{n\to\infty} \left(1 + \tfrac{1}{2} + \tfrac{1}{3} - \tfrac{1}{n+1} - \tfrac{1}{n+2} - \tfrac{1}{n+3}\right) = 1 + \tfrac{1}{2} + \tfrac{1}{3} = \tfrac{11}{6}$. Converges

31. $0.\overline{2} = \frac{2}{10} + \frac{2}{10^2} + \cdots$ is a geometric series with $a = \frac{2}{10}$ and $r = \frac{1}{10}$. It converges to $\frac{a}{1-r} = \frac{2/10}{1-1/10} = \frac{2}{9}$.

33. $3.\overline{417} = 3 + \frac{417}{10^3} + \frac{417}{10^6} + \cdots$. Now $\frac{417}{10^3} + \frac{417}{10^6} + \cdots$ is a geometric series with $a = \frac{417}{10^3}$ and $r = \frac{1}{10^3}$. It converges to

$\frac{a}{1-r} = \frac{417/10^3}{1-1/10^3} = \frac{417/10^3}{999/10^3} = \frac{417}{999}$. Thus, $3.\overline{417} = 3 + \frac{417}{999} = \frac{3414}{999} = \frac{1138}{333}$.

35. $\sum_{n=1}^{\infty} \frac{x^n}{3^n} = \sum_{n=1}^{\infty} \left(\frac{x}{3}\right)^n$ is a geometric series with $r = \frac{x}{3}$, so the series converges $\Leftrightarrow$ $|r| < 1$ $\Leftrightarrow$ $\frac{|x|}{3} < 1$ $\Leftrightarrow$ $|x| < 3$;

that is, $-3 < x < 3$. In that case, the sum of the series is $\frac{a}{1-r} = \frac{x/3}{1-x/3} = \frac{x/3}{1-x/3} \cdot \frac{3}{3} = \frac{x}{3-x}$.

37. $\sum_{n=0}^{\infty} \frac{\cos^n x}{2^n}$ is a geometric series with first term 1 and ratio $r = \frac{\cos x}{2}$, so it converges $\Leftrightarrow$ $|r| < 1$. But $|r| = \frac{|\cos x|}{2} \leq \frac{1}{2}$ for all x. Thus, the series converges for all real values of x and the sum of the series is $\frac{1}{1 - (\cos x)/2} = \frac{2}{2 - \cos x}$.

39. After defining f, We use `convert(f,parfrac);` in Maple, `Apart` in Mathematica, or `Expand Rational` and `Simplify` in Derive to find that the general term is $\frac{1}{(4n+1)(4n-3)} = -\frac{1/4}{4n+1} + \frac{1/4}{4n-3}$. So the nth partial sum is

$$s_n = \sum_{k=1}^{n} \left(-\frac{1/4}{4k+1} + \frac{1/4}{4k-3} \right) = \frac{1}{4} \sum_{k=1}^{n} \left(\frac{1}{4k-3} - \frac{1}{4k+1} \right)$$

$$= \frac{1}{4} \left[\left(1 - \frac{1}{5}\right) + \left(\frac{1}{5} - \frac{1}{9}\right) + \left(\frac{1}{9} - \frac{1}{13}\right) + \cdots + \left(\frac{1}{4n-3} - \frac{1}{4n+1}\right) \right] = \frac{1}{4}\left(1 - \frac{1}{4n+1}\right)$$

The series converges to $\lim_{n\to\infty} s_n = \frac{1}{4}$. This can be confirmed by directly computing the sum using `sum(f,1..infinity);` (in Maple), `Sum[f,{n,1,Infinity}]` (in Mathematica), or `Calculus Sum` (from 1 to ∞) and `Simplify` (in Derive).

41. For $n = 1$, $a_1 = 0$ since $s_1 = 0$. For $n > 1$,

$$a_n = s_n - s_{n-1} = \frac{n-1}{n+1} - \frac{(n-1)-1}{(n-1)+1} = \frac{(n-1)n - (n+1)(n-2)}{(n+1)n} = \frac{2}{n(n+1)}$$

Also, $\sum_{n=1}^{\infty} a_n = \lim_{n\to\infty} s_n = \lim_{n\to\infty} \frac{1 - 1/n}{1 + 1/n} = 1$.

43. (a) The first step in the chain occurs when the local government spends D dollars. The people who receive it spend a fraction c of those D dollars, that is, Dc dollars. Those who receive the Dc dollars spend a fraction c of it, that is, Dc^2 dollars. Continuing in this way, we see that the total spending after n transactions is

$$S_n = D + Dc + Dc^2 + \cdots + Dc^{n-1} = \frac{D(1 - c^n)}{1 - c} \text{ by (3).}$$

(b) $$\lim_{n\to\infty} S_n = \lim_{n\to\infty} \frac{D(1 - c^n)}{1 - c} = \frac{D}{1 - c} \lim_{n\to\infty} (1 - c^n) = \frac{D}{1 - c} \quad [\text{since } 0 < c < 1 \Rightarrow \lim_{n\to\infty} c^n = 0]$$

$$= \frac{D}{s} \quad [\text{since } c + s = 1] = kD \quad [\text{since } k = 1/s]$$

If $c = 0.8$, then $s = 1 - c = 0.2$ and the multiplier is $k = 1/s = 5$.

45. $\sum_{n=2}^{\infty} (1 + c)^{-n}$ is a geometric series with $a = (1 + c)^{-2}$ and $r = (1 + c)^{-1}$, so the series converges when $\left|(1 + c)^{-1}\right| < 1$ $\Leftrightarrow$ $|1 + c| > 1$ $\Leftrightarrow$ $1 + c > 1$ or $1 + c < -1$ $\Leftrightarrow$ $c > 0$ or $c < -2$. We calculate the sum of the series and set it equal to 2: $\frac{(1 + c)^{-2}}{1 - (1 + c)^{-1}} = 2$ $\Leftrightarrow$ $\left(\frac{1}{1 + c}\right)^2 = 2 - 2\left(\frac{1}{1 + c}\right)$ $\Leftrightarrow$ $1 = 2(1 + c)^2 - 2(1 + c)$ $\Leftrightarrow$ $2c^2 + 2c - 1 = 0$ $\Leftrightarrow$ $c = \frac{-2 \pm \sqrt{12}}{4} = \frac{\pm\sqrt{3} - 1}{2}$. However, the negative root is inadmissible because $-2 < \frac{-\sqrt{3} - 1}{2} < 0$. So $c = \frac{\sqrt{3} - 1}{2}$.

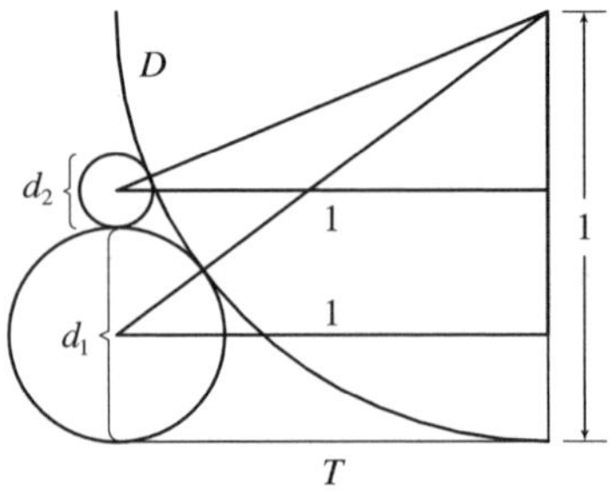

47. Let d_n be the diameter of C_n. We draw lines from the centres of the C_i to the centre of D (or C), and using the Pythagorean Theorem, we can write

$1^2 + \left(1 - \frac{1}{2}d_1\right)^2 = \left(1 + \frac{1}{2}d_1\right)^2 \quad \Leftrightarrow$

$1 = \left(1 + \frac{1}{2}d_1\right)^2 - \left(1 - \frac{1}{2}d_1\right)^2 = 2d_1$ [difference of squares] $\Rightarrow \quad d_1 = \frac{1}{2}$.

Similarly,

$1 = \left(1 + \frac{1}{2}d_2\right)^2 - \left(1 - d_1 - \frac{1}{2}d_2\right)^2 = 2d_2 + 2d_1 - d_1^2 - d_1 d_2$
$= (2 - d_1)(d_1 + d_2) \quad \Leftrightarrow$

$d_2 = \dfrac{1}{2 - d_1} - d_1 = \dfrac{(1 - d_1)^2}{2 - d_1}$, $1 = \left(1 + \frac{1}{2}d_3\right)^2 - \left(1 - d_1 - d_2 - \frac{1}{2}d_3\right)^2 \quad \Leftrightarrow \quad d_3 = \dfrac{[1 - (d_1 + d_2)]^2}{2 - (d_1 + d_2)}$, and in general,

$d_{n+1} = \dfrac{\left(1 - \sum_{i=1}^{n} d_i\right)^2}{2 - \sum_{i=1}^{n} d_i}$. If we actually calculate d_2 and d_3 from the formulas above, we find that they are $\dfrac{1}{6} = \dfrac{1}{2 \cdot 3}$ and

$\dfrac{1}{12} = \dfrac{1}{3 \cdot 4}$ respectively, so we suspect that in general, $d_n = \dfrac{1}{n(n+1)}$. To prove this, we use induction: Assume that for all

$k \leq n$, $d_k = \dfrac{1}{k(k+1)} = \dfrac{1}{k} - \dfrac{1}{k+1}$. Then $\sum\limits_{i=1}^{n} d_i = 1 - \dfrac{1}{n+1} = \dfrac{n}{n+1}$ [telescoping sum]. Substituting this into our

formula for d_{n+1}, we get $d_{n+1} = \dfrac{\left[1 - \dfrac{n}{n+1}\right]^2}{2 - \left(\dfrac{n}{n+1}\right)} = \dfrac{\dfrac{1}{(n+1)^2}}{\dfrac{n+2}{n+1}} = \dfrac{1}{(n+1)(n+2)}$, and the induction is complete.

Now, we observe that the partial sums $\sum_{i=1}^{n} d_i$ of the diameters of the circles approach 1 as $n \to \infty$; that is,

$\sum\limits_{n=1}^{\infty} a_n = \sum\limits_{n=1}^{\infty} \dfrac{1}{n(n+1)} = 1$, which is what we wanted to prove.

49. The series $1 - 1 + 1 - 1 + 1 - 1 + \cdots$ diverges (geometric series with $r = -1$) so we cannot say that $0 = 1 - 1 + 1 - 1 + 1 - 1 + \cdots$.

51. Suppose on the contrary that $\sum(a_n + b_n)$ converges. Then $\sum(a_n + b_n)$ and $\sum a_n$ are convergent series. So by Theorem 8, $\sum[(a_n + b_n) - a_n]$ would also be convergent. But $\sum[(a_n + b_n) - a_n] = \sum b_n$, a contradiction, since $\sum b_n$ is given to be divergent.

53. The partial sums $\{s_n\}$ form an increasing sequence, since $s_n - s_{n-1} = a_n > 0$ for all n. Also, the sequence $\{s_n\}$ is bounded since $s_n \leq 1000$ for all n. So by Theorem 8.1.7, the sequence of partial sums converges, that is, the series $\sum a_n$ is convergent.

55. (a) At the first step, only the interval $\left(\frac{1}{3}, \frac{2}{3}\right)$ (length $\frac{1}{3}$) is removed. At the second step, we remove the intervals $\left(\frac{1}{9}, \frac{2}{9}\right)$ and $\left(\frac{7}{9}, \frac{8}{9}\right)$, which have a total length of $2 \cdot \left(\frac{1}{3}\right)^2$. At the third step, we remove 2^2 intervals, each of length $\left(\frac{1}{3}\right)^3$. In general, at the nth step we remove 2^{n-1} intervals, each of length $\left(\frac{1}{3}\right)^n$, for a length of $2^{n-1} \cdot \left(\frac{1}{3}\right)^n = \frac{1}{3}\left(\frac{2}{3}\right)^{n-1}$. Thus, the total length of all removed intervals is $\sum\limits_{n=1}^{\infty} \frac{1}{3}\left(\frac{2}{3}\right)^{n-1} = \frac{1/3}{1 - 2/3} = 1$ $\left[\text{geometric series with } a = \frac{1}{3} \text{ and } r = \frac{2}{3}\right]$. Notice that at the nth step, the leftmost interval that is removed is $\left(\left(\frac{1}{3}\right)^n, \left(\frac{2}{3}\right)^n\right)$, so we never remove 0, and 0 is in the Cantor set. Also, the rightmost interval removed is $\left(1 - \left(\frac{2}{3}\right)^n, 1 - \left(\frac{1}{3}\right)^n\right)$, so 1 is never removed. Some other numbers in the Cantor set are $\frac{1}{3}, \frac{2}{3}, \frac{1}{9}, \frac{2}{9}, \frac{7}{9}$, and $\frac{8}{9}$.

(b) The area removed at the first step is $\frac{1}{9}$; at the second step, $8 \cdot \left(\frac{1}{9}\right)^2$; at the third step, $(8)^2 \cdot \left(\frac{1}{9}\right)^3$. In general, the area removed at the nth step is $(8)^{n-1}\left(\frac{1}{9}\right)^n = \frac{1}{9}\left(\frac{8}{9}\right)^{n-1}$, so the total area of all removed squares is

$$\sum_{n=1}^{\infty} \frac{1}{9}\left(\frac{8}{9}\right)^{n-1} = \frac{1/9}{1 - 8/9} = 1.$$

57. (a) For $\sum_{n=1}^{\infty} \frac{n}{(n+1)!}$, $s_1 = \frac{1}{1 \cdot 2} = \frac{1}{2}$, $s_2 = \frac{1}{2} + \frac{2}{1 \cdot 2 \cdot 3} = \frac{5}{6}$, $s_3 = \frac{5}{6} + \frac{3}{1 \cdot 2 \cdot 3 \cdot 4} = \frac{23}{24}$,

$s_4 = \frac{23}{24} + \frac{4}{1 \cdot 2 \cdot 3 \cdot 4 \cdot 5} = \frac{119}{120}$. The denominators are $(n+1)!$, so a guess would be $s_n = \frac{(n+1)! - 1}{(n+1)!}$.

(b) For $n = 1$, $s_1 = \frac{1}{2} = \frac{2! - 1}{2!}$, so the formula holds for $n = 1$. Assume $s_k = \frac{(k+1)! - 1}{(k+1)!}$. Then

$$s_{k+1} = \frac{(k+1)! - 1}{(k+1)!} + \frac{k+1}{(k+2)!} = \frac{(k+1)! - 1}{(k+1)!} + \frac{k+1}{(k+1)!(k+2)}$$

$$= \frac{(k+2)! - (k+2) + k + 1}{(k+2)!} = \frac{(k+2)! - 1}{(k+2)!}$$

Thus, the formula is true for $n = k + 1$. So by induction, the guess is correct.

(c) $\lim\limits_{n\to\infty} s_n = \lim\limits_{n\to\infty} \frac{(n+1)! - 1}{(n+1)!} = \lim\limits_{n\to\infty} \left[1 - \frac{1}{(n+1)!}\right] = 1$ and so $\sum_{n=1}^{\infty} \frac{n}{(n+1)!} = 1$.

8.3 The Integral and Comparison Tests; Estimating Sums

1. The picture shows that $a_2 = \frac{1}{2^{1.3}} < \int_1^2 \frac{1}{x^{1.3}}\,dx$,

$a_3 = \frac{1}{3^{1.3}} < \int_2^3 \frac{1}{x^{1.3}}\,dx$, and so on, so $\sum_{n=2}^{\infty} \frac{1}{n^{1.3}} < \int_1^{\infty} \frac{1}{x^{1.3}}\,dx$. The integral converges by (5.10.2) with $p = 1.3 > 1$, so the series converges.

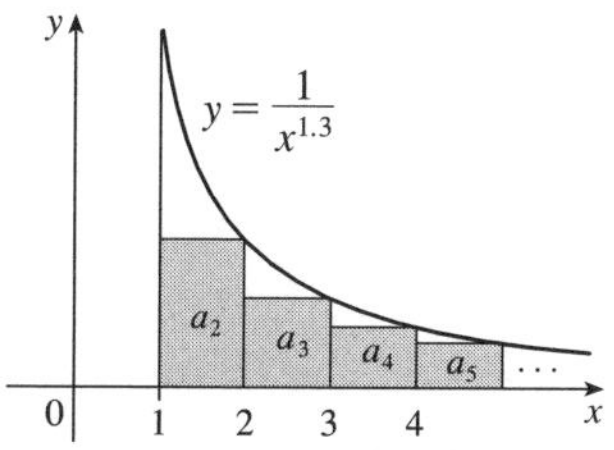

3. (a) We cannot say anything about $\sum a_n$. If $a_n > b_n$ for all n and $\sum b_n$ is convergent, then $\sum a_n$ could be convergent or divergent. (See the note after Example 4.)

(b) If $a_n < b_n$ for all n, then $\sum a_n$ is convergent. [This is part (i) of the Comparison Test.]

5. $\sum_{n=1}^{\infty} n^b$ is a p-series with $p = -b$. $\sum_{n=1}^{\infty} b^n$ is a geometric series. By (1), the p-series is convergent if $p > 1$. In this case, $\sum_{n=1}^{\infty} n^b = \sum_{n=1}^{\infty} (1/n^{-b})$, so $-b > 1 \iff b < -1$ are the values for which the series converge. A geometric series $\sum_{n=1}^{\infty} ar^{n-1}$ converges if $|r| < 1$, so $\sum_{n=1}^{\infty} b^n$ converges if $|b| < 1 \iff -1 < b < 1$.

7. The function $f(x) = 1/x^4$ is continuous, positive, and decreasing on $[1, \infty)$, so the Integral Test applies.

$\int_1^{\infty} \frac{1}{x^4}\,dx = \lim\limits_{t\to\infty} \int_1^t x^{-4}\,dx = \lim\limits_{t\to\infty} \left[\frac{x^{-3}}{-3}\right]_1^t = \lim\limits_{t\to\infty} \left(-\frac{1}{3t^3} + \frac{1}{3}\right) = \frac{1}{3}$. Since this improper integral is convergent, the series $\sum_{n=1}^{\infty} \frac{1}{n^4}$ is also convergent by the Integral Test.

9. $\frac{1}{n^2 + n + 1} < \frac{1}{n^2}$ for all $n \geq 1$, so $\sum_{n=1}^{\infty} \frac{1}{n^2 + n + 1}$ converges by comparison with $\sum_{n=1}^{\infty} \frac{1}{n^2}$, which converges because it is a p-series with $p = 2 > 1$.

11. $1 + \frac{1}{8} + \frac{1}{27} + \frac{1}{64} + \frac{1}{125} + \cdots = \sum_{n=1}^{\infty} \frac{1}{n^3}$. This is a p-series with $p = 3 > 1$, so it converges by (1).

13. $f(x) = xe^{-x}$ is continuous and positive on $[1, \infty)$. $f'(x) = -xe^{-x} + e^{-x} = e^{-x}(1 - x) < 0$ for $x > 1$, so f is decreasing on $[1, \infty)$. Thus, the Integral Test applies.

$$\int_1^\infty xe^{-x}\,dx = \lim_{b\to\infty} \int_1^b xe^{-x}\,dx = \lim_{b\to\infty} \left[-xe^{-x} - e^{-x}\right]_1^b \quad \text{[by parts]}$$

$$= \lim_{b\to\infty} \left[-be^{-b} - e^{-b} + e^{-1} + e^{-1}\right] = 2/e$$

since $\lim_{b\to\infty} be^{-b} = \lim_{b\to\infty} (b/e^b) \overset{\text{H}}{=} \lim_{b\to\infty} (1/e^b) = 0$ and $\lim_{b\to\infty} e^{-b} = 0$. Thus, $\sum_{n=1}^{\infty} ne^{-n}$ converges.

15. $f(x) = \dfrac{1}{x \ln x}$ is continuous and positive on $[2, \infty)$, and also decreasing since $f'(x) = -\dfrac{1 + \ln x}{x^2 (\ln x)^2} < 0$ for $x > 2$, so we can use the Integral Test. $\displaystyle\int_2^\infty \frac{1}{x \ln x}\,dx = \lim_{t\to\infty} [\ln(\ln x)]_2^t = \lim_{t\to\infty} [\ln(\ln t) - \ln(\ln 2)] = \infty$, so the series $\displaystyle\sum_{n=2}^{\infty} \frac{1}{n \ln n}$ diverges.

17. $\dfrac{\cos^2 n}{n^2 + 1} \le \dfrac{1}{n^2 + 1} < \dfrac{1}{n^2}$, so the series $\displaystyle\sum_{n=1}^{\infty} \frac{\cos^2 n}{n^2 + 1}$ converges by comparison with the p-series $\displaystyle\sum_{n=1}^{\infty} \frac{1}{n^2}$ $(p = 2 > 1)$.

19. $\dfrac{n - 1}{n\,4^n}$ is positive for $n > 1$ and $\dfrac{n - 1}{n\,4^n} < \dfrac{n}{n\,4^n} = \dfrac{1}{4^n} = \left(\dfrac{1}{4}\right)^n$, so $\displaystyle\sum_{n=1}^{\infty} \frac{n - 1}{n\,4^n}$ converges by comparison with the convergent geometric series $\displaystyle\sum_{n=1}^{\infty} \left(\frac{1}{4}\right)^n$.

21. Use the Limit Comparison Test with $a_n = \dfrac{1}{\sqrt{n^2 + 1}}$ and $b_n = \dfrac{1}{n}$:

$\displaystyle\lim_{n\to\infty} \frac{a_n}{b_n} = \lim_{n\to\infty} \frac{n}{\sqrt{n^2 + 1}} = \lim_{n\to\infty} \frac{1}{\sqrt{1 + (1/n^2)}} = 1 > 0$. Since the harmonic series $\displaystyle\sum_{n=1}^{\infty} \frac{1}{n}$ diverges, so does $\displaystyle\sum_{n=1}^{\infty} \frac{1}{\sqrt{n^2 + 1}}$.

23. $\dfrac{2 + (-1)^n}{n\sqrt{n}} \le \dfrac{3}{n\sqrt{n}}$, and $\displaystyle\sum_{n=1}^{\infty} \frac{3}{n\sqrt{n}}$ converges because it is a constant multiple of the convergent p-series $\displaystyle\sum_{n=1}^{\infty} \frac{1}{n\sqrt{n}}$ $\left(p = \frac{3}{2} > 1\right)$, so the given series converges by the Comparison Test.

25. Use the Limit Comparison Test with $a_n = \sin\left(\dfrac{1}{n}\right)$ and $b_n = \dfrac{1}{n}$. Then $\sum a_n$ and $\sum b_n$ are series with positive terms and

$\displaystyle\lim_{n\to\infty} \frac{a_n}{b_n} = \lim_{n\to\infty} \frac{\sin(1/n)}{1/n} = \lim_{\theta\to 0} \frac{\sin\theta}{\theta} = 1 > 0$. Since $\displaystyle\sum_{n=1}^{\infty} b_n$ is the divergent harmonic series, $\displaystyle\sum_{n=1}^{\infty} \sin(1/n)$ also diverges.

[Note that we could also use l'Hospital's Rule to evaluate the limit:

$\displaystyle\lim_{x\to\infty} \frac{\sin(1/x)}{1/x} \overset{\text{H}}{=} \lim_{x\to\infty} \frac{\cos(1/x) \cdot (-1/x^2)}{-1/x^2} = \lim_{x\to\infty} \cos\frac{1}{x} = \cos 0 = 1.$]

27. We have already shown (in Exercise 15) that when $p = 1$ the series $\displaystyle\sum_{n=2}^{\infty} \frac{1}{n(\ln n)^p}$ diverges, so assume that $p \ne 1$.

$f(x) = \dfrac{1}{x(\ln x)^p}$ is continuous and positive on $[2, \infty)$, and $f'(x) = -\dfrac{p + \ln x}{x^2 (\ln x)^{p+1}} < 0$ if $x > e^{-p}$, so that f is eventually decreasing and we can use the Integral Test.

$$\int_2^\infty \frac{1}{x(\ln x)^p}\,dx = \lim_{t\to\infty} \left[\frac{(\ln x)^{1-p}}{1 - p}\right]_2^t \quad [\text{for } p \ne 1] \quad = \lim_{t\to\infty} \left[\frac{(\ln t)^{1-p}}{1 - p}\right] - \frac{(\ln 2)^{1-p}}{1 - p}$$

This limit exists whenever $1 - p < 0 \;\Leftrightarrow\; p > 1$, so the series converges for $p > 1$.

29. (a) $f(x) = \dfrac{1}{x^2}$ is positive and continuous and $f'(x) = -\dfrac{2}{x^3}$ is negative for $x > 0$, and so the Integral Test applies.

$\sum\limits_{n=1}^{\infty} \dfrac{1}{n^2} \approx s_{10} = \dfrac{1}{1^2} + \dfrac{1}{2^2} + \dfrac{1}{3^2} + \cdots + \dfrac{1}{10^2} \approx 1.549\,768.$

$R_{10} \le \displaystyle\int_{10}^{\infty} \frac{1}{x^2}\,dx = \lim_{t\to\infty}\left[\frac{-1}{x}\right]_{10}^{t} = \lim_{t\to\infty}\left(-\frac{1}{t} + \frac{1}{10}\right) = \frac{1}{10}$, so the error is at most 0.1.

(b) $s_{10} + \displaystyle\int_{11}^{\infty} \frac{1}{x^2}\,dx \le s \le s_{10} + \int_{10}^{\infty} \frac{1}{x^2}\,dx \quad\Rightarrow\quad s_{10} + \tfrac{1}{11} \le s \le s_{10} + \tfrac{1}{10} \quad\Rightarrow$

$1.549\,768 + 0.090\,909 = 1.640\,677 \le s \le 1.549\,768 + 0.1 = 1.649\,768$, so we get $s \approx 1.64522$ (the average of $1.640\,677$ and $1.649\,768$) with error ≤ 0.005 (the maximum of $1.649\,768 - 1.64522$ and $1.64522 - 1.640\,677$, rounded up).

(c) $R_n \le \displaystyle\int_{n}^{\infty} \frac{1}{x^2}\,dx = \frac{1}{n}$. So $R_n < 0.001$ if $\dfrac{1}{n} < \dfrac{1}{1000} \quad\Leftrightarrow\quad n > 1000$.

31. $f(x) = x^{-3/2}$ is positive and continuous and $f'(x) = -\frac{3}{2}x^{-5/2}$ is negative for $x > 0$, so the Integral Test applies. From the end of Example 7, we see that the error is at most half the length of the interval. From (4), the interval is $\left(s_n + \int_{n+1}^{\infty} f(x)\,dx,\, s_n + \int_n^{\infty} f(x)\,dx\right)$, so its length is $\int_n^{\infty} f(x)\,dx - \int_{n+1}^{\infty} f(x)\,dx = \int_n^{n+1} f(x)\,dx$. Thus, we need n such that $0.01 > \dfrac{1}{2}\displaystyle\int_n^{n+1} x^{-3/2}\,dx = \frac{1}{2}\left[\frac{-2}{\sqrt{x}}\right]_n^{n+1} = \frac{1}{\sqrt{n}} - \frac{1}{\sqrt{n+1}} \quad\Leftrightarrow\quad n > 13.08$ (use a graphing calculator to solve $1/\sqrt{x} - 1/\sqrt{x+1} < 0.01$). Again from the end of Example 7, we approximate s by the midpoint of this interval. In general, the midpoint is $\frac{1}{2}\left[\left(s_n + \int_{n+1}^{\infty} f(x)\,dx\right) + \left(s_n + \int_n^{\infty} f(x)\,dx\right)\right] = s_n + \frac{1}{2}\left(\int_{n+1}^{\infty} f(x)\,dx + \int_n^{\infty} f(x)\,dx\right)$. So using $n = 14$, we have $s \approx s_{14} + \frac{1}{2}\left(\int_{14}^{\infty} x^{-3/2}\,dx + \int_{15}^{\infty} x^{-3/2}\,dx\right) \approx 2.0872 + \frac{1}{\sqrt{14}} + \frac{1}{\sqrt{15}} \approx 2.6127 \approx 2.61$. Any larger value of n will also work. For instance, $s \approx s_{30} + \frac{1}{\sqrt{30}} + \frac{1}{\sqrt{31}} \approx 2.6124$.

33. $\sum\limits_{n=1}^{10} \dfrac{1}{n^4 + n^2} = \dfrac{1}{2} + \dfrac{1}{20} + \dfrac{1}{90} + \cdots + \dfrac{1}{10\,100} \approx 0.567\,975$. Now $\dfrac{1}{n^4 + n^2} < \dfrac{1}{n^4}$, so using the reasoning and notation of Example 8, the error is $R_{10} \le T_{10} = \sum\limits_{n=11}^{\infty} \dfrac{1}{n^4} \le \displaystyle\int_{10}^{\infty} \frac{dx}{x^4} = \lim_{t\to\infty}\left[-\frac{x^{-3}}{3}\right]_{10}^{t} = \frac{1}{3000} = 0.000\overline{3}$.

35. (a) From the figure, $a_2 + a_3 + \cdots + a_n \le \int_1^n f(x)\,dx$, so with

$f(x) = \dfrac{1}{x}$, $\dfrac{1}{2} + \dfrac{1}{3} + \dfrac{1}{4} + \cdots + \dfrac{1}{n} \le \displaystyle\int_1^n \frac{1}{x}\,dx = \ln n$. Thus,

$s_n = 1 + \dfrac{1}{2} + \dfrac{1}{3} + \dfrac{1}{4} + \cdots + \dfrac{1}{n} \le 1 + \ln n$.

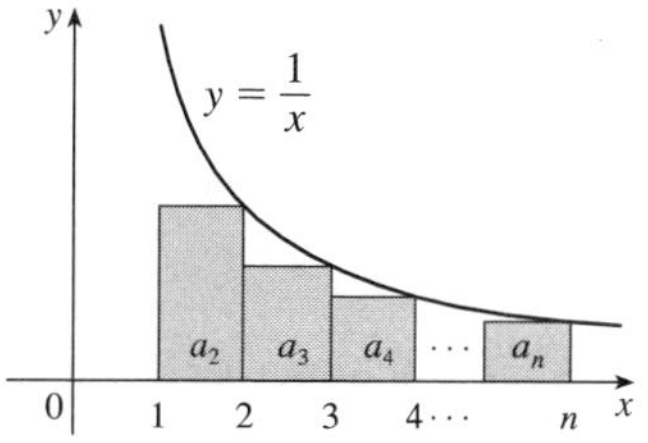

(b) By part (a), $s_{10^6} \le 1 + \ln 10^6 \approx 14.82 < 15$ and $s_{10^9} \le 1 + \ln 10^9 \approx 21.72 < 22$.

37. Since $\dfrac{d_n}{10^n} \le \dfrac{9}{10^n}$ for each n, and since $\sum\limits_{n=1}^{\infty} \dfrac{9}{10^n}$ is a convergent geometric series $\left(|r| = \frac{1}{10} < 1\right)$, $0.d_1d_2d_3\ldots = \sum\limits_{n=1}^{\infty} \dfrac{d_n}{10^n}$ will always converge by the Comparison Test.

39. Yes. Since $\sum a_n$ is a convergent series with positive terms, $\lim\limits_{n\to\infty} a_n = 0$ by Theorem 8.2.6, and $\sum b_n = \sum \sin(a_n)$ is a series with positive terms (for large enough n). We have $\lim\limits_{n\to\infty} \dfrac{b_n}{a_n} = \lim\limits_{n\to\infty} \dfrac{\sin(a_n)}{a_n} = 1 > 0$ by Theorem 3.4.2. Thus, $\sum b_n$ is also convergent by the Limit Comparison Test.

8.4 Other Convergence Tests

1. (a) An alternating series is a series whose terms are alternately positive and negative.

(b) An alternating series $\sum_{n=1}^{\infty}(-1)^{n-1}b_n$ converges if $0 < b_{n+1} \le b_n$ for all n and $\lim\limits_{n\to\infty} b_n = 0$. (This is the Alternating Series Test.)

(c) The error involved in using the partial sum s_n as an approximation to the total sum s is the remainder $R_n = s - s_n$ and the size of the error is smaller than b_{n+1}; that is, $|R_n| \le b_{n+1}$. (This is the Alternating Series Estimation Theorem.)

3. $\frac{4}{7} - \frac{4}{8} + \frac{4}{9} - \frac{4}{10} + \frac{4}{11} - \cdots = \sum\limits_{n=1}^{\infty}(-1)^{n-1}\frac{4}{n+6}$. Now $b_n = \frac{4}{n+6} > 0$, $\{b_n\}$ is decreasing, and $\lim\limits_{n\to\infty} b_n = 0$, so the series converges by the Alternating Series Test.

5. $b_n = \frac{1}{\sqrt{n}} > 0$, $\{b_n\}$ is decreasing, and $\lim\limits_{n\to\infty} b_n = 0$, so the series $\sum\limits_{n=1}^{\infty}\frac{(-1)^{n-1}}{\sqrt{n}}$ converges by the Alternating Series Test.

7. $\sum\limits_{n=1}^{\infty} a_n = \sum\limits_{n=1}^{\infty}(-1)^n\frac{3n-1}{2n+1} = \sum\limits_{n=1}^{\infty}(-1)^n b_n$. Now $\lim\limits_{n\to\infty} b_n = \lim\limits_{n\to\infty}\frac{3-1/n}{2+1/n} = \frac{3}{2} \ne 0$. Since $\lim\limits_{n\to\infty} a_n \ne 0$ (in fact the limit does not exist), the series diverges by the Test for Divergence.

9. $\sum\limits_{n=1}^{\infty}\frac{(-1)^{n-1}}{n} = 1 - \frac{1}{2} + \frac{1}{3} - \frac{1}{4} + \cdots + \frac{1}{49} - \frac{1}{50} + \frac{1}{51} - \frac{1}{52} + \cdots$. The 50th partial sum of this series is an underestimate, since $\sum\limits_{n=1}^{\infty}\frac{(-1)^{n-1}}{n} = s_{50} + \left(\frac{1}{51} - \frac{1}{52}\right) + \left(\frac{1}{53} - \frac{1}{54}\right) + \cdots$, and the terms in parentheses are all positive. The result can be seen geometrically in Figure 1.

11. If $p > 0$, $\frac{1}{(n+1)^p} \le \frac{1}{n^p}$ ($\{1/n^p\}$ is decreasing) and $\lim\limits_{n\to\infty}\frac{1}{n^p} = 0$, so the series converges by the Alternating Series Test. If $p \le 0$, $\lim\limits_{n\to\infty}\frac{(-1)^{n-1}}{n^p}$ does not exist, so the series diverges by the Test for Divergence. Thus, $\sum\limits_{n=1}^{\infty}\frac{(-1)^{n-1}}{n^p}$ converges $\Leftrightarrow\ p > 0$.

13. The series $\sum\limits_{n=1}^{\infty}\frac{(-1)^{n+1}}{n^6}$ satisfies (i) of the Alternating Series Test because $\frac{1}{(n+1)^6} < \frac{1}{n^6}$ and (ii) $\lim\limits_{n\to\infty}\frac{1}{n^6} = 0$, so the series is convergent. Now $b_5 = \frac{1}{5^6} = 0.000\,064 > 0.00005$ and $b_6 = \frac{1}{6^6} \approx 0.00002 < 0.00005$, so by the Alternating Series Estimation Theorem, $n = 5$. (That is, since the 6th term is less than the desired error, we need to add the first 5 terms to get the sum to the desired accuracy.)

15. The graph gives us an estimate for the sum of the series $\sum\limits_{n=1}^{\infty}\frac{(-1)^{n-1}}{(2n-1)!}$ of 0.84.

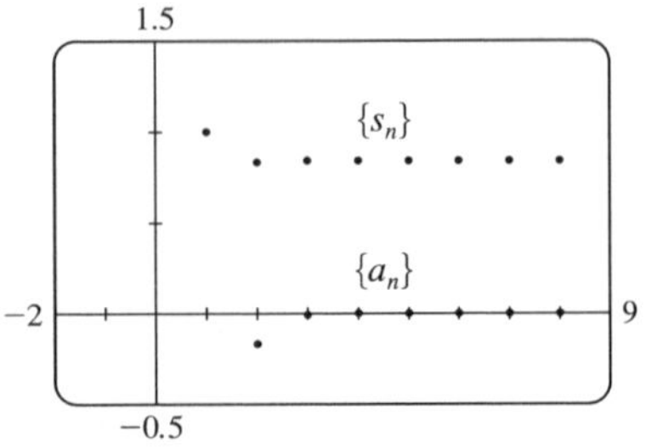

$$b_5 = \frac{1}{(2\cdot 5-1)!} = \frac{1}{362\,880} \approx 0.000\,003, \text{ so}$$

$$\sum_{n=1}^{\infty}\frac{(-1)^{n-1}}{(2n-1)!} \approx s_4 = \sum_{n=1}^{4}\frac{(-1)^{n-1}}{(2n-1)!} = 1 - \frac{1}{6} + \frac{1}{120} - \frac{1}{5040} \approx 0.841\,468.$$

Adding b_5 to s_4 does not change the fourth decimal place of s_4, so the sum of the series, correct to four decimal places, is 0.8415.

17. $b_7 = \dfrac{7^2}{10^7} = 0.000\,004\,9$, so

$$\sum_{n=1}^{\infty} \frac{(-1)^{n-1} n^2}{10^n} \approx s_6 = \sum_{n=1}^{6} \frac{(-1)^{n-1} n^2}{10^n} = \tfrac{1}{10} - \tfrac{4}{100} + \tfrac{9}{1000} - \tfrac{16}{10\,000} + \tfrac{25}{100\,000} - \tfrac{36}{1\,000\,000} = 0.067\,614.$$ Adding b_7 to s_6 does not change the fourth decimal place of s_6, so the sum of the series, correct to four decimal places, is 0.0676.

19. Using the Ratio Test, $\lim\limits_{n\to\infty} \left|\dfrac{a_{n+1}}{a_n}\right| = \lim\limits_{n\to\infty} \left|\dfrac{(-3)^{n+1}/(n+1)^3}{(-3)^n/n^3}\right| = \lim\limits_{n\to\infty} \left|\dfrac{(-3)n^3}{(n+1)^3}\right| = 3 \lim\limits_{n\to\infty} \left(\dfrac{n}{n+1}\right)^3 = 3 > 1,$ so the series diverges.

21. $\sum\limits_{n=0}^{\infty} \dfrac{(-10)^n}{n!}$. Using the Ratio Test, $\lim\limits_{n\to\infty} \left|\dfrac{a_{n+1}}{a_n}\right| = \lim\limits_{n\to\infty} \left|\dfrac{(-10)^{n+1}}{(n+1)!} \cdot \dfrac{n!}{(-10)^n}\right| = \lim\limits_{n\to\infty} \left|\dfrac{-10}{n+1}\right| = 0 < 1$, so the series is absolutely convergent.

23. Consider the series whose terms are the absolute values of the terms of the given series. $\sum\limits_{n=1}^{\infty} \left|\dfrac{(-1)^{n-1}}{\sqrt{n}}\right| = \sum\limits_{n=1}^{\infty} \dfrac{1}{n^{1/2}}$, which is a divergent p-series ($p = \frac{1}{2} \le 1$). Thus, $\sum\limits_{n=1}^{\infty} \dfrac{(-1)^{n-1}}{\sqrt{n}}$ is *not* absolutely convergent.

25. $$\lim_{n\to\infty} \left|\frac{a_{n+1}}{a_n}\right| = \lim_{n\to\infty} \left[\frac{10^{n+1}}{(n+2)4^{2(n+1)+1}} \cdot \frac{(n+1)4^{2n+1}}{10^n}\right] = \lim_{n\to\infty} \left[\frac{10^{n+1}}{(n+2)4^{2n+3}} \cdot \frac{(n+1)4^{2n+1}}{10^n}\right]$$
$$= \lim_{n\to\infty} \left(\frac{10}{4^2} \cdot \frac{n+1}{n+2}\right) = \frac{5}{8} < 1,$$

so the series is absolutely convergent by the Ratio Test. Since the terms of this series are positive, absolute convergence is the same as convergence.

27. $\left|\dfrac{\sin 2n}{n^2}\right| \le \dfrac{1}{n^2}$ and $\sum\limits_{n=1}^{\infty} \dfrac{1}{n^2}$ converges (p-series, $p = 2 > 1$), so $\sum\limits_{n=1}^{\infty} \dfrac{\sin 2n}{n^2}$ converges absolutely by the Comparison Test.

29. Use the Ratio Test with the series

$$1 - \frac{1\cdot 3}{3!} + \frac{1\cdot 3\cdot 5}{5!} - \frac{1\cdot 3\cdot 5\cdot 7}{7!} + \cdots + (-1)^{n-1}\frac{1\cdot 3\cdot 5\cdot\cdots\cdot(2n-1)}{(2n-1)!} + \cdots = \sum_{n=1}^{\infty} (-1)^{n-1}\frac{1\cdot 3\cdot 5\cdot\cdots\cdot(2n-1)}{(2n-1)!}.$$

$$\lim_{n\to\infty} \left|\frac{a_{n+1}}{a_n}\right| = \lim_{n\to\infty} \left|\frac{(-1)^n\cdot 1\cdot 3\cdot 5\cdot\cdots\cdot(2n-1)[2(n+1)-1]}{[2(n+1)-1]!} \cdot \frac{(2n-1)!}{(-1)^{n-1}\cdot 1\cdot 3\cdot 5\cdot\cdots\cdot(2n-1)}\right|$$
$$= \lim_{n\to\infty} \left|\frac{(-1)(2n+1)(2n-1)!}{(2n+1)(2n)(2n-1)!}\right| = \lim_{n\to\infty} \frac{1}{2n} = 0 < 1,$$

so the given series is absolutely convergent and therefore convergent.

31. By the recursive definition, $\lim\limits_{n\to\infty} \left|\dfrac{a_{n+1}}{a_n}\right| = \lim\limits_{n\to\infty} \left|\dfrac{5n+1}{4n+3}\right| = \dfrac{5}{4} > 1$, so the series diverges by the Ratio Test.

33. (a) $\lim\limits_{n\to\infty} \left|\dfrac{1/(n+1)^3}{1/n^3}\right| = \lim\limits_{n\to\infty} \dfrac{n^3}{(n+1)^3} = \lim\limits_{n\to\infty} \dfrac{1}{(1+1/n)^3} = 1.$ Inconclusive

(b) $\lim\limits_{n\to\infty} \left|\dfrac{(n+1)}{2^{n+1}} \cdot \dfrac{2^n}{n}\right| = \lim\limits_{n\to\infty} \dfrac{n+1}{2n} = \lim\limits_{n\to\infty} \left(\dfrac{1}{2} + \dfrac{1}{2n}\right) = \dfrac{1}{2}.$ Conclusive (convergent)

(c) $\lim\limits_{n\to\infty} \left|\dfrac{(-3)^n}{\sqrt{n+1}} \cdot \dfrac{\sqrt{n}}{(-3)^{n-1}}\right| = 3 \lim\limits_{n\to\infty} \sqrt{\dfrac{n}{n+1}} = 3 \lim\limits_{n\to\infty} \sqrt{\dfrac{1}{1+1/n}} = 3.$ Conclusive (divergent)

(d) $\lim\limits_{n\to\infty} \left|\dfrac{\sqrt{n+1}}{1+(n+1)^2} \cdot \dfrac{1+n^2}{\sqrt{n}}\right| = \lim\limits_{n\to\infty} \left[\sqrt{1+\dfrac{1}{n}} \cdot \dfrac{1/n^2+1}{1/n^2+(1+1/n)^2}\right] = 1.$ Inconclusive

35. (a) $\lim_{n\to\infty}\left|\frac{a_{n+1}}{a_n}\right| = \lim_{n\to\infty}\left|\frac{x^{n+1}}{(n+1)!}\cdot\frac{n!}{x^n}\right| = \lim_{n\to\infty}\left|\frac{x}{n+1}\right| = |x|\lim_{n\to\infty}\frac{1}{n+1} = |x|\cdot 0 = 0 < 1$, so by the Ratio Test the series $\sum_{n=0}^{\infty}\frac{x^n}{n!}$ converges for all x.

(b) Since the series of part (a) always converges, we must have $\lim_{n\to\infty}\frac{x^n}{n!} = 0$ by Theorem 8.2.6.

8.5 Power Series

1. A power series is a series of the form $\sum_{n=0}^{\infty} c_n x^n = c_0 + c_1 x + c_2 x^2 + c_3 x^3 + \cdots$, where x is a variable and the c_n's are constants called the coefficients of the series.
More generally, a series of the form $\sum_{n=0}^{\infty} c_n (x-a)^n = c_0 + c_1(x-a) + c_2(x-a)^2 + \cdots$ is called a power series in $(x-a)$ or a power series centred at a or a power series about a, where a is a constant.

3. If $a_n = \frac{x^n}{\sqrt{n}}$, then $\lim_{n\to\infty}\left|\frac{a_{n+1}}{a_n}\right| = \lim_{n\to\infty}\left|\frac{x^{n+1}}{\sqrt{n+1}}\cdot\frac{\sqrt{n}}{x^n}\right| = \lim_{n\to\infty}\left|\frac{x}{\sqrt{n+1}/\sqrt{n}}\right| = \lim_{n\to\infty}\frac{|x|}{\sqrt{1+1/n}} = |x|$.

By the Ratio Test, the series $\sum_{n=1}^{\infty}\frac{x^n}{\sqrt{n}}$ converges when $|x| < 1$, so the radius of convergence $R = 1$. Now we'll check the endpoints, that is, $x = \pm 1$. When $x = 1$, the series $\sum_{n=1}^{\infty}\frac{1}{\sqrt{n}}$ diverges because it is a p-series with $p = \frac{1}{2} \le 1$. When $x = -1$, the series $\sum_{n=1}^{\infty}\frac{(-1)^n}{\sqrt{n}}$ converges by the Alternating Series Test. Thus, the interval of convergence is $I = [-1, 1)$.

5. If $a_n = \frac{(-1)^{n-1}x^n}{n^3}$, then

$$\lim_{n\to\infty}\left|\frac{a_{n+1}}{a_n}\right| = \lim_{n\to\infty}\left|\frac{(-1)^n x^{n+1}}{(n+1)^3}\cdot\frac{n^3}{(-1)^{n-1}x^n}\right| = \lim_{n\to\infty}\left|\frac{(-1)xn^3}{(n+1)^3}\right|$$
$$= \lim_{n\to\infty}\left[\left(\frac{n}{n+1}\right)^3|x|\right] = 1^3\cdot|x| = |x|$$

By the Ratio Test, the series $\sum_{n=1}^{\infty}\frac{(-1)^{n-1}x^n}{n^3}$ converges when $|x| < 1$, so the radius of convergence $R = 1$. Now we'll check the endpoints, that is, $x = \pm 1$. When $x = 1$, the series $\sum_{n=1}^{\infty}\frac{(-1)^{n-1}}{n^3}$ converges by the Alternating Series Test. When $x = -1$, the series $\sum_{n=1}^{\infty}\frac{(-1)^{n-1}(-1)^n}{n^3} = -\sum_{n=1}^{\infty}\frac{1}{n^3}$ converges because it is a constant multiple of a convergent p-series $(p = 3 > 1)$. Thus, the interval of convergence is $I = [-1, 1]$.

7. If $a_n = \frac{x^n}{n!}$, then $\lim_{n\to\infty}\left|\frac{a_{n+1}}{a_n}\right| = \lim_{n\to\infty}\left|\frac{x^{n+1}}{(n+1)!}\cdot\frac{n!}{x^n}\right| = \lim_{n\to\infty}\left|\frac{x}{n+1}\right| = |x|\lim_{n\to\infty}\frac{1}{n+1} = |x|\cdot 0 = 0 < 1$ for *all* real x.
So, by the Ratio Test, $R = \infty$, and $I = (-\infty, \infty)$.

9. If $a_n = \frac{3^n x^n}{(n+1)^2}$, then $\lim_{n\to\infty}\left|\frac{a_{n+1}}{a_n}\right| = \lim_{n\to\infty}\left|\frac{3^{n+1}x^{n+1}}{(n+2)^2}\cdot\frac{(n+1)^2}{3^n x^n}\right| = 3|x|\lim_{n\to\infty}\left(\frac{n+1}{n+2}\right)^2 = 3|x|\cdot 1 = 3|x|$.
By the Ratio Test, the series converges when $3|x| < 1 \quad\Leftrightarrow\quad |x| < \frac{1}{3}$, so $R = \frac{1}{3}$. When $x = \frac{1}{3}$, $\sum_{n=0}^{\infty}\frac{3^n x^n}{(n+1)^2} = \sum_{n=0}^{\infty}\frac{1}{(n+1)^2} = \sum_{n=1}^{\infty}\frac{1}{n^2}$, which is a convergent p-series $(p = 2 > 1)$. When $x = -\frac{1}{3}$, $\sum_{n=0}^{\infty}\frac{3^n x^n}{(n+1)^2} = \sum_{n=0}^{\infty}\frac{(-1)^n}{(n+1)^2}$, which converges by the Alternating Series Test. Thus, $I = \left[-\frac{1}{3}, \frac{1}{3}\right]$.

11. If $a_n = (-1)^n \dfrac{x^n}{4^n \ln n}$, then $\lim\limits_{n\to\infty} \left|\dfrac{a_{n+1}}{a_n}\right| = \lim\limits_{n\to\infty} \left|\dfrac{x^{n+1}}{4^{n+1}\ln(n+1)} \cdot \dfrac{4^n \ln n}{x^n}\right| = \dfrac{|x|}{4} \lim\limits_{n\to\infty} \dfrac{\ln n}{\ln(n+1)} = \dfrac{|x|}{4} \cdot 1$

[by l'Hospital's Rule] $= \dfrac{|x|}{4}$. By the Ratio Test, the series converges when $\dfrac{|x|}{4} < 1 \iff |x| < 4$, so $R = 4$. When

$x = -4$, $\sum\limits_{n=2}^{\infty} (-1)^n \dfrac{x^n}{4^n \ln n} = \sum\limits_{n=2}^{\infty} \dfrac{[(-1)(-4)]^n}{4^n \ln n} = \sum\limits_{n=2}^{\infty} \dfrac{1}{\ln n}$. Since $\ln n < n$ for $n \geq 2$, $\dfrac{1}{\ln n} > \dfrac{1}{n}$ and $\sum\limits_{n=2}^{\infty} \dfrac{1}{n}$ is the

divergent harmonic series (without the $n = 1$ term), $\sum\limits_{n=2}^{\infty} \dfrac{1}{\ln n}$ is divergent by the Comparison Test. When $x = 4$,

$\sum\limits_{n=2}^{\infty} (-1)^n \dfrac{x^n}{4^n \ln n} = \sum\limits_{n=2}^{\infty} (-1)^n \dfrac{1}{\ln n}$, which converges by the Alternating Series Test. Thus, $I = (-4, 4]$.

13. If $a_n = (-1)^n \dfrac{(x+2)^n}{n\,2^n}$, then $\lim\limits_{n\to\infty} \left|\dfrac{a_{n+1}}{a_n}\right| = \lim\limits_{n\to\infty} \left[\dfrac{|x+2|^{n+1}}{(n+1)2^{n+1}} \cdot \dfrac{n2^n}{|x+2|^n}\right] = \lim\limits_{n\to\infty} \dfrac{n}{n+1} \cdot \dfrac{|x+2|}{2} = \dfrac{|x+2|}{2}$. By the

Ratio Test, the series converges when $\dfrac{|x+2|}{2} < 1 \iff |x+2| < 2$ [so $R = 2$] $\iff -2 < x+2 < 2 \iff$

$-4 < x < 0$. When $x = -4$, the series becomes $\sum\limits_{n=1}^{\infty} (-1)^n \dfrac{(-2)^n}{n2^n} = \sum\limits_{n=1}^{\infty} \dfrac{2^n}{n\,2^n} = \sum\limits_{n=1}^{\infty} \dfrac{1}{n}$, which is the divergent harmonic

series. When $x = 0$, the series is $\sum\limits_{n=1}^{\infty} \dfrac{(-1)^n}{n}$, the alternating harmonic series, which converges by the Alternating Series Test.

Thus, $I = (-4, 0]$.

15. $a_n = \dfrac{n}{b^n}(x-a)^n$, where $b > 0$.

$$\lim_{n\to\infty} \left|\frac{a_{n+1}}{a_n}\right| = \lim_{n\to\infty} \frac{(n+1)\,|x-a|^{n+1}}{b^{n+1}} \cdot \frac{b^n}{n\,|x-a|^n} = \lim_{n\to\infty} \left(1 + \frac{1}{n}\right) \frac{|x-a|}{b} = \frac{|x-a|}{b}.$$

By the Ratio Test, the series converges when $\dfrac{|x-a|}{b} < 1 \iff |x-a| < b$ [so $R = b$] $\iff -b < x - a < b \iff$

$a - b < x < a + b$. When $|x-a| = b$, $\lim\limits_{n\to\infty} |a_n| = \lim\limits_{n\to\infty} n = \infty$, so the series diverges. Thus, $I = (a-b, a+b)$.

17. If $a_n = n!(2x-1)^n$, then $\lim\limits_{n\to\infty} \left|\dfrac{a_{n+1}}{a_n}\right| = \lim\limits_{n\to\infty} \left|\dfrac{(n+1)!(2x-1)^{n+1}}{n!(2x-1)^n}\right| = \lim\limits_{n\to\infty} (n+1)\,|2x-1| \to \infty$ as $n \to \infty$ for all

$x \neq \frac{1}{2}$. Since the series diverges for all $x \neq \frac{1}{2}$, $R = 0$ and $I = \{\frac{1}{2}\}$.

19. (a) We are given that the power series $\sum_{n=0}^{\infty} c_n x^n$ is convergent for $x = 4$. So by Theorem 3, it must converge for at least $-4 < x \leq 4$. In particular, it converges when $x = -2$; that is, $\sum_{n=0}^{\infty} c_n (-2)^n$ is convergent.

(b) It does not follow that $\sum_{n=0}^{\infty} c_n (-4)^n$ is necessarily convergent. [See the comments after Theorem 3 about convergence at the endpoint of an interval. An example is $c_n = (-1)^n/(n4^n)$.]

21. If $a_n = \dfrac{(n!)^k}{(kn)!} x^n$, then

$$\lim_{n\to\infty} \left|\frac{a_{n+1}}{a_n}\right| = \lim_{n\to\infty} \frac{[(n+1)!]^k\,(kn)!}{(n!)^k\,[k(n+1)]!}\,|x| = \lim_{n\to\infty} \frac{(n+1)^k}{(kn+k)(kn+k-1)\cdots(kn+2)(kn+1)}\,|x|$$

$$= \lim_{n\to\infty} \left[\frac{(n+1)}{(kn+1)} \frac{(n+1)}{(kn+2)} \cdots \frac{(n+1)}{(kn+k)}\right] |x|$$

$$= \lim_{n\to\infty} \left[\frac{n+1}{kn+1}\right] \lim_{n\to\infty} \left[\frac{n+1}{kn+2}\right] \cdots \lim_{n\to\infty} \left[\frac{n+1}{kn+k}\right] |x|$$

$= \left(\dfrac{1}{k}\right)^k |x| < 1 \iff |x| < k^k$ for convergence, and the radius of convergence is $R = k^k$.

23. (a) If $a_n = \dfrac{(-1)^n\, x^{2n+1}}{n!(n+1)!\, 2^{2n+1}}$, then

$$\lim_{n\to\infty}\left|\frac{a_{n+1}}{a_n}\right| = \lim_{n\to\infty}\left|\frac{x^{2n+3}}{(n+1)!(n+2)!\,2^{2n+3}}\cdot\frac{n!(n+1)!\,2^{2n+1}}{x^{2n+1}}\right| = \left(\frac{x}{2}\right)^2\lim_{n\to\infty}\frac{1}{(n+1)(n+2)} = 0 \text{ for all } x.$$

So $J_1(x)$ converges for all x and its domain is $(-\infty, \infty)$.

(b), (c) The initial terms of $J_1(x)$ up to $n = 5$ are $a_0 = \dfrac{x}{2}$,

$a_1 = -\dfrac{x^3}{16}$, $a_2 = \dfrac{x^5}{384}$, $a_3 = -\dfrac{x^7}{18\,432}$, $a_4 = \dfrac{x^9}{1\,474\,560}$, and

$a_5 = -\dfrac{x^{11}}{176\,947\,200}$. The partial sums seem to approximate $J_1(x)$ well near the origin, but as $|x|$ increases, we need to take a large number of terms to get a good approximation.

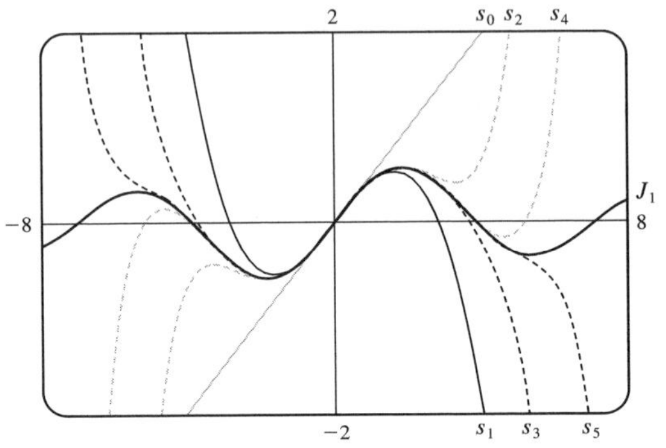

25. $s_{2n-1} = 1 + 2x + x^2 + 2x^3 + x^4 + 2x^5 + \cdots + x^{2n-2} + 2x^{2n-1}$

$= 1(1+2x) + x^2(1+2x) + x^4(1+2x) + \cdots + x^{2n-2}(1+2x) = (1+2x)(1 + x^2 + x^4 + \cdots + x^{2n-2})$

$= (1+2x)\dfrac{1-x^{2n}}{1-x^2}$ [by (8.2.3)] with $r = x^2$] $\to \dfrac{1+2x}{1-x^2}$ as $n \to \infty$ [by (8.2.4)], when $|x| < 1$.

Also $s_{2n} = s_{2n-1} + x^{2n} \to \dfrac{1+2x}{1-x^2}$ since $x^{2n} \to 0$ for $|x| < 1$. Therefore, $s_n \to \dfrac{1+2x}{1-x^2}$ since s_{2n} and s_{2n-1} both approach $\dfrac{1+2x}{1-x^2}$ as $n \to \infty$. Thus, the interval of convergence is $(-1, 1)$ and $f(x) = \dfrac{1+2x}{1-x^2}$.

27. For $2 < x < 3$, $\sum c_n x^n$ diverges and $\sum d_n x^n$ converges. By Exercise 8.2.51, $\sum (c_n + d_n)\, x^n$ diverges. Since both series converge for $|x| < 2$, the radius of convergence of $\sum (c_n + d_n)\, x^n$ is 2.

8.6 Representations of Functions as Power Series

1. If $f(x) = \sum\limits_{n=0}^{\infty} c_n x^n$ has radius of convergence 10, then $f'(x) = \sum\limits_{n=1}^{\infty} n c_n x^{n-1}$ also has radius of convergence 10 by Theorem 2.

3. Our goal is to write the function in the form $\dfrac{1}{1-r}$, and then use Equation (1) to represent the function as a sum of a power series. $f(x) = \dfrac{1}{1+x} = \dfrac{1}{1-(-x)} = \sum\limits_{n=0}^{\infty}(-x)^n = \sum\limits_{n=0}^{\infty}(-1)^n x^n$ with $|-x| < 1 \;\Leftrightarrow\; |x| < 1$, so $R = 1$ and $I = (-1, 1)$.

5. Replacing x with x^3 in (1) gives $f(x) = \dfrac{1}{1-x^3} = \sum\limits_{n=0}^{\infty}(x^3)^n = \sum\limits_{n=0}^{\infty} x^{3n}$. The series converges when $|x^3| < 1 \;\Leftrightarrow\; |x|^3 < 1 \;\Leftrightarrow\; |x| < \sqrt[3]{1} \;\Leftrightarrow\; |x| < 1$. Thus, $R = 1$ and $I = (-1, 1)$.

7. $f(x) = \dfrac{1}{x-5} = -\dfrac{1}{5}\left(\dfrac{1}{1-x/5}\right) = -\dfrac{1}{5}\sum\limits_{n=0}^{\infty}\left(\dfrac{x}{5}\right)^n$ or equivalently, $-\sum\limits_{n=0}^{\infty}\dfrac{1}{5^{n+1}}x^n$. The series converges when $\left|\dfrac{x}{5}\right| < 1$; that is, when $|x| < 5$, so $I = (-5, 5)$.

9. If the constant term in the denominator is something other than 1, factor it out of the binomial to obtain a 1.

$f(x) = \dfrac{1}{4+x^2} = \dfrac{1}{4}\left(\dfrac{1}{1+x^2/4}\right) = \dfrac{1}{4}\left(\dfrac{1}{1-(-x^2/4)}\right) = \dfrac{1}{4}\sum\limits_{n=0}^{\infty}\left(-\dfrac{x^2}{4}\right)^n = \sum\limits_{n=0}^{\infty}\dfrac{(-1)^n x^{2n}}{4^{n+1}}$. The series converges when $\left|-\dfrac{x^2}{4}\right| < 1 \quad\Leftrightarrow\quad x^2 < 4 \quad\Leftrightarrow\quad |x| < 2$, so $R = 2$ and $I = (-2, 2)$.

11. (a) $f(x) = \dfrac{1}{(1+x)^2} = \dfrac{d}{dx}\left(\dfrac{-1}{1+x}\right) = -\dfrac{d}{dx}\left[\sum\limits_{n=0}^{\infty}(-1)^n x^n\right]$ [from Exercise 3]

$= \sum\limits_{n=1}^{\infty}(-1)^{n+1}nx^{n-1}$ [from Theorem 2(i)] $= \sum\limits_{n=0}^{\infty}(-1)^n(n+1)x^n$ with $R = 1$.

In the last step, note that we *decreased* the initial value of the summation variable n by 1, and then *increased* each occurrence of n in the term by 1 [also note that $(-1)^{n+2} = (-1)^n$].

(b) $f(x) = \dfrac{1}{(1+x)^3} = -\dfrac{1}{2}\dfrac{d}{dx}\left[\dfrac{1}{(1+x)^2}\right] = -\dfrac{1}{2}\dfrac{d}{dx}\left[\sum\limits_{n=0}^{\infty}(-1)^n(n+1)x^n\right]$ [from part (a)]

$= -\frac{1}{2}\sum\limits_{n=1}^{\infty}(-1)^n(n+1)nx^{n-1} = \frac{1}{2}\sum\limits_{n=0}^{\infty}(-1)^n(n+2)(n+1)x^n$ with $R = 1$.

(c) $f(x) = \dfrac{x^2}{(1+x)^3} = x^2 \cdot \dfrac{1}{(1+x)^3} = x^2 \cdot \dfrac{1}{2}\sum\limits_{n=0}^{\infty}(-1)^n(n+2)(n+1)x^n$ [from part (b)]

$= \dfrac{1}{2}\sum\limits_{n=0}^{\infty}(-1)^n(n+2)(n+1)x^{n+2}$

To write the power series with x^n rather than x^{n+2}, we will *decrease* each occurrence of n in the term by 2 and *increase* the initial value of the summation variable by 2. This gives us $\dfrac{1}{2}\sum\limits_{n=2}^{\infty}(-1)^n(n)(n-1)x^n$ with $R = 1$.

13. $f(x) = \ln(5-x) = -\displaystyle\int\frac{dx}{5-x} = -\frac{1}{5}\int\frac{dx}{1-x/5} = -\frac{1}{5}\int\left[\sum_{n=0}^{\infty}\left(\frac{x}{5}\right)^n\right]dx = C - \frac{1}{5}\sum_{n=0}^{\infty}\frac{x^{n+1}}{5^n(n+1)} = C - \sum_{n=1}^{\infty}\frac{x^n}{n\,5^n}$

Putting $x = 0$, we get $C = \ln 5$. The series converges for $|x/5| < 1 \quad\Leftrightarrow\quad |x| < 5$, so $R = 5$.

15. $\dfrac{1}{2-x} = \dfrac{1}{2(1-x/2)} = \dfrac{1}{2}\sum\limits_{n=0}^{\infty}\left(\dfrac{x}{2}\right)^n = \sum\limits_{n=0}^{\infty}\dfrac{1}{2^{n+1}}x^n$ for $\left|\dfrac{x}{2}\right| < 1 \quad\Leftrightarrow\quad |x| < 2$. Now

$\dfrac{1}{(x-2)^2} = \dfrac{d}{dx}\left(\dfrac{1}{2-x}\right) = \dfrac{d}{dx}\left(\sum\limits_{n=0}^{\infty}\dfrac{1}{2^{n+1}}x^n\right) = \sum\limits_{n=1}^{\infty}\dfrac{n}{2^{n+1}}x^{n-1} = \sum\limits_{n=0}^{\infty}\dfrac{n+1}{2^{n+2}}x^n$. So

$f(x) = \dfrac{x^3}{(x-2)^2} = x^3\sum\limits_{n=0}^{\infty}\dfrac{n+1}{2^{n+2}}x^n = \sum\limits_{n=0}^{\infty}\dfrac{n+1}{2^{n+2}}x^{n+3}$ or $\sum\limits_{n=3}^{\infty}\dfrac{n-2}{2^{n-1}}x^n$ for $|x| < 2$. Thus, $R = 2$ and $I = (-2, 2)$.

17. $f(x) = \ln(3 + x) = \int \frac{dx}{3 + x} = \frac{1}{3}\int \frac{dx}{1 + x/3} = \frac{1}{3}\int \frac{dx}{1 - (-x/3)} = \frac{1}{3}\int \sum_{n=0}^{\infty}\left(-\frac{x}{3}\right)^n dx$

$= C + \frac{1}{3}\sum_{n=0}^{\infty}\frac{(-1)^n}{(n+1)3^n}x^{n+1} = \ln 3 + \frac{1}{3}\sum_{n=1}^{\infty}\frac{(-1)^{n-1}}{n3^{n-1}}x^n \quad [C = f(0) = \ln 3]$

$= \ln 3 + \sum_{n=1}^{\infty}\frac{(-1)^{n-1}}{n\,3^n}x^n$. The series converges when $|-x/3| < 1 \;\Leftrightarrow\; |x| < 3$, so $R = 3$.

The terms of the series are $a_0 = \ln 3, a_1 = \frac{x}{3}, a_2 = -\frac{x^2}{18}, a_3 = \frac{x^3}{81}, a_4 = -\frac{x^4}{324}, a_5 = \frac{x^5}{1215}, \ldots$.

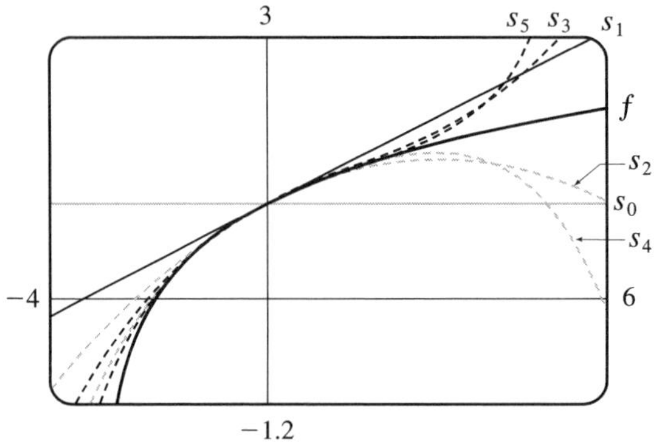

As n increases, $s_n(x)$ approximates f better on the interval of convergence, which is $(-3, 3)$.

19. $f(x) = \ln\left(\frac{1+x}{1-x}\right) = \ln(1 + x) - \ln(1 - x) = \int \frac{dx}{1+x} + \int \frac{dx}{1-x} = \int \frac{dx}{1-(-x)} + \int \frac{dx}{1-x}$

$= \int\left[\sum_{n=0}^{\infty}(-1)^n x^n + \sum_{n=0}^{\infty}x^n\right]dx = \int\left[(1 - x + x^2 - x^3 + x^4 - \cdots) + (1 + x + x^2 + x^3 + x^4 + \cdots)\right]dx$

$= \int (2 + 2x^2 + 2x^4 + \cdots)\,dx = \int \sum_{n=0}^{\infty}2x^{2n}\,dx = C + \sum_{n=0}^{\infty}\frac{2x^{2n+1}}{2n+1}$

But $f(0) = \ln\frac{1}{1} = 0$, so $C = 0$ and we have $f(x) = \sum_{n=0}^{\infty}\frac{2x^{2n+1}}{2n+1}$ with $R = 1$. If $x = \pm 1$, then $f(x) = \pm 2\sum_{n=0}^{\infty}\frac{1}{2n+1}$,

which both diverge by the Limit Comparison Test with $b_n = \frac{1}{n}$.

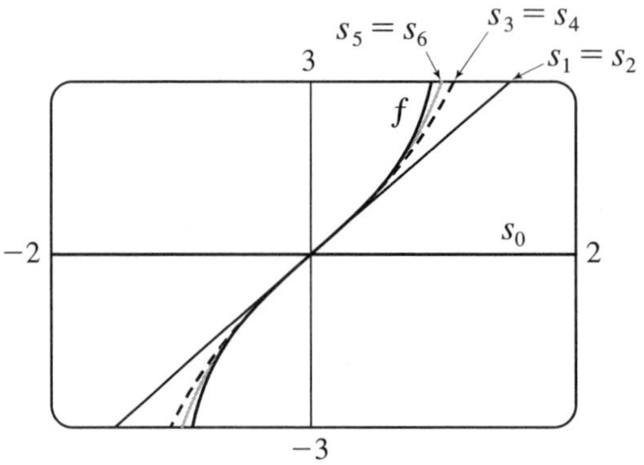

As n increases, $s_n(x)$ approximates f better on the interval of convergence, which is $(-1, 1)$.

21. $\dfrac{t}{1-t^8} = t\cdot\dfrac{1}{1-t^8} = t\sum\limits_{n=0}^{\infty}(t^8)^n = \sum\limits_{n=0}^{\infty}t^{8n+1} \quad\Rightarrow\quad \displaystyle\int\frac{t}{1-t^8}\,dt = C+\sum_{n=0}^{\infty}\frac{t^{8n+2}}{8n+2}$. The series for $\dfrac{1}{1-t^8}$ converges when $|t^8|<1 \quad\Leftrightarrow\quad |t|<1$, so $R=1$ for that series and also the series for $t/(1-t^8)$. By Theorem 2, the series for $\displaystyle\int\frac{t}{1-t^8}\,dt$ also has $R=1$.

23. By Example 7, $\tan^{-1}x = \sum\limits_{n=0}^{\infty}(-1)^n\dfrac{x^{2n+1}}{2n+1}$ with $R=1$, so

$$x-\tan^{-1}x = x-\left(x-\frac{x^3}{3}+\frac{x^5}{5}-\frac{x^7}{7}+\cdots\right) = \frac{x^3}{3}-\frac{x^5}{5}+\frac{x^7}{7}-\cdots = \sum_{n=1}^{\infty}(-1)^{n+1}\frac{x^{2n+1}}{2n+1} \text{ and}$$

$$\frac{x-\tan^{-1}x}{x^3} = \sum_{n=1}^{\infty}(-1)^{n+1}\frac{x^{2n-2}}{2n+1}, \text{ so}$$

$$\int\frac{x-\tan^{-1}x}{x^3}\,dx = C+\sum_{n=1}^{\infty}(-1)^{n+1}\frac{x^{2n-1}}{(2n+1)(2n-1)} = C+\sum_{n=1}^{\infty}(-1)^{n+1}\frac{x^{2n-1}}{4n^2-1}.$$ By Theorem 2, $R=1$.

25. $\dfrac{1}{1+x^5} = \dfrac{1}{1-(-x^5)} = \sum\limits_{n=0}^{\infty}(-x^5)^n = \sum\limits_{n=0}^{\infty}(-1)^n x^{5n} \quad\Rightarrow$

$$\int\frac{1}{1+x^5}\,dx = \int\sum_{n=0}^{\infty}(-1)^n x^{5n}\,dx = C+\sum_{n=0}^{\infty}(-1)^n\frac{x^{5n+1}}{5n+1}.$$ Thus,

$I = \displaystyle\int_0^{0.2}\frac{1}{1+x^5}\,dx = \left[x-\frac{x^6}{6}+\frac{x^{11}}{11}-\cdots\right]_0^{0.2} = 0.2-\frac{(0.2)^6}{6}+\frac{(0.2)^{11}}{11}-\cdots$. The series is alternating, so if we use the first two terms, the error is at most $(0.2)^{11}/11\approx 1.9\times10^{-9}$. So $I\approx 0.2-(0.2)^6/6\approx 0.199\,989$ to six decimal places.

27. We substitute $3x$ for x in Example 7, and find that

$$\begin{aligned}\int x\arctan(3x)\,dx &= \int x\sum_{n=0}^{\infty}(-1)^n\frac{(3x)^{2n+1}}{2n+1}\,dx = \int\sum_{n=0}^{\infty}(-1)^n\frac{3^{2n+1}x^{2n+2}}{2n+1}\,dx\\ &= C+\sum_{n=0}^{\infty}(-1)^n\frac{3^{2n+1}x^{2n+3}}{(2n+1)(2n+3)}\end{aligned}$$

$$\begin{aligned}\text{So}\int_0^{0.1}x\arctan(3x)\,dx &= \left[\frac{3x^3}{1\cdot3}-\frac{3^3x^5}{3\cdot5}+\frac{3^5x^7}{5\cdot7}-\frac{3^7x^9}{7\cdot9}+\cdots\right]_0^{0.1}\\ &= \frac{1}{10^3}-\frac{9}{5\times10^5}+\frac{243}{35\times10^7}-\frac{2187}{63\times10^9}+\cdots.\end{aligned}$$

The series is alternating, so if we use three terms, the error is at most $\dfrac{2187}{63\times10^9}\approx 3.5\times10^{-8}$. So

$\displaystyle\int_0^{0.1}x\arctan(3x)\,dx \approx \frac{1}{10^3}-\frac{9}{5\times10^5}+\frac{243}{35\times10^7}\approx 0.000\,983$ to six decimal places.

29. Using the result of Example 6, $\ln(1-x) = -\sum\limits_{n=1}^{\infty}\dfrac{x^n}{n}$, with $x=-0.1$, we have

$\ln 1.1 = \ln[1-(-0.1)] = 0.1-\dfrac{0.01}{2}+\dfrac{0.001}{3}-\dfrac{0.0001}{4}+\dfrac{0.00001}{5}-\cdots$. The series is alternating, so if we use only the first four terms, the error is at most $\dfrac{0.00001}{5} = 0.000\,002$. So $\ln 1.1\approx 0.1-\dfrac{0.01}{2}+\dfrac{0.001}{3}-\dfrac{0.0001}{4}\approx 0.09531$.

31. (a) $J_0(x) = \sum\limits_{n=0}^{\infty} \dfrac{(-1)^n x^{2n}}{2^{2n}(n!)^2}$, $J_0'(x) = \sum\limits_{n=1}^{\infty} \dfrac{(-1)^n 2nx^{2n-1}}{2^{2n}(n!)^2}$, and $J_0''(x) = \sum\limits_{n=1}^{\infty} \dfrac{(-1)^n 2n(2n-1)x^{2n-2}}{2^{2n}(n!)^2}$, so

$$\begin{aligned} x^2 J_0''(x) + xJ_0'(x) + x^2 J_0(x) &= \sum_{n=1}^{\infty} \frac{(-1)^n 2n(2n-1)x^{2n}}{2^{2n}(n!)^2} + \sum_{n=1}^{\infty} \frac{(-1)^n 2nx^{2n}}{2^{2n}(n!)^2} + \sum_{n=0}^{\infty} \frac{(-1)^n x^{2n+2}}{2^{2n}(n!)^2} \\ &= \sum_{n=1}^{\infty} \frac{(-1)^n 2n(2n-1)x^{2n}}{2^{2n}(n!)^2} + \sum_{n=1}^{\infty} \frac{(-1)^n 2nx^{2n}}{2^{2n}(n!)^2} + \sum_{n=1}^{\infty} \frac{(-1)^{n-1} x^{2n}}{2^{2n-2}\left[(n-1)!\right]^2} \\ &= \sum_{n=1}^{\infty} \frac{(-1)^n 2n(2n-1)x^{2n}}{2^{2n}(n!)^2} + \sum_{n=1}^{\infty} \frac{(-1)^n 2nx^{2n}}{2^{2n}(n!)^2} + \sum_{n=1}^{\infty} \frac{(-1)^n(-1)^{-1}2^2n^2x^{2n}}{2^{2n}(n!)^2} \\ &= \sum_{n=1}^{\infty} (-1)^n \left[\frac{2n(2n-1) + 2n - 2^2n^2}{2^{2n}(n!)^2}\right] x^{2n} \\ &= \sum_{n=1}^{\infty} (-1)^n \left[\frac{4n^2 - 2n + 2n - 4n^2}{2^{2n}(n!)^2}\right] x^{2n} = 0 \end{aligned}$$

(b) $$\int_0^1 J_0(x)\,dx = \int_0^1 \left[\sum_{n=0}^{\infty} \frac{(-1)^n x^{2n}}{2^{2n}(n!)^2}\right] dx = \int_0^1 \left(1 - \frac{x^2}{4} + \frac{x^4}{64} - \frac{x^6}{2304} + \cdots\right) dx$$

$$= \left[x - \frac{x^3}{3\cdot 4} + \frac{x^5}{5\cdot 64} - \frac{x^7}{7\cdot 2304} + \cdots\right]_0^1 = 1 - \frac{1}{12} + \frac{1}{320} - \frac{1}{16\,128} + \cdots$$

Since $\frac{1}{16\,128} \approx 0.000\,062$, it follows from The Alternating Series Estimation Theorem that, correct to three decimal places, $\int_0^1 J_0(x)\,dx \approx 1 - \frac{1}{12} + \frac{1}{320} \approx 0.920$.

33. (a) $f(x) = \sum\limits_{n=0}^{\infty} \dfrac{x^n}{n!} \;\Rightarrow\; f'(x) = \sum\limits_{n=1}^{\infty} \dfrac{nx^{n-1}}{n!} = \sum\limits_{n=1}^{\infty} \dfrac{x^{n-1}}{(n-1)!} = \sum\limits_{n=0}^{\infty} \dfrac{x^n}{n!} = f(x)$

(b) By Theorem 7.4.2, the only solution to the differential equation $df(x)/dx = f(x)$ is $f(x) = Ke^x$, but $f(0) = 1$, so $K = 1$ and $f(x) = e^x$.

Or: We could solve the equation $df(x)/dx = f(x)$ as a separable differential equation.

35. If $a_n = \dfrac{x^n}{n^2}$, then by the Ratio Test, $\lim\limits_{n\to\infty} \left|\dfrac{a_{n+1}}{a_n}\right| = \lim\limits_{n\to\infty} \left|\dfrac{x^{n+1}}{(n+1)^2} \cdot \dfrac{n^2}{x^n}\right| = |x| \lim\limits_{n\to\infty} \left(\dfrac{n}{n+1}\right)^2 = |x| < 1$ for convergence, so $R = 1$. When $x = \pm 1$, $\sum\limits_{n=1}^{\infty} \left|\dfrac{x^n}{n^2}\right| = \sum\limits_{n=1}^{\infty} \dfrac{1}{n^2}$ which is a convergent p-series ($p = 2 > 1$), so the interval of convergence for f is $[-1, 1]$. By Theorem 2, the radii of convergence of f' and f'' are both 1, so we need only check the endpoints. $f(x) = \sum\limits_{n=1}^{\infty} \dfrac{x^n}{n^2} \;\Rightarrow\; f'(x) = \sum\limits_{n=1}^{\infty} \dfrac{nx^{n-1}}{n^2} = \sum\limits_{n=0}^{\infty} \dfrac{x^n}{n+1}$, and this series diverges for $x = 1$ (harmonic series) and converges for $x = -1$ (Alternating Series Test), so the interval of convergence is $[-1, 1)$. $f''(x) = \sum\limits_{n=1}^{\infty} \dfrac{nx^{n-1}}{n+1}$ diverges at both 1 and -1 (Test for Divergence) since $\lim\limits_{n\to\infty} \dfrac{n}{n+1} = 1 \neq 0$, so its interval of convergence is $(-1, 1)$.

37. By Example 7, $\tan^{-1} x = \sum\limits_{n=0}^{\infty} (-1)^n \dfrac{x^{2n+1}}{2n+1}$ for $|x| < 1$. In particular, for $x = \dfrac{1}{\sqrt{3}}$, we have $\dfrac{\pi}{6} = \tan^{-1}\left(\dfrac{1}{\sqrt{3}}\right) = \sum\limits_{n=0}^{\infty} (-1)^n \dfrac{\left(1/\sqrt{3}\right)^{2n+1}}{2n+1} = \sum\limits_{n=0}^{\infty} (-1)^n \left(\dfrac{1}{3}\right)^n \dfrac{1}{\sqrt{3}} \dfrac{1}{2n+1}$, so

$$\pi = \frac{6}{\sqrt{3}} \sum_{n=0}^{\infty} \frac{(-1)^n}{(2n+1)3^n} = 2\sqrt{3} \sum_{n=0}^{\infty} \frac{(-1)^n}{(2n+1)3^n}.$$

8.7 Taylor and Maclaurin Series

1. Using Theorem 5 with $\sum_{n=0}^{\infty} b_n(x-5)^n$, $b_n = \dfrac{f^{(n)}(a)}{n!}$, so $b_8 = \dfrac{f^{(8)}(5)}{8!}$.

3. Since $f^{(n)}(0) = (n+1)!$, Equation 7 gives the Maclaurin series

$\sum_{n=0}^{\infty} \dfrac{f^{(n)}(0)}{n!}x^n = \sum_{n=0}^{\infty} \dfrac{(n+1)!}{n!}x^n = \sum_{n=0}^{\infty}(n+1)x^n$. Applying the Ratio Test with $a_n = (n+1)x^n$ gives us

$\lim_{n\to\infty}\left|\dfrac{a_{n+1}}{a_n}\right| = \lim_{n\to\infty}\left|\dfrac{(n+2)x^{n+1}}{(n+1)x^n}\right| = |x|\lim_{n\to\infty}\dfrac{n+2}{n+1} = |x|\cdot 1 = |x|$. For convergence, we must have $|x| < 1$, so the radius of convergence $R = 1$.

5.

n	$f^{(n)}(x)$	$f^{(n)}(0)$
0	$\cos x$	1
1	$-\sin x$	0
2	$-\cos x$	-1
3	$\sin x$	0
4	$\cos x$	1
⋮	⋮	⋮

We use Equation 7 with $f(x) = \cos x$.

$$\cos x = f(0) + f'(0)x + \frac{f''(0)}{2!}x^2 + \frac{f^{(3)}(0)}{3!}x^3 + \frac{f^{(4)}(0)}{4!}x^4 + \cdots$$

$$= 1 - \frac{x^2}{2!} + \frac{x^4}{4!} - \cdots = \sum_{n=0}^{\infty}\frac{(-1)^n x^{2n}}{(2n)!}$$

If $a_n = \dfrac{(-1)^n x^{2n}}{(2n)!}$, then

$$\lim_{n\to\infty}\left|\frac{a_{n+1}}{a_n}\right| = \lim_{n\to\infty}\left|\frac{x^{2n+2}}{(2n+2)!}\cdot\frac{(2n)!}{x^{2n}}\right| = x^2\lim_{n\to\infty}\frac{1}{(2n+2)(2n+1)} = 0 < 1 \text{ for all } x. \text{ So } R = \infty \text{ (Ratio Test).}$$

7.

n	$f^{(n)}(x)$	$f^{(n)}(0)$
0	$(1+x)^{-3}$	1
1	$-3(1+x)^{-4}$	-3
2	$12(1+x)^{-5}$	12
3	$-60(1+x)^{-6}$	-60
4	$360(1+x)^{-7}$	360
⋮	⋮	⋮

$$(1+x)^{-3} = f(0) + f'(0)x + \frac{f''(0)}{2!}x^2 + \frac{f'''(0)}{3!}x^3 + \frac{f^{(4)}(0)}{4!}x^4 + \cdots$$

$$= 1 - 3x + \frac{4\cdot 3}{2!}x^2 - \frac{5\cdot 4\cdot 3}{3!}x^3 + \frac{6\cdot 5\cdot 4\cdot 3}{4!}x^4 - \cdots$$

$$= 1 - 3x + \frac{4\cdot 3\cdot 2}{2\cdot 2!}x^2 - \frac{5\cdot 4\cdot 3\cdot 2}{2\cdot 3!}x^3 + \frac{6\cdot 5\cdot 4\cdot 3\cdot 2}{2\cdot 4!}x^4 - \cdots$$

$$= \sum_{n=0}^{\infty}\frac{(-1)^n(n+2)!\,x^n}{2(n!)} = \sum_{n=0}^{\infty}\frac{(-1)^n(n+2)(n+1)x^n}{2}$$

$$\lim_{n\to\infty}\left|\frac{a_{n+1}}{a_n}\right| = \lim_{n\to\infty}\left|\frac{(n+3)(n+2)x^{n+1}}{2}\cdot\frac{2}{(n+2)(n+1)x^n}\right| = |x|\lim_{n\to\infty}\frac{n+3}{n+1} = |x| < 1 \text{ for convergence,}$$

so $R = 1$ (Ratio Test).

9.

n	$f^{(n)}(x)$	$f^{(n)}(2)$
0	$1+x+x^2$	7
1	$1+2x$	5
2	2	2
3	0	0
4	0	0
⋮	⋮	⋮

$$f(x) = 7 + 5(x-2) + \frac{2}{2!}(x-2)^2 + \sum_{n=3}^{\infty}\frac{0}{n!}(x-2)^n$$

$$= 7 + 5(x-2) + (x-2)^2$$

Since $a_n = 0$ for large n, $R = \infty$.

11. Clearly, $f^{(n)}(x) = e^x$, so $f^{(n)}(3) = e^3$ and $e^x = \sum_{n=0}^{\infty} \frac{e^3}{n!}(x-3)^n$. If $a_n = \frac{e^3}{n!}(x-3)^n$, then

$$\lim_{n\to\infty}\left|\frac{a_{n+1}}{a_n}\right| = \lim_{n\to\infty}\left|\frac{e^3(x-3)^{n+1}}{(n+1)!}\cdot\frac{n!}{e^3(x-3)^n}\right| = \lim_{n\to\infty}\frac{|x-3|}{n+1} = 0 < 1 \text{ for all } x, \text{ so } R = \infty.$$

13.

n	$f^{(n)}(x)$	$f^{(n)}(\pi)$
0	$\cos x$	-1
1	$-\sin x$	0
2	$-\cos x$	1
3	$\sin x$	0
4	$\cos x$	-1
$\vdots$	$\vdots$	$\vdots$

$$\begin{aligned}\cos x &= \sum_{k=0}^{\infty}\frac{f^{(k)}(\pi)}{k!}(x-\pi)^k \\ &= -1 + \frac{(x-\pi)^2}{2!} - \frac{(x-\pi)^4}{4!} + \frac{(x-\pi)^6}{6!} - \cdots \\ &= \sum_{n=0}^{\infty}(-1)^{n+1}\frac{(x-\pi)^{2n}}{(2n)!}\end{aligned}$$

$$\lim_{n\to\infty}\left|\frac{a_{n+1}}{a_n}\right| = \lim_{n\to\infty}\left[\frac{|x-\pi|^{2n+2}}{(2n+2)!}\cdot\frac{(2n)!}{|x-\pi|^{2n}}\right] = \lim_{n\to\infty}\frac{|x-\pi|^2}{(2n+2)(2n+1)} = 0 < 1 \text{ for all } x, \text{ so } R = \infty.$$

15.

n	$f^{(n)}(x)$	$f^{(n)}(9)$
0	$x^{-1/2}$	$\frac{1}{3}$
1	$-\frac{1}{2}x^{-3/2}$	$-\frac{1}{2}\cdot\frac{1}{3^3}$
2	$\frac{3}{4}x^{-5/2}$	$-\frac{1}{2}\cdot\left(-\frac{3}{2}\right)\cdot\frac{1}{3^5}$
3	$-\frac{15}{8}x^{-7/2}$	$-\frac{1}{2}\cdot\left(-\frac{3}{2}\right)\cdot\left(-\frac{5}{2}\right)\cdot\frac{1}{3^7}$
$\vdots$	$\vdots$	$\vdots$

$$\begin{aligned}\frac{1}{\sqrt{x}} &= \frac{1}{3} - \frac{1}{2\cdot 3^3}(x-9) + \frac{3}{2^2\cdot 3^5}\frac{(x-9)^2}{2!} - \frac{3\cdot 5}{2^3\cdot 3^7}\frac{(x-9)^3}{3!} + \cdots \\ &= \sum_{n=0}^{\infty}(-1)^n\frac{1\cdot 3\cdot 5\cdot\cdots\cdot(2n-1)}{2^n\cdot 3^{2n+1}\cdot n!}(x-9)^n.\end{aligned}$$

$$\begin{aligned}\lim_{n\to\infty}\left|\frac{a_{n+1}}{a_n}\right| &= \lim_{n\to\infty}\left[\frac{1\cdot 3\cdot 5\cdot\cdots\cdot(2n-1)[2(n+1)-1]\,|x-9|^{n+1}}{2^{n+1}\cdot 3^{[2(n+1)+1]}\cdot(n+1)!}\cdot\frac{2^n\cdot 3^{2n+1}\cdot n!}{1\cdot 3\cdot 5\cdot\cdots\cdot(2n-1)\,|x-9|^n}\right] \\ &= \lim_{n\to\infty}\left[\frac{(2n+1)\,|x-9|}{2\cdot 3^2(n+1)}\right] = \frac{1}{9}|x-9| < 1\end{aligned}$$

for convergence, so $|x-9| < 9$ and $R = 9$.

17. If $f(x) = \cos x$, then $f^{(n+1)}(x) = \pm\sin x$ or $\pm\cos x$. In each case, $\left|f^{(n+1)}(x)\right| \le 1$, so by Formula 9 with $a = 0$ and $M = 1$, $|R_n(x)| \le \frac{1}{(n+1)!}|x|^{n+1}$. Thus, $|R_n(x)| \to 0$ as $n \to \infty$ by Equation 10. So $\lim_{n\to\infty} R_n(x) = 0$ and, by Theorem 8, the series in Exercise 5 represents $\cos x$ for all x.

19. $\cos x = \sum_{n=0}^{\infty}(-1)^n\frac{x^{2n}}{(2n)!} \Rightarrow f(x) = \cos(\pi x) = \sum_{n=0}^{\infty}\frac{(-1)^n(\pi x)^{2n}}{(2n)!} = \sum_{n=0}^{\infty}\frac{(-1)^n\pi^{2n}x^{2n}}{(2n)!}$, $R = \infty$

21. $\tan^{-1} x = \sum\limits_{n=0}^{\infty} (-1)^n \dfrac{x^{2n+1}}{2n+1} \quad \Rightarrow \quad f(x) = x \tan^{-1} x = x \sum\limits_{n=0}^{\infty} (-1)^n \dfrac{x^{2n+1}}{2n+1} = \sum\limits_{n=0}^{\infty} (-1)^n \dfrac{x^{2n+2}}{2n+1}, \quad R = 1$

23. $e^x = \sum\limits_{n=0}^{\infty} \dfrac{x^n}{n!} \quad \Rightarrow \quad f(x) = x^2 e^{-x} = x^2 \sum\limits_{n=0}^{\infty} \dfrac{(-x)^n}{n!} = \sum\limits_{n=0}^{\infty} \dfrac{(-1)^n \, x^{n+2}}{n!}, \quad R = \infty$

25. $\sin^2 x = \dfrac{1}{2}(1 - \cos 2x) = \dfrac{1}{2}\left[1 - \sum\limits_{n=0}^{\infty} \dfrac{(-1)^n (2x)^{2n}}{(2n)!}\right] = \dfrac{1}{2}\left[1 - 1 - \sum\limits_{n=1}^{\infty} \dfrac{(-1)^n (2x)^{2n}}{(2n)!}\right]$

$= \sum\limits_{n=1}^{\infty} \dfrac{(-1)^{n+1} 2^{2n-1} x^{2n}}{(2n)!}, \quad R = \infty$

27.

n	$f^{(n)}(x)$	$f^{(n)}(0)$
0	$(1+x)^{1/2}$	1
1	$\frac{1}{2}(1+x)^{-1/2}$	$\frac{1}{2}$
2	$-\frac{1}{4}(1+x)^{-3/2}$	$-\frac{1}{4}$
3	$\frac{3}{8}(1+x)^{-5/2}$	$\frac{3}{8}$
4	$-\frac{15}{16}(1+x)^{-7/2}$	$-\frac{15}{16}$
$\vdots$	$\vdots$	$\vdots$

So $f^{(n)}(0) = \dfrac{(-1)^{n-1}\, 1 \cdot 3 \cdot 5 \cdot \cdots \cdot (2n-3)}{2^n}$ for $n \geq 2$, and $\sqrt{1+x} = 1 + \dfrac{x}{2} + \sum\limits_{n=2}^{\infty} \dfrac{(-1)^{n-1} 1 \cdot 3 \cdot 5 \cdot \cdots \cdot (2n-3)}{2^n n!} x^n$.

If $a_n = \dfrac{(-1)^{n-1} 1 \cdot 3 \cdot 5 \cdot \cdots \cdot (2n-3)}{2^n n!} x^n$, then

$$\lim_{n\to\infty} \left|\frac{a_{n+1}}{a_n}\right| = \lim_{n\to\infty} \left| \frac{1 \cdot 3 \cdot 5 \cdot \cdots \cdot (2n-3)(2n-1) x^{n+1}}{2^{n+1}(n+1)!} \cdot \frac{2^n n!}{1 \cdot 3 \cdot 5 \cdot \cdots \cdot (2n-3) x^n} \right|$$

$$= \frac{|x|}{2} \lim_{n\to\infty} \frac{2n-1}{n+1} = \frac{|x|}{2} \cdot 2 = |x| < 1 \text{ for convergence, so } R = 1.$$

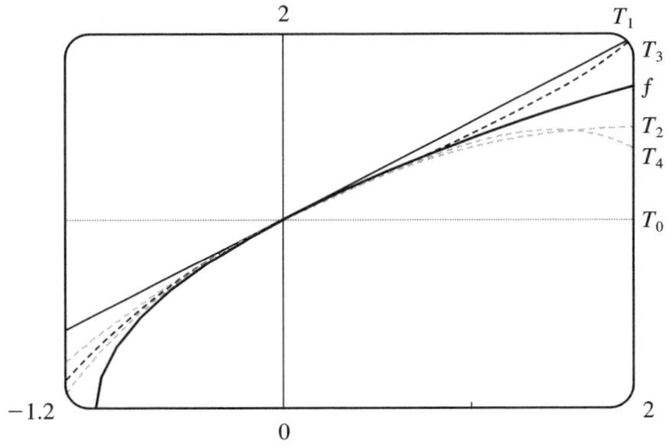

Notice that, as n increases, $T_n(x)$ becomes a better approximation to $f(x)$ for $-1 < x < 1$.

29. $\cos x = \sum_{n=0}^{\infty} (-1)^n \dfrac{x^{2n}}{(2n)!} \Rightarrow f(x) = \cos(x^2) = \sum_{n=0}^{\infty} \dfrac{(-1)^n (x^2)^{2n}}{(2n)!} = \sum_{n=0}^{\infty} \dfrac{(-1)^n x^{4n}}{(2n)!}$, $R = \infty$

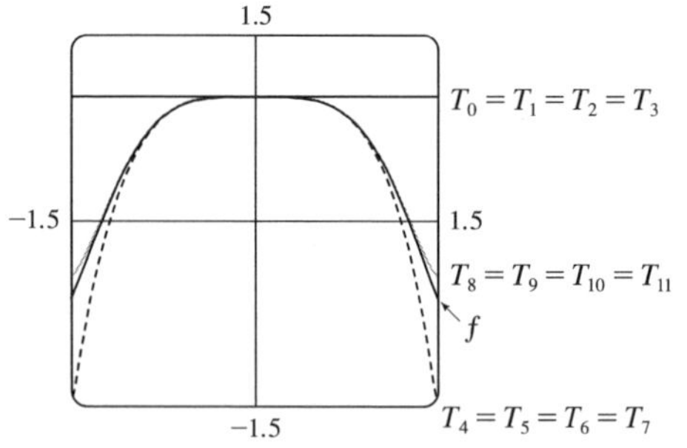

Notice that, as n increases, $T_n(x)$ becomes a better approximation to $f(x)$.

31. $e^x = \sum_{n=0}^{\infty} \dfrac{x^n}{n!}$, so $e^{-0.2} = \sum_{n=0}^{\infty} \dfrac{(-0.2)^n}{n!} = 1 - 0.2 + \dfrac{1}{2!}(0.2)^2 - \dfrac{1}{3!}(0.2)^3 + \dfrac{1}{4!}(0.2)^4 - \dfrac{1}{5!}(0.2)^5 + \dfrac{1}{6!}(0.2)^6 - \cdots$.

But $\dfrac{1}{6!}(0.2)^6 = 8.\overline{8} \times 10^{-8}$, so by the Alternating Series Estimation Theorem, $e^{-0.2} \approx \sum_{n=0}^{5} \dfrac{(-0.2)^n}{n!} \approx 0.81873$, correct to five decimal places.

33. $\cos x \overset{(16)}{=} \sum_{n=0}^{\infty} (-1)^n \dfrac{x^{2n}}{(2n)!} \Rightarrow \cos(x^3) = \sum_{n=0}^{\infty} (-1)^n \dfrac{(x^3)^{2n}}{(2n)!} = \sum_{n=0}^{\infty} (-1)^n \dfrac{x^{6n}}{(2n)!} \Rightarrow$

$x\cos(x^3) = \sum_{n=0}^{\infty} (-1)^n \dfrac{x^{6n+1}}{(2n)!} \Rightarrow \int x\cos(x^3)\,dx = C + \sum_{n=0}^{\infty} (-1)^n \dfrac{x^{6n+2}}{(6n+2)(2n)!}$, with $R = \infty$.

35. Using the series from Exercise 27 and substituting x^3 for x, we get

$$\int \sqrt{x^3+1}\,dx = \int \left[1 + \frac{x^3}{2} + \sum_{n=2}^{\infty} \frac{(-1)^{n-1}\, 1 \cdot 3 \cdot 5 \cdot \cdots \cdot (2n-3)}{2^n n!} x^{3n}\right] dx$$
$$= C + x + \frac{x^4}{8} + \sum_{n=2}^{\infty} \frac{(-1)^{n-1}\, 1 \cdot 3 \cdot 5 \cdot \cdots \cdot (2n-3)}{2^n n!(3n+1)} x^{3n+1}$$

37. $\int \sin(x^2)\,dx = \int \sum_{n=0}^{\infty} (-1)^n \dfrac{(x^2)^{2n+1}}{(2n+1)!}\,dx = \int \sum_{n=0}^{\infty} \dfrac{(-1)^n x^{4n+2}}{(2n+1)!}\,dx = C + \sum_{n=0}^{\infty} \dfrac{(-1)^n x^{4n+3}}{(4n+3)(2n+1)!}$.

Thus, $\int_0^1 \sin(x^2)\,dx = \sum_{n=0}^{\infty} \left[\dfrac{(-1)^n x^{4n+3}}{(4n+3)(2n+1)!}\right]_0^1 = \sum_{n=0}^{\infty} \dfrac{(-1)^n}{(4n+3)(2n+1)!}$ and $|c_3| = \dfrac{1}{75\,600} < 0.000\,014$, so by the Alternating Series Estimation Theorem, we have $\int_0^1 \sin(x^2)\,dx \approx \sum_{n=0}^{2} \dfrac{(-1)^n}{(4n+3)(2n+1)!} = \dfrac{1}{3} - \dfrac{1}{42} + \dfrac{1}{1320} \approx 0.310$ (correct to three decimal places).

39. We first find a series representation for $f(x) = (1+x)^{-1/2}$, and then substitute.

n	$f^{(n)}(x)$	$f^{(n)}(0)$
0	$(1+x)^{-1/2}$	1
1	$-\frac{1}{2}(1+x)^{-3/2}$	$-\frac{1}{2}$
2	$\frac{3}{4}(1+x)^{-5/2}$	$\frac{3}{4}$
3	$-\frac{15}{8}(1+x)^{-7/2}$	$-\frac{15}{8}$
⋮	⋮	⋮

$\dfrac{1}{\sqrt{1+x}} = 1 - \dfrac{x}{2} + \dfrac{3}{4}\left(\dfrac{x^2}{2!}\right) - \dfrac{15}{8}\left(\dfrac{x^3}{3!}\right) + \cdots \Rightarrow \dfrac{1}{\sqrt{1+x^3}} = 1 - \dfrac{1}{2}x^3 + \dfrac{3}{8}x^6 - \dfrac{5}{16}x^9 + \cdots \Rightarrow$

$\displaystyle\int_0^{0.1} \frac{dx}{\sqrt{1+x^3}} = \left[x - \frac{1}{8}x^4 + \frac{3}{56}x^7 - \frac{1}{32}x^{10} + \cdots\right]_0^{0.1} \approx (0.1) - \frac{1}{8}(0.1)^4$, by the Alternating Series Estimation Theorem, since $\frac{3}{56}(0.1)^7 \approx 0.000\,000\,005\,4 < 10^{-8}$, which is the maximum desired error. Therefore,

$\displaystyle\int_0^{0.1} \frac{dx}{\sqrt{1+x^3}} \approx 0.099\,987\,50.$

41. $\displaystyle\lim_{x\to 0} \frac{x - \tan^{-1} x}{x^3} = \lim_{x\to 0} \frac{x - \left(x - \frac{1}{3}x^3 + \frac{1}{5}x^5 - \frac{1}{7}x^7 + \cdots\right)}{x^3} = \lim_{x\to 0} \frac{\frac{1}{3}x^3 - \frac{1}{5}x^5 + \frac{1}{7}x^7 - \cdots}{x^3}$

$\displaystyle = \lim_{x\to 0}\left(\tfrac{1}{3} - \tfrac{1}{5}x^2 + \tfrac{1}{7}x^4 - \cdots\right) = \tfrac{1}{3}$

since power series are continuous functions.

43. $\displaystyle\lim_{x\to 0} \frac{\sin x - x + \frac{1}{6}x^3}{x^5} = \lim_{x\to 0} \frac{\left(x - \frac{1}{3!}x^3 + \frac{1}{5!}x^5 - \frac{1}{7!}x^7 + \cdots\right) - x + \frac{1}{6}x^3}{x^5}$

$\displaystyle = \lim_{x\to 0} \frac{\frac{1}{5!}x^5 - \frac{1}{7!}x^7 + \cdots}{x^5} = \lim_{x\to 0}\left(\frac{1}{5!} - \frac{x^2}{7!} + \frac{x^4}{9!} - \cdots\right) = \frac{1}{5!} = \frac{1}{120}$

since power series are continuous functions.

45. As in Example 8(a), we have $e^{-x^2} = 1 - \dfrac{x^2}{1!} + \dfrac{x^4}{2!} - \dfrac{x^6}{3!} + \cdots$ and we know that $\cos x = 1 - \dfrac{x^2}{2!} + \dfrac{x^4}{4!} - \cdots$ from Equation 16. Therefore, $e^{-x^2}\cos x = \left(1 - x^2 + \frac{1}{2}x^4 - \cdots\right)\left(1 - \frac{1}{2}x^2 + \frac{1}{24}x^4 - \cdots\right)$. Writing only the terms with degree ≤ 4, we get $e^{-x^2}\cos x = 1 - \frac{1}{2}x^2 + \frac{1}{24}x^4 - x^2 + \frac{1}{2}x^4 + \frac{1}{2}x^4 + \cdots = 1 - \frac{3}{2}x^2 + \frac{25}{24}x^4 + \cdots$.

47. $\dfrac{x}{\sin x} \overset{(15)}{=} \dfrac{x}{x - \frac{1}{6}x^3 + \frac{1}{120}x^5 - \cdots}$.

$$\begin{array}{r|l}
 & 1 + \frac{1}{6}x^2 + \frac{7}{360}x^4 + \cdots \\
\hline
x - \frac{1}{6}x^3 + \frac{1}{120}x^5 - \cdots & x \\
 & x - \frac{1}{6}x^3 + \frac{1}{120}x^5 - \cdots \\
\hline
 & \frac{1}{6}x^3 - \frac{1}{120}x^5 + \cdots \\
 & \frac{1}{6}x^3 - \frac{1}{36}x^5 + \cdots \\
\hline
 & \frac{7}{360}x^5 + \cdots \\
 & \frac{7}{360}x^5 + \cdots \\
\hline
 & \cdots
\end{array}$$

From the long division above, $\dfrac{x}{\sin x} = 1 + \frac{1}{6}x^2 + \frac{7}{360}x^4 + \cdots$.

49. $\displaystyle\sum_{n=0}^{\infty} (-1)^n \frac{x^{4n}}{n!} = \sum_{n=0}^{\infty} \frac{\left(-x^4\right)^n}{n!} = e^{-x^4}$, by (11).

51. $\displaystyle\sum_{n=0}^{\infty} \frac{(-1)^n \pi^{2n+1}}{4^{2n+1}(2n+1)!} = \sum_{n=0}^{\infty} \frac{(-1)^n \left(\frac{\pi}{4}\right)^{2n+1}}{(2n+1)!} = \sin\frac{\pi}{4} = \frac{1}{\sqrt{2}}$, by (15).

53. $\displaystyle 3 + \frac{9}{2!} + \frac{27}{3!} + \frac{81}{4!} + \cdots = \frac{3^1}{1!} + \frac{3^2}{2!} + \frac{3^3}{3!} + \frac{3^4}{4!} + \cdots = \sum_{n=1}^{\infty} \frac{3^n}{n!} = \sum_{n=0}^{\infty} \frac{3^n}{n!} - 1 = e^3 - 1$, by (11).

55. Assume that $|f'''(x)| \le M$, so $f'''(x) \le M$ for $a \le x \le a+d$. Now $\int_a^x f'''(t)\,dt \le \int_a^x M\,dt \Rightarrow$

$f''(x) - f''(a) \le M(x-a) \Rightarrow f''(x) \le f''(a) + M(x-a)$. Thus, $\int_a^x f''(t)\,dt \le \int_a^x [f''(a) + M(t-a)]\,dt \Rightarrow$

$f'(x) - f'(a) \le f''(a)(x-a) + \frac{1}{2}M(x-a)^2 \Rightarrow f'(x) \le f'(a) + f''(a)(x-a) + \frac{1}{2}M(x-a)^2 \Rightarrow$

$\int_a^x f'(t)\,dt \le \int_a^x \left[f'(a) + f''(a)(t-a) + \frac{1}{2}M(t-a)^2\right] dt \Rightarrow$

$f(x) - f(a) \le f'(a)(x-a) + \frac{1}{2}f''(a)(x-a)^2 + \frac{1}{6}M(x-a)^3$. So

$f(x) - f(a) - f'(a)(x-a) - \frac{1}{2}f''(a)(x-a)^2 \le \frac{1}{6}M(x-a)^3$. But

$R_2(x) = f(x) - T_2(x) = f(x) - f(a) - f'(a)(x-a) - \frac{1}{2}f''(a)(x-a)^2$, so $R_2(x) \le \frac{1}{6}M(x-a)^3$.

A similar argument using $f'''(x) \ge -M$ shows that $R_2(x) \ge -\frac{1}{6}M(x-a)^3$. So $|R_2(x_2)| \le \frac{1}{6}M\,|x-a|^3$.

Although we have assumed that $x > a$, a similar calculation shows that this inequality is also true if $x < a$.

8.8 The Binomial Series

1. The general binomial series in (2) is

$$(1+x)^k = \sum_{n=0}^{\infty} \binom{k}{n} x^n = 1 + kx + \frac{k(k-1)}{2!}x^2 + \frac{k(k-1)(k-2)}{3!}x^3 + \cdots.$$

$$\begin{aligned}(1+x)^{1/2} &= \sum_{n=0}^{\infty} \binom{\frac{1}{2}}{n} x^n = 1 + \left(\tfrac{1}{2}\right)x + \frac{\left(\frac{1}{2}\right)\left(-\frac{1}{2}\right)}{2!}x^2 + \frac{\left(\frac{1}{2}\right)\left(-\frac{1}{2}\right)\left(-\frac{3}{2}\right)}{3!}x^3 + \cdots \\ &= 1 + \frac{x}{2} - \frac{x^2}{2^2 \cdot 2!} + \frac{1 \cdot 3 \cdot x^3}{2^3 \cdot 3!} - \frac{1 \cdot 3 \cdot 5 \cdot x^4}{2^4 \cdot 4!} + \cdots \\ &= 1 + \frac{x}{2} + \sum_{n=2}^{\infty} \frac{(-1)^{n-1}\, 1 \cdot 3 \cdot 5 \cdot \cdots \cdot (2n-3)x^n}{2^n \cdot n!} \quad \text{for } |x| < 1, \quad \text{so } R = 1.\end{aligned}$$

3. $\dfrac{1}{(2+x)^3} = \dfrac{1}{[2(1+x/2)]^3} = \dfrac{1}{8}\left(1 + \dfrac{x}{2}\right)^{-3} = \dfrac{1}{8}\displaystyle\sum_{n=0}^{\infty} \binom{-3}{n}\left(\frac{x}{2}\right)^n$. The binomial coefficient is

$$\begin{aligned}\binom{-3}{n} &= \frac{(-3)(-4)(-5)\cdot \cdots \cdot (-3-n+1)}{n!} = \frac{(-3)(-4)(-5)\cdot \cdots \cdot [-(n+2)]}{n!} \\ &= \frac{(-1)^n \cdot 2 \cdot 3 \cdot 4 \cdot 5 \cdot \cdots \cdot (n+1)(n+2)}{2 \cdot n!} = \frac{(-1)^n (n+1)(n+2)}{2}\end{aligned}$$

Thus, $\dfrac{1}{(2+x)^3} = \dfrac{1}{8}\displaystyle\sum_{n=0}^{\infty} \frac{(-1)^n (n+1)(n+2)}{2}\frac{x^n}{2^n} = \sum_{n=0}^{\infty} \frac{(-1)^n (n+1)(n+2)x^n}{2^{n+4}}$ for $\left|\dfrac{x}{2}\right| < 1 \Leftrightarrow$

$|x| < 2$, so $R = 2$.

5. We must write the binomial in the form (1+ expression), so we'll factor out a 4.

$$\frac{x}{\sqrt{4+x^2}} = \frac{x}{\sqrt{4(1+x^2/4)}} = \frac{x}{2\sqrt{1+x^2/4}} = \frac{x}{2}\left(1+\frac{x^2}{4}\right)^{-1/2} = \frac{x}{2}\sum_{n=0}^{\infty}\binom{-\frac{1}{2}}{n}\left(\frac{x^2}{4}\right)^n$$

$$= \frac{x}{2}\left[1 + \left(-\tfrac{1}{2}\right)\frac{x^2}{4} + \frac{\left(-\frac{1}{2}\right)\left(-\frac{3}{2}\right)}{2!}\left(\frac{x^2}{4}\right)^2 + \frac{\left(-\frac{1}{2}\right)\left(-\frac{3}{2}\right)\left(-\frac{5}{2}\right)}{3!}\left(\frac{x^2}{4}\right)^3 + \cdots\right]$$

$$= \frac{x}{2} + \frac{x}{2}\sum_{n=1}^{\infty}(-1)^n\frac{1\cdot 3\cdot 5\cdot\cdots\cdot(2n-1)}{2^n\cdot 4^n\cdot n!}x^{2n}$$

$$= \frac{x}{2} + \sum_{n=1}^{\infty}(-1)^n\frac{1\cdot 3\cdot 5\cdot\cdots\cdot(2n-1)}{n!\,2^{3n+1}}x^{2n+1} \text{ and } \frac{x^2}{4} < 1 \;\Leftrightarrow\; \frac{|x|}{2} < 1 \;\Leftrightarrow\; |x| < 2, \quad \text{so } R = 2.$$

7. $$(1+2x)^{3/4} = 1 + \frac{3}{4}(2x) + \frac{\left(\frac{3}{4}\right)\left(-\frac{1}{4}\right)}{2!}(2x)^2 + \frac{\left(\frac{3}{4}\right)\left(-\frac{1}{4}\right)\left(-\frac{5}{4}\right)}{3!}(2x)^3 + \cdots$$

$$= 1 + \frac{3}{2}x + 3\sum_{n=2}^{\infty}(-1)^{n+1}\frac{1\cdot 5\cdot 9\cdot\cdots\cdot(4n-7)}{4^n\cdot n!}\cdot 2^n x^n$$

$$= 1 + \frac{3}{2}x + 3\sum_{n=2}^{\infty}(-1)^{n+1}\frac{1\cdot 5\cdot 9\cdot\cdots\cdot(4n-7)}{2^n\cdot n!}x^n \text{ and } |2x| < 1 \;\Leftrightarrow\; |x| < \frac{1}{2}, \quad \text{so } R = \frac{1}{2}.$$

The three Taylor polynomials are $T_1(x) = 1 + \frac{3}{2}x$, $T_2(x) = 1 + \frac{3}{2}x - \frac{3}{8}x^2$, and $T_3(x) = 1 + \frac{3}{2}x - \frac{3}{8}x^2 + \frac{5}{16}x^3$.

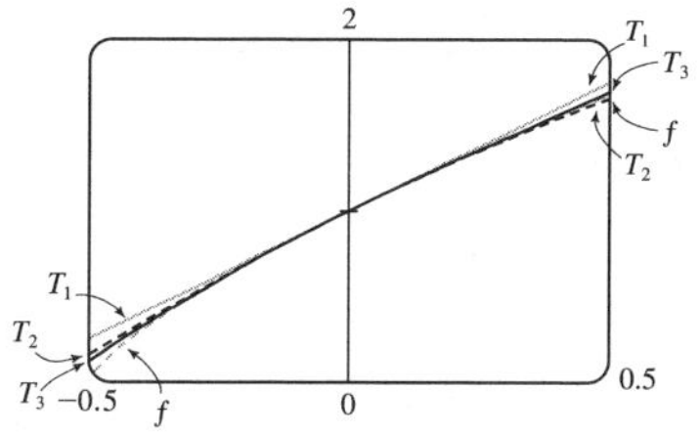

9. (a) $$1/\sqrt{1-x^2} = \left[1+(-x^2)\right]^{-1/2} = 1 + \left(-\tfrac{1}{2}\right)(-x^2) + \frac{\left(-\frac{1}{2}\right)\left(-\frac{3}{2}\right)}{2!}(-x^2)^2 + \frac{\left(-\frac{1}{2}\right)\left(-\frac{3}{2}\right)\left(-\frac{5}{2}\right)}{3!}(-x^2)^3 + \cdots$$

$$= 1 + \sum_{n=1}^{\infty}\frac{1\cdot 3\cdot 5\cdot\cdots\cdot(2n-1)}{2^n\cdot n!}x^{2n}$$

(b) $$\sin^{-1}x = \int\frac{1}{\sqrt{1-x^2}}\,dx = C + x + \sum_{n=1}^{\infty}\frac{1\cdot 3\cdot 5\cdot\cdots\cdot(2n-1)}{(2n+1)2^n\cdot n!}x^{2n+1}$$

$$= x + \sum_{n=1}^{\infty}\frac{1\cdot 3\cdot 5\cdot\cdots\cdot(2n-1)}{(2n+1)2^n\cdot n!}x^{2n+1} \quad \text{since } 0 = \sin^{-1}0 = C.$$

11. (a) $$[1+(-x)]^{-2} = 1 + (-2)(-x) + \frac{(-2)(-3)}{2!}(-x)^2 + \frac{(-2)(-3)(-4)}{3!}(-x)^3 + \cdots$$

$$= 1 + 2x + 3x^2 + 4x^3 + \cdots = \sum_{n=0}^{\infty}(n+1)x^n,$$

so $$\frac{x}{(1-x)^2} = x\sum_{n=0}^{\infty}(n+1)x^n = \sum_{n=0}^{\infty}(n+1)x^{n+1} = \sum_{n=1}^{\infty}nx^n.$$

(b) With $x = \frac{1}{2}$ in part (a), we have $\displaystyle\sum_{n=1}^{\infty}n\left(\tfrac{1}{2}\right)^n = \sum_{n=1}^{\infty}\frac{n}{2^n} = \frac{\frac{1}{2}}{\left(1-\frac{1}{2}\right)^2} = \frac{\frac{1}{2}}{\frac{1}{4}} = 2.$

13. (a) $\left(1+x^2\right)^{1/2} = 1 + \left(\frac{1}{2}\right)x^2 + \frac{\left(\frac{1}{2}\right)\left(-\frac{1}{2}\right)}{2!}\left(x^2\right)^2 + \frac{\left(\frac{1}{2}\right)\left(-\frac{1}{2}\right)\left(-\frac{3}{2}\right)}{3!}\left(x^2\right)^3 + \cdots$

$$= 1 + \frac{x^2}{2} + \sum_{n=2}^{\infty} \frac{(-1)^{n-1}\, 1\cdot 3\cdot 5\cdot \cdots \cdot (2n-3)}{2^n \cdot n!} x^{2n}$$

(b) The coefficient of x^{10} (corresponding to $n=5$) in the above Maclaurin series is $\frac{f^{(10)}(0)}{10!}$, so

$$\frac{f^{(10)}(0)}{10!} = \frac{(-1)^4 \cdot 1\cdot 3\cdot 5\cdot 7}{2^5\cdot 5!} \quad\Rightarrow\quad f^{(10)}(0) = 10!\left(\frac{1\cdot 3\cdot 5\cdot 7}{2^5\cdot 5!}\right) = 99\,225.$$

15. (a) $g(x) = \sum_{n=0}^{\infty}\binom{k}{n}x^n \quad\Rightarrow\quad g'(x) = \sum_{n=1}^{\infty}\binom{k}{n}nx^{n-1}$, so

$$(1+x)g'(x) = (1+x)\sum_{n=1}^{\infty}\binom{k}{n}nx^{n-1} = \sum_{n=1}^{\infty}\binom{k}{n}nx^{n-1} + \sum_{n=1}^{\infty}\binom{k}{n}nx^{n}$$

$$= \sum_{n=0}^{\infty}\binom{k}{n+1}(n+1)x^n + \sum_{n=0}^{\infty}\binom{k}{n}nx^n \qquad \left[\begin{matrix}\text{Replace } n \text{ with } n+1\\ \text{in the first series}\end{matrix}\right]$$

$$= \sum_{n=0}^{\infty}(n+1)\frac{k(k-1)(k-2)\cdots(k-n+1)(k-n)}{(n+1)!}x^n + \sum_{n=0}^{\infty}\left[(n)\frac{k(k-1)(k-2)\cdots(k-n+1)}{n!}\right]x^n$$

$$= \sum_{n=0}^{\infty}\frac{(n+1)k(k-1)(k-2)\cdots(k-n+1)}{(n+1)!}\left[(k-n)+n\right]x^n$$

$$= k\sum_{n=0}^{\infty}\frac{k(k-1)(k-2)\cdots(k-n+1)}{n!}x^n = k\sum_{n=0}^{\infty}\binom{k}{n}x^n = kg(x)$$

Thus, $g'(x) = \dfrac{kg(x)}{1+x}$.

(b) $h(x) = (1+x)^{-k}g(x) \quad\Rightarrow$

$$h'(x) = -k(1+x)^{-k-1}g(x) + (1+x)^{-k}g'(x) \quad \text{[Product Rule]}$$

$$= -k(1+x)^{-k-1}g(x) + (1+x)^{-k}\frac{kg(x)}{1+x} \quad \text{[from part (a)]}$$

$$= -k(1+x)^{-k-1}g(x) + k(1+x)^{-k-1}g(x) = 0$$

(c) From part (b) we see that $h(x)$ must be constant for $x\in(-1,1)$, so $h(x)=h(0)=1$ for $x\in(-1,1)$.

Thus, $h(x) = 1 = (1+x)^{-k}g(x) \quad\Leftrightarrow\quad g(x) = (1+x)^k$ for $x\in(-1,1)$.

8.9 Applications of Taylor Polynomials

1. (a)

n	$f^{(n)}(x)$	$f^{(n)}(0)$	$T_n(x)$
0	$\cos x$	1	1
1	$-\sin x$	0	1
2	$-\cos x$	-1	$1 - \frac{1}{2}x^2$
3	$\sin x$	0	$1 - \frac{1}{2}x^2$
4	$\cos x$	1	$1 - \frac{1}{2}x^2 + \frac{1}{24}x^4$
5	$-\sin x$	0	$1 - \frac{1}{2}x^2 + \frac{1}{24}x^4$
6	$-\cos x$	-1	$1 - \frac{1}{2}x^2 + \frac{1}{24}x^4 - \frac{1}{720}x^6$

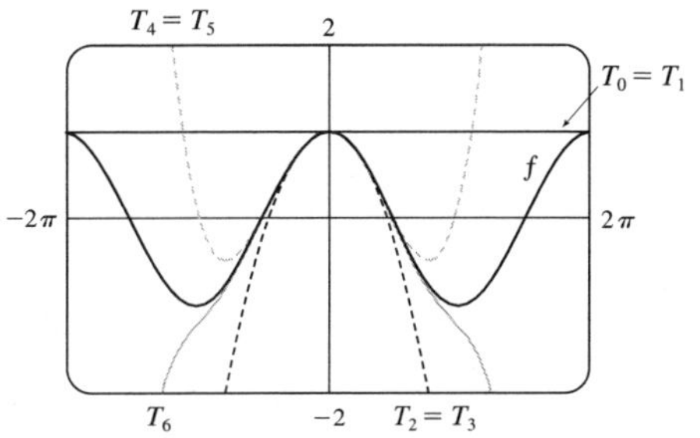

(b)

x	f	$T_0 = T_1$	$T_2 = T_3$	$T_4 = T_5$	T_6
$\frac{\pi}{4}$	0.7071	1	0.6916	0.7074	0.7071
$\frac{\pi}{2}$	0	1	-0.2337	0.0200	-0.0009
π	-1	1	-3.9348	0.1239	-1.2114

(c) As n increases, $T_n(x)$ is a good approximation to $f(x)$ on a larger and larger interval.

3.

n	$f^{(n)}(x)$	$f^{(n)}\left(\frac{\pi}{6}\right)$
0	$\sin x$	$\frac{1}{2}$
1	$\cos x$	$\frac{\sqrt{3}}{2}$
2	$-\sin x$	$-\frac{1}{2}$
3	$-\cos x$	$-\frac{\sqrt{3}}{2}$

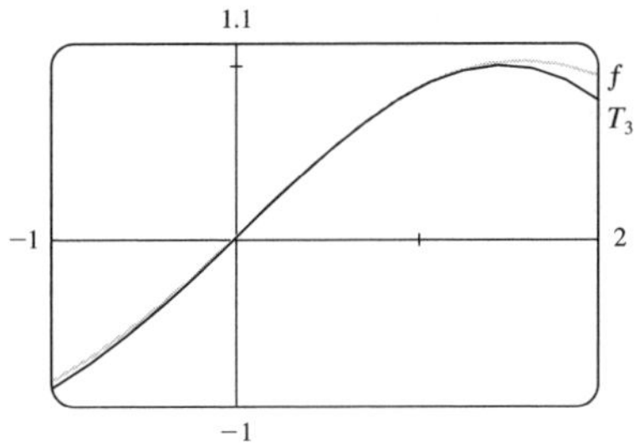

$$T_3(x) = \sum_{n=0}^{3} \frac{f^{(n)}\left(\frac{\pi}{6}\right)}{n!}\left(x - \frac{\pi}{6}\right)^n = \frac{1}{2} + \frac{\sqrt{3}}{2}\left(x - \frac{\pi}{6}\right) - \frac{1}{4}\left(x - \frac{\pi}{6}\right)^2 - \frac{\sqrt{3}}{12}\left(x - \frac{\pi}{6}\right)^3$$

5.

n	$f^{(n)}(x)$	$f^{(n)}(0)$
0	$\arcsin x$	0
1	$1/\sqrt{1-x^2}$	1
2	$x/(1-x^2)^{3/2}$	0
3	$(2x^2+1)/(1-x^2)^{5/2}$	1

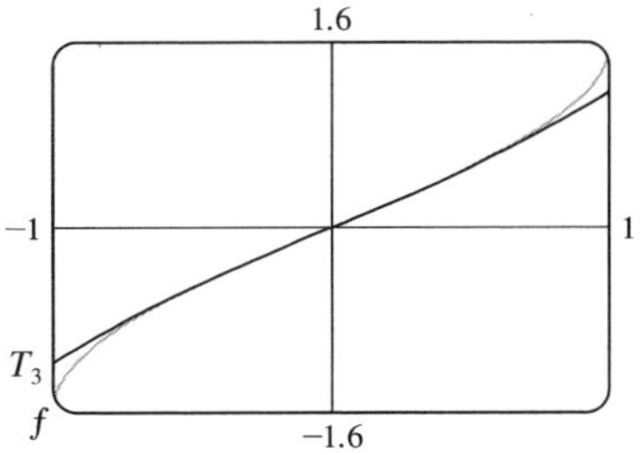

$$T_3(x) = \sum_{n=0}^{3} \frac{f^{(n)}(0)}{n!}x^n = x + \frac{x^3}{6}$$

7.

n	$f^{(n)}(x)$	$f^{(n)}(0)$
0	xe^{-2x}	0
1	$(1-2x)e^{-2x}$	1
2	$4(x-1)e^{-2x}$	-4
3	$4(3-2x)e^{-2x}$	12

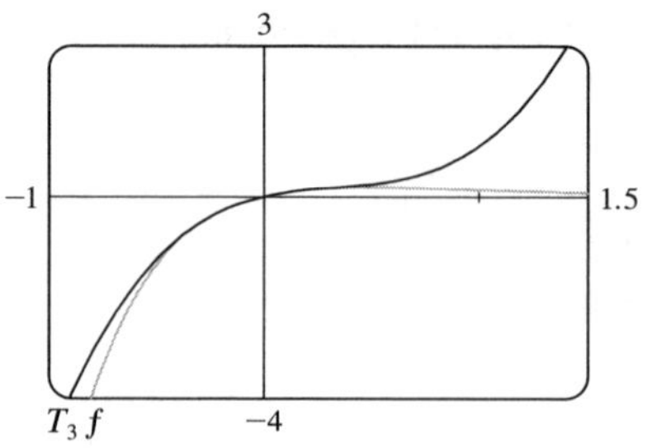

$$T_3(x) = \sum_{n=0}^{3} \frac{f^{(n)}(0)}{n!} x^n = \tfrac{0}{1} \cdot 1 + \tfrac{1}{1}x^1 + \tfrac{-4}{2}x^2 + \tfrac{12}{6}x^3 = x - 2x^2 + 2x^3$$

9. In Maple, we can find the Taylor polynomials by the following method: first define `f:=sec(x);` and then set `T2:=convert(taylor(f,x=0,3),polynom);`, `T4:=convert(taylor(f,x=0,5),polynom);`, etc. (The third argument in the `taylor` function is one more than the degree of the desired polynomial). We must `convert` to the type `polynom` because the output of the `taylor` function contains an error term which we do not want. In Mathematica, we use `Tn:=Normal[Series[f,{x,0,n}]]`, with `n=2, 4`, etc. Note that in Mathematica, the "degree" argument is the same as the degree of the desired polynomial. In Derive, author $\sec x$, then enter `Calculus,Taylor,8,0;` and then simplify the expression. The eighth Taylor polynomial is

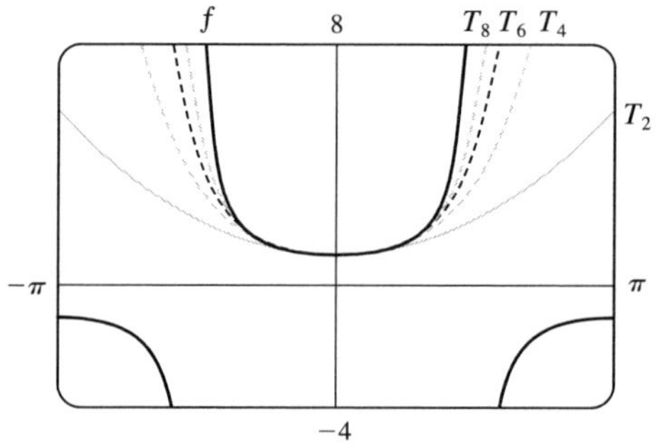

$T_8(x) = 1 + \frac{1}{2}x^2 + \frac{5}{24}x^4 + \frac{61}{720}x^6 + \frac{277}{8064}x^8$.

11.

n	$f^{(n)}(x)$	$f^{(n)}(4)$
0	$\sqrt{x}$	2
1	$\frac{1}{2}x^{-1/2}$	$\frac{1}{4}$
2	$-\frac{1}{4}x^{-3/2}$	$-\frac{1}{32}$
3	$\frac{3}{8}x^{-5/2}$	

(a) $f(x) = \sqrt{x} \approx T_2(x) = 2 + \frac{1}{4}(x-4) - \frac{1/32}{2!}(x-4)^2 = 2 + \frac{1}{4}(x-4) - \frac{1}{64}(x-4)^2$

(b) $|R_2(x)| \le \dfrac{M}{3!}|x-4|^3$, where $|f'''(x)| \le M$. Now $4 \le x \le 4.2 \;\Rightarrow\; |x-4| \le 0.2 \;\Rightarrow\; |x-4|^3 \le 0.008$.

Since $f'''(x)$ is decreasing on $[4, 4.2]$, we can take $M = |f'''(4)| = \frac{3}{8}4^{-5/2} = \frac{3}{256}$, so

$|R_2(x)| \le \frac{3/256}{6}(0.008) = \frac{0.008}{512} = 0.000\,015\,625$.

(c)

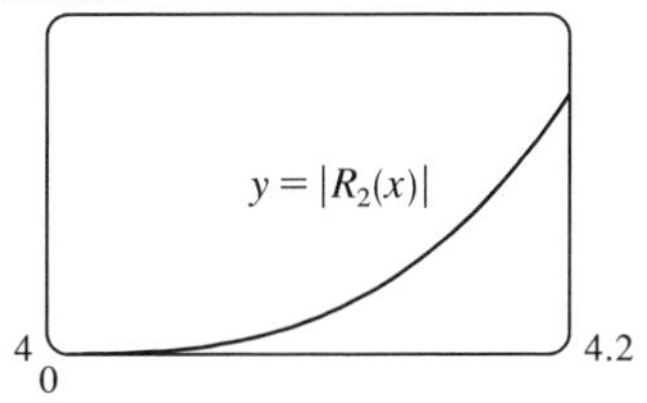

From the graph of $|R_2(x)| = |\sqrt{x} - T_2(x)|$, it seems that the error is less than 1.52×10^{-5} on $[4, 4.2]$.

13.

n	$f^{(n)}(x)$	$f^{(n)}(1)$
0	$x^{2/3}$	1
1	$\frac{2}{3}x^{-1/3}$	$\frac{2}{3}$
2	$-\frac{2}{9}x^{-4/3}$	$-\frac{2}{9}$
3	$\frac{8}{27}x^{-7/3}$	$\frac{8}{27}$
4	$-\frac{56}{81}x^{-10/3}$	

(a) $f(x) = x^{2/3} \approx T_3(x) = 1 + \frac{2}{3}(x-1) - \frac{2/9}{2!}(x-1)^2 + \frac{8/27}{3!}(x-1)^3 = 1 + \frac{2}{3}(x-1) - \frac{1}{9}(x-1)^2 + \frac{4}{81}(x-1)^3$

(b) $|R_3(x)| \le \frac{M}{4!}|x-1|^4$, where $\left|f^{(4)}(x)\right| \le M$. Now $0.8 \le x \le 1.2 \;\Rightarrow\; |x-1| \le 0.2 \;\Rightarrow\; |x-1|^4 \le 0.0016$.

Since $\left|f^{(4)}(x)\right|$ is decreasing on $[0.8, 1.2]$, we can take $M = \left|f^{(4)}(0.8)\right| = \frac{56}{81}(0.8)^{-10/3}$, so

$$|R_3(x)| \le \frac{\frac{56}{81}(0.8)^{-10/3}}{24}(0.0016) \approx 0.000\,096\,97.$$

(c)

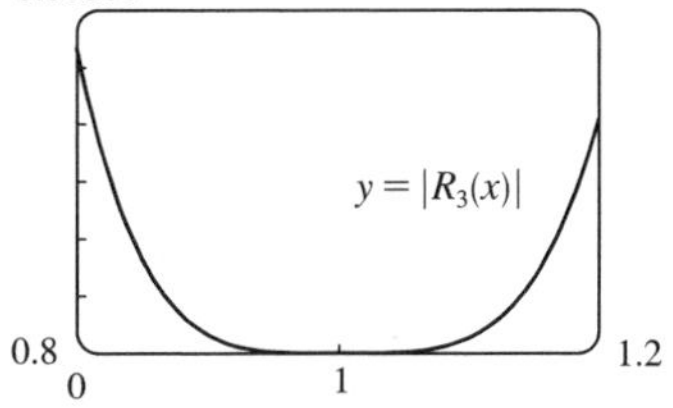

From the graph of $|R_3(x)| = \left|x^{2/3} - T_3(x)\right|$, it seems that the error is less than $0.000\,053\,3$ on $[0.8, 1.2]$.

15.

n	$f^{(n)}(x)$	$f^{(n)}(0)$
0	e^{x^2}	1
1	$e^{x^2}(2x)$	0
2	$e^{x^2}(2+4x^2)$	2
3	$e^{x^2}(12x+8x^3)$	0
4	$e^{x^2}(12+48x^2+16x^4)$	

(a) $f(x) = e^{x^2} \approx T_3(x) = 1 + \frac{2}{2!}x^2 = 1 + x^2$

(b) $|R_3(x)| \le \dfrac{M}{4!}|x|^4$, where $\left|f^{(4)}(x)\right| \le M$. Now $0 \le x \le 0.1 \;\Rightarrow\; x^4 \le (0.1)^4$, and letting $x = 0.1$ gives

$$|R_3(x)| \le \frac{e^{0.01}(12 + 0.48 + 0.0016)}{24}(0.1)^4 \approx 0.00006.$$

(c)

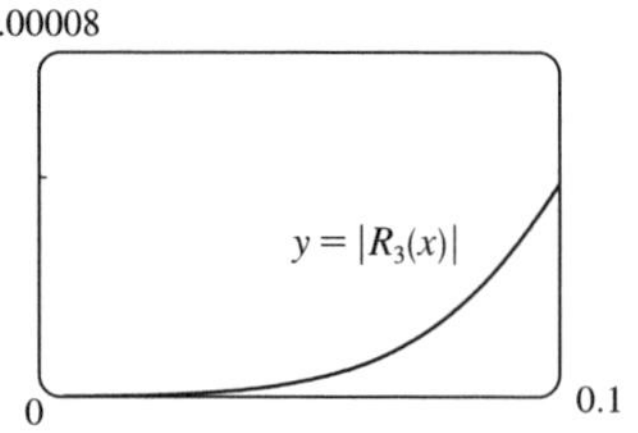

From the graph of $|R_3(x)| = \left| e^{x^2} - (1 + x^2) \right|$, it appears that the error is less than 0.000 051 on $[0, 0.1]$.

17.

n	$f^{(n)}(x)$	$f^{(n)}(0)$
0	$x \sin x$	0
1	$\sin x + x \cos x$	0
2	$2 \cos x - x \sin x$	2
3	$-3 \sin x - x \cos x$	0
4	$-4 \cos x + x \sin x$	-4
5	$5 \sin x + x \cos x$	

(a) $f(x) = x \sin x \approx T_4(x) = \dfrac{2}{2!}(x-0)^2 + \dfrac{-4}{4!}(x-0)^4 = x^2 - \dfrac{1}{6}x^4$

(b) $|R_4(x)| \le \dfrac{M}{5!}|x|^5$, where $\left| f^{(5)}(x) \right| \le M$. Now $-1 \le x \le 1 \;\Rightarrow\; |x| \le 1$, and a graph of $f^{(5)}(x)$ shows that $\left| f^{(5)}(x) \right| \le 5$ for $-1 \le x \le 1$. Thus, we can take $M = 5$ and get $|R_4(x)| \le \dfrac{5}{5!} \cdot 1^5 = \dfrac{1}{24} = 0.041\overline{6}$.

(c)

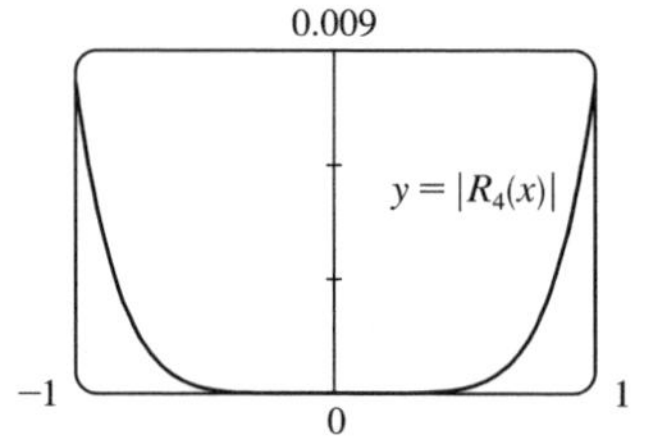

From the graph of $|R_4(x)| = |x \sin x - T_4(x)|$, it seems that the error is less than 0.0082 on $[-1, 1]$.

19. From Exercise 3, $\sin x = \frac{1}{2} + \frac{\sqrt{3}}{2}\left(x - \frac{\pi}{6}\right) - \frac{1}{4}\left(x - \frac{\pi}{6}\right)^2 - \frac{\sqrt{3}}{12}\left(x - \frac{\pi}{6}\right)^3 + R_3(x)$, where $|R_3(x)| \le \dfrac{M}{4!}\left|x - \frac{\pi}{6}\right|^4$ with $\left| f^{(4)}(x) \right| = |\sin x| \le M = 1$. Now $x = 35° = (30° + 5°) = \left(\frac{\pi}{6} + \frac{\pi}{36}\right)$ radians, so the error is $\left| R_3\left(\frac{\pi}{36}\right) \right| \le \dfrac{\left(\frac{\pi}{36}\right)^4}{4!} < 0.000\,003$. Therefore, to five decimal places, $\sin 35° \approx \frac{1}{2} + \frac{\sqrt{3}}{2}\left(\frac{\pi}{36}\right) - \frac{1}{4}\left(\frac{\pi}{36}\right)^2 - \frac{\sqrt{3}}{12}\left(\frac{\pi}{36}\right)^3 \approx 0.57358$.

21. All derivatives of e^x are e^x, so $|R_n(x)| \le \dfrac{e^x}{(n+1)!}|x|^{n+1}$, where $0 < x < 0.1$. Letting $x = 0.1$, $R_n(0.1) \le \dfrac{e^{0.1}}{(n+1)!}(0.1)^{n+1} < 0.00001$, and by trial and error we find that $n = 3$ satisfies this inequality since $R_3(0.1) < 0.000\,004\,6$. Thus, by adding the four terms of the Maclaurin series for e^x corresponding to $n = 0, 1, 2$, and 3, we can estimate $e^{0.1}$ to within 0.00001. (In fact, this sum is $1.1051\overline{6}$ and $e^{0.1} \approx 1.10517$.)

23. $\sin x = x - \frac{1}{3!}x^3 + \frac{1}{5!}x^5 - \cdots$. By the Alternating Series Estimation Theorem, the error in the approximation $\sin x = x - \frac{1}{3!}x^3$ is less than $\left|\frac{1}{5!}x^5\right| < 0.01 \Leftrightarrow |x^5| < 120(0.01) \Leftrightarrow$ $|x| < (1.2)^{1/5} \approx 1.037$. The curves $y = x - \frac{1}{6}x^3$ and $y = \sin x - 0.01$ intersect at $x \approx 1.043$, so the graph confirms our estimate. Since both the sine function and the given approximation are odd functions, we need to check the estimate only for $x > 0$. Thus, the desired range of values for x is $-1.037 < x < 1.037$.

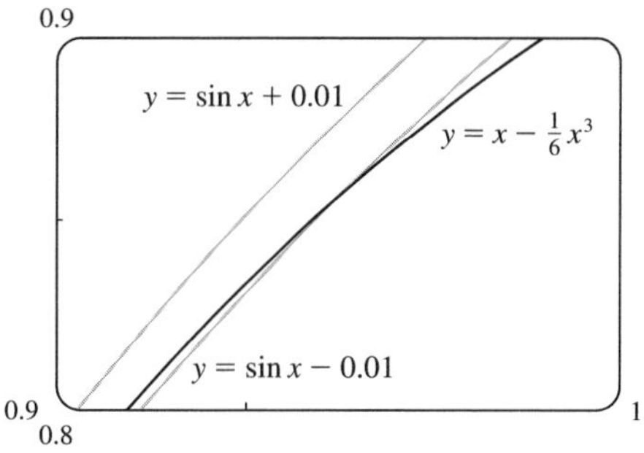

25. Let $s(t)$ be the position function of the car, and for convenience set $s(0) = 0$. The velocity of the car is $v(t) = s'(t)$ and the acceleration is $a(t) = s''(t)$, so the second degree Taylor polynomial is $T_2(t) = s(0) + v(0)t + \frac{a(0)}{2}t^2 = 20t + t^2$. We estimate the distance travelled during the next second to be $s(1) \approx T_2(1) = 20 + 1 = 21$ m. The function $T_2(t)$ would not be accurate over a full minute, since the car could not possibly maintain an acceleration of 2 m/s^2 for that long (if it did, its final speed would be $140\text{ m/s} = 504\text{ km/h}$!)

27. $E = \frac{q}{D^2} - \frac{q}{(D+d)^2} = \frac{q}{D^2} - \frac{q}{D^2(1+d/D)^2} = \frac{q}{D^2}\left[1 - \left(1 + \frac{d}{D}\right)^{-2}\right]$.

We use the Binomial Series to expand $(1 + d/D)^{-2}$:

$$E = \frac{q}{D^2}\left[1 - \left(1 - 2\left(\frac{d}{D}\right) + \frac{2\cdot 3}{2!}\left(\frac{d}{D}\right)^2 - \frac{2\cdot 3\cdot 4}{3!}\left(\frac{d}{D}\right)^3 + \cdots\right)\right]$$

$$= \frac{q}{D^2}\left[2\left(\frac{d}{D}\right) - 3\left(\frac{d}{D}\right)^2 + 4\left(\frac{d}{D}\right)^3 - \cdots\right] \approx \frac{q}{D^2}\cdot 2\left(\frac{d}{D}\right) = 2qd\cdot\frac{1}{D^3}$$

when D is much larger than d; that is, when P is far away from the dipole.

29. (a) L is the length of the arc subtended by the angle θ, so $L = R\theta \Rightarrow$ $\theta = L/R$. Now $\sec\theta = (R+C)/R \Rightarrow R\sec\theta = R + C \Rightarrow$ $C = R\sec\theta - R = R\sec(L/R) - R$.

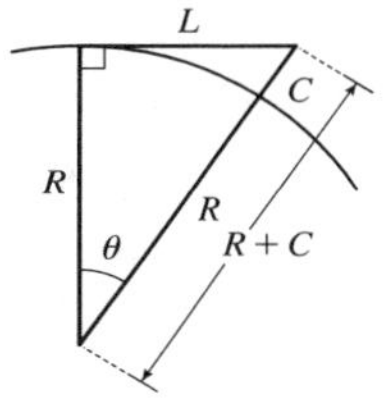

(b) From Exercise 9, $\sec x \approx T_4(x) = 1 + \frac{1}{2}x^2 + \frac{5}{24}x^4$. By part (a),

$$C \approx R\left[1 + \frac{1}{2}\left(\frac{L}{R}\right)^2 + \frac{5}{24}\left(\frac{L}{R}\right)^4\right] - R = R + \frac{1}{2}R\cdot\frac{L^2}{R^2} + \frac{5}{24}R\cdot\frac{L^4}{R^4} - R = \frac{L^2}{2R} + \frac{5L^4}{24R^3}.$$

(c) Taking $L = 100$ km and $R = 6370$ km, the formula in part (a) says that

$C = R\sec(L/R) - R = 6370\sec(100/6370) - 6370 \approx 0.785\,009\,965\,44$ km. The formula in part (b) says that

$$C \approx \frac{L^2}{2R} + \frac{5L^4}{24R^3} = \frac{100^2}{2\cdot 6370} + \frac{5\cdot 100^4}{24\cdot 6370^3} \approx 0.785\,009\,957\,36 \text{ km}.$$

The difference between these two results is only 0.000 000 008 08 km, or 0.000 008 08 m!

31. Using $f(x) = T_n(x) + R_n(x)$ with $n = 1$ and $x = r$, we have $f(r) = T_1(r) + R_1(r)$, where T_1 is the first-degree Taylor polynomial of f at a. Because $a = x_n$, $f(r) = f(x_n) + f'(x_n)(r - x_n) + R_1(r)$. But r is a root of f, so $f(r) = 0$ and we have $0 = f(x_n) + f'(x_n)(r - x_n) + R_1(r)$. Taking the first two terms to the left side gives us

$f'(x_n)(x_n - r) - f(x_n) = R_1(r)$. Dividing by $f'(x_n)$, we get $x_n - r - \dfrac{f(x_n)}{f'(x_n)} = \dfrac{R_1(r)}{f'(x_n)}$. By the formula for Newton's method, the left side of the preceding equation is $x_{n+1} - r$, so $|x_{n+1} - r| = \left|\dfrac{R_1(r)}{f'(x_n)}\right|$. Taylor's Inequality gives us

$|R_1(r)| \le \dfrac{|f''(r)|}{2!}|r - x_n|^2$. Combining this inequality with the facts $|f''(x)| \le M$ and $|f'(x)| \ge K$ gives us

$|x_{n+1} - r| \le \dfrac{M}{2K}|x_n - r|^2$.

8 Review

CONCEPT CHECK

1. (a) See Definition 8.1.1.

(b) See Definition 8.2.2.

(c) The terms of the sequence $\{a_n\}$ approach 3 as n becomes large.

(d) By adding sufficiently many terms of the series, we can make the partial sums as close to 3 as we like.

2. (a) See the definition on page 563.

(b) A sequence is monotonic if it is either increasing or decreasing.

(c) By Theorem 8.1.7, every bounded, monotonic sequence is convergent.

3. (a) See (4) in Section 8.2.

(b) The p-series $\sum\limits_{n=1}^{\infty} \dfrac{1}{n^p}$ is convergent if $p > 1$.

4. If $\sum a_n = 3$, then $\lim\limits_{n\to\infty} a_n = 0$ and $\lim\limits_{n\to\infty} s_n = 3$.

5. (a) See the Test for Divergence on page 572.

(b) See the Integral Test on page 578.

(c) See the Comparison Test on page 580.

(d) See the Limit Comparison Test on page 582.

(e) See the Alternating Series Test on page 587.

(f) See the Ratio Test on page 591.

6. (a) A series $\sum a_n$ is called *absolutely convergent* if the series of absolute values $\sum |a_n|$ is convergent.

(b) If a series $\sum a_n$ is absolutely convergent, then it is convergent.

7. (a) Use (4) in Section 8.3.

(b) See Example 8 in Section 8.3.

(c) By adding terms until you reach the desired accuracy given by the Alternating Series Estimation Theorem on page 588.

8. (a) $\sum_{n=0}^{\infty} c_n(x - a)^n$

(b) Given the power series $\sum_{n=0}^{\infty} c_n(x - a)^n$, the radius of convergence is:

(i) 0 if the series converges only when $x = a$

(ii) ∞ if the series converges for all x, or

(iii) a positive number R such that the series converges if $|x - a| < R$ and diverges if $|x - a| > R$.

(c) The interval of convergence of a power series is the interval that consists of all values of x for which the series converges. Corresponding to the cases in part (b), the interval of convergence is: (i) the single point $\{a\}$, (ii) all real numbers, that is, the real number line $(-\infty, \infty)$, or (iii) an interval with endpoints $a - R$ and $a + R$ which can contain neither, either, or both of the endpoints. In this case, we must test the series for convergence at each endpoint to determine the interval of convergence.

9. (a), (b) See Theorem 8.6.2.

10. (a) $T_n(x) = \sum\limits_{i=0}^{n} \dfrac{f^{(i)}(a)}{i!}(x-a)^i$

(b) $\sum\limits_{n=0}^{\infty} \dfrac{f^{(n)}(a)}{n!}(x-a)^n$

(c) $\sum\limits_{n=0}^{\infty} \dfrac{f^{(n)}(0)}{n!}x^n$ [$a = 0$ in part (b)]

(d) See Theorem 8.7.8.

(e) See Taylor's Inequality (8.7.9).

11. (a) – (e) See the table on page 612.

12. See the Binomial Series (8.8.2) for the expansion. The radius of convergence for the binomial series is 1.

TRUE-FALSE QUIZ

1. False. See Note 2 after Theorem 8.2.6.

3. True. If $\lim\limits_{n\to\infty} a_n = L$, then given any $\varepsilon > 0$, we can find a positive integer N such that $|a_n - L| < \varepsilon$ whenever $n > N$. If $n > N$, then $2n + 1 > N$ and $|a_{2n+1} - L| < \varepsilon$. Thus, $\lim\limits_{n\to\infty} a_{2n+1} = L$.

5. False. For example, take $c_n = (-1)^n/(n6^n)$.

7. False, since $\lim\limits_{n\to\infty}\left|\dfrac{a_{n+1}}{a_n}\right| = \lim\limits_{n\to\infty}\left|\dfrac{1}{(n+1)^3}\cdot\dfrac{n^3}{1}\right| = \lim\limits_{n\to\infty}\left|\dfrac{n^3}{(n+1)^3}\cdot\dfrac{1/n^3}{1/n^3}\right| = \lim\limits_{n\to\infty}\dfrac{1}{(1+1/n)^3} = 1.$

9. False. See the note after Example 4 in Section 8.3.

11. True. See (6) in Section 8.1.

13. True. By Theorem 8.7.5 the coefficient of x^3 is $\dfrac{f'''(0)}{3!} = \dfrac{1}{3} \Rightarrow f'''(0) = 2$.

Or: Use Theorem 8.6.2 to differentiate f three times.

15. False. For example, let $a_n = b_n = (-1)^n$. Then $\{a_n\}$ and $\{b_n\}$ are divergent, but $a_n b_n = 1$, so $\{a_n b_n\}$ is convergent.

17. True by Theorem 8.4.1. $\left[\sum (-1)^n a_n\right.$ is absolutely convergent and hence convergent.$\left.\right]$

EXERCISES

1. $\left\{\dfrac{2+n^3}{1+2n^3}\right\}$ converges since $\lim\limits_{n\to\infty}\dfrac{2+n^3}{1+2n^3} = \lim\limits_{n\to\infty}\dfrac{2/n^3+1}{1/n^3+2} = \dfrac{1}{2}$.

3. $\lim\limits_{n\to\infty} a_n = \lim\limits_{n\to\infty}\dfrac{n^3}{1+n^2} = \lim\limits_{n\to\infty}\dfrac{n}{1/n^2+1} = \infty$, so the sequence diverges.

5. $|a_n| = \left|\dfrac{n\sin n}{n^2+1}\right| \le \dfrac{n}{n^2+1} < \dfrac{1}{n}$, so $|a_n| \to 0$ as $n \to \infty$. Thus, $\lim\limits_{n\to\infty} a_n = 0$. The sequence $\{a_n\}$ is convergent.

7. $\left\{\left(1+\dfrac{3}{n}\right)^{4n}\right\}$ is convergent. Let $y=\left(1+\dfrac{3}{x}\right)^{4x}$. Then

$$\lim_{x\to\infty}\ln y=\lim_{x\to\infty}4x\ln(1+3/x)=\lim_{x\to\infty}\frac{\ln(1+3/x)}{1/(4x)}\overset{\text{H}}{=}\lim_{x\to\infty}\frac{\dfrac{1}{1+3/x}\left(-\dfrac{3}{x^2}\right)}{-1/(4x^2)}=\lim_{x\to\infty}\frac{12}{1+3/x}=12$$

so $\displaystyle\lim_{x\to\infty}y=\lim_{n\to\infty}\left(1+\frac{3}{n}\right)^{4n}=e^{12}$.

Or: Use Exercise 4.5.38.

9. $\dfrac{n}{n^3+1}<\dfrac{n}{n^3}=\dfrac{1}{n^2}$, so $\displaystyle\sum_{n=1}^{\infty}\frac{n}{n^3+1}$ converges by the Comparison Test with the convergent p-series $\displaystyle\sum_{n=1}^{\infty}\frac{1}{n^2}$ $(p=2>1)$.

11. $\displaystyle\lim_{n\to\infty}\left|\frac{a_{n+1}}{a_n}\right|=\lim_{n\to\infty}\left[\frac{(n+1)^3}{5^{n+1}}\cdot\frac{5^n}{n^3}\right]=\lim_{n\to\infty}\left(1+\frac{1}{n}\right)^3\cdot\frac{1}{5}=\frac{1}{5}<1$, so $\displaystyle\sum_{n=1}^{\infty}\frac{n^3}{5^n}$ converges by the Ratio Test.

13. Let $f(x)=\dfrac{1}{x\sqrt{\ln x}}$. Then f is continuous, positive, and decreasing on $[2,\infty)$, so the Integral Test applies.

$$\int_2^{\infty}f(x)\,dx=\lim_{t\to\infty}\int_2^t\frac{1}{x\sqrt{\ln x}}\,dx\quad\left[\begin{aligned}u&=\ln x,\\ du&=\frac{1}{x}\,dx\end{aligned}\right]\quad=\lim_{t\to\infty}\int_{\ln 2}^{\ln t}u^{-1/2}\,du$$

$$=\lim_{t\to\infty}\left[2\sqrt{u}\right]_{\ln 2}^{\ln t}=\lim_{t\to\infty}\left(2\sqrt{\ln t}-2\sqrt{\ln 2}\right)=\infty,\text{ so the series }\sum_{n=2}^{\infty}\frac{1}{n\sqrt{\ln n}}\text{ diverges.}$$

15. $b_n=\dfrac{\sqrt{n}}{n+1}>0$, $\{b_n\}$ is decreasing, and $\displaystyle\lim_{n\to\infty}b_n=0$, so the series $\displaystyle\sum_{n=1}^{\infty}(-1)^{n-1}\frac{\sqrt{n}}{n+1}$ converges by the Alternating Series Test.

17. $\displaystyle\lim_{n\to\infty}\left|\frac{a_{n+1}}{a_n}\right|=\lim_{n\to\infty}\frac{1\cdot3\cdot5\cdot\cdots\cdot(2n-1)(2n+1)}{5^{n+1}(n+1)!}\cdot\frac{5^n n!}{1\cdot3\cdot5\cdot\cdots\cdot(2n-1)}=\lim_{n\to\infty}\frac{2n+1}{5(n+1)}=\frac{2}{5}<1$, so the series converges by the Ratio Test.

19. $\dfrac{2^{2n+1}}{5^n}=\dfrac{2^{2n}\cdot2^1}{5^n}=\dfrac{(2^2)^n\cdot2}{5^n}=2\left(\dfrac{4}{5}\right)^n$, so $\displaystyle\sum_{n=1}^{\infty}\frac{2^{2n+1}}{5^n}=2\sum_{n=1}^{\infty}\left(\frac{4}{5}\right)^n$ is a geometric series with $a=\dfrac{8}{5}$ and $r=\dfrac{4}{5}$. Since $|r|=\dfrac{4}{5}<1$, the series converges to $\dfrac{a}{1-r}=\dfrac{8/5}{1-4/5}=\dfrac{8/5}{1/5}=8$.

21.
$$\begin{aligned}\sum_{n=1}^{\infty}\left[\tan^{-1}(n+1)-\tan^{-1}n\right]&=\lim_{n\to\infty}s_n\\&=\lim_{n\to\infty}\left[(\tan^{-1}2-\tan^{-1}1)+(\tan^{-1}3-\tan^{-1}2)+\cdots\right.\\&\qquad\left.+(\tan^{-1}(n+1)-\tan^{-1}n)\right]\\&=\lim_{n\to\infty}\left[\tan^{-1}(n+1)-\tan^{-1}1\right]=\tfrac{\pi}{2}-\tfrac{\pi}{4}=\tfrac{\pi}{4}\end{aligned}$$

23. $1.2345345345\ldots=1.2+0.0\overline{345}=\dfrac{12}{10}+\dfrac{345/10\,000}{1-1/1000}=\dfrac{12}{10}+\dfrac{345}{9990}=\dfrac{4111}{3330}$

25. $\displaystyle\sum_{n=1}^{\infty}\frac{(-1)^{n+1}}{n^5}=1-\frac{1}{32}+\frac{1}{243}-\frac{1}{1024}+\frac{1}{3125}-\frac{1}{7776}+\frac{1}{16\,807}-\frac{1}{32\,768}+\cdots$.

Since $b_8=\dfrac{1}{8^5}=\dfrac{1}{32\,768}<0.000\,031$, $\displaystyle\sum_{n=1}^{\infty}\frac{(-1)^{n+1}}{n^5}\approx\sum_{n=1}^{7}\frac{(-1)^{n+1}}{n^5}\approx0.9721$.

27. $\displaystyle\sum_{n=1}^{\infty}\frac{1}{2+5^n}\approx\sum_{n=1}^{8}\frac{1}{2+5^n}\approx0.189\,762\,24$. To estimate the error, note that $\dfrac{1}{2+5^n}<\dfrac{1}{5^n}$, so the remainder term is

$R_8=\displaystyle\sum_{n=9}^{\infty}\frac{1}{2+5^n}<\sum_{n=9}^{\infty}\frac{1}{5^n}=\frac{1/5^9}{1-1/5}=6.4\times10^{-7}$ [geometric series with $a=\frac{1}{5^9}$ and $r=\frac{1}{5}$].

29. Use the Limit Comparison Test. $\lim\limits_{n\to\infty}\left|\dfrac{\left(\frac{n+1}{n}\right)a_n}{a_n}\right| = \lim\limits_{n\to\infty}\dfrac{n+1}{n} = \lim\limits_{n\to\infty}\left(1+\dfrac{1}{n}\right) = 1 > 0.$

Since $\sum |a_n|$ is convergent, so is $\sum\left|\left(\dfrac{n+1}{n}\right)a_n\right|$, by the Limit Comparison Test.

31. $\lim\limits_{n\to\infty}\left|\dfrac{a_{n+1}}{a_n}\right| = \lim\limits_{n\to\infty}\left[\dfrac{|x+2|^{n+1}}{(n+1)\,4^{n+1}}\cdot\dfrac{n\,4^n}{|x+2|^n}\right] = \lim\limits_{n\to\infty}\left[\dfrac{n}{n+1}\,\dfrac{|x+2|}{4}\right] = \dfrac{|x+2|}{4} < 1 \;\Leftrightarrow\; |x+2| < 4$, so $R = 4$.

$|x+2| < 4 \;\Leftrightarrow\; -4 < x+2 < 4 \;\Leftrightarrow\; -6 < x < 2$. If $x = -6$, then the series $\sum\limits_{n=1}^{\infty}\dfrac{(x+2)^n}{n4^n}$ becomes $\sum\limits_{n=1}^{\infty}\dfrac{(-4)^n}{n4^n} = \sum\limits_{n=1}^{\infty}\dfrac{(-1)^n}{n}$, the alternating harmonic series, which converges by the Alternating Series Test. When $x = 2$, the series becomes the harmonic series $\sum\limits_{n=1}^{\infty}\dfrac{1}{n}$, which diverges. Thus, $I = [-6, 2)$.

33. $\lim\limits_{n\to\infty}\left|\dfrac{a_{n+1}}{a_n}\right| = \lim\limits_{n\to\infty}\left|\dfrac{2^{n+1}(x-3)^{n+1}}{\sqrt{n+4}}\cdot\dfrac{\sqrt{n+3}}{2^n(x-3)^n}\right| = 2\,|x-3|\lim\limits_{n\to\infty}\sqrt{\dfrac{n+3}{n+4}} = 2\,|x-3| < 1 \;\Leftrightarrow\; |x-3| < \frac{1}{2}$, so $R = \frac{1}{2}$. $|x-3| < \frac{1}{2} \;\Leftrightarrow\; -\frac{1}{2} < x-3 < \frac{1}{2} \;\Leftrightarrow\; \frac{5}{2} < x < \frac{7}{2}$. For $x = \frac{7}{2}$, the series $\sum\limits_{n=1}^{\infty}\dfrac{2^n(x-3)^n}{\sqrt{n+3}}$ becomes $\sum\limits_{n=0}^{\infty}\dfrac{1}{\sqrt{n+3}} = \sum\limits_{n=3}^{\infty}\dfrac{1}{n^{1/2}}$, which diverges ($p = \frac{1}{2} \le 1$), but for $x = \frac{5}{2}$, we get $\sum\limits_{n=0}^{\infty}\dfrac{(-1)^n}{\sqrt{n+3}}$, which is a convergent alternating series, so $I = \left[\frac{5}{2}, \frac{7}{2}\right)$.

35.

n	$f^{(n)}(x)$	$f^{(n)}\left(\frac{\pi}{6}\right)$
0	$\sin x$	$\frac{1}{2}$
1	$\cos x$	$\frac{\sqrt{3}}{2}$
2	$-\sin x$	$-\frac{1}{2}$
3	$-\cos x$	$-\frac{\sqrt{3}}{2}$
4	$\sin x$	$\frac{1}{2}$
$\vdots$	$\vdots$	$\vdots$

$$\begin{aligned}\sin x &= f\left(\frac{\pi}{6}\right) + f'\left(\frac{\pi}{6}\right)\left(x-\frac{\pi}{6}\right) + \frac{f''\left(\frac{\pi}{6}\right)}{2!}\left(x-\frac{\pi}{6}\right)^2 + \frac{f^{(3)}\left(\frac{\pi}{6}\right)}{3!}\left(x-\frac{\pi}{6}\right)^3 + \frac{f^{(4)}\left(\frac{\pi}{6}\right)}{4!}\left(x-\frac{\pi}{6}\right)^4 + \cdots \\ &= \frac{1}{2}\left[1 - \frac{1}{2!}\left(x-\frac{\pi}{6}\right)^2 + \frac{1}{4!}\left(x-\frac{\pi}{6}\right)^4 - \cdots\right] + \frac{\sqrt{3}}{2}\left[\left(x-\frac{\pi}{6}\right) - \frac{1}{3!}\left(x-\frac{\pi}{6}\right)^3 + \cdots\right] \\ &= \frac{1}{2}\sum_{n=0}^{\infty}(-1)^n\frac{1}{(2n)!}\left(x-\frac{\pi}{6}\right)^{2n} + \frac{\sqrt{3}}{2}\sum_{n=0}^{\infty}(-1)^n\frac{1}{(2n+1)!}\left(x-\frac{\pi}{6}\right)^{2n+1}\end{aligned}$$

or

$$\frac{1}{2}\sum_{n=0}^{\infty}(-1)^n\left[\frac{1}{(2n)!}\left(x-\frac{\pi}{6}\right)^{2n} + \frac{\sqrt{3}}{(2n+1)!}\left(x-\frac{\pi}{6}\right)^{2n+1}\right]$$

37. $\dfrac{1}{1+x} = \dfrac{1}{1-(-x)} = \sum\limits_{n=0}^{\infty}(-x)^n = \sum\limits_{n=0}^{\infty}(-1)^n x^n$ for $|x| < 1 \;\Rightarrow\; \dfrac{x^2}{1+x} = \sum\limits_{n=0}^{\infty}(-1)^n x^{n+2}$ with $R = 1$.

39. $\dfrac{1}{1-x} = \sum\limits_{n=0}^{\infty}x^n$ for $|x| < 1 \;\Rightarrow\; \ln(1-x) = -\displaystyle\int\frac{dx}{1-x} = -\int\sum_{n=0}^{\infty}x^n\,dx = C - \sum_{n=0}^{\infty}\frac{x^{n+1}}{n+1}.$

$\ln(1-0) = C - 0 \;\Rightarrow\; C = 0 \;\Rightarrow\; \ln(1-x) = -\sum\limits_{n=0}^{\infty}\dfrac{x^{n+1}}{n+1} = \sum\limits_{n=1}^{\infty}\dfrac{-x^n}{n}$ with $R = 1$.

41. $\sin x = \sum\limits_{n=0}^{\infty} \dfrac{(-1)^n x^{2n+1}}{(2n+1)!} \Rightarrow \sin(x^4) = \sum\limits_{n=0}^{\infty} \dfrac{(-1)^n (x^4)^{2n+1}}{(2n+1)!} = \sum\limits_{n=0}^{\infty} \dfrac{(-1)^n x^{8n+4}}{(2n+1)!}$ for all x, so the radius of convergence is ∞.

43. $f(x) = \dfrac{1}{\sqrt[4]{16-x}} = \dfrac{1}{\sqrt[4]{16(1-x/16)}} = \dfrac{1}{\sqrt[4]{16}\left(1-\frac{1}{16}x\right)^{1/4}} = \frac{1}{2}\left(1-\frac{1}{16}x\right)^{-1/4}$

$$= \frac{1}{2}\left[1+\left(-\frac{1}{4}\right)\left(-\frac{x}{16}\right)+\frac{\left(-\frac{1}{4}\right)\left(-\frac{5}{4}\right)}{2!}\left(-\frac{x}{16}\right)^2+\frac{\left(-\frac{1}{4}\right)\left(-\frac{5}{4}\right)\left(-\frac{9}{4}\right)}{3!}\left(-\frac{x}{16}\right)^3+\cdots\right]$$

$$= \frac{1}{2}+\sum_{n=1}^{\infty}\frac{1\cdot 5\cdot 9\cdot\cdots\cdot(4n-3)}{2\cdot 4^n\cdot n!\cdot 16^n}x^n = \frac{1}{2}+\sum_{n=1}^{\infty}\frac{1\cdot 5\cdot 9\cdot\cdots\cdot(4n-3)}{2^{6n+1}\,n!}x^n$$

for $\left|-\dfrac{x}{16}\right| < 1 \Leftrightarrow |x| < 16$, so $R = 16$.

45. $e^x = \sum\limits_{n=0}^{\infty} \dfrac{x^n}{n!}$, so $\dfrac{e^x}{x} = \dfrac{1}{x}\sum\limits_{n=0}^{\infty}\dfrac{x^n}{n!} = \sum\limits_{n=0}^{\infty}\dfrac{x^{n-1}}{n!} = x^{-1}+\sum\limits_{n=1}^{\infty}\dfrac{x^{n-1}}{n!} = \dfrac{1}{x}+\sum\limits_{n=1}^{\infty}\dfrac{x^{n-1}}{n!}$ and

$\displaystyle\int \frac{e^x}{x}\,dx = C + \ln|x| + \sum_{n=1}^{\infty}\frac{x^n}{n\cdot n!}$.

47. (a)

n	$f^{(n)}(x)$	$f^{(n)}(1)$
0	$x^{1/2}$	1
1	$\frac{1}{2}x^{-1/2}$	$\frac{1}{2}$
2	$-\frac{1}{4}x^{-3/2}$	$-\frac{1}{4}$
3	$\frac{3}{8}x^{-5/2}$	$\frac{3}{8}$
4	$-\frac{15}{16}x^{-7/2}$	$-\frac{15}{16}$
$\vdots$	$\vdots$	$\vdots$

$$\sqrt{x} \approx T_3(x) = 1+\frac{1/2}{1!}(x-1)-\frac{1/4}{2!}(x-1)^2+\frac{3/8}{3!}(x-1)^3$$
$$= 1+\tfrac{1}{2}(x-1)-\tfrac{1}{8}(x-1)^2+\tfrac{1}{16}(x-1)^3$$

(b)

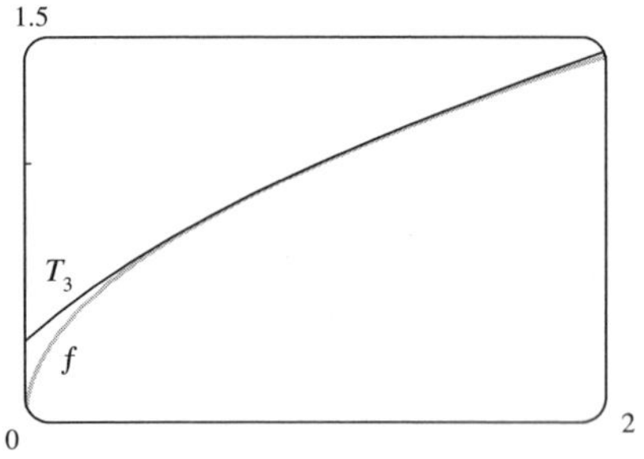

(c) $|R_3(x)| \le \dfrac{M}{4!}|x-1|^4$, where $\left|f^{(4)}(x)\right| \le M$ with $f^{(4)}(x) = -\frac{15}{16}x^{-7/2}$. Now $0.9 \le x \le 1.1 \Rightarrow -0.1 \le x-1 \le 0.1 \Rightarrow (x-1)^4 \le (0.1)^4$, and letting $x = 0.9$ gives $M = \dfrac{15}{16(0.9)^{7/2}}$, so

$$|R_3(x)| \le \frac{15}{16(0.9)^{7/2}\,4!}(0.1)^4 \approx 0.000\,005\,648$$
$$\approx 0.000\,006 = 6\times 10^{-6}$$

(d)

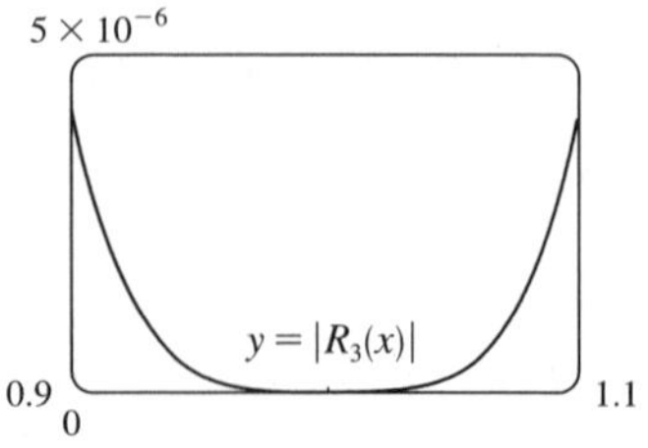

From the graph of $|R_3(x)| = |\sqrt{x} - T_3(x)|$, it appears that the error is less than 5×10^{-6} on $[0.9, 1.1]$.

49. $\sin x = \sum\limits_{n=0}^{\infty} (-1)^n \dfrac{x^{2n+1}}{(2n+1)!} = x - \dfrac{x^3}{3!} + \dfrac{x^5}{5!} - \dfrac{x^7}{7!} + \cdots$, so $\sin x - x = -\dfrac{x^3}{3!} + \dfrac{x^5}{5!} - \dfrac{x^7}{7!} + \cdots$ and

$\dfrac{\sin x - x}{x^3} = -\dfrac{1}{3!} + \dfrac{x^2}{5!} - \dfrac{x^4}{7!} + \cdots$. Thus, $\lim\limits_{x \to 0} \dfrac{\sin x - x}{x^3} = \lim\limits_{x \to 0} \left(-\dfrac{1}{6} + \dfrac{x^2}{120} - \dfrac{x^4}{5040} + \cdots \right) = -\dfrac{1}{6}$.

51. (a) From Formula 14a in Appendix C, with $x = y = \theta$, we get $\tan 2\theta = \dfrac{2\tan\theta}{1 - \tan^2\theta}$, so $\cot 2\theta = \dfrac{1 - \tan^2\theta}{2\tan\theta}$ $\Rightarrow$

$2\cot 2\theta = \dfrac{1 - \tan^2\theta}{\tan\theta} = \cot\theta - \tan\theta$. Replacing θ by $\frac{1}{2}x$, we get $2\cot x = \cot\frac{1}{2}x - \tan\frac{1}{2}x$, or $\tan\frac{1}{2}x = \cot\frac{1}{2}x - 2\cot x$.

(b) From part (a) with $\dfrac{x}{2^{n-1}}$ in place of x, $\tan\dfrac{x}{2^n} = \cot\dfrac{x}{2^n} - 2\cot\dfrac{x}{2^{n-1}}$, so the nth partial sum of $\sum\limits_{n=1}^{\infty} \dfrac{1}{2^n}\tan\dfrac{x}{2^n}$ is

$$\begin{aligned} s_n &= \frac{\tan(x/2)}{2} + \frac{\tan(x/4)}{4} + \frac{\tan(x/8)}{8} + \cdots + \frac{\tan(x/2^n)}{2^n} \\ &= \left[\frac{\cot(x/2)}{2} - \cot x\right] + \left[\frac{\cot(x/4)}{4} - \frac{\cot(x/2)}{2}\right] + \left[\frac{\cot(x/8)}{8} - \frac{\cot(x/4)}{4}\right] + \cdots \\ &\qquad + \left[\frac{\cot(x/2^n)}{2^n} - \frac{\cot(x/2^{n-1})}{2^{n-1}}\right] = -\cot x + \frac{\cot(x/2^n)}{2^n} \quad \text{[telescoping sum]} \end{aligned}$$

Now $\dfrac{\cot(x/2^n)}{2^n} = \dfrac{\cos(x/2^n)}{2^n \sin(x/2^n)} = \dfrac{\cos(x/2^n)}{x} \cdot \dfrac{x/2^n}{\sin(x/2^n)} \to \dfrac{1}{x} \cdot 1 = \dfrac{1}{x}$ as $n \to \infty$ since $x/2^n \to 0$ for $x \neq 0$. Therefore, if $x \neq 0$ and $x \neq k\pi$ where k is any integer, then

$$\sum_{n=1}^{\infty} \frac{1}{2^n}\tan\frac{x}{2^n} = \lim_{n\to\infty} s_n = \lim_{n\to\infty}\left(-\cot x + \frac{1}{2^n}\cot\frac{x}{2^n}\right) = -\cot x + \frac{1}{x}$$

If $x = 0$, then all terms in the series are 0, so the sum is 0.

FOCUS ON PROBLEM SOLVING

1. It would be far too much work to compute 15 derivatives of f. The key idea is to remember that $f^{(n)}(0)$ occurs in the coefficient of x^n in the Maclaurin series of f. We start with the Maclaurin series for sin: $\sin x = x - \dfrac{x^3}{3!} + \dfrac{x^5}{5!} - \cdots$

Then $\sin(x^3) = x^3 - \dfrac{x^9}{3!} + \dfrac{x^{15}}{5!} - \cdots$, and so the coefficient of x^{15} is $\dfrac{f^{(15)}(0)}{15!} = \dfrac{1}{5!}$. Therefore,

$f^{(15)}(0) = \dfrac{15!}{5!} = 6 \cdot 7 \cdot 8 \cdot 9 \cdot 10 \cdot 11 \cdot 12 \cdot 13 \cdot 14 \cdot 15 = 10\,897\,286\,400.$

3. (a) At each stage, each side is replaced by four shorter sides, each of length $\frac{1}{3}$ of the side length at the preceding stage. Writing s_0 and ℓ_0 for the number of sides and the length of the side of the initial triangle, we generate the table at right. In general, we have $s_n = 3 \cdot 4^n$ and $\ell_n = \left(\frac{1}{3}\right)^n$, so the length of the perimeter at the nth stage of construction is $p_n = s_n \ell_n = 3 \cdot 4^n \cdot \left(\frac{1}{3}\right)^n = 3 \cdot \left(\frac{4}{3}\right)^n$.

$s_0 = 3$	$\ell_0 = 1$
$s_1 = 3 \cdot 4$	$\ell_1 = 1/3$
$s_2 = 3 \cdot 4^2$	$\ell_2 = 1/3^2$
$s_3 = 3 \cdot 4^3$	$\ell_3 = 1/3^3$
$\vdots$	$\vdots$

(b) $p_n = \dfrac{4^n}{3^{n-1}} = 4\left(\dfrac{4}{3}\right)^{n-1}$. Since $\frac{4}{3} > 1$, $p_n \to \infty$ as $n \to \infty$.

(c) The area of each of the small triangles added at a given stage is one-ninth of the area of the triangle added at the preceding stage. Let a be the area of the original triangle. Then the area a_n of each of the small triangles added at stage n is $a_n = a \cdot \dfrac{1}{9^n} = \dfrac{a}{9^n}$. Since a small triangle is added to each side at every stage, it follows that the total area A_n added to the figure at the nth stage is $A_n = s_{n-1} \cdot a_n = 3 \cdot 4^{n-1} \cdot \dfrac{a}{9^n} = a \cdot \dfrac{4^{n-1}}{3^{2n-1}}$. Then the total area enclosed by the snowflake curve is $A = a + A_1 + A_2 + A_3 + \cdots = a + a \cdot \dfrac{1}{3} + a \cdot \dfrac{4}{3^3} + a \cdot \dfrac{4^2}{3^5} + a \cdot \dfrac{4^3}{3^7} + \cdots$. After the first term, this is a geometric series with common ratio $\dfrac{4}{9}$, so $A = a + \dfrac{a/3}{1 - \frac{4}{9}} = a + \dfrac{a}{3} \cdot \dfrac{9}{5} = \dfrac{8a}{5}$. But the area of the original equilateral triangle with side 1 is $a = \dfrac{1}{2} \cdot 1 \cdot \sin\dfrac{\pi}{3} = \dfrac{\sqrt{3}}{4}$. So the area enclosed by the snowflake curve is $\dfrac{8}{5} \cdot \dfrac{\sqrt{3}}{4} = \dfrac{2\sqrt{3}}{5}$.

5.
$$\begin{aligned}\ln\left(1 - \frac{1}{n^2}\right) &= \ln\left(\frac{n^2-1}{n^2}\right) = \ln\frac{(n+1)(n-1)}{n^2} = \ln[(n+1)(n-1)] - \ln n^2\\ &= \ln(n+1) + \ln(n-1) - 2\ln n = \ln(n-1) - \ln n - \ln n + \ln(n+1)\\ &= \ln\frac{n-1}{n} - [\ln n - \ln(n+1)] = \ln\frac{n-1}{n} - \ln\frac{n}{n+1}.\end{aligned}$$

Let $s_k = \displaystyle\sum_{n=2}^{k} \ln\left(1 - \frac{1}{n^2}\right) = \sum_{n=2}^{k}\left(\ln\frac{n-1}{n} - \ln\frac{n}{n+1}\right)$ for $k \geq 2$. Then

$s_k = \left(\ln\dfrac{1}{2} - \ln\dfrac{2}{3}\right) + \left(\ln\dfrac{2}{3} - \ln\dfrac{3}{4}\right) + \cdots + \left(\ln\dfrac{k-1}{k} - \ln\dfrac{k}{k+1}\right) = \ln\dfrac{1}{2} - \ln\dfrac{k}{k+1}$, so

$\displaystyle\sum_{n=2}^{\infty} \ln\left(1 - \frac{1}{n^2}\right) = \lim_{k\to\infty} s_k = \lim_{k\to\infty}\left(\ln\frac{1}{2} - \ln\frac{k}{k+1}\right) = \ln\frac{1}{2} - \ln 1 = \ln 1 - \ln 2 - \ln 1 = -\ln 2.$

7. $u = 1 + \frac{x^3}{3!} + \frac{x^6}{6!} + \frac{x^9}{9!} + \cdots$, $v = x + \frac{x^4}{4!} + \frac{x^7}{7!} + \frac{x^{10}}{10!} + \cdots$, $w = \frac{x^2}{2!} + \frac{x^5}{5!} + \frac{x^8}{8!} + \cdots$.

Use the Ratio Test to show that the series for u, v, and w have positive radii of convergence (∞ in each case), so Theorem 8.6.2 applies, and hence, we may differentiate each of these series:

$$\frac{du}{dx} = \frac{3x^2}{3!} + \frac{6x^5}{6!} + \frac{9x^8}{9!} + \cdots = \frac{x^2}{2!} + \frac{x^5}{5!} + \frac{x^8}{8!} + \cdots = w$$

Similarly, $\frac{dv}{dx} = 1 + \frac{x^3}{3!} + \frac{x^6}{6!} + \frac{x^9}{9!} + \cdots = u$, and $\frac{dw}{dx} = x + \frac{x^4}{4!} + \frac{x^7}{7!} + \frac{x^{10}}{10!} + \cdots = v$.

So $u' = w$, $v' = u$, and $w' = v$. Now differentiate the left hand side of the desired equation:

$$\begin{aligned}\frac{d}{dx}\left(u^3 + v^3 + w^3 - 3uvw\right) &= 3u^2u' + 3v^2v' + 3w^2w' - 3(u'vw + uv'w + uvw') \\ &= 3u^2w + 3v^2u + 3w^2v - 3(vw^2 + u^2w + uv^2) = 0 \quad \Rightarrow\end{aligned}$$

$u^3 + v^3 + w^3 - 3uvw = C$. To find the value of the constant C, we put $x = 0$ in the last equation and get $1^3 + 0^3 + 0^3 - 3(1 \cdot 0 \cdot 0) = C \;\Rightarrow\; C = 1$, so $u^3 + v^3 + w^3 - 3uvw = 1$.

9. If L is the length of a side of the equilateral triangle, then the area is $A = \frac{1}{2}L \cdot \frac{\sqrt{3}}{2}L = \frac{\sqrt{3}}{4}L^2$ and so $L^2 = \frac{4}{\sqrt{3}}A$.

Let r be the radius of one of the circles. When there are n rows of circles, the figure shows that

$L = \sqrt{3}r + r + (n-2)(2r) + r + \sqrt{3}\,r = r\left(2n - 2 + 2\sqrt{3}\,\right)$, so $r = \dfrac{L}{2\left(n + \sqrt{3} - 1\right)}$.

The number of circles is $1 + 2 + \cdots + n = \dfrac{n(n+1)}{2}$, and so the total area of the circles is

$$\begin{aligned}A_n &= \frac{n(n+1)}{2}\pi r^2 = \frac{n(n+1)}{2}\pi\frac{L^2}{4\left(n+\sqrt{3}-1\right)^2} = \frac{n(n+1)}{2}\pi\frac{4A/\sqrt{3}}{4\left(n+\sqrt{3}-1\right)^2} \\ &= \frac{n(n+1)}{\left(n+\sqrt{3}-1\right)^2}\frac{\pi A}{2\sqrt{3}} \quad \Rightarrow\end{aligned}$$

$$\begin{aligned}\frac{A_n}{A} &= \frac{n(n+1)}{\left(n+\sqrt{3}-1\right)^2}\frac{\pi}{2\sqrt{3}} \\ &= \frac{1 + 1/n}{\left[1 + \left(\sqrt{3}-1\right)/n\right]^2}\frac{\pi}{2\sqrt{3}} \to \frac{\pi}{2\sqrt{3}} \text{ as } n \to \infty\end{aligned}$$

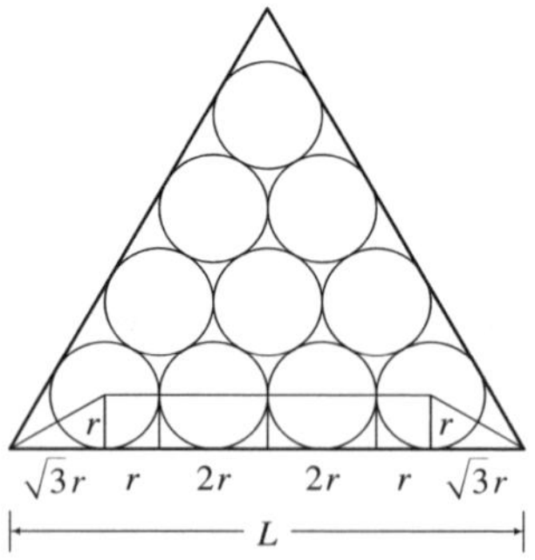

11. As in Section 8.6 we have to integrate the function x^x by integrating series. Writing $x^x = \left(e^{\ln x}\right)^x = e^{x\ln x}$ and using the Maclaurin series for e^x, we have $x^x = \left(e^{\ln x}\right)^x = e^{x\ln x} = \sum\limits_{n=0}^{\infty}\frac{(x\ln x)^n}{n!} = \sum\limits_{n=0}^{\infty}\frac{x^n(\ln x)^n}{n!}$. As with power series, we can integrate this series term-by-term: $\int_0^1 x^x\,dx = \sum\limits_{n=0}^{\infty}\int_0^1 \frac{x^n(\ln x)^n}{n!}\,dx = \sum\limits_{n=0}^{\infty}\frac{1}{n!}\int_0^1 x^n(\ln x)^n\,dx$. We integrate by parts with $u = (\ln x)^n$, $dv = x^n\,dx$, so $du = \dfrac{n(\ln x)^{n-1}}{x}\,dx$ and $v = \dfrac{x^{n+1}}{n+1}$:

$$\int_0^1 x^n (\ln x)^n\,dx = \lim_{t\to 0^+}\int_t^1 x^n (\ln x)^n\,dx = \lim_{t\to 0^+}\left[\frac{x^{n+1}}{n+1}(\ln x)^n\right]_t^1 - \lim_{t\to 0^+}\int_t^1 \frac{n}{n+1}x^n(\ln x)^{n-1}\,dx$$
$$= 0 - \frac{n}{n+1}\int_0^1 x^n (\ln x)^{n-1}\,dx$$

(where l'Hospital's Rule was used to help evaluate the first limit). Further integration by parts gives

$\displaystyle\int_0^1 x^n (\ln x)^k\,dx = -\frac{k}{n+1}\int_0^1 x^n(\ln x)^{k-1}\,dx$ and, combining these steps, we get

$$\int_0^1 x^n (\ln x)^n\,dx = \frac{(-1)^n\,n!}{(n+1)^n}\int_0^1 x^n\,dx = \frac{(-1)^n\,n!}{(n+1)^{n+1}} \quad\Rightarrow$$

$$\int_0^1 x^x\,dx = \sum_{n=0}^{\infty}\frac{1}{n!}\int_0^1 x^n(\ln x)^n\,dx = \sum_{n=0}^{\infty}\frac{1}{n!}\frac{(-1)^n\,n!}{(n+1)^{n+1}} = \sum_{n=0}^{\infty}\frac{(-1)^n}{(n+1)^{n+1}} = \sum_{n=1}^{\infty}\frac{(-1)^{n-1}}{n^n}.$$

13. Let $f(x) = \sum\limits_{m=0}^{\infty} c_m x^m$ and $g(x) = e^{f(x)} = \sum\limits_{n=0}^{\infty} d_n x^n$. Then $g'(x) = \sum\limits_{n=0}^{\infty} n d_n x^{n-1}$, so nd_n occurs as the coefficient of x^{n-1}. But also

$$g'(x) = e^{f(x)}f'(x) = \left(\sum_{n=0}^{\infty} d_n x^n\right)\left(\sum_{m=1}^{\infty} m c_m x^{m-1}\right)$$
$$= \left(d_0 + d_1 x + d_2 x^2 + \cdots + d_{n-1}x^{n-1} + \cdots\right)\left(c_1 + 2c_2 x + 3c_3 x^2 + \cdots + nc_n x^{n-1} + \cdots\right)$$

so the coefficient of x^{n-1} is $c_1 d_{n-1} + 2c_2 d_{n-2} + 3c_3 d_{n-3} + \cdots + nc_n d_0 = \sum\limits_{i=1}^{n} ic_i d_{n-i}$. Therefore, $nd_n = \sum\limits_{i=1}^{n} ic_i d_{n-i}$.

15. Call the series S. We group the terms according to the number of digits in their denominators:

$$S = \underbrace{\left(\tfrac{1}{1}+\tfrac{1}{2}+\cdots+\tfrac{1}{8}+\tfrac{1}{9}\right)}_{g_1} + \underbrace{\left(\tfrac{1}{11}+\cdots+\tfrac{1}{99}\right)}_{g_2} + \underbrace{\left(\tfrac{1}{111}+\cdots+\tfrac{1}{999}\right)}_{g_3} + \cdots$$

Now in the group g_n, since we have 9 choices for each of the n digits in the denominator, there are 9^n terms. Furthermore, each term in g_n is less than $\frac{1}{10^{n-1}}$ [except for the first term in g_1]. So $g_n < 9^n\cdot\frac{1}{10^{n-1}} = 9\left(\frac{9}{10}\right)^{n-1}$. Now $\sum_{n=1}^{\infty} 9\left(\frac{9}{10}\right)^{n-1}$ is a geometric series with $a = 9$ and $r = \frac{9}{10} < 1$. Therefore, by the Comparison Test, $S = \sum_{n=1}^{\infty} g_n < \sum_{n=1}^{\infty} 9\left(\frac{9}{10}\right)^{n-1} = \frac{9}{1-9/10} = 90$.

9 ☐ VECTORS AND THE GEOMETRY OF SPACE

9.1 Three-Dimensional Coordinate Systems

1. We start at the origin, which has coordinates $(0, 0, 0)$. First we move 4 units along the positive x-axis, affecting only the x-coordinate, bringing us to the point $(4, 0, 0)$. We then move 3 units straight downward, in the negative z-direction. Thus only the z-coordinate is affected, and we arrive at $(4, 0, -3)$.

3. The distance from a point to the xz-plane is the absolute value of the y-coordinate of the point. $Q(-5, -1, 4)$ has the y-coordinate with the smallest absolute value, so Q is the point closest to the xz-plane. $R(0, 3, 8)$ must lie in the yz-plane since the distance from R to the yz-plane, given by the x-coordinate of R, is 0.

5. The equation $x + y = 2$ represents the set of all points in $\mathbb{R}^3$ whose x- and y-coordinates have a sum of 2, or equivalently where $y = 2 - x$. This is the set $\{(x, 2 - x, z) \mid x \in \mathbb{R}, z \in \mathbb{R}\}$ which is a vertical plane that intersects the xy-plane in the line $y = 2 - x$, $z = 0$.

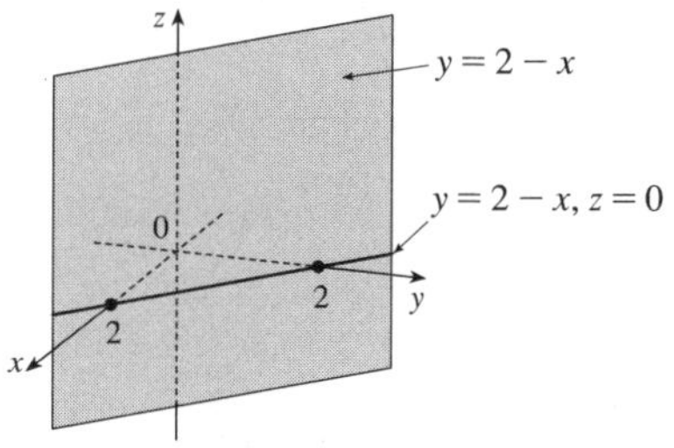

7. (a) We can find the lengths of the sides of the triangle by using the distance formula between pairs of vertices:

$$|PQ| = \sqrt{(7-3)^2 + [0-(-2)]^2 + [1-(-3)]^2} = \sqrt{16+4+16} = 6$$

$$|QR| = \sqrt{(1-7)^2 + (2-0)^2 + (1-1)^2} = \sqrt{36+4+0} = \sqrt{40} = 2\sqrt{10}$$

$$|RP| = \sqrt{(3-1)^2 + (-2-2)^2 + (-3-1)^2} = \sqrt{4+16+16} = 6$$

The longest side is QR, but the Pythagorean Theorem is not satisfied: $|PQ|^2 + |RP|^2 \neq |QR|^2$. Thus PQR is not a right triangle. PQR is isosceles, as two sides have the same length.

(b) Compute the lengths of the sides of the triangle by using the distance formula between pairs of vertices:

$$|PQ| = \sqrt{(4-2)^2 + [1-(-1)]^2 + (1-0)^2} = \sqrt{4+4+1} = 3$$

$$|QR| = \sqrt{(4-4)^2 + (-5-1)^2 + (4-1)^2} = \sqrt{0+36+9} = \sqrt{45} = 3\sqrt{5}$$

$$|RP| = \sqrt{(2-4)^2 + [-1-(-5)]^2 + (0-4)^2} = \sqrt{4+16+16} = 6$$

Since the Pythagorean Theorem is satisfied by $|PQ|^2 + |RP|^2 = |QR|^2$, PQR is a right triangle. PQR is not isosceles, as no two sides have the same length.

9. (a) First we find the distances between points:

$$|AB| = \sqrt{(7-5)^2 + (9-1)^2 + (-1-3)^2} = \sqrt{84} = 2\sqrt{21}$$

$$|BC| = \sqrt{(1-7)^2 + (-15-9)^2 + [11-(-1)]^2} = \sqrt{756} = 6\sqrt{21}$$

$$|AC| = \sqrt{(1-5)^2 + (-15-1)^2 + (11-3)^2} = \sqrt{336} = 4\sqrt{21}$$

In order for the points to lie on a straight line, the sum of the two shortest distances must be equal to the longest distance. Since $|AB| + |AC| = |BC|$, the three points lie on a straight line.

(b) The distances between points are

$$|KL| = \sqrt{(1-0)^2 + (2-3)^2 + [-2-(-4)]^2} = \sqrt{6}$$

$$|LM| = \sqrt{(3-1)^2 + (0-2)^2 + [1-(-2)]^2} = \sqrt{17}$$

$$|KM| = \sqrt{(3-0)^2 + (0-3)^2 + [1-(-4)]^2} = \sqrt{43}$$

Since $\sqrt{6} + \sqrt{17} \neq \sqrt{43}$, the three points do not lie on a straight line.

11. The radius of the sphere is the distance between $(4, 3, -1)$ and $(3, 8, 1)$:

$r = \sqrt{(3-4)^2 + (8-3)^2 + [1-(-1)]^2} = \sqrt{30}$. Thus, an equation of the sphere is $(x-3)^2 + (y-8)^2 + (z-1)^2 = 30$.

Another solution: An equation of a sphere having centre $(3, 8, 1)$ is of the form $(x-3)^2 + (y-8)^2 + (z-1)^2 = r^2$. Since the sphere passes through the point $(4, 3, -1)$, we let $x = 4$, $y = 3$, and $z = -1$ to obtain $1 + 25 + 4 = r^2$, so an equation is $(x-3)^2 + (y-8)^2 + (z-1)^2 = 30$.

13. Completing squares in the equation $x^2 + y^2 + z^2 - 6x + 4y - 2z = 11$ gives

$(x^2 - 6x + 9) + (y^2 + 4y + 4) + (z^2 - 2z + 1) = 11 + 9 + 4 + 1 \quad \Rightarrow$

$(x-3)^2 + (y+2)^2 + (z-1)^2 = 25$ which we recognize as an equation of a sphere with centre $(3, -2, 1)$ and radius 5.

15. (a) If the midpoint of the line segment from $P_1(x_1, y_1, z_1)$ to $P_2(x_2, y_2, z_2)$ is $Q = \left(\frac{x_1 + x_2}{2}, \frac{y_1 + y_2}{2}, \frac{z_1 + z_2}{2}\right)$, then the distances $|P_1Q|$ and $|QP_2|$ are equal, and each is half of $|P_1P_2|$. We verify that this is the case:

$$|P_1P_2| = \sqrt{(x_2 - x_1)^2 + (y_2 - y_1)^2 + (z_2 - z_1)^2}$$

$$\begin{aligned}
|P_1Q| &= \sqrt{\left[\tfrac{1}{2}(x_1 + x_2) - x_1\right]^2 + \left[\tfrac{1}{2}(y_1 + y_2) - y_1\right]^2 + \left[\tfrac{1}{2}(z_1 + z_2) - z_1\right]^2} \\
&= \sqrt{\left(\tfrac{1}{2}x_2 - \tfrac{1}{2}x_1\right)^2 + \left(\tfrac{1}{2}y_2 - \tfrac{1}{2}y_1\right)^2 + \left(\tfrac{1}{2}z_2 - \tfrac{1}{2}z_1\right)^2} \\
&= \sqrt{\left(\tfrac{1}{2}\right)^2\left[(x_2 - x_1)^2 + (y_2 - y_1)^2 + (z_2 - z_1)^2\right]} \\
&= \tfrac{1}{2}\sqrt{(x_2 - x_1)^2 + (y_2 - y_1)^2 + (z_2 - z_1)^2} \\
&= \tfrac{1}{2}\,|P_1P_2|
\end{aligned}$$

$$\begin{aligned}|QP_2| &= \sqrt{\left[x_2 - \tfrac{1}{2}(x_1 + x_2)\right]^2 + \left[y_2 - \tfrac{1}{2}(y_1 + y_2)\right]^2 + \left[z_2 - \tfrac{1}{2}(z_1 + z_2)\right]^2} \\ &= \sqrt{\left(\tfrac{1}{2}x_2 - \tfrac{1}{2}x_1\right)^2 + \left(\tfrac{1}{2}y_2 - \tfrac{1}{2}y_1\right)^2 + \left(\tfrac{1}{2}z_2 - \tfrac{1}{2}z_1\right)^2} \\ &= \sqrt{\left(\tfrac{1}{2}\right)^2\left[(x_2 - x_1)^2 + (y_2 - y_1)^2 + (z_2 - z_1)^2\right]} \\ &= \tfrac{1}{2}\sqrt{(x_2 - x_1)^2 + (y_2 - y_1)^2 + (z_2 - z_1)^2} \\ &= \tfrac{1}{2}\,|P_1P_2|\end{aligned}$$

So Q is indeed the midpoint of P_1P_2.

(b) By part (a), the midpoints of sides AB, BC and CA are $P_1\left(-\frac{1}{2}, 1, 4\right)$, $P_2\left(1, \frac{1}{2}, 5\right)$ and $P_3\left(\frac{5}{2}, \frac{3}{2}, 4\right)$. (Recall that a median of a triangle is a line segment from a vertex to the midpoint of the opposite side.) Then the lengths of the medians are:

$$\begin{aligned}|AP_2| &= \sqrt{0^2 + \left(\tfrac{1}{2} - 2\right)^2 + (5 - 3)^2} = \sqrt{\tfrac{9}{4} + 4} = \sqrt{\tfrac{25}{4}} = \tfrac{5}{2} \\ |BP_3| &= \sqrt{\left(\tfrac{5}{2} + 2\right)^2 + \left(\tfrac{3}{2}\right)^2 + (4 - 5)^2} = \sqrt{\tfrac{81}{4} + \tfrac{9}{4} + 1} = \sqrt{\tfrac{94}{4}} = \tfrac{1}{2}\sqrt{94} \\ |CP_1| &= \sqrt{\left(-\tfrac{1}{2} - 4\right)^2 + (1 - 1)^2 + (4 - 5)^2} = \sqrt{\tfrac{81}{4} + 1} = \tfrac{1}{2}\sqrt{85}\end{aligned}$$

17. (a) Since the sphere touches the xy-plane, its radius is the distance from its centre, $(2, -3, 6)$, to the xy-plane, namely 6. Therefore $r = 6$ and an equation of the sphere is $(x - 2)^2 + (y + 3)^2 + (z - 6)^2 = 6^2 = 36$.

(b) The radius of this sphere is the distance from its centre $(2, -3, 6)$ to the yz-plane, which is 2. Therefore, an equation is $(x - 2)^2 + (y + 3)^2 + (z - 6)^2 = 4$.

(c) Here the radius is the distance from the centre $(2, -3, 6)$ to the xz-plane, which is 3. Therefore, an equation is $(x - 2)^2 + (y + 3)^2 + (z - 6)^2 = 9$.

19. The equation $y = -4$ represents a plane parallel to the xz-plane and 4 units to the left of it.

21. The inequality $x > 3$ represents a half-space consisting of all points in front of the plane $x = 3$.

23. The inequality $0 \leq z \leq 6$ represents all points on or between the horizontal planes $z = 0$ (the xy-plane) and $z = 6$.

25. The inequality $x^2 + y^2 + z^2 > 1$ is equivalent to $\sqrt{x^2 + y^2 + z^2} > 1$, so the region consists of those points whose distance from the origin is greater than 1. This is the set of all points outside the sphere with radius 1 and centre $(0, 0, 0)$.

27. Here $x^2 + z^2 \leq 9$ or equivalently $\sqrt{x^2 + z^2} \leq 3$ which describes the set of all points in $\mathbb{R}^3$ whose distance from the y-axis is at most 3. Thus, the inequality represents the region consisting of all points on or inside a circular cylinder of radius 3 with axis the y-axis.

29. This describes all points with negative y-coordinates, that is, $y < 0$.

31. This describes a region all of whose points have a distance to the origin which is greater than r, but smaller than R. So inequalities describing the region are $r < \sqrt{x^2 + y^2 + z^2} < R$, or $r^2 < x^2 + y^2 + z^2 < R^2$.

33. (a) To find the x- and y-coordinates of the point P, we project it onto L_2 and project the resulting point Q onto the x- and y-axes. To find the z-coordinate, we project P onto either the xz-plane or the yz-plane (using our knowledge of its x- or y-coordinate) and then project the resulting point onto the z-axis. (Or, we could draw a line parallel to QO from P to the z-axis.) The coordinates of P are $(2, 1, 4)$.

(b) A is the intersection of L_1 and L_2, B is directly below the y-intercept of L_2, and C is directly above the x-intercept of L_2.

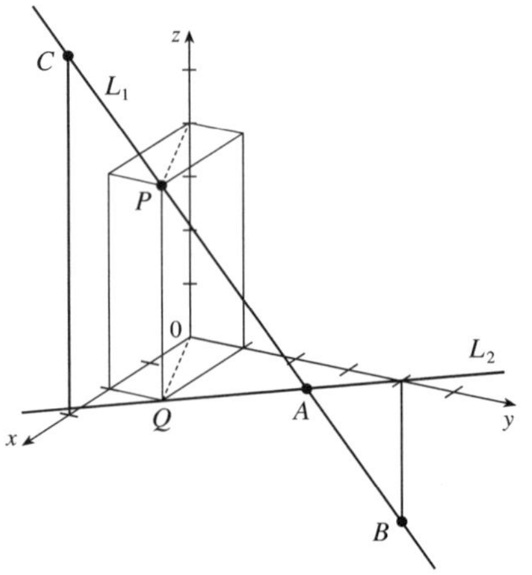

35. We need to find a set of points $\{P(x,y,z) \mid |AP| = |BP|\}$.

$\sqrt{(x+1)^2 + (y-5)^2 + (z-3)^2} = \sqrt{(x-6)^2 + (y-2)^2 + (z+2)^2} \quad \Rightarrow$

$(x+1)^2 + (y-5) + (z-3)^2 = (x-6)^2 + (y-2)^2 + (z+2)^2 \quad \Rightarrow$

$x^2 + 2x + 1 + y^2 - 10y + 25 + z^2 - 6z + 9 = x^2 - 12x + 36 + y^2 - 4y + 4 + z^2 + 4z + 4 \quad \Rightarrow \quad 14x - 6y - 10z = 9.$

Thus the set of points is a plane perpendicular to the line segment joining A and B (since this plane must contain the perpendicular bisector of the line segment AB).

9.2 Vectors

1. (a) The cost of a theatre ticket is a scalar, because it has only magnitude.

(b) The current in a river is a vector, because it has both magnitude (the speed of the current) and direction at any given location.

(c) If we assume that the initial path is linear, the initial flight path from Houston to Dallas is a vector, because it has both magnitude (distance) and direction.

(d) The population of the world is a scalar, because it has only magnitude.

3. Vectors are equal when they share the same length and direction (but not necessarily location). Using the symmetry of the parallelogram as a guide, we see that $\overrightarrow{AB} = \overrightarrow{DC}$, $\overrightarrow{DA} = \overrightarrow{CB}$, $\overrightarrow{DE} = \overrightarrow{EB}$, and $\overrightarrow{EA} = \overrightarrow{CE}$.

5. (a)

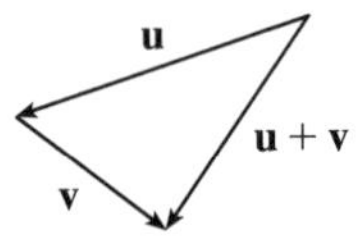

(b)

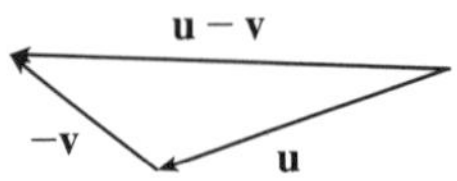

(c)

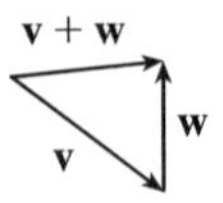

(d)

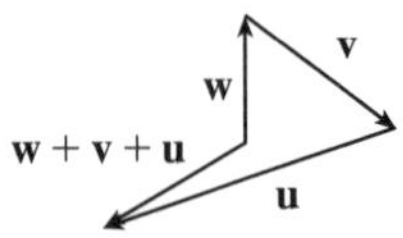

7. $\mathbf{a} = \langle -2-2, 1-3 \rangle = \langle -4, -2 \rangle$

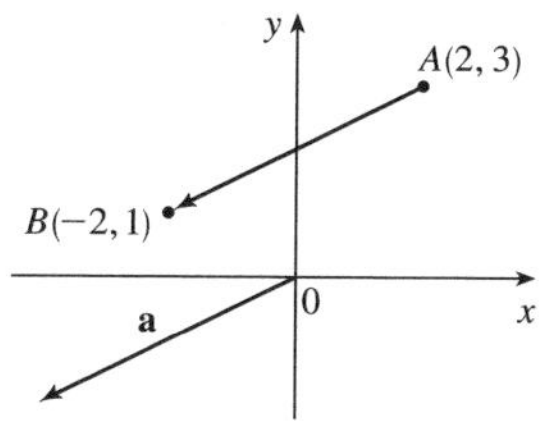

9. $\mathbf{a} = \langle 2-0, 3-3, -1-1 \rangle = \langle 2, 0, -2 \rangle$

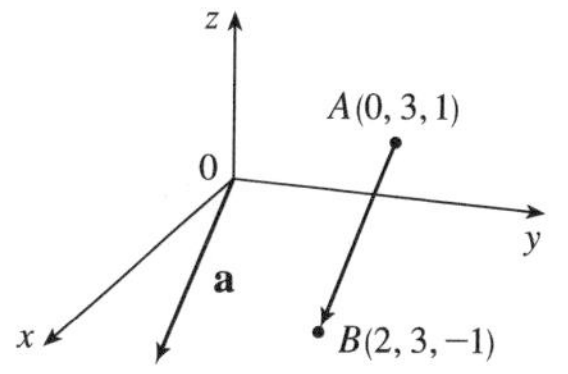

11. $\langle 3, -1 \rangle + \langle -2, 4 \rangle = \langle 3 + (-2), -1 + 4 \rangle$
$= \langle 1, 3 \rangle$

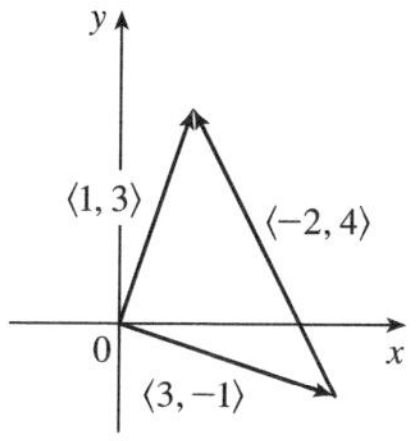

13. $\langle 0, 1, 2 \rangle + \langle 0, 0, -3 \rangle = \langle 0 + 0, 1 + 0, 2 + (-3) \rangle$
$= \langle 0, 1, -1 \rangle$

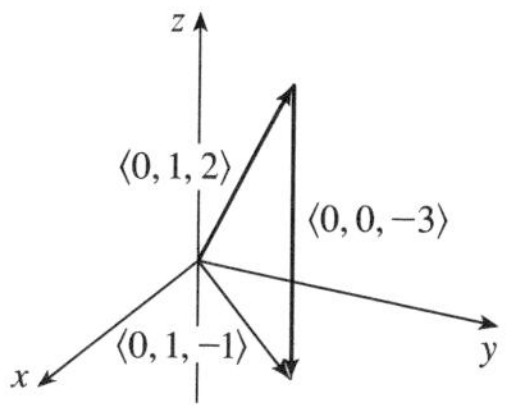

15. $\mathbf{a} + \mathbf{b} = \langle 5, -12 \rangle + \langle -3, -6 \rangle = \langle 5 + (-3), -12 + (-6) \rangle = \langle 2, -18 \rangle$

$2\mathbf{a} + 3\mathbf{b} = \langle 10, -24 \rangle + \langle -9, -18 \rangle = \langle 1, -42 \rangle$

$|\mathbf{a}| = \sqrt{5^2 + (-12)^2} = \sqrt{169} = 13$

$|\mathbf{a} - \mathbf{b}| = |\langle 5 - (-3), -12 - (-6) \rangle| = |\langle 8, -6 \rangle| = \sqrt{8^2 + (-6)^2} = \sqrt{100} = 10$

17. $\mathbf{a} + \mathbf{b} = (\mathbf{i} + 2\mathbf{j} - 3\mathbf{k}) + (-2\mathbf{i} - \mathbf{j} + 5\mathbf{k}) = -\mathbf{i} + \mathbf{j} + 2\mathbf{k}$

$2\mathbf{a} + 3\mathbf{b} = 2(\mathbf{i} + 2\mathbf{j} - 3\mathbf{k}) + 3(-2\mathbf{i} - \mathbf{j} + 5\mathbf{k}) = 2\mathbf{i} + 4\mathbf{j} - 6\mathbf{k} - 6\mathbf{i} - 3\mathbf{j} + 15\mathbf{k} = -4\mathbf{i} + \mathbf{j} + 9\mathbf{k}$

$|\mathbf{a}| = \sqrt{1^2 + 2^2 + (-3)^2} = \sqrt{14}$

$|\mathbf{a} - \mathbf{b}| = |(\mathbf{i} + 2\mathbf{j} - 3\mathbf{k}) - (-2\mathbf{i} - \mathbf{j} + 5\mathbf{k})| = |3\mathbf{i} + 3\mathbf{j} - 8\mathbf{k}| = \sqrt{3^2 + 3^2 + (-8)^2} = \sqrt{82}$

19. The vector $8\mathbf{i} - \mathbf{j} + 4\mathbf{k}$ has length $|8\mathbf{i} - \mathbf{j} + 4\mathbf{k}| = \sqrt{8^2 + (-1)^2 + 4^2} = \sqrt{81} = 9$, so by Equation 4 the unit vector with the same direction is $\frac{1}{9}(8\mathbf{i} - \mathbf{j} + 4\mathbf{k}) = \frac{8}{9}\mathbf{i} - \frac{1}{9}\mathbf{j} + \frac{4}{9}\mathbf{k}$.

21. From the figure, we see that the x-component of $\mathbf{v}$ is $v_1 = |\mathbf{v}|\cos(\pi/3) = 4 \cdot \frac{1}{2} = 2$ and the y-component is $v_2 = |\mathbf{v}|\sin(\pi/3) = 4 \cdot \frac{\sqrt{3}}{2} = 2\sqrt{3}$. Thus $\mathbf{v} = \langle v_1, v_2 \rangle = \langle 2, 2\sqrt{3} \rangle$.

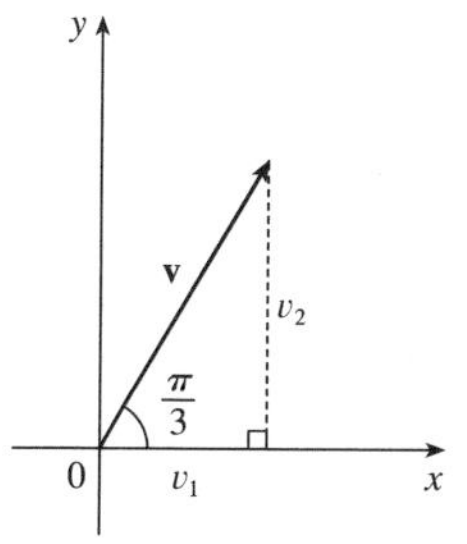

23. $|\mathbf{F}_1| = 10$ N and $|\mathbf{F}_2| = 12$ N.

$\mathbf{F}_1 = -|\mathbf{F}_1|\cos 45°\,\mathbf{i} + |\mathbf{F}_1|\sin 45°\,\mathbf{j} = -10\cos 45°\,\mathbf{i} + 10\sin 45°\,\mathbf{j} = -5\sqrt{2}\,\mathbf{i} + 5\sqrt{2}\,\mathbf{j}$

$\mathbf{F}_2 = |\mathbf{F}_2|\cos 30°\,\mathbf{i} + |\mathbf{F}_2|\sin 30°\,\mathbf{j} = 12\cos 30°\,\mathbf{i} + 12\sin 30°\,\mathbf{j} = 6\sqrt{3}\,\mathbf{i} + 6\,\mathbf{j}$

$\mathbf{F} = \mathbf{F}_1 + \mathbf{F}_2 = \left(6\sqrt{3} - 5\sqrt{2}\right)\mathbf{i} + \left(6 + 5\sqrt{2}\right)\mathbf{j} \approx 3.32\,\mathbf{i} + 13.07\,\mathbf{j}$

$|\mathbf{F}| \approx \sqrt{(3.32)^2 + (13.07)^2} \approx 13.5$ N. $\quad \tan\theta = \dfrac{6 + 5\sqrt{2}}{6\sqrt{3} - 5\sqrt{2}} \quad \Rightarrow \quad \theta = \tan^{-1}\dfrac{6 + 5\sqrt{2}}{6\sqrt{3} - 5\sqrt{2}} \approx 76°.$

25. With respect to the water's surface, the woman's velocity is the vector sum of the velocity of the ship with respect to the water, and the woman's velocity with respect to the ship. If we let north be the positive y-direction, then $\mathbf{v} = \langle 0, 35\rangle + \langle -5, 0\rangle = \langle -5, 35\rangle$. The woman's speed is $|\mathbf{v}| = \sqrt{(-5)^2 + 35^2} = \sqrt{1250} \approx 35.4$ km/h. The vector $\mathbf{v}$ makes an angle θ with the east, where $\theta = \cos^{-1}\left(-5/\sqrt{1250}\right) \approx 98°$. Therefore, the woman's direction is about N$(98 - 90)°$ W $=$ N8° W.

27. Let $\mathbf{T}_1$ and $\mathbf{T}_2$ represent the tension vectors in each side of the clothesline as shown in the figure. $\mathbf{T}_1$ and $\mathbf{T}_2$ have equal vertical components and opposite horizontal components, so we can write

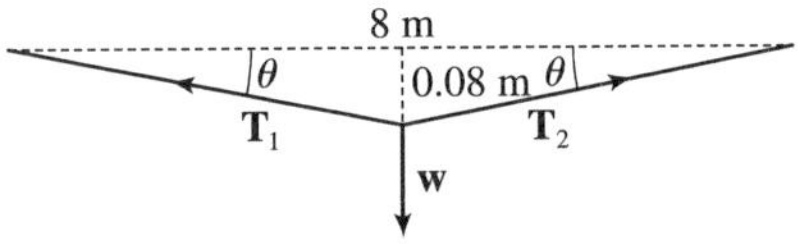

$\mathbf{T}_1 = -a\,\mathbf{i} + b\,\mathbf{j}$ and $\mathbf{T}_2 = a\,\mathbf{i} + b\,\mathbf{j}$ $(a, b > 0)$. By similar triangles, $\dfrac{b}{a} = \dfrac{0.08}{4} \quad \Rightarrow \quad a = 50b$. The force due to gravity acting on the shirt has magnitude $0.8g \approx (0.8)(9.8) = 7.84$ N, hence we have $\mathbf{w} = -7.84\,\mathbf{j}$. The resultant $\mathbf{T}_1 + \mathbf{T}_2$ of the tensile forces counterbalances $\mathbf{w}$, so $\mathbf{T}_1 + \mathbf{T}_2 = -\mathbf{w} \quad \Rightarrow \quad (-a\,\mathbf{i} + b\,\mathbf{j}) + (a\,\mathbf{i} + b\,\mathbf{j}) = 7.84\,\mathbf{j} \quad \Rightarrow$
$(-50b\,\mathbf{i} + b\,\mathbf{j}) + (50b\,\mathbf{i} + b\,\mathbf{j}) = 2b\,\mathbf{j} = 7.84\,\mathbf{j} \quad \Rightarrow \quad b = \frac{7.84}{2} = 3.92$ and $a = 50b = 196$. Thus the tensions are $\mathbf{T}_1 = -a\,\mathbf{i} + b\,\mathbf{j} = -196\,\mathbf{i} + 3.92\,\mathbf{j}$ and $\mathbf{T}_2 = a\,\mathbf{i} + b\,\mathbf{j} = 196\,\mathbf{i} + 3.92\,\mathbf{j}$.

Alternatively, we can find the value of θ and proceed as in Example 7.

29. (a), (b)

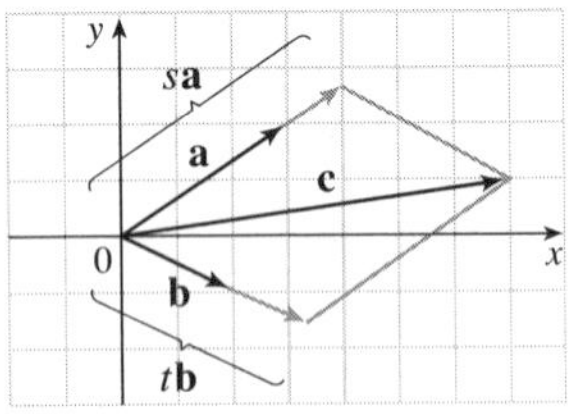

(c) From the sketch, we estimate that $s \approx 1.3$ and $t \approx 1.6$.

(d) $\mathbf{c} = s\,\mathbf{a} + t\,\mathbf{b} \quad \Leftrightarrow \quad 7 = 3s + 2t$ and $1 = 2s - t$.

Solving these equations gives $s = \frac{9}{7}$ and $t = \frac{11}{7}$.

31.

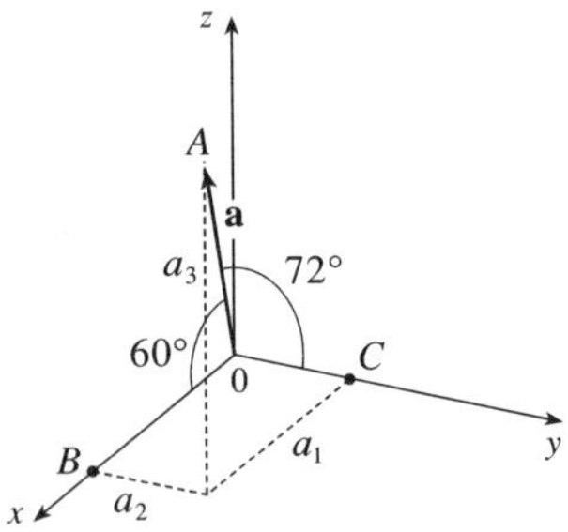

Let $\mathbf{a} = \langle a_1, a_2, a_3\rangle$, as shown in the figure. Since $|\mathbf{a}| = 1$ and triangle ABO is a right triangle, we have $\cos 60° = \dfrac{a_1}{1} \quad \Rightarrow \quad a_1 = \cos 60°$. Similarly, triangle ACO is a right triangle, so $a_2 = \cos 72°$. Finally, since $|\mathbf{a}| = 1$ we have

$\sqrt{(\cos 60°)^2 + (\cos 72°)^2 + a_3^2} = 1 \quad \Rightarrow$

$a_3^2 = 1 - (\cos 60°)^2 - (\cos 72°)^2 \quad \Rightarrow$

$a_3 = \sqrt{1 - (\cos 60°)^2 - (\cos 72°)^2}$. Thus

$\mathbf{a} = \left\langle \cos 60°, \cos 72°, \sqrt{1 - (\cos 60°)^2 - (\cos 72°)^2} \right\rangle \approx \langle 0.50, 0.31, 0.81\rangle.$

33. $|\mathbf{r} - \mathbf{r}_0|$ is the distance between the points (x, y, z) and (x_0, y_0, z_0), so the set of points is a sphere with radius 1 and centre (x_0, y_0, z_0).

Alternate method: $|\mathbf{r} - \mathbf{r}_0| = 1 \quad \Leftrightarrow \quad \sqrt{(x - x_0)^2 + (y - y_0)^2 + (z - z_0)^2} = 1 \quad \Leftrightarrow$

$(x - x_0)^2 + (y - y_0)^2 + (z - z_0)^2 = 1$, which is the equation of a sphere with radius 1 and centre (x_0, y_0, z_0).

35.
$$\begin{aligned} \mathbf{a} + (\mathbf{b} + \mathbf{c}) &= \langle a_1, a_2 \rangle + (\langle b_1, b_2 \rangle + \langle c_1, c_2 \rangle) = \langle a_1, a_2 \rangle + \langle b_1 + c_1, b_2 + c_2 \rangle \\ &= \langle a_1 + b_1 + c_1, a_2 + b_2 + c_2 \rangle = \langle (a_1 + b_1) + c_1, (a_2 + b_2) + c_2 \rangle \\ &= \langle a_1 + b_1, a_2 + b_2 \rangle + \langle c_1, c_2 \rangle = (\langle a_1, a_2 \rangle + \langle b_1, b_2 \rangle) + \langle c_1, c_2 \rangle \\ &= (\mathbf{a} + \mathbf{b}) + \mathbf{c} \end{aligned}$$

37. Consider triangle ABC, where D and E are the midpoints of AB and BC. We know that $\overrightarrow{AB} + \overrightarrow{BC} = \overrightarrow{AC}$ (1) and $\overrightarrow{DB} + \overrightarrow{BE} = \overrightarrow{DE}$ (2). However, $\overrightarrow{DB} = \frac{1}{2}\overrightarrow{AB}$, and $\overrightarrow{BE} = \frac{1}{2}\overrightarrow{BC}$. Substituting these expressions for $\overrightarrow{DB}$ and $\overrightarrow{BE}$ into (2) gives $\frac{1}{2}\overrightarrow{AB} + \frac{1}{2}\overrightarrow{BC} = \overrightarrow{DE}$. Comparing this with (1) gives $\overrightarrow{DE} = \frac{1}{2}\overrightarrow{AC}$. Therefore $\overrightarrow{AC}$ and $\overrightarrow{DE}$ are parallel and $\left|\overrightarrow{DE}\right| = \frac{1}{2}\left|\overrightarrow{AC}\right|$.

9.3 The Dot Product

1. (a) $\mathbf{a} \cdot \mathbf{b}$ is a scalar, and the dot product is defined only for vectors, so $(\mathbf{a} \cdot \mathbf{b}) \cdot \mathbf{c}$ has no meaning.

(b) $(\mathbf{a} \cdot \mathbf{b})\,\mathbf{c}$ is a scalar multiple of a vector, so it does have meaning.

(c) Both $|\mathbf{a}|$ and $\mathbf{b} \cdot \mathbf{c}$ are scalars, so $|\mathbf{a}|\,(\mathbf{b} \cdot \mathbf{c})$ is an ordinary product of real numbers, and has meaning.

(d) Both $\mathbf{a}$ and $\mathbf{b} + \mathbf{c}$ are vectors, so the dot product $\mathbf{a} \cdot (\mathbf{b} + \mathbf{c})$ has meaning.

(e) $\mathbf{a} \cdot \mathbf{b}$ is a scalar, but $\mathbf{c}$ is a vector, and so the two quantities cannot be added and this expression has no meaning.

(f) $|\mathbf{a}|$ is a scalar, and the dot product is defined only for vectors, so $|\mathbf{a}| \cdot (\mathbf{b} + \mathbf{c})$ has no meaning.

3. $\mathbf{a} \cdot \mathbf{b} = |\mathbf{a}|\,|\mathbf{b}| \cos\theta = (6)(5)\cos\frac{2\pi}{3} = 30\left(-\frac{1}{2}\right) = -15$

5. $\mathbf{a} \cdot \mathbf{b} = \left\langle 4, 1, \frac{1}{4} \right\rangle \cdot \langle 6, -3, -8 \rangle = 4(6) + 1(-3) + \frac{1}{4}(-8) = 24 - 3 - 2 = 19$

7. $\mathbf{a} \cdot \mathbf{b} = (\mathbf{i} - 2\,\mathbf{j} + 3\,\mathbf{k}) \cdot (5\,\mathbf{i} + 9\,\mathbf{k}) = (1)(5) + (-2)(0) + (3)(9) = 32$

9. $\mathbf{u}$, $\mathbf{v}$, and $\mathbf{w}$ are all unit vectors, so the triangle is an equilateral triangle. Thus the angle between $\mathbf{u}$ and $\mathbf{v}$ is 60° and $\mathbf{u} \cdot \mathbf{v} = |\mathbf{u}|\,|\mathbf{v}| \cos 60^\circ = (1)(1)\left(\frac{1}{2}\right) = \frac{1}{2}$. If $\mathbf{w}$ is moved so it has the same initial point as $\mathbf{u}$, we can see that the angle between them is 120° and we have $\mathbf{u} \cdot \mathbf{w} = |\mathbf{u}|\,|\mathbf{w}| \cos 120^\circ = (1)(1)\left(-\frac{1}{2}\right) = -\frac{1}{2}$.

11. (a) $\mathbf{i} \cdot \mathbf{j} = \langle 1, 0, 0 \rangle \cdot \langle 0, 1, 0 \rangle = (1)(0) + (0)(1) + (0)(0) = 0$. Similarly $\mathbf{j} \cdot \mathbf{k} = (0)(0) + (1)(0) + (0)(1) = 0$ and $\mathbf{k} \cdot \mathbf{i} = (0)(1) + (0)(0) + (1)(0) = 0$.

Another method: Because $\mathbf{i}$, $\mathbf{j}$, and $\mathbf{k}$ are mutually perpendicular, the cosine factor in each dot product is $\cos\frac{\pi}{2} = 0$.

(b) By Property 1 of the dot product, $\mathbf{i} \cdot \mathbf{i} = |\mathbf{i}|^2 = 1^2 = 1$ since $\mathbf{i}$ is a unit vector. Similarly, $\mathbf{j} \cdot \mathbf{j} = |\mathbf{j}|^2 = 1$ and $\mathbf{k} \cdot \mathbf{k} = |\mathbf{k}|^2 = 1$.

13. $|\mathbf{a}| = \sqrt{3^2 + 4^2} = 5$, $|\mathbf{b}| = \sqrt{5^2 + 12^2} = 13$, and $\mathbf{a} \cdot \mathbf{b} = (3)(5) + (4)(12) = 63$. From the definition of the dot product, we have $\cos\theta = \dfrac{\mathbf{a} \cdot \mathbf{b}}{|\mathbf{a}|\,|\mathbf{b}|} = \dfrac{63}{5 \cdot 13} = \dfrac{63}{65}$. So the angle between $\mathbf{a}$ and $\mathbf{b}$ is $\theta = \cos^{-1}\left(\frac{63}{65}\right) \approx 14°$.

15. $|\mathbf{a}| = \sqrt{0^2 + 1^2 + 1^2} = \sqrt{2}$, $|\mathbf{b}| = \sqrt{1^2 + 2^2 + (-3)^2} = \sqrt{14}$, and $\mathbf{a} \cdot \mathbf{b} = (0)(1) + (1)(2) + (1)(-3) = -1$. Then $\cos\theta = \dfrac{\mathbf{a} \cdot \mathbf{b}}{|\mathbf{a}|\,|\mathbf{b}|} = \dfrac{-1}{\sqrt{2} \cdot \sqrt{14}} = \dfrac{-1}{2\sqrt{7}}$ and $\theta = \cos^{-1}\left(-\dfrac{1}{2\sqrt{7}}\right) \approx 101°$.

17. (a) $\mathbf{a} \cdot \mathbf{b} = (-5)(6) + (3)(-8) + (7)(2) = -40 \neq 0$, so $\mathbf{a}$ and $\mathbf{b}$ are not orthogonal. Also, since $\mathbf{a}$ is not a scalar multiple of $\mathbf{b}$, $\mathbf{a}$ and $\mathbf{b}$ are not parallel.

(b) $\mathbf{a} \cdot \mathbf{b} = (4)(-3) + (6)(2) = 0$, so $\mathbf{a}$ and $\mathbf{b}$ are orthogonal (and not parallel).

(c) $\mathbf{a} \cdot \mathbf{b} = (-1)(3) + (2)(4) + (5)(-1) = 0$, so $\mathbf{a}$ and $\mathbf{b}$ are orthogonal (and not parallel).

(d) Because $\mathbf{a} = -\frac{2}{3}\,\mathbf{b}$, $\mathbf{a}$ and $\mathbf{b}$ are parallel.

19. $\overrightarrow{QP} = \langle -1, -3, 2\rangle$, $\overrightarrow{QR} = \langle 4, -2, -1\rangle$, and $\overrightarrow{QP} \cdot \overrightarrow{QR} = -4 + 6 - 2 = 0$. Thus $\overrightarrow{QP}$ and $\overrightarrow{QR}$ are orthogonal, so the angle of the triangle at vertex Q is a right angle.

21. Let $\mathbf{a} = a_1\,\mathbf{i} + a_2\,\mathbf{j} + a_3\,\mathbf{k}$ be a vector orthogonal to both $\mathbf{i} + \mathbf{j}$ and $\mathbf{i} + \mathbf{k}$. Then $\mathbf{a} \cdot (\mathbf{i} + \mathbf{j}) = 0 \quad\Leftrightarrow\quad a_1 + a_2 = 0$ and $\mathbf{a} \cdot (\mathbf{i} + \mathbf{k}) = 0 \quad\Leftrightarrow\quad a_1 + a_3 = 0$, so $a_1 = -a_2 = -a_3$. Furthermore $\mathbf{a}$ is to be a unit vector, so $1 = a_1^2 + a_2^2 + a_3^2 = 3a_1^2$ implies $a_1 = \pm\frac{1}{\sqrt{3}}$. Thus $\mathbf{a} = \frac{1}{\sqrt{3}}\,\mathbf{i} - \frac{1}{\sqrt{3}}\,\mathbf{j} - \frac{1}{\sqrt{3}}\,\mathbf{k}$ and $\mathbf{a} = -\frac{1}{\sqrt{3}}\,\mathbf{i} + \frac{1}{\sqrt{3}}\,\mathbf{j} + \frac{1}{\sqrt{3}}\,\mathbf{k}$ are two such unit vectors.

23. $|\mathbf{a}| = \sqrt{3^2 + (-4)^2} = 5$. The scalar projection of $\mathbf{b}$ onto $\mathbf{a}$ is $\operatorname{comp}_{\mathbf{a}}\mathbf{b} = \dfrac{\mathbf{a} \cdot \mathbf{b}}{|\mathbf{a}|} = \dfrac{3 \cdot 5 + (-4) \cdot 0}{5} = 3$ and the vector projection of $\mathbf{b}$ onto $\mathbf{a}$ is $\operatorname{proj}_{\mathbf{a}}\mathbf{b} = \left(\dfrac{\mathbf{a} \cdot \mathbf{b}}{|\mathbf{a}|}\right)\dfrac{\mathbf{a}}{|\mathbf{a}|} = 3 \cdot \frac{1}{5}\langle 3, -4\rangle = \left\langle \frac{9}{5}, -\frac{12}{5}\right\rangle$.

25. $|\mathbf{a}| = \sqrt{9 + 36 + 4} = 7$ so the scalar projection of $\mathbf{b}$ onto $\mathbf{a}$ is $\operatorname{comp}_{\mathbf{a}}\mathbf{b} = \dfrac{\mathbf{a} \cdot \mathbf{b}}{|\mathbf{a}|} = \frac{1}{7}(3 + 12 - 6) = \frac{9}{7}$. The vector projection of $\mathbf{b}$ onto $\mathbf{a}$ is $\operatorname{proj}_{\mathbf{a}}\mathbf{b} = \dfrac{9}{7}\dfrac{\mathbf{a}}{|\mathbf{a}|} = \frac{9}{7} \cdot \frac{1}{7}\langle 3, 6, -2\rangle = \frac{9}{49}\langle 3, 6, -2\rangle = \left\langle \frac{27}{49}, \frac{54}{49}, -\frac{18}{49}\right\rangle$.

27.
$$\begin{aligned}(\operatorname{orth}_{\mathbf{a}}\mathbf{b}) \cdot \mathbf{a} &= (\mathbf{b} - \operatorname{proj}_{\mathbf{a}}\mathbf{b}) \cdot \mathbf{a} = \mathbf{b} \cdot \mathbf{a} - (\operatorname{proj}_{\mathbf{a}}\mathbf{b}) \cdot \mathbf{a} = \mathbf{b} \cdot \mathbf{a} - \frac{\mathbf{a} \cdot \mathbf{b}}{|\mathbf{a}|^2}\,\mathbf{a} \cdot \mathbf{a} \\ &= \mathbf{b} \cdot \mathbf{a} - \frac{\mathbf{a} \cdot \mathbf{b}}{|\mathbf{a}|^2}\,|\mathbf{a}|^2 = \mathbf{b} \cdot \mathbf{a} - \mathbf{a} \cdot \mathbf{b} = 0\end{aligned}$$

So they are orthogonal by (2).

29. $\operatorname{comp}_{\mathbf{a}}\mathbf{b} = \dfrac{\mathbf{a} \cdot \mathbf{b}}{|\mathbf{a}|} = 2 \quad\Leftrightarrow\quad \mathbf{a} \cdot \mathbf{b} = 2\,|\mathbf{a}| = 2\sqrt{10}$. If $\mathbf{b} = \langle b_1, b_2, b_3\rangle$, then we need $3b_1 + 0b_2 - 1b_3 = 2\sqrt{10}$. One possible solution is obtained by taking $b_1 = 0$, $b_2 = 0$, $b_3 = -2\sqrt{10}$. In general, $\mathbf{b} = \left\langle s, t, 3s - 2\sqrt{10}\right\rangle$, $s, t \in \mathbb{R}$.

31. Here $\mathbf{D} = (4 - 2)\,\mathbf{i} + (9 - 3)\,\mathbf{j} + (15 - 0)\,\mathbf{k} = 2\,\mathbf{i} + 6\,\mathbf{j} + 15\,\mathbf{k}$, so $W = \mathbf{F} \cdot \mathbf{D} = (10\mathbf{i} + 18\mathbf{j} - 6\mathbf{k}) \cdot (2\mathbf{i} + 6\mathbf{j} + 15\mathbf{k}) = 20 + 108 - 90 = 38$ joules.

33.
$$\begin{aligned}W &= |\mathbf{F}|\,|\mathbf{D}|\cos\theta \\ &= (140)(4)\cos 20° \\ &= 560\cos 20° \\ &\approx 526\text{ J}\end{aligned}$$

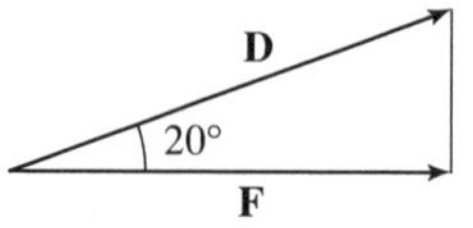

35. First note that $\mathbf{n} = \langle a, b\rangle$ is perpendicular to the line, because if $Q_1 = (a_1, b_1)$ and $Q_2 = (a_2, b_2)$ lie on the line, then $\mathbf{n}\cdot\overrightarrow{Q_1Q_2} = aa_2 - aa_1 + bb_2 - bb_1 = 0$, since $aa_2 + bb_2 = -c = aa_1 + bb_1$ from the equation of the line.

Let $P_2 = (x_2, y_2)$ lie on the line. Then the distance from P_1 to the line is the absolute value of the scalar projection of $\overrightarrow{P_1P_2}$ onto $\mathbf{n}$. $\text{comp}_{\mathbf{n}}\left(\overrightarrow{P_1P_2}\right) = \dfrac{|\mathbf{n}\cdot\langle x_2 - x_1, y_2 - y_1\rangle|}{|\mathbf{n}|} = \dfrac{|ax_2 - ax_1 + by_2 - by_1|}{\sqrt{a^2+b^2}} = \dfrac{|ax_1 + by_1 + c|}{\sqrt{a^2+b^2}}$ since $ax_2 + by_2 = -c$. The required distance is $\dfrac{|3\cdot -2 + -4\cdot 3 + 5|}{\sqrt{3^2+4^2}} = \dfrac{13}{5}$.

37. For convenience, consider the unit cube positioned so that its back left corner is at the origin, and its edges lie along the coordinate axes. The diagonal of the cube that begins at the origin and ends at $(1,1,1)$ has vector representation $\langle 1,1,1\rangle$. The angle θ between this vector and the vector of the edge which also begins at the origin and runs along the x-axis [that is, $\langle 1,0,0\rangle$] is given by $\cos\theta = \dfrac{\langle 1,1,1\rangle\cdot\langle 1,0,0\rangle}{|\langle 1,1,1\rangle|\,|\langle 1,0,0\rangle|} = \dfrac{1}{\sqrt{3}} \quad\Rightarrow\quad \theta = \cos^{-1}\!\left(\frac{1}{\sqrt{3}}\right) \approx 55^\circ$.

39. Consider the H-C-H combination consisting of the sole carbon atom and the two hydrogen atoms that are at $(1,0,0)$ and $(0,1,0)$ (or any H-C-H combination, for that matter). Vector representations of the line segments emanating from the carbon atom and extending to these two hydrogen atoms are $\left\langle 1-\frac12, 0-\frac12, 0-\frac12\right\rangle = \left\langle \frac12, -\frac12, -\frac12\right\rangle$ and $\left\langle 0-\frac12, 1-\frac12, 0-\frac12\right\rangle = \left\langle -\frac12, \frac12, -\frac12\right\rangle$. The bond angle, θ, is therefore given by

$$\cos\theta = \frac{\left\langle \frac12, -\frac12, -\frac12\right\rangle\cdot\left\langle -\frac12, \frac12, -\frac12\right\rangle}{\left|\left\langle \frac12, -\frac12, -\frac12\right\rangle\right|\left|\left\langle -\frac12, \frac12, -\frac12\right\rangle\right|} = \frac{-\frac14 - \frac14 + \frac14}{\sqrt{\frac34}\sqrt{\frac34}} = -\frac13 \quad\Rightarrow\quad \theta = \cos^{-1}\left(-\tfrac13\right) \approx 109.5^\circ.$$

41. If $c = 0$ then $c\mathbf{a} = \mathbf{0}$, so $(c\mathbf{a})\cdot\mathbf{b} = \mathbf{0}\cdot\mathbf{b} = 0$ by Property 5. Similarly, $\mathbf{a}\cdot(c\mathbf{b}) = \mathbf{a}\cdot\mathbf{0} = 0$, and $c(\mathbf{a}\cdot\mathbf{b}) = 0(|\mathbf{a}|\,|\mathbf{b}|\cos\theta) = 0$, thus $(c\mathbf{a})\cdot\mathbf{b} = c(\mathbf{a}\cdot\mathbf{b}) = \mathbf{a}\cdot(c\mathbf{b})$. If $c > 0$, the angle θ between $\mathbf{a}$ and $\mathbf{b}$ coincides with the angle between $c\mathbf{a}$ and $\mathbf{b}$, so by definition of the dot product, $(c\mathbf{a})\cdot\mathbf{b} = |c\mathbf{a}|\,|\mathbf{b}|\cos\theta = |c|\,|\mathbf{a}|\,|\mathbf{b}|\cos\theta = c\,|\mathbf{a}|\,|\mathbf{b}|\cos\theta$. Similarly, $\mathbf{a}\cdot(c\mathbf{b}) = |\mathbf{a}|\,|c\mathbf{b}|\cos\theta = |\mathbf{a}|\,|c|\,|\mathbf{b}|\cos\theta = c\,|\mathbf{a}|\,|\mathbf{b}|\cos\theta$, and $c(\mathbf{a}\cdot\mathbf{b}) = c\,|\mathbf{a}|\,|\mathbf{b}|\cos\theta$. Thus, $(c\mathbf{a})\cdot\mathbf{b} = c(\mathbf{a}\cdot\mathbf{b}) = \mathbf{a}\cdot(c\mathbf{b})$. The case for $c < 0$ is similar. Using components, let $\mathbf{a} = \langle a_1, a_2, a_3\rangle$ and $\mathbf{b} = \langle b_1, b_2, b_3\rangle$. Then

$$\begin{aligned}(c\mathbf{a})\cdot\mathbf{b} &= \langle ca_1, ca_2, ca_3\rangle\cdot\langle b_1, b_2, b_3\rangle = (ca_1)b_1 + (ca_2)b_2 + (ca_3)b_3\\ &= c(a_1b_1 + a_2b_2 + a_3b_3) = c(\mathbf{a}\cdot\mathbf{b})\\ &= a_1(cb_1) + a_2(cb_2) + a_3(cb_3) = \langle a_1, a_2, a_3\rangle\cdot\langle cb_1, cb_2, cb_3\rangle = \mathbf{a}\cdot(c\mathbf{b})\end{aligned}$$

43. $|\mathbf{a}\cdot\mathbf{b}| = \big|\,|\mathbf{a}|\,|\mathbf{b}|\cos\theta\big| = |\mathbf{a}|\,|\mathbf{b}|\,|\cos\theta|$. Since $|\cos\theta| \le 1$, $|\mathbf{a}\cdot\mathbf{b}| = |\mathbf{a}|\,|\mathbf{b}|\,|\cos\theta| \le |\mathbf{a}|\,|\mathbf{b}|$.

Note: We have equality in the case of $\cos\theta = \pm 1$, so $\theta = 0$ or $\theta = \pi$, thus equality when $\mathbf{a}$ and $\mathbf{b}$ are parallel.

45. (a)

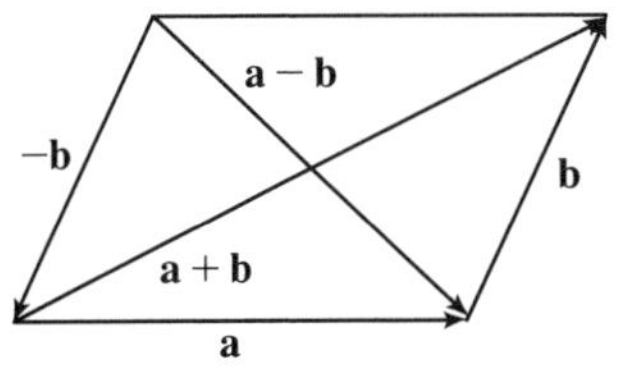

The Parallelogram Law states that the sum of the squares of the lengths of the diagonals of a parallelogram equals the sum of the squares of its (four) sides.

(b) $|\mathbf{a}+\mathbf{b}|^2 = (\mathbf{a}+\mathbf{b})\cdot(\mathbf{a}+\mathbf{b}) = |\mathbf{a}|^2 + 2(\mathbf{a}\cdot\mathbf{b}) + |\mathbf{b}|^2$ and $|\mathbf{a}-\mathbf{b}|^2 = (\mathbf{a}-\mathbf{b})\cdot(\mathbf{a}-\mathbf{b}) = |\mathbf{a}|^2 - 2(\mathbf{a}\cdot\mathbf{b}) + |\mathbf{b}|^2$. Adding these two equations gives $|\mathbf{a}+\mathbf{b}|^2 + |\mathbf{a}-\mathbf{b}|^2 = 2\,|\mathbf{a}|^2 + 2\,|\mathbf{b}|^2$.

9.4 The Cross Product

1. (a) Since $\mathbf{b}\times\mathbf{c}$ is a vector, the dot product $\mathbf{a}\cdot(\mathbf{b}\times\mathbf{c})$ is meaningful and is a scalar.

(b) $\mathbf{b}\cdot\mathbf{c}$ is a scalar, so $\mathbf{a}\times(\mathbf{b}\cdot\mathbf{c})$ is meaningless, as the cross product is defined only for two *vectors*.

(c) Since $\mathbf{b}\times\mathbf{c}$ is a vector, the cross product $\mathbf{a}\times(\mathbf{b}\times\mathbf{c})$ is meaningful and results in another vector.

(d) $\mathbf{a}\cdot\mathbf{b}$ is a scalar, so the cross product $(\mathbf{a}\cdot\mathbf{b})\times\mathbf{c}$ is meaningless.

(e) Since $(\mathbf{a}\cdot\mathbf{b})$ and $(\mathbf{c}\cdot\mathbf{d})$ are both scalars, the cross product $(\mathbf{a}\cdot\mathbf{b})\times(\mathbf{c}\cdot\mathbf{d})$ is meaningless.

(f) $\mathbf{a}\times\mathbf{b}$ and $\mathbf{c}\times\mathbf{d}$ are both vectors, so the dot product $(\mathbf{a}\times\mathbf{b})\cdot(\mathbf{c}\times\mathbf{d})$ is meaningful and is a scalar.

3. If we sketch $\mathbf{u}$ and $\mathbf{v}$ starting from the same initial point, we see that the angle between them is $30°$, so

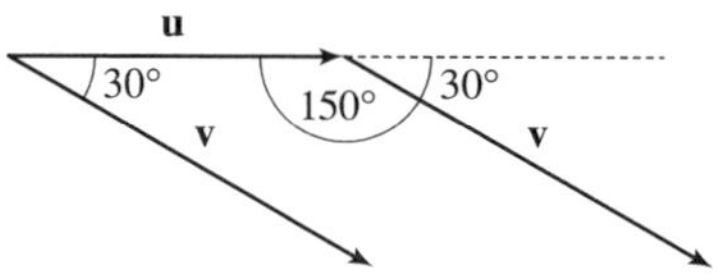

$$|\mathbf{u}\times\mathbf{v}| = |\mathbf{u}|\,|\mathbf{v}|\sin 30° = (6)(8)\left(\tfrac{1}{2}\right) = 24$$

By the right-hand rule, $\mathbf{u}\times\mathbf{v}$ is directed into the page.

5. The magnitude of the torque is $|\boldsymbol{\tau}| = |\mathbf{r}\times\mathbf{F}| = |\mathbf{r}|\,|\mathbf{F}|\sin\theta = (0.18\text{ m})(60\text{ N})\sin(70+10)° = 10.8\sin 80° \approx 10.6\text{ N}\cdot\text{m}$.

7. $$\mathbf{a}\times\mathbf{b} = \begin{vmatrix} \mathbf{i} & \mathbf{j} & \mathbf{k} \\ 1 & -1 & 0 \\ 3 & 2 & 1 \end{vmatrix} = \begin{vmatrix} -1 & 0 \\ 2 & 1 \end{vmatrix}\mathbf{i} - \begin{vmatrix} 1 & 0 \\ 3 & 1 \end{vmatrix}\mathbf{j} + \begin{vmatrix} 1 & -1 \\ 3 & 2 \end{vmatrix}\mathbf{k} = (-1-0)\,\mathbf{i} - (1-0)\,\mathbf{j} + [2-(-3)]\,\mathbf{k} = -\mathbf{i} - \mathbf{j} + 5\,\mathbf{k}$$

Since $(\mathbf{a}\times\mathbf{b})\cdot\mathbf{a} = \langle -1,-1,5\rangle\cdot\langle 1,-1,0\rangle = -1+1+0 = 0$, $\mathbf{a}\times\mathbf{b}$ is orthogonal to $\mathbf{a}$.

Since $(\mathbf{a}\times\mathbf{b})\cdot\mathbf{b} = \langle -1,-1,5\rangle\cdot\langle 3,2,1\rangle = -3-2+5 = 0$, $\mathbf{a}\times\mathbf{b}$ is orthogonal to $\mathbf{b}$.

9. $$\mathbf{a}\times\mathbf{b} = \begin{vmatrix} \mathbf{i} & \mathbf{j} & \mathbf{k} \\ t & t^2 & t^3 \\ 1 & 2t & 3t^2 \end{vmatrix} = \begin{vmatrix} t^2 & t^3 \\ 2t & 3t^2 \end{vmatrix}\mathbf{i} - \begin{vmatrix} t & t^3 \\ 1 & 3t^2 \end{vmatrix}\mathbf{j} + \begin{vmatrix} t & t^2 \\ 1 & 2t \end{vmatrix}\mathbf{k}$$

$$= (3t^4 - 2t^4)\,\mathbf{i} - (3t^3 - t^3)\,\mathbf{j} + (2t^2 - t^2)\,\mathbf{k} = t^4\,\mathbf{i} - 2t^3\,\mathbf{j} + t^2\,\mathbf{k}$$

Since $(\mathbf{a}\times\mathbf{b})\cdot\mathbf{a} = \langle t^4, -2t^3, t^2\rangle\cdot\langle t, t^2, t^3\rangle = t^5 - 2t^5 + t^5 = 0$, $\mathbf{a}\times\mathbf{b}$ is orthogonal to $\mathbf{a}$.

Since $(\mathbf{a}\times\mathbf{b})\cdot\mathbf{b} = \langle t^4, -2t^3, t^2\rangle\cdot\langle 1, 2t, 3t^2\rangle = t^4 - 4t^4 + 3t^4 = 0$, $\mathbf{a}\times\mathbf{b}$ is orthogonal to $\mathbf{b}$.

11. $$\mathbf{a}\times\mathbf{b} = \begin{vmatrix} \mathbf{i} & \mathbf{j} & \mathbf{k} \\ 3 & 2 & 4 \\ 1 & -2 & -3 \end{vmatrix} = \begin{vmatrix} 2 & 4 \\ -2 & -3 \end{vmatrix}\mathbf{i} - \begin{vmatrix} 3 & 4 \\ 1 & -3 \end{vmatrix}\mathbf{j} + \begin{vmatrix} 3 & 2 \\ 1 & -2 \end{vmatrix}\mathbf{k}$$

$$= [-6-(-8)]\,\mathbf{i} - (-9-4)\,\mathbf{j} + (-6-2)\,\mathbf{k} = 2\,\mathbf{i} + 13\,\mathbf{j} - 8\,\mathbf{k}$$

Since $(\mathbf{a}\times\mathbf{b})\cdot\mathbf{a} = (2\,\mathbf{i} + 13\,\mathbf{j} - 8\,\mathbf{k})\cdot(3\,\mathbf{i} + 2\,\mathbf{j} + 4\,\mathbf{k}) = 6 + 26 - 32 = 0$, $\mathbf{a}\times\mathbf{b}$ is orthogonal to $\mathbf{a}$.

Since $(\mathbf{a}\times\mathbf{b})\cdot\mathbf{b} = (2\,\mathbf{i} + 13\,\mathbf{j} - 8\,\mathbf{k})\cdot(\mathbf{i} - 2\,\mathbf{j} - 3\,\mathbf{k}) = 2 - 26 + 24 = 0$, $\mathbf{a}\times\mathbf{b}$ is orthogonal to $\mathbf{b}$.

13. We know that the cross product of two vectors is orthogonal to both. So we calculate

$$\langle 2,0,-3\rangle\times\langle -1,4,2\rangle = \begin{vmatrix} \mathbf{i} & \mathbf{j} & \mathbf{k} \\ 2 & 0 & -3 \\ -1 & 4 & 2 \end{vmatrix} = \begin{vmatrix} 0 & -3 \\ 4 & 2 \end{vmatrix}\mathbf{i} - \begin{vmatrix} 2 & -3 \\ -1 & 2 \end{vmatrix}\mathbf{j} + \begin{vmatrix} 2 & 0 \\ -1 & 4 \end{vmatrix}\mathbf{k} = 12\,\mathbf{i} - \mathbf{j} + 8\,\mathbf{k}$$

So two unit vectors orthogonal to both are $\pm\dfrac{\langle 12,-1,8\rangle}{\sqrt{144+1+64}} = \pm\dfrac{\langle 12,-1,8\rangle}{\sqrt{209}}$, that is, $\left\langle \frac{12}{\sqrt{209}}, -\frac{1}{\sqrt{209}}, \frac{8}{\sqrt{209}} \right\rangle$ and $\left\langle -\frac{12}{\sqrt{209}}, \frac{1}{\sqrt{209}}, -\frac{8}{\sqrt{209}} \right\rangle$.

15. By plotting the vertices, we can see that the parallelogram is determined by the vectors $\overrightarrow{AB} = \langle 2,3\rangle$ and $\overrightarrow{AD} = \langle 4,-2\rangle$. We know that the area of the parallelogram determined by two vectors is equal to the length of the cross product of these vectors. In order to compute the cross product, we consider the vector $\overrightarrow{AB}$ as the three-dimensional vector $\langle 2,3,0\rangle$ (and similarly for $\overrightarrow{AD}$), and then the area of parallelogram $ABCD$ is

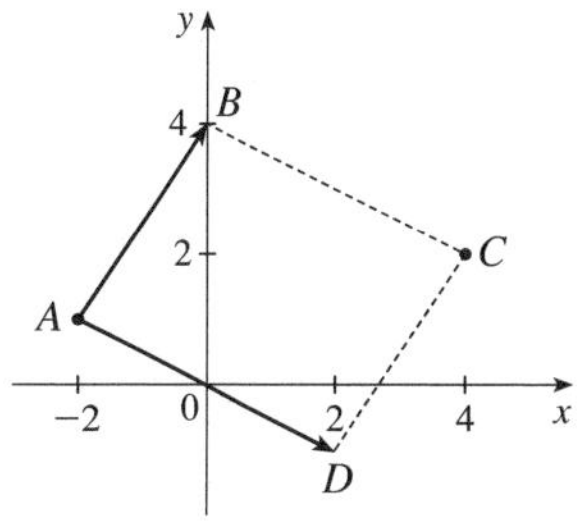

$$\left|\overrightarrow{AB}\times\overrightarrow{AD}\right| = \left\| \begin{matrix} \mathbf{i} & \mathbf{j} & \mathbf{k} \\ 2 & 3 & 0 \\ 4 & -2 & 0 \end{matrix} \right\| = |(0)\,\mathbf{i} - (0)\,\mathbf{j} + (-4-12)\,\mathbf{k}| = |-16\,\mathbf{k}| = 16$$

17. (a) $\overrightarrow{PQ} = \langle 4,3,-2\rangle$ and $\overrightarrow{PR} = \langle 5,5,1\rangle$, so a vector orthogonal to the plane through P, Q, and R is $\overrightarrow{PQ}\times\overrightarrow{PR} = \langle (3)(1)-(-2)(5), (-2)(5)-(4)(1), (4)(5)-(3)(5)\rangle = \langle 13,-14,5\rangle$ (or any scalar mutiple thereof).

(b) The area of the parallelogram determined by $\overrightarrow{PQ}$ and $\overrightarrow{PR}$ is $\left|\overrightarrow{PQ}\times\overrightarrow{PR}\right| = |\langle 13,-14,5\rangle| = \sqrt{13^2+(-14)^2+5^2} = \sqrt{390}$, so the area of triangle PQR is $\frac{1}{2}\sqrt{390}$.

19. Using the notation of (1), $\mathbf{r} = \langle 0,0.3,0\rangle$ and $\mathbf{F}$ has direction $\langle 0,3,-4\rangle$. The angle θ between them can be determined by $\cos\theta = \dfrac{\langle 0,0.3,0\rangle\cdot\langle 0,3,-4\rangle}{|\langle 0,0.3,0\rangle|\,|\langle 0,3,-4\rangle|} \Rightarrow \cos\theta = \dfrac{0.9}{(0.3)(5)} \Rightarrow \cos\theta = 0.6 \Rightarrow \theta\approx 53.1^\circ$. Then $|\boldsymbol{\tau}| = |\mathbf{r}|\,|\mathbf{F}|\sin\theta \Rightarrow 100 = 0.3\,|\mathbf{F}|\sin 53.1^\circ \Rightarrow |\mathbf{F}|\approx 417$ N.

21. We know that the volume of the parallelepiped determined by $\mathbf{a}$, $\mathbf{b}$, and $\mathbf{c}$ is the magnitude of their scalar triple product, which is

$$\mathbf{a}\cdot(\mathbf{b}\times\mathbf{c}) = \begin{vmatrix} 6 & 3 & -1 \\ 0 & 1 & 2 \\ 4 & -2 & 5 \end{vmatrix} = 6\begin{vmatrix} 1 & 2 \\ -2 & 5 \end{vmatrix} - 3\begin{vmatrix} 0 & 2 \\ 4 & 5 \end{vmatrix} + (-1)\begin{vmatrix} 0 & 1 \\ 4 & -2 \end{vmatrix}$$
$$= 6(5+4) - 3(0-8) - (0-4) = 82$$

Thus the volume of the parallelepiped is 82 cubic units.

23. $\mathbf{a} = \overrightarrow{PQ} = \langle 2,1,1\rangle$, $\mathbf{b} = \overrightarrow{PR} = \langle 1,-1,2\rangle$, and $\mathbf{c} = \overrightarrow{PS} = \langle 0,-2,3\rangle$.

$$\mathbf{a}\cdot(\mathbf{b}\times\mathbf{c}) = \begin{vmatrix} 2 & 1 & 1 \\ 1 & -1 & 2 \\ 0 & -2 & 3 \end{vmatrix} = 2\begin{vmatrix} -1 & 2 \\ -2 & 3 \end{vmatrix} - 1\begin{vmatrix} 1 & 2 \\ 0 & 3 \end{vmatrix} + 1\begin{vmatrix} 1 & -1 \\ 0 & -2 \end{vmatrix} = 2-3-2 = -3,$$

so the volume of the parallelepiped is 3 cubic units.

25. $\mathbf{u}\cdot(\mathbf{v}\times\mathbf{w}) = \begin{vmatrix} 1 & 5 & -2 \\ 3 & -1 & 0 \\ 5 & 9 & -4 \end{vmatrix} = 1\begin{vmatrix} -1 & 0 \\ 9 & -4 \end{vmatrix} - 5\begin{vmatrix} 3 & 0 \\ 5 & -4 \end{vmatrix} + (-2)\begin{vmatrix} 3 & -1 \\ 5 & 9 \end{vmatrix} = 4 + 60 - 64 = 0$, which says that the volume of the parallelepiped determined by $\mathbf{u}$, $\mathbf{v}$ and $\mathbf{w}$ is 0, and thus these three vectors are coplanar.

27. (a)

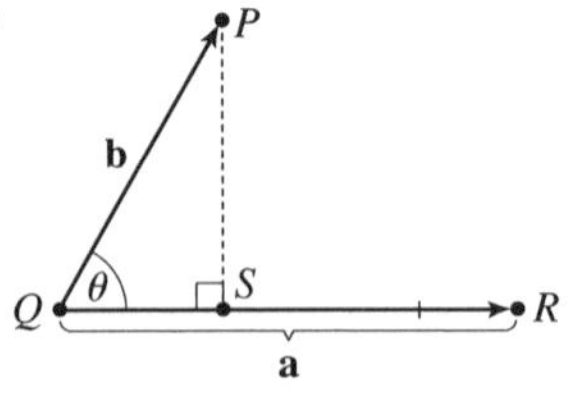

The distance between a point and a line is the length of the perpendicular from the point to the line, here $\left|\overrightarrow{PS}\right| = d$. But referring to triangle PQS, $d = \left|\overrightarrow{PS}\right| = \left|\overrightarrow{QP}\right| \sin\theta = |\mathbf{b}|\sin\theta$. But θ is the angle between $\overrightarrow{QP} = \mathbf{b}$ and $\overrightarrow{QR} = \mathbf{a}$. Thus by the definition of the cross product, $\sin\theta = \dfrac{|\mathbf{a}\times\mathbf{b}|}{|\mathbf{a}|\,|\mathbf{b}|}$ and so

$$d = |\mathbf{b}|\sin\theta = \frac{|\mathbf{b}|\,|\mathbf{a}\times\mathbf{b}|}{|\mathbf{a}|\,|\mathbf{b}|} = \frac{|\mathbf{a}\times\mathbf{b}|}{|\mathbf{a}|}.$$

(b) $\mathbf{a} = \overrightarrow{QR} = \langle -1, -2, -1\rangle$ and $\mathbf{b} = \overrightarrow{QP} = \langle 1, -5, -7\rangle$. Then

$\mathbf{a}\times\mathbf{b} = \langle(-2)(-7) - (-1)(-5), (-1)(1) - (-1)(-7), (-1)(-5) - (-2)(1)\rangle = \langle 9, -8, 7\rangle$.

Thus the distance is $d = \dfrac{|\mathbf{a}\times\mathbf{b}|}{|\mathbf{a}|} = \frac{1}{\sqrt{6}}\sqrt{81 + 64 + 49} = \sqrt{\frac{194}{6}} = \sqrt{\frac{97}{3}}$.

29.

$$\begin{aligned}(\mathbf{a}-\mathbf{b})\times(\mathbf{a}+\mathbf{b}) &= (\mathbf{a}-\mathbf{b})\times\mathbf{a} + (\mathbf{a}-\mathbf{b})\times\mathbf{b} && \text{by Property 3 of the cross product}\\ &= \mathbf{a}\times\mathbf{a} + (-\mathbf{b})\times\mathbf{a} + \mathbf{a}\times\mathbf{b} + (-\mathbf{b})\times\mathbf{b} && \text{by Property 4}\\ &= (\mathbf{a}\times\mathbf{a}) - (\mathbf{b}\times\mathbf{a}) + (\mathbf{a}\times\mathbf{b}) - (\mathbf{b}\times\mathbf{b}) && \text{by Property 2 (with } c = -1)\\ &= \mathbf{0} - (\mathbf{b}\times\mathbf{a}) + (\mathbf{a}\times\mathbf{b}) - \mathbf{0} && \text{by the margin note on page 658}\\ &= (\mathbf{a}\times\mathbf{b}) + (\mathbf{a}\times\mathbf{b}) && \text{by Property 1}\\ &= 2(\mathbf{a}\times\mathbf{b})\end{aligned}$$

31. $\mathbf{a}\times(\mathbf{b}\times\mathbf{c}) + \mathbf{b}\times(\mathbf{c}\times\mathbf{a}) + \mathbf{c}\times(\mathbf{a}\times\mathbf{b})$

$$= [(\mathbf{a}\cdot\mathbf{c})\mathbf{b} - (\mathbf{a}\cdot\mathbf{b})\mathbf{c}] + [(\mathbf{b}\cdot\mathbf{a})\mathbf{c} - (\mathbf{b}\cdot\mathbf{c})\mathbf{a}] + [(\mathbf{c}\cdot\mathbf{b})\mathbf{a} - (\mathbf{c}\cdot\mathbf{a})\mathbf{b}] \qquad \text{by Exercise 30}$$

$$= (\mathbf{a}\cdot\mathbf{c})\mathbf{b} - (\mathbf{a}\cdot\mathbf{b})\mathbf{c} + (\mathbf{a}\cdot\mathbf{b})\mathbf{c} - (\mathbf{b}\cdot\mathbf{c})\mathbf{a} + (\mathbf{b}\cdot\mathbf{c})\mathbf{a} - (\mathbf{a}\cdot\mathbf{c})\mathbf{b} = \mathbf{0}$$

33. (a) No. If $\mathbf{a}\cdot\mathbf{b} = \mathbf{a}\cdot\mathbf{c}$, then $\mathbf{a}\cdot(\mathbf{b}-\mathbf{c}) = 0$, so $\mathbf{a}$ is perpendicular to $\mathbf{b}-\mathbf{c}$, which can happen if $\mathbf{b}\neq\mathbf{c}$. For example, let $\mathbf{a} = \langle 1, 1, 1\rangle$, $\mathbf{b} = \langle 1, 0, 0\rangle$ and $\mathbf{c} = \langle 0, 1, 0\rangle$.

(b) No. If $\mathbf{a}\times\mathbf{b} = \mathbf{a}\times\mathbf{c}$ then $\mathbf{a}\times(\mathbf{b}-\mathbf{c}) = \mathbf{0}$, which implies that $\mathbf{a}$ is parallel to $\mathbf{b}-\mathbf{c}$, which of course can happen if $\mathbf{b}\neq\mathbf{c}$.

(c) Yes. Since $\mathbf{a}\cdot\mathbf{c} = \mathbf{a}\cdot\mathbf{b}$, $\mathbf{a}$ is perpendicular to $\mathbf{b}-\mathbf{c}$, by part (a). From part (b), $\mathbf{a}$ is also parallel to $\mathbf{b}-\mathbf{c}$. Thus since $\mathbf{a}\neq\mathbf{0}$ but is both parallel and perpendicular to $\mathbf{b}-\mathbf{c}$, we have $\mathbf{b}-\mathbf{c} = \mathbf{0}$, so $\mathbf{b} = \mathbf{c}$.

9.5 Equations of Lines and Planes

1. (a) True; each of the first two lines has a direction vector parallel to the direction vector of the third line, so these vectors are each scalar multiples of the third direction vector. Then the first two direction vectors are also scalar multiples of each other, so these vectors, and hence the two lines, are parallel.

(b) False; for example, the x- and y-axes are both perpendicular to the z-axis, yet the x- and y-axes are not parallel.

(c) True; each of the first two planes has a normal vector parallel to the normal vector of the third plane, so these two normal vectors are parallel to each other and the planes are parallel.

(d) False; for example, the xy- and yz-planes are not parallel, yet they are both perpendicular to the xz-plane.

(e) False; the x- and y-axes are not parallel, yet they are both parallel to the plane $z = 1$.

(f) True; if each line is perpendicular to a plane, then the lines' direction vectors are both parallel to a normal vector for the plane. Thus, the direction vectors are parallel to each other and the lines are parallel.

(g) False; the planes $y = 1$ and $z = 1$ are not parallel, yet they are both parallel to the x-axis.

(h) True; if each plane is perpendicular to a line, then any normal vector for each plane is parallel to a direction vector for the line. Thus, the normal vectors are parallel to each other and the planes are parallel.

(i) True; see Figure 9 and the accompanying discussion.

(j) False; they can be skew, as in Example 3.

(k) True. Consider any normal vector for the plane and any direction vector for the line. If the normal vector is perpendicular to the direction vector, the line and plane are parallel. Otherwise, the vectors meet at an angle θ, $0^\circ \leq \theta < 90^\circ$, and the line will intersect the plane at an angle $90^\circ - \theta$.

3. For this line, we have $\mathbf{r}_0 = -2\,\mathbf{i} + 4\,\mathbf{j} + 10\,\mathbf{k}$ and $\mathbf{v} = 3\,\mathbf{i} + \mathbf{j} - 8\,\mathbf{k}$, so a vector equation is $\mathbf{r} = \mathbf{r}_0 + t\,\mathbf{v} = (-2\,\mathbf{i} + 4\,\mathbf{j} + 10\,\mathbf{k}) + t(3\,\mathbf{i} + \mathbf{j} - 8\,\mathbf{k}) = (-2 + 3t)\,\mathbf{i} + (4 + t)\,\mathbf{j} + (10 - 8t)\,\mathbf{k}$ and parametric equations are $x = -2 + 3t$, $y = 4 + t$, $z = 10 - 8t$.

5. A line perpendicular to the given plane has the same direction as a normal vector to the plane, such as $\mathbf{n} = \langle 1, 3, 1 \rangle$. So $\mathbf{r}_0 = \mathbf{i} + 6\,\mathbf{k}$, and we can take $\mathbf{v} = \mathbf{i} + 3\,\mathbf{j} + \mathbf{k}$. Then a vector equation is $\mathbf{r} = (\mathbf{i} + 6\,\mathbf{k}) + t(\mathbf{i} + 3\,\mathbf{j} + \mathbf{k}) = (1 + t)\,\mathbf{i} + 3t\,\mathbf{j} + (6 + t)\,\mathbf{k}$, and parametric equations are $x = 1 + t$, $y = 3t$, $z = 6 + t$.

7. $\mathbf{v} = \left\langle 2 - 0, 1 - \frac{1}{2}, -3 - 1 \right\rangle = \left\langle 2, \frac{1}{2}, -4 \right\rangle$, and letting $P_0 = (2, 1, -3)$, parametric equations are $x = 2 + 2t$, $y = 1 + \frac{1}{2}t$, $z = -3 - 4t$, while symmetric equations are $\dfrac{x-2}{2} = \dfrac{y-1}{1/2} = \dfrac{z+3}{-4}$ or $\dfrac{x-2}{2} = 2y - 2 = \dfrac{z+3}{-4}$.

9. The line has direction $\mathbf{v} = \langle 1, 2, 1 \rangle$. Letting $P_0 = (1, -1, 1)$, parametric equations are $x = 1 + t$, $y = -1 + 2t$, $z = 1 + t$ and symmetric equations are $x - 1 = \dfrac{y+1}{2} = z - 1$.

11. Direction vectors of the lines are $\mathbf{v}_1 = \langle -2 - (-4), 0 - (-6), -3 - 1 \rangle = \langle 2, 6, -4 \rangle$ and $\mathbf{v}_2 = \langle 5 - 10, 3 - 18, 14 - 4 \rangle = \langle -5, -15, 10 \rangle$, and since $\mathbf{v}_2 = -\frac{5}{2}\mathbf{v}_1$, the direction vectors and thus the lines are parallel.

13. (a) A direction vector of the line with parametric equations $x = 1 + 2t$, $y = 3t$, $z = 5 - 7t$ is $\mathbf{v} = \langle 2, 3, -7 \rangle$ and the desired parallel line must also have $\mathbf{v}$ as a direction vector. Here $P_0 = (0, 2, -1)$, so symmetric equations for the line are $\dfrac{x}{2} = \dfrac{y-2}{3} = \dfrac{z+1}{-7}$.

(b) The line intersects the xy-plane when $z = 0$, so we need $\dfrac{x}{2} = \dfrac{y-2}{3} = \dfrac{1}{-7}$ or $x = -\frac{2}{7}$, $y = \frac{11}{7}$. Thus the point of intersection with the xy-plane is $\left(-\frac{2}{7}, \frac{11}{7}, 0\right)$. Similarly for the yz-plane, we need $x = 0 \;\Leftrightarrow\; 0 = \dfrac{y-2}{3} = \dfrac{z+1}{-7} \;\Leftrightarrow\; y = 2$, $z = -1$. Thus the line intersects the yz-plane at $(0, 2, -1)$. For the xz-plane, we need $y = 0 \;\Leftrightarrow\; \dfrac{x}{2} = -\dfrac{2}{3} = \dfrac{z+1}{-7} \;\Leftrightarrow\; x = -\frac{4}{3}$, $z = \frac{11}{3}$. So the line intersects the xz-plane at $\left(-\frac{4}{3}, 0, \frac{11}{3}\right)$.

15. From Equation 4, the line segment from $\mathbf{r}_0 = 2\,\mathbf{i} - \mathbf{j} + 4\,\mathbf{k}$ to $\mathbf{r}_1 = 4\,\mathbf{i} + 6\,\mathbf{j} + \mathbf{k}$ is $\mathbf{r}(t) = (1 - t)\,\mathbf{r}_0 + t\,\mathbf{r}_1 = (1 - t)(2\,\mathbf{i} - \mathbf{j} + 4\,\mathbf{k}) + t(4\,\mathbf{i} + 6\,\mathbf{j} + \mathbf{k}) = (2\,\mathbf{i} - \mathbf{j} + 4\,\mathbf{k}) + t(2\,\mathbf{i} + 7\,\mathbf{j} - 3\,\mathbf{k})$, $0 \leq t \leq 1$.

17. Since the direction vectors are $\mathbf{v}_1 = \langle -6, 9, -3 \rangle$ and $\mathbf{v}_2 = \langle 2, -3, 1 \rangle$, we have $\mathbf{v}_1 = -3\mathbf{v}_2$ so the lines are parallel.

19. Since the direction vectors $\langle 1, 2, 3\rangle$ and $\langle -4, -3, 2\rangle$ are not scalar multiples of each other, the lines are not parallel, so we check to see if the lines intersect. The parametric equations of the lines are L_1: $x = t$, $y = 1 + 2t$, $z = 2 + 3t$ and L_2: $x = 3 - 4s$, $y = 2 - 3s$, $z = 1 + 2s$. For the lines to intersect, we must be able to find one value of t and one value of s that produce the same point from the respective parametric equations. Thus we need to satisfy the following three equations: $t = 3 - 4s$, $1 + 2t = 2 - 3s$, $2 + 3t = 1 + 2s$. Solving the first two equations we get $t = -1$, $s = 1$ and checking, we see that these values don't satisfy the third equation. Thus the lines aren't parallel and don't intersect, so they must be skew lines.

21. Since the plane is perpendicular to the vector $\langle -2, 1, 5\rangle$, we can take $\langle -2, 1, 5\rangle$ as a normal vector to the plane. $(6, 3, 2)$ is a point on the plane, so setting $a = -2$, $b = 1$, $c = 5$ and $x_0 = 6$, $y_0 = 3$, $z_0 = 2$ in Equation 7 gives $-2(x - 6) + 1(y - 3) + 5(z - 2) = 0$ or $-2x + y + 5z = 1$ to be an equation of the plane.

23. Since the two planes are parallel, they will have the same normal vectors. So we can take $\mathbf{n} = \langle 2, -1, 3\rangle$, and an equation of the plane is $2(x - 0) - 1(y - 0) + 3(z - 0) = 0$ or $2x - y + 3z = 0$.

25. Here the vectors $\mathbf{a} = \langle 1 - 0, 0 - 1, 1 - 1\rangle = \langle 1, -1, 0\rangle$ and $\mathbf{b} = \langle 1 - 0, 1 - 1, 0 - 1\rangle = \langle 1, 0, -1\rangle$ lie in the plane, so $\mathbf{a} \times \mathbf{b}$ is a normal vector to the plane. Thus, we can take $\mathbf{n} = \mathbf{a} \times \mathbf{b} = \langle 1 - 0, 0 + 1, 0 + 1\rangle = \langle 1, 1, 1\rangle$. If P_0 is the point $(0, 1, 1)$, an equation of the plane is $1(x - 0) + 1(y - 1) + 1(z - 1) = 0$ or $x + y + z = 2$.

27. If we first find two nonparallel vectors in the plane, their cross product will be a normal vector to the plane. Since the given line lies in the plane, its direction vector $\mathbf{a} = \langle -2, 5, 4\rangle$ is one vector in the plane. We can verify that the given point $(6, 0, -2)$ does not lie on this line, so to find another nonparallel vector $\mathbf{b}$ which lies in the plane, we can pick any point on the line and find a vector connecting the points. If we put $t = 0$, we see that $(4, 3, 7)$ is on the line, so $\mathbf{b} = \langle 6 - 4, 0 - 3, -2 - 7\rangle = \langle 2, -3, -9\rangle$ and $\mathbf{n} = \mathbf{a} \times \mathbf{b} = \langle -45 + 12, 8 - 18, 6 - 10\rangle = \langle -33, -10, -4\rangle$. Thus, an equation of the plane is $-33(x - 6) - 10(y - 0) - 4[z - (-2)] = 0$ or $33x + 10y + 4z = 190$.

29. A direction vector for the line of intersection is $\mathbf{a} = \mathbf{n}_1 \times \mathbf{n}_2 = \langle 1, 1, -1\rangle \times \langle 2, -1, 3\rangle = \langle 2, -5, -3\rangle$, and $\mathbf{a}$ is parallel to the desired plane. Another vector parallel to the plane is the vector connecting any point on the line of intersection to the given point $(-1, 2, 1)$ in the plane. Setting $x = 0$, the equations of the planes reduce to $y - z = 2$ and $-y + 3z = 1$ with simultaneous solution $y = \frac{7}{2}$ and $z = \frac{3}{2}$. So a point on the line is $\left(0, \frac{7}{2}, \frac{3}{2}\right)$ and another vector parallel to the plane is $\left\langle -1, -\frac{3}{2}, -\frac{1}{2}\right\rangle$. Then a normal vector to the plane is $\mathbf{n} = \langle 2, -5, -3\rangle \times \left\langle -1, -\frac{3}{2}, -\frac{1}{2}\right\rangle = \langle -2, 4, -8\rangle$ and an equation of the plane is $-2(x + 1) + 4(y - 2) - 8(z - 1) = 0$ or $x - 2y + 4z = -1$.

31. Substituting the parametric equations of the line into the equation of the plane gives $2x + y - z + 5 = 2(1 + 2t) + (-1) - t + 5 = 0 \Rightarrow 3t + 6 = 0 \Rightarrow t = -2$. Therefore, the point of intersection is given by $x = 1 + 2(-2) = -3$, $y = -1$ and $z = -2$, that is, the point $(-3, -1, -2)$.

33. Normal vectors for the planes are $\mathbf{n}_1 = \langle 1, 1, 1\rangle$ and $\mathbf{n}_2 = \langle 1, -1, 1\rangle$. The normals are not parallel, so neither are the planes. Furthermore, $\mathbf{n}_1 \cdot \mathbf{n}_2 = 1 - 1 + 1 = 1 \neq 0$, so the planes aren't perpendicular. The angle between them is given by

$$\cos\theta = \frac{\mathbf{n}_1 \cdot \mathbf{n}_2}{|\mathbf{n}_1|\,|\mathbf{n}_2|} = \frac{1}{\sqrt{3}\,\sqrt{3}} = \frac{1}{3} \quad \Rightarrow \quad \theta = \cos^{-1}\left(\tfrac{1}{3}\right) \approx 70.5^\circ.$$

35. The normals are $\mathbf{n}_1 = \langle 1, -4, 2\rangle$ and $\mathbf{n}_2 = \langle 2, -8, 4\rangle$. Since $\mathbf{n}_2 = 2\mathbf{n}_1$, the normals (and thus the planes) are parallel.

37. (a) To find a point on the line of intersection, set one of the variables equal to a constant, say $z = 0$. (This will only work if the line of intersection crosses the xy-plane; otherwise, try setting x or y equal to 0.) Then the equations of the planes reduce to $x + y = 2$ and $3x - 4y = 6$. Solving these two equations gives $x = 2$, $y = 0$. So a point on the line of intersection is $(2, 0, 0)$. The direction of the line is $\mathbf{v} = \mathbf{n}_1 \times \mathbf{n}_2 = \langle 5 - 4, -3 - 5, -4 - 3 \rangle = \langle 1, -8, -7 \rangle$, and symmetric equations for the line are $x - 2 = \dfrac{y}{-8} = \dfrac{z}{-7}$.

(b) The angle between the planes satisfies $\cos\theta = \dfrac{\mathbf{n}_1 \cdot \mathbf{n}_2}{|\mathbf{n}_1|\,|\mathbf{n}_2|} = \dfrac{3 - 4 - 5}{\sqrt{3}\,\sqrt{50}} = -\dfrac{\sqrt{6}}{5}$. Therefore $\theta = \cos^{-1}\!\left(-\frac{\sqrt{6}}{5}\right) \approx 119^\circ$ (or 61°).

39. The plane contains the points $(a, 0, 0)$, $(0, b, 0)$ and $(0, 0, c)$. Thus the vectors $\mathbf{a} = \langle -a, b, 0 \rangle$ and $\mathbf{b} = \langle -a, 0, c \rangle$ lie in the plane, and $\mathbf{n} = \mathbf{a} \times \mathbf{b} = \langle bc - 0, 0 + ac, 0 + ab \rangle = \langle bc, ac, ab \rangle$ is a normal vector to the plane. The equation of the plane is therefore $bcx + acy + abz = abc + 0 + 0$ or $bcx + acy + abz = abc$. Notice that if $a \neq 0$, $b \neq 0$ and $c \neq 0$ then we can rewrite the equation as $\dfrac{x}{a} + \dfrac{y}{b} + \dfrac{z}{c} = 1$. This is a good equation to remember!

41. Two vectors which are perpendicular to the required line are the normal of the given plane, $\langle 1, 1, 1 \rangle$, and a direction vector for the given line, $\langle 1, -1, 2 \rangle$. So a direction vector for the required line is $\langle 1, 1, 1 \rangle \times \langle 1, -1, 2 \rangle = \langle 3, -1, -2 \rangle$. Thus L is given by $\langle x, y, z \rangle = \langle 0, 1, 2 \rangle + t\langle 3, -1, -2 \rangle$, or in parametric form, $x = 3t$, $y = 1 - t$, $z = 2 - 2t$.

43. Let P_i have normal vector $\mathbf{n}_i$. Then $\mathbf{n}_1 = \langle 4, -2, 6 \rangle$, $\mathbf{n}_2 = \langle 4, -2, -2 \rangle$, $\mathbf{n}_3 = \langle -6, 3, -9 \rangle$, $\mathbf{n}_4 = \langle 2, -1, -1 \rangle$. Now $\mathbf{n}_1 = -\frac{2}{3}\mathbf{n}_3$, so $\mathbf{n}_1$ and $\mathbf{n}_3$ are parallel, and hence P_1 and P_3 are parallel; similarly P_2 and P_4 are parallel because $\mathbf{n}_2 = 2\mathbf{n}_4$. However, $\mathbf{n}_1$ and $\mathbf{n}_2$ are not parallel. $\left(0, 0, \frac{1}{2}\right)$ lies on P_1, but not on P_3, so they are not the same plane, but both P_2 and P_4 contain the point $(0, 0, -3)$, so these two planes are identical.

45. Let $Q = (2, 2, 0)$ and $R = (3, -1, 5)$, points on the line corresponding to $t = 0$ and $t = 1$.

Let $P = (1, 2, 3)$. Then $\mathbf{a} = \overrightarrow{QR} = \langle 1, -3, 5 \rangle$, $\mathbf{b} = \overrightarrow{QP} = \langle -1, 0, 3 \rangle$. The distance is

$$d = \frac{|\mathbf{a} \times \mathbf{b}|}{|\mathbf{a}|} = \frac{|\langle 1, -3, 5 \rangle \times \langle -1, 0, 3 \rangle|}{|\langle 1, -3, 5 \rangle|} = \frac{|\langle -9, -8, -3 \rangle|}{|\langle 1, -3, 5 \rangle|} = \frac{\sqrt{9^2 + 8^2 + 3^2}}{\sqrt{1^2 + 3^2 + 5^2}} = \frac{\sqrt{154}}{\sqrt{35}} = \sqrt{\frac{22}{5}}.$$

47. By Equation 9, the distance is $D = \dfrac{1}{\sqrt{1 + 4 + 4}}\,[(1)(2) + (-2)(8) + (-2)(5) - 1] = \dfrac{25}{3}$.

49. Put $y = z = 0$ in the equation of the first plane to get the point $(-1, 0, 0)$ on the plane. Because the planes are parallel, the distance D between them is the distance from $(-1, 0, 0)$ to the second plane. By Equation 9,

$$D = \frac{|3(-1) + 6(0) - 3(0) - 4|}{\sqrt{3^2 + 6^2 + (-3)^2}} = \frac{7}{3\sqrt{6}} \text{ or } \frac{7\sqrt{6}}{18}.$$

51. The distance between two parallel planes is the same as the distance between a point on one of the planes and the other plane. Let $P_0 = (x_0, y_0, z_0)$ be a point on the plane given by $ax + by + cz + d_1 = 0$. Then $ax_0 + by_0 + cz_0 + d_1 = 0$ and the distance between P_0 and the plane given by $ax + by + cz + d_2 = 0$ is, from Equation 9,

$$D = \frac{|ax_0 + by_0 + cz_0 + d_2|}{\sqrt{a^2 + b^2 + c^2}} = \frac{|-d_1 + d_2|}{\sqrt{a^2 + b^2 + c^2}} = \frac{|d_1 - d_2|}{\sqrt{a^2 + b^2 + c^2}}.$$

53. L_1: $x = y = z \Rightarrow x = y$ (1). L_2: $x + 1 = y/2 = z/3 \Rightarrow x + 1 = y/2$ (2). The solution of (1) and (2) is $x = y = -2$. However, when $x = -2$, $x = z \Rightarrow z = -2$, but $x + 1 = z/3 \Rightarrow z = -3$, a contradiction. Hence the lines do not intersect. For L_1, $\mathbf{v}_1 = \langle 1, 1, 1 \rangle$, and for L_2, $\mathbf{v}_2 = \langle 1, 2, 3 \rangle$, so the lines are not parallel. Thus the lines are skew lines. If two lines are skew, they can be viewed as lying in two parallel planes and so the distance between the skew lines would be the same as the distance between these parallel planes. The common normal vector to the planes must be perpendicular to both $\langle 1, 1, 1 \rangle$ and $\langle 1, 2, 3 \rangle$, the direction vectors of the two lines. So set $\mathbf{n} = \langle 1, 1, 1 \rangle \times \langle 1, 2, 3 \rangle = \langle 3 - 2, -3 + 1, 2 - 1 \rangle = \langle 1, -2, 1 \rangle$. From above, we know that $(-2, -2, -2)$ and $(-2, -2, -3)$ are points of L_1 and L_2 respectively. So in the notation of Equation 8, $1(-2) - 2(-2) + 1(-2) + d_1 = 0 \Rightarrow d_1 = 0$ and $1(-2) - 2(-2) + 1(-3) + d_2 = 0 \Rightarrow d_2 = 1$.

By Exercise 51, the distance between these two skew lines is $D = \dfrac{|0 - 1|}{\sqrt{1 + 4 + 1}} = \dfrac{1}{\sqrt{6}}$.

Alternate solution (without reference to planes): A vector which is perpendicular to both of the lines is $\mathbf{n} = \langle 1, 1, 1 \rangle \times \langle 1, 2, 3 \rangle = \langle 1, -2, 1 \rangle$. Pick any point on each of the lines, say $(-2, -2, -2)$ and $(-2, -2, -3)$, and form the vector $\mathbf{b} = \langle 0, 0, 1 \rangle$ connecting the two points. The distance between the two skew lines is the absolute value of the scalar projection of $\mathbf{b}$ along $\mathbf{n}$, that is, $D = \dfrac{|\mathbf{n} \cdot \mathbf{b}|}{|\mathbf{n}|} = \dfrac{|1 \cdot 0 - 2 \cdot 0 + 1 \cdot 1|}{\sqrt{1 + 4 + 1}} = \dfrac{1}{\sqrt{6}}$.

55. If $a \neq 0$, then $ax + by + cz + d = 0 \Rightarrow a(x + d/a) + b(y - 0) + c(z - 0) = 0$ which by (7) is the scalar equation of the plane through the point $(-d/a, 0, 0)$ with normal vector $\langle a, b, c \rangle$. Similarly, if $b \neq 0$ (or if $c \neq 0$) the equation of the plane can be rewritten as $a(x - 0) + b(y + d/b) + c(z - 0) = 0$ [or as $a(x - 0) + b(y - 0) + c(z + d/c) = 0$] which by (7) is the scalar equation of a plane through the point $(0, -d/b, 0)$ [or the point $(0, 0, -d/c)$] with normal vector $\langle a, b, c \rangle$.

9.6 Functions and Surfaces

1. (a) According to Table 1, $f(80, 15) = 7.7$, which means that if an 80-km/h wind has been blowing in the open sea for 15 hours, it will create waves with estimated heights of 7.7 metres.

(b) $h = f(60, t)$ means we fix v at 60 and allow t to vary, resulting in a function of one variable. Thus here, $h = f(60, t)$ gives the wave heights produced by 60-km/h winds blowing for t hours. From the table (look at the row corresponding to $v = 60$), the function increases but at a declining rate as t increases. In fact, the function values appear to be approaching a limiting value of approximately 5.9, which suggests that 60-km/h winds cannot produce waves higher than about 5.9 metres.

(c) $h = f(v, 30)$ means we fix t at 30, again giving a function of one variable. So, $h = f(v, 30)$ gives the wave heights produced by winds of speed v blowing for 30 hours. From the table (look at the column corresponding to $t = 30$), the function appears to increase at an increasing rate, with no apparent limiting value. This suggests that faster winds (lasting 30 hours) always create higher waves.

3. (a) $f(x, y) = x^2 e^{3xy}$, so $f(2, 0) = 2^2 e^{3(2)(0)} = 4(1) = 4$.

(b) Since both x^2 and the exponential function are defined everywhere, $x^2 e^{3xy}$ is defined for all choices of values for x and y. Thus, the domain of f is $\mathbb{R}^2$.

(c) Because the range of $g(x, y) = 3xy$ is $\mathbb{R}$, and the range of e^x is $(0, \infty)$, the range of $e^{g(x,y)} = e^{3xy}$ is $(0, \infty)$. The range of x^2 is $[0, \infty)$, so the range of the product $x^2 e^{3xy}$ is $[0, \infty)$.

5. $\sqrt{y - x^2}$ is defined only when $y - x^2 \geq 0$, or $y \geq x^2$. In addition, f is not defined if $1 - x^2 = 0 \Rightarrow x = \pm 1$. Thus the domain of f is $\{(x, y) \mid y \geq x^2, x \neq \pm 1\}$.

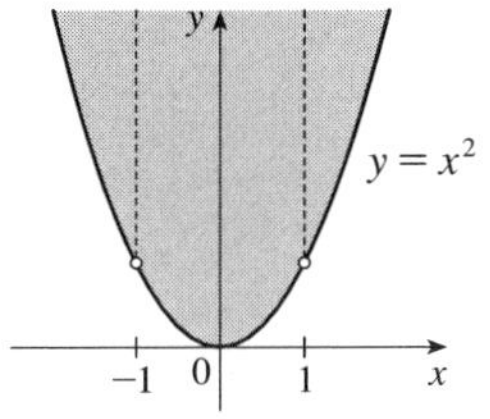

7. $\sqrt{1 - x^2}$ is defined only when $1 - x^2 \geq 0$, or $x^2 \leq 1 \Leftrightarrow -1 \leq x \leq 1$, and $\sqrt{1 - y^2}$ is defined only when $1 - y^2 \geq 0$, or $y^2 \leq 1 \Leftrightarrow -1 \leq y \leq 1$. Thus the domain of f is $\{(x, y) \mid -1 \leq x \leq 1, -1 \leq y \leq 1\}$.

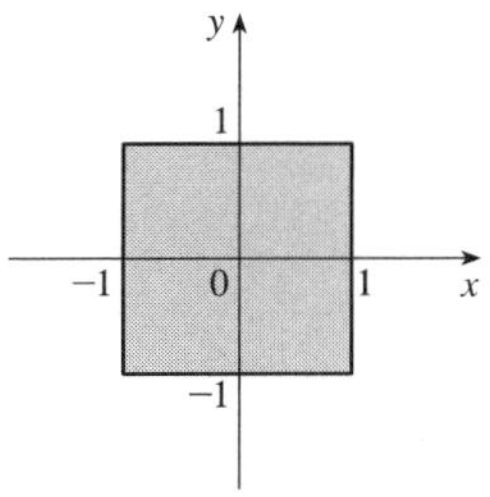

9. $z = 3$, a horizontal plane through the point $(0, 0, 3)$.

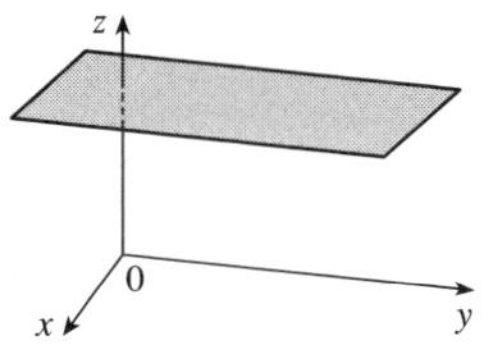

11. $z = 1 - x - y$ or $x + y + z = 1$, a plane with intercepts 1, 1, and 1.

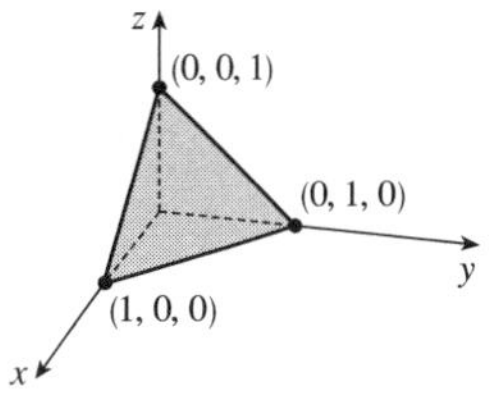

13. $z = y^2 + 1$, a parabolic cylinder.

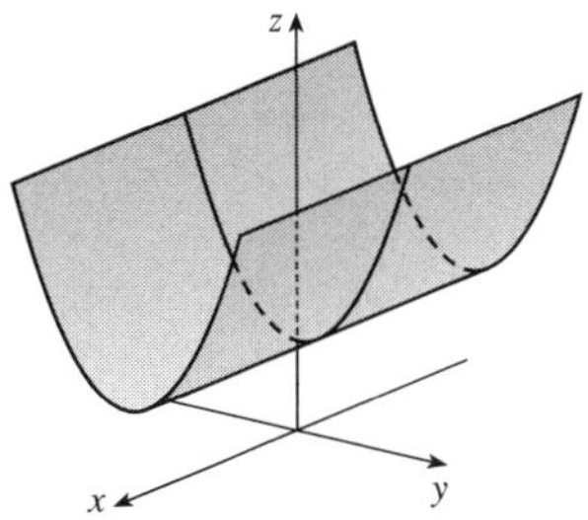

15. All six graphs have different traces in the planes $x = 0$ and $y = 0$, so we investigate these for each function.

(a) $f(x, y) = |x| + |y|$. The trace in $x = 0$ is $z = |y|$, and in $y = 0$ is $z = |x|$, so it must be graph VI.

(b) $f(x, y) = |xy|$. The trace in $x = 0$ is $z = 0$, and in $y = 0$ is $z = 0$, so it must be graph V.

(c) $f(x, y) = \dfrac{1}{1 + x^2 + y^2}$. The trace in $x = 0$ is $z = \dfrac{1}{1 + y^2}$, and in $y = 0$ is $z = \dfrac{1}{1 + x^2}$. In addition, we can see that f is close to 0 for large values of x and y, so this is graph I.

(d) $f(x, y) = (x^2 - y^2)^2$. The trace in $x = 0$ is $z = y^4$, and in $y = 0$ is $z = x^4$. Both graph II and graph IV seem plausible; notice the trace in $z = 0$ is $0 = (x^2 - y^2)^2 \Rightarrow y = \pm x$, so it must be graph IV.

(e) $f(x,y) = (x-y)^2$. The trace in $x=0$ is $z=y^2$, and in $y=0$ is $z=x^2$. Both graph II and graph IV seem plausible; notice the trace in $z=0$ is $0=(x-y)^2 \quad\Rightarrow\quad y=x$, so it must be graph II.

(f) $f(x,y) = \sin(|x|+|y|)$. The trace in $x=0$ is $z=\sin|y|$, and in $y=0$ is $z=\sin|x|$. In addition, notice that the oscillating nature of the graph is characteristic of trigonometric functions. So this is graph III.

17. The equation of the graph is $z=\sqrt{4x^2+y^2}$ or equivalently $4x^2+y^2=z^2$, $z\ge 0$. Traces in $x=k$ are $z^2-y^2=4k^2$, $z\ge 0$, a family of hyperbolas where we have only the upper branch. Traces in $y=k$ are $z^2-4x^2=k^2$, $z\ge 0$, again a family of half-hyperbolas. Traces in $z=k$, $k\ge 0$, are $4x^2+y^2=k^2$ or $x^2+\dfrac{y^2}{4}=\dfrac{k^2}{4}$, a family of ellipses. Note that the original equation can be written as $x^2+\dfrac{y^2}{4}=\dfrac{z^2}{4}$, $z\ge 0$, which we recognize as the upper half of an elliptical cone.

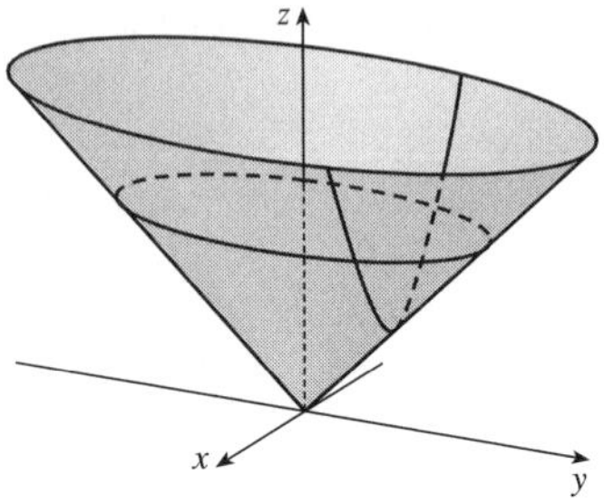

19. $y=z^2-x^2$. The traces in $x=k$ are the parabolas $y=z^2-k^2$; the traces in $y=k$ are $k=z^2-x^2$, which are hyperbolas (note the hyperbolas are oriented differently for $k>0$ than for $k<0$); and the traces in $z=k$ are the parabolas $y=k^2-x^2$. Thus, $\dfrac{y}{1}=\dfrac{z^2}{1^2}-\dfrac{x^2}{1^2}$ is a hyperbolic paraboloid.

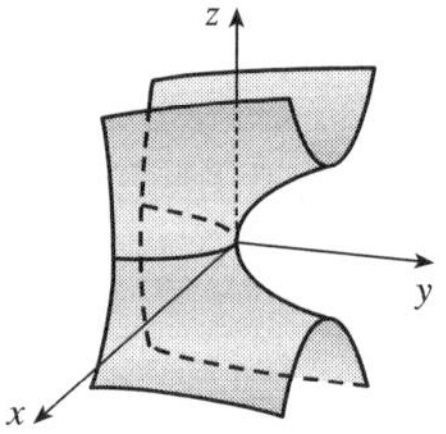

21. Completing squares in y and z gives $4x^2+(y-2)^2+4(z-3)^2=4$ or $x^2+\dfrac{(y-2)^2}{4}+(z-3)^2=1$, an ellipsoid with centre $(0,2,3)$.

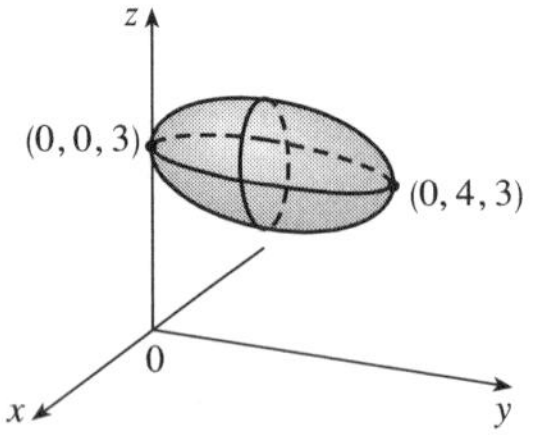

23. (a) In $\mathbb{R}^2$, $x^2+y^2=1$ represents a circle of radius 1 centred at the origin.

(b) In $\mathbb{R}^3$, the equation doesn't involve z, which means that any horizontal plane $z=k$ intersects the surface in a circle $x^2+y^2=1$, $z=k$. Thus the surface is a circular cylinder, made up of infinitely many shifted copies of the circle $x^2+y^2=1$, with axis the z-axis.

(c) In $\mathbb{R}^3$, $x^2+z^2=1$ also represents a circular cylinder of radius 1, this time with axis the y-axis.

25. (a) The traces of $x^2+y^2-z^2=1$ in $x=k$ are $y^2-z^2=1-k^2$, a family of hyperbolas. (Note that the hyperbolas are oriented differently for $-1<k<1$ than for $k<-1$ or $k>1$.) The traces in $y=k$ are $x^2-z^2=1-k^2$, a similar family of hyperbolas. The traces in $z=k$ are $x^2+y^2=1+k^2$, a family of circles. For $k=0$, the trace in the xy-plane, the circle is of radius 1. As $|k|$ increases, so does the radius of the circle. This behaviour, combined with the hyperbolic vertical traces, gives the graph of the hyperboloid of one sheet in Table 2.

(b) The shape of the surface is unchanged, but the hyperboloid is rotated so that its axis is the y-axis. Traces in $y = k$ are circles, while traces in $x = k$ and $z = k$ are hyperbolas.

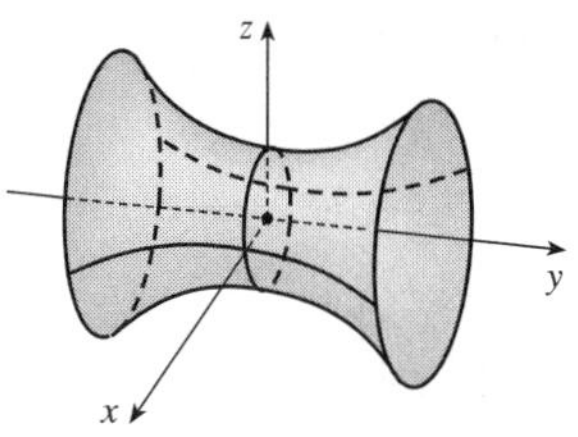

(c) Completing the square in y gives $x^2 + (y + 1)^2 - z^2 = 1$. The surface is a hyperboloid identical to the one in part (a) but shifted one unit in the negative y-direction.

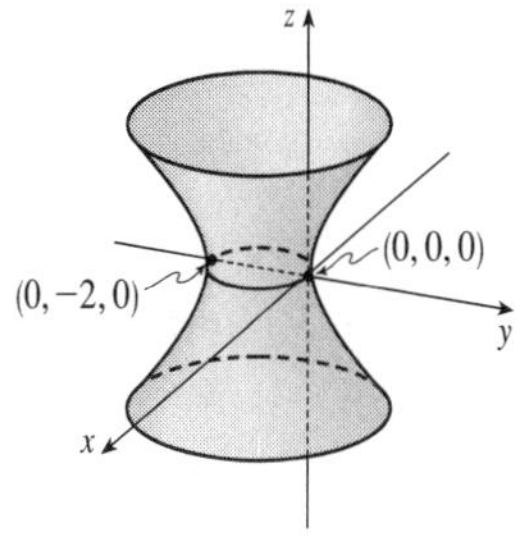

27. $f(x, y) = 3x - x^4 - 4y^2 - 10xy$

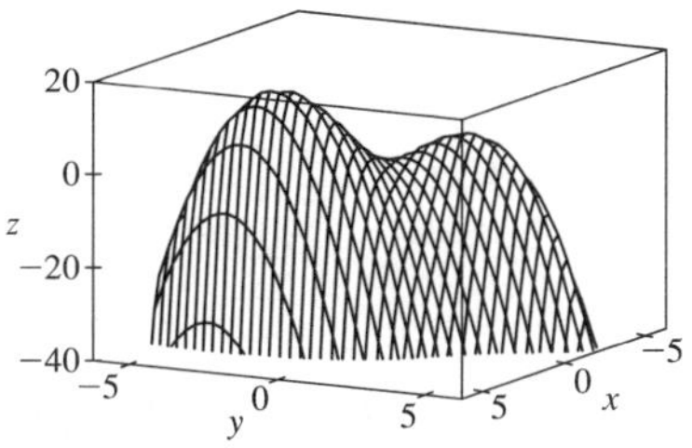

Three-dimensional view

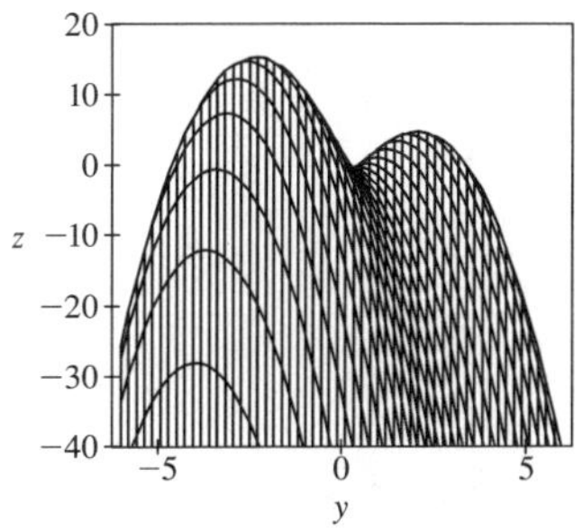

Front view

It does appear that the function has a maximum value, at the higher of the two "hilltops." From the front view graph, the maximum value appears to be approximately 15. Both hilltops could be considered local maximum points, as the values of f there are larger than at the neighboring points. There does not appear to be any local minimum point; although the valley shape between the two peaks looks like a minimum of some kind, some neighboring points have lower function values.

29.

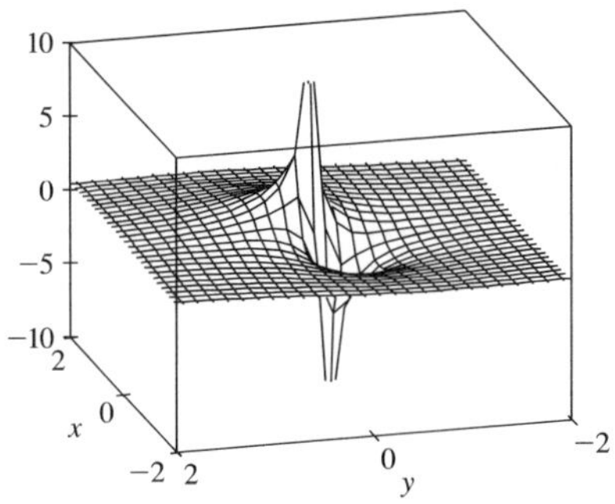

$f(x, y) = \dfrac{x + y}{x^2 + y^2}$. As both x and y become large, the function values appear to approach 0, regardless of which direction is considered. As (x, y) approaches the origin, the graph exhibits asymptotic behaviour. From some directions, $f(x, y) \to \infty$, while in others $f(x, y) \to -\infty$. (These are the vertical spikes visible in the graph.) If the graph is examined carefully, however, one can see that $f(x, y)$ approaches 0 along the line $y = -x$.

31.

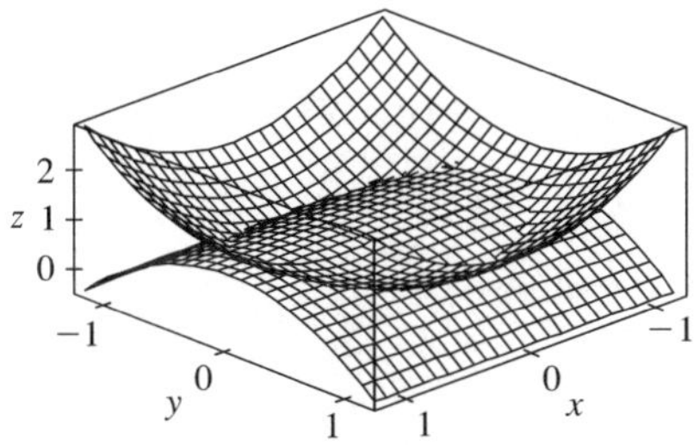

The curve of intersection looks like a bent ellipse. The projection of this curve onto the xy-plane is the set of points $(x, y, 0)$ which satisfy $x^2 + y^2 = 1 - y^2 \quad \Leftrightarrow \quad x^2 + 2y^2 = 1 \quad \Leftrightarrow$

$$x^2 + \frac{y^2}{\left(1/\sqrt{2}\right)^2} = 1.$$ This is an equation of an ellipse.

33. If (a, b, c) satisfies $z = y^2 - x^2$, then $c = b^2 - a^2$. L_1: $x = a + t$, $y = b + t$, $z = c + 2(b - a)t$,
L_2: $x = a + t$, $y = b - t$, $z = c - 2(b + a)t$. Substitute the parametric equations of L_1 into the equation of the hyperbolic paraboloid in order to find the points of intersection: $z = y^2 - x^2 \Rightarrow$
$c + 2(b - a)t = (b + t)^2 - (a + t)^2 = b^2 - a^2 + 2(b - a)t \Rightarrow c = b^2 - a^2$. As this is true for all values of t, L_1 lies on $z = y^2 - x^2$. Performing similar operations with L_2 gives: $z = y^2 - x^2 \Rightarrow$
$c - 2(b + a)t = (b - t)^2 - (a + t)^2 = b^2 - a^2 - 2(b + a)t \Rightarrow c = b^2 - a^2$. This tells us that all of L_2 also lies on $z = y^2 - x^2$.

9.7 Cylindrical and Spherical Coordinates

1. See Figure 2 and the accompanying discussion on page 685; see the paragraph preceding Example 2 on page 686.

3. (a)

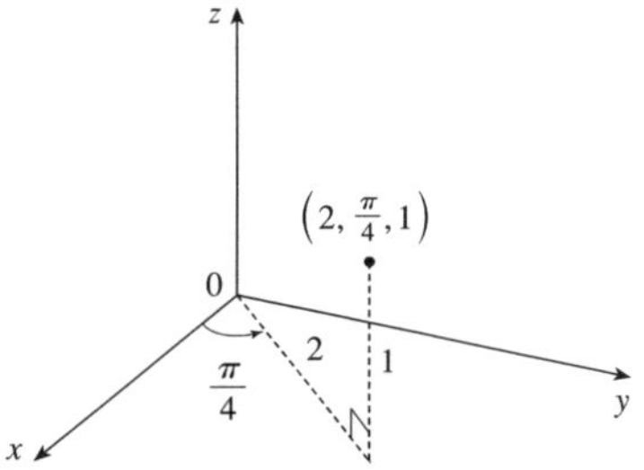

$x = 2\cos\frac{\pi}{4} = \sqrt{2}$, $y = 2\sin\frac{\pi}{4} = \sqrt{2}$, $z = 1$, so the point is $(\sqrt{2}, \sqrt{2}, 1)$ in rectangular coordinates.

(b)

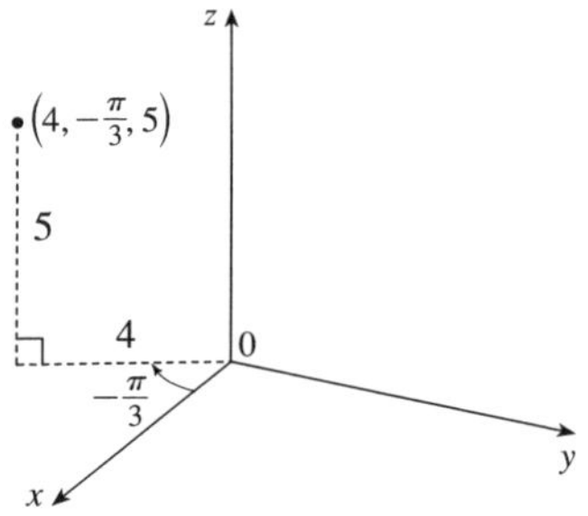

$x = 4\cos\left(-\frac{\pi}{3}\right) = 2$, $y = 4\sin\left(-\frac{\pi}{3}\right) = -2\sqrt{3}$, and $z = 5$, so the point is $(2, -2\sqrt{3}, 5)$ in rectangular coordinates.

5. (a) $r^2 = x^2 + y^2 = 1^2 + (-1)^2 = 2$ so $r = \sqrt{2}$; $\tan\theta = \dfrac{y}{x} = \dfrac{-1}{1} = -1$ and the point $(1, -1)$ is in the fourth quadrant of the xy-plane, so $\theta = \frac{7\pi}{4} + 2n\pi$; $z = 4$. Thus, one set of cylindrical coordinates is $\left(\sqrt{2}, \frac{7\pi}{4}, 4\right)$.

(b) $r^2 = (-1)^2 + \left(-\sqrt{3}\right)^2 = 4$ so $r = 2$; $\tan\theta = \frac{-\sqrt{3}}{-1} = \sqrt{3}$ and the point $\left(-1, -\sqrt{3}\right)$ is in the third quadrant of the xy-plane, so $\theta = \frac{4\pi}{3} + 2n\pi$; $z = 2$. Thus, one set of cylindrical coordinates is $\left(2, \frac{4\pi}{3}, 2\right)$.

7. (a)

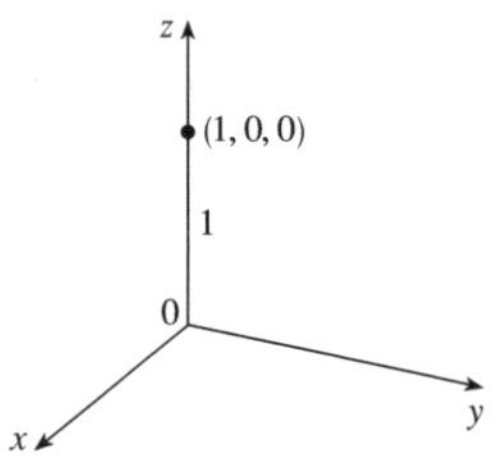

$x = \rho\sin\phi\cos\theta = (1)\sin 0\cos 0 = 0$,
$y = \rho\sin\phi\sin\theta = (1)\sin 0\sin 0 = 0$, and
$z = \rho\cos\phi = (1)\cos 0 = 1$ so the point is
$(0, 0, 1)$ in rectangular coordinates.

(b)

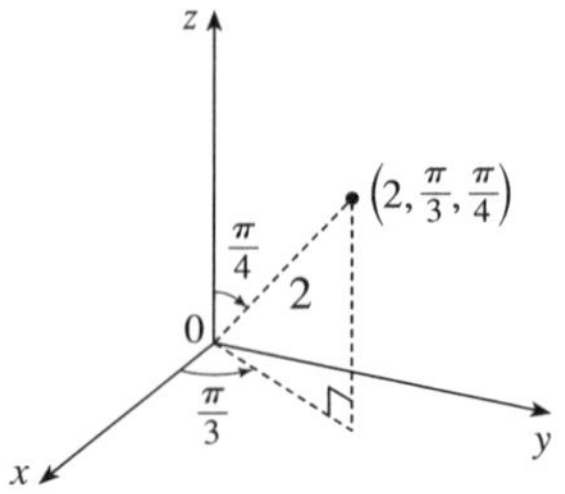

$x = 2\sin\frac{\pi}{4}\cos\frac{\pi}{3} = \frac{\sqrt{2}}{2}$, $y = 2\sin\frac{\pi}{4}\sin\frac{\pi}{3} = \frac{\sqrt{6}}{2}$,
$z = 2\cos\frac{\pi}{4} = \sqrt{2}$ so the point is $\left(\frac{\sqrt{2}}{2}, \frac{\sqrt{6}}{2}, \sqrt{2}\right)$ in rectangular coordinates.

9. (a) $\rho = \sqrt{x^2 + y^2 + z^2} = \sqrt{1 + 3 + 12} = 4$, $\cos\phi = \dfrac{z}{\rho} = \dfrac{2\sqrt{3}}{4} = \dfrac{\sqrt{3}}{2} \quad\Rightarrow\quad \phi = \dfrac{\pi}{6}$, and

$\cos\theta = \dfrac{x}{\rho\sin\phi} = \dfrac{1}{4\sin(\pi/6)} = \dfrac{1}{2} \quad\Rightarrow\quad \theta = \dfrac{\pi}{3}$ (since $y > 0$). Thus spherical coordinates are $\left(4, \dfrac{\pi}{3}, \dfrac{\pi}{6}\right)$.

(b) $\rho = \sqrt{0 + 1 + 1} = \sqrt{2}$, $\cos\phi = \dfrac{-1}{\sqrt{2}} \quad\Rightarrow\quad \phi = \dfrac{3\pi}{4}$, and $\cos\theta = \dfrac{0}{\sqrt{2}\sin(3\pi/4)} = 0 \quad\Rightarrow\quad \theta = \dfrac{3\pi}{2}$

(since $y < 0$). Thus spherical coordinates are $\left(\sqrt{2}, \dfrac{3\pi}{2}, \dfrac{3\pi}{4}\right)$.

11. Since $r = 3$, $x^2 + y^2 = 9$ and the surface is a circular cylinder with radius 3 and axis the z-axis.

13. Since $\phi = \frac{\pi}{3}$, the surface is the top half of the right circular cone with vertex at the origin and axis the positive z-axis.

15. $z = r^2 = x^2 + y^2$, so the surface is a circular paraboloid with vertex at the origin and axis the positive z-axis.

17. $r = 2\cos\theta \quad\Rightarrow\quad r^2 = x^2 + y^2 = 2r\cos\theta = 2x \quad\Leftrightarrow\quad (x-1)^2 + y^2 = 1$, which is the equation of a circular cylinder with radius 1, whose axis is the vertical line $x = 1$, $y = 0$, $z = z$.

19. Since $r^2 + z^2 = 25$ and $r^2 = x^2 + y^2$, we have $x^2 + y^2 + z^2 = 25$, a sphere with radius 5 and centre at the origin.

21. (a) $x^2 + y^2 = r^2$, so the equation becomes $z = r^2$.

(b) $x = \rho\sin\phi\cos\theta$, $y = \rho\sin\phi\sin\theta$, and $z = \rho\cos\phi$, so the equation becomes

$\rho\cos\phi = (\rho\sin\phi\cos\theta)^2 + (\rho\sin\phi\sin\theta)^2$ or $\rho\cos\phi = \rho^2\sin^2\phi$ or $\rho\sin^2\phi = \cos\phi$.

23. (a) $r^2 = 2r\sin\theta$ or $r = 2\sin\theta$.

(b) $\rho^2\sin^2\phi\,(\cos^2\theta + \sin^2\theta) = 2\rho\sin\phi\sin\theta$ or $\rho\sin^2\phi = 2\sin\phi\sin\theta$ or $\rho\sin\phi = 2\sin\theta$.

25.

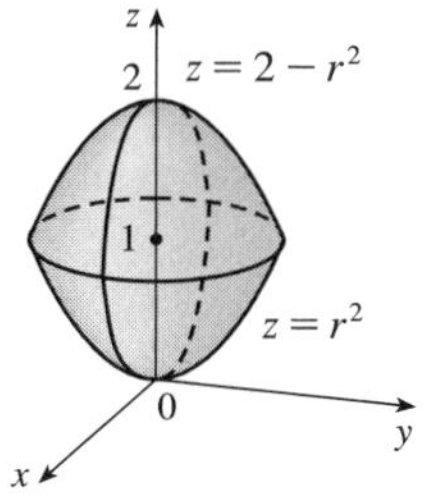

$z = r^2 = x^2 + y^2$ is a circular paraboloid with vertex $(0, 0, 0)$, opening upward. $z = 2 - r^2 \quad\Rightarrow\quad z - 2 = -(x^2 + y^2)$ is a circular paraboloid with vertex $(0, 0, 2)$ opening downward. Thus $r^2 \le z \le 2 - r^2$ is the solid region enclosed by these two surfaces.

27.

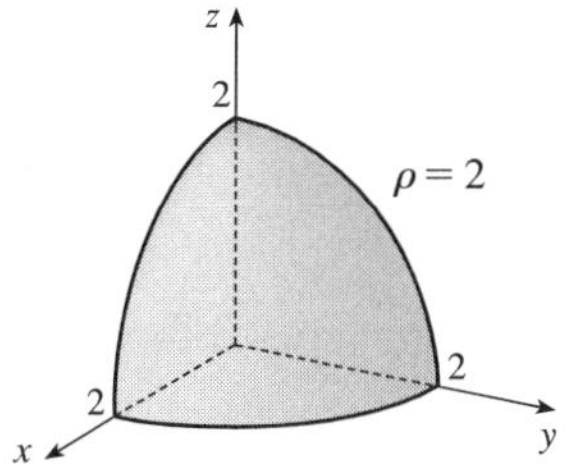

$\rho = 2$ represents a sphere of radius 2, centred at the origin, so $\rho \le 2$ is this sphere and its interior. $0 \le \phi \le \frac{\pi}{2}$ restricts the solid to that portion of the region that lies on or above the xy-plane, and $0 \le \theta \le \frac{\pi}{2}$ further restricts the solid to the first octant. Thus the solid is the portion in the first octant of the solid ball centred at the origin with radius 2.

29. $-\frac{\pi}{2} \le \theta \le \frac{\pi}{2}$ restricts the solid to the 4 octants in which x is positive.

$\rho = \sec\phi \quad\Rightarrow\quad \rho\cos\phi = z = 1$, which is the equation of a horizontal plane.

$0 \le \phi \le \frac{\pi}{6}$ describes a cone, opening upward. So the solid lies above the cone $\phi = \frac{\pi}{6}$ and below the plane $z = 1$.

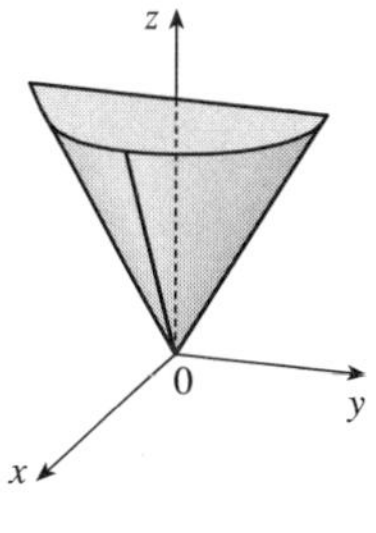

31. We can position the cylindrical shell vertically so that its axis coincides with the z-axis and its base lies in the xy-plane. If we use centimetres as the unit of measurement, then cylindrical coordinates conveniently describe the shell as $6 \le r \le 7$, $0 \le \theta \le 2\pi$, $0 \le z \le 20$.

33. $z \ge \sqrt{x^2 + y^2}$ because the solid lies above the cone. Squaring both sides of this inequality gives $z^2 \ge x^2 + y^2 \Rightarrow 2z^2 \ge x^2 + y^2 + z^2 = \rho^2 \Rightarrow z^2 = \rho^2 \cos^2 \phi \ge \frac{1}{2}\rho^2 \Rightarrow \cos^2 \phi \ge \frac{1}{2}$. The cone opens upward so that the inequality is $\cos \phi \ge \frac{1}{\sqrt{2}}$, or equivalently $0 \le \phi \le \frac{\pi}{4}$. In spherical coordinates the sphere $z = x^2 + y^2 + z^2$ is $\rho \cos \phi = \rho^2 \Rightarrow \rho = \cos \phi$. $0 \le \rho \le \cos \phi$ because the solid lies below the sphere. The solid can therefore be described as the region in spherical coordinates satisfying $0 \le \rho \le \cos \phi$, $0 \le \phi \le \frac{\pi}{4}$.

35. In cylindrical coordinates, the equation of the cylinder is $r = 3$, $0 \le z \le 10$. The hemisphere is the upper part of the sphere radius 3, centre $(0, 0, 10)$, equation $r^2 + (z - 10)^2 = 3^2$, $z \ge 10$. In Maple, we can use either the `coords=cylindrical` option in a regular `plot` command, or the `plots[cylinderplot]` command. In Mathematica, we can use `ParametricPlot3d`.

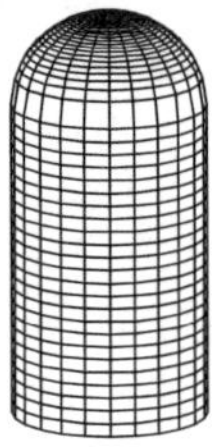

9 Review

CONCEPT CHECK

1. A scalar is a real number, while a vector is a quantity that has both a real-valued magnitude and a direction.

2. To add two vectors geometrically, we can use either the Triangle Law or the Parallelogram Law, as illustrated in Figures 3 and 4 in Section 9.2. (See also the definition of vector addition on page 643.) Algebraically, we add the corresponding components of the vectors.

3. For $c > 0$, $c\,\mathbf{a}$ is a vector with the same direction as $\mathbf{a}$ and length c times the length of $\mathbf{a}$. If $c < 0$, $c\mathbf{a}$ points in the opposite direction as $\mathbf{a}$ and has length $|c|$ times the length of $\mathbf{a}$. (See Figures 7 and 15 in Section 9.2.) Algebraically, to find $c\,\mathbf{a}$ we multiply each component of $\mathbf{a}$ by c.

4. See (1) in Section 9.2.

5. See the definition on page 651 and the boxed equation on page 653.

6. The dot product can be used to determine the work done moving an object given the force and displacement vectors. The dot product can also be used to find the angle between two vectors and the scalar projection of one vector onto another. In particular, the dot product can determine if two vectors are orthogonal.

7. See the boxed equations on page 655 as well as Figures 5 and 6 and the accompanying discussion on pages 654–55.

8. See the definition on page 658; use either (2) or (4) in Section 9.4.

9. The cross product can be used to determine torque if the force and position vectors are known. In addition, the cross product can be used to create a vector orthogonal to two given vectors as well as to determine if two vectors are parallel. The cross product can also be used to find the area of a parallelogram determined by two vectors.

10. (a) The area of the parallelogram determined by $\mathbf{a}$ and $\mathbf{b}$ is the length of the cross product: $|\mathbf{a} \times \mathbf{b}|$.

(b) The volume of the parallelepiped determined by $\mathbf{a}$, $\mathbf{b}$, and $\mathbf{c}$ is the magnitude of their scalar triple product: $|\mathbf{a} \cdot (\mathbf{b} \times \mathbf{c})|$.

11. If an equation of the plane is known, it can be written as $ax + by + cz + d = 0$. A normal vector, which is perpendicular to the plane, is $\langle a, b, c \rangle$ (or any scalar multiple of $\langle a, b, c \rangle$). If an equation is not known, we can use points on the plane to find two non-parallel vectors which lie in the plane. The cross product of these vectors is a vector perpendicular to the plane.

12. The angle between two intersecting planes is defined as the acute angle between their normal vectors. We can find this angle using the definition of the dot product on page 651.

13. See (1), (2), and (3) in Section 9.5.

14. See (5), (6), and (7) in Section 9.5.

15. (a) Two (nonzero) vectors are parallel if and only if one is a scalar multiple of the other. In addition, two nonzero vectors are parallel if and only if their cross product is $\mathbf{0}$.

(b) Two vectors are perpendicular if and only if their dot product is 0.

(c) Two planes are parallel if and only if their normal vectors are parallel.

16. (a) Determine the vectors $\overrightarrow{PQ} = \langle a_1, a_2, a_3 \rangle$ and $\overrightarrow{PR} = \langle b_1, b_2, b_3 \rangle$. If there is a scalar t such that $\langle a_1, a_2, a_3 \rangle = t \langle b_1, b_2, b_3 \rangle$, then the vectors are parallel and the points must all lie on the same line.

Alternatively, if $\overrightarrow{PQ} \times \overrightarrow{PR} = \mathbf{0}$, then $\overrightarrow{PQ}$ and $\overrightarrow{PR}$ are parallel, so P, Q, and R are collinear.

Thirdly, an algebraic method is to determine an equation of the line joining two of the points, and then check whether or not the third point satisfies this equation.

(b) Find the vectors $\overrightarrow{PQ} = \mathbf{a}$, $\overrightarrow{PR} = \mathbf{b}$, $\overrightarrow{PS} = \mathbf{c}$. $\mathbf{a} \times \mathbf{b}$ is normal to the plane formed by P, Q and R, and so S lies on this plane if $\mathbf{a} \times \mathbf{b}$ and $\mathbf{c}$ are orthogonal, that is, if $(\mathbf{a} \times \mathbf{b}) \cdot \mathbf{c} = 0$. (Or use the reasoning in Example 6 in Section 9.4.) Alternatively, find an equation for the plane determined by three of the points and check whether or not the fourth point satisfies this equation.

17. (a) See Exercise 9.4.27.

(b) See Example 8 in Section 9.5.

(c) See Example 10 in Section 9.5.

18. One method of graphing a function of two variables is to first find traces (see Example 6 in Section 9.6 and the discussion preceding it).

19. See Table 2 in Section 9.6.

20. (a) See (1) and the discussion accompanying Figure 4 in Section 9.7.

(b) See (3) and Figures 7–9, and the accompanying discussion, in Section 9.7.

TRUE-FALSE QUIZ

1. True, by Property 2 of the dot product. (See page 654.)

3. True. If θ is the angle between $\mathbf{u}$ and $\mathbf{v}$, then by the definition of the cross product, $|\mathbf{u} \times \mathbf{v}| = |\mathbf{u}|\,|\mathbf{v}| \sin\theta = |\mathbf{v}|\,|\mathbf{u}| \sin\theta = |\mathbf{v} \times \mathbf{u}|$.

(Or, by Properties 1 and 2 of the cross product, $|\mathbf{u} \times \mathbf{v}| = |-\mathbf{v} \times \mathbf{u}| = |-1|\,|\mathbf{v} \times \mathbf{u}| = |\mathbf{v} \times \mathbf{u}|$.)

5. Property 2 of the cross product tells us that this is true.

7. This is true by (6) in Section 9.4.

9. This is true because $\mathbf{u} \times \mathbf{v}$ is orthogonal to $\mathbf{u}$ (see page 658), and the dot product of two orthogonal vectors is 0.

11. If $|\mathbf{u}| = 1$, $|\mathbf{v}| = 1$ and θ is the angle between these two vectors (so $0 \leq \theta \leq \pi$), then by the definition of the cross product, $|\mathbf{u} \times \mathbf{v}| = |\mathbf{u}|\,|\mathbf{v}| \sin\theta = \sin\theta$, which is equal to 1 if and only if $\theta = \frac{\pi}{2}$ (that is, if and only if the two vectors are orthogonal). Therefore, the assertion that the cross product of two unit vectors is a unit vector is false.

13. This is false. In $\mathbb{R}^2$, $x^2 + y^2 = 1$ represents a circle, but $\{(x, y, z) \mid x^2 + y^2 = 1\}$ represents a *three-dimensional surface*, namely, a circular cylinder with axis the z-axis.

15. False. For example, $\mathbf{i} \cdot \mathbf{j} = \mathbf{0}$ but $\mathbf{i} \neq \mathbf{0}$ and $\mathbf{j} \neq \mathbf{0}$.

EXERCISES

1. (a) The radius of the sphere is the distance between the points $(-1, 2, 1)$ and $(6, -2, 3)$, namely $\sqrt{[6 - (-1)]^2 + (-2 - 2)^2 + (3 - 1)^2} = \sqrt{69}$. By the formula for an equation of a sphere (see page 640), an equation of the sphere with centre $(-1, 2, 1)$ and radius $\sqrt{69}$ is $(x + 1)^2 + (y - 2)^2 + (z - 1)^2 = 69$.

(b) The intersection of this sphere with the yz-plane is the set of points on the sphere whose x-coordinate is 0. Putting $x = 0$ into the equation, we have $(y - 2)^2 + (z - 1)^2 = 68, x = 0$ which represents a circle in the yz-plane with centre $(0, 2, 1)$ and radius $\sqrt{68}$.

(c) Completing squares gives $(x - 4)^2 + (y + 1)^2 + (z + 3)^2 = -1 + 16 + 1 + 9 = 25$. Thus, the sphere is centred at $(4, -1, -3)$ and has radius 5.

3. $\mathbf{u} \cdot \mathbf{v} = |\mathbf{u}|\,|\mathbf{v}| \cos 45^\circ = (2)(3)\frac{\sqrt{2}}{2} = 3\sqrt{2}$. $|\mathbf{u} \times \mathbf{v}| = |\mathbf{u}|\,|\mathbf{v}| \sin 45^\circ = (2)(3)\frac{\sqrt{2}}{2} = 3\sqrt{2}$. By the right-hand rule, $\mathbf{u} \times \mathbf{v}$ is directed out of the page.

5. For the two vectors to be orthogonal, we need $\langle 3, 2, x\rangle \cdot \langle 2x, 4, x\rangle = 0 \quad \Leftrightarrow \quad (3)(2x) + (2)(4) + (x)(x) = 0 \quad \Leftrightarrow$ $x^2 + 6x + 8 = 0 \quad \Leftrightarrow \quad (x + 2)(x + 4) = 0 \quad \Leftrightarrow \quad x = -2$ or $x = -4$.

7. (a) $(\mathbf{u} \times \mathbf{v}) \cdot \mathbf{w} = \mathbf{u} \cdot (\mathbf{v} \times \mathbf{w}) = 2$

(b) $\mathbf{u} \cdot (\mathbf{w} \times \mathbf{v}) = \mathbf{u} \cdot [-(\mathbf{v} \times \mathbf{w})] = -\mathbf{u} \cdot (\mathbf{v} \times \mathbf{w}) = -2$

(c) $\mathbf{v} \cdot (\mathbf{u} \times \mathbf{w}) = (\mathbf{v} \times \mathbf{u}) \cdot \mathbf{w} = -(\mathbf{u} \times \mathbf{v}) \cdot \mathbf{w} = -2$

(d) $(\mathbf{u} \times \mathbf{v}) \cdot \mathbf{v} = \mathbf{u} \cdot (\mathbf{v} \times \mathbf{v}) = \mathbf{u} \cdot \mathbf{0} = 0$

9. For simplicity, consider a unit cube positioned with its back left corner at the origin. Vector representations of the diagonals joining the points $(0, 0, 0)$ to $(1, 1, 1)$ and $(1, 0, 0)$ to $(0, 1, 1)$ are $\langle 1, 1, 1\rangle$ and $\langle -1, 1, 1\rangle$. Let θ be the angle between these two vectors. $\langle 1, 1, 1\rangle \cdot \langle -1, 1, 1\rangle = -1 + 1 + 1 = 1 = |\langle 1, 1, 1\rangle|\,|\langle -1, 1, 1\rangle| \cos\theta = 3\cos\theta \quad \Rightarrow \quad \cos\theta = \frac{1}{3} \quad \Rightarrow$ $\theta = \cos^{-1}\left(\frac{1}{3}\right) \approx 71^\circ$.

11. $\overrightarrow{AB} = \langle 1, 0, -1\rangle$, $\overrightarrow{AC} = \langle 0, 4, 3\rangle$, so

(a) a vector perpendicular to the plane is $\overrightarrow{AB} \times \overrightarrow{AC} = \langle 0 + 4, -(3 + 0), 4 - 0\rangle = \langle 4, -3, 4\rangle$.

(b) $\frac{1}{2}\left|\overrightarrow{AB} \times \overrightarrow{AC}\right| = \frac{1}{2}\sqrt{16 + 9 + 16} = \frac{\sqrt{41}}{2}$.

13. Let F_1 be the magnitude of the force directed 20° away from the direction of shore, and let F_2 be the magnitude of the other force. Separating these forces into components parallel to the direction of the resultant force and perpendicular to it gives $F_1 \cos 20^\circ + F_2 \cos 30^\circ = 255$ **(1)**, and $F_1 \sin 20^\circ - F_2 \sin 30^\circ = 0 \quad \Rightarrow \quad F_1 = F_2 \dfrac{\sin 30^\circ}{\sin 20^\circ}$ **(2)**. Substituting **(2)** into **(1)** gives $F_2(\sin 30^\circ \cot 20^\circ + \cos 30^\circ) = 255 \quad \Rightarrow \quad F_2 \approx 114$ N. Substituting this into **(2)** gives $F_1 \approx 166$ N.

15. The line has direction $\mathbf{v} = \langle -3, 2, 3 \rangle$. Letting $P_0 = (4, -1, 2)$, parametric equations are $x = 4 - 3t$, $y = -1 + 2t$, $z = 2 + 3t$.

17. A direction vector for the line is a normal vector for the plane, $\mathbf{n} = \langle 2, -1, 5 \rangle$, and parametric equations for the line are $x = -2 + 2t$, $y = 2 - t$, $z = 4 + 5t$.

19. Here the vectors $\mathbf{a} = \langle 4 - 3, 0 - (-1), 2 - 1 \rangle = \langle 1, 1, 1 \rangle$ and $\mathbf{b} = \langle 6 - 3, 3 - (-1), 1 - 1 \rangle = \langle 3, 4, 0 \rangle$ lie in the plane, so $\mathbf{n} = \mathbf{a} \times \mathbf{b} = \langle -4, 3, 1 \rangle$ is a normal vector to the plane and an equation of the plane is $-4(x - 3) + 3(y - (-1)) + 1(z - 1) = 0$ or $-4x + 3y + z = -14$.

21. $\mathbf{n}_1 = \langle 1, 0, -1 \rangle$ and $\mathbf{n}_2 = \langle 0, 1, 2 \rangle$. Setting $z = 0$, it is easy to see that $(1, 3, 0)$ is a point on the line of intersection of $x - z = 1$ and $y + 2z = 3$. The direction of this line is $\mathbf{v}_1 = \mathbf{n}_1 \times \mathbf{n}_2 = \langle 1, -2, 1 \rangle$. A second vector parallel to the desired plane is $\mathbf{v}_2 = \langle 1, 1, -2 \rangle$, since it is perpendicular to $x + y - 2z = 1$. Therefore, the normal of the plane in question is $\mathbf{n} = \mathbf{v}_1 \times \mathbf{v}_2 = \langle 4 - 1, 1 + 2, 1 + 2 \rangle = 3 \langle 1, 1, 1 \rangle$. Taking $(x_0, y_0, z_0) = (1, 3, 0)$, the equation we are looking for is $(x - 1) + (y - 3) + z = 0 \quad \Leftrightarrow \quad x + y + z = 4$.

23. Since the direction vectors $\langle 2, 3, 4 \rangle$ and $\langle 6, -1, 2 \rangle$ aren't parallel, neither are the lines. For the lines to intersect, the three equations $1 + 2t = -1 + 6s$, $2 + 3t = 3 - s$, $3 + 4t = -5 + 2s$ must be satisfied simultaneously. Solving the first two equations gives $t = \frac{1}{5}$, $s = \frac{2}{5}$ and checking we see these values don't satisfy the third equation. Thus the lines aren't parallel and they don't intersect, so they must be skew.

25. By Exercise 9.5.51, $D = \dfrac{|2 - 24|}{\sqrt{26}} = \dfrac{22}{\sqrt{26}}$.

27. $\ln(x - y^2)$ is defined only when $x - y^2 > 0$, or $x > y^2$, and x is defined for all real numbers, so the domain of the product $x \ln(x - y^2)$ is $\{(x, y) \mid x > y^2\}$.

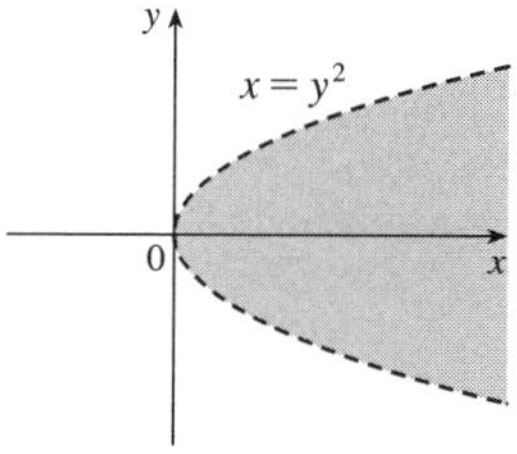

29. The graph is the plane $z = 6 - 2x - 3y \quad \Rightarrow \quad 2x + 3y + z = 6$. The intercepts with the coordinate axes are $(3, 0, 0)$, $(0, 2, 0)$, and $(0, 0, 6)$ which enable us to sketch the portion of the plane that lies in the first octant.

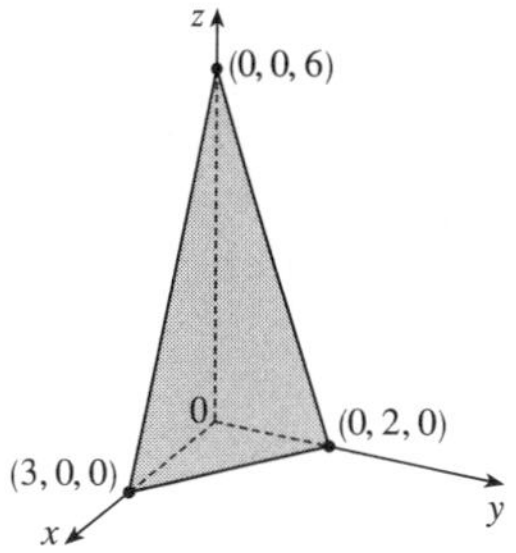

31. The equation is $z = 4 - x^2 - 4y^2$. The traces in $x = k$ are $z = 4 - k^2 - 4y^2$, a family of parabolas opening downward, as are the traces in $y = k$, $z = 4 - 4k^2 - x^2$. The traces in $z = k$ are $x^2 + 4y^2 = 4 - k$, a family of ellipses, so the surface is an elliptic paraboloid.

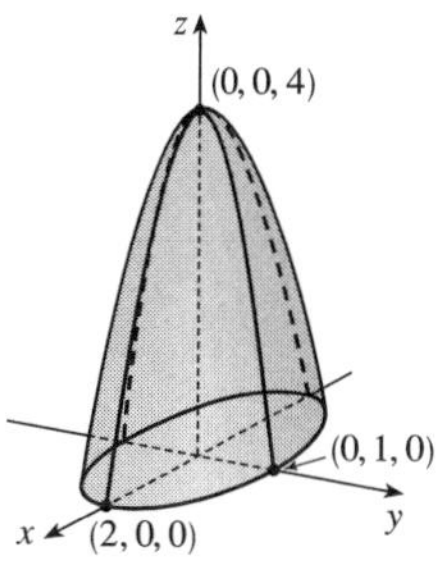

33. An equivalent equation is $\dfrac{x^2}{(1/2)^2} + y^2 + z^2 = 1$, an ellipsoid centred at the origin with intercepts $\pm\frac{1}{2}$, ± 1, and ± 1.

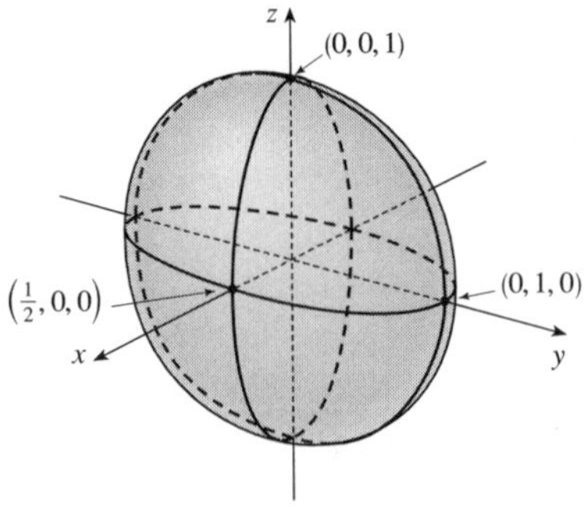

35. $y^2 + z^2 = 1$ is the equation of a circular cylinder with axis the x-axis.

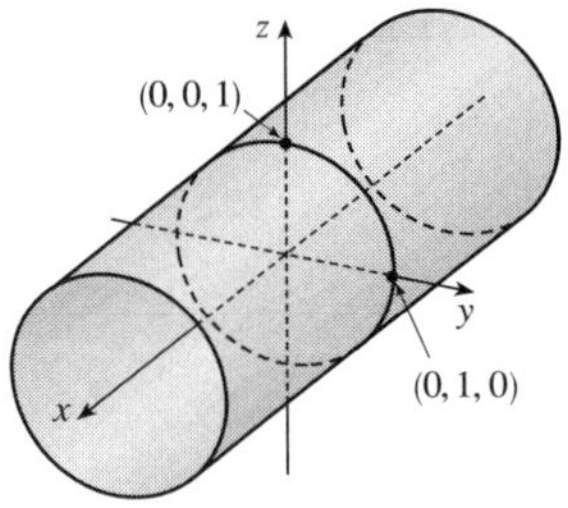

37. $x = r\cos\theta = 2\sqrt{3}\cos\frac{\pi}{3} = 2\sqrt{3}\cdot\frac{1}{2} = \sqrt{3}$, $y = r\sin\theta = 2\sqrt{3}\sin\frac{\pi}{3} = 2\sqrt{3}\cdot\frac{\sqrt{3}}{2} = 3$, $z = 2$, so in rectangular coordinates the point is $\left(\sqrt{3}, 3, 2\right)$. $\rho = \sqrt{r^2 + z^2} = \sqrt{12 + 4} = 4$, $\theta = \frac{\pi}{3}$, and $\cos\phi = \frac{z}{\rho} = \frac{1}{2}$, so $\phi = \frac{\pi}{3}$ and spherical coordinates are $\left(4, \frac{\pi}{3}, \frac{\pi}{3}\right)$.

39. $x = \rho\sin\phi\cos\theta = 8\sin\frac{\pi}{6}\cos\frac{\pi}{4} = 8\cdot\frac{1}{2}\cdot\frac{\sqrt{2}}{2} = 2\sqrt{2}$, $y = \rho\sin\phi\sin\theta = 8\sin\frac{\pi}{6}\sin\frac{\pi}{4} = 2\sqrt{2}$, and $z = \rho\cos\phi = 8\cos\frac{\pi}{6} = 8\cdot\frac{\sqrt{3}}{2} = 4\sqrt{3}$. Thus rectangular coordinates for the point are $\left(2\sqrt{2}, 2\sqrt{2}, 4\sqrt{3}\right)$.
$r^2 = x^2 + y^2 = 8 + 8 = 16 \quad\Rightarrow\quad r = 4$, $\theta = \frac{\pi}{4}$, and $z = 4\sqrt{3}$, so cylindrical coordinates are $\left(4, \frac{\pi}{4}, 4\sqrt{3}\right)$.

41. $x^2 + y^2 + z^2 = 4$. In cylindrical coordinates, this becomes $r^2 + z^2 = 4$. In spherical coordinates, it becomes $\rho^2 = 4$ or $\rho = 2$.

43. The resulting surface is a circular paraboloid with equation $z = 4x^2 + 4y^2$. Changing to cylindrical coordinates we have $z = 4\left(x^2 + y^2\right) = 4r^2$.

FOCUS ON PROBLEM SOLVING

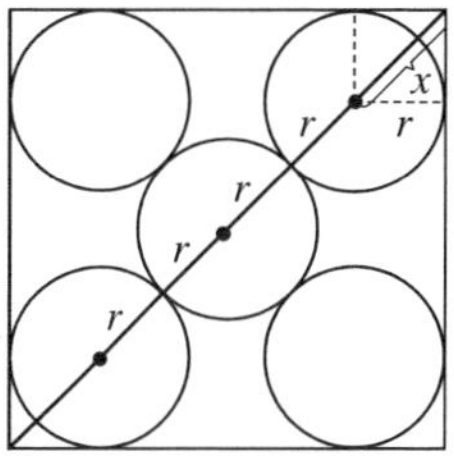

1. Since three-dimensional situations are often difficult to visualize and work with, let us first try to find an analogous problem in two dimensions. The analogue of a cube is a square and the analogue of a sphere is a circle. Thus a similar problem in two dimensions is the following: if five circles with the same radius r are contained in a square of side 1 m so that the circles touch each other and four of the circles touch two sides of the square, find r.

The diagonal of the square is $\sqrt{2}$. The diagonal is also $4r + 2x$. But x is the diagonal of a smaller square of side r. Therefore $x = \sqrt{2}\,r \quad \Rightarrow \quad \sqrt{2} = 4r + 2x = 4r + 2\sqrt{2}\,r = \left(4 + 2\sqrt{2}\right)r \quad \Rightarrow \quad r = \frac{\sqrt{2}}{4+2\sqrt{2}}$.

Let's use these ideas to solve the original three-dimensional problem. The diagonal of the cube is $\sqrt{1^2 + 1^2 + 1^2} = \sqrt{3}$. The diagonal of the cube is also $4r + 2x$ where x is the diagonal of a smaller cube with edge r. Therefore

$x = \sqrt{r^2 + r^2 + r^2} = \sqrt{3}\,r \quad \Rightarrow \quad \sqrt{3} = 4r + 2x = 4r + 2\sqrt{3}\,r = \left(4 + 2\sqrt{3}\right)r$. Thus $r = \dfrac{\sqrt{3}}{4 + 2\sqrt{3}} = \dfrac{2\sqrt{3} - 3}{2}$.

The radius of each ball is $\left(\sqrt{3} - \frac{3}{2}\right)$ m.

3. (a) We find the line of intersection L as in Example 9.5.7(b). Observe that the point $(-1, c, c)$ lies on both planes. Now since L lies in both planes, it is perpendicular to both of the normal vectors $\mathbf{n}_1$ and $\mathbf{n}_2$, and thus parallel to their cross product

$\mathbf{n}_1 \times \mathbf{n}_2 = \begin{vmatrix} \mathbf{i} & \mathbf{j} & \mathbf{k} \\ c & 1 & 1 \\ 1 & -c & c \end{vmatrix} = \left\langle 2c, -c^2 + 1, -c^2 - 1 \right\rangle$. So symmetric equations of L can be written as

$\dfrac{x+1}{-2c} = \dfrac{y-c}{c^2-1} = \dfrac{z-c}{c^2+1}$, provided that $c \neq 0, \pm 1$.

If $c = 0$, then the two planes are given by $y + z = 0$ and $x = -1$, so symmetric equations of L are $x = -1$, $y = -z$. If $c = -1$, then the two planes are given by $-x + y + z = -1$ and $x + y + z = -1$, and they intersect in the line $x = 0$, $y = -z - 1$. If $c = 1$, then the two planes are given by $x + y + z = 1$ and $x - y + z = 1$, and they intersect in the line $y = 0$, $x = 1 - z$.

(b) If we set $z = t$ in the symmetric equations and solve for x and y separately, we get $x + 1 = \dfrac{(t-c)(-2c)}{c^2+1}$,

$y - c = \dfrac{(t-c)(c^2-1)}{c^2+1} \quad \Rightarrow \quad x = \dfrac{-2ct + (c^2-1)}{c^2+1}$, $y = \dfrac{(c^2-1)t + 2c}{c^2+1}$. Eliminating c from these equations, we have $x^2 + y^2 = t^2 + 1$. So the curve traced out by L in the plane $z = t$ is a circle with centre at $(0, 0, t)$ and radius $\sqrt{t^2+1}$.

(c) The area of a horizontal cross-section of the solid is $A(z) = \pi(z^2 + 1)$, so $V = \int_0^1 A(z)dz = \pi\left[\frac{1}{3}z^3 + z\right]_0^1 = \frac{4\pi}{3}$.

5. (a) When $\theta = \theta_s$, the block is not moving, so the sum of the forces on the block must be $\mathbf{0}$, thus $\mathbf{N} + \mathbf{F} + \mathbf{W} = \mathbf{0}$. This relationship is illustrated geometrically in the figure. Since the vectors form a right triangle, we have

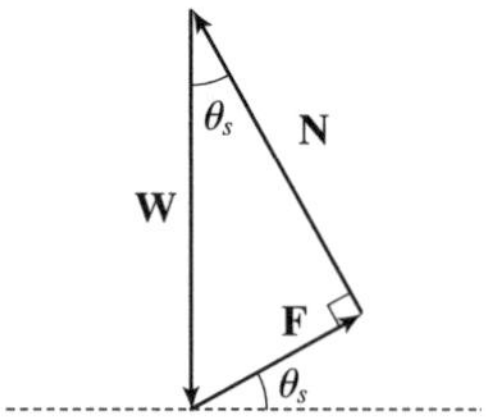

$$\tan(\theta_s) = \frac{|\mathbf{F}|}{|\mathbf{N}|} = \frac{\mu_s n}{n} = \mu_s.$$

(b) We place the block at the origin and sketch the force vectors acting on the block, including the additional horizontal force $\mathbf{H}$, with initial points at the origin. We then rotate this system so that $\mathbf{F}$ lies along the positive x-axis and the inclined plane is parallel to the x-axis.

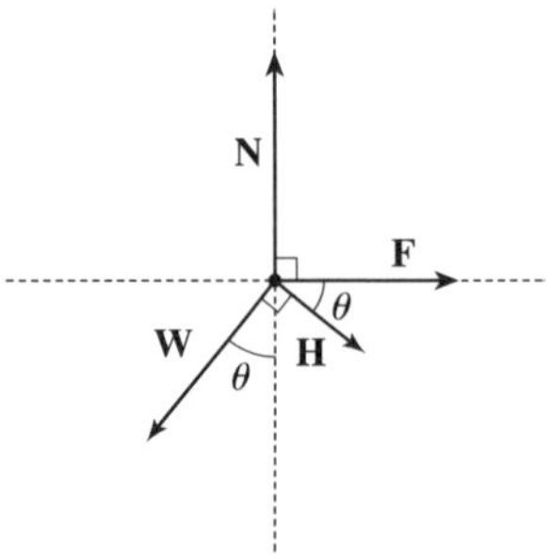

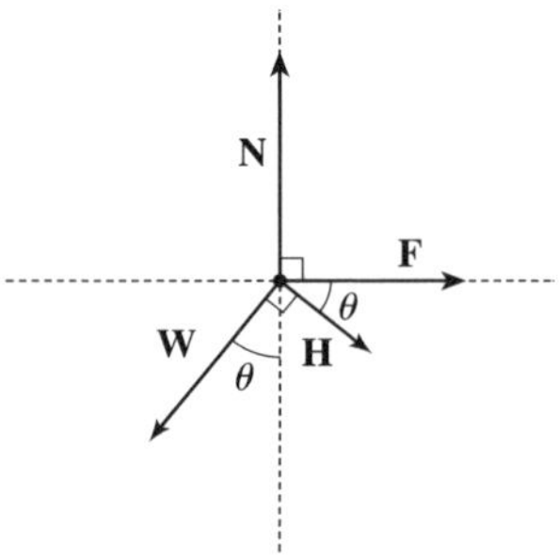

$|\mathbf{F}|$ is maximal, so $|\mathbf{F}| = \mu_s n$ for $\theta > \theta_s$. Then the vectors, in terms of components parallel and perpendicular to the inclined plane, are

$$\mathbf{N} = n\,\mathbf{j} \qquad \mathbf{F} = (\mu_s n)\,\mathbf{i}$$

$$\mathbf{W} = (-mg\sin\theta)\,\mathbf{i} + (-mg\cos\theta)\,\mathbf{j} \qquad \mathbf{H} = (h_{\min}\cos\theta)\,\mathbf{i} + (-h_{\min}\sin\theta)\,\mathbf{j}$$

Equating components, we have

$$\mu_s n - mg\sin\theta + h_{\min}\cos\theta = 0 \quad\Rightarrow\quad h_{\min}\cos\theta + \mu_s n = mg\sin\theta \qquad \textbf{(1)}$$

$$n - mg\cos\theta - h_{\min}\sin\theta = 0 \quad\Rightarrow\quad h_{\min}\sin\theta + mg\cos\theta = n \qquad \textbf{(2)}$$

(c) Since **(2)** is solved for n, we substitute into **(1)**:

$$h_{\min}\cos\theta + \mu_s(h_{\min}\sin\theta + mg\cos\theta) = mg\sin\theta \quad\Rightarrow$$

$$h_{\min}\cos\theta + h_{\min}\mu_s\sin\theta = mg\sin\theta - mg\mu_s\cos\theta \quad\Rightarrow$$

$$h_{\min} = mg\left(\frac{\sin\theta - \mu_s\cos\theta}{\cos\theta + \mu_s\sin\theta}\right) = mg\left(\frac{\tan\theta - \mu_s}{1 + \mu_s\tan\theta}\right)$$

From part (a) we know $\mu_s = \tan\theta_s$, so this becomes $h_{\min} = mg\left(\dfrac{\tan\theta - \tan\theta_s}{1 + \tan\theta_s\tan\theta}\right)$ and using a trigonometric identity, this is $mg\tan(\theta - \theta_s)$ as desired.

Note for $\theta = \theta_s$, $h_{\min} = mg\tan 0 = 0$, which makes sense since the block is at rest for θ_s, thus no additional force $\mathbf{H}$ is necessary to prevent it from moving. As θ increases, the factor $\tan(\theta - \theta_s)$, and hence the value of $h_{\min}$, increases slowly for small values of $\theta - \theta_s$ but much more rapidly as $\theta - \theta_s$ becomes significant. This seems reasonable, as the steeper the inclined plane, the less the horizontal components of the various forces affect the movement of the block, so we would need a much larger magnitude of horizontal force to keep the block motionless. If we allow $\theta \to 90°$, corresponding to the inclined plane being placed vertically, the value of $h_{\min}$ is quite large; this is to be expected, as it takes a great amount of horizontal force to keep an object from moving vertically. In fact, without friction (so $\theta_s = 0$), we would have $\theta \to 90° \quad \Rightarrow \quad h_{\min} \to \infty$, and it would be impossible to keep the block from slipping.

(d) Since $h_{\max}$ is the largest value of h that keeps the block from slipping, the force of friction is keeping the block from moving *up* the inclined plane; thus, $\mathbf{F}$ is directed *down* the plane. Our system of forces is similar to that in part (b), then, except that we have $\mathbf{F} = -(\mu_s n)\,\mathbf{i}$. (Note that $|\mathbf{F}|$ is again maximal.) Following our procedure in parts (b) and (c), we equate components:

$$-\mu_s n - mg\sin\theta + h_{\max}\cos\theta = 0 \quad \Rightarrow \quad h_{\max}\cos\theta - \mu_s n = mg\sin\theta$$

$$n - mg\cos\theta - h_{\max}\sin\theta = 0 \quad \Rightarrow \quad h_{\max}\sin\theta + mg\cos\theta = n$$

Then substituting,

$$h_{\max}\cos\theta - \mu_s(h_{\max}\sin\theta + mg\cos\theta) = mg\sin\theta \quad \Rightarrow$$

$$h_{\max}\cos\theta - h_{\max}\mu_s\sin\theta = mg\sin\theta + mg\mu_s\cos\theta \quad \Rightarrow$$

$$h_{\max} = mg\left(\frac{\sin\theta + \mu_s\cos\theta}{\cos\theta - \mu_s\sin\theta}\right) = mg\left(\frac{\tan\theta + \mu_s}{1 - \mu_s\tan\theta}\right)$$

$$= mg\left(\frac{\tan\theta + \tan\theta_s}{1 - \tan\theta_s\tan\theta}\right) = mg\tan(\theta + \theta_s)$$

We would expect $h_{\max}$ to increase as θ increases, with similar behavior as we established for $h_{\min}$, but with $h_{\max}$ values always larger than $h_{\min}$. We can see that this is the case if we graph $h_{\max}$ as a function of θ, as the curve is the graph of $h_{\min}$ translated $2\theta_s$ to the left, so the equation does seem reasonable. Notice that the equation predicts $h_{\max} \to \infty$ as $\theta \to (90° - \theta_s)$. In fact, as $h_{\max}$ increases, the normal force increases as well. When $(90° - \theta_s) \le \theta \le 90°$, the horizontal force is completely counteracted by the sum of the normal and frictional forces, so no part of the horizontal force contributes to moving the block up the plane no matter how large its magnitude.

10 □ VECTOR FUNCTIONS

10.1 Vector Functions and Space Curves

1. The component functions t^2, $\sqrt{t-1}$, and $\sqrt{5-t}$ are all defined when $t-1 \geq 0 \Rightarrow t \geq 1$ and $5-t \geq 0 \Rightarrow t \leq 5$, so the domain of $\mathbf{r}(t)$ is $[1, 5]$.

3. $\lim\limits_{t\to 0^+} \cos t = \cos 0 = 1$, $\lim\limits_{t\to 0^+} \sin t = \sin 0 = 0$, $\lim\limits_{t\to 0^+} t\ln t = \lim\limits_{t\to 0^+} \dfrac{\ln t}{1/t} = \lim\limits_{t\to 0^+} \dfrac{1/t}{-1/t^2} = \lim\limits_{t\to 0^+} -t = 0$ [by l'Hospital's Rule]. Thus $\lim\limits_{t\to 0^+} \langle \cos t, \sin t, t\ln t\rangle = \left\langle \lim\limits_{t\to 0^+} \cos t, \lim\limits_{t\to 0^+} \sin t, \lim\limits_{t\to 0^+} t\ln t \right\rangle = \langle 1, 0, 0\rangle$.

5. The corresponding parametric equations for this curve are $x = \sin t$, $y = t$. We can make a table of values, or we can eliminate the parameter: $t = y \Rightarrow x = \sin y$, with $y \in \mathbb{R}$. By comparing different values of t, we find the direction in which t increases as indicated in the graph.

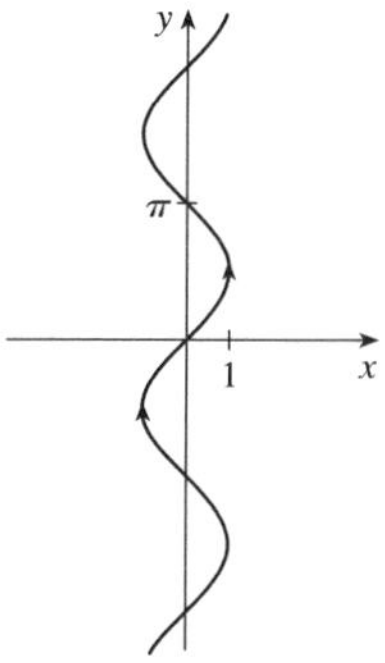

7. The corresponding parametric equations are $x = t$, $y = \cos 2t$, $z = \sin 2t$. Note that $y^2 + z^2 = \cos^2 2t + \sin^2 2t = 1$, so the curve lies on the circular cylinder $y^2 + z^2 = 1$. Since $x = t$, the curve is a helix.

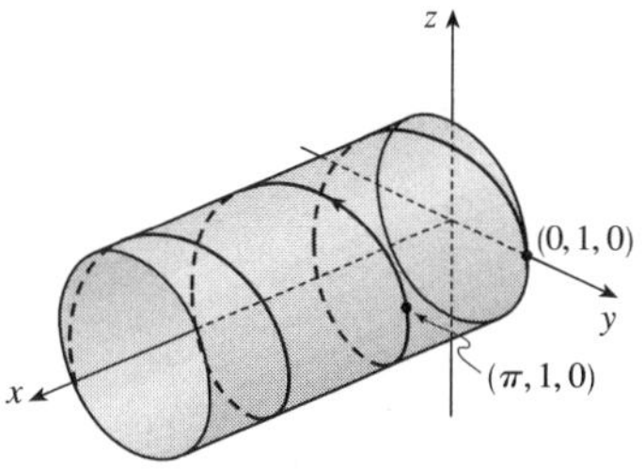

9. The parametric equations give $x^2 + z^2 = \sin^2 t + \cos^2 t = 1$, $y = 3$, which is a circle of radius 1, centre $(0, 3, 0)$ in the plane $y = 3$.

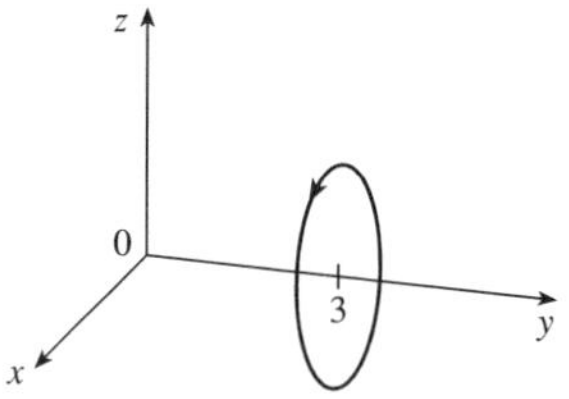

11. The parametric equations are $x = t^2$, $y = t^4$, $z = t^6$. These are positive for $t \neq 0$ and 0 when $t = 0$. So the curve lies entirely in the first quadrant. The projection of the graph onto the xy-plane is $y = x^2$, $y > 0$, a half parabola. On the xz-plane $z = x^3$, $z > 0$, a half cubic, and the yz-plane, $y^3 = z^2$.

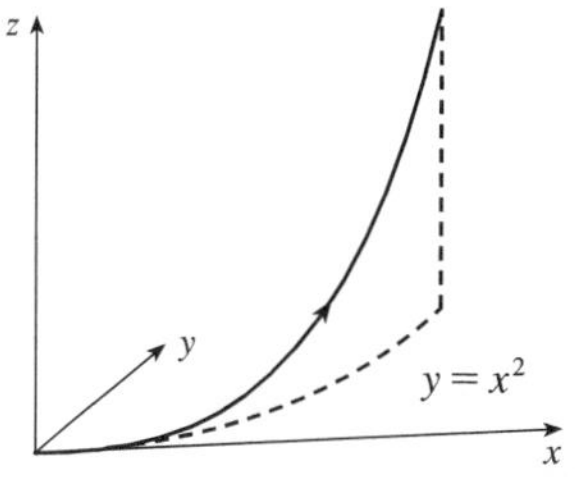

13. Taking $\mathbf{r}_0 = \langle 0, 0, 0\rangle$ and $\mathbf{r}_1 = \langle 1, 2, 3\rangle$, we have from Equation 9.5.4
$\mathbf{r}(t) = (1-t)\,\mathbf{r}_0 + t\,\mathbf{r}_1 = (1-t)\,\langle 0,0,0\rangle + t\,\langle 1,2,3\rangle$, $0 \le t \le 1$ or $\mathbf{r}(t) = \langle t, 2t, 3t\rangle$, $0 \le t \le 1$.
Parametric equations are $x = t$, $y = 2t$, $z = 3t$, $0 \le t \le 1$.

15. Taking $\mathbf{r}_0 = \langle 1, -1, 2\rangle$ and $\mathbf{r}_1 = \langle 4, 1, 7\rangle$, we have $\mathbf{r}(t) = (1-t)\,\mathbf{r}_0 + t\,\mathbf{r}_1 = (1-t)\,\langle 1,-1,2\rangle + t\,\langle 4,1,7\rangle$, $0 \le t \le 1$ or $\mathbf{r}(t) = \langle 1+3t, -1+2t, 2+5t\rangle$, $0 \le t \le 1$. Parametric equations are $x = 1+3t$, $y = -1+2t$, $z = 2+5t$, $0 \le t \le 1$.

17. $x = \cos 4t$, $y = t$, $z = \sin 4t$. At any point (x, y, z) on the curve, $x^2 + z^2 = \cos^2 4t + \sin^2 4t = 1$. So the curve lies on a circular cylinder with axis the y-axis. Since $y = t$, this is a helix. So the graph is VI.

19. $x = t$, $y = 1/(1+t^2)$, $z = t^2$. Note that y and z are positive for all t. The curve passes through $(0, 1, 0)$ when $t = 0$. As $t \to \infty$, $(x, y, z) \to (\infty, 0, \infty)$, and as $t \to -\infty$, $(x, y, z) \to (-\infty, 0, \infty)$. So the graph is IV.

21. $x = \cos t$, $y = \sin t$, $z = \sin 5t$. $x^2 + y^2 = \cos^2 t + \sin^2 t = 1$, so the curve lies on a circular cylinder with axis the z-axis. Each of x, y and z is periodic, and at $t = 0$ and $t = 2\pi$ the curve passes through the same point, so the curve repeats itself and the graph is V.

23. If $x = t\cos t$, $y = t\sin t$, and $z = t$, then
$x^2 + y^2 = t^2\cos^2 t + t^2\sin^2 t = t^2 = z^2$, so the curve lies on the cone $z^2 = x^2 + y^2$. Since $z = t$, the curve is a spiral on this cone.

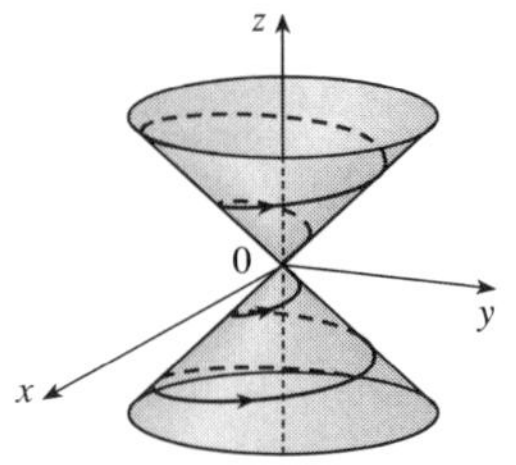

25. Parametric equations for the curve are $x = t$, $y = 0$, $z = 2t - t^2$. Substituting into the equation of the paraboloid gives
$2t - t^2 = t^2 \;\Rightarrow\; 2t = 2t^2 \;\Rightarrow\; t = 0, 1$. Since $\mathbf{r}(0) = \mathbf{0}$ and $\mathbf{r}(1) = \mathbf{i} + \mathbf{k}$, the points of intersection are $(0, 0, 0)$ and $(1, 0, 1)$.

27. $\mathbf{r}(t) = \langle t^2, \sqrt{t-1}, \sqrt{5-t}\,\rangle$

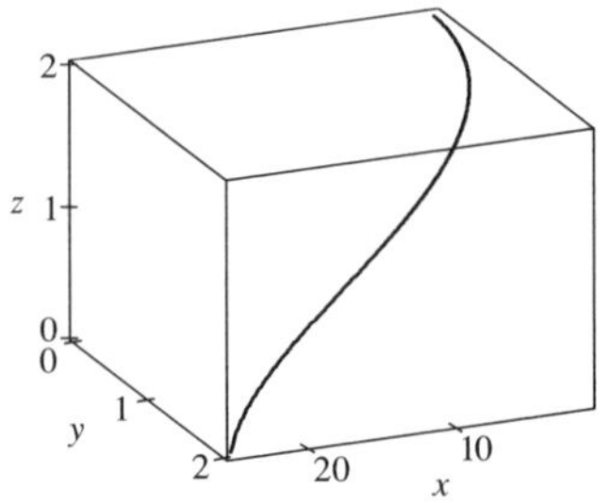

29.

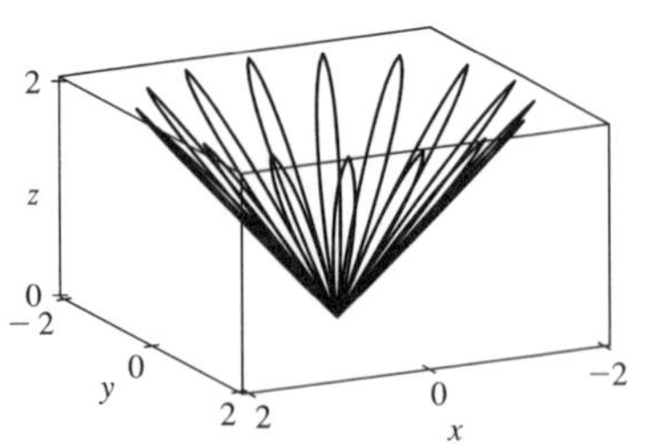

$x = (1 + \cos 16t)\cos t$, $y = (1 + \cos 16t)\sin t$, $z = 1 + \cos 16t$. At any point on the graph,

$$x^2 + y^2 = (1 + \cos 16t)^2\cos^2 t + (1 + \cos 16t)^2\sin^2 t$$
$$= (1 + \cos 16t)^2 = z^2,$$ so the graph lies on the cone $x^2 + y^2 = z^2$.

From the graph at left, we see that this curve looks like the projection of a leaved two-dimensional curve onto a cone.

31. If $t = -1$, then $x = 1$, $y = 4$, $z = 0$, so the curve passes through the point $(1, 4, 0)$. If $t = 3$, then $x = 9$, $y = -8$, $z = 28$, so the curve passes through the point $(9, -8, 28)$. For the point $(4, 7, -6)$ to be on the curve, we require $y = 1 - 3t = 7 \;\Rightarrow\; t = -2$. But then $z = 1 + (-2)^3 = -7 \ne -6$, so $(4, 7, -6)$ is not on the curve.

33. Both equations are solved for z, so we can substitute to eliminate z: $\sqrt{x^2 + y^2} = 1 + y \;\Rightarrow\; x^2 + y^2 = 1 + 2y + y^2 \;\Rightarrow\; x^2 = 1 + 2y \;\Rightarrow\; y = \frac{1}{2}(x^2 - 1)$. We can form parametric equations for the curve C of intersection by choosing a parameter $x = t$, then $y = \frac{1}{2}(t^2 - 1)$ and $z = 1 + y = 1 + \frac{1}{2}(t^2 - 1) = \frac{1}{2}(t^2 + 1)$. Thus a vector function representing C is $\mathbf{r}(t) = t\,\mathbf{i} + \frac{1}{2}(t^2 - 1)\,\mathbf{j} + \frac{1}{2}(t^2 + 1)\,\mathbf{k}$.

35. 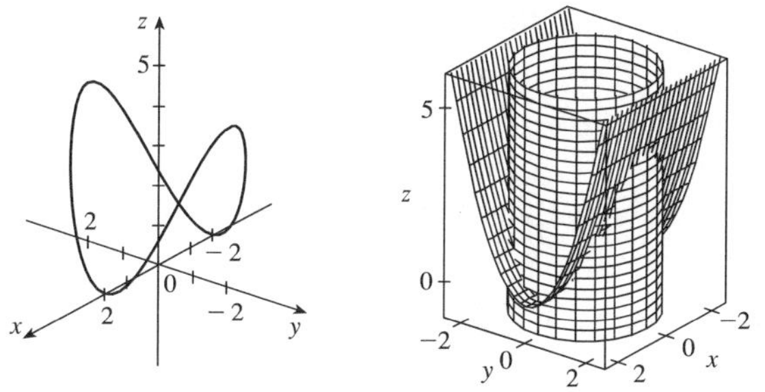

The projection of the curve C of intersection onto the xy-plane is the circle $x^2 + y^2 = 4$, $z = 0$. Then we can write $x = 2\cos t$, $y = 2\sin t$, $0 \le t \le 2\pi$. Since C also lies on the surface $z = x^2$, we have $z = x^2 = (2\cos t)^2 = 4\cos^2 t$. Then parametric equations for C are $x = 2\cos t$, $y = 2\sin t$, $z = 4\cos^2 t$, $0 \le t \le 2\pi$.

37. For the particles to collide, we require $\mathbf{r}_1(t) = \mathbf{r}_2(t) \iff \langle t^2, 7t - 12, t^2 \rangle = \langle 4t - 3, t^2, 5t - 6 \rangle$. Equating components gives $t^2 = 4t - 3$, $7t - 12 = t^2$, and $t^2 = 5t - 6$. From the first equation, $t^2 - 4t + 3 = 0 \iff (t-3)(t-1) = 0$ so $t = 1$ or $t = 3$. $t = 1$ does not satisfy the other two equations, but $t = 3$ does. The particles collide when $t = 3$, at the point $(9, 9, 9)$.

39. Let $\mathbf{u}(t) = \langle u_1(t), u_2(t), u_3(t) \rangle$ and $\mathbf{v}(t) = \langle v_1(t), v_2(t), v_3(t) \rangle$. In each part of this problem the basic procedure is to use Equation 1 and then analyze the individual component functions using the limit properties we have already developed for real-valued functions.

(a) $\lim\limits_{t\to a} \mathbf{u}(t) + \lim\limits_{t\to a} \mathbf{v}(t) = \left\langle \lim\limits_{t\to a} u_1(t), \lim\limits_{t\to a} u_2(t), \lim\limits_{t\to a} u_3(t) \right\rangle + \left\langle \lim\limits_{t\to a} v_1(t), \lim\limits_{t\to a} v_2(t), \lim\limits_{t\to a} v_3(t) \right\rangle$ and the limits of these component functions must each exist since the vector functions both possess limits as $t \to a$. Then adding the two vectors and using the addition property of limits for real-valued functions, we have that

$$\begin{aligned}
\lim_{t\to a} \mathbf{u}(t) + \lim_{t\to a} \mathbf{v}(t) &= \left\langle \lim_{t\to a} u_1(t) + \lim_{t\to a} v_1(t), \lim_{t\to a} u_2(t) + \lim_{t\to a} v_2(t), \lim_{t\to a} u_3(t) + \lim_{t\to a} v_3(t) \right\rangle \\
&= \left\langle \lim_{t\to a} [u_1(t) + v_1(t)], \lim_{t\to a} [u_2(t) + v_2(t)], \lim_{t\to a} [u_3(t) + v_3(t)] \right\rangle \\
&= \lim_{t\to a} \langle u_1(t) + v_1(t), u_2(t) + v_2(t), u_3(t) + v_3(t) \rangle \qquad \text{[using (1) backward]} \\
&= \lim_{t\to a} [\mathbf{u}(t) + \mathbf{v}(t)]
\end{aligned}$$

(b)
$$\begin{aligned}
\lim_{t\to a} c\mathbf{u}(t) &= \lim_{t\to a} \langle cu_1(t), cu_2(t), cu_3(t) \rangle = \left\langle \lim_{t\to a} cu_1(t), \lim_{t\to a} cu_2(t), \lim_{t\to a} cu_3(t) \right\rangle \\
&= \left\langle c\lim_{t\to a} u_1(t), c\lim_{t\to a} u_2(t), c\lim_{t\to a} u_3(t) \right\rangle = c\left\langle \lim_{t\to a} u_1(t), \lim_{t\to a} u_2(t), \lim_{t\to a} u_3(t) \right\rangle \\
&= c\lim_{t\to a} \langle u_1(t), u_2(t), u_3(t) \rangle = c\lim_{t\to a} \mathbf{u}(t)
\end{aligned}$$

(c)
$$\begin{aligned}
\lim_{t\to a} \mathbf{u}(t) \cdot \lim_{t\to a} \mathbf{v}(t) &= \left\langle \lim_{t\to a} u_1(t), \lim_{t\to a} u_2(t), \lim_{t\to a} u_3(t) \right\rangle \cdot \left\langle \lim_{t\to a} v_1(t), \lim_{t\to a} v_2(t), \lim_{t\to a} v_3(t) \right\rangle \\
&= \left[\lim_{t\to a} u_1(t)\right]\left[\lim_{t\to a} v_1(t)\right] + \left[\lim_{t\to a} u_2(t)\right]\left[\lim_{t\to a} v_2(t)\right] + \left[\lim_{t\to a} u_3(t)\right]\left[\lim_{t\to a} v_3(t)\right] \\
&= \lim_{t\to a} u_1(t)v_1(t) + \lim_{t\to a} u_2(t)v_2(t) + \lim_{t\to a} u_3(t)v_3(t) \\
&= \lim_{t\to a} [u_1(t)v_1(t) + u_2(t)v_2(t) + u_3(t)v_3(t)] = \lim_{t\to a} [\mathbf{u}(t) \cdot \mathbf{v}(t)]
\end{aligned}$$

(d) $\lim_{t\to a}\mathbf{u}(t)\times\lim_{t\to a}\mathbf{v}(t)=\left\langle\lim_{t\to a}u_1(t),\lim_{t\to a}u_2(t),\lim_{t\to a}u_3(t)\right\rangle\times\left\langle\lim_{t\to a}v_1(t),\lim_{t\to a}v_2(t),\lim_{t\to a}v_3(t)\right\rangle$

$$=\left\langle\left[\lim_{t\to a}u_2(t)\right]\left[\lim_{t\to a}v_3(t)\right]-\left[\lim_{t\to a}u_3(t)\right]\left[\lim_{t\to a}v_2(t)\right],\right.$$

$$\left[\lim_{t\to a}u_3(t)\right]\left[\lim_{t\to a}v_1(t)\right]-\left[\lim_{t\to a}u_1(t)\right]\left[\lim_{t\to a}v_3(t)\right],$$

$$\left.\left[\lim_{t\to a}u_1(t)\right]\left[\lim_{t\to a}v_2(t)\right]-\left[\lim_{t\to a}u_2(t)\right]\left[\lim_{t\to a}v_1(t)\right]\right\rangle$$

$$=\left\langle\lim_{t\to a}\left[u_2(t)v_3(t)-u_3(t)v_2(t)\right],\lim_{t\to a}\left[u_3(t)v_1(t)-u_1(t)v_3(t)\right],\right.$$

$$\left.\lim_{t\to a}\left[u_1(t)v_2(t)-u_2(t)v_1(t)\right]\right\rangle$$

$$=\lim_{t\to a}\left\langle u_2(t)v_3(t)-u_3(t)v_2(t),u_3(t)\,v_1(t)-u_1(t)v_3(t),u_1(t)v_2(t)-u_2(t)v_1(t)\right\rangle$$

$$=\lim_{t\to a}\left[\mathbf{u}(t)\times\mathbf{v}(t)\right]$$

10.2 Derivatives and Integrals of Vector Functions

1. (a)

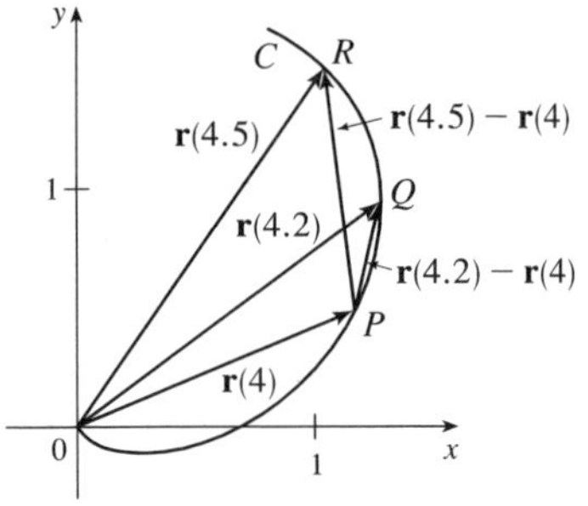

(b) $\dfrac{\mathbf{r}(4.5)-\mathbf{r}(4)}{0.5}=2[\mathbf{r}(4.5)-\mathbf{r}(4)]$, so we draw a vector in the same direction but with twice the length of the vector $\mathbf{r}(4.5)-\mathbf{r}(4)$. $\dfrac{\mathbf{r}(4.2)-\mathbf{r}(4)}{0.2}=5[\mathbf{r}(4.2)-\mathbf{r}(4)]$, so we draw a vector in the same direction but with 5 times the length of the vector $\mathbf{r}(4.2)-\mathbf{r}(4)$.

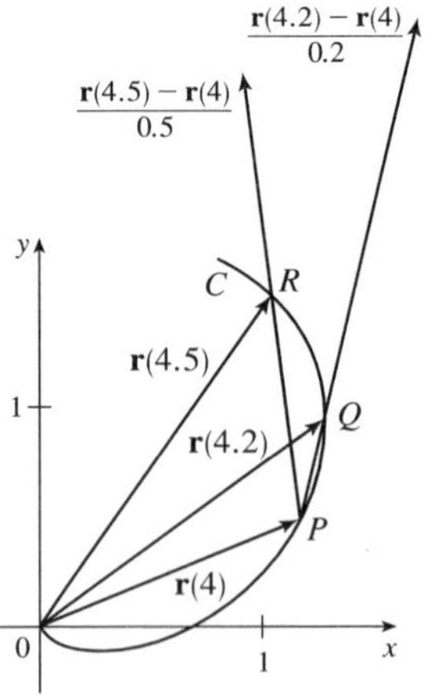

(c) By Definition 1, $\mathbf{r}'(4)=\lim_{h\to 0}\dfrac{\mathbf{r}(4+h)-\mathbf{r}(4)}{h}$.

$\mathbf{T}(4)=\dfrac{\mathbf{r}'(4)}{|\mathbf{r}'(4)|}$.

(d) $\mathbf{T}(4)$ is a unit vector in the same direction as $\mathbf{r}'(4)$, that is, parallel to the tangent line to the curve at $\mathbf{r}(4)$ with length 1.

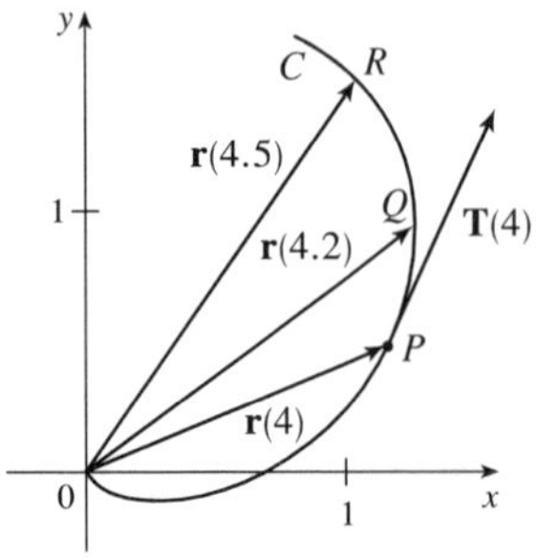

3. (a), (c)

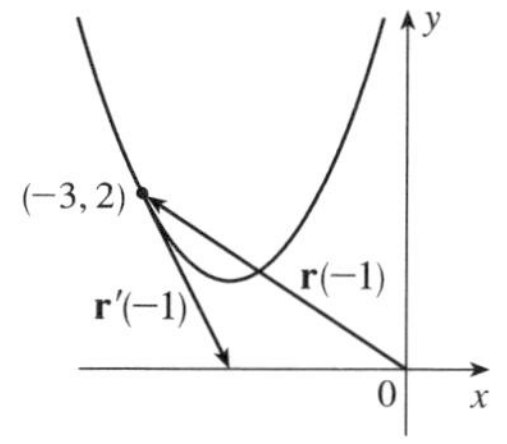

(b) $\mathbf{r}'(t) = \langle 1, 2t \rangle$

5. $x^{-2} = e^{-2t} = y$, so $y = 1/x^2$, $x > 0$

(a), (c)

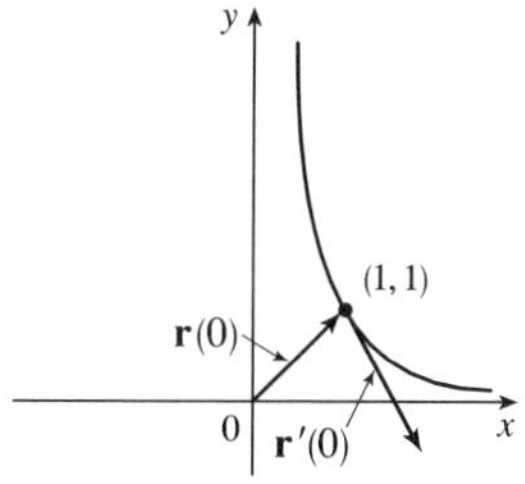

(b) $\mathbf{r}'(t) = e^t\,\mathbf{i} - 2e^{-2t}\,\mathbf{j}$

7. (a), (c)

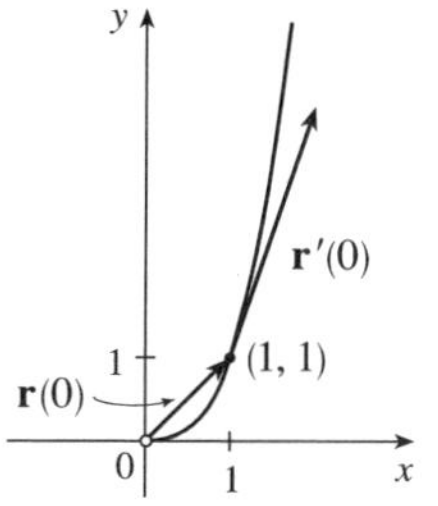

(b) $\mathbf{r}'(t) = e^t\,\mathbf{i} + 3e^{3t}\,\mathbf{j}$

9. $\mathbf{r}'(t) = \left\langle \dfrac{d}{dt}\left[t^2\right], \dfrac{d}{dt}\left[1-t\right], \dfrac{d}{dt}\left[\sqrt{t}\right] \right\rangle = \left\langle 2t, -1, \dfrac{1}{2\sqrt{t}} \right\rangle$

11. $\mathbf{r}(t) = e^{t^2}\,\mathbf{i} - \mathbf{j} + \ln(1+3t)\,\mathbf{k} \quad\Rightarrow\quad \mathbf{r}'(t) = 2te^{t^2}\,\mathbf{i} + \dfrac{3}{1+3t}\,\mathbf{k}$

13. $\mathbf{r}'(t) = \mathbf{0} + \mathbf{b} + 2t\,\mathbf{c} = \mathbf{b} + 2t\,\mathbf{c}$ by Formulas 1 and 3 of Theorem 3.

15. $\mathbf{r}'(t) = -\sin t\,\mathbf{i} + 3\,\mathbf{j} + 4\cos 2t\,\mathbf{k} \quad\Rightarrow\quad \mathbf{r}'(0) = 3\,\mathbf{j} + 4\,\mathbf{k}$. Thus

$$\mathbf{T}(0) = \frac{\mathbf{r}'(0)}{|\mathbf{r}'(0)|} = \frac{1}{\sqrt{0^2+3^2+4^2}}(3\,\mathbf{j}+4\,\mathbf{k}) = \tfrac{1}{5}(3\,\mathbf{j}+4\,\mathbf{k}) = \tfrac{3}{5}\,\mathbf{j} + \tfrac{4}{5}\,\mathbf{k}.$$

17. $\mathbf{r}(t) = \langle t, t^2, t^3 \rangle \quad\Rightarrow\quad \mathbf{r}'(t) = \langle 1, 2t, 3t^2 \rangle$. Then $\mathbf{r}'(1) = \langle 1, 2, 3 \rangle$ and $|\mathbf{r}'(1)| = \sqrt{1^2+2^2+3^2} = \sqrt{14}$, so

$$\mathbf{T}(1) = \frac{\mathbf{r}'(1)}{|\mathbf{r}'(1)|} = \tfrac{1}{\sqrt{14}}\langle 1, 2, 3 \rangle = \left\langle \tfrac{1}{\sqrt{14}}, \tfrac{2}{\sqrt{14}}, \tfrac{3}{\sqrt{14}} \right\rangle. \quad \mathbf{r}''(t) = \langle 0, 2, 6t \rangle, \text{ so}$$

$$\mathbf{r}'(t)\times\mathbf{r}''(t) = \begin{vmatrix} \mathbf{i} & \mathbf{j} & \mathbf{k} \\ 1 & 2t & 3t^2 \\ 0 & 2 & 6t \end{vmatrix} = \begin{vmatrix} 2t & 3t^2 \\ 2 & 6t \end{vmatrix}\mathbf{i} - \begin{vmatrix} 1 & 3t^2 \\ 0 & 6t \end{vmatrix}\mathbf{j} + \begin{vmatrix} 1 & 2t \\ 0 & 2 \end{vmatrix}\mathbf{k}$$

$$= (12t^2 - 6t^2)\,\mathbf{i} - (6t - 0)\,\mathbf{j} + (2 - 0)\,\mathbf{k} = \langle 6t^2, -6t, 2 \rangle.$$

19. The vector equation for the curve is $\mathbf{r}(t) = \langle t^5, t^4, t^3 \rangle$, so $\mathbf{r}'(t) = \langle 5t^4, 4t^3, 3t^2 \rangle$. The point $(1, 1, 1)$ corresponds to $t = 1$, so the tangent vector there is $\mathbf{r}'(1) = \langle 5, 4, 3 \rangle$. Thus, the tangent line goes through the point $(1, 1, 1)$ and is parallel to the vector $\langle 5, 4, 3 \rangle$. Parametric equations are $x = 1 + 5t$, $y = 1 + 4t$, $z = 1 + 3t$.

21. The vector equation for the curve is $\mathbf{r}(t) = \langle e^{-t}\cos t, e^{-t}\sin t, e^{-t} \rangle$, so

$$\mathbf{r}'(t) = \langle e^{-t}(-\sin t) + (\cos t)(-e^{-t}), e^{-t}\cos t + (\sin t)(-e^{-t}), (-e^{-t}) \rangle$$
$$= \langle -e^{-t}(\cos t + \sin t), e^{-t}(\cos t - \sin t), -e^{-t} \rangle.$$

The point $(1, 0, 1)$ corresponds to $t = 0$, so the tangent vector there is

$\mathbf{r}'(0) = \langle -e^0(\cos 0 + \sin 0), e^0(\cos 0 - \sin 0), -e^0 \rangle = \langle -1, 1, -1 \rangle$. Thus, the tangent line is parallel to the vector $\langle -1, 1, -1 \rangle$ and parametric equations are $x = 1 + (-1)t = 1 - t$, $y = 0 + 1\cdot t = t$, $z = 1 + (-1)t = 1 - t$.

23. $\mathbf{r}(t) = \langle t, e^{-t}, 2t - t^2 \rangle \Rightarrow \mathbf{r}'(t) = \langle 1, -e^{-t}, 2 - 2t \rangle$. At $(0, 1, 0)$, $t = 0$ and $\mathbf{r}'(0) = \langle 1, -1, 2 \rangle$. Thus, parametric equations of the tangent line are $x = t$, $y = 1 - t$, $z = 2t$.

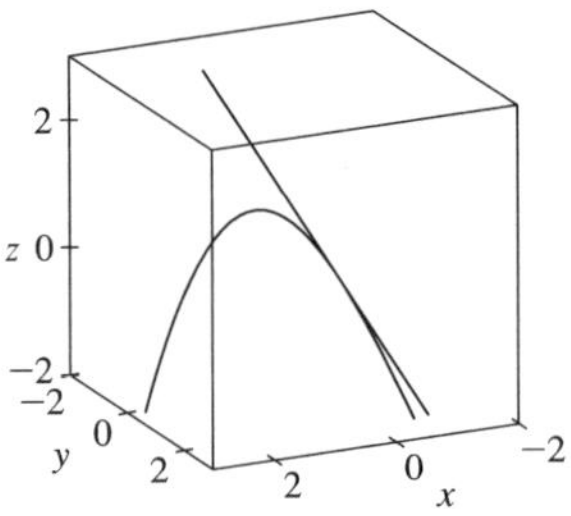

25. (a) $\mathbf{r}(t) = \langle t^3, t^4, t^5 \rangle \Rightarrow \mathbf{r}'(t) = \langle 3t^2, 4t^3, 5t^4 \rangle$, and since $\mathbf{r}'(0) = \langle 0, 0, 0 \rangle = \mathbf{0}$, the curve is not smooth.

(b) $\mathbf{r}(t) = \langle t^3 + t, t^4, t^5 \rangle \Rightarrow \mathbf{r}'(t) = \langle 3t^2 + 1, 4t^3, 5t^4 \rangle$. $\mathbf{r}'(t)$ is continuous since its component functions are continuous. Also, $\mathbf{r}'(t) \neq \mathbf{0}$, as the y- and z-components are 0 only for $t = 0$, but $\mathbf{r}'(0) = \langle 1, 0, 0 \rangle \neq \mathbf{0}$. Thus, the curve is smooth.

(c) $\mathbf{r}(t) = \langle \cos^3 t, \sin^3 t \rangle \Rightarrow \mathbf{r}'(t) = \langle -3\cos^2 t \sin t, 3\sin^2 t \cos t \rangle$. Since $\mathbf{r}'(0) = \langle -3\cos^2 0 \sin 0, 3\sin^2 0 \cos 0 \rangle = \langle 0, 0 \rangle = \mathbf{0}$, the curve is not smooth.

27. The angle of intersection of the two curves is the angle between the two tangent vectors to the curves at the point of intersection. Since $\mathbf{r}_1'(t) = \langle 1, 2t, 3t^2 \rangle$ and $t = 0$ at $(0, 0, 0)$, $\mathbf{r}_1'(0) = \langle 1, 0, 0 \rangle$ is a tangent vector to $\mathbf{r}_1$ at $(0, 0, 0)$. Similarly, $\mathbf{r}_2'(t) = \langle \cos t, 2\cos 2t, 1 \rangle$ and since $\mathbf{r}_2(0) = \langle 0, 0, 0 \rangle$, $\mathbf{r}_2'(0) = \langle 1, 2, 1 \rangle$ is a tangent vector to $\mathbf{r}_2$ at $(0, 0, 0)$. If θ is the angle between these two tangent vectors, then $\cos\theta = \frac{1}{\sqrt{1}\sqrt{6}} \langle 1, 0, 0 \rangle \cdot \langle 1, 2, 1 \rangle = \frac{1}{\sqrt{6}}$ and $\theta = \cos^{-1}\left(\frac{1}{\sqrt{6}}\right) \approx 66^\circ$.

29.
$$\begin{aligned}\int_0^1 (16t^3\,\mathbf{i} - 9t^2\,\mathbf{j} + 25t^4\,\mathbf{k})\,dt &= \left(\int_0^1 16t^3\,dt\right)\mathbf{i} - \left(\int_0^1 9t^2\,dt\right)\mathbf{j} + \left(\int_0^1 25t^4\,dt\right)\mathbf{k} \\ &= \left[4t^4\right]_0^1\,\mathbf{i} - \left[3t^3\right]_0^1\,\mathbf{j} + \left[5t^5\right]_0^1\,\mathbf{k} = 4\,\mathbf{i} - 3\,\mathbf{j} + 5\,\mathbf{k}\end{aligned}$$

31.
$$\begin{aligned}&\int_0^{\pi/2} (3\sin^2 t\cos t\,\mathbf{i} + 3\sin t\cos^2 t\,\mathbf{j} + 2\sin t\cos t\,\mathbf{k})\,dt \\ &= \left(\int_0^{\pi/2} 3\sin^2 t\cos t\,dt\right)\mathbf{i} + \left(\int_0^{\pi/2} 3\sin t\cos^2 t\,dt\right)\mathbf{j} + \left(\int_0^{\pi/2} 2\sin t\cos t\,dt\right)\mathbf{k} \\ &= \left[\sin^3 t\right]_0^{\pi/2}\,\mathbf{i} + \left[-\cos^3 t\right]_0^{\pi/2}\,\mathbf{j} + \left[\sin^2 t\right]_0^{\pi/2}\,\mathbf{k} = (1-0)\,\mathbf{i} + (0+1)\,\mathbf{j} + (1-0)\,\mathbf{k} = \mathbf{i} + \mathbf{j} + \mathbf{k}\end{aligned}$$

33.
$$\begin{aligned}\int (e^t\,\mathbf{i} + 2t\,\mathbf{j} + \ln t\,\mathbf{k})\,dt &= \left(\int e^t\,dt\right)\mathbf{i} + \left(\int 2t\,dt\right)\mathbf{j} + \left(\int \ln t\,dt\right)\mathbf{k} \\ &= e^t\,\mathbf{i} + t^2\,\mathbf{j} + (t\ln t - t)\,\mathbf{k} + \mathbf{C}, \text{ where } \mathbf{C} \text{ is a vector constant of integration.}\end{aligned}$$

35. $\mathbf{r}'(t) = 2t\,\mathbf{i} + 3t^2\,\mathbf{j} + \sqrt{t}\,\mathbf{k} \Rightarrow \mathbf{r}(t) = t^2\,\mathbf{i} + t^3\,\mathbf{j} + \frac{2}{3}t^{3/2}\,\mathbf{k} + \mathbf{C}$, where $\mathbf{C}$ is a constant vector. But $\mathbf{i} + \mathbf{j} = \mathbf{r}(1) = \mathbf{i} + \mathbf{j} + \frac{2}{3}\mathbf{k} + \mathbf{C}$. Thus $\mathbf{C} = -\frac{2}{3}\mathbf{k}$ and $\mathbf{r}(t) = t^2\,\mathbf{i} + t^3\,\mathbf{j} + \left(\frac{2}{3}t^{3/2} - \frac{2}{3}\right)\mathbf{k}$.

For Exercises 37–40, let $\mathbf{u}(t) = \langle u_1(t), u_2(t), u_3(t) \rangle$ and $\mathbf{v}(t) = \langle v_1(t), v_2(t), v_3(t) \rangle$. In each of these exercises, the procedure is to apply Theorem 2 so that the corresponding properties of derivatives of real-valued functions can be used.

37.
$$\begin{aligned}\frac{d}{dt}\left[\mathbf{u}(t) + \mathbf{v}(t)\right] &= \frac{d}{dt}\langle u_1(t) + v_1(t), u_2(t) + v_2(t), u_3(t) + v_3(t) \rangle \\ &= \left\langle \frac{d}{dt}\left[u_1(t) + v_1(t)\right], \frac{d}{dt}\left[u_2(t) + v_2(t)\right], \frac{d}{dt}\left[u_3(t) + v_3(t)\right] \right\rangle \\ &= \langle u_1'(t) + v_1'(t), u_2'(t) + v_2'(t), u_3'(t) + v_3'(t) \rangle \\ &= \langle u_1'(t), u_2'(t), u_3'(t) \rangle + \langle v_1'(t), v_2'(t), v_3'(t) \rangle = \mathbf{u}'(t) + \mathbf{v}'(t).\end{aligned}$$

39. $$\begin{aligned}\frac{d}{dt}[\mathbf{u}(t)\times\mathbf{v}(t)] &= \frac{d}{dt}\langle u_2(t)v_3(t)-u_3(t)v_2(t), u_3(t)v_1(t)-u_1(t)v_3(t), u_1(t)v_2(t)-u_2(t)v_1(t)\rangle \\ &= \langle u_2'v_3(t)+u_2(t)v_3'(t)-u_3'(t)v_2(t)-u_3(t)v_2'(t), \\ &\qquad u_3'(t)v_1(t)+u_3(t)v_1'(t)-u_1'(t)v_3(t)-u_1(t)v_3'(t), \\ &\qquad u_1'(t)v_2(t)+u_1(t)v_2'(t)-u_2'(t)v_1(t)-u_2(t)v_1'(t)\rangle \\ &= \langle u_2'(t)v_3(t)-u_3'(t)v_2(t), u_3'(t)v_1(t)-u_1'(t)v_3(t), u_1'(t)v_2(t)-u_2'(t)v_1(t)\rangle \\ &\qquad + \langle u_2(t)v_3'(t)-u_3(t)v_2'(t), u_3(t)v_1'(t)-u_1(t)v_3'(t), u_1(t)v_2'(t)-u_2(t)v_1'(t)\rangle \\ &= \mathbf{u}'(t)\times\mathbf{v}(t)+\mathbf{u}(t)\times\mathbf{v}'(t)\end{aligned}$$

Alternate solution: Let $\mathbf{r}(t)=\mathbf{u}(t)\times\mathbf{v}(t)$. Then

$$\begin{aligned}\mathbf{r}(t+h)-\mathbf{r}(t) &= [\mathbf{u}(t+h)\times\mathbf{v}(t+h)]-[\mathbf{u}(t)\times\mathbf{v}(t)] \\ &= [\mathbf{u}(t+h)\times\mathbf{v}(t+h)]-[\mathbf{u}(t)\times\mathbf{v}(t)]+[\mathbf{u}(t+h)\times\mathbf{v}(t)]-[\mathbf{u}(t+h)\times\mathbf{v}(t)] \\ &= \mathbf{u}(t+h)\times[\mathbf{v}(t+h)-\mathbf{v}(t)]+[\mathbf{u}(t+h)-\mathbf{u}(t)]\times\mathbf{v}(t)\end{aligned}$$

(Be careful of the order of the cross product.)
Dividing through by h and taking the limit as $h\to 0$ we have

$$\begin{aligned}\mathbf{r}'(t) &= \lim_{h\to 0}\frac{\mathbf{u}(t+h)\times[\mathbf{v}(t+h)-\mathbf{v}(t)]}{h}+\lim_{h\to 0}\frac{[\mathbf{u}(t+h)-\mathbf{u}(t)]\times\mathbf{v}(t)}{h} \\ &= \mathbf{u}(t)\times\mathbf{v}'(t)+\mathbf{u}'(t)\times\mathbf{v}(t)\end{aligned}$$

by Exercise 10.1.39(a) and Definition 1.

41. $$\begin{aligned}\frac{d}{dt}[\mathbf{u}(t)\cdot\mathbf{v}(t)] &= \mathbf{u}'(t)\cdot\mathbf{v}(t)+\mathbf{u}(t)\cdot\mathbf{v}'(t) \qquad \text{[by Formula 4 of Theorem 3]} \\ &= (-4t\,\mathbf{j}+9t^2\,\mathbf{k})\cdot(t\,\mathbf{i}+\cos t\,\mathbf{j}+\sin t\,\mathbf{k})+(\mathbf{i}-2t^2\,\mathbf{j}+3t^3\,\mathbf{k})\cdot(\mathbf{i}-\sin t\,\mathbf{j}+\cos t\,\mathbf{k}) \\ &= -4t\cos t+9t^2\sin t+1+2t^2\sin t+3t^3\cos t \\ &= 1-4t\cos t+11t^2\sin t+3t^3\cos t\end{aligned}$$

43. $\frac{d}{dt}[\mathbf{r}(t)\times\mathbf{r}'(t)]=\mathbf{r}'(t)\times\mathbf{r}'(t)+\mathbf{r}(t)\times\mathbf{r}''(t)$ by Formula 5 of Theorem 3. But $\mathbf{r}'(t)\times\mathbf{r}'(t)=\mathbf{0}$ [by the margin note on page 658]. Thus, $\frac{d}{dt}[\mathbf{r}(t)\times\mathbf{r}'(t)]=\mathbf{r}(t)\times\mathbf{r}''(t)$.

45. $\frac{d}{dt}|\mathbf{r}(t)|=\frac{d}{dt}[\mathbf{r}(t)\cdot\mathbf{r}(t)]^{1/2}=\frac{1}{2}[\mathbf{r}(t)\cdot\mathbf{r}(t)]^{-1/2}[2\mathbf{r}(t)\cdot\mathbf{r}'(t)]=\frac{1}{|\mathbf{r}(t)|}\mathbf{r}(t)\cdot\mathbf{r}'(t)$

47. Since $\mathbf{u}(t)=\mathbf{r}(t)\cdot[\mathbf{r}'(t)\times\mathbf{r}''(t)]$,

$$\begin{aligned}\mathbf{u}'(t) &= \mathbf{r}'(t)\cdot[\mathbf{r}'(t)\times\mathbf{r}''(t)]+\mathbf{r}(t)\cdot\frac{d}{dt}[\mathbf{r}'(t)\times\mathbf{r}''(t)] \\ &= 0+\mathbf{r}(t)\cdot[\mathbf{r}''(t)\times\mathbf{r}''(t)+\mathbf{r}'(t)\times\mathbf{r}'''(t)] \qquad [\text{since } \mathbf{r}'(t)\perp\mathbf{r}'(t)\times\mathbf{r}''(t)] \\ &= \mathbf{r}(t)\cdot[\mathbf{r}'(t)\times\mathbf{r}'''(t)] \qquad [\text{since } \mathbf{r}''(t)\times\mathbf{r}''(t)=\mathbf{0}]\end{aligned}$$

10.3 Arc Length and Curvature

1. $\mathbf{r}'(t) = \langle 2\cos t, 5, -2\sin t\rangle \Rightarrow |\mathbf{r}'(t)| = \sqrt{(2\cos t)^2 + 5^2 + (-2\sin t)^2} = \sqrt{29}$. Then using Formula 3, we have $L = \int_{-10}^{10} |\mathbf{r}'(t)|\,dt = \int_{-10}^{10} \sqrt{29}\,dt = \sqrt{29}\,t\Big]_{-10}^{10} = 20\sqrt{29}$.

3. $\mathbf{r}'(t) = 2t\,\mathbf{j} + 3t^2\,\mathbf{k} \Rightarrow |\mathbf{r}'(t)| = \sqrt{4t^2 + 9t^4} = t\sqrt{4 + 9t^2}$ (since $t \geq 0$). Then $L = \int_0^1 |\mathbf{r}'(t)|\,dt = \int_0^1 t\sqrt{4 + 9t^2}\,dt = \frac{1}{18} \cdot \frac{2}{3}(4 + 9t^2)^{3/2}\Big]_0^1 = \frac{1}{27}(13^{3/2} - 4^{3/2}) = \frac{1}{27}(13^{3/2} - 8)$.

5. The point $(2, 4, 8)$ corresponds to $t = 2$, so by Equation 2, $L = \int_0^2 \sqrt{(1)^2 + (2t)^2 + (3t^2)^2}\,dt$. If $f(t) = \sqrt{1 + 4t^2 + 9t^4}$, then Simpson's Rule gives $L \approx \dfrac{2-0}{10 \cdot 3}\left[f(0) + 4f(0.2) + 2f(0.4) + \cdots + 4f(1.8) + f(2)\right] \approx 9.5706$.

7. $\mathbf{r}'(t) = 2\,\mathbf{i} - 3\,\mathbf{j} + 4\,\mathbf{k}$ and $\frac{ds}{dt} = |\mathbf{r}'(t)| = \sqrt{4 + 9 + 16} = \sqrt{29}$. Then $s = s(t) = \int_0^t |\mathbf{r}'(u)|\,du = \int_0^t \sqrt{29}\,du = \sqrt{29}\,t$. Therefore, $t = \frac{1}{\sqrt{29}}s$, and substituting for t in the original equation, we have

$\mathbf{r}(t(s)) = \frac{2}{\sqrt{29}}s\,\mathbf{i} + \left(1 - \frac{3}{\sqrt{29}}s\right)\mathbf{j} + \left(5 + \frac{4}{\sqrt{29}}s\right)\mathbf{k}$.

9. Here $\mathbf{r}(t) = \langle 3\sin t, 4t, 3\cos t\rangle$, so $\mathbf{r}'(t) = \langle 3\cos t, 4, -3\sin t\rangle$ and $|\mathbf{r}'(t)| = \sqrt{9\cos^2 t + 16 + 9\sin^2 t} = \sqrt{25} = 5$. The point $(0, 0, 3)$ corresponds to $t = 0$, so the arc length function beginning at $(0, 0, 3)$ and measuring in the positive direction is given by $s(t) = \int_0^t |\mathbf{r}'(u)|\,du = \int_0^t 5\,du = 5t$. $s(t) = 5 \Rightarrow 5t = 5 \Rightarrow t = 1$, thus your location after moving 5 units along the curve is $(3\sin 1, 4, 3\cos 1)$.

11. (a) $\mathbf{r}'(t) = \langle 2\cos t, 5, -2\sin t\rangle \Rightarrow |\mathbf{r}'(t)| = \sqrt{4\cos^2 t + 25 + 4\sin^2 t} = \sqrt{29}$. Then

$$\mathbf{T}(t) = \frac{\mathbf{r}'(t)}{|\mathbf{r}'(t)|} = \frac{1}{\sqrt{29}}\langle 2\cos t, 5, -2\sin t\rangle \text{ or } \left\langle \frac{2}{\sqrt{29}}\cos t, \frac{5}{\sqrt{29}}, -\frac{2}{\sqrt{29}}\sin t\right\rangle.$$

$\mathbf{T}'(t) = \frac{1}{\sqrt{29}}\langle -2\sin t, 0, -2\cos t\rangle \Rightarrow |\mathbf{T}'(t)| = \frac{1}{\sqrt{29}}\sqrt{4\sin^2 t + 0 + 4\cos^2 t} = \frac{2}{\sqrt{29}}$. Thus

$$\mathbf{N}(t) = \frac{\mathbf{T}'(t)}{|\mathbf{T}'(t)|} = \frac{1/\sqrt{29}}{2/\sqrt{29}}\langle -2\sin t, 0, -2\cos t\rangle = \langle -\sin t, 0, -\cos t\rangle.$$

(b) $\kappa(t) = \dfrac{|\mathbf{T}'(t)|}{|\mathbf{r}'(t)|} = \dfrac{2/\sqrt{29}}{\sqrt{29}} = \dfrac{2}{29}$

13. (a) $\mathbf{r}'(t) = \langle t^2, 2t, 2\rangle \Rightarrow |\mathbf{r}'(t)| = \sqrt{t^4 + 4t^2 + 4} = \sqrt{(t^2 + 2)^2} = t^2 + 2$. Then

$$\mathbf{T}(t) = \frac{\mathbf{r}'(t)}{|\mathbf{r}'(t)|} = \frac{1}{t^2 + 2}\langle t^2, 2t, 2\rangle.$$

$$\begin{aligned}\mathbf{T}'(t) &= \frac{-2t}{(t^2+2)^2}\langle t^2, 2t, 2\rangle + \frac{1}{t^2+2}\langle 2t, 2, 0\rangle \quad \text{[by Formula 3 of Theorem 10.2.3]}\\ &= \frac{1}{(t^2+2)^2}\langle -2t^3, -4t^2, -4t\rangle + \frac{1}{(t^2+2)^2}\langle 2t^3 + 4t, 2t^2 + 4, 0\rangle\\ &= \frac{1}{(t^2+2)^2}\langle 4t, 4 - 2t^2, -4t\rangle\end{aligned}$$

$$\begin{aligned}|\mathbf{T}'(t)| &= \frac{1}{(t^2+2)^2}\sqrt{16t^2 + (16 - 16t^2 + 4t^4) + 16t^2}\\ &= \frac{1}{(t^2+2)^2}\sqrt{4t^4 + 16t^2 + 16}\\ &= \frac{1}{(t^2+2)^2}\sqrt{4(t^2+2)^2} = \frac{2(t^2+2)}{(t^2+2)^2} = \frac{2}{t^2+2}\end{aligned}$$

Thus $\mathbf{N}(t) = \dfrac{\mathbf{T}'(t)}{|\mathbf{T}'(t)|} = \dfrac{1/(t^2+2)^2}{2/(t^2+2)}\langle 4t, 4 - 2t^2, -4t\rangle = \dfrac{1}{t^2+2}\langle 2t, 2 - t^2, -2t\rangle$.

(b) $\kappa(t) = \dfrac{|\mathbf{T}'(t)|}{|\mathbf{r}'(t)|} = \dfrac{2/(t^2+2)}{t^2+2} = \dfrac{2}{(t^2+2)^2}$

15. $\mathbf{r}'(t) = 2t\,\mathbf{i} + \mathbf{k}$, $\mathbf{r}''(t) = 2\,\mathbf{i}$, $|\mathbf{r}'(t)| = \sqrt{(2t)^2 + 0^2 + 1^2} = \sqrt{4t^2+1}$, $\mathbf{r}'(t) \times \mathbf{r}''(t) = 2\,\mathbf{j}$, $|\mathbf{r}'(t) \times \mathbf{r}''(t)| = 2$.

Then $\kappa(t) = \dfrac{|\mathbf{r}'(t) \times \mathbf{r}''(t)|}{|\mathbf{r}'(t)|^3} = \dfrac{2}{\left(\sqrt{4t^2+1}\right)^3} = \dfrac{2}{(4t^2+1)^{3/2}}$.

17. $\mathbf{r}'(t) = 3\,\mathbf{i} + 4\cos t\,\mathbf{j} - 4\sin t\,\mathbf{k}$, $\mathbf{r}''(t) = -4\sin t\,\mathbf{j} - 4\cos t\,\mathbf{k}$, $|\mathbf{r}'(t)| = \sqrt{9 + 16\cos^2 t + 16\sin^2 t} = \sqrt{9+16} = 5$,

$\mathbf{r}'(t) \times \mathbf{r}''(t) = -16\,\mathbf{i} + 12\cos t\,\mathbf{j} - 12\sin t\,\mathbf{k}$, $|\mathbf{r}'(t) \times \mathbf{r}''(t)| = \sqrt{256 + 144\cos^2 t + 144\sin^2 t} = \sqrt{400} = 20$.

Then $\kappa(t) = \dfrac{|\mathbf{r}'(t) \times \mathbf{r}''(t)|}{|\mathbf{r}'(t)|^3} = \dfrac{20}{5^3} = \dfrac{4}{25}$.

19. $\mathbf{r}'(t) = \langle 1, 2t, 3t^2\rangle$. The point $(1,1,1)$ corresponds to $t = 1$, and

$\mathbf{r}'(1) = \langle 1,2,3\rangle \;\Rightarrow\; |\mathbf{r}'(1)| = \sqrt{1+4+9} = \sqrt{14}$. $\mathbf{r}''(t) = \langle 0,2,6t\rangle \;\Rightarrow\; \mathbf{r}''(1) = \langle 0,2,6\rangle$.

$\mathbf{r}'(1) \times \mathbf{r}''(1) = \langle 6,-6,2\rangle$, so $|\mathbf{r}'(1) \times \mathbf{r}''(1)| = \sqrt{36+36+4} = \sqrt{76}$. Then

$\kappa(1) = \dfrac{|\mathbf{r}'(1) \times \mathbf{r}''(1)|}{|\mathbf{r}'(1)|^3} = \dfrac{\sqrt{76}}{\sqrt{14}^3} = \dfrac{1}{7}\sqrt{\dfrac{19}{14}}$.

21. $f(x) = xe^x$, $f'(x) = xe^x + e^x$, $f''(x) = xe^x + 2e^x$,

$\kappa(x) = \dfrac{|f''(x)|}{\left[1 + (f'(x))^2\right]^{3/2}} = \dfrac{|xe^x + 2e^x|}{[1 + (xe^x + e^x)^2]^{3/2}} = \dfrac{|x+2|\,e^x}{[1 + (xe^x + e^x)^2]^{3/2}}$

23. $f(x) = 4x^{5/2}$, $f'(x) = 10x^{3/2}$, $f''(x) = 15x^{1/2}$, $\kappa(x) = \dfrac{|f''(x)|}{[1 + (f'(x))^2]^{3/2}} = \dfrac{\left|15x^{1/2}\right|}{[1 + (10x^{3/2})^2]^{3/2}} = \dfrac{15\sqrt{x}}{(1 + 100x^3)^{3/2}}$

25. Since $y' = y'' = e^x$, the curvature is $\kappa(x) = \dfrac{|y''(x)|}{[1 + (y'(x))^2]^{3/2}} = \dfrac{e^x}{(1+e^{2x})^{3/2}} = e^x(1+e^{2x})^{-3/2}$.

To find the maximum curvature, we first find the critical numbers of $\kappa(x)$:

$\kappa'(x) = e^x(1+e^{2x})^{-3/2} + e^x\left(-\frac{3}{2}\right)(1+e^{2x})^{-5/2}(2e^{2x}) = e^x\dfrac{1 + e^{2x} - 3e^{2x}}{(1+e^{2x})^{5/2}} = e^x\dfrac{1 - 2e^{2x}}{(1+e^{2x})^{5/2}}$.

$\kappa'(x) = 0$ when $1 - 2e^{2x} = 0$, so $e^{2x} = \frac{1}{2}$ or $x = -\frac{1}{2}\ln 2$. And since $1 - 2e^{2x} > 0$ for $x < -\frac{1}{2}\ln 2$ and $1 - 2e^{2x} < 0$ for $x > -\frac{1}{2}\ln 2$, the maximum curvature is attained at the point $\left(-\frac{1}{2}\ln 2, e^{(-\ln 2)/2}\right) = \left(-\frac{1}{2}\ln 2, \frac{1}{\sqrt{2}}\right)$. Since $\lim\limits_{x\to\infty} e^x(1+e^{2x})^{-3/2} = 0$, $\kappa(x)$ approaches 0 as $x \to \infty$.

27. (a) C appears to be changing direction more quickly at P than Q, so we would expect the curvature to be greater at P.

(b) First we sketch approximate osculating circles at P and Q. Using the axes scale as a guide, we measure the radius of the osculating circle at P to be approximately 0.8 units, thus $\rho = \dfrac{1}{\kappa} \;\Rightarrow\; \kappa = \dfrac{1}{\rho} \approx \dfrac{1}{0.8} \approx 1.3$. Similarly, we estimate the radius of the osculating circle at Q to be 1.4 units, so $\kappa = \dfrac{1}{\rho} \approx \dfrac{1}{1.4} \approx 0.7$.

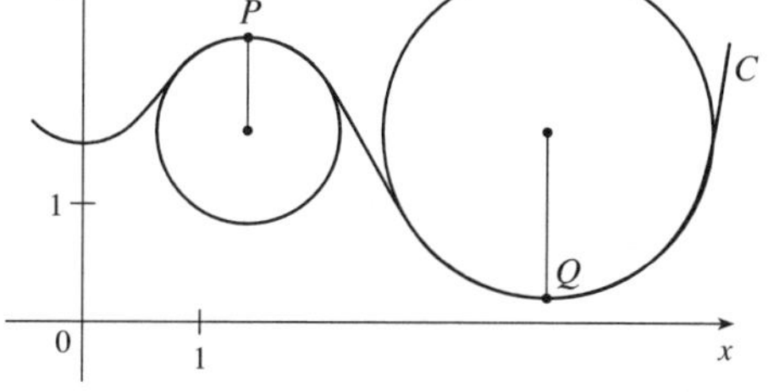

29. $y = x^{-2} \;\Rightarrow\; y' = -2x^{-3}$, $y'' = 6x^{-4}$, and $\kappa(x) = \dfrac{|y''|}{\left[1+(y')^2\right]^{3/2}} = \dfrac{|6x^{-4}|}{\left[1+(-2x^{-3})^2\right]^{3/2}} = \dfrac{6}{x^4\left(1+4x^{-6}\right)^{3/2}}$. The appearance of the two humps in this graph is perhaps a little surprising, but it is explained by the fact that $y = x^{-2}$ increases asymptotically at the origin from both directions, and so its graph has very little bend there. (Note that $\kappa(0)$ is undefined.)

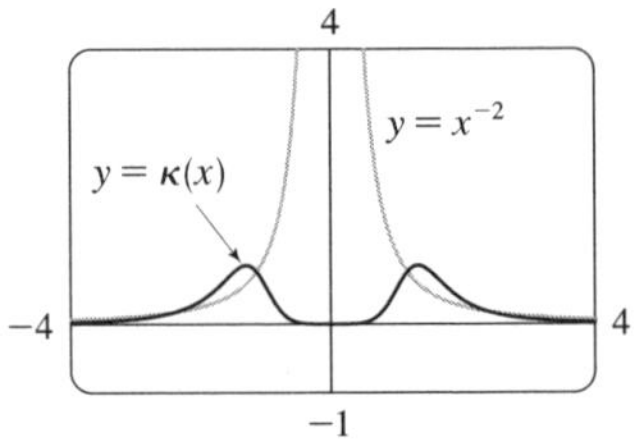

31. Notice that the curve b has two inflection points at which the graph appears almost straight. We would expect the curvature to be 0 or nearly 0 at these values, but the curve a isn't near 0 there. Thus, a must be the graph of $y = f(x)$ rather than the graph of curvature, and b is the graph of $y = \kappa(x)$.

33. Using a CAS, we find (after simplifying)

$\kappa(t) = \dfrac{6\sqrt{4\cos^2 t - 12\cos t + 13}}{(17 - 12\cos t)^{3/2}}$. (To compute cross products in Maple, use the `Linalg` package and the `crossprod(a,b)` command; in Mathematica, use `Cross[a,b]`.) Curvature is largest at integer multiples of 2π.

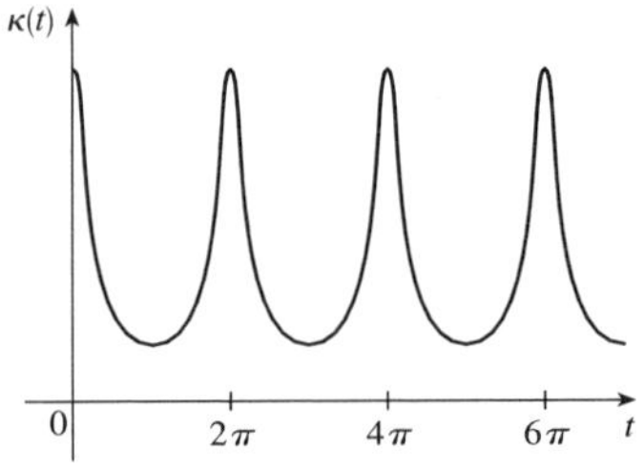

35. $x = e^t \cos t \;\Rightarrow\; \dot{x} = e^t(\cos t - \sin t) \;\Rightarrow\; \ddot{x} = e^t(-\sin t - \cos t) + e^t(\cos t - \sin t) = -2e^t \sin t$,

$y = e^t \sin t \;\Rightarrow\; \dot{y} = e^t(\cos t + \sin t) \;\Rightarrow\; \ddot{y} = e^t(-\sin t + \cos t) + e^t(\cos t + \sin t) = 2e^t \cos t$. Then

$$\kappa(t) = \frac{|\dot{x}\ddot{y} - \dot{y}\ddot{x}|}{(\dot{x}^2 + \dot{y}^2)^{3/2}} = \frac{\left|e^t(\cos t - \sin t)(2e^t\cos t) - e^t(\cos t + \sin t)(-2e^t \sin t)\right|}{\left([e^t(\cos t - \sin t)]^2 + [e^t(\cos t + \sin t)]^2\right)^{3/2}}$$

$$= \frac{\left|2e^{2t}(\cos^2 t - \sin t\cos t + \sin t\cos t + \sin^2 t)\right|}{\left[e^{2t}(\cos^2 t - 2\cos t\sin t + \sin^2 t + \cos^2 t + 2\cos t\sin t + \sin^2 t)\right]^{3/2}}$$

$$= \frac{\left|2e^{2t}(1)\right|}{\left[e^{2t}(1+1)\right]^{3/2}} = \frac{2e^{2t}}{e^{3t}(2)^{3/2}} = \frac{1}{\sqrt{2}\,e^t}$$

37. $\left(1, \frac{2}{3}, 1\right)$ corresponds to $t = 1$. $\mathbf{T}(t) = \dfrac{\mathbf{r}'(t)}{|\mathbf{r}'(t)|} = \dfrac{\langle 2t, 2t^2, 1\rangle}{\sqrt{4t^2 + 4t^4 + 1}} = \dfrac{\langle 2t, 2t^2, 1\rangle}{2t^2 + 1}$, so $\mathbf{T}(1) = \left\langle \frac{2}{3}, \frac{2}{3}, \frac{1}{3}\right\rangle$.

$$\mathbf{T}'(t) = -4t(2t^2+1)^{-2}\langle 2t, 2t^2, 1\rangle + (2t^2+1)^{-1}\langle 2, 4t, 0\rangle \qquad \text{[by Formula 3 of Theorem 10.2.3]}$$

$$= (2t^2+1)^{-2}\left\langle -8t^2 + 4t^2 + 2, -8t^3 + 8t^3 + 4t, -4t\right\rangle = 2(2t^2+1)^{-2}\left\langle 1 - 2t^2, 2t, -2t\right\rangle$$

$$\mathbf{N}(t) = \frac{\mathbf{T}'(t)}{|\mathbf{T}'(t)|} = \frac{2(2t^2+1)^{-2}\langle 1-2t^2, 2t, -2t\rangle}{2(2t^2+1)^{-2}\sqrt{(1-2t^2)^2 + (2t)^2 + (-2t)^2}} = \frac{\langle 1-2t^2, 2t, -2t\rangle}{\sqrt{1 - 4t^2 + 4t^4 + 8t^2}} = \frac{\langle 1-2t^2, 2t, -2t\rangle}{1 + 2t^2}$$

$\mathbf{N}(1) = \left\langle -\frac{1}{3}, \frac{2}{3}, -\frac{2}{3}\right\rangle$ and $\mathbf{B}(1) = \mathbf{T}(1) \times \mathbf{N}(1) = \left\langle -\frac{4}{9} - \frac{2}{9}, -\left(-\frac{4}{9} + \frac{1}{9}\right), \frac{4}{9} + \frac{2}{9}\right\rangle = \left\langle -\frac{2}{3}, \frac{1}{3}, \frac{2}{3}\right\rangle$.

39. $(0, \pi, -2)$ corresponds to $t = \pi$. $\mathbf{r}(t) = \langle 2\sin 3t, t, 2\cos 3t\rangle \Rightarrow$

$$\mathbf{T}(t) = \frac{\mathbf{r}'(t)}{|\mathbf{r}'(t)|} = \frac{\langle 6\cos 3t, 1, -6\sin 3t\rangle}{\sqrt{36\cos^2 3t + 1 + 36\sin^2 3t}} = \frac{1}{\sqrt{37}}\langle 6\cos 3t, 1, -6\sin 3t\rangle.$$

$\mathbf{T}(\pi) = \frac{1}{\sqrt{37}}\langle -6, 1, 0\rangle$ is a normal vector for the normal plane, and so $\langle -6, 1, 0\rangle$ is also normal. Thus an equation for the plane is $-6(x-0) + 1(y-\pi) + 0(z+2) = 0$ or $y - 6x = \pi$.

$$\mathbf{T}'(t) = \frac{1}{\sqrt{37}}\langle -18\sin 3t, 0, -18\cos 3t\rangle \Rightarrow |\mathbf{T}'(t)| = \frac{\sqrt{18^2\sin^2 3t + 18^2\cos^2 3t}}{\sqrt{37}} = \frac{18}{\sqrt{37}} \Rightarrow$$

$\mathbf{N}(t) = \dfrac{\mathbf{T}'(t)}{|\mathbf{T}'(t)|} = \langle -\sin 3t, 0, -\cos 3t\rangle$. So $\mathbf{N}(\pi) = \langle 0, 0, 1\rangle$ and $\mathbf{B}(\pi) = \frac{1}{\sqrt{37}}\langle -6, 1, 0\rangle \times \langle 0, 0, 1\rangle = \frac{1}{\sqrt{37}}\langle 1, 6, 0\rangle$.

Since $\mathbf{B}(\pi)$ is a normal to the osculating plane, so is $\langle 1, 6, 0\rangle$ and an equation for the plane is $1(x-0) + 6(y-\pi) + 0(z+2) = 0$ or $x + 6y = 6\pi$.

41. The ellipse is given by the parametric equations $x = 2\cos t$, $y = 3\sin t$, so using the result from Exercise 34,

$$\kappa(t) = \frac{|\dot{x}\ddot{y} - \ddot{x}\dot{y}|}{(\dot{x}^2 + \dot{y}^2)^{3/2}} = \frac{|(-2\sin t)(-3\sin t) - (3\cos t)(-2\cos t)|}{(4\sin^2 t + 9\cos^2 t)^{3/2}} = \frac{6}{(4\sin^2 t + 9\cos^2 t)^{3/2}}$$

At $(2, 0)$, $t = 0$. Now $\kappa(0) = \frac{6}{27} = \frac{2}{9}$, so the radius of the osculating circle is $1/\kappa(0) = \frac{9}{2}$ and its centre is $\left(-\frac{5}{2}, 0\right)$. Its equation is therefore $\left(x + \frac{5}{2}\right)^2 + y^2 = \frac{81}{4}$. At $(0, 3)$, $t = \frac{\pi}{2}$, and $\kappa\left(\frac{\pi}{2}\right) = \frac{6}{8} = \frac{3}{4}$. So the radius of the osculating circle is $\frac{4}{3}$ and its centre is $\left(0, \frac{5}{3}\right)$. Hence its equation is $x^2 + \left(y - \frac{5}{3}\right)^2 = \frac{16}{9}$.

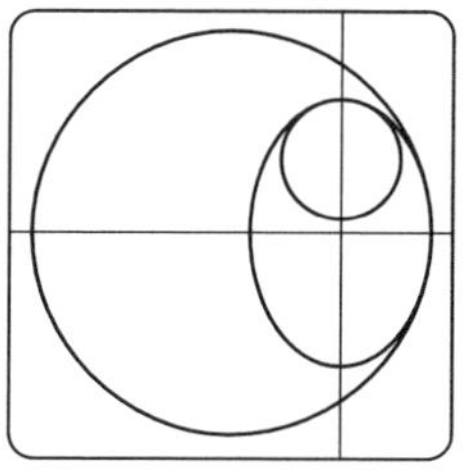

43. The tangent vector is normal to the normal plane, and the vector $\langle 6, 6, -8\rangle$ is normal to the given plane. [$\parallel$ means "is parallel to".] But $\mathbf{T}(t) \parallel \mathbf{r}'(t)$ and $\langle 6, 6, -8\rangle \parallel \langle 3, 3, -4\rangle$, so we need to find t such that $\mathbf{r}'(t) \parallel \langle 3, 3, -4\rangle$. $\mathbf{r}(t) = \langle t^3, 3t, t^4\rangle \Rightarrow \mathbf{r}'(t) = \langle 3t^2, 3, 4t^3\rangle \parallel \langle 3, 3, -4\rangle$ when $t = -1$. So the planes are parallel at the point $((-1)^3, 3(-1), (-1)^4) = (-1, -3, 1)$.

45. $\kappa = \left|\dfrac{d\mathbf{T}}{ds}\right| = \left|\dfrac{d\mathbf{T}/dt}{ds/dt}\right| = \dfrac{|d\mathbf{T}/dt|}{ds/dt}$ and $\mathbf{N} = \dfrac{d\mathbf{T}/dt}{|d\mathbf{T}/dt|}$, so $\kappa\mathbf{N} = \dfrac{\left|\dfrac{d\mathbf{T}}{dt}\right| \dfrac{d\mathbf{T}}{dt}}{\left|\dfrac{d\mathbf{T}}{dt}\right| \dfrac{ds}{dt}} = \dfrac{d\mathbf{T}/dt}{ds/dt} = \dfrac{d\mathbf{T}}{ds}$ by the Chain Rule.

47. (a) $|\mathbf{B}| = 1 \Rightarrow \mathbf{B}\cdot\mathbf{B} = 1 \Rightarrow \dfrac{d}{ds}(\mathbf{B}\cdot\mathbf{B}) = 0 \Rightarrow 2\dfrac{d\mathbf{B}}{ds}\cdot\mathbf{B} = 0 \Rightarrow \dfrac{d\mathbf{B}}{ds} \perp \mathbf{B}$

(b) $\mathbf{B} = \mathbf{T}\times\mathbf{N} \Rightarrow$

$$\frac{d\mathbf{B}}{ds} = \frac{d}{ds}(\mathbf{T}\times\mathbf{N}) = \frac{d}{dt}(\mathbf{T}\times\mathbf{N})\frac{1}{ds/dt} = \frac{d}{dt}(\mathbf{T}\times\mathbf{N})\frac{1}{|\mathbf{r}'(t)|}$$

$$= [(\mathbf{T}'\times\mathbf{N}) + (\mathbf{T}\times\mathbf{N}')]\frac{1}{|\mathbf{r}'(t)|} = \left[\left(\mathbf{T}'\times\frac{\mathbf{T}'}{|\mathbf{T}'|}\right) + (\mathbf{T}\times\mathbf{N}')\right]\frac{1}{|\mathbf{r}'(t)|} = \frac{\mathbf{T}\times\mathbf{N}'}{|\mathbf{r}'(t)|}$$

$$\Rightarrow \frac{d\mathbf{B}}{ds} \perp \mathbf{T}$$

(c) $\mathbf{B} = \mathbf{T} \times \mathbf{N} \Rightarrow \mathbf{T} \perp \mathbf{N}, \mathbf{B} \perp \mathbf{T}$ and $\mathbf{B} \perp \mathbf{N}$. So $\mathbf{B}$, $\mathbf{T}$ and $\mathbf{N}$ form an orthogonal set of vectors in the three-dimensional space $\mathbb{R}^3$. From parts (a) and (b), $d\mathbf{B}/ds$ is perpendicular to both $\mathbf{B}$ and $\mathbf{T}$, so $d\mathbf{B}/ds$ is parallel to $\mathbf{N}$. Therefore, $d\mathbf{B}/ds = -\tau(s)\mathbf{N}$, where $\tau(s)$ is a scalar.

(d) Since $\mathbf{B} = \mathbf{T} \times \mathbf{N}$, $\mathbf{T} \perp \mathbf{N}$ and both $\mathbf{T}$ and $\mathbf{N}$ are unit vectors, $\mathbf{B}$ is a unit vector mutually perpendicular to both $\mathbf{T}$ and $\mathbf{N}$. For a plane curve, $\mathbf{T}$ and $\mathbf{N}$ always lie in the plane of the curve, so that $\mathbf{B}$ is a constant unit vector always perpendicular to the plane. Thus $d\mathbf{B}/ds = \mathbf{0}$, but $d\mathbf{B}/ds = -\tau(s)\mathbf{N}$ and $\mathbf{N} \neq \mathbf{0}$, so $\tau(s) = 0$.

49. (a) $\mathbf{r}' = s'\,\mathbf{T} \Rightarrow \mathbf{r}'' = s''\,\mathbf{T} + s'\,\mathbf{T}' = s''\,\mathbf{T} + s'\,\dfrac{d\mathbf{T}}{ds}s' = s''\,\mathbf{T} + \kappa(s')^2\,\mathbf{N}$ by the first Serret-Frenet formula.

(b) Using part (a), we have

$$\begin{aligned}\mathbf{r}' \times \mathbf{r}'' &= (s'\,\mathbf{T}) \times [s''\,\mathbf{T} + \kappa(s')^2\,\mathbf{N}] \\ &= [(s'\,\mathbf{T}) \times (s''\,\mathbf{T})] + [(s'\mathbf{T}) \times (\kappa(s')^2\,\mathbf{N})] \qquad \text{[by Property 3 of the cross product]} \\ &= (s's'')(\mathbf{T} \times \mathbf{T}) + \kappa(s')^3(\mathbf{T} \times \mathbf{N}) = \mathbf{0} + \kappa(s')^3\,\mathbf{B} = \kappa(s')^3\,\mathbf{B}\end{aligned}$$

(c) Using part (a), we have

$$\begin{aligned}\mathbf{r}''' &= [s''\,\mathbf{T} + \kappa(s')^2\,\mathbf{N}]' = s'''\,\mathbf{T} + s''\,\mathbf{T}' + \kappa'(s')^2\,\mathbf{N} + 2\kappa s's''\,\mathbf{N} + \kappa(s')^2\,\mathbf{N}' \\ &= s'''\,\mathbf{T} + s''\frac{d\mathbf{T}}{ds}\,s' + \kappa'(s')^2\,\mathbf{N} + 2\kappa s's''\,\mathbf{N} + \kappa(s')^2\,\frac{d\mathbf{N}}{ds}\,s' \\ &= s'''\,\mathbf{T} + s''s'\kappa\,\mathbf{N} + \kappa'(s')^2\,\mathbf{N} + 2\kappa s's''\,\mathbf{N} + \kappa(s')^3(-\kappa\,\mathbf{T} + \tau\,\mathbf{B}) \qquad \text{[by the second formula]} \\ &= [s''' - \kappa^2(s')^3]\,\mathbf{T} + [3\kappa s's'' + \kappa'(s')^2]\,\mathbf{N} + \kappa\tau(s')^3\,\mathbf{B}\end{aligned}$$

(d) Using parts (b) and (c) and the facts that $\mathbf{B} \cdot \mathbf{T} = 0$, $\mathbf{B} \cdot \mathbf{N} = 0$, and $\mathbf{B} \cdot \mathbf{B} = 1$, we get

$$\frac{(\mathbf{r}' \times \mathbf{r}'') \cdot \mathbf{r}'''}{|\mathbf{r}' \times \mathbf{r}''|^2} = \frac{\kappa(s')^3\,\mathbf{B} \cdot \{[s''' - \kappa^2(s')^3]\,\mathbf{T} + [3\kappa s's'' + \kappa'(s')^2]\,\mathbf{N} + \kappa\tau(s')^3\,\mathbf{B}\}}{|\kappa(s')^3\,\mathbf{B}|^2} = \frac{\kappa(s')^3\kappa\tau(s')^3}{[\kappa(s')^3]^2} = \tau$$

51. For one helix, the vector equation is $\mathbf{r}(t) = \langle 10\cos t, 10\sin t, 34t/(2\pi)\rangle$ (measuring in angstroms), because the radius of each helix is 10 angstroms, and z increases by 34 angstroms for each increase of 2π in t. Using the arc length formula, letting t go from 0 to $2.9 \times 10^8 \times 2\pi$, we find the approximate length of each helix to be

$$\begin{aligned}L &= \int_0^{2.9\times10^8\times2\pi} |\mathbf{r}'(t)|\,dt = \int_0^{2.9\times10^8\times2\pi} \sqrt{(-10\sin t)^2 + (10\cos t)^2 + \left(\tfrac{34}{2\pi}\right)^2}\,dt \\ &= \sqrt{100 + \left(\tfrac{34}{2\pi}\right)^2}\,t\,\Big]_0^{2.9\times10^8\times2\pi} = 2.9 \times 10^8 \times 2\pi\sqrt{100 + \left(\tfrac{34}{2\pi}\right)^2} \\ &\approx 2.07 \times 10^{10}\ \text{Å} \text{ — more than two metres!}\end{aligned}$$

10.4 Motion in Space: Velocity and Acceleration

1. (a) If $\mathbf{r}(t) = x(t)\,\mathbf{i} + y\,(t)\;\mathbf{j} + z(t)\,\mathbf{k}$ is the position vector of the particle at time t, then the average velocity over the time interval $[0, 1]$ is

$\mathbf{v}_{\text{ave}} = \dfrac{\mathbf{r}(1) - \mathbf{r}(0)}{1 - 0} = \dfrac{(4.5\,\mathbf{i} + 6.0\,\mathbf{j} + 3.0\,\mathbf{k}) - (2.7\,\mathbf{i} + 9.8\,\mathbf{j} + 3.7\,\mathbf{k})}{1} = 1.8\,\mathbf{i} - 3.8\,\mathbf{j} - 0.7\,\mathbf{k}$. Similarly, over the other intervals we have

$$[0.5, 1]: \quad \mathbf{v}_{\text{ave}} = \frac{\mathbf{r}(1) - \mathbf{r}(0.5)}{1 - 0.5} = \frac{(4.5\,\mathbf{i} + 6.0\,\mathbf{j} + 3.0\,\mathbf{k}) - (3.5\,\mathbf{i} + 7.2\,\mathbf{j} + 3.3\,\mathbf{k})}{0.5} = 2.0\,\mathbf{i} - 2.4\,\mathbf{j} - 0.6\,\mathbf{k}$$

$$[1, 2]: \quad \mathbf{v}_{\text{ave}} = \frac{\mathbf{r}(2) - \mathbf{r}(1)}{2 - 1} = \frac{(7.3\,\mathbf{i} + 7.8\,\mathbf{j} + 2.7\,\mathbf{k}) - (4.5\,\mathbf{i} + 6.0\,\mathbf{j} + 3.0\,\mathbf{k})}{1} = 2.8\,\mathbf{i} + 1.8\,\mathbf{j} - 0.3\,\mathbf{k}$$

$$[1, 1.5]: \quad \mathbf{v}_{\text{ave}} = \frac{\mathbf{r}(1.5) - \mathbf{r}(1)}{1.5 - 1} = \frac{(5.9\,\mathbf{i} + 6.4\,\mathbf{j} + 2.8\,\mathbf{k}) - (4.5\,\mathbf{i} + 6.0\,\mathbf{j} + 3.0\,\mathbf{k})}{0.5} = 2.8\,\mathbf{i} + 0.8\,\mathbf{j} - 0.4\,\mathbf{k}$$

(b) We can estimate the velocity at $t = 1$ by averaging the average velocities over the time intervals $[0.5, 1]$ and $[1, 1.5]$: $\mathbf{v}(1) \approx \frac{1}{2}[(2\,\mathbf{i} - 2.4\,\mathbf{j} - 0.6\,\mathbf{k}) + (2.8\,\mathbf{i} + 0.8\,\mathbf{j} - 0.4\,\mathbf{k})] = 2.4\,\mathbf{i} - 0.8\,\mathbf{j} - 0.5\,\mathbf{k}$. Then the speed is $|\mathbf{v}(1)| \approx \sqrt{(2.4)^2 + (-0.8)^2 + (-0.5)^2} \approx 2.58$.

3. $\mathbf{r}(t) = \langle t^2 - 1, t\rangle \;\Rightarrow$ At $t = 1$:

$\mathbf{v}(t) = \mathbf{r}'(t) = \langle 2t, 1\rangle$, $\quad \mathbf{v}(1) = \langle 2, 1\rangle$

$\mathbf{a}(t) = \mathbf{r}''(t) = \langle 2, 0\rangle$, $\quad \mathbf{a}(1) = \langle 2, 0\rangle$

$|\mathbf{v}(t)| = \sqrt{4t^2 + 1}$

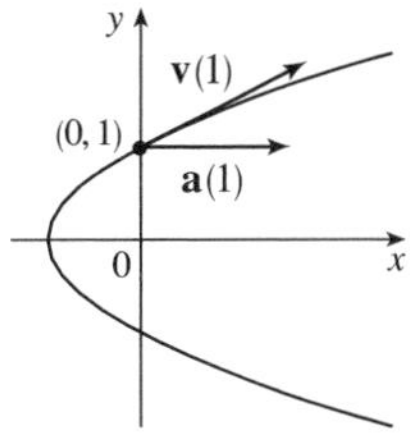

5. $\mathbf{r}(t) = e^t\,\mathbf{i} + e^{-t}\,\mathbf{j} \;\Rightarrow$ At $t = 0$:

$\mathbf{v}(t) = e^t\,\mathbf{i} - e^{-t}\,\mathbf{j}$, $\quad \mathbf{v}(0) = \mathbf{i} - \mathbf{j}$,

$\mathbf{a}(t) = e^t\,\mathbf{i} + e^{-t}\,\mathbf{j}$ $\quad \mathbf{a}(0) = \mathbf{i} + \mathbf{j}$

$|\mathbf{v}(t)| = \sqrt{e^{2t} + e^{-2t}} = e^{-t}\sqrt{e^{4t} + 1}$

Since $x = e^t$, $t = \ln x$ and $y = e^{-t} = e^{-\ln x} = 1/x$, and $x > 0$, $y > 0$.

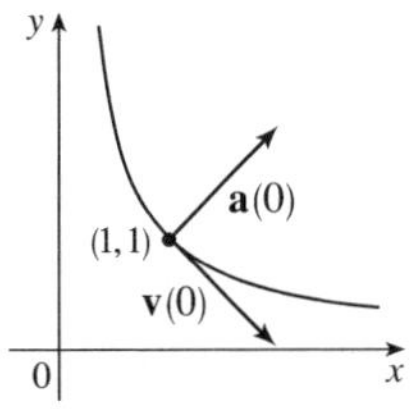

7. $\mathbf{r}(t) = t\,\mathbf{i} + t^2\,\mathbf{j} + 2\,\mathbf{k} \;\Rightarrow$

$\mathbf{v}(t) = \mathbf{i} + 2t\,\mathbf{j}$, $\mathbf{v}(1) = \mathbf{i} + 2\,\mathbf{j}$

$\mathbf{a}(t) = 2\,\mathbf{j}$, $\mathbf{a}(1) = 2\,\mathbf{j}$

$|\mathbf{v}(t)| = \sqrt{1 + 4t^2}$

Here $x = t$, $y = t^2 \;\Rightarrow\; y = x^2$ and $z = 2$, so the path of the particle is a parabola in the plane $z = 2$.

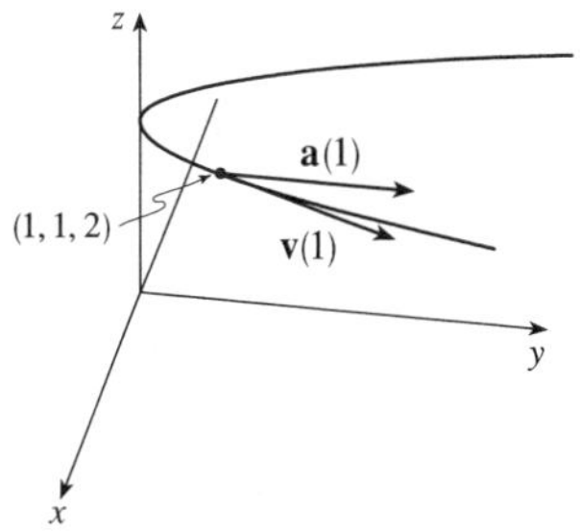

9. $\mathbf{r}(t) = \langle t^2 + 1, t^3, t^2 - 1\rangle \;\Rightarrow\; \mathbf{v}(t) = \mathbf{r}'(t) = \langle 2t, 3t^2, 2t\rangle$, $\mathbf{a}(t) = \mathbf{v}'(t) = \langle 2, 6t, 2\rangle$,

$|\mathbf{v}(t)| = \sqrt{(2t)^2 + (3t^2)^2 + (2t)^2} = \sqrt{9t^4 + 8t^2} = |t|\,\sqrt{9t^2 + 8}$.

11. $\mathbf{r}(t) = \sqrt{2}\,t\,\mathbf{i} + e^t\,\mathbf{j} + e^{-t}\,\mathbf{k} \;\Rightarrow\; \mathbf{v}(t) = \mathbf{r}'(t) = \sqrt{2}\,\mathbf{i} + e^t\,\mathbf{j} - e^{-t}\,\mathbf{k}$, $\mathbf{a}(t) = \mathbf{v}'(t) = e^t\,\mathbf{j} + e^{-t}\,\mathbf{k}$,

$|\mathbf{v}(t)| = \sqrt{2 + e^{2t} + e^{-2t}} = \sqrt{(e^t + e^{-t})^2} = e^t + e^{-t}$.

13. $\mathbf{a}(t) = \mathbf{i} + 2\,\mathbf{j} \;\Rightarrow\; \mathbf{v}(t) = \int \mathbf{a}(t)\,dt = \int (\mathbf{i} + 2\mathbf{j})\,dt = t\,\mathbf{i} + 2t\,\mathbf{j} + \mathbf{C}$ and $\mathbf{k} = \mathbf{v}\,(0) = \mathbf{C}$, so $\mathbf{C} = \mathbf{k}$ and $\mathbf{v}(t) = t\,\mathbf{i} + 2t\,\mathbf{j} + \mathbf{k}$. $\mathbf{r}(t) = \int \mathbf{v}(t)\,dt = \int (t\,\mathbf{i} + 2t\,\mathbf{j} + \mathbf{k})\,dt = \frac{1}{2}t^2\,\mathbf{i} + t^2\,\mathbf{j} + t\,\mathbf{k} + \mathbf{D}$. But $\mathbf{i} = \mathbf{r}\,(0) = \mathbf{D}$, so $\mathbf{D} = \mathbf{i}$ and $\mathbf{r}(t) = \left(\frac{1}{2}t^2 + 1\right)\mathbf{i} + t^2\,\mathbf{j} + t\,\mathbf{k}$.

15. (a) $\mathbf{a}(t) = 2t\,\mathbf{i} + \sin t\,\mathbf{j} + \cos 2t\,\mathbf{k} \;\Rightarrow$

$\mathbf{v}(t) = \int (2t\,\mathbf{i} + \sin t\,\mathbf{j} + \cos 2t\,\mathbf{k})\,dt = t^2\,\mathbf{i} - \cos t\,\mathbf{j} + \frac{1}{2}\sin 2t\,\mathbf{k} + \mathbf{C}$

and $\mathbf{i} = \mathbf{v}\,(0) = -\mathbf{j} + \mathbf{C}$, so $\mathbf{C} = \mathbf{i} + \mathbf{j}$ and

$\mathbf{v}(t) = (t^2 + 1)\,\mathbf{i} + (1 - \cos t)\,\mathbf{j} + \frac{1}{2}\sin 2t\,\mathbf{k}$.

$\mathbf{r}(t) = \int [(t^2 + 1)\;\mathbf{i} + (1 - \cos t)\,\mathbf{j} + \frac{1}{2}\sin 2t\,\mathbf{k}]dt$

$= \left(\frac{1}{3}t^3 + t\right)\,\mathbf{i} + (t - \sin t)\,\mathbf{j} - \frac{1}{4}\cos 2t\,\mathbf{k} + \mathbf{D}$

(b)

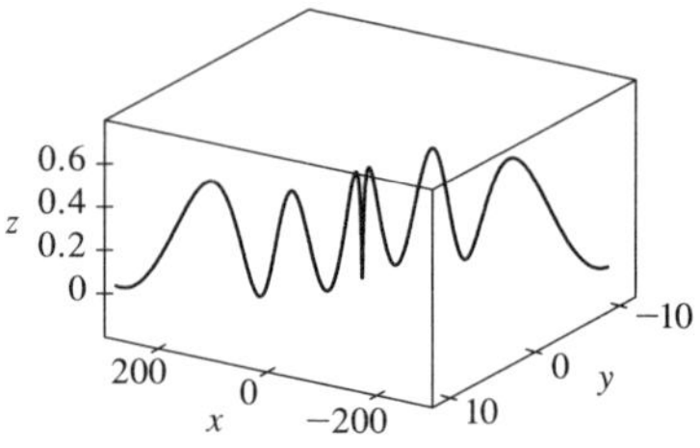

But $\mathbf{j} = \mathbf{r}\,(0) = -\frac{1}{4}\,\mathbf{k} + \mathbf{D}$, so $\mathbf{D} = \mathbf{j} + \frac{1}{4}\,\mathbf{k}$ and $\mathbf{r}(t) = \left(\frac{1}{3}t^3 + t\right)\mathbf{i} + (t - \sin t + 1)\mathbf{j} + \left(\frac{1}{4} - \frac{1}{4}\cos 2t\right)\mathbf{k}$.

17. $\mathbf{r}(t) = \left\langle t^2, 5t, t^2 - 16t\right\rangle \;\Rightarrow\; \mathbf{v}(t) = \langle 2t, 5, 2t - 16\rangle$, $|\mathbf{v}(t)| = \sqrt{4t^2 + 25 + 4t^2 - 64t + 256} = \sqrt{8t^2 - 64t + 281}$ and $\dfrac{d}{dt}\,|\mathbf{v}(t)| = \frac{1}{2}(8t^2 - 64t + 281)^{-1/2}(16t - 64)$. This is zero if and only if the numerator is zero, that is, $16t - 64 = 0$ or $t = 4$. Since $\dfrac{d}{dt}\,|\mathbf{v}(t)| < 0$ for $t < 4$ and $\dfrac{d}{dt}\,|\mathbf{v}(t)| > 0$ for $t > 4$, the minimum speed of $\sqrt{153}$ is attained at $t = 4$ units of time.

19. $|\mathbf{F}(t)| = 20$ N in the direction of the positive z-axis, so $\mathbf{F}(t) = 20\,\mathbf{k}$. Also $m = 4$ kg, $\mathbf{r}(0) = \mathbf{0}$ and $\mathbf{v}(0) = \mathbf{i} - \mathbf{j}$. Since $20\mathbf{k} = \mathbf{F}(t) = 4\,\mathbf{a}(t)$, $\mathbf{a}(t) = 5\,\mathbf{k}$. Then $\mathbf{v}(t) = 5t\,\mathbf{k} + \mathbf{c}_1$ where $\mathbf{c}_1 = \mathbf{i} - \mathbf{j}$ so $\mathbf{v}(t) = \mathbf{i} - \mathbf{j} + 5t\,\mathbf{k}$ and the speed is $|\mathbf{v}(t)| = \sqrt{1 + 1 + 25t^2} = \sqrt{25t^2 + 2}$. Also $\mathbf{r}(t) = t\,\mathbf{i} - t\,\mathbf{j} + \frac{5}{2}t^2\,\mathbf{k} + \mathbf{c}_2$ and $\mathbf{0} = \mathbf{r}(0)$, so $\mathbf{c}_2 = \mathbf{0}$ and $\mathbf{r}(t) = t\,\mathbf{i} - t\,\mathbf{j} + \frac{5}{2}t^2\,\mathbf{k}$.

21. $|\mathbf{v}(0)| = 500$ m/s and since the angle of elevation is $30°$, the direction of the velocity is $\frac{1}{2}\left(\sqrt{3}\,\mathbf{i} + \mathbf{j}\right)$. Thus $\mathbf{v}(0) = 250\left(\sqrt{3}\,\mathbf{i} + \mathbf{j}\right)$ and if we set up the axes so the projectile starts at the origin, then $\mathbf{r}(0) = \mathbf{0}$. Ignoring air resistance, the only force is that due to gravity, so $\mathbf{F}(t) = -mg\,\mathbf{j}$ where $g \approx 9.8$ m/s^2. Thus $\mathbf{a}(t) = -g\,\mathbf{j}$ and $\mathbf{v}(t) = -gt\,\mathbf{j} + \mathbf{c}_1$. But $250\left(\sqrt{3}\,\mathbf{i} + \mathbf{j}\right) = \mathbf{v}(0) = \mathbf{c}_1$, so $\mathbf{v}(t) = 250\,\sqrt{3}\,\mathbf{i} + (250 - gt)\,\mathbf{j}$ and $\mathbf{r}(t) = 250\,\sqrt{3}\,t\,\mathbf{i} + \left(250t - \frac{1}{2}gt^2\right)\mathbf{j} + \mathbf{c}_2$ where $\mathbf{0} = \mathbf{r}(0) = \mathbf{c}_2$. Thus $\mathbf{r}(t) = 250\,\sqrt{3}\,t\,\mathbf{i} + \left(250t - \frac{1}{2}gt^2\right)\mathbf{j}$.

(a) Setting $250t - \frac{1}{2}gt^2 = 0$ gives $t = 0$ or $t = \frac{500}{g} \approx 51.0$ s. So the range is $250\,\sqrt{3} \cdot \frac{500}{g} \approx 22$ km.

(b) $0 = \dfrac{d}{dt}\left(250t - \frac{1}{2}gt^2\right) = 250 - gt$ implies that the maximum height is attained when $t = 250/g \approx 25.5$ s. Thus, the maximum height is $(250)(250/g) - g(250/g)^2\,\frac{1}{2} = (250)^2/(2g) \approx 3.2$ km.

(c) From part (a), impact occurs at $t = 500/g \approx 51.0$. Thus, the velocity at impact is $\mathbf{v}(500/g) = 250\,\sqrt{3}\,\mathbf{i} + [250 - g(500/g)]\,\mathbf{j} = 250\,\sqrt{3}\,\mathbf{i} - 250\,\mathbf{j}$ and the speed is $|\mathbf{v}(500/g)| = 250\,\sqrt{3 + 1} = 500$ m/s.

23. As in Example 5, $\mathbf{r}(t) = (v_0 \cos 45\,°)t\,\mathbf{i} + \left[(v_0 \sin 45\,°)t - \frac{1}{2}gt^2\right]\mathbf{j} = \frac{1}{2}\left[v_0\sqrt{2}\,t\,\mathbf{i} + \left(v_0\sqrt{2}\,t - gt^2\right)\mathbf{j}\right]$. Then the ball lands at $t = \dfrac{v_0\sqrt{2}}{g}$ s. Now since it lands 90 m away, $90 = \frac{1}{2}v_0\,\sqrt{2}\,\dfrac{v_0\sqrt{2}}{g}$ or $v_0^2 = 90g$ and the initial velocity is $v_0 = \sqrt{90g} \approx 30$ m/s.

25. Let α be the angle of elevation. Then $v_0 = 150$ m/s and from Example 5, the horizontal distance travelled by the projectile is $d = \dfrac{v_0^2 \sin 2\alpha}{g}$. Thus $\dfrac{150^2 \sin 2\alpha}{g} = 800 \;\Rightarrow\; \sin 2\alpha = \dfrac{800g}{150^2} \approx 0.3484 \;\Rightarrow\; 2\alpha \approx 20.4^\circ$ or $180 - 20.4 = 159.6^\circ$.

Two angles of elevation then are $\alpha \approx 10.2^\circ$ and $\alpha \approx 79.8^\circ$.

27. Place the catapult at the origin and assume the catapult is 100 metres from the city, so the city lies between $(100, 0)$ and $(600, 0)$. The initial speed is $v_0 = 80$ m/s and let θ be the angle the catapult is set at. As in Example 5, the trajectory of the catapulted rock is given by $\mathbf{r}(t) = (80\cos\theta)t\,\mathbf{i} + \left[(80\sin\theta)t - 4.9t^2\right]\mathbf{j}$. The top of the near city wall is at $(100, 15)$ which the rock will hit when $(80\cos\theta)\,t = 100 \;\Rightarrow\; t = \dfrac{5}{4\cos\theta}$ and $(80\sin\theta)t - 4.9t^2 = 15 \;\Rightarrow$

$80\sin\theta \cdot \dfrac{5}{4\cos\theta} - 4.9\left(\dfrac{5}{4\cos\theta}\right)^2 = 15 \;\Rightarrow\; 100\tan\theta - 7.65625\sec^2\theta = 15$. Replacing $\sec^2\theta$ with $\tan^2\theta + 1$ gives

$7.65625\tan^2\theta - 100\tan\theta + 22.62625 = 0$. Using the quadratic formula, we have $\tan\theta \approx 0.230\,324,\ 12.8309 \;\Rightarrow$ $\theta \approx 13.0^\circ, 85.5^\circ$. So for $13.0^\circ < \theta < 85.5^\circ$, the rock will land beyond the near city wall. The base of the far wall is located at $(600, 0)$ which the rock hits if $(80\cos\theta)t = 600 \;\Rightarrow\; t = \dfrac{15}{2\cos\theta}$ and $(80\sin\theta)t - 4.9t^2 = 0 \;\Rightarrow$

$80\sin\theta \cdot \dfrac{15}{2\cos\theta} - 4.9\left(\dfrac{15}{2\cos\theta}\right)^2 = 0 \;\Rightarrow\; 600\tan\theta - 275.625\sec^2\theta = 0 \;\Rightarrow$

$275.625\tan^2\theta - 600\tan\theta + 275.625 = 0$. Solutions are $\tan\theta \approx 0.658\,678,\ 1.51819 \;\Rightarrow\; \theta \approx 33.4^\circ, 56.6^\circ$. Thus the rock lands beyond the enclosed city ground for $33.4^\circ < \theta < 56.6^\circ$, and the angles that allow the rock to land on city ground are $13.0^\circ < \theta < 33.4^\circ$, $56.6^\circ < \theta < 85.5^\circ$. If you consider that the rock can hit the far wall and bounce back into the city, we calculate the angles that cause the rock to hit the top of the wall at $(600, 15)$: $(80\cos\theta)t = 600 \;\Rightarrow\; t = \dfrac{15}{2\cos\theta}$ and

$(80\sin\theta)t - 4.9t^2 = 15 \;\Rightarrow\; 600\tan\theta - 275.625\sec^2\theta = 15 \;\Rightarrow\; 275.625\tan^2\theta - 600\tan\theta + 290.625 = 0$.

Solutions are $\tan\theta \approx 0.727\,506,\ 1.44936 \;\Rightarrow\; \theta \approx 36.0^\circ, 55.4^\circ$, so the catapult should be set with angle θ where $13.0^\circ < \theta < 36.0^\circ$, $55.4^\circ < \theta < 85.5^\circ$.

29. (a) After t seconds, the boat will be $5t$ metres west of point A. The velocity of the water at that location is $\frac{3}{400}(5t)(40 - 5t)\,\mathbf{j}$. The velocity of the boat in still water is $5\,\mathbf{i}$, so the resultant velocity of the boat is $\mathbf{v}(t) = 5\,\mathbf{i} + \frac{3}{400}(5t)(40 - 5t)\,\mathbf{j} = 5\mathbf{i} + \left(\frac{3}{2}t - \frac{3}{16}t^2\right)\mathbf{j}$.

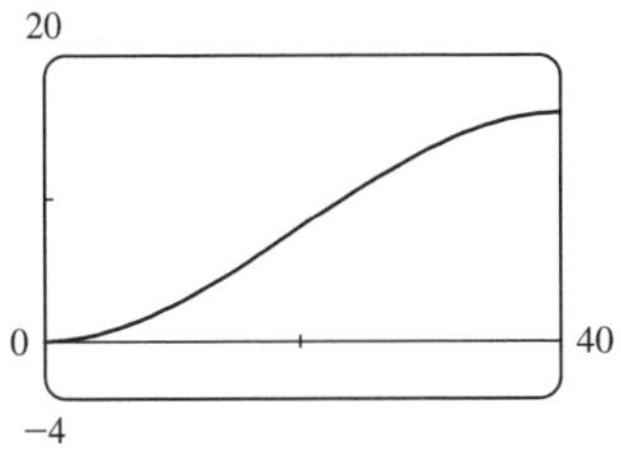

Integrating, we obtain $\mathbf{r}(t) = 5t\,\mathbf{i} + \left(\frac{3}{4}t^2 - \frac{1}{16}t^3\right)\mathbf{j} + \mathbf{C}$.

If we place the origin at A (and consider $\mathbf{j}$ to coincide with the northern direction) then $\mathbf{r}(0) = \mathbf{0} \;\Rightarrow\; \mathbf{C} = \mathbf{0}$ and we have $\mathbf{r}(t) = 5t\,\mathbf{i} + \left(\frac{3}{4}t^2 - \frac{1}{16}t^3\right)\mathbf{j}$. The boat reaches the east bank after 8 s, and it is located at $\mathbf{r}(8) = 5(8)\mathbf{i} + \left(\frac{3}{4}(8)^2 - \frac{1}{16}(8)^3\right)\mathbf{j} = 40\,\mathbf{i} + 16\,\mathbf{j}$. Thus the boat is 16 m downstream.

(b) Let α be the angle north of east that the boat heads. Then the velocity of the boat in still water is given by $5(\cos\alpha)\,\mathbf{i} + 5(\sin\alpha)\,\mathbf{j}$. At t seconds, the boat is $5(\cos\alpha)t$ metres from the west bank, at which point the velocity of the water is $\frac{3}{400}[5(\cos\alpha)t][40 - 5(\cos\alpha)t]\,\mathbf{j}$. The resultant velocity of the boat is given by

$$\begin{aligned}\mathbf{v}(t) &= 5(\cos\alpha)\,\mathbf{i} + \left[5\sin\alpha + \tfrac{3}{400}(5t\cos\alpha)(40 - 5t\cos\alpha)\right]\mathbf{j} \\ &= (5\cos\alpha)\,\mathbf{i} + \left(5\sin\alpha + \tfrac{3}{2}t\cos\alpha - \tfrac{3}{16}t^2\cos^2\alpha\right)\mathbf{j}\end{aligned}$$

Integrating, $\mathbf{r}(t) = (5t\cos\alpha)\,\mathbf{i} + \left(5t\sin\alpha + \frac{3}{4}t^2\cos\alpha - \frac{1}{16}t^3\cos^2\alpha\right)\mathbf{j}$ (where we have again placed the origin at A). The boat will reach the east bank when $5t\cos\alpha = 40 \;\Rightarrow\; t = \dfrac{40}{5\cos\alpha} = \dfrac{8}{\cos\alpha}$.

In order to land at point $B(40, 0)$ we need $5t\sin\alpha + \frac{3}{4}t^2\cos\alpha - \frac{1}{16}t^3\cos^2\alpha = 0 \;\Rightarrow$

$$5\left(\frac{8}{\cos\alpha}\right)\sin\alpha + \frac{3}{4}\left(\frac{8}{\cos\alpha}\right)^2\cos\alpha - \frac{1}{16}\left(\frac{8}{\cos\alpha}\right)^3\cos^2\alpha = 0 \;\Rightarrow\; \frac{1}{\cos\alpha}(40\sin\alpha + 48 - 32) = 0 \;\Rightarrow$$

$40\sin\alpha + 16 = 0 \;\Rightarrow\; \sin\alpha = -\frac{2}{5}$. Thus $\alpha = \sin^{-1}\left(-\frac{2}{5}\right) \approx -23.6^\circ$, so the boat should head 23.6° south of east (upstream).

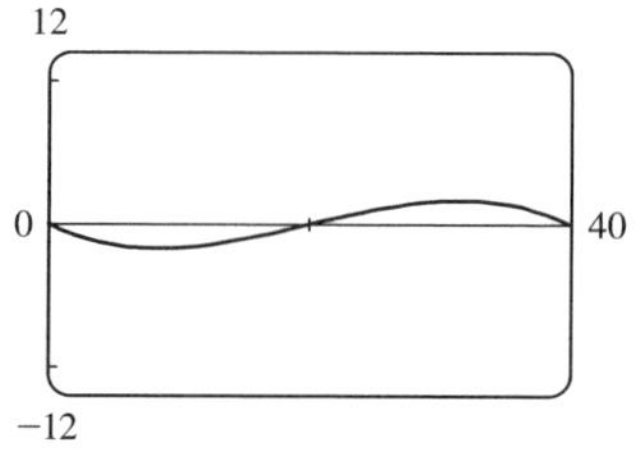

The path does seem realistic. The boat initially heads upstream to counteract the effect of the current. Near the centre of the river, the current is stronger and the boat is pushed downstream. When the boat nears the eastern bank, the current is slower and the boat is able to progress upstream to arrive at point B.

31. $\mathbf{r}(t) = (3t - t^3)\,\mathbf{i} + 3t^2\,\mathbf{j} \;\Rightarrow\; \mathbf{r}'(t) = (3 - 3t^2)\,\mathbf{i} + 6t\,\mathbf{j}$,

$|\mathbf{r}'(t)| = \sqrt{(3 - 3t^2)^2 + (6t)^2} = \sqrt{9 + 18t^2 + 9t^4} = \sqrt{(3 - 3t^2)^2} = 3 + 3t^2$,

$\mathbf{r}''(t) = -6t\,\mathbf{i} + 6\,\mathbf{j}$, $\mathbf{r}'(t) \times \mathbf{r}''(t) = (18 + 18t^2)\,\mathbf{k}$. Then Equation 9 gives

$$a_T = \frac{\mathbf{r}'(t)\cdot\mathbf{r}''(t)}{|\mathbf{r}'(t)|} = \frac{(3 - 3t^2)(-6t) + (6t)(6)}{3 + 3t^2} = \frac{18t + 18t^3}{3 + 3t^2} = \frac{18t(1 + t^2)}{3(1 + t^2)} = 6t \quad \left[\text{or by Equation 8, } a_T = v' = \frac{d}{dt}\left[3 + 3t^2\right] = 6t\right]$$

and Equation 10 gives $a_N = \dfrac{|\mathbf{r}'(t) \times \mathbf{r}''(t)|}{|\mathbf{r}'(t)|} = \dfrac{18 + 18t^2}{3 + 3t^2} = \dfrac{18(1 + t^2)}{3(1 + t^2)} = 6$.

33. $\mathbf{r}(t) = \cos t\,\mathbf{i} + \sin t\,\mathbf{j} + t\,\mathbf{k} \;\Rightarrow\; \mathbf{r}'(t) = -\sin t\,\mathbf{i} + \cos t\,\mathbf{j} + \mathbf{k}$, $|\mathbf{r}'(t)| = \sqrt{\sin^2 t + \cos^2 t + 1} = \sqrt{2}$,

$\mathbf{r}''(t) = -\cos t\,\mathbf{i} - \sin t\,\mathbf{j}$, $\mathbf{r}'(t) \times \mathbf{r}''(t) = \sin t\,\mathbf{i} - \cos t\,\mathbf{j} + \mathbf{k}$. Then $a_T = \dfrac{\mathbf{r}'(t)\cdot\mathbf{r}''(t)}{|\mathbf{r}'(t)|} = \dfrac{\sin t\cos t - \sin t\cos t}{\sqrt{2}} = 0$

and $a_N = \dfrac{|\mathbf{r}'(t) \times \mathbf{r}''(t)|}{|\mathbf{r}'(t)|} = \dfrac{\sqrt{\sin^2 t + \cos^2 t + 1}}{\sqrt{2}} = \dfrac{\sqrt{2}}{\sqrt{2}} = 1$.

35. The tangential component of $\mathbf{a}$ is the length of the projection of $\mathbf{a}$ onto $\mathbf{T}$, so we sketch the scalar projection of $\mathbf{a}$ in the tangential direction to the curve and estimate its length to be 4.5 (using the fact that $\mathbf{a}$ has length 10 as a guide). Similarly, the normal component of $\mathbf{a}$ is the length of the projection of $\mathbf{a}$ onto $\mathbf{N}$, so we sketch the scalar projection of $\mathbf{a}$ in the normal direction to the curve and estimate its length to be 9.0. Thus $a_T \approx 4.5\text{ cm/s}^2$ and $a_N \approx 9.0\text{ cm/s}^2$.

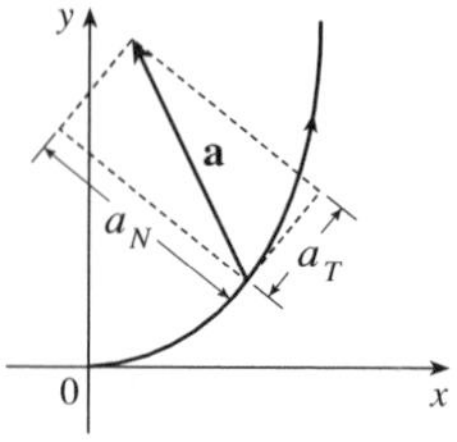

37. If the engines are turned off at time t, then the spacecraft will continue to travel in the direction of $\mathbf{v}(t)$, so we need a t such that for some scalar $s > 0$, $\mathbf{r}(t) + s\,\mathbf{v}(t) = \langle 6, 4, 9 \rangle$. $\mathbf{v}(t) = \mathbf{r}'(t) = \mathbf{i} + \frac{1}{t}\mathbf{j} + \frac{8t}{(t^2+1)^2}\mathbf{k} \Rightarrow$

$$\mathbf{r}(t) + s\,\mathbf{v}(t) = \left\langle 3 + t + s, 2 + \ln t + \frac{s}{t}, 7 - \frac{4}{t^2+1} + \frac{8st}{(t^2+1)^2} \right\rangle \Rightarrow 3 + t + s = 6 \Rightarrow s = 3 - t,$$

so $7 - \frac{4}{t^2+1} + \frac{8(3-t)t}{(t^2+1)^2} = 9 \Leftrightarrow \frac{24t - 12t^2 - 4}{(t^2+1)^2} = 2 \Leftrightarrow t^4 + 8t^2 - 12t + 3 = 0$. It is easily seen that $t = 1$ is a root of this polynomial. Also $2 + \ln 1 + \frac{3-1}{1} = 4$, so $t = 1$ is the desired solution.

10.5 Parametric Surfaces

1. $\mathbf{r}(u, v) = (u + v)\,\mathbf{i} + (3 - v)\,\mathbf{j} + (1 + 4u + 5v)\,\mathbf{k} = \langle 0, 3, 1 \rangle + u\,\langle 1, 0, 4 \rangle + v\,\langle 1, -1, 5 \rangle$. From Example 3, we recognize this as a vector equation of a plane through the point $(0, 3, 1)$ and containing vectors $\mathbf{a} = \langle 1, 0, 4 \rangle$ and $\mathbf{b} = \langle 1, -1, 5 \rangle$. If we wish to find a more conventional equation for the plane, a normal vector to the plane is $\mathbf{a} \times \mathbf{b} = \begin{vmatrix} \mathbf{i} & \mathbf{j} & \mathbf{k} \\ 1 & 0 & 4 \\ 1 & -1 & 5 \end{vmatrix} = 4\,\mathbf{i} - \mathbf{j} - \mathbf{k}$ and an equation of the plane is $4(x - 0) - (y - 3) - (z - 1) = 0$ or $4x - y - z = -4$.

3. $\mathbf{r}(s, t) = \langle s, t, t^2 - s^2 \rangle$, so the corresponding parametric equations for the surface are $x = s$, $y = t$, $z = t^2 - s^2$. For any point (x, y, z) on the surface, we have $z = y^2 - x^2$. With no restrictions on the parameters, the surface is $z = y^2 - x^2$, which we recognize as a hyperbolic paraboloid.

5. $\mathbf{r}(u, v) = \langle u^2 + 1, v^3 + 1, u + v \rangle$, $-1 \le u \le 1$, $-1 \le v \le 1$.

The surface has parametric equations $x = u^2 + 1$, $y = v^3 + 1$, $z = u + v$, $-1 \le u \le 1$, $-1 \le v \le 1$. If we keep u constant at u_0, $x = u_0^2 + 1$, a constant, so the corresponding grid curves must be the curves parallel to the yz-plane. If v is constant, we have $y = v_0^3 + 1$, a constant, so these grid curves are the curves parallel to the xz-plane.

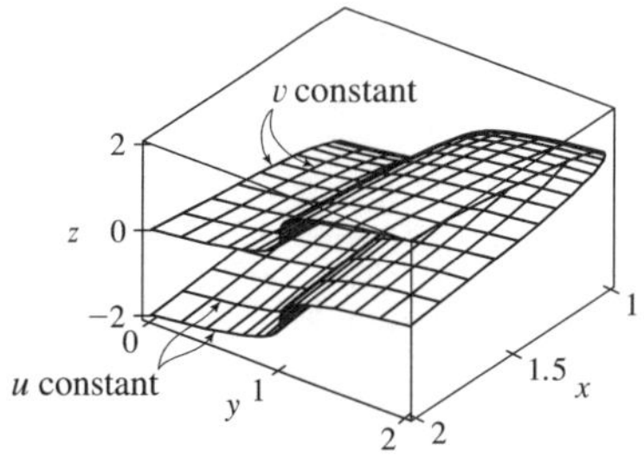

7. $\mathbf{r}(u, v) = \langle \cos^3 u \cos^3 v, \sin^3 u \cos^3 v, \sin^3 v \rangle$.

The surface has parametric equations $x = \cos^3 u \cos^3 v$, $y = \sin^3 u \cos^3 v$, $z = \sin^3 v$, $0 \le u \le \pi$, $0 \le v \le 2\pi$. Note that if $v = v_0$ is constant then $z = \sin^3 v_0$ is constant, so the corresponding grid curves must be the curves parallel to the xy-plane. The vertically oriented grid curves, then, correspond to $u = u_0$ being held constant, giving $x = \cos^3 u_0 \cos^3 v$, $y = \sin^3 u_0 \cos^3 v$, $z = \sin^3 v$. These curves lie in vertical planes that contain the z-axis.

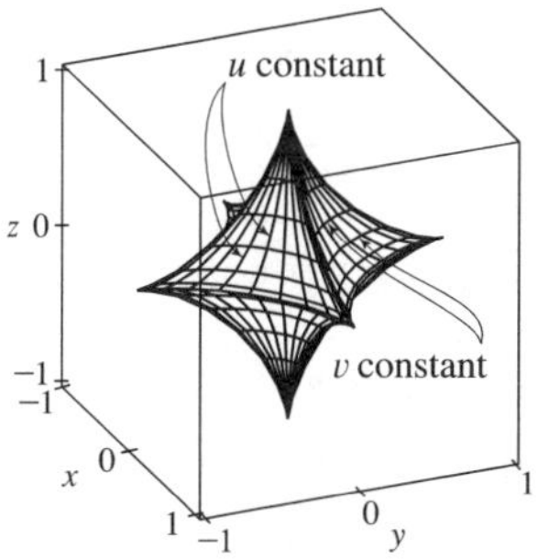

9. $x = \cos u \sin 2v$, $y = \sin u \sin 2v$, $z = \sin v$.

The complete graph of the surface is given by the parametric domain $0 \le u \le \pi, 0 \le v \le 2\pi$. Note that if $v = v_0$ is constant, the parametric equations become $x = \cos u \sin 2v_0$, $y = \sin u \sin 2v_0$, $z = \sin v_0$ which represent a circle of radius $\sin 2v_0$ in the plane $z = \sin v_0$. So the circular grid curves we see lying horizontally are the grid curves which have v constant. The vertical grid curves, then, correspond to $u = u_0$ being held constant, giving $x = \cos u_0 \sin 2v$ and $y = \sin u_0 \sin 2v$ with $z = \sin v$ which has a "figure-eight" shape.

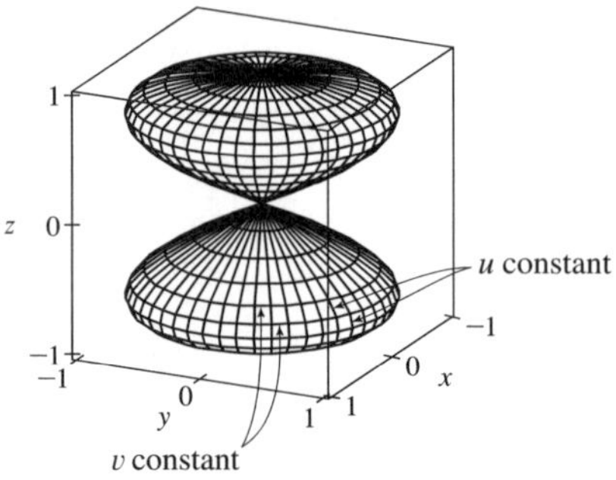

11. $\mathbf{r}(u, v) = \cos v\,\mathbf{i} + \sin v\,\mathbf{j} + u\,\mathbf{k}$. The parametric equations for the surface are $x = \cos v$, $y = \sin v$, $z = u$. Then $x^2 + y^2 = \cos^2 v + \sin^2 v = 1$ and $z = u$ with no restriction on u, so we have a circular cylinder, graph IV. The grid curves with u constant are the horizontal circles we see in the plane $z = u$. If v is constant, both x and y are constant with z free to vary, so the corresponding grid curves are the lines on the cylinder parallel to the z-axis.

13. $\mathbf{r}(u, v) = u \cos v\,\mathbf{i} + u \sin v\,\mathbf{j} + v\,\mathbf{k}$. The parametric equations for the surface are $x = u \cos v$, $y = u \sin v$, $z = v$. We look at the grid curves first; if we fix v, then x and y parametrize a straight line in the plane $z = v$ which intersects the z-axis. If u is held constant, the projection onto the xy-plane is circular; with $z = v$, each grid curve is a helix. The surface is a spiraling ramp, graph I.

15. $x = (u - \sin u) \cos v$, $y = (1 - \cos u) \sin v$, $z = u$. If u is held constant, x and y give an equation of an ellipse in the plane $z = u$, thus the grid curves are horizontally oriented ellipses. Note that when $u = 0$, the "ellipse" is the single point $(0, 0, 0)$, and when $u = \pi$, we have $y = 0$ while x ranges from $-\pi$ to π, a line segment parallel to the x-axis in the plane $z = \pi$. This is the upper "seam" we see in graph II. When v is held constant, $z = u$ is free to vary, so the corresponding grid curves are the curves we see running up and down along the surface.

17. From Example 3, parametric equations for the plane through the point $(1, 2, -3)$ that contains the vectors $\mathbf{a} = \langle 1, 1, -1 \rangle$ and $\mathbf{b} = \langle 1, -1, 1 \rangle$ are $x = 1 + u(1) + v(1) = 1 + u + v$, $y = 2 + u(1) + v(-1) = 2 + u - v$, $z = -3 + u(-1) + v(1) = -3 - u + v$.

19. Solving the equation for y gives $y^2 = 1 - x^2 + z^2 \quad \Rightarrow \quad y = \sqrt{1 - x^2 + z^2}$. (We choose the positive root since we want the part of the hyperboloid that corresponds to $y \ge 0$.) If we let x and z be the parameters, parametric equations are $x = x$, $z = z$, $y = \sqrt{1 - x^2 + z^2}$.

21. Since the cone intersects the sphere in the circle $x^2 + y^2 = 2$, $z = \sqrt{2}$ and we want the portion of the sphere above this, we can parametrize the surface as $x = x$, $y = y$, $z = \sqrt{4 - x^2 - y^2}$ where $x^2 + y^2 \le 2$.

Alternate solution: Using spherical coordinates, $x = 2 \sin \phi \cos \theta$, $y = 2 \sin \phi \sin \theta$, $z = 2 \cos \phi$ where $0 \le \phi \le \frac{\pi}{4}$ and $0 \le \theta \le 2\pi$.

23. Parametric equations are $x = x$, $y = 4 \cos \theta$, $z = 4 \sin \theta$, $0 \le x \le 5$, $0 \le \theta \le 2\pi$.

25. The surface appears to be a portion of a circular cylinder of radius 3 with axis the x-axis. An equation of the cylinder is $y^2 + z^2 = 9$, and we can impose the restrictions $0 \le x \le 5$, $y \le 0$ to obtain the portion shown.

To graph the surface on a CAS, we can use parametric equations $x = u$, $y = 3 \cos v$, $z = 3 \sin v$ with the parameter domain $0 \le u \le 5$, $\frac{\pi}{2} \le v \le \frac{3\pi}{2}$. Alternatively, we can regard x and z as parameters. Then parametric equations are $x = x$, $z = z$, $y = -\sqrt{9 - z^2}$, where $0 \le x \le 5$ and $-3 \le z \le 3$.

27. Using Equations 3, we have the parametrization $x = x$, $y = e^{-x}\cos\theta$, $z = e^{-x}\sin\theta$, $0 \le x \le 3$, $0 \le \theta \le 2\pi$.

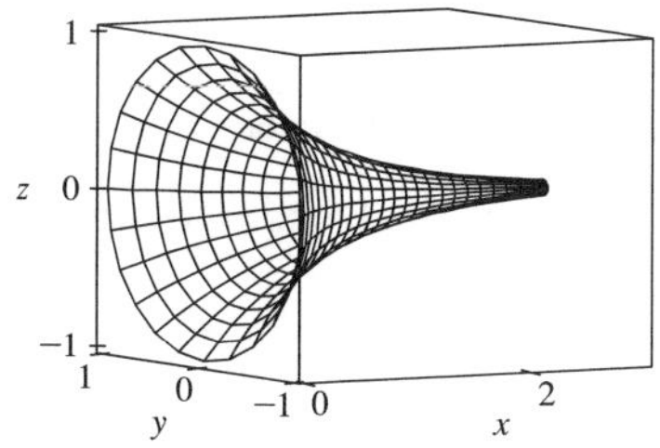

29. (a) $x = a\sin u\cos v$, $y = b\sin u\sin v$, $z = c\cos u$ $\Rightarrow$

$$\frac{x^2}{a^2} + \frac{y^2}{b^2} + \frac{z^2}{c^2} = (\sin u\cos v)^2 + (\sin u\sin v)^2 + (\cos u)^2$$
$$= \sin^2 u + \cos^2 u = 1$$

and since the ranges of u and v are sufficient to generate the entire graph, the parametric equations represent an ellipsoid.

(b)

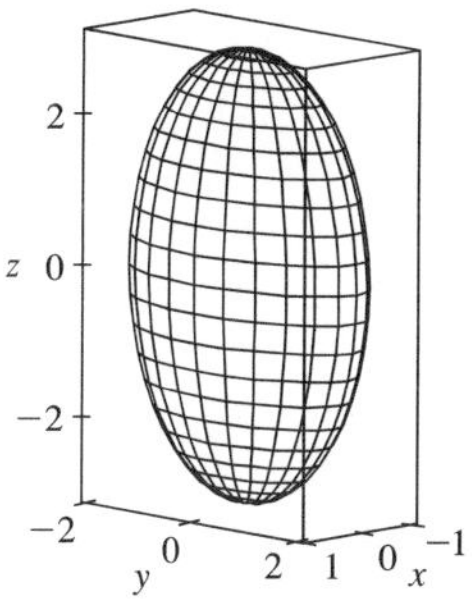

31. (a) Replacing $\cos u$ by $\sin u$ and $\sin u$ by $\cos u$ gives parametric equations $x = (2 + \sin v)\sin u$, $y = (2 + \sin v)\cos u$, $z = u + \cos v$. From the graph, it appears that the direction of the spiral is reversed. We can verify this observation by noting that the projection of the spiral grid curves onto the xy-plane, given by $x = (2 + \sin v)\sin u$, $y = (2 + \sin v)\cos u$, $z = 0$, draws a circle in the clockwise direction for each value of v. The original equations, on the other hand, give circular projections drawn in the counterclockwise direction. The equation for z is identical in both surfaces, so as z increases, these grid curves spiral up in opposite directions for the two surfaces.

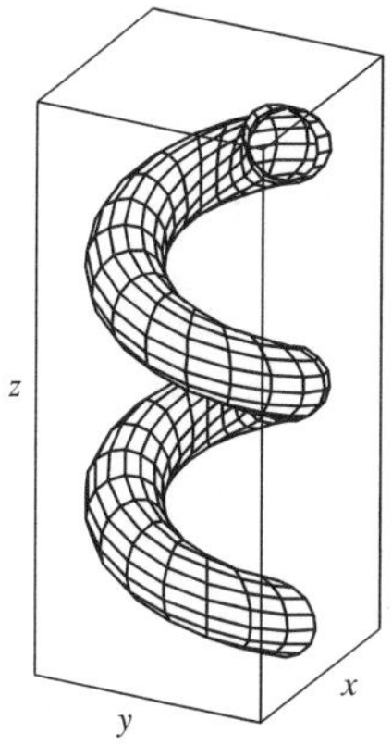

(b) Replacing $\cos u$ by $\cos 2u$ and $\sin u$ by $\sin 2u$ gives parametric equations $x = (2 + \sin v)\cos 2u$, $y = (2 + \sin v)\sin 2u$, $z = u + \cos v$. From the graph, it appears that the number of coils in the surface doubles within the same parametric domain. We can verify this observation by noting that the projection of the spiral grid curves onto the xy-plane, given by $x = (2 + \sin v)\cos 2u$, $y = (2 + \sin v)\sin 2u$, $z = 0$ (where v is constant), complete circular revolutions for $0 \le u \le \pi$ while the original surface requires $0 \le u \le 2\pi$ for a complete revolution. Thus, the new surface winds around twice as fast as the original surface, and since the equation for z is identical in both surfaces, we observe twice as many circular coils in the same z-interval.

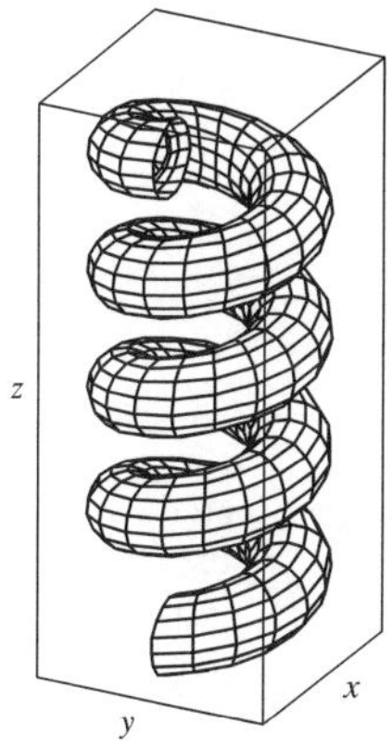

10 Review

CONCEPT CHECK

1. A vector function is a function whose domain is a set of real numbers and whose range is a set of vectors. To find the derivative or integral, we can differentiate or integrate each component of the vector function.

2. The tip of the moving vector $\mathbf{r}(t)$ of a continuous vector function traces out a space curve.

3. (a) A curve represented by the vector function $\mathbf{r}(t)$ is smooth if $\mathbf{r}'(t)$ is continuous and $\mathbf{r}'(t) \neq \mathbf{0}$ on its parametric domain (except possibly at the endpoints).

   (b) The tangent vector to a smooth curve at a point P with position vector $\mathbf{r}(t)$ is the vector $\mathbf{r}'(t)$. The tangent line at P is the line through P parallel to the tangent vector $\mathbf{r}'(t)$. The unit tangent vector is $\mathbf{T}(t) = \dfrac{\mathbf{r}'(t)}{|\mathbf{r}'(t)|}$.

4. (a)–(f) See Theorem 10.2.3.

5. Use Formula 10.3.2, or equivalently 10.3.3.

6. (a) The curvature of a curve is $\kappa = \left|\dfrac{d\mathbf{T}}{ds}\right|$ where $\mathbf{T}$ is the unit tangent vector.

   (b) $\kappa(t) = \left|\dfrac{\mathbf{T}'(t)}{\mathbf{r}'(t)}\right|$ (c) $\kappa(t) = \dfrac{|\mathbf{r}'(t) \times \mathbf{r}''(t)|}{|\mathbf{r}'(t)|^3}$ (d) $\kappa(x) = \dfrac{|f''(x)|}{[1 + (f'(x))^2]^{3/2}}$

7. (a) The unit normal vector: $\mathbf{N}(t) = \dfrac{\mathbf{T}'(t)}{|\mathbf{T}'(t)|}$. The binormal vector: $\mathbf{B}(t) = \mathbf{T}(t) \times \mathbf{N}(t)$.

   (b) See the discussion at the bottom of page 713.

8. (a) If $\mathbf{r}(t)$ is the position vector of the particle on the space curve, the velocity $\mathbf{v}(t) = \mathbf{r}'(t)$, the speed is given by $|\mathbf{v}(t)|$, and the acceleration $\mathbf{a}(t) = \mathbf{v}'(t) = \mathbf{r}''(t)$.

   (b) $\mathbf{a} = a_T\mathbf{T} + a_N\mathbf{N}$ where $a_T = v'$ and $a_N = \kappa v^2$.

9. See the statement of Kepler's Laws on page 722.

10. See the discussion on pages 728 and 729.

TRUE-FALSE QUIZ

1. True. If we reparametrize the curve by replacing $u = t^3$, we have $\mathbf{r}(u) = u\,\mathbf{i} + 2u\,\mathbf{j} + 3u\,\mathbf{k}$, which is a line through the origin with direction vector $\mathbf{i} + 2\,\mathbf{j} + 3\,\mathbf{k}$.

3. False. $\mathbf{r}'(t) = \langle -\sin t, 2t, 4t^3\rangle$, and since $\mathbf{r}'(0) = \langle 0, 0, 0\rangle = \mathbf{0}$, the curve is not smooth.

5. False. By Formula 5 of Theorem 10.2.3, $\dfrac{d}{dt}\,[\mathbf{u}(t) \times \mathbf{v}(t)] = \mathbf{u}'(t) \times \mathbf{v}(t) + \mathbf{u}(t) \times \mathbf{v}'(t)$.

7. False. κ is the magnitude of the rate of change of the unit tangent vector $\mathbf{T}$ with respect to arc length s, not with respect to t.

9. True. See the discussion at the bottom of page 713.

EXERCISES

1. (a) The corresponding parametric equations for the curve are $x = t$, $y = \cos \pi t$, $z = \sin \pi t$. Since $y^2 + z^2 = 1$, the curve is contained in a circular cylinder with axis the x-axis. Since $x = t$, the curve is a helix.

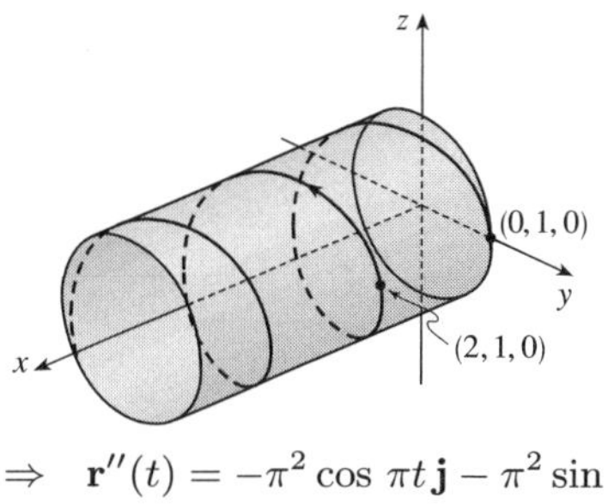

(b) $\mathbf{r}(t) = t\,\mathbf{i} + \cos \pi t\,\mathbf{j} + \sin \pi t\,\mathbf{k} \;\Rightarrow\; \mathbf{r}'(t) = \mathbf{i} - \pi \sin \pi t\,\mathbf{j} + \pi \cos \pi t\,\mathbf{k} \;\Rightarrow\; \mathbf{r}''(t) = -\pi^2 \cos \pi t\,\mathbf{j} - \pi^2 \sin \pi t\,\mathbf{k}$

3. The projection of the curve C of intersection onto the xy-plane is the circle $x^2 + y^2 = 16$, $z = 0$. So we can write $x = 4\cos t$, $y = 4\sin t$, $0 \le t \le 2\pi$. From the equation of the plane, we have $z = 5 - x = 5 - 4\cos t$, so parametric equations for C are $x = 4\cos t$, $y = 4\sin t$, $z = 5 - 4\cos t$, $0 \le t \le 2\pi$, and the corresponding vector function is $\mathbf{r}(t) = 4\cos t\,\mathbf{i} + 4\sin t\,\mathbf{j} + (5 - 4\cos t)\,\mathbf{k}$, $0 \le t \le 2\pi$.

5.
$$\begin{aligned}\int_0^1 (t^2\,\mathbf{i} + t\cos \pi t\,\mathbf{j} + \sin \pi t\,\mathbf{k})\,dt &= \left(\int_0^1 t^2\,dt\right)\mathbf{i} + \left(\int_0^1 t\cos \pi t\,dt\right)\mathbf{j} + \left(\int_0^1 \sin \pi t\,dt\right)\mathbf{k} \\ &= \left[\tfrac{1}{3}t^3\right]_0^1 \mathbf{i} + \left(\tfrac{t}{\pi}\sin \pi t\Big]_0^1 - \int_0^1 \tfrac{1}{\pi}\sin \pi t\,dt\right)\mathbf{j} + \left[-\tfrac{1}{\pi}\cos \pi t\right]_0^1 \mathbf{k} \\ &= \tfrac{1}{3}\,\mathbf{i} + \left[\tfrac{1}{\pi^2}\cos \pi t\right]_0^1 \mathbf{j} + \tfrac{2}{\pi}\,\mathbf{k} = \tfrac{1}{3}\,\mathbf{i} - \tfrac{2}{\pi^2}\,\mathbf{j} + \tfrac{2}{\pi}\,\mathbf{k}\end{aligned}$$

where we integrated by parts in the y-component.

7. $\mathbf{r}(t) = \langle t^2, t^3, t^4 \rangle \;\Rightarrow\; \mathbf{r}'(t) = \langle 2t, 3t^2, 4t^3 \rangle \;\Rightarrow\; |\mathbf{r}'(t)| = \sqrt{4t^2 + 9t^4 + 16t^6}$ and $L = \int_0^3 |\mathbf{r}'(t)|\,dt = \int_0^3 \sqrt{4t^2 + 9t^4 + 16t^6}\,dt$. Using Simpson's Rule with $f(t) = \sqrt{4t^2 + 9t^4 + 16t^6}$ and $n = 6$ we have $\Delta t = \frac{3-0}{6} = \frac{1}{2}$ and

$$\begin{aligned}L &\approx \tfrac{\Delta t}{3}\left[f(0) + 4f\left(\tfrac{1}{2}\right) + 2f(1) + 4f\left(\tfrac{3}{2}\right) + 2f(2) + 4f\left(\tfrac{5}{2}\right) + f(3)\right] \\ &= \tfrac{1}{6}\Bigg[\sqrt{0+0+0} + 4\cdot\sqrt{4\left(\tfrac{1}{2}\right)^2 + 9\left(\tfrac{1}{2}\right)^4 + 16\left(\tfrac{1}{2}\right)^6} + 2\cdot\sqrt{4(1)^2 + 9(1)^4 + 16(1)^6} \\ &\qquad + 4\cdot\sqrt{4\left(\tfrac{3}{2}\right)^2 + 9\left(\tfrac{3}{2}\right)^4 + 16\left(\tfrac{3}{2}\right)^6} + 2\cdot\sqrt{4(2)^2 + 9(2)^4 + 16(2)^6} \\ &\qquad + 4\cdot\sqrt{4\left(\tfrac{5}{2}\right)^2 + 9\left(\tfrac{5}{2}\right)^4 + 16\left(\tfrac{5}{2}\right)^6} + \sqrt{4(3)^2 + 9(3)^4 + 16(3)^6}\,\Bigg] \\ &\approx 86.631\end{aligned}$$

9. The angle of intersection of the two curves, θ, is the angle between their respective tangents at the point of intersection. For both curves the point $(1, 0, 0)$ occurs when $t = 0$. $\mathbf{r}_1'(t) = -\sin t\,\mathbf{i} + \cos t\,\mathbf{j} + \mathbf{k} \;\Rightarrow$ $\mathbf{r}_1'(0) = \mathbf{j} + \mathbf{k}$ and $\mathbf{r}_2'(t) = \mathbf{i} + 2t\,\mathbf{j} + 3t^2\,\mathbf{k} \;\Rightarrow\; \mathbf{r}_2'(0) = \mathbf{i}$. $\mathbf{r}_1'(0) \cdot \mathbf{r}_2'(0) = (\mathbf{j} + \mathbf{k}) \cdot \mathbf{i} = 0$. Therefore, the curves intersect in a right angle, that is, $\theta = \frac{\pi}{2}$.

11. (a) $\mathbf{T}(t) = \dfrac{\mathbf{r}'(t)}{|\mathbf{r}'(t)|} = \dfrac{\langle t^2, t, 1\rangle}{|\langle t^2, t, 1\rangle|} = \dfrac{\langle t^2, t, 1\rangle}{\sqrt{t^4 + t^2 + 1}}$

(b) $\mathbf{T}'(t) = -\frac{1}{2}(t^4 + t^2 + 1)^{-3/2}(4t^3 + 2t)\,\langle t^2, t, 1\rangle + (t^4 + t^2 + 1)^{-1/2}\langle 2t, 1, 0\rangle$

$$= \frac{-2t^3 - t}{(t^4 + t^2 + 1)^{3/2}}\,\langle t^2, t, 1\rangle + \frac{1}{(t^4 + t^2 + 1)^{1/2}}\,\langle 2t, 1, 0\rangle$$

$$= \frac{\langle -2t^5 - t^3, -2t^4 - t^2, -2t^3 - t\rangle + \langle 2t^5 + 2t^3 + 2t, t^4 + t^2 + 1, 0\rangle}{(t^4 + t^2 + 1)^{3/2}}$$

$$= \frac{\langle 2t, -t^4 + 1, -2t^3 - t\rangle}{(t^4 + t^2 + 1)^{3/2}}$$

$|\mathbf{T}'(t)| = \dfrac{\sqrt{4t^2 + t^8 - 2t^4 + 1 + 4t^6 + 4t^4 + t^2}}{(t^4 + t^2 + 1)^{3/2}} = \dfrac{\sqrt{t^8 + 4t^6 + 2t^4 + 5t^2}}{(t^4 + t^2 + 1)^{3/2}}$, and $\mathbf{N}(t) = \dfrac{\langle 2t, 1 - t^4, -2t^3 - t\rangle}{\sqrt{t^8 + 4t^6 + 2t^4 + 5t^2}}$.

(c) $\kappa(t) = \dfrac{|\mathbf{T}'(t)|}{|\mathbf{r}'(t)|} = \dfrac{\sqrt{t^8 + 4t^6 + 2t^4 + 5t^2}}{(t^4 + t^2 + 1)^2}$

13. $y' = 4x^3$, $y'' = 12x^2$ and $\kappa(x) = \dfrac{|y''|}{[1 + (y')^2]^{3/2}} = \dfrac{|12x^2|}{(1 + 16x^6)^{3/2}}$, so $\kappa(1) = \dfrac{12}{17^{3/2}}$.

15. $\mathbf{r}(t) = \langle \sin 2t, t, \cos 2t\rangle \;\Rightarrow\; \mathbf{r}'(t) = \langle 2\cos 2t, 1, -2\sin 2t\rangle \;\Rightarrow\; \mathbf{T}(t) = \frac{1}{\sqrt{5}}\,\langle 2\cos 2t, 1, -2\sin 2t\rangle \;\Rightarrow$

$\mathbf{T}'(t) = \frac{1}{\sqrt{5}}\,\langle -4\sin 2t, 0, -4\cos 2t\rangle \;\Rightarrow\; \mathbf{N}(t) = \langle -\sin 2t, 0, -\cos 2t\rangle$. So $\mathbf{N} = \mathbf{N}(\pi) = \langle 0, 0, -1\rangle$ and

$\mathbf{B} = \mathbf{T} \times \mathbf{N} = \frac{1}{\sqrt{5}}\,\langle -1, 2, 0\rangle$. So a normal to the osculating plane is $\langle -1, 2, 0\rangle$ and an equation is

$-1(x - 0) + 2(y - \pi) + 0(z - 1) = 0$ or $x - 2y + 2\pi = 0$.

17. $\mathbf{r}(t) = t\ln t\,\mathbf{i} + t\,\mathbf{j} + e^{-t}\,\mathbf{k}$, $\mathbf{v}(t) = \mathbf{r}'(t) = (1 + \ln t)\,\mathbf{i} + \mathbf{j} - e^{-t}\,\mathbf{k}$,

$|\mathbf{v}(t)| = \sqrt{(1 + \ln t)^2 + 1^2 + (-e^{-t})^2} = \sqrt{2 + 2\ln t + (\ln t)^2 + e^{-2t}}$, $\mathbf{a}(t) = \mathbf{v}'(t) = \frac{1}{t}\,\mathbf{i} + e^{-t}\,\mathbf{k}$.

19. We set up the axes so that the shot leaves the athlete's hand 2 m above the origin. Then we are given $\mathbf{r}(0) = 2\mathbf{j}$, $|\mathbf{v}(0)| = 13$ m/s, and $\mathbf{v}(0)$ has direction given by a 45° angle of elevation. Then a unit vector in the direction of $\mathbf{v}(0)$ is $\frac{1}{\sqrt{2}}(\mathbf{i} + \mathbf{j}) \;\Rightarrow\; \mathbf{v}(0) = \frac{13}{\sqrt{2}}(\mathbf{i} + \mathbf{j})$. Assuming air resistance is negligible, the only external force is due to gravity, so as in Example 10.4.5 we have $\mathbf{a} = -g\,\mathbf{j}$ where here $g \approx 9.8\ \text{m/s}^2$. Since $\mathbf{v}'(t) = \mathbf{a}(t)$, we integrate, giving $\mathbf{v}(t) = -gt\,\mathbf{j} + \mathbf{C}$ where $\mathbf{C} = \mathbf{v}(0) = \frac{13}{\sqrt{2}}(\mathbf{i} + \mathbf{j}) \;\Rightarrow\; \mathbf{v}(t) = \frac{13}{\sqrt{2}}\,\mathbf{i} + \left(\frac{13}{\sqrt{2}} - gt\right)\mathbf{j}$. Since $\mathbf{r}'(t) = \mathbf{v}(t)$ we integrate again, so

$\mathbf{r}(t) = \frac{13}{\sqrt{2}}t\,\mathbf{i} + \left(\frac{13}{\sqrt{2}}t - \frac{1}{2}gt^2\right)\mathbf{j} + \mathbf{D}$. But $\mathbf{D} = \mathbf{r}(0) = 2\,\mathbf{j} \;\Rightarrow\; \mathbf{r}(t) = \frac{13}{\sqrt{2}}t\,\mathbf{i} + \left(\frac{13}{\sqrt{2}}t - \frac{1}{2}gt^2 + 2\right)\mathbf{j}$.

(a) At 2 seconds, the shot is at $\mathbf{r}(2) = \frac{13}{\sqrt{2}}(2)\,\mathbf{i} + \left(\frac{13}{\sqrt{2}}(2) - \frac{1}{2}g(2)^2 + 2\right)\mathbf{j} \approx 18.4\,\mathbf{i} + 0.8\,\mathbf{j}$, so the shot is about 0.8 m above the ground, at a horizontal distance of 18.4 m from the athlete.

(b) The shot reaches its maximum height when the vertical component of velocity is 0: $\frac{13}{\sqrt{2}} - gt = 0 \;\Rightarrow$ $t = \dfrac{13}{\sqrt{2}\,g} \approx 0.94$ s. Then $\mathbf{r}(0.94) \approx 8.6\,\mathbf{i} + 6.3\,\mathbf{j}$, so the maximum height is approximately 6.3 m.

(c) The shot hits the ground when the vertical component of $\mathbf{r}(t)$ is 0, so $\frac{13}{\sqrt{2}}t - \frac{1}{2}gt^2 + 2 = 0 \;\Rightarrow\; -4.9t^2 + \frac{13}{\sqrt{2}}t + 2 = 0$ $\Rightarrow\; t \approx 2.07$ s. $\mathbf{r}(2.07) \approx 19.1\,\mathbf{i} - 0\,\mathbf{j}$, thus the shot lands approximately 19.1 m from the athlete.

21. From Example 4 in Section 10.5, a parametric representation of the sphere $x^2 + y^2 + z^2 = 4$ is $x = 2\sin\phi\cos\theta$, $y = 2\sin\phi\sin\theta$, $z = 2\cos\phi$ with $0 \le \theta \le 2\pi$ and $0 \le \phi \le \pi$. We can restrict the surface to that portion between the planes $z = 1$ and $z = -1$ by restricting $-1 \le z \le 1 \;\Rightarrow\; -1 \le 2\cos\phi \le 1 \;\Rightarrow\; \frac{\pi}{3} \le \phi \le \frac{2\pi}{3}$.

23. By the Fundamental Theorem of Calculus, $\mathbf{r}'(t) = \left\langle \sin(\pi t^2/2), \cos(\pi t^2/2) \right\rangle$, $|\mathbf{r}'(t)| = 1$ and so $\mathbf{T}(t) = \mathbf{r}'(t)$. Thus $\mathbf{T}'(t) = \pi t \left\langle \cos(\pi t^2/2), -\sin(\pi t^2/2) \right\rangle$ and the curvature is $\kappa = |\mathbf{T}'(t)| / |\mathbf{r}'(t)| = \sqrt{(\pi t)^2(1)}/1 = \pi\,|t|$.

FOCUS ON PROBLEM SOLVING

1. (a) $\mathbf{r}(t) = R\cos\omega t\,\mathbf{i} + R\sin\omega t\,\mathbf{j} \;\Rightarrow\; \mathbf{v} = \mathbf{r}'(t) = -\omega R\sin\omega t\,\mathbf{i} + \omega R\cos\omega t\,\mathbf{j}$, so $\mathbf{r} = R(\cos\omega t\,\mathbf{i} + \sin\omega t\,\mathbf{j})$ and $\mathbf{v} = \omega R(-\sin\omega t\,\mathbf{i} + \cos\omega t\,\mathbf{j})$. $\mathbf{v}\cdot\mathbf{r} = \omega R^2(-\cos\omega t\sin\omega t + \sin\omega t\cos\omega t) = 0$, so $\mathbf{v}\perp\mathbf{r}$. Since $\mathbf{r}$ points along a radius of the circle, and $\mathbf{v}\perp\mathbf{r}$, $\mathbf{v}$ is tangent to the circle. Because it is a velocity vector, $\mathbf{v}$ points in the direction of motion.

(b) In (a), we wrote $\mathbf{v}$ in the form $\omega R\,\mathbf{u}$, where $\mathbf{u}$ is the unit vector $-\sin\omega t\,\mathbf{i} + \cos\omega t\,\mathbf{j}$. Clearly $|\mathbf{v}| = \omega R\,|\mathbf{u}| = \omega R$. At speed ωR, the particle completes one revolution, a distance $2\pi R$, in time $T = \dfrac{2\pi R}{\omega R} = \dfrac{2\pi}{\omega}$.

(c) $\mathbf{a} = \dfrac{d\mathbf{v}}{dt} = -\omega^2 R\cos\omega t\,\mathbf{i} - \omega^2 R\sin\omega t\,\mathbf{j} = -\omega^2 R(\cos\omega t\,\mathbf{i} + \sin\omega t\,\mathbf{j})$, so $\mathbf{a} = -\omega^2\mathbf{r}$. This shows that $\mathbf{a}$ is proportional to $\mathbf{r}$ and points in the opposite direction (toward the origin). Also, $|\mathbf{a}| = \omega^2\,|\mathbf{r}| = \omega^2 R$.

(d) By Newton's Second Law (see Section 10.4), $\mathbf{F} = m\mathbf{a}$, so $|\mathbf{F}| = m\,|\mathbf{a}| = mR\omega^2 = \dfrac{m\,(\omega R)^2}{R} = \dfrac{m\,|\mathbf{v}|^2}{R}$.

3. (a) The projectile reaches maximum height when $0 = \dfrac{dy}{dt} = \dfrac{d}{dt}\left[(v_0\sin\alpha)t - \frac{1}{2}gt^2\right] = v_0\sin\alpha - gt$; that is, when

$t = \dfrac{v_0\sin\alpha}{g}$ and $y = (v_0\sin\alpha)\left(\dfrac{v_0\sin\alpha}{g}\right) - \dfrac{1}{2}g\left(\dfrac{v_0\sin\alpha}{g}\right)^2 = \dfrac{v_0^2\sin^2\alpha}{2g}$. This is the maximum height attained when the projectile is fired with an angle of elevation α. This maximum height is largest when $\alpha = \frac{\pi}{2}$. In that case, $\sin\alpha = 1$ and the maximum height is $\dfrac{v_0^2}{2g}$.

(b) Let $R = v_0^2/g$. We are asked to consider the parabola $x^2 + 2Ry - R^2 = 0$ which can be rewritten as $y = -\dfrac{1}{2R}x^2 + \dfrac{R}{2}$. The points on or inside this parabola are those for which $-R \le x \le R$ and $0 \le y \le \dfrac{-1}{2R}x^2 + \dfrac{R}{2}$. When the projectile is fired at angle of elevation α, the points (x, y) along its path satisfy the relations $x = (v_0\cos\alpha)\,t$ and $y = (v_0\sin\alpha)t - \frac{1}{2}gt^2$, where $0 \le t \le (2v_0\sin\alpha)/g$ (as in Example 10.4.5). Thus

$|x| \le \left|v_0\cos\alpha\left(\dfrac{2v_0\sin\alpha}{g}\right)\right| = \left|\dfrac{v_0^2}{g}\sin 2\alpha\right| \le \left|\dfrac{v_0^2}{g}\right| = |R|$. This shows that $-R \le x \le R$.

For t in the specified range, we also have $y = t\left(v_0\sin\alpha - \frac{1}{2}gt\right) = \frac{1}{2}gt\left(\dfrac{2v_0\sin\alpha}{g} - t\right) \ge 0$ and

$y = (v_0\sin\alpha)\dfrac{x}{v_0\cos\alpha} - \dfrac{g}{2}\left(\dfrac{x}{v_0\cos\alpha}\right)^2 = (\tan\alpha)\,x - \dfrac{g}{2v_0^2\cos^2\alpha}x^2 = -\dfrac{1}{2R\cos^2\alpha}x^2 + (\tan\alpha)\,x$. Thus

$$\begin{aligned} y - \left(\frac{-1}{2R}x^2 + \frac{R}{2}\right) &= \frac{-1}{2R\cos^2\alpha}x^2 + \frac{1}{2R}x^2 + (\tan\alpha)\,x - \frac{R}{2} \\ &= \frac{x^2}{2R}\left(1 - \frac{1}{\cos^2\alpha}\right) + (\tan\alpha)\,x - \frac{R}{2} = \frac{x^2(1-\sec^2\alpha) + 2R(\tan\alpha)\,x - R^2}{2R} \\ &= \frac{-(\tan^2\alpha)\,x^2 + 2R(\tan\alpha)\,x - R^2}{2R} = \frac{-\left[(\tan\alpha)\,x - R\right]^2}{2R} \le 0 \end{aligned}$$

[continued]

We have shown that every target that can be hit by the projectile lies on or inside the parabola $y = -\frac{1}{2R}x^2 + \frac{R}{2}$.

Now let (a,b) be any point on or inside the parabola $y = -\frac{1}{2R}x^2 + \frac{R}{2}$. Then $-R \le a \le R$ and $0 \le b \le -\frac{1}{2R}a^2 + \frac{R}{2}$.

We seek an angle α such that (a,b) lies in the path of the projectile; that is, we wish to find an angle α such that

$b = -\frac{1}{2R\cos^2\alpha}a^2 + (\tan\alpha)\,a$ or equivalently $b = \frac{-1}{2R}(\tan^2\alpha + 1)a^2 + (\tan\alpha)\,a$. Rearranging this equation we get

$\frac{a^2}{2R}\tan^2\alpha - a\tan\alpha + \left(\frac{a^2}{2R} + b\right) = 0$ or $a^2(\tan\alpha)^2 - 2aR(\tan\alpha) + (a^2 + 2bR) = 0$ $(*)$. This quadratic equation

for $\tan\alpha$ has real solutions exactly when the discriminant is nonnegative. Now $B^2 - 4AC \ge 0 \;\Leftrightarrow$

$(-2aR)^2 - 4a^2(a^2 + 2bR) \ge 0 \;\Leftrightarrow\; 4a^2(R^2 - a^2 - 2bR) \ge 0 \;\Leftrightarrow\; -a^2 - 2bR + R^2 \ge 0 \;\Leftrightarrow$

$b \le \frac{1}{2R}(R^2 - a^2) \;\Leftrightarrow\; b \le \frac{-1}{2R}a^2 + \frac{R}{2}$. This condition is satisfied since (a,b) is on or inside the parabola

$y = -\frac{1}{2R}x^2 + \frac{R}{2}$. It follows that (a,b) lies in the path of the projectile when $\tan\alpha$ satisfies $(*)$, that is, when

$$\tan\alpha = \frac{2aR \pm \sqrt{4a^2(R^2 - a^2 - 2bR)}}{2a^2} = \frac{R \pm \sqrt{R^2 - 2bR - a^2}}{a}.$$

(c)

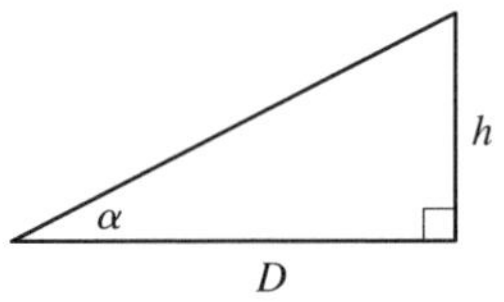

If the gun is pointed at a target with height h at a distance D downrange, then $\tan\alpha = h/D$. When the projectile reaches a distance D downrange (remember we are assuming that it doesn't hit the ground first), we have $D = x = (v_0\cos\alpha)t$,

so $t = \frac{D}{v_0\cos\alpha}$ and $y = (v_0\sin\alpha)t - \frac{1}{2}gt^2 = D\tan\alpha - \frac{gD^2}{2v_0^2\cos^2\alpha}$.

Meanwhile, the target, whose x-coordinate is also D, has fallen from height h to height

$h - \frac{1}{2}gt^2 = D\tan\alpha - \frac{gD^2}{2v_0^2\cos^2\alpha}$. Thus the projectile hits the target.

5. (a) $\mathbf{a} = -g\,\mathbf{j} \;\Rightarrow\; \mathbf{v} = \mathbf{v}_0 - gt\,\mathbf{j} = 0.5\,\mathbf{i} - gt\,\mathbf{j} \;\Rightarrow\; \mathbf{s} = \mathbf{s}_0 + 0.5t\,\mathbf{i} - \frac{1}{2}gt^2\,\mathbf{j} = 1.2\,\mathbf{j} + 0.5t\,\mathbf{i} - \frac{1}{2}gt^2\,\mathbf{j} \;\Rightarrow$

$\mathbf{s} = 0.5t\,\mathbf{i} + \left(1.2 - \frac{1}{2}gt^2\right)\mathbf{j}$. Therefore $y = 0$ when $t = \sqrt{2.4/g}$ seconds. At that instant, the ball is

$0.5\sqrt{2.4/g} \approx 0.25$ m to the right of the table top. Its coordinates (relative to an origin on the floor directly under the table's edge) are $(0.25, 0)$. At impact, the velocity is $\mathbf{v} = 0.5\,\mathbf{i} - \sqrt{2.4g}\,\mathbf{j}$, so the speed is

$|\mathbf{v}| = \sqrt{0.25 + 2.4g} \approx 4.9$ m/s.

(b) The slope of the curve when $t = \sqrt{\frac{2.4}{g}}$ is $\frac{dy}{dx} = \frac{dy/dt}{dx/dt} = \frac{-gt}{0.5} = -2g\sqrt{\frac{2.4}{g}} = \frac{-\sqrt{9.6g}}{1}$. Thus $\cot\theta = \frac{\sqrt{9.6g}}{1}$

and $\theta \approx 5.8863^\circ \approx 5.9^\circ$.

(c) From (a), $|\mathbf{v}| = \sqrt{0.25 + 2.4g}$. So the ball rebounds with speed $0.8\sqrt{0.25 + 2.4g} \approx 3.90$ m/s at angle of inclination

$90^\circ - \theta \approx 84.1137^\circ$. By Example 10.4.5, the horizontal distance travelled between bounces is $d = \frac{v_0^2\sin 2\alpha}{g}$, where

$v_0 \approx 3.90$ m/s and $\alpha \approx 84.1137^\circ$. Therefore, $d \approx 0.317$ m. So the ball strikes the floor at about

$0.5\sqrt{2.4/g} + 0.317 \approx 0.56$ m to the right of the table's edge.

7. The trajectory of the projectile is given by $\mathbf{r}(t) = (v\cos\alpha)t\,\mathbf{i} + \left[(v\sin\alpha)t - \frac{1}{2}gt^2\right]\mathbf{j}$, so $\mathbf{v}(t) = \mathbf{r}'(t) = v\cos\alpha\,\mathbf{i} + (v\sin\alpha - gt)\,\mathbf{j}$ and

$$\begin{aligned}|\mathbf{v}(t)| &= \sqrt{(v\cos\alpha)^2 + (v\sin\alpha - gt)^2} = \sqrt{v^2 - (2vg\sin\alpha)\,t + g^2t^2}\\ &= \sqrt{g^2\left(t^2 - \frac{2v}{g}(\sin\alpha)\,t + \frac{v^2}{g^2}\right)} = g\sqrt{\left(t - \frac{v}{g}\sin\alpha\right)^2 + \frac{v^2}{g^2} - \frac{v^2}{g^2}\sin^2\alpha}\\ &= g\sqrt{\left(t - \frac{v}{g}\sin\alpha\right)^2 + \frac{v^2}{g^2}\cos^2\alpha}\end{aligned}$$

The projectile hits the ground when $(v\sin\alpha)t - \frac{1}{2}gt^2 = 0 \;\Rightarrow\; t = \frac{2v}{g}\sin\alpha$, so the distance travelled by the projectile is

$$\begin{aligned}L(\alpha) &= \int_0^{(2v/g)\sin\alpha} |\mathbf{v}(t)|\,dt = \int_0^{(2v/g)\sin\alpha} g\sqrt{\left(t - \frac{v}{g}\sin\alpha\right)^2 + \frac{v^2}{g^2}\cos^2\alpha\,dt}\\ &= g\left[\frac{t - (v/g)\sin\alpha}{2}\sqrt{\left(t - \frac{v}{g}\sin\alpha\right)^2 + \left(\frac{v}{g}\cos\alpha\right)^2}\right.\\ &\qquad\left. + \frac{[(v/g)\cos\alpha]^2}{2}\ln\left(t - \frac{v}{g}\sin\alpha + \sqrt{\left(t - \frac{v}{g}\sin\alpha\right)^2 + \left(\frac{v}{g}\cos\alpha\right)^2}\right)\right]_0^{(2v/g)\sin\alpha}\end{aligned}$$

[using Formula 21 in the Table of Integrals]

$$\begin{aligned}&= \frac{g}{2}\left[\frac{v}{g}\sin\alpha\sqrt{\left(\frac{v}{g}\sin\alpha\right)^2 + \left(\frac{v}{g}\cos\alpha\right)^2} + \left(\frac{v}{g}\cos\alpha\right)^2\ln\left(\frac{v}{g}\sin\alpha + \sqrt{\left(\frac{v}{g}\sin\alpha\right)^2 + \left(\frac{v}{g}\cos\alpha\right)^2}\right)\right.\\ &\qquad\left. + \frac{v}{g}\sin\alpha\sqrt{\left(\frac{v}{g}\sin\alpha\right)^2 + \left(\frac{v}{g}\cos\alpha\right)^2} - \left(\frac{v}{g}\cos\alpha\right)^2\ln\left(-\frac{v}{g}\sin\alpha + \sqrt{\left(\frac{v}{g}\sin\alpha\right)^2 + \left(\frac{v}{g}\cos\alpha\right)^2}\right)\right]\\ &= \frac{g}{2}\left[\frac{v}{g}\sin\alpha\cdot\frac{v}{g} + \frac{v^2}{g^2}\cos^2\alpha\ln\left(\frac{v}{g}\sin\alpha + \frac{v}{g}\right) + \frac{v}{g}\sin\alpha\cdot\frac{v}{g} - \frac{v^2}{g^2}\cos^2\alpha\ln\left(-\frac{v}{g}\sin\alpha + \frac{v}{g}\right)\right]\\ &= \frac{v^2}{g}\sin\alpha + \frac{v^2}{2g}\cos^2\alpha\ln\left(\frac{(v/g)\sin\alpha + v/g}{-(v/g)\sin\alpha + v/g}\right) = \frac{v^2}{g}\sin\alpha + \frac{v^2}{2g}\cos^2\alpha\ln\left(\frac{1+\sin\alpha}{1-\sin\alpha}\right)\end{aligned}$$

We want to maximize $L(\alpha)$ for $0 \le \alpha \le \pi/2$.

$$\begin{aligned}L'(\alpha) &= \frac{v^2}{g}\cos\alpha + \frac{v^2}{2g}\left[\cos^2\alpha\cdot\frac{1-\sin\alpha}{1+\sin\alpha}\cdot\frac{2\cos\alpha}{(1-\sin\alpha)^2} - 2\cos\alpha\sin\alpha\ln\left(\frac{1+\sin\alpha}{1-\sin\alpha}\right)\right]\\ &= \frac{v^2}{g}\cos\alpha + \frac{v^2}{2g}\left[\cos^2\alpha\cdot\frac{2}{\cos\alpha} - 2\cos\alpha\sin\alpha\ln\left(\frac{1+\sin\alpha}{1-\sin\alpha}\right)\right]\\ &= \frac{v^2}{g}\cos\alpha + \frac{v^2}{g}\cos\alpha\left[1 - \sin\alpha\ln\left(\frac{1+\sin\alpha}{1-\sin\alpha}\right)\right] = \frac{v^2}{g}\cos\alpha\left[2 - \sin\alpha\ln\left(\frac{1+\sin\alpha}{1-\sin\alpha}\right)\right]\end{aligned}$$

$L(\alpha)$ has critical points for $0 < \alpha < \pi/2$ when $L'(\alpha) = 0 \;\Rightarrow\; 2 - \sin\alpha\ln\left(\frac{1+\sin\alpha}{1-\sin\alpha}\right) = 0$ (since $\cos\alpha \ne 0$).

Solving by graphing (or using a CAS) gives $\alpha \approx 0.9855$. Compare values at the critical point and the endpoints: $L(0) = 0$, $L(\pi/2) = v^2/g$, and $L(0.9855) \approx 1.20v^2/g$. Thus the distance travelled by the projectile is maximized for $\alpha \approx 0.9855$ or $\approx 56^\circ$.

11 □ PARTIAL DERIVATIVES

11.1 Functions of Several Variables

1. (a) From Table 1, $f(-15, 40) = -27$, which means that if the temperature is $-15°\text{C}$ and the wind speed is 40 km/h, then the air would feel equivalent to approximately $-27°\text{C}$ without wind.

(b) The question is asking: when the temperature is $-20°\text{C}$, what wind speed gives a wind-chill index of $-30°\text{C}$? From Table 1, the speed is 20 km/h.

(c) The question is asking: when the wind speed is 20 km/h, what temperature gives a wind-chill index of $-49°\text{C}$? From Table 1, the temperature is $-35°\text{C}$.

(d) The function $W = f(-5, v)$ means that we fix T at -5 and allow v to vary, resulting in a function of one variable. In other words, the function gives wind-chill index values for different wind speeds when the temperature is $-5°\text{C}$. From Table 1 (look at the row corresponding to $T = -5$), the function decreases and appears to approach a constant value as v increases.

(e) The function $W = f(T, 50)$ means that we fix v at 50 and allow T to vary, again giving a function of one variable. In other words, the function gives wind-chill index values for different temperatures when the wind speed is 50 km/h . From Table 1 (look at the column corresponding to $v = 50$), the function increases almost linearly as T increases.

3. If the amounts of labour and capital are both doubled, we replace L, K in the function with $2L, 2K$, giving

$$P(2L, 2K) = 1.01(2L)^{0.75}(2K)^{0.25} = 1.01(2^{0.75})(2^{0.25})L^{0.75}K^{0.25} = (2^1)1.01L^{0.75}K^{0.25} = 2P(L, K)$$

Thus, the production is doubled. It is also true for the general case $P(L, K) = bL^{\alpha}K^{1-\alpha}$:
$P(2L, 2K) = b(2L)^{\alpha}(2K)^{1-\alpha} = b(2^{\alpha})(2^{1-\alpha})L^{\alpha}K^{1-\alpha} = (2^{\alpha+1-\alpha})bL^{\alpha}K^{1-\alpha} = 2P(L, K)$.

5. $\ln(9 - x^2 - 9y^2)$ is defined only when $9 - x^2 - 9y^2 > 0$, or $\frac{1}{9}x^2 + y^2 < 1$. So the domain of f is $\left\{(x, y) \mid \frac{1}{9}x^2 + y^2 < 1\right\}$, the interior of an ellipse.

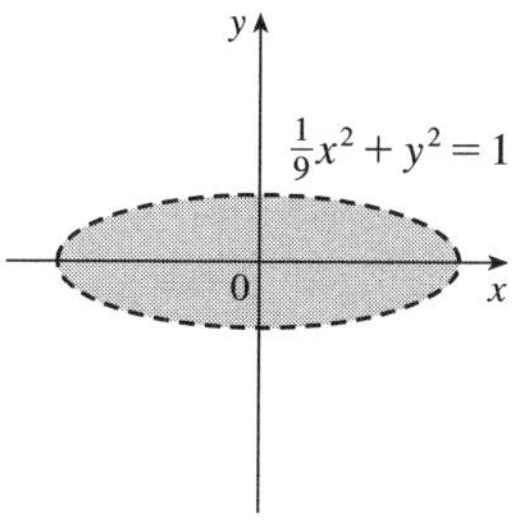

7. (a) $f(2, -1, 6) = e^{\sqrt{6-2^2-(-1)^2}} = e^{\sqrt{1}} = e$.

(b) $e^{\sqrt{z-x^2-y^2}}$ is defined when $z - x^2 - y^2 \geq 0 \;\Rightarrow\; z \geq x^2 + y^2$. Thus the domain of f is $\left\{(x, y, z) \mid z \geq x^2 + y^2\right\}$.

(c) Since $\sqrt{z - x^2 - y^2} \geq 0$, we have $e^{\sqrt{z-x^2-y^2}} \geq 1$. Thus the range of f is $[1, \infty)$.

9. The point $(-3, 3)$ lies between the level curves with z-values 50 and 60. Since the point is a little closer to the level curve with $z = 60$, we estimate that $f(-3, 3) \approx 56$. The point $(3, -2)$ appears to be just about halfway between the level curves with z-values 30 and 40, so we estimate $f(3, -2) \approx 35$. The graph rises as we approach the origin, gradually from above, steeply from below.

11. Near A, the level curves are very close together, indicating that the terrain is quite steep. At B, the level curves are much further apart, so we would expect the terrain to be much less steep than near A, perhaps almost flat.

13.

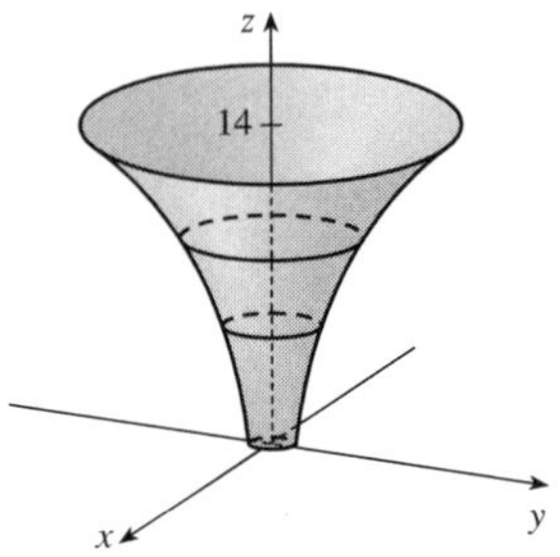

15. The level curves are $(y - 2x)^2 = k$ or $y = 2x \pm \sqrt{k}$, $k \geq 0$, a family of pairs of parallel lines.

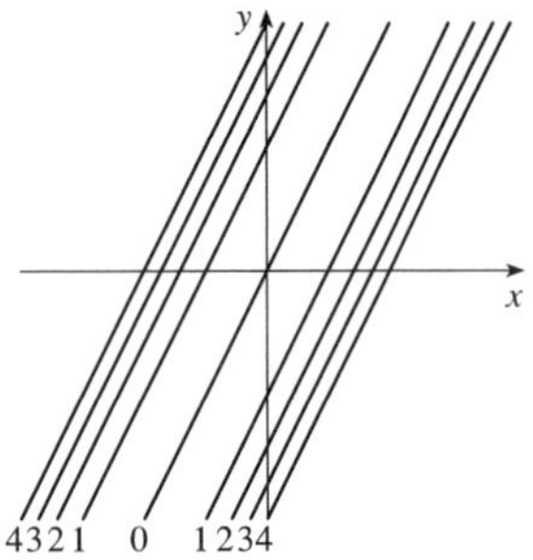

17. The level curves are $y - \ln x = k$ or $y = \ln x + k$.

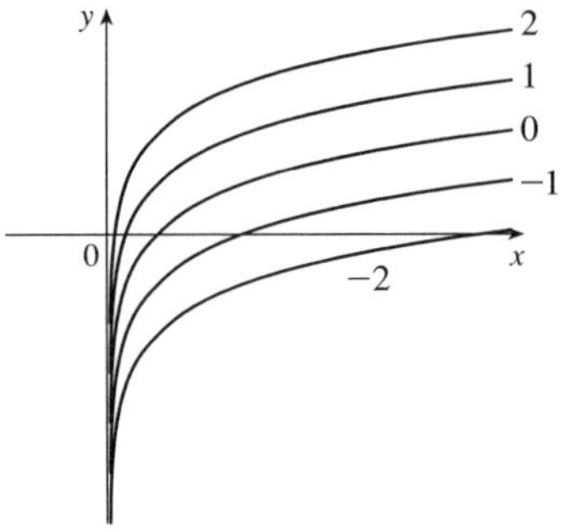

19. The level curves are $ye^x = k$ or $y = ke^{-x}$, a family of exponential curves.

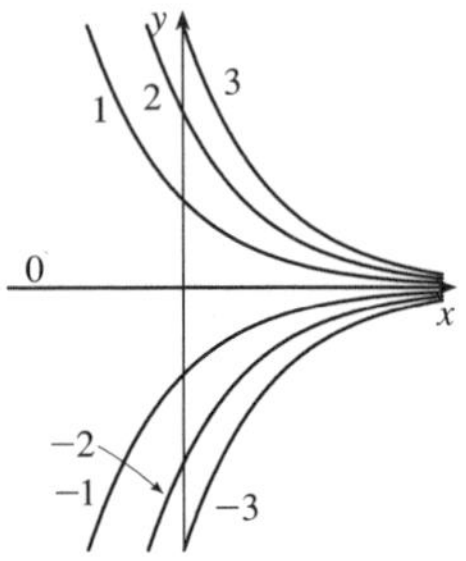

21. The level curves are $\sqrt{y^2 - x^2} = k$ or $y^2 - x^2 = k^2$, $k \geq 0$. When $k = 0$ the level curve is the pair of lines $y = \pm x$. For $k > 0$, the level curves are hyperbolas with axis the y-axis.

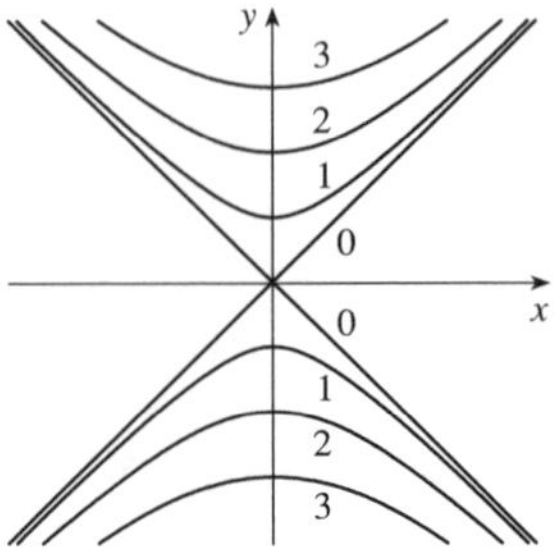

23. The contour map consists of the level curves $k = x^2 + 9y^2$, a family of ellipses with major axis the x-axis. (Or, if $k = 0$, the origin.)

The graph of $f(x, y)$ is the surface $z = x^2 + 9y^2$, an elliptic paraboloid.

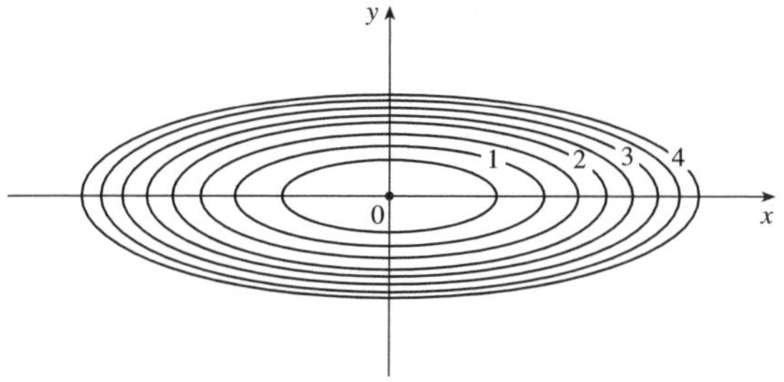

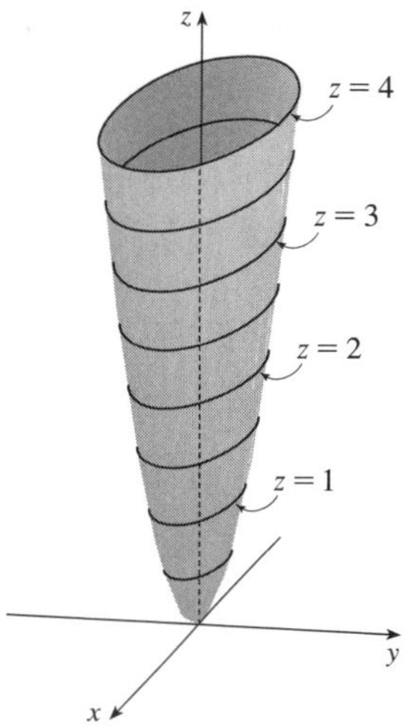

If we visualize lifting each ellipse $k = x^2 + 9y^2$ of the contour map to the plane $z = k$, we have horizontal traces that indicate the shape of the graph of f.

25. The isothermals are given by $k = 100/(1 + x^2 + 2y^2)$ or $x^2 + 2y^2 = (100 - k)/k$ $(0 < k \leq 100)$, a family of ellipses.

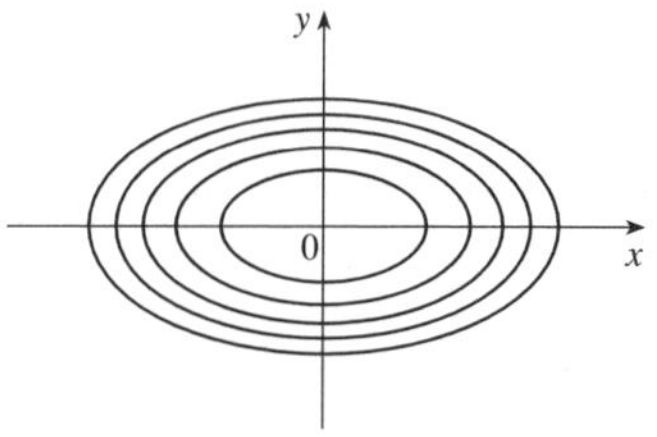

27. $f(x, y) = e^x \cos y$

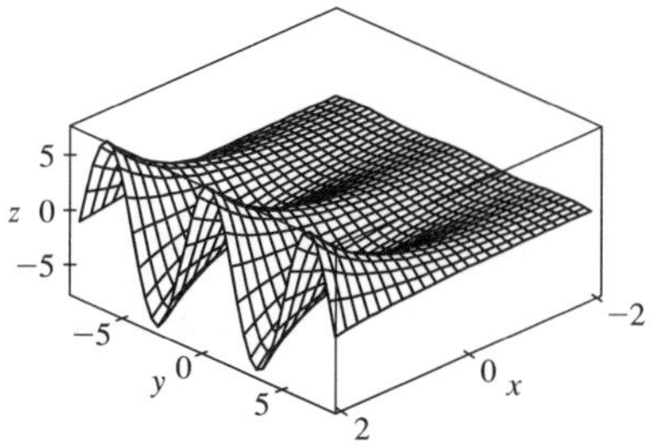

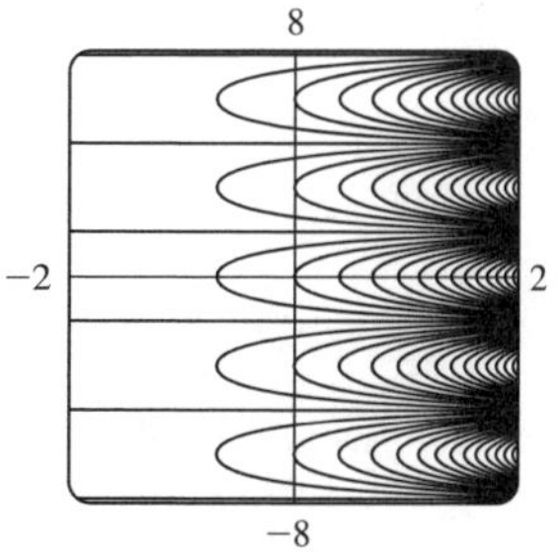

Traces parallel to the yz-plane (such as the left-front trace in the first graph above) are cosine curves. The amplitudes of these curves decrease as x decreases.

29. $f(x, y) = xy^2 - x^3$

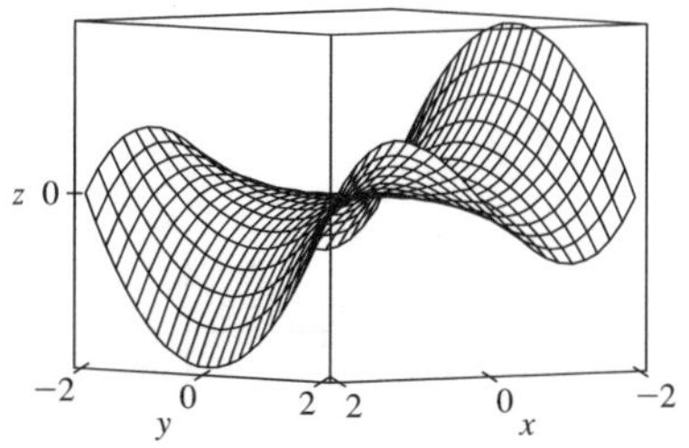

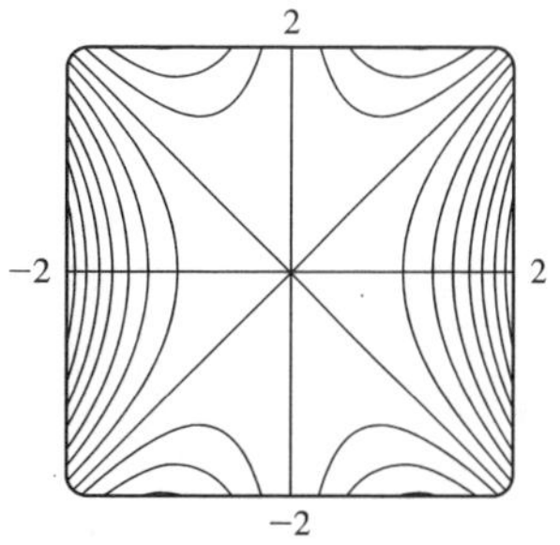

The traces parallel to the yz-plane (such as the left-front trace in the graph above) are parabolas; those parallel to the xz-plane (such as the right-front tracc) arc cubic curves. The surface is called a monkey saddle because a monkey sitting on the surface near the origin has places for both legs and tail to rest.

31. (a) C (b) II

Reasons: This function is periodic in both x and y, and the function is the same when x is interchanged with y, so its graph is symmetric about the plane $y = x$. In addition, the function is 0 along the x- and y-axes. These conditions are satisfied only by C and II.

33. (a) F (b) I

Reasons: This function is periodic in both x and y but is constant along the lines $y = x + k$, a condition satisfied only by F and I.

35. (a) B (b) VI

Reasons: This function is 0 along the lines $x = \pm 1$ and $y = \pm 1$. The only contour map in which this could occur is VI. Also note that the trace in the xz-plane is the parabola $z = 1 - x^2$ and the trace in the yz-plane is the parabola $z = 1 - y^2$, so the graph is B.

37. $k = x + 3y + 5z$ is a family of parallel planes with normal vector $\langle 1, 3, 5 \rangle$.

39. $k = x^2 - y^2 + z^2$ are the equations of the level surfaces. For $k = 0$, the surface is a right circular cone with vertex the origin and axis the y-axis. For $k > 0$, we have a family of hyperboloids of one sheet with axis the y-axis. For $k < 0$, we have a family of hyperboloids of two sheets with axis the y-axis.

41. (a) The graph of g is the graph of f shifted upward 2 units.

(b) The graph of g is the graph of f stretched vertically by a factor of 2.

(c) The graph of g is the graph of f reflected about the xy-plane.

(d) The graph of $g(x, y) = -f(x, y) + 2$ is the graph of f reflected about the xy-plane and then shifted upward 2 units.

43. $f(x, y) = e^{cx^2 + y^2}$. First, if $c = 0$, the graph is the cylindrical surface $z = e^{y^2}$ (whose level curves are parallel lines). When $c > 0$, the vertical trace above the y-axis remains fixed while the sides of the surface in the x-direction "curl" upward, giving the graph a shape resembling an elliptic paraboloid. The level curves of the surface are ellipses centred at the origin.

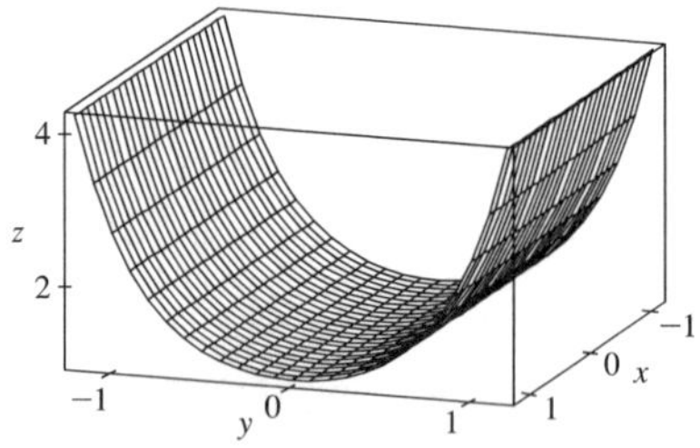

$c = 0$

For $0 < c < 1$, the ellipses have major axis the x-axis and the eccentricity increases as $c \to 0$.

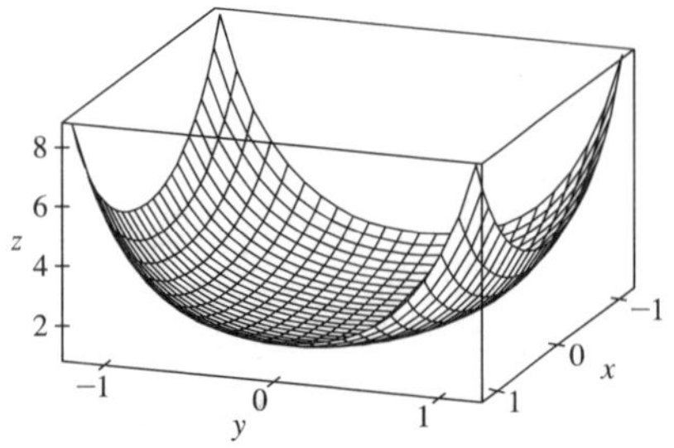

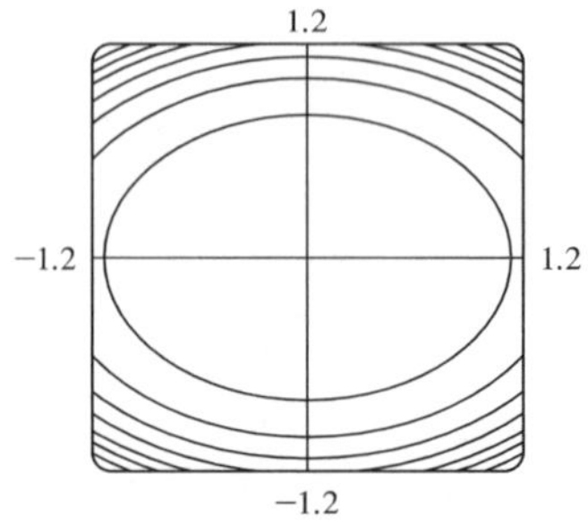

$c = 0.5$ (level curves in increments of 1)

For $c = 1$ the level curves are circles centred at the origin.

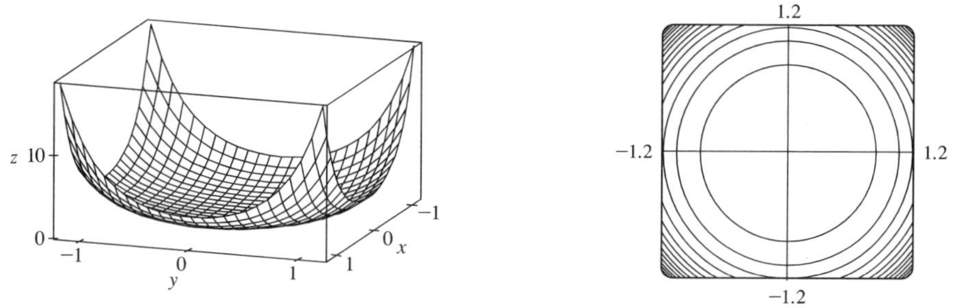

$c = 1$ (level curves in increments of 1)

When $c > 1$, the level curves are ellipses with major axis the y-axis, and the eccentricity increases as c increases.

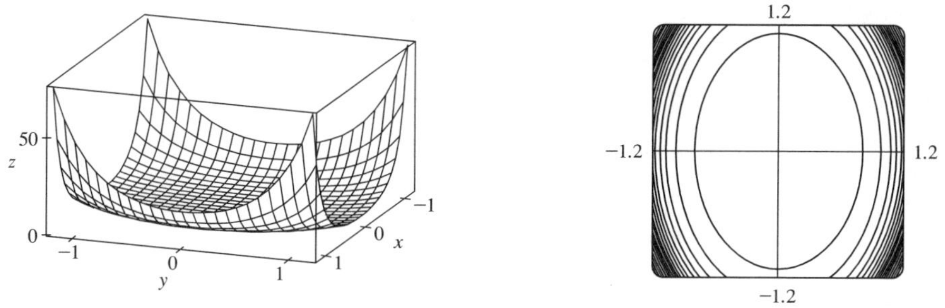

$c = 2$ (level curves in increments of 4)

For values of $c < 0$, the sides of the surface in the x-direction curl downward and approach the xy-plane (while the vertical trace $x = 0$ remains fixed), giving a saddle-shaped appearance to the graph near the point $(0, 0, 1)$. The level curves consist of a family of hyperbolas. As c decreases, the surface becomes flatter in the x-direction and the surface's approach to the curve in the trace $x = 0$ becomes steeper, as the graphs demonstrate.

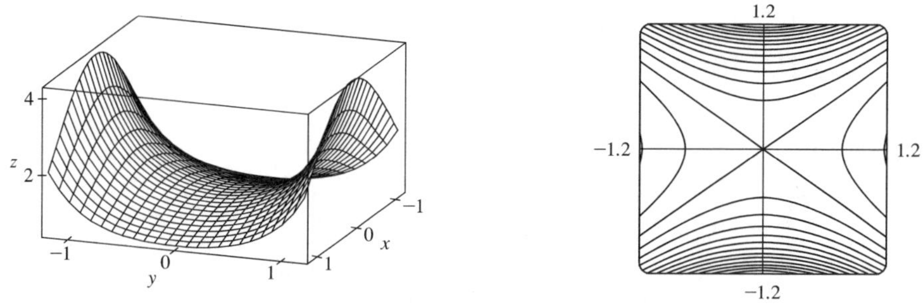

$c = -0.5$ (level curves in increments of 0.25)

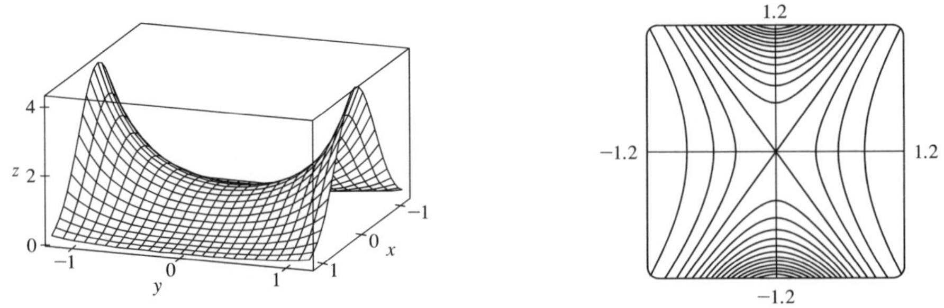

$c = -2$ (level curves in increments of 0.25)

45. (a) $P = bL^{\alpha}K^{1-\alpha} \quad\Rightarrow\quad \dfrac{P}{K} = bL^{\alpha}K^{-\alpha} \quad\Rightarrow\quad \dfrac{P}{K} = b\left(\dfrac{L}{K}\right)^{\alpha} \quad\Rightarrow\quad \ln\dfrac{P}{K} = \ln\left(b\left(\dfrac{L}{K}\right)^{\alpha}\right) \quad\Rightarrow$

$$\ln\frac{P}{K} = \ln b + \alpha\ln\left(\frac{L}{K}\right)$$

(b) We list the values for $\ln(L/K)$ and $\ln(P/K)$ for the years 1899–1922. (Historically, these values were rounded to 2 decimal places.)

Year	$x = \ln(L/K)$	$y = \ln(P/K)$
1899	0	0
1900	−0.02	−0.06
1901	−0.04	−0.02
1902	−0.04	0
1903	−0.07	−0.05
1904	−0.13	−0.12
1905	−0.18	−0.04
1906	−0.20	−0.07
1907	−0.23	−0.15
1908	−0.41	−0.38
1909	−0.33	−0.24
1910	−0.35	−0.27

Year	$x = \ln(L/K)$	$y = \ln(P/K)$
1911	−0.38	−0.34
1912	−0.38	−0.24
1913	−0.41	−0.25
1914	−0.47	−0.37
1915	−0.53	−0.34
1916	−0.49	−0.28
1917	−0.53	−0.39
1918	−0.60	−0.50
1919	−0.68	−0.57
1920	−0.74	−0.57
1921	−1.05	−0.85
1922	−0.98	−0.59

After entering the (x, y) pairs into a calculator or CAS, the resulting least squares regression line through the points is approximately $y = 0.75136x + 0.01053$, which we round to $y = 0.75x + 0.01$.

(c) Comparing the regression line from part (b) to the equation $y = \ln b + \alpha x$ with $x = \ln(L/K)$ and $y = \ln(P/K)$, we have $\alpha = 0.75$ and $\ln b = 0.01 \quad\Rightarrow\quad b = e^{0.01} \approx 1.01$. Thus, the Cobb-Douglas production function is $P = bL^{\alpha}K^{1-\alpha} = 1.01L^{0.75}K^{0.25}$.

11.2 Limits and Continuity

1. In general, we can't say anything about $f(3,1)$! $\displaystyle\lim_{(x,y)\to(3,1)} f(x,y) = 6$ means that the values of $f(x,y)$ approach 6 as (x,y) approaches, but is not equal to, $(3,1)$. If f is continuous, we know that $\displaystyle\lim_{(x,y)\to(a,b)} f(x,y) = f(a,b)$, so $\displaystyle\lim_{(x,y)\to(3,1)} f(x,y) = f(3,1) = 6$.

3. We make a table of values of $f(x,y)=\dfrac{x^2y^3+x^3y^2-5}{2-xy}$ for a set of (x,y) points near the origin.

x \ y	-0.2	-0.1	-0.05	0	0.05	0.1	0.2
-0.2	-2.551	-2.525	-2.513	-2.500	-2.488	-2.475	-2.451
-0.1	-2.525	-2.513	-2.506	-2.500	-2.494	-2.488	-2.475
-0.05	-2.513	-2.506	-2.503	-2.500	-2.497	-2.494	-2.488
0	-2.500	-2.500	-2.500		-2.500	-2.500	-2.500
0.05	-2.488	-2.494	-2.497	-2.500	-2.503	-2.506	-2.513
0.1	-2.475	-2.488	-2.494	-2.500	-2.506	-2.513	-2.525
0.2	-2.451	-2.475	-2.488	-2.500	-2.513	-2.525	-2.551

As the table shows, the values of $f(x,y)$ seem to approach -2.5 as (x,y) approaches the origin from a variety of different directions. This suggests that $\lim\limits_{(x,y)\to(0,0)} f(x,y)=-2.5$.

Since f is a rational function, it is continuous on its domain. f is defined at $(0,0)$, so we can use direct substitution to establish that $\lim\limits_{(x,y)\to(0,0)} f(x,y)=\dfrac{0^20^3+0^30^2-5}{2-0\cdot 0}=-\dfrac{5}{2}$, verifying our guess.

5. $f(x,y)=x^5+4x^3y-5xy^2$ is a polynomial, and hence continuous, so
$\lim\limits_{(x,y)\to(5,-2)} f(x,y)=f(5,-2)=5^5+4(5)^3(-2)-5(5)(-2)^2=2025$.

7. $f(x,y)=y^4/(x^4+3y^4)$. First approach $(0,0)$ along the x-axis. Then $f(x,0)=0/x^4=0$ for $x\neq 0$, so $f(x,y)\to 0$. Now approach $(0,0)$ along the y-axis. Then for $y\neq 0$, $f(0,y)=y^4/3y^4=1/3$, so $f(x,y)\to 1/3$. Since f has two different limits along two different lines, the limit does not exist.

9. $f(x,y)=(xy\cos y)/(3x^2+y^2)$. On the x-axis, $f(x,0)=0$ for $x\neq 0$, so $f(x,y)\to 0$ as $(x,y)\to(0,0)$ along the x-axis. Approaching $(0,0)$ along the line $y=x$, $f(x,x)=(x^2\cos x)/4x^2=\frac{1}{4}\cos x$ for $x\neq 0$, so $f(x,y)\to\frac{1}{4}$ along this line. Thus the limit does not exist.

11. $f(x,y)=\dfrac{xy}{\sqrt{x^2+y^2}}$. We can see that the limit along any line through $(0,0)$ is 0, as well as along other paths through $(0,0)$ such as $x=y^2$ and $y=x^2$. So we suspect that the limit exists and equals 0; we use the Squeeze Theorem to prove our assertion. $0\leq\left|\dfrac{xy}{\sqrt{x^2+y^2}}\right|\leq|x|$ since $|y|\leq\sqrt{x^2+y^2}$, and $|x|\to 0$ as $(x,y)\to(0,0)$. So $\lim\limits_{(x,y)\to(0,0)} f(x,y)=0$.

13. Let $f(x,y)=\dfrac{2x^2y}{x^4+y^2}$. Then $f(x,0)=0$ for $x\neq 0$, so $f(x,y)\to 0$ as $(x,y)\to(0,0)$ along the x-axis. But $f(x,x^2)=\dfrac{2x^4}{2x^4}=1$ for $x\neq 0$, so $f(x,y)\to 1$ as $(x,y)\to(0,0)$ along the parabola $y=x^2$. Thus the limit doesn't exist.

15.
$$\lim_{(x,y)\to(0,0)}\frac{x^2+y^2}{\sqrt{x^2+y^2+1}-1}=\lim_{(x,y)\to(0,0)}\frac{x^2+y^2}{\sqrt{x^2+y^2+1}-1}\cdot\frac{\sqrt{x^2+y^2+1}+1}{\sqrt{x^2+y^2+1}+1}$$
$$=\lim_{(x,y)\to(0,0)}\frac{(x^2+y^2)\left(\sqrt{x^2+y^2+1}+1\right)}{x^2+y^2}=\lim_{(x,y)\to(0,0)}\left(\sqrt{x^2+y^2+1}+1\right)=2$$

17. $f(x,y,z)=\dfrac{xy+yz^2+xz^2}{x^2+y^2+z^4}$. Then $f(x,0,0)=0/x^2=0$ for $x\neq 0$, so as $(x,y,z)\to(0,0,0)$ along the x-axis, $f(x,y,z)\to 0$. But $f(x,x,0)=x^2/(2x^2)=\frac{1}{2}$ for $x\neq 0$, so as $(x,y,z)\to(0,0,0)$ along the line $y=x$, $z=0$, $f(x,y,z)\to\frac{1}{2}$. Thus the limit doesn't exist.

19.

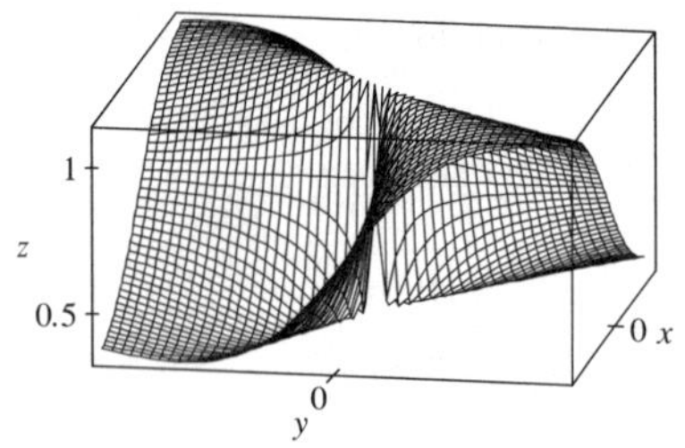

From the ridges on the graph, we see that as $(x, y) \to (0, 0)$ along the lines under the two ridges, $f(x, y)$ approaches different values. So the limit does not exist.

21. $h(x, y) = g(f(x, y)) = (2x + 3y - 6)^2 + \sqrt{2x + 3y - 6}$. Since f is a polynomial, it is continuous on $\mathbb{R}^2$ and g is continuous on its domain $\{t \mid t \geq 0\}$. Thus h is continuous on its domain.

$D = \{(x, y) \mid 2x + 3y - 6 \geq 0\} = \{(x, y) \mid y \geq -\frac{2}{3}x + 2\}$, which consists of all points on or above the line $y = -\frac{2}{3}x + 2$.

23.

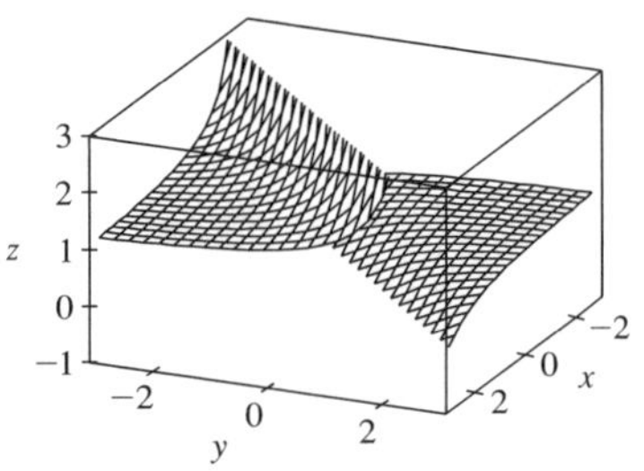

From the graph, it appears that f is discontinuous along the line $y = x$. If we consider $f(x, y) = e^{1/(x-y)}$ as a composition of functions, $g(x, y) = 1/(x - y)$ is a rational function and therefore continuous except where $x - y = 0 \Rightarrow y = x$. Since the function $h(t) = e^t$ is continuous everywhere, the composition $h(g(x, y)) = e^{1/(x-y)} = f(x, y)$ is continuous except along the line $y = x$, as we suspected.

25. $F(x, y) = \dfrac{1}{x^2 - y}$ is a rational function and thus is continuous on its domain $\{(x, y) \mid x^2 - y \neq 0\} = \{(x, y) \mid y \neq x^2\}$, so F is continuous on $\mathbb{R}^2$ except the parabola $y = x^2$.

27. $F(x, y) = \arctan(x + \sqrt{y}) = g(f(x, y))$ where $f(x, y) = x + \sqrt{y}$, continuous on its domain $\{(x, y) \mid y \geq 0\}$, and $g(t) = \arctan t$ is continuous everywhere. Thus F is continuous on its domain $\{(x, y) \mid y \geq 0\}$.

29. $f(x, y, z) = \dfrac{xyz}{x^2 + y^2 - z}$ is a rational function and thus is continuous on its domain $\{(x, y, z) \mid x^2 + y^2 - z \neq 0\} = \{(x, y, z) \mid z \neq x^2 + y^2\}$, so f is continuous on $\mathbb{R}^3$ except on the circular paraboloid $z = x^2 + y^2$.

31. $f(x, y) = \begin{cases} \dfrac{x^2y^3}{2x^2 + y^2} & \text{if } (x, y) \neq (0, 0) \\ 1 & \text{if } (x, y) = (0, 0) \end{cases}$ The first piece of f is a rational function defined everywhere except at the origin, so f is continuous on $\mathbb{R}^2$ except possibly at the origin. Since $x^2 \leq 2x^2 + y^2$, we have $\left|x^2y^3/(2x^2 + y^2)\right| \leq \left|y^3\right|$. We know that $\left|y^3\right| \to 0$ as $(x, y) \to (0, 0)$. So, by the Squeeze Theorem, $\displaystyle\lim_{(x,y)\to(0,0)} f(x, y) = \lim_{(x,y)\to(0,0)} \frac{x^2y^3}{2x^2 + y^2} = 0$. But $f(0, 0) = 1$, so f is discontinuous at $(0, 0)$. Therefore, f is continuous on the set $\{(x, y) \mid (x, y) \neq (0, 0)\}$.

33. $\displaystyle\lim_{(x,y)\to(0,0)} \frac{x^3 + y^3}{x^2 + y^2} = \lim_{r\to 0+} \frac{(r\cos\theta)^3 + (r\sin\theta)^3}{r^2} = \lim_{r\to 0+} (r\cos^3\theta + r\sin^3\theta) = 0$

35. $\displaystyle\lim_{(x,y,z)\to(0,0,0)} \frac{xyz}{x^2 + y^2 + z^2} = \lim_{\rho\to 0+} \frac{(\rho\sin\phi\cos\theta)(\rho\sin\phi\sin\theta)(\rho\cos\phi)}{\rho^2} = \lim_{\rho\to 0+} (\rho\sin^2\phi\cos\phi\sin\theta\cos\theta) = 0$

11.3 Partial Derivatives

1. (a) $\partial T/\partial x$ represents the rate of change of T when we fix y and t and consider T as a function of the single variable x, which describes how quickly the temperature changes when longitude changes but latitude and time are constant. $\partial T/\partial y$ represents the rate of change of T when we fix x and t and consider T as a function of y, which describes how quickly the temperature changes when latitude changes but longitude and time are constant. $\partial T/\partial t$ represents the rate of change of T when we fix x and y and consider T as a function of t, which describes how quickly the temperature changes over time for a constant longitude and latitude.

(b) $f_x(158, 21, 9)$ represents the rate of change of temperature at longitude 158°W, latitude 21°N at 9:00 A.M. when only longitude varies. Since the air is warmer to the west than to the east, increasing longitude results in an increased air temperature, so we would expect $f_x(158, 21, 9)$ to be positive. $f_y(158, 21, 9)$ represents the rate of change of temperature at the same time and location when only latitude varies. Since the air is warmer to the south and cooler to the north, increasing latitude results in a decreased air temperature, so we would expect $f_y(158, 21, 9)$ to be negative. $f_t(158, 21, 9)$ represents the rate of change of temperature at the same time and location when only time varies. Since typically air temperature increases from the morning to the afternoon as the sun warms it, we would expect $f_t(158, 21, 9)$ to be positive.

3. (a) By Definition 4, $f_T(-15, 30) = \lim\limits_{h\to 0} \dfrac{f(-15+h, 30) - f(-15, 30)}{h}$, which we can approximate by considering $h = 5$ and $h = -5$ and using the values given in the table:

$f_T(-15, 30) \approx \dfrac{f(-10, 30) - f(-15, 30)}{5} = \dfrac{-20 - (-26)}{5} = \dfrac{6}{5} = 1.2$,

$f_T(-15, 30) \approx \dfrac{f(-20, 30) - f(-15, 30)}{-5} = \dfrac{-33 - (-26)}{-5} = \dfrac{-7}{-5} = 1.4$. Averaging these values, we estimate $f_T(-15, 30)$ to be approximately 1.3. Thus, when the actual temperature is $-15°$C and the wind speed is 30 km/h, the apparent temperature rises by about $1.3°$C for every degree that the actual temperature rises.

Similarly, $f_v(-15, 30) = \lim\limits_{h\to 0} \dfrac{f(-15, 30+h) - f(-15, 30)}{h}$ which we can approximate by considering $h = 10$ and $h = -10$: $f_v(-15, 30) \approx \dfrac{f(-15, 40) - f(-15, 30)}{10} = \dfrac{-27 - (-26)}{10} = \dfrac{-1}{10} = -0.1$,

$f_v(-15, 30) \approx \dfrac{f(-15, 20) - f(-15, 30)}{-10} = \dfrac{-24 - (-26)}{-10} = \dfrac{2}{-10} = -0.2$. Averaging these values, we estimate $f_v(-15, 30)$ to be approximately -0.15. Thus, when the actual temperature is $-15°$C and the wind speed is 30 km/h, the apparent temperature decreases by about $0.15°$C for every km/h that the wind speed increases.

(b) For a fixed wind speed v, the values of the wind-chill index W increase as temperature T increases (look at a column of the table), so $\dfrac{\partial W}{\partial T}$ is positive. For a fixed temperature T, the values of W decrease (or remain constant) as v increases (look at a row of the table), so $\dfrac{\partial W}{\partial v}$ is negative (or perhaps 0).

(c) For fixed values of T, the function values $f(T, v)$ appear to become constant (or nearly constant) as v increases, so the corresponding rate of change is 0 or near 0 as v increases. This suggests that $\lim\limits_{v\to\infty} (\partial W/\partial v) = 0$.

5. (a) If we start at $(1, 2)$ and move in the positive x-direction, the graph of f increases. Thus $f_x(1, 2)$ is positive.

(b) If we start at $(1, 2)$ and move in the positive y-direction, the graph of f decreases. Thus $f_y(1, 2)$ is negative.

7. First of all, if we start at the point $(3, -3)$ and move in the positive y-direction, we see that both b and c decrease, while a increases. Both b and c have a low point at about $(3, -1.5)$, while a is 0 at this point. So a is definitely the graph of f_y, and one of b and c is the graph of f. To see which is which, we start at the point $(-3, -1.5)$ and move in the positive x-direction. b traces out a line with negative slope, while c traces out a parabola opening downward. This tells us that b is the x-derivative of c. So c is the graph of f, b is the graph of f_x, and a is the graph of f_y.

9. $f(x, y) = 16 - 4x^2 - y^2 \Rightarrow f_x(x, y) = -8x$ and $f_y(x, y) = -2y \Rightarrow f_x(1, 2) = -8$ and $f_y(1, 2) = -4$. The graph of f is the paraboloid $z = 16 - 4x^2 - y^2$ and the vertical plane $y = 2$ intersects it in the parabola $z = 12 - 4x^2$, $y = 2$ (the curve C_1 in the first figure).

The slope of the tangent line to this parabola at $(1, 2, 8)$ is $f_x(1, 2) = -8$. Similarly the plane $x = 1$ intersects the paraboloid in the parabola $z = 12 - y^2$, $x = 1$ (the curve C_2 in the second figure) and the slope of the tangent line at $(1, 2, 8)$ is $f_y(1, 2) = -4$.

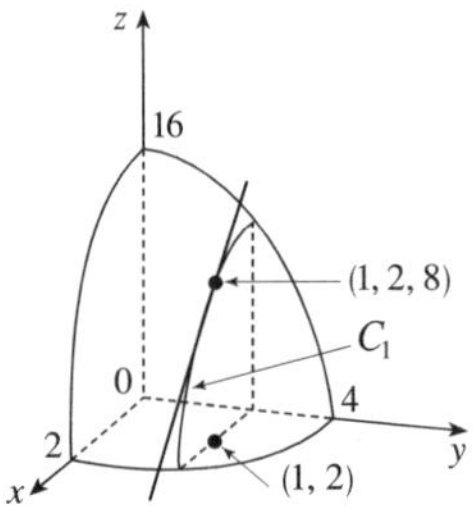

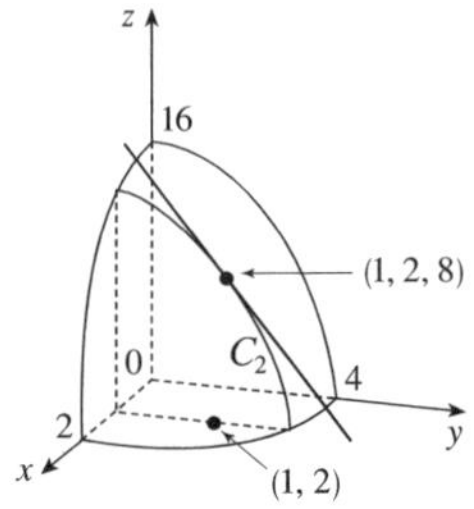

11. $f(x, y) = x^2 + y^2 + x^2y \Rightarrow f_x = 2x + 2xy$, $f_y = 2y + x^2$

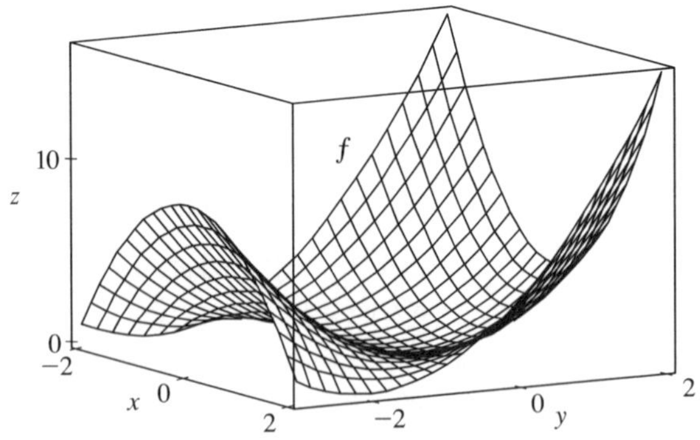

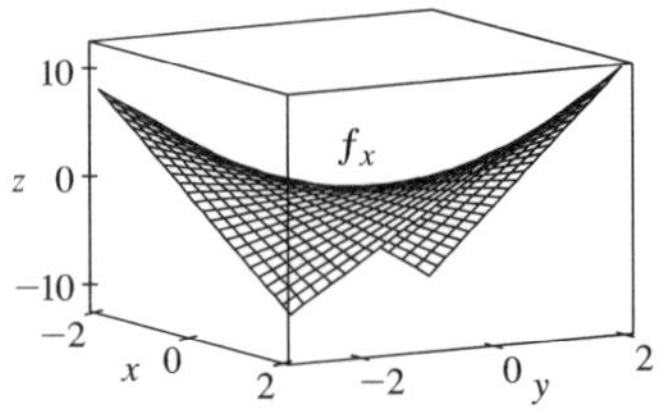

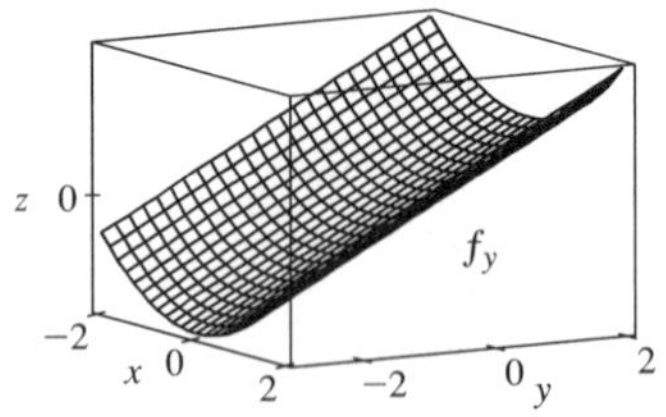

Note that the traces of f in planes parallel to the xz-plane are parabolas which open downward for $y < -1$ and upward for $y > -1$, and the traces of f_x in these planes are straight lines, which have negative slopes for $y < -1$ and positive slopes for $y > -1$. The traces of f in planes parallel to the yz-plane are parabolas which always open upward, and the traces of f_y in these planes are straight lines with positive slopes.

13. $f(x, y) = 3x - 2y^4 \Rightarrow f_x(x, y) = 3 - 0 = 3$, $f_y(x, y) = 0 - 8y^3 = -8y^3$

15. $z = xe^{3y} \Rightarrow \dfrac{\partial z}{\partial x} = e^{3y}$, $\dfrac{\partial z}{\partial y} = 3xe^{3y}$

17. $f(x,y) = \dfrac{x-y}{x+y} \quad\Rightarrow\quad f_x(x,y) = \dfrac{(1)(x+y)-(x-y)(1)}{(x+y)^2} = \dfrac{2y}{(x+y)^2}$,

$f_y(x,y) = \dfrac{(-1)(x+y)-(x-y)(1)}{(x+y)^2} = -\dfrac{2x}{(x+y)^2}$

19. $w = \sin\alpha\cos\beta \quad\Rightarrow\quad \dfrac{\partial w}{\partial \alpha} = \cos\alpha\cos\beta,\ \dfrac{\partial w}{\partial \beta} = -\sin\alpha\sin\beta$

21. $f(u,v) = \tan^{-1}\left(\dfrac{u}{v}\right) \quad\Rightarrow\quad f_u(u,v) = \dfrac{1}{1+(u/v)^2}\left(\dfrac{1}{v}\right) = \dfrac{1}{v}\left(\dfrac{v^2}{u^2+v^2}\right) = \dfrac{v}{u^2+v^2}$,

$f_v(u,v) = \dfrac{1}{1+(u/v)^2}\left(-\dfrac{u}{v^2}\right) = -\dfrac{u}{v^2}\left(\dfrac{v^2}{u^2+v^2}\right) = -\dfrac{u}{u^2+v^2}$

23. $z = \ln\left(x+\sqrt{x^2+y^2}\right) \quad\Rightarrow$

$$\frac{\partial z}{\partial x} = \frac{1}{x+\sqrt{x^2+y^2}}\left[1+\tfrac{1}{2}(x^2+y^2)^{-1/2}(2x)\right] = \frac{\left(\sqrt{x^2+y^2}+x\right)\Big/\sqrt{x^2+y^2}}{\left(x+\sqrt{x^2+y^2}\right)} = \frac{1}{\sqrt{x^2+y^2}},$$

$$\frac{\partial z}{\partial y} = \frac{1}{x+\sqrt{x^2+y^2}}\left(\frac{1}{2}\right)(x^2+y^2)^{-1/2}(2y) = \frac{y}{x\sqrt{x^2+y^2}+x^2+y^2}$$

25. $f(x,y,z) = xy^2z^3+3yz \quad\Rightarrow\quad f_x(x,y,z) = y^2z^3,\ f_y(x,y,z) = 2xyz^3+3z,\ f_z(x,y,z) = 3xy^2z^2+3y$

27. $w = \ln(x+2y+3z) \quad\Rightarrow\quad \dfrac{\partial w}{\partial x} = \dfrac{1}{x+2y+3z},\ \dfrac{\partial w}{\partial y} = \dfrac{2}{x+2y+3z},\ \dfrac{\partial w}{\partial z} = \dfrac{3}{x+2y+3z}$

29. $u = xe^{-t}\sin\theta \quad\Rightarrow\quad \dfrac{\partial u}{\partial x} = e^{-t}\sin\theta,\ \dfrac{\partial u}{\partial t} = -xe^{-t}\sin\theta,\ \dfrac{\partial u}{\partial \theta} = xe^{-t}\cos\theta$

31. $f(x,y,z,t) = xyz^2\tan(yt) \quad\Rightarrow\quad f_x(x,y,z,t) = yz^2\tan(yt)$,

$f_y(x,y,z,t) = xyz^2\cdot\sec^2(yt)\cdot t + xz^2\tan(yt) = xyz^2t\sec^2(yt)+xz^2\tan(yt)$,

$f_z(x,y,z,t) = 2xyz\tan(yt),\ f_t(x,y,z,t) = xyz^2\sec^2(yt)\cdot y = xy^2z^2\sec^2(yt)$.

33. $u = \sqrt{x_1^2+x_2^2+\cdots+x_n^2}$. For each $i = 1,\ldots,n$, $u_{x_i} = \frac{1}{2}\left(x_1^2+x_2^2+\cdots+x_n^2\right)^{-1/2}(2x_i) = \dfrac{x_i}{\sqrt{x_1^2+x_2^2+\cdots+x_n^2}}$.

35. $f(x,y) = \sqrt{x^2+y^2} \quad\Rightarrow\quad f_x(x,y) = \frac{1}{2}(x^2+y^2)^{-1/2}(2x) = \dfrac{x}{\sqrt{x^2+y^2}}$, so $f_x(3,4) = \dfrac{3}{\sqrt{3^2+4^2}} = \dfrac{3}{5}$.

37. $f(x,y,z) = \dfrac{x}{y+z} = x(y+z)^{-1} \quad\Rightarrow\quad f_z(x,y,z) = x(-1)(y+z)^{-2} = -\dfrac{x}{(y+z)^2}$, so

$f_z(3,2,1) = -\dfrac{3}{(2+1)^2} = -\dfrac{1}{3}$.

39. $f(x,y) = xy^2-x^3y \quad\Rightarrow$

$$f_x(x,y) = \lim_{h\to 0}\frac{f(x+h,y)-f(x,y)}{h} = \lim_{h\to 0}\frac{(x+h)y^2-(x+h)^3y-(xy^2-x^3y)}{h}$$

$$= \lim_{h\to 0}\frac{h(y^2-3x^2y-3xyh-yh^2)}{h} = \lim_{h\to 0}(y^2-3x^2y-3xyh-yh^2) = y^2-3x^2y$$

$$f_y(x,y) = \lim_{h\to 0}\frac{f(x,y+h)-f(x,y)}{h} = \lim_{h\to 0}\frac{x(y+h)^2-x^3(y+h)-(xy^2-x^3y)}{h}$$

$$= \lim_{h\to 0}\frac{h(2xy+xh-x^3)}{h} = \lim_{h\to 0}(2xy+xh-x^3) = 2xy-x^3$$

41. $x^2+y^2+z^2=3xyz \;\Rightarrow\; \dfrac{\partial}{\partial x}\left(x^2+y^2+z^2\right)=\dfrac{\partial}{\partial x}(3xyz) \;\Rightarrow\; 2x+0+2z\,\dfrac{\partial z}{\partial x}=3y\left(x\,\dfrac{\partial z}{\partial x}+z\cdot 1\right) \;\Leftrightarrow$

$2z\,\dfrac{\partial z}{\partial x}-3xy\,\dfrac{\partial z}{\partial x}=3yz-2x \;\Leftrightarrow\; (2z-3xy)\,\dfrac{\partial z}{\partial x}=3yz-2x$, so $\dfrac{\partial z}{\partial x}=\dfrac{3yz-2x}{2z-3xy}$.

$\dfrac{\partial}{\partial y}\left(x^2+y^2+z^2\right)=\dfrac{\partial}{\partial y}(3xyz) \;\Rightarrow\; 0+2y+2z\,\dfrac{\partial z}{\partial y}=3x\left(y\,\dfrac{\partial z}{\partial y}+z\cdot 1\right) \;\Leftrightarrow\; 2z\,\dfrac{\partial z}{\partial y}-3xy\,\dfrac{\partial z}{\partial y}=3xz-2y \;\Leftrightarrow$

$(2z-3xy)\,\dfrac{\partial z}{\partial y}=3xz-2y$, so $\dfrac{\partial z}{\partial y}=\dfrac{3xz-2y}{2z-3xy}$.

43. $x-z=\arctan(yz) \;\Rightarrow\; \dfrac{\partial}{\partial x}(x-z)=\dfrac{\partial}{\partial x}(\arctan(yz)) \;\Rightarrow\; 1-\dfrac{\partial z}{\partial x}=\dfrac{1}{1+(yz)^2}\cdot y\,\dfrac{\partial z}{\partial x} \;\Leftrightarrow$

$1=\left(\dfrac{y}{1+y^2z^2}+1\right)\dfrac{\partial z}{\partial x} \;\Leftrightarrow\; 1=\left(\dfrac{y+1+y^2z^2}{1+y^2z^2}\right)\dfrac{\partial z}{\partial x}$, so $\dfrac{\partial z}{\partial x}=\dfrac{1+y^2z^2}{1+y+y^2z^2}$.

$\dfrac{\partial}{\partial y}(x-z)=\dfrac{\partial}{\partial y}(\arctan(yz)) \;\Rightarrow\; 0-\dfrac{\partial z}{\partial y}=\dfrac{1}{1+(yz)^2}\cdot\left(y\,\dfrac{\partial z}{\partial y}+z\cdot 1\right) \;\Leftrightarrow$

$-\dfrac{z}{1+y^2z^2}=\left(\dfrac{y}{1+y^2z^2}+1\right)\dfrac{\partial z}{\partial y} \;\Leftrightarrow\; -\dfrac{z}{1+y^2z^2}=\left(\dfrac{y+1+y^2z^2}{1+y^2z^2}\right)\dfrac{\partial z}{\partial y} \;\Leftrightarrow\; \dfrac{\partial z}{\partial y}=-\dfrac{z}{1+y+y^2z^2}$.

45. (a) $z=f(x)+g(y) \;\Rightarrow\; \dfrac{\partial z}{\partial x}=f'(x),\ \dfrac{\partial z}{\partial y}=g'(y)$

(b) $z=f(x+y)$. Let $u=x+y$. Then $\dfrac{\partial z}{\partial x}=\dfrac{df}{du}\dfrac{\partial u}{\partial x}=\dfrac{df}{du}(1)=f'(u)=f'(x+y)$,

$\dfrac{\partial z}{\partial y}=\dfrac{df}{du}\dfrac{\partial u}{\partial y}=\dfrac{df}{du}(1)=f'(u)=f'(x+y)$.

47. $f(x,y)=x^4-3x^2y^3 \;\Rightarrow\; f_x(x,y)=4x^3-6xy^3,\ f_y(x,y)=-9x^2y^2$. Then $f_{xx}(x,y)=12x^2-6y^3$,

$f_{xy}(x,y)=-18xy^2$, $f_{yx}(x,y)=-18xy^2$, and $f_{yy}(x,y)=-18x^2y$.

49. $z=\dfrac{x}{x+y}=x(x+y)^{-1} \;\Rightarrow\; z_x=\dfrac{1(x+y)-1(x)}{(x+y)^2}=\dfrac{y}{(x+y)^2},\ z_y=x(-1)(x+y)^{-2}=-\dfrac{x}{(x+y)^2}$. Then

$z_{xx}=y(-2)(x+y)^{-3}=-\dfrac{2y}{(x+y)^3},\ z_{xy}=\dfrac{1(x+y)^2-y(2)(x+y)}{[(x+y)^2]^2}=\dfrac{x+y-2y}{(x+y)^3}=\dfrac{x-y}{(x+y)^3}$,

$z_{yx}=-\dfrac{1(x+y)^2-x(2)(x+y)}{[(x+y)^2]^2}=-\dfrac{-x^2+xy+y^2}{(x+y)^2}=\dfrac{(x+y)(x-y)}{(x+y)^2}=\dfrac{x-y}{(x+y)^3}$, and

$z_{yy}=-x(-2)(x+y)^{-3}=\dfrac{2x}{(x+y)^3}$.

51. $u=e^{-s}\sin t \;\Rightarrow\; u_s=-e^{-s}\sin t,\ u_t=e^{-s}\cos t$. Then $u_{ss}=e^{-s}\sin t,\ u_{st}=-e^{-s}\cos t,\ u_{ts}=-e^{-s}\cos t$,

and $u_{tt}=-e^{-s}\sin t$.

53. $u=x\sin(x+2y) \;\Rightarrow\; u_x=x\cdot\cos(x+2y)(1)+\sin(x+2y)\cdot 1=x\cos(x+2y)+\sin(x+2y)$,

$u_{xy}=x(-\sin(x+2y)(2))+\cos(x+2y)(2)=2\cos(x+2y)-2x\sin(x+2y)$

and $u_y=x\cos(x+2y)(2)=2x\cos(x+2y)$,

$u_{yx}=2x\cdot(-\sin(x+2y)(1))+\cos(x+2y)\cdot 2=2\cos(x+2y)-2x\sin(x+2y)$. Thus $u_{xy}=u_{yx}$.

55. $f(x,y)=3xy^4+x^3y^2 \;\Rightarrow\; f_x=3y^4+3x^2y^2,\ f_{xx}=6xy^2,\ f_{xxy}=12xy$ and $f_y=12xy^3+2x^3y$,

$f_{yy}=36xy^2+2x^3,\ f_{yyy}=72xy$.

57. $f(x, y, z) = \cos(4x + 3y + 2z) \Rightarrow$

$f_x = -\sin(4x + 3y + 2z)(4) = -4\sin(4x + 3y + 2z)$,

$f_{xy} = -4\cos(4x + 3y + 2z)(3) = -12\cos(4x + 3y + 2z)$,

$f_{xyz} = -12(-\sin(4x + 3y + 2z))(2) = 24\sin(4x + 3y + 2z)$ and

$f_y = -\sin(4x + 3y + 2z)(3) = -3\sin(4x + 3y + 2z)$,

$f_{yz} = -3\cos(4x + 3y + 2z)(2) = -6\cos(4x + 3y + 2z)$,

$f_{yzz} = -6(-\sin(4x + 3y + 2z))(2) = 12\sin(4x + 3y + 2z)$.

59. $u = e^{r\theta}\sin\theta \Rightarrow \dfrac{\partial u}{\partial\theta} = e^{r\theta}\cos\theta + \sin\theta \cdot e^{r\theta}(r) = e^{r\theta}(\cos\theta + r\sin\theta)$,

$\dfrac{\partial^2 u}{\partial r\,\partial\theta} = e^{r\theta}(\sin\theta) + (\cos\theta + r\sin\theta)\,e^{r\theta}(\theta) = e^{r\theta}(\sin\theta + \theta\cos\theta + r\theta\sin\theta)$,

$\dfrac{\partial^3 u}{\partial r^2\,\partial\theta} = e^{r\theta}(\theta\sin\theta) + (\sin\theta + \theta\cos\theta + r\theta\sin\theta)\cdot e^{r\theta}(\theta) = \theta e^{r\theta}(2\sin\theta + \theta\cos\theta + r\theta\sin\theta)$.

61. By Definition 4, $f_x(3,2) = \lim\limits_{h\to 0}\dfrac{f(3+h,2) - f(3,2)}{h}$ which we can approximate by considering $h = 0.5$ and $h = -0.5$: $f_x(3,2) \approx \dfrac{f(3.5,2) - f(3,2)}{0.5} = \dfrac{22.4 - 17.5}{0.5} = 9.8$,

$f_x(3,2) \approx \dfrac{f(2.5,2) - f(3,2)}{-0.5} = \dfrac{10.2 - 17.5}{-0.5} = 14.6$. Averaging these values, we estimate $f_x(3,2)$ to be approximately 12.2. Similarly, $f_x(3,2.2) = \lim\limits_{h\to 0}\dfrac{f(3+h,2.2) - f(3,2.2)}{h}$ which we can approximate by considering $h = 0.5$ and $h = -0.5$: $f_x(3,2.2) \approx \dfrac{f(3.5,2.2) - f(3,2.2)}{0.5} = \dfrac{26.1 - 15.9}{0.5} = 20.4$,

$f_x(3,2.2) \approx \dfrac{f(2.5,2.2) - f(3,2.2)}{-0.5} = \dfrac{9.3 - 15.9}{-0.5} = 13.2$. Averaging these values, we have $f_x(3,2.2) \approx 16.8$.

To estimate $f_{xy}(3,2)$, we first need an estimate for $f_x(3,1.8)$: $f_x(3,1.8) \approx \dfrac{f(3.5,1.8) - f(3,1.8)}{0.5} = \dfrac{20.0 - 18.1}{0.5} = 3.8$,

$f_x(3,1.8) \approx \dfrac{f(2.5,1.8) - f(3,1.8)}{-0.5} = \dfrac{12.5 - 18.1}{-0.5} = 11.2$. Averaging these values, we get $f_x(3,1.8) \approx 7.5$.

Now $f_{xy}(x,y) = \dfrac{\partial}{\partial y}[f_x(x,y)]$ and $f_x(x,y)$ is itself a function of 2 variables, so Definition 4 says that

$$f_{xy}(x,y) = \frac{\partial}{\partial y}[f_x(x,y)] = \lim_{h\to 0}\frac{f_x(x,y+h) - f_x(x,y)}{h} \quad\Rightarrow\quad f_{xy}(3,2) = \lim_{h\to 0}\frac{f_x(3,2+h) - f_x(3,2)}{h}.$$

We can estimate this value using our previous work with $h = 0.2$ and $h = -0.2$:

$f_{xy}(3,2) \approx \dfrac{f_x(3,2.2) - f_x(3,2)}{0.2} = \dfrac{16.8 - 12.2}{0.2} = 23$, $f_{xy}(3,2) \approx \dfrac{f_x(3,1.8) - f_x(3,2)}{-0.2} = \dfrac{7.5 - 12.2}{-0.2} = 23.5$.

Averaging these values, we estimate $f_{xy}(3,2)$ to be approximately 23.25.

63. $u = e^{-\alpha^2 k^2 t}\sin kx \Rightarrow u_x = ke^{-\alpha^2 k^2 t}\cos kx$, $u_{xx} = -k^2 e^{-\alpha^2 k^2 t}\sin kx$, and $u_t = -\alpha^2 k^2 e^{-\alpha^2 k^2 t}\sin kx$.

Thus $\alpha^2 u_{xx} = u_t$.

65. $u = \dfrac{1}{\sqrt{x^2+y^2+z^2}} \Rightarrow u_x = \left(-\frac{1}{2}\right)(x^2+y^2+z^2)^{-3/2}(2x) = -x(x^2+y^2+z^2)^{-3/2}$ and

$$u_{xx} = -(x^2+y^2+z^2)^{-3/2} - x\left(-\tfrac{3}{2}\right)(x^2+y^2+z^2)^{-5/2}(2x) = \frac{2x^2-y^2-z^2}{(x^2+y^2+z^2)^{5/2}}.$$

By symmetry, $u_{yy} = \dfrac{2y^2-x^2-z^2}{(x^2+y^2+z^2)^{5/2}}$ and $u_{zz} = \dfrac{2z^2-x^2-y^2}{(x^2+y^2+z^2)^{5/2}}$.

Thus $u_{xx}+u_{yy}+u_{zz} = \dfrac{2x^2-y^2-z^2+2y^2-x^2-z^2+2z^2-x^2-y^2}{(x^2+y^2+z^2)^{5/2}} = 0$.

67. Let $v = x+at$, $w = x-at$. Then $u_t = \dfrac{\partial[f(v)+g(w)]}{\partial t} = \dfrac{df(v)}{dv}\dfrac{\partial v}{\partial t} + \dfrac{dg(w)}{dw}\dfrac{\partial w}{\partial t} = af'(v) - ag'(w)$ and

$u_{tt} = \dfrac{\partial[af'(v)-ag'(w)]}{\partial t} = a[af''(v)+ag''(w)] = a^2[f''(v)+g''(w)]$. Similarly, by using the Chain Rule we have

$u_x = f'(v)+g'(w)$ and $u_{xx} = f''(v)+g''(w)$. Thus $u_{tt} = a^2 u_{xx}$.

69. If we fix $K = K_0$, $P(L,K_0)$ is a function of a single variable L, and $\dfrac{dP}{dL} = \alpha\dfrac{P}{L}$ is a separable differential equation. Then

$\dfrac{dP}{P} = \alpha\dfrac{dL}{L} \Rightarrow \displaystyle\int\frac{dP}{P} = \int\alpha\frac{dL}{L} \Rightarrow \ln|P| = \alpha\ln|L| + C(K_0)$, where $C(K_0)$ can depend on K_0. Then

$|P| = e^{\alpha\ln|L|+C(K_0)}$, and since $P>0$ and $L>0$, we have $P = e^{\alpha\ln L}e^{C(K_0)} = e^{C(K_0)}e^{\ln L^\alpha} = C_1(K_0)L^\alpha$ where

$C_1(K_0) = e^{C(K_0)}$.

71. By the Chain Rule, taking the partial derivative of both sides with respect to R_1 gives

$\dfrac{\partial R^{-1}}{\partial R}\dfrac{\partial R}{\partial R_1} = \dfrac{\partial[(1/R_1)+(1/R_2)+(1/R_3)]}{\partial R_1}$ or $-R^{-2}\dfrac{\partial R}{\partial R_1} = -R_1^{-2}$. Thus $\dfrac{\partial R}{\partial R_1} = \dfrac{R^2}{R_1^2}$.

73. By Exercise 72, $PV = mRT \Rightarrow P = \dfrac{mRT}{V}$, so $\dfrac{\partial P}{\partial T} = \dfrac{mR}{V}$. Also, $PV = mRT \Rightarrow V = \dfrac{mRT}{P}$

and $\dfrac{\partial V}{\partial T} = \dfrac{mR}{P}$. Since $T = \dfrac{PV}{mR}$, we have $T\dfrac{\partial P}{\partial T}\dfrac{\partial V}{\partial T} = \dfrac{PV}{mR}\cdot\dfrac{mR}{V}\cdot\dfrac{mR}{P} = mR$.

75. $\dfrac{\partial K}{\partial m} = \frac{1}{2}v^2$, $\dfrac{\partial K}{\partial v} = mv$, $\dfrac{\partial^2 K}{\partial v^2} = m$. Thus $\dfrac{\partial K}{\partial m}\cdot\dfrac{\partial^2 K}{\partial v^2} = \frac{1}{2}v^2 m = K$.

77. $f_x(x,y) = x+4y \Rightarrow f_{xy}(x,y) = 4$ and $f_y(x,y) = 3x-y \Rightarrow f_{yx}(x,y) = 3$. Since f_{xy} and f_{yx} are continuous everywhere but $f_{xy}(x,y) \neq f_{yx}(x,y)$, Clairaut's Theorem implies that such a function $f(x,y)$ does not exist.

79. By the geometry of partial derivatives, the slope of the tangent line is $f_x(1,2)$. By implicit differentiation of $4x^2+2y^2+z^2 = 16$, we get $8x + 2z\,(\partial z/\partial x) = 0 \Rightarrow \partial z/\partial x = -4x/z$, so when $x=1$ and $z=2$ we have $\partial z/\partial x = -2$. So the slope is $f_x(1,2) = -2$. Thus the tangent line is given by $z-2 = -2(x-1)$, $y=2$. Taking the parameter to be $t = x-1$, we can write parametric equations for this line: $x = 1+t$, $y=2$, $z = 2-2t$.

81. Let $g(x) = f(x,0) = x(x^2)^{-3/2}e^0 = x\,|x|^{-3}$. But we are using the point $(1,0)$, so near $(1,0)$, $g(x) = x^{-2}$. Then $g'(x) = -2x^{-3}$ and $g'(1) = -2$, so using (1) we have $f_x(1,0) = g'(1) = -2$.

83. (a)

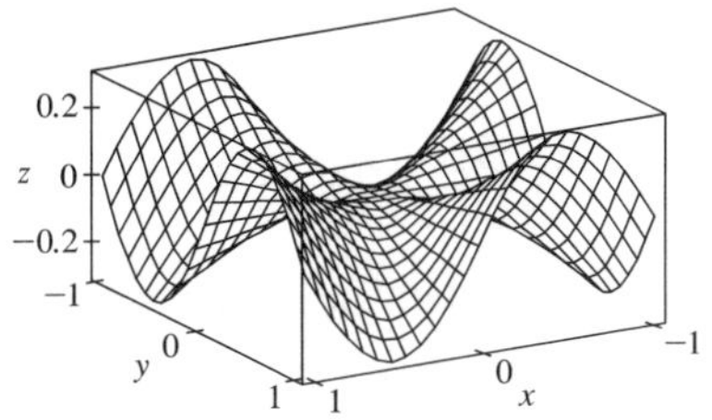

(b) For $(x,y) \neq (0,0)$, $f_x(x,y) = \dfrac{(3x^2y - y^3)(x^2+y^2) - (x^3y - xy^3)(2x)}{(x^2+y^2)^2} = \dfrac{x^4y + 4x^2y^3 - y^5}{(x^2+y^2)^2}$, and by symmetry

$$f_y(x,y) = \frac{x^5 - 4x^3y^2 - xy^4}{(x^2+y^2)^2}.$$

(c) $f_x(0,0) = \lim\limits_{h\to 0} \dfrac{f(h,0) - f(0,0)}{h} = \lim\limits_{h\to 0} \dfrac{(0/h^2) - 0}{h} = 0$ and $f_y(0,0) = \lim\limits_{h\to 0} \dfrac{f(0,h) - f(0,0)}{h} = 0.$

(d) By (3), $f_{xy}(0,0) = \dfrac{\partial f_x}{\partial y} = \lim\limits_{h\to 0} \dfrac{f_x(0,h) - f_x(0,0)}{h} = \lim\limits_{h\to 0} \dfrac{(-h^5 - 0)/h^4}{h} = -1$ while by (2),

$$f_{yx}(0,0) = \frac{\partial f_y}{\partial x} = \lim_{h\to 0} \frac{f_y(h,0) - f_y(0,0)}{h} = \lim_{h\to 0} \frac{h^5/h^4}{h} = 1.$$

(e) For $(x,y) \neq (0,0)$, we use a CAS to compute

$$f_{xy}(x,y) = \frac{x^6 + 9x^4y^2 - 9x^2y^4 - y^6}{(x^2+y^2)^3}.$$

Now as $(x,y) \to (0,0)$ along the x-axis, $f_{xy}(x,y) \to 1$ while as $(x,y) \to (0,0)$ along the y-axis, $f_{xy}(x,y) \to -1$. Thus f_{xy} isn't continuous at $(0,0)$ and Clairaut's Theorem doesn't apply, so there is no contradiction. The graphs of f_{xy} and f_{yx} are identical except at the origin, where we observe the discontinuity.

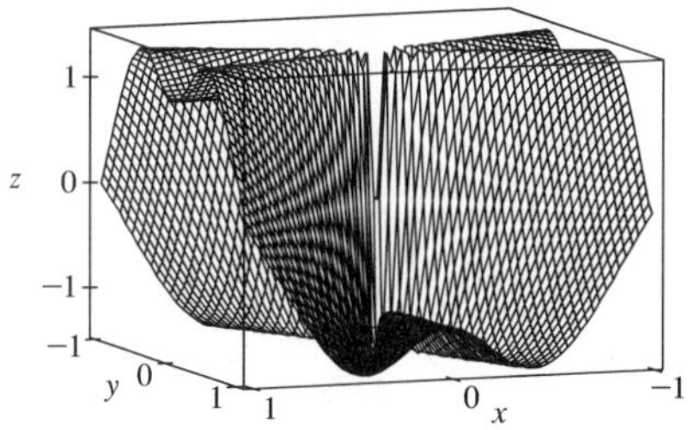

11.4 Tangent Planes and Linear Approximations

1. $z = f(x,y) = 4x^2 - y^2 + 2y \;\Rightarrow\; f_x(x,y) = 8x$, $f_y(x,y) = -2y + 2$, so $f_x(-1,2) = -8$, $f_y(-1,2) = -2$. By Equation 2, an equation of the tangent plane is $z - 4 = f_x(-1,2)[x - (-1)] + f_y(-1,2)(y-2) \;\Rightarrow$ $z - 4 = -8(x+1) - 2(y-2)$ or $z = -8x - 2y$.

3. $z = f(x,y) = y\cos(x-y) \;\Rightarrow\; f_x = y(-\sin(x-y)(1)) = -y\sin(x-y)$, $f_y = y(-\sin(x-y)(-1)) + \cos(x-y) = y\sin(x-y) + \cos(x-y)$, so $f_x(2,2) = -2\sin(0) = 0$, $f_y(2,2) = 2\sin(0) + \cos(0) = 1$ and an equation of the tangent plane is $z - 2 = 0(x-2) + 1(y-2)$ or $z = y$.

5. $z = f(x,y) = x^2 + xy + 3y^2$, so $f_x(x,y) = 2x + y \;\Rightarrow\; f_x(1,1) = 3$, $f_y(x,y) = x + 6y \;\Rightarrow\; f_y(1,1) = 7$ and an equation of the tangent plane is $z - 5 = 3(x-1) + 7(y-1)$ or $z = 3x + 7y - 5$. After zooming in, the surface and the tangent plane become almost indistinguishable. (Here, the tangent plane is below the surface.) If we zoom in further, the surface and the tangent plane will appear to coincide.

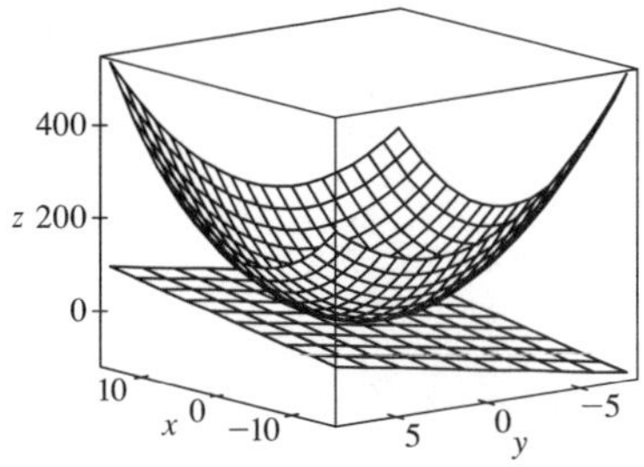

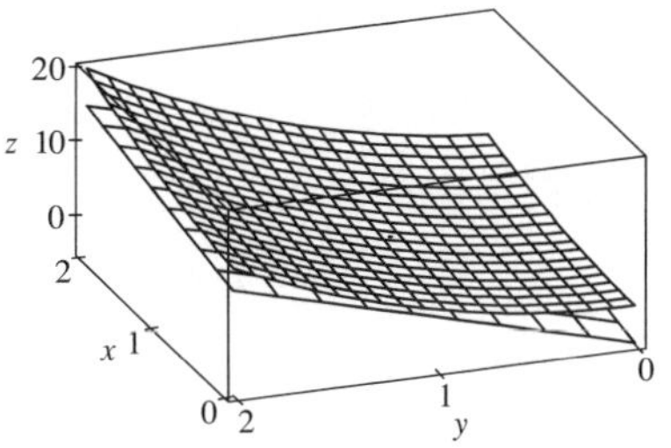

7. $f(x,y) = \dfrac{xy\sin(x-y)}{1+x^2+y^2}$. A CAS gives $f_x(x,y) = \dfrac{y\sin(x-y)+xy\cos(x-y)}{1+x^2+y^2} - \dfrac{2x^2y\sin(x-y)}{(1+x^2+y^2)^2}$ and

$f_y(x,y) = \dfrac{x\sin(x-y)-xy\cos(x-y)}{1+x^2+y^2} - \dfrac{2xy^2\sin(x-y)}{(1+x^2+y^2)^2}$. We use the CAS to evaluate these at $(1,1)$, and then substitute the results into Equation 2 to compute an equation of the tangent plane: $z = \frac{1}{3}x - \frac{1}{3}y$. The surface and tangent plane are shown in the first graph below. After zooming in, the surface and the tangent plane become almost indistinguishable, as shown in the second graph. (Here, the tangent plane is shown with fewer traces than the surface.) If we zoom in further, the surface and the tangent plane will appear to coincide.

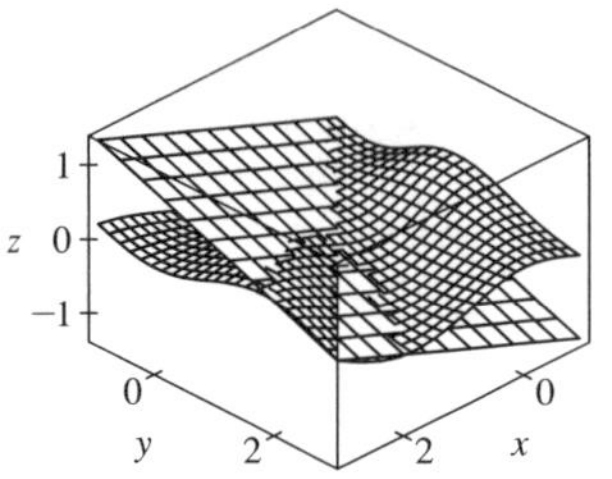

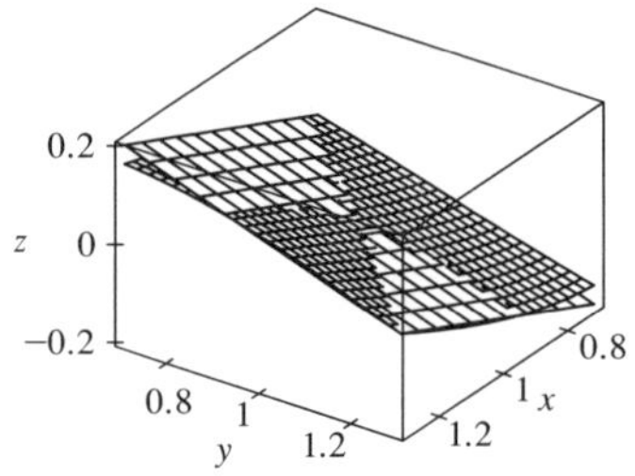

9. $f(x,y) = x\sqrt{y}$. The partial derivatives are $f_x(x,y) = \sqrt{y}$ and $f_y(x,y) = \dfrac{x}{2\sqrt{y}}$, so $f_x(1,4) = 2$ and $f_y(1,4) = \frac{1}{4}$. Both f_x and f_y are continuous functions for $y > 0$, so by Theorem 8, f is differentiable at $(1,4)$. By Equation 3, the linearization of f at $(1,4)$ is given by $L(x,y) = f(1,4) + f_x(1,4)(x-1) + f_y(1,4)(y-4) = 2 + 2(x-1) + \frac{1}{4}(y-4) = 2x + \frac{1}{4}y - 1$.

11. $f(x,y) = \tan^{-1}(x+2y)$. The partial derivatives are $f_x(x,y) = \dfrac{1}{1+(x+2y)^2}$ and $f_y(x,y) = \dfrac{2}{1+(x+2y)^2}$, so $f_x(1,0) = \frac{1}{2}$ and $f_y(1,0) = 1$. Both f_x and f_y are continuous functions, so f is differentiable at $(1,0)$, and the linearization of f at $(1,0)$ is $L(x,y) = f(1,0) + f_x(1,0)(x-1) + f_y(1,0)(y-0) = \frac{\pi}{4} + \frac{1}{2}(x-1) + 1(y) = \frac{1}{2}x + y + \frac{\pi}{4} - \frac{1}{2}$.

13. $f(x,y) = \sqrt{20-x^2-7y^2} \;\Rightarrow\; f_x(x,y) = -\dfrac{x}{\sqrt{20-x^2-7y^2}}$ and $f_y(x,y) = -\dfrac{7y}{\sqrt{20-x^2-7y^2}}$, so $f_x(2,1) = -\frac{2}{3}$ and $f_y(2,1) = -\frac{7}{3}$. Then the linear approximation of f at $(2,1)$ is given by $f(x,y) \approx f(2,1) + f_x(2,1)(x-2) + f_y(2,1)(y-1) = 3 - \frac{2}{3}(x-2) - \frac{7}{3}(y-1) = -\frac{2}{3}x - \frac{7}{3}y + \frac{20}{3}$. Thus $f(1.95, 1.08) \approx -\frac{2}{3}(1.95) - \frac{7}{3}(1.08) + \frac{20}{3} = 2.84\overline{6}$.

15. $f(x,y,z) = \sqrt{x^2+y^2+z^2} \;\Rightarrow\; f_x(x,y,z) = \dfrac{x}{\sqrt{x^2+y^2+z^2}}$, $f_y(x,y,z) = \dfrac{y}{\sqrt{x^2+y^2+z^2}}$, and $f_z(x,y,z) = \dfrac{z}{\sqrt{x^2+y^2+z^2}}$, so $f_x(3,2,6) = \frac{3}{7}$, $f_y(3,2,6) = \frac{2}{7}$, and $f_z(3,2,6) = \frac{6}{7}$. Then the linear approximation of f at $(3,2,6)$ is given by

$$\begin{aligned} f(x,y,z) &\approx f(3,2,6) + f_x(3,2,6)(x-3) + f_y(3,2,6)(y-2) + f_z(3,2,6)(z-6) \\ &= 7 + \tfrac{3}{7}(x-3) + \tfrac{2}{7}(y-2) + \tfrac{6}{7}(z-6) = \tfrac{3}{7}x + \tfrac{2}{7}y + \tfrac{6}{7}z \end{aligned}$$

Thus $\sqrt{(3.02)^2 + (1.97)^2 + (5.99)^2} = f(3.02, 1.97, 5.99) \approx \frac{3}{7}(3.02) + \frac{2}{7}(1.97) + \frac{6}{7}(5.99) \approx 6.9914$.

17. From the table, $f(32,65)=43$. To estimate $f_T(32,65)$ and $f_H(32,65)$ we follow the procedure used in Section 11.3. Since $f_T(32,65)=\lim\limits_{h\to 0}\dfrac{f(32+h,65)-f(32,65)}{h}$, we approximate this quantity with $h=\pm 2$ and use the values given in the table: $f_T(32,65)\approx\dfrac{f(34,65)-f(32,65)}{2}=\dfrac{48-43}{2}=2.5$, $f_T(32,65)\approx\dfrac{f(30,65)-f(32,65)}{-2}=\dfrac{40-43}{-2}=1.5$.

Averaging these values gives $f_T(32,65)\approx 2$. Similarly, $f_H(32,65)=\lim\limits_{h\to 0}\dfrac{f(32,65+h)-f(32,65)}{h}$, so we use $h=\pm 5$:

$f_H(32,65)\approx\dfrac{f(32,70)-f(32,65)}{5}=\dfrac{45-43}{5}=0.4$, $f_H(32,65)\approx\dfrac{f(32,60)-f(32,65)}{-5}=\dfrac{42-43}{-5}=0.2$.

Averaging these values gives $f_H(32,65)\approx 0.3$. The linear approximation, then, is

$$\begin{aligned} f(T,H) &\approx f(32,65)+f_T(32,65)(T-32)+f_H(32,65)(H-65) \\ &\approx 43+2(T-32)+0.3(H-65) \qquad [\text{or } 2T+0.3H-40.5] \end{aligned}$$

Thus when $T=33$ and $H=63$, $f(33,63)\approx 43+2(33-32)+0.3(63-65)=44.4$, so we estimate the humidex to be approximately $44.4°\text{C}$.

19. $z=x^3\ln(y^2)\quad\Rightarrow\quad dz=\dfrac{\partial z}{\partial x}dx+\dfrac{\partial z}{\partial y}dy=3x^2\ln(y^2)\,dx+x^3\cdot\dfrac{1}{y^2}(2y)\,dy=3x^2\ln(y^2)\,dx+\dfrac{2x^3}{y}dy.$

21. $R=\alpha\beta^2\cos\gamma\quad\Rightarrow\quad dR=\dfrac{\partial R}{\partial\alpha}\,d\alpha+\dfrac{\partial R}{\partial\beta}\,d\beta+\dfrac{\partial R}{\partial\gamma}\,d\gamma=\beta^2\cos\gamma\,d\alpha+2\alpha\beta\cos\gamma\,d\beta-\alpha\beta^2\sin\gamma\,d\gamma$

23. $dx=\Delta x=0.05$, $dy=\Delta y=0.1$, $z=5x^2+y^2$, $z_x=10x$, $z_y=2y$. Thus when $x=1$ and $y=2$,
$dz=z_x(1,2)\,dx+z_y(1,2)\,dy=(10)(0.05)+(4)(0.1)=0.9$ while
$\Delta z=f(1.05,2.1)-f(1,2)=5(1.05)^2+(2.1)^2-5-4=0.9225$.

25. $dA=\dfrac{\partial A}{\partial x}\,dx+\dfrac{\partial A}{\partial y}\,dy=y\,dx+x\,dy$ and $|\Delta x|\le 0.1$, $|\Delta y|\le 0.1$. We use $dx=0.1$, $dy=0.1$ with $x=30$, $y=24$; then the maximum error in the area is about $dA=24(0.1)+30(0.1)=5.4\text{ cm}^2$.

27. The volume of a can is $V=\pi r^2h$ and $\Delta V\approx dV$ is an estimate of the amount of tin. Here $dV=2\pi rh\,dr+\pi r^2\,dh$, so put $dr=0.04$, $dh=0.08$ (0.04 on top, 0.04 on bottom) and then $\Delta V\approx dV=2\pi(48)(0.04)+\pi(16)(0.08)\approx 16.08\text{ cm}^3$. Thus the amount of tin is about 16 cm^3.

29. The errors in measurement are at most 2%, so $\left|\dfrac{\Delta w}{w}\right|\le 0.02$ and $\left|\dfrac{\Delta h}{h}\right|\le 0.02$. The relative error in the calculated surface area is

$$\begin{aligned} \frac{\Delta S}{S}\approx\frac{dS}{S}=\frac{S_w\,dw+S_h\,dh}{S}&=\frac{72.09(0.425w^{0.425-1})h^{0.725}\,dw+72.09w^{0.425}(0.725h^{0.725-1})\,dh}{72.09w^{0.425}h^{0.725}} \\ &=0.425\frac{dw}{w}+0.725\frac{dh}{h} \end{aligned}$$

To estimate the maximum relative error, we use

$$\frac{dw}{w}=\left|\frac{\Delta w}{w}\right|=0.02 \text{ and } \frac{dh}{h}=\left|\frac{\Delta h}{h}\right|=0.02 \quad\Rightarrow\quad \frac{dS}{S}=0.425\,(0.02)+0.725\,(0.02)=0.023.$$

Thus the maximum *percentage* error is approximately 2.3%.

31. First we find $\dfrac{\partial R}{\partial R_1}$ implicitly by taking partial derivatives of both sides with respect to R_1:

$\dfrac{\partial}{\partial R_1}\left(\dfrac{1}{R}\right) = \dfrac{\partial\left[(1/R_1)+(1/R_2)+(1/R_3)\right]}{\partial R_1} \;\Rightarrow\; -R^{-2}\dfrac{\partial R}{\partial R_1} = -R_1^{-2} \;\Rightarrow\; \dfrac{\partial R}{\partial R_1} = \dfrac{R^2}{R_1^2}$. Then by symmetry,

$\dfrac{\partial R}{\partial R_2} = \dfrac{R^2}{R_2^2}$, $\dfrac{\partial R}{\partial R_3} = \dfrac{R^2}{R_3^2}$. When $R_1 = 25$, $R_2 = 40$ and $R_3 = 50$, $\dfrac{1}{R} = \dfrac{17}{200} \;\Leftrightarrow\; R = \frac{200}{17}\ \Omega$. Since the possible error for each R_i is 0.5%, the maximum error of R is attained by setting $\Delta R_i = 0.005R_i$. So

$$\Delta R \approx dR = \frac{\partial R}{\partial R_1}\Delta R_1 + \frac{\partial R}{\partial R_2}\Delta R_2 + \frac{\partial R}{\partial R_3}\Delta R_3 = (0.005)R^2\left(\frac{1}{R_1}+\frac{1}{R_2}+\frac{1}{R_3}\right) = (0.005)R = \tfrac{1}{17} \approx 0.059\ \Omega.$$

33. $\mathbf{r}(u,v) = (u+v)\,\mathbf{i} + 3u^2\,\mathbf{j} + (u-v)\,\mathbf{k}$.

$\mathbf{r}_u = \mathbf{i} + 6u\,\mathbf{j} + \mathbf{k}$ and $\mathbf{r}_v = \mathbf{i} - \mathbf{k}$, so

$\mathbf{r}_u \times \mathbf{r}_v = -6u\,\mathbf{i} + 2\,\mathbf{j} - 6u\,\mathbf{k}$. Since the point $(2,3,0)$ corresponds to $u = 1$, $v = 1$, a normal vector to the surface at $(2,3,0)$ is $-6\,\mathbf{i} + 2\,\mathbf{j} - 6\,\mathbf{k}$, and an equation of the tangent plane is $-6x + 2y - 6z = -6$ or $3x - y + 3z = 3$.

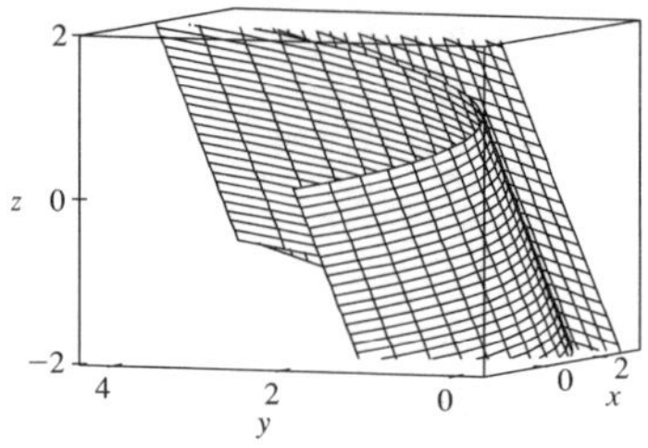

35. $\mathbf{r}(u,v) = u^2\,\mathbf{i} + 2u\sin v\,\mathbf{j} + u\cos v\,\mathbf{k} \;\Rightarrow\; \mathbf{r}(1,0) = (1,0,1)$.

$\mathbf{r}_u = 2u\,\mathbf{i} + 2\sin v\,\mathbf{j} + \cos v\,\mathbf{k}$ and $\mathbf{r}_v = 2u\cos v\,\mathbf{j} - u\sin v\,\mathbf{k}$,

so a normal vector to the surface at the point $(1,0,1)$ is

$\mathbf{r}_u(1,0) \times \mathbf{r}_v(1,0) = (2\,\mathbf{i} + \mathbf{k}) \times (2\,\mathbf{j}) = -2\,\mathbf{i} + 4\,\mathbf{k}$.

Thus an equation of the tangent plane at $(1,0,1)$ is

$-2(x-1) + 0(y-0) + 4(z-1) = 0$ or $-x + 2z = 1$.

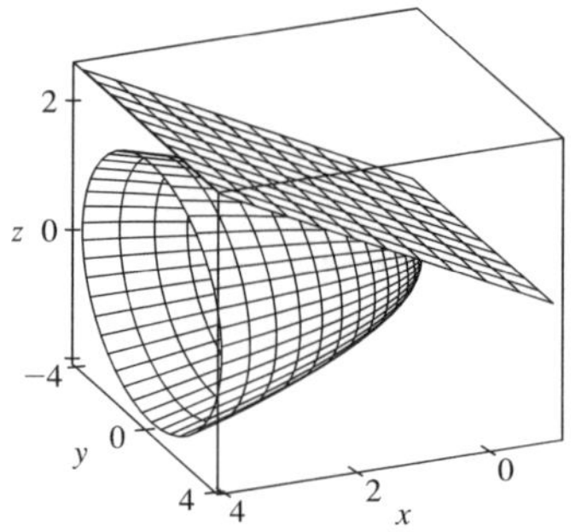

37. $\mathbf{r}(u,v) = u\,\mathbf{i} + \ln(uv)\,\mathbf{j} + v\,\mathbf{k} \;\Rightarrow\; \mathbf{r}_u(u,v) = \mathbf{i} + \frac{1}{u}\,\mathbf{j}$,

$\mathbf{r}_v(u,v) = \frac{1}{v}\,\mathbf{j} + \mathbf{k}$. $\mathbf{r}(1,1) = \mathbf{i} + \mathbf{k}$, so the point corresponding to $u = 1$, $v = 1$ is $(1,0,1)$. A normal vector for the tangent plane is $\mathbf{r}_u(1,1) \times \mathbf{r}_v(1,1) = (\mathbf{i} + \mathbf{j}) \times (\mathbf{j} + \mathbf{k}) = \mathbf{i} - \mathbf{j} + \mathbf{k}$, so an equation of the tangent plane is $(x-1) - (y-0) + (z-1) = 0$ or $x - y + z = 2$.

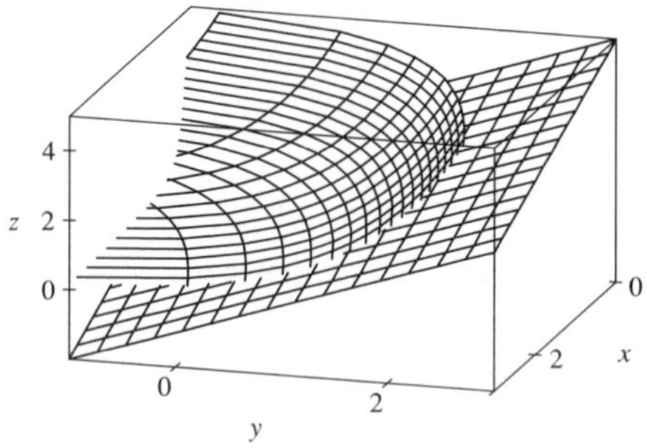

39. $\Delta z = f(a+\Delta x, b+\Delta y) - f(a,b) = (a+\Delta x)^2 + (b+\Delta y)^2 - (a^2+b^2)$

$\quad = a^2 + 2a\,\Delta x + (\Delta x)^2 + b^2 + 2b\,\Delta y + (\Delta y)^2 - a^2 - b^2 = 2a\,\Delta x + (\Delta x)^2 + 2b\,\Delta y + (\Delta y)^2$

But $f_x(a,b) = 2a$ and $f_y(a,b) = 2b$ and so $\Delta z = f_x(a,b)\,\Delta x + f_y(a,b)\,\Delta y + \Delta x\,\Delta x + \Delta y\,\Delta y$, which is Definition 7 with $\varepsilon_1 = \Delta x$ and $\varepsilon_2 = \Delta y$. Hence f is differentiable.

41. To show that f is continuous at (a, b) we need to show that $\lim_{(x,y)\to(a,b)} f(x, y) = f(a, b)$ or equivalently $\lim_{(\Delta x,\Delta y)\to(0,0)} f(a + \Delta x, b + \Delta y) = f(a, b)$. Since f is differentiable at (a, b), $f(a + \Delta x, b + \Delta y) - f(a, b) = \Delta z = f_x(a, b)\,\Delta x + f_y(a, b)\,\Delta y + \varepsilon_1\,\Delta x + \varepsilon_2\,\Delta y$, where ϵ_1 and $\epsilon_2 \to 0$ as $(\Delta x, \Delta y) \to (0, 0)$. Thus $f(a + \Delta x, b + \Delta y) = f(a, b) + f_x(a, b)\,\Delta x + f_y(a, b)\,\Delta y + \varepsilon_1\,\Delta x + \varepsilon_2\,\Delta y$. Taking the limit of both sides as $(\Delta x, \Delta y) \to (0, 0)$ gives $\lim_{(\Delta x,\Delta y)\to(0,0)} f(a + \Delta x, b + \Delta y) = f(a, b)$. Thus f is continuous at (a, b).

11.5 The Chain Rule

1. $z = \sin x \cos y,\ x = \pi t,\ y = \sqrt{t} \ \Rightarrow$

$$\frac{dz}{dt} = \frac{\partial z}{\partial x}\frac{dx}{dt} + \frac{\partial z}{\partial y}\frac{dy}{dt} = \cos x \cos y \cdot \pi + \sin x\,(-\sin y) \cdot \tfrac{1}{2}t^{-1/2} = \pi \cos x \cos y - \frac{1}{2\sqrt{t}} \sin x \sin y$$

3. $w = xe^{y/z},\ x = t^2,\ y = 1 - t,\ z = 1 + 2t \ \Rightarrow$

$$\frac{dw}{dt} = \frac{\partial w}{\partial x}\frac{dx}{dt} + \frac{\partial w}{\partial y}\frac{dy}{dt} + \frac{\partial w}{\partial z}\frac{dz}{dt} = e^{y/z} \cdot 2t + xe^{y/z}\left(\frac{1}{z}\right) \cdot (-1) + xe^{y/z}\left(-\frac{y}{z^2}\right) \cdot 2 = e^{y/z}\left(2t - \frac{x}{z} - \frac{2xy}{z^2}\right)$$

5. $z = x^2 + xy + y^2,\ x = s + t,\ y = st \ \Rightarrow$

$$\frac{\partial z}{\partial s} = \frac{\partial z}{\partial x}\frac{\partial x}{\partial s} + \frac{\partial z}{\partial y}\frac{\partial y}{\partial s} = (2x + y)(1) + (x + 2y)(t) = 2x + y + xt + 2yt$$

$$\frac{\partial z}{\partial t} = \frac{\partial z}{\partial x}\frac{\partial x}{\partial t} + \frac{\partial z}{\partial y}\frac{\partial y}{\partial t} = (2x + y)(1) + (x + 2y)(s) = 2x + y + xs + 2ys$$

7. $z = e^r \cos\theta,\ r = st,\ \theta = \sqrt{s^2 + t^2} \ \Rightarrow$

$$\frac{\partial z}{\partial s} = \frac{\partial z}{\partial r}\frac{\partial r}{\partial s} + \frac{\partial z}{\partial \theta}\frac{\partial \theta}{\partial s} = e^r \cos\theta \cdot t + e^r(-\sin\theta) \cdot \tfrac{1}{2}(s^2 + t^2)^{-1/2}(2s)$$
$$= te^r \cos\theta - e^r \sin\theta \cdot \frac{s}{\sqrt{s^2 + t^2}} = e^r\left(t\cos\theta - \frac{s}{\sqrt{s^2 + t^2}}\sin\theta\right)$$

$$\frac{\partial z}{\partial t} = \frac{\partial z}{\partial r}\frac{\partial r}{\partial t} + \frac{\partial z}{\partial \theta}\frac{\partial \theta}{\partial t} = e^r \cos\theta \cdot s + e^r(-\sin\theta) \cdot \tfrac{1}{2}(s^2 + t^2)^{-1/2}(2t)$$
$$= se^r \cos\theta - e^r \sin\theta \cdot \frac{t}{\sqrt{s^2 + t^2}} = e^r\left(s\cos\theta - \frac{t}{\sqrt{s^2 + t^2}}\sin\theta\right)$$

9. When $t = 3$, $x = g(3) = 2$ and $y = h(3) = 7$. By the Chain Rule (2),

$$\frac{dz}{dt} = \frac{\partial f}{\partial x}\frac{dx}{dt} + \frac{\partial f}{\partial y}\frac{dy}{dt} = f_x(2, 7)g'(3) + f_y(2, 7)\,h'(3) = (6)(5) + (-8)(-4) = 62.$$

11. $g(u, v) = f(x(u, v), y(u, v))$ where $x = e^u + \sin v$, $y = e^u + \cos v \ \Rightarrow \ \dfrac{\partial x}{\partial u} = e^u$, $\dfrac{\partial x}{\partial v} = \cos v$, $\dfrac{\partial y}{\partial u} = e^u$, $\dfrac{\partial y}{\partial v} = -\sin v$. By the Chain Rule (3), $\dfrac{\partial g}{\partial u} = \dfrac{\partial f}{\partial x}\dfrac{\partial x}{\partial u} + \dfrac{\partial f}{\partial y}\dfrac{\partial y}{\partial u}$. Then

$$g_u(0, 0) = f_x(x(0, 0), y(0, 0))\,x_u(0, 0) + f_y(x(0, 0), y(0, 0))\,y_u(0, 0)$$
$$= f_x(1, 2)(e^0) + f_y(1, 2)(e^0) = 2(1) + 5(1) = 7$$

Similarly $\dfrac{\partial g}{\partial v} = \dfrac{\partial f}{\partial x}\dfrac{\partial x}{\partial v} + \dfrac{\partial f}{\partial y}\dfrac{\partial y}{\partial v}$. Then

$$g_v(0, 0) = f_x(x(0, 0), y(0, 0))\,x_v(0, 0) + f_y(x(0, 0), y(0, 0))\,y_v(0, 0)$$
$$= f_x(1, 2)(\cos 0) + f_y(1, 2)(-\sin 0) = 2(1) + 5(0) = 2$$

13.

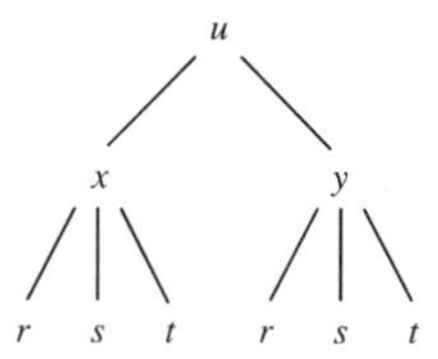

$u = f(x, y)$, $x = x(r, s, t)$, $y = y(r, s, t)$ $\Rightarrow$

$$\frac{\partial u}{\partial r} = \frac{\partial u}{\partial x}\frac{\partial x}{\partial r} + \frac{\partial u}{\partial y}\frac{\partial y}{\partial r},\ \frac{\partial u}{\partial s} = \frac{\partial u}{\partial x}\frac{\partial x}{\partial s} + \frac{\partial u}{\partial y}\frac{\partial y}{\partial s},$$

$$\frac{\partial u}{\partial t} = \frac{\partial u}{\partial x}\frac{\partial x}{\partial t} + \frac{\partial u}{\partial y}\frac{\partial y}{\partial t}$$

15.

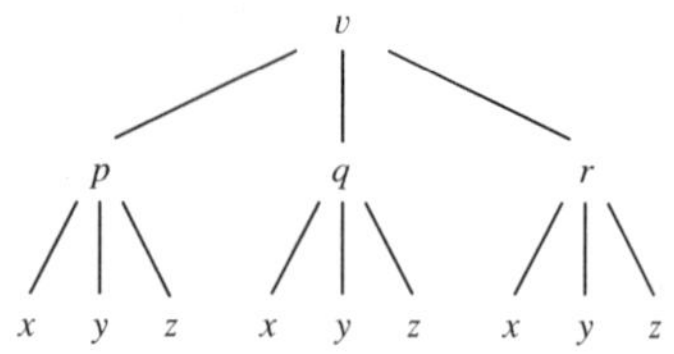

$v = f(p, q, r)$, $p = p(x, y, z)$, $q = q(x, y, z)$, $r = r(x, y, z)$ $\Rightarrow$

$$\frac{\partial v}{\partial x} = \frac{\partial v}{\partial p}\frac{\partial p}{\partial x} + \frac{\partial v}{\partial q}\frac{\partial q}{\partial x} + \frac{\partial v}{\partial r}\frac{\partial r}{\partial x},\ \frac{\partial v}{\partial y} = \frac{\partial v}{\partial p}\frac{\partial p}{\partial y} + \frac{\partial v}{\partial q}\frac{\partial q}{\partial y} + \frac{\partial v}{\partial r}\frac{\partial r}{\partial y},$$

$$\frac{\partial v}{\partial z} = \frac{\partial v}{\partial p}\frac{\partial p}{\partial z} + \frac{\partial v}{\partial q}\frac{\partial q}{\partial z} + \frac{\partial v}{\partial r}\frac{\partial r}{\partial z}$$

17. $z = x^2 + xy^3$, $x = uv^2 + w^3$, $y = u + ve^w$ $\Rightarrow$ $\dfrac{\partial z}{\partial u} = \dfrac{\partial z}{\partial x}\dfrac{\partial x}{\partial u} + \dfrac{\partial z}{\partial y}\dfrac{\partial y}{\partial u} = (2x + y^3)(v^2) + (3xy^2)(1)$,

$$\frac{\partial z}{\partial v} = \frac{\partial z}{\partial x}\frac{\partial x}{\partial v} + \frac{\partial z}{\partial y}\frac{\partial y}{\partial v} = (2x + y^3)(2uv) + (3xy^2)(e^w),\ \frac{\partial z}{\partial w} = \frac{\partial z}{\partial x}\frac{\partial x}{\partial w} + \frac{\partial z}{\partial y}\frac{\partial y}{\partial w} = (2x + y^3)(3w^2) + (3xy^2)(ve^w).$$

When $u = 2$, $v = 1$, and $w = 0$, we have $x = 2$, $y = 3$, so $\dfrac{\partial z}{\partial u} = (31)(1) + (54)(1) = 85$, $\dfrac{\partial z}{\partial v} = (31)\,(4) + (54)(1) = 178$,

$\dfrac{\partial z}{\partial w} = (31)(0) + (54)(1) = 54$.

19. $R = \ln(u^2 + v^2 + w^2)$, $u = x + 2y$, $v = 2x - y$, $w = 2xy$ $\Rightarrow$

$$\frac{\partial R}{\partial x} = \frac{\partial R}{\partial u}\frac{\partial u}{\partial x} + \frac{\partial R}{\partial v}\frac{\partial v}{\partial x} + \frac{\partial R}{\partial w}\frac{\partial w}{\partial x} = \frac{2u}{u^2 + v^2 + w^2}(1) + \frac{2v}{u^2 + v^2 + w^2}(2) + \frac{2w}{u^2 + v^2 + w^2}(2y)$$
$$= \frac{2u + 4v + 4wy}{u^2 + v^2 + w^2},$$

$$\frac{\partial R}{\partial y} = \frac{\partial R}{\partial u}\frac{\partial u}{\partial y} + \frac{\partial R}{\partial v}\frac{\partial v}{\partial y} + \frac{\partial R}{\partial w}\frac{\partial w}{\partial y} = \frac{2u}{u^2 + v^2 + w^2}(2) + \frac{2v}{u^2 + v^2 + w^2}(-1) + \frac{2w}{u^2 + v^2 + w^2}(2x)$$
$$= \frac{4u - 2v + 4wx}{u^2 + v^2 + w^2}.$$

When $x = y = 1$ we have $u = 3$, $v = 1$, and $w = 2$, so $\dfrac{\partial R}{\partial x} = \dfrac{9}{7}$ and $\dfrac{\partial R}{\partial y} = \dfrac{9}{7}$.

21. $u = x^2 + yz$, $x = pr\cos\theta$, $y = pr\sin\theta$, $z = p + r$ $\Rightarrow$

$$\frac{\partial u}{\partial p} = \frac{\partial u}{\partial x}\frac{\partial x}{\partial p} + \frac{\partial u}{\partial y}\frac{\partial y}{\partial p} + \frac{\partial u}{\partial z}\frac{\partial z}{\partial p} = (2x)(r\cos\theta) + (z)(r\sin\theta) + (y)(1) = 2xr\cos\theta + zr\sin\theta + y,$$

$$\frac{\partial u}{\partial r} = \frac{\partial u}{\partial x}\frac{\partial x}{\partial r} + \frac{\partial u}{\partial y}\frac{\partial y}{\partial r} + \frac{\partial u}{\partial z}\frac{\partial z}{\partial r} = (2x)(p\cos\theta) + (z)(p\sin\theta) + (y)(1) = 2xp\cos\theta + zp\sin\theta + y,$$

$$\frac{\partial u}{\partial \theta} = \frac{\partial u}{\partial x}\frac{\partial x}{\partial \theta} + \frac{\partial u}{\partial y}\frac{\partial y}{\partial \theta} + \frac{\partial u}{\partial z}\frac{\partial z}{\partial \theta} = (2x)(-pr\sin\theta) + (z)(pr\cos\theta) + (y)(0) = -2xpr\sin\theta + zpr\cos\theta.$$

When $p = 2$, $r = 3$, and $\theta = 0$ we have $x = 6$, $y = 0$, and $z = 5$, so $\dfrac{\partial u}{\partial p} = 36$, $\dfrac{\partial u}{\partial r} = 24$, and $\dfrac{\partial u}{\partial \theta} = 30$.

23. $\sqrt{xy} = 1 + x^2y$, so let $F(x, y) = (xy)^{1/2} - 1 - x^2y = 0$. Then by Equation 6

$$\frac{dy}{dx} = -\frac{F_x}{F_y} = -\frac{\frac{1}{2}(xy)^{-1/2}(y) - 2xy}{\frac{1}{2}(xy)^{-1/2}(x) - x^2} = -\frac{y - 4xy\sqrt{xy}}{x - 2x^2\sqrt{xy}} = \frac{4(xy)^{3/2} - y}{x - 2x^2\sqrt{xy}}.$$

25. $xy^2 + yz^2 + zx^2 = 3$, so let $F(x, y) = xy^2 + yz^2 + zx^2 - 3 = 0$.

Then $\dfrac{\partial z}{\partial x} = -\dfrac{F_x}{F_z} = -\dfrac{y^2 + 2zx}{2yz + x^2}$ and $\dfrac{\partial z}{\partial y} = -\dfrac{F_y}{F_z} = -\dfrac{2xy + z^2}{2yz + x^2}$.

27. Let $F(x, y, z) = xe^y + yz + ze^x = 0$. Then $\dfrac{\partial z}{\partial x} = -\dfrac{F_x}{F_z} = -\dfrac{e^y + ze^x}{y + e^x}$, $\dfrac{\partial z}{\partial y} = -\dfrac{F_y}{F_z} = -\dfrac{xe^y + z}{y + e^x}$.

29. Since x and y are each functions of t, $T(x, y)$ is a function of t, so by the Chain Rule, $\dfrac{dT}{dt} = \dfrac{\partial T}{\partial x}\dfrac{dx}{dt} + \dfrac{\partial T}{\partial y}\dfrac{dy}{dt}$. After 3 seconds, $x = \sqrt{1+t} = \sqrt{1+3} = 2$, $y = 2 + \frac{1}{3}t = 2 + \frac{1}{3}(3) = 3$, $\dfrac{dx}{dt} = \dfrac{1}{2\sqrt{1+t}} = \dfrac{1}{2\sqrt{1+3}} = \dfrac{1}{4}$, and $\dfrac{dy}{dt} = \dfrac{1}{3}$.

Then $\dfrac{dT}{dt} = T_x(2,3)\,\dfrac{dx}{dt} + T_y(2,3)\,\dfrac{dy}{dt} = 4\left(\frac{1}{4}\right) + 3\left(\frac{1}{3}\right) = 2$. Thus the temperature is rising at a rate of $2°\text{C/s}$.

31. $C = 1449.2 + 4.6T - 0.055T^2 + 0.00029T^3 + 0.016D$, so $\dfrac{\partial C}{\partial T} = 4.6 - 0.11T + 0.00087T^2$ and $\dfrac{\partial C}{\partial D} = 0.016$.

According to the graph, the diver is experiencing a temperature of approximately $12.5°\text{C}$ at $t = 20$ minutes, so

$\dfrac{\partial C}{\partial T} = 4.6 - 0.11(12.5) + 0.00087(12.5)^2 \approx 3.36$. By sketching tangent lines at $t = 20$ to the graphs given, we estimate

$\dfrac{dD}{dt} \approx \dfrac{1}{2}$ and $\dfrac{dT}{dt} \approx -\dfrac{1}{10}$. Then, by the Chain Rule, $\dfrac{dC}{dt} = \dfrac{\partial C}{\partial T}\dfrac{dT}{dt} + \dfrac{\partial C}{\partial D}\dfrac{dD}{dt} \approx (3.36)\left(-\frac{1}{10}\right) + (0.016)\left(\frac{1}{2}\right) \approx -0.33$.

Thus the speed of sound experienced by the diver is decreasing at a rate of approximately 0.33 m/s per minute.

33. (a) $V = \ell wh$, so by the Chain Rule,

$$\begin{aligned}\frac{dV}{dt} &= \frac{\partial V}{\partial \ell}\frac{d\ell}{dt} + \frac{\partial V}{\partial w}\frac{dw}{dt} + \frac{\partial V}{\partial h}\frac{dh}{dt} = wh\,\frac{d\ell}{dt} + \ell h\,\frac{dw}{dt} + \ell w\,\frac{dh}{dt}\\ &= 2\cdot 2\cdot 2 + 1\cdot 2\cdot 2 + 1\cdot 2\cdot(-3) = 6\ \text{m}^3/\text{s}\end{aligned}$$

(b) $S = 2(\ell w + \ell h + wh)$, so by the Chain Rule,

$$\begin{aligned}\frac{dS}{dt} &= \frac{\partial S}{\partial \ell}\frac{d\ell}{dt} + \frac{\partial S}{\partial w}\frac{dw}{dt} + \frac{\partial S}{\partial h}\frac{dh}{dt} = 2(w+h)\,\frac{d\ell}{dt} + 2(\ell+h)\,\frac{dw}{dt} + 2(\ell+w)\,\frac{dh}{dt}\\ &= 2(2+2)2 + 2(1+2)2 + 2(1+2)(-3) = 10\ \text{m}^2/\text{s}\end{aligned}$$

(c) $L^2 = \ell^2 + w^2 + h^2 \;\Rightarrow\; 2L\,\dfrac{dL}{dt} = 2\ell\,\dfrac{d\ell}{dt} + 2w\,\dfrac{dw}{dt} + 2h\,\dfrac{dh}{dt} = 2(1)(2) + 2(2)(2) + 2(2)(-3) = 0 \;\Rightarrow$ $dL/dt = 0$ m/s.

35. $\dfrac{dP}{dt} = 0.05$, $\dfrac{dT}{dt} = 0.15$, $V = 8.31\dfrac{T}{P}$ and $\dfrac{dV}{dt} = \dfrac{8.31}{P}\dfrac{dT}{dt} - 8.31\dfrac{T}{P^2}\dfrac{dP}{dt}$. Thus when $P = 20$ and $T = 320$,

$\dfrac{dV}{dt} = 8.31\left[\dfrac{0.15}{20} - \dfrac{(0.05)(320)}{400}\right] \approx -0.27$ L/s.

37. (a) By the Chain Rule, $\dfrac{\partial z}{\partial r} = \dfrac{\partial z}{\partial x}\cos\theta + \dfrac{\partial z}{\partial y}\sin\theta$, $\dfrac{\partial z}{\partial \theta} = \dfrac{\partial z}{\partial x}(-r\sin\theta) + \dfrac{\partial z}{\partial y}\,r\cos\theta$.

(b) $\left(\dfrac{\partial z}{\partial r}\right)^2 = \left(\dfrac{\partial z}{\partial x}\right)^2\cos^2\theta + 2\,\dfrac{\partial z}{\partial x}\dfrac{\partial z}{\partial y}\cos\theta\sin\theta + \left(\dfrac{\partial z}{\partial y}\right)^2\sin^2\theta$,

$\left(\dfrac{\partial z}{\partial \theta}\right)^2 = \left(\dfrac{\partial z}{\partial x}\right)^2 r^2\sin^2\theta - 2\,\dfrac{\partial z}{\partial x}\dfrac{\partial z}{\partial y}\,r^2\cos\theta\sin\theta + \left(\dfrac{\partial z}{\partial y}\right)^2 r^2\cos^2\theta$. Thus

$\left(\dfrac{\partial z}{\partial r}\right)^2 + \dfrac{1}{r^2}\left(\dfrac{\partial z}{\partial \theta}\right)^2 = \left[\left(\dfrac{\partial z}{\partial x}\right)^2 + \left(\dfrac{\partial z}{\partial y}\right)^2\right](\cos^2\theta + \sin^2\theta) = \left(\dfrac{\partial z}{\partial x}\right)^2 + \left(\dfrac{\partial z}{\partial y}\right)^2$.

39. Let $u = x - y$. Then $\dfrac{\partial z}{\partial x} = \dfrac{dz}{du}\dfrac{\partial u}{\partial x} = \dfrac{dz}{du}$ and $\dfrac{\partial z}{\partial y} = \dfrac{dz}{du}(-1)$. Thus $\dfrac{\partial z}{\partial x} + \dfrac{\partial z}{\partial y} = 0$.

41. Let $u = x + at$, $v = x - at$. Then $z = f(u) + g(v)$, so $\partial z/\partial u = f'(u)$ and $\partial z/\partial v = g'(v)$.

Thus $\dfrac{\partial z}{\partial t} = \dfrac{\partial z}{\partial u}\dfrac{\partial u}{\partial t} + \dfrac{\partial z}{\partial v}\dfrac{\partial v}{\partial t} = af'(u) - ag'(v)$ and

$$\frac{\partial^2 z}{\partial t^2} = a\frac{\partial}{\partial t}\left[f'(u) - g'(v)\right] = a\left(\frac{df'(u)}{du}\frac{\partial u}{\partial t} - \frac{dg'(v)}{dv}\frac{\partial v}{\partial t}\right) = a^2 f''(u) + a^2 g''(v).$$

Similarly $\dfrac{\partial z}{\partial x} = f'(u) + g'(v)$ and $\dfrac{\partial^2 z}{\partial x^2} = f''(u) + g''(v)$. Thus $\dfrac{\partial^2 z}{\partial t^2} = a^2\dfrac{\partial^2 z}{\partial x^2}$.

43. $\dfrac{\partial z}{\partial s} = \dfrac{\partial z}{\partial x}2s + \dfrac{\partial z}{\partial y}2r$. Then

$$\begin{aligned}
\frac{\partial^2 z}{\partial r\,\partial s} &= \frac{\partial}{\partial r}\left(\frac{\partial z}{\partial x}2s\right) + \frac{\partial}{\partial r}\left(\frac{\partial z}{\partial y}2r\right) \\
&= \frac{\partial^2 z}{\partial x^2}\frac{\partial x}{\partial r}2s + \frac{\partial}{\partial y}\left(\frac{\partial z}{\partial x}\right)\frac{\partial y}{\partial r}2s + \frac{\partial z}{\partial x}\frac{\partial}{\partial r}2s + \frac{\partial^2 z}{\partial y^2}\frac{\partial y}{\partial r}2r + \frac{\partial}{\partial x}\left(\frac{\partial z}{\partial y}\right)\frac{\partial x}{\partial r}2r + \frac{\partial z}{\partial y}2 \\
&= 4rs\frac{\partial^2 z}{\partial x^2} + \frac{\partial^2 z}{\partial y\,\partial x}4s^2 + 0 + 4rs\frac{\partial^2 z}{\partial y^2} + \frac{\partial^2 z}{\partial x\,\partial y}4r^2 + 2\frac{\partial z}{\partial y}
\end{aligned}$$

By the continuity of the partials, $\dfrac{\partial^2 z}{\partial r\partial s} = 4rs\dfrac{\partial^2 z}{\partial x^2} + 4rs\dfrac{\partial^2 z}{\partial y^2} + (4r^2 + 4s^2)\dfrac{\partial^2 z}{\partial x\,\partial y} + 2\dfrac{\partial z}{\partial y}$.

45. $\dfrac{\partial z}{\partial r} = \dfrac{\partial z}{\partial x}\cos\theta + \dfrac{\partial z}{\partial y}\sin\theta$ and $\dfrac{\partial z}{\partial \theta} = -\dfrac{\partial z}{\partial x}r\sin\theta + \dfrac{\partial z}{\partial y}r\cos\theta$. Then

$$\begin{aligned}
\frac{\partial^2 z}{\partial r^2} &= \cos\theta\left(\frac{\partial^2 z}{\partial x^2}\cos\theta + \frac{\partial^2 z}{\partial y\,\partial x}\sin\theta\right) + \sin\theta\left(\frac{\partial^2 z}{\partial y^2}\sin\theta + \frac{\partial^2 z}{\partial x\,\partial y}\cos\theta\right) \\
&= \cos^2\theta\frac{\partial^2 z}{\partial x^2} + 2\cos\theta\sin\theta\frac{\partial^2 z}{\partial x\,\partial y} + \sin^2\theta\frac{\partial^2 z}{\partial y^2}
\end{aligned}$$

and

$$\begin{aligned}
\frac{\partial^2 z}{\partial \theta^2} &= -r\cos\theta\frac{\partial z}{\partial x} + (-r\sin\theta)\left(\frac{\partial^2 z}{\partial x^2}(-r\sin\theta) + \frac{\partial^2 z}{\partial y\,\partial x}r\cos\theta\right) \\
&\qquad -r\sin\theta\frac{\partial z}{\partial y} + r\cos\theta\left(\frac{\partial^2 z}{\partial y^2}r\cos\theta + \frac{\partial^2 z}{\partial x\,\partial y}(-r\sin\theta)\right) \\
&= -r\cos\theta\frac{\partial z}{\partial x} - r\sin\theta\frac{\partial z}{\partial y} + r^2\sin^2\theta\frac{\partial^2 z}{\partial x^2} - 2r^2\cos\theta\sin\theta\frac{\partial^2 z}{\partial x\,\partial y} + r^2\cos^2\theta\frac{\partial^2 z}{\partial y^2}
\end{aligned}$$

Thus

$$\begin{aligned}
\frac{\partial^2 z}{\partial r^2} + \frac{1}{r^2}\frac{\partial^2 z}{\partial \theta^2} + \frac{1}{r}\frac{\partial z}{\partial r} &= (\cos^2\theta + \sin^2\theta)\frac{\partial^2 z}{\partial x^2} + (\sin^2\theta + \cos^2\theta)\frac{\partial^2 z}{\partial y^2} - \frac{1}{r}\cos\theta\frac{\partial z}{\partial x} \\
&\qquad - \frac{1}{r}\sin\theta\frac{\partial z}{\partial y} + \frac{1}{r}\left(\cos\theta\frac{\partial z}{\partial x} + \sin\theta\frac{\partial z}{\partial y}\right) \\
&= \frac{\partial^2 z}{\partial x^2} + \frac{\partial^2 z}{\partial y^2} \text{ as desired.}
\end{aligned}$$

47. $F(x, y, z) = 0$ is assumed to define z as a function of x and y, that is, $z = f(x, y)$. So by (7), $\dfrac{\partial z}{\partial x} = -\dfrac{F_x}{F_z}$ since $F_z \neq 0$. Similarly, it is assumed that $F(x, y, z) = 0$ defines x as a function of y and z, that is $x = h(x, z)$. Then $F(h(y, z), y, z) = 0$ and by the Chain Rule, $F_x\dfrac{\partial x}{\partial y} + F_y\dfrac{\partial y}{\partial y} + F_z\dfrac{\partial z}{\partial y} = 0$. But $\dfrac{\partial z}{\partial y} = 0$ and $\dfrac{\partial y}{\partial y} = 1$, so $F_x\dfrac{\partial x}{\partial y} + F_y = 0 \;\Rightarrow\; \dfrac{\partial x}{\partial y} = -\dfrac{F_y}{F_x}$.

A similar calculation shows that $\dfrac{\partial y}{\partial z} = -\dfrac{F_z}{F_y}$. Thus $\dfrac{\partial z}{\partial x}\dfrac{\partial x}{\partial y}\dfrac{\partial y}{\partial z} = \left(-\dfrac{F_x}{F_z}\right)\left(-\dfrac{F_y}{F_x}\right)\left(-\dfrac{F_z}{F_y}\right) = -1$.

11.6 Directional Derivatives and the Gradient Vector

1. First we draw a line passing through Alice Springs and Adelaide. We can approximate the directional derivative at Alice Springs in the direction of Adelaide by the average rate of change of pressure between the points where this line intersects the contour lines closest to Alice Springs. In the direction of Adelaide, the pressure changes from 1008 hPa to 1012 hPa. We estimate the distance between these two points to be approximately 500 km, so the rate of change of pressure in the direction given is approximately $\frac{1012-1008}{500} = 0.008$ hPa/km.

3. $D_{\mathbf{u}}\, f(-20,30) = \nabla f\,(-20,30)\cdot \mathbf{u} = f_T(-20,30)\left(\frac{1}{\sqrt{2}}\right) + f_v(-20,30)\left(\frac{1}{\sqrt{2}}\right)$.

$f_T(-20,30) = \lim\limits_{h\to 0} \dfrac{f(-20+h,30)-f(-20,30)}{h}$, so we can approximate $f_T(-20,30)$ by considering $h=\pm 5$ and using the values given in the table: $f_T(-20,30) \approx \dfrac{f(-15,30)-f(-20,30)}{5} = \dfrac{-26-(-33)}{5} = 1.4$,

$f_T(-20,30) \approx \dfrac{f(-25,30)-f(-20,30)}{-5} = \dfrac{-39-(-33)}{-5} = 1.2$. Averaging these values gives $f_T(-20,30)\approx 1.3$.

Similarly, $f_v(-20,30) = \lim\limits_{h\to 0} \dfrac{f(-20,30+h)-f(-20,30)}{h}$, so we can approximate $f_v(-20,30)$ with $h=\pm 10$:

$f_v(-20,30) \approx \dfrac{f(-20,40)-f(-20,30)}{10} = \dfrac{-34-(-33)}{10} = -0.1$,

$f_v(-20,30) \approx \dfrac{f(-20,20)-f(-20,30)}{-10} = \dfrac{-30-(-33)}{-10} = -0.3$. Averaging these values gives $f_v(-20,30)\approx -0.2$.

Then $D_{\mathbf{u}}f(-20,30) \approx 1.3\left(\frac{1}{\sqrt{2}}\right) + (-0.2)\left(\frac{1}{\sqrt{2}}\right) \approx 0.778$.

5. $f(x,y) = \sqrt{5x-4y} \;\Rightarrow\; f_x(x,y) = \frac{1}{2}(5x-4y)^{-1/2}(5) = \dfrac{5}{2\sqrt{5x-4y}}$ and

$f_y(x,y) = \frac{1}{2}(5x-4y)^{-1/2}(-4) = -\dfrac{2}{\sqrt{5x-4y}}$. If $\mathbf{u}$ is a unit vector in the direction of $\theta = -\frac{\pi}{6}$, then from Equation 6,

$D_{\mathbf{u}}\, f(4,1) = f_x(4,1)\cos\left(-\frac{\pi}{6}\right) + f_y(4,1)\sin\left(-\frac{\pi}{6}\right) = \frac{5}{8}\cdot\frac{\sqrt{3}}{2} + \left(-\frac{1}{2}\right)\left(-\frac{1}{2}\right) = \frac{5\sqrt{3}}{16} + \frac{1}{4}$.

7. $f(x,y) = 5xy^2 - 4x^3y$

(a) $\nabla f(x,y) = \langle f_x(x,y), f_y(x,y)\rangle = \langle 5y^2 - 12x^2y, 10xy - 4x^3\rangle$

(b) $\nabla f(1,2) = \langle 5(2)^2 - 12(1)^2(2), 10(1)(2) - 4(1)^3\rangle = \langle -4, 16\rangle$

(c) By Equation 9, $D_{\mathbf{u}}\, f(1,2) = \nabla f(1,2)\cdot\mathbf{u} = \langle -4,16\rangle\cdot\left\langle \frac{5}{13},\frac{12}{13}\right\rangle = (-4)\left(\frac{5}{13}\right) + (16)\left(\frac{12}{13}\right) = \frac{172}{13}$.

9. $f(x,y,z) = xe^{2yz}$

(a) $\nabla f(x,y,z) = \langle f_x(x,y,z), f_y(x,y,z), f_z(x,y,z)\rangle = \langle e^{2yz}, 2xze^{2yz}, 2xye^{2yz}\rangle$

(b) $\nabla f(3,0,2) = \langle 1, 12, 0\rangle$

(c) By Equation 14, $D_{\mathbf{u}}f(3,0,2) = \nabla f(3,0,2)\cdot\mathbf{u} = \langle 1,12,0\rangle\cdot\left\langle \frac{2}{3},-\frac{2}{3},\frac{1}{3}\right\rangle = \frac{2}{3} - \frac{24}{3} + 0 = -\frac{22}{3}$.

11. $f(x,y) = 1 + 2x\sqrt{y} \;\Rightarrow\; \nabla f(x,y) = \left\langle 2\sqrt{y}, 2x\cdot\frac{1}{2}y^{-1/2}\right\rangle = \langle 2\sqrt{y}, x/\sqrt{y}\rangle$, $\nabla f(3,4) = \left\langle 4, \frac{3}{2}\right\rangle$, and a unit vector in the direction of $\mathbf{v}$ is $\mathbf{u} = \frac{1}{\sqrt{4^2+(-3)^2}}\langle 4,-3\rangle = \left\langle \frac{4}{5}, -\frac{3}{5}\right\rangle$, so $D_{\mathbf{u}}\, f(3,4) = \nabla f(3,4)\cdot\mathbf{u} = \left\langle 4,\frac{3}{2}\right\rangle\cdot\left\langle\frac{4}{5},-\frac{3}{5}\right\rangle = \frac{23}{10}$.

13. $g(s,t) = s^2 e^t \;\Rightarrow\; \nabla g(s,t) = 2se^t\,\mathbf{i} + s^2e^t\,\mathbf{j}$, $\nabla g(2,0) = 4\,\mathbf{i} + 4\,\mathbf{j}$, and a unit vector in the direction of $\mathbf{v}$ is

$\mathbf{u} = \frac{1}{\sqrt{2}}(\mathbf{i}+\mathbf{j})$, so $D_{\mathbf{u}}\, g(2,0) = \nabla g(2,0)\cdot\mathbf{u} = (4\,\mathbf{i}+4\,\mathbf{j})\cdot\frac{1}{\sqrt{2}}(\mathbf{i}+\mathbf{j}) = \frac{8}{\sqrt{2}} = 4\sqrt{2}$.

15. $g(x,y,z) = (x+2y+3z)^{3/2} \;\Rightarrow$

$$\nabla g(x,y,z) = \left\langle \tfrac{3}{2}(x+2y+3z)^{1/2}(1), \tfrac{3}{2}(x+2y+3z)^{1/2}(2), \tfrac{3}{2}(x+2y+3z)^{1/2}(3)\right\rangle$$
$$= \left\langle \tfrac{3}{2}\sqrt{x+2y+3z}, 3\sqrt{x+2y+3z}, \tfrac{9}{2}\sqrt{x+2y+3z}\,\right\rangle, \nabla g(1,1,2) = \left\langle \tfrac{9}{2}, 9, \tfrac{27}{2}\right\rangle,$$

and a unit vector in the direction of $\mathbf{v} = 2\,\mathbf{j} - \mathbf{k}$ is $\mathbf{u} = \frac{2}{\sqrt{5}}\,\mathbf{j} - \frac{1}{\sqrt{5}}\,\mathbf{k}$, so

$D_{\mathbf{u}}\, g(1,1,2) = \left\langle \frac{9}{2}, 9, \frac{27}{2}\right\rangle \cdot \left\langle 0, \frac{2}{\sqrt{5}}, -\frac{1}{\sqrt{5}}\right\rangle = \frac{18}{\sqrt{5}} - \frac{27}{2\sqrt{5}} = \frac{9}{2\sqrt{5}}$.

17. $f(x,y) = \sqrt{xy} \;\Rightarrow\; \nabla f(x,y) = \left\langle \frac{1}{2}(xy)^{-1/2}(y), \frac{1}{2}(xy)^{-1/2}(x)\right\rangle = \left\langle \dfrac{y}{2\sqrt{xy}}, \dfrac{x}{2\sqrt{xy}}\right\rangle$, so $\nabla f(2,8) = \left\langle 1, \frac{1}{4}\right\rangle$.

The unit vector in the direction of $\overrightarrow{PQ} = \langle 5-2, 4-8\rangle = \langle 3,-4\rangle$ is $\mathbf{u} = \left\langle \frac{3}{5}, -\frac{4}{5}\right\rangle$, so

$D_{\mathbf{u}}\, f(2,8) = \nabla f(2,8)\cdot\mathbf{u} = \left\langle 1, \frac{1}{4}\right\rangle \cdot \left\langle \frac{3}{5}, -\frac{4}{5}\right\rangle = \frac{2}{5}$.

19. $f(x,y) = y^2/x = y^2x^{-1} \;\Rightarrow\; \nabla f(x,y) = \langle -y^2x^{-2}, 2yx^{-1}\rangle = \langle -y^2/x^2, 2y/x\rangle$.

$\nabla f(2,4) = \langle -4, 4\rangle$, or equivalently $\langle -1, 1\rangle$, is the direction of maximum rate of change, and the maximum rate is $|\nabla f(2,4)| = \sqrt{16+16} = 4\sqrt{2}$.

21. $f(x,y,z) = \ln(xy^2z^3) \;\Rightarrow\; \nabla f(x,y,z) = \left\langle \dfrac{y^2z^3}{xy^2z^3}, \dfrac{2xyz^3}{xy^2z^3}, \dfrac{3xy^2z^2}{xy^2z^3}\right\rangle = \left\langle \dfrac{1}{x}, \dfrac{2}{y}, \dfrac{3}{z}\right\rangle$.

$\nabla f(1,-2,-3) = \langle 1,-1,-1\rangle$ is the direction of maximum rate of change and the maximum rate is $|\nabla f(1,-2,-3)| = \sqrt{3}$.

23. (a) As in the proof of Theorem 15, $D_{\mathbf{u}}\, f = |\nabla f|\cos\theta$. Since the minimum value of $\cos\theta$ is -1 occurring when $\theta = \pi$, the minimum value of $D_{\mathbf{u}}\, f$ is $-|\nabla f|$ occurring when $\theta = \pi$, that is when $\mathbf{u}$ is in the opposite direction of ∇f (assuming $\nabla f \neq \mathbf{0}$).

(b) $f(x,y) = x^4y - x^2y^3 \;\Rightarrow\; \nabla f(x,y) = \langle 4x^3y - 2xy^3, x^4 - 3x^2y^2\rangle$, so f decreases fastest at the point $(2,-3)$ in the direction $-\nabla f(2,-3) = -\langle 12,-92\rangle = \langle -12, 92\rangle$.

25. The direction of fastest change is $\nabla f(x,y) = (2x-2)\,\mathbf{i} + (2y-4)\,\mathbf{j}$, so we need to find all points (x,y) where $\nabla f(x,y)$ is parallel to $\mathbf{i}+\mathbf{j} \;\Leftrightarrow\; (2x-2)\,\mathbf{i} + (2y-4)\,\mathbf{j} = k\,(\mathbf{i}+\mathbf{j}) \;\Leftrightarrow\; k = 2x-2$ and $k = 2y-4$. Then $2x-2 = 2y-4 \;\Rightarrow$ $y = x+1$, so the direction of fastest change is $\mathbf{i}+\mathbf{j}$ at all points on the line $y = x+1$.

27. $T = \dfrac{k}{\sqrt{x^2+y^2+z^2}}$ and $120 = T(1,2,2) = \dfrac{k}{3}$ so $k = 360$.

(a) $\mathbf{u} = \dfrac{\langle 1,-1,1\rangle}{\sqrt{3}}$,

$$D_{\mathbf{u}}T(1,2,2) = \nabla T(1,2,2)\cdot\mathbf{u} = \left[-360\left(x^2+y^2+z^2\right)^{-3/2}\langle x,y,z\rangle\right]_{(1,2,2)}\cdot\mathbf{u} = -\tfrac{40}{3}\langle 1,2,2\rangle\cdot\tfrac{1}{\sqrt{3}}\langle 1,-1,1\rangle = -\tfrac{40}{3\sqrt{3}}$$

(b) From (a), $\nabla T = -360\left(x^2+y^2+z^2\right)^{-3/2}\langle x,y,z\rangle$, and since $\langle x,y,z\rangle$ is the position vector of the point (x,y,z), the vector $-\langle x,y,z\rangle$, and thus ∇T, always points toward the origin.

29. $\nabla V(x,y,z) = \langle 10x - 3y + yz, xz - 3x, xy\rangle$, $\nabla V(3,4,5) = \langle 38,6,12\rangle$

(a) $D_{\mathbf{u}}\, V(3,4,5) = \langle 38,6,12\rangle\cdot\frac{1}{\sqrt{3}}\langle 1,1,-1\rangle = \frac{32}{\sqrt{3}}$

(b) $\nabla V(3,4,5) = \langle 38,6,12\rangle$ or equivalently $\langle 19,3,6\rangle$.

(c) $|\nabla V(3,4,5)| = \sqrt{38^2+6^2+12^2} = \sqrt{1624} = 2\sqrt{406}$

31. A unit vector in the direction of $\overrightarrow{AB}$ is $\mathbf{i}$ and a unit vector in the direction of $\overrightarrow{AC}$ is $\mathbf{j}$. Thus $D_{\overrightarrow{AB}}\, f(1,3) = f_x(1,3) = 3$ and

$D_{\overrightarrow{AC}}\, f(1,3) = f_y(1,3) = 26$. Therefore $\nabla f(1,3) = \langle f_x(1,3), f_y(1,3)\rangle = \langle 3, 26\rangle$, and by definition,

$D_{\overrightarrow{AD}}\, f(1,3) = \nabla f \cdot \mathbf{u}$ where $\mathbf{u}$ is a unit vector in the direction of $\overrightarrow{AD}$, which is $\left\langle \frac{5}{13}, \frac{12}{13}\right\rangle$. Therefore,

$D_{\overrightarrow{AD}}\, f\,(1,3) = \langle 3, 26\rangle \cdot \left\langle \frac{5}{13}, \frac{12}{13}\right\rangle = 3 \cdot \frac{5}{13} + 26 \cdot \frac{12}{13} = \frac{327}{13}$.

33. (a) $$\nabla(au+bv) = \left\langle \frac{\partial(au+bv)}{\partial x}, \frac{\partial(au+bv)}{\partial y}\right\rangle = \left\langle a\,\frac{\partial u}{\partial x} + b\,\frac{\partial v}{\partial x}, a\,\frac{\partial u}{\partial y} + b\,\frac{\partial v}{\partial y}\right\rangle = a\left\langle \frac{\partial u}{\partial x}, \frac{\partial u}{\partial y}\right\rangle + b\left\langle \frac{\partial v}{\partial x}, \frac{\partial v}{\partial y}\right\rangle$$

$$= a\,\nabla u + b\,\nabla v$$

(b) $$\nabla(uv) = \left\langle \frac{\partial(uv)}{\partial x}, \frac{\partial(uv)}{\partial y}\right\rangle = \left\langle v\,\frac{\partial u}{\partial x} + u\,\frac{\partial v}{\partial x}, v\,\frac{\partial u}{\partial y} + u\,\frac{\partial v}{\partial y}\right\rangle = v\left\langle \frac{\partial u}{\partial x}, \frac{\partial u}{\partial y}\right\rangle + u\left\langle \frac{\partial v}{\partial x}, \frac{\partial v}{\partial y}\right\rangle = v\,\nabla u + u\,\nabla v$$

(c) $$\nabla\left(\frac{u}{v}\right) = \left\langle \frac{\partial(u/v)}{\partial x}, \frac{\partial(u/v)}{\partial y}\right\rangle = \left\langle \frac{v\,\frac{\partial u}{\partial x} - u\,\frac{\partial v}{\partial x}}{v^2}, \frac{v\,\frac{\partial u}{\partial y} - u\,\frac{\partial v}{\partial y}}{v^2}\right\rangle = \frac{v\left\langle \frac{\partial u}{\partial x}, \frac{\partial u}{\partial y}\right\rangle - u\left\langle \frac{\partial v}{\partial x}, \frac{\partial v}{\partial y}\right\rangle}{v^2}$$

$$= \frac{v\,\nabla u - u\,\nabla v}{v^2}$$

(d) $$\nabla u^n = \left\langle \frac{\partial(u^n)}{\partial x}, \frac{\partial(u^n)}{\partial y}\right\rangle = \left\langle nu^{n-1}\,\frac{\partial u}{\partial x}, nu^{n-1}\,\frac{\partial u}{\partial y}\right\rangle = nu^{n-1}\,\nabla u.$$

35. Let $F(x,y,z) = x^2 - 2y^2 + z^2 + yz$. Then $x^2 - 2y^2 + z^2 + yz = 2$ is a level surface of F

and $\nabla F(x,y,z) = \langle 2x, -4y+z, 2z+y\rangle$.

(a) $\nabla F(2,1,-1) = \langle 4, -5, -1\rangle$ is a normal vector for the tangent plane at $(2,1,-1)$, so an equation of the tangent plane is $4(x-2) - 5(y-1) - 1(z+1) = 0$ or $4x - 5y - z = 4$.

(b) The normal line has direction $\langle 4, -5, -1\rangle$, so parametric equations are $x = 2 + 4t$, $y = 1 - 5t$, $z = -1 - t$, and symmetric equations are $\dfrac{x-2}{4} = \dfrac{y-1}{-5} = \dfrac{z+1}{-1}$.

37. $F(x,y,z) = -z + xe^y \cos z \quad \Rightarrow \quad \nabla F(x,y,z) = \langle e^y \cos z, xe^y \cos z, -1 - xe^y \sin z\rangle$, $\nabla F(1,0,0) = \langle 1, 1, -1\rangle$

(a) $1(x-1) + 1(y-0) - 1(z-0) = 0$ or $x + y - z = 1$

(b) $\dfrac{x-1}{1} = \dfrac{y-0}{1} = \dfrac{z-0}{-1}$ or $x - 1 = y = -z$

39. $F(x,y,z) = xy + yz + zx$, $\nabla F(x,y,z) = \langle y+z, x+z, y+x\rangle$, $\nabla F(1,1,1) = \langle 2,2,2\rangle$, so an equation of the tangent plane is $2x + 2y + 2z = 6$ or $x + y + z = 3$, and the normal line is given by $x - 1 = y - 1 = z - 1$ or $x = y = z$.

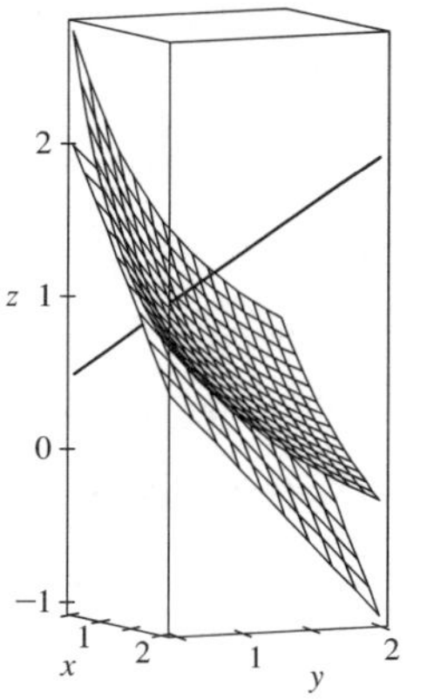

41. $\nabla f(x,y) = \langle 2x, 8y\rangle$, $\nabla f(2,1) = \langle 4, 8\rangle$. The tangent line has equation $\nabla f(2,1)\cdot\langle x-2, y-1\rangle = 0 \Rightarrow$ $4(x-2)+8(y-1)=0$, which simplifies to $x+2y=4$.

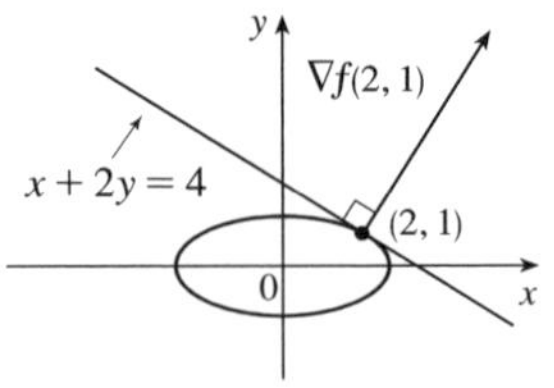

43. $\nabla F(x_0,y_0,z_0) = \left\langle \dfrac{2x_0}{a^2}, \dfrac{2y_0}{b^2}, \dfrac{2z_0}{c^2}\right\rangle$. Thus an equation of the tangent plane at (x_0,y_0,z_0) is

$\dfrac{2x_0}{a^2}x + \dfrac{2y_0}{b^2}y + \dfrac{2z_0}{c^2}z = 2\left(\dfrac{x_0^2}{a^2}+\dfrac{y_0^2}{b^2}+\dfrac{z_0^2}{c^2}\right) = 2(1) = 2$ since (x_0,y_0,z_0) is a point on the ellipsoid. Hence

$\dfrac{x_0}{a^2}x + \dfrac{y_0}{b^2}y + \dfrac{z_0}{c^2}z = 1$ is an equation of the tangent plane.

45. $\nabla f(x_0,y_0,z_0) = \langle 2x_0, -2y_0, 4z_0\rangle$ and the given line has direction numbers 2, 4, 6, so $\langle 2x_0,-2y_0,4z_0\rangle = k\langle 2,4,6\rangle$ or $x_0 = k$, $y_0=-2k$ and $z_0 = \frac{3}{2}k$. But $x_0^2 - y_0^2 + 2z_0^2 = 1$ or $\left(1-4+\frac{9}{2}\right)k^2 = 1$, so $k = \pm\sqrt{\frac{2}{3}} = \pm\frac{\sqrt{6}}{3}$ and there are two such points: $\left(\pm\frac{\sqrt{6}}{3}, \mp\frac{2\sqrt{6}}{3}, \pm\frac{\sqrt{6}}{2}\right)$.

47. Let (x_0,y_0,z_0) be a point on the surface. Then an equation of the tangent plane at the point is

$\dfrac{x}{2\sqrt{x_0}} + \dfrac{y}{2\sqrt{y_0}} + \dfrac{z}{2\sqrt{z_0}} = \dfrac{\sqrt{x_0}+\sqrt{y_0}+\sqrt{z_0}}{2}$. But $\sqrt{x_0}+\sqrt{y_0}+\sqrt{z_0} = \sqrt{c}$, so the equation is

$\dfrac{x}{\sqrt{x_0}} + \dfrac{y}{\sqrt{y_0}} + \dfrac{z}{\sqrt{z_0}} = \sqrt{c}$. The x-, y-, and z-intercepts are $\sqrt{cx_0}$, $\sqrt{cy_0}$ and $\sqrt{cz_0}$ respectively. (The x-intercept is found by setting $y=z=0$ and solving the resulting equation for x, and the y- and z-intercepts are found similarly.) So the sum of the intercepts is $\sqrt{c}\left(\sqrt{x_0}+\sqrt{y_0}+\sqrt{z_0}\right) = c$, a constant.

49. If $f(x,y,z) = z - x^2 - y^2$ and $g(x,y,z) = 4x^2+y^2+z^2$, then the tangent line is perpendicular to both ∇f and ∇g at $(-1,1,2)$. The vector $\mathbf{v} = \nabla f \times \nabla g$ will therefore be parallel to the tangent line. We have: $\nabla f(x,y,z) = \langle -2x,-2y,1\rangle$ $\Rightarrow$ $\nabla f(-1,1,2) = \langle 2,-2,1\rangle$, and $\nabla g(x,y,z) = \langle 8x, 2y, 2z\rangle \Rightarrow \nabla g(-1,1,2) = \langle -8,2,4\rangle$. Hence

$$\mathbf{v} = \nabla f\times\nabla g = \begin{vmatrix} \mathbf{i} & \mathbf{j} & \mathbf{k} \\ 2 & -2 & 1 \\ -8 & 2 & 4 \end{vmatrix} = -10\,\mathbf{i} - 16\,\mathbf{j} - 12\,\mathbf{k}.$$ Parametric equations are: $x = -1-10t$, $y = 1-16t$, $z = 2-12t$.

51. (a) The direction of the normal line of F is given by ∇F, and that of G by ∇G. Assuming that $\nabla F \neq 0 \neq \nabla G$, the two normal lines are perpendicular at P if $\nabla F\cdot\nabla G = 0$ at P $\Leftrightarrow$ $\langle \partial F/\partial x, \partial F/\partial y, \partial F/\partial z\rangle \cdot \langle \partial G/\partial x, \partial G/\partial y, \partial G/\partial z\rangle = 0$ at P $\Leftrightarrow$ $F_xG_x + F_yG_y + F_zG_z = 0$ at P.

(b) Here $F = x^2+y^2-z^2$ and $G = x^2+y^2+z^2-r^2$, so

$\nabla F\cdot\nabla G = \langle 2x,2y,-2z\rangle\cdot\langle 2x,2y,2z\rangle = 4x^2+4y^2-4z^2 = 4F = 0$, since the point (x,y,z) lies on the graph of $F=0$. To see that this is true without using calculus, note that $G=0$ is the equation of a sphere centred at the origin and $F=0$ is the equation of a right circular cone with vertex at the origin (which is generated by lines through the origin). At any point of intersection, the sphere's normal line (which passes through the origin) lies on the cone, and thus is perpendicular to the cone's normal line. So the surfaces with equations $F=0$ and $G=0$ are everywhere orthogonal.

53. Let $\mathbf{u} = \langle a, b \rangle$ and $\mathbf{v} = \langle c, d \rangle$. Then we know that at the given point, $D_{\mathbf{u}} f = \nabla f \cdot \mathbf{u} = af_x + bf_y$ and $D_{\mathbf{v}} f = \nabla f \cdot \mathbf{v} = cf_x + df_y$. But these are just two linear equations in the two unknowns f_x and f_y, and since $\mathbf{u}$ and $\mathbf{v}$ are not parallel, we can solve the equations to find $\nabla f = \langle f_x, f_y \rangle$ at the given point. In fact,

$$\nabla f = \left\langle \frac{d\,D_{\mathbf{u}} f - b\,D_{\mathbf{v}} f}{ad - bc}, \frac{a\,D_{\mathbf{v}} f - c\,D_{\mathbf{u}} f}{ad - bc} \right\rangle.$$

11.7 Maximum and Minimum Values

1. (a) First we compute $D(1, 1) = f_{xx}(1, 1)\, f_{yy}(1, 1) - [f_{xy}(1, 1)]^2 = (4)(2) - (1)^2 = 7$. Since $D(1, 1) > 0$ and $f_{xx}(1, 1) > 0$, f has a local minimum at $(1, 1)$ by the Second Derivatives Test.

(b) $D(1, 1) = f_{xx}(1, 1)\, f_{yy}(1, 1) - [f_{xy}(1, 1)]^2 = (4)(2) - (3)^2 = -1$. Since $D(1, 1) < 0$, f has a saddle point at $(1, 1)$ by the Second Derivatives Test.

3. In the figure, a point at approximately $(1, 1)$ is enclosed by level curves which are oval in shape and indicate that as we move away from the point in any direction the values of f are increasing. Hence we would expect a local minimum at or near $(1, 1)$. The level curves near $(0, 0)$ resemble hyperbolas, and as we move away from the origin, the values of f increase in some directions and decrease in others, so we would expect to find a saddle point there.

To verify our predictions, we have $f(x, y) = 4 + x^3 + y^3 - 3xy \;\Rightarrow\; f_x(x, y) = 3x^2 - 3y$, $f_y(x, y) = 3y^2 - 3x$. We have critical points where these partial derivatives are equal to 0: $3x^2 - 3y = 0$, $3y^2 - 3x = 0$. Substituting $y = x^2$ from the first equation into the second equation gives $3(x^2)^2 - 3x = 0 \;\Rightarrow\; 3x(x^3 - 1) = 0 \;\Rightarrow\; x = 0$ or $x = 1$. Then we have two critical points, $(0, 0)$ and $(1, 1)$. The second partial derivatives are $f_{xx}(x, y) = 6x$, $f_{xy}(x, y) = -3$, and $f_{yy}(x, y) = 6y$, so $D(x, y) = f_{xx}(x, y)\, f_{yy}(x, y) - [f_{xy}(x, y)]^2 = (6x)(6y) - (-3)^2 = 36xy - 9$. Then $D(0, 0) = 36(0)(0) - 9 = -9$, and $D(1, 1) = 36(1)(1) - 9 = 27$. Since $D(0, 0) < 0$, f has a saddle point at $(0, 0)$ by the Second Derivatives Test. Since $D(1, 1) > 0$ and $f_{xx}(1, 1) > 0$, f has a local minimum at $(1, 1)$.

5. $f(x, y) = 9 - 2x + 4y - x^2 - 4y^2 \;\Rightarrow\; f_x = -2 - 2x$, $f_y = 4 - 8y$, $f_{xx} = -2$, $f_{xy} = 0$, $f_{yy} = -8$. Then $f_x = 0$ and $f_y = 0$ imply $x = -1$ and $y = \frac{1}{2}$, and the only critical point is $\left(-1, \frac{1}{2}\right)$. $D(x, y) = f_{xx} f_{yy} - (f_{xy})^2 = (-2)(-8) - 0^2 = 16$, and since $D\left(-1, \frac{1}{2}\right) = 16 > 0$ and $f_{xx}\left(-1, \frac{1}{2}\right) = -2 < 0$, $f\left(-1, \frac{1}{2}\right) = 11$ is a local maximum by the Second Derivatives Test.

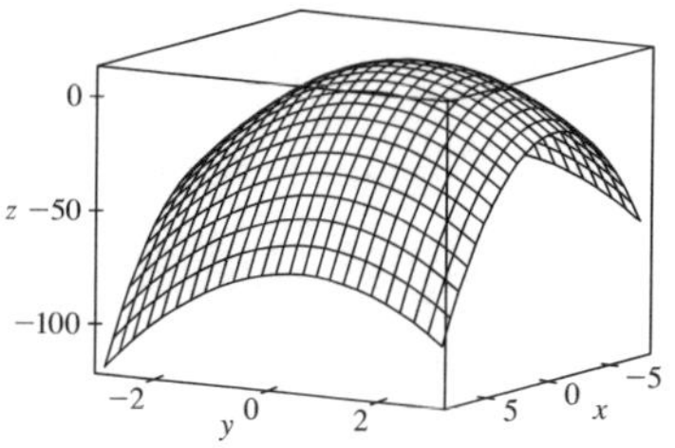

7. $f(x, y) = x^2 + y^2 + x^2 y + 4 \;\Rightarrow\; f_x = 2x + 2xy$, $f_y = 2y + x^2$, $f_{xx} = 2 + 2y$, $f_{yy} = 2$, $f_{xy} = 2x$. Then $f_y = 0$ implies $y = -\frac{1}{2}x^2$, substituting into $f_x = 0$ gives $2x - x^3 = 0$ so $x = 0$ or $x = \pm\sqrt{2}$. Thus the critical points are $(0, 0)$, $\left(\sqrt{2}, -1\right)$ and $\left(-\sqrt{2}, -1\right)$. Now $D(0, 0) = 4$, $D\left(\sqrt{2}, -1\right) = -8 = Ds\left(-\sqrt{2}, -1\right)$, $f_{xx}(0, 0) = 2$, $f_{xx}\left(\pm\sqrt{2}, -1\right) = 0$. Thus $f(0, 0) = 4$ is a local minimum and $\left(\pm\sqrt{2}, -1\right)$ are saddle points.

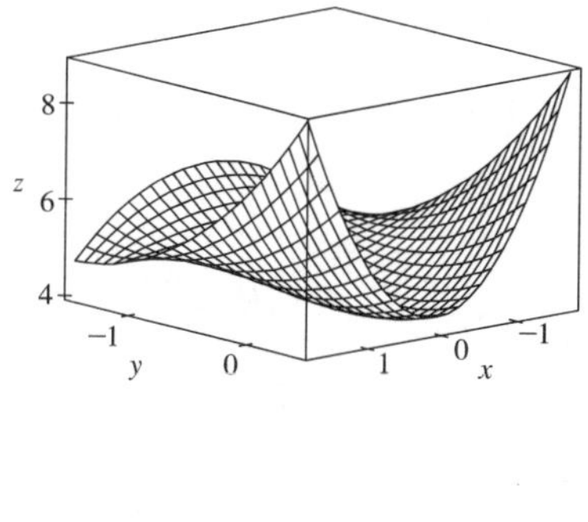

9. $f(x,y) = xy - 2x - y \quad \Rightarrow \quad f_x = y - 2,\ f_y = x - 1,$

$f_{xx} = f_{yy} = 0,\ f_{xy} = 1$ and the only critical point is $(1,2)$. Now

$D(1,2) = -1$, so $(1,2)$ is a saddle point and f has no local

maximum or minimum.

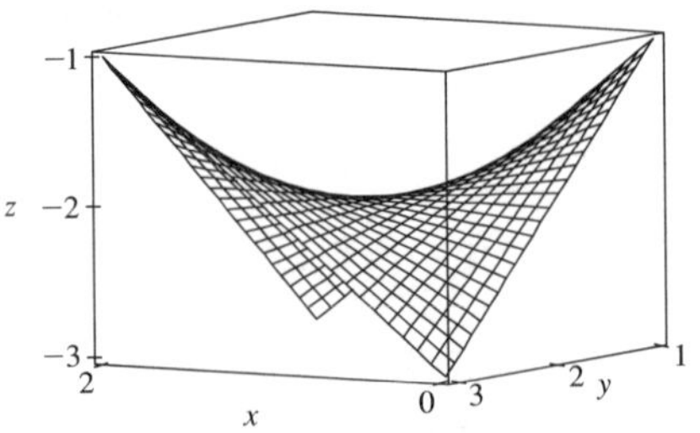

11. $f(x,y) = e^x \cos y \quad \Rightarrow \quad f_x = e^x \cos y,\ f_y = -e^x \sin y$. Now $f_x = 0$

implies $\cos y = 0$ or $y = \frac{\pi}{2} + n\pi$ for n an integer. But $\sin\left(\frac{\pi}{2} + n\pi\right) \neq 0$,

so there are no critical points.

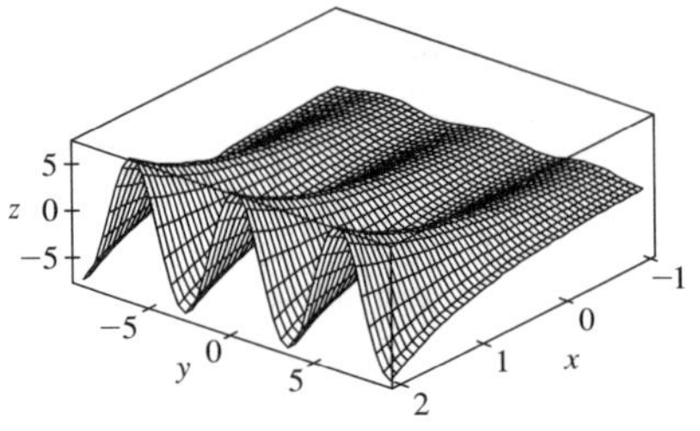

13. $f(x,y) = x \sin y \quad \Rightarrow \quad f_x = \sin y,\ f_y = x \cos y,\ f_{xx} = 0,\ f_{yy} = -x$

$\sin y$ and $f_{xy} = \cos y$. Then $f_x = 0$ if and only if $y = n\pi$, n an integer,

and substituting into $f_y = 0$ requires $x = 0$ for each of these y-values.

Thus the critical points are $(0, n\pi)$, n an integer. But

$D(0, n\pi) = -\cos^2(n\pi) < 0$ so each critical point is a saddle point.

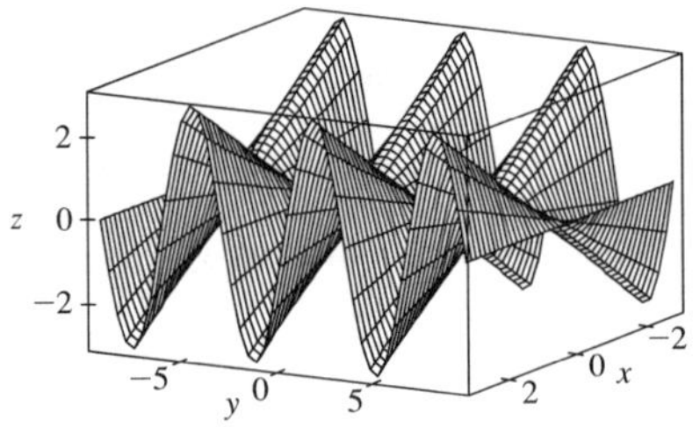

15. $f(x,y) = (x^2 + y^2)e^{y^2 - x^2} \quad \Rightarrow \quad f_x = (x^2 + y^2)e^{y^2 - x^2}(-2x) + 2xe^{y^2 - x^2} = 2xe^{y^2 - x^2}(1 - x^2 - y^2),$

$f_y = (x^2 + y^2)e^{y^2 - x^2}(2y) + 2ye^{y^2 - x^2} = 2ye^{y^2 - x^2}(1 + x^2 + y^2),$

$f_{xx} = 2xe^{y^2 - x^2}(-2x) + (1 - x^2 - y^2)\left(2x\left(-2xe^{y^2 - x^2}\right) + 2e^{y^2 - x^2}\right) = 2e^{y^2 - x^2}((1 - x^2 - y^2)(1 - 2x^2) - 2x^2),$

$f_{xy} = 2xe^{y^2 - x^2}(-2y) + 2x(2y)e^{y^2 - x^2}(1 - x^2 - y^2) = -4xye^{y^2 - x^2}(x^2 + y^2),$

$f_{yy} = 2ye^{y^2 - x^2}(2y) + (1 + x^2 + y^2)\left(2y\left(2ye^{y^2 - x^2}\right) + 2e^{y^2 - x^2}\right) = 2e^{y^2 - x^2}((1 + x^2 + y^2)(1 + 2y^2) + 2y^2).$

$f_y = 0$ implies $y = 0$, and substituting into $f_x = 0$ gives

$2xe^{-x^2}(1 - x^2) = 0 \quad \Rightarrow \quad x = 0$ or $x = \pm 1$.

Thus the critical points are $(0,0)$ and $(\pm 1, 0)$. $D(0,0) = (2)(2) - 0 > 0$

and $f_{xx}(0,0) = 2 > 0$, so $f(0,0) = 0$ is a local minimum.

$D(\pm 1, 0) = (-4e^{-1})(4e^{-1}) - 0 < 0$ so $(\pm 1, 0)$ are saddle points.

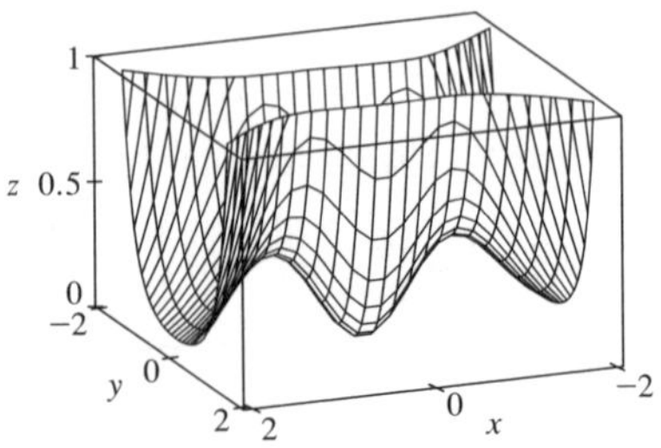

17. $f(x,y) = 3x^2y + y^3 - 3x^2 - 3y^2 + 2$

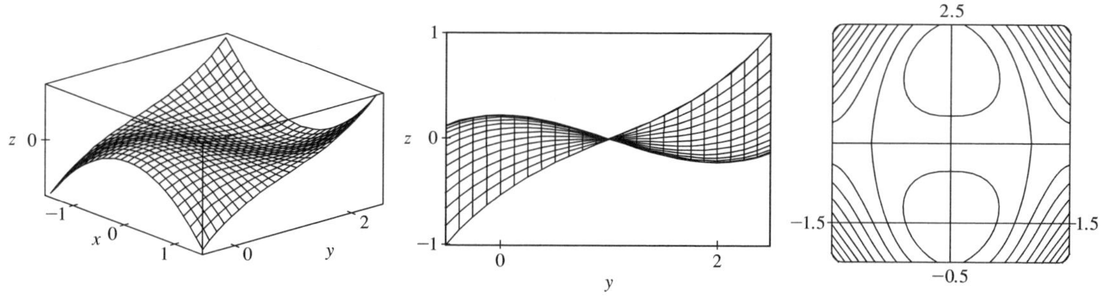

From the graphs, it appears that f has a local maximum $f(0,0) \approx 2$ and a local minimum $f(0,2) \approx -2$. There appear to be saddle points near $(\pm 1, 1)$.

$f_x = 6xy - 6x$, $f_y = 3x^2 + 3y^2 - 6y$. Then $f_x = 0$ implies $x = 0$ or $y = 1$ and when $x = 0$, $f_y = 0$ implies $y = 0$ or $y = 2$; when $y = 1$, $f_y = 0$ implies $x^2 = 1$ or $x = \pm 1$. Thus the critical points are $(0,0)$, $(0,2)$, $(\pm 1, 1)$. Now $f_{xx} = 6y - 6$, $f_{yy} = 6y - 6$ and $f_{xy} = 6x$, so $D(0,0) = D(0,2) = 36 > 0$ while $D(\pm 1, 1) = -36 < 0$ and $f_{xx}(0,0) = -6$, $f_{xx}(0,2) = 6$. Hence $(\pm 1, 1)$ are saddle points while $f(0,0) = 2$ is a local maximum and $f(0,2) = -2$ is a local minimum.

19. $f(x,y) = \sin x + \sin y + \sin(x+y)$, $0 \le x \le 2\pi$, $0 \le y \le 2\pi$

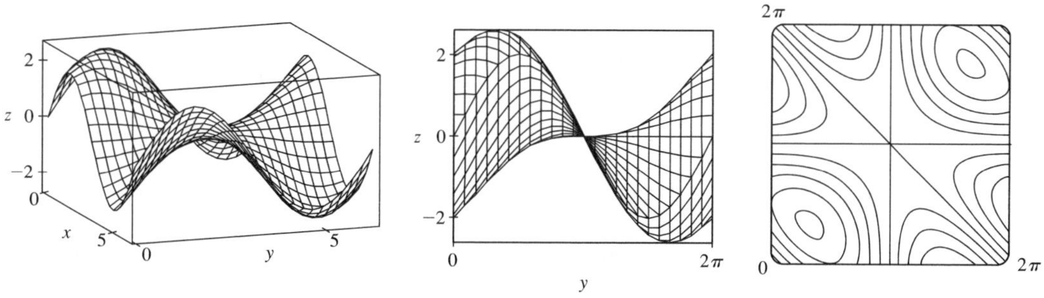

From the graphs it appears that f has a local maximum at about $(1,1)$ with value approximately 2.6, a local minimum at about $(5,5)$ with value approximately -2.6, and a saddle point at about $(3,3)$.

$f_x = \cos x + \cos(x+y)$, $f_y = \cos y + \cos(x+y)$, $f_{xx} = -\sin x - \sin(x+y)$, $f_{yy} = -\sin y - \sin(x+y)$, $f_{xy} = -\sin(x+y)$. Setting $f_x = 0$ and $f_y = 0$ and subtracting gives $\cos x - \cos y = 0$ or $\cos x = \cos y$. Thus $x = y$ or $x = 2\pi - y$. If $x = y$, $f_x = 0$ becomes $\cos x + \cos 2x = 0$ or $2\cos^2 x + \cos x - 1 = 0$, a quadratic in $\cos x$. Thus $\cos x = -1$ or $\frac{1}{2}$ and $x = \pi$, $\frac{\pi}{3}$, or $\frac{5\pi}{3}$, yielding the critical points (π,π), $\left(\frac{\pi}{3}, \frac{\pi}{3}\right)$ and $\left(\frac{5\pi}{3}, \frac{5\pi}{3}\right)$. Similarly if $x = 2\pi - y$, $f_x = 0$ becomes $(\cos x) + 1 = 0$ and the resulting critical point is (π, π). Now $D(x,y) = \sin x \sin y + \sin x \sin(x+y)$ $+ \sin y \sin(x+y)$. So $D(\pi,\pi) = 0$ and the Second Derivatives Test doesn't apply. $D\left(\frac{\pi}{3}, \frac{\pi}{3}\right) = \frac{9}{4} > 0$ and $f_{xx}\left(\frac{\pi}{3}, \frac{\pi}{3}\right) < 0$ so $f\left(\frac{\pi}{3}, \frac{\pi}{3}\right) = \frac{3\sqrt{3}}{2}$ is a local maximum while $D\left(\frac{5\pi}{3}, \frac{5\pi}{3}\right) = \frac{9}{4} > 0$ and $f_{xx}\left(\frac{5\pi}{3}, \frac{5\pi}{3}\right) > 0$, so $f\left(\frac{5\pi}{3}, \frac{5\pi}{3}\right) = -\frac{3\sqrt{3}}{2}$ is a local minimum.

21. $f(x,y) = x^4 - 5x^2 + y^2 + 3x + 2 \;\Rightarrow\; f_x(x,y) = 4x^3 - 10x + 3$ and $f_y(x,y) = 2y$. $f_y = 0 \;\Rightarrow\; y = 0$, and the graph of f_x shows that the roots of $f_x = 0$ are approximately $x = -1.714$, 0.312 and 1.402. (Alternatively, we could have used a calculator or a CAS to find these roots.) So to three decimal places, the critical points are $(-1.714, 0)$, $(1.402, 0)$, and $(0.312, 0)$. Now since $f_{xx} = 12x^2 - 10$, $f_{xy} = 0$, $f_{yy} = 2$, and $D = 24x^2 - 20$, we have $D(-1.714, 0) > 0$, $f_{xx}(-1.714, 0) > 0$, $D(1.402, 0) > 0$, $f_{xx}(1.402, 0) > 0$, and $D(0.312, 0) < 0$. Therefore $f(-1.714, 0) \approx -9.200$ and $f(1.402, 0) \approx 0.242$ are local minima, and $(0.312, 0)$ is a saddle point. The lowest point on the graph is approximately $(-1.714, 0, -9.200)$.

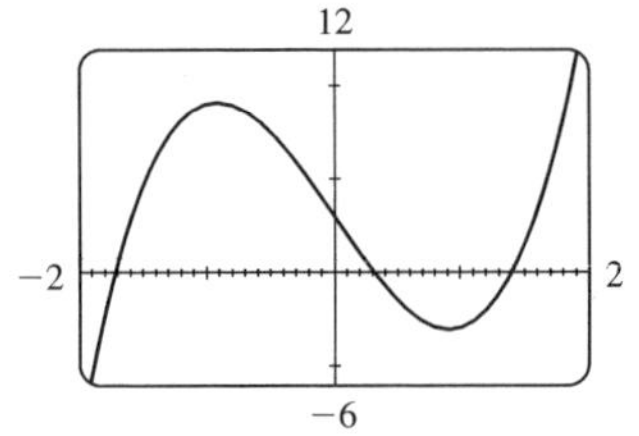

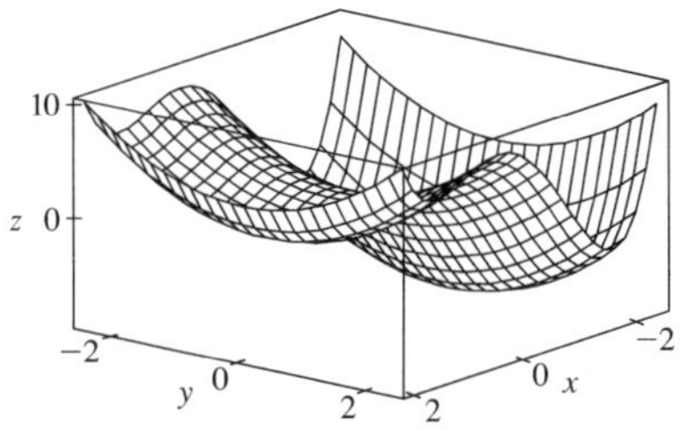

23. $f(x,y) = 2x + 4x^2 - y^2 + 2xy^2 - x^4 - y^4 \;\Rightarrow\; f_x(x,y) = 2 + 8x + 2y^2 - 4x^3$, $f_y(x,y) = -2y + 4xy - 4y^3$. Now $f_y = 0 \;\Leftrightarrow\; 2y(2y^2 - 2x + 1) = 0 \;\Leftrightarrow\; y = 0$ or $y^2 = x - \frac{1}{2}$. The first of these implies that $f_x = -4x^3 + 8x + 2$, and the second implies that $f_x = 2 + 8x + 2(x - \frac{1}{2}) - 4x^3 = -4x^3 + 10x + 1$. From the graphs, we see that the first possibility for f_x has roots at approximately -1.267, -0.259, and 1.526, and the second has a root at approximately 1.629 (the negative roots do not give critical points, since $y^2 = x - \frac{1}{2}$ must be positive). So to three decimal places, f has critical points at $(-1.267, 0)$, $(-0.259, 0)$, $(1.526, 0)$, and $(1.629, \pm 1.063)$. Now since $f_{xx} = 8 - 12x^2$, $f_{xy} = 4y$, $f_{yy} = 4x - 12y^2$, and $D = (8 - 12x^2)(4x - 12y^2) - 16y^2$, we have $D(-1.267, 0) > 0$, $f_{xx}(-1.267, 0) > 0$, $D(-0.259, 0) < 0$, $D(1.526, 0) < 0$, $D(1.629, \pm 1.063) > 0$, and $f_{xx}(1.629, \pm 1.063) < 0$. Therefore, to three decimal places, $f(-1.267, 0) \approx 1.310$ and $f(1.629, \pm 1.063) \approx 8.105$ are local maxima, and $(-0.259, 0)$ and $(1.526, 0)$ are saddle points. The highest points on the graph are approximately $(1.629, \pm 1.063, 8.105)$.

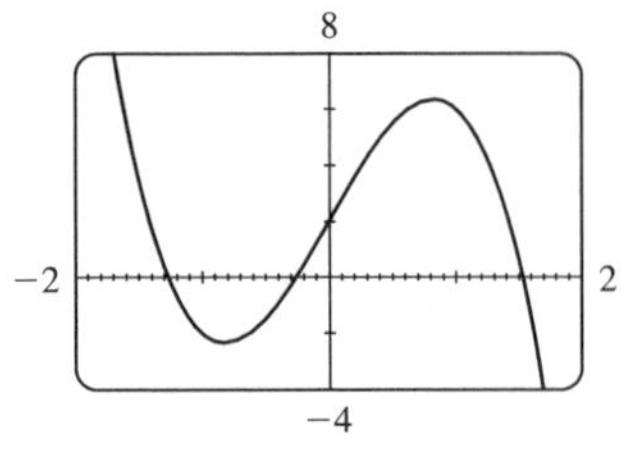

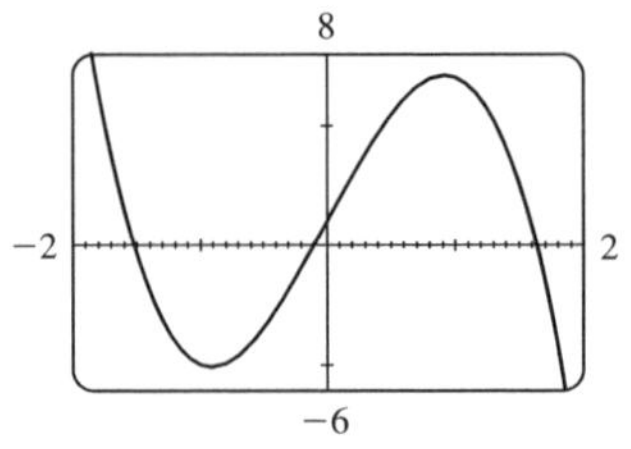

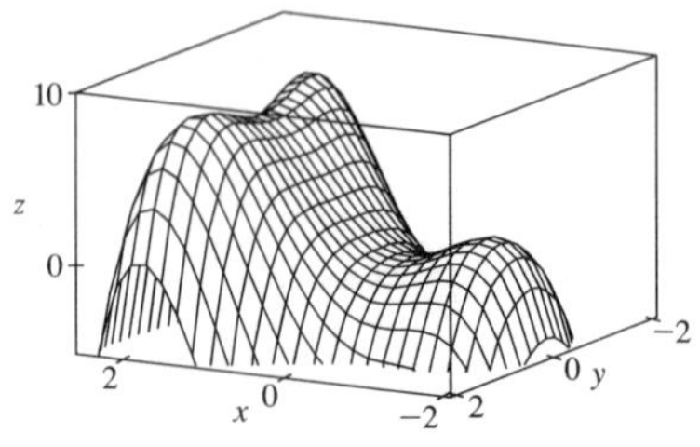

25. Since f is a polynomial it is continuous on D, so an absolute maximum and minimum exist. Here $f_x = 4$, $f_y = -5$ so there are no critical points inside D. Thus the absolute extrema must both occur on the boundary. Along L_1, $x = 0$ and $f(0, y) = 1 - 5y$ for $0 \le y \le 3$, a decreasing function in y, so the maximum value is $f(0, 0) = 1$ and the minimum value is $f(0, 3) = -14$. Along L_2, $y = 0$ and $f(x, 0) = 1 + 4x$ for $0 \le x \le 2$, an increasing function in x, so the minimum value is $f(0, 0) = 1$ and the maximum value is $f(2, 0) = 9$. Along L_3, $y = -\frac{3}{2}x + 3$ and $f\left(x, -\frac{3}{2}x + 3\right) = \frac{23}{2}x - 14$ for $0 \le x \le 2$, an increasing function in x, so the minimum value is $f(0, 3) = -14$ and the maximum value is $f(2, 0) = 9$. Thus the absolue maximum of f on D is $f(2, 0) = 9$ and the absolute minimum is $f(0, 3) = -14$.

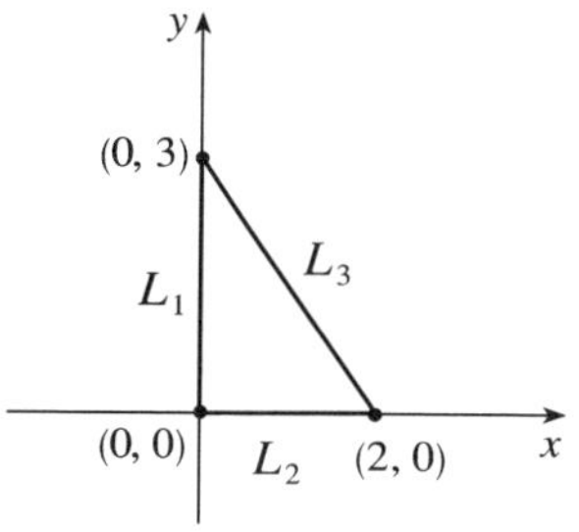

27. $f_x(x, y) = 2x + 2xy$, $f_y(x, y) = 2y + x^2$, and setting $f_x = f_y = 0$ gives $(0, 0)$ as the only critical point in D, with $f(0, 0) = 4$.
On L_1: $y = -1$, $f(x, -1) = 5$, a constant.
On L_2: $x = 1$, $f(1, y) = y^2 + y + 5$, a quadratic in y which attains its maximum at $(1, 1)$, $f(1, 1) = 7$ and its minimum at $\left(1, -\frac{1}{2}\right)$, $f\left(1, -\frac{1}{2}\right) = \frac{19}{4}$.
On L_3: $f(x, 1) = 2x^2 + 5$ which attains its maximum at $(-1, 1)$ and $(1, 1)$ with $f(\pm 1, 1) = 7$ and its minimum at $(0, 1)$, $f(0, 1) = 5$.

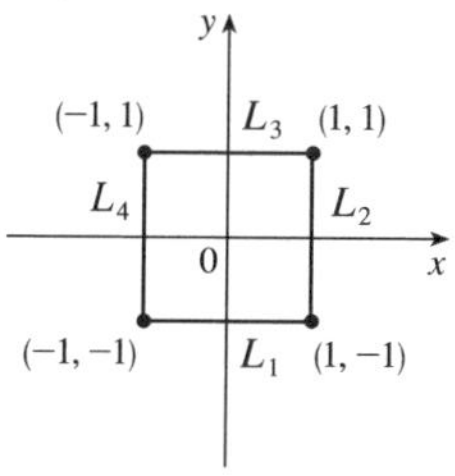

On L_4: $f(-1, y) = y^2 + y + 5$ with maximum at $(-1, 1)$, $f(-1, 1) = 7$ and minimum at $\left(-1, -\frac{1}{2}\right)$, $f\left(-1, -\frac{1}{2}\right) = \frac{19}{4}$.
Thus the absolute maximum is attained at both $(\pm 1, 1)$ with $f(\pm 1, 1) = 7$ and the absolute minimum on D is attained at $(0, 0)$ with $f(0, 0) = 4$.

29. $f(x, y) = x^4 + y^4 - 4xy + 2$ is a polynomial and hence continuous on D, so it has an absolute maximum and minimum on D. In Exercise 7, we found the critical points of f; only $(1, 1)$ with $f(1, 1) = 0$ is inside D. On L_1: $y = 0$, $f(x, 0) = x^4 + 2$, $0 \le x \le 3$, a polynomial in x which attains its maximum at $x = 3$, $f(3, 0) = 83$, and its minimum at $x = 0$, $f(0, 0) = 2$.
On L_2: $x = 3$, $f(3, y) = y^4 - 12y + 83$, $0 \le y \le 2$, a polynomial in y which attains its minimum at $y = \sqrt[3]{3}$, $f\left(3, \sqrt[3]{3}\right) = 83 - 9\sqrt[3]{3} \approx 70.0$, and its maximum at $y = 0$, $f(3, 0) = 83$.

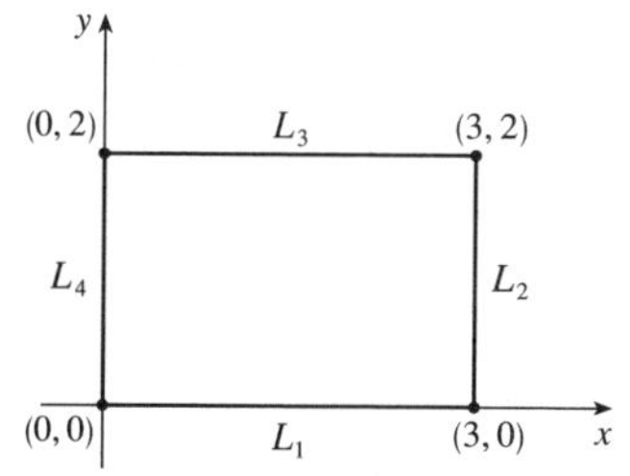

On L_3: $y = 2$, $f(x, 2) = x^4 - 8x + 18$, $0 \le x \le 3$, a polynomial in x which attains its minimum at $x = \sqrt[3]{2}$, $f\left(\sqrt[3]{2}, 2\right) = 18 - 6\sqrt[3]{2} \approx 10.4$, and its maximum at $x = 3$, $f(3, 2) = 75$. On L_4: $x = 0$, $f(0, y) = y^4 + 2$, $0 \le y \le 2$, a polynomial in y which attains its maximum at $y = 2$, $f(0, 2) = 18$, and its minimum at $y = 0$, $f(0, 0) = 2$. Thus the absolute maximum of f on D is $f(3, 0) = 83$ and the absolute minimum is $f(1, 1) = 0$.

31. $f(x,y) = -(x^2-1)^2 - (x^2y-x-1)^2 \;\Rightarrow\; f_x(x,y) = -2(x^2-1)(2x) - 2(x^2y-x-1)(2xy-1)$ and $f_y(x,y) = -2(x^2y-x-1)x^2$. Setting $f_y(x,y) = 0$ gives either $x=0$ or $x^2y-x-1=0$.

There are no critical points for $x=0$, since $f_x(0,y) = -2$, so we set $x^2y-x-1=0 \;\Leftrightarrow\; y = \dfrac{x+1}{x^2}$ $(x \neq 0)$, so

$$f_x\left(x, \frac{x+1}{x^2}\right) = -2(x^2-1)(2x) - 2\left(x^2\,\frac{x+1}{x^2} - x - 1\right)\left(2x\,\frac{x+1}{x^2} - 1\right) = -4x(x^2-1).\text{ Therefore}$$

$f_x(x,y) = f_y(x,y) = 0$ at the points $(1,2)$ and $(-1,0)$. To classify these critical points, we calculate

$f_{xx}(x,y) = -12x^2 - 12x^2y^2 + 12xy + 4y + 2$, $f_{yy}(x,y) = -2x^4$,

and $f_{xy}(x,y) = -8x^3y + 6x^2 + 4x$. In order to use the Second Derivatives Test we calculate

$D(-1,0) = f_{xx}(-1,0)\,f_{yy}(-1,0) - [f_{xy}(-1,0)]^2 = 16 > 0$,

$f_{xx}(-1,0) = -10 < 0$, $D(1,2) = 16 > 0$, and $f_{xx}(1,2) = -26 < 0$, so both $(-1,0)$ and $(1,2)$ give local maxima.

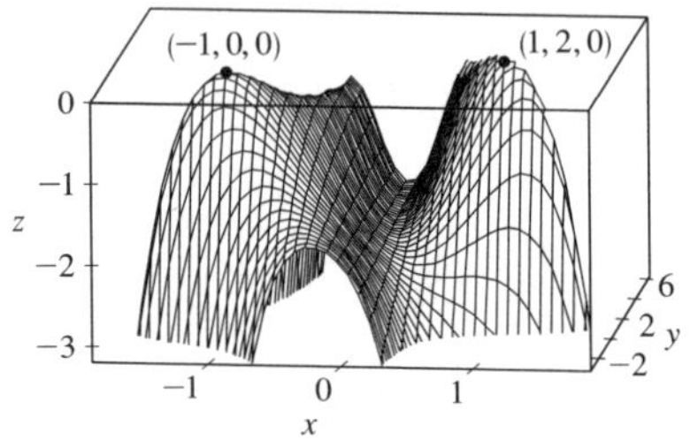

33. Let d be the distance from $(2,1,-1)$ to any point (x,y,z) on the plane $x+y-z=1$, so

$d = \sqrt{(x-2)^2 + (y-1)^2 + (z+1)^2}$ where $z = x+y-1$, and we minimize

$d^2 = f(x,y) = (x-2)^2 + (y-1)^2 + (x+y)^2$. Then $f_x(x,y) = 2(x-2) + 2(x+y) = 4x+2y-4$,

$f_y(x,y) = 2(y-1) + 2(x+y) = 2x+4y-2$. Solving $4x+2y-4=0$ and $2x+4y-2=0$ simultaneously gives $x=1$, $y=0$. An absolute minimum exists (since there is a minimum distance from the point to the plane) and it must occur at a critical point, so the shortest distance occurs for $x=1$, $y=0$ for which $d = \sqrt{(1-2)^2 + (0-1)^2 + (0+1)^2} = \sqrt{3}$.

35. Let d be the distance from the point $(4,2,0)$ to any point (x,y,z) on the cone, so $d = \sqrt{(x-4)^2 + (y-2)^2 + z^2}$ where $z^2 = x^2 + y^2$, and we minimize $d^2 = (x-4)^2 + (y-2)^2 + x^2 + y^2 = f(x,y)$. Then

$f_x(x,y) = 2(x-4) + 2x = 4x - 8$, $f_y(x,y) = 2(y-2) + 2y = 4y - 4$, and the critical points occur when

$f_x = 0 \;\Rightarrow\; x = 2$, $f_y = 0 \;\Rightarrow\; y = 1$. Thus the only critical point is $(2,1)$. An absolute minimum exists (since there is a minimum distance from the cone to the point) which must occur at a critical point, so the points on the cone closest to $(4,2,0)$ are $\left(2, 1, \pm\sqrt{5}\right)$.

37. $x+y+z=100$, so maximize $f(x,y) = xy(100-x-y)$. $f_x = 100y - 2xy - y^2$, $f_y = 100x - x^2 - 2xy$, $f_{xx} = -2y$, $f_{yy} = -2x$, $f_{xy} = 100 - 2x - 2y$. Then $f_x = 0$ implies $y = 0$ or $y = 100 - 2x$. Substituting $y = 0$ into $f_y = 0$ gives $x = 0$ or $x = 100$ and substituting $y = 100 - 2x$ into $f_y = 0$ gives $3x^2 - 100x = 0$ so $x = 0$ or $\frac{100}{3}$. Thus the critical points are $(0,0)$, $(100,0)$, $(0,100)$ and $\left(\frac{100}{3}, \frac{100}{3}\right)$.

$D(0,0) = D(100,0) = D(0,100) = -10{,}000$ while $D\left(\frac{100}{3}, \frac{100}{3}\right) = \frac{10{,}000}{3}$ and $f_{xx}\left(\frac{100}{3}, \frac{100}{3}\right) = -\frac{200}{3} < 0$. Thus $(0,0)$, $(100,0)$ and $(0,100)$ are saddle points whereas $f\left(\frac{100}{3}, \frac{100}{3}\right)$ is a local maximum. Thus the numbers are $x = y = z = \frac{100}{3}$.

39. Maximize $f(x,y) = xy\left(36 - 9x^2 - 36y^2\right)^{1/2}/\,2$ with (x,y,z) in first octant. Then

$$f_x = \frac{y(36 - 9x^2 - 36y^2)^{1/2}}{2} + \frac{-9x^2y(36 - 9x^2 - 36y^2)^{-1/2}}{2} = \frac{(36y - 18x^2y - 36y^3)}{2(36 - 9x^2 - 36y^2)^{1/2}} \text{ and}$$

$f_y = \dfrac{36x - 9x^3 - 72xy^2}{2(36 - 9x^2 - 36y^2)^{1/2}}$. Setting $f_x = 0$ gives $y = 0$ or $y^2 = \dfrac{2 - x^2}{2}$ but $y > 0$, so only the latter solution applies.

Substituting this y into $f_y = 0$ gives $x^2 = \frac{4}{3}$ or $x = \frac{2}{\sqrt{3}}$, $y = \frac{1}{\sqrt{3}}$ and then $z^2 = (36 - 12 - 12)/4 = 3$. The fact that this gives a maximum volume follows from the geometry. This maximum volume is

$V = (2x)(2y)(2z) = 8\left(\frac{2}{\sqrt{3}}\right)\left(\frac{1}{\sqrt{3}}\right)\left(\sqrt{3}\right) = \frac{16}{\sqrt{3}}$.

41. Maximize $f(x,y) = \dfrac{xy}{3}\,(6 - x - 2y)$, then the maximum volume is $V = xyz$. $f_x = \frac{1}{3}(6y - 2xy - y^2) = \frac{1}{3}y(6 - 2x - 2y)$ and $f_y = \frac{1}{3}x\,(6 - x - 4y)$. Setting $f_x = 0$ and $f_y = 0$ gives the critical point $(2,1)$ which geometrically must yield a maximum. Thus the volume of the largest such box is $V = (2)(1)\left(\frac{2}{3}\right) = \frac{4}{3}$.

43. Let the dimensions be x, y, and z; then $4x + 4y + 4z = c$ and the volume is

$V = xyz = xy\left(\frac{1}{4}c - x - y\right) = \frac{1}{4}cxy - x^2y - xy^2$, $x > 0$, $y > 0$. Then $V_x = \frac{1}{4}cy - 2xy - y^2$ and $V_y = \frac{1}{4}cx - x^2 - 2xy$, so $V_x = 0 = V_y$ when $2x + y = \frac{1}{4}c$ and $x + 2y = \frac{1}{4}c$. Solving, we get $x = \frac{1}{12}c$, $y = \frac{1}{12}c$ and $z = \frac{1}{4}c - x - y = \frac{1}{12}c$. From the geometrical nature of the problem, this critical point must give an absolute maximum. Thus the box is a cube with edge length $\frac{1}{12}c$.

45. Let the dimensions be x, y and z, then minimize $xy + 2(xz + yz)$ if $xyz = 32\,000\text{ m}^3$. Then $f(x,y) = xy + [64\,000(x + y)/xy] = xy + 64\,000(x^{-1} + y^{-1})$, $f_x = y - 64\,000x^{-2}$, $f_y = x - 64\,000y^{-2}$. And $f_x = 0$ implies $y = 64\,000/x^2$; substituting into $f_y = 0$ implies $x^3 = 64\,000$ or $x = 40$ and then $y = 40$. Now $D(x,y) = [(2)(64\,000)]^2x^{-3}y^{-3} - 1 > 0$ for $(40,40)$ and $f_{xx}(40,40) > 0$ so this is indeed a minimum. Thus the dimensions of the box are $x = y = 40$ cm, $z = 20$ cm.

47. Let x, y, z be the dimensions of the rectangular box. Then the volume of the box is xyz and $L = \sqrt{x^2 + y^2 + z^2} \;\Rightarrow\; L^2 = x^2 + y^2 + z^2 \;\Rightarrow\; z = \sqrt{L^2 - x^2 - y^2}$. Substituting, we have volume $V(x,y) = xy\,\sqrt{L^2 - x^2 - y^2}$, $x, y > 0$.

$$V_x = xy \cdot \tfrac{1}{2}(L^2 - x^2 - y^2)^{-1/2}(-2x) + y\,\sqrt{L^2 - x^2 - y^2} = y\,\sqrt{L^2 - x^2 - y^2} - \frac{x^2y}{\sqrt{L^2 - x^2 - y^2}},$$

$$V_y = x\,\sqrt{L^2 - x^2 - y^2} - \frac{xy^2}{\sqrt{L^2 - x^2 - y^2}},$$

$V_x = 0$ implies $y(L^2 - x^2 - y^2) = x^2y \;\Rightarrow\; y(L^2 - 2x^2 - y^2) = 0 \;\Rightarrow\; 2x^2 + y^2 = L^2$ (since $y > 0$), and $V_y = 0$ implies $x(L^2 - x^2 - y^2) = xy^2 \;\Rightarrow\; x(L^2 - x^2 - 2y^2) = 0 \;\Rightarrow\; x^2 + 2y^2 = L^2$ (since $x > 0$). Substituting $y^2 = L^2 - 2x^2$ into $x^2 + 2y^2 = L^2$ gives $x^2 + 2L^2 - 4x^2 = L^2 \;\Rightarrow\; 3x^2 = L^2 \;\Rightarrow\; x = L/\sqrt{3}$ (since $x > 0$) and then $y = \sqrt{L^2 - 2\left(L/\sqrt{3}\right)^2} = L/\sqrt{3}$. So the only critical point is $\left(L/\sqrt{3}, L/\sqrt{3}\right)$ which, from the geometrical nature of the problem, must give an absolute maximum. Thus the maximum volume is

$V\left(L/\sqrt{3}, L/\sqrt{3}\right) = \left(L/\sqrt{3}\right)^2\sqrt{L^2 - \left(L/\sqrt{3}\right)^2 - \left(L/\sqrt{3}\right)^2} = L^3/\left(3\sqrt{3}\right)$ cubic units.

49. Note that here the variables are m and b, and $f(m,b) = \sum_{i=1}^{n} [y_i - (mx_i + b)]^2$. Then $f_m = \sum_{i=1}^{n} -2x_i[y_i - (mx_i + b)] = 0$ implies $\sum_{i=1}^{n} (x_i y_i - mx_i^2 - bx_i) = 0$ or $\sum_{i=1}^{n} x_i y_i = m \sum_{i=1}^{n} x_i^2 + b \sum_{i=1}^{n} x_i$ and $f_b = \sum_{i=1}^{n} -2[y_i - (mx_i + b)] = 0$ implies $\sum_{i=1}^{n} y_i = m \sum_{i=1}^{n} x_i + \sum_{i=1}^{n} b = m\left(\sum_{i=1}^{n} x_i\right) + nb$. Thus we have the two desired equations. Now $f_{mm} = \sum_{i=1}^{n} 2x_i^2$, $f_{bb} = \sum_{i=1}^{n} 2 = 2n$ and $f_{mb} = \sum_{i=1}^{n} 2x_i$. And $f_{mm}(m,b) > 0$ always and $D(m,b) = 4n\left(\sum_{i=1}^{n} x_i^2\right) - 4\left(\sum_{i=1}^{n} x_i\right)^2 = 4\left[n\left(\sum_{i=1}^{n} x_i^2\right) - \left(\sum_{i=1}^{n} x_i\right)^2\right] > 0$ always so the solutions of these two equations do indeed minimize $\sum_{i=1}^{n} d_i^2$.

11.8 Lagrange Multipliers

1. At the extreme values of f, the level curves of f just touch the curve $g(x,y) = 8$ with a common tangent line. (See Figure 1 and the accompanying discussion.) We can observe several such occurrences on the contour map, but the level curve $f(x,y) = c$ with the largest value of c which still intersects the curve $g(x,y) = 8$ is approximately $c = 59$, and the smallest value of c corresponding to a level curve which intersects $g(x,y) = 8$ appears to be $c = 30$. Thus we estimate the maximum value of f subject to the constraint $g(x,y) = 8$ to be about 59 and the minimum to be 30.

3. $f(x,y) = x^2 + y^2$, $g(x,y) = xy = 1$, and $\nabla f = \lambda \nabla g \Rightarrow \langle 2x, 2y \rangle = \langle \lambda y, \lambda x \rangle$, so $2x = \lambda y$, $2y = \lambda x$, and $xy = 1$. From the last equation, $x \neq 0$ and $y \neq 0$, so $2x = \lambda y \Rightarrow \lambda = 2x/y$. Substituting, we have $2y = (2x/y)\,x \Rightarrow y^2 = x^2 \Rightarrow y = \pm x$. But $xy = 1$, so $x = y = \pm 1$ and the possible points for the extreme values of f are $(1,1)$ and $(-1,-1)$. Here there is no maximum value, since the constraint $xy = 1$ allows x or y to become arbitrarily large, and hence $f(x,y) = x^2 + y^2$ can be made arbitrarily large. The minimum value is $f(1,1) = f(-1,-1) = 2$.

5. $f(x,y) = x^2 y$, $g(x,y) = x^2 + 2y^2 = 6 \Rightarrow \nabla f = \langle 2xy, x^2 \rangle$, $\lambda \nabla g = \langle 2\lambda x, 4\lambda y \rangle$. Then $2xy = 2\lambda x$ implies $x = 0$ or $\lambda = y$. If $x = 0$, then $x^2 = 4\lambda y$ implies $\lambda = 0$ or $y = 0$. However, if $y = 0$ then $g(x,y) = 0$, a contradiction. So $\lambda = 0$ and then $g(x,y) = 6 \Rightarrow y = \pm\sqrt{3}$. If $\lambda = y$, then $x^2 = 4\lambda y$ implies $x^2 = 4y^2$, and so $g(x,y) = 6 \Rightarrow 4y^2 + 2y^2 = 6 \Rightarrow y^2 = 1 \Rightarrow y = \pm 1$. Thus f has possible extreme values at the points $(0, \pm\sqrt{3})$, $(\pm 2, 1)$, and $(\pm 2, -1)$. After evaluating f at these points, we find the maximum value to be $f(\pm 2, 1) = 4$ and the minimum to be $f(\pm 2, -1) = -4$.

7. $f(x,y,z) = 2x + 6y + 10z$, $g(x,y,z) = x^2 + y^2 + z^2 = 35 \Rightarrow \nabla f = \langle 2, 6, 10 \rangle$, $\lambda \nabla g = \langle 2\lambda x, 2\lambda y, 2\lambda z \rangle$. Then $2\lambda x = 2$, $2\lambda y = 6$, $2\lambda z = 10$ imply $x = \frac{1}{\lambda}$, $y = \frac{3}{\lambda}$, and $z = \frac{5}{\lambda}$. But $35 = x^2 + y^2 + z^2 = \left(\frac{1}{\lambda}\right)^2 + \left(\frac{3}{\lambda}\right)^2 + \left(\frac{5}{\lambda}\right)^2 \Rightarrow 35 = \frac{35}{\lambda^2} \Rightarrow \lambda = \pm 1$, so f has possible extreme values at the points $(1,3,5)$, $(-1,-3,-5)$. The maximum value of f on $x^2 + y^2 + z^2 = 35$ is $f(1,3,5) = 70$, and the minimum is $f(-1,-3,-5) = -70$.

9. $f(x,y,z) = xyz$, $g(x,y,z) = x^2 + 2y^2 + 3z^2 = 6 \Rightarrow \nabla f = \langle yz, xz, xy\rangle$, $\lambda \nabla g = \langle 2\lambda x, 4\lambda y, 6\lambda z\rangle$. Then $\nabla f = \lambda \nabla g$ implies $\lambda = (yz)/(2x) = (xz)/(4y) = (xy)/(6z)$ or $x^2 = 2y^2$ and $z^2 = \frac{2}{3}y^2$. Thus $x^2 + 2y^2 + 3z^2 = 6$ implies $6y^2 = 6$ or $y = \pm 1$. Then the possible points are $\left(\sqrt{2}, \pm 1, \sqrt{\frac{2}{3}}\right)$, $\left(\sqrt{2}, \pm 1, -\sqrt{\frac{2}{3}}\right)$, $\left(-\sqrt{2}, \pm 1, \sqrt{\frac{2}{3}}\right)$, $\left(-\sqrt{2}, \pm 1, -\sqrt{\frac{2}{3}}\right)$. The maximum value of f on the ellipsoid is $\frac{2}{\sqrt{3}}$, occurring when all coordinates are positive or exactly two are negative and the minimum is $-\frac{2}{\sqrt{3}}$ occurring when 1 or 3 of the coordinates are negative.

11. $f(x,y,z) = x^2 + y^2 + z^2$, $g(x,y,z) = x^4 + y^4 + z^4 = 1 \Rightarrow \nabla f = \langle 2x, 2y, 2z\rangle$, $\lambda\nabla g = \langle 4\lambda x^3, 4\lambda y^3, 4\lambda z^3\rangle$.

Case 1: If $x \neq 0$, $y \neq 0$ and $z \neq 0$, then $\nabla f = \lambda \nabla g$ implies $\lambda = 1/(2x^2) = 1/(2y^2) = 1/(2z^2)$ or $x^2 = y^2 = z^2$ and $3x^4 = 1$ or $x = \pm\frac{1}{\sqrt[4]{3}}$ giving the points $\left(\pm\frac{1}{\sqrt[4]{3}}, \frac{1}{\sqrt[4]{3}}, \frac{1}{\sqrt[4]{3}}\right)$, $\left(\pm\frac{1}{\sqrt[4]{3}}, -\frac{1}{\sqrt[4]{3}}, \frac{1}{\sqrt[4]{3}}\right)$, $\left(\pm\frac{1}{\sqrt[4]{3}}, \frac{1}{\sqrt[4]{3}}, -\frac{1}{\sqrt[4]{3}}\right)$, $\left(\pm\frac{1}{\sqrt[4]{3}}, -\frac{1}{\sqrt[4]{3}}, -\frac{1}{\sqrt[4]{3}}\right)$ all with an f-value of $\sqrt{3}$.

Case 2: If one of the variables equals zero and the other two are not zero, then the squares of the two nonzero coordinates are equal with common value $\frac{1}{\sqrt{2}}$ and corresponding f value of $\sqrt{2}$.

Case 3: If exactly two of the variables are zero, then the third variable has value ± 1 with the corresponding f value of 1. Thus on $x^4 + y^4 + z^4 = 1$, the maximum value of f is $\sqrt{3}$ and the minimum value is 1.

13. $f(x,y,z,t) = x + y + z + t$, $g(x,y,z,t) = x^2 + y^2 + z^2 + t^2 = 1 \Rightarrow \langle 1,1,1,1\rangle = \langle 2\lambda x, 2\lambda y, 2\lambda z, 2\lambda t\rangle$, so $\lambda = 1/(2x) = 1/(2y) = 1/(2z) = 1/(2t)$ and $x = y = z = t$. But $x^2 + y^2 + z^2 + t^2 = 1$, so the possible points are $\left(\pm\frac{1}{2}, \pm\frac{1}{2}, \pm\frac{1}{2}, \pm\frac{1}{2}\right)$. Thus the maximum value of f is $f\left(\frac{1}{2}, \frac{1}{2}, \frac{1}{2}, \frac{1}{2}\right) = 2$ and the minimum value is $f\left(-\frac{1}{2}, -\frac{1}{2}, -\frac{1}{2}, -\frac{1}{2}\right) = -2$.

15. $f(x,y,z) = x + 2y$, $g(x,y,z) = x + y + z = 1$, $h(x,y,z) = y^2 + z^2 = 4 \Rightarrow \nabla f = \langle 1, 2, 0\rangle$, $\lambda\nabla g = \langle \lambda, \lambda, \lambda\rangle$ and $\mu\nabla h = \langle 0, 2\mu y, 2\mu z\rangle$. Then $1 = \lambda$, $2 = \lambda + 2\mu y$ and $0 = \lambda + 2\mu z$ so $\mu y = \frac{1}{2} = -\mu z$ or $y = 1/(2\mu)$, $z = -1/(2\mu)$. Thus $x + y + z = 1$ implies $x = 1$ and $y^2 + z^2 = 4$ implies $\mu = \pm\frac{1}{2\sqrt{2}}$. Then the possible points are $\left(1, \pm\sqrt{2}, \mp\sqrt{2}\right)$ and the maximum value is $f\left(1, \sqrt{2}, -\sqrt{2}\right) = 1 + 2\sqrt{2}$ and the minimum value is $f\left(1, -\sqrt{2}, \sqrt{2}\right) = 1 - 2\sqrt{2}$.

17. $f(x,y,z) = yz + xy$, $g(x,y,z) = xy = 1$, $h(x,y,z) = y^2 + z^2 = 1 \Rightarrow \nabla f = \langle y, x + z, y\rangle$, $\lambda\nabla g = \langle \lambda y, \lambda x, 0\rangle$, $\mu\nabla h = \langle 0, 2\mu y, 2\mu z\rangle$. Then $y = \lambda y$ implies $\lambda = 1$ [$y \neq 0$ since $g(x,y,z) = 1$], $x + z = \lambda x + 2\mu y$ and $y = 2\mu z$. Thus $\mu = z/(2y) = y/(2y)$ or $y^2 = z^2$, and so $y^2 + z^2 = 1$ implies $y = \pm\frac{1}{\sqrt{2}}$, $z = \pm\frac{1}{\sqrt{2}}$. Then $xy = 1$ implies $x = \pm\sqrt{2}$ and the possible points are $\left(\pm\sqrt{2}, \pm\frac{1}{\sqrt{2}}, \frac{1}{\sqrt{2}}\right)$, $\left(\pm\sqrt{2}, \pm\frac{1}{\sqrt{2}}, -\frac{1}{\sqrt{2}}\right)$. Hence the maximum of f subject to the constraints is $f\left(\pm\sqrt{2}, \pm\frac{1}{\sqrt{2}}, \pm\frac{1}{\sqrt{2}}\right) = \frac{3}{2}$ and the minimum is $f\left(\pm\sqrt{2}, \pm\frac{1}{\sqrt{2}}, \mp\frac{1}{\sqrt{2}}\right) = \frac{1}{2}$.

Note: Since $xy = 1$ is one of the constraints we could have solved the problem by solving $f(y,z) = yz + 1$ subject to $y^2 + z^2 = 1$.

19. $f(x,y) = e^{-xy}$. For the interior of the region, we find the critical points: $f_x = -ye^{-xy}$, $f_y = -xe^{-xy}$, so the only critical point is $(0,0)$, and $f(0,0) = 1$. For the boundary, we use Lagrange multipliers. $g(x,y) = x^2 + 4y^2 = 1 \Rightarrow$ $\lambda\nabla g = \langle 2\lambda x, 8\lambda y\rangle$, so setting $\nabla f = \lambda\nabla g$ we get $-ye^{-xy} = 2\lambda x$ and $-xe^{-xy} = 8\lambda y$. The first of these gives $e^{-xy} = -2\lambda x/y$, and then the second gives $-x(-2\lambda x/y) = 8\lambda y \Rightarrow x^2 = 4y^2$. Solving this last equation with the constraint $x^2 + 4y^2 = 1$ gives $x = \pm\frac{1}{\sqrt{2}}$ and $y = \pm\frac{1}{2\sqrt{2}}$. Now $f\left(\pm\frac{1}{\sqrt{2}}, \mp\frac{1}{2\sqrt{2}}\right) = e^{1/4} \approx 1.284$ and $f\left(\pm\frac{1}{\sqrt{2}}, \pm\frac{1}{2\sqrt{2}}\right) = e^{-1/4} \approx 0.779$. The former are the maxima on the region and the latter are the minima.

21. $P(L,K) = bL^{\alpha}K^{1-\alpha}$, $g(L,K) = mL + nK = p \Rightarrow \nabla P = \left\langle \alpha bL^{\alpha-1}K^{1-\alpha}, (1-\alpha)bL^{\alpha}K^{-\alpha}\right\rangle$, $\lambda\nabla g = \langle \lambda m, \lambda n\rangle$. Then $\alpha b(K/L)^{1-\alpha} = \lambda m$ and $(1-\alpha)b(L/K)^{\alpha} = \lambda n$ and $mL + nK = p$, so $\alpha b(K/L)^{1-\alpha}/m = (1-\alpha)b(L/K)^{\alpha}/n$ or $n\alpha/[m(1-\alpha)] = (L/K)^{\alpha}(L/K)^{1-\alpha}$ or $L = Kn\alpha/[m(1-\alpha)]$. Substituting into $mL + nK = p$ gives $K = (1-\alpha)p/n$ and $L = \alpha p/m$ for the maximum production.

23. Let the sides of the rectangle be x and y. Then $f(x,y) = xy$, $g(x,y) = 2x + 2y = p \Rightarrow \nabla f(x,y) = \langle y, x\rangle$, $\lambda\nabla g = \langle 2\lambda, 2\lambda\rangle$. Then $\lambda = \frac{1}{2}y = \frac{1}{2}x$ implies $x = y$ and the rectangle with maximum area is a square with side length $\frac{1}{4}p$.

25. Let $f(x,y,z) = d^2 = (x-2)^2 + (y-1)^2 + (z+1)^2$, then we want to minimize f subject to the constraint $g(x,y,z) = x + y - z = 1$. $\nabla f = \lambda\nabla g \Rightarrow \langle 2(x-2), 2(y-1), 2(z+1)\rangle = \lambda\langle 1, 1, -1\rangle$, so $x = (\lambda+4)/2$, $y = (\lambda+2)/2$, $z = -(\lambda+2)/2$. Substituting into the constraint equation gives $\dfrac{\lambda+4}{2} + \dfrac{\lambda+2}{2} + \dfrac{\lambda+2}{2} = 1 \Rightarrow$ $3\lambda + 8 = 2 \Rightarrow \lambda = -2$, so $x = 1$, $y = 0$, and $z = 0$. This must correspond to a minimum, so the shortest distance is $d = \sqrt{(1-2)^2 + (0-1)^2 + (0+1)^2} = \sqrt{3}$.

27. Let $f(x,y,z) = d^2 = (x-4)^2 + (y-2)^2 + z^2$. Then we want to minimize f subject to the constraint $g(x,y,z) = x^2 + y^2 - z^2 = 0$. $\nabla f = \lambda\nabla g \Rightarrow \langle 2(x-4), 2(y-2), 2z\rangle = \langle 2\lambda x, 2\lambda y, -2\lambda z\rangle$, so $x - 4 = \lambda x$, $y - 2 = \lambda y$, and $z = -\lambda z$. From the last equation we have $z + \lambda z = 0 \Rightarrow z(1+\lambda) = 0$, so either $z = 0$ or $\lambda = -1$. But from the constraint equation we have $z = 0 \Rightarrow x^2 + y^2 = 0 \Rightarrow x = y = 0$ which is not possible from the first two equations. So $\lambda = -1$ and $x - 4 = \lambda x \Rightarrow x = 2$, $y - 2 = \lambda y \Rightarrow y = 1$, and $x^2 + y^2 - z^2 = 0 \Rightarrow$ $4 + 1 - z^2 = 0 \Rightarrow z = \pm\sqrt{5}$. This must correspond to a minimum, so the points on the cone closest to $(4,2,0)$ are $(2, 1, \pm\sqrt{5})$.

29. $f(x,y,z) = xyz$, $g(x,y,z) = x + y + z = 100 \Rightarrow \nabla f = \langle yz, xz, xy\rangle = \lambda\nabla g = \langle \lambda, \lambda, \lambda\rangle$. Then $\lambda = yz = xz = xy$ implies $x = y = z = \frac{100}{3}$.

31. If the dimensions are $2x$, $2y$ and $2z$, then $f(x,y,z) = 8xyz$ and $g(x,y,z) = 9x^2 + 36y^2 + 4z^2 = 36 \Rightarrow$ $\nabla f = \langle 8yz, 8xz, 8xy\rangle = \lambda\nabla g = \langle 18\lambda x, 72\lambda y, 8\lambda z\rangle$. Thus $18\lambda x = 8yz$, $72\lambda y = 8xz$, $8\lambda z = 8xy$ so $x^2 = 4y^2$, $z^2 = 9y^2$ and $36y^2 + 36y^2 + 36y^2 = 36$ or $y = \frac{1}{\sqrt{3}}$ $(y > 0)$. Thus the volume of the largest such box is $8\left(\frac{1}{\sqrt{3}}\right)\left(\frac{2}{\sqrt{3}}\right)\left(\frac{3}{\sqrt{3}}\right) = 16/\sqrt{3}$.

33. $f(x,y,z) = xyz$, $g(x,y,z) = x + 2y + 3z = 6 \Rightarrow \nabla f = \langle yz, xz, xy\rangle = \lambda\nabla g = \langle \lambda, 2\lambda, 3\lambda\rangle$. Then $\lambda = yz = \frac{1}{2}xz = \frac{1}{3}xy$ implies $x = 2y$, $z = \frac{2}{3}y$. But $2y + 2y + 2y = 6$ so $y = 1$, $x = 2$, $z = \frac{2}{3}$ and the volume is $V = \frac{4}{3}$.

35. $f(x,y,z)=xyz$, $g(x,y,z)=4(x+y+z)=c \;\Rightarrow\; \nabla f=\langle yz,xz,xy\rangle$, $\lambda\nabla g=\langle 4\lambda,4\lambda,4\lambda\rangle$. Thus $4\lambda=yz=xz=xy$ or $x=y=z=\frac{1}{12}c$ are the dimensions giving the maximum volume.

37. If the dimensions of the box are given by x, y, and z, then we need to find the maximum value of $f(x,y,z)=xyz$ $(x,y,z>0)$ subject to the constraint $L=\sqrt{x^2+y^2+z^2}$ or $g(x,y,z)=x^2+y^2+z^2=L^2$. $\nabla f=\lambda\nabla g \;\Rightarrow$ $\langle yz,xz,xy\rangle=\lambda\langle 2x,2y,2z\rangle$, so $yz=2\lambda x \;\Rightarrow\; \lambda=\dfrac{yz}{2x}$, $xz=2\lambda y \;\Rightarrow\; \lambda=\dfrac{xz}{2y}$, and $xy=2\lambda z \;\Rightarrow\; \lambda=\dfrac{xy}{2z}$. Thus $\lambda=\dfrac{yz}{2x}=\dfrac{xz}{2y} \;\Rightarrow\; x^2=y^2$ [since $z\neq 0$] $\Rightarrow\; x=y$ and $\lambda=\dfrac{yz}{2x}=\dfrac{xy}{2z} \;\Rightarrow\; x=z$ [since $y\neq 0$]. Substituting into the constraint equation gives $x^2+x^2+x^2=L^2 \;\Rightarrow\; x^2=L^2/3 \;\Rightarrow\; x=L/\sqrt{3}=y=z$ and the maximum volume is $\left(L/\sqrt{3}\right)^3=L^3/\left(3\sqrt{3}\right)$.

39. We need to find the extreme values of $f(x,y,z)=x^2+y^2+z^2$ subject to the two constraints $g(x,y,z)=x+y+2z=2$ and $h(x,y,z)=x^2+y^2-z=0$. $\nabla f=\langle 2x,2y,2z\rangle$, $\lambda\nabla g=\langle\lambda,\lambda,2\lambda\rangle$ and $\mu\nabla h=\langle 2\mu x,2\mu y,-\mu\rangle$. Thus we need (1) $2x=\lambda+2\mu x$, (2) $2y=\lambda+2\mu y$, (3) $2z=2\lambda-\mu$, (4) $x+y+2z=2$, and (5) $x^2+y^2-z=0$. From (1) and (2), $2(x-y)=2\mu(x-y)$, so if $x\neq y$, $\mu=1$. Putting this in (3) gives $2z=2\lambda-1$ or $\lambda=z+\frac{1}{2}$, but putting $\mu=1$ into (1) says $\lambda=0$. Hence $z+\frac{1}{2}=0$ or $z=-\frac{1}{2}$. Then (4) and (5) become $x+y-3=0$ and $x^2+y^2+\frac{1}{2}=0$. The last equation cannot be true, so this case gives no solution. So we must have $x=y$. Then (4) and (5) become $2x+2z=2$ and $2x^2-z=0$ which imply $z=1-x$ and $z=2x^2$. Thus $2x^2=1-x$ or $2x^2+x-1=(2x-1)(x+1)=0$ so $x=\frac{1}{2}$ or $x=-1$. The two points to check are $\left(\frac{1}{2},\frac{1}{2},\frac{1}{2}\right)$ and $(-1,-1,2)$: $f\left(\frac{1}{2},\frac{1}{2},\frac{1}{2}\right)=\frac{3}{4}$ and $f(-1,-1,2)=6$. Thus $\left(\frac{1}{2},\frac{1}{2},\frac{1}{2}\right)$ is the point on the ellipse nearest the origin and $(-1,-1,2)$ is the one furthest from the origin.

41. $f(x,y,z)=ye^{x-z}$, $g(x,y,z)=9x^2+4y^2+36z^2=36$, $h(x,y,z)=xy+yz=1$. $\nabla f=\lambda\nabla g+\mu\nabla h \;\Rightarrow$ $\left\langle ye^{x-z},e^{x-z},-ye^{x-z}\right\rangle=\lambda\langle 18x,8y,72z\rangle+\mu\langle y,x+z,y\rangle$, so $ye^{x-z}=18\lambda x+\mu y$, $e^{x-z}=8\lambda y+\mu(x+z)$, $-ye^{x-z}=72\lambda z+\mu y$, $9x^2+4y^2+36z^2=36$, $xy+yz=1$. Using a CAS to solve these 5 equations simultaneously for x, y, z, λ, and μ (in Maple, use the `allvalues` command), we get 4 real-valued solutions:

$x\approx 0.222\,444$,	$y\approx -2.157\,012$,	$z\approx -0.686\,049$,	$\lambda\approx -0.200\,401$,	$\mu\approx 2.108\,584$
$x\approx -1.951\,921$,	$y\approx -0.545\,867$,	$z\approx 0.119\,973$,	$\lambda\approx 0.003\,141$,	$\mu\approx -0.076\,238$
$x\approx 0.155\,142$,	$y\approx 0.904\,622$,	$z\approx 0.950\,293$,	$\lambda\approx -0.012\,447$,	$\mu\approx 0.489\,938$
$x\approx 1.138\,731$,	$y\approx 1.768\,057$,	$z\approx -0.573\,138$,	$\lambda\approx 0.317\,141$,	$\mu\approx 1.862\,675$

Substituting these values into f gives $f(0.222\,444,-2.157\,012,-0.686\,049)\approx -5.3506$, $f(-1.951\,921,-0.545\,867,0.119\,973)\approx -0.0688$, $f(0.155\,142,0.904\,622,0.950\,293)\approx 0.4084$, $f(1.138\,731,1.768\,057,-0.573\,138)\approx 9.7938$. Thus the maximum is approximately 9.7938, and the mininum is approximately -5.3506.

43. (a) We wish to maximize $f(x_1, x_2, \ldots, x_n) = \sqrt[n]{x_1 x_2 \cdots x_n}$ subject to

$g(x_1, x_2, \ldots, x_n) = x_1 + x_2 + \cdots + x_n = c$ and $x_i > 0$.

$$\nabla f = \left\langle \tfrac{1}{n}(x_1 x_2 \cdots x_n)^{\frac{1}{n}-1}(x_2 \cdots x_n), \tfrac{1}{n}(x_1 x_2 \cdots x_n)^{\frac{1}{n}-1}(x_1 x_3 \cdots x_n), \ldots, \tfrac{1}{n}(x_1 x_2 \cdots x_n)^{\frac{1}{n}-1}(x_1 \cdots x_{n-1}) \right\rangle$$

and $\lambda \nabla g = \langle \lambda, \lambda, \ldots, \lambda \rangle$, so we need to solve the system of equations

$$\frac{1}{n}(x_1 x_2 \cdots x_n)^{\frac{1}{n}-1}(x_2 \cdots x_n) = \lambda \quad \Rightarrow \quad x_1^{1/n} x_2^{1/n} \cdots x_n^{1/n} = n\lambda x_1$$

$$\frac{1}{n}(x_1 x_2 \cdots x_n)^{\frac{1}{n}-1}(x_1 x_3 \cdots x_n) = \lambda \quad \Rightarrow \quad x_1^{1/n} x_2^{1/n} \cdots x_n^{1/n} = n\lambda x_2$$

$$\vdots$$

$$\frac{1}{n}(x_1 x_2 \cdots x_n)^{\frac{1}{n}-1}(x_1 \cdots x_{n-1}) = \lambda \quad \Rightarrow \quad x_1^{1/n} x_2^{1/n} \cdots x_n^{1/n} = n\lambda x_n$$

This implies $n\lambda x_1 = n\lambda x_2 = \cdots = n\lambda x_n$. Note $\lambda \neq 0$, otherwise we can't have all $x_i > 0$. Thus $x_1 = x_2 = \cdots = x_n$. But $x_1 + x_2 + \cdots + x_n = c \Rightarrow nx_1 = c \Rightarrow x_1 = \frac{c}{n} = x_2 = x_3 = \cdots = x_n$. Then the only point where f can have an extreme value is $\left(\frac{c}{n}, \frac{c}{n}, \ldots, \frac{c}{n}\right)$. Since we can choose values for $(x_1, x_2, \ldots, x_n)$ that make f as close to zero (but not equal) as we like, f has no minimum value. Thus the maximum value is

$$f\left(\frac{c}{n}, \frac{c}{n}, \ldots, \frac{c}{n}\right) = \sqrt[n]{\frac{c}{n} \cdot \frac{c}{n} \cdot \cdots \cdot \frac{c}{n}} = \frac{c}{n}.$$

(b) From part (a), $\frac{c}{n}$ is the maximum value of f. Thus $f(x_1, x_2, \ldots, x_n) = \sqrt[n]{x_1 x_2 \cdots x_n} \leq \frac{c}{n}$. But $x_1 + x_2 + \cdots + x_n = c$, so $\sqrt[n]{x_1 x_2 \cdots x_n} \leq \frac{x_1 + x_2 + \cdots + x_n}{n}$. These two means are equal when f attains its maximum value $\frac{c}{n}$, but this can occur only at the point $\left(\frac{c}{n}, \frac{c}{n}, \ldots, \frac{c}{n}\right)$ we found in part (a). So the means are equal only when $x_1 = x_2 = x_3 = \cdots = x_n = \frac{c}{n}$.

11 Review

CONCEPT CHECK

1. (a) A function f of two variables is a rule that assigns to each ordered pair (x, y) of real numbers in its domain a unique real number denoted by $f(x, y)$.

(b) One way to visualize a function of two variables is by graphing it, resulting in the surface $z = f(x, y)$. Another method for visualizing a function of two variables is a contour map. The contour map consists of level curves of the function which are horizontal traces of the graph of the function projected onto the xy-plane.

2. A function f of three variables is a rule that assigns to each ordered triple (x, y, z) in its domain a unique real number $f(x, y, z)$. We can visualize a function of three variables by examining its level surfaces $f(x, y, z) = k$, where k is a constant.

3. $\lim\limits_{(x,y)\to(a,b)} f(x, y) = L$ means the values of $f(x, y)$ approach the number L as the point (x, y) approaches the point (a, b) along any path that is within the domain of f. We can show that a limit at a point does not exist by finding two different paths approaching the point along which $f(x, y)$ has different limits.

4. (a) See Definition 11.2.3.

(b) If f is continuous on $\mathbb{R}^2$, its graph will appear as a surface without holes or breaks.

5. (a) See (2) and (3) in Section 11.3.

(b) See "Interpretations of Partial Derivatives" on page 759.

(c) To find f_x, regard y as a constant and differentiate $f(x,y)$ with respect to x. To find f_y, regard x as a constant and differentiate $f(x,y)$ with respect to y.

6. See the statement of Clairaut's Theorem on page 763.

7. (a) See (2) in Section 11.4.

(b) See (19) and the preceding discussion in Section 11.6.

(c) See "Tangent Planes to Parametric Surfaces" on page 777.

8. See (3) and (4) and the accompanying discussion in Section 11.4. We can interpret the linearization of f at (a,b) geometrically as the linear function whose graph is the tangent plane to the graph of f at (a,b). Thus it is the linear function which best approximates f near (a,b).

9. (a) See Definition 11.4.7.

(b) Use Theorem 11.4.8.

10. See (10) and the associated discussion in Section 11.4.

11. See (2) and (3) in Section 11.5.

12. See (7) and the preceding discussion in Section 11.5.

13. (a) See Definition 11.6.2. We can interpret it as the rate of change of f at (x_0, y_0) in the direction of $\mathbf{u}$. Geometrically, if P is the point $(x_0, y_0, f(x_0, y_0))$ on the graph of f and C is the curve of intersection of the graph of f with the vertical plane that passes through P in the direction $\mathbf{u}$, the directional derivative of f at (x_0, y_0) in the direction of $\mathbf{u}$ is the slope of the tangent line to C at P. (See Figure 5 in Section 11.6.)

(b) See Theorem 11.6.3.

14. (a) See (8) and (13) in Section 11.6.

(b) $D_\mathbf{u} f(x,y) = \nabla f(x,y) \cdot \mathbf{u}$ or $D_\mathbf{u} f(x,y,z) = \nabla f(x,y,z) \cdot \mathbf{u}$

(c) The gradient vector of a function points in the direction of maximum rate of increase of the function. On a graph of the function, the gradient points in the direction of steepest ascent.

15. (a) f has a local maximum at (a,b) if $f(x,y) \le f(a,b)$ when (x,y) is near (a,b).

(b) f has an absolute maximum at (a,b) if $f(x,y) \le f(a,b)$ for all points (x,y) in the domain of f.

(c) f has a local minimum at (a,b) if $f(x,y) \ge f(a,b)$ when (x,y) is near (a,b).

(d) f has an absolute minimum at (a,b) if $f(x,y) \ge f(a,b)$ for all points (x,y) in the domain of f.

(e) f has a saddle point at (a,b) if $f(a,b)$ is a local maximum in one direction but a local minimum in another.

16. (a) By Theorem 11.7.2, if f has a local maximum at (a,b) and the first-order partial derivatives of f exist there, then $f_x(a,b) = 0$ and $f_y(a,b) = 0$.

(b) A critical point of f is a point (a,b) such that $f_x(a,b) = 0$ and $f_y(a,b) = 0$ or one of these partial derivatives does not exist.

17. See (3) in Section 11.7.

18. (a) See Figure 11 and the accompanying discussion in Section 11.7.

(b) See Theorem 11.7.8.

(c) See the procedure outlined in (9) in Section 11.7.

19. See the discussion beginning on page 813; see "Two Constraints" on page 817.

TRUE-FALSE QUIZ

1. True. $f_y(a,b) = \lim\limits_{h\to 0} \dfrac{f(a,b+h) - f(a,b)}{h}$ from Equation 11.3.3. Let $h = y - b$. As $h \to 0$, $y \to b$. Then by substituting, we get $f_y(a,b) = \lim\limits_{y\to b} \dfrac{f(a,y) - f(a,b)}{y - b}$.

3. False. $f_{xy} = \dfrac{\partial^2 f}{\partial y\,\partial x}$.

5. False. See Example 11.2.3.

7. True. If f has a local minimum and f is differentiable at (a,b) then by Theorem 11.7.2, $f_x(a,b) = 0$ and $f_y(a,b) = 0$, so $\nabla f(a,b) = \langle f_x(a,b), f_y(a,b)\rangle = \langle 0,0\rangle = \mathbf{0}$.

9. False. $\nabla f(x,y) = \langle 0, 1/y\rangle$.

11. True. $\nabla f = \langle \cos x, \cos y\rangle$, so $|\nabla f| = \sqrt{\cos^2 x + \cos^2 y}$. But $|\cos\theta| \le 1$, so $|\nabla f| \le \sqrt{2}$. Now $D_{\mathbf{u}}\, f(x,y) = \nabla f \cdot \mathbf{u} = |\nabla f|\,|\mathbf{u}| \cos\theta$, but $\mathbf{u}$ is a unit vector, so $|D_{\mathbf{u}}\, f(x,y)| \le \sqrt{2}\cdot 1 \cdot 1 = \sqrt{2}$.

EXERCISES

1. $\ln(x+y+1)$ is defined only when $x+y+1>0 \;\Rightarrow\; y > -x-1$, so the domain of f is $\{(x,y) \mid y > -x-1\}$, all those points above the line $y = -x-1$.

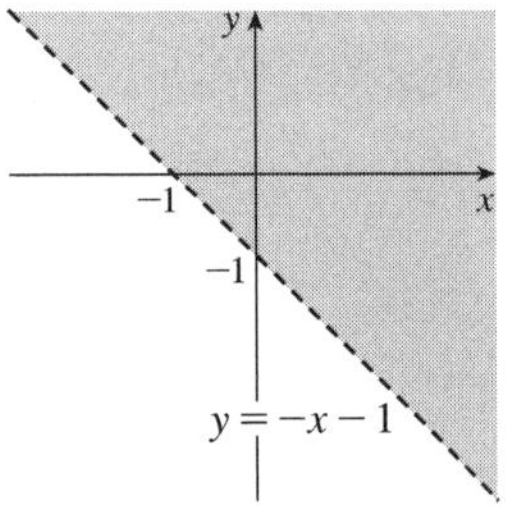

3. $z = f(x,y) = 1 - y^2$, a parabolic cylinder.

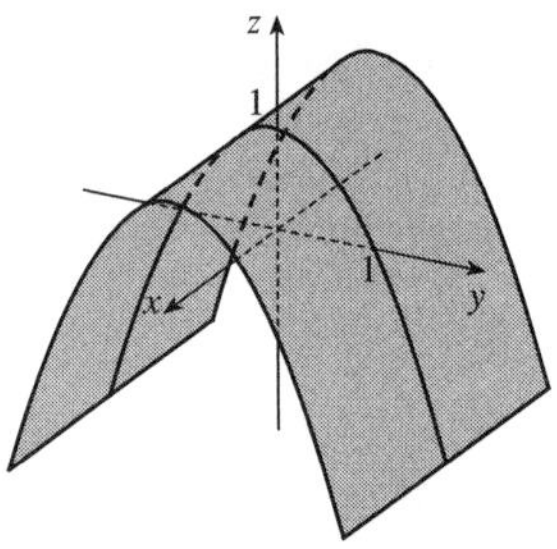

5. The level curves are $\sqrt{4x^2 + y^2} = k$ or $4x^2 + y^2 = k^2$, $k \ge 0$, a family of ellipses.

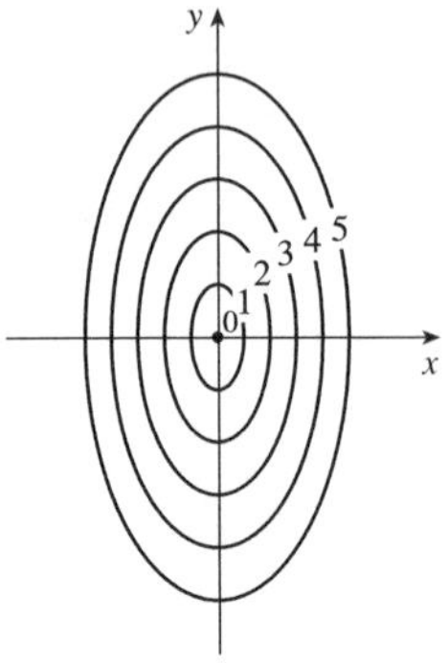

7.

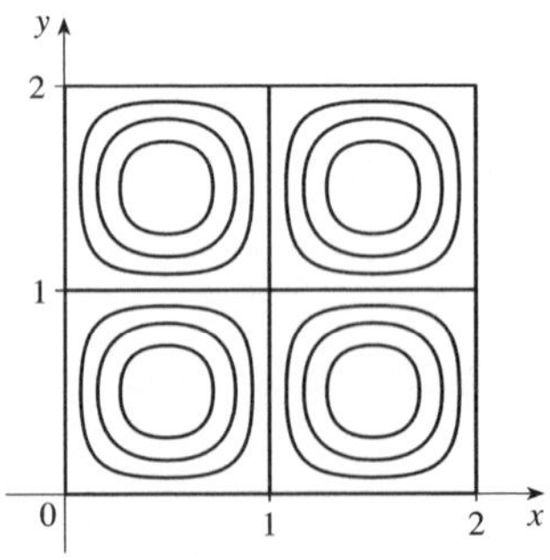

9. f is a rational function, so it is continuous on its domain. Since f is defined at $(1,1)$, we use direct substitution to evaluate the limit: $\displaystyle\lim_{(x,y)\to(1,1)} \frac{2xy}{x^2+2y^2} = \frac{2(1)(1)}{1^2+2(1)^2} = \frac{2}{3}$.

11. (a) $T_x(6,4) = \displaystyle\lim_{h\to 0} \frac{T(6+h,4)-T(6,4)}{h}$, so we can approximate $T_x(6,4)$ by considering $h=\pm 2$ and using the values given in the table: $T_x(6,4) \approx \dfrac{T(8,4)-T(6,4)}{2} = \dfrac{86-80}{2} = 3$,

$T_x(6,4) \approx \dfrac{T(4,4)-T(6,4)}{-2} = \dfrac{72-80}{-2} = 4$. Averaging these values, we estimate $T_x(6,4)$ to be approximately 3.5°C/m. Similarly, $T_y(6,4) = \displaystyle\lim_{h\to 0} \frac{T(6,4+h)-T(6,4)}{h}$, which we can approximate with $h=\pm 2$:

$T_y(6,4) \approx \dfrac{T(6,6)-T(6,4)}{2} = \dfrac{75-80}{2} = -2.5$, $T_y(6,4) \approx \dfrac{T(6,2)-T(6,4)}{-2} = \dfrac{87-80}{-2} = -3.5$. Averaging these values, we estimate $T_y(6,4)$ to be approximately -3.0°C/m.

(b) Here $\mathbf{u} = \left\langle \frac{1}{\sqrt{2}}, \frac{1}{\sqrt{2}} \right\rangle$, so by Equation 11.6.9, $D_{\mathbf{u}}\, T(6,4) = \nabla T(6,4)\cdot \mathbf{u} = T_x(6,4)\,\frac{1}{\sqrt{2}} + T_y(6,4)\,\frac{1}{\sqrt{2}}$. Using our estimates from part (a), we have $D_{\mathbf{u}}\, T(6,4) \approx (3.5)\,\frac{1}{\sqrt{2}} + (-3.0)\,\frac{1}{\sqrt{2}} = \frac{1}{2\sqrt{2}} \approx 0.35$. This means that as we move through the point $(6,4)$ in the direction of $\mathbf{u}$, the temperature increases at a rate of approximately 0.35°C/m.

Alternatively, we can use Definition 11.6.2: $D_{\mathbf{u}}\, T(6,4) = \displaystyle\lim_{h\to 0} \frac{T\left(6+h\,\frac{1}{\sqrt{2}}, 4+h\,\frac{1}{\sqrt{2}}\right) - T(6,4)}{h}$,

which we can estimate with $h = \pm 2\sqrt{2}$. Then $D_{\mathbf{u}}\, T(6,4) \approx \dfrac{T(8,6)-T(6,4)}{2\sqrt{2}} = \dfrac{80-80}{2\sqrt{2}} = 0$,

$D_{\mathbf{u}}\, T(6,4) \approx \dfrac{T(4,2)-T(6,4)}{-2\sqrt{2}} = \dfrac{74-80}{-2\sqrt{2}} = \dfrac{3}{\sqrt{2}}$. Averaging these values, we have $D_{\mathbf{u}}\, T(6,4) \approx \frac{3}{2\sqrt{2}} \approx 1.1^\circ\text{C/m}$.

(c) $T_{xy}(x,y) = \dfrac{\partial}{\partial y}\,[T_x(x,y)] = \displaystyle\lim_{h\to 0} \frac{T_x(x,y+h)-T_x(x,y)}{h}$, so $T_{xy}(6,4) = \displaystyle\lim_{h\to 0} \frac{T_x(6,4+h)-T_x(6,4)}{h}$ which we can estimate with $h=\pm 2$. We have $T_x(6,4)\approx 3.5$ from part (a), but we will also need values for $T_x(6,6)$ and $T_x(6,2)$. If we use $h=\pm 2$ and the values given in the table, we have

$T_x(6,6) \approx \dfrac{T(8,6)-T(6,6)}{2} = \dfrac{80-75}{2} = 2.5$, $T_x(6,6) \approx \dfrac{T(4,6)-T(6,6)}{-2} = \dfrac{68-75}{-2} = 3.5$.

Averaging these values, we estimate $T_x(6,6) \approx 3.0$. Similarly,

$T_x(6,2) \approx \dfrac{T(8,2)-T_x(6,2)}{2} = \dfrac{90-87}{2} = 1.5$, $T_x(6,2) \approx \dfrac{T(4,2)-T(6,2)}{-2} = \dfrac{74-87}{-2} = 6.5$.

Averaging these values, we estimate $T_x(6,2) \approx 4.0$. Finally, we estimate $T_{xy}(6,4)$:

$T_{xy}(6,4) \approx \dfrac{T_x(6,6)-T_x(6,4)}{2} = \dfrac{3.0-3.5}{2} = -0.25$, $T_{xy}(6,4) \approx \dfrac{T_x(6,2)-T_x(6,4)}{-2} = \dfrac{4.0-3.5}{-2} = -0.25$.

Averaging these values, we have $T_{xy}(6,4) \approx -0.25$.

13. $f(x,y) = \sqrt{2x+y^2} \quad\Rightarrow\quad f_x = \frac{1}{2}(2x+y^2)^{-1/2}(2) = \dfrac{1}{\sqrt{2x+y^2}}$, $f_y = \frac{1}{2}(2x+y^2)^{-1/2}(2y) = \dfrac{y}{\sqrt{2x+y^2}}$

15. $g(u,v) = u\tan^{-1} v \quad\Rightarrow\quad g_u = \tan^{-1} v$, $g_v = \dfrac{u}{1+v^2}$

17. $T(p,q,r) = p\ln(q+e^r) \quad\Rightarrow\quad T_p = \ln(q+e^r)$, $T_q = \dfrac{p}{q+e^r}$, $T_r = \dfrac{pe^r}{q+e^r}$

19. $f(x,y) = 4x^3 - xy^2 \quad\Rightarrow\quad f_x = 12x^2 - y^2$, $f_y = -2xy$, $f_{xx} = 24x$, $f_{yy} = -2x$, and $f_{xy} = f_{yx} = -2y$.

21. $f(x,y,z) = x^k y^l z^m \quad\Rightarrow\quad f_x = kx^{k-1}y^l z^m$, $f_y = lx^k y^{l-1} z^m$, $f_z = mx^k y^l z^{m-1}$, $f_{xx} = k(k-1)x^{k-2}y^l z^m$, $f_{yy} = l(l-1)x^k y^{l-2} z^m$, $f_{zz} = m(m-1)x^k y^l z^{m-2}$, $f_{xy} = f_{yx} = klx^{k-1}y^{l-1}z^m$, $f_{xz} = f_{zx} = kmx^{k-1}y^l z^{m-1}$, and $f_{yz} = f_{zy} = lmx^k y^{l-1} z^{m-1}$.

23. $z = xy + xe^{y/x} \;\Rightarrow\; \dfrac{\partial z}{\partial x} = y - \dfrac{y}{x}e^{y/x} + e^{y/x},\ \dfrac{\partial z}{\partial y} = x + e^{y/x}$ and

$$\begin{aligned} x\frac{\partial z}{\partial x} + y\frac{\partial z}{\partial y} &= x\left(y - \frac{y}{x}e^{y/x} + e^{y/x}\right) + y\left(x + e^{y/x}\right) \\ &= xy - ye^{y/x} + xe^{y/x} + xy + ye^{y/x} = xy + xy + xe^{y/x} = xy + z \end{aligned}$$

25. (a) $z_x = 6x + 2 \;\Rightarrow\; z_x(1,-2) = 8$ and $z_y = -2y \;\Rightarrow\; z_y(1,-2) = 4$, so an equation of the tangent plane is $z - 1 = 8(x-1) + 4(y+2)$ or $z = 8x + 4y + 1$.

(b) A normal vector to the tangent plane (and the surface) at $(1,-2,1)$ is $\langle 8, 4, -1\rangle$. Then parametric equations for the normal line there are $x = 1 + 8t$, $y = -2 + 4t$, $z = 1 - t$, and symmetric equations are $\dfrac{x-1}{8} = \dfrac{y+2}{4} = \dfrac{z-1}{-1}$.

27. (a) Let $F(x,y,z) = x^2 + 2y^2 - 3z^2$. Then $F_x = 2x$, $F_y = 4y$, $F_z = -6z$, so $F_x(2,-1,1) = 4$, $F_y(2,-1,1) = -4$, $F_z(2,-1,1) = -6$. From Equation 11.6.19, an equation of the tangent plane is $4(x-2) - 4(y+1) - 6(z-1) = 0$ or equivalently $2x - 2y - 3z = 3$.

(b) From Equations 11.6.20, symmetric equations for the normal line are $\dfrac{x-2}{4} = \dfrac{y+1}{-4} = \dfrac{z-1}{-6}$.

29. (a) $\mathbf{r}(u,v) = (u+v)\,\mathbf{i} + u^2\,\mathbf{j} + v^2\,\mathbf{k}$ and the point $(3,4,1)$ corresponds to $u = 2$, $v = 1$. Then $\mathbf{r}_u = \mathbf{i} + 2u\,\mathbf{j} \;\Rightarrow$ $\mathbf{r}_u(2,1) = \mathbf{i} + 4\,\mathbf{j}$ and $\mathbf{r}_v = \mathbf{i} + 2v\,\mathbf{k} \;\Rightarrow\; \mathbf{r}_v(2,1) = \mathbf{i} + 2\,\mathbf{j}$. A normal vector to the surface at $(3,4,1)$ is $\mathbf{r}_u \times \mathbf{r}_v = 8\,\mathbf{i} - 2\,\mathbf{j} - 4\,\mathbf{k}$, so an equation of the tangent plane there is $8(x-3) - 2(y-4) - 4(z-1) = 0$ or equivalently $4x - y - 2z = 6$.

(b) A direction vector for the normal line through $(3,4,1)$ is $8\,\mathbf{i} - 2\,\mathbf{j} - 4\,\mathbf{k}$, so a vector equation is $\mathbf{r}(t) = (3\,\mathbf{i} + 4\,\mathbf{j} + \mathbf{k}) + t\,(8\,\mathbf{i} - 2\,\mathbf{j} - 4\,\mathbf{k})$, and the corresponding parametric equations are $x = 3 + 8t$, $y = 4 - 2t$, $z = 1 - 4t$.

31. The hyperboloid is a level surface of the function $F(x,y,z) = x^2 + 4y^2 - z^2$, so a normal vector to the surface at (x_0, y_0, z_0) is $\nabla F(x_0, y_0, z_0) = \langle 2x_0, 8y_0, -2z_0\rangle$. A normal vector for the plane $2x + 2y + z = 5$ is $\langle 2, 2, 1\rangle$. For the planes to be parallel, we need the normal vectors to be parallel, so $\langle 2x_0, 8y_0, -2z_0\rangle = k\,\langle 2, 2, 1\rangle$, or $x_0 = k$, $y_0 = \frac{1}{4}k$, and $z_0 = -\frac{1}{2}k$. But $x_0^2 + 4y_0^2 - z_0^2 = 4 \;\Rightarrow\; k^2 + \frac{1}{4}k^2 - \frac{1}{4}k^2 = 4 \;\Rightarrow\; k^2 = 4 \;\Rightarrow\; k = \pm 2$. So there are two such points: $\left(2, \frac{1}{2}, -1\right)$ and $\left(-2, -\frac{1}{2}, 1\right)$.

33. $f(x,y,z) = x^3\sqrt{y^2 + z^2} \;\Rightarrow\; f_x(x,y,z) = 3x^2\sqrt{y^2+z^2}$, $f_y(x,y,z) = \dfrac{yx^3}{\sqrt{y^2+z^2}}$, and $f_z(x,y,z) = \dfrac{zx^3}{\sqrt{y^2+z^2}}$, so $f(2,3,4) = 8(5) = 40$, $f_x(2,3,4) = 3(4)\sqrt{25} = 60$, $f_y(2,3,4) = \frac{3(8)}{\sqrt{25}} = \frac{24}{5}$, and $f_z(2,3,4) = \frac{4(8)}{\sqrt{25}} = \frac{32}{5}$. Then the linear approximation of f at $(2,3,4)$ is

$$\begin{aligned} f(x,y,z) &\approx f(2,3,4) + f_x(2,3,4)(x-2) + f_y(2,3,4)(y-3) + f_z(2,3,4)(z-4) \\ &= 40 + 60(x-2) + \tfrac{24}{5}(y-3) + \tfrac{32}{5}(z-4) = 60x + \tfrac{24}{5}y + \tfrac{32}{5}z - 120 \end{aligned}$$

Then $(1.98)^3\sqrt{(3.01)^2 + (3.97)^2} = f(1.98, 3.01, 3.97) \approx 60(1.98) + \frac{24}{5}(3.01) + \frac{32}{5}(3.97) - 120 = 38.656$.

35. $\dfrac{du}{dp} = \dfrac{\partial u}{\partial x}\dfrac{dx}{dp} + \dfrac{\partial u}{\partial y}\dfrac{dy}{dp} + \dfrac{\partial u}{\partial z}\dfrac{dz}{dp} = 2xy^3(1 + 6p) + 3x^2y^2(pe^p + e^p) + 4z^3(p\cos p + \sin p)$

37. By the Chain Rule, $\dfrac{\partial z}{\partial s} = \dfrac{\partial z}{\partial x}\dfrac{\partial x}{\partial s} + \dfrac{\partial z}{\partial y}\dfrac{\partial y}{\partial s}$. When $s = 1$ and $t = 2$, $x = g(1,2) = 3$ and $y = h(1,2) = 6$, so

$\dfrac{\partial z}{\partial s} = f_x(3,6)g_s(1,2) + f_y(3,6)\,h_s(1,2) = (7)(-1) + (8)(-5) = -47$. Similarly, $\dfrac{\partial z}{\partial t} = \dfrac{\partial z}{\partial x}\dfrac{\partial x}{\partial t} + \dfrac{\partial z}{\partial y}\dfrac{\partial y}{\partial t}$, so

$\dfrac{\partial z}{\partial t} = f_x(3,6)g_t(1,2) + f_y(3,6)\,h_t(1,2) = (7)(4) + (8)(10) = 108$.

39. $\dfrac{\partial z}{\partial x} = 2xf'(x^2 - y^2)$, $\dfrac{\partial z}{\partial y} = 1 - 2yf'(x^2 - y^2)$ $\left[\text{where } f' = \dfrac{df}{d(x^2 - y^2)}\right]$. Then

$y\,\dfrac{\partial z}{\partial x} + x\,\dfrac{\partial z}{\partial y} = 2xyf'(x^2 - y^2) + x - 2xyf'(x^2 - y^2) = x$.

41. $\dfrac{\partial z}{\partial x} = \dfrac{\partial z}{\partial u}\,y + \dfrac{\partial z}{\partial v}\dfrac{-y}{x^2}$ and

$$\frac{\partial^2 z}{\partial x^2} = y\,\frac{\partial}{\partial x}\left(\frac{\partial z}{\partial u}\right) + \frac{2y}{x^3}\frac{\partial z}{\partial v} + \frac{-y}{x^2}\frac{\partial}{\partial x}\left(\frac{\partial z}{\partial v}\right) = \frac{2y}{x^3}\frac{\partial z}{\partial v} + y\left(\frac{\partial^2 z}{\partial u^2}\,y + \frac{\partial^2 z}{\partial v\,\partial u}\frac{-y}{x^2}\right) + \frac{-y}{x^2}\left(\frac{\partial^2 z}{\partial v^2}\frac{-y}{x^2} + \frac{\partial^2 z}{\partial u\,\partial v}\,y\right)$$

$$= \frac{2y}{x^3}\frac{\partial z}{\partial v} + y^2\,\frac{\partial^2 z}{\partial u^2} - \frac{2y^2}{x^2}\frac{\partial^2 z}{\partial u\,\partial v} + \frac{y^2}{x^4}\frac{\partial^2 z}{\partial v^2}$$

Also $\dfrac{\partial z}{\partial y} = x\,\dfrac{\partial z}{\partial u} + \dfrac{1}{x}\dfrac{\partial z}{\partial v}$ and

$$\frac{\partial^2 z}{\partial y^2} = x\,\frac{\partial}{\partial y}\left(\frac{\partial z}{\partial u}\right) + \frac{1}{x}\frac{\partial}{\partial y}\left(\frac{\partial z}{\partial v}\right) = x\left(\frac{\partial^2 z}{\partial u^2}\,x + \frac{\partial^2 z}{\partial v\,\partial u}\frac{1}{x}\right) + \frac{1}{x}\left(\frac{\partial^2 z}{\partial v^2}\frac{1}{x} + \frac{\partial^2 z}{\partial u\,\partial v}\,x\right)$$

$$= x^2\,\frac{\partial^2 z}{\partial u^2} + 2\,\frac{\partial^2 z}{\partial u\,\partial v} + \frac{1}{x^2}\frac{\partial^2 z}{\partial v^2}$$

Thus

$$x^2\,\frac{\partial^2 z}{\partial x^2} - y^2\,\frac{\partial^2 z}{\partial y^2} = \frac{2y}{x}\frac{\partial z}{\partial v} + x^2y^2\,\frac{\partial^2 z}{\partial u^2} - 2y^2\,\frac{\partial^2 z}{\partial u\,\partial v} + \frac{y^2}{x^2}\frac{\partial^2 z}{\partial v^2} - x^2y^2\,\frac{\partial^2 z}{\partial u^2} - 2y^2\,\frac{\partial^2 z}{\partial u\,\partial v} - \frac{y^2}{x^2}\frac{\partial^2 z}{\partial v^2}$$

$$= \frac{2y}{x}\frac{\partial z}{\partial v} - 4y^2\,\frac{\partial^2 z}{\partial u\,\partial v} = 2v\,\frac{\partial z}{\partial v} - 4uv\,\frac{\partial^2 z}{\partial u\,\partial v}$$

since $y = xv = \dfrac{uv}{y}$ or $y^2 = uv$.

43. $\nabla f = \left\langle z^2\sqrt{y}\,e^{x\sqrt{y}}, \dfrac{xz^2e^{x\sqrt{y}}}{2\sqrt{y}}, 2ze^{x\sqrt{y}}\right\rangle = ze^{x\sqrt{y}}\left\langle z\sqrt{y}, \dfrac{xz}{2\sqrt{y}}, 2\right\rangle$

45. $\nabla f = \langle 1/\sqrt{x}, -2y\rangle$, $\nabla f(1,5) = \langle 1, -10\rangle$, $\mathbf{u} = \frac{1}{5}\langle 3, -4\rangle$. Then $D_{\mathbf{u}}\,f(1,5) = \frac{43}{5}$.

47. $\nabla f = \langle 2xy, x^2 + 1/(2\sqrt{y})\rangle$, $|\nabla f(2,1)| = \left|\langle 4, \frac{9}{2}\rangle\right|$. Thus the maximum rate of change of f at $(2,1)$ is $\frac{\sqrt{145}}{2}$ in the direction $\langle 4, \frac{9}{2}\rangle$.

49. First we draw a line passing through Homestead and the eye of the hurricane. We can approximate the directional derivative at Homestead in the direction of the eye of the hurricane by the average rate of change of wind speed between the points where this line intersects the contour lines closest to Homestead. In the direction of the eye of the hurricane, the wind speed changes from 45 to 50 knots. We estimate the distance between these two points to be approximately 8 miles, so the rate of change of wind speed in the direction given is approximately $\frac{50-45}{8} = \frac{5}{8} = 0.625$ knot/mi.

51. $f(x,y) = x^2 - xy + y^2 + 9x - 6y + 10 \quad \Rightarrow \quad f_x = 2x - y + 9$, $f_y = -x + 2y - 6$, $f_{xx} = 2 = f_{yy}$, $f_{xy} = -1$. Then $f_x = 0$ and $f_y = 0$ imply $y = 1$, $x = -4$. Thus the only critical point is $(-4, 1)$ and $f_{xx}(-4,1) > 0$, $D(-4,1) = 3 > 0$, so $f(-4,1) = -11$ is a local minimum.

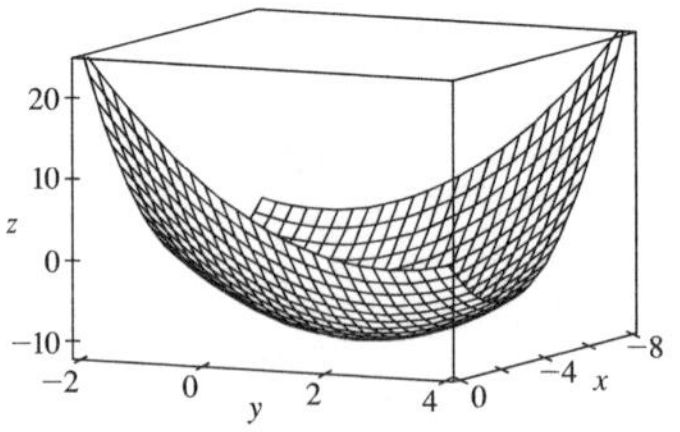

53. $f(x,y) = 3xy - x^2y - xy^2 \quad \Rightarrow \quad f_x = 3y - 2xy - y^2$, $f_y = 3x - x^2 - 2xy$, $f_{xx} = -2y$, $f_{yy} = -2x$, $f_{xy} = 3 - 2x - 2y$. Then $f_x = 0$ implies $y(3 - 2x - y) = 0$ so $y = 0$ or $y = 3 - 2x$. Substituting into $f_y = 0$ implies $x(3 - x) = 0$ or $3x(-1 + x) = 0$. Hence the critical points are $(0,0)$, $(3,0)$, $(0,3)$ and $(1,1)$. $D(0,0) = D(3,0) = D(0,3) = -9 < 0$ so $(0,0)$, $(3,0)$, and $(0,3)$ are saddle points. $D(1,1) = 3 > 0$ and $f_{xx}(1,1) = -2 < 0$, so $f(1,1) = 1$ is a local maximum.

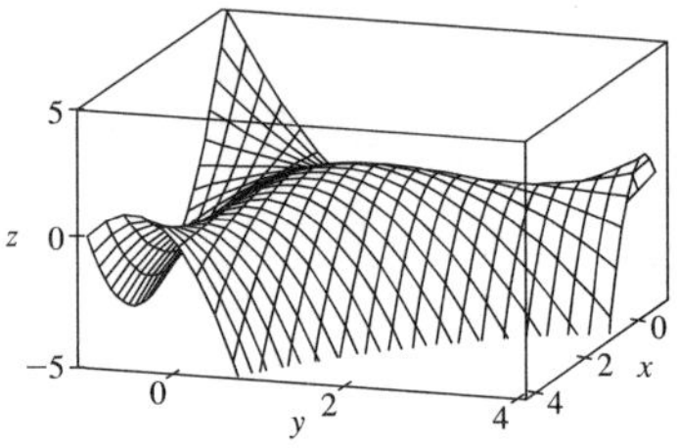

55. First solve inside D. Here $f_x = 4y^2 - 2xy^2 - y^3$, $f_y = 8xy - 2x^2y - 3xy^2$. Then $f_x = 0$ implies $y = 0$ or $y = 4 - 2x$, but $y = 0$ isn't inside D. Substituting $y = 4 - 2x$ into $f_y = 0$ implies $x = 0$, $x = 2$ or $x = 1$, but $x = 0$ isn't inside D, and when $x = 2$, $y = 0$ but $(2,0)$ isn't inside D. Thus the only critical point inside D is $(1,2)$ and $f(1,2) = 4$. Secondly we consider the boundary of D. On L_1, $f(x,0) = 0$ and so $f = 0$ on L_1. On L_2, $x = -y + 6$ and $f(-y+6, y) = y^2(6-y)(-2) = -2(6y^2 - y^3)$ which has critical points at $y = 0$ and $y = 4$. Then $f(6,0) = 0$ while $f(2,4) = -64$. On L_3, $f(0,y) = 0$, so $f = 0$ on L_3. Thus on D the absolute maximum of f is $f(1,2) = 4$ while the absolute minimum is $f(2,4) = -64$.

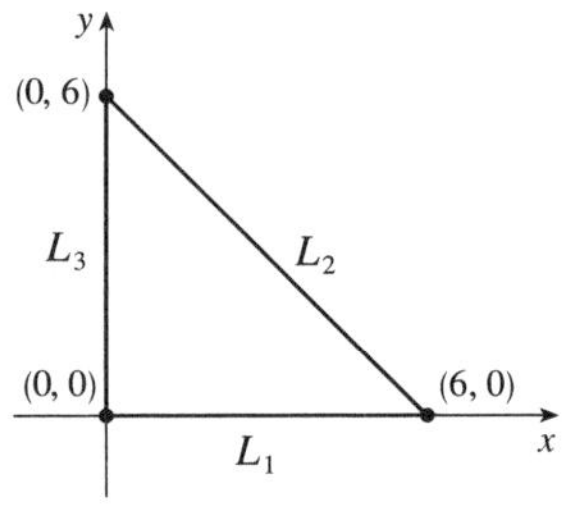

57. $f(x,y) = x^3 - 3x + y^4 - 2y^2$

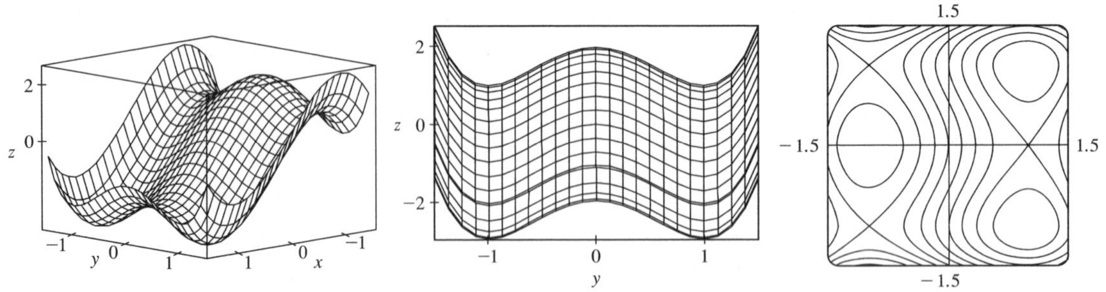

From the graphs, it appears that f has a local maximum $f(-1,0) \approx 2$, local minima $f(1,\pm 1) \approx -3$, and saddle points at $(-1, \pm 1)$ and $(1, 0)$.

To find the exact quantities, we calculate $f_x = 3x^2 - 3 = 0 \quad \Leftrightarrow \quad x = \pm 1$ and $f_y = 4y^3 - 4y = 0 \quad \Leftrightarrow$ $y = 0, \pm 1$, giving the critical points estimated above. Also $f_{xx} = 6x$, $f_{xy} = 0$, $f_{yy} = 12y^2 - 4$, so using the Second Derivatives Test, $D(-1,0) = 24 > 0$ and $f_{xx}(-1,0) = -6 < 0$ indicating a local maximum $f(-1,0) = 2$; $D(1,\pm 1) = 48 > 0$ and $f_{xx}(1,\pm 1) = 6 > 0$ indicating local minima $f(1,\pm 1) = -3$; and $D(-1,\pm 1) = -48$ and $D(1,0) = -24$, indicating saddle points.

59. $f(x, y) = x^2 y$, $g(x, y) = x^2 + y^2 = 1 \Rightarrow \nabla f = \langle 2xy, x^2 \rangle = \lambda \nabla g = \langle 2\lambda x, 2\lambda y \rangle$. Then $2xy = 2\lambda x$ and $x^2 = 2\lambda y$ imply $\lambda = x^2/(2y)$ and $\lambda = y$ if $x \neq 0$ and $y \neq 0$. Hence $x^2 = 2y^2$. Then $x^2 + y^2 = 1$ implies $3y^2 = 1$ so $y = \pm\frac{1}{\sqrt{3}}$ and $x = \pm\sqrt{\frac{2}{3}}$. [Note if $x = 0$ then $x^2 = 2\lambda y$ implies $y = 0$ and $f(0, 0) = 0$.] Thus the possible points are $\left(\pm\sqrt{\frac{2}{3}}, \pm\frac{1}{\sqrt{3}}\right)$ and the absolute maxima are $f\left(\pm\sqrt{\frac{2}{3}}, \frac{1}{\sqrt{3}}\right) = \frac{2}{3\sqrt{3}}$ while the absolute minima are $f\left(\pm\sqrt{\frac{2}{3}}, -\frac{1}{\sqrt{3}}\right) = -\frac{2}{3\sqrt{3}}$.

61. $f(x, y, z) = xyz$, $g(x, y, z) = x^2 + y^2 + z^2 = 3$. $\nabla f = \lambda \nabla g \Rightarrow \langle yz, xz, xy \rangle = \lambda \langle 2x, 2y, 2z \rangle$. If any of x, y, or z is zero, then $x = y = z = 0$ which contradicts $x^2 + y^2 + z^2 = 3$. Then $\lambda = \dfrac{yz}{2x} = \dfrac{xz}{2y} = \dfrac{xy}{2z} \Rightarrow 2y^2 z = 2x^2 z \Rightarrow y^2 = x^2$, and similarly $2yz^2 = 2x^2 y \Rightarrow z^2 = x^2$. Substituting into the constraint equation gives $x^2 + x^2 + x^2 = 3 \Rightarrow x^2 = 1 = y^2 = z^2$. Thus the possible points are $(1, 1, \pm 1)$, $(1, -1, \pm 1)$, $(-1, 1, \pm 1)$, $(-1, -1, \pm 1)$. The absolute maximum is $f(1, 1, 1) = f(1, -1, -1) = f(-1, 1, -1) = f(-1, -1, 1) = 1$ and the absolute minimum is $f(1, 1, -1) = f(1, -1, 1) = f(-1, 1, 1) = f(-1, -1, -1) = -1$.

63. $f(x, y, z) = x^2 + y^2 + z^2$, $g(x, y, z) = xy^2 z^3 = 2 \Rightarrow \nabla f = \langle 2x, 2y, 2z \rangle = \lambda \nabla g = \langle \lambda y^2 z^3, 2\lambda xyz^3, 3\lambda xy^2 z^2 \rangle$. Since $xy^2 z^3 = 2$, $x \neq 0$, $y \neq 0$ and $z \neq 0$, so (1) $2x = \lambda y^2 z^3$, (2) $1 = \lambda xz^3$, (3) $2 = 3\lambda xy^2 z$. Then (2) and (3) imply $\dfrac{1}{xz^3} = \dfrac{2}{3xy^2 z}$ or $y^2 = \frac{2}{3}z^2$ so $y = \pm z\sqrt{\frac{2}{3}}$. Similarly (1) and (3) imply $\dfrac{2x}{y^2 z^3} = \dfrac{2}{3xy^2 z}$ or $3x^2 = z^2$ so $x = \pm\frac{1}{\sqrt{3}}z$. But $xy^2 z^3 = 2$ so x and z must have the same sign, that is, $x = \frac{1}{\sqrt{3}}z$. Thus $g(x, y, z) = 2$ implies $\frac{1}{\sqrt{3}}z\left(\frac{2}{3}z^2\right)z^3 = 2$ or $z = \pm 3^{1/4}$ and the possible points are $(\pm 3^{-1/4}, 3^{-1/4}\sqrt{2}, \pm 3^{1/4})$, $(\pm 3^{-1/4}, -3^{-1/4}\sqrt{2}, \pm 3^{1/4})$. However at each of these points f takes on the same value, $2\sqrt{3}$. But $(2, 1, 1)$ also satisfies $g(x, y, z) = 2$ and $f(2, 1, 1) = 6 > 2\sqrt{3}$. Thus f has an absolute minimum value of $2\sqrt{3}$ and no absolute maximum subject to the constraint $xy^2 z^3 = 2$.

Alternate solution: $g(x, y, z) = xy^2 z^3 = 2$ implies $y^2 = \dfrac{2}{xz^3}$, so minimize $f(x, z) = x^2 + \dfrac{2}{xz^3} + z^2$. Then $f_x = 2x - \dfrac{2}{x^2 z^3}$, $f_z = -\dfrac{6}{xz^4} + 2z$, $f_{xx} = 2 + \dfrac{4}{x^3 z^3}$, $f_{zz} = \dfrac{24}{xz^5} + 2$ and $f_{xz} = \dfrac{6}{x^2 z^4}$. Now $f_x = 0$ implies $2x^3 z^3 - 2 = 0$ or $z = 1/x$. Substituting into $f_y = 0$ implies $-6x^3 + 2x^{-1} = 0$ or $x = \frac{1}{\sqrt[4]{3}}$, so the two critical points are $\left(\pm\frac{1}{\sqrt[4]{3}}, \pm\sqrt[4]{3}\right)$. Then $D\left(\pm\frac{1}{\sqrt[4]{3}}, \pm\sqrt[4]{3}\right) = (2 + 4)\left(2 + \frac{24}{3}\right) - \left(\frac{6}{\sqrt{3}}\right)^2 > 0$ and $f_{xx}\left(\pm\frac{1}{\sqrt[4]{3}}, \pm\sqrt[4]{3}\right) = 6 > 0$, so each point is a minimum. Finally, $y^2 = \dfrac{2}{xz^3}$, so the four points closest to the origin are $\left(\pm\frac{1}{\sqrt[4]{3}}, \frac{\sqrt{2}}{\sqrt[4]{3}}, \pm\sqrt[4]{3}\right)$, $\left(\pm\frac{1}{\sqrt[4]{3}}, -\frac{\sqrt{2}}{\sqrt[4]{3}}, \pm\sqrt[4]{3}\right)$.

65.

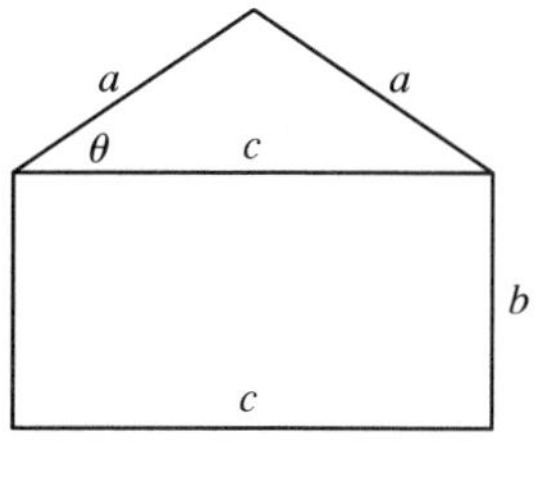

The area of the triangle is $\frac{1}{2}ca\sin\theta$ and the area of the rectangle is bc. Thus, the area of the whole object is $f(a, b, c) = \frac{1}{2}ca\sin\theta + bc$. The perimeter of the object is $g(a, b, c) = 2a + 2b + c = P$. To simplify $\sin\theta$ in terms of a, b, and c notice that $a^2\sin^2\theta + \left(\frac{1}{2}c\right)^2 = a^2 \Rightarrow \sin\theta = \dfrac{1}{2a}\sqrt{4a^2 - c^2}$. Thus $f(a, b, c) = \dfrac{c}{4}\sqrt{4a^2 - c^2} + bc$. (Instead of using θ, we could just have used the Pythagorean Theorem.) As a result, by Lagrange's method, we must find a, b, c, and λ by solving $\nabla f = \lambda \nabla g$ which gives the following equations: (1) $ca(4a^2 - c^2)^{-1/2} = 2\lambda$, (2) $c = 2\lambda$, (3) $\frac{1}{4}(4a^2 - c^2)^{1/2} - \frac{1}{4}c^2(4a^2 - c^2)^{-1/2} + b = \lambda$, and (4) $2a + 2b + c = P$. From (2), $\lambda = \frac{1}{2}c$ and so (1) produces $ca(4a^2 - c^2)^{-1/2} = c \Rightarrow (4a^2 - c^2)^{1/2} = a \Rightarrow$ $4a^2 - c^2 = a^2 \Rightarrow$ (5) $c = \sqrt{3}\,a$. Similarly, since $(4a^2 - c^2)^{1/2} = a$ and $\lambda = \frac{1}{2}c$, (3) gives $\frac{a}{4} - \frac{c^2}{4a} + b = \frac{c}{2}$, so from (5), $\frac{a}{4} - \frac{3a}{4} + b = \frac{\sqrt{3}\,a}{2} \Rightarrow -\frac{a}{2} - \frac{\sqrt{3}\,a}{2} = -b \Rightarrow$ (6) $b = \frac{a}{2}(1 + \sqrt{3})$. Substituting (5) and (6) into (4) we get:

$2a + a(1 + \sqrt{3}) + \sqrt{3}\,a = P \Rightarrow 3a + 2\sqrt{3}\,a = P \Rightarrow a = \dfrac{P}{3 + 2\sqrt{3}} = \dfrac{2\sqrt{3} - 3}{3}P$ and thus

$b = \dfrac{(2\sqrt{3} - 3)(1 + \sqrt{3})}{6}P = \dfrac{3 - \sqrt{3}}{6}P$ and $c = (2 - \sqrt{3})\,P$.C

FOCUS ON PROBLEM SOLVING

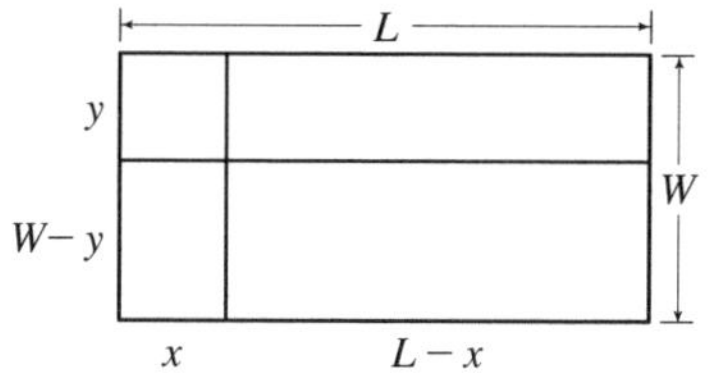

1. The areas of the smaller rectangles are $A_1 = xy$, $A_2 = (L - x)y$, $A_3 = (L - x)(W - y)$, $A_4 = x(W - y)$. For $0 \le x \le L$, $0 \le y \le W$, let

$$f(x, y) = A_1^2 + A_2^2 + A_3^2 + A_4^2$$
$$= x^2y^2 + (L - x)^2y^2 + (L - x)^2(W - y)^2 + x^2(W - y)^2$$
$$= [x^2 + (L - x)^2][y^2 + (W - y)^2]$$

Then we need to find the maximum and minimum values of $f(x, y)$. Here $f_x(x, y) = [2x - 2(L - x)][y^2 + (W - y)^2] = 0 \;\Rightarrow\; 4x - 2L = 0$ or $x = \frac{1}{2}L$, and $f_y(x, y) = [x^2 + (L - x)^2][2y - 2(W - y)] = 0 \;\Rightarrow\; 4y - 2W = 0$ or $y = W/2$. Also $f_{xx} = 4[y^2 + (W - y)^2]$, $f_{yy} = 4[x^2 + (L - x)^2]$, and $f_{xy} = (4x - 2L)(4y - 2W)$. Then $D = 16[y^2 + (W - y)^2][x^2 + (L - x)^2] - (4x - 2L)^2(4y - 2W)^2$. Thus when $x = \frac{1}{2}L$ and $y = \frac{1}{2}W$, $D > 0$ and $f_{xx} = 2W^2 > 0$. Thus a minimum of f occurs at $\left(\frac{1}{2}L, \frac{1}{2}W\right)$ and this minimum value is $f\left(\frac{1}{2}L, \frac{1}{2}W\right) = \frac{1}{4}L^2W^2$. There are no other critical points, so the maximum must occur on the boundary. Now along the width of the rectangle let $g(y) = f(0, y) = f(L, y) = L^2[y^2 + (W - y)^2]$, $0 \le y \le W$. Then $g'(y) = L^2[2y - 2(W - y)] = 0 \;\Leftrightarrow\; y = \frac{1}{2}W$. And $g\left(\frac{1}{2}\right) = \frac{1}{2}L^2W^2$. Checking the endpoints, we get $g(0) = g(W) = L^2W^2$. Along the length of the rectangle let $h(x) = f(x, 0) = f(x, W) = W^2[x^2 + (L - x)^2]$, $0 \le x \le L$. By symmetry $h'(x) = 0 \;\Leftrightarrow\; x = \frac{1}{2}L$ and $h\left(\frac{1}{2}L\right) = \frac{1}{2}L^2W^2$. At the endpoints we have $h(0) = h(L) = L^2W^2$. Therefore L^2W^2 is the maximum value of f. This maximum value of f occurs when the "cutting" lines correspond to sides of the rectangle.

3. (a) The area of a trapezoid is $\frac{1}{2}h(b_1 + b_2)$, where h is the height (the distance between the two parallel sides) and b_1, b_2 are the lengths of the bases (the parallel sides). From the figure in the text, we see that $h = x\sin\theta$, $b_1 = w - 2x$, and $b_2 = w - 2x + 2x\cos\theta$. Therefore the cross-sectional area of the rain gutter is

$$A(x, \theta) = \tfrac{1}{2}x\sin\theta\,[(w - 2x) + (w - 2x + 2x\cos\theta)]$$
$$= (x\sin\theta)(w - 2x + x\cos\theta)$$
$$= wx\sin\theta - 2x^2\sin\theta + x^2\sin\theta\cos\theta,\;\; 0 < x \le \tfrac{1}{2}w, 0 < \theta \le \tfrac{\pi}{2}$$

We look for the critical points of A: $\partial A/\partial x = w\sin\theta - 4x\sin\theta + 2x\sin\theta\cos\theta$ and $\partial A/\partial\theta = wx\cos\theta - 2x^2\cos\theta + x^2(\cos^2\theta - \sin^2\theta)$, so $\partial A/\partial x = 0 \;\Leftrightarrow\; \sin\theta\,(w - 4x + 2x\cos\theta) = 0 \;\Leftrightarrow\;$ $\cos\theta = \dfrac{4x - w}{2x} = 2 - \dfrac{w}{2x}$ $\;(0 < \theta \le \frac{\pi}{2} \;\Rightarrow\; \sin\theta > 0)$. If, in addition, $\partial A/\partial\theta = 0$, then

$$0 = wx\cos\theta - 2x^2\cos\theta + x^2(2\cos^2\theta - 1)$$
$$= wx\left(2 - \frac{w}{2x}\right) - 2x^2\left(2 - \frac{w}{2x}\right) + x^2\left[2\left(2 - \frac{w}{2x}\right)^2 - 1\right]$$
$$= 2wx - \tfrac{1}{2}w^2 - 4x^2 + wx + x^2\left[8 - \frac{4w}{x} + \frac{w^2}{2x^2} - 1\right] = -wx + 3x^2 = x(3x - w)$$

Since $x > 0$, we must have $x = \frac{1}{3}w$, in which case $\cos\theta = \frac{1}{2}$, so $\theta = \frac{\pi}{3}$, $\sin\theta = \frac{\sqrt{3}}{2}$, $k = \frac{\sqrt{3}}{6}w$, $b_1 = \frac{1}{3}w$, $b_2 = \frac{2}{3}w$, and $A = \frac{\sqrt{3}}{12}w^2$. As in Example 11.7.6, we can argue from the physical nature of this problem that we have found a local maximum of A. Now checking the boundary of A, let

$g(\theta) = A(w/2, \theta) = \frac{1}{2}w^2\sin\theta - \frac{1}{2}w^2\sin\theta + \frac{1}{4}w^2\sin\theta\cos\theta = \frac{1}{8}w^2\sin 2\theta$, $0 < \theta \leq \frac{\pi}{2}$. Clearly g is maximized when $\sin 2\theta = 1$ in which case $A = \frac{1}{8}w^2$. Also along the line $\theta = \frac{\pi}{2}$, let $h(x) = A\left(x, \frac{\pi}{2}\right) = wx - 2x^2$, $0 < x < \frac{1}{2}w \Rightarrow$ $h'(x) = w - 4x = 0 \Leftrightarrow x = \frac{1}{4}w$, and $h\left(\frac{1}{4}w\right) = w\left(\frac{1}{4}w\right) - 2\left(\frac{1}{4}w\right)^2 = \frac{1}{8}w^2$. Since $\frac{1}{8}w^2 < \frac{\sqrt{3}}{12}w^2$, we conclude that the local maximum found earlier was an absolute maximum.

(b) If the metal were bent into a semi-circular gutter of radius r, we would have $w = \pi r$ and $A = \frac{1}{2}\pi r^2 = \frac{1}{2}\pi\left(\frac{w}{\pi}\right)^2 = \frac{w^2}{2\pi}$.

Since $\dfrac{w^2}{2\pi} > \dfrac{\sqrt{3}\,w^2}{12}$, it *would* be better to bend the metal into a gutter with a semicircular cross-section.

5. Let $g(x,y) = xf\left(\frac{y}{x}\right)$. Then $g_x(x,y) = f\left(\frac{y}{x}\right) + xf'\left(\frac{y}{x}\right)\left(-\frac{y}{x^2}\right) = f\left(\frac{y}{x}\right) - \frac{y}{x}f'\left(\frac{y}{x}\right)$ and

$g_y(x,y) = xf'\left(\frac{y}{x}\right)\left(\frac{1}{x}\right) = f'\left(\frac{y}{x}\right)$. Thus the tangent plane at (x_0, y_0, z_0) on the surface has equation

$$z - x_0 f\left(\frac{y_0}{x_0}\right) = \left[f\left(\frac{y_0}{x_0}\right) - y_0 x_0^{-1} f'\left(\frac{y_0}{x_0}\right)\right](x - x_0) + f'\left(\frac{y_0}{x_0}\right)(y - y_0) \Rightarrow$$

$\left[f\left(\frac{y_0}{x_0}\right) - y_0 x_0^{-1} f'\left(\frac{y_0}{x_0}\right)\right]x + \left[f'\left(\frac{y_0}{x_0}\right)\right]y - z = 0$. But any plane whose equation is of the form $ax + by + cz = 0$ passes through the origin. Thus the origin is the common point of intersection.

7. (a) $x = r\cos\theta$, $y = r\sin\theta$, $z = z$. Then $\dfrac{\partial u}{\partial r} = \dfrac{\partial u}{\partial x}\dfrac{\partial x}{\partial r} + \dfrac{\partial u}{\partial y}\dfrac{\partial y}{\partial r} + \dfrac{\partial u}{\partial z}\dfrac{\partial z}{\partial r} = \dfrac{\partial u}{\partial x}\cos\theta + \dfrac{\partial u}{\partial y}\sin\theta$ and

$$\begin{aligned}\frac{\partial^2 u}{\partial r^2} &= \cos\theta\left[\frac{\partial^2 u}{\partial x^2}\frac{\partial x}{\partial r} + \frac{\partial^2 u}{\partial y\,\partial x}\frac{\partial y}{\partial r} + \frac{\partial^2 u}{\partial z\,\partial x}\frac{\partial z}{\partial r}\right] + \sin\theta\left[\frac{\partial^2 u}{\partial y^2}\frac{\partial y}{\partial r} + \frac{\partial^2 u}{\partial x\,\partial y}\frac{\partial x}{\partial r} + \frac{\partial^2 u}{\partial z\,\partial y}\frac{\partial z}{\partial r}\right] \\ &= \frac{\partial^2 u}{\partial x^2}\cos^2\theta + \frac{\partial^2 u}{\partial y^2}\sin^2\theta + 2\frac{\partial^2 u}{\partial y\,\partial x}\cos\theta\sin\theta\end{aligned}$$

Similarly $\dfrac{\partial u}{\partial \theta} = -\dfrac{\partial u}{\partial x}r\sin\theta + \dfrac{\partial u}{\partial y}r\cos\theta$ and

$$\frac{\partial^2 u}{\partial \theta^2} = \frac{\partial^2 u}{\partial x^2}r^2\sin^2\theta + \frac{\partial^2 u}{\partial y^2}r^2\cos^2\theta - 2\frac{\partial^2 u}{\partial y\,\partial x}r^2\sin\theta\cos\theta - \frac{\partial u}{\partial x}r\cos\theta - \frac{\partial u}{\partial y}r\sin\theta. \text{ So}$$

$$\begin{aligned}&\frac{\partial^2 u}{\partial r^2} + \frac{1}{r}\frac{\partial u}{\partial r} + \frac{1}{r^2}\frac{\partial^2 u}{\partial \theta^2} + \frac{\partial^2 u}{\partial z^2} \\ &= \frac{\partial^2 u}{\partial x^2}\cos^2\theta + \frac{\partial^2 u}{\partial y^2}\sin^2\theta + 2\frac{\partial^2 u}{\partial y\,\partial x}\cos\theta\sin\theta + \frac{\partial u}{\partial x}\frac{\cos\theta}{r} + \frac{\partial u}{\partial y}\frac{\sin\theta}{r} \\ &\quad + \frac{\partial^2 u}{\partial x^2}\sin^2\theta + \frac{\partial^2 u}{\partial y^2}\cos^2\theta - 2\frac{\partial^2 u}{\partial y\,\partial x}\sin\theta\cos\theta - \frac{\partial u}{\partial x}\frac{\cos\theta}{r} - \frac{\partial u}{\partial y}\frac{\sin\theta}{r} + \frac{\partial^2 u}{\partial z^2} \\ &= \frac{\partial^2 u}{\partial x^2} + \frac{\partial^2 u}{\partial y^2} + \frac{\partial^2 u}{\partial z^2}\end{aligned}$$

(b) $x = \rho\sin\phi\cos\theta$, $y = \rho\sin\phi\sin\theta$, $z = \rho\cos\phi$. Then

$$\frac{\partial u}{\partial \rho} = \frac{\partial u}{\partial x}\frac{\partial x}{\partial \rho} + \frac{\partial u}{\partial y}\frac{\partial y}{\partial \rho} + \frac{\partial u}{\partial z}\frac{\partial z}{\partial \rho} = \frac{\partial u}{\partial x}\sin\phi\cos\theta + \frac{\partial u}{\partial y}\sin\phi\sin\theta + \frac{\partial u}{\partial z}\cos\phi, \text{ and}$$

$$\begin{aligned}\frac{\partial^2 u}{\partial \rho^2} &= \sin\phi\cos\theta\left[\frac{\partial^2 u}{\partial x^2}\frac{\partial x}{\partial \rho} + \frac{\partial^2 u}{\partial y\,\partial x}\frac{\partial y}{\partial \rho} + \frac{\partial^2 u}{\partial z\,\partial x}\frac{\partial z}{\partial \rho}\right] \\ &\quad + \sin\phi\sin\theta\left[\frac{\partial^2 u}{\partial y^2}\frac{\partial y}{\partial \rho} + \frac{\partial^2 u}{\partial x\,\partial y}\frac{\partial x}{\partial \rho} + \frac{\partial^2 u}{\partial z\,\partial y}\frac{\partial z}{\partial \rho}\right] \\ &\quad + \cos\phi\left[\frac{\partial^2 u}{\partial z^2}\frac{\partial z}{\partial \rho} + \frac{\partial^2 u}{\partial x\,\partial z}\frac{\partial x}{\partial \rho} + \frac{\partial^2 u}{\partial y\,\partial z}\frac{\partial y}{\partial \rho}\right] \\ &= 2\frac{\partial^2 u}{\partial y\,\partial x}\sin^2\phi\sin\theta\cos\theta + 2\frac{\partial^2 u}{\partial z\,\partial x}\sin\phi\cos\phi\cos\theta + 2\frac{\partial^2 u}{\partial y\,\partial z}\sin\phi\cos\phi\sin\theta \\ &\quad + \frac{\partial^2 u}{\partial x^2}\sin^2\phi\cos^2\theta + \frac{\partial^2 u}{\partial y^2}\sin^2\phi\sin^2\theta + \frac{\partial^2 u}{\partial z^2}\cos^2\phi\end{aligned}$$

Similarly $\dfrac{\partial u}{\partial \phi} = \dfrac{\partial u}{\partial x}\rho\cos\phi\cos\theta + \dfrac{\partial u}{\partial y}\rho\cos\phi\sin\theta - \dfrac{\partial u}{\partial z}\rho\sin\phi$, and

$$\begin{aligned}\frac{\partial^2 u}{\partial \phi^2} &= 2\frac{\partial^2 u}{\partial y\,\partial x}\rho^2\cos^2\phi\sin\theta\cos\theta - 2\frac{\partial^2 u}{\partial x\,\partial z}\rho^2\sin\phi\cos\phi\cos\theta \\ &\quad - 2\frac{\partial^2 u}{\partial y\,\partial z}\rho^2\sin\phi\cos\phi\sin\theta + \frac{\partial^2 u}{\partial x^2}\rho^2\cos^2\phi\cos^2\theta + \frac{\partial^2 u}{\partial y^2}\rho^2\cos^2\phi\sin^2\theta \\ &\quad + \frac{\partial^2 u}{\partial z^2}\rho^2\sin^2\phi - \frac{\partial u}{\partial x}\rho\sin\phi\cos\theta - \frac{\partial u}{\partial y}\rho\sin\phi\sin\theta - \frac{\partial u}{\partial z}\rho\cos\phi\end{aligned}$$

And $\dfrac{\partial u}{\partial \theta} = -\dfrac{\partial u}{\partial x}\rho\sin\phi\sin\theta + \dfrac{\partial u}{\partial y}\rho\sin\phi\cos\theta$, while

$$\begin{aligned}\frac{\partial^2 u}{\partial \theta^2} &= -2\frac{\partial^2 u}{\partial y\,\partial x}\rho^2\sin^2\phi\cos\theta\sin\theta + \frac{\partial^2 u}{\partial x^2}\rho^2\sin^2\phi\sin^2\theta \\ &\quad + \frac{\partial^2 u}{\partial y^2}\rho^2\sin^2\phi\cos^2\theta - \frac{\partial u}{\partial x}\rho\sin\phi\cos\theta - \frac{\partial u}{\partial y}\rho\sin\phi\sin\theta\end{aligned}$$

Therefore

$$\begin{aligned}&\frac{\partial^2 u}{\partial \rho^2} + \frac{2}{\rho}\frac{\partial u}{\partial \rho} + \frac{\cot\phi}{\rho^2}\frac{\partial u}{\partial \phi} + \frac{1}{\rho^2}\frac{\partial^2 u}{\partial \phi^2} + \frac{1}{\rho^2\sin^2\phi}\frac{\partial^2 u}{\partial \theta^2} \\ &= \frac{\partial^2 u}{\partial x^2}\left[(\sin^2\phi\cos^2\theta) + (\cos^2\phi\cos^2\theta) + \sin^2\theta\right] \\ &\quad + \frac{\partial^2 u}{\partial y^2}\left[(\sin^2\phi\sin^2\theta) + (\cos^2\phi\sin^2\theta) + \cos^2\theta\right] + \frac{\partial^2 u}{\partial z^2}\left[\cos^2\phi + \sin^2\phi\right] \\ &\quad + \frac{\partial u}{\partial x}\left[\frac{2\sin^2\phi\cos\theta + \cos^2\phi\cos\theta - \sin^2\phi\cos\theta - \cos\theta}{\rho\sin\phi}\right] \\ &\quad + \frac{\partial u}{\partial y}\left[\frac{2\sin^2\phi\sin\theta + \cos^2\phi\sin\theta - \sin^2\phi\sin\theta - \sin\theta}{\rho\sin\phi}\right]\end{aligned}$$

But $2\sin^2\phi\cos\theta + \cos^2\phi\cos\theta - \sin^2\phi\cos\theta - \cos\theta = (\sin^2\phi + \cos^2\phi - 1)\cos\theta = 0$ and similarly the coefficient of $\partial u/\partial y$ is 0. Also $\sin^2\phi\cos^2\theta + \cos^2\phi\cos^2\theta + \sin^2\theta = \cos^2\theta\,(\sin^2\phi + \cos^2\phi) + \sin^2\theta = 1$, and similarly the coefficient of $\partial^2 u/\partial y^2$ is 1. So Laplace's Equation in spherical coordinates is as stated.

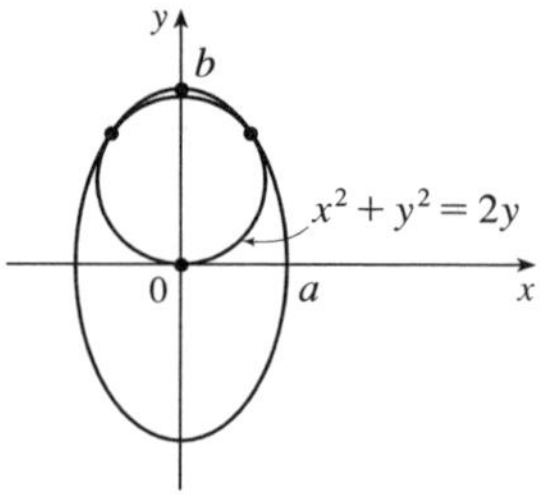

9. Since we are minimizing the area of the ellipse, and the circle lies above the x-axis, the ellipse will intersect the circle for only one value of y. This y-value must satisfy both the equation of the circle and the equation of the ellipse. Now $\frac{x^2}{a^2} + \frac{y^2}{b^2} = 1 \Rightarrow x^2 = \frac{a^2}{b^2}(b^2 - y^2)$. Substituting into the equation of the circle gives $\frac{a^2}{b^2}(b^2 - y^2) + y^2 - 2y = 0 \Rightarrow \left(\frac{b^2 - a^2}{b^2}\right)y^2 - 2y + a^2 = 0$.

In order for there to be only one solution to this quadratic equation, the discriminant must be 0, so $4 - 4a^2\,\frac{b^2 - a^2}{b^2} = 0 \Rightarrow$ $b^2 - a^2b^2 + a^4 = 0$. The area of the ellipse is $A(a, b) = \pi ab$, and we minimize this function subject to the constraint $g(a, b) = b^2 - a^2b^2 + a^4 = 0$.

Now $\nabla A = \lambda \nabla g \Leftrightarrow \pi b = \lambda(4a^3 - 2ab^2)$, $\pi a = \lambda(2b - 2ba^2) \Rightarrow$ (1) $\lambda = \frac{\pi b}{2a(2a^2 - b^2)}$, (2) $\lambda = \frac{\pi a}{2b(1 - a^2)}$, (3) $b^2 - a^2b^2 + a^4 = 0$. Comparing (1) and (2) gives $\frac{\pi b}{2a(2a^2 - b^2)} = \frac{\pi a}{2b(1 - a^2)} \Rightarrow$ $2\pi b^2 = 4\pi a^4 \Leftrightarrow a^2 = \frac{1}{\sqrt{2}}\,b$. Substitute this into (3) to get $b = \frac{3}{\sqrt{2}} \Rightarrow a = \sqrt{\frac{3}{2}}$.

12 □ MULTIPLE INTEGRALS

12.1 Double Integrals over Rectangles

1. (a) The subrectangles are shown in the figure.

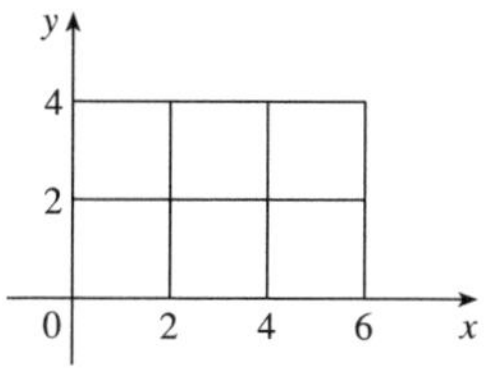

The surface is the graph of $f(x, y) = xy$ and $\Delta A = 4$, so we estimate

$$V \approx \sum_{i=1}^{3} \sum_{j=1}^{2} f(x_i, y_j)\,\Delta A$$
$$= f(2,2)\,\Delta A + f(2,4)\,\Delta A + f(4,2)\,\Delta A + f(4,4)\,\Delta A + f(6,2)\,\Delta A + f(6,4)\,\Delta A$$
$$= 4(4) + 8(4) + 8(4) + 16(4) + 12(4) + 24(4) = 288$$

(b) $V \approx \sum_{i=1}^{3} \sum_{j=1}^{2} f(\overline{x}_i, \overline{y}_j)\,\Delta A = f(1,1)\,\Delta A + f(1,3)\,\Delta A + f(3,1)\,\Delta A + f(3,3)\,\Delta A + f(5,1)\,\Delta A + f(5,3)\,\Delta A$

$$= 1(4) + 3(4) + 3(4) + 9(4) + 5(4) + 15(4) = 144$$

3. (a) The subrectangles are shown in the figure. Since $\Delta A = \pi^2/4$, we estimate

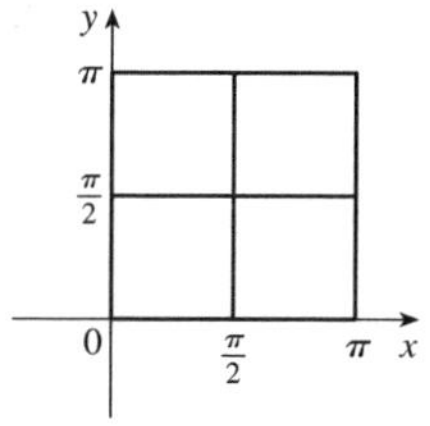

$$\iint_R \sin(x+y)\,dA \approx \sum_{i=1}^{2} \sum_{j=1}^{2} f(x_{ij}^*, y_{ij}^*)\,\Delta A$$
$$= f(0,0)\,\Delta A + f(0, \tfrac{\pi}{2})\,\Delta A + f(\tfrac{\pi}{2}, 0)\,\Delta A + f(\tfrac{\pi}{2}, \tfrac{\pi}{2})\,\Delta A$$
$$= 0\left(\tfrac{\pi^2}{4}\right) + 1\left(\tfrac{\pi^2}{4}\right) + 1\left(\tfrac{\pi^2}{4}\right) + 0\left(\tfrac{\pi^2}{4}\right) = \tfrac{\pi^2}{2} \approx 4.935$$

(b) $\iint_R \sin(x+y)\,dA \approx \sum_{i=1}^{2} \sum_{j=1}^{2} f(\overline{x}_i, \overline{y}_j)\,\Delta A$

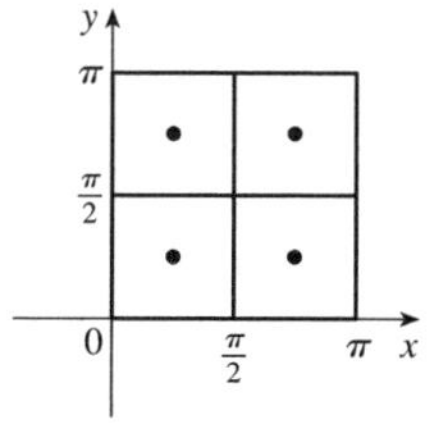

$$= f(\tfrac{\pi}{4}, \tfrac{\pi}{4})\,\Delta A + f(\tfrac{\pi}{4}, \tfrac{3\pi}{4})\,\Delta A + f(\tfrac{3\pi}{4}, \tfrac{\pi}{4})\,\Delta A + f(\tfrac{3\pi}{4}, \tfrac{3\pi}{4})\,\Delta A$$
$$= 1\left(\tfrac{\pi^2}{4}\right) + 0\left(\tfrac{\pi^2}{4}\right) + 0\left(\tfrac{\pi^2}{4}\right) + (-1)\left(\tfrac{\pi^2}{4}\right) = 0$$

5. (a) Each subrectangle and its midpoint are shown in the figure. The area of each subrectangle is $\Delta A = 2$, so we evaluate f at each midpoint and estimate

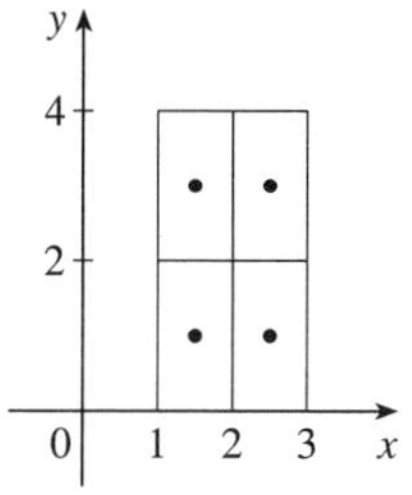

$$\iint_R f(x,y)\,dA \approx \sum_{i=1}^{2} \sum_{j=1}^{2} f(\overline{x}_i, \overline{y}_j)\,\Delta A$$
$$= f(1.5, 1)\,\Delta A + f(1.5, 3)\,\Delta A + f(2.5, 1)\,\Delta A + f(2.5, 3)\,\Delta A$$
$$= 1(2) + (-8)(2) + 5(2) + (-1)(2) = -6$$

(b) The subrectangles are shown in the figure. In each subrectangle, the sample point furthest from the origin is the upper right corner, and the area of each subrectangle is $\Delta A = \frac{1}{2}$. Thus we estimate

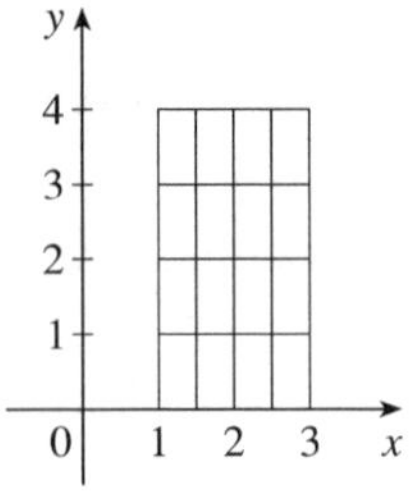

$$\iint_R f(x,y)\,dA \approx \sum_{i=1}^{4}\sum_{j=1}^{4} f(x_i,y_j)\,\Delta A$$

$$= f(1.5,1)\,\Delta A + f(1.5,2)\,\Delta A + f(1.5,3)\,\Delta A + f(1.5,4)\,\Delta A$$
$$+ f(2,1)\,\Delta A + f(2,2)\,\Delta A + f(2,3)\,\Delta A + f(2,4)\,\Delta A$$
$$+ f(2.5,1)\,\Delta A + f(2.5,2)\,\Delta A + f(2.5,3)\,\Delta A + f(2.5,4)\,\Delta A$$
$$+ f(3,1)\,\Delta A + f(3,2)\,\Delta A + f(3,3)\,\Delta A + f(3,4)\,\Delta A$$
$$= 1(\tfrac{1}{2}) + (-4)(\tfrac{1}{2}) + (-8)(\tfrac{1}{2}) + (-6)(\tfrac{1}{2}) + 3(\tfrac{1}{2}) + 0(\tfrac{1}{2}) + (-5)(\tfrac{1}{2}) + (-8)(\tfrac{1}{2})$$
$$+ 5(\tfrac{1}{2}) + 3(\tfrac{1}{2}) + (-1)(\tfrac{1}{2}) + (-4)(\tfrac{1}{2}) + 8(\tfrac{1}{2}) + 6(\tfrac{1}{2}) + 3(\tfrac{1}{2}) + 0(\tfrac{1}{2})$$
$$= -3.5$$

7. The values of $f(x,y) = \sqrt{52 - x^2 - y^2}$ get smaller as we move further from the origin, so on any of the subrectangles in the problem, the function will have its largest value at the lower left corner of the subrectangle and its smallest value at the upper right corner, and any other value will lie between these two. So using these subrectangles we have $U < V < L$. (Note that this is true no matter how R is divided into subrectangles.)

9. (a) With $m = n = 2$, we have $\Delta A = 4$. Using the contour map to estimate the value of f at the centre of each subrectangle, we have

$$\iint_R f(x,y)\,dA \approx \sum_{i=1}^{2}\sum_{j=1}^{2} f(\overline{x}_i,\overline{y}_j)\,\Delta A = \Delta A[f(1,1) + f(1,3) + f(3,1) + f(3,3)]$$
$$\approx 4(27 + 4 + 14 + 17) = 248$$

(b) $f_{\text{ave}} = \frac{1}{A(R)}\iint_R f(x,y)\,dA \approx \frac{1}{16}(248) = 15.5$

11. $z = 3 > 0$, so we can interpret the integral as the volume of the solid S that lies below the plane $z = 3$ and above the rectangle $[-2,2] \times [1,6]$. S is a rectangular solid, thus $\iint_R 3\,dA = 4 \cdot 5 \cdot 3 = 60$.

13. $z = f(x,y) = 4 - 2y \geq 0$ for $0 \leq y \leq 1$. Thus the integral represents the volume of that part of the rectangular solid $[0,1] \times [0,1] \times [0,4]$ which lies below the plane $z = 4 - 2y$. So

$$\iint_R (4 - 2y)\,dA = (1)(1)(2) + \tfrac{1}{2}(1)(1)(2) = 3$$

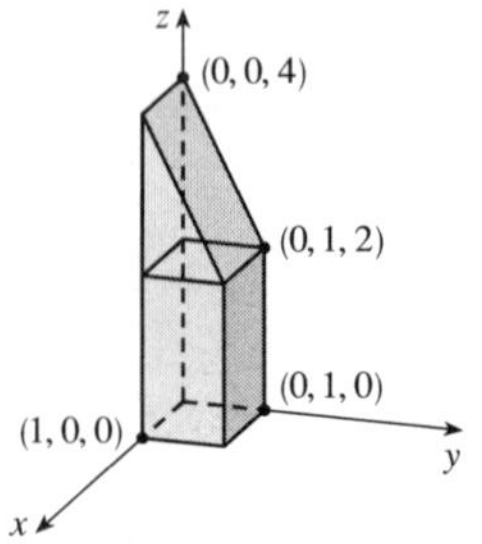

15. To calculate the estimates using a programmable calculator, we can use an algorithm similar to that of Exercise 5.1.7. In Maple, we can define the function $f(x,y) = \sqrt{1 + xe^{-y}}$ (calling it `f`), load the `student` package, and then use the command

```
middlesum(middlesum(f,x=0..1,m),
                         y=0..1,m);
```

to get the estimate with $n = m^2$ squares of equal size. Mathematica has no special Riemann sum command, but we can define `f` and then use nested `Sum` commands to calculate the estimates.

n	estimate
1	1.141 606
4	1.143 191
16	1.143 535
64	1.143 617
256	1.143 637
1024	1.143 642

17. If we divide R into mn subrectangles, $\iint_R k\,dA \approx \sum\limits_{i=1}^{m}\sum\limits_{j=1}^{n} f\left(x_{ij}^*, y_{ij}^*\right)\Delta A$ for any choice of sample points $\left(x_{ij}^*, y_{ij}^*\right)$. But $f\left(x_{ij}^*, y_{ij}^*\right) = k$ always and $\sum\limits_{i=1}^{m}\sum\limits_{j=1}^{n}\Delta A =$ area of $R = (b-a)(d-c)$. Thus, no matter how we choose the sample points, $\sum\limits_{i=1}^{m}\sum\limits_{j=1}^{n} f\left(x_{ij}^*, y_{ij}^*\right)\Delta A = k\sum\limits_{i=1}^{m}\sum\limits_{j=1}^{n}\Delta A = k(b-a)(d-c)$ and so

$$\begin{aligned}\iint_R k\,dA &= \lim_{m,n\to\infty}\sum_{i=1}^{m}\sum_{j=1}^{n} f\left(x_{ij}^*, y_{ij}^*\right)\Delta A = \lim_{m,n\to\infty} k\sum_{i=1}^{m}\sum_{j=1}^{n}\Delta A \\ &= \lim_{m,n\to\infty} k(b-a)(d-c) = k(b-a)(d-c)\end{aligned}$$

12.2 Iterated Integrals

1. $\int_0^3 (2x + 3x^2y)\,dx = \left[x^2 + x^3y\right]_{x=0}^{x=3} = (9 + 27y) - (0+0) = 9 + 27y$,

$\int_0^4 (2x + 3x^2y)\,dy = \left[2xy + 3x^2\,\dfrac{y^2}{2}\right]_{y=0}^{y=4} = \left(8x + 3x^2\cdot\dfrac{16}{2}\right) - (0+0) = 8x + 24x^2$

3. $\int_1^3\int_0^1 (1 + 4xy)\,dx\,dy = \int_1^3 \left[x + 2x^2y\right]_{x=0}^{x=1} dy = \int_1^3 (1 + 2y)\,dy = \left[y + y^2\right]_1^3 = (3+9) - (1+1) = 10$

5.
$$\begin{aligned}\int_0^3\int_0^1 \sqrt{x+y}\,dx\,dy &= \int_0^3 \left[\tfrac{2}{3}(x+y)^{3/2}\right]_{x=0}^{x=1} dy = \tfrac{2}{3}\int_0^3 \left[(1+y)^{3/2} - y^{3/2}\right] dy \\ &= \tfrac{2}{3}\left[\tfrac{2}{5}(1+y)^{5/2} - \tfrac{2}{5}y^{5/2}\right]_0^3 = \tfrac{4}{15}\left[32 - 3^{5/2} - 1\right] = \tfrac{4}{15}\left(31 - 9\sqrt{3}\right)\end{aligned}$$

7.
$$\begin{aligned}\int_0^2\int_0^1 (2x+y)^8\,dx\,dy &= \int_0^2 \left[\frac{1}{2}\frac{(2x+y)^9}{9}\right]_{x=0}^{x=1} dy \qquad \left[\text{substitute } u = 2x + y \;\Rightarrow\; dx = \tfrac{1}{2}du\right] \\ &= \frac{1}{18}\int_0^2 \left[(2+y)^9 - (0+y)^9\right] dy = \frac{1}{18}\left[\frac{(2+y)^{10}}{10} - \frac{y^{10}}{10}\right]_0^2 \\ &= \tfrac{1}{180}\left[(4^{10} - 2^{10}) - (2^{10} - 0^{10})\right] = \tfrac{1\,046\,528}{180} = \tfrac{261\,632}{45}\end{aligned}$$

9.
$$\begin{aligned}\int_1^4\int_1^2 \left(\frac{x}{y} + \frac{y}{x}\right) dy\,dx &= \int_1^4 \left[x\ln|y| + \frac{1}{x}\cdot\frac{1}{2}y^2\right]_{y=1}^{y=2} dx = \int_1^4 \left(x\ln 2 + \frac{3}{2x}\right) dx = \left[\frac{1}{2}x^2\ln 2 + \frac{3}{2}\ln|x|\right]_1^4 \\ &= 8\ln 2 + \tfrac{3}{2}\ln 4 - \tfrac{1}{2}\ln 2 = \tfrac{15}{2}\ln 2 + 3\ln 4^{1/2} = \tfrac{21}{2}\ln 2\end{aligned}$$

11. $\int_0^{\ln 2}\int_0^{\ln 5} e^{2x-y}\,dx\,dy = \left(\int_0^{\ln 5} e^{2x}\,dx\right)\left(\int_0^{\ln 2} e^{-y}\,dy\right) = \left[\tfrac{1}{2}e^{2x}\right]_0^{\ln 5}\left[-e^{-y}\right]_0^{\ln 2} = \left(\tfrac{25}{2} - \tfrac{1}{2}\right)\left(-\tfrac{1}{2} + 1\right) = 6$

13. $\iint_R \frac{xy^2}{x^2+1}\,dA = \int_0^1 \int_{-3}^3 \frac{xy^2}{x^2+1}\,dy\,dx = \int_0^1 \frac{x}{x^2+1}\,dx \int_{-3}^3 y^2\,dy$

$= \left[\frac{1}{2}\ln(x^2+1)\right]_0^1 \left[\frac{1}{3}y^3\right]_{-3}^3 = \frac{1}{2}(\ln 2 - \ln 1)\cdot\frac{1}{3}(27+27) = 9\ln 2$

15. $\int_0^{\pi/6}\int_0^{\pi/3} x\sin(x+y)\,dy\,dx$

$= \int_0^{\pi/6}\left[-x\cos(x+y)\right]_{y=0}^{y=\pi/3}\,dx = \int_0^{\pi/6}\left[x\cos x - x\cos\left(x+\frac{\pi}{3}\right)\right]dx$

$= x\left[\sin x - \sin\left(x+\frac{\pi}{3}\right)\right]_0^{\pi/6} - \int_0^{\pi/6}\left[\sin x - \sin\left(x+\frac{\pi}{3}\right)\right]dx$ [by integrating by parts separately for each term]

$= \frac{\pi}{6}\left[\frac{1}{2}-1\right] - \left[-\cos x + \cos\left(x+\frac{\pi}{3}\right)\right]_0^{\pi/6} = -\frac{\pi}{12} - \left[-\frac{\sqrt{3}}{2} + 0 - \left(-1+\frac{1}{2}\right)\right] = \frac{\sqrt{3}-1}{2} - \frac{\pi}{12}$

17. $\iint_R xye^{x^2y}\,dA = \int_0^2\int_0^1 xye^{x^2y}\,dx\,dy = \int_0^2\left[\frac{1}{2}e^{x^2y}\right]_{x=0}^{x=1}dy = \frac{1}{2}\int_0^2(e^y-1)\,dy$

$= \frac{1}{2}\left[e^y - y\right]_0^2 = \frac{1}{2}[(e^2-2)-(1-0)] = \frac{1}{2}(e^2-3)$

19. $z = f(x,y) = 4 - x - 2y \geq 0$ for $0 \leq x \leq 1$ and $0 \leq y \leq 1$. So the solid is the region in the first octant which lies below the plane $z = 4 - x - 2y$ and above $[0,1]\times[0,1]$.

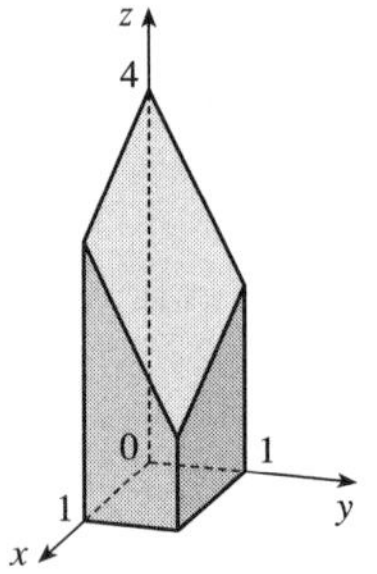

21. $V = \iint_R(12-3x-2y)\,dA = \int_{-2}^3\int_0^1(12-3x-2y)\,dx\,dy = \int_{-2}^3\left[12x - \frac{3}{2}x^2 - 2xy\right]_{x=0}^{x=1}dy$

$= \int_{-2}^3\left(\frac{21}{2}-2y\right)dy = \left[\frac{21}{2}y - y^2\right]_{-2}^3 = \frac{95}{2}$

23. $V = \int_{-2}^2\int_{-1}^1\left(1-\frac{1}{4}x^2-\frac{1}{9}y^2\right)dx\,dy = 4\int_0^2\int_0^1\left(1-\frac{1}{4}x^2-\frac{1}{9}y^2\right)dx\,dy$

$= 4\int_0^2\left[x - \frac{1}{12}x^3 - \frac{1}{9}y^2x\right]_{x=0}^{x=1}dy = 4\int_0^2\left(\frac{11}{12}-\frac{1}{9}y^2\right)dy = 4\left[\frac{11}{12}y - \frac{1}{27}y^3\right]_0^2 = 4\cdot\frac{83}{54} = \frac{166}{27}$

25. Here we need the volume of the solid lying under the surface $z = x\sqrt{x^2+y}$ and above the square $R = [0,1]\times[0,1]$ in the xy-plane.

$$V = \int_0^1\int_0^1 x\sqrt{x^2+y}\,dx\,dy = \int_0^1\frac{1}{3}\left[(x^2+y)^{3/2}\right]_{x=0}^{x=1}dy = \frac{1}{3}\int_0^1\left[(1+y)^{3/2}-y^{3/2}\right]dy$$

$$= \frac{1}{3}\cdot\frac{2}{5}\left[(1+y)^{5/2}-y^{5/2}\right]_0^1 = \frac{4}{15}(2\sqrt{2}-1)$$

27. In the first octant, $z \geq 0 \quad\Rightarrow\quad y \leq 3$, so

$$V = \int_0^3\int_0^2(9-y^2)\,dx\,dy = \int_0^3\left[9x - y^2x\right]_{x=0}^{x=2}dy = \int_0^3(18-2y^2)\,dy = \left[18y - \frac{2}{3}y^3\right]_0^3 = 36$$

29. In Maple, we can calculate the integral by defining the integrand as `f` and then using the command `int(int(f,x=0..1),y=0..1);`. In Mathematica, we can use the command `Integrate[Integrate[f,{x,0,1}],{y,0,1}]`. We find that $\iint_R x^5y^3e^{xy}\,dA = 21e - 57 \approx 0.0839$. We can use `plot3d` (in Maple) or `Plot3d` (in Mathematica) to graph the function.

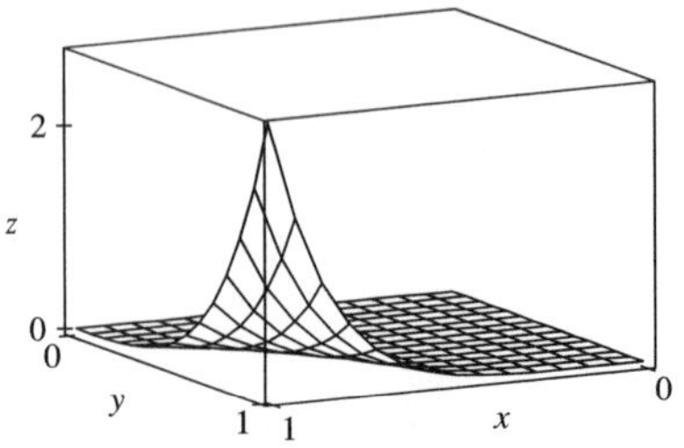

31. R is the rectangle $[-1,1]\times[0,5]$. Thus, $A(R)=2\cdot 5=10$ and

$$f_{\text{ave}}=\frac{1}{A(R)}\iint_R f(x,y)\,dA=\tfrac{1}{10}\int_0^5\int_{-1}^1 x^2y\,dx\,dy=\tfrac{1}{10}\int_0^5\left[\tfrac{1}{3}x^3y\right]_{x=-1}^{x=1}dy=\tfrac{1}{10}\int_0^5\tfrac{2}{3}y\,dy=\tfrac{1}{10}\left[\tfrac{1}{3}y^2\right]_0^5=\tfrac{5}{6}.$$

33. Let $f(x,y)=\dfrac{x-y}{(x+y)^3}$. Then a CAS gives $\int_0^1\int_0^1 f(x,y)\,dy\,dx=\frac{1}{2}$ and $\int_0^1\int_0^1 f(x,y)\,dx\,dy=-\frac{1}{2}$.

To explain the seeming violation of Fubini's Theorem, note that f has an infinite discontinuity at $(0,0)$ and thus does not satisfy the conditions of Fubini's Theorem. In fact, both iterated integrals involve improper integrals which diverge at their lower limits of integration.

12.3 Double Integrals over General Regions

1. $\int_0^1\int_0^{x^2}(x+2y)\,dy\,dx=\int_0^1\left[xy+y^2\right]_{y=0}^{y=x^2}dx=\int_0^1\left[x(x^2)+(x^2)^2-0-0\right]dx$

$=\int_0^1(x^3+x^4)\,dx=\left[\frac{1}{4}x^4+\frac{1}{5}x^5\right]_0^1=\frac{9}{20}$

3. $\int_0^1\int_y^{e^y}\sqrt{x}\,dx\,dy=\int_0^1\left[\frac{2}{3}x^{3/2}\right]_{x=y}^{x=e^y}dy=\frac{2}{3}\int_0^1(e^{3y/2}-y^{3/2})\,dy=\frac{2}{3}\left[\frac{2}{3}e^{3y/2}-\frac{2}{5}y^{5/2}\right]_0^1$

$=\frac{2}{3}\left(\frac{2}{3}e^{3/2}-\frac{2}{5}-\frac{2}{3}e^0+0\right)=\frac{4}{9}e^{3/2}-\frac{32}{45}$

5. $\int_0^{\pi/2}\int_0^{\cos\theta}e^{\sin\theta}\,dr\,d\theta=\int_0^{\pi/2}\left[re^{\sin\theta}\right]_{r=0}^{r=\cos\theta}d\theta=\int_0^{\pi/2}(\cos\theta)\,e^{\sin\theta}\,d\theta=e^{\sin\theta}\Big]_0^{\pi/2}=e^{\sin(\pi/2)}-e^0=e-1$

7. $\iint_D x^3y^2\,dA=\int_0^2\int_{-x}^{x}x^3y^2\,dy\,dx=\int_0^2\left[\frac{1}{3}x^3y^3\right]_{y=-x}^{y=x}dx=\frac{1}{3}\int_0^2 2x^6\,dx=\frac{2}{3}\left[\frac{1}{7}x^7\right]_0^2=\frac{2}{21}\left[2^7-0\right]=\frac{256}{21}$

9. $\displaystyle\int_0^1\int_0^{\sqrt{x}}\frac{2y}{x^2+1}\,dy\,dx=\int_0^1\left[\frac{y^2}{x^2+1}\right]_{y=0}^{y=\sqrt{x}}dx=\int_0^1\frac{x}{x^2+1}\,dx=\tfrac{1}{2}\ln\left|x^2+1\right|\Big]_0^1=\tfrac{1}{2}(\ln 2-\ln 1)=\tfrac{1}{2}\ln 2$

11. $\int_0^1\int_0^{x^2}x\cos y\,dy\,dx=\int_0^1\left[x\sin y\right]_{y=0}^{y=x^2}dx=\int_0^1 x\sin x^2\,dx=-\frac{1}{2}\cos x^2\Big]_0^1=\frac{1}{2}(1-\cos 1)$

13.

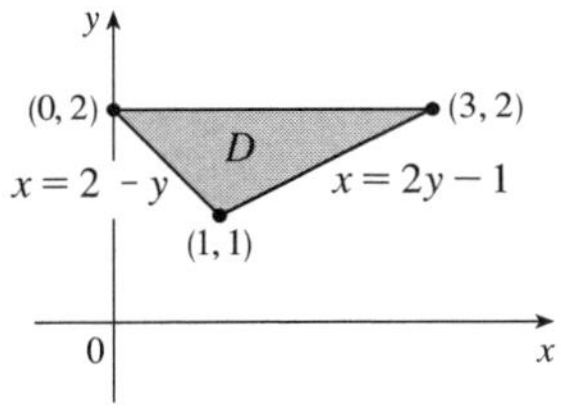

$$\int_1^2\int_{2-y}^{2y-1}y^3\,dx\,dy=\int_1^2\left[xy^3\right]_{x=2-y}^{x=2y-1}dy$$

$=\int_1^2\left[(2y-1)-(2-y)\right]y^3\,dy$

$=\int_1^2(3y^4-3y^3)\,dy=\left[\frac{3}{5}y^5-\frac{3}{4}y^4\right]_1^2$

$=\frac{96}{5}-12-\frac{3}{5}+\frac{3}{4}=\frac{147}{20}$

15.

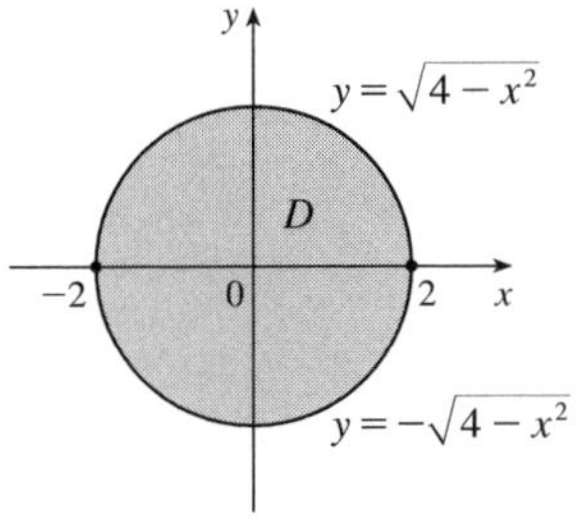

$$\int_{-2}^2\int_{-\sqrt{4-x^2}}^{\sqrt{4-x^2}}(2x-y)\,dy\,dx$$

$$=\int_{-2}^2\left[2xy-\frac{1}{2}y^2\right]_{y=-\sqrt{4-x^2}}^{y=\sqrt{4-x^2}}dx$$

$=\int_{-2}^2\left[2x\sqrt{4-x^2}-\frac{1}{2}(4-x^2)+2x\sqrt{4-x^2}+\frac{1}{2}(4-x^2)\right]dx$

$=\int_{-2}^2 4x\sqrt{4-x^2}\,dx=-\frac{4}{3}(4-x^2)^{3/2}\Big]_{-2}^2=0$

(Or, note that $4x\sqrt{4-x^2}$ is an odd function, so $\int_{-2}^2 4x\sqrt{4-x^2}\,dx=0$.)

17.

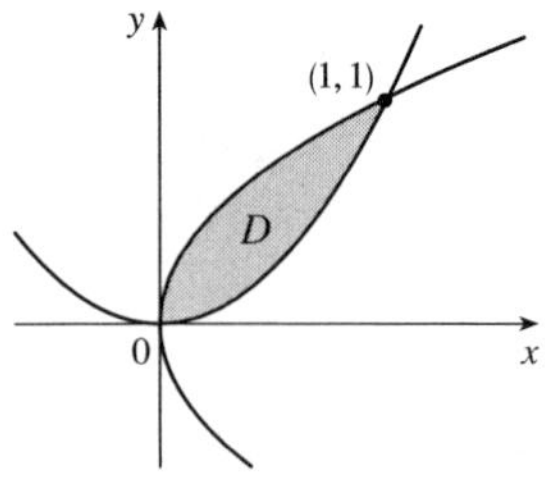

$$V = \int_0^1 \int_{x^2}^{\sqrt{x}} (x^2 + y^2)\, dy\, dx = \int_0^1 \left[\left(x^2 y + \frac{y^3}{3} \right) \right]_{y = x^2}^{y = \sqrt{x}} dx$$
$$= \int_0^1 \left(x^{5/2} - x^4 + \tfrac{1}{3} x^{3/2} - \tfrac{1}{3} x^6 \right) dx$$
$$= \left[\tfrac{2}{7} x^{7/2} - \tfrac{1}{5} x^5 + \tfrac{2}{15} x^{5/2} - \tfrac{1}{21} x^7 \right]_0^1$$
$$= \tfrac{18}{105} = \tfrac{6}{35}$$

19.

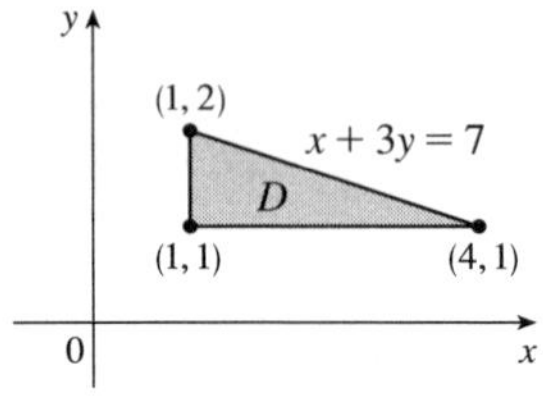

$$V = \int_1^2 \int_1^{7 - 3y} xy\, dx\, dy = \int_1^2 \left[\tfrac{1}{2} x^2 y \right]_{x = 1}^{x = 7 - 3y} dy$$
$$= \tfrac{1}{2} \int_1^2 (48y - 42y^2 + 9y^3)\, dy$$
$$= \tfrac{1}{2} \left[24y^2 - 14y^3 + \tfrac{9}{4} y^4 \right]_1^2 = \tfrac{31}{8}$$

21.

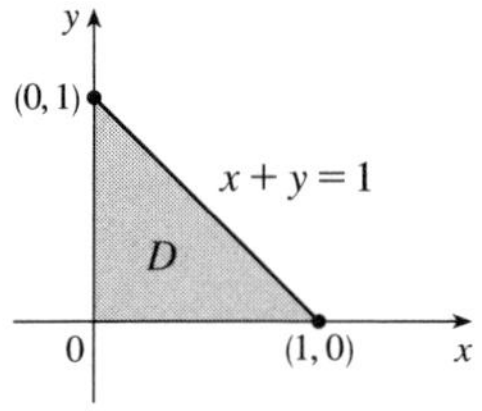

$$V = \int_0^1 \int_0^{1 - x} (1 - x - y)\, dy\, dx = \int_0^1 \left[y - xy - \frac{y^2}{2} \right]_{y = 0}^{y = 1 - x} dx$$
$$= \int_0^1 \left[(1 - x)^2 - \tfrac{1}{2}(1 - x)^2 \right] dx$$
$$= \int_0^1 \tfrac{1}{2}(1 - x)^2\, dx = \left[-\tfrac{1}{6}(1 - x)^3 \right]_0^1 = \tfrac{1}{6}$$

23.

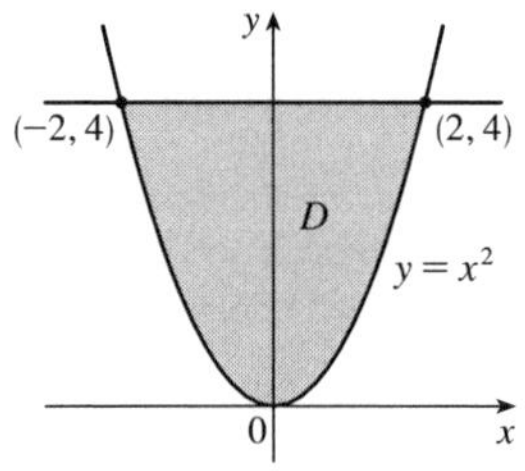

$$V = \int_{-2}^2 \int_{x^2}^4 x^2\, dy\, dx$$
$$= \int_{-2}^2 x^2 \left[y \right]_{y = x^2}^{y = 4} dx = \int_{-2}^2 (4x^2 - x^4)\, dx$$
$$= \left[\tfrac{4}{3} x^3 - \tfrac{1}{5} x^5 \right]_{-2}^2 = \tfrac{32}{3} - \tfrac{32}{5} + \tfrac{32}{3} - \tfrac{32}{5} = \tfrac{128}{15}$$

25.

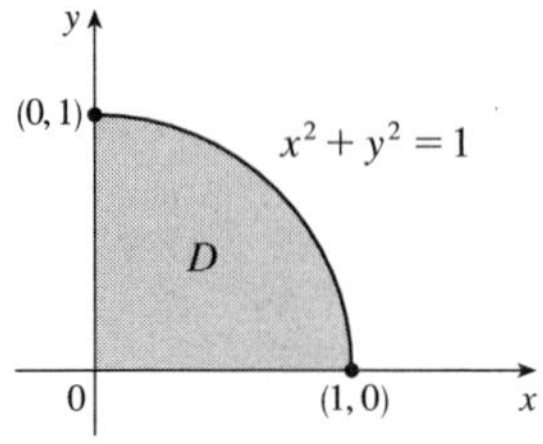

$$V = \int_0^1 \int_0^{\sqrt{1 - x^2}} y\, dy\, dx = \int_0^1 \left[\frac{y^2}{2} \right]_{y = 0}^{y = \sqrt{1 - x^2}} dx$$
$$= \int_0^1 \frac{1 - x^2}{2}\, dx = \tfrac{1}{2} \left[x - \tfrac{1}{3} x^3 \right]_0^1 = \tfrac{1}{3}$$

27.

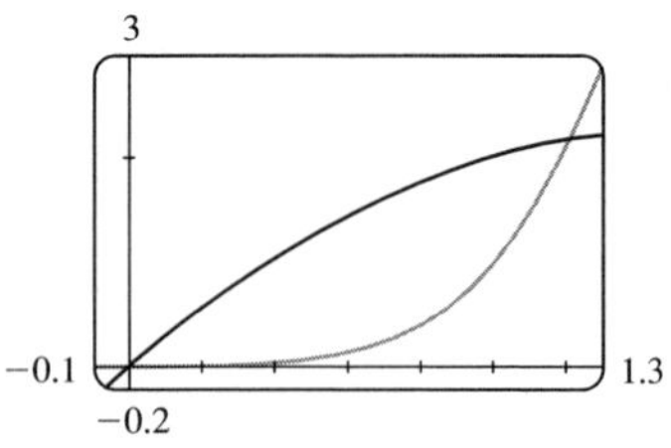

From the graph, it appears that the two curves intersect at $x = 0$ and at $x = a \approx 1.213$. Thus the desired integral is

$$\iint_D x\, dA = \int_0^a \int_{x^4}^{3x - x^2} x\, dy\, dx = \int_0^a \left[xy \right]_{y = x^4}^{y = 3x - x^2} dx$$
$$= \int_0^a (3x^2 - x^3 - x^5)\, dx = \left[x^3 - \tfrac{1}{4} x^4 - \tfrac{1}{6} x^6 \right]_0^a$$
$$\approx 0.713$$

29. The two bounding curves $y = 1 - x^2$ and $y = x^2 - 1$ intersect at $(\pm 1, 0)$ with $1 - x^2 \geq x^2 - 1$ on $[-1, 1]$. Within this region, the plane $z = 2x + 2y + 10$ is above the plane $z = 2 - x - y$, so

$$V = \int_{-1}^{1} \int_{x^2-1}^{1-x^2} (2x + 2y + 10)\, dy\, dx - \int_{-1}^{1} \int_{x^2-1}^{1-x^2} (2 - x - y)\, dy\, dx$$

$$= \int_{-1}^{1} \int_{x^2-1}^{1-x^2} (2x + 2y + 10 - (2 - x - y))\, dy\, dx = \int_{-1}^{1} \int_{x^2-1}^{1-x^2} (3x + 3y + 8)\, dy\, dx$$

$$= \int_{-1}^{1} \left[3xy + \tfrac{3}{2}y^2 + 8y\right]_{y=x^2-1}^{y=1-x^2} dx$$

$$= \int_{-1}^{1} \left[3x(1 - x^2) + \tfrac{3}{2}(1 - x^2)^2 + 8(1 - x^2) - 3x(x^2 - 1) - \tfrac{3}{2}(x^2 - 1)^2 - 8(x^2 - 1)\right] dx$$

$$= \int_{-1}^{1} (-6x^3 - 16x^2 + 6x + 16)\, dx = \left[-\tfrac{3}{2}x^4 - \tfrac{16}{3}x^3 + 3x^2 + 16x\right]_{-1}^{1}$$

$$= -\tfrac{3}{2} - \tfrac{16}{3} + 3 + 16 + \tfrac{3}{2} - \tfrac{16}{3} - 3 + 16 = \tfrac{64}{3}$$

31. The two surfaces intersect in the circle $x^2 + y^2 = 1$, $z = 0$ and the region of integration is the disc D: $x^2 + y^2 \leq 1$. Using a CAS, the volume is $\iint_D (1 - x^2 - y^2)\, dA = \displaystyle\int_{-1}^{1} \int_{-\sqrt{1-x^2}}^{\sqrt{1-x^2}} (1 - x^2 - y^2)\, dy\, dx = \frac{\pi}{2}$.

33.

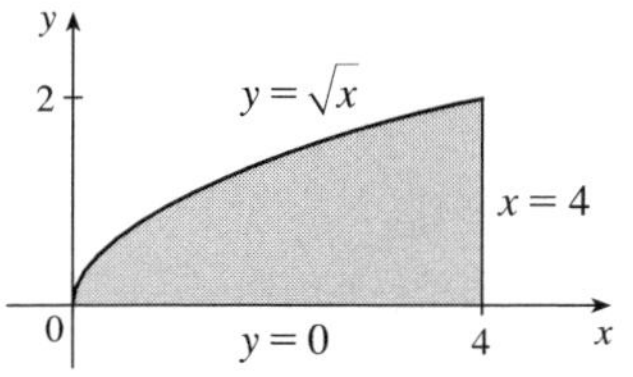

Because the region of integration is

$D = \{(x, y) \mid 0 \leq y \leq \sqrt{x}, 0 \leq x \leq 4\} = \{(x, y) \mid y^2 \leq x \leq 4, 0 \leq y \leq 2\}$

we have $\int_0^4 \int_0^{\sqrt{x}} f(x, y)\, dy\, dx = \iint_D f(x, y)\, dA = \int_0^2 \int_{y^2}^4 f(x, y)\, dx\, dy$.

35.

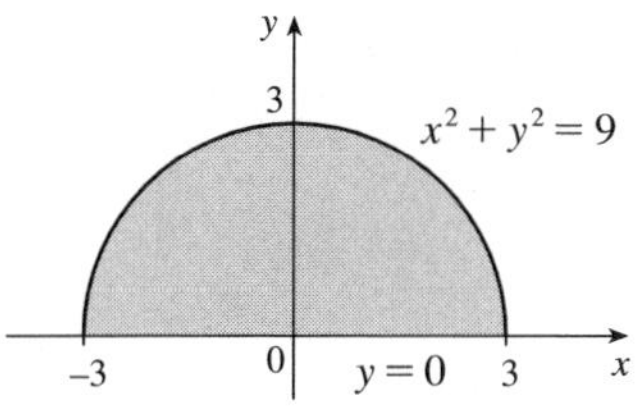

Because the region of integration is

$$D = \left\{(x, y) \mid -\sqrt{9 - y^2} \leq x \leq \sqrt{9 - y^2}, 0 \leq y \leq 3\right\}$$
$$= \{(x, y) \mid 0 \leq y \leq \sqrt{9 - x^2}, -3 \leq x \leq 3\}$$

we have

$$\int_0^3 \int_{-\sqrt{9-y^2}}^{\sqrt{9-y^2}} f(x, y)\, dx\, dy = \iint_D f(x, y)\, dA$$
$$= \int_{-3}^{3} \int_0^{\sqrt{9-x^2}} f(x, y)\, dy\, dx.$$

37.

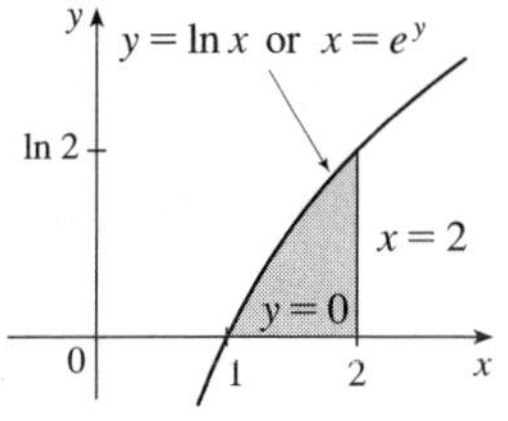

Because the region of integration is

$D = \{(x, y) \mid 0 \leq y \leq \ln x, 1 \leq x \leq 2\} = \{(x, y) \mid e^y \leq x \leq 2, 0 \leq y \leq \ln 2\}$

we have

$$\int_1^2 \int_0^{\ln x} f(x, y)\, dy\, dx = \iint_D f(x, y)\, dA = \int_0^{\ln 2} \int_{e^y}^{2} f(x, y)\, dx\, dy.$$

39.

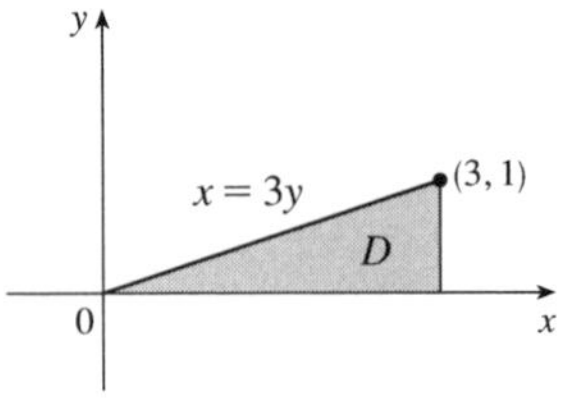

$$\int_0^1 \int_{3y}^3 e^{x^2}\,dx\,dy = \int_0^3 \int_0^{x/3} e^{x^2}\,dy\,dx$$
$$= \int_0^3 \left[e^{x^2} y\right]_{y=0}^{y=x/3} dx = \int_0^3 \left(\frac{x}{3}\right) e^{x^2}\,dx$$
$$= \tfrac{1}{6}\, e^{x^2}\Big]_0^3 = \frac{e^9 - 1}{6}$$

41.

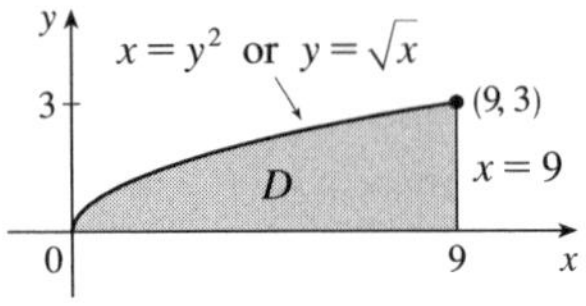

$$\int_0^3 \int_{y^2}^9 y\cos(x^2)\,dx\,dy = \int_0^9 \int_0^{\sqrt{x}} y\cos(x^2)\,dy\,dx$$
$$= \int_0^9 \cos(x^2)\left[\frac{y^2}{2}\right]_{y=0}^{y=\sqrt{x}} dx = \int_0^9 \tfrac{1}{2} x\cos(x^2)\,dx$$
$$= \tfrac{1}{4}\sin(x^2)\Big]_0^9 = \tfrac{1}{4}\sin 81$$

43.

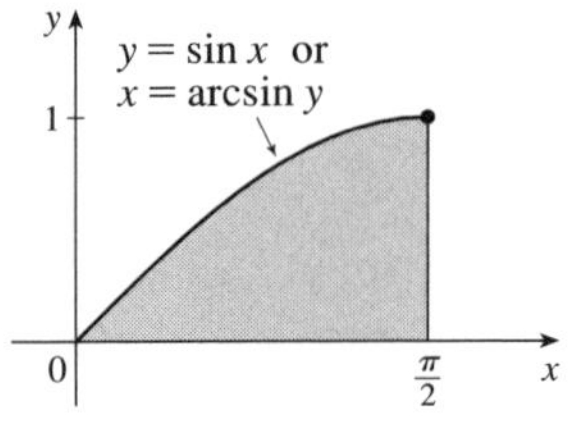

$$\int_0^1 \int_{\arcsin y}^{\pi/2} \cos x\,\sqrt{1+\cos^2 x}\,dx\,dy = \int_0^{\pi/2} \int_0^{\sin x} \cos x\,\sqrt{1+\cos^2 x}\,dy\,dx$$
$$= \int_0^{\pi/2} \cos x\,\sqrt{1+\cos^2 x}\,\left[y\right]_{y=0}^{y=\sin x}\,dx$$
$$= \int_0^{\pi/2} \cos x\,\sqrt{1+\cos^2 x}\,\sin x\,dx$$

[Let $u = \cos x$, $du = -\sin x\,dx$, $dx = du/(-\sin x)$]

$$= \int_1^0 -u\sqrt{1+u^2}\,du = -\tfrac{1}{3}\left(1+u^2\right)^{3/2}\Big]_1^0 = \tfrac{1}{3}\left(\sqrt{8}-1\right) = \tfrac{1}{3}\left(2\sqrt{2}-1\right)$$

45. $D = \{(x,y) \mid 0 \le x \le 1,\ -x+1 \le y \le 1\} \cup \{(x,y) \mid -1 \le x \le 0, x+1 \le y \le 1\}$
$\cup \{(x,y) \mid 0 \le x \le 1,\ -1 \le y \le x-1\} \cup \{(x,y) \mid -1 \le x \le 0,\ -1 \le y \le -x-1\}$,

all type I.

$$\iint_D x^2\,dA = \int_0^1 \int_{1-x}^1 x^2\,dy\,dx + \int_{-1}^0 \int_{x+1}^1 x^2\,dy\,dx + \int_0^1 \int_{-1}^{x-1} x^2\,dy\,dx + \int_{-1}^0 \int_{-1}^{-x-1} x^2\,dy\,dx$$
$$= 4\int_0^1 \int_{1-x}^1 x^2\,dy\,dx \qquad \text{[by symmetry of the regions and because } f(x,y) = x^2 \ge 0\text{]}$$
$$= 4\int_0^1 x^3\,dx = 4\left[\tfrac{1}{4}x^4\right]_0^1 = 1$$

47. For $D = [0,1] \times [0,1]$, $0 \le \sqrt{x^3+y^3} \le \sqrt{2}$ and $A(D) = 1$, so $0 \le \iint_D \sqrt{x^3+y^3}\,dA \le \sqrt{2}$.

49. Since $m \le f(x,y) \le M$, $\iint_D m\,dA \le \iint_D f(x,y)\,dA \le \iint_D M\,dA$ by (8) $\Rightarrow$
$m\iint_D 1\,dA \le \iint_D f(x,y)\,dA \le M\iint_D 1\,dA$ by (7) $\Rightarrow$ $mA(D) \le \iint_D f(x,y)\,dA \le MA(D)$ by (10).

51. $\iint_D (x^2\tan x + y^3 + 4)\,dA = \iint_D x^2\tan x\,dA + \iint_D y^3\,dA + \iint_D 4\,dA$. But $x^2\tan x$ is an odd function of x and D is symmetric with respect to the y-axis, so $\iint_D x^2\tan x\,dA = 0$. Similarly, y^3 is an odd function of y and D is symmetric with respect to the x-axis, so $\iint_D y^3\,dA = 0$. Thus $\iint_D (x^2\tan x + y^3 + 4)\,dA = 4\iint_D dA = 4(\text{area of } D) = 4\cdot\pi\left(\sqrt{2}\right)^2 = 8\pi$.

53. Since $\sqrt{1-x^2-y^2} \ge 0$, we can interpret $\iint_D \sqrt{1-x^2-y^2}\,dA$ as the volume of the solid that lies below the graph of $z = \sqrt{1-x^2-y^2}$ and above the region D in the xy-plane. $z = \sqrt{1-x^2-y^2}$ is equivalent to $x^2+y^2+z^2 = 1$, $z \ge 0$ which meets the xy-plane in the circle $x^2+y^2 = 1$, the boundary of D. Thus, the solid is an upper hemisphere of radius 1 which has volume $\tfrac{1}{2}\left[\tfrac{4}{3}\pi(1)^3\right] = \tfrac{2}{3}\pi$.

12.4 Double Integrals in Polar Coordinates

1. The region R is more easily described by polar coordinates: $R = \{(r, \theta) \mid 0 \le r \le 2, 0 \le \theta \le 2\pi\}$.
Thus $\iint_R f(x, y)\, dA = \int_0^{2\pi} \int_0^2 f(r\cos\theta, r\sin\theta)\, r\, dr\, d\theta$.

3. The region R is more easily described by rectangular coordinates: $R = \{(x, y) \mid -2 \le x \le 2, x \le y \le 2\}$.
Thus $\iint_R f(x, y)\, dA = \int_{-2}^{2} \int_x^2 f(x, y)\, dy\, dx$.

5. The region R is more easily described by polar coordinates: $R = \{(r, \theta) \mid 2 \le r \le 5, 0 \le \theta \le 2\pi\}$.
Thus $\iint_R f(x, y)\, dA = \int_0^{2\pi} \int_2^5 f(r\cos\theta, r\sin\theta)\, r\, dr\, d\theta$.

7. The integral $\int_\pi^{2\pi} \int_4^7 r\, dr\, d\theta$ represents the area of the region $R = \{(r, \theta) \mid 4 \le r \le 7, \pi \le \theta \le 2\pi\}$, the lower half of a ring.

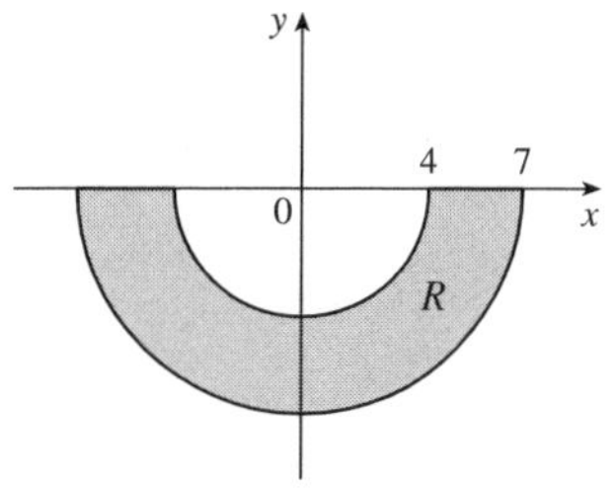

$$\int_\pi^{2\pi} \int_4^7 r\, dr\, d\theta = \left(\int_\pi^{2\pi} d\theta\right)\left(\int_4^7 r\, dr\right)$$
$$= \left[\theta\right]_\pi^{2\pi} \left[\tfrac{1}{2}r^2\right]_4^7 = \pi \cdot \tfrac{1}{2}(49 - 16) = \frac{33\pi}{2}$$

9. The disc D can be described in polar coordinates as $D = \{(r, \theta) \mid 0 \le r \le 3, 0 \le \theta \le 2\pi\}$. Then
$$\iint_D xy\, dA = \int_0^{2\pi} \int_0^3 (r\cos\theta)(r\sin\theta)\, r\, dr\, d\theta = \left(\int_0^{2\pi} \sin\theta\cos\theta\, d\theta\right)\left(\int_0^3 r^3\, dr\right) = \left[\tfrac{1}{2}\sin^2\theta\right]_0^{2\pi} \left[\tfrac{1}{4}r^4\right]_0^3 = 0.$$

11. $\iint_R \cos(x^2 + y^2)\, dA = \int_0^\pi \int_0^3 \cos(r^2)\, r\, dr\, d\theta = \left(\int_0^\pi d\theta\right)\left(\int_0^3 r\cos(r^2)\, dr\right)$
$$= \left[\theta\right]_0^\pi \left[\tfrac{1}{2}\sin(r^2)\right]_0^3 = \pi \cdot \tfrac{1}{2}(\sin 9 - \sin 0) = \tfrac{\pi}{2}\sin 9$$

13. R is the region shown in the figure, and can be described by $R = \{(r, \theta) \mid 0 \le \theta \le \pi/4, 1 \le r \le 2\}$. Thus
$\iint_R \arctan(y/x)\, dA = \int_0^{\pi/4} \int_1^2 \arctan(\tan\theta)\, r\, dr\, d\theta$ since $y/x = \tan\theta$.
Also, $\arctan(\tan\theta) = \theta$ for $0 \le \theta \le \pi/4$, so the integral becomes
$$\int_0^{\pi/4} \int_1^2 \theta\, r\, dr\, d\theta = \int_0^{\pi/4} \theta\, d\theta \int_1^2 r\, dr = \left[\tfrac{1}{2}\theta^2\right]_0^{\pi/4} \left[\tfrac{1}{2}r^2\right]_1^2 = \tfrac{\pi^2}{32} \cdot \tfrac{3}{2} = \tfrac{3}{64}\pi^2.$$

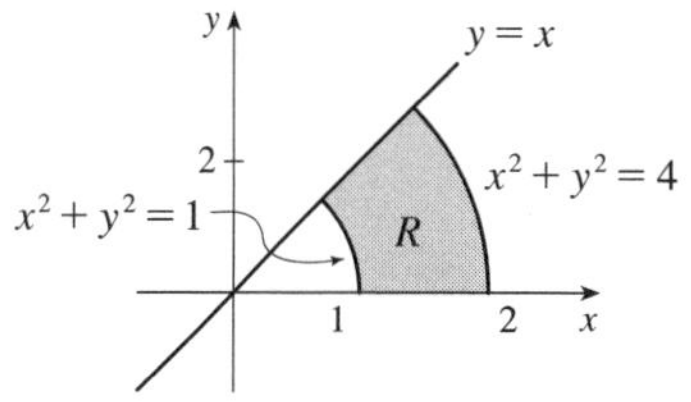

15. $V = \iint_{x^2+y^2 \le 4} \sqrt{x^2 + y^2}\, dA = \int_0^{2\pi} \int_0^2 \sqrt{r^2}\, r\, dr\, d\theta = \int_0^{2\pi} d\theta \int_0^2 r^2\, dr = \left[\theta\right]_0^{2\pi} \left[\tfrac{1}{3}r^3\right]_0^2 = 2\pi\left(\tfrac{8}{3}\right) = \tfrac{16}{3}\pi$

17. By symmetry,
$$V = 2 \iint_{x^2+y^2 \le a^2} \sqrt{a^2 - x^2 - y^2}\, dA = 2\int_0^{2\pi} \int_0^a \sqrt{a^2 - r^2}\, r\, dr\, d\theta = 2\int_0^{2\pi} d\theta \int_0^a r\sqrt{a^2 - r^2}\, dr$$
$$= 2\left[\theta\right]_0^{2\pi} \left[-\tfrac{1}{3}(a^2 - r^2)^{3/2}\right]_0^a = 2(2\pi)\left(0 + \tfrac{1}{3}a^3\right) = \tfrac{4\pi}{3}a^3$$

19. The cone $z = \sqrt{x^2 + y^2}$ intersects the sphere $x^2 + y^2 + z^2 = 1$ when $x^2 + y^2 + \left(\sqrt{x^2 + y^2}\right)^2 = 1$ or $x^2 + y^2 = \frac{1}{2}$. So
$$V = \iint_{x^2+y^2 \le 1/2} \left(\sqrt{1 - x^2 - y^2} - \sqrt{x^2 + y^2}\right) dA = \int_0^{2\pi} \int_0^{1/\sqrt{2}} \left(\sqrt{1 - r^2} - r\right) r\, dr\, d\theta$$
$$= \int_0^{2\pi} d\theta \int_0^{1/\sqrt{2}} \left(r\sqrt{1 - r^2} - r^2\right) dr = \left[\theta\right]_0^{2\pi} \left[-\tfrac{1}{3}(1 - r^2)^{3/2} - \tfrac{1}{3}r^3\right]_0^{1/\sqrt{2}}$$
$$= 2\pi\left(-\tfrac{1}{3}\right)\left(\tfrac{1}{\sqrt{2}} - 1\right) = \tfrac{\pi}{3}\left(2 - \sqrt{2}\right)$$

21. The given solid is the region inside the cylinder $x^2 + y^2 = 4$ between the surfaces $z = \sqrt{64 - 4x^2 - 4y^2}$ and $z = -\sqrt{64 - 4x^2 - 4y^2}$. So

$$V = \iint\limits_{x^2+y^2 \le 4} \left[\sqrt{64 - 4x^2 - 4y^2} - \left(-\sqrt{64 - 4x^2 - 4y^2}\right)\right] dA = \iint\limits_{x^2+y^2 \le 4} 2\sqrt{64 - 4x^2 - 4y^2}\, dA$$

$$= 4\int_0^{2\pi}\int_0^2 \sqrt{16 - r^2}\, r\, dr\, d\theta = 4\int_0^{2\pi} d\theta \int_0^2 r\sqrt{16 - r^2}\, dr = 4\left[\,\theta\,\right]_0^{2\pi}\left[-\tfrac{1}{3}(16 - r^2)^{3/2}\right]_0^2$$

$$= 8\pi\left(-\tfrac{1}{3}\right)(12^{3/2} - 16^{2/3}) = \tfrac{8\pi}{3}\left(64 - 24\sqrt{3}\right)$$

23. One loop is given by the region $D = \{(r, \theta) \mid -\pi/6 \le \theta \le \pi/6, 0 \le r \le \cos 3\theta\}$, so the area is

$$\iint_D dA = \int_{-\pi/6}^{\pi/6}\int_0^{\cos 3\theta} r\, dr\, d\theta = \int_{-\pi/6}^{\pi/6}\left[\frac{1}{2}r^2\right]_{r=0}^{r=\cos 3\theta} d\theta$$

$$= \int_{-\pi/6}^{\pi/6} \frac{1}{2}\cos^2 3\theta\, d\theta = 2\int_0^{\pi/6} \frac{1}{2}\left(\frac{1 + \cos 6\theta}{2}\right) d\theta$$

$$= \frac{1}{2}\left[\theta + \frac{1}{6}\sin 6\theta\right]_0^{\pi/6} = \frac{\pi}{12}$$

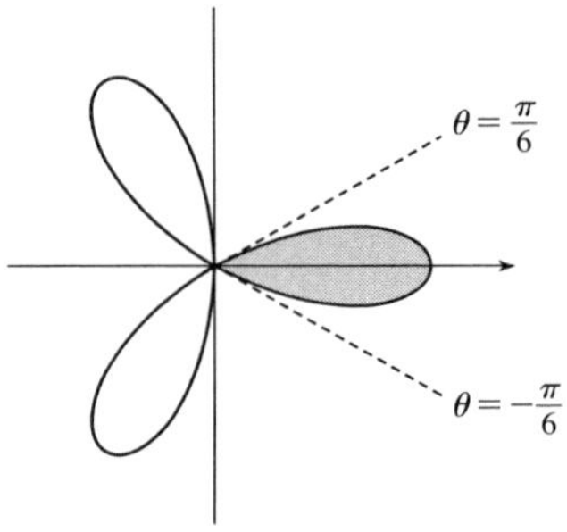

25.

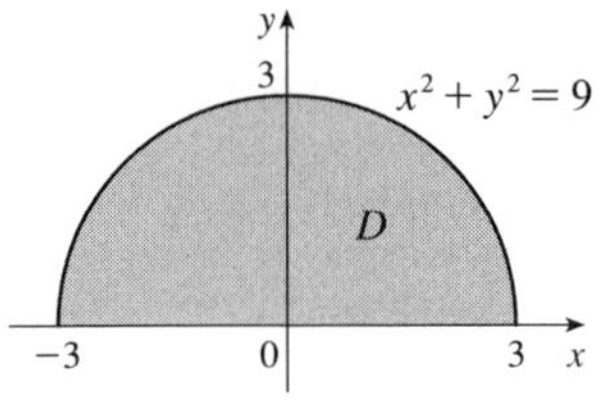

$$\int_{-3}^3\int_0^{\sqrt{9-x^2}} \sin\left(x^2 + y^2\right) dy\, dx = \int_0^\pi \int_0^3 \sin\left(r^2\right) r\, dr\, d\theta$$

$$= \int_0^\pi d\theta \int_0^3 r\sin\left(r^2\right) dr$$

$$= [\theta]_0^\pi \left[-\tfrac{1}{2}\cos\left(r^2\right)\right]_0^3$$

$$= \pi\left(-\tfrac{1}{2}\right)(\cos 9 - 1) = \tfrac{\pi}{2}(1 - \cos 9)$$

27.

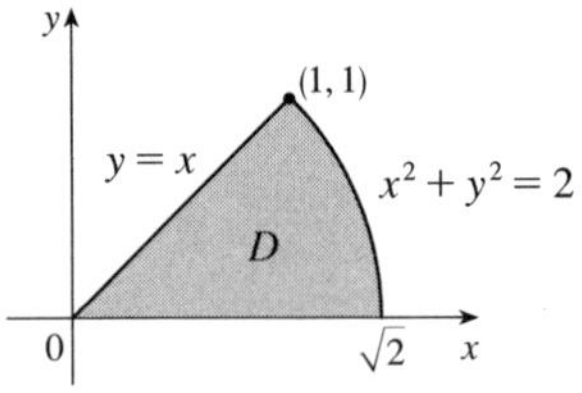

$$\int_0^{\pi/4}\int_0^{\sqrt{2}} (r\cos\theta + r\sin\theta)\, r\, dr\, d\theta = \int_0^{\pi/4} (\cos\theta + \sin\theta)\, d\theta \int_0^{\sqrt{2}} r^2\, dr$$

$$= [\sin\theta - \cos\theta]_0^{\pi/4}\left[\tfrac{1}{3}r^3\right]_0^{\sqrt{2}}$$

$$= \left[\tfrac{\sqrt{2}}{2} - \tfrac{\sqrt{2}}{2} - 0 + 1\right]\cdot \tfrac{1}{3}\left(2\sqrt{2} - 0\right) = \tfrac{2\sqrt{2}}{3}$$

29. The surface of the water in the pool is a circular disc D with radius 5 m. If we place D on coordinate axes with the origin at the centre of D and define $f(x, y)$ to be the depth of the water at (x, y), then the volume of water in the pool is the volume of the solid that lies above $D = \left\{(x, y) \mid x^2 + y^2 \le 25\right\}$ and below the graph of $f(x, y)$. We can associate north with the positive y-direction, so we are given that the depth is constant in the x-direction and the depth increases linearly in the y-direction from $f(0, -5) = 1$ to $f(0, 5) = 2$. The trace in the yz-plane is a line segment from $(0, -5, 1)$ to $(0, 5, 2)$. The slope of this line is $\frac{2-1}{5-(-5)} = \frac{1}{10}$, so an equation of the line is $z - 2 = \frac{1}{10}(y - 5) \Rightarrow z = \frac{1}{10}y + \frac{3}{2}$. Since $f(x, y)$ is independent of x, $f(x, y) = \frac{1}{10}y + \frac{3}{2}$. Thus the volume is given by $\iint_D f(x, y)\, dA$, which is most conveniently evaluated using polar coordinates. Then $D = \{(r, \theta) \mid 0 \le r \le 5, 0 \le \theta \le 2\pi\}$ and substituting $x = r\cos\theta$, $y = r\sin\theta$ the integral becomes

$$\int_0^{2\pi}\int_0^5 \left(\tfrac{1}{10}r\sin\theta + \tfrac{3}{2}\right) r\, dr\, d\theta = \int_0^{2\pi}\left[\tfrac{1}{30}r^3\sin\theta + \tfrac{3}{4}r^2\right]_{r=0}^{r=5} d\theta = \int_0^{2\pi}\left(\tfrac{25}{6}\sin\theta + \tfrac{75}{4}\right) d\theta$$

$$= \left[-\tfrac{25}{6}\cos\theta + \tfrac{75}{4}\theta\right]_0^{2\pi} = \tfrac{75}{2}\pi = 37.5\pi$$

Thus the pool contains $37.5\pi \approx 118\ \text{m}^3$ of water.

31. $\displaystyle\int_{1/\sqrt{2}}^{1}\int_{\sqrt{1-x^2}}^{x} xy\,dy\,dx + \int_{1}^{\sqrt{2}}\int_{0}^{x} xy\,dy\,dx + \int_{\sqrt{2}}^{2}\int_{0}^{\sqrt{4-x^2}} xy\,dy\,dx$

$$= \int_0^{\pi/4}\int_1^2 r^3\cos\theta\sin\theta\,dr\,d\theta = \int_0^{\pi/4}\left[\frac{r^4}{4}\cos\theta\sin\theta\right]_{r=1}^{r=2} d\theta$$

$$= \frac{15}{4}\int_0^{\pi/4}\sin\theta\cos\theta\,d\theta = \frac{15}{4}\left[\frac{\sin^2\theta}{2}\right]_0^{\pi/4} = \frac{15}{16}$$

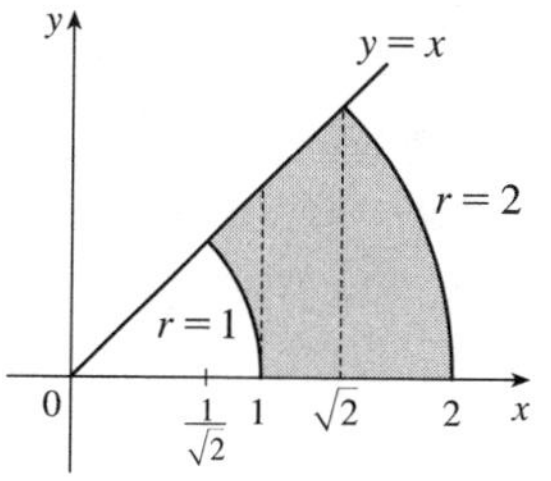

33. (a) We integrate by parts with $u = x$ and $dv = xe^{-x^2}\,dx$. Then $du = dx$ and $v = -\frac{1}{2}e^{-x^2}$, so

$$\int_0^\infty x^2 e^{-x^2}\,dx = \lim_{t\to\infty}\int_0^t x^2 e^{-x^2}\,dx = \lim_{t\to\infty}\left(-\tfrac{1}{2}xe^{-x^2}\Big]_0^t + \int_0^t \tfrac{1}{2}e^{-x^2}\,dx\right)$$

$$= \lim_{t\to\infty}\left(-\tfrac{1}{2}te^{-t^2}\right) + \tfrac{1}{2}\int_0^\infty e^{-x^2}\,dx = 0 + \tfrac{1}{2}\int_0^\infty e^{-x^2}\,dx \qquad \text{[by l'Hospital's Rule]}$$

$$= \tfrac{1}{4}\int_{-\infty}^\infty e^{-x^2}\,dx \qquad \text{[since } e^{-x^2} \text{ is an even function]}$$

$$= \tfrac{1}{4}\sqrt{\pi} \qquad \text{[by Exercise 32(c)]}$$

(b) Let $u = \sqrt{x}$. Then $u^2 = x \;\Rightarrow\; dx = 2u\,du \;\Rightarrow$

$$\int_0^\infty \sqrt{x}e^{-x}\,dx = \lim_{t\to\infty}\int_0^t \sqrt{x}\,e^{-x}\,dx = \lim_{t\to\infty}\int_0^{\sqrt{t}} ue^{-u^2}2u\,du = 2\int_0^\infty u^2e^{-u^2}\,du = 2\left(\tfrac{1}{4}\sqrt{\pi}\right) \quad \text{[by part(a)]}$$

$$= \tfrac{1}{2}\sqrt{\pi}$$

12.5 Applications of Double Integrals

1. $Q = \iint_D \sigma(x,y)\,dA = \int_1^3\int_0^2 (2xy + y^2)\,dy\,dx = \int_1^3 \left[xy^2 + \frac{1}{3}y^3\right]_{y=0}^{y=2} dx$

$= \int_1^3 \left(4x + \frac{8}{3}\right) dx = \left[2x^2 + \frac{8}{3}x\right]_1^3 = 16 + \frac{16}{3} = \frac{64}{3}$ C

3. $m = \iint_D \rho(x,y)\,dA = \int_0^2\int_{-1}^1 xy^2\,dy\,dx = \int_0^2 x\,dx \int_{-1}^1 y^2\,dy = \left[\frac{1}{2}x^2\right]_0^2 \left[\frac{1}{3}y^3\right]_{-1}^1 = 2\cdot\frac{2}{3} = \frac{4}{3}$,

$\overline{x} = \frac{1}{m}\iint_D x\rho(x,y)\,dA = \frac{3}{4}\int_0^2\int_{-1}^1 x^2y^2\,dy\,dx = \frac{3}{4}\int_0^2 x^2\,dx\int_{-1}^1 y^2\,dy = \frac{3}{4}\left[\frac{1}{3}x^3\right]_0^2\left[\frac{1}{3}y^3\right]_{-1}^1 = \frac{3}{4}\cdot\frac{8}{3}\cdot\frac{2}{3} = \frac{4}{3}$,

$\overline{y} = \frac{1}{m}\iint_D y\rho(x,y)\,dA = \frac{3}{4}\int_0^2\int_{-1}^1 xy^3\,dy\,dx = \frac{3}{4}\int_0^2 x\,dx\int_{-1}^1 y^3\,dy = \frac{3}{4}\left[\frac{1}{2}x^2\right]_0^2\left[\frac{1}{4}y^4\right]_{-1}^1 = \frac{3}{4}\cdot 2\cdot 0 = 0.$

Hence, $(\overline{x},\overline{y}) = \left(\frac{4}{3}, 0\right)$.

5. $m = \int_0^2\int_{x/2}^{3-x}(x+y)\,dy\,dx = \int_0^2\left[xy + \frac{1}{2}y^2\right]_{y=x/2}^{y=3-x} dx = \int_0^2\left[x\left(3 - \frac{3}{2}x\right) + \frac{1}{2}(3-x)^2 - \frac{1}{8}x^2\right] dx$

$= \int_0^2\left(-\frac{9}{8}x^2 + \frac{9}{2}\right) dx = \left[-\frac{9}{8}\left(\frac{1}{3}x^3\right) + \frac{9}{2}x\right]_0^2 = 6$,

$M_y = \int_0^2\int_{x/2}^{3-x}(x^2 + xy)\,dy\,dx = \int_0^2\left[x^2y + \frac{1}{2}xy^2\right]_{y=x/2}^{y=3-x} dx = \int_0^2\left(\frac{9}{2}x - \frac{9}{8}x^3\right) dx = \frac{9}{2}$, and

$M_x = \int_0^2\int_{x/2}^{3-y}(xy + y^2)\,dy\,dx = \int_0^2\left[\frac{1}{2}xy^2 + \frac{1}{3}y^3\right]_{y=x/2}^{y=3-x} dx = \int_0^2\left(9 - \frac{9}{2}x\right) dx = 9.$

Hence $m = 6$, $(\overline{x},\overline{y}) = \left(\dfrac{M_y}{m}, \dfrac{M_x}{m}\right) = \left(\dfrac{3}{4}, \dfrac{3}{2}\right)$.

7. $m = \int_0^1\int_0^{e^x} y\,dy\,dx = \int_0^1\left[\frac{1}{2}y^2\right]_{y=0}^{y=e^x} dx = \frac{1}{2}\int_0^1 e^{2x}\,dx = \frac{1}{4}e^{2x}\Big]_0^1 = \frac{1}{4}(e^2 - 1)$,

$M_y = \int_0^1\int_0^{e^x} xy\,dy\,dx = \frac{1}{2}\int_0^1 xe^{2x}\,dx = \frac{1}{2}\left[\frac{1}{2}xe^{2x} - \frac{1}{4}e^{2x}\right]_0^1 = \frac{1}{8}(e^2 + 1)$, and

$M_x = \int_0^1\int_0^{e^x} y^2\,dy\,dx = \int_0^1\left[\frac{1}{3}y^3\right]_{y=0}^{y=e^x} dx = \frac{1}{3}\int_0^1 e^{3x}\,dx = \frac{1}{3}\left[\frac{1}{3}e^{3x}\right]_0^1 = \frac{1}{9}(e^3 - 1).$

Hence $m = \frac{1}{4}(e^2 - 1)$, $(\overline{x},\overline{y}) = \left(\dfrac{\frac{1}{8}(e^2+1)}{\frac{1}{4}(e^2-1)}, \dfrac{\frac{1}{9}(e^3-1)}{\frac{1}{4}(e^2-1)}\right) = \left(\dfrac{e^2+1}{2(e^2-1)}, \dfrac{4(e^3-1)}{9(e^2-1)}\right).$

9. 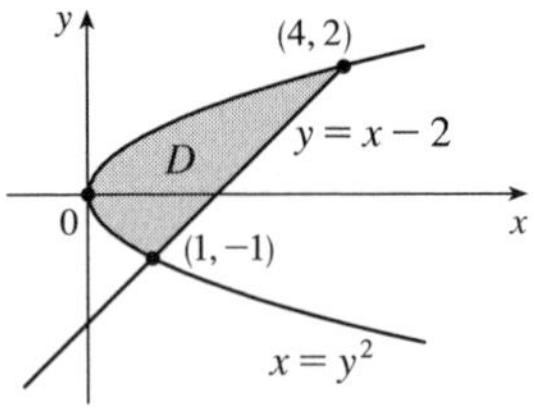

$m = \int_{-1}^{2}\int_{y^2}^{y+2} 3\,dx\,dy = \int_{-1}^{2}(3y+6-3y^2)\,dy = \frac{27}{2}$,

$$M_y = \int_{-1}^{2}\int_{y^2}^{y+2} 3x\,dx\,dy = \int_{-1}^{2}\tfrac{3}{2}\left[(y+2)^2 - y^4\right]dy$$
$$= \left[\tfrac{1}{2}(y+2)^3 - \tfrac{3}{10}y^5\right]_{-1}^{2} = \tfrac{108}{5}$$

and

$$M_x = \int_{-1}^{2}\int_{y^2}^{y+2} 3y\,dx\,dy = \int_{-1}^{2}(3y^2+6y-3y^3)\,dy$$
$$= \left[y^3 + 3y^2 - \tfrac{3}{4}y^4\right]_{-1}^{2} = \tfrac{27}{4}$$

Hence $m = \frac{27}{2}$, $(\overline{x}, \overline{y}) = \left(\frac{8}{5}, \frac{1}{2}\right)$.

11. $\rho(x,y) = ky = kr\sin\theta$, $m = \int_0^{\pi/2}\int_0^1 kr^2\sin\theta\,dr\,d\theta = \frac{1}{3}k\int_0^{\pi/2}\sin\theta\,d\theta = \frac{1}{3}k\left[-\cos\theta\right]_0^{\pi/2} = \frac{1}{3}k$,

$M_y = \int_0^{\pi/2}\int_0^1 kr^3\sin\theta\cos\theta\,dr\,d\theta = \frac{1}{4}k\int_0^{\pi/2}\sin\theta\cos\theta\,d\theta = \frac{1}{8}k\left[-\cos 2\theta\right]_0^{\pi/2} = \frac{1}{8}k$,

$M_x = \int_0^{\pi/2}\int_0^1 kr^3\sin^2\theta\,dr\,d\theta = \frac{1}{4}k\int_0^{\pi/2}\sin^2\theta\,d\theta = \frac{1}{8}k\left[\theta + \sin 2\theta\right]_0^{\pi/2} = \frac{\pi}{16}k$.

Hence $(\overline{x}, \overline{y}) = \left(\frac{3}{8}, \frac{3\pi}{16}\right)$.

13. Placing the vertex opposite the hypotenuse at $(0,0)$, $\rho(x,y) = k(x^2+y^2)$. Then

$m = \int_0^a\int_0^{a-x} k(x^2+y^2)\,dy\,dx = k\int_0^a\left[ax^2 - x^3 + \frac{1}{3}(a-x)^3\right]dx = k\left[\frac{1}{3}ax^3 - \frac{1}{4}x^4 - \frac{1}{12}(a-x)^4\right]_0^a = \frac{1}{6}ka^4$.

By symmetry,

$$M_y = M_x = \int_0^a\int_0^{a-x} ky(x^2+y^2)\,dy\,dx = k\int_0^a\left[\tfrac{1}{2}(a-x)^2x^2 + \tfrac{1}{4}(a-x)^4\right]dx$$
$$= k\left[\tfrac{1}{6}a^2x^3 - \tfrac{1}{4}ax^4 + \tfrac{1}{10}x^5 - \tfrac{1}{20}(a-x)^5\right]_0^a = \tfrac{1}{15}ka^5$$

Hence $(\overline{x}, \overline{y}) = \left(\frac{2}{5}a, \frac{2}{5}a\right)$.

15. $I_x = \iint_D y^2\rho(x,y)\,dA = \int_0^1\int_0^{e^x} y^2\cdot y\,dy\,dx = \int_0^1\left[\frac{1}{4}y^4\right]_{y=0}^{y=e^x}dx = \frac{1}{4}\int_0^1 e^{4x}\,dx = \frac{1}{4}\left[\frac{1}{4}e^{4x}\right]_0^1 = \frac{1}{16}(e^4-1)$,

$I_y = \iint_D x^2\rho(x,y)\,dA = \int_0^1\int_0^{e^x} x^2y\,dy\,dx = \int_0^1 x^2\left[\frac{1}{2}y^2\right]_{y=0}^{y=e^x}dx = \frac{1}{2}\int_0^1 x^2e^{2x}\,dx$

$= \frac{1}{2}\left[\left(\frac{1}{2}x^2 - \frac{1}{2}x + \frac{1}{4}\right)e^{2x}\right]_0^1$ [integrate by parts twice]

$= \frac{1}{8}(e^2-1)$,

and $I_0 = I_x + I_y = \frac{1}{16}(e^4-1) + \frac{1}{8}(e^2-1) = \frac{1}{16}(e^4+2e^2-3)$.

17. As in Exercise 13, we place the vertex opposite the hypotenuse at $(0,0)$ and the equal sides along the positive axes.

$$I_x = \int_0^a\int_0^{a-x} y^2k(x^2+y^2)\,dy\,dx = k\int_0^a\int_0^{a-x}(x^2y^2+y^4)\,dy\,dx$$
$$= k\int_0^a\left[\tfrac{1}{3}x^2y^3 + \tfrac{1}{5}y^5\right]_{y=0}^{y=a-x}dx = k\int_0^a\left[\tfrac{1}{3}x^2(a-x)^3 + \tfrac{1}{5}(a-x)^5\right]dx$$
$$= k\left[\tfrac{1}{3}\left(\tfrac{1}{3}a^3x^3 - \tfrac{3}{4}a^2x^4 + \tfrac{3}{5}ax^5 - \tfrac{1}{6}x^6\right) - \tfrac{1}{30}(a-x)^6\right]_0^a = \tfrac{7}{180}ka^6,$$

$$I_y = \int_0^a\int_0^{a-x} x^2k(x^2+y^2)\,dy\,dx = k\int_0^a\int_0^{a-x}(x^4+x^2y^2)\,dy\,dx$$
$$= k\int_0^a\left[x^4y + \tfrac{1}{3}x^2y^3\right]_{y=0}^{y=a-x}dx = k\int_0^a\left[x^4(a-x) + \tfrac{1}{3}x^2(a-x)^3\right]dx$$
$$= k\left[\tfrac{1}{5}ax^5 - \tfrac{1}{6}x^6 + \tfrac{1}{3}\left(\tfrac{1}{3}a^3x^3 - \tfrac{3}{4}a^2x^4 + \tfrac{3}{5}ax^5 - \tfrac{1}{6}x^6\right)\right]_0^a = \tfrac{7}{180}ka^6,$$

and $I_0 = I_x + I_y = \frac{7}{90}ka^6$.

19. Using a CAS, we find $m = \iint_D \rho(x,y)\,dA = \int_0^\pi \int_0^{\sin x} xy\,dy\,dx = \frac{\pi^2}{8}$. Then

$$\overline{x} = \frac{1}{m}\iint_D x\rho(x,y)\,dA = \frac{8}{\pi^2}\int_0^\pi \int_0^{\sin x} x^2 y\,dy\,dx = \frac{2\pi}{3} - \frac{1}{\pi} \text{ and}$$

$$\overline{y} = \frac{1}{m}\iint_D y\rho(x,y)\,dA = \frac{8}{\pi^2}\int_0^\pi \int_0^{\sin x} xy^2\,dy\,dx = \frac{16}{9\pi}, \text{ so } (\overline{x},\overline{y}) = \left(\frac{2\pi}{3} - \frac{1}{\pi}, \frac{16}{9\pi}\right).$$

The moments of inertia are $I_x = \iint_D y^2\rho(x,y)\,dA = \int_0^\pi \int_0^{\sin x} xy^3\,dy\,dx = \frac{3\pi^2}{64}$,

$I_y = \iint_D x^2\rho(x,y)\,dA = \int_0^\pi \int_0^{\sin x} x^3 y\,dy\,dx = \frac{\pi^2}{16}(\pi^2 - 3)$, and $I_0 = I_x + I_y = \frac{\pi^2}{64}(4\pi^2 - 9)$.

21. (a) $f(x,y)$ is a joint density function, so we know $\iint_{\mathbb{R}^2} f(x,y)\,dA = 1$. Since $f(x,y) = 0$ outside the rectangle $[0,1] \times [0,2]$, we can say

$$\iint_{\mathbb{R}^2} f(x,y)\,dA = \int_{-\infty}^{\infty}\int_{-\infty}^{\infty} f(x,y)\,dy\,dx = \int_0^1\int_0^2 Cx(1+y)\,dy\,dx$$

$$= C\int_0^1 x\left[y + \tfrac{1}{2}y^2\right]_{y=0}^{y=2}\,dx = C\int_0^1 4x\,dx = C\Big[2x^2\Big]_0^1 = 2C$$

Then $2C = 1 \;\Rightarrow\; C = \frac{1}{2}$.

(b) $P(X \le 1, Y \le 1) = \int_{-\infty}^1\int_{-\infty}^1 f(x,y)\,dy\,dx = \int_0^1\int_0^1 \frac{1}{2}x(1+y)\,dy\,dx$

$$= \int_0^1 \tfrac{1}{2}x\left[y + \tfrac{1}{2}y^2\right]_{y=0}^{y=1}\,dx = \int_0^1 \tfrac{1}{2}x\left(\tfrac{3}{2}\right)dx = \tfrac{3}{4}\left[\tfrac{1}{2}x^2\right]_0^1 = \tfrac{3}{8} \text{ or } 0.375$$

(c) $P(X + Y \le 1) = P((X,Y) \in D)$ where D is the triangular region shown in the figure. Thus

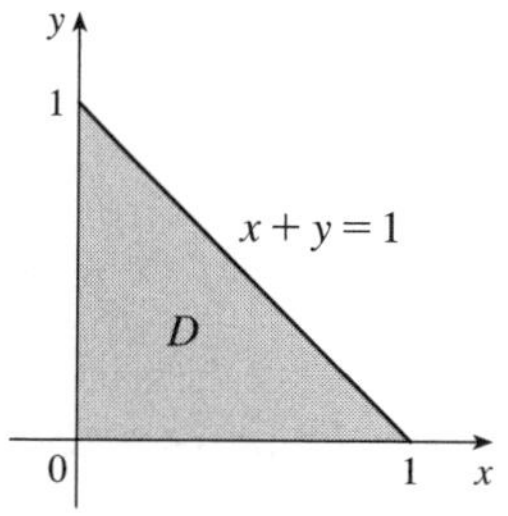

$$P(X + Y \le 1) = \iint_D f(x,y)\,dA = \int_0^1\int_0^{1-x} \tfrac{1}{2}x(1+y)\,dy\,dx$$

$$= \int_0^1 \tfrac{1}{2}x\left[y + \tfrac{1}{2}y^2\right]_{y=0}^{y=1-x}\,dx = \int_0^1 \tfrac{1}{2}x\left(\tfrac{1}{2}x^2 - 2x + \tfrac{3}{2}\right)dx$$

$$= \tfrac{1}{4}\int_0^1 (x^3 - 4x^2 + 3x)\,dx = \tfrac{1}{4}\left[\tfrac{x^4}{4} - 4\tfrac{x^3}{3} + 3\tfrac{x^2}{2}\right]_0^1$$

$$= \tfrac{5}{48} \approx 0.1042$$

23. (a) $f(x,y) \ge 0$, so f is a joint density function if $\iint_{\mathbb{R}^2} f(x,y)\,dA = 1$. Here, $f(x,y) = 0$ outside the first quadrant, so

$$\iint_{\mathbb{R}^2} f(x,y)\,dA = \int_0^\infty\int_0^\infty 0.1e^{-(0.5x+0.2y)}\,dy\,dx = 0.1\int_0^\infty\int_0^\infty e^{-0.5x}e^{-0.2y}\,dy\,dx$$

$$= 0.1\int_0^\infty e^{-0.5x}\,dx\int_0^\infty e^{-0.2y}\,dy = 0.1\lim_{t\to\infty}\int_0^t e^{-0.5x}\,dx\lim_{t\to\infty}\int_0^t e^{-0.2y}\,dy$$

$$= 0.1\lim_{t\to\infty}\left[-2e^{-0.5x}\right]_0^t\lim_{t\to\infty}\left[-5e^{-0.2y}\right]_0^t = 0.1\lim_{t\to\infty}\left[-2(e^{-0.5t} - 1)\right]\lim_{t\to\infty}\left[-5(e^{-0.2t} - 1)\right]$$

$$= (0.1)\cdot(-2)(0-1)\cdot(-5)(0-1) = 1$$

Thus $f(x,y)$ is a joint density function.

(b) (i) No restriction is placed on X, so

$$P(Y \geq 1) = \int_{-\infty}^{\infty} \int_1^{\infty} f(x,y)\,dy\,dx = \int_0^{\infty} \int_1^{\infty} 0.1e^{-(0.5x+0.2y)}\,dy\,dx$$
$$= 0.1 \int_0^{\infty} e^{-0.5x}\,dx \int_1^{\infty} e^{-0.2y}\,dy = 0.1 \lim_{t\to\infty} \int_0^t e^{-0.5x}\,dx \lim_{t\to\infty} \int_1^t e^{-0.2y}\,dy$$
$$= 0.1 \lim_{t\to\infty} \left[-2e^{-0.5x}\right]_0^t \lim_{t\to\infty} \left[-5e^{-0.2y}\right]_1^t = 0.1 \lim_{t\to\infty} \left[-2(e^{-0.5t}-1)\right] \lim_{t\to\infty} \left[-5(e^{-0.2t}-e^{-0.2})\right]$$
$$= (0.1)\cdot(-2)(0-1)\cdot(-5)(0-e^{-0.2}) = e^{-0.2} \approx 0.8187$$

(ii) $$P(X \leq 2, Y \leq 4) = \int_{-\infty}^{2} \int_{-\infty}^{4} f(x,y)\,dy\,dx = \int_0^2 \int_0^4 0.1e^{-(0.5x+0.2y)}\,dy\,dx$$
$$= 0.1 \int_0^2 e^{-0.5x}\,dx \int_0^4 e^{-0.2y}\,dy = 0.1\left[-2e^{-0.5x}\right]_0^2 \left[-5e^{-0.2y}\right]_0^4$$
$$= (0.1)\cdot(-2)(e^{-1}-1)\cdot(-5)(e^{-0.8}-1)$$
$$= (e^{-1}-1)(e^{-0.8}-1) = 1 + e^{-1.8} - e^{-0.8} - e^{-1} \approx 0.3481$$

(c) The expected value of X is given by

$$\mu_1 = \iint_{\mathbb{R}^2} x\,f(x,y)\,dA = \int_0^{\infty} \int_0^{\infty} x\left[0.1e^{-(0.5x+0.2y)}\right] dy\,dx$$
$$= 0.1 \int_0^{\infty} xe^{-0.5x}\,dx \int_0^{\infty} e^{-0.2y}\,dy = 0.1 \lim_{t\to\infty} \int_0^t xe^{-0.5x}\,dx \lim_{t\to\infty} \int_0^t e^{-0.2y}\,dy$$

To evaluate the first integral, we integrate by parts with $u = x$ and $dv = e^{-0.5x}\,dx$ (or we can use Formula 96 in the Table of Integrals): $\int xe^{-0.5x}\,dx = -2xe^{-0.5x} - \int -2e^{-0.5x}\,dx = -2xe^{-0.5x} - 4e^{-0.5x} = -2(x+2)e^{-0.5x}$. Thus

$$\mu_1 = 0.1 \lim_{t\to\infty} \left[-2(x+2)e^{-0.5x}\right]_0^t \lim_{t\to\infty} \left[-5e^{-0.2y}\right]_0^t$$
$$= 0.1 \lim_{t\to\infty} (-2)\left[(t+2)e^{-0.5t} - 2\right] \lim_{t\to\infty} (-5)\left[e^{-0.2t} - 1\right]$$
$$= 0.1(-2)\left(\lim_{t\to\infty} \frac{t+2}{e^{0.5t}} - 2\right)(-5)(-1) = 2 \qquad \text{[by l'Hospital's Rule]}$$

The expected value of Y is given by

$$\mu_2 = \iint_{\mathbb{R}^2} y\,f(x,y)\,dA = \int_0^{\infty} \int_0^{\infty} y\left[0.1e^{-(0.5+0.2y)}\right] dy\,dx$$
$$= 0.1 \int_0^{\infty} e^{-0.5x}\,dx \int_0^{\infty} ye^{-0.2y}\,dy = 0.1 \lim_{t\to\infty} \int_0^t e^{-0.5x}\,dx \lim_{t\to\infty} \int_0^t ye^{-0.2y}\,dy$$

To evaluate the second integral, we integrate by parts with $u = y$ and $dv = e^{-0.2y}\,dy$ (or again we can use Formula 96 in the Table of Integrals) which gives $\int ye^{-0.2y}\,dy = -5ye^{-0.2y} + \int 5e^{-0.2y}\,dy = -5(y+5)e^{-0.2y}$. Then

$$\mu_2 = 0.1 \lim_{t\to\infty} \left[-2e^{-0.5x}\right]_0^t \lim_{t\to\infty} \left[-5(y+5)e^{-0.2y}\right]_0^t$$
$$= 0.1 \lim_{t\to\infty} \left[-2(e^{-0.5t}-1)\right] \lim_{t\to\infty} \left(-5\left[(t+5)e^{-0.2t} - 5\right]\right)$$
$$= 0.1(-2)(-1)\cdot(-5)\left(\lim_{t\to\infty} \frac{t+5}{e^{0.2t}} - 5\right) = 5 \qquad \text{[by l'Hospital's Rule]}$$

25. (a) The random variables X and Y are normally distributed with $\mu_1 = 45$, $\mu_2 = 20$, $\sigma_1 = 0.5$, and $\sigma_2 = 0.1$. The individual density functions for X and Y, then, are $f_1(x) = \dfrac{1}{0.5\sqrt{2\pi}} e^{-(x-45)^2/0.5}$ and $f_2(y) = \dfrac{1}{0.1\sqrt{2\pi}} e^{-(y-20)^2/0.02}$. Since X and Y are independent, the joint density function is the product

$$f(x,y) = f_1(x)f_2(y) = \frac{1}{0.5\sqrt{2\pi}} e^{-(x-45)^2/0.5} \frac{1}{0.1\sqrt{2\pi}} e^{-(y-20)^2/0.02} = \frac{10}{\pi} e^{-2(x-45)^2 - 50(y-20)^2}.$$

Then $P(40 \leq X \leq 50, 20 \leq Y \leq 25) = \int_{40}^{50} \int_{20}^{25} f(x,y)\,dy\,dx = \frac{10}{\pi} \int_{40}^{50} \int_{20}^{25} e^{-2(x-45)^2 - 50(y-20)^2}\,dy\,dx$.

Using a CAS or calculator to evaluate the integral, we get $P(40 \leq X \leq 50, 20 \leq Y \leq 25) \approx 0.500$.

(b) $P(4(X-45)^2+100(Y-20)^2 \le 2) = \iint_D \frac{10}{\pi} e^{-2(x-45)^2-50(y-20)^2}\, dA$, where D is the region enclosed by the ellipse $4(x-45)^2+100(y-20)^2=2$. Solving for y gives $y = 20 \pm \frac{1}{10}\sqrt{2-4(x-45)^2}$, the upper and lower halves of the ellipse, and these two halves meet where $y=20$ [since the ellipse is centred at $(45,20)$] $\Rightarrow$ $4(x-45)^2=2$ $\Rightarrow$ $x = 45 \pm \frac{1}{\sqrt{2}}$. Thus

$$\iint_D \frac{10}{\pi} e^{-2(x-45)^2-50(y-20)^2}\, dA = \frac{10}{\pi}\int_{45-1/\sqrt{2}}^{45+1/\sqrt{2}} \int_{20-\frac{1}{10}\sqrt{2-4(x-45)^2}}^{20+\frac{1}{10}\sqrt{2-4(x-45)^2}} e^{-2(x-45)^2-50(y-20)^2}\, dy\, dx.$$

Using a CAS or calculator to evaluate the integral, we get $P(4(X-45)^2+100(Y-20)^2 \le 2) \approx 0.632$.

27. (a) If $f(P,A)$ is the probability that an individual at A will be infected by an individual at P, and $k\,dA$ is the number of infected individuals in an element of area dA, then $f(P,A)k\,dA$ is the number of infections that should result from exposure of the individual at A to infected people in the element of area dA. Integration over D gives the number of infections of the person at A due to all the infected people in D. In rectangular coordinates (with the origin at the city's centre), the exposure of a person at A is

$$E = \iint_D kf(P,A)\, dA = k\iint_D \frac{20-d(P,A)}{20}\, dA = k\iint_D \left[1 - \frac{\sqrt{(x-x_0)^2+(y-y_0)^2}}{20}\right] dx\, dy$$

(b) If $A=(0,0)$, then

$$E = k\iint_D \left[1 - \frac{1}{20}\sqrt{x^2+y^2}\right] dx\, dy$$

$$= k\int_0^{2\pi}\int_0^{10}\left(1-\frac{r}{20}\right) r\, dr\, d\theta = 2\pi k\left[\frac{r^2}{2}-\frac{r^3}{60}\right]_0^{10}$$

$$= 2\pi k\left(50 - \tfrac{50}{3}\right) = \tfrac{200}{3}\pi k \approx 209k$$

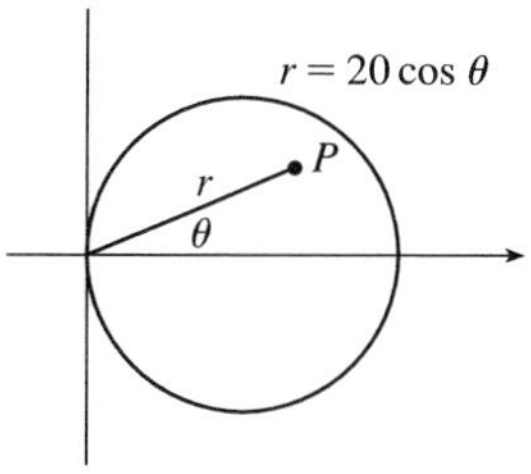

For A at the edge of the city, it is convenient to use a polar coordinate system centred at A. Then the polar equation for the circular boundary of the city becomes $r = 20\cos\theta$ instead of $r=10$, and the distance from A to a point P in the city is again r (see the figure). So

$$E = k\int_{-\pi/2}^{\pi/2}\int_0^{20\cos\theta}\left(1-\frac{r}{20}\right) r\, dr\, d\theta = k\int_{-\pi/2}^{\pi/2}\left[\frac{r^2}{2}-\frac{r^3}{60}\right]_{r=0}^{r=20\cos\theta} d\theta$$

$$= k\int_{-\pi/2}^{\pi/2}\left(200\cos^2\theta - \tfrac{400}{3}\cos^3\theta\right) d\theta = 200k\int_{-\pi/2}^{\pi/2}\left[\tfrac{1}{2}+\tfrac{1}{2}\cos 2\theta - \tfrac{2}{3}\left(1-\sin^2\theta\right)\cos\theta\right] d\theta$$

$$= 200k\left[\tfrac{1}{2}\theta + \tfrac{1}{4}\sin 2\theta - \tfrac{2}{3}\sin\theta + \tfrac{2}{3}\cdot\tfrac{1}{3}\sin^3\theta\right]_{-\pi/2}^{\pi/2} = 200k\left[\tfrac{\pi}{4}+0-\tfrac{2}{3}+\tfrac{2}{9}+\tfrac{\pi}{4}+0-\tfrac{2}{3}+\tfrac{2}{9}\right]$$

$$= 200k\left(\tfrac{\pi}{2}-\tfrac{8}{9}\right) \approx 136k$$

Therefore the risk of infection is much lower at the edge of the city than in the middle, so it is better to live at the edge.

12.6 Surface Area

1. Here $z = f(x,y) = 2 + 3x + 4y$ and D is the rectangle $[0,5] \times [1,4]$, so by Formula 6 the area of the surface is

$$A(S) = \iint_D \sqrt{1 + \left(\frac{\partial z}{\partial x}\right)^2 + \left(\frac{\partial z}{\partial y}\right)^2}\, dA = \iint_D \sqrt{1 + 3^2 + 4^2}\, dA = \sqrt{26} \iint_D dA = \sqrt{26}\, A(D)$$
$$= \sqrt{26}\,(5)(3) = 15\sqrt{26}$$

3. $z = f(x,y) = 6 - 3x - 2y$ which intersects the xy-plane in the line $3x + 2y = 6$, so D is the triangular region given by $\{(x,y) \mid 0 \le x \le 2, 0 \le y \le 3 - \frac{3}{2}x\}$. Thus

$A(S) = \iint_D \sqrt{1 + (-3)^2 + (-2)^2}\, dA = \sqrt{14} \iint_D dA = \sqrt{14}\, A(D) = \sqrt{14}\left(\frac{1}{2} \cdot 2 \cdot 3\right) = 3\sqrt{14}.$

5. $z = f(x,y) = y^2 - x^2$ with $1 \le x^2 + y^2 \le 4$. Then

$$A(S) = \iint_D \sqrt{1 + 4x^2 + 4y^2}\, dA = \int_0^{2\pi}\int_1^2 \sqrt{1 + 4r^2}\, r\, dr\, d\theta = \int_0^{2\pi} d\theta \int_1^2 r\sqrt{1 + 4r^2}\, dr$$
$$= \Big[\,\theta\,\Big]_0^{2\pi} \left[\tfrac{1}{12}(1 + 4r^2)^{3/2}\right]_1^2 = \tfrac{\pi}{6}\left(17\sqrt{17} - 5\sqrt{5}\right)$$

7. $\mathbf{r}_u = \langle 2u, v, 0\rangle$, $\mathbf{r}_v = \langle 0, u, v\rangle$, and $\mathbf{r}_u \times \mathbf{r}_v = \langle v^2, -2uv, 2u^2\rangle$. Then

$$A(S) = \iint_D |\mathbf{r}_u \times \mathbf{r}_v|\, dA = \int_0^1\int_0^2 \sqrt{v^4 + 4u^2v^2 + 4u^4}\, dv\, du$$
$$= \int_0^1\int_0^2 \sqrt{(v^2 + 2u^2)^2}\, dv\, du = \int_0^1\int_0^2 (v^2 + 2u^2)\, dv\, du$$
$$= \int_0^1 \left[\tfrac{1}{3}v^3 + 2u^2v\right]_{v=0}^{v=2} du = \int_0^1 \left(\tfrac{8}{3} + 4u^2\right) du = \left[\tfrac{8}{3}u + \tfrac{4}{3}u^3\right]_0^1 = 4$$

9. A parametric representation of the surface is $x = x$, $y = 4x + z^2$, $z = z$ with $0 \le x \le 1$, $0 \le z \le 1$.

Hence $\mathbf{r}_x \times \mathbf{r}_z = (\mathbf{i} + 4\mathbf{j}) \times (2z\,\mathbf{j} + \mathbf{k}) = 4\,\mathbf{i} - \mathbf{j} + 2z\,\mathbf{k}$.

Note: In general, if $y = f(x,z)$ then $\mathbf{r}_x \times \mathbf{r}_z = \dfrac{\partial f}{\partial x}\,\mathbf{i} - \mathbf{j} + \dfrac{\partial f}{\partial z}\,\mathbf{k}$ and $A(S) = \displaystyle\iint_D \sqrt{1 + \left(\frac{\partial f}{\partial x}\right)^2 + \left(\frac{\partial f}{\partial z}\right)^2}\, dA$. Then

$$A(S) = \int_0^1\int_0^1 \sqrt{17 + 4z^2}\, dx\, dz = \int_0^1 \sqrt{17 + 4z^2}\, dz$$
$$= \tfrac{1}{2}\left(z\sqrt{17 + 4z^2} + \tfrac{17}{2}\ln\left|2z + \sqrt{4z^2 + 17}\right|\right)\Big]_0^1 = \tfrac{\sqrt{21}}{2} + \tfrac{17}{4}\left[\ln\left(2 + \sqrt{21}\right) - \ln\sqrt{17}\right]$$

11. $z = f(x,y) = xy$ with $0 \le x^2 + y^2 \le 1$, so $f_x = y$, $f_y = x$ $\Rightarrow$

$$A(S) = \iint_D \sqrt{1 + y^2 + x^2}\, dA = \int_0^{2\pi}\int_0^1 \sqrt{r^2 + 1}\, r\, dr\, d\theta = \int_0^{2\pi} \left[\tfrac{1}{3}(r^2 + 1)^{3/2}\right]_{r=0}^{r=1} d\theta$$
$$= \int_0^{2\pi} \tfrac{1}{3}\left(2\sqrt{2} - 1\right) d\theta = \tfrac{2\pi}{3}\left(2\sqrt{2} - 1\right)$$

13. $z = f(x,y) = e^{-x^2 - y^2}$, $f_x = -2xe^{-x^2-y^2}$, $f_y = -2ye^{-x^2-y^2}$. Then

$$A(S) = \iint\limits_{x^2+y^2\le 4} \sqrt{1 + (-2xe^{-x^2-y^2})^2 + (-2ye^{-x^2-y^2})^2}\, dA = \iint\limits_{x^2+y^2\le 4} \sqrt{1 + 4(x^2 + y^2)e^{-2(x^2+y^2)}}\, dA.$$

Converting to polar coordinates we have

$$A(S) = \int_0^{2\pi}\int_0^2 \sqrt{1+4r^2e^{-2r^2}}\, r\,dr\,d\theta = \int_0^{2\pi} d\theta \int_0^2 r\sqrt{1+4r^2e^{-2r^2}}\,dr$$

$$= 2\pi \int_0^2 r\sqrt{1+4r^2e^{-2r^2}}\,dr \approx 13.9783 \text{ using a calculator.}$$

15. (a) $A(S) = \iint_D \sqrt{1+\left(\frac{\partial z}{\partial x}\right)^2+\left(\frac{\partial z}{\partial y}\right)^2}\,dA = \int_0^6\int_0^4 \sqrt{1+\frac{4x^2+4y^2}{(1+x^2+y^2)^4}}\,dy\,dx$. Using the Midpoint Rule with

$f(x,y) = \sqrt{1+\frac{4x^2+4y^2}{(1+x^2+y^2)^4}}$, $m=3$, $n=2$ we have

$$A(S) \approx \sum_{i=1}^{3}\sum_{j=1}^{2} f(\overline{x}_i,\overline{y}_j)\,\Delta A = 4\,[f(1,1)+f(1,3)+f(3,1)+f(3,3)+f(5,1)+f(5,3)] \approx 24.2055$$

(b) Using a CAS we have $A(S) = \int_0^6\int_0^4 \sqrt{1+\frac{4x^2+4y^2}{(1+x^2+y^2)^4}}\,dy\,dx \approx 24.2476$. This agrees with the estimate in part (a) to the first decimal place.

17. $\mathbf{r}(u,v) = \langle \cos^3 u\cos^3 v, \sin^3 u\cos^3 v, \sin^3 v\rangle$, so $\mathbf{r}_u = \langle -3\cos^2 u\sin u\cos^3 v, 3\sin^2 u\cos u\cos^3 v, 0\rangle$,

$\mathbf{r}_v = \langle -3\cos^3 u\cos^2 v\sin v, -3\sin^3 u\cos^2 v\sin v, 3\sin^2 v\cos v\rangle$, and

$\mathbf{r}_u\times\mathbf{r}_v = \langle 9\cos u\sin^2 u\cos^4 v\sin^2 v, 9\cos^2 u\sin u\cos^4 v\sin^2 v, 9\cos^2 u\sin^2 u\cos^5 v\sin v\rangle$. Then

$$|\mathbf{r}_u\times\mathbf{r}_v| = 9\sqrt{\cos^2 u\sin^4 u\cos^8 v\sin^4 v+\cos^4 u\sin^2 u\cos^8 v\sin^4 v+\cos^4 u\sin^4 u\cos^{10} v\sin^2 v}$$

$$= 9\sqrt{\cos^2 u\sin^2 u\cos^8 v\sin^2 v\,(\sin^2 v+\cos^2 u\sin^2 u\cos^2 v)}$$

$$= 9\cos^4 v\,|\cos u\sin u\sin v|\,\sqrt{\sin^2 v+\cos^2 u\sin^2 u\cos^2 v}$$

Using a CAS, we have $A(S) = \int_0^{\pi}\int_0^{2\pi} 9\cos^4 v\,|\cos u\sin u\sin v|\,\sqrt{\sin^2 v+\cos^2 u\sin^2 u\cos^2 v}\,dv\,du \approx 4.4506$.

19. $z = 1+2x+3y+4y^2$, so

$$A(S) = \iint_D \sqrt{1+\left(\frac{\partial z}{\partial x}\right)^2+\left(\frac{\partial z}{\partial y}\right)^2}\,dA = \int_1^4\int_0^1 \sqrt{1+4+(3+8y)^2}\,dy\,dx = \int_1^4\int_0^1 \sqrt{14+48y+64y^2}\,dy\,dx.$$

Using a CAS, we have

$$\int_1^4\int_0^1 \sqrt{14+48y+64y^2}\,dy\,dx = \tfrac{45}{8}\sqrt{14}+\tfrac{15}{16}\ln\left(11\sqrt{5}+3\sqrt{14}\sqrt{5}\right)-\tfrac{15}{16}\ln\left(3\sqrt{5}+\sqrt{14}\sqrt{5}\right)$$

or $\frac{45}{8}\sqrt{14}+\frac{15}{16}\ln\frac{11\sqrt{5}+3\sqrt{70}}{3\sqrt{5}+\sqrt{70}}$.

21. (a) $x = a\sin u\cos v$, $y = b\sin u\sin v$, $z = c\cos u$ $\Rightarrow$

$$\frac{x^2}{a^2}+\frac{y^2}{b^2}+\frac{z^2}{c^2} = (\sin u\cos v)^2+(\sin u\sin v)^2+(\cos u)^2$$

$$= \sin^2 u+\cos^2 u = 1$$

and since the ranges of u and v are sufficient to generate the entire graph, the parametric equations represent an ellipsoid.

(b)

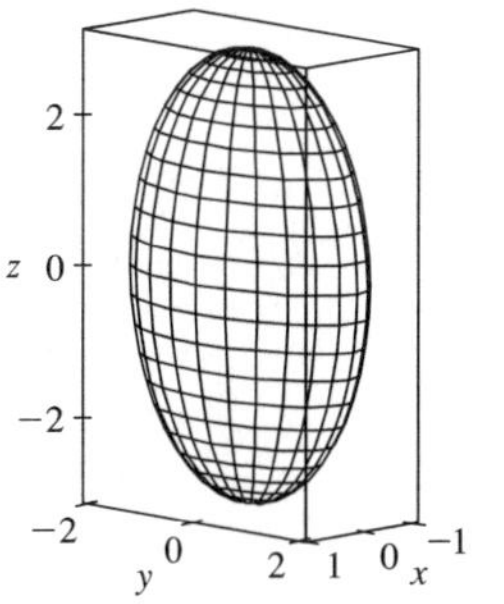

(c) From the parametric equations (with $a=1$, $b=2$, and $c=3$), we calculate $\mathbf{r}_u = \cos u\cos v\,\mathbf{i}+2\cos u\sin v\,\mathbf{j}-3\sin u\,\mathbf{k}$ and

$\mathbf{r}_v = -\sin u\sin v\,\mathbf{i}+2\sin u\cos v\,\mathbf{j}$. So $\mathbf{r}_u\times\mathbf{r}_v = 6\sin^2 u\cos v\,\mathbf{i}+3\sin^2 u\sin v\,\mathbf{j}+2\sin u\cos u\,\mathbf{k}$, and the surface area is given by $A(S) = \int_0^{2\pi}\int_0^{\pi} |\mathbf{r}_u\times\mathbf{r}_v|\,du\,dv = \int_0^{2\pi}\int_0^{\pi}\sqrt{36\sin^4 u\cos^2 v+9\sin^4 u\sin^2 v+4\cos^2 u\sin^2 u}\,du\,dv$.

23. To find the region D: $z = x^2 + y^2$ implies $z + z^2 = 4z$ or $z^2 - 3z = 0$. Thus $z = 0$ or $z = 3$ are the planes where the surfaces intersect. But $x^2 + y^2 + z^2 = 4z$ implies $x^2 + y^2 + (z-2)^2 = 4$, so $z = 3$ intersects the upper hemisphere. Thus $(z-2)^2 = 4 - x^2 - y^2$ or $z = 2 + \sqrt{4 - x^2 - y^2}$. Therefore D is the region inside the circle $x^2 + y^2 + (3-2)^2 = 4$, that is, $D = \{(x, y) \mid x^2 + y^2 \le 3\}$.

$$\begin{aligned} A(S) &= \iint_D \sqrt{1 + [(-x)(4 - x^2 - y^2)^{-1/2}]^2 + [(-y)(4 - x^2 - y^2)^{-1/2}]^2}\, dA \\ &= \int_0^{2\pi} \int_0^{\sqrt{3}} \sqrt{1 + \frac{r^2}{4 - r^2}}\, r\, dr\, d\theta = \int_0^{2\pi} \int_0^{\sqrt{3}} \frac{2r\, dr}{\sqrt{4 - r^2}}\, d\theta = \int_0^{2\pi} \Big[-2(4 - r^2)^{1/2}\Big]_{r=0}^{r=\sqrt{3}}\, d\theta \\ &= \int_0^{2\pi} (-2 + 4)\, d\theta = 2\theta\Big]_0^{2\pi} = 4\pi \end{aligned}$$

25. If we revolve the curve $y = f(x)$, $a \le x \le b$ about the x-axis, where $f(x) \ge 0$, then from Equations 10.5.3 we know we can parametrize the surface using $x = x$, $y = f(x)\cos\theta$, and $z = f(x)\sin\theta$, where $a \le x \le b$ and $0 \le \theta \le 2\pi$. Thus we can say the surface is represented by $\mathbf{r}(x, \theta) = x\,\mathbf{i} + f(x)\cos\theta\,\mathbf{j} + f(x)\sin\theta\,\mathbf{k}$, with $a \le x \le b$ and $0 \le \theta \le 2\pi$. Then by (4), the surface area is given by $A(S) = \iint_D |\mathbf{r}_x \times \mathbf{r}_\theta|\, dA$ where D is the rectangular parameter region $[a, b] \times [0, 2\pi]$. Here, $\mathbf{r}_x(x, \theta) = \mathbf{i} + f'(x)\cos\theta\,\mathbf{j} + f'(x)\sin\theta\,\mathbf{k}$ and $\mathbf{r}_\theta(x) = -f(x)\sin\theta\,\mathbf{j} + f(x)\cos\theta\,\mathbf{k}$. So

$$\mathbf{r}_x \times \mathbf{r}_\theta = \begin{vmatrix} \mathbf{i} & \mathbf{j} & \mathbf{k} \\ 1 & f'(x)\cos\theta & f'(x)\sin\theta \\ 0 & -f(x)\sin\theta & f(x)\cos\theta \end{vmatrix} = \left[f(x)f'(x)\cos^2\theta + f(x)f'(x)\sin^2\theta\right]\mathbf{i} - f(x)\cos\theta\,\mathbf{j} - f(x)\sin\theta\,\mathbf{k}$$

$$= f(x)f'(x)\mathbf{i} - f(x)\cos\theta\,\mathbf{j} - f(x)\sin\theta\,\mathbf{k} \text{ and}$$

$$\begin{aligned} |\mathbf{r}_x \times \mathbf{r}_\theta| &= \sqrt{[f(x)f'(x)]^2 + [f(x)]^2\cos^2\theta + [f(x)]^2\sin^2\theta} \\ &= \sqrt{[f(x)]^2\left([f'(x)]^2 + 1\right)} = f(x)\sqrt{1 + [f'(x)]^2} \text{ [since } f(x) \ge 0\text{]. Thus} \end{aligned}$$

$$\begin{aligned} A(S) &= \iint_D |\mathbf{r}_x \times \mathbf{r}_\theta|\, dA = \int_a^b \int_0^{2\pi} f(x)\sqrt{1 + [f'(x)]^2}\, d\theta\, dx \\ &= \int_a^b f(x)\sqrt{1 + [f'(x)]^2}\, [\theta]_0^{2\pi}\, dx = 2\pi \int_a^b f(x)\sqrt{1 + [f'(x)]^2}\, dx \end{aligned}$$

27. $y = \sqrt{x} \quad\Rightarrow\quad 1 + \left(\dfrac{dy}{dx}\right)^2 = 1 + \left(\dfrac{1}{2\sqrt{x}}\right)^2 = 1 + \dfrac{1}{4x}$. So

$$\begin{aligned} S &= \int_4^9 2\pi y \sqrt{1 + \left(\frac{dy}{dx}\right)^2}\, dx = \int_4^9 2\pi\sqrt{x}\sqrt{1 + \frac{1}{4x}}\, dx = 2\pi \int_4^9 \left(x + \tfrac{1}{4}\right) dx \\ &= 2\pi\left[\tfrac{2}{3}\left(x + \tfrac{1}{4}\right)^{3/2}\right]_4^9 = \tfrac{4\pi}{3}\left[\tfrac{1}{8}(4x + 1)^{3/2}\right]_4^9 = \tfrac{\pi}{6}\left(37\sqrt{37} - 17\sqrt{17}\right) \end{aligned}$$

12.7 Triple Integrals

1. $\iiint_B xyz^2\,dV = \int_0^1\int_{-1}^2\int_0^3 xyz^2\,dz\,dx\,dy = \int_0^1\int_{-1}^2 xy\left[\frac{1}{3}z^3\right]_{z=0}^{z=3} dx\,dy = \int_0^1\int_{-1}^2 9xy\,dx\,dy$

$= \int_0^1\left[\frac{9}{2}x^2y\right]_{x=-1}^{x=2} dy = \int_0^1 \frac{27}{2}\,y\,dy = \frac{27}{4}y^2\Big]_0^1 = \frac{27}{4}$

3. $\int_0^1\int_0^z\int_0^{x+z} 6xz\,dy\,dx\,dz = \int_0^1\int_0^z \left[6xyz\right]_{y=0}^{y=x+z} dx\,dz = \int_0^1\int_0^z 6xz(x+z)\,dx\,dz$

$= \int_0^1\left[2x^3z + 3x^2z^2\right]_{x=0}^{x=z} dz = \int_0^1(2z^4+3z^4)\,dz = \int_0^1 5z^4\,dz = z^5\Big]_0^1 = 1$

5. $\int_0^3\int_0^1\int_0^{\sqrt{1-z^2}} ze^y\,dx\,dz\,dy = \int_0^3\int_0^1\left[xze^y\right]_{x=0}^{x=\sqrt{1-z^2}} dz\,dy = \int_0^3\int_0^1 ze^y\sqrt{1-z^2}\,dz\,dy$

$= \int_0^3\left[-\frac{1}{3}(1-z^2)^{3/2}e^y\right]_{z=0}^{z=1} dy = \int_0^3 \frac{1}{3}e^y\,dy = \frac{1}{3}e^y\Big]_0^3 = \frac{1}{3}(e^3-1)$

7. $\iiint_E 2x\,dV = \int_0^2\int_0^{\sqrt{4-y^2}}\int_0^y 2x\,dz\,dx\,dy = \int_0^2\int_0^{\sqrt{4-y^2}}\left[2xz\right]_{z=0}^{z=y} dx\,dy = \int_0^2\int_0^{\sqrt{4-y^2}} 2xy\,dx\,dy$

$= \int_0^2\left[x^2y\right]_{x=0}^{x=\sqrt{4-y^2}} dy = \int_0^2(4-y^2)y\,dy = \left[2y^2 - \frac{1}{4}y^4\right]_0^2 = 4$

9. Here $E = \{(x,y,z) \mid 0 \le x \le 1, 0 \le y \le \sqrt{x}, 0 \le z \le 1+x+y\}$, so

$$\begin{aligned}\iiint_E 6xy\,dV &= \int_0^1\int_0^{\sqrt{x}}\int_0^{1+x+y} 6xy\,dz\,dy\,dx = \int_0^1\int_0^{\sqrt{x}}\left[6xyz\right]_{z=0}^{z=1+x+y} dy\,dx\\ &= \int_0^1\int_0^{\sqrt{x}} 6xy(1+x+y)\,dy\,dx = \int_0^1\left[3xy^2 + 3x^2y^2 + 2xy^3\right]_{y=0}^{y=\sqrt{x}} dx\\ &= \int_0^1(3x^2+3x^3+2x^{5/2})\,dx = \left[x^3 + \frac{3}{4}x^4 + \frac{4}{7}x^{7/2}\right]_0^1 = \frac{65}{28}\end{aligned}$$

11. Here E is the region that lies below the plane with x-, y-, and z-intercepts 1, 2, and 3 respectively, that is, below the plane $2z+6x+3y=6$ and above the region in the xy-plane bounded by the lines $x=0$, $y=0$ and $6x+3y=6$. So

$$\begin{aligned}\iiint_E xy\,dV &= \int_0^1\int_0^{2-2x}\int_0^{3-3x-3y/2} xy\,dz\,dy\,dx = \int_0^1\int_0^{2-2x}\left(3xy - 3x^2y - \frac{3}{2}xy^2\right) dy\,dx\\ &= \int_0^1\left[\frac{3}{2}xy^2 - \frac{3}{2}x^2y^2 - \frac{1}{2}xy^3\right]_{y=0}^{y=2-2x} dx = \int_0^1(2x-6x^2+6x^3-2x^4)\,dx\\ &= \left[x^2 - 2x^3 + \frac{3}{2}x^4 - \frac{2}{5}x^5\right]_0^1 = \frac{1}{10}\end{aligned}$$

13.

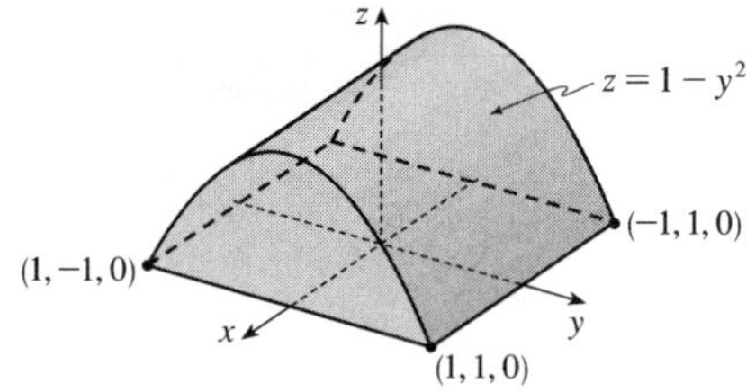

E is the region below the parabolic cylinder $z = 1-y^2$ and above the square $[-1,1]\times[-1,1]$ in the xy-plane.

$$\begin{aligned}\iiint_E x^2e^y\,dV &= \int_{-1}^1\int_{-1}^1\int_0^{1-y^2} x^2e^y\,dz\,dy\,dx\\ &= \int_{-1}^1\int_{-1}^1 x^2e^y(1-y^2)\,dy\,dx\\ &= \int_{-1}^1 x^2\,dx\int_{-1}^1(e^y - y^2e^y)\,dy\\ &= \left[\frac{1}{3}x^3\right]_{-1}^1\left[e^y - (y^2-2y+2)e^y\right]_{-1}^1\\ &\qquad\text{[integrate by parts twice]}\\ &= \frac{1}{3}(2)[e - e - e^{-1} + 5e^{-1}] = \frac{8}{3e}\end{aligned}$$

15.

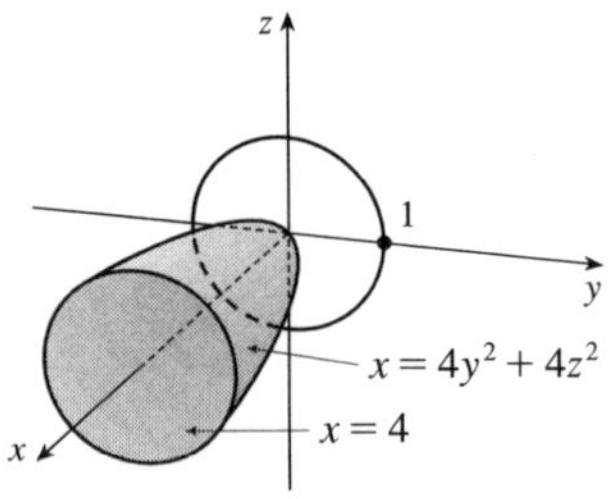

The projection E on the yz-plane is the disc $y^2 + z^2 \le 1$.
Using polar coordinates $y = r\cos\theta$ and $z = r\sin\theta$, we get

$$\iiint_E x\,dV = \iint_D \left[\int_{4y^2+4z^2}^{4} x\,dx\right] dA$$
$$= \tfrac{1}{2}\iint_D \left[4^2 - (4y^2+4z^2)^2\right] dA = 8\int_0^{2\pi}\int_0^1 (1-r^4)\,r\,dr\,d\theta$$
$$= 8\int_0^{2\pi} d\theta \int_0^1 (r - r^5)\,dr = 8(2\pi)\left[\tfrac{1}{2}r^2 - \tfrac{1}{6}r^6\right]_0^1 = \tfrac{16\pi}{3}$$

17. The plane $2x + y + z = 4$ intersects the xy-plane when
$2x + y + 0 = 4 \;\Rightarrow\; y = 4 - 2x$, so
$E = \{(x,y,z) \mid 0 \le x \le 2, 0 \le y \le 4 - 2x, 0 \le z \le 4 - 2x - y\}$ and

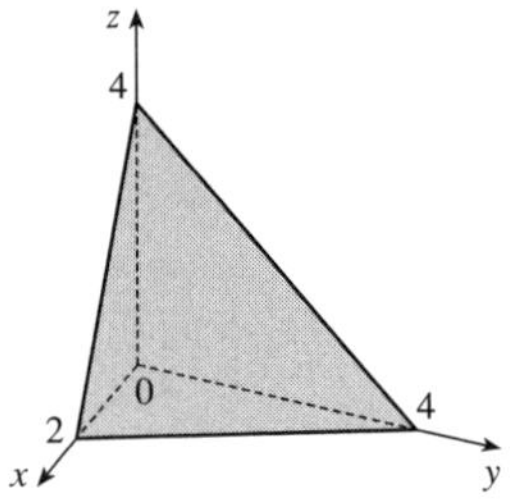

$$V = \int_0^2 \int_0^{4-2x} \int_0^{4-2x-y} dz\,dy\,dx = \int_0^2 \int_0^{4-2x} (4 - 2x - y)\,dy\,dx$$
$$= \int_0^2 \left[4y - 2xy - \tfrac{1}{2}y^2\right]_{y=0}^{y=4-2x} dx$$
$$= \int_0^2 \left[4(4-2x) - 2x(4-2x) - \tfrac{1}{2}(4-2x)^2\right] dx$$
$$= \int_0^2 (2x^2 - 8x + 8)\,dx = \left[\tfrac{2}{3}x^3 - 4x^2 + 8x\right]_0^2 = \tfrac{16}{3}$$

19.

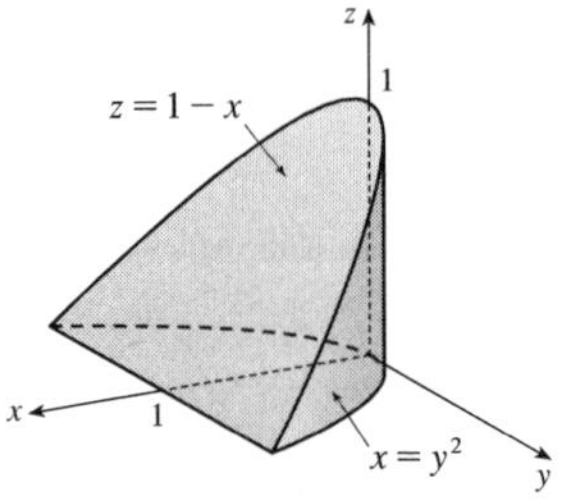

$$V = \int_0^1 \int_{-\sqrt{x}}^{\sqrt{x}} \int_0^{1-x} dz\,dy\,dx = \int_0^1 \int_{-\sqrt{x}}^{\sqrt{x}} (1-x)\,dy\,dx$$
$$= \int_0^1 2\sqrt{x}\,(1-x)\,dx = \int_0^1 2\left(\sqrt{x} - x^{3/2}\right) dx$$
$$= 2\left[\tfrac{2}{3}x^{3/2} - \tfrac{2}{5}x^{5/2}\right]_0^1 = 2\left(\tfrac{2}{3} - \tfrac{2}{5}\right) = \tfrac{8}{15}$$

21. (a) The wedge can be described as the region

$$D = \{(x,y,z) \mid y^2 + z^2 \le 1, 0 \le x \le 1, 0 \le y \le x\}$$
$$= \left\{(x,y,z) \mid 0 \le x \le 1, 0 \le y \le x, 0 \le z \le \sqrt{1-y^2}\right\}$$

So the integral expressing the volume of the wedge is

$$\iiint_D dV = \int_0^1 \int_0^x \int_0^{\sqrt{1-y^2}} dz\,dy\,dx.$$

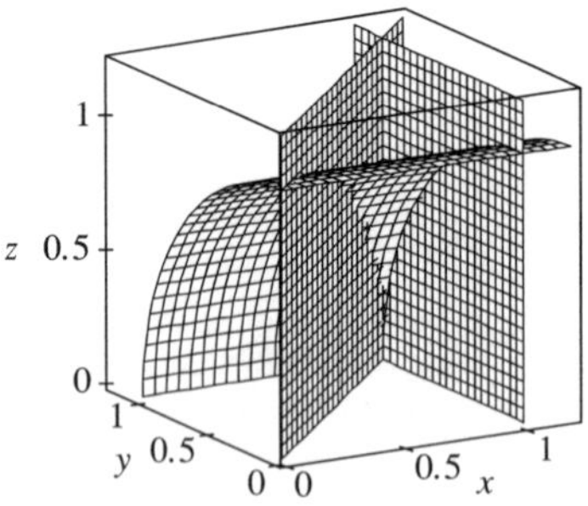

(b) A CAS gives $\int_0^1 \int_0^x \int_0^{\sqrt{1-y^2}} dz\,dy\,dx = \frac{\pi}{4} - \frac{1}{3}$.

(Or use Formulas 30 and 87 from the Table of Integrals.)

23. Here $f(x, y, z) = \dfrac{1}{\ln(1 + x + y + z)}$ and $\Delta V = 2 \cdot 4 \cdot 2 = 16$, so the Midpoint Rule gives

$$\begin{aligned}\iiint_B f(x,y,z)\,dV &\approx \sum_{i=1}^{l}\sum_{j=1}^{m}\sum_{k=1}^{n} f(\overline{x}_i, \overline{y}_j, \overline{z}_k)\,\Delta V \\ &= 16\,[f(1,2,1) + f(1,2,3) + f(1,6,1) + f(1,6,3) \\ &\qquad + f(3,2,1) + f(3,2,3) + f(3,6,1) + f(3,6,3)] \\ &= 16\left[\frac{1}{\ln 5} + \frac{1}{\ln 7} + \frac{1}{\ln 9} + \frac{1}{\ln 11} + \frac{1}{\ln 7} + \frac{1}{\ln 9} + \frac{1}{\ln 11} + \frac{1}{\ln 13}\right] \approx 60.533\end{aligned}$$

25. $E = \{(x, y, z) \mid 0 \le x \le 1, 0 \le z \le 1 - x, 0 \le y \le 2 - 2z\}$, the solid bounded by the three coordinate planes and the planes $z = 1 - x$, $y = 2 - 2z$.

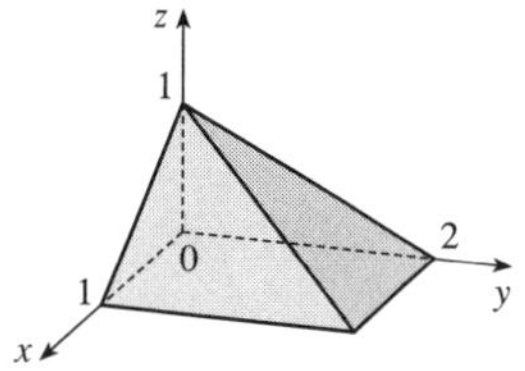

27.

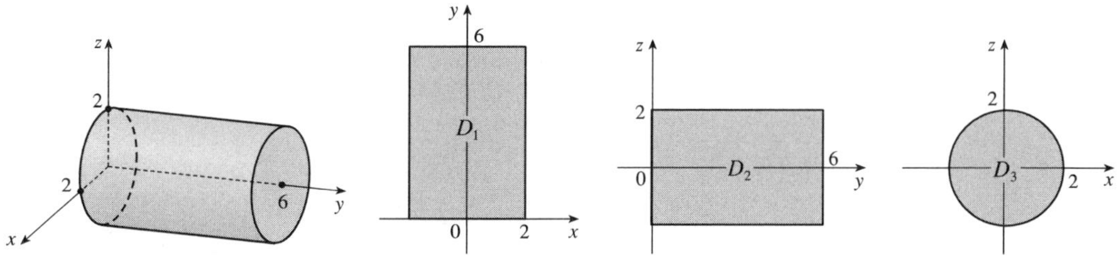

If D_1, D_2, D_3 are the projections of E on the xy-, yz-, and xz-planes, then

$$D_1 = \{(x, y) \mid -2 \le x \le 2, 0 \le y \le 6\}$$

$$D_2 = \{(y, z) \mid -2 \le z \le 2, 0 \le y \le 6\}$$

$$D_3 = \{(x, z) \mid x^2 + z^2 \le 4\}$$

Therefore

$$\begin{aligned}E &= \left\{(x,y,z) \mid -\sqrt{4-x^2} \le z \le \sqrt{4-x^2},\ -2 \le x \le 2, 0 \le y \le 6\right\} \\ &= \left\{(x,y,z) \mid -\sqrt{4-z^2} \le x \le \sqrt{4-z^2},\ -2 \le z \le 2, 0 \le y \le 6\right\}\end{aligned}$$

$$\begin{aligned}\iiint_E f(x,y,z)\,dV &= \int_{-2}^{2}\int_0^6\int_{-\sqrt{4-x^2}}^{\sqrt{4-x^2}} f(x,y,z)\,dz\,dy\,dx = \int_0^6\int_{-2}^{2}\int_{-\sqrt{4-x^2}}^{\sqrt{4-x^2}} f(x,y,z)\,dz\,dx\,dy \\ &= \int_0^6\int_{-2}^{2}\int_{-\sqrt{4-z^2}}^{\sqrt{4-z^2}} f(x,y,z)\,dx\,dz\,dy = \int_{-2}^{2}\int_0^6\int_{-\sqrt{4-z^2}}^{\sqrt{4-z^2}} f(x,y,z)\,dx\,dy\,dz \\ &= \int_{-2}^{2}\int_{-\sqrt{4-x^2}}^{\sqrt{4-x^2}}\int_0^6 f(x,y,z)\,dy\,dz\,dx = \int_{-2}^{2}\int_{-\sqrt{4-z^2}}^{\sqrt{4-z^2}}\int_0^6 f(x,y,z)\,dy\,dx\,dz\end{aligned}$$

29.

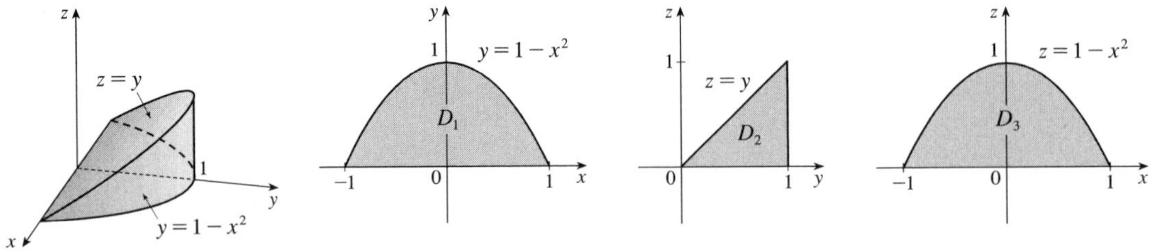

If D_1, D_2, and D_3 are the projections of E on the xy-, yz-, and xz-planes, then

$$D_1 = \{(x,y) \mid -1 \le x \le 1, 0 \le y \le 1 - x^2\} = \left\{(x,y) \mid 0 \le y \le 1, -\sqrt{1-y} \le x \le \sqrt{1-y}\right\},$$

$$D_2 = \{(y,z) \mid 0 \le y \le 1, 0 \le z \le y\} = \{(y,z) \mid 0 \le z \le 1, z \le y \le 1\}, \text{ and}$$

$$D_3 = \{(x,z) \mid -1 \le x \le 1, 0 \le z \le 1 - x^2\} = \left\{(x,z) \mid 0 \le z \le 1, -\sqrt{1-z} \le x \le \sqrt{1-z}\right\}$$

Therefore

$$\begin{aligned} E &= \{(x,y,z) \mid -1 \le x \le 1, 0 \le y \le 1 - x^2, 0 \le z \le y\} \\ &= \left\{(x,y,z) \mid 0 \le y \le 1,\ -\sqrt{1-y} \le x \le \sqrt{1-y}, 0 \le z \le y\right\} \\ &= \left\{(x,y,z) \mid 0 \le y \le 1, 0 \le z \le y,\ -\sqrt{1-y} \le x \le \sqrt{1-y}\right\} \\ &= \left\{(x,y,z) \mid 0 \le z \le 1, z \le y \le 1,\ -\sqrt{1-y} \le x \le \sqrt{1-y}\right\} \\ &= \{(x,y,z) \mid -1 \le x \le 1, 0 \le z \le 1 - x^2, z \le y \le 1 - x^2\} \\ &= \left\{(x,y,z) \mid 0 \le z \le 1,\ -\sqrt{1-z} \le x \le \sqrt{1-z}, z \le y \le 1 - x^2\right\} \end{aligned}$$

Then

$$\begin{aligned} \iiint_E f(x,y,z)\,dV &= \int_{-1}^{1}\int_0^{1-x^2}\int_0^y f(x,y,z)\,dz\,dy\,dx = \int_0^1\int_{-\sqrt{1-y}}^{\sqrt{1-y}}\int_0^y f(x,y,z)\,dz\,dx\,dy \\ &= \int_0^1\int_0^y\int_{-\sqrt{1-y}}^{\sqrt{1-y}} f(x,y,z)\,dx\,dz\,dy = \int_0^1\int_z^1\int_{-\sqrt{1-y}}^{\sqrt{1-y}} f(x,y,z)\,dx\,dy\,dz \\ &= \int_{-1}^{1}\int_0^{1-x^2}\int_z^{1-x^2} f(x,y,z)\,dy\,dz\,dx = \int_0^1\int_{-\sqrt{1-z}}^{\sqrt{1-z}}\int_z^{1-x^2} f(x,y,z)\,dy\,dx\,dz \end{aligned}$$

31.

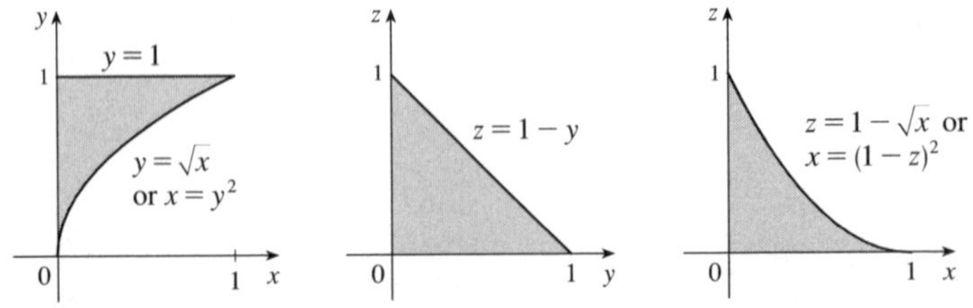

The diagrams show the projections of E on the xy-, yz-, and xz-planes. Therefore

$$\begin{aligned} \int_0^1\int_{\sqrt{x}}^1\int_0^{1-y} f(x,y,z)\,dz\,dy\,dx &= \int_0^1\int_0^{y^2}\int_0^{1-y} f(x,y,z)\,dz\,dx\,dy = \int_0^1\int_0^{1-z}\int_0^{y^2} f(x,y,z)\,dx\,dy\,dz \\ &= \int_0^1\int_0^{1-y}\int_0^{y^2} f(x,y,z)\,dx\,dz\,dy = \int_0^1\int_0^{1-\sqrt{x}}\int_{\sqrt{x}}^{1-z} f(x,y,z)\,dy\,dz\,dx \\ &= \int_0^1\int_0^{(1-z)^2}\int_{\sqrt{x}}^{1-z} f(x,y,z)\,dy\,dx\,dz \end{aligned}$$

33.

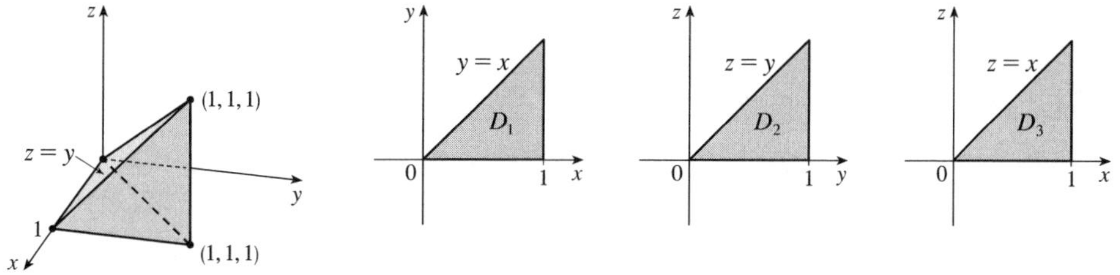

$\int_0^1 \int_y^1 \int_0^y f(x,y,z)\,dz\,dx\,dy = \iiint_E f(x,y,z)\,dV$ where $E = \{(x,y,z) \mid 0 \le z \le y, y \le x \le 1, 0 \le y \le 1\}$.

If D_1, D_2, and D_3 are the projections of E on the xy-, yz- and xz-planes then

$$D_1 = \{(x,y) \mid 0 \le y \le 1, y \le x \le 1\} = \{(x,y) \mid 0 \le x \le 1, 0 \le y \le x\},$$

$$D_2 = \{(y,z) \mid 0 \le y \le 1, 0 \le z \le y\} = \{(y,z) \mid 0 \le z \le 1, z \le y \le 1\}, \text{ and}$$

$$D_3 = \{(x,z) \mid 0 \le x \le 1, 0 \le z \le x\} = \{(x,z) \mid 0 \le z \le 1, z \le x \le 1\}.$$

Thus we also have

$$\begin{aligned} E &= \{(x,y,z) \mid 0 \le x \le 1, 0 \le y \le x, 0 \le z \le y\} = \{(x,y,z) \mid 0 \le y \le 1, 0 \le z \le y, y \le x \le 1\} \\ &= \{(x,y,z) \mid 0 \le z \le 1, z \le y \le 1, y \le x \le 1\} = \{(x,y,z) \mid 0 \le x \le 1, 0 \le z \le x, z \le y \le x\} \\ &= \{(x,y,z) \mid 0 \le z \le 1, z \le x \le 1, z \le y \le x\}. \end{aligned}$$

Then

$$\begin{aligned} \int_0^1 \int_y^1 \int_0^y f(x,y,z)\,dz\,dx\,dy &= \int_0^1 \int_0^x \int_0^y f(x,y,z)\,dz\,dy\,dx = \int_0^1 \int_0^y \int_y^1 f(x,y,z)\,dx\,dz\,dy \\ &= \int_0^1 \int_z^1 \int_y^1 f(x,y,z)\,dx\,dy\,dz = \int_0^1 \int_0^x \int_z^x f(x,y,z)\,dy\,dz\,dx \\ &= \int_0^1 \int_z^1 \int_z^x f(x,y,z)\,dy\,dx\,dz \end{aligned}$$

35. $m = \iiint_E \rho(x,y,z)\,dV = \int_0^1 \int_0^{\sqrt{x}} \int_0^{1+x+y} 2\,dz\,dy\,dx = \int_0^1 \int_0^{\sqrt{x}} 2(1+x+y)\,dy\,dx$

$= \int_0^1 \left[2y + 2xy + y^2\right]_{y=0}^{y=\sqrt{x}} dx = \int_0^1 \left(2\sqrt{x} + 2x^{3/2} + x\right) dx = \left[\frac{4}{3}x^{3/2} + \frac{4}{5}x^{5/2} + \frac{1}{2}x^2\right]_0^1 = \frac{79}{30}$

$M_{yz} = \iiint_E x\rho(x,y,z)\,dV = \int_0^1 \int_0^{\sqrt{x}} \int_0^{1+x+y} 2x\,dz\,dy\,dx = \int_0^1 \int_0^{\sqrt{x}} 2x(1+x+y)\,dy\,dx$

$= \int_0^1 \left[2xy + 2x^2y + xy^2\right]_{y=0}^{y=\sqrt{x}} dx = \int_0^1 (2x^{3/2} + 2x^{5/2} + x^2)\,dx = \left[\frac{4}{5}x^{5/2} + \frac{4}{7}x^{7/2} + \frac{1}{3}x^3\right]_0^1 = \frac{179}{105}$

$M_{xz} = \iiint_E y\rho(x,y,z)\,dV = \int_0^1 \int_0^{\sqrt{x}} \int_0^{1+x+y} 2y\,dz\,dy\,dx = \int_0^1 \int_0^{\sqrt{x}} 2y(1+x+y)\,dy\,dx$

$= \int_0^1 \left[y^2 + xy^2 + \frac{2}{3}y^3\right]_{y=0}^{y=\sqrt{x}} dx = \int_0^1 \left(x + x^2 + \frac{2}{3}x^{3/2}\right) dx = \left[\frac{1}{2}x^2 + \frac{1}{3}x^3 + \frac{4}{15}x^{5/2}\right]_0^1 = \frac{11}{10}$

$M_{xy} = \iiint_E z\rho(x,y,z)\,dV = \int_0^1 \int_0^{\sqrt{x}} \int_0^{1+x+y} 2z\,dz\,dy\,dx = \int_0^1 \int_0^{\sqrt{x}} \left[z^2\right]_{z=0}^{z=1+x+y} dy\,dx = \int_0^1 \int_0^{\sqrt{x}} (1+x+y)^2\,dy\,dx$

$= \int_0^1 \int_0^{\sqrt{x}} (1 + 2x + 2y + 2xy + x^2 + y^2)\,dy\,dx = \int_0^1 \left[y + 2xy + y^2 + xy^2 + x^2y + \frac{1}{3}y^3\right]_{y=0}^{y=\sqrt{x}} dx$

$= \int_0^1 \left(\sqrt{x} + \frac{7}{3}x^{3/2} + x + x^2 + x^{5/2}\right) dx = \left[\frac{2}{3}x^{3/2} + \frac{14}{15}x^{5/2} + \frac{1}{2}x^2 + \frac{1}{3}x^3 + \frac{2}{7}x^{7/2}\right]_0^1 = \frac{571}{210}$

Thus the mass is $\frac{79}{30}$ and the centre of mass is $(\overline{x}, \overline{y}, \overline{z}) = \left(\frac{M_{yz}}{m}, \frac{M_{xz}}{m}, \frac{M_{xy}}{m}\right) = \left(\frac{358}{553}, \frac{33}{79}, \frac{571}{553}\right)$.

37. $m = \int_0^a \int_0^a \int_0^a (x^2 + y^2 + z^2)\,dx\,dy\,dz = \int_0^a \int_0^a \left[\frac{1}{3}x^3 + xy^2 + xz^2\right]_{x=0}^{x=a} dy\,dz = \int_0^a \int_0^a \left(\frac{1}{3}a^3 + ay^2 + az^2\right) dy\,dz$

$= \int_0^a \left[\frac{1}{3}a^3 y + \frac{1}{3}ay^3 + ayz^2\right]_{y=0}^{y=a} dz = \int_0^a \left(\frac{2}{3}a^4 + a^2z^2\right) dz = \left[\frac{2}{3}a^4 z + \frac{1}{3}a^2z^3\right]_0^a = \frac{2}{3}a^5 + \frac{1}{3}a^5 = a^5$

$M_{yz} = \int_0^a \int_0^a \int_0^a \left[x^3 + x(y^2 + z^2)\right] dx\,dy\,dz = \int_0^a \int_0^a \left[\frac{1}{4}a^4 + \frac{1}{2}a^2(y^2 + z^2)\right] dy\,dz$

$= \int_0^a \left(\frac{1}{4}a^5 + \frac{1}{6}a^5 + \frac{1}{2}a^3z^2\right) dz = \frac{1}{4}a^6 + \frac{1}{3}a^6 = \frac{7}{12}a^6 = M_{xz} = M_{xy}$ by symmetry of E and $\rho(x, y, z)$

Hence $(\overline{x}, \overline{y}, \overline{z}) = \left(\frac{7}{12}a, \frac{7}{12}a, \frac{7}{12}a\right)$.

39. (a) $m = \int_{-3}^{3} \int_{-\sqrt{9-x^2}}^{\sqrt{9-x^2}} \int_1^{5-y} \sqrt{x^2 + y^2}\,dz\,dy\,dx$

(b) $(\overline{x}, \overline{y}, \overline{z}) = \left(\dfrac{M_{yz}}{m}, \dfrac{M_{xz}}{m}, \dfrac{M_{xy}}{m}\right)$ where

$M_{yz} = \int_{-3}^{3} \int_{-\sqrt{9-x^2}}^{\sqrt{9-x^2}} \int_1^{5-y} x\sqrt{x^2 + y^2}\,dz\,dy\,dx$, $M_{xz} = \int_{-3}^{3} \int_{-\sqrt{9-x^2}}^{\sqrt{9-x^2}} \int_1^{5-y} y\sqrt{x^2 + y^2}\,dz\,dy\,dx$, and

$M_{xy} = \int_{-3}^{3} \int_{-\sqrt{9-x^2}}^{\sqrt{9-x^2}} \int_1^{5-y} z\sqrt{x^2 + y^2}\,dz\,dy\,dx$.

(c) $I_z = \int_{-3}^{3} \int_{-\sqrt{9-x^2}}^{\sqrt{9-x^2}} \int_1^{5-y} (x^2 + y^2)\sqrt{x^2 + y^2}\,dz\,dy\,dx = \int_{-3}^{3} \int_{-\sqrt{9-x^2}}^{\sqrt{9-x^2}} \int_1^{5-y} (x^2 + y^2)^{3/2}\,dz\,dy\,dx$

41. (a) $m = \int_0^1 \int_0^{\sqrt{1-x^2}} \int_0^y (1 + x + y + z)\,dz\,dy\,dx = \frac{3\pi}{32} + \frac{11}{24}$

(b) $(\overline{x}, \overline{y}, \overline{z}) = \Big(m^{-1} \int_0^1 \int_0^{\sqrt{1-x^2}} \int_0^y x(1 + x + y + z)\,dz\,dy\,dx,$

$m^{-1} \int_0^1 \int_0^{\sqrt{1-x^2}} \int_0^y y(1 + x + y + z)\,dz\,dy\,dx,$

$m^{-1} \int_0^1 \int_0^{\sqrt{1-x^2}} \int_0^y z(1 + x + y + z)\,dz\,dy\,dx\Big)$

$$= \left(\frac{28}{9\pi + 44}, \frac{30\pi + 128}{45\pi + 220}, \frac{45\pi + 208}{135\pi + 660}\right)$$

(c) $I_z = \displaystyle\int_0^1 \int_0^{\sqrt{1-x^2}} \int_0^y (x^2 + y^2)(1 + x + y + z)\,dz\,dy\,dx = \frac{68 + 15\pi}{240}$

43. $I_x = \int_0^L \int_0^L \int_0^L k(y^2 + z^2)\,dz\,dy\,dx = k\int_0^L \int_0^L \left(Ly^2 + \frac{1}{3}L^3\right) dy\,dx = k\int_0^L \frac{2}{3}L^4\,dx = \frac{2}{3}kL^5$.

By symmetry, $I_x = I_y = I_z = \frac{2}{3}kL^5$.

45. (a) $f(x, y, z)$ is a joint density function, so we know $\iiint_{\mathbb{R}^3} f(x, y, z)\,dV = 1$. Here we have

$$\iiint_{\mathbb{R}^3} f(x, y, z)\,dV = \int_{-\infty}^{\infty} \int_{-\infty}^{\infty} \int_{-\infty}^{\infty} f(x, y, z)\,dz\,dy\,dx = \int_0^2 \int_0^2 \int_0^2 Cxyz\,dz\,dy\,dx$$

$$= C\int_0^2 x\,dx \int_0^2 y\,dy \int_0^2 z\,dz = C\left[\frac{x^2}{2}\right]_0^2 \left[\frac{y^2}{2}\right]_0^2 \left[\frac{z^2}{2}\right]_0^2 = 8C$$

Then we must have $8C = 1 \;\Rightarrow\; C = \frac{1}{8}$.

(b) $P(X \leq 1, Y \leq 1, Z \leq 1) = \int_{-\infty}^{1} \int_{-\infty}^{1} \int_{-\infty}^{1} f(x, y, z)\,dz\,dy\,dx = \int_0^1 \int_0^1 \int_0^1 \frac{1}{8}xyz\,dz\,dy\,dx$

$$= \frac{1}{8}\int_0^1 x\,dx \int_0^1 y\,dy \int_0^1 z\,dz = \frac{1}{8}\left[\frac{x^2}{2}\right]_0^1 \left[\frac{y^2}{2}\right]_0^1 \left[\frac{z^2}{2}\right]_0^1 = \frac{1}{8}\left(\frac{1}{2}\right)^3 = \frac{1}{64}$$

(c) $P(X+Y+Z \le 1) = P((X,Y,Z) \in E)$ where E is the solid region in the first octant bounded by the coordinate planes and the plane $x+y+z=1$. The plane $x+y+z=1$ meets the xy-plane in the line $x+y=1$, so we have

$$P(X+Y+Z \le 1) = \iiint_E f(x,y,z)\,dV = \int_0^1 \int_0^{1-x} \int_0^{1-x-y} \tfrac{1}{8}xyz\,dz\,dy\,dx$$

$$= \tfrac{1}{8}\int_0^1 \int_0^{1-x} xy\left[\tfrac{1}{2}z^2\right]_{z=0}^{z=1-x-y} dy\,dx = \tfrac{1}{16}\int_0^1 \int_0^{1-x} xy(1-x-y)^2\,dy\,dx$$

$$= \tfrac{1}{16}\int_0^1 \int_0^{1-x} \left[(x^3-2x^2+x)y + (2x^2-2x)y^2 + xy^3\right] dy\,dx$$

$$= \tfrac{1}{16}\int_0^1 \left[(x^3-2x^2+x)\tfrac{1}{2}y^2 + (2x^2-2x)\tfrac{1}{3}y^3 + x\left(\tfrac{1}{4}y^4\right)\right]_{y=0}^{y=1-x} dx$$

$$= \tfrac{1}{192}\int_0^1 (x-4x^2+6x^3-4x^4+x^5)\,dx = \tfrac{1}{192}\left(\tfrac{1}{30}\right) = \tfrac{1}{5760}$$

47. $V(E) = L^3$,

$$f_{\text{ave}} = \frac{1}{L^3}\int_0^L \int_0^L \int_0^L xyz\,dx\,dy\,dz = \frac{1}{L^3}\int_0^L x\,dx \int_0^L y\,dy \int_0^L z\,dz$$

$$= \frac{1}{L^3}\left[\frac{x^2}{2}\right]_0^L \left[\frac{y^2}{2}\right]_0^L \left[\frac{z^2}{2}\right]_0^L = \frac{1}{L^3}\frac{L^2}{2}\frac{L^2}{2}\frac{L^2}{2} = \frac{L^3}{8}$$

49. The triple integral will attain its maximum when the integrand $1-x^2-2y^2-3z^2$ is positive in the region E and negative everywhere else. For if E contains some region F where the integrand is negative, the integral could be increased by excluding F from E, and if E fails to contain some part G of the region where the integrand is positive, the integral could be increased by including G in E. So we require that $x^2+2y^2+3z^2 \le 1$. This describes the region bounded by the ellipsoid $x^2+2y^2+3z^2=1$.

12.8 Triple Integrals in Cylindrical and Spherical Coordinates

1.

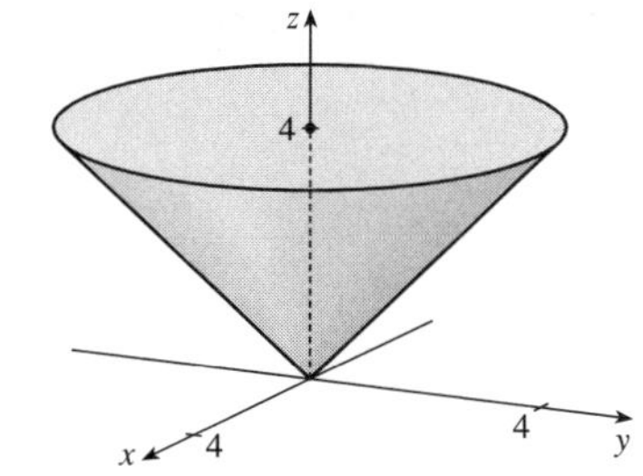

The region of integration is given in cylindrical coordinates by $E = \{(r,\theta,z) \mid 0 \le \theta \le 2\pi, 0 \le r \le 4, r \le z \le 4\}$. This represents the solid region bounded below by the cone $z=r$ and above by the horizontal plane $z=4$.

$$\int_0^4 \int_0^{2\pi} \int_r^4 r\,dz\,d\theta\,dr = \int_0^4 \int_0^{2\pi} \left[rz\right]_{z=r}^{z=4} d\theta\,dr = \int_0^4 \int_0^{2\pi} r(4-r)\,d\theta\,dr$$

$$= \int_0^4 (4r-r^2)\,dr \int_0^{2\pi} d\theta = \left[2r^2 - \tfrac{1}{3}r^3\right]_0^4 \left[\theta\right]_0^{2\pi}$$

$$= \left(32 - \tfrac{64}{3}\right)(2\pi) = \tfrac{64\pi}{3}$$

3.

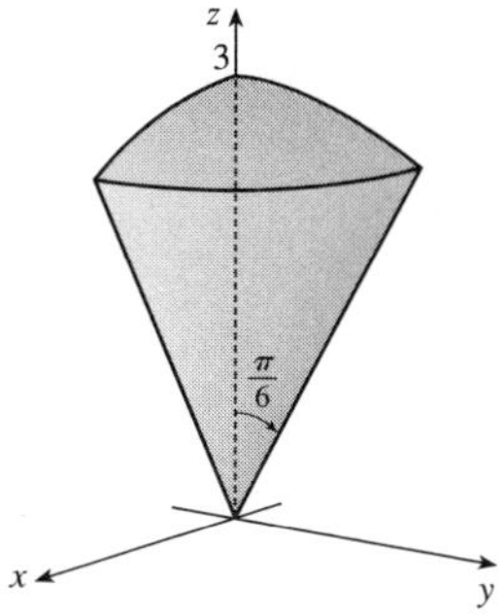

The region of integration is given in spherical coordinates by $E = \{(\rho,\theta,\phi) \mid 0 \le \rho \le 3, 0 \le \theta \le \pi/2, 0 \le \phi \le \pi/6\}$. This represents the solid region in the first octant bounded above by the sphere $\rho = 3$ and below by the cone $\phi = \pi/6$.

$$\int_0^{\pi/6} \int_0^{\pi/2} \int_0^3 \rho^2 \sin\phi\,d\rho\,d\theta\,d\phi = \int_0^{\pi/6} \sin\phi\,d\phi \int_0^{\pi/2} d\theta \int_0^3 \rho^2\,d\rho$$

$$= \left[-\cos\phi\right]_0^{\pi/6} \left[\theta\right]_0^{\pi/2} \left[\tfrac{1}{3}\rho^3\right]_0^3$$

$$= \left(1 - \frac{\sqrt{3}}{2}\right)\left(\frac{\pi}{2}\right)(9) = \frac{9\pi}{4}\left(2-\sqrt{3}\right)$$

5. The solid E is most conveniently described if we use cylindrical coordinates:

$E = \{(r, \theta, z) \mid 0 \le \theta \le \frac{\pi}{2}, 0 \le r \le 3, 0 \le z \le 2\}$. Then $\iiint_E f(x, y, z)\, dV = \int_0^{\pi/2} \int_0^3 \int_0^2 f(r\cos\theta, r\sin\theta, z)\, r\, dz\, dr\, d\theta$.

7. In cylindrical coordinates, E is given by $\{(r, \theta, z) \mid 0 \le \theta \le 2\pi, 0 \le r \le 4, -5 \le z \le 4\}$. So

$$\iiint_E \sqrt{x^2 + y^2}\, dV = \int_0^{2\pi} \int_0^4 \int_{-5}^4 \sqrt{r^2}\, r\, dz\, dr\, d\theta = \int_0^{2\pi} d\theta \int_0^4 r^2\, dr \int_{-5}^4 dz$$
$$= \left[\theta\right]_0^{2\pi} \left[\tfrac{1}{3}r^3\right]_0^4 \left[z\right]_{-5}^4 = (2\pi)\left(\tfrac{64}{3}\right)(9) = 384\pi$$

9. In cylindrical coordinates E is bounded by the paraboloid $z = 1 + r^2$, the cylinder $r^2 = 5$ or $r = \sqrt{5}$, and the xy-plane, so E is given by $\{(r, \theta, z) \mid 0 \le \theta \le 2\pi, 0 \le r \le \sqrt{5}, 0 \le z \le 1 + r^2\}$. Thus

$$\iiint_E e^z\, dV = \int_0^{2\pi} \int_0^{\sqrt{5}} \int_0^{1+r^2} e^z\, r\, dz\, dr\, d\theta = \int_0^{2\pi} \int_0^{\sqrt{5}} r\left[e^z\right]_{z=0}^{z=1+r^2} dr\, d\theta = \int_0^{2\pi} \int_0^{\sqrt{5}} r(e^{1+r^2} - 1)\, dr\, d\theta$$
$$= \int_0^{2\pi} d\theta \int_0^{\sqrt{5}} \left(re^{1+r^2} - r\right) dr = 2\pi \left[\tfrac{1}{2}e^{1+r^2} - \tfrac{1}{2}r^2\right]_0^{\sqrt{5}} = \pi(e^6 - e - 5)$$

11. In cylindrical coordinates, E is bounded by the cylinder $r = 1$, the plane $z = 0$, and the cone $z = 2r$. So $E = \{(r, \theta, z) \mid 0 \le \theta \le 2\pi, 0 \le r \le 1, 0 \le z \le 2r\}$ and

$$\iiint_E x^2\, dV = \int_0^{2\pi} \int_0^1 \int_0^{2r} r^2 \cos^2\theta\, r\, dz\, dr\, d\theta = \int_0^{2\pi} \int_0^1 \left[r^3 \cos^2\theta\, z\right]_{z=0}^{z=2r} dr\, d\theta$$
$$= \int_0^{2\pi} \int_0^1 2r^4 \cos^2\theta\, dr\, d\theta = \int_0^{2\pi} \left[\tfrac{2}{5}r^5 \cos^2\theta\right]_{r=0}^{r=1} d\theta = \tfrac{2}{5} \int_0^{2\pi} \cos^2\theta\, d\theta$$
$$= \frac{2}{5} \int_0^{2\pi} \frac{1 + \cos 2\theta}{2}\, d\theta = \frac{1}{5}\left[\theta + \frac{1}{2}\sin 2\theta\right]_0^{2\pi} = \frac{2\pi}{5}$$

13. (a) The paraboloids intersect when $x^2 + y^2 = 36 - 3x^2 - 3y^2 \Rightarrow x^2 + y^2 = 9$, so the region of integration is $D = \{(x, y) \mid x^2 + y^2 \le 9\}$. Then, in cylindrical coordinates, $E = \{(r, \theta, z) \mid r^2 \le z \le 36 - 3r^2, 0 \le r \le 3, 0 \le \theta \le 2\pi\}$ and

$$V = \int_0^{2\pi} \int_0^3 \int_{r^2}^{36 - 3r^2} r\, dz\, dr\, d\theta = \int_0^{2\pi} \int_0^3 (36r - 4r^3)\, dr\, d\theta$$
$$= \int_0^{2\pi} \left[18r^2 - r^4\right]_{r=0}^{r=3} d\theta = \int_0^{2\pi} 81\, d\theta = 162\pi$$

(b) For constant density K, $m = KV = 162\pi K$ from part (a). Since the region is homogeneous and symmetric, $M_{yz} = M_{xz} = 0$ and

$$M_{xy} = \int_0^{2\pi} \int_0^3 \int_{r^2}^{36-3r^2} (zK)\, r\, dz\, dr\, d\theta = K \int_0^{2\pi} \int_0^3 r\left[\tfrac{1}{2}z^2\right]_{z=r^2}^{z=36-3r^2} dr\, d\theta$$
$$= \tfrac{K}{2} \int_0^{2\pi} \int_0^3 r((36 - 3r^2)^2 - r^4)\, dr\, d\theta = \tfrac{K}{2} \int_0^{2\pi} d\theta \int_0^3 (8r^5 - 216r^3 + 1296r)\, dr$$
$$= \tfrac{K}{2}(2\pi)\left[\tfrac{8}{6}r^6 - \tfrac{216}{4}r^4 + \tfrac{1296}{2}r^2\right]_0^3 = \pi K(2430) = 2430\pi K$$

Thus $(\bar{x}, \bar{y}, \bar{z}) = \left(\dfrac{M_{yz}}{m}, \dfrac{M_{xz}}{m}, \dfrac{M_{xy}}{m}\right) = \left(0, 0, \frac{2430\pi K}{162\pi K}\right) = (0, 0, 15)$.

15. The paraboloid $z = 4x^2 + 4y^2$ intersects the plane $z = a$ when $a = 4x^2 + 4y^2$ or $x^2 + y^2 = \frac{1}{4}a$. So, in cylindrical coordinates, $E = \{(r, \theta, z) \mid 0 \le r \le \frac{1}{2}\sqrt{a}, 0 \le \theta \le 2\pi, 4r^2 \le z \le a\}$. Thus

$$m = \int_0^{2\pi} \int_0^{\sqrt{a}/2} \int_{4r^2}^a Kr\, dz\, dr\, d\theta = K \int_0^{2\pi} \int_0^{\sqrt{a}/2} (ar - 4r^3)\, dr\, d\theta$$
$$= K \int_0^{2\pi} \left[\tfrac{1}{2}ar^2 - r^4\right]_{r=0}^{r=\sqrt{a}/2} d\theta = K \int_0^{2\pi} \tfrac{1}{16}a^2\, d\theta = \tfrac{1}{8}a^2\pi K$$

Since the region is homogeneous and symmetric, $M_{yz} = M_{xz} = 0$ and

$$M_{xy} = \int_0^{2\pi} \int_0^{\sqrt{a}/2} \int_{4r^2}^a Krz\, dz\, dr\, d\theta = K \int_0^{2\pi} \int_0^{\sqrt{a}/2} \left(\tfrac{1}{2}a^2 r - 8r^5\right) dr\, d\theta$$
$$= K \int_0^{2\pi} \left[\tfrac{1}{4}a^2r^2 - \tfrac{4}{3}r^6\right]_{r=0}^{r=\sqrt{a}/2} d\theta = K \int_0^{2\pi} \tfrac{1}{24}a^3\, d\theta = \tfrac{1}{12}a^3\pi K$$

Hence $(\bar{x}, \bar{y}, \bar{z}) = \left(0, 0, \frac{2}{3}a\right)$.

17. In spherical coordinates, B is represented by $\{(\rho,\theta,\phi)\,|\,0\le\rho\le 1, 0\le\theta\le 2\pi, 0\le\phi\le\pi\}$. Thus

$$\iiint_B (x^2+y^2+z^2)\,dV = \int_0^\pi\int_0^{2\pi}\int_0^1 (\rho^2)\rho^2\sin\phi\,d\rho\,d\theta\,d\phi = \int_0^\pi \sin\phi\,d\phi\int_0^{2\pi} d\theta\int_0^1 \rho^4\,d\rho$$
$$= \big[-\cos\phi\big]_0^\pi\,\big[\,\theta\,\big]_0^{2\pi}\,\big[\tfrac15\rho^5\big]_0^1 = (2)(2\pi)\big(\tfrac15\big) = \tfrac{4\pi}{5}$$

19. In spherical coordinates, E is represented by $\{(\rho,\theta,\phi)\,|\,1\le\rho\le 2, 0\le\theta\le\frac{\pi}{2}, 0\le\phi\le\frac{\pi}{2}\}$. Thus

$$\iiint_E z\,dV = \int_0^{\pi/2}\int_0^{\pi/2}\int_1^2 (\rho\cos\phi)\,\rho^2\sin\phi\,d\rho\,d\theta\,d\phi$$
$$= \int_0^{\pi/2}\cos\phi\sin\phi\,d\phi\int_0^{\pi/2}d\theta\int_1^2\rho^3\,d\rho = \big[\tfrac12\sin^2\phi\big]_0^{\pi/2}\,\big[\,\theta\,\big]_0^{\pi/2}\,\big[\tfrac14\rho^4\big]_1^2$$
$$= \big(\tfrac12\big)\big(\tfrac{\pi}{2}\big)\big(\tfrac{15}{4}\big) = \tfrac{15\pi}{16}$$

21. $\iiint_E x^2\,dV = \int_0^\pi\int_0^\pi\int_3^4 (\rho\sin\phi\cos\theta)^2\,\rho^2\sin\phi\,d\rho\,d\phi\,d\theta = \int_0^\pi\cos^2\theta\,d\theta\int_0^\pi\sin^3\phi\,d\phi\int_3^4\rho^4\,d\rho$

$$= \big[\tfrac12\theta+\tfrac14\sin 2\theta\big]_0^\pi\,\big[-\tfrac13(2+\sin^2\phi)\cos\phi\big]_0^\pi\,\big[\tfrac15\rho^5\big]_3^4 = \big(\tfrac{\pi}{2}\big)\big(\tfrac23+\tfrac23\big)\tfrac15(4^5-3^5) = \tfrac{1562}{15}\pi$$

23. (a) Since $\rho = 4\cos\phi$ implies $\rho^2 = 4\rho\cos\phi$, the equation is that of a sphere of radius 2 with centre at $(0,0,2)$. Thus

$$V = \int_0^{2\pi}\int_0^{\pi/3}\int_0^{4\cos\phi}\rho^2\sin\phi\,d\rho\,d\phi\,d\theta = \int_0^{2\pi}\int_0^{\pi/3}\big[\tfrac13\rho^3\big]_{\rho=0}^{\rho=4\cos\phi}\sin\phi\,d\phi\,d\theta = \int_0^{2\pi}\int_0^{\pi/3}\big(\tfrac{64}{3}\cos^3\phi\big)\sin\phi\,d\phi\,d\theta$$
$$= \int_0^{2\pi}\big[-\tfrac{16}{3}\cos^4\phi\big]_{\phi=0}^{\phi=\pi/3}\,d\theta = \int_0^{2\pi}-\tfrac{16}{3}\big(\tfrac{1}{16}-1\big)\,d\theta = 5\theta\Big]_0^{2\pi} = 10\pi$$

(b) By the symmetry of the problem $M_{yz} = M_{xz} = 0$. Then

$$M_{xy} = \int_0^{2\pi}\int_0^{\pi/3}\int_0^{4\cos\phi}\rho^3\cos\phi\sin\phi\,d\rho\,d\phi\,d\theta = \int_0^{2\pi}\int_0^{\pi/3}\cos\phi\sin\phi\,\big(64\cos^4\phi\big)\,d\phi\,d\theta$$
$$= \int_0^{2\pi}64\big[-\tfrac16\cos^6\phi\big]_{\phi=0}^{\phi=\pi/3}\,d\theta = \int_0^{2\pi}\tfrac{21}{2}\,d\theta = 21\pi$$

Hence $(\overline{x},\overline{y},\overline{z}) = (0,0,2.1)$.

25. (a) The density function is $\rho(x,y,z) = K$, a constant, and by the symmetry of the problem $M_{xz} = M_{yz} = 0$. Then $M_{xy} = \int_0^{2\pi}\int_0^{\pi/2}\int_0^a K\rho^3\sin\phi\cos\phi\,d\rho\,d\phi\,d\theta = \frac12\pi Ka^4\int_0^{\pi/2}\sin\phi\cos\phi\,d\phi = \frac18\pi Ka^4$. But the mass is K(volume of the hemisphere) $= \frac23\pi Ka^3$, so the centroid is $\big(0,0,\frac38 a\big)$.

(b) Place the centre of the base at $(0,0,0)$; the density function is $\rho(x,y,z) = K$. By symmetry, the moments of inertia about any two such diameters will be equal, so we just need to find I_x:

$$I_x = \int_0^{2\pi}\int_0^{\pi/2}\int_0^a \big(K\rho^2\sin\phi\big)\,\rho^2\,\big(\sin^2\phi\sin^2\theta+\cos^2\phi\big)\,d\rho\,d\phi\,d\theta$$
$$= K\int_0^{2\pi}\int_0^{\pi/2}\big(\sin^3\phi\sin^2\theta+\sin\phi\cos^2\phi\big)\big(\tfrac15 a^5\big)\,d\phi\,d\theta$$
$$= \tfrac15 Ka^5\int_0^{2\pi}\big[\sin^2\theta\,\big(-\cos\phi+\tfrac13\cos^3\phi\big)+\big(-\tfrac13\cos^3\phi\big)\big]_{\phi=0}^{\phi=\pi/2}\,d\theta = \tfrac15 Ka^5\int_0^{2\pi}\big[\tfrac23\sin^2\theta+\tfrac13\big]\,d\theta$$
$$= \tfrac15 Ka^5\big[\tfrac23\big(\tfrac12\theta-\tfrac14\sin 2\theta\big)+\tfrac13\theta\big]_0^{2\pi} = \tfrac15 Ka^5\big[\tfrac23(\pi-0)+\tfrac13(2\pi-0)\big] = \tfrac{4}{15}Ka^5\pi$$

27. In spherical coordinates $z = \sqrt{x^2+y^2}$ becomes $\cos\phi = \sin\phi$ or $\phi = \frac{\pi}{4}$. Then

$V = \int_0^{2\pi}\int_0^{\pi/4}\int_0^1 \rho^2\sin\phi\,d\rho\,d\phi\,d\theta = \int_0^{2\pi}d\theta\int_0^{\pi/4}\sin\phi\,d\phi\int_0^1\rho^2\,d\rho = \frac13\pi\big(2-\sqrt2\big)$,

$M_{xy} = \int_0^{2\pi}\int_0^{\pi/4}\int_0^1 \rho^3\sin\phi\cos\phi\,d\rho\,d\phi\,d\theta = 2\pi\big[-\frac14\cos 2\phi\big]_0^{\pi/4}\big(\frac14\big) = \frac{\pi}{8}$ and by symmetry $M_{yz} = M_{xz} = 0$.

Hence $(\overline{x},\overline{y},\overline{z}) = \left(0,0,\dfrac{3}{8(2-\sqrt2)}\right)$.

29. In cylindrical coordinates the paraboloid is given by $z = r^2$ and the plane by $z = 2r\sin\theta$ and they intersect in the circle $r = 2\sin\theta$. Then $\iiint_E z\,dV = \int_0^\pi \int_0^{2\sin\theta} \int_{r^2}^{2r\sin\theta} rz\,dz\,dr\,d\theta = \frac{5\pi}{6}$ [using a CAS].

31. The region of integration is the region above the cone $z = \sqrt{x^2+y^2}$, or $z = r$, and below the plane $z = 2$. Also, we have $-2 \le y \le 2$ with $-\sqrt{4-y^2} \le x \le \sqrt{4-y^2}$ which describes a circle of radius 2 in the xy-plane centred at $(0,0)$. Thus,

$$\int_{-2}^{2}\int_{-\sqrt{4-y^2}}^{\sqrt{4-y^2}}\int_{\sqrt{x^2+y^2}}^{2} xz\,dz\,dx\,dy = \int_0^{2\pi}\int_0^2\int_r^2 (r\cos\theta)\,z\,r\,dz\,dr\,d\theta = \int_0^{2\pi}\int_0^2\int_r^2 r^2(\cos\theta)\,z\,dz\,dr\,d\theta$$

$$= \int_0^{2\pi}\int_0^2 r^2(\cos\theta)\left[\tfrac{1}{2}z^2\right]_{z=r}^{z=2} dr\,d\theta = \tfrac{1}{2}\int_0^{2\pi}\int_0^2 r^2(\cos\theta)(4-r^2)\,dr\,d\theta$$

$$= \tfrac{1}{2}\int_0^{2\pi}\cos\theta\,d\theta\int_0^2\left(4r^2-r^4\right)dr = \tfrac{1}{2}\left[\sin\theta\right]_0^{2\pi}\left[\tfrac{4}{3}r^3-\tfrac{1}{5}r^5\right]_0^2 = 0$$

33. The region E of integration is the region above the cone $z = \sqrt{x^2+y^2}$ and below the sphere $x^2+y^2+z^2 = 2$ in the first octant. Because E is in the first octant we have $0 \le \theta \le \frac{\pi}{2}$. The cone has equation $\phi = \frac{\pi}{4}$ (as in Example 4), so $0 \le \phi \le \frac{\pi}{4}$, and $0 \le \rho \le \sqrt{2}$. So the integral becomes

$$\int_0^{\pi/4}\int_0^{\pi/2}\int_0^{\sqrt{2}} (\rho\sin\phi\cos\theta)\,(\rho\sin\phi\sin\theta)\,\rho^2\sin\phi\,d\rho\,d\theta\,d\phi$$

$$= \int_0^{\pi/4}\sin^3\phi\,d\phi\int_0^{\pi/2}\sin\theta\cos\theta\,d\theta\int_0^{\sqrt{2}}\rho^4\,d\rho = \left(\int_0^{\pi/4}\left(1-\cos^2\phi\right)\sin\phi\,d\phi\right)\left[\tfrac{1}{2}\sin^2\theta\right]_0^{\pi/2}\left[\tfrac{1}{5}\rho^5\right]_0^{\sqrt{2}}$$

$$= \left[\tfrac{1}{3}\cos^3\phi-\cos\phi\right]_0^{\pi/4}\cdot\tfrac{1}{2}\cdot\tfrac{1}{5}\left(\sqrt{2}\right)^5 = \left[\tfrac{\sqrt{2}}{12}-\tfrac{\sqrt{2}}{2}-\left(\tfrac{1}{3}-1\right)\right]\cdot\tfrac{2\sqrt{2}}{5} = \tfrac{4\sqrt{2}-5}{15}$$

35. If E is the solid enclosed by the surface $\rho = 1 + \frac{1}{5}\sin 6\theta\sin 5\phi$, it can be described in spherical coordinates as $E = \left\{(\rho,\theta,\phi) \mid 0 \le \rho \le 1 + \frac{1}{5}\sin 6\theta\sin 5\phi, 0 \le \theta \le 2\pi, 0 \le \phi \le \pi\right\}$. Its volume is given by $V(E) = \iiint_E dV = \int_0^\pi\int_0^{2\pi}\int_0^{1+(\sin 6\theta\sin 5\phi)/5}\rho^2\sin\phi\,d\rho\,d\theta\,d\phi = \frac{136\pi}{99}$ [using a CAS].

37. (a) The mountain comprises a solid conical region C. The work done in lifting a small volume of material ΔV with density $g(P)$ to a height $h(P)$ above sea level is $h(P)g(P)\,\Delta V$. Summing over the whole mountain we get $W = \iiint_C h(P)g(P)\,\Delta V$.

(b) Here C is a solid right circular cone with radius $R = 19\,000$ m, height $H = 3800$ m, and density $g(P) = 3200\text{ kg/m}^3$ at all points P in C. We use cylindrical coordinates:

$$W = \int_0^{2\pi}\int_0^H\int_0^{R(1-z/H)} z\cdot 3200r\,dr\,dz\,d\theta = 2\pi\int_0^H 3200z\left[\tfrac{1}{2}r^2\right]_{r=0}^{r=R(1-z/H)}dz$$

$$= 6400\pi\int_0^H z\,\frac{R^2}{2}\left(1-\frac{z}{H}\right)^2 dz = 3200\pi R^2\int_0^H\left(z-\frac{2z^2}{H}+\frac{z^3}{H^2}\right)dz$$

$$= 3200\pi R^2\left[\frac{z^2}{2}-\frac{2z^3}{3H}+\frac{z^4}{4H^2}\right]_0^H = 3200\pi R^2\left(\frac{H^2}{2}-\frac{2H^2}{3}+\frac{H^2}{4}\right)$$

$$= 3200\pi R^2\left(\tfrac{1}{12}H^2\right) = \tfrac{800}{3}\pi R^2H^2 = \tfrac{800}{3}\pi(19\,000)^2(3800)^2$$

$$\approx 4.4\times 10^{18}\text{ J}$$

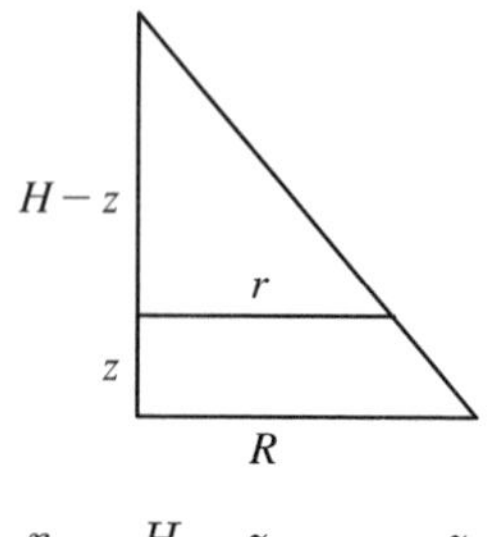

$$\frac{r}{R} = \frac{H-z}{H} = 1-\frac{z}{H}$$

12.9 Change of Variables in Multiple Integrals

1. $x = u + 4v$, $y = 3u - 2v$. The Jacobian is $\dfrac{\partial(x,y)}{\partial(u,v)} = \begin{vmatrix} \partial x/\partial u & \partial x/\partial v \\ \partial y/\partial u & \partial y/\partial v \end{vmatrix} = \begin{vmatrix} 1 & 4 \\ 3 & -2 \end{vmatrix} = 1(-2) - 4(3) = -14.$

3. $$\frac{\partial(x,y)}{\partial(u,v)} = \begin{vmatrix} \dfrac{\partial x}{\partial u} & \dfrac{\partial x}{\partial v} \\ \dfrac{\partial y}{\partial u} & \dfrac{\partial y}{\partial v} \end{vmatrix} = \begin{vmatrix} \dfrac{v}{(u+v)^2} & -\dfrac{u}{(u+v)^2} \\ -\dfrac{v}{(u-v)^2} & \dfrac{u}{(u-v)^2} \end{vmatrix} = \frac{uv}{(u+v)^2(u-v)^2} - \frac{uv}{(u+v)^2(u-v)^2} = 0$$

5. $$\frac{\partial(x,y,z)}{\partial(u,v,w)} = \begin{vmatrix} \partial x/\partial u & \partial x/\partial v & \partial x/\partial w \\ \partial y/\partial u & \partial y/\partial v & \partial y/\partial w \\ \partial z/\partial u & \partial z/\partial v & \partial z/\partial w \end{vmatrix} = \begin{vmatrix} v & u & 0 \\ 0 & w & v \\ w & 0 & u \end{vmatrix} = v\begin{vmatrix} w & v \\ 0 & u \end{vmatrix} - u\begin{vmatrix} 0 & v \\ w & u \end{vmatrix} + 0\begin{vmatrix} 0 & w \\ w & 0 \end{vmatrix}$$
$$= v(uw - 0) - u(0 - vw) = 2uvw$$

7. The transformation maps the boundary of S to the boundary of the image R, so we first look at side S_1 in the uv-plane. S_1 is described by $v = 0$ $(0 \le u \le 3)$, so $x = 2u + 3v = 2u$ and $y = u - v = u$. Eliminating u, we have $x = 2y$, $0 \le x \le 6$. S_2 is the line segment $u = 3$, $0 \le v \le 2$, so $x = 6 + 3v$ and $y = 3 - v$. Then $v = 3 - y \;\Rightarrow\; x = 6 + 3(3 - y) = 15 - 3y$, $6 \le x \le 12$. S_3 is the line segment $v = 2$, $0 \le u \le 3$, so $x = 2u + 6$ and $y = u - 2$, giving $u = y + 2 \;\Rightarrow\; x = 2y + 10$, $6 \le x \le 12$. Finally, S_4 is the segment $u = 0$, $0 \le v \le 2$, so $x = 3v$ and $y = -v \;\Rightarrow\; x = -3y$, $0 \le x \le 6$. The image of set S is the region R shown in the xy-plane, a parallelogram bounded by these four segments.

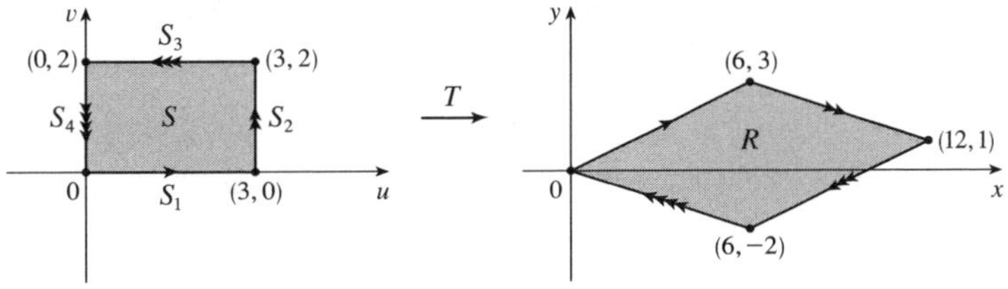

9. S_1 is the line segment $u = v$, $0 \le u \le 1$, so $y = v = u$ and $x = u^2 = y^2$. Since $0 \le u \le 1$, the image is the portion of the parabola $x = y^2$, $0 \le y \le 1$. S_2 is the segment $v = 1$, $0 \le u \le 1$, thus $y = v = 1$ and $x = u^2$, so $0 \le x \le 1$. The image is the line segment $y = 1$, $0 \le x \le 1$. S_3 is the segment $u = 0$, $0 \le v \le 1$, so $x = u^2 = 0$ and $y = v \;\Rightarrow\; 0 \le y \le 1$. The image is the segment $x = 0$, $0 \le y \le 1$. Thus, the image of S is the region R in the first quadrant bounded by the parabola $x = y^2$, the y-axis, and the line $y = 1$.

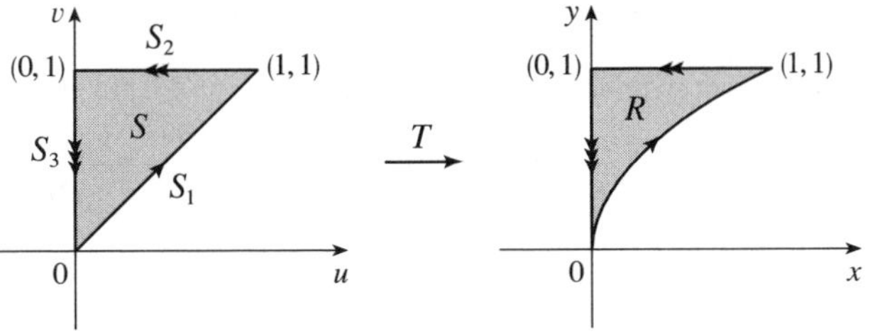

11. $\dfrac{\partial(x,y)}{\partial(u,v)} = \begin{vmatrix} 2 & 1 \\ 1 & 2 \end{vmatrix} = 3$ and $x - 3y = (2u + v) - 3(u + 2v) = -u - 5v$. To find the region S in the uv-plane that corresponds to R we first find the corresponding boundary under the given transformation. The line through $(0,0)$ and $(2,1)$ is $y = \frac{1}{2}x$ which is the image of $u + 2v = \frac{1}{2}(2u + v) \;\Rightarrow\; v = 0$; the line through $(2,1)$ and $(1,2)$ is $x + y = 3$ which is the image of $(2u + v) + (u + 2v) = 3 \;\Rightarrow\; u + v = 1$; the line through $(0,0)$ and $(1,2)$ is $y = 2x$ which is the image of $u + 2v = 2(2u + v) \;\Rightarrow\; u = 0$. Thus S is the triangle $0 \le v \le 1 - u$, $0 \le u \le 1$ in the uv-plane and

$$\iint_R (x - 3y)\,dA = \int_0^1 \int_0^{1-u} (-u - 5v)\,|3|\,dv\,du = -3\int_0^1 \left[uv + \tfrac{5}{2}v^2\right]_{v=0}^{v=1-u} du$$
$$= -3\int_0^1 \left(u - u^2 + \tfrac{5}{2}(1-u)^2\right) du = -3\left[\tfrac{1}{2}u^2 - \tfrac{1}{3}u^3 - \tfrac{5}{6}(1-u)^3\right]_0^1 = -3\left(\tfrac{1}{2} - \tfrac{1}{3} + \tfrac{5}{6}\right) = -3$$

13. $\dfrac{\partial(x,y)}{\partial(u,v)} = \begin{vmatrix} 2 & 0 \\ 0 & 3 \end{vmatrix} = 6$, $x^2 = 4u^2$ and the planar ellipse $9x^2 + 4y^2 \le 36$ is the image of the disc $u^2 + v^2 \le 1$. Thus

$$\iint_R x^2\,dA = \iint\limits_{u^2+v^2\le 1} (4u^2)(6)\,du\,dv = \int_0^{2\pi}\int_0^1 (24r^2\cos^2\theta)\,r\,dr\,d\theta = 24\int_0^{2\pi}\cos^2\theta\,d\theta\int_0^1 r^3\,dr$$
$$= 24\left[\tfrac{1}{2}x + \tfrac{1}{4}\sin 2x\right]_0^{2\pi}\left[\tfrac{1}{4}r^4\right]_0^1 = 24(\pi)\left(\tfrac{1}{4}\right) = 6\pi$$

15. $\dfrac{\partial(x,y)}{\partial(u,v)} = \begin{vmatrix} 1/v & -u/v^2 \\ 0 & 1 \end{vmatrix} = \dfrac{1}{v}$, $xy = u$, $y = x$ is the image of the parabola $v^2 = u$, $y = 3x$ is the image of the parabola $v^2 = 3u$, and the hyperbolas $xy = 1$, $xy = 3$ are the images of the lines $u = 1$ and $u = 3$ respectively. Thus

$$\iint_R xy\,dA = \int_1^3\int_{\sqrt{u}}^{\sqrt{3u}} u\left(\frac{1}{v}\right)dv\,du = \int_1^3 u\left(\ln\sqrt{3u} - \ln\sqrt{u}\right)du = \int_1^3 u\ln\sqrt{3}\,du = 4\ln\sqrt{3} = 2\ln 3.$$

17. (a) $\dfrac{\partial(x,y,z)}{\partial(u,v,w)} = \begin{vmatrix} a & 0 & 0 \\ 0 & b & 0 \\ 0 & 0 & c \end{vmatrix} = abc$ and since $u = \dfrac{x}{a}$, $v = \dfrac{y}{b}$, $w = \dfrac{z}{c}$ the solid enclosed by the ellipsoid is the image of the ball $u^2 + v^2 + w^2 \le 1$. So

$$\iiint_E dV = \iiint\limits_{u^2+v^2+w^2\le 1} abc\,du\,dv\,dw = (abc)(\text{volume of the ball}) = \tfrac{4}{3}\pi abc$$

(b) If we approximate the surface of the Earth by the ellipsoid $\dfrac{x^2}{6378^2} + \dfrac{y^2}{6378^2} + \dfrac{z^2}{6356^2} = 1$, then we can estimate the volume of the Earth by finding the volume of the solid E enclosed by the ellipsoid. From part (a), this is $\iiint_E dV = \frac{4}{3}\pi(6378)(6378)(6356) \approx 1.083 \times 10^{12}\ \text{km}^3$.

19. Letting $u = x - 2y$ and $v = 3x - y$, we have $x = \frac{1}{5}(2v - u)$ and $y = \frac{1}{5}(v - 3u)$. Then $\dfrac{\partial(x,y)}{\partial(u,v)} = \begin{vmatrix} -1/5 & 2/5 \\ -3/5 & 1/5 \end{vmatrix} = \dfrac{1}{5}$ and R is the image of the rectangle enclosed by the lines $u = 0$, $u = 4$, $v = 1$, and $v = 8$. Thus

$$\iint_R \frac{x-2y}{3x-y}\,dA = \int_0^4\int_1^8 \frac{u}{v}\left|\frac{1}{5}\right|dv\,du = \frac{1}{5}\int_0^4 u\,du\int_1^8 \frac{1}{v}\,dv = \tfrac{1}{5}\left[\tfrac{1}{2}u^2\right]_0^4\left[\ln|v|\right]_1^8 = \tfrac{8}{5}\ln 8.$$

21. Letting $u = y - x$, $v = y + x$, we have $y = \frac{1}{2}(u + v)$, $x = \frac{1}{2}(v - u)$. Then $\dfrac{\partial(x,y)}{\partial(u,v)} = \begin{vmatrix} -1/2 & 1/2 \\ 1/2 & 1/2 \end{vmatrix} = -\dfrac{1}{2}$ and R is the image of the trapezoidal region with vertices $(-1,1)$, $(-2,2)$, $(2,2)$, and $(1,1)$. Thus

$$\iint_R \cos\frac{y-x}{y+x}\,dA = \int_1^2\int_{-v}^{v}\cos\frac{u}{v}\left|-\frac{1}{2}\right|du\,dv = \frac{1}{2}\int_1^2\left[v\sin\frac{u}{v}\right]_{u=-v}^{u=v}dv$$
$$= \tfrac{1}{2}\int_1^2 2v\sin(1)\,dv = \tfrac{3}{2}\sin 1$$

23. Let $u = x + y$ and $v = -x + y$. Then $u + v = 2y \;\Rightarrow\; y = \frac{1}{2}(u + v)$ and

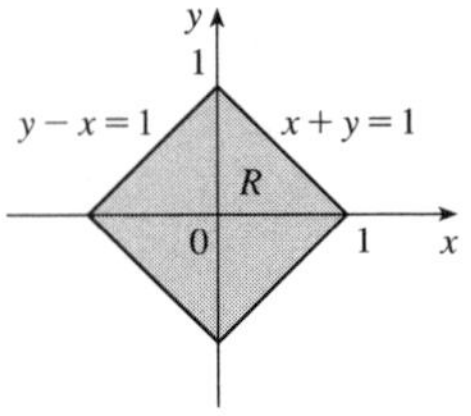

$u - v = 2x \;\Rightarrow\; x = \frac{1}{2}(u - v)$. $\dfrac{\partial(x,y)}{\partial(u,v)} = \begin{vmatrix} 1/2 & -1/2 \\ 1/2 & 1/2 \end{vmatrix} = \dfrac{1}{2}$.

Now $|u| = |x + y| \le |x| + |y| \le 1 \;\Rightarrow\; -1 \le u \le 1$, and $|v| = |-x + y| \le |x| + |y| \le 1 \;\Rightarrow\; -1 \le v \le 1$. R is the image of the square region with vertices $(1,1)$, $(1,-1)$, $(-1,-1)$, and $(-1,1)$. So $\iint_R e^{x+y}\,dA = \frac{1}{2}\int_{-1}^1\int_{-1}^1 e^u\,du\,dv = \frac{1}{2}\left[e^u\right]_{-1}^1\left[v\right]_{-1}^1 = e - e^{-1}$.

12 Review

CONCEPT CHECK

1. (a) A double Riemann sum of f is $\sum_{i=1}^{m} \sum_{j=1}^{n} f(x_{ij}^*, y_{ij}^*)\,\Delta A$, where ΔA is the area of each subrectangle and (x_{ij}^*, y_{ij}^*) is a sample point in each subrectangle. If $f(x, y) \geq 0$, this sum represents an approximation to the volume of the solid that lies above the rectangle R and below the graph of f.

   (b) $\iint_R f(x, y)\,dA = \lim_{m,n\to\infty} \sum_{i=1}^{m} \sum_{j=1}^{n} f(x_{ij}^*, y_{ij}^*)\,\Delta A$

   (c) If $f(x, y) \geq 0$, $\iint_R f(x, y)\,dA$ represents the volume of the solid that lies above the rectangle R and below the surface $z = f(x, y)$. If f takes on both positive and negative values, $\iint_R f(x, y)\,dA$ is the difference of the volume above R but below the surface $z = f(x, y)$ and the volume below R but above the surface $z = f(x, y)$.

   (d) We usually evaluate $\iint_R f(x, y)\,dA$ as an iterated integral according to Fubini's Theorem (see Theorem 12.2.4).

   (e) The Midpoint Rule for Double Integrals says that we approximate the double integral $\iint_R f(x, y)\,dA$ by the double Riemann sum $\sum_{i=1}^{m} \sum_{j=1}^{n} f(\overline{x}_i, \overline{y}_j)\,\Delta A$ where the sample points $(\overline{x}_i, \overline{y}_j)$ are the centres of the subrectangles.

   (f) $f_{\text{ave}} = \dfrac{1}{A(R)} \displaystyle\iint_R f(x, y)\,dA$ where $A(R)$ is the area of R.

2. (a) See (1) and (2) and the accompanying discussion in Section 12.3.

   (b) See (3) and the accompanying discussion in Section 12.3.

   (c) See (5) and the preceding discussion in Section 12.3.

   (d) See (6)–(11) in Section 12.3.

3. We may want to change from rectangular to polar coordinates in a double integral if the region R of integration is more easily described in polar coordinates. To accomplish this, we use $\iint_R f(x, y)\,dA = \int_\alpha^\beta \int_a^b f(r\cos\theta, r\sin\theta)\,r\,dr\,d\theta$ where R is given by $0 \leq a \leq r \leq b$, $\alpha \leq \theta \leq \beta$.

4. (a) $m = \iint_D \rho(x, y)\,dA$

   (b) $M_x = \iint_D y\rho(x, y)\,dA$, $M_y = \iint_D x\rho(x, y)\,dA$

   (c) The centre of mass is $(\overline{x}, \overline{y})$ where $\overline{x} = \dfrac{M_y}{m}$ and $\overline{y} = \dfrac{M_x}{m}$.

   (d) $I_x = \iint_D y^2\rho(x, y)\,dA$, $I_y = \iint_D x^2\rho(x, y)\,dA$, $I_0 = \iint_D (x^2 + y^2)\rho(x, y)\,dA$

5. (a) $P(a \leq X \leq b, c \leq Y \leq d) = \int_a^b \int_c^d f(x, y)\,dy\,dx$

   (b) $f(x, y) \geq 0$ and $\iint_{\mathbb{R}^2} f(x, y)\,dA = 1$.

   (c) The expected value of X is $\mu_1 = \iint_{\mathbb{R}^2} x f(x, y)\,dA$; the expected value of Y is $\mu_2 = \iint_{\mathbb{R}^2} y f(x, y)\,dA$.

6. (a) $A(S) = \iint_D |\mathbf{r}_u \times \mathbf{r}_v|\,dA$

   (b) $A(S) = \displaystyle\iint_D \sqrt{1 + \left(\frac{\partial z}{\partial x}\right)^2 + \left(\frac{\partial z}{\partial y}\right)^2}\,dA$

   (c) $A(S) = 2\pi \int_a^b f(x)\sqrt{1 + [f'(x)]^2}\,dx$

7. (a) $\iiint_B f(x, y, z)\,dV = \lim_{l,m,n\to\infty} \sum_{i=1}^{l} \sum_{j=1}^{m} \sum_{k=1}^{n} f(x_{ijk}^*, y_{ijk}^*, z_{ijk}^*)\,\Delta V$

   (b) We usually evaluate $\iiint_B f(x, y, z)\,dV$ as an iterated integral according to Fubini's Theorem for Triple Integrals (see Theorem 12.7.4).

   (c) See the paragraph following Example 12.7.1.

(d) See (5) and (6) and the accompanying discussion in Section 12.7.

(e) See (10) and the accompanying discussion in Section 12.7.

(f) See (11) and the preceding discussion in Section 12.7.

8. (a) $m = \iiint_E \rho(x, y, z)\, dV$

(b) $M_{yz} = \iiint_E x\rho(x, y, z)\, dV$, $M_{xz} = \iiint_E y\rho(x, y, z)\, dV$, $M_{xy} = \iiint_E z\rho(x, y, z)\, dV$.

(c) The centre of mass is $(\overline{x}, \overline{y}, \overline{z})$ where $\overline{x} = \dfrac{M_{yz}}{m}$, $\overline{y} = \dfrac{M_{xz}}{m}$, and $\overline{z} = \dfrac{M_{xy}}{m}$.

(d) $I_x = \iiint_E (y^2 + z^2)\rho(x, y, z)\, dV$, $I_y = \iiint_E (x^2 + z^2)\rho(x, y, z)\, dV$, $I_z = \iiint_E (x^2 + y^2)\rho(x, y, z)\, dV$.

9. (a) See Formula 12.8.2 and the accompanying discussion.

(b) See Formula 12.8.4 and the accompanying discussion.

(c) We may want to change from rectangular to cylindrical or spherical coordinates in a triple integral if the region E of integration is more easily described in cylindrical or spherical coordinates or if the triple integral is easier to evaluate using cylindrical or spherical coordinates.

10. (a) $\dfrac{\partial(x, y)}{\partial(u, v)} = \begin{vmatrix} \partial x/\partial u & \partial x/\partial v \\ \partial y/\partial u & \partial y/\partial v \end{vmatrix} = \dfrac{\partial x}{\partial u}\dfrac{\partial y}{\partial v} - \dfrac{\partial x}{\partial v}\dfrac{\partial y}{\partial u}$

(b) See (9) and the accompanying discussion in Section 12.9.

(c) See (13) and the accompanying discussion in Section 12.9.

TRUE-FALSE QUIZ

1. This is true by Fubini's Theorem.

3. True. See the discussion following Example 4 on page 841.

5. True:

$$\begin{aligned} \iint_D \sqrt{4 - x^2 - y^2}\, dA &= \text{the volume under the surface } x^2 + y^2 + z^2 = 4 \text{ and above the } xy\text{-plane} \\ &= \tfrac{1}{2}\left(\text{the volume of the sphere } x^2 + y^2 + z^2 = 4\right) = \tfrac{1}{2} \cdot \tfrac{4}{3}\pi(2)^3 = \tfrac{16}{3}\pi \end{aligned}$$

7. The volume enclosed by the cone $z = \sqrt{x^2 + y^2}$ and the plane $z = 2$ is, in cylindrical coordinates, $V = \int_0^{2\pi} \int_0^2 \int_r^2 r\, dz\, dr\, d\theta \neq \int_0^{2\pi} \int_0^2 \int_r^2 dz\, dr\, d\theta$, so the assertion is false.

EXERCISES

1. As shown in the contour map, we divide R into 9 equally sized subsquares, each with area $\Delta A = 1$. Then we approximate $\iint_R f(x, y)\, dA$ by a Riemann sum with $m = n = 3$ and the sample points the upper right corners of each square, so

$$\begin{aligned} \iint_R f(x, y)\, dA &\approx \sum_{i=1}^{3} \sum_{j=1}^{3} f(x_i, y_j)\, \Delta A \\ &= \Delta A\,[f(1, 1) + f(1, 2) + f(1, 3) + f(2, 1) + f(2, 2) + f(2, 3) + f(3, 1) + f(3, 2) + f(3, 3)] \end{aligned}$$

Using the contour lines to estimate the function values, we have

$$\iint_R f(x, y)\, dA \approx 1[2.7 + 4.7 + 8.0 + 4.7 + 6.7 + 10.0 + 6.7 + 8.6 + 11.9] \approx 64.0$$

3. $\int_1^2 \int_0^2 (y + 2xe^y)\, dx\, dy = \int_1^2 \left[xy + x^2e^y\right]_{x=0}^{x=2} dy = \int_1^2 (2y + 4e^y)\, dy = \left[y^2 + 4e^y\right]_1^2$
$= 4 + 4e^2 - 1 - 4e = 4e^2 - 4e + 3$

5. $\int_0^1 \int_0^x \cos(x^2)\, dy\, dx = \int_0^1 \left[\cos(x^2)y\right]_{y=0}^{y=x} dx = \int_0^1 x\cos(x^2)\, dx = \frac{1}{2}\sin(x^2)\Big]_0^1 = \frac{1}{2}\sin 1$

7. $\int_0^\pi \int_0^1 \int_0^{\sqrt{1-y^2}} y \sin x\, dz\, dy\, dx = \int_0^\pi \int_0^1 \left[(y\sin x)z\right]_{z=0}^{z=\sqrt{1-y^2}} dy\, dx = \int_0^\pi \int_0^1 y\sqrt{1-y^2}\sin x\, dy\, dx$

$$= \int_0^\pi \left[-\tfrac{1}{3}(1-y^2)^{3/2}\sin x\right]_{y=0}^{y=1} dx = \int_0^\pi \tfrac{1}{3}\sin x\, dx = -\tfrac{1}{3}\cos x\Big]_0^\pi = \tfrac{2}{3}$$

9. The region R is more easily described by polar coordinates: $R = \{(r,\theta) \mid 2 \le r \le 4, 0 \le \theta \le \pi\}$. Thus $\iint_R f(x,y)\, dA = \int_0^\pi \int_2^4 f(r\cos\theta, r\sin\theta)\, r\, dr\, d\theta$.

11.

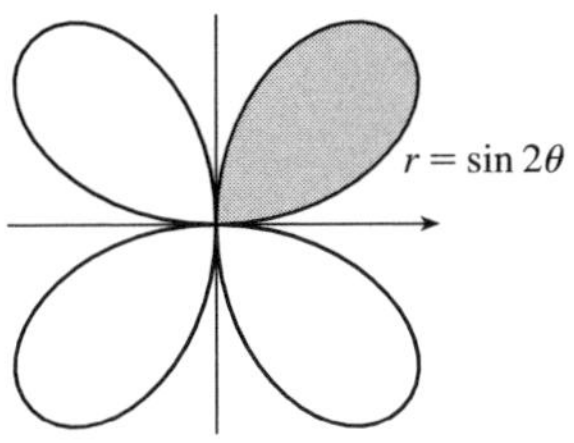

The region whose area is given by $\int_0^{\pi/2} \int_0^{\sin 2\theta} r\, dr\, d\theta$ is $\{(r,\theta) \mid 0 \le \theta \le \frac{\pi}{2}, 0 \le r \le \sin 2\theta\}$, which is the region contained in the loop in the first quadrant of the four-leaved rose $r = \sin 2\theta$.

13.

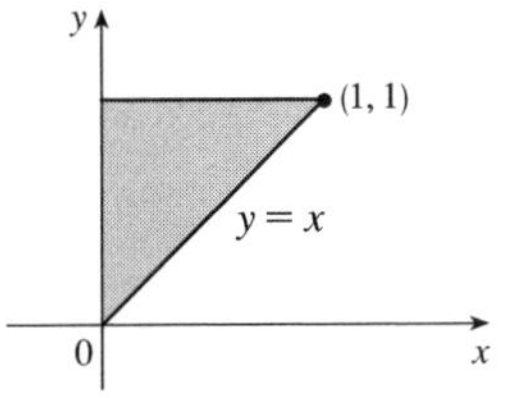

$$\int_0^1 \int_x^1 \cos(y^2)\, dy\, dx = \int_0^1 \int_0^y \cos(y^2)\, dx\, dy$$
$$= \int_0^1 \cos(y^2)\left[x\right]_{x=0}^{x=y} dy = \int_0^1 y\cos(y^2)\, dy$$
$$= \left[\tfrac{1}{2}\sin(y^2)\right]_0^1 = \tfrac{1}{2}\sin 1$$

15. $\iint_R ye^{xy}\, dA = \int_0^3 \int_0^2 ye^{xy}\, dx\, dy = \int_0^3 \left[e^{xy}\right]_{x=0}^{x=2} dy = \int_0^3 (e^{2y} - 1)\, dy = \left[\frac{1}{2}e^{2y} - y\right]_0^3$
$= \frac{1}{2}e^6 - 3 - \frac{1}{2} = \frac{1}{2}e^6 - \frac{7}{2}$

17.

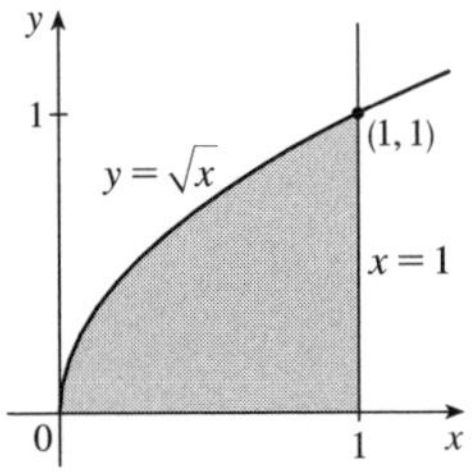

$$\iint_D \frac{y}{1+x^2}\, dA = \int_0^1 \int_0^{\sqrt{x}} \frac{y}{1+x^2}\, dy\, dx = \int_0^1 \frac{1}{1+x^2}\left[\tfrac{1}{2}y^2\right]_{y=0}^{y=\sqrt{x}} dx$$
$$= \frac{1}{2}\int_0^1 \frac{x}{1+x^2}\, dx = \left[\tfrac{1}{4}\ln(1+x^2)\right]_0^1 = \tfrac{1}{4}\ln 2$$

19.

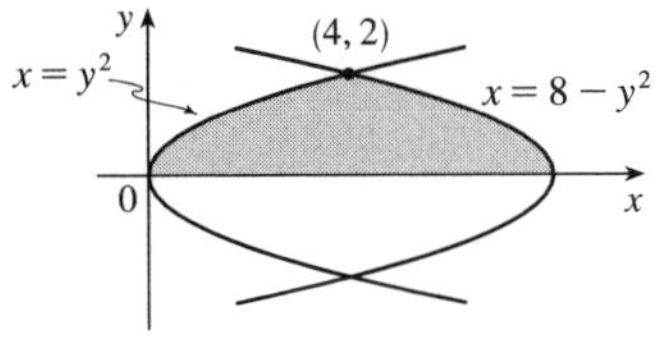

$$\iint_D y\, dA = \int_0^2 \int_{y^2}^{8-y^2} y\, dx\, dy$$
$$= \int_0^2 y\left[x\right]_{x=y^2}^{x=8-y^2} dy = \int_0^2 y(8 - y^2 - y^2)\, dy$$
$$= \int_0^2 (8y - 2y^3)\, dy = \left[4y^2 - \tfrac{1}{2}y^4\right]_0^2 = 8$$

21.

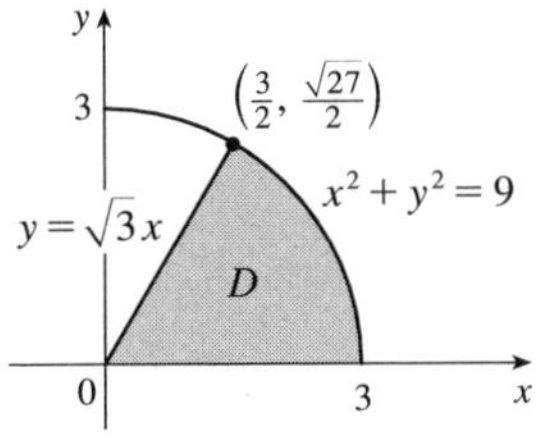

$$\iint_D (x^2+y^2)^{3/2}\, dA = \int_0^{\pi/3} \int_0^3 (r^2)^{3/2} r\, dr\, d\theta$$
$$= \int_0^{\pi/3} d\theta \int_0^3 r^4\, dr = \left[\theta\right]_0^{\pi/3} \left[\frac{1}{5}r^5\right]_0^3$$
$$= \frac{\pi}{3}\frac{3^5}{5} = \frac{81\pi}{5}$$

23. $\iiint_E xy\,dV = \int_0^3 \int_0^x \int_0^{x+y} xy\,dz\,dy\,dx = \int_0^3 \int_0^x xy\left[z\right]_{z=0}^{z=x+y} dy\,dx = \int_0^3 \int_0^x xy(x+y)\,dy\,dx$
$= \int_0^3 \int_0^x (x^2y + xy^2)\,dy\,dx = \int_0^3 \left[\frac{1}{2}x^2y^2 + \frac{1}{3}xy^3\right]_{y=0}^{y=x} dx = \int_0^3 \left(\frac{1}{2}x^4 + \frac{1}{3}x^4\right) dx$
$= \frac{5}{6}\int_0^3 x^4\,dx = \left[\frac{1}{6}x^5\right]_0^3 = \frac{81}{2} = 40.5$

25. $\iiint_E y^2z^2\,dV = \int_{-1}^1 \int_{-\sqrt{1-y^2}}^{\sqrt{1-y^2}} \int_0^{1-y^2-z^2} y^2z^2\,dx\,dz\,dy = \int_{-1}^1 \int_{-\sqrt{1-y^2}}^{\sqrt{1-y^2}} y^2z^2(1-y^2-z^2)\,dz\,dy$
$= \int_0^{2\pi} \int_0^1 (r^2\cos^2\theta)(r^2\sin^2\theta)(1-r^2)\,r\,dr\,d\theta = \int_0^{2\pi} \int_0^1 \frac{1}{4}\sin^2 2\theta(r^5 - r^7)\,dr\,d\theta$
$= \int_0^{2\pi} \frac{1}{8}(1-\cos 4\theta)\left[\frac{1}{6}r^6 - \frac{1}{8}r^8\right]_{r=0}^{r=1} d\theta = \frac{1}{192}\left[\theta - \frac{1}{4}\sin 4\theta\right]_0^{2\pi} = \frac{2\pi}{192} = \frac{\pi}{96}$

27. $\iiint_E yz\,dV = \int_{-2}^2 \int_0^{\sqrt{4-x^2}} \int_0^y yz\,dz\,dy\,dx = \int_{-2}^2 \int_0^{\sqrt{4-x^2}} \frac{1}{2}y^3 dy\,dx = \int_0^\pi \int_0^2 \frac{1}{2}r^3(\sin^3\theta)\,r\,dr\,d\theta$
$= \frac{16}{5}\int_0^\pi \sin^3\theta\,d\theta = \frac{16}{5}\left[-\cos\theta + \frac{1}{3}\cos^3\theta\right]_0^\pi = \frac{64}{15}$

29. $V = \int_0^2 \int_1^4 (x^2 + 4y^2)\,dy\,dx = \int_0^2 \left[x^2y + \frac{4}{3}y^3\right]_{y=1}^{y=4} dx = \int_0^2 (3x^2 + 84)\,dx = 176$

31.

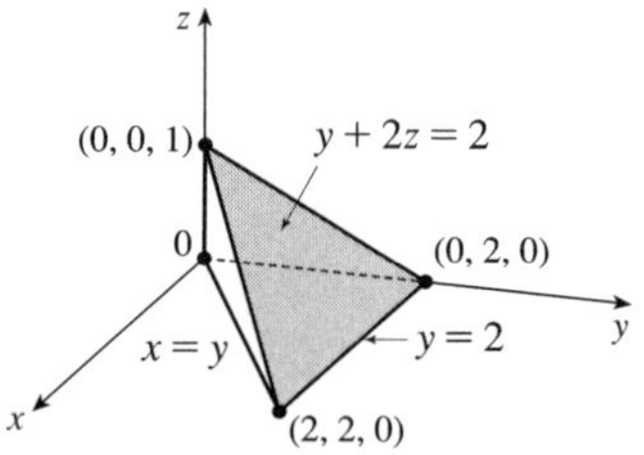

$V = \int_0^2 \int_0^y \int_0^{(2-y)/2} dz\,dx\,dy$
$= \int_0^2 \int_0^y \left(1 - \frac{1}{2}y\right) dx\,dy$
$= \int_0^2 \left(y - \frac{1}{2}y^2\right) dy = \frac{2}{3}$

33. Using the wedge above the plane $z = 0$ and below the plane $z = mx$ and noting that we have the same volume for $m < 0$ as for $m > 0$ (so use $m > 0$), we have

$V = 2\int_0^{a/3} \int_0^{\sqrt{a^2-9y^2}} mx\,dx\,dy = 2\int_0^{a/3} \frac{1}{2}m(a^2 - 9y^2)\,dy = m\left[a^2y - 3y^3\right]_0^{a/3} = m\left(\frac{1}{3}a^3 - \frac{1}{9}a^3\right) = \frac{2}{9}ma^3.$

35. (a) $m = \int_0^1 \int_0^{1-y^2} y\,dx\,dy = \int_0^1 (y - y^3)\,dy = \frac{1}{2} - \frac{1}{4} = \frac{1}{4}$

(b) $M_y = \int_0^1 \int_0^{1-y^2} xy\,dx\,dy = \int_0^1 \frac{1}{2}y(1-y^2)^2\,dy = -\frac{1}{12}(1-y^2)^3\Big]_0^1 = \frac{1}{12}$,

$M_x = \int_0^1 \int_0^{1-y^2} y^2\,dx\,dy = \int_0^1 (y^2 - y^4)\,dy = \frac{2}{15}$. Hence $(\overline{x}, \overline{y}) = \left(\frac{1}{3}, \frac{8}{15}\right)$.

(c) $I_x = \int_0^1 \int_0^{1-y^2} y^3\,dx\,dy = \int_0^1 (y^3 - y^5)\,dy = \frac{1}{12}$,

$I_y = \int_0^1 \int_0^{1-y^2} yx^2\,dx\,dy = \int_0^1 \frac{1}{3}y(1-y^2)^3\,dy = -\frac{1}{24}(1-y^2)^4\Big]_0^1 = \frac{1}{24}$,

$I_0 = I_x + I_y = \frac{1}{8}$, $\overline{\overline{y}}^2 = \frac{1/12}{1/4} = \frac{1}{3} \Rightarrow \overline{\overline{y}} = \frac{1}{\sqrt{3}}$, and $\overline{\overline{x}}^2 = \frac{1/24}{1/4} = \frac{1}{6} \Rightarrow \overline{\overline{x}} = \frac{1}{\sqrt{6}}$.

37. (a) The equation of the cone with the suggested orientation is $(h - z) = \frac{h}{a}\sqrt{x^2 + y^2}$, $0 \le z \le h$. Then $V = \frac{1}{3}\pi a^2 h$ is the volume of one frustum of a cone; by symmetry $M_{yz} = M_{xz} = 0$; and

$$M_{xy} = \iint\limits_{x^2+y^2 \le a^2} \int_0^{h-(h/a)\sqrt{x^2+y^2}} z\,dz\,dA = \int_0^{2\pi} \int_0^a \int_0^{(h/a)(a-r)} rz\,dz\,dr\,d\theta = \pi \int_0^a r\frac{h^2}{a^2}(a-r)^2\,dr$$

$$= \frac{\pi h^2}{a^2} \int_0^a (a^2r - 2ar^2 + r^3)\,dr = \frac{\pi h^2}{a^2}\left(\frac{a^4}{2} - \frac{2a^4}{3} + \frac{a^4}{4}\right) = \frac{\pi h^2 a^2}{12}$$

Hence the centroid is $(\overline{x}, \overline{y}, \overline{z}) = \left(0, 0, \frac{1}{4}h\right)$.

(b) $$I_z = \int_0^{2\pi} \int_0^a \int_0^{(h/a)(a-r)} r^3\,dz\,dr\,d\theta = 2\pi \int_0^a \frac{h}{a}(ar^3 - r^4)\,dr = \frac{2\pi h}{a}\left(\frac{a^5}{4} - \frac{a^5}{5}\right) = \frac{\pi a^4 h}{10}$$

39. Let D represent the given triangle; then D can be described as the area enclosed by the x- and y-axes and the line $y = 2 - 2x$, or equivalently $D = \{(x, y) \mid 0 \le x \le 1, 0 \le y \le 2 - 2x\}$. We want to find the surface area of the part of the graph of $z = x^2 + y$ that lies over D, so using Equation 12.6.6 we have

$$A(S) = \iint_D \sqrt{1 + \left(\frac{\partial z}{\partial x}\right)^2 + \left(\frac{\partial z}{\partial y}\right)^2}\, dA = \iint_D \sqrt{1 + (2x)^2 + (1)^2}\, dA = \int_0^1 \int_0^{2-2x} \sqrt{2 + 4x^2}\, dy\, dx$$

$$= \int_0^1 \sqrt{2 + 4x^2}\, \big[\, y \,\big]_{y=0}^{y=2-2x}\, dx = \int_0^1 (2 - 2x)\sqrt{2 + 4x^2}\, dx = \int_0^1 2\sqrt{2 + 4x^2}\, dx - \int_0^1 2x\sqrt{2 + 4x^2}\, dx$$

Using Formula 21 in the Table of Integrals with $a = \sqrt{2}$, $u = 2x$, and $du = 2\, dx$, we have $\int 2\sqrt{2 + 4x^2}\, dx = x\sqrt{2 + 4x^2} + \ln\big(2x + \sqrt{2 + 4x^2}\,\big)$. If we substitute $u = 2 + 4x^2$ in the second integral, then $du = 8x\, dx$ and $\int 2x\sqrt{2 + 4x^2}\, dx = \frac{1}{4}\int \sqrt{u}\, du = \frac{1}{4} \cdot \frac{2}{3}u^{3/2} = \frac{1}{6}(2 + 4x^2)^{3/2}$. Thus

$$A(S) = \left[x\sqrt{2 + 4x^2} + \ln\left(2x + \sqrt{2 + 4x^2}\right) - \tfrac{1}{6}(2 + 4x^2)^{3/2}\right]_0^1$$

$$= \sqrt{6} + \ln\left(2 + \sqrt{6}\right) - \tfrac{1}{6}(6)^{3/2} - \ln\sqrt{2} + \tfrac{\sqrt{2}}{3} = \ln\tfrac{2 + \sqrt{6}}{\sqrt{2}} + \tfrac{\sqrt{2}}{3}$$

$$= \ln\left(\sqrt{2} + \sqrt{3}\right) + \tfrac{\sqrt{2}}{3} \approx 1.6176$$

41.

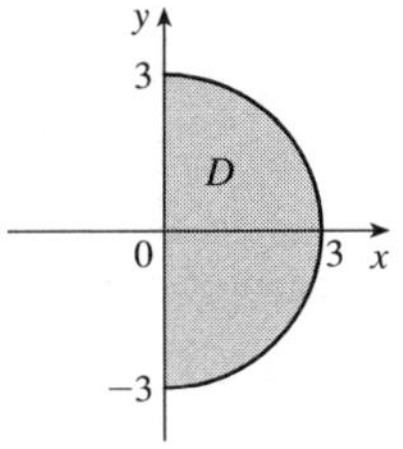

$$\int_0^3 \int_{-\sqrt{9-x^2}}^{\sqrt{9-x^2}} (x^3 + xy^2)\, dy\, dx = \int_0^3 \int_{-\sqrt{9-x^2}}^{\sqrt{9-x^2}} x(x^2 + y^2)\, dy\, dx$$

$$= \int_{-\pi/2}^{\pi/2} \int_0^3 (r\cos\theta)(r^2)\, r\, dr\, d\theta$$

$$= \int_{-\pi/2}^{\pi/2} \cos\theta\, d\theta \int_0^3 r^4\, dr$$

$$= \big[\sin\theta\big]_{-\pi/2}^{\pi/2} \left[\tfrac{1}{5}r^5\right]_0^3 = 2 \cdot \tfrac{1}{5}(243) = \tfrac{486}{5} = 97.2$$

43. From the graph, it appears that $1 - x^2 = e^x$ at $x = a \approx -0.71$ and at $x = 0$, with $1 - x^2 > e^x$ on $(a, 0)$. So the desired integral is

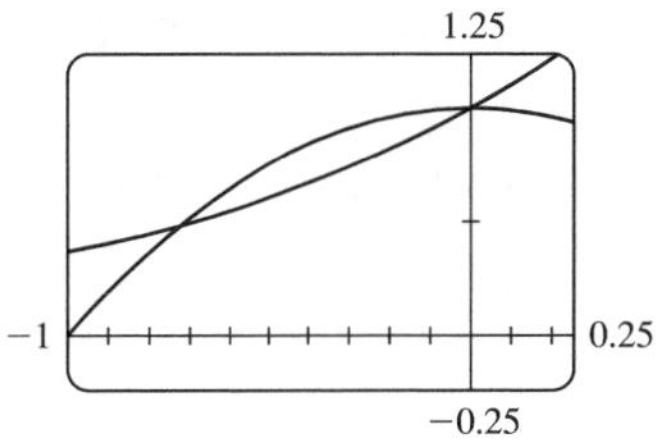

$$\iint_D y^2\, dA \approx \int_a^0 \int_{e^x}^{1-x^2} y^2\, dy\, dx = \int_a^0 \left[\tfrac{1}{3}y^3\right]_{e^x}^{1-x^2} dx$$

$$= \tfrac{1}{3}\int_a^0 [(1 - x^2)^3 - e^{3x}]\, dx = \tfrac{1}{3}\int_a^0 (1 - 3x^2 + 3x^4 - x^6 - e^{3x})\, dx$$

$$= \tfrac{1}{3}\left[x - x^3 + \tfrac{3}{5}x^5 - \tfrac{1}{7}x^7 - \tfrac{1}{3}e^{3x}\right]_a^0 \approx 0.0512$$

45. (a) $f(x, y)$ is a joint density function, so we know that $\iint_{\mathbb{R}^2} f(x, y)\, dA = 1$. Since $f(x, y) = 0$ outside the rectangle $[0, 3] \times [0, 2]$, we can say

$$\iint_{\mathbb{R}^2} f(x, y)\, dA = \int_{-\infty}^{\infty}\int_{-\infty}^{\infty} f(x, y)\, dy\, dx = \int_0^3 \int_0^2 C(x + y)\, dy\, dx$$

$$= C\int_0^3 \left[xy + \tfrac{1}{2}y^2\right]_{y=0}^{y=2} dx = C\int_0^3 (2x + 2)\, dx = C\big[x^2 + 2x\big]_0^3 = 15C$$

Then $15C = 1 \;\Rightarrow\; C = \frac{1}{15}$.

(b) $P(X \le 2, Y \ge 1) = \int_{-\infty}^2 \int_1^{\infty} f(x, y)\, dy\, dx = \int_0^2 \int_1^2 \frac{1}{15}(x, y)\, dy\, dx = \frac{1}{15}\int_0^2 \left[xy + \frac{1}{2}y^2\right]_{y=1}^{y=2} dx$

$$= \tfrac{1}{15}\int_0^2 \left(x + \tfrac{3}{2}\right) dx = \tfrac{1}{15}\left[\tfrac{1}{2}x^2 + \tfrac{3}{2}x\right]_0^2 = \tfrac{1}{3}$$

(c) $P(X + Y \le 1) = P((X, Y) \in D)$ where D is the triangular region shown in the figure. Thus

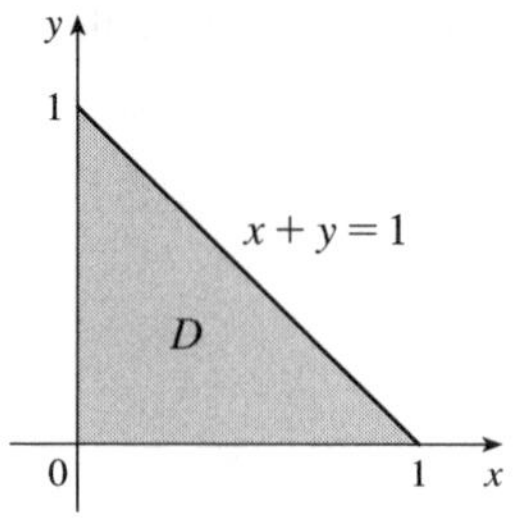

$$P(X + Y \le 1) = \iint_D f(x, y)\, dA = \int_0^1 \int_0^{1-x} \tfrac{1}{15}(x + y)\, dy\, dx$$

$$= \tfrac{1}{15} \int_0^1 \left[xy + \tfrac{1}{2}y^2\right]_{y=0}^{y=1-x} dx$$

$$= \tfrac{1}{15} \int_0^1 \left[x(1 - x) + \tfrac{1}{2}(1 - x)^2\right] dx$$

$$= \tfrac{1}{30} \int_0^1 (1 - x^2)\, dx = \tfrac{1}{30}\left[x - \tfrac{1}{3}x^3\right]_0^1 = \tfrac{1}{45}$$

47.

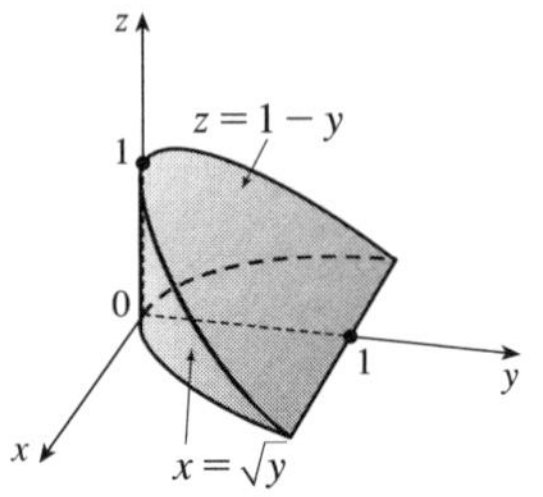

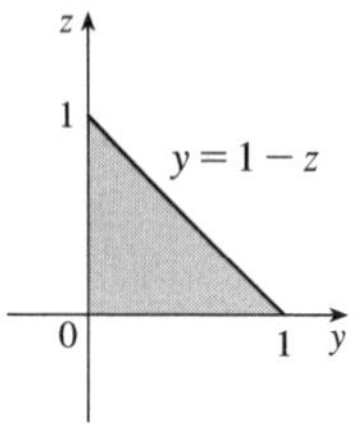

$$\int_{-1}^1 \int_{x^2}^1 \int_0^{1-y} f(x, y, z)\, dz\, dy\, dx = \int_0^1 \int_0^{1-z} \int_{-\sqrt{y}}^{\sqrt{y}} f(x, y, z)\, dx\, dy\, dz$$

49. Since $u = x - y$ and $v = x + y$, $x = \frac{1}{2}(u + v)$ and $y = \frac{1}{2}(v - u)$.

Thus $\dfrac{\partial(x, y)}{\partial(u, v)} = \begin{vmatrix} 1/2 & 1/2 \\ -1/2 & 1/2 \end{vmatrix} = \dfrac{1}{2}$ and $\displaystyle\iint_R \frac{x - y}{x + y}\, dA = \int_2^4 \int_{-2}^0 \frac{u}{v}\left(\frac{1}{2}\right) du\, dv = -\int_2^4 \frac{dv}{v} = -\ln 2.$

51. Let $u = y - x$ and $v = y + x$ so $x = y - u = (v - x) - u \;\Rightarrow\; x = \frac{1}{2}(v - u)$ and $y = v - \frac{1}{2}(v - u) = \frac{1}{2}(v + u)$.

$\left|\dfrac{\partial(x, y)}{\partial(u, v)}\right| = \left|\dfrac{\partial x}{\partial u}\dfrac{\partial y}{\partial v} - \dfrac{\partial x}{\partial v}\dfrac{\partial y}{\partial u}\right| = \left|-\frac{1}{2}\left(\frac{1}{2}\right) - \frac{1}{2}\left(\frac{1}{2}\right)\right| = \left|-\frac{1}{2}\right| = \frac{1}{2}$. R is the image under this transformation of the square with vertices $(u, v) = (0, 0)$, $(-2, 0)$, $(0, 2)$, and $(-2, 2)$. So

$$\iint_R xy\, dA = \int_0^2 \int_{-2}^0 \frac{v^2 - u^2}{4}\left(\frac{1}{2}\right) du\, dv = \tfrac{1}{8} \int_0^2 \left[v^2 u - \tfrac{1}{3}u^3\right]_{u=-2}^{u=0} dv = \tfrac{1}{8} \int_0^2 \left(2v^2 - \tfrac{8}{3}\right) dv = \tfrac{1}{8}\left[\tfrac{2}{3}v^3 - \tfrac{8}{3}v\right]_0^2 = 0$$

This result could have been anticipated by symmetry, since the integrand is an odd function of y and R is symmetric about the x-axis.

FOCUS ON PROBLEM SOLVING

1. 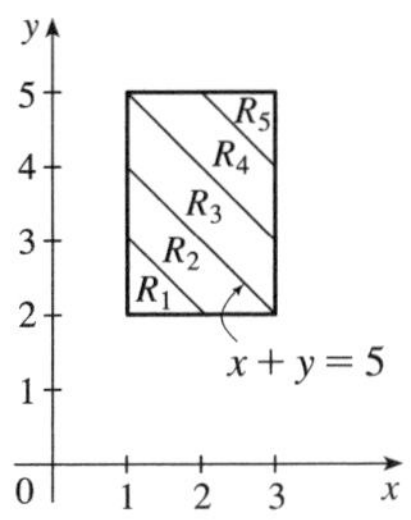

Let $R = \bigcup_{i=1}^{5} R_i$, where

$R_i = \{(x, y) \mid x + y \geq i + 2, x + y < i + 3, 1 \leq x \leq 3, 2 \leq y \leq 5\}$.

$\iint_R [\![x+y]\!]\, dA = \sum_{i=1}^{5} \iint_{R_i} [\![x+y]\!]\, dA = \sum_{i=1}^{5} [\![x+y]\!] \iint_{R_i} dA$, since

$[\![x+y]\!] = \text{constant} = i + 2$ for $(x, y) \in R_i$. Therefore

$$\begin{aligned}\iint_R [\![x+y]\!]\, dA &= \textstyle\sum_{i=1}^{5} (i+2)\,[A(R_i)] \\ &= 3A(R_1) + 4A(R_2) + 5A(R_3) + 6A(R_4) + 7A(R_5) \\ &= 3\left(\tfrac{1}{2}\right) + 4\left(\tfrac{3}{2}\right) + 5(2) + 6\left(\tfrac{3}{2}\right) + 7\left(\tfrac{1}{2}\right) = 30\end{aligned}$$

3. $f_{\text{ave}} = \dfrac{1}{b-a}\displaystyle\int_a^b f(x)\,dx = \dfrac{1}{1-0}\int_0^1 \left[\int_x^1 \cos(t^2)\,dt\right] dx$

$= \int_0^1 \int_x^1 \cos(t^2)\,dt\,dx$

$= \int_0^1 \int_0^t \cos(t^2)\,dx\,dt$ [changing the order of integration]

$= \int_0^1 t\cos(t^2)\,dt = \frac{1}{2}\sin(t^2)\big]_0^1 = \frac{1}{2}\sin 1$

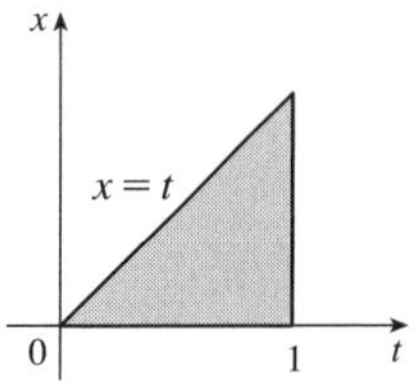

5. Since $|xy| < 1$, except at $(1, 1)$, the formula for the sum of a geometric series gives $\dfrac{1}{1-xy} = \displaystyle\sum_{n=0}^{\infty} (xy)^n$, so

$$\begin{aligned}\int_0^1\int_0^1 \frac{1}{1-xy}\,dx\,dy &= \int_0^1\int_0^1 \sum_{n=0}^{\infty} (xy)^n\,dx\,dy = \sum_{n=0}^{\infty}\int_0^1\int_0^1 (xy)^n\,dx\,dy \\ &= \sum_{n=0}^{\infty}\left[\int_0^1 x^n\,dx\right]\left[\int_0^1 y^n\,dy\right] = \sum_{n=0}^{\infty}\frac{1}{n+1}\cdot\frac{1}{n+1} \\ &= \sum_{n=0}^{\infty}\frac{1}{(n+1)^2} = \frac{1}{1^2} + \frac{1}{2^2} + \frac{1}{3^2} + \cdots = \sum_{n=1}^{\infty}\frac{1}{n^2}\end{aligned}$$

7. (a) Since $|xyz| < 1$ except at $(1, 1, 1)$, the formula for the sum of a geometric series gives $\dfrac{1}{1-xyz} = \displaystyle\sum_{n=0}^{\infty}(xyz)^n$,

so

$$\begin{aligned}\int_0^1\int_0^1\int_0^1 \frac{1}{1-xyz}\,dx\,dy\,dz &= \int_0^1\int_0^1\int_0^1 \sum_{n=0}^{\infty}(xyz)^n\,dx\,dy\,dz \\ &= \sum_{n=0}^{\infty}\int_0^1\int_0^1\int_0^1 (xyz)^n\,dx\,dy\,dz \\ &= \sum_{n=0}^{\infty}\left[\int_0^1 x^n\,dx\right]\left[\int_0^1 y^n\,dy\right]\left[\int_0^1 z^n\,dz\right] \\ &= \sum_{n=0}^{\infty}\frac{1}{n+1}\cdot\frac{1}{n+1}\cdot\frac{1}{n+1} \\ &= \sum_{n=0}^{\infty}\frac{1}{(n+1)^3} = \frac{1}{1^3} + \frac{1}{2^3} + \frac{1}{3^3} + \cdots = \sum_{n=1}^{\infty}\frac{1}{n^3}\end{aligned}$$

(b) Since $|-xyz| < 1$, except at $(1,1,1)$, the formula for the sum of a geometric series gives $\dfrac{1}{1+xyz} = \sum_{n=0}^{\infty}(-xyz)^n$,

so

$$\int_0^1\int_0^1\int_0^1 \frac{1}{1+xyz}\,dx\,dy\,dz = \int_0^1\int_0^1\int_0^1 \sum_{n=0}^{\infty}(-xyz)^n\,dx\,dy\,dz$$

$$= \sum_{n=0}^{\infty}\int_0^1\int_0^1\int_0^1 (-xyz)^n\,dx\,dy\,dz$$

$$= \sum_{n=0}^{\infty}(-1)^n\left[\int_0^1 x^n\,dx\right]\left[\int_0^1 y^n\,dy\right]\left[\int_0^1 z^n\,dz\right]$$

$$= \sum_{n=0}^{\infty}(-1)^n\frac{1}{n+1}\cdot\frac{1}{n+1}\cdot\frac{1}{n+1}$$

$$= \sum_{n=0}^{\infty}\frac{(-1)^n}{(n+1)^3} = \frac{1}{1^3} - \frac{1}{2^3} + \frac{1}{3^3} - \cdots = \sum_{n=1}^{\infty}\frac{(-1)^{n-1}}{n^3}$$

To evaluate this sum, we first write out a few terms: $s = 1 - \frac{1}{2^3} + \frac{1}{3^3} - \frac{1}{4^3} + \frac{1}{5^3} - \frac{1}{6^3} \approx 0.8998$. Notice that $a_7 = \frac{1}{7^3} < 0.003$. By the Alternating Series Estimation Theorem from Section 8.4, we have $|s - s_6| \le a_7 < 0.003$. This error of 0.003 will not affect the second decimal place, so we have $s \approx 0.90$.

9. $\int_0^x\int_0^y\int_0^z f(t)\,dt\,dz\,dy = \iiint_E f(t)\,dV$, where $E = \{(t,z,y) \mid 0 \le t \le z, 0 \le z \le y, 0 \le y \le x\}$.

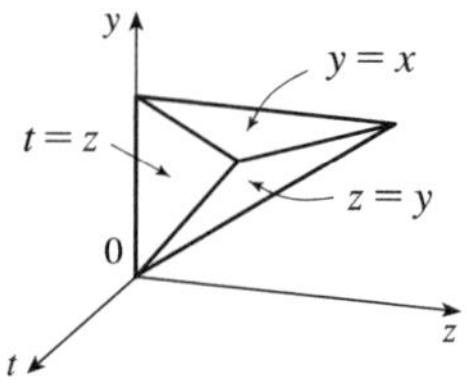

If we let D be the projection of E on the yt-plane then $D = \{(y,t) \mid 0 \le t \le x, t \le y \le x\}$. And we see from the diagram that $E = \{(t,z,y) \mid t \le z \le y, t \le y \le x, 0 \le t \le x\}$. So

$$\int_0^x\int_0^y\int_0^z f(t)\,dt\,dz\,dy = \int_0^x\int_t^x\int_t^y f(t)\,dz\,dy\,dt = \int_0^x\left[\int_t^x (y-t)\,f(t)\,dy\right]dt$$

$$= \int_0^x \left[\left(\tfrac12 y^2 - ty\right)f(t)\right]_{y=t}^{y=x}dt = \int_0^x \left[\tfrac12 x^2 - tx - \tfrac12 t^2 + t^2\right]f(t)\,dt$$

$$= \int_0^x \left[\tfrac12 x^2 - tx + \tfrac12 t^2\right]f(t)\,dt = \int_0^x \left(\tfrac12 x^2 - 2tx + t^2\right)f(t)\,dt$$

$$= \tfrac12\int_0^x (x-t)^2\,f(t)\,dt$$

13 ☐ VECTOR CALCULUS

13.1 Vector Fields

1. $\mathbf{F}(x, y) = \frac{1}{2}(\mathbf{i} + \mathbf{j})$

   All vectors in this field are identical, with length $\frac{1}{\sqrt{2}}$ and direction parallel to the line $y = x$.

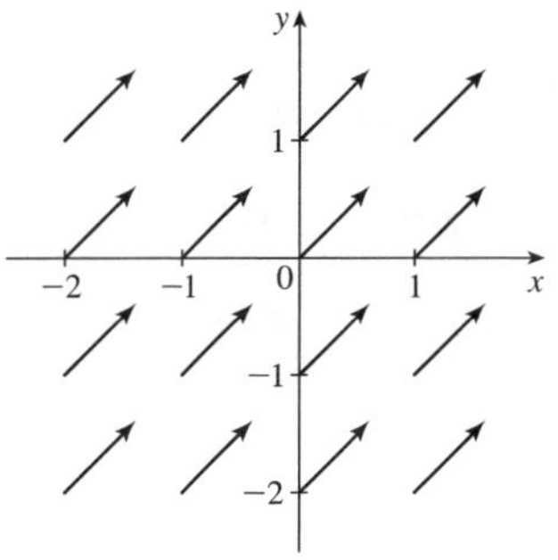

3. $\mathbf{F}(x, y) = y\,\mathbf{i} + \frac{1}{2}\,\mathbf{j}$

   The length of the vector $y\,\mathbf{i} + \frac{1}{2}\,\mathbf{j}$ is $\sqrt{y^2 + \frac{1}{4}}$. Vectors are tangent to parabolas opening about the x-axis.

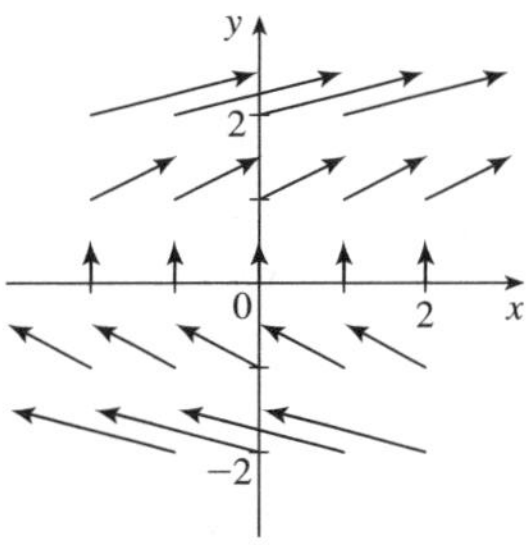

5. $\mathbf{F}(x, y) = \dfrac{y\,\mathbf{i} + x\,\mathbf{j}}{\sqrt{x^2 + y^2}}$

   The length of the vector $\dfrac{y\,\mathbf{i} + x\,\mathbf{j}}{\sqrt{x^2 + y^2}}$ is 1.

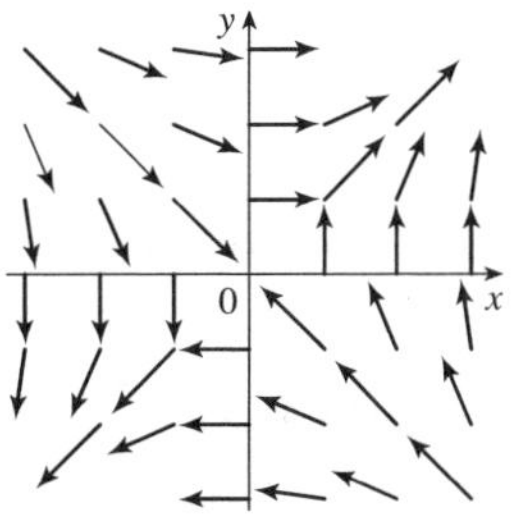

7. $\mathbf{F}(x, y, z) = \mathbf{k}$

   All vectors in this field are parallel to the z-axis and have length 1.

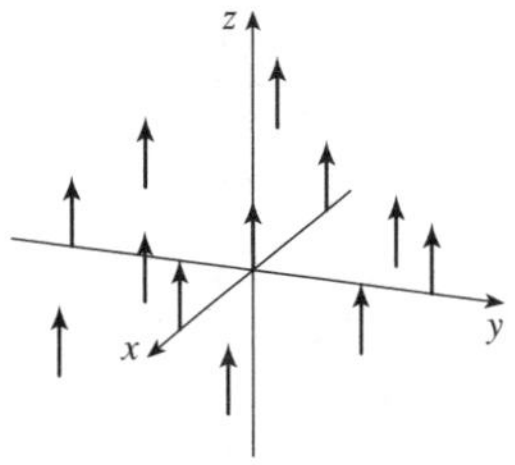

9. $\mathbf{F}(x, y, z) = x\,\mathbf{k}$

   At each point (x, y, z), $\mathbf{F}(x, y, z)$ is a vector of length $|x|$. For $x > 0$, all point in the direction of the positive z-axis, while for $x < 0$, all are in the direction of the negative z-axis. In each plane $x = k$, all the vectors are identical.

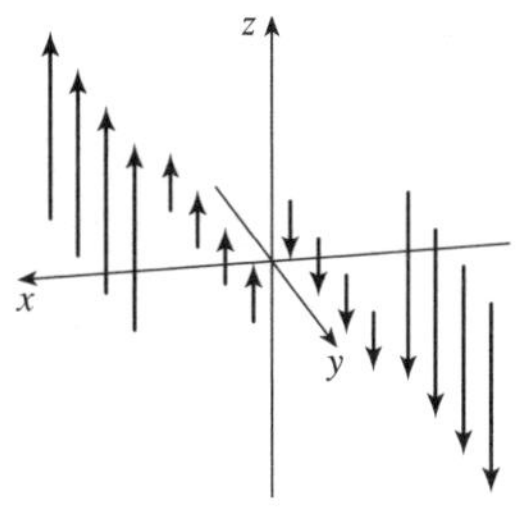

11. $\mathbf{F}(x, y) = \langle y, x \rangle$ corresponds to graph II. In the first quadrant all the vectors have positive x- and y-components, in the second quadrant all vectors have positive x-components and negative y-components, in the third quadrant all vectors have negative x- and y-components, and in the fourth quadrant all vectors have negative x-components and positive y-components. In addition, the vectors get shorter as we approach the origin.

13. $\mathbf{F}(x, y) = \langle x - 2, x + 1 \rangle$ corresponds to graph I since the vectors are independent of y (vectors along vertical lines are identical) and, as we move to the right, both the x- and the y-components get larger.

15. $\mathbf{F}(x, y, z) = \mathbf{i} + 2\,\mathbf{j} + 3\,\mathbf{k}$ corresponds to graph IV, since all vectors have identical length and direction.

17. $\mathbf{F}(x, y, z) = x\,\mathbf{i} + y\,\mathbf{j} + 3\,\mathbf{k}$ corresponds to graph III; the projection of each vector onto the xy-plane is $x\,\mathbf{i} + y\,\mathbf{j}$, which points away from the origin, and the vectors point generally upward because their z-components are all 3.

19.

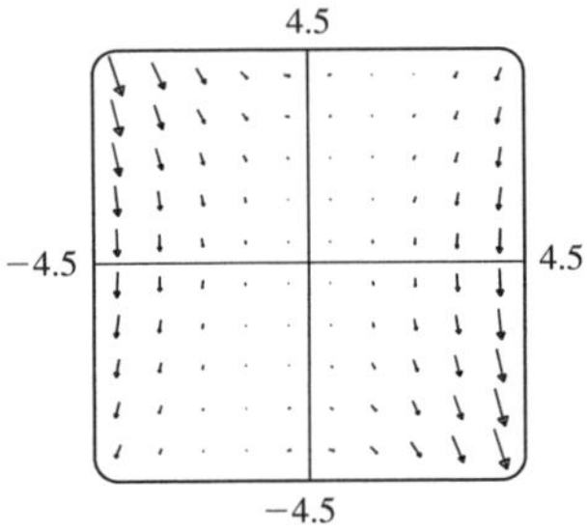

The vector field seems to have very short vectors near the line $y = 2x$. For $\mathbf{F}(x, y) = \langle 0, 0 \rangle$ we must have $y^2 - 2xy = 0$ and $3xy - 6x^2 = 0$. The first equation holds if $y = 0$ or $y = 2x$, and the second holds if $x = 0$ or $y = 2x$. So both equations hold [and thus $\mathbf{F}(x, y) = \mathbf{0}$] along the line $y = 2x$.

21. $\nabla f(x, y) = f_x(x, y)\mathbf{i} + f_y\,(x, y)\,\mathbf{j} = \dfrac{1}{x + 2y}\,\mathbf{i} + \dfrac{2}{x + 2y}\,\mathbf{j}$

23. $\nabla f(x, y, z) = f_x(x, y, z)\,\mathbf{i} + f_y(x, y, z)\,\mathbf{j} + f_z(x, y, z)\,\mathbf{k} = \dfrac{x}{\sqrt{x^2 + y^2 + z^2}}\,\mathbf{i} + \dfrac{y}{\sqrt{x^2 + y^2 + z^2}}\,\mathbf{j} + \dfrac{z}{\sqrt{x^2 + y^2 + z^2}}\,\mathbf{k}$

25. $f(x, y) = xy - 2x \quad \Rightarrow \quad \nabla f(x, y) = (y - 2)\,\mathbf{i} + x\,\mathbf{j}$.
The length of $\nabla f(x, y)$ is $\sqrt{(y - 2)^2 + x^2}$ and $\nabla f(x, y)$ terminates on the line $y = x + 2$ at the point $(x + y - 2, x + y)$.

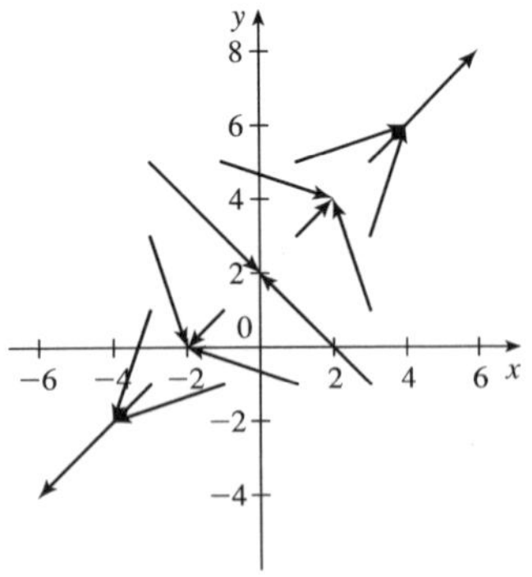

27. We graph ∇f along with a contour map of f.

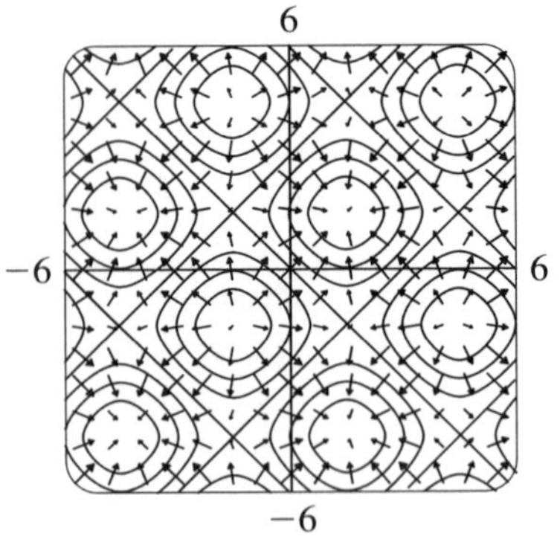

The graph shows that the gradient vectors are perpendicular to the level curves. Also, the gradient vectors point in the direction in which f is increasing and are longer where the level curves are closer together.

29. $f(x,y)=xy \Rightarrow \nabla f(x,y)=y\,\mathbf{i}+x\,\mathbf{j}$. In the first quadrant, both components of each vector are positive, while in the third quadrant both components are negative. However, in the second quadrant each vector's x-component is positive while its y-component is negative (and vice versa in the fourth quadrant). Thus, ∇f is graph IV.

31. $f(x,y)=x^2+y^2 \Rightarrow \nabla f(x,y)=2x\,\mathbf{i}+2y\,\mathbf{j}$. Thus, each vector $\nabla f(x,y)$ has the same direction and twice the length of the position vector of the point (x,y), so the vectors all point directly away from the origin and their lengths increase as we move away from the origin. Hence, ∇f is graph II.

33. At $t=3$ the particle is at $(2,1)$ so its velocity is $\mathbf{V}(2,1)=\langle 4,3\rangle$. After 0.01 units of time, the particle's change in location should be approximately $0.01\,\mathbf{V}(2,1)=0.01\,\langle 4,3\rangle=\langle 0.04,0.03\rangle$, so the particle should be approximately at the point $(2.04,1.03)$.

35. (a) We sketch the vector field $\mathbf{F}(x,y)=x\,\mathbf{i}-y\,\mathbf{j}$ along with several approximate flow lines. The flow lines appear to be hyperbolas with shape similar to the graph of $y=\pm 1/x$, so we might guess that the flow lines have equations $y=C/x$.

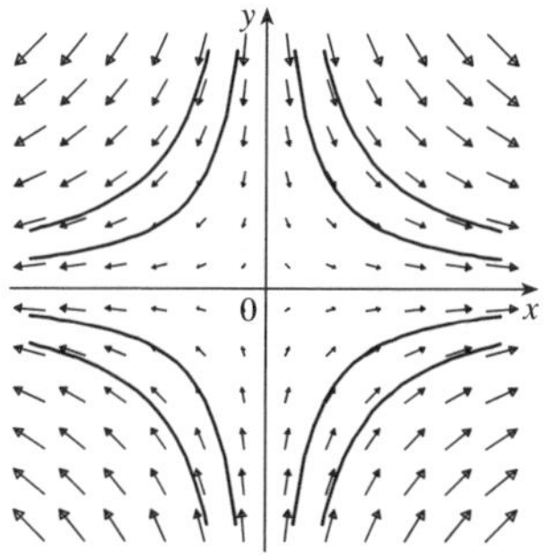

(b) If $x=x(t)$ and $y=y(t)$ are parametric equations of a flow line, then the velocity vector of the flow line at the point (x,y) is $x'(t)\,\mathbf{i}+y'(t)\,\mathbf{j}$. Since the velocity vectors coincide with the vectors in the vector field, we have $x'(t)\,\mathbf{i}+y'(t)\,\mathbf{j}=x\,\mathbf{i}-y\,\mathbf{j} \Rightarrow dx/dt=x,\ dy/dt=-y$. To solve these differential equations, we know $dx/dt=x \Rightarrow dx/x=dt \Rightarrow \ln|x|=t+C \Rightarrow x=\pm e^{t+C}=Ae^t$ for some constant A, and $dy/dt=-y \Rightarrow dy/y=-dt \Rightarrow \ln|y|=-t+K \Rightarrow y=\pm e^{-t+K}=Be^{-t}$ for some constant B. Therefore $xy=Ae^tBe^{-t}=AB=$ constant. If the flow line passes through $(1,1)$ then $(1)\,(1)=$ constant $=1 \Rightarrow xy=1 \Rightarrow y=1/x,\ x>0$.

13.2 Line Integrals

1. $x=t^2$ and $y=t$, $0\le t\le 2$, so by Formula 3

$$\int_C y\,ds=\int_0^2 t\sqrt{\left(\frac{dx}{dt}\right)^2+\left(\frac{dy}{dt}\right)^2}\,dt=\int_0^2 t\,\sqrt{(2t)^2+(1)^2}\,dt$$

$$=\int_0^2 t\,\sqrt{4t^2+1}\,dt=\tfrac{1}{12}\left(4t^2+1\right)^{3/2}\Big]_0^2=\tfrac{1}{12}\left(17\sqrt{17}-1\right)$$

3. Parametric equations for C are $x=4\cos t$, $y=4\sin t$, $-\frac{\pi}{2}\le t\le\frac{\pi}{2}$. Then

$$\int_C xy^4\,ds=\int_{-\pi/2}^{\pi/2}(4\cos t)(4\sin t)^4\sqrt{(-4\sin t)^2+(4\cos t)^2}\,dt=\int_{-\pi/2}^{\pi/2}4^5\cos t\sin^4 t\sqrt{16\left(\sin^2 t+\cos^2 t\right)}\,dt$$

$$=4^5\int_{-\pi/2}^{\pi/2}\left(\sin^4 t\cos t\right)(4)\,dt=(4)^6\left[\tfrac{1}{5}\sin^5 t\right]_{-\pi/2}^{\pi/2}=\tfrac{2\cdot 4^6}{5}=1638.4$$

5. 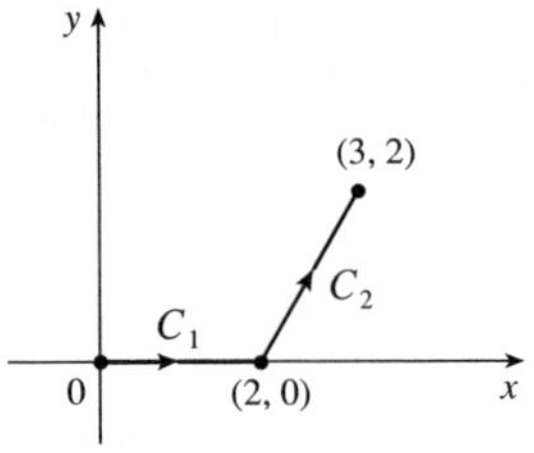

$C = C_1 + C_2$

On C_1: $x = x$, $y = 0 \Rightarrow dy = 0\,dx$, $0 \le x \le 2$.

On C_2: $x = x$, $y = 2x - 4 \Rightarrow dy = 2\,dx$, $2 \le x \le 3$.

Then

$$\begin{aligned}\int_C xy\,dx + (x-y)\,dy &= \int_{C_1} xy\,dx + (x-y)\,dy + \int_{C_2} xy\,dx + (x-y)\,dy \\ &= \int_0^2 (0+0)\,dx + \int_2^3 [(2x^2 - 4x) + (-x+4)(2)]\,dx \\ &= \int_2^3 (2x^2 - 6x + 8)\,dx = \tfrac{17}{3}\end{aligned}$$

7. $x = 4\sin t$, $y = 4\cos t$, $z = 3t$, $0 \le t \le \frac{\pi}{2}$. Then by Formula 9,

$$\begin{aligned}\int_C xy^3\,ds &= \int_0^{\pi/2} (4\sin t)(4\cos t)^3 \sqrt{\left(\tfrac{dx}{dt}\right)^2 + \left(\tfrac{dy}{dt}\right)^2 + \left(\tfrac{dz}{dt}\right)^2}\,dt \\ &= \int_0^{\pi/2} 4^4 \cos^3 t \sin t\,\sqrt{(4\cos t)^2 + (-4\sin t)^2 + (3)^2}\,dt \\ &= \int_0^{\pi/2} 256\cos^3 t \sin t\,\sqrt{16(\cos^2 t + \sin^2 t) + 9}\,dt \\ &= 1280 \int_0^{\pi/2} \cos^3 t \sin t\,dt = -320\cos^4 t\Big]_0^{\pi/2} = 320\end{aligned}$$

9. Parametric equations for C are $x = t$, $y = 2t$, $z = 3t$, $0 \le t \le 1$. Then

$$\int_C xe^{yz}\,ds = \int_0^1 te^{(2t)(3t)}\sqrt{1^2 + 2^2 + 3^2}\,dt = \sqrt{14}\int_0^1 te^{6t^2}\,dt = \sqrt{14}\left[\tfrac{1}{12}e^{6t^2}\right]_0^1 = \tfrac{\sqrt{14}}{12}(e^6 - 1).$$

11. $\int_C x^2 y\sqrt{z}\,dz = \int_0^1 (t^3)^2(t)\sqrt{t^2}\cdot 2t\,dt = \int_0^1 2t^9\,dt = \frac{1}{5}t^{10}\Big]_0^1 = \frac{1}{5}$

13. 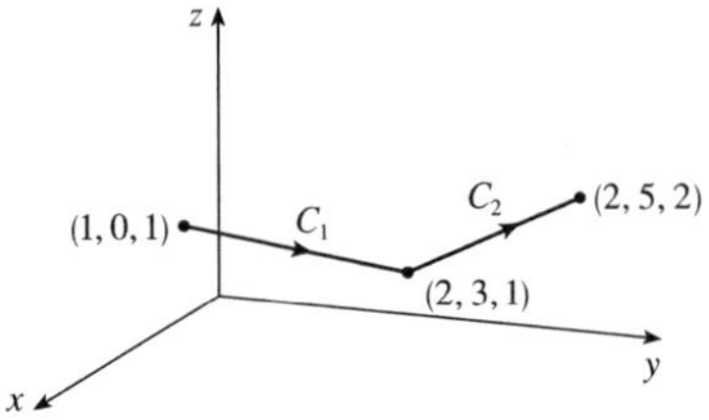

On C_1: $x = 1 + t \Rightarrow dx = dt$, $y = 3t \Rightarrow dy = 3\,dt$, $z = 1$
$\Rightarrow dz = 0\,dt$, $0 \le t \le 1$.

On C_2: $x = 2 \Rightarrow dx = 0\,dt$, $y = 3 + 2t \Rightarrow$
$dy = 2\,dt$, $z = 1 + t \Rightarrow dz = dt$, $0 \le t \le 1$.

Then $\int_C (x + yz)\,dx + 2x\,dy + xyz\,dz$

$$\begin{aligned}&= \int_{C_1} (x+yz)\,dx + 2x\,dy + xyz\,dz + \int_{C_2} (x+yz)\,dx + 2x\,dy + xyz\,dz \\ &= \int_0^1 (1 + t + (3t)(1))\,dt + 2(1+t)\cdot 3\,dt + (1+t)(3t)(1)\cdot 0\,dt \\ &\qquad + \int_0^1 (2 + (3+2t)(1+t))\cdot 0\,dt + 2(2)\cdot 2\,dt + (2)(3+2t)(1+t)\,dt \\ &= \int_0^1 (10t + 7)\,dt + \int_0^1 (4t^2 + 10t + 14)\,dt \\ &= \left[5t^2 + 7t\right]_0^1 + \left[\tfrac{4}{3}t^3 + 5t^2 + 14t\right]_0^1 = 12 + \tfrac{61}{3} = \tfrac{97}{3}\end{aligned}$$

15. (a) Along the line $x = -3$, the vectors of $\mathbf{F}$ have positive y-components, so since the path goes upward, the integrand $\mathbf{F}\cdot\mathbf{T}$ is always positive. Therefore $\int_{C_1} \mathbf{F}\cdot d\mathbf{r} = \int_{C_1} \mathbf{F}\cdot\mathbf{T}\,ds$ is positive.

(b) All of the (nonzero) field vectors along the circle with radius 3 are pointed in the clockwise direction, that is, opposite the direction to the path. So $\mathbf{F}\cdot\mathbf{T}$ is negative, and therefore $\int_{C_2} \mathbf{F}\cdot d\mathbf{r} = \int_{C_2} \mathbf{F}\cdot\mathbf{T}\,ds$ is negative.

17. $\mathbf{r}(t) = t^2\,\mathbf{i} - t^3\mathbf{j}$, so $\mathbf{F}(\mathbf{r}(t)) = (t^2)^2(-t^3)^3\mathbf{i} - (-t^3)\sqrt{t^2}\,\mathbf{j} = -t^{13}\,\mathbf{i} + t^4\,\mathbf{j}$ and $\mathbf{r}'(t) = 2t\,\mathbf{i} - 3t^2\,\mathbf{j}$.

Thus $\int_C \mathbf{F}\cdot d\mathbf{r} = \int_0^1 \mathbf{F}(\mathbf{r}(t))\cdot\mathbf{r}'(t)\,dt = \int_0^1 (-2t^{14} - 3t^6)dt = \left[-\frac{2}{15}t^{15} - \frac{3}{7}t^7\right]_0^1 = -\frac{59}{105}$.

19. $$\begin{aligned}\int_C \mathbf{F}\cdot d\mathbf{r} &= \int_0^1 \langle \sin t^3, \cos(-t^2), t^4\rangle \cdot \langle 3t^2, -2t, 1\rangle\,dt \\ &= \int_0^1 (3t^2 \sin t^3 - 2t\cos t^2 + t^4)\,dt = \left[-\cos t^3 - \sin t^2 + \tfrac{1}{5}t^5\right]_0^1 = \tfrac{6}{5} - \cos 1 - \sin 1\end{aligned}$$

21. We graph $\mathbf{F}(x,y) = (x-y)\,\mathbf{i} + xy\,\mathbf{j}$ and the curve C. We see that most of the vectors starting on C point in roughly the same direction as C, so for these portions of C the tangential component $\mathbf{F}\cdot\mathbf{T}$ is positive. Although some vectors in the third quadrant which start on C point in roughly the opposite direction, and hence give negative tangential components, it seems reasonable that the effect of these portions of C is outweighed by the positive tangential components. Thus, we would expect $\int_C \mathbf{F}\cdot d\mathbf{r} = \int_C \mathbf{F}\cdot\mathbf{T}\,ds$ to be positive.

To verify, we evaluate $\int_C \mathbf{F}\cdot d\mathbf{r}$. The curve C can be represented by $\mathbf{r}(t) = 2\cos t\,\mathbf{i} + 2\sin t\,\mathbf{j}$, $0 \le t \le \frac{3\pi}{2}$, so $\mathbf{F}(\mathbf{r}(t)) = (2\cos t - 2\sin t)\,\mathbf{i} + 4\cos t\sin t\,\mathbf{j}$ and $\mathbf{r}'(t) = -2\sin t\,\mathbf{i} + 2\cos t\,\mathbf{j}$. Then

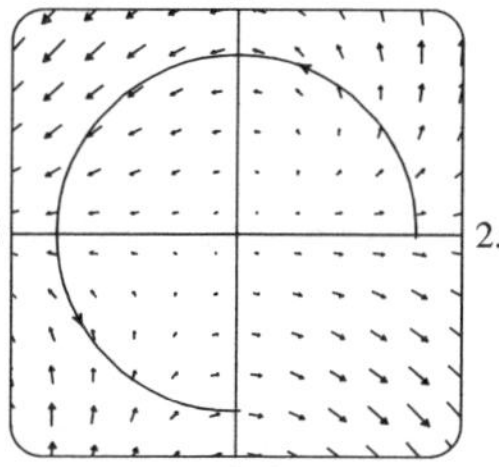

$$\begin{aligned}\int_C \mathbf{F}\cdot d\mathbf{r} &= \int_0^{3\pi/2} \mathbf{F}(\mathbf{r}(t))\cdot\mathbf{r}'(t)\,dt\\ &= \int_0^{3\pi/2} [-2\sin t(2\cos t - 2\sin t) + 2\cos t(4\cos t\sin t)]\,dt\\ &= 4\int_0^{3\pi/2} (\sin^2 t - \sin t\cos t + 2\sin t\cos^2 t)\,dt\\ &= 3\pi + \tfrac{2}{3} \quad \text{[using a CAS]}\end{aligned}$$

23. (a) $\int_C \mathbf{F}\cdot d\mathbf{r} = \int_0^1 \left\langle e^{t^2-1}, t^5\right\rangle\cdot\langle 2t, 3t^2\rangle\,dt = \int_0^1 \left(2te^{t^2-1} + 3t^7\right)dt = \left[e^{t^2-1} + \frac{3}{8}t^8\right]_0^1 = \frac{11}{8} - 1/e$

(b) $\mathbf{r}(0) = \mathbf{0}$, $\mathbf{F}(\mathbf{r}(0)) = \langle e^{-1}, 0\rangle$;

$\mathbf{r}\left(\frac{1}{\sqrt{2}}\right) = \left\langle \frac{1}{2}, \frac{1}{2\sqrt{2}}\right\rangle$, $\mathbf{F}\left(\mathbf{r}\left(\frac{1}{\sqrt{2}}\right)\right) = \left\langle e^{-1/2}, \frac{1}{4\sqrt{2}}\right\rangle$;

$\mathbf{r}(1) = \langle 1,1\rangle$, $\mathbf{F}(\mathbf{r}(1)) = \langle 1,1\rangle$.

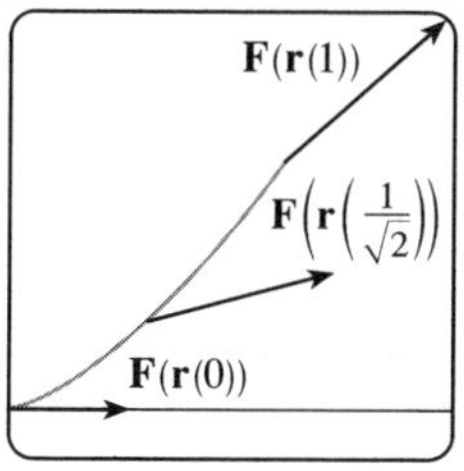

In order to generate the graph with Maple, we use the PLOT command (not to be confused with the plot command) to define each of the vectors. For example,

```
v1:=PLOT(CURVES([[0,0],[evalf(1/exp(1)),0]]));
```

generates the vector from the vector field at the point $(0,0)$ (but without an arrowhead) and gives it the name v1. To show everything on the same screen, we use the display command. In Mathematica, we use ListPlot (with the PlotJoined -> True option) to generate the vectors, and then Show to show everything on the same screen.

25. The part of the astroid that lies in the quadrant is parametrized by $x = \cos^3 t$, $y = \sin^3 t$, $0 \le t \le \frac{\pi}{2}$.

Now $\dfrac{dx}{dt} = 3\cos^2 t\,(-\sin t)$ and $\dfrac{dy}{dt} = 3\sin^2 t\cos t$, so

$$\sqrt{\left(\frac{dx}{dt}\right)^2 + \left(\frac{dy}{dt}\right)^2} = \sqrt{9\cos^4 t\sin^2 t + 9\sin^4 t\cos^2 t} = 3\cos t\sin t\,\sqrt{\cos^2 t + \sin^2 t} = 3\cos t\sin t.$$

Therefore $\int_C x^3y^5\,ds = \int_0^{\pi/2} \cos^9 t\sin^{15} t\,(3\cos t\sin t)\,dt = \frac{945}{16\,777\,216}\pi$.

27. We use the parametrization $x = 2\cos t$, $y = 2\sin t$, $-\frac{\pi}{2} \le t \le \frac{\pi}{2}$. Then

$$ds = \sqrt{\left(\frac{dx}{dt}\right)^2 + \left(\frac{dy}{dt}\right)^2}\,dt = \sqrt{(-2\sin t)^2 + (2\cos t)^2}\,dt = 2\,dt, \text{ so } m = \int_C k\,ds = 2k\int_{-\pi/2}^{\pi/2} dt = 2k(\pi),$$

$$\overline{x} = \frac{1}{2\pi k}\int_C xk\,ds = \frac{1}{2\pi}\int_{-\pi/2}^{\pi/2}(2\cos t)2\,dt = \frac{1}{2\pi}\left[4\sin t\right]_{-\pi/2}^{\pi/2} = \frac{4}{\pi},\ \overline{y} = \frac{1}{2\pi k}\int_C yk\,ds = \frac{1}{2\pi}\int_{-\pi/2}^{\pi/2}(2\sin t)2\,dt = 0.$$

Hence $(\overline{x},\overline{y}) = \left(\frac{4}{\pi}, 0\right)$.

29. (a) $\overline{x} = \dfrac{1}{m}\displaystyle\int_C x\rho(x,y,z)\,ds$, $\overline{y} = \dfrac{1}{m}\displaystyle\int_C y\rho(x,y,z)\,ds$, $\overline{z} = \dfrac{1}{m}\displaystyle\int_C z\rho(x,y,z)\,ds$ where $m = \int_C \rho(x,y,z)\,ds$.

(b) $m = \int_C k\,ds = k\int_0^{2\pi}\sqrt{4\sin^2 t + 4\cos^2 t + 9}\,dt = k\sqrt{13}\int_0^{2\pi} dt = 2\pi k\sqrt{13}$,

$$\overline{x} = \frac{1}{2\pi k\sqrt{13}}\int_0^{2\pi} 2k\sqrt{13}\sin t\,dt = 0,\ \overline{y} = \frac{1}{2\pi k\sqrt{13}}\int_0^{2\pi} 2k\sqrt{13}\cos t\,dt = 0,$$

$$\overline{z} = \frac{1}{2\pi k\sqrt{13}}\int_0^{2\pi}\left(k\sqrt{13}\right)(3t)\,dt = \frac{3}{2\pi}\left(2\pi^2\right) = 3\pi.$$ Hence $(\overline{x},\overline{y},\overline{z}) = (0,0,3\pi)$.

31. From Example 3, $\rho(x,y) = k(1-y)$, $x = \cos t$, $y = \sin t$, and $ds = dt$, $0 \le t \le \pi \;\Rightarrow$

$$I_x = \int_C y^2\rho(x,y)\,ds = \int_0^\pi \sin^2 t\,[k(1-\sin t)]\,dt = k\int_0^\pi(\sin^2 t - \sin^3 t)\,dt$$

$$= \tfrac{1}{2}k\int_0^\pi(1-\cos 2t)\,dt - k\int_0^\pi\left(1-\cos^2 t\right)\sin t\,dt \quad \left[\begin{array}{c}\text{Let } u = \cos t,\ du = -\sin t\,dt \\ \text{in the second integral}\end{array}\right]$$

$$= k\left[\tfrac{\pi}{2} + \int_1^{-1}(1-u^2)\,du\right] = k\left(\tfrac{\pi}{2} - \tfrac{4}{3}\right)$$

$$I_y = \int_C x^2\rho(x,y)\,ds = k\int_0^\pi\cos^2 t\,(1-\sin t)\,dt = \tfrac{k}{2}\int_0^\pi(1+\cos 2t)\,dt - k\int_0^\pi\cos^2 t\sin t\,dt$$

$= k\left(\frac{\pi}{2} - \frac{2}{3}\right)$, using the same substitution as above.

33. $W = \int_C \mathbf{F}\cdot d\mathbf{r} = \int_0^{2\pi}\langle t-\sin t, 3-\cos t\rangle\cdot\langle 1-\cos t, \sin t\rangle\,dt$

$= \int_0^{2\pi}(t - t\cos t - \sin t + \sin t\cos t + 3\sin t - \sin t\cos t)\,dt$

$= \int_0^{2\pi}(t - t\cos t + 2\sin t)\,dt = \left[\frac{1}{2}t^2 - (t\sin t + \cos t) - 2\cos t\right]_0^{2\pi}$ [by integrating by parts in the second term]

$= 2\pi^2$

35. $\mathbf{r}(t) = \langle 1+2t, 4t, 2t\rangle$, $0 \le t \le 1$,

$$W = \int_C \mathbf{F}\cdot d\mathbf{r} = \int_0^1\langle 6t, 1+4t, 1+6t\rangle\cdot\langle 2,4,2\rangle\,dt = \int_0^1(12t + 4(1+4t) + 2(1+6t))\,dt$$

$$= \int_0^1(40t+6)\,dt = \left[20t^2 + 6t\right]_0^1 = 26$$

37. Let $\mathbf{F} = 185\,\mathbf{k}$. To parametrize the staircase, let

$x = 20\cos t$, $y = 20\sin t$, $z = \frac{90}{6\pi}t = \frac{15}{\pi}t$, $0 \le t \le 6\pi \;\Rightarrow$

$$W = \int_C \mathbf{F}\cdot d\mathbf{r} = \int_0^{6\pi}\langle 0,0,185\rangle\cdot\left\langle -20\sin t, 20\cos t, \tfrac{15}{\pi}\right\rangle dt = (185)\tfrac{15}{\pi}\int_0^{6\pi}dt = (185)(90) \approx 1.67\times 10^4 \text{ ft-lb}$$

39. (a) $\mathbf{r}(t) = \langle\cos t, \sin t\rangle$, $0 \le t \le 2\pi$, and let $\mathbf{F} = \langle a, b\rangle$. Then

$$W = \int_C \mathbf{F}\cdot d\mathbf{r} = \int_0^{2\pi}\langle a,b\rangle\cdot\langle -\sin t, \cos t\rangle\,dt = \int_0^{2\pi}(-a\sin t + b\cos t)\,dt = \left[a\cos t + b\sin t\right]_0^{2\pi}$$

$$= a + 0 - a + 0 = 0$$

(b) Yes. $\mathbf{F}(x,y) = k\,\mathbf{x} = \langle kx, ky\rangle$ and

$$W = \int_C \mathbf{F}\cdot d\mathbf{r} = \int_0^{2\pi}\langle k\cos t, k\sin t\rangle\cdot\langle -\sin t, \cos t\rangle\,dt = \int_0^{2\pi}(-k\sin t\cos t + k\sin t\cos t)\,dt = \int_0^{2\pi}0\,dt = 0$$

41. The work done in moving the object is $\int_C \mathbf{F}\cdot d\mathbf{r} = \int_C \mathbf{F}\cdot\mathbf{T}\,ds$. We can approximate this integral by dividing C into 7 segments of equal length $\Delta s = 2$ and approximating $\mathbf{F}\cdot\mathbf{T}$, that is, the tangential component of force, at a point (x_i^*, y_i^*) on each segment. Since C is composed of straight line segments, $\mathbf{F}\cdot\mathbf{T}$ is the scalar projection of each force vector onto C. If we choose (x_i^*, y_i^*) to be the point on the segment closest to the origin, then the work done is

$\int_C \mathbf{F}\cdot\mathbf{T}\,ds \approx \sum\limits_{i=1}^{7}[\mathbf{F}(x_i^*, y_i^*)\cdot\mathbf{T}(x_i^*, y_i^*)]\,\Delta s = [2+2+2+2+1+1+1](2) = 22$. Thus, we estimate the work done to be approximately 22 J.

13.3 The Fundamental Theorem for Line Integrals

1. C appears to be a smooth curve, and since ∇f is continuous, we know f is differentiable. Then Theorem 2 says that the value of $\int_C \nabla f \cdot d\mathbf{r}$ is simply the difference of the values of f at the terminal and initial points of C. From the graph, this is $50 - 10 = 40$.

3. $\partial(6x + 5y)/\partial y = 5 = \partial(5x + 4y)/\partial x$ and the domain of $\mathbf{F}$ is $\mathbb{R}^2$ which is open and simply-connected, so by Theorem 6 $\mathbf{F}$ is conservative. Thus, there exists a function f such that $\nabla f = \mathbf{F}$, that is, $f_x(x, y) = 6x + 5y$ and $f_y(x, y) = 5x + 4y$. But $f_x(x, y) = 6x + 5y$ implies $f(x, y) = 3x^2 + 5xy + g(y)$ and differentiating both sides of this equation with respect to y gives $f_y(x, y) = 5x + g'(y)$. Thus $5x + 4y = 5x + g'(y)$ so $g'(y) = 4y$ and $g(y) = 2y^2 + K$ where K is a constant. Hence $f(x, y) = 3x^2 + 5xy + 2y^2 + K$ is a potential function for $\mathbf{F}$.

5. $\partial(xe^y)/\partial y = xe^y$, $\partial(ye^x)/\partial x = ye^x$. Since these are not equal, $\mathbf{F}$ is not conservative.

7. $\partial(2x\cos y - y\cos x)/\partial y = -2x\sin y - \cos x = \partial(-x^2\sin y - \sin x)/\partial x$ and the domain of $\mathbf{F}$ is $\mathbb{R}^2$. Hence $\mathbf{F}$ is conservative so there exists a function f such that $\nabla f = \mathbf{F}$. Then $f_x(x, y) = 2x\cos y - y\cos x$ implies $f(x, y) = x^2\cos y - y\sin x + g(y)$ and $f_y(x, y) = -x^2\sin y - \sin x + g'(y)$. But $f_y(x, y) = -x^2\sin y - \sin x$ so $g'(y) = 0 \;\Rightarrow\; g(y) = K$. Then $f(x, y) = x^2\cos y - y\sin x + K$ is a potential function for $\mathbf{F}$.

9. $\partial(ye^x + \sin y)/\partial y = e^x + \cos y = \partial(e^x + x\cos y)/\partial x$ and the domain of $\mathbf{F}$ is $\mathbb{R}^2$. Hence $\mathbf{F}$ is conservative so there exists a function f such that $\nabla f = \mathbf{F}$. Then $f_x(x, y) = ye^x + \sin y$ implies $f(x, y) = ye^x + x\sin y + g(y)$ and $f_y(x, y) = e^x + x\cos y + g'(y)$. But $f_y(x, y) = e^x + x\cos y$ so $g(y) = K$ and $f(x, y) = ye^x + x\sin y + K$ is a potential function for $\mathbf{F}$.

11. (a) $\mathbf{F}$ has continuous first-order partial derivatives and $\dfrac{\partial}{\partial y}2xy = 2x = \dfrac{\partial}{\partial x}(x^2)$ on $\mathbb{R}^2$, which is open and simply-connected. Thus, $\mathbf{F}$ is conservative by Theorem 6. Then we know that the line integral of $\mathbf{F}$ is independent of path; in particular, the value of $\int_C \mathbf{F} \cdot d\mathbf{r}$ depends only on the endpoints of C. Since all three curves have the same initial and terminal points, $\int_C \mathbf{F} \cdot d\mathbf{r}$ will have the same value for each curve.

(b) We first find a potential function f, so that $\nabla f = \mathbf{F}$. We know $f_x(x, y) = 2xy$ and $f_y(x, y) = x^2$. Integrating $f_x(x, y)$ with respect to x, we have $f(x, y) = x^2y + g(y)$. Differentiating both sides with respect to y gives $f_y(x, y) = x^2 + g'(y)$, so we must have $x^2 + g'(y) = x^2 \;\Rightarrow\; g'(y) = 0 \;\Rightarrow\; g(y) = K$, a constant. Thus $f(x, y) = x^2y + K$. All three curves start at $(1, 2)$ and end at $(3, 2)$, so by Theorem 2, $\int_C \mathbf{F} \cdot d\mathbf{r} = f(3, 2) - f(1, 2) = 18 - 2 = 16$ for each curve.

13. (a) $f_x(x, y) = x^3y^4$ implies $f(x, y) = \frac{1}{4}x^4y^4 + g(y)$ and $f_y(x, y) = x^4y^3 + g'(y)$. But $f_y(x, y) = x^4y^3$ so $g'(y) = 0 \;\Rightarrow\; g(y) = K$, a constant. We can take $K = 0$, so $f(x, y) = \frac{1}{4}x^4y^4$.

(b) The initial point of C is $\mathbf{r}(0) = (0, 1)$ and the terminal point is $\mathbf{r}(1) = (1, 2)$, so $\int_C \mathbf{F} \cdot d\mathbf{r} = f(1, 2) - f(0, 1) = 4 - 0 = 4$.

15. (a) $f_x(x, y, z) = yz$ implies $f(x, y, z) = xyz + g(y, z)$ and so $f_y(x, y, z) = xz + g_y(y, z)$. But $f_y(x, y, z) = xz$ so $g_y(y, z) = 0 \;\Rightarrow\; g(y, z) = h(z)$. Thus $f(x, y, z) = xyz + h(z)$ and $f_z(x, y, z) = xy + h'(z)$. But $f_z(x, y, z) = xy + 2z$, so $h'(z) = 2z \;\Rightarrow\; h(z) = z^2 + K$. Hence $f(x, y, z) = xyz + z^2$ (taking $K = 0$).

(b) $\int_C \mathbf{F} \cdot d\mathbf{r} = f(4, 6, 3) - f(1, 0, -2) = 81 - 4 = 77$.

17. (a) $f_x(x,y,z) = y^2 \cos z$ implies $f(x,y,z) = xy^2 \cos z + g(y,z)$ and so $f_y(x,y,z) = 2xy \cos z + g_y(y,z)$. But $f_y(x,y,z) = 2xy \cos z$ so $g_y(y,z) = 0 \;\Rightarrow\; g(y,z) = h(z)$. Thus $f(x,y,z) = xy^2 \cos z + h(z)$ and $f_z(x,y,z) = -xy^2 \sin z + h'(z)$. But $f_z(x,y,z) = -xy^2 \sin z$, so $h'(z) = 0 \;\Rightarrow\; h(z) = K$. Hence $f(x,y,z) = xy^2 \cos z$ (taking $K = 0$).

(b) $\mathbf{r}(0) = \langle 0,0,0\rangle$, $\mathbf{r}(\pi) = \langle \pi^2, 0, \pi\rangle$ so $\int_C \mathbf{F} \cdot d\mathbf{r} = f(\pi^2, 0, \pi) - f(0,0,0) = 0 - 0 = 0$.

19. Here $\mathbf{F}(x,y) = (2x \sin y)\,\mathbf{i} + (x^2 \cos y - 3y^2)\,\mathbf{j}$. Then $f(x,y) = x^2 \sin y - y^3$ is a potential function for $\mathbf{F}$, that is, $\nabla f = \mathbf{F}$ so $\mathbf{F}$ is conservative and thus its line integral is independent of path. Hence

$\int_C 2x \sin y\, dx + (x^2 \cos y - 3y^2)\, dy = \int_C \mathbf{F} \cdot d\mathbf{r} = f(5,1) - f(-1,0) = 25 \sin 1 - 1$.

21. $\mathbf{F}(x,y) = 2y^{3/2}\,\mathbf{i} + 3x\sqrt{y}\,\mathbf{j}$, $W = \int_C \mathbf{F} \cdot d\mathbf{r}$. Since $\partial(2y^{3/2})/\partial y = 3\sqrt{y} = \partial(3x\sqrt{y})/\partial x$, there exists a function f such that $\nabla f = \mathbf{F}$. In fact, $f_x(x,y) = 2y^{3/2} \;\Rightarrow\; f(x,y) = 2xy^{3/2} + g(y) \;\Rightarrow\; f_y(x,y) = 3xy^{1/2} + g'(y)$. But $f_y(x,y) = 3x\sqrt{y}$ so $g'(y) = 0$ or $g(y) = K$. We can take $K = 0 \;\Rightarrow\; f(x,y) = 2xy^{3/2}$. Thus

$W = \int_C \mathbf{F} \cdot d\mathbf{r} = f(2,4) - f(1,1) = 2(2)(8) - 2(1) = 30$.

23. We know that if the vector field (call it $\mathbf{F}$) is conservative, then around any closed path C, $\int_C \mathbf{F} \cdot d\mathbf{r} = 0$. But take C to be a circle centred at the origin, oriented counterclockwise. All of the field vectors that start on C are roughly in the direction of motion along C, so the integral around C will be positive. Therefore the field is not conservative.

25.

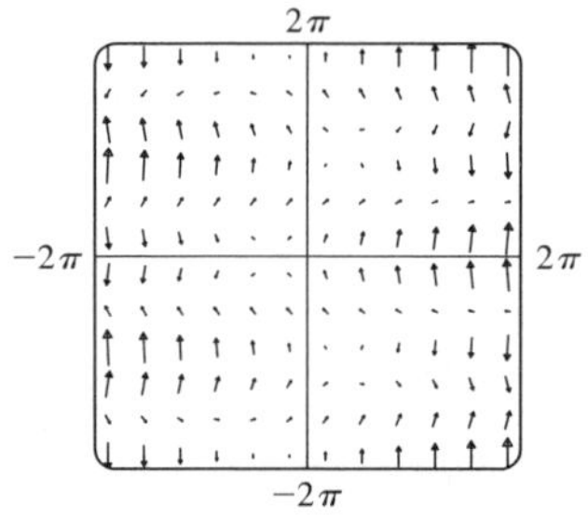

From the graph, it appears that $\mathbf{F}$ is conservative, since around all closed paths, the number and size of the field vectors pointing in directions similar to that of the path seem to be roughly the same as the number and size of the vectors pointing in the opposite direction. To check, we calculate

$\dfrac{\partial}{\partial y}(\sin y) = \cos y = \dfrac{\partial}{\partial x}(1 + x \cos y)$. Thus $\mathbf{F}$ is conservative, by Theorem 6.

27. Since $\mathbf{F}$ is conservative, there exists a function f such that $\mathbf{F} = \nabla f$, that is, $P = f_x$, $Q = f_y$, and $R = f_z$. Since P, Q and R have continuous first order partial derivatives, Clairaut's Theorem says that $\partial P/\partial y = f_{xy} = f_{yx} = \partial Q/\partial x$, $\partial P/\partial z = f_{xz} = f_{zx} = \partial R/\partial x$, and $\partial Q/\partial z = f_{yz} = f_{zy} = \partial R/\partial y$.

29. $D = \{(x,y) \mid x > 0, y > 0\}$ = the first quadrant (excluding the axes).

(a) D is open because around every point in D we can put a disc that lies in D.

(b) D is connected because the straight line segment joining any two points in D lies in D.

(c) D is simply-connected because it's connected and has no holes.

31. $D = \{(x,y) \mid 1 < x^2 + y^2 < 4\}$ = the annular region between the circles with centre $(0,0)$ and radii 1 and 2.

(a) D is open.

(b) D is connected.

(c) D is not simply-connected. For example, $x^2 + y^2 = (1.5)^2$ is simple and closed and lies within D but encloses points that are not in D. (Or we can say, D has a hole, so is not simply-connected.)

33. (a) $P = -\dfrac{y}{x^2+y^2}$, $\dfrac{\partial P}{\partial y} = \dfrac{y^2-x^2}{(x^2+y^2)^2}$ and $Q = \dfrac{x}{x^2+y^2}$, $\dfrac{\partial Q}{\partial x} = \dfrac{y^2-x^2}{(x^2+y^2)^2}$. Thus $\dfrac{\partial P}{\partial y} = \dfrac{\partial Q}{\partial x}$.

(b) C_1: $x = \cos t$, $y = \sin t$, $0 \le t \le \pi$, C_2: $x = \cos t$, $y = \sin t$, $t = 2\pi$ to $t = \pi$. Then

$$\int_{C_1} \mathbf{F} \cdot d\mathbf{r} = \int_0^{\pi} \frac{(-\sin t)(-\sin t) + (\cos t)(\cos t)}{\cos^2 t + \sin^2 t}\,dt = \int_0^{\pi} dt = \pi \text{ and } \int_{C_2} \mathbf{F} \cdot d\mathbf{r} = \int_{2\pi}^{\pi} dt = -\pi$$

Since these aren't equal, the line integral of $\mathbf{F}$ isn't independent of path. (Or notice that $\int_{C_3} \mathbf{F} \cdot d\mathbf{r} = \int_0^{2\pi} dt = 2\pi$ where C_3 is the circle $x^2 + y^2 = 1$, and apply the contrapositive of Theorem 3.) This doesn't contradict Theorem 6, since the domain of $\mathbf{F}$, which is $\mathbb{R}^2$ except the origin, isn't simply-connected.

13.4 Green's Theorem

1. (a)

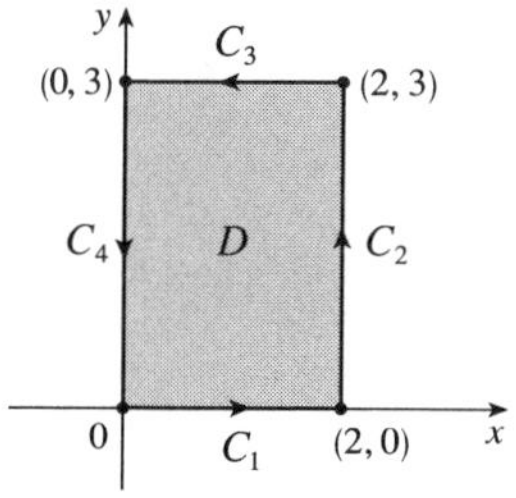

C_1: $x = t \;\Rightarrow\; dx = dt$, $y = 0 \;\Rightarrow\; dy = 0\,dt$, $0 \le t \le 2$.

C_2: $x = 2 \;\Rightarrow\; dx = 0\,dt$, $y = t \;\Rightarrow\; dy = dt$, $0 \le t \le 3$.

C_3: $x = 2 - t \;\Rightarrow\; dx = -dt$, $y = 3 \;\Rightarrow\; dy = 0\,dt$, $0 \le t \le 2$.

C_4: $x = 0 \;\Rightarrow\; dx = 0\,dt$, $y = 3 - t \;\Rightarrow\; dy = -dt$, $0 \le t \le 3$.

Thus $\oint_C xy^2\,dx + x^3\,dy = \displaystyle\oint_{C_1+C_2+C_3+C_4} xy^2\,dx + x^3\,dy$

$$= \int_0^2 0\,dt + \int_0^3 8\,dt + \int_0^2 -9(2-t)\,dt + \int_0^3 0\,dt$$

$$= 0 + 24 - 18 + 0 = 6$$

(b) $\oint_C xy^2\,dx + x^3\,dy = \iint_D \left[\frac{\partial}{\partial x}(x^3) - \frac{\partial}{\partial y}(xy^2)\right] dA = \int_0^2 \int_0^3 (3x^2 - 2xy)\,dy\,dx = \int_0^2 (9x^2 - 9x)\,dx = 24 - 18 = 6$

3. (a)

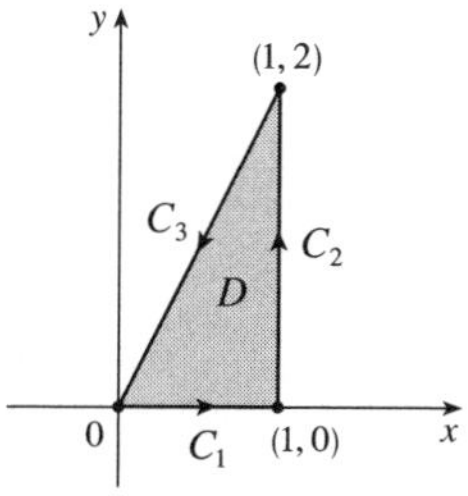

C_1: $x = t \;\Rightarrow\; dx = dt$, $y = 0 \;\Rightarrow\; dy = 0\,dt$, $0 \le t \le 1$.

C_2: $x = 1 \;\Rightarrow\; dx = 0\,dt$, $y = t \;\Rightarrow\; dy = dt$, $0 \le t \le 2$.

C_3: $x = 1 - t \;\Rightarrow\; dx = -dt$, $y = 2 - 2t \;\Rightarrow\; dy = -2\,dt$, $0 \le t \le 1$.

Thus

$$\oint_C xy\,dx + x^2y^3\,dy = \oint_{C_1+C_2+C_3} xy\,dx + x^2y^3\,dy$$

$$= \int_0^1 0\,dt + \int_0^2 t^3\,dt + \int_0^1 \left[-(1-t)(2-2t) - 2(1-t)^2(2-2t)^3\right] dt$$

$$= 0 + \left[\tfrac{1}{4}t^4\right]_0^2 + \left[\tfrac{2}{3}(1-t)^3 + \tfrac{8}{3}(1-t)^6\right]_0^1 = 4 - \tfrac{10}{3} = \tfrac{2}{3}$$

(b) $\oint_C xy\,dx + x^2y^3\,dy = \iint_D \left[\frac{\partial}{\partial x}(x^2y^3) - \frac{\partial}{\partial y}(xy)\right] dA = \int_0^1 \int_0^{2x} (2xy^3 - x)\,dy\,dx$

$$= \int_0^1 \left[\tfrac{1}{2}xy^4 - xy\right]_{y=0}^{y=2x} dx = \int_0^1 (8x^5 - 2x^2)\,dx = \tfrac{4}{3} - \tfrac{2}{3} = \tfrac{2}{3}$$

5. We can parametrize C as $x = \cos\theta$, $y = \sin\theta$, $0 \le \theta \le 2\pi$. Then the line integral is

$\oint_C P\,dx + Q\,dy = \int_0^{2\pi} \cos^4\theta \sin^5\theta\,(-\sin\theta)\,d\theta + \int_0^{2\pi} (-\cos^7\theta \sin^6\theta)\cos\theta\,d\theta = -\frac{29\pi}{1024}$, according to a CAS.

The double integral is

$$\iint_D \left(\frac{\partial Q}{\partial x} - \frac{\partial P}{\partial y}\right) dA = \int_{-1}^{1} \int_{-\sqrt{1-x^2}}^{\sqrt{1-x^2}} (-7x^6y^6 - 5x^4y^4)\,dy\,dx = -\frac{29\pi}{1024},$$

verifying Green's Theorem in this case.

7. The region D enclosed by C is $[0,1]\times[0,1]$, so

$$\int_C e^y\,dx + 2xe^y\,dy = \iint_D\left[\frac{\partial}{\partial x}(2xe^y) - \frac{\partial}{\partial y}(e^y)\right]dA = \int_0^1\int_0^1 (2e^y - e^y)\,dy\,dx$$
$$= \int_0^1 dx\int_0^1 e^y\,dy = (1)(e^1 - e^0) = e - 1$$

9. $$\int_C \left(y + e^{\sqrt{x}}\right)dx + (2x + \cos y^2)\,dy = \iint_D\left[\frac{\partial}{\partial x}(2x + \cos y^2) - \frac{\partial}{\partial y}\left(y + e^{\sqrt{x}}\right)\right]dA$$
$$= \int_0^1\int_{y^2}^{\sqrt{y}}(2-1)\,dx\,dy = \int_0^1 (y^{1/2} - y^2)\,dy = \tfrac{1}{3}$$

11. $$\int_C y^3\,dx - x^3\,dy = \iint_D\left[\frac{\partial}{\partial x}(-x^3) - \frac{\partial}{\partial y}(y^3)\right]dA = \iint_D(-3x^2 - 3y^2)\,dA = \int_0^{2\pi}\int_0^2(-3r^2)\,r\,dr\,d\theta$$
$$= -3\int_0^{2\pi}d\theta\int_0^2 r^3\,dr = -3(2\pi)(4) = -24\pi$$

13. $\mathbf{F}(x,y) = \left\langle \sqrt{x} + y^3, x^2 + \sqrt{y}\right\rangle$ and the region D enclosed by C is given by $\{(x,y)\mid 0\le x\le\pi, 0\le y\le\sin x\}$. C is traversed clockwise, so $-C$ gives the positive orientation.

$$\int_C \mathbf{F}\cdot d\mathbf{r} = -\int_{-C}(\sqrt{x}+y^3)\,dx + (x^2+\sqrt{y})\,dy = -\iint_D\left[\frac{\partial}{\partial x}(x^2+\sqrt{y}) - \frac{\partial}{\partial y}(\sqrt{x}+y^3)\right]dA$$
$$= -\int_0^\pi\int_0^{\sin x}(2x - 3y^2)\,dy\,dx = -\int_0^\pi\left[2xy - y^3\right]_{y=0}^{y=\sin x}dx$$
$$= -\int_0^\pi(2x\sin x - \sin^3 x)\,dx = -\int_0^\pi(2x\sin x - (1-\cos^2 x)\sin x)\,dx$$
$$= -\left[2\sin x - 2x\cos x + \cos x - \tfrac{1}{3}\cos^3 x\right]_0^\pi \qquad \text{[integrate by parts in the first term]}$$
$$= -\left(2\pi - 2 + \tfrac{2}{3}\right) = \tfrac{4}{3} - 2\pi$$

15. $\mathbf{F}(x,y) = \left\langle e^x + x^2y, e^y - xy^2\right\rangle$ and the region D enclosed by C is the disc $x^2+y^2\le 25$. C is traversed clockwise, so $-C$ gives the positive orientation.

$$\int_C \mathbf{F}\cdot d\mathbf{r} = -\int_{-C}(e^x + x^2y)\,dx + (e^y - xy^2)\,dy = -\iint_D\left[\frac{\partial}{\partial x}(e^y - xy^2) - \frac{\partial}{\partial y}(e^x + x^2y)\right]dA$$
$$= -\iint_D(-y^2 - x^2)\,dA = \iint_D(x^2+y^2)\,dA = \int_0^{2\pi}\int_0^5(r^2)\,r\,dr\,d\theta = \int_0^{2\pi}d\theta\int_0^5 r^3\,dr = 2\pi\left[\tfrac{1}{4}r^4\right]_0^5 = \tfrac{625}{2}\pi$$

17. By Green's Theorem, $W = \int_C \mathbf{F}\cdot d\mathbf{r} = \int_C x(x+y)\,dx + xy^2\,dy = \iint_D(y^2 - x)\,dy\,dx$ where C is the path described in the question and D is the triangle bounded by C. So

$$W = \int_0^1\int_0^{1-x}(y^2 - x)\,dy\,dx = \int_0^1\left[\tfrac{1}{3}y^3 - xy\right]_{y=0}^{y=1-x}dx = \int_0^1\left(\tfrac{1}{3}(1-x)^3 - x(1-x)\right)dx$$
$$= \left[-\tfrac{1}{12}(1-x)^4 - \tfrac{1}{2}x^2 + \tfrac{1}{3}x^3\right]_0^1 = \left(-\tfrac{1}{2} + \tfrac{1}{3}\right) - \left(-\tfrac{1}{12}\right) = -\tfrac{1}{12}$$

19. Let C_1 be the arch of the cycloid from $(0,0)$ to $(2\pi,0)$, which corresponds to $0\le t\le 2\pi$, and let C_2 be the segment from $(2\pi,0)$ to $(0,0)$, so C_2 is given by $x = 2\pi - t$, $y = 0$, $0\le t\le 2\pi$. Then $C = C_1\cup C_2$ is traversed clockwise, so $-C$ is oriented positively. Thus $-C$ encloses the area under one arch of the cycloid and from (5) we have

$$A = -\oint_{-C} y\,dx = \int_{C_1} y\,dx + \int_{C_2} y\,dx = \int_0^{2\pi}(1-\cos t)(1-\cos t)\,dt + \int_0^{2\pi}0\,(-dt)$$
$$= \int_0^{2\pi}(1 - 2\cos t + \cos^2 t)\,dt + 0 = \left[t - 2\sin t + \tfrac{1}{2}t + \tfrac{1}{4}\sin 2t\right]_0^{2\pi} = 3\pi$$

21. (a) Using Equation 13.2.8, we write parametric equations of the line segment as $x = (1-t)x_1 + tx_2$, $y = (1-t)y_1 + ty_2$, $0\le t\le 1$. Then $dx = (x_2 - x_1)\,dt$ and $dy = (y_2 - y_1)\,dt$, so

$$\int_C x\,dy - y\,dx = \int_0^1[(1-t)x_1 + tx_2](y_2 - y_1)\,dt + [(1-t)y_1 + ty_2](x_2 - x_1)\,dt$$
$$= \int_0^1(x_1(y_2-y_1) - y_1(x_2-x_1) + t[(y_2-y_1)(x_2-x_1) - (x_2-x_1)(y_2-y_1)])\,dt$$
$$= \int_0^1(x_1y_2 - x_2y_1)\,dt = x_1y_2 - x_2y_1$$

(b) We apply Green's Theorem to the path $C = C_1 \cup C_2 \cup \cdots \cup C_n$, where C_i is the line segment that joins (x_i, y_i) to (x_{i+1}, y_{i+1}) for $i = 1, 2, \ldots, n-1$, and C_n is the line segment that joins (x_n, y_n) to (x_1, y_1). From (5), $\frac{1}{2}\int_C x\,dy - y\,dx = \iint_D dA$, where D is the polygon bounded by C. Therefore

$$\text{area of polygon} = A(D) = \iint_D dA = \tfrac{1}{2}\int_C x\,dy - y\,dx$$
$$= \tfrac{1}{2}\left(\int_{C_1} x\,dy - y\,dx + \int_{C_2} x\,dy - y\,dx + \cdots + \int_{C_{n-1}} x\,dy - y\,dx + \int_{C_n} x\,dy - y\,dx\right)$$

To evaluate these integrals we use the formula from (a) to get
$A(D) = \frac{1}{2}[(x_1y_2 - x_2y_1) + (x_2y_3 - x_3y_2) + \cdots + (x_{n-1}y_n - x_ny_{n-1}) + (x_ny_1 - x_1y_n)]$.

(c) $A = \frac{1}{2}[(0\cdot 1 - 2\cdot 0) + (2\cdot 3 - 1\cdot 1) + (1\cdot 2 - 0\cdot 3) + (0\cdot 1 - (-1)\cdot 2) + (-1\cdot 0 - 0\cdot 1)]$
$= \frac{1}{2}(0 + 5 + 2 + 2) = \frac{9}{2}$

23. Here $A = \frac{1}{2}(1)(1) = \frac{1}{2}$ and $C = C_1 + C_2 + C_3$, where C_1: $x = x$, $y = 0$, $0 \le x \le 1$;
C_2: $x = x$, $y = 1 - x$, $x = 1$ to $x = 0$; and C_3: $x = 0$, $y = 1$ to $y = 0$. Then
$\overline{x} = \frac{1}{2A}\int_C x^2\,dy = \int_{C_1} x^2\,dy + \int_{C_2} x^2\,dy + \int_{C_3} x^2\,dy = 0 + \int_1^0 (x^2)(-dx) + 0 = \frac{1}{3}$. Similarly,
$\overline{y} = -\frac{1}{2A}\int_C y^2\,dx = \int_{C_1} y^2\,dx + \int_{C_2} y^2\,dx + \int_{C_3} y^2\,dx = 0 + \int_1^0 (1-x)^2(-dx) + 0 = \frac{1}{3}$.
Therefore $(\overline{x}, \overline{y}) = \left(\frac{1}{3}, \frac{1}{3}\right)$.

25. By Green's Theorem, $-\frac{1}{3}\rho\oint_C y^3\,dx = -\frac{1}{3}\rho\iint_D(-3y^2)\,dA = \iint_D y^2\rho\,dA = I_x$ and
$\frac{1}{3}\rho\oint_C x^3\,dy = \frac{1}{3}\rho\iint_D(3x^2)\,dA = \iint_D x^2\rho\,dA = I_y$.

27. Since C is a simple closed path which doesn't pass through or enclose the origin, there exists an open region that doesn't contain the origin but does contain D. Thus $P = -y/(x^2 + y^2)$ and $Q = x/(x^2 + y^2)$ have continuous partial derivatives on this open region containing D and we can apply Green's Theorem. But by Exercise 13.3.33(a), $\partial P/\partial y = \partial Q/\partial x$, so $\oint_C \mathbf{F}\cdot d\mathbf{r} = \iint_D 0\,dA = 0$.

29. Using the first part of (5), we have that $\iint_R dx\,dy = A(R) = \int_{\partial R} x\,dy$. But $x = g(u,v)$, and $dy = \dfrac{\partial h}{\partial u}\,du + \dfrac{\partial h}{\partial v}\,dv$, and we orient ∂S by taking the positive direction to be that which corresponds, under the mapping, to the positive direction along ∂R, so

$$\int_{\partial R} x\,dy = \int_{\partial S} g(u,v)\left(\frac{\partial h}{\partial u}\,du + \frac{\partial h}{\partial v}\,dv\right) = \int_{\partial S} g(u,v)\,\frac{\partial h}{\partial u}\,du + g(u,v)\,\frac{\partial h}{\partial v}\,dv$$
$$= \pm\iint_S \left[\frac{\partial}{\partial u}\left(g(u,v)\,\frac{\partial h}{\partial v}\right) - \frac{\partial}{\partial v}\left(g(u,v)\,\frac{\partial h}{\partial u}\right)\right] dA \quad \text{[using Green's Theorem in the } uv\text{-plane]}$$
$$= \pm\iint_S \left(\frac{\partial g}{\partial u}\frac{\partial h}{\partial v} + g(u,v)\,\frac{\partial^2 h}{\partial u\,\partial v} - \frac{\partial g}{\partial v}\frac{\partial h}{\partial u} - g(u,v)\,\frac{\partial^2 h}{\partial v\,\partial u}\right) dA \qquad \text{[using the Chain Rule]}$$
$$= \pm\iint_S \left(\frac{\partial x}{\partial u}\frac{\partial y}{\partial v} - \frac{\partial x}{\partial v}\frac{\partial y}{\partial u}\right) dA \text{ [by the equality of mixed partials]} = \pm\iint_S \frac{\partial(x,y)}{\partial(u,v)}\,du\,dv$$

The sign is chosen to be positive if the orientation that we gave to ∂S corresponds to the usual positive orientation, and it is negative otherwise. In either case, since $A(R)$ is positive, the sign chosen must be the same as the sign of $\dfrac{\partial(x,y)}{\partial(u,v)}$. Therefore

$$A(R) = \iint_R dx\,dy = \iint_S \left|\frac{\partial(x,y)}{\partial(u,v)}\right| du\,dv.$$

13.5 Curl and Divergence

1. (a) $\operatorname{curl}\mathbf{F} = \nabla\times\mathbf{F} = \begin{vmatrix} \mathbf{i} & \mathbf{j} & \mathbf{k} \\ \partial/\partial x & \partial/\partial y & \partial/\partial z \\ xyz & 0 & -x^2y \end{vmatrix} = (-x^2 - 0)\,\mathbf{i} - (-2xy - xy)\,\mathbf{j} + (0 - xz)\,\mathbf{k}$

$= -x^2\,\mathbf{i} + 3xy\,\mathbf{j} - xz\,\mathbf{k}$

(b) $\operatorname{div}\mathbf{F} = \nabla\cdot\mathbf{F} = \dfrac{\partial}{\partial x}(xyz) + \dfrac{\partial}{\partial y}(0) + \dfrac{\partial}{\partial z}(-x^2y) = yz + 0 + 0 = yz$

3. (a) $\operatorname{curl}\mathbf{F} = \nabla\times\mathbf{F} = \begin{vmatrix} \mathbf{i} & \mathbf{j} & \mathbf{k} \\ \partial/\partial x & \partial/\partial y & \partial/\partial z \\ 1 & x+yz & xy-\sqrt{z} \end{vmatrix} = (x-y)\,\mathbf{i} - (y-0)\,\mathbf{j} + (1-0)\,\mathbf{k}$

$= (x-y)\,\mathbf{i} - y\,\mathbf{j} + \mathbf{k}$

(b) $\operatorname{div}\mathbf{F} = \nabla\cdot\mathbf{F} = \dfrac{\partial}{\partial x}(1) + \dfrac{\partial}{\partial y}(x+yz) + \dfrac{\partial}{\partial z}(xy - \sqrt{z}) = z - \dfrac{1}{2\sqrt{z}}$

5. (a) $\operatorname{curl}\mathbf{F} = \nabla\times\mathbf{F} = \begin{vmatrix} \mathbf{i} & \mathbf{j} & \mathbf{k} \\ \partial/\partial x & \partial/\partial y & \partial/\partial z \\ e^x\sin y & e^x\cos y & z \end{vmatrix} = (0-0)\,\mathbf{i} - (0-0)\,\mathbf{j} + (e^x\cos y - e^x\cos y)\,\mathbf{k} = \mathbf{0}$

(b) $\operatorname{div}\mathbf{F} = \nabla\cdot\mathbf{F} = \dfrac{\partial}{\partial x}(e^x\sin y) + \dfrac{\partial}{\partial y}(e^x\cos y) + \dfrac{\partial}{\partial z}(z) = e^x\sin y - e^x\sin y + 1 = 1$

7. If the vector field is $\mathbf{F} = P\,\mathbf{i} + Q\,\mathbf{j} + R\,\mathbf{k}$, then we know $R = 0$. In addition, the x-component of each vector of $\mathbf{F}$ is 0, so $P = 0$, hence $\dfrac{\partial P}{\partial x} = \dfrac{\partial P}{\partial y} = \dfrac{\partial P}{\partial z} = \dfrac{\partial R}{\partial x} = \dfrac{\partial R}{\partial y} = \dfrac{\partial R}{\partial z} = 0$. Q decreases as y increases, so $\dfrac{\partial Q}{\partial y} < 0$, but Q doesn't change in the x- or z-directions, so $\dfrac{\partial Q}{\partial x} = \dfrac{\partial Q}{\partial z} = 0$.

(a) $\operatorname{div}\mathbf{F} = \dfrac{\partial P}{\partial x} + \dfrac{\partial Q}{\partial y} + \dfrac{\partial R}{\partial z} = 0 + \dfrac{\partial Q}{\partial y} + 0 < 0$

(b) $\operatorname{curl}\mathbf{F} = \left(\dfrac{\partial R}{\partial y} - \dfrac{\partial Q}{\partial z}\right)\mathbf{i} + \left(\dfrac{\partial P}{\partial z} - \dfrac{\partial R}{\partial x}\right)\mathbf{j} + \left(\dfrac{\partial Q}{\partial x} - \dfrac{\partial P}{\partial y}\right)\mathbf{k} = (0-0)\,\mathbf{i} + (0-0)\,\mathbf{j} + (0-0)\,\mathbf{k} = \mathbf{0}$

9. If the vector field is $\mathbf{F} = P\,\mathbf{i} + Q\,\mathbf{j} + R\,\mathbf{k}$, then we know $R = 0$. In addition, the y-component of each vector of $\mathbf{F}$ is 0, so $Q = 0$, hence $\dfrac{\partial Q}{\partial x} = \dfrac{\partial Q}{\partial y} = \dfrac{\partial Q}{\partial z} = \dfrac{\partial R}{\partial x} = \dfrac{\partial R}{\partial y} = \dfrac{\partial R}{\partial z} = 0$. P increases as y increases, so $\dfrac{\partial P}{\partial y} > 0$, but P doesn't change in the x- or z-directions, so $\dfrac{\partial P}{\partial x} = \dfrac{\partial P}{\partial z} = 0$.

(a) $\operatorname{div}\mathbf{F} = \dfrac{\partial P}{\partial x} + \dfrac{\partial Q}{\partial y} + \dfrac{\partial R}{\partial z} = 0 + 0 + 0 = 0$

(b) $\operatorname{curl}\mathbf{F} = \left(\dfrac{\partial R}{\partial y} - \dfrac{\partial Q}{\partial z}\right)\mathbf{i} + \left(\dfrac{\partial P}{\partial z} - \dfrac{\partial R}{\partial x}\right)\mathbf{j} + \left(\dfrac{\partial Q}{\partial x} - \dfrac{\partial P}{\partial y}\right)\mathbf{k}$

$= (0-0)\,\mathbf{i} + (0-0)\,\mathbf{j} + \left(0 - \dfrac{\partial P}{\partial y}\right)\mathbf{k} = -\dfrac{\partial P}{\partial y}\mathbf{k}$

Since $\dfrac{\partial P}{\partial y} > 0$, $-\dfrac{\partial P}{\partial y}\mathbf{k}$ is a vector pointing in the negative z-direction.

11. $\operatorname{curl}\mathbf{F} = \nabla \times \mathbf{F} = \begin{vmatrix} \mathbf{i} & \mathbf{j} & \mathbf{k} \\ \partial/\partial x & \partial/\partial y & \partial/\partial z \\ yz & xz & xy \end{vmatrix} = (x - x)\,\mathbf{i} - (y - y)\,\mathbf{j} + (z - z)\,\mathbf{k} = \mathbf{0}$

and $\mathbf{F}$ is defined on all of $\mathbb{R}^3$ with component functions which have continuous partial derivatives, so by Theorem 4, $\mathbf{F}$ is conservative. Thus, there exists a function f such that $\mathbf{F} = \nabla f$. Then $f_x(x, y, z) = yz$ implies $f(x, y, z) = xyz + g(y, z)$ and $f_y(x, y, z) = xz + g_y(y, z)$. But $f_y(x, y, z) = xz$, so $g(y, z) = h(z)$ and $f(x, y, z) = xyz + h(z)$. Thus $f_z(x, y, z) = xy + h'(z)$ but $f_z(x, y, z) = xy$ so $h(z) = K$, a constant. Hence a potential function for $\mathbf{F}$ is $f(x, y, z) = xyz + K$.

13. $\operatorname{curl}\mathbf{F} = \nabla \times \mathbf{F} = \begin{vmatrix} \mathbf{i} & \mathbf{j} & \mathbf{k} \\ \partial/\partial x & \partial/\partial y & \partial/\partial z \\ 2xy & x^2 + 2yz & y^2 \end{vmatrix} = (2y - 2y)\,\mathbf{i} - (0 - 0)\,\mathbf{j} + (2x - 2x)\,\mathbf{k} = \mathbf{0}$, $\mathbf{F}$ is defined on all of $\mathbb{R}^3$,

and the partial derivatives of the component functions are continuous, so $\mathbf{F}$ is conservative. Thus there exists a function f such that $\nabla f = \mathbf{F}$. Then $f_x(x, y, z) = 2xy$ implies $f(x, y, z) = x^2y + g(y, z)$ and $f_y(x, y, z) = x^2 + g_y(y, z)$. But $f_y(x, y, z) = x^2 + 2yz$, so $g(y, z) = y^2z + h(z)$ and $f(x, y, z) = x^2y + y^2z + h(z)$. Thus $f_z(x, y, z) = y^2 + h'(z)$ but $f_z(x, y, z) = y^2$ so $h(z) = K$ and $f(x, y, z) = x^2y + y^2z + K$.

15. $\operatorname{curl}\mathbf{F} = \nabla \times \mathbf{F} = \begin{vmatrix} \mathbf{i} & \mathbf{j} & \mathbf{k} \\ \partial/\partial x & \partial/\partial y & \partial/\partial z \\ e^x & e^z & e^y \end{vmatrix} = (e^y - e^z)\,\mathbf{i} - (0 - 0)\,\mathbf{j} + (0 - 0)\,\mathbf{k} \neq \mathbf{0}$, so $\mathbf{F}$ isn't conservative.

17. No. Assume there is such a $\mathbf{G}$. Then $\operatorname{div}(\operatorname{curl}\mathbf{G}) = y^2 + z^2 + x^2 \neq 0$, which contradicts Theorem 11.

19. $\operatorname{curl}\mathbf{F} = \begin{vmatrix} \mathbf{i} & \mathbf{j} & \mathbf{k} \\ \partial/\partial x & \partial/\partial y & \partial/\partial z \\ f(x) & g(y) & h(z) \end{vmatrix} = (0 - 0)\,\mathbf{i} + (0 - 0)\,\mathbf{j} + (0 - 0)\,\mathbf{k} = \mathbf{0}$.

Hence $\mathbf{F} = f(x)\,\mathbf{i} + g(y)\,\mathbf{j} + h(z)\,\mathbf{k}$ is irrotational.

For Exercises 21–27, let $\mathbf{F}(x, y, z) = P_1\,\mathbf{i} + Q_1\,\mathbf{j} + R_1\,\mathbf{k}$ and $\mathbf{G}(x, y, z) = P_2\,\mathbf{i} + Q_2\,\mathbf{j} + R_2\,\mathbf{k}$.

21. $\operatorname{div}(\mathbf{F} + \mathbf{G}) = \operatorname{div}\langle P_1 + P_2, Q_1 + Q_2, R_1 + R_2\rangle = \dfrac{\partial(P_1 + P_2)}{\partial x} + \dfrac{\partial(Q_1 + Q_2)}{\partial y} + \dfrac{\partial(R_1 + R_2)}{\partial z}$

$$= \frac{\partial P_1}{\partial x} + \frac{\partial P_2}{\partial x} + \frac{\partial Q_1}{\partial y} + \frac{\partial Q_2}{\partial y} + \frac{\partial R_1}{\partial z} + \frac{\partial R_2}{\partial z} = \left(\frac{\partial P_1}{\partial x} + \frac{\partial Q_1}{\partial y} + \frac{\partial R_1}{\partial z}\right) + \left(\frac{\partial P_2}{\partial x} + \frac{\partial Q_2}{\partial y} + \frac{\partial R_2}{\partial z}\right)$$

$$= \operatorname{div}\langle P_1, Q_1, R_1\rangle + \operatorname{div}\langle P_2, Q_2, R_2\rangle = \operatorname{div}\mathbf{F} + \operatorname{div}\mathbf{G}$$

23. $\operatorname{div}(f\mathbf{F}) = \operatorname{div}(f\,\langle P_1, Q_1, R_1\rangle) = \operatorname{div}\langle fP_1, fQ_1, fR_1\rangle = \dfrac{\partial(fP_1)}{\partial x} + \dfrac{\partial(fQ_1)}{\partial y} + \dfrac{\partial(fR_1)}{\partial z}$

$$= \left(f\,\frac{\partial P_1}{\partial x} + P_1\,\frac{\partial f}{\partial x}\right) + \left(f\,\frac{\partial Q_1}{\partial y} + Q_1\,\frac{\partial f}{\partial y}\right) + \left(f\,\frac{\partial R_1}{\partial z} + R_1\,\frac{\partial f}{\partial z}\right)$$

$$= f\left(\frac{\partial P_1}{\partial x} + \frac{\partial Q_1}{\partial y} + \frac{\partial R_1}{\partial z}\right) + \langle P_1, Q_1, R_1\rangle \cdot \left\langle \frac{\partial f}{\partial x}, \frac{\partial f}{\partial y}, \frac{\partial f}{\partial z}\right\rangle = f\operatorname{div}\mathbf{F} + \mathbf{F} \cdot \nabla f$$

25. $\operatorname{div}(\mathbf{F}\times\mathbf{G}) = \nabla\cdot(\mathbf{F}\times\mathbf{G}) = \begin{vmatrix} \partial/\partial x & \partial/\partial y & \partial/\partial z \\ P_1 & Q_1 & R_1 \\ P_2 & Q_2 & R_2 \end{vmatrix} = \dfrac{\partial}{\partial x}\begin{vmatrix} Q_1 & R_1 \\ Q_2 & R_2 \end{vmatrix} - \dfrac{\partial}{\partial y}\begin{vmatrix} P_1 & R_1 \\ P_2 & R_2 \end{vmatrix} + \dfrac{\partial}{\partial z}\begin{vmatrix} P_1 & Q_1 \\ P_2 & Q_2 \end{vmatrix}$

$$= \left[Q_1\frac{\partial R_2}{\partial x} + R_2\frac{\partial Q_1}{\partial x} - Q_2\frac{\partial R_1}{\partial x} - R_1\frac{\partial Q_2}{\partial x}\right]$$

$$- \left[P_1\frac{\partial R_2}{\partial y} + R_2\frac{\partial P_1}{\partial y} - P_2\frac{\partial R_1}{\partial y} - R_1\frac{\partial P_2}{\partial y}\right]$$

$$+ \left[P_1\frac{\partial Q_2}{\partial z} + Q_2\frac{\partial P_1}{\partial z} - P_2\frac{\partial Q_1}{\partial z} - Q_1\frac{\partial P_2}{\partial z}\right]$$

$$= \left[P_2\left(\frac{\partial R_1}{\partial y} - \frac{\partial Q_1}{\partial z}\right) + Q_2\left(\frac{\partial P_1}{\partial z} - \frac{\partial R_1}{\partial x}\right) + R_2\left(\frac{\partial Q_1}{\partial x} - \frac{\partial P_1}{\partial y}\right)\right]$$

$$- \left[P_1\left(\frac{\partial R_2}{\partial y} - \frac{\partial Q_2}{\partial z}\right) + Q_1\left(\frac{\partial P_2}{\partial z} - \frac{\partial R_2}{\partial x}\right) + R_1\left(\frac{\partial Q_2}{\partial x} - \frac{\partial P_2}{\partial y}\right)\right]$$

$$= \mathbf{G}\cdot\operatorname{curl}\mathbf{F} - \mathbf{F}\cdot\operatorname{curl}\mathbf{G}$$

27. $\operatorname{curl}(\operatorname{curl}\mathbf{F}) = \nabla\times(\nabla\times\mathbf{F}) = \begin{vmatrix} \mathbf{i} & \mathbf{j} & \mathbf{k} \\ \partial/\partial x & \partial/\partial y & \partial/\partial z \\ \partial R_1/\partial y - \partial Q_1/\partial z & \partial P_1/\partial z - \partial R_1/\partial x & \partial Q_1/\partial x - \partial P_1/\partial y \end{vmatrix}$

$$= \left(\frac{\partial^2 Q_1}{\partial y\partial x} - \frac{\partial^2 P_1}{\partial y^2} - \frac{\partial^2 P_1}{\partial z^2} + \frac{\partial^2 R_1}{\partial z\partial x}\right)\mathbf{i} + \left(\frac{\partial^2 R_1}{\partial z\partial y} - \frac{\partial^2 Q_1}{\partial z^2} - \frac{\partial^2 Q_1}{\partial x^2} + \frac{\partial^2 P_1}{\partial x\partial y}\right)\mathbf{j}$$

$$+ \left(\frac{\partial^2 P_1}{\partial x\partial z} - \frac{\partial^2 R_1}{\partial x^2} - \frac{\partial^2 R_1}{\partial y^2} + \frac{\partial^2 Q_1}{\partial y\partial z}\right)\mathbf{k}$$

Now let's consider $\operatorname{grad}(\operatorname{div}\mathbf{F}) - \nabla^2\mathbf{F}$ and compare with the above.

(Note that $\nabla^2\mathbf{F}$ is defined on page 944.)

$$\operatorname{grad}(\operatorname{div}\mathbf{F}) - \nabla^2\mathbf{F} = \left[\left(\frac{\partial^2 P_1}{\partial x^2} + \frac{\partial^2 Q_1}{\partial x\partial y} + \frac{\partial^2 R_1}{\partial x\partial z}\right)\mathbf{i} + \left(\frac{\partial^2 P_1}{\partial y\partial x} + \frac{\partial^2 Q_1}{\partial y^2} + \frac{\partial^2 R_1}{\partial y\partial z}\right)\mathbf{j}\right.$$

$$\left. + \left(\frac{\partial^2 P_1}{\partial z\partial x} + \frac{\partial^2 Q_1}{\partial z\partial y} + \frac{\partial^2 R_1}{\partial z^2}\right)\mathbf{k}\right]$$

$$- \left[\left(\frac{\partial^2 P_1}{\partial x^2} + \frac{\partial^2 P_1}{\partial y^2} + \frac{\partial^2 P_1}{\partial z^2}\right)\mathbf{i} + \left(\frac{\partial^2 Q_1}{\partial x^2} + \frac{\partial^2 Q_1}{\partial y^2} + \frac{\partial^2 Q_1}{\partial z^2}\right)\mathbf{j}\right.$$

$$\left. + \left(\frac{\partial^2 R_1}{\partial x^2} + \frac{\partial^2 R_1}{\partial y^2} + \frac{\partial^2 R_1}{\partial z^2}\right)\mathbf{k}\right]$$

$$= \left(\frac{\partial^2 Q_1}{\partial x\partial y} + \frac{\partial^2 R_1}{\partial x\partial z} - \frac{\partial^2 P_1}{\partial y^2} - \frac{\partial^2 P_1}{\partial z^2}\right)\mathbf{i} + \left(\frac{\partial^2 P_1}{\partial y\partial x} + \frac{\partial^2 R_1}{\partial y\partial z} - \frac{\partial^2 Q_1}{\partial x^2} - \frac{\partial^2 Q_1}{\partial z^2}\right)\mathbf{j}$$

$$+ \left(\frac{\partial^2 P_1}{\partial z\partial x} + \frac{\partial^2 Q_1}{\partial z\partial y} - \frac{\partial^2 R_1}{\partial x^2} - \frac{\partial^2 R_2}{\partial y^2}\right)\mathbf{k}$$

Then applying Clairaut's Theorem to reverse the order of differentiation in the second partial derivatives as needed and comparing, we have $\operatorname{curl}\operatorname{curl}\mathbf{F} = \operatorname{grad}\operatorname{div}\mathbf{F} - \nabla^2\mathbf{F}$ as desired.

29. (a) $\nabla r = \nabla\sqrt{x^2+y^2+z^2} = \dfrac{x}{\sqrt{x^2+y^2+z^2}}\,\mathbf{i} + \dfrac{y}{\sqrt{x^2+y^2+z^2}}\,\mathbf{j} + \dfrac{z}{\sqrt{x^2+y^2+z^2}}\,\mathbf{k} = \dfrac{x\,\mathbf{i}+y\,\mathbf{j}+z\,\mathbf{k}}{\sqrt{x^2+y^2+z^2}} = \dfrac{\mathbf{r}}{r}$

(b) $\nabla\times\mathbf{r} = \begin{vmatrix} \mathbf{i} & \mathbf{j} & \mathbf{k} \\ \dfrac{\partial}{\partial x} & \dfrac{\partial}{\partial y} & \dfrac{\partial}{\partial z} \\ x & y & z \end{vmatrix} = \left[\dfrac{\partial}{\partial y}(z) - \dfrac{\partial}{\partial z}(y)\right]\mathbf{i} + \left[\dfrac{\partial}{\partial z}(x) - \dfrac{\partial}{\partial x}(z)\right]\mathbf{j} + \left[\dfrac{\partial}{\partial x}(y) - \dfrac{\partial}{\partial y}(x)\right]\mathbf{k} = \mathbf{0}$

(c) $\nabla\left(\dfrac{1}{r}\right) = \nabla\left(\dfrac{1}{\sqrt{x^2+y^2+z^2}}\right)$

$$= \frac{-\dfrac{1}{2\sqrt{x^2+y^2+z^2}}(2x)}{x^2+y^2+z^2}\,\mathbf{i} - \frac{\dfrac{1}{2\sqrt{x^2+y^2+z^2}}(2y)}{x^2+y^2+z^2}\,\mathbf{j} - \frac{\dfrac{1}{2\sqrt{x^2+y^2+z^2}}(2z)}{x^2+y^2+z^2}\,\mathbf{k}$$

$$= -\frac{x\,\mathbf{i}+y\,\mathbf{j}+z\,\mathbf{k}}{(x^2+y^2+z^2)^{3/2}} = -\frac{\mathbf{r}}{r^3}$$

(d) $\nabla\ln r = \nabla\ln(x^2+y^2+z^2)^{1/2} = \frac{1}{2}\nabla\ln(x^2+y^2+z^2)$

$$= \frac{x}{x^2+y^2+z^2}\,\mathbf{i} + \frac{y}{x^2+y^2+z^2}\,\mathbf{j} + \frac{z}{x^2+y^2+z^2}\,\mathbf{k} = \frac{x\,\mathbf{i}+y\,\mathbf{j}+z\,\mathbf{k}}{x^2+y^2+z^2} = \frac{\mathbf{r}}{r^2}$$

31. By (13), $\oint_C f(\nabla g)\cdot\mathbf{n}\,ds = \iint_D \operatorname{div}(f\nabla g)\,dA = \iint_D [f\operatorname{div}(\nabla g) + \nabla g\cdot\nabla f]\,dA$ by Exercise 23. But $\operatorname{div}(\nabla g) = \nabla^2 g$. Hence $\iint_D f\nabla^2 g\,dA = \oint_C f(\nabla g)\cdot\mathbf{n}\,ds - \iint_D \nabla g\cdot\nabla f\,dA$.

33. Let $f(x,y) = 1$. Then $\nabla f = \mathbf{0}$ and Green's first identity (see Exercise 31) says $\iint_D \nabla^2 g\,dA = \oint_C (\nabla g)\cdot\mathbf{n}\,ds - \iint_D \mathbf{0}\cdot\nabla g\,dA \;\Rightarrow\; \iint_D \nabla^2 g\,dA = \oint_C \nabla g\cdot\mathbf{n}\,ds$. But g is harmonic on D, so $\nabla^2 g = 0 \;\Rightarrow\; \oint_C \nabla g\cdot\mathbf{n}\,ds = 0$ and $\oint_C D_{\mathbf{n}} g\,ds = \oint_C (\nabla g\cdot\mathbf{n})\,ds = 0$.

35. (a) We know that $\omega = v/d$, and from the diagram $\sin\theta = d/r \;\Rightarrow\; v = d\omega = (\sin\theta)r\omega = |\mathbf{w}\times\mathbf{r}|$. But $\mathbf{v}$ is perpendicular to both $\mathbf{w}$ and $\mathbf{r}$, so that $\mathbf{v} = \mathbf{w}\times\mathbf{r}$.

(b) From (a), $\mathbf{v} = \mathbf{w}\times\mathbf{r} = \begin{vmatrix} \mathbf{i} & \mathbf{j} & \mathbf{k} \\ 0 & 0 & \omega \\ x & y & z \end{vmatrix} = (0\cdot z - \omega y)\,\mathbf{i} + (\omega x - 0\cdot z)\,\mathbf{j} + (0\cdot y - x\cdot 0)\,\mathbf{k} = -\omega y\,\mathbf{i} + \omega x\,\mathbf{j}$

(c) $\operatorname{curl}\mathbf{v} = \nabla\times\mathbf{v} = \begin{vmatrix} \mathbf{i} & \mathbf{j} & \mathbf{k} \\ \partial/\partial x & \partial/\partial y & \partial/\partial z \\ -\omega y & \omega x & 0 \end{vmatrix}$

$$= \left[\frac{\partial}{\partial y}(0) - \frac{\partial}{\partial z}(\omega x)\right]\mathbf{i} + \left[\frac{\partial}{\partial z}(-\omega y) - \frac{\partial}{\partial x}(0)\right]\mathbf{j} + \left[\frac{\partial}{\partial x}(\omega x) - \frac{\partial}{\partial y}(-\omega y)\right]\mathbf{k}$$

$$= [\omega - (-\omega)]\,\mathbf{k} = 2\omega\,\mathbf{k} = 2\mathbf{w}$$

37. For any continuous function f on $\mathbb{R}^3$, define a vector field $\mathbf{G}(x,y,z) = \langle g(x,y,z), 0, 0\rangle$ where $g(x,y,z) = \int_0^x f(t,y,z)\,dt$. Then $\operatorname{div}\mathbf{G} = \dfrac{\partial}{\partial x}(g(x,y,z)) + \dfrac{\partial}{\partial y}(0) + \dfrac{\partial}{\partial z}(0) = \dfrac{\partial}{\partial x}\int_0^x f(t,y,z)\,dt = f(x,y,z)$ by the Fundamental Theorem of Calculus. Thus every continuous function f on $\mathbb{R}^3$ is the divergence of some vector field.

13.6 Surface Integrals

1. Each face of the cube has surface area $2^2 = 4$, and the points P_{ij}^* are the points where the cube intersects the coordinate axes. Here, $f(x, y, z) = \sqrt{x^2 + 2y^2 + 3z^2}$, so by Definition 1,

$$\begin{aligned}\textstyle\iint_S f(x,y,z)\,dS &\approx [f(1,0,0)](4) + [f(-1,0,0)](4) + [f(0,1,0)](4) + [f(0,-1,0)](4)\\ &\quad + [f(0,0,1)](4) + [f(0,0,-1)](4)\\ &= 4\left(1 + 1 + 2\sqrt{2} + 2\sqrt{3}\right) = 8\left(1 + \sqrt{2} + \sqrt{3}\right) \approx 33.170\end{aligned}$$

3. We can use the xz- and yz-planes to divide H into four patches of equal size, each with surface area equal to $\frac{1}{8}$ the surface area of a sphere with radius $\sqrt{50}$, so $\Delta S = \frac{1}{8}(4)\pi\left(\sqrt{50}\right)^2 = 25\pi$. Then $(\pm 3, \pm 4, 5)$ are sample points in the four patches, and using a Riemann sum as in Definition 1, we have

$$\begin{aligned}\textstyle\iint_H f(x,y,z)\,dS &\approx f(3,4,5)\,\Delta S + f(3,-4,5)\,\Delta S + f(-3,4,5)\,\Delta S + f(-3,-4,5)\,\Delta S\\ &= (7 + 8 + 9 + 12)(25\pi) = 900\pi \approx 2827\end{aligned}$$

5. $\mathbf{r}(u,v) = u^2\,\mathbf{i} + u\sin v\,\mathbf{j} + u\cos v\,\mathbf{k}$, $0 \le u \le 1$, $0 \le v \le \pi/2$ and
$\mathbf{r}_u \times \mathbf{r}_v = (2u\,\mathbf{i} + \sin v\,\mathbf{j} + \cos v\,\mathbf{k}) \times (u\cos v\,\mathbf{j} - u\sin v\,\mathbf{k}) = -u\,\mathbf{i} + 2u^2\sin v\,\mathbf{j} + 2u^2\cos v\,\mathbf{k}$ and
$|\mathbf{r}_u \times \mathbf{r}_v| = \sqrt{u^2 + 4u^4\sin^2 v + 4u^4\cos^2 v} = \sqrt{u^2 + 4u^4(\sin^2 v + \cos^2 v)} = u\sqrt{1 + 4u^2}$ (since $u \ge 0$). Then

$$\begin{aligned}\textstyle\iint_S yz\,dS &= \textstyle\int_0^{\pi/2}\int_0^1 (u\sin v)(u\cos v)\cdot u\sqrt{1+4u^2}\,du\,dv = \int_0^1 u^3\sqrt{1+4u^2}\,du\int_0^{\pi/2}\sin v\cos v\,dv\\ &\qquad\left[\text{let } t = 1 + 4u^2 \;\Rightarrow\; u^2 = \tfrac{1}{4}(t-1) \text{ and } \tfrac{1}{8}\,dt = u\,du\right]\\ &= \textstyle\int_1^5 \frac{1}{8}\cdot\frac{1}{4}(t-1)\sqrt{t}\,dt\int_0^{\pi/2}\sin v\cos v\,dv = \frac{1}{32}\int_1^5\left(t^{3/2} - \sqrt{t}\right)dt\int_0^{\pi/2}\sin v\cos v\,dv\\ &= \tfrac{1}{32}\left[\tfrac{2}{5}t^{5/2} - \tfrac{2}{3}t^{3/2}\right]_1^5\left[\tfrac{1}{2}\sin^2 v\right]_0^{\pi/2} = \tfrac{1}{32}\left(\tfrac{2}{5}(5)^{5/2} - \tfrac{2}{3}(5)^{3/2} - \tfrac{2}{5} + \tfrac{2}{3}\right)\cdot\tfrac{1}{2}(1-0)\\ &= \tfrac{5}{48}\sqrt{5} + \tfrac{1}{240}\end{aligned}$$

7. $z = 1 + 2x + 3y$ so $\dfrac{\partial z}{\partial x} = 2$ and $\dfrac{\partial z}{\partial y} = 3$. Then by Formula 4,

$$\begin{aligned}\iint_S x^2yz\,dS &= \iint_D x^2yz\sqrt{\left(\frac{\partial z}{\partial x}\right)^2 + \left(\frac{\partial z}{\partial y}\right)^2 + 1}\,dA = \textstyle\int_0^3\int_0^2 x^2y(1+2x+3y)\sqrt{4+9+1}\,dy\,dx\\ &= \textstyle\sqrt{14}\int_0^3\int_0^2(x^2y + 2x^3y + 3x^2y^2)\,dy\,dx = \sqrt{14}\int_0^3\left[\frac{1}{2}x^2y^2 + x^3y^2 + x^2y^3\right]_{y=0}^{y=2}dx\\ &= \textstyle\sqrt{14}\int_0^3(10x^2 + 4x^3)\,dx = \sqrt{14}\left[\frac{10}{3}x^3 + x^4\right]_0^3 = 171\sqrt{14}\end{aligned}$$

9. S is the part of the plane $z = 1 - x - y$ over the region $D = \{(x,y) \mid 0 \le x \le 1, 0 \le y \le 1 - x\}$. Thus

$$\begin{aligned}\textstyle\iint_S yz\,dS &= \textstyle\iint_D y(1-x-y)\sqrt{(-1)^2 + (-1)^2 + 1}\,dA = \sqrt{3}\int_0^1\int_0^{1-x}(y - xy - y^2)\,dy\,dx\\ &= \textstyle\sqrt{3}\int_0^1\left[\frac{1}{2}y^2 - \frac{1}{2}xy^2 - \frac{1}{3}y^3\right]_{y=0}^{y=1-x}dx = \sqrt{3}\int_0^1\frac{1}{6}(1-x)^3\,dx = -\frac{\sqrt{3}}{24}(1-x)^4\Big]_0^1 = \frac{\sqrt{3}}{24}\end{aligned}$$

11. S is the portion of the cone $z^2 = x^2 + y^2$ for $1 \le z \le 3$, or equivalently, S is the part of the surface $z = \sqrt{x^2 + y^2}$ over the region $D = \{(x,y) \mid 1 \le x^2 + y^2 \le 9\}$. Thus

$$\begin{aligned}\iint_S x^2z^2\,dS &= \iint_D x^2(x^2+y^2)\sqrt{\left(\frac{x}{\sqrt{x^2+y^2}}\right)^2 + \left(\frac{y}{\sqrt{x^2+y^2}}\right)^2 + 1}\,dA\\ &= \iint_D x^2(x^2+y^2)\sqrt{\frac{x^2+y^2}{x^2+y^2} + 1}\,dA = \iint_D \sqrt{2}\,x^2(x^2+y^2)\,dA = \sqrt{2}\int_0^{2\pi}\int_1^3 (r\cos\theta)^2(r^2)\,r\,dr\,d\theta\\ &= \textstyle\sqrt{2}\int_0^{2\pi}\cos^2\theta\,d\theta\int_1^3 r^5\,dr = \sqrt{2}\left[\frac{1}{2}\theta + \frac{1}{4}\sin 2\theta\right]_0^{2\pi}\left[\frac{1}{6}r^6\right]_1^3 = \sqrt{2}\,(\pi)\cdot\frac{1}{6}(3^6 - 1) = \dfrac{364\sqrt{2}}{3}\,\pi\end{aligned}$$

13. Using x and z as parameters, we have $\mathbf{r}(x,z) = x\,\mathbf{i} + (x^2+z^2)\,\mathbf{j} + z\,\mathbf{k}$, $x^2+z^2 \le 4$. Then $\mathbf{r}_x \times \mathbf{r}_z = (\mathbf{i} + 2x\,\mathbf{j}) \times (2z\,\mathbf{j} + \mathbf{k}) = 2x\,\mathbf{i} - \mathbf{j} + 2z\,\mathbf{k}$ and $|\mathbf{r}_x \times \mathbf{r}_z| = \sqrt{4x^2+1+4z^2} = \sqrt{1+4(x^2+z^2)}$. Thus

$$\iint_S y\,dS = \iint_{x^2+z^2\le 4} (x^2+z^2)\sqrt{1+4(x^2+z^2)}\,dA = \int_0^{2\pi}\int_0^2 r^2\sqrt{1+4r^2}\,r\,dr\,d\theta$$

$$= \int_0^{2\pi} d\theta \int_0^2 r^2\sqrt{1+4r^2}\,r\,dr = 2\pi\int_0^2 r^2\sqrt{1+4r^2}\,r\,dr$$

$$[\text{let } u = 1+4r^2 \;\Rightarrow\; r^2 = \tfrac{1}{4}(u-1) \text{ and } \tfrac{1}{8}du = r\,dr]$$

$$= 2\pi\int_1^{17} \tfrac{1}{4}(u-1)\sqrt{u}\cdot\tfrac{1}{8}du = \tfrac{1}{16}\pi\int_1^{17}(u^{3/2}-u^{1/2})\,du$$

$$= \tfrac{1}{16}\pi\left[\tfrac{2}{5}u^{5/2} - \tfrac{2}{3}u^{3/2}\right]_1^{17} = \tfrac{1}{16}\pi\left[\tfrac{2}{5}(17)^{5/2} - \tfrac{2}{3}(17)^{3/2} - \tfrac{2}{5} + \tfrac{2}{3}\right] = \frac{\pi}{60}(391\sqrt{17}+1)$$

15. Using spherical coordinates and Example 12.6.1 we have $\mathbf{r}(\phi,\theta) = 2\sin\phi\cos\theta\,\mathbf{i} + 2\sin\phi\sin\theta\,\mathbf{j} + 2\cos\phi\,\mathbf{k}$ and $|\mathbf{r}_\phi \times \mathbf{r}_\theta| = 4\sin\phi$. Then $\iint_S (x^2z + y^2z)\,dS = \int_0^{2\pi}\int_0^{\pi/2}(4\sin^2\phi)(2\cos\phi)(4\sin\phi)\,d\phi\,d\theta = 16\pi\sin^4\phi\Big]_0^{\pi/2} = 16\pi$.

17. Using cylindrical coordinates, we have $\mathbf{r}(\theta,z) = 3\cos\theta\,\mathbf{i} + 3\sin\theta\,\mathbf{j} + z\,\mathbf{k}$, $0 \le \theta \le 2\pi$, $0 \le z \le 2$, and $|\mathbf{r}_\theta \times \mathbf{r}_z| = 3$.

$$\iint_S (x^2y + z^2)\,dS = \int_0^{2\pi}\int_0^2 (27\cos^2\theta\sin\theta + z^2)\,3\,dz\,d\theta = \int_0^{2\pi}(162\cos^2\theta\sin\theta + 8)\,d\theta = 16\pi$$

19. $\mathbf{F}(x,y,z) = xy\,\mathbf{i} + yz\,\mathbf{j} + zx\,\mathbf{k}$, $z = g(x,y) = 4 - x^2 - y^2$, and D is the square $[0,1]\times[0,1]$, so by Equation 10

$$\iint_S \mathbf{F}\cdot d\mathbf{S} = \iint_D [-xy(-2x) - yz(-2y) + zx]\,dA = \int_0^1\int_0^1 [2x^2y + 2y^2(4-x^2-y^2) + x(4-x^2-y^2)]\,dy\,dx$$

$$= \int_0^1 \left(\tfrac{1}{3}x^2 + \tfrac{11}{3}x - x^3 + \tfrac{34}{15}\right)dx = \tfrac{713}{180}$$

21. $\mathbf{F}(x,y,z) = xze^y\,\mathbf{i} - xze^y\,\mathbf{j} + z\,\mathbf{k}$, $z = g(x,y) = 1 - x - y$, and $D = \{(x,y) \mid 0 \le x \le 1, 0 \le y \le 1-x\}$. Since S has downward orientation, we have

$$\iint_S \mathbf{F}\cdot d\mathbf{S} = -\iint_D [-xze^y(-1) - (-xze^y)(-1) + z]\,dA = -\int_0^1\int_0^{1-x}(1-x-y)\,dy\,dx$$

$$= -\int_0^1 \left(\tfrac{1}{2}x^2 - x + \tfrac{1}{2}\right)dx = -\tfrac{1}{6}$$

23. $\mathbf{F}(x,y,z) = x\,\mathbf{i} - z\,\mathbf{j} + y\,\mathbf{k}$, $z = g(x,y) = \sqrt{4-x^2-y^2}$ and D is the quarter disc $\{(x,y) \mid 0 \le x \le 2, 0 \le y \le \sqrt{4-x^2}\}$. S has downward orientation, so by Formula 10,

$$\iint_S \mathbf{F}\cdot d\mathbf{S} = -\iint_D \left[-x\cdot\tfrac{1}{2}(4-x^2-y^2)^{-1/2}(-2x) - (-z)\cdot\tfrac{1}{2}(4-x^2-y^2)^{-1/2}(-2y) + y\right]dA$$

$$= -\iint_D \left(\frac{x^2}{\sqrt{4-x^2-y^2}} - \sqrt{4-x^2-y^2}\cdot\frac{y}{\sqrt{4-x^2-y^2}} + y\right)dA$$

$$= -\iint_D x^2(4-(x^2+y^2))^{-1/2}\,dA = -\int_0^{\pi/2}\int_0^2 (r\cos\theta)^2(4-r^2)^{-1/2}\,r\,dr\,d\theta$$

$$= -\int_0^{\pi/2}\cos^2\theta\,d\theta\int_0^2 r^3(4-r^2)^{-1/2}\,dr \quad [\text{let } u = 4-r^2 \;\Rightarrow\; r^2 = 4-u \text{ and } -\tfrac{1}{2}\,du = r\,dr]$$

$$= -\int_0^{\pi/2}\left(\tfrac{1}{2} + \tfrac{1}{2}\cos 2\theta\right)d\theta\int_4^0 -\tfrac{1}{2}(4-u)(u)^{-1/2}\,du$$

$$= -\left[\tfrac{1}{2}\theta + \tfrac{1}{4}\sin 2\theta\right]_0^{\pi/2}\left(-\tfrac{1}{2}\right)\left[8\sqrt{u} - \tfrac{2}{3}u^{3/2}\right]_4^0 = -\tfrac{\pi}{4}\left(-\tfrac{1}{2}\right)\left(-16 + \tfrac{16}{3}\right) = -\tfrac{4}{3}\pi$$

25. Let S_1 be the paraboloid $y = x^2 + z^2$, $0 \le y \le 1$ and S_2 the disc $x^2 + z^2 \le 1$, $y = 1$. Since S is a closed surface, we use the outward orientation. On S_1: $\mathbf{F}(\mathbf{r}(x,z)) = (x^2+z^2)\,\mathbf{j} - z\,\mathbf{k}$ and $\mathbf{r}_x \times \mathbf{r}_z = 2x\,\mathbf{i} - \mathbf{j} + 2z\,\mathbf{k}$ (since the $\mathbf{j}$-component must be negative on S_1). Then

$$\iint_{S_1}\mathbf{F}\cdot d\mathbf{S} = \iint_{x^2+z^2\le 1}[-(x^2+z^2) - 2z^2]\,dA = -\int_0^{2\pi}\int_0^1 (r^2 + 2r^2\cos^2\theta)\,r\,dr\,d\theta$$

$$= -\int_0^{2\pi}\tfrac{1}{4}(1+2\cos^2\theta)\,d\theta = -\left(\tfrac{\pi}{2} + \tfrac{\pi}{2}\right) = -\pi$$

On S_2: $\mathbf{F}(\mathbf{r}(x,z)) = \mathbf{j} - z\,\mathbf{k}$ and $\mathbf{r}_z \times \mathbf{r}_x = \mathbf{j}$. Then $\iint_{S_2}\mathbf{F}\cdot d\mathbf{S} = \iint_{x^2+z^2\le 1}(1)\,dA = \pi$. Hence $\iint_S \mathbf{F}\cdot d\mathbf{S} = -\pi + \pi = 0$.

27. Here S consists of four surfaces: S_1, the top surface (a portion of the circular cylinder $y^2+z^2=1$); S_2, the bottom surface (a portion of the xy-plane); S_3, the front half-disc in the plane $x=2$, and S_4, the back half-disc in the plane $x=0$.

On S_1: The surface is $z=\sqrt{1-y^2}$ for $0\le x\le 2$, $-1\le y\le 1$ with upward orientation, so

$$\iint_{S_1}\mathbf{F}\cdot d\mathbf{S}=\int_0^2\int_{-1}^1\left[-x^2\,(0)-y^2\left(-\frac{y}{\sqrt{1-y^2}}\right)+z^2\right]dy\,dx=\int_0^2\int_{-1}^1\left(\frac{y^3}{\sqrt{1-y^2}}+1-y^2\right)dy\,dx$$

$$=\int_0^2\left[-\sqrt{1-y^2}+\tfrac{1}{3}(1-y^2)^{3/2}+y-\tfrac{1}{3}y^3\right]_{y=-1}^{y=1}dx=\int_0^2\tfrac{4}{3}\,dx=\tfrac{8}{3}$$

On S_2: The surface is $z=0$ with downward orientation, so

$$\iint_{S_2}\mathbf{F}\cdot d\mathbf{S}=\int_0^2\int_{-1}^1(-z^2)\,dy\,dx=\int_0^2\int_{-1}^1(0)\,dy\,dx=0$$

On S_3: The surface is $x=2$ for $-1\le y\le 1$, $0\le z\le\sqrt{1-y^2}$, oriented in the positive x-direction. Regarding y and z as parameters, we have $\mathbf{r}_y\times\mathbf{r}_z=\mathbf{i}$ and

$$\iint_{S_3}\mathbf{F}\cdot d\mathbf{S}=\int_{-1}^1\int_0^{\sqrt{1-y^2}}x^2\,dz\,dy=\int_{-1}^1\int_0^{\sqrt{1-y^2}}4\,dz\,dy=4A\,(S_3)=2\pi$$

On S_4: The surface is $x=0$ for $-1\le y\le 1$, $0\le z\le\sqrt{1-y^2}$, oriented in the negative x-direction. Regarding y and z as parameters, we use $-\,(\mathbf{r}_y\times\mathbf{r}_z)=-\mathbf{i}$ and

$$\iint_{S_4}\mathbf{F}\cdot d\mathbf{S}=\int_{-1}^1\int_0^{\sqrt{1-y^2}}x^2\,dz\,dy=\int_{-1}^1\int_0^{\sqrt{1-y^2}}(0)\,dz\,dy=0$$

Thus $\iint_S\mathbf{F}\cdot d\mathbf{S}=\frac{8}{3}+0+2\pi+0=2\pi+\frac{8}{3}$.

29. We use Formula 4 with $z=3-2x^2-y^2\;\Rightarrow\;\partial z/\partial x=-4x,\ \partial z/\partial y=-2y$. The boundaries of the region $3-2x^2-y^2\ge 0$ are $-\sqrt{\frac{3}{2}}\le x\le\sqrt{\frac{3}{2}}$ and $-\sqrt{3-2x^2}\le y\le\sqrt{3-2x^2}$, so we use a CAS (with precision reduced to seven or fewer digits; otherwise the calculation takes a very long time) to calculate

$$\iint_S x^2y^2z^2\,dS=\int_{-\sqrt{3/2}}^{\sqrt{3/2}}\int_{-\sqrt{3-2x^2}}^{\sqrt{3-2x^2}}x^2y^2(3-2x^2-y^2)^2\sqrt{16x^2+4y^2+1}\,dy\,dx\approx 3.4895$$

31. If S is given by $y=h(x,z)$, then S is also the level surface $f(x,y,z)=y-h(x,z)=0$.
$\mathbf{n}=\dfrac{\nabla f(x,y,z)}{|\nabla f(x,y,z)|}=\dfrac{-h_x\,\mathbf{i}+\mathbf{j}-h_z\,\mathbf{k}}{\sqrt{h_x^2+1+h_z^2}}$, and $-\mathbf{n}$ is the unit normal that points to the left. Now we proceed as in the derivation of (10), using Formula 4 to evaluate

$$\iint_S\mathbf{F}\cdot d\mathbf{S}=\iint_S\mathbf{F}\cdot\mathbf{n}\,dS=\iint_D(P\,\mathbf{i}+Q\,\mathbf{j}+R\,\mathbf{k})\,\frac{\dfrac{\partial h}{\partial x}\mathbf{i}-\mathbf{j}+\dfrac{\partial h}{\partial z}\mathbf{k}}{\sqrt{\left(\dfrac{\partial h}{\partial x}\right)^2+1+\left(\dfrac{\partial h}{\partial z}\right)^2}}\sqrt{\left(\frac{\partial h}{\partial x}\right)^2+1+\left(\frac{\partial h}{\partial z}\right)^2}\,dA$$

where D is the projection of $f(x,y,z)$ onto the xz-plane. Therefore

$$\iint_S\mathbf{F}\cdot d\mathbf{S}=\iint_D\left(P\frac{\partial h}{\partial x}-Q+R\frac{\partial h}{\partial z}\right)dA$$

33. $m=\iint_S K\,dS=K\cdot 4\pi\left(\frac{1}{2}a^2\right)=2\pi a^2K$; by symmetry $M_{xz}=M_{yz}=0$, and
$M_{xy}=\iint_S zK\,dS=K\int_0^{2\pi}\int_0^{\pi/2}(a\cos\phi)(a^2\sin\phi)\,d\phi\,d\theta=2\pi Ka^3\left[-\frac{1}{4}\cos 2\phi\right]_0^{\pi/2}=\pi Ka^3$.
Hence $(\overline{x},\overline{y},\overline{z})=\left(0,0,\frac{1}{2}a\right)$.

35. (a) $I_z=\iint_S(x^2+y^2)\rho(x,y,z)\,dS$

(b) $I_z=\iint_S(x^2+y^2)\left(10-\sqrt{x^2+y^2}\right)dS=\iint\limits_{1\le x^2+y^2\le 16}(x^2+y^2)\left(10-\sqrt{x^2+y^2}\right)\sqrt{2}\,dA$

$$=\int_0^{2\pi}\int_1^4\sqrt{2}\,(10r^3-r^4)\,dr\,d\theta=2\sqrt{2}\,\pi\left(\tfrac{4329}{10}\right)=\tfrac{4329}{5}\sqrt{2}\,\pi$$

37. The rate of flow through the cylinder is the flux $\iint_S \rho\mathbf{v}\cdot\mathbf{n}\,dS = \iint_S \rho\mathbf{v}\cdot d\mathbf{S}$. We use the parametric representation $\mathbf{r}(u,v) = 2\cos u\,\mathbf{i} + 2\sin u\,\mathbf{j} + v\,\mathbf{k}$ for S, where $0 \le u \le 2\pi$, $0 \le v \le 1$, so $\mathbf{r}_u = -2\sin u\,\mathbf{i} + 2\cos u\,\mathbf{j}$, $\mathbf{r}_v = \mathbf{k}$, and the outward orientation is given by $\mathbf{r}_u \times \mathbf{r}_v = 2\cos u\,\mathbf{i} + 2\sin u\,\mathbf{j}$. Then

$$\iint_S \rho\mathbf{v}\cdot d\mathbf{S} = \rho\int_0^{2\pi}\int_0^1 \left(v\,\mathbf{i} + 4\sin^2 u\,\mathbf{j} + 4\cos^2 u\,\mathbf{k}\right)\cdot(2\cos u\,\mathbf{i} + 2\sin u\,\mathbf{j})\,dv\,du$$

$$= \rho\int_0^{2\pi}\int_0^1 \left(2v\cos u + 8\sin^3 u\right)dv\,du = \rho\int_0^{2\pi}\left(\cos u + 8\sin^3 u\right)du$$

$$= \rho\left[\sin u + 8\left(-\tfrac{1}{3}\right)(2+\sin^2 u)\cos u\right]_0^{2\pi} = 0\text{ kg/s}$$

39. S consists of the hemisphere S_1 given by $z = \sqrt{a^2 - x^2 - y^2}$ and the disc S_2 given by $0 \le x^2 + y^2 \le a^2$, $z = 0$. On S_1: $\mathbf{E} = a\sin\phi\cos\theta\,\mathbf{i} + a\sin\phi\sin\theta\,\mathbf{j} + 2a\cos\phi\,\mathbf{k}$, $\mathbf{T}_\phi \times \mathbf{T}_\theta = a^2\sin^2\phi\cos\theta\,\mathbf{i} + a^2\sin^2\phi\sin\theta\,\mathbf{j} + a^2\sin\phi\cos\phi\,\mathbf{k}$. Thus

$$\iint_{S_1}\mathbf{E}\cdot d\mathbf{S} = \int_0^{2\pi}\int_0^{\pi/2}(a^3\sin^3\phi + 2a^3\sin\phi\cos^2\phi)\,d\phi\,d\theta$$

$$= \int_0^{2\pi}\int_0^{\pi/2}(a^3\sin\phi + a^3\sin\phi\cos^2\phi)\,d\phi\,d\theta = (2\pi)a^3\left(1+\tfrac{1}{3}\right) = \tfrac{8}{3}\pi a^3$$

On S_2: $\mathbf{E} = x\,\mathbf{i} + y\,\mathbf{j}$, and $\mathbf{r}_y \times \mathbf{r}_x = -\mathbf{k}$ so $\iint_{S_2}\mathbf{E}\cdot d\mathbf{S} = 0$. Hence the total charge is $q = \varepsilon_0\iint_S \mathbf{E}\cdot d\mathbf{S} = \frac{8}{3}\pi a^3\varepsilon_0$.

41. $K\nabla u = 6.5(4y\,\mathbf{j} + 4z\,\mathbf{k})$. S is given by $\mathbf{r}(x,\theta) = x\,\mathbf{i} + \sqrt{6}\,\cos\theta\,\mathbf{j} + \sqrt{6}\,\sin\theta\,\mathbf{k}$ and since we want the inward heat flow, we use $\mathbf{r}_x \times \mathbf{r}_\theta = -\sqrt{6}\,\cos\theta\,\mathbf{j} - \sqrt{6}\,\sin\theta\,\mathbf{k}$. Then the rate of heat flow inward is given by $\iint_S(-K\,\nabla u)\cdot d\mathbf{S} = \int_0^{2\pi}\int_0^4 -(6.5)(-24)\,dx\,d\theta = (2\pi)(156)(4) = 1248\pi$.

43. Let S be a sphere of radius a centred at the origin. Then $|\mathbf{r}| = a$ and $\mathbf{F}(\mathbf{r}) = c\mathbf{r}/|\mathbf{r}|^3 = \left(c/a^3\right)(x\,\mathbf{i} + y\,\mathbf{j} + z\,\mathbf{k})$. A parametric representation for S is $\mathbf{r}(\phi,\theta) = a\sin\phi\cos\theta\,\mathbf{i} + a\sin\phi\sin\theta\,\mathbf{j} + a\cos\phi\,\mathbf{k}$, $0 \le \phi \le \pi$, $0 \le \theta \le 2\pi$. Then $\mathbf{r}_\phi = a\cos\phi\cos\theta\,\mathbf{i} + a\cos\phi\sin\theta\,\mathbf{j} - a\sin\phi\,\mathbf{k}$, $\mathbf{r}_\theta = -a\sin\phi\sin\theta\,\mathbf{i} + a\sin\phi\cos\theta\,\mathbf{j}$, and the outward orientation is given by $\mathbf{r}_\phi \times \mathbf{r}_\theta = a^2\sin^2\phi\cos\theta\,\mathbf{i} + a^2\sin^2\phi\sin\theta\,\mathbf{j} + a^2\sin\phi\cos\phi\,\mathbf{k}$. The flux of $\mathbf{F}$ across S is

$$\iint_S \mathbf{F}\cdot d\mathbf{S} = \int_0^\pi\int_0^{2\pi}\frac{c}{a^3}\,(a\sin\phi\cos\theta\,\mathbf{i} + a\sin\phi\sin\theta\,\mathbf{j} + a\cos\phi\,\mathbf{k})$$

$$\cdot\left(a^2\sin^2\phi\cos\theta\,\mathbf{i} + a^2\sin^2\phi\sin\theta\,\mathbf{j} + a^2\sin\phi\cos\phi\,\mathbf{k}\right)d\theta\,d\phi$$

$$= \frac{c}{a^3}\int_0^\pi\int_0^{2\pi}a^3\left(\sin^3\phi + \sin\phi\cos^2\phi\right)d\theta\,d\phi = c\int_0^\pi\int_0^{2\pi}\sin\phi\,d\theta\,d\phi = 4\pi c$$

Thus the flux does not depend on the radius a.

13.7 Stokes' Theorem

1. Both H and P are oriented piecewise-smooth surfaces that are bounded by the simple, closed, smooth curve $x^2 + y^2 = 4$, $z = 0$ (which we can take to be oriented positively for both surfaces). Then H and P satisfy the hypotheses of Stokes' Theorem, so by (3) we know $\iint_H \operatorname{curl}\mathbf{F}\cdot d\mathbf{S} = \int_C \mathbf{F}\cdot d\mathbf{r} = \iint_P \operatorname{curl}\mathbf{F}\cdot d\mathbf{S}$ (where C is the boundary curve).

3. The boundary curve C is the circle $x^2 + y^2 = 4$, $z = 0$ oriented in the counterclockwise direction. The vector equation is $\mathbf{r}(t) = 2\cos t\,\mathbf{i} + 2\sin t\,\mathbf{j}$, $0 \le t \le 2\pi$, so $\mathbf{r}'(t) = -2\sin t\,\mathbf{i} + 2\cos t\,\mathbf{j}$ and $\mathbf{F}(\mathbf{r}(t)) = (2\cos t)^2 e^{(2\sin t)(0)}\,\mathbf{i} + (2\sin t)^2 e^{(2\cos t)(0)}\,\mathbf{j} + (0)^2 e^{(2\cos t)(2\sin t)}\,\mathbf{k} = 4\cos^2 t\,\mathbf{i} + 4\sin^2 t\,\mathbf{j}$. Then, by Stokes' Theorem,

$$\iint_S \operatorname{curl}\mathbf{F}\cdot d\mathbf{S} = \int_C \mathbf{F}\cdot d\mathbf{r} = \int_0^{2\pi}\mathbf{F}(\mathbf{r}(t))\cdot\mathbf{r}'(t)\,dt = \int_0^{2\pi}(-8\cos^2 t\sin t + 8\sin^2 t\cos t)\,dt$$

$$= 8\left[\tfrac{1}{3}\cos^3 t + \tfrac{1}{3}\sin^3 t\right]_0^{2\pi} = 0$$

5. C is the square in the plane $z = -1$. By (3), $\iint_{S_1} \operatorname{curl} \mathbf{F} \cdot d\mathbf{S} = \oint_C \mathbf{F} \cdot d\mathbf{r} = \iint_{S_2} \operatorname{curl} \mathbf{F} \cdot d\mathbf{S}$ where S_1 is the original cube without the bottom and S_2 is the bottom face of the cube. $\operatorname{curl} \mathbf{F} = x^2 z\,\mathbf{i} + (xy - 2xyz)\,\mathbf{j} + (y - xz)\,\mathbf{k}$. For S_2, we choose $\mathbf{n} = \mathbf{k}$ so that C has the same orientation for both surfaces. Then $\operatorname{curl} \mathbf{F} \cdot \mathbf{n} = y - xz = x + y$ on S_2, where $z = -1$. Thus $\iint_{S_2} \operatorname{curl} \mathbf{F} \cdot d\mathbf{S} = \int_{-1}^{1} \int_{-1}^{1} (x + y)\, dx\, dy = 0$ so $\iint_{S_1} \operatorname{curl} \mathbf{F} \cdot d\mathbf{S} = 0$.

7. $\operatorname{curl} \mathbf{F} = -2z\,\mathbf{i} - 2x\,\mathbf{j} - 2y\,\mathbf{k}$ and we take the surface S to be the planar region enclosed by C, so S is the portion of the plane $x + y + z = 1$ over $D = \{(x, y) \mid 0 \le x \le 1, 0 \le y \le 1 - x\}$. Since C is oriented counterclockwise, we orient S upward. Using Equation 13.6.10, we have $z = g(x, y) = 1 - x - y$, $P = -2z$, $Q = -2x$, $R = -2y$, and

$$\begin{aligned}\int_C \mathbf{F} \cdot d\mathbf{r} &= \iint_S \operatorname{curl} \mathbf{F} \cdot d\mathbf{S} = \iint_D [-(-2z)(-1) - (-2x)(-1) + (-2y)]\, dA \\ &= \int_0^1 \int_0^{1-x} (-2)\, dy\, dx = -2 \int_0^1 (1 - x)\, dx = -1\end{aligned}$$

9. $\operatorname{curl} \mathbf{F} = (xe^{xy} - 2x)\,\mathbf{i} - (ye^{xy} - y)\,\mathbf{j} + (2z - z)\,\mathbf{k}$ and we take S to be the disc $x^2 + y^2 \le 16$, $z = 5$. Since C is oriented counterclockwise (from above), we orient S upward. Then $\mathbf{n} = \mathbf{k}$ and $\operatorname{curl} \mathbf{F} \cdot \mathbf{n} = 2z - z$ on S, where $z = 5$. Thus

$$\oint \mathbf{F} \cdot d\mathbf{r} = \iint_S \operatorname{curl} \mathbf{F} \cdot \mathbf{n}\, dS = \iint_S (2z - z)\, dS = \iint_S (10 - 5)\, dS = 5(\text{area of } S) = 5(\pi \cdot 4^2) = 80\pi$$

11. (a) The curve of intersection is an ellipse in the plane $x + y + z = 1$ with unit normal $\mathbf{n} = \frac{1}{\sqrt{3}}(\mathbf{i} + \mathbf{j} + \mathbf{k})$, $\operatorname{curl} \mathbf{F} = x^2\,\mathbf{j} + y^2\,\mathbf{k}$, and $\operatorname{curl} \mathbf{F} \cdot \mathbf{n} = \frac{1}{\sqrt{3}}(x^2 + y^2)$. Then

$$\oint_C \mathbf{F} \cdot d\mathbf{r} = \iint_S \tfrac{1}{\sqrt{3}}(x^2 + y^2)\, dS = \iint_{x^2 + y^2 \le 9} (x^2 + y^2)\, dx\, dy = \int_0^{2\pi} \int_0^3 r^3\, dr\, d\theta = 2\pi\left(\tfrac{81}{4}\right) = \tfrac{81\pi}{2}$$

(b)

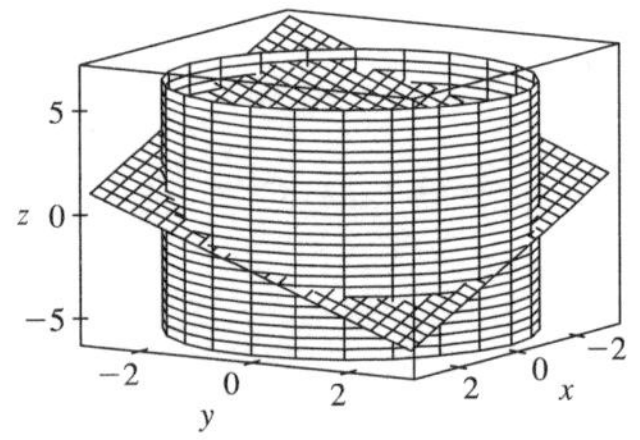

(c) One possible parametrization is $x = 3\cos t$, $y = 3\sin t$, $z = 1 - 3\cos t - 3\sin t$, $0 \le t \le 2\pi$.

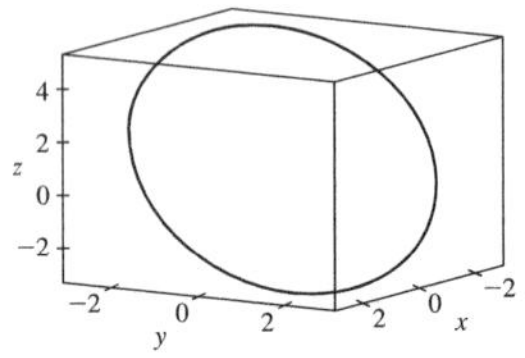

13. The boundary curve C is the circle $x^2 + y^2 = 1$, $z = 1$ oriented in the counterclockwise direction as viewed from above. We can parametrize C by $\mathbf{r}(t) = \cos t\,\mathbf{i} + \sin t\,\mathbf{j} + \mathbf{k}$, $0 \le t \le 2\pi$, and then $\mathbf{r}'(t) = -\sin t\,\mathbf{i} + \cos t\,\mathbf{j}$. Thus $\mathbf{F}(\mathbf{r}(t)) = \sin^2 t\,\mathbf{i} + \cos t\,\mathbf{j} + \mathbf{k}$, $\mathbf{F}(\mathbf{r}(t)) \cdot \mathbf{r}'(t) = \cos^2 t - \sin^3 t$, and

$$\begin{aligned}\oint_C \mathbf{F} \cdot d\mathbf{r} &= \int_0^{2\pi} (\cos^2 t - \sin^3 t)\, dt = \int_0^{2\pi} \tfrac{1}{2}(1 + \cos 2t)\, dt - \int_0^{2\pi} (1 - \cos^2 t) \sin t\, dt \\ &= \tfrac{1}{2}\left[t + \tfrac{1}{2}\sin 2t\right]_0^{2\pi} - \left[-\cos t + \tfrac{1}{3}\cos^3 t\right]_0^{2\pi} = \pi\end{aligned}$$

Now $\operatorname{curl} \mathbf{F} = (1 - 2y)\,\mathbf{k}$, and the projection D of S on the xy-plane is the disc $x^2 + y^2 \le 1$, so by Equation 13.6.10 with $z = g(x, y) = x^2 + y^2$ we have

$$\iint_S \operatorname{curl} \mathbf{F} \cdot d\mathbf{S} = \iint_D (1 - 2y)\, dA = \int_0^{2\pi} \int_0^1 (1 - 2r\sin\theta)\, r\, dr\, d\theta = \int_0^{2\pi} \left(\tfrac{1}{2} - \tfrac{2}{3}\sin\theta\right) d\theta = \pi$$

15. The boundary curve C is the circle $x^2 + z^2 = 1$, $y = 0$ oriented in the counterclockwise direction as viewed from the positive y-axis. Then C can be described by $\mathbf{r}(t) = \cos t\,\mathbf{i} - \sin t\,\mathbf{k}$, $0 \le t \le 2\pi$, and $\mathbf{r}'(t) = -\sin t\,\mathbf{i} - \cos t\,\mathbf{k}$. Thus $\mathbf{F}(\mathbf{r}(t)) = -\sin t\,\mathbf{j} + \cos t\,\mathbf{k}$, $\mathbf{F}(\mathbf{r}(t)) \cdot \mathbf{r}'(t) = -\cos^2 t$, and $\oint_C \mathbf{F} \cdot d\mathbf{r} = \int_0^{2\pi} -\cos^2 t\,dt = -\frac{1}{2}t - \frac{1}{4}\sin 2t\Big]_0^{2\pi} = -\pi$.

Now curl $\mathbf{F} = -\mathbf{i} - \mathbf{j} - \mathbf{k}$, and S can be parametrized (see Example 12.6.1) by

$\mathbf{r}(\phi, \theta) = \sin\phi\cos\theta\,\mathbf{i} + \sin\phi\sin\theta\,\mathbf{j} + \cos\phi\,\mathbf{k}$, $0 \le \theta \le \pi$, $0 \le \phi \le \pi$. Then

$\mathbf{r}_\phi \times \mathbf{r}_\theta = \sin^2\phi\cos\theta\,\mathbf{i} + \sin^2\phi\sin\theta\,\mathbf{j} + \sin\phi\cos\phi\,\mathbf{k}$ and

$$\begin{aligned}\iint_S \text{curl}\,\mathbf{F} \cdot d\mathbf{S} &= \iint_{x^2+z^2\le 1} \text{curl}\,\mathbf{F} \cdot (\mathbf{r}_\phi \times \mathbf{r}_\theta)\,dA \\ &= \int_0^\pi \int_0^\pi (-\sin^2\phi\cos\theta - \sin^2\phi\sin\theta - \sin\phi\cos\phi)\,d\theta\,d\phi \\ &= \int_0^\pi (-2\sin^2\phi - \pi\sin\phi\cos\phi)\,d\phi = \left[\tfrac{1}{2}\sin 2\phi - \phi - \tfrac{\pi}{2}\sin^2\phi\right]_0^\pi = -\pi\end{aligned}$$

17. It is easier to use Stokes' Theorem than to compute the work directly. Let S be the planar region enclosed by the path of the particle, so S is the portion of the plane $z = \frac{1}{2}y$ for $0 \le x \le 1$, $0 \le y \le 2$, with upward orientation.

curl $\mathbf{F} = 8y\,\mathbf{i} + 2z\,\mathbf{j} + 2y\,\mathbf{k}$ and

$$\begin{aligned}\oint_C \mathbf{F} \cdot d\mathbf{r} &= \iint_S \text{curl}\,\mathbf{F} \cdot d\mathbf{S} = \iint_D \left[-8y\,(0) - 2z\left(\tfrac{1}{2}\right) + 2y\right] dA = \int_0^1 \int_0^2 \left(2y - \tfrac{1}{2}y\right) dy\,dx \\ &= \int_0^1 \int_0^2 \tfrac{3}{2}y\,dy\,dx = \int_0^1 \left[\tfrac{3}{4}y^2\right]_{y=0}^{y=2} dx = \int_0^1 3\,dx = 3\end{aligned}$$

19. Assume S is centred at the origin with radius a and let H_1 and H_2 be the upper and lower hemispheres, respectively, of S. Then $\iint_S \text{curl}\,\mathbf{F} \cdot d\mathbf{S} = \iint_{H_1} \text{curl}\,\mathbf{F} \cdot d\mathbf{S} + \iint_{H_2} \text{curl}\,\mathbf{F} \cdot d\mathbf{S} = \oint_{C_1} \mathbf{F} \cdot d\mathbf{r} + \oint_{C_2} \mathbf{F} \cdot d\mathbf{r}$ by Stokes' Theorem. But C_1 is the circle $x^2 + y^2 = a^2$ oriented in the counterclockwise direction while C_2 is the same circle oriented in the clockwise direction. Hence $\oint_{C_2} \mathbf{F} \cdot d\mathbf{r} = -\oint_{C_1} \mathbf{F} \cdot d\mathbf{r}$ so $\iint_S \text{curl}\,\mathbf{F} \cdot d\mathbf{S} = 0$ as desired.

13.8 The Divergence Theorem

1. div $\mathbf{F} = 3 + x + 2x = 3 + 3x$, so

$\iiint_E \text{div}\,\mathbf{F}\,dV = \int_0^1 \int_0^1 \int_0^1 (3x + 3)\,dx\,dy\,dz = \frac{9}{2}$ (notice the triple integral is three times the volume of the cube plus three times $\overline{x}$).

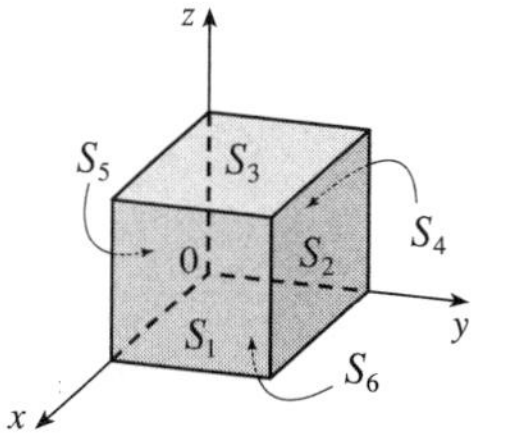

To compute $\iint_S \mathbf{F} \cdot d\mathbf{S}$, on S_1: $\mathbf{n} = \mathbf{i}$, $\mathbf{F} = 3\,\mathbf{i} + y\,\mathbf{j} + 2z\,\mathbf{k}$, and $\iint_{S_1} \mathbf{F} \cdot d\mathbf{S} = \iint_{S_1} 3\,dS = 3$;

S_2: $\mathbf{F} = 3x\,\mathbf{i} + x\,\mathbf{j} + 2xz\,\mathbf{k}$, $\mathbf{n} = \mathbf{j}$ and $\iint_{S_2} \mathbf{F} \cdot d\mathbf{S} = \iint_{S_2} x\,dS = \frac{1}{2}$;

S_3: $\mathbf{F} = 3x\,\mathbf{i} + xy\,\mathbf{j} + 2x\,\mathbf{k}$, $\mathbf{n} = \mathbf{k}$ and $\iint_{S_3} \mathbf{F} \cdot d\mathbf{S} = \iint_{S_3} 2x\,dS = 1$;

S_4: $\mathbf{F} = \mathbf{0}$, $\iint_{S_4} \mathbf{F} \cdot d\mathbf{S} = 0$; S_5: $\mathbf{F} = 3x\,\mathbf{i} + 2x\,\mathbf{k}$, $\mathbf{n} = -\mathbf{j}$ and $\iint_{S_5} \mathbf{F} \cdot d\mathbf{S} = \iint_{S_5} 0\,dS = 0$;

S_6: $\mathbf{F} = 3x\,\mathbf{i} + xy\,\mathbf{j}$, $\mathbf{n} = -\mathbf{k}$ and $\iint_{S_6} \mathbf{F} \cdot d\mathbf{S} = \iint_{S_6} 0\,dS = 0$. Thus $\iint_S \mathbf{F} \cdot d\mathbf{S} = \frac{9}{2}$.

3. div $\mathbf{F} = x + y + z$, so

$$\iiint_E \text{div}\,\mathbf{F}\,dV = \int_0^{2\pi}\int_0^1\int_0^1 (r\cos\theta + r\sin\theta + z)\,r\,dz\,dr\,d\theta = \int_0^{2\pi}\int_0^1 \left(r^2\cos\theta + r^2\sin\theta + \tfrac{1}{2}r\right)dr\,d\theta$$
$$= \int_0^{2\pi}\left(\tfrac{1}{3}\cos\theta + \tfrac{1}{3}\sin\theta + \tfrac{1}{4}\right)d\theta = \tfrac{1}{4}(2\pi) = \tfrac{\pi}{2}$$

Let S_1 be the top of the cylinder, S_2 the bottom, and S_3 the vertical edge. On S_1, $z = 1$, $\mathbf{n} = \mathbf{k}$, and $\mathbf{F} = xy\,\mathbf{i} + y\,\mathbf{j} + x\,\mathbf{k}$, so $\iint_{S_1}\mathbf{F}\cdot d\mathbf{S} = \iint_{S_1}\mathbf{F}\cdot\mathbf{n}\,dS = \iint_{S_1} x\,dS = \int_0^{2\pi}\int_0^1 (r\cos\theta)\,r\,dr\,d\theta = \left[\sin\theta\right]_0^{2\pi}\left[\tfrac{1}{3}r^3\right]_0^1 = 0$. On S_2, $z = 0$, $\mathbf{n} = -\mathbf{k}$, and $\mathbf{F} = xy\,\mathbf{i}$ so $\iint_{S_2}\mathbf{F}\cdot d\mathbf{S} = \iint_{S_2} 0\,dS = 0$. S_3 is given by $\mathbf{r}(\theta, z) = \cos\theta\,\mathbf{i} + \sin\theta\,\mathbf{j} + z\,\mathbf{k}$, $0 \le \theta \le 2\pi$, $0 \le z \le 1$. Then $\mathbf{r}_\theta \times \mathbf{r}_z = \cos\theta\,\mathbf{i} + \sin\theta\,\mathbf{j}$ and

$$\iint_{S_3}\mathbf{F}\cdot d\mathbf{S} = \iint_D \mathbf{F}\cdot(\mathbf{r}_\theta\times\mathbf{r}_z)\,dA = \int_0^{2\pi}\int_0^1 (\cos^2\theta\sin\theta + z\sin^2\theta)\,dz\,d\theta$$
$$= \int_0^{2\pi}\left(\cos^2\theta\sin\theta + \tfrac{1}{2}\sin^2\theta\right)d\theta = \left[-\tfrac{1}{3}\cos^3\theta + \tfrac{1}{4}\left(\theta - \tfrac{1}{2}\sin 2\theta\right)\right]_0^{2\pi} = \tfrac{\pi}{2}$$

Thus $\iint_S \mathbf{F}\cdot d\mathbf{S} = 0 + 0 + \frac{\pi}{2} = \frac{\pi}{2}$.

5. div $\mathbf{F} = \frac{\partial}{\partial x}(e^x\sin y) + \frac{\partial}{\partial y}(e^x\cos y) + \frac{\partial}{\partial z}(yz^2) = e^x\sin y - e^x\sin y + 2yz = 2yz$, so by the Divergence Theorem,

$$\iint_S \mathbf{F}\cdot d\mathbf{S} = \iiint_E \text{div}\,\mathbf{F}\,dV = \int_0^1\int_0^1\int_0^2 2yz\,dz\,dy\,dx = 2\int_0^1 dx\int_0^1 y\,dy\int_0^2 z\,dz$$
$$= 2\left[x\right]_0^1\left[\tfrac{1}{2}y^2\right]_0^1\left[\tfrac{1}{2}z^2\right]_0^2 = 2$$

7. div $\mathbf{F} = 3y^2 + 0 + 3z^2$, so using cylindrical coordinates with $y = r\cos\theta$, $z = r\sin\theta$, $x = x$ we have

$$\iint_S \mathbf{F}\cdot d\mathbf{S} = \iiint_E (3y^2 + 3z^2)\,dV = \int_0^{2\pi}\int_0^1\int_{-1}^2 (3r^2\cos^2\theta + 3r^2\sin^2\theta)\,r\,dx\,dr\,d\theta$$
$$= 3\int_0^{2\pi} d\theta\int_0^1 r^3\,dr\int_{-1}^2 dx = 3(2\pi)\left(\tfrac{1}{4}\right)(3) = \tfrac{9\pi}{2}$$

9. div $\mathbf{F} = y\sin z + 0 - y\sin z = 0$, so by the Divergence Theorem, $\iint_S \mathbf{F}\cdot d\mathbf{S} = \iiint_E 0\,dV = 0$.

11. div $\mathbf{F} = y^2 + 0 + x^2 = x^2 + y^2$ so

$$\iint_S \mathbf{F}\cdot d\mathbf{S} = \iiint_E (x^2 + y^2)\,dV = \int_0^{2\pi}\int_0^2\int_{r^2}^4 r^2\cdot r\,dz\,dr\,d\theta = \int_0^{2\pi}\int_0^2 r^3(4 - r^2)\,dr\,d\theta$$
$$= \int_0^{2\pi} d\theta\int_0^2 (4r^3 - r^5)\,dr = 2\pi\left[r^4 - \tfrac{1}{6}r^6\right]_0^2 = \tfrac{32}{3}\pi$$

13. div $\mathbf{F} = 12x^2z + 12y^2z + 12z^3$ so

$$\iint_S \mathbf{F}\cdot d\mathbf{S} = \iiint_E 12z(x^2 + y^2 + z^2)\,dV = \int_0^{2\pi}\int_0^{\pi}\int_0^R 12(\rho\cos\phi)(\rho^2)\rho^2\sin\phi\,d\rho\,d\phi\,d\theta$$
$$= 12\int_0^{2\pi} d\theta\int_0^{\pi}\sin\phi\cos\phi\,d\phi\int_0^R \rho^5\,d\rho = 12(2\pi)\left[\tfrac{1}{2}\sin^2\phi\right]_0^{\pi}\left[\tfrac{1}{6}\rho^6\right]_0^R = 0$$

15. $\iint_S \mathbf{F}\cdot d\mathbf{S} = \iiint_E \sqrt{3 - x^2}\,dV = \int_{-1}^1\int_{-1}^1\int_0^{2 - x^4 - y^4}\sqrt{3 - x^2}\,dz\,dy\,dx = \frac{341}{60}\sqrt{2} + \frac{81}{20}\sin^{-1}\left(\frac{\sqrt{3}}{3}\right)$

17. For S_1 we have $\mathbf{n} = -\mathbf{k}$, so $\mathbf{F}\cdot\mathbf{n} = \mathbf{F}\cdot(-\mathbf{k}) = -x^2z - y^2 = -y^2$ (since $z = 0$ on S_1). So if D is the unit disc, we get $\iint_{S_1}\mathbf{F}\cdot d\mathbf{S} = \iint_{S_1}\mathbf{F}\cdot\mathbf{n}\,dS = \iint_D(-y^2)\,dA = -\int_0^{2\pi}\int_0^1 r^2(\sin^2\theta)\,r\,dr\,d\theta = -\frac{1}{4}\pi$. Now since S_2 is closed, we can use the Divergence Theorem. Since div $\mathbf{F} = \frac{\partial}{\partial x}(z^2x) + \frac{\partial}{\partial y}\left(\frac{1}{3}y^3 + \tan z\right) + \frac{\partial}{\partial z}(x^2z + y^2) = z^2 + y^2 + x^2$, we use spherical coordinates to get $\iint_{S_2}\mathbf{F}\cdot d\mathbf{S} = \iiint_E \text{div}\,\mathbf{F}\,dV = \int_0^{2\pi}\int_0^{\pi/2}\int_0^1 \rho^2\cdot\rho^2\sin\phi\,d\rho\,d\phi\,d\theta = \frac{2}{5}\pi$. Finally $\iint_S \mathbf{F}\cdot d\mathbf{S} = \iint_{S_2}\mathbf{F}\cdot d\mathbf{S} - \iint_{S_1}\mathbf{F}\cdot d\mathbf{S} = \frac{2}{5}\pi - \left(-\frac{1}{4}\pi\right) = \frac{13}{20}\pi$.

19. The vectors that end near P_1 are longer than the vectors that start near P_1, so the net flow is inward near P_1 and $\operatorname{div}\mathbf{F}(P_1)$ is negative. The vectors that end near P_2 are shorter than the vectors that start near P_2, so the net flow is outward near P_2 and $\operatorname{div}\mathbf{F}(P_2)$ is positive.

21.

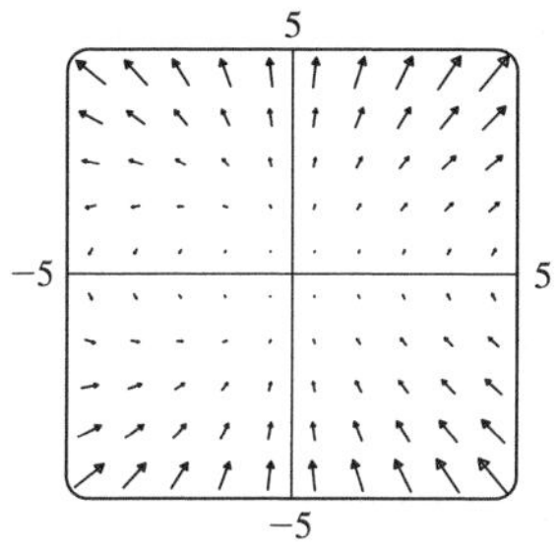

From the graph it appears that for points above the x-axis, vectors starting near a particular point are longer than vectors ending there, so divergence is positive. The opposite is true at points below the x-axis, where divergence is negative.

$\mathbf{F}(x,y) = \langle xy, x+y^2\rangle \quad\Rightarrow$

$\operatorname{div}\mathbf{F} = \frac{\partial}{\partial x}(xy) + \frac{\partial}{\partial y}(x+y^2) = y + 2y = 3y$. Thus $\operatorname{div}\mathbf{F} > 0$ for $y > 0$, and $\operatorname{div}\mathbf{F} < 0$ for $y < 0$.

23. Since $\dfrac{\mathbf{x}}{|\mathbf{x}|^3} = \dfrac{x\,\mathbf{i} + y\,\mathbf{j} + z\,\mathbf{k}}{(x^2+y^2+z^2)^{3/2}}$ and $\dfrac{\partial}{\partial x}\left(\dfrac{x}{(x^2+y^2+z^2)^{3/2}}\right) = \dfrac{(x^2+y^2+z^2) - 3x^2}{(x^2+y^2+z^2)^{5/2}}$ with similar expressions for $\dfrac{\partial}{\partial y}\left(\dfrac{y}{(x^2+y^2+z^2)^{3/2}}\right)$ and $\dfrac{\partial}{\partial z}\left(\dfrac{z}{(x^2+y^2+z^2)^{3/2}}\right)$, we have

$$\operatorname{div}\left(\frac{\mathbf{x}}{|\mathbf{x}|^3}\right) = \frac{3(x^2+y^2+z^2) - 3(x^2+y^2+z^2)}{(x^2+y^2+z^2)^{5/2}} = 0, \text{ except at } (0,0,0) \text{ where it is undefined.}$$

25. $\iint_S \mathbf{a}\cdot\mathbf{n}\,dS = \iiint_E \operatorname{div}\mathbf{a}\,dV = 0$ since $\operatorname{div}\mathbf{a} = 0$.

27. $\iint_S \operatorname{curl}\mathbf{F}\cdot d\mathbf{S} = \iiint_E \operatorname{div}(\operatorname{curl}\mathbf{F})\,dV = 0$ by Theorem 13.5.11.

29. $\iint_S (f\nabla g)\cdot\mathbf{n}\,dS = \iiint_E \operatorname{div}(f\nabla g)\,dV = \iiint_E (f\nabla^2 g + \nabla g\cdot\nabla f)\,dV$ by Exercise 13.5.23.

31. If $\mathbf{c} = c_1\,\mathbf{i} + c_2\,\mathbf{j} + c_3\,\mathbf{k}$ is an arbitrary constant vector, we define $\mathbf{F} = f\mathbf{c} = fc_1\,\mathbf{i} + fc_2\,\mathbf{j} + fc_3\,\mathbf{k}$. Then $\operatorname{div}\mathbf{F} = \operatorname{div} f\mathbf{c} = \dfrac{\partial f}{\partial x}c_1 + \dfrac{\partial f}{\partial y}c_2 + \dfrac{\partial f}{\partial z}c_3 = \nabla f\cdot\mathbf{c}$ and the Divergence Theorem says $\iint_S \mathbf{F}\cdot d\mathbf{S} = \iiint_E \operatorname{div}\mathbf{F}\,dV \;\Rightarrow$ $\iint_S \mathbf{F}\cdot\mathbf{n}\,dS = \iiint_E \nabla f\cdot\mathbf{c}\,dV$. In particular, if $\mathbf{c} = \mathbf{i}$ then $\iint_S f\mathbf{i}\cdot\mathbf{n}\,dS = \iiint_E \nabla f\cdot\mathbf{i}\,dV \;\Rightarrow$ $\displaystyle\iint_S fn_1\,dS = \iiint_E \frac{\partial f}{\partial x}\,dV$ (where $\mathbf{n} = n_1\,\mathbf{i} + n_2\,\mathbf{j} + n_3\,\mathbf{k}$). Similarly, if $\mathbf{c} = \mathbf{j}$ we have $\displaystyle\iint_S fn_2\,dS = \iiint_E \frac{\partial f}{\partial y}\,dV$, and $\mathbf{c} = \mathbf{k}$ gives $\displaystyle\iint_S fn_3\,dS = \iiint_E \frac{\partial f}{\partial z}\,dV$. Then

$$\begin{aligned}
\iint_S f\mathbf{n}\,dS &= \left(\iint_S fn_1\,dS\right)\mathbf{i} + \left(\iint_S fn_2\,dS\right)\mathbf{j} + \left(\iint_S fn_3\,dS\right)\mathbf{k} \\
&= \left(\iiint_E \frac{\partial f}{\partial x}\,dV\right)\mathbf{i} + \left(\iiint_E \frac{\partial f}{\partial y}\,dV\right)\mathbf{j} + \left(\iiint_E \frac{\partial f}{\partial z}\,dV\right)\mathbf{k} \\
&= \iiint_E \left(\frac{\partial f}{\partial x}\mathbf{i} + \frac{\partial f}{\partial y}\mathbf{j} + \frac{\partial f}{\partial z}\mathbf{k}\right)dV = \iiint_E \nabla f\,dV
\end{aligned}$$

as desired.

13 Review

CONCEPT CHECK

1. See Definitions 1 and 2 in Section 13.1. A vector field can represent, for example, the wind velocity at any location in space, the speed and direction of the ocean current at any location, or the force vectors of Earth's gravitational field at a location in space.

2. (a) A conservative vector field $\mathbf{F}$ is a vector field which is the gradient of some scalar function f.

(b) The function f in part (a) is called a potential function for $\mathbf{F}$, that is, $\mathbf{F} = \nabla f$.

3. (a) See Definition 13.2.2.

(b) We normally evaluate the line integral using Formula 13.2.3.

(c) The mass is $m = \int_C \rho\,(x, y)\; ds$, and the centre of mass is $(\overline{x}, \overline{y})$ where $\overline{x} = \frac{1}{m}\int_C x\rho\,(x, y)\; ds$, $\overline{y} = \frac{1}{m}\int_C y\rho\,(x, y)\; ds$.

(d) See (5) and (6) in Section 13.2 for plane curves; we have similar definitions when C is a space curve [see the equation preceding (10) in Section 13.2].

(e) For plane curves, see Equations 13.2.7. We have similar results for space curves [see the equation preceding (10) in Section 13.2].

4. (a) See Definition 13.2.13.

(b) If $\mathbf{F}$ is a force field, $\int_C \mathbf{F} \cdot d\mathbf{r}$ represents the work done by $\mathbf{F}$ in moving a particle along the curve C.

(c) $\int_C \mathbf{F} \cdot d\mathbf{r} = \int_C P\,dx + Q\,dy + R\,dz$

5. See Theorem 13.3.2.

6. (a) $\int_C \mathbf{F} \cdot d\mathbf{r}$ is independent of path if the line integral has the same value for any two curves that have the same initial and terminal points.

(b) See Theorem 13.3.4.

7. See the statement of Green's Theorem on page 933.

8. See Equations 13.4.5.

9. (a) $\operatorname{curl} \mathbf{F} = \left(\dfrac{\partial R}{\partial y} - \dfrac{\partial Q}{\partial z}\right)\mathbf{i} + \left(\dfrac{\partial P}{\partial z} - \dfrac{\partial R}{\partial x}\right)\mathbf{j} + \left(\dfrac{\partial Q}{\partial x} - \dfrac{\partial P}{\partial y}\right)\mathbf{k} = \nabla \times \mathbf{F}$

(b) $\operatorname{div} \mathbf{F} = \dfrac{\partial P}{\partial x} + \dfrac{\partial Q}{\partial y} + \dfrac{\partial R}{\partial z} = \nabla \cdot \mathbf{F}$

(c) For $\operatorname{curl} \mathbf{F}$, see the discussion accompanying Figure 1 on page 943 as well as Figure 6 and the accompanying discussion on page 963. For $\operatorname{div} \mathbf{F}$, see the discussion following Example 5 on page 944 as well as the discussion preceding (8) on page 970.

10. See Theorem 13.3.6; see Theorem 13.5.4.

11. (a) See (1) in Section 13.6.

(b) We normally evaluate the surface integral using Formula 13.6.2.

(c) See Formula 13.6.4.

(d) The mass is $m = \iint_S \rho(x, y, z)\, dS$ and the centre of mass is $(\overline{x}, \overline{y}, \overline{z})$ where $\overline{x} = \frac{1}{m}\iint_S x\rho(x, y, z)\, dS$, $\overline{y} = \frac{1}{m}\iint_S y\rho(x, y, z)\, dS$, $\overline{z} = \frac{1}{m}\iint_S z\rho(x, y, z)\, dS$.

12. (a) See Figures 6 and 7 and the accompanying discussion in Section 13.6. A Möbius strip is a nonorientable surface; see Figures 4 and 5 and the accompanying discussion on page 952.

(b) See Definition 13.6.8.

(c) See Formula 13.6.9.

(d) See Formula 13.6.10.

13. See the statement of Stokes' Theorem on page 959.

14. See the statement of the Divergence Theorem on page 966.

15. In each theorem, we have an integral of a "derivative" over a region on the left side, while the right side involves the values of the original function only on the boundary of the region.

TRUE-FALSE QUIZ

1. False; div $\mathbf{F}$ is a scalar field.

3. True, by Theorem 13.5.3 and the fact that div $\mathbf{0} = 0$.

5. False. See Exercise 13.3.33. (But the assertion is true if D is simply-connected; see Theorem 13.3.6.)

7. True. Apply the Divergence Theorem and use the fact that div $\mathbf{F} = 0$.

EXERCISES

1. (a) Vectors starting on C point in roughly the direction opposite to C, so the tangential component $\mathbf{F} \cdot \mathbf{T}$ is negative. Thus $\int_C \mathbf{F} \cdot d\mathbf{r} = \int_C \mathbf{F} \cdot \mathbf{T}\, ds$ is negative.

(b) The vectors that end near P are shorter than the vectors that start near P, so the net flow is outward near P and div $\mathbf{F}\,(P)$ is positive.

3. $$\int_C yz\cos x\, ds = \int_0^\pi (3\cos t)\,(3\sin t)\cos t\,\sqrt{(1)^2 + (-3\sin t)^2 + (3\cos t)^2}\, dt = \int_0^\pi \left(9\cos^2 t\sin t\right)\sqrt{10}\, dt$$
$$= 9\sqrt{10}\left(-\tfrac{1}{3}\cos^3 t\right)\Big]_0^\pi = -3\sqrt{10}\,(-2) = 6\sqrt{10}$$

5. $$\int_C y^3\, dx + x^2\, dy = \int_{-1}^1 \left[y^3(-2y) + \left(1 - y^2\right)^2\right] dy = \int_{-1}^1 \left(-y^4 - 2y^2 + 1\right) dy$$
$$= \left[-\tfrac{1}{5}y^5 - \tfrac{2}{3}y^3 + y\right]_{-1}^1 = -\tfrac{1}{5} - \tfrac{2}{3} + 1 - \tfrac{1}{5} - \tfrac{2}{3} + 1 = \tfrac{4}{15}$$

7. C: $x = 1 + 2t \quad\Rightarrow\quad dx = 2\, dt,\ y = 4t \quad\Rightarrow\quad dy = 4\, dt,\ z = -1 + 3t \quad\Rightarrow\quad dz = 3\, dt,\ 0 \le t \le 1.$

$$\int_C xy\, dx + y^2\, dy + yz\, dz = \int_0^1 \left[(1 + 2t)(4t)(2) + (4t)^2(4) + (4t)(-1 + 3t)(3)\right] dt$$
$$= \int_0^1 (116t^2 - 4t)\, dt = \left[\tfrac{116}{3}t^3 - 2t^2\right]_0^1 = \tfrac{116}{3} - 2 = \tfrac{110}{3}$$

9. $\mathbf{F}(\mathbf{r}(t)) = e^{-t}\,\mathbf{i} + t^2(-t)\,\mathbf{j} + (t^2 + t^3)\,\mathbf{k}$, $\mathbf{r}'(t) = 2t\,\mathbf{i} + 3t^2\,\mathbf{j} - \mathbf{k}$ and

$$\int_C \mathbf{F} \cdot d\mathbf{r} = \int_0^1 (2te^{-t} - 3t^5 - (t^2 + t^3))\, dt = \left[-2te^{-t} - 2e^{-t} - \tfrac{1}{2}t^6 - \tfrac{1}{3}t^3 - \tfrac{1}{4}t^4\right]_0^1 = \tfrac{11}{12} - \tfrac{4}{e}.$$

11. $\frac{\partial}{\partial y}\left[(1+xy)e^{xy}\right] = 2xe^{xy} + x^2ye^{xy} = \frac{\partial}{\partial x}\left[e^y + x^2e^{xy}\right]$ and the domain of $\mathbf{F}$ is $\mathbb{R}^2$, so $\mathbf{F}$ is conservative. Thus there exists a function f such that $\mathbf{F} = \nabla f$. Then $f_y(x,y) = e^y + x^2e^{xy}$ implies $f(x,y) = e^y + xe^{xy} + g(x)$ and then $f_x(x,y) = xye^{xy} + e^{xy} + g'(x) = (1+xy)e^{xy} + g'(x)$. But $f_x(x,y) = (1+xy)e^{xy}$, so $g'(x) = 0 \;\Rightarrow\; g(x) = K$. Thus $f(x,y) = e^y + xe^{xy} + K$ is a potential function for $\mathbf{F}$.

13. Since $\frac{\partial}{\partial y}\left(4x^3y^2 - 2xy^3\right) = 8x^3y - 6xy^2 = \frac{\partial}{\partial x}\left(2x^4y - 3x^2y^2 + 4y^3\right)$ and the domain of $\mathbf{F}$ is $\mathbb{R}^2$, $\mathbf{F}$ is conservative. Furthermore $f(x,y) = x^4y^2 - x^2y^3 + y^4$ is a potential function for $\mathbf{F}$. $t = 0$ corresponds to the point $(0,1)$ and $t = 1$ corresponds to $(1,1)$, so $\int_C \mathbf{F}\cdot d\mathbf{r} = f(1,1) - f(0,1) = 1 - 1 = 0$.

15.

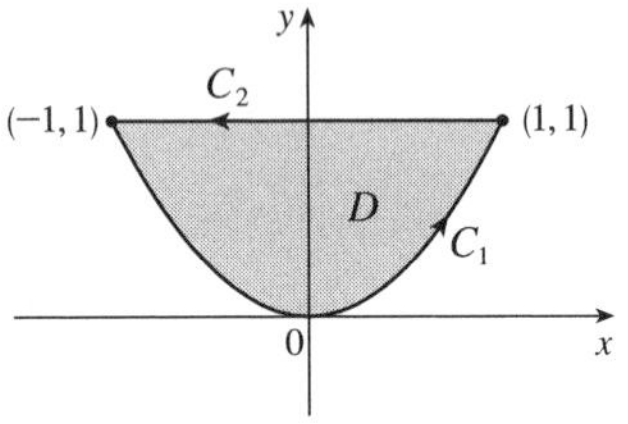

C_1: $\mathbf{r}(t) = t\,\mathbf{i} + t^2\,\mathbf{j}$, $-1 \le t \le 1$;

C_2: $\mathbf{r}(t) = -t\,\mathbf{i} + \mathbf{j}$, $-1 \le t \le 1$.

Then

$$\int_C xy^2\,dx - x^2y\,dy = \int_{-1}^{1}(t^5 - 2t^5)\,dt + \int_{-1}^{1} t\,dt$$
$$= \left[-\tfrac{1}{6}t^6\right]_{-1}^{1} + \left[\tfrac{1}{2}t^2\right]_{-1}^{1} = 0$$

Using Green's Theorem, we have

$$\int_C xy^2\,dx - x^2y\,dy = \iint_D \left[\frac{\partial}{\partial x}\left(-x^2y\right) - \frac{\partial}{\partial y}\left(xy^2\right)\right] dA = \iint_D (-2xy - 2xy)\,dA = \int_{-1}^{1}\int_{x^2}^{1} -4xy\,dy\,dx$$
$$= \int_{-1}^{1}\left[-2xy^2\right]_{y=x^2}^{y=1} dx = \int_{-1}^{1}\left(2x^5 - 2x\right)dx = \left[\tfrac{1}{3}x^6 - x^2\right]_{-1}^{1} = 0$$

17. $\int_C x^2y\,dx - xy^2\,dy = \iint\limits_{x^2+y^2\le 4}\left[\frac{\partial}{\partial x}\left(-xy^2\right) - \frac{\partial}{\partial y}\left(x^2y\right)\right] dA = \iint\limits_{x^2+y^2\le 4}\left(-y^2 - x^2\right) dA = -\int_0^{2\pi}\int_0^2 r^3\,dr\,d\theta = -8\pi$

19. If we assume there is such a vector field $\mathbf{G}$, then $\operatorname{div}(\operatorname{curl}\mathbf{G}) = 2 + 3z - 2xz$. But $\operatorname{div}(\operatorname{curl}\mathbf{F}) = 0$ for all vector fields $\mathbf{F}$. Thus such a $\mathbf{G}$ cannot exist.

21. For any piecewise-smooth simple closed plane curve C bounding a region D, we can apply Green's Theorem to $\mathbf{F}(x,y) = f(x)\,\mathbf{i} + g(y)\,\mathbf{j}$ to get $\int_C f(x)\,dx + g(y)\,dy = \iint_D\left[\frac{\partial}{\partial x}g(y) - \frac{\partial}{\partial y}f(x)\right] dA = \iint_D 0\,dA = 0$.

23. $\nabla^2 f = 0$ means that $\dfrac{\partial^2 f}{\partial x^2} + \dfrac{\partial^2 f}{\partial y^2} = 0$. Now if $\mathbf{F} = f_y\,\mathbf{i} - f_x\,\mathbf{j}$ and C is any closed path in D, then applying Green's Theorem, we get

$$\int_C \mathbf{F}\cdot d\mathbf{r} = \int_C f_y\,dx - f_x\,dy = \iint_D\left[\frac{\partial}{\partial x}(-f_x) - \frac{\partial}{\partial y}(f_y)\right] dA = -\iint_D (f_{xx} + f_{yy})\,dA = -\iint_D 0\,dA = 0$$

Therefore the line integral is independent of path, by Theorem 13.3.3.

25. $z = f(x,y) = x^2 + y^2$ with $0 \le x^2 + y^2 \le 4$ so $\mathbf{r}_x \times \mathbf{r}_y = -2x\,\mathbf{i} - 2y\,\mathbf{j} + \mathbf{k}$ (using upward orientation). Then

$$\iint_S z\,dS = \iint\limits_{x^2+y^2\le 4}\left(x^2 + y^2\right)\sqrt{4x^2 + 4y^2 + 1}\,dA = \int_0^{2\pi}\int_0^2 r^3\sqrt{1 + 4r^2}\,dr\,d\theta = \tfrac{1}{60}\pi\left(391\sqrt{17} + 1\right)$$

(Substitute $u = 1 + 4r^2$ and use tables.)

27. Since the sphere bounds a simple solid region, the Divergence Theorem applies and

$\iint_S \mathbf{F}\cdot d\mathbf{S} = \iiint_E (z-2)\,dV = \iiint_E z\,dV - 2\iiint_E dV = m\bar{z} - 2\left(\frac{4}{3}\pi 2^3\right) = -\frac{64}{3}\pi.$

Alternate solution: $\mathbf{F}(\mathbf{r}(\phi,\theta)) = 4\sin\phi\cos\theta\cos\phi\,\mathbf{i} - 4\sin\phi\sin\theta\,\mathbf{j} + 6\sin\phi\cos\theta\,\mathbf{k}$,

$\mathbf{r}_\phi \times \mathbf{r}_\theta = 4\sin^2\phi\cos\theta\,\mathbf{i} + 4\sin^2\phi\sin\theta\,\mathbf{j} + 4\sin\phi\cos\phi\,\mathbf{k}$, and

$\mathbf{F}\cdot(\mathbf{r}_\phi\times\mathbf{r}_\theta) = 16\sin^3\phi\cos^2\theta\cos\phi - 16\sin^3\phi\sin^2\theta + 24\sin^2\phi\cos\phi\cos\theta$. Then

$$\iint_S F\cdot dS = \int_0^{2\pi}\int_0^{\pi}(16\sin^3\phi\cos\phi\cos^2\theta - 16\sin^3\phi\sin^2\theta + 24\sin^2\phi\cos\phi\cos\theta)\,d\phi\,d\theta$$
$$= \int_0^{2\pi}\tfrac{4}{3}(-16\sin^2\theta)\,d\theta = -\tfrac{64}{3}\pi$$

29. Since $\operatorname{curl}\mathbf{F} = \mathbf{0}$, $\iint_S(\operatorname{curl}\mathbf{F})\cdot d\mathbf{S} = 0$. We parametrize C: $\mathbf{r}(t) = \cos t\,\mathbf{i} + \sin t\,\mathbf{j}$, $0\le t\le 2\pi$ and

$\oint_C \mathbf{F}\cdot d\mathbf{r} = \int_0^{2\pi}(-\cos^2 t\sin t + \sin^2 t\cos t)\,dt = \frac{1}{3}\cos^3 t + \frac{1}{3}\sin^3 t\Big]_0^{2\pi} = 0.$

31. The surface is given by $x+y+z=1$ or $z = 1-x-y$, $0\le x\le 1$, $0\le y\le 1-x$ and $\mathbf{r}_x\times\mathbf{r}_y = \mathbf{i}+\mathbf{j}+\mathbf{k}$. Then

$$\oint_C \mathbf{F}\cdot d\mathbf{r} = \iint_S \operatorname{curl}\mathbf{F}\cdot d\mathbf{S} = \iint_D(-y\,\mathbf{i} - z\,\mathbf{j} - x\,\mathbf{k})\cdot(\mathbf{i}+\mathbf{j}+\mathbf{k})\,dA = \iint_D(-1)\,dA = -(\text{area of } D) = -\tfrac{1}{2}$$

33. $\iiint_E \operatorname{div}\mathbf{F}\,dV = \iiint_{x^2+y^2+z^2\le 1} 3\,dV = 3(\text{volume of sphere}) = 4\pi$. Then

$\mathbf{F}(\mathbf{r}(\phi,\theta))\cdot(\mathbf{r}_\phi\times\mathbf{r}_\theta) = \sin^3\phi\cos^2\theta + \sin^3\phi\sin^2\theta + \sin\phi\cos^2\phi = \sin\phi$ and

$\iint_S \mathbf{F}\cdot d\mathbf{S} = \int_0^{2\pi}\int_0^{\pi}\sin\phi\,d\phi\,d\theta = (2\pi)(2) = 4\pi.$

35. Because $\operatorname{curl}\mathbf{F} = \mathbf{0}$, $\mathbf{F}$ is conservative, and if $f(x,y,z) = x^3yz - 3xy + z^2$, then $\nabla f = \mathbf{F}$. Hence

$\int_C \mathbf{F}\cdot d\mathbf{r} = \int_C \nabla f\cdot d\mathbf{r} = f(0,3,0) - f(0,0,2) = 0 - 4 = -4.$

37. By the Divergence Theorem, $\iint_S \mathbf{F}\cdot\mathbf{n}\,dS = \iiint_E \operatorname{div}\mathbf{F}\,dV = 3(\text{volume of } E) = 3(8-1) = 21.$

FOCUS ON PROBLEM SOLVING

1. Let S_1 be the portion of $\Omega(S)$ between $S(a)$ and S, and let ∂S_1 be its boundary. Also let S_L be the lateral surface of S_1 [that is, the surface of S_1 except S and $S(a)$]. Applying the Divergence Theorem we have $\iint_{\partial S_1} \frac{\mathbf{r}\cdot\mathbf{n}}{r^3}\,dS = \iiint_{S_1} \nabla\cdot\frac{\mathbf{r}}{r^3}\,dV$.

But

$$\begin{aligned}\nabla\cdot\frac{\mathbf{r}}{r^3} &= \left\langle \frac{\partial}{\partial x}, \frac{\partial}{\partial y}, \frac{\partial}{\partial z}\right\rangle \cdot \left\langle \frac{x}{(x^2+y^2+z^2)^{3/2}}, \frac{y}{(x^2+y^2+z^2)^{3/2}}, \frac{z}{(x^2+y^2+z^2)^{3/2}}\right\rangle \\ &= \frac{(x^2+y^2+z^2-3x^2)+(x^2+y^2+z^2-3y^2)+(x^2+y^2+z^2-3z^2)}{(x^2+y^2+z^2)^{5/2}} = 0\end{aligned}$$

$\Rightarrow \iint_{\partial S_1} \frac{\mathbf{r}\cdot\mathbf{n}}{r^3}\,dS = \iiint_{S_1} 0\,dV = 0$. On the other hand, notice that for the surfaces of ∂S_1 other than $S(a)$ and S, $\mathbf{r}\cdot\mathbf{n} = 0 \;\Rightarrow$

$$\begin{aligned}0 &= \iint_{\partial S_1} \frac{\mathbf{r}\cdot\mathbf{n}}{r^3}\,dS = \iint_S \frac{\mathbf{r}\cdot\mathbf{n}}{r^3}\,dS + \iint_{S(a)} \frac{\mathbf{r}\cdot\mathbf{n}}{r^3}\,dS + \iint_{S_L} \frac{\mathbf{r}\cdot\mathbf{n}}{r^3}\,dS \\ &= \iint_S \frac{\mathbf{r}\cdot\mathbf{n}}{r^3}\,dS + \iint_{S(a)} \frac{\mathbf{r}\cdot\mathbf{n}}{r^3}\,dS\end{aligned}$$

$\Rightarrow \iint_S \frac{\mathbf{r}\cdot\mathbf{n}}{r^3}\,dS = -\iint_{S(a)} \frac{\mathbf{r}\cdot\mathbf{n}}{r^3}\,dS$. Notice that on $S(a)$, $r = a \;\Rightarrow\; \mathbf{n} = -\frac{\mathbf{r}}{r} = -\frac{\mathbf{r}}{a}$ and $\mathbf{r}\cdot\mathbf{r} = r^2 = a^2$, so that $-\iint_{S(a)} \frac{\mathbf{r}\cdot\mathbf{n}}{r^3}\,dS = \iint_{S(a)} \frac{\mathbf{r}\cdot\mathbf{r}}{a^4}\,dS = \iint_{S(a)} \frac{a^2}{a^4}\,dS = \frac{1}{a^2}\iint_{S(a)} dS = \frac{\text{area of } S(a)}{a^2} = |\Omega(S)|$. Therefore $|\Omega(S)| = \iint_S \frac{\mathbf{r}\cdot\mathbf{n}}{r^3}\,dS$.

3. Let $\mathbf{F} = \mathbf{a}\times\mathbf{r} = \langle a_1, a_2, a_3\rangle \times \langle x, y, z\rangle = \langle a_2 z - a_3 y, a_3 x - a_1 z, a_1 y - a_2 x\rangle$. Then $\operatorname{curl}\mathbf{F} = \langle 2a_1, 2a_2, 2a_3\rangle = 2\mathbf{a}$, and

$$\iint_S 2\mathbf{a}\cdot d\mathbf{S} = \iint_S \operatorname{curl}\mathbf{F}\cdot d\mathbf{S} = \int_C \mathbf{F}\cdot d\mathbf{r} = \int_C (\mathbf{a}\times\mathbf{r})\cdot d\mathbf{r}$$

by Stokes' Theorem.

5. The given line integral $\frac{1}{2}\int_C (bz - cy)\,dx + (cx - az)\,dy + (ay - bx)\,dz$ can be expressed as $\int_C \mathbf{F}\cdot d\mathbf{r}$ if we define the vector field $\mathbf{F}$ by $\mathbf{F}(x,y,z) = P\,\mathbf{i} + Q\,\mathbf{j} + R\,\mathbf{k} = \frac{1}{2}(bz - cy)\,\mathbf{i} + \frac{1}{2}(cx - az)\,\mathbf{j} + \frac{1}{2}(ay - bx)\,\mathbf{k}$. Then define S to be the planar interior of C, so S is an oriented, smooth surface. Stokes' Theorem says $\int_C \mathbf{F}\cdot d\mathbf{r} = \iint_S \operatorname{curl}\mathbf{F}\cdot d\mathbf{S} = \iint_S \operatorname{curl}\mathbf{F}\cdot\mathbf{n}\,dS$.

Now

$$\begin{aligned}\operatorname{curl}\mathbf{F} &= \left(\frac{\partial R}{\partial y} - \frac{\partial Q}{\partial z}\right)\mathbf{i} + \left(\frac{\partial P}{\partial z} - \frac{\partial R}{\partial x}\right)\mathbf{j} + \left(\frac{\partial Q}{\partial x} - \frac{\partial P}{\partial y}\right)\mathbf{k} \\ &= \left(\tfrac{1}{2}a + \tfrac{1}{2}a\right)\mathbf{i} + \left(\tfrac{1}{2}b + \tfrac{1}{2}b\right)\mathbf{j} + \left(\tfrac{1}{2}c + \tfrac{1}{2}c\right)\mathbf{k} = a\,\mathbf{i} + b\,\mathbf{j} + c\,\mathbf{k} = \mathbf{n}\end{aligned}$$

so $\operatorname{curl}\mathbf{F}\cdot\mathbf{n} = \mathbf{n}\cdot\mathbf{n} = |\mathbf{n}|^2 = 1$, hence $\iint_S \operatorname{curl}\mathbf{F}\cdot\mathbf{n}\,dS = \iint_S dS$ which is simply the surface area of S. Thus, $\int_C \mathbf{F}\cdot d\mathbf{r} = \frac{1}{2}\int_C (bz - cy)\,dx + (cx - az)\,dy + (ay - bx)\,dz$ is the plane area enclosed by C.

☐ APPENDIXES

A INTERVALS, INEQUALITIES, AND ABSOLUTE VALUES

1. $|5-23| = |-18| = 18$

3. $\left|\sqrt{5}-5\right| = -\left(\sqrt{5}-5\right) = 5-\sqrt{5}$ because $\sqrt{5}-5<0$.

5. If $x<2$, $x-2<0$, so $|x-2| = -(x-2) = 2-x$.

7. $|x+1| = \begin{cases} x+1 & \text{if } x+1\geq 0 \\ -(x+1) & \text{if } x+1<0 \end{cases} = \begin{cases} x+1 & \text{if } x\geq -1 \\ -x-1 & \text{if } x<-1 \end{cases}$

9. $\left|x^2+1\right| = x^2+1$ (since $x^2+1\geq 0$ for all x).

11. $2x+7>3 \quad\Leftrightarrow\quad 2x>-4 \quad\Leftrightarrow\quad x>-2$, so $x\in(-2,\infty)$.

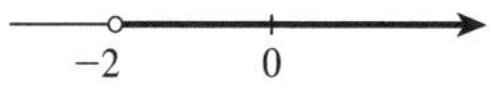

13. $1-x\leq 2 \quad\Leftrightarrow\quad -x\leq 1 \quad\Leftrightarrow\quad x\geq -1$, so $x\in[-1,\infty)$.

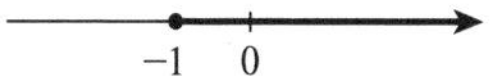

15. $0\leq 1-x<1 \quad\Leftrightarrow\quad -1\leq -x<0 \quad\Leftrightarrow\quad 1\geq x>0$, so $x\in(0,1]$.

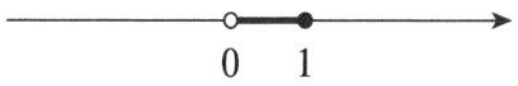

17. $(x-1)(x-2)>0$. *Case 1:* (both factors are positive, so their product is positive)

$x-1>0 \quad\Leftrightarrow\quad x>1$, and $x-2>0 \quad\Leftrightarrow\quad x>2$, so $x\in(2,\infty)$.

Case 2: (both factors are negative, so their product is positive)

$x-1<0 \quad\Leftrightarrow\quad x<1$, and $x-2<0 \quad\Leftrightarrow\quad x<2$, so $x\in(-\infty,1)$.

Thus, the solution set is $(-\infty,1)\cup(2,\infty)$.

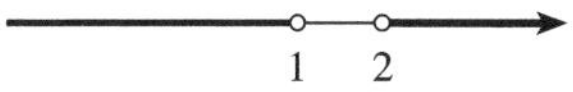

19. $x^2<3 \quad\Leftrightarrow\quad x^2-3<0 \quad\Leftrightarrow\quad \left(x-\sqrt{3}\right)\left(x+\sqrt{3}\right)<0$. *Case 1:* $x>\sqrt{3}$ and $x<-\sqrt{3}$, which is impossible. *Case 2:* $x<\sqrt{3}$ and $x>-\sqrt{3}$. Thus, the solution set is $\left(-\sqrt{3},\sqrt{3}\right)$.

Another method: $x^2<3 \quad\Leftrightarrow\quad |x|<\sqrt{3} \quad\Leftrightarrow\quad -\sqrt{3}<x<\sqrt{3}$.

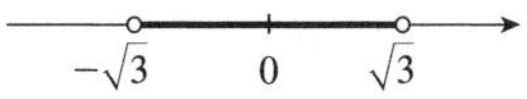

21. $x^3-x^2\leq 0 \quad\Leftrightarrow\quad x^2(x-1)\leq 0$. Since $x^2\geq 0$ for all x, the inequality is satisfied when $x-1\leq 0 \quad\Leftrightarrow\quad x\leq 1$. Thus, the solution set is $(-\infty,1]$.

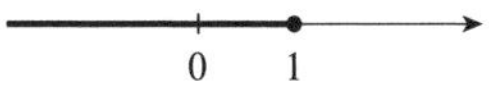

23. $x^3>x \quad\Leftrightarrow\quad x^3-x>0 \quad\Leftrightarrow\quad x\left(x^2-1\right)>0 \quad\Leftrightarrow\quad x(x-1)(x+1)>0$. Construct a chart:

Interval	x	$x-1$	$x+1$	$x(x-1)(x+1)$
$x<-1$	$-$	$-$	$-$	$-$
$-1<x<0$	$-$	$-$	$+$	$+$
$0<x<1$	$+$	$-$	$+$	$-$
$x>1$	$+$	$+$	$+$	$+$

Since $x^3>x$ when the last column is positive, the solution set is $(-1,0)\cup(1,\infty)$.

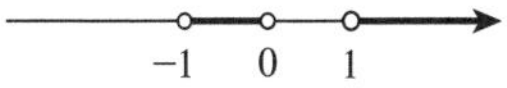

25. $1/x < 4$. This is clearly true for $x < 0$. So suppose $x > 0$. then $1/x < 4 \iff 1 < 4x \iff \frac{1}{4} < x$. Thus, the solution set is $(-\infty, 0) \cup \left(\frac{1}{4}, \infty\right)$.

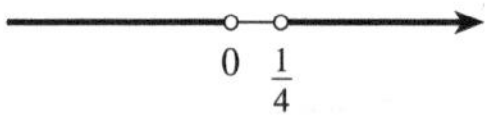

27. $C = \frac{5}{9}(F - 32) \Rightarrow F = \frac{9}{5}C + 32$. So $50 \leq F \leq 95 \Rightarrow 50 \leq \frac{9}{5}C + 32 \leq 95 \Rightarrow 18 \leq \frac{9}{5}C \leq 63 \Rightarrow$ $10 \leq C \leq 35$. So the interval is $[10, 35]$.

29. (a) Let T represent the temperature in degrees Celsius and h the height in km. $T = 20$ when $h = 0$ and T decreases by $10°$C for every km ($1°$C for each 100-m rise). Thus, $T = 20 - 10h$ when $0 \leq h \leq 12$.

(b) From part (a), $T = 20 - 10h \Rightarrow 10h = 20 - T \Rightarrow h = 2 - T/10$. So $0 \leq h \leq 5 \Rightarrow 0 \leq 2 - T/10 \leq 5 \Rightarrow$ $-2 \leq -T/10 \leq 3 \Rightarrow -20 \leq -T \leq 30 \Rightarrow 20 \geq T \geq -30 \Rightarrow -30 \leq T \leq 20$. Thus, the range of temperatures (in °C) to be expected is $[-30, 20]$.

31. $|x + 3| = |2x + 1| \iff$ either $x + 3 = 2x + 1$ or $x + 3 = -(2x + 1)$. In the first case, $x = 2$, and in the second case, $x + 3 = -2x - 1 \iff 3x = -4 \iff x = -\frac{4}{3}$. So the solutions are $-\frac{4}{3}$ and 2.

33. By Property 5 of absolute values, $|x| < 3 \iff -3 < x < 3$, so $x \in (-3, 3)$.

35. $|x - 4| < 1 \iff -1 < x - 4 < 1 \iff 3 < x < 5$, so $x \in (3, 5)$.

37. $|x + 5| \geq 2 \iff x + 5 \geq 2$ or $x + 5 \leq -2 \iff x \geq -3$ or $x \leq -7$, so $x \in (-\infty, -7] \cup [-3, \infty)$.

39. $|2x - 3| \leq 0.4 \iff -0.4 \leq 2x - 3 \leq 0.4 \iff 2.6 \leq 2x \leq 3.4 \iff 1.3 \leq x \leq 1.7$, so $x \in [1.3, 1.7]$.

41. $a(bx - c) \geq bc \iff bx - c \geq \dfrac{bc}{a} \iff bx \geq \dfrac{bc}{a} + c = \dfrac{bc + ac}{a} \iff x \geq \dfrac{bc + ac}{ab}$

43. $|ab| = \sqrt{(ab)^2} = \sqrt{a^2 b^2} = \sqrt{a^2}\sqrt{b^2} = |a|\,|b|$

B COORDINATE GEOMETRY

1. Use the distance formula with $P_1(x_1, y_1) = (1, 1)$ and $P_2(x_2, y_2) = (4, 5)$ to get

$$|P_1 P_2| = \sqrt{(4 - 1)^2 + (5 - 1)^2} = \sqrt{3^2 + 4^2} = \sqrt{25} = 5$$

3. With $P(-3, 3)$ and $Q(-1, -6)$, the slope m of the line through P and Q is $m = \dfrac{-6 - 3}{-1 - (-3)} = -\dfrac{9}{2}$.

5. Using $A(-2, 9)$, $B(4, 6)$, $C(1, 0)$, and $D(-5, 3)$, we have

$|AB| = \sqrt{[4 - (-2)]^2 + (6 - 9)^2} = \sqrt{6^2 + (-3)^2} = \sqrt{45} = \sqrt{9}\sqrt{5} = 3\sqrt{5}$,

$|BC| = \sqrt{(1 - 4)^2 + (0 - 6)^2} = \sqrt{(-3)^2 + (-6)^2} = \sqrt{45} = \sqrt{9}\sqrt{5} = 3\sqrt{5}$,

$|CD| = \sqrt{(-5 - 1)^2 + (3 - 0)^2} = \sqrt{(-6)^2 + 3^2} = \sqrt{45} = \sqrt{9}\sqrt{5} = 3\sqrt{5}$, and

$|DA| = \sqrt{[-2 - (-5)]^2 + (9 - 3)^2} = \sqrt{3^2 + 6^2} = \sqrt{45} = \sqrt{9}\sqrt{5} = 3\sqrt{5}$. So all sides are of equal length and we have a rhombus. Moreover, $m_{AB} = \dfrac{6 - 9}{4 - (-2)} = -\dfrac{1}{2}$, $m_{BC} = \dfrac{0 - 6}{1 - 4} = 2$, $m_{CD} = \dfrac{3 - 0}{-5 - 1} = -\dfrac{1}{2}$, and

$m_{DA} = \dfrac{9 - 3}{-2 - (-5)} = 2$, so the sides are perpendicular. Thus, A, B, C, and D are vertices of a square.

7. The graph of the equation $x = 3$ is a vertical line with x-intercept 3. The line does not have a slope.

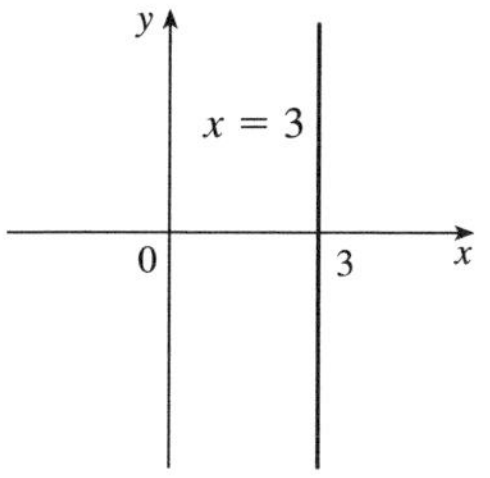

9. $xy = 0 \;\Leftrightarrow\; x = 0$ or $y = 0$. The graph consists of the coordinate axes.

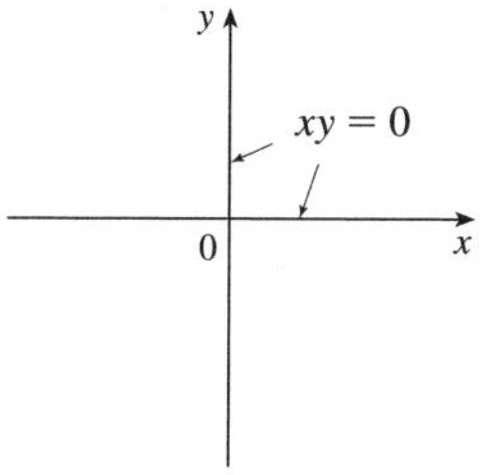

11. By the point-slope form of the equation of a line, an equation of the line through $(2, -3)$ with slope 6 is $y - (-3) = 6(x - 2)$ or $y = 6x - 15$.

13. The slope of the line through $(2, 1)$ and $(1, 6)$ is $m = \dfrac{6 - 1}{1 - 2} = -5$, so an equation of the line is $y - 1 = -5(x - 2)$ or $y = -5x + 11$.

15. By the slope-intercept form of the equation of a line, an equation of the line is $y = 3x - 2$.

17. Since the line passes through $(1, 0)$ and $(0, -3)$, its slope is $m = \dfrac{-3 - 0}{0 - 1} = 3$, so an equation is $y = 3x - 3$.

Another method: From Exercise 46, $\dfrac{x}{1} + \dfrac{y}{-3} = 1 \;\Rightarrow\; -3x + y = -3 \;\Rightarrow\; y = 3x - 3$.

19. The line is parallel to the x-axis, so it is horizontal and must have the form $y = k$. Since it goes through the point $(x, y) = (4, 5)$, the equation is $y = 5$.

21. Putting the line $x + 2y = 6$ into its slope-intercept form gives us $y = -\frac{1}{2}x + 3$, so we see that this line has slope $-\frac{1}{2}$. Thus, we want the line of slope $-\frac{1}{2}$ that passes through the point $(1, -6)$: $y - (-6) = -\frac{1}{2}(x - 1) \;\Leftrightarrow\; y = -\frac{1}{2}x - \frac{11}{2}$.

23. $2x + 5y + 8 = 0 \;\Leftrightarrow\; y = -\frac{2}{5}x - \frac{8}{5}$. Since this line has slope $-\frac{2}{5}$, a line perpendicular to it would have slope $\frac{5}{2}$, so the required line is $y - (-2) = \frac{5}{2}\,[x - (-1)] \;\Leftrightarrow\; y = \frac{5}{2}x + \frac{1}{2}$.

25. $x + 3y = 0 \;\Leftrightarrow\; y = -\frac{1}{3}x$, so the slope is $-\frac{1}{3}$ and the y-intercept is 0.

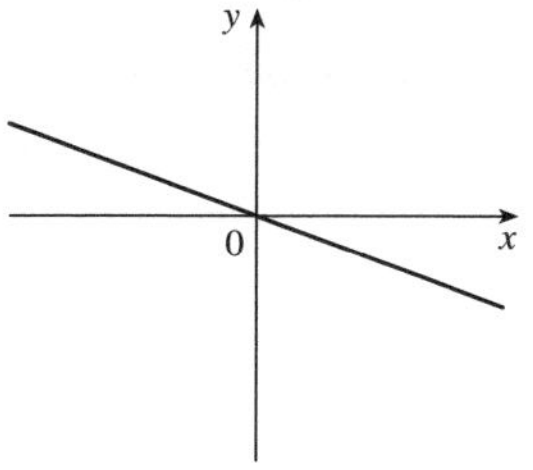

27. $3x - 4y = 12 \;\Leftrightarrow\; y = \frac{3}{4}x - 3$, so the slope is $\frac{3}{4}$ and the y-intercept is -3.

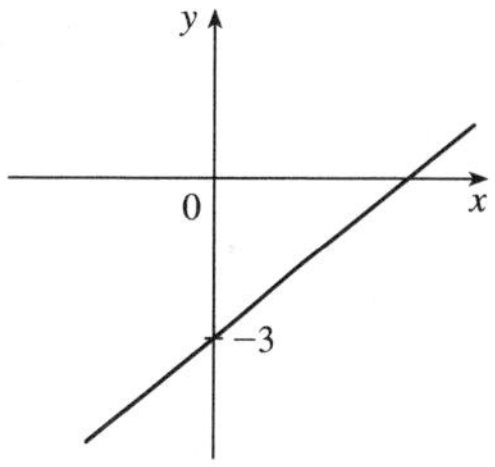

29. $\{(x, y) \mid x < 0\}$

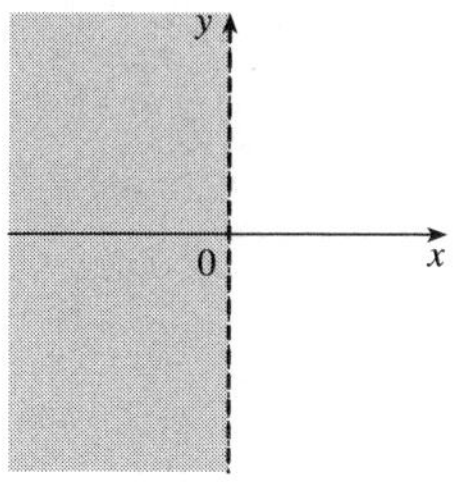

31. $\left\{(x, y) \,\middle|\, |x| \le 2\right\} = \{(x, y) \mid -2 \le x \le 2\}$

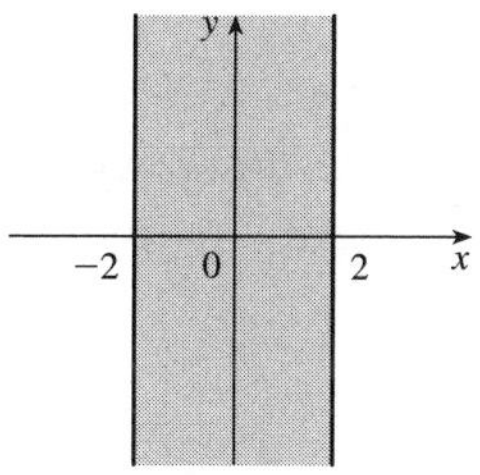

33. $\{(x, y) \mid 0 \le y \le 4, x \le 2\}$

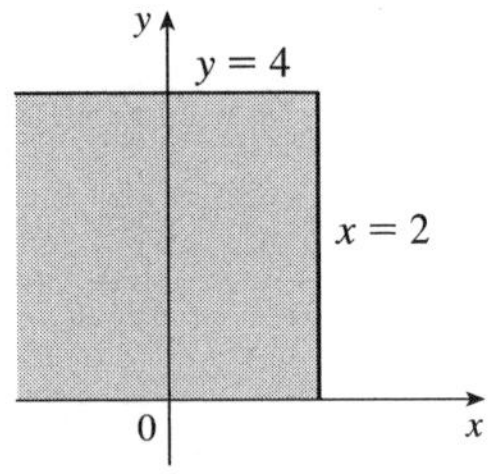

35. $\{(x, y) \mid 1 + x \le y \le 1 - 2x\}$

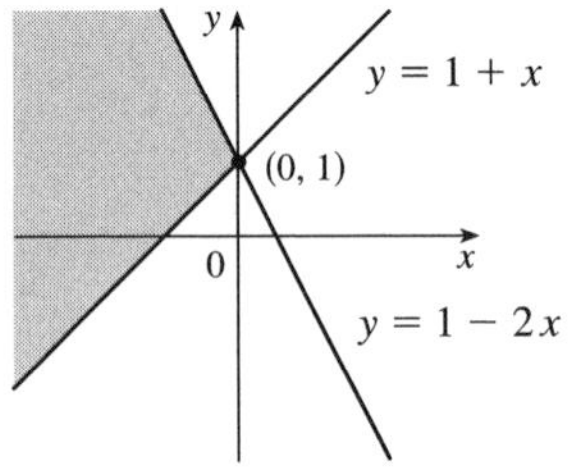

37. An equation of the circle with centre $(3, -1)$ and radius 5 is $(x-3)^2 + (y+1)^2 = 5^2 = 25$.

39. $x^2 + y^2 - 4x + 10y + 13 = 0 \quad \Leftrightarrow \quad x^2 - 4x + y^2 + 10y = -13 \quad \Leftrightarrow$
$(x^2 - 4x + 4) + (y^2 + 10y + 25) = -13 + 4 + 25 = 16 \quad \Leftrightarrow \quad (x-2)^2 + (y+5)^2 = 4^2$. Thus, we have a circle with centre $(2, -5)$ and radius 4.

41. $2x - y = 4 \quad \Leftrightarrow \quad y = 2x - 4 \quad \Rightarrow \quad m_1 = 2$ and $6x - 2y = 10 \quad \Leftrightarrow \quad 2y = 6x - 10 \quad \Leftrightarrow \quad y = 3x - 5 \quad \Rightarrow \quad m_2 = 3$. Since $m_1 \ne m_2$, the two lines are not parallel. To find the point of intersection: $2x - 4 = 3x - 5 \quad \Leftrightarrow \quad x = 1 \quad \Rightarrow$ $y = -2$. Thus, the point of intersection is $(1, -2)$.

43. Let M be the point $\left(\dfrac{x_1 + x_2}{2}, \dfrac{y_1 + y_2}{2}\right)$. Then

$|MP_1|^2 = \left(x_1 - \dfrac{x_1 + x_2}{2}\right)^2 + \left(y_1 - \dfrac{y_1 + y_2}{2}\right)^2 = \left(\dfrac{x_1 - x_2}{2}\right)^2 + \left(\dfrac{y_1 - y_2}{2}\right)^2$ and

$|MP_2|^2 = \left(x_2 - \dfrac{x_1 + x_2}{2}\right)^2 + \left(y_2 - \dfrac{y_1 + y_2}{2}\right)^2 = \left(\dfrac{x_2 - x_1}{2}\right)^2 + \left(\dfrac{y_2 - y_1}{2}\right)^2$. Hence, $|MP_1| = |MP_2|$; that is, M is equidistant from P_1 and P_2.

45. With $A(1, 4)$ and $B(7, -2)$, the slope of segment AB is $\frac{-2-4}{7-1} = -1$, so its perpendicular bisector has slope 1. The midpoint of AB is $\left(\frac{1+7}{2}, \frac{4+(-2)}{2}\right) = (4, 1)$, so an equation of the perpendicular bisector is $y - 1 = 1(x - 4)$ or $y = x - 3$.

47. If $P(x, y)$ is any point on the parabola, then the distance from P to the focus is $|PF| = \sqrt{x^2 + (y-p)^2}$ and the distance from P to the directrix is $|y + p|$. (Figure 14 in the text illustrates the case where $p > 0$.) The defining property of a parabola is that these distances are equal: $\sqrt{x^2 + (y-p)^2} = |y + p|$. We get an equivalent equation by squaring and simplifying: $x^2 + (y-p)^2 = |y+p|^2 = (y+p)^2 \quad \Leftrightarrow \quad x^2 + y^2 - 2py + p^2 = y^2 + 2py + p^2 \quad \Leftrightarrow \quad x^2 = 4py$. Thus, an equation of a parabola with focus $(0, p)$ and directrix $y = -p$ is $x^2 = 4py$.

49. See Figure 20 in the text. $P(x, y)$ is a point on the ellipse when $|PF_1| + |PF_2| = 2a$; that is,

$\sqrt{(x+c)^2 + y^2} + \sqrt{(x-c)^2 + y^2} = 2a$ or $\sqrt{(x+c)^2 + y^2} = 2a - \sqrt{(x+c)^2 + y^2}$. Squaring both sides, we have

$x^2 - 2cx + c^2 + y^2 = 4a^2 - 4a\sqrt{(x+c)^2 + y^2} + x^2 + 2cx + c^2 + y^2$, which simplifies to $a\sqrt{(x+c)^2 + y^2} = a^2 + cx$.

We square again: $a^2\left(x^2 + 2cx + c^2 + y^2\right) = a^4 + 2a^2cx + c^2x^2$, which becomes $(a^2 - c^2)x^2 + a^2y^2 = a^2\left(a^2 - c^2\right)$.

From triangle F_1F_2P in Figure 20, we see that $2c < 2a$, so $c < a$ and, therefore, $a^2 - c^2 > 0$. For convenience, let $b^2 = a^2 - c^2$. Then the equation of the ellipse becomes $b^2x^2 + a^2y^2 = a^2b^2$ or, if both sides are divided by a^2b^2,

$\dfrac{x^2}{a^2} + \dfrac{y^2}{b^2} = 1$.

51. From Figure 23 in the text, $|PF_1| - |PF_2| = \pm 2a \quad\Leftrightarrow\quad \sqrt{(x+c)^2+y^2} - \sqrt{(x-c)^2+y^2} = \pm 2a \quad\Leftrightarrow$
$\sqrt{(x+c)^2+y^2} = \sqrt{(x-c)^2+y^2} \pm 2a \quad\Leftrightarrow$
$(x+c)^2+y^2 = (x-c)^2+y^2+4a^2 \pm 4a\sqrt{(x-c)^2+y^2} \quad\Leftrightarrow\quad 4cx - 4a^2 = \pm 4a\sqrt{(x-c)^2+y^2} \quad\Leftrightarrow$
$c^2x^2 - 2a^2cx + a^4 = a^2(x^2 - 2cx + c^2 + y^2) \quad\Leftrightarrow\quad (c^2-a^2)\,x^2 - a^2y^2 = a^2(c^2-a^2) \quad\Leftrightarrow$
$b^2x^2 - a^2y^2 = a^2b^2$ [where $b^2 = c^2 - a^2$] $\quad\Leftrightarrow\quad \dfrac{x^2}{a^2} - \dfrac{y^2}{b^2} = 1.$

53. 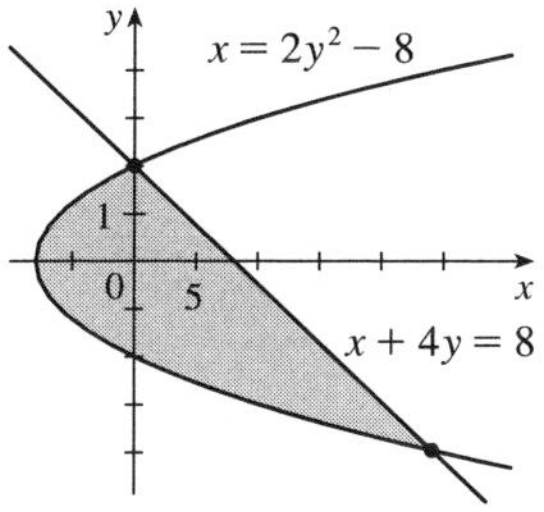

$x + 4y = 8$, $x = 2y^2 - 8$. Substitute x from the second equation into the first: $(2y^2-8)+4y = 8 \quad\Leftrightarrow\quad 2y^2+4y-16 = 0 \quad\Leftrightarrow$
$y^2+2y-8=0 \quad\Leftrightarrow\quad (y+4)(y-2)=0 \quad\Leftrightarrow\quad y=-4$ or 2.
So the points of intersection are $(24,-4)$ and $(0,2)$.

55. Differentiating implicitly, $\dfrac{x^2}{a^2} + \dfrac{y^2}{b^2} = 1 \quad\Rightarrow\quad \dfrac{2x}{a^2} + \dfrac{2yy'}{b^2} = 0 \quad\Rightarrow\quad y' = -\dfrac{b^2x}{a^2y}$ $(y \neq 0)$. Thus, the slope of the tangent line at P is $-\dfrac{b^2x_1}{a^2y_1}$. The slope of F_1P is $\dfrac{y_1}{x_1+c}$ and of F_2P is $\dfrac{y_1}{x_1-c}$. By the formula in Problem 15 on text page 260, we have

$$\tan\alpha = \frac{\dfrac{y_1}{x_1+c} + \dfrac{b^2x_1}{a^2y_1}}{1 - \dfrac{b^2x_1y_1}{a^2y_1(x_1+c)}} = \frac{a^2y_1^2 + b^2x_1(x_1+c)}{a^2y_1(x_1+c) - b^2x_1y_1}$$
$$= \frac{a^2b^2 + b^2cx_1}{c^2x_1y_1 + a^2cy_1} \qquad \left[\begin{array}{c}\text{using } b^2x_1^2 + a^2y_1^2 = a^2b^2 \\ \text{and } a^2 - b^2 = c^2\end{array}\right]$$
$$= \frac{b^2(cx_1+a^2)}{cy_1(cx_1+a^2)}$$
$$= \frac{b^2}{cy_1}$$

and

$$\tan\beta = \frac{-\dfrac{b^2x_1}{a^2y_1} - \dfrac{y_1}{x_1-c}}{1 - \dfrac{b^2x_1y_1}{a^2y_1(x_1-c)}} = \frac{-a^2y_1^2 - b^2x_1(x_1-c)}{a^2y_1\,(x_1-c) - b^2x_1y_1}$$
$$= \frac{-a^2b^2 + b^2cx_1}{c^2x_1y_1 - a^2cy_1} = \frac{b^2(cx_1-a^2)}{cy_1(cx_1-a^2)}$$
$$= \frac{b^2}{cy_1}$$

Thus, $\alpha = \beta$.

C TRIGONOMETRY

1. (a) $210^\circ = 210\left(\frac{\pi}{180}\right) = \frac{7\pi}{6}$ rad (b) $9^\circ = 9\left(\frac{\pi}{180}\right) = \frac{\pi}{20}$ rad

3. (a) 4π rad $= 4\pi\left(\frac{180}{\pi}\right) = 720^\circ$ (b) $-\frac{3\pi}{8}$ rad $= -\frac{3\pi}{8}\left(\frac{180}{\pi}\right) = -67.5^\circ$

5. Using Formula 3, $a = r\theta = 36 \cdot \frac{\pi}{12} = 3\pi$ cm.

7. Using Formula 3, $\theta = a/r = \frac{1}{1.5} = \frac{2}{3}$ rad $= \frac{2}{3}\left(\frac{180}{\pi}\right) = \left(\frac{120}{\pi}\right)^\circ \approx 38.2^\circ$.

9. (a)

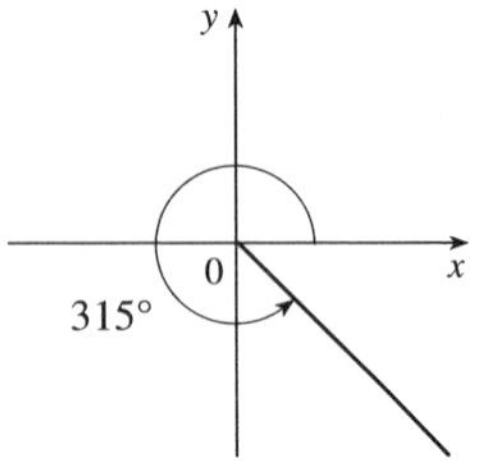

(b)

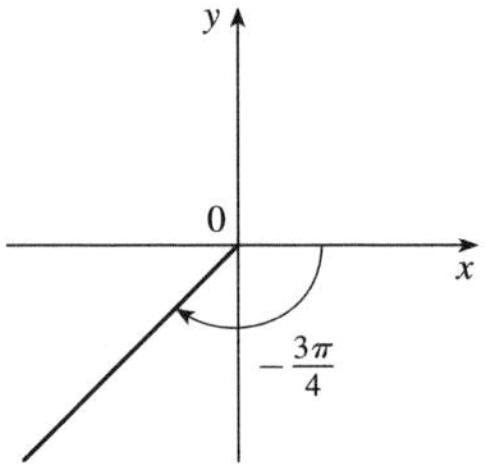

11.

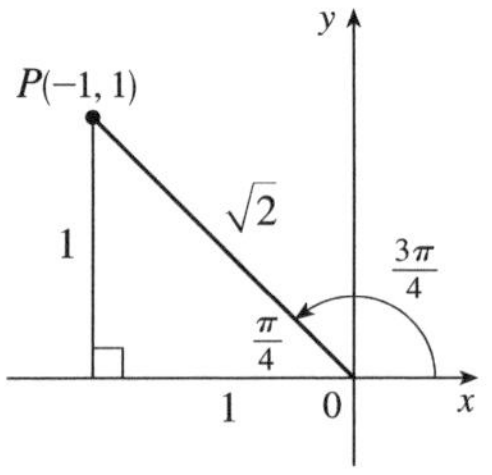

From the diagram we see that a point on the terminal side is $P(-1, 1)$. Therefore, taking $x = -1$, $y = 1$, $r = \sqrt{2}$ in the definitions of the trigonometric ratios, we have $\sin\frac{3\pi}{4} = \frac{1}{\sqrt{2}}$, $\cos\frac{3\pi}{4} = -\frac{1}{\sqrt{2}}$, $\tan\frac{3\pi}{4} = -1$, $\csc\frac{3\pi}{4} = \sqrt{2}$, $\sec\frac{3\pi}{4} = -\sqrt{2}$, and $\cot\frac{3\pi}{4} = -1$.

13. $\sin\theta = y/r = \frac{3}{5} \Rightarrow y = 3$, $r = 5$, and $x = \sqrt{r^2 - y^2} = 4$ (since $0 < \theta < \frac{\pi}{2}$). Therefore taking $x = 4$, $y = 3$, $r = 5$ in the definitions of the trigonometric ratios, we have $\cos\theta = \frac{4}{5}$, $\tan\theta = \frac{3}{4}$, $\csc\theta = \frac{5}{3}$, $\sec\theta = \frac{5}{4}$, and $\cot\theta = \frac{4}{3}$.

15. $\sin 35^\circ = \dfrac{x}{10} \Rightarrow x = 10\sin 35^\circ \approx 5.73576$ cm

17. $\tan\dfrac{2\pi}{5} = \dfrac{x}{8} \Rightarrow x = 8\tan\dfrac{2\pi}{5} \approx 24.62147$ cm

19.

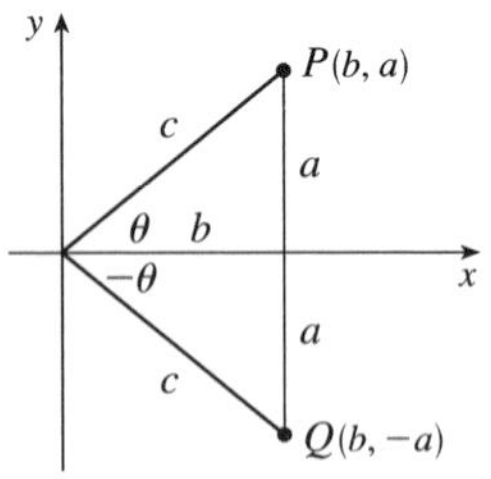

(a) From the diagram we see that $\sin\theta = \dfrac{y}{r} = \dfrac{a}{c}$, and

$$\sin(-\theta) = \frac{-a}{c} = -\frac{a}{c} = -\sin\theta.$$

(b) Again from the diagram we see that $\cos\theta = \dfrac{x}{r} = \dfrac{b}{c} = \cos(-\theta)$.

21. Using (12a), we have $\sin\left(\frac{\pi}{2} + x\right) = \sin\frac{\pi}{2}\cos x + \cos\frac{\pi}{2}\sin x = 1 \cdot \cos x + 0 \cdot \sin x = \cos x$.

23. Using (6), we have $\sin\theta\cot\theta = \sin\theta \cdot \dfrac{\cos\theta}{\sin\theta} = \cos\theta$.

25. Using (14a), we have $\tan 2\theta = \tan(\theta + \theta) = \dfrac{\tan\theta + \tan\theta}{1 - \tan\theta\tan\theta} = \dfrac{2\tan\theta}{1 - \tan^2\theta}$.

27. Since $\sin x = \frac{1}{3}$ we can label the opposite side as having length 1, the hypotenuse as having length 3, and use the Pythagorean Theorem to get that the adjacent side has length $\sqrt{8}$. Then, from the diagram, $\cos x = \frac{\sqrt{8}}{3}$. Similarly we have that $\sin y = \frac{3}{5}$. Now use (12a):

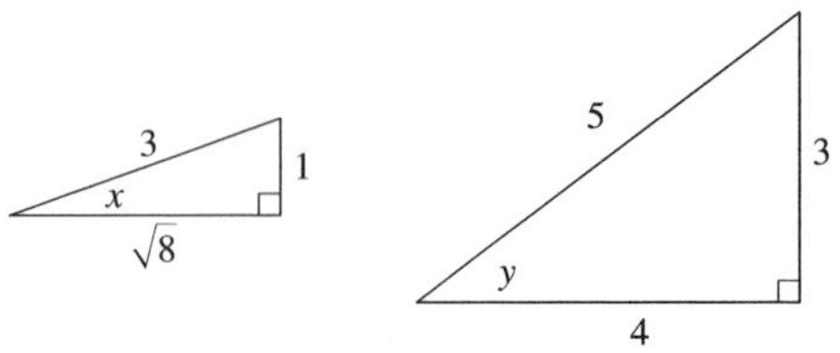

$\sin(x + y) = \sin x\cos y + \cos x\,\sin y = \frac{1}{3} \cdot \frac{4}{5} + \frac{\sqrt{8}}{3} \cdot \frac{3}{5} = \frac{4}{15} + \frac{3\sqrt{8}}{15} = \frac{4 + 6\sqrt{2}}{15}$.

29. $2\cos x - 1 = 0 \quad \Leftrightarrow \quad \cos x = \frac{1}{2} \quad \Rightarrow \quad x = \frac{\pi}{3}, \frac{5\pi}{3}$ for $x \in [0, 2\pi]$.

31. Using (15a), we have $\sin 2x = \cos x \quad \Leftrightarrow \quad 2\sin x \cos x - \cos x = 0 \quad \Leftrightarrow \quad \cos x(2\sin x - 1) = 0 \quad \Leftrightarrow \quad \cos x = 0$ or $2\sin x - 1 = 0 \quad \Rightarrow \quad x = \frac{\pi}{2}, \frac{3\pi}{2}$ or $\sin x = \frac{1}{2} \quad \Rightarrow \quad x = \frac{\pi}{6}$ or $\frac{5\pi}{6}$. Therefore, the solutions are $x = \frac{\pi}{6}, \frac{\pi}{2}, \frac{5\pi}{6}, \frac{3\pi}{2}$.

33. We know that $\sin x = \frac{1}{2}$ when $x = \frac{\pi}{6}$ or $\frac{5\pi}{6}$, and from Figure 13(a), we see that $\sin x \le \frac{1}{2} \quad \Rightarrow \quad 0 \le x \le \frac{\pi}{6}$ or $\frac{5\pi}{6} \le x \le 2\pi$ for $x \in [0, 2\pi]$.

35. $\tan x = -1$ when $x = \frac{3\pi}{4}, \frac{7\pi}{4}$, and $\tan x = 1$ when $x = \frac{\pi}{4}$ or $\frac{5\pi}{4}$. From Figure 14 we see that $-1 < \tan x < 1 \quad \Rightarrow$ $0 \le x < \frac{\pi}{4}, \frac{3\pi}{4} < x < \frac{5\pi}{4}$, and $\frac{7\pi}{4} < x \le 2\pi$.

37. $y = \cos\left(x - \frac{\pi}{3}\right)$. We start with the graph of $y = \cos x$ and shift it $\frac{\pi}{3}$ units to the right.

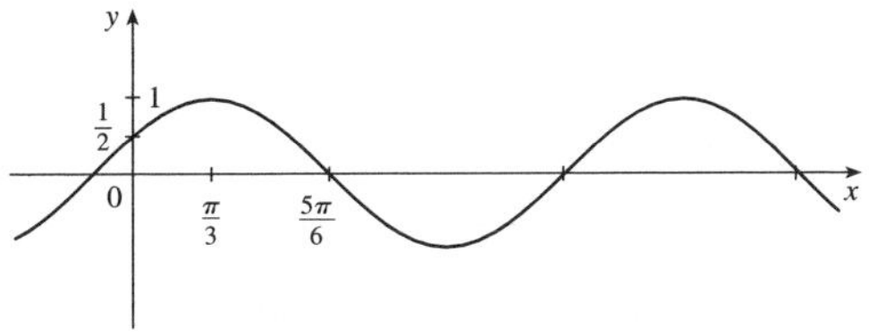

39. $y = \frac{1}{3}\tan\left(x - \frac{\pi}{2}\right)$. We start with the graph of $y = \tan x$, shift it $\frac{\pi}{2}$ units to the right and compress it to $\frac{1}{3}$ of its original vertical size.

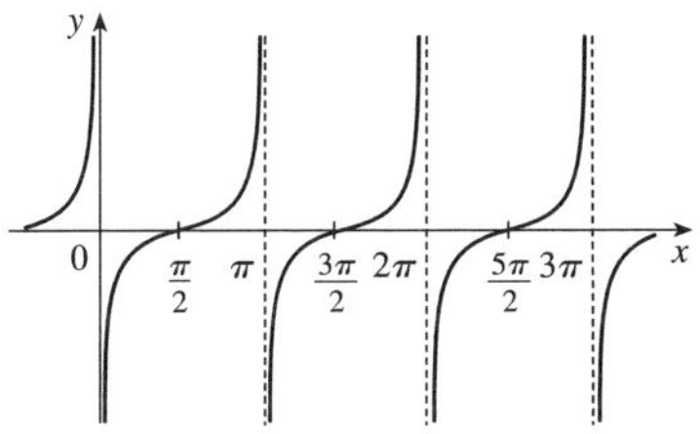

41. (a) $\sin^{-1}(0.5) = \frac{\pi}{6}$ since $\sin\frac{\pi}{6} = 0.5$ and $-\frac{\pi}{2} \le \frac{\pi}{6} \le \frac{\pi}{2}$.

(b) $\arctan(-1) = -\frac{\pi}{4}$ since $\tan\left(-\frac{\pi}{4}\right) = -1$ and $-\frac{\pi}{4}$ is in $\left(-\frac{\pi}{2}, \frac{\pi}{2}\right)$.

43. (a) $\sin\left(\sin^{-1}(0.7)\right) = 0.7$ since 0.7 is in $[-1, 1]$.

(b) $\arcsin\left(\sin\frac{5\pi}{4}\right) = \arcsin\left(-\frac{1}{\sqrt{2}}\right) = -\frac{\pi}{4}$

45. Let $y = \sin^{-1} x$. Then $-\frac{\pi}{2} \le y \le \frac{\pi}{2} \quad \Rightarrow \quad \cos y \ge 0$, so $\cos\left(\sin^{-1} x\right) = \cos y = \sqrt{1 - \sin^2 y} = \sqrt{1 - x^2}$.

47. $g(x) = \sin^{-1}(3x + 1)$.

Domain $(g) = \{x \mid -1 \le 3x + 1 \le 1\} = \{x \mid -2 \le 3x \le 0\} = \left\{x \mid -\frac{2}{3} \le x \le 0\right\} = \left[-\frac{2}{3}, 0\right]$.

Range $(g) = \left\{y \mid -\frac{\pi}{2} \le y \le \frac{\pi}{2}\right\} = \left[-\frac{\pi}{2}, \frac{\pi}{2}\right]$.

49. From the figure in the text, we see that $x = b\cos\theta$, $y = b\sin\theta$, and from the distance formula we have that the distance c from (x, y) to $(a, 0)$ is $c = \sqrt{(x - a)^2 + (y - 0)^2} \quad \Rightarrow$

$$\begin{aligned} c^2 &= (b\cos\theta - a)^2 + (b\sin\theta)^2 = b^2\cos^2\theta - 2ab\cos\theta + a^2 + b^2\sin^2\theta \\ &= a^2 + b^2\left(\cos^2\theta + \sin^2\theta\right) - 2ab\cos\theta = a^2 + b^2 - 2ab\cos\theta \quad \text{[by (7)]} \end{aligned}$$

51. Using the Law of Cosines, we have $c^2 = 1^2 + 1^2 - 2(1)(1)\cos(\alpha - \beta) = 2\left[1 - \cos(\alpha - \beta)\right]$. Now, using the distance formula, $c^2 = |AB|^2 = (\cos\alpha - \cos\beta)^2 + (\sin\alpha - \sin\beta)^2$. Equating these two expressions for c^2, we get $2\left[1 - \cos(\alpha - \beta)\right] = \cos^2\alpha + \sin^2\alpha + \cos^2\beta + \sin^2\beta - 2\cos\alpha\cos\beta - 2\sin\alpha\sin\beta \quad \Rightarrow$ $1 - \cos(\alpha - \beta) = 1 - \cos\alpha\cos\beta - \sin\alpha\sin\beta \quad \Rightarrow \quad \cos(\alpha - \beta) = \cos\alpha\cos\beta + \sin\alpha\sin\beta$.

53. In Exercise 52 we used the subtraction formula for cosine to prove the addition formula for cosine. Using that formula with $x = \frac{\pi}{2} - \alpha$, $y = \beta$, we get $\cos\left[\left(\frac{\pi}{2} - \alpha\right) + \beta\right] = \cos\left(\frac{\pi}{2} - \alpha\right)\cos\beta - \sin\left(\frac{\pi}{2} - \alpha\right)\sin\beta \quad \Rightarrow$ $\cos\left[\frac{\pi}{2} - (\alpha - \beta)\right] = \cos\left(\frac{\pi}{2} - \alpha\right)\cos\beta - \sin\left(\frac{\pi}{2} - \alpha\right)\sin\beta$. Now we use the identities given in the problem, $\cos\left(\frac{\pi}{2} - \theta\right) = \sin\theta$ and $\sin\left(\frac{\pi}{2} - \theta\right) = \cos\theta$, to get $\sin(\alpha - \beta) = \sin\alpha\cos\beta - \cos\alpha\sin\beta$.

D PRECISE DEFINITIONS OF LIMITS

1. On the left side of $x = 2$, we need $|x - 2| < \left|\frac{10}{7} - 2\right| = \frac{4}{7}$. On the right side, we need $|x - 2| < \left|\frac{10}{3} - 2\right| = \frac{4}{3}$. For both of these conditions to be satisfied at once, we need the more restrictive of the two to hold, that is, $|x - 2| < \frac{4}{7}$. So we can choose $\delta = \frac{4}{7}$, or any smaller positive number.

3. $\left|\sqrt{4x+1} - 3\right| < 0.5 \quad \Leftrightarrow \quad 2.5 < \sqrt{4x+1} < 3.5$. We plot the three parts of this inequality on the same screen and identify the x-coordinates of the points of intersection using the cursor. It appears that the inequality holds for $1.3125 \le x \le 2.8125$. Since $|2 - 1.3125| = 0.6875$ and $|2 - 2.8125| = 0.8125$, we choose $0 < \delta < \min\{0.6875, 0.8125\} = 0.6875$.

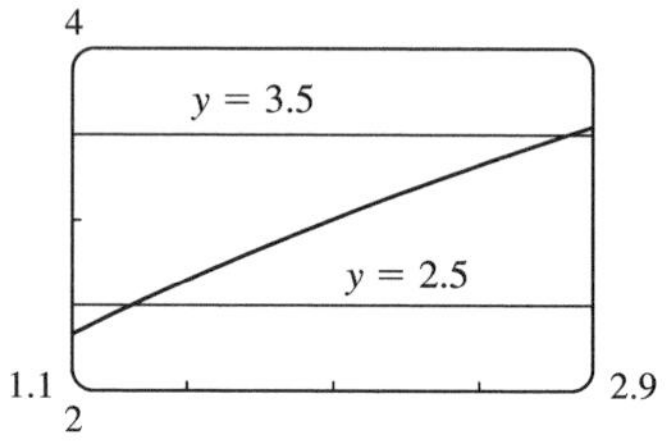

5. For $\varepsilon = 1$, the definition of a limit requires that we find δ such that $\left|(4 + x - 3x^3) - 2\right| < 1 \quad \Leftrightarrow \quad 1 < 4 + x - 3x^3 < 3$ whenever $0 < |x - 1| < \delta$. If we plot the graphs of $y = 1$, $y = 4 + x - 3x^3$ and $y = 3$ on the same screen, we see that we need $0.86 \le x \le 1.11$. So since $|1 - 0.86| = 0.14$ and $|1 - 1.11| = 0.11$, we choose $\delta = 0.11$ (or any smaller positive number). For $\varepsilon = 0.1$, we must find δ such that $\left|(4 + x - 3x^3) - 2\right| < 0.1 \quad \Leftrightarrow \quad 1.9 < 4 + x - 3x^3 < 2.1$ whenever $0 < |x - 1| < \delta$. From the graph, we see that we need $0.988 \le x \le 1.012$. So since $|1 - 0.988| = 0.012$ and $|1 - 1.012| = 0.012$, we choose $\delta = 0.012$ (or any smaller positive number) for the inequality to hold.

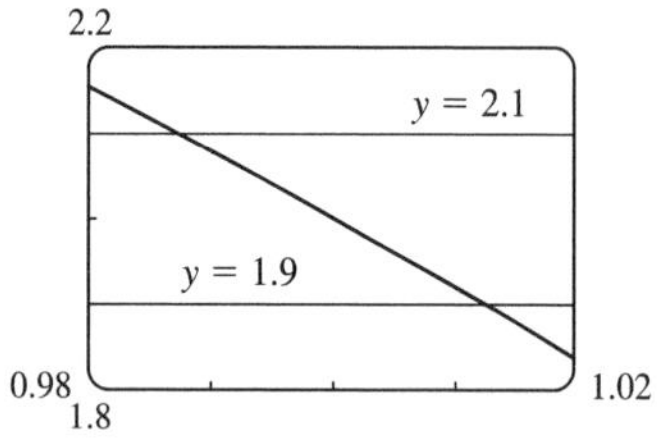

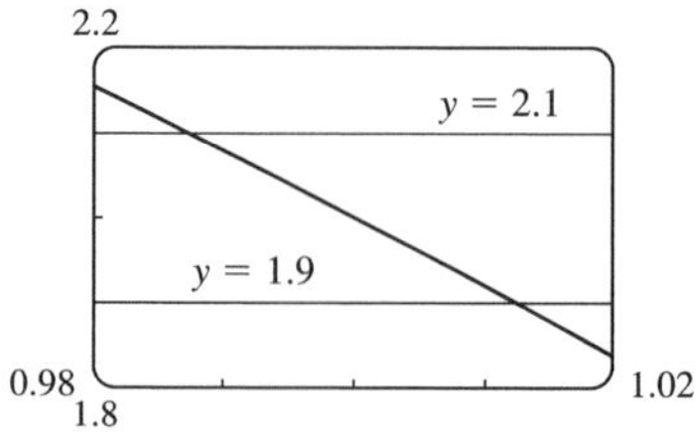

7. Given $\varepsilon > 0$, we need $\delta > 0$ such that if $|x| < \delta$ then $\left|x^3 - 0\right| < \varepsilon \quad \Leftrightarrow \quad |x|^3 < \varepsilon \quad \Leftrightarrow \quad |x| < \sqrt[3]{\varepsilon}$. Take $\delta = \sqrt[3]{\varepsilon}$. Then $|x - 0| < \delta \quad \Rightarrow \quad \left|x^3 - 0\right| < \delta^3 = \varepsilon$. Thus, $\lim\limits_{x \to 0} x^3 = 0$ by the definition of a limit.

9. Given $\varepsilon > 0$, we need $\delta > 0$ such that if $|x - 2| < \delta$, then

$\left|(3x - 2) - 4\right| < \varepsilon \quad \Leftrightarrow \quad |3x - 6| < \varepsilon \quad \Leftrightarrow \quad 3\,|x - 2| < \varepsilon \quad \Leftrightarrow$

$|x - 2| < \varepsilon/3$. So if we choose $\delta = \varepsilon/3$, then $|x - 2| < \delta \quad \Rightarrow$
$\left|(3x - 2) - 4\right| < \varepsilon$. Thus, $\lim\limits_{x \to 2}(3x - 2) = 4$ by the definition of a limit.

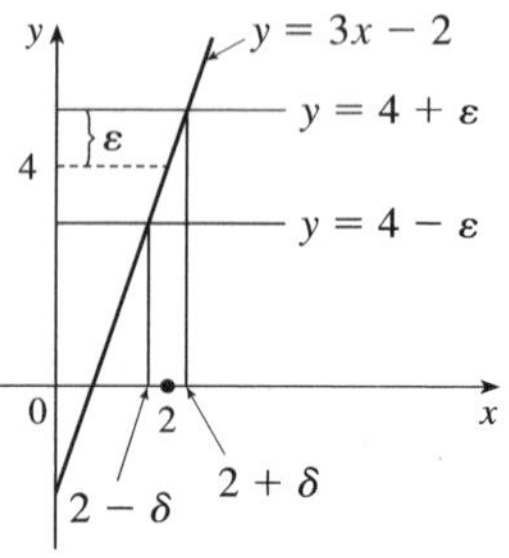

11. (a) $A = \pi r^2$ and $A = 1000\text{ cm}^2 \quad \Rightarrow \quad \pi r^2 = 1000 \quad \Rightarrow \quad r^2 = \frac{1000}{\pi} \quad \Rightarrow$

$r = \sqrt{\frac{1000}{\pi}} \quad (r > 0) \quad \approx 17.8412\text{ cm}$.

(b) $|A - 1000| \leq 5 \quad \Rightarrow \quad -5 \leq \pi r^2 - 1000 \leq 5 \quad \Rightarrow \quad 1000 - 5 \leq \pi r^2 \leq 1000 + 5 \quad \Rightarrow$

$\sqrt{\frac{995}{\pi}} \leq r \leq \sqrt{\frac{1005}{\pi}} \quad \Rightarrow \quad 17.7966 \leq r \leq 17.8858.$ $\sqrt{\frac{1000}{\pi}} - \sqrt{\frac{995}{\pi}} \approx 0.04466$ and $\sqrt{\frac{1005}{\pi}} - \sqrt{\frac{1000}{\pi}} \approx 0.04455.$

So if the machinist gets the radius within 0.0445 cm of 17.8412, the area will be within 5 cm^2 of 1000.

(c) x is the radius, $f(x)$ is the area, a is the target radius given in part (a), L is the target area (1000), ε is the tolerance in the area (5), and δ is the tolerance in the radius given in part (b).

13. $\left| \dfrac{6x^2 + 5x - 3}{2x^2 - 1} - 3 \right| < 0.2 \quad \Leftrightarrow \quad 2.8 < \dfrac{6x^2 + 5x - 3}{2x^2 - 1} < 3.2$. So we graph the three parts of this inequality on the same screen, and find that the curve $y = \dfrac{6x^2 + 5x - 3}{2x^2 - 1}$ seems to lie between the lines $y = 2.8$ and $y = 3.2$ whenever $x > 12.5$. So we can choose $N = 13$ (or any larger number), so that the inequality holds whenever $x \geq N$.

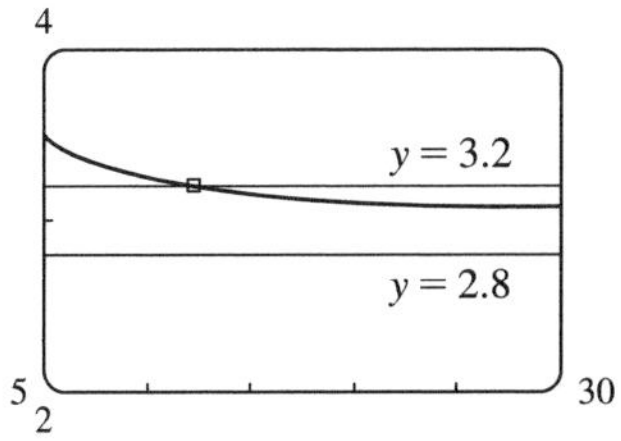

15. (a) $1/x^2 < 0.0001 \quad \Leftrightarrow \quad x^2 > 1/0.0001 = 10\,000 \quad \Leftrightarrow \quad x > 100 \quad (x > 0)$

(b) If $\varepsilon > 0$ is given, then $1/x^2 < \varepsilon \quad \Leftrightarrow \quad x^2 > 1/\varepsilon \quad \Leftrightarrow \quad x > 1/\sqrt{\varepsilon}$. Let $N = 1/\sqrt{\varepsilon}$. Then $x > N \quad \Rightarrow \quad x > \dfrac{1}{\sqrt{\varepsilon}} \quad \Rightarrow$

$\left| \dfrac{1}{x^2} - 0 \right| = \dfrac{1}{x^2} < \varepsilon$, so $\displaystyle\lim_{x \to \infty} \frac{1}{x^2} = 0.$

17. (a)

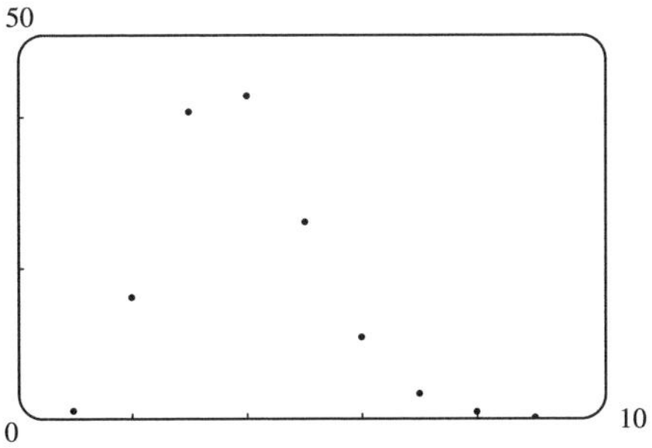

From the graph, it appears that the sequence $\left\{ \dfrac{n^5}{n!} \right\}$ converges to 0, that is, $\displaystyle\lim_{n \to \infty} \frac{n^5}{n!} = 0.$

(b)

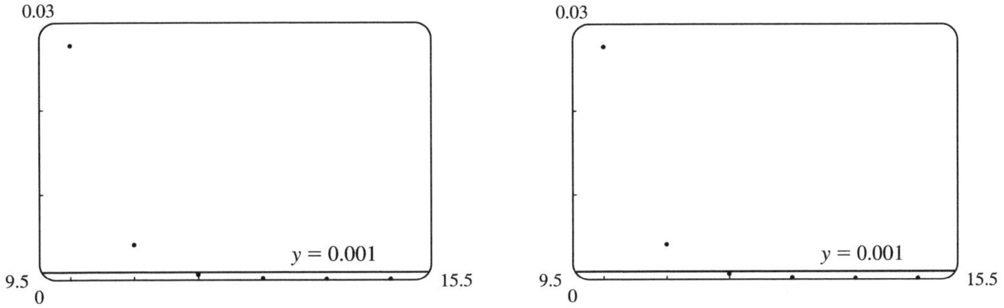

From the first graph, it seems that the smallest possible value of N corresponding to $\varepsilon = 0.1$ is 9, since $n^5/n! < 0.1$ whenever $n \geq 10$, but $9^5/9! > 0.1$. From the second graph, it seems that for $\varepsilon = 0.001$, the smallest possible value for N is 11 since $n^5/n! < 0.001$ whenever $n \geq 12$.

19. If $\displaystyle\lim_{n \to \infty} |a_n| = 0$, then $\displaystyle\lim_{n \to \infty} (-|a_n|) = 0$, and since $-|a_n| \leq a_n \leq |a_n|$, we have that $\displaystyle\lim_{n \to \infty} a_n = 0$ by the Squeeze Theorem.

21. (*Note:* This exercise does not appear in *Single Variable Calculus.*)

Let $\varepsilon > 0$. We want to find $\delta > 0$ such that

$$\left|\frac{xy}{\sqrt{x^2+y^2}} - 0\right| < \varepsilon \qquad \text{whenever} \qquad 0 < \sqrt{x^2+y^2} < \delta$$

that is,

$$\frac{|xy|}{\sqrt{x^2+y^2}} < \varepsilon \qquad \text{whenever} \qquad 0 < \sqrt{x^2+y^2} < \delta$$

But $|x| = \sqrt{x^2} \leq \sqrt{x^2+y^2}$ and $|y| = \sqrt{y^2} \leq \sqrt{x^2+y^2}$, so

$$\frac{|xy|}{\sqrt{x^2+y^2}} \leq \frac{\left(\sqrt{x^2+y^2}\right)^2}{\sqrt{x^2+y^2}} = \sqrt{x^2+y^2}$$

Thus, if we choose $\delta = \varepsilon$ and let $0 < \sqrt{x^2+y^2} < \delta$, then

$$\left|\frac{xy}{\sqrt{x^2+y^2}} - 0\right| \leq \sqrt{x^2+y^2} < \delta = \varepsilon$$

Hence, by Definition 1 (Definition 5 in the full version of the text),

$$\lim_{(x,y)\to(0,0)} \frac{xy}{\sqrt{x^2+y^2}} = 0$$

F SIGMA NOTATION

1. $\sum_{i=1}^{5} \sqrt{i} = \sqrt{1} + \sqrt{2} + \sqrt{3} + \sqrt{4} + \sqrt{5}$

3. $\sum_{i=4}^{6} 3^i = 3^4 + 3^5 + 3^6$

5. $\sum_{k=0}^{4} \frac{2k-1}{2k+1} = -1 + \frac{1}{3} + \frac{3}{5} + \frac{5}{7} + \frac{7}{9}$

7. $\sum_{i=1}^{n} i^{10} = 1^{10} + 2^{10} + 3^{10} + \cdots + n^{10}$

9. $\sum_{j=0}^{n-1} (-1)^j = 1 - 1 + 1 - 1 + \cdots + (-1)^{n-1}$

11. $1 + 2 + 3 + 4 + \cdots + 10 = \sum_{i=1}^{10} i$

13. $\frac{1}{2} + \frac{2}{3} + \frac{3}{4} + \frac{4}{5} + \cdots + \frac{19}{20} = \sum_{i=1}^{19} \frac{i}{i+1}$

15. $2 + 4 + 6 + 8 + \cdots + 2n = \sum_{i=1}^{n} 2i$

17. $1 + 2 + 4 + 8 + 16 + 32 = \sum_{i=0}^{5} 2^i$

19. $x + x^2 + x^3 + \cdots + x^n = \sum_{i=1}^{n} x^i$

21. $\sum_{i=4}^{8} (3i-2) = [3(4)-2] + [3(5)-2] + [3(6)-2] + [3(7)-2] + [3(8)-2] = 10 + 13 + 16 + 19 + 22 = 80$

23. $\sum_{j=1}^{6} 3^{j+1} = 3^2 + 3^3 + 3^4 + 3^5 + 3^6 + 3^7 = 9 + 27 + 81 + 243 + 729 + 2187 = 3276$

(For a more general method, see Exercise 47.)

25. $\sum_{n=1}^{20} (-1)^n = -1 + 1 - 1 + 1 - 1 + 1 - 1 + 1 - 1 + 1 - 1 + 1 - 1 + 1 - 1 + 1 - 1 + 1 - 1 + 1 = 0$

27. $\sum_{i=0}^{4}(2^i+i^2)=(1+0)+(2+1)+(4+4)+(8+9)+(16+16)=61$

29. $\sum_{i=1}^{n}2i=2\sum_{i=1}^{n}i=2\cdot\dfrac{n(n+1)}{2}$ [by Theorem 3(c)] $=n(n+1)$

31. $$\sum_{i=1}^{n}(i^2+3i+4)=\sum_{i=1}^{n}i^2+3\sum_{i=1}^{n}i+\sum_{i=1}^{n}4=\frac{n(n+1)(2n+1)}{6}+\frac{3n(n+1)}{2}+4n$$
$$=\tfrac{1}{6}\left[(2n^3+3n^2+n)+(9n^2+9n)+24n\right]=\tfrac{1}{6}\left(2n^3+12n^2+34n\right)$$
$$=\tfrac{1}{3}n(n^2+6n+17)$$

33. $$\sum_{i=1}^{n}(i+1)(i+2)=\sum_{i=1}^{n}(i^2+3i+2)=\sum_{i=1}^{n}i^2+3\sum_{i=1}^{n}i+\sum_{i=1}^{n}2$$
$$=\frac{n(n+1)(2n+1)}{6}+\frac{3n(n+1)}{2}+2n=\frac{n(n+1)}{6}[(2n+1)+9]+2n$$
$$=\frac{n(n+1)}{3}(n+5)+2n=\frac{n}{3}[(n+1)(n+5)+6]=\frac{n}{3}\left(n^2+6n+11\right)$$

35. $$\sum_{i=1}^{n}(i^3-i-2)=\sum_{i=1}^{n}i^3-\sum_{i=1}^{n}i-\sum_{i=1}^{n}2=\left[\frac{n(n+1)}{2}\right]^2-\frac{n(n+1)}{2}-2n$$
$$=\tfrac{1}{4}n(n+1)\left[n(n+1)-2\right]-2n=\tfrac{1}{4}n(n+1)(n+2)(n-1)-2n$$
$$=\tfrac{1}{4}n\left[(n+1)(n-1)(n+2)-8\right]=\tfrac{1}{4}n\left[(n^2-1)(n+2)-8\right]=\tfrac{1}{4}n(n^3+2n^2-n-10)$$

37. By Theorem 2(a) and Example 3, $\sum_{i=1}^{n}c=c\sum_{i=1}^{n}1=cn$.

39. $$\sum_{i=1}^{n}\left[(i+1)^4-i^4\right]=(2^4-1^4)+(3^4-2^4)+(4^4-3^4)+\cdots+\left[(n+1)^4-n^4\right]$$
$$=(n+1)^4-1^4=n^4+4n^3+6n^2+4n$$

On the other hand,

$$\sum_{i=1}^{n}\left[(i+1)^4-i^4\right]=\sum_{i=1}^{n}(4i^3+6i^2+4i+1)=4\sum_{i=1}^{n}i^3+6\sum_{i=1}^{n}i^2+4\sum_{i=1}^{n}i+\sum_{i=1}^{n}1$$
$$=4S+n(n+1)(2n+1)+2n(n+1)+n\quad\left[\text{where } S=\sum_{i=1}^{n}i^3\right]$$
$$=4S+2n^3+3n^2+n+2n^2+2n+n=4S+2n^3+5n^2+4n$$

Thus, $n^4+4n^3+6n^2+4n=4S+2n^3+5n^2+4n$, from which it follows that

$4S=n^4+2n^3+n^2=n^2(n^2+2n+1)=n^2(n+1)^2$ and $S=\left[\dfrac{n(n+1)}{2}\right]^2$.

41. (a) $\sum_{i=1}^{n}\left[i^4-(i-1)^4\right]=(1^4-0^4)+(2^4-1^4)+(3^4-2^4)+\cdots+\left[n^4-(n-1)^4\right]=n^4-0=n^4$

(b) $\sum_{i=1}^{100}(5^i-5^{i-1})=(5^1-5^0)+(5^2-5^1)+(5^3-5^2)+\cdots+(5^{100}-5^{99})=5^{100}-5^0=5^{100}-1$

(c) $\sum_{i=3}^{99}\left(\dfrac{1}{i}-\dfrac{1}{i+1}\right)=\left(\dfrac{1}{3}-\dfrac{1}{4}\right)+\left(\dfrac{1}{4}-\dfrac{1}{5}\right)+\left(\dfrac{1}{5}-\dfrac{1}{6}\right)+\cdots+\left(\dfrac{1}{99}-\dfrac{1}{100}\right)=\dfrac{1}{3}-\dfrac{1}{100}=\dfrac{97}{300}$

(d) $\sum_{i=1}^{n}(a_i-a_{i-1})=(a_1-a_0)+(a_2-a_1)+(a_3-a_2)+\cdots+(a_n-a_{n-1})=a_n-a_0$

43. $$\lim_{n\to\infty}\sum_{i=1}^{n}\frac{1}{n}\left(\frac{i}{n}\right)^2=\lim_{n\to\infty}\frac{1}{n^3}\sum_{i=1}^{n}i^2=\lim_{n\to\infty}\frac{1}{n^3}\frac{n(n+1)(2n+1)}{6}=\lim_{n\to\infty}\frac{1}{6}\left(1+\frac{1}{n}\right)\left(2+\frac{1}{n}\right)$$
$$=\tfrac{1}{6}(1)(2)=\tfrac{1}{3}$$

45. $\displaystyle \lim_{n\to\infty} \sum_{i=1}^{n} \frac{2}{n}\left[\left(\frac{2i}{n}\right)^3 + 5\left(\frac{2i}{n}\right)\right] = \lim_{n\to\infty} \sum_{i=1}^{n} \left[\frac{16}{n^4} i^3 + \frac{20}{n^2} i\right] = \lim_{n\to\infty} \left[\frac{16}{n^4} \sum_{i=1}^{n} i^3 + \frac{20}{n^2} \sum_{i=1}^{n} i\right]$

$$= \lim_{n\to\infty} \left[\frac{16}{n^4} \frac{n^2(n+1)^2}{4} + \frac{20}{n^2} \frac{n(n+1)}{2}\right] = \lim_{n\to\infty} \left[\frac{4(n+1)^2}{n^2} + \frac{10n(n+1)}{n^2}\right]$$

$$= \lim_{n\to\infty} \left[4\left(1+\frac{1}{n}\right)^2 + 10\left(1+\frac{1}{n}\right)\right] = 4\cdot 1 + 10 \cdot 1 = 14$$

47. Let $S = \sum_{i=1}^{n} ar^{i-1} = a + ar + ar^2 + \cdots + ar^{n-1}$. Multiplying both sides by r gives us $rS = ar + ar^2 + \cdots + ar^{n-1} + ar^n$. Subtracting the first equation from the second, we find $(r-1)S = ar^n - a = a(r^n - 1)$, so $S = \dfrac{a(r^n-1)}{r-1}$ (since $r \neq 1$).

49. $\displaystyle \sum_{i=1}^{n} (2i + 2^i) = 2\sum_{i=1}^{n} i + \sum_{i=1}^{n} 2 \cdot 2^{i-1} = 2\frac{n(n+1)}{2} + \frac{2(2^n-1)}{2-1} = 2^{n+1} + n^2 + n - 2.$

For the first sum we have used Theorem 3(c), and for the second, Exercise 47 with $a = r = 2$.

G INTEGRATION OF RATIONAL FUNCTIONS BY PARTIAL FRACTIONS

1. (a) $\dfrac{2x}{(x+3)(3x+1)} = \dfrac{A}{x+3} + \dfrac{B}{3x+1}$

(b) $\dfrac{1}{x^3+2x^2+x} = \dfrac{1}{x(x^2+2x+1)} = \dfrac{1}{x(x+1)^2} = \dfrac{A}{x} + \dfrac{B}{x+1} + \dfrac{C}{(x+1)^2}$

3. (a) $\dfrac{2}{x^2+3x-4} = \dfrac{2}{(x+4)(x-1)} = \dfrac{A}{x+4} + \dfrac{B}{x-1}$

(b) $x^2 + x + 1$ is irreducible, so $\dfrac{x^2}{(x-1)(x^2+x+1)} = \dfrac{A}{x-1} + \dfrac{Bx+C}{x^2+x+1}$.

5. (a) $\dfrac{x^4}{x^4-1} = \dfrac{(x^4-1)+1}{x^4-1} = 1 + \dfrac{1}{x^4-1}$ [or use long division] $= 1 + \dfrac{1}{(x^2-1)(x^2+1)}$

$$= 1 + \frac{1}{(x-1)(x+1)(x^2+1)} = 1 + \frac{A}{x-1} + \frac{B}{x+1} + \frac{Cx+D}{x^2+1}$$

(b) $\dfrac{t^4+t^2+1}{(t^2+1)(t^2+4)^2} = \dfrac{At+B}{t^2+1} + \dfrac{Ct+D}{t^2+4} + \dfrac{Et+F}{(t^2+4)^2}$

7. $\displaystyle \int \frac{x}{x-6}\,dx = \int \frac{(x-6)+6}{x-6}\,dx = \int \left(1 + \frac{6}{x-6}\right) dx = x + 6\ln|x-6| + C$

9. $\dfrac{x-9}{(x+5)(x-2)} = \dfrac{A}{x+5} + \dfrac{B}{x-2}$. Multiply both sides by $(x+5)(x-2)$ to get $x - 9 = A(x-2) + B(x+5)$. Substituting 2 for x gives $-7 = 7B \Leftrightarrow B = -1$. Substituting -5 for x gives $-14 = -7A \Leftrightarrow A = 2$. Thus,

$$\int \frac{x-9}{(x+5)(x-2)}\,dx = \int \left(\frac{2}{x+5} + \frac{-1}{x-2}\right) dx = 2\ln|x+5| - \ln|x-2| + C$$

11. $\dfrac{1}{x^2-1} = \dfrac{1}{(x+1)(x-1)} = \dfrac{A}{x+1} + \dfrac{B}{x-1}$. Multiply both sides by $(x+1)(x-1)$ to get $1 = A(x-1) + B(x+1)$. Substituting 1 for x gives $1 = 2B \Leftrightarrow B = \frac{1}{2}$. Substituting -1 for x gives $1 = -2A \Leftrightarrow A = -\frac{1}{2}$. Thus,

$$\int_2^3 \frac{1}{x^2-1}\,dx = \int_2^3 \left(\frac{-1/2}{x+1} + \frac{1/2}{x-1}\right) dx = \left[-\tfrac{1}{2}\ln|x+1| + \tfrac{1}{2}\ln|x-1|\right]_2^3$$

$$= \left(-\tfrac{1}{2}\ln 4 + \tfrac{1}{2}\ln 2\right) - \left(-\tfrac{1}{2}\ln 3 + \tfrac{1}{2}\ln 1\right) = \tfrac{1}{2}(\ln 2 + \ln 3 - \ln 4) \quad \left[\text{or } \tfrac{1}{2}\ln\tfrac{3}{2}\right]$$

13. $\displaystyle \int \frac{ax}{x^2-bx}\,dx = \int \frac{ax}{x(x-b)}\,dx = \int \frac{a}{x-b}\,dx = a\ln|x-b| + C$

15. $\dfrac{2x+3}{(x+1)^2} = \dfrac{A}{x+1} + \dfrac{B}{(x+1)^2} \Rightarrow 2x+3 = A(x+1)+B$. Take $x=-1$ to get $B=1$, and equate coefficients of x to get $A=2$. Now

$$\int_0^1 \frac{2x+3}{(x+1)^2}\,dx = \int_0^1 \left[\frac{2}{x+1} + \frac{1}{(x+1)^2}\right] dx = \left[2\ln(x+1) - \frac{1}{x+1}\right]_0^1$$
$$= 2\ln 2 - \tfrac{1}{2} - (2\ln 1 - 1) = 2\ln 2 + \tfrac{1}{2}$$

17. $\dfrac{4y^2-7y-12}{y(y+2)(y-3)} = \dfrac{A}{y} + \dfrac{B}{y+2} + \dfrac{C}{y-3} \Rightarrow 4y^2-7y-12 = A(y+2)(y-3) + By(y-3) + Cy(y+2)$. Setting $y=0$ gives $-12=-6A$, so $A=2$. Setting $y=-2$ gives $18=10B$, so $B=\frac{9}{5}$. Setting $y=3$ gives $3=15C$, so $C=\frac{1}{5}$. Now

$$\int_1^2 \frac{4y^2-7y-12}{y(y+2)(y-3)}\,dy = \int_1^2 \left(\frac{2}{y} + \frac{9/5}{y+2} + \frac{1/5}{y-3}\right) dy = \left[2\ln|y| + \tfrac{9}{5}\ln|y+2| + \tfrac{1}{5}\ln|y-3|\right]_1^2$$
$$= 2\ln 2 + \tfrac{9}{5}\ln 4 + \tfrac{1}{5}\ln 1 - 2\ln 1 - \tfrac{9}{5}\ln 3 - \tfrac{1}{5}\ln 2$$
$$= 2\ln 2 + \tfrac{18}{5}\ln 2 - \tfrac{1}{5}\ln 2 - \tfrac{9}{5}\ln 3 = \tfrac{27}{5}\ln 2 - \tfrac{9}{5}\ln 3 = \tfrac{9}{5}(3\ln 2 - \ln 3) = \tfrac{9}{5}\ln\tfrac{8}{3}$$

19. $\dfrac{1}{(x+5)^2(x-1)} = \dfrac{A}{x+5} + \dfrac{B}{(x+5)^2} + \dfrac{C}{x-1} \Rightarrow 1 = A(x+5)(x-1) + B(x-1) + C(x+5)^2$. Setting $x=-5$ gives $1=-6B$, so $B=-\frac{1}{6}$. Setting $x=1$ gives $1=36C$, so $C=\frac{1}{36}$. Setting $x=-2$ gives $1 = A(3)(-3) + B(-3) + C(3^2) = -9A - 3B + 9C = -9A + \frac{1}{2} + \frac{1}{4} = -9A + \frac{3}{4}$, so $9A = -\frac{1}{4}$ and $A = -\frac{1}{36}$. Now

$$\int \frac{1}{(x+5)^2(x-1)}\,dx = \int \left[\frac{-1/36}{x+5} - \frac{1/6}{(x+5)^2} + \frac{1/36}{x-1}\right] dx$$
$$= -\frac{1}{36}\ln|x+5| + \frac{1}{6(x+5)} + \frac{1}{36}\ln|x-1| + C$$

21. $\dfrac{5x^2+3x-2}{x^3+2x^2} = \dfrac{5x^2+3x-2}{x^2(x+2)} = \dfrac{A}{x} + \dfrac{B}{x^2} + \dfrac{C}{x+2}$. Multiply by $x^2(x+2)$ to get $5x^2+3x-2 = Ax(x+2) + B(x+2) + Cx^2$. Set $x=-2$ to get $C=3$, and take $x=0$ to get $B=-1$. Equating the coefficients of x^2 gives $5 = A + C \Rightarrow A = 2$. So

$$\int \frac{5x^2+3x-2}{x^3+2x^2}\,dx = \int \left(\frac{2}{x} - \frac{1}{x^2} + \frac{3}{x+2}\right) dx = 2\ln|x| + \frac{1}{x} + 3\ln|x+2| + C.$$

23. $\dfrac{10}{(x-1)(x^2+9)} = \dfrac{A}{x-1} + \dfrac{Bx+C}{x^2+9}$. Multiply both sides by $(x-1)(x^2+9)$ to get $10 = A(x^2+9) + (Bx+C)(x-1)$ $(*)$. Substituting 1 for x gives $10 = 10A \Leftrightarrow A=1$. Substituting 0 for x gives $10 = 9A - C \Rightarrow C = 9(1) - 10 = -1$.
The coefficients of the x^2-terms in $(*)$ must be equal, so $0 = A + B \Rightarrow B = -1$. Thus,

$$\int \frac{10}{(x-1)(x^2+9)}\,dx = \int \left(\frac{1}{x-1} + \frac{-x-1}{x^2+9}\right) dx = \int \left(\frac{1}{x-1} - \frac{x}{x^2+9} - \frac{1}{x^2+9}\right) dx$$
$$= \ln|x-1| - \tfrac{1}{2}\ln(x^2+9) \text{ [let } u = x^2+9\text{]} - \tfrac{1}{3}\tan^{-1}\left(\tfrac{x}{3}\right) \text{ [Formula 10]} + C$$

25. $\dfrac{x^3+x^2+2x+1}{(x^2+1)(x^2+2)} = \dfrac{Ax+B}{x^2+1} + \dfrac{Cx+D}{x^2+2}$. Multiply both sides by $(x^2+1)(x^2+2)$ to get
$x^3+x^2+2x+1 = (Ax+B)(x^2+2) + (Cx+D)(x^2+1) \Leftrightarrow$
$x^3+x^2+2x+1 = (Ax^3+Bx^2+2Ax+2B) + (Cx^3+Dx^2+Cx+D) \Leftrightarrow$
$x^3+x^2+2x+1 = (A+C)x^3 + (B+D)x^2 + (2A+C)x + (2B+D)$. Comparing coefficients gives us the following system of equations:

$$A + C = 1 \quad \textbf{(1)} \qquad B + D = 1 \quad \textbf{(2)}$$
$$2A + C = 2 \quad \textbf{(3)} \qquad 2B + D = 1 \quad \textbf{(4)}$$

Subtracting equation **(1)** from equation **(3)** gives us $A = 1$, so $C = 0$. Subtracting equation **(2)** from equation **(4)** gives us $B = 0$, so $D = 1$. Thus, $I = \int \frac{x^3 + x^2 + 2x + 1}{(x^2 + 1)(x^2 + 2)}\,dx = \int \left(\frac{x}{x^2 + 1} + \frac{1}{x^2 + 2}\right) dx$. For $\int \frac{x}{x^2 + 1}\,dx$, let $u = x^2 + 1$ so $du = 2x\,dx$ and then $\int \frac{x}{x^2 + 1}\,dx = \frac{1}{2}\int \frac{1}{u}\,du = \frac{1}{2}\ln|u| + C = \frac{1}{2}\ln(x^2 + 1) + C$. For $\int \frac{1}{x^2 + 2}\,dx$, use Formula 10 with $a = \sqrt{2}$. So $\int \frac{1}{x^2 + 2}\,dx = \int \frac{1}{x^2 + (\sqrt{2})^2}\,dx = \frac{1}{\sqrt{2}}\tan^{-1}\frac{x}{\sqrt{2}} + C$.

Thus, $I = \frac{1}{2}\ln(x^2 + 1) + \frac{1}{\sqrt{2}}\tan^{-1}\frac{x}{\sqrt{2}} + C$.

27. $\displaystyle\int \frac{x + 4}{x^2 + 2x + 5}\,dx = \int \frac{x + 1}{x^2 + 2x + 5}\,dx + \int \frac{3}{x^2 + 2x + 5}\,dx = \frac{1}{2}\int \frac{(2x + 2)\,dx}{x^2 + 2x + 5} + \int \frac{3\,dx}{(x + 1)^2 + 4}$

$$= \frac{1}{2}\ln\left|x^2 + 2x + 5\right| + 3\int \frac{2\,du}{4(u^2 + 1)} \qquad \begin{bmatrix}\text{where } x + 1 = 2u, \\ \text{and } dx = 2\,du\end{bmatrix}$$

$$= \frac{1}{2}\ln(x^2 + 2x + 5) + \frac{3}{2}\tan^{-1}u + C = \frac{1}{2}\ln(x^2 + 2x + 5) + \frac{3}{2}\tan^{-1}\left(\frac{x + 1}{2}\right) + C$$

29. $\displaystyle\frac{1}{x^3 - 1} = \frac{1}{(x - 1)(x^2 + x + 1)} = \frac{A}{x - 1} + \frac{Bx + C}{x^2 + x + 1} \quad\Rightarrow\quad 1 = A(x^2 + x + 1) + (Bx + C)(x - 1).$

Take $x = 1$ to get $A = \frac{1}{3}$. Equating coefficients of x^2 and then comparing the constant terms, we get $0 = \frac{1}{3} + B$, $1 = \frac{1}{3} - C$, so $B = -\frac{1}{3}$, $C = -\frac{2}{3}$ $\Rightarrow$

$$\int \frac{1}{x^3 - 1}\,dx = \int \frac{\frac{1}{3}}{x - 1}\,dx + \int \frac{-\frac{1}{3}x - \frac{2}{3}}{x^2 + x + 1}\,dx = \tfrac{1}{3}\ln|x - 1| - \frac{1}{3}\int \frac{x + 2}{x^2 + x + 1}\,dx$$

$$= \tfrac{1}{3}\ln|x - 1| - \frac{1}{3}\int \frac{x + 1/2}{x^2 + x + 1}\,dx - \frac{1}{3}\int \frac{(3/2)\,dx}{(x + 1/2)^2 + 3/4}$$

$$= \tfrac{1}{3}\ln|x - 1| - \tfrac{1}{6}\ln(x^2 + x + 1) - \tfrac{1}{2}\left(\tfrac{2}{\sqrt{3}}\right)\tan^{-1}\left(\frac{x + \frac{1}{2}}{\sqrt{3}/2}\right) + K$$

$$= \tfrac{1}{3}\ln|x - 1| - \tfrac{1}{6}\ln(x^2 + x + 1) - \tfrac{1}{\sqrt{3}}\tan^{-1}\left(\tfrac{1}{\sqrt{3}}(2x + 1)\right) + K$$

31. $\displaystyle\frac{1}{x^4 - x^2} = \frac{1}{x^2(x - 1)(x + 1)} = \frac{A}{x} + \frac{B}{x^2} + \frac{C}{x - 1} + \frac{D}{x + 1}$. Multiply by $x^2(x - 1)(x + 1)$ to get $1 = Ax(x - 1)(x + 1) + B(x - 1)(x + 1) + Cx^2(x + 1) + Dx^2(x - 1)$. Setting $x = 1$ gives $C = \frac{1}{2}$, taking $x = -1$ gives $D = -\frac{1}{2}$. Equating the coefficients of x^3 gives $0 = A + C + D = A$. Finally, setting $x = 0$ yields $B = -1$. Now $\displaystyle\int \frac{dx}{x^4 - x^2} = \int \left[\frac{-1}{x^2} + \frac{1/2}{x - 1} - \frac{1/2}{x + 1}\right] dx = \frac{1}{x} + \tfrac{1}{2}\ln\left|\frac{x - 1}{x + 1}\right| + C.$

33. $\displaystyle\int \frac{x - 3}{(x^2 + 2x + 4)^2}\,dx = \int \frac{x - 3}{\left[(x + 1)^2 + 3\right]^2}\,dx = \int \frac{u - 4}{(u^2 + 3)^2}\,du \qquad [\text{with } u = x + 1]$

$$= \int \frac{u\,du}{(u^2 + 3)^2} - 4\int \frac{du}{(u^2 + 3)^2} = \frac{1}{2}\int \frac{dv}{v^2} - 4\int \frac{\sqrt{3}\sec^2\theta\,d\theta}{9\sec^4\theta} \qquad \begin{bmatrix} v = u^2 + 3 \text{ in the first integral;} \\ u = \sqrt{3}\tan\theta \text{ in the second}\end{bmatrix}$$

$$= \frac{-1}{2v} - \frac{4\sqrt{3}}{9}\int \cos^2\theta\,d\theta = \frac{-1}{2(u^2 + 3)} - \frac{2\sqrt{3}}{9}(\theta + \sin\theta\cos\theta) + C$$

$$= \frac{-1}{2(x^2 + 2x + 4)} - \frac{2\sqrt{3}}{9}\left[\tan^{-1}\left(\frac{x + 1}{\sqrt{3}}\right) + \frac{\sqrt{3}(x + 1)}{x^2 + 2x + 4}\right] + C$$

$$= \frac{-1}{2(x^2 + 2x + 4)} - \frac{2\sqrt{3}}{9}\tan^{-1}\left(\frac{x + 1}{\sqrt{3}}\right) - \frac{2(x + 1)}{3(x^2 + 2x + 4)} + C$$

35. Let $u = \sqrt{x}$, so $u^2 = x$ and $dx = 2u\,du$. Thus,

$$\int_9^{16} \frac{\sqrt{x}}{x-4}\,dx = \int_3^4 \frac{u}{u^2-4} 2u\,du = 2\int_3^4 \frac{u^2}{u^2-4}\,du = 2\int_3^4 \left(1 + \frac{4}{u^2-4}\right) du \qquad \text{[by long division]}$$

$$= 2 + 8\int_3^4 \frac{du}{(u+2)(u-2)} \quad (*)$$

Multiply $\dfrac{1}{(u+2)(u-2)} = \dfrac{A}{u+2} + \dfrac{B}{u-2}$ by $(u+2)(u-2)$ to get $1 = A(u-2) + B(u+2)$. Equating coefficients we get $A + B = 0$ and $-2A + 2B = 1$. Solving gives us $B = \frac{1}{4}$ and $A = -\frac{1}{4}$, so $\dfrac{1}{(u+2)(u-2)} = \dfrac{-1/4}{u+2} + \dfrac{1/4}{u-2}$ and $(*)$ is

$$2 + 8\int_3^4 \left(\frac{-1/4}{u+2} + \frac{1/4}{u-2}\right) du = 2 + 8\Big[-\tfrac{1}{4}\ln|u+2| + \tfrac{1}{4}\ln|u-2|\Big]_3^4$$

$$= 2 + \Big[2\ln|u-2| - 2\ln|u+2|\Big]_3^4 = 2 + 2\left[\ln\left|\frac{u-2}{u+2}\right|\right]_3^4$$

$$= 2 + 2\left(\ln\tfrac{2}{6} - \ln\tfrac{1}{5}\right) = 2 + 2\ln\tfrac{2/6}{1/5}$$

$$= 2 + 2\ln\tfrac{5}{3} \text{ or } 2 + \ln\left(\tfrac{5}{3}\right)^2 = 2 + \ln\tfrac{25}{9}$$

37. Let $u = e^x$. Then $x = \ln u$, $dx = \dfrac{du}{u} \Rightarrow$

$$\int \frac{e^{2x}\,dx}{e^{2x} + 3e^x + 2} = \int \frac{u^2\,(du/u)}{u^2 + 3u + 2} = \int \frac{u\,du}{(u+1)(u+2)} = \int \left[\frac{-1}{u+1} + \frac{2}{u+2}\right] du$$

$$= 2\ln|u+2| - \ln|u+1| + C = \ln\left[(e^x+2)^2/(e^x+1)\right] + C$$

39.

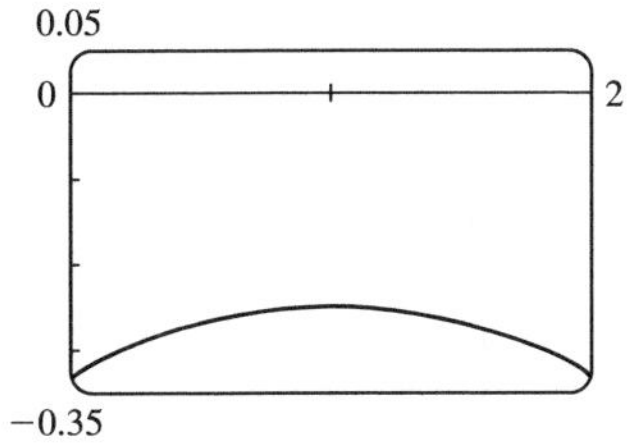

From the graph, we see that the integral will be negative, and we guess that the area is about the same as that of a rectangle with width 2 and height 0.3, so we estimate the integral to be $-(2 \cdot 0.3) = -0.6$. Now

$$\frac{1}{x^2 - 2x - 3} = \frac{1}{(x-3)(x+1)} = \frac{A}{x-3} + \frac{B}{x+1} \Leftrightarrow$$

$1 = (A+B)x + A - 3B$, so $A = -B$ and $A - 3B = 1 \Leftrightarrow A = \frac{1}{4}$ and $B = -\frac{1}{4}$, so the integral becomes

$$\int_0^2 \frac{dx}{x^2-2x-3} = \frac{1}{4}\int_0^2 \frac{dx}{x-3} - \frac{1}{4}\int_0^2 \frac{dx}{x+1} = \tfrac{1}{4}\Big[\ln|x-3| - \ln|x+1|\Big]_0^2$$

$$= \frac{1}{4}\left[\ln\left|\frac{x-3}{x+1}\right|\right]_0^2 = \tfrac{1}{4}\left(\ln\tfrac{1}{3} - \ln 3\right) = -\tfrac{1}{2}\ln 3 \approx -0.55$$

41. $\dfrac{P+S}{P\,[(r-1)P-S]} = \dfrac{A}{P} + \dfrac{B}{(r-1)P-S} \Rightarrow P + S = A\,[(r-1)P - S] + BP = [(r-1)A + B]\,P - AS \Rightarrow$ $(r-1)A + B = 1$, $-A = 1 \Rightarrow A = -1$, $B = r$. Now

$$t = \int \frac{P+S}{P\,[(r-1)P-S]}\,dP = \int \left[\frac{-1}{P} + \frac{r}{(r-1)P-S}\right] dP = -\int \frac{dP}{P} + \frac{r}{r-1}\int \frac{r-1}{(r-1)P-S}\,dP$$

so $t = -\ln P + \dfrac{r}{r-1}\ln|(r-1)P - S| + C$. Here $r = 0.10$ and $S = 900$, so

$$t = -\ln P + \tfrac{0.1}{-0.9}\ln|-0.9P - 900| + C = -\ln P - \tfrac{1}{9}\ln(|-1|\,|0.9P + 900|)$$
$$= -\ln P - \tfrac{1}{9}\ln(0.9P + 900) + C$$

When $t = 0$, $P = 10\,000$, so $0 = -\ln 10\,000 - \frac{1}{9}\ln(9900) + C$. Thus, $C = \ln 10\,000 + \frac{1}{9}\ln 9900$ $[\approx 10.2326]$, so our equation becomes

$$t = \ln 10\,000 - \ln P + \tfrac{1}{9}\ln 9900 - \tfrac{1}{9}\ln(0.9P + 900) = \ln\frac{10\,000}{P} + \frac{1}{9}\ln\frac{9900}{0.9P + 900}$$
$$= \ln\frac{10\,000}{P} + \frac{1}{9}\ln\frac{1100}{0.1P + 100} = \ln\frac{10\,000}{P} + \frac{1}{9}\ln\frac{11\,000}{P + 1000}$$

43. (a) In Maple, we define $f(x)$, and then use `convert(f,parfrac,x);` to obtain

$$f(x) = \frac{24\,110/4879}{5x + 2} - \frac{668/323}{2x + 1} - \frac{9438/80\,155}{3x - 7} + \frac{(22\,098x + 48\,935)/260\,015}{x^2 + x + 5}$$

In Mathematica, we use the command `Apart`, and in Derive, we use `Expand`.

(b) $$\int f(x)\,dx = \tfrac{24\,110}{4879}\cdot\tfrac{1}{5}\ln|5x + 2| - \tfrac{668}{323}\cdot\tfrac{1}{2}\ln|2x + 1| - \tfrac{9438}{80\,155}\cdot\tfrac{1}{3}\ln|3x - 7|$$
$$+ \frac{1}{260\,015}\int\frac{22\,098\left(x + \frac{1}{2}\right) + 37\,886}{\left(x + \frac{1}{2}\right)^2 + \frac{19}{4}}\,dx + C$$
$$= \tfrac{24\,110}{4879}\cdot\tfrac{1}{5}\ln|5x + 2| - \tfrac{668}{323}\cdot\tfrac{1}{2}\ln|2x + 1| - \tfrac{9438}{80\,155}\cdot\tfrac{1}{3}\ln|3x - 7|$$
$$+ \tfrac{1}{260\,015}\left[22\,098\cdot\tfrac{1}{2}\ln\left(x^2 + x + 5\right) + 37\,886\cdot\sqrt{\tfrac{4}{19}}\tan^{-1}\left(\tfrac{1}{\sqrt{19/4}}\left(x + \tfrac{1}{2}\right)\right)\right] + C$$
$$= \tfrac{4822}{4879}\ln|5x + 2| - \tfrac{334}{323}\ln|2x + 1| - \tfrac{3146}{80\,155}\ln|3x - 7| + \tfrac{11\,049}{260\,015}\ln\left(x^2 + x + 5\right)$$
$$+ \tfrac{75\,772}{260\,015\sqrt{19}}\tan^{-1}\left[\tfrac{1}{\sqrt{19}}(2x + 1)\right] + C$$

Using a CAS, we get

$$\frac{4822\ln(5x + 2)}{4879} - \frac{334\ln(2x + 1)}{323} - \frac{3146\ln(3x - 7)}{80\,155}$$
$$+ \frac{11\,049\ln\left(x^2 + x + 5\right)}{260\,015} + \frac{3988\sqrt{19}}{260\,015}\tan^{-1}\left[\frac{\sqrt{19}}{19}(2x + 1)\right]$$

The main difference in this answer is that the absolute value signs and the constant of integration have been omitted. Also, the fractions have been reduced and the denominators rationalized.

45. There are only finitely many values of x where $Q(x) = 0$ (assuming that Q is not the zero polynomial). At all other values of x, $F(x)/Q(x) = G(x)/Q(x)$, so $F(x) = G(x)$. In other words, the values of F and G agree at all except perhaps finitely many values of x. By continuity of F and G, the polynomials F and G must agree at those values of x too.

More explicitly: if a is a value of x such that $Q(a) = 0$, then $Q(x) \neq 0$ for all x sufficiently close to a. Thus,

$$F(a) = \lim_{x\to a} F(x) \quad [\text{by continuity of } F] \quad = \lim_{x\to a} G(x) \quad [\text{whenever } Q(x) \neq 0]$$
$$= G(a) \quad [\text{by continuity of } G]$$

H POLAR COORDINATES

H.1 Curves in Polar Coordinates

1. (a) By adding 2π to $\frac{\pi}{2}$, we obtain the point $\left(1, \frac{5\pi}{2}\right)$. The direction opposite $\frac{\pi}{2}$ is $\frac{3\pi}{2}$, so $\left(-1, \frac{3\pi}{2}\right)$ is a point that satisfies the $r < 0$ requirement.

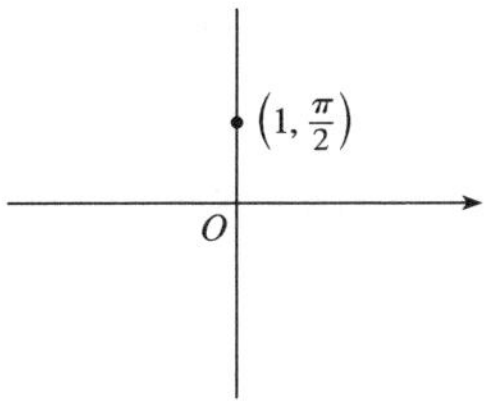

(b) $\left(-2, \frac{\pi}{4}\right)$

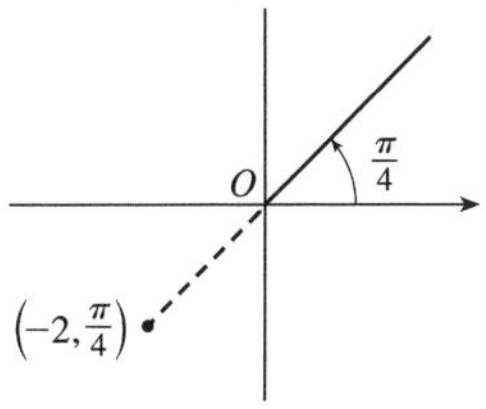

$\left(2, \frac{5\pi}{4}\right), \left(-2, \frac{9\pi}{4}\right)$

(c) $(3, 2)$

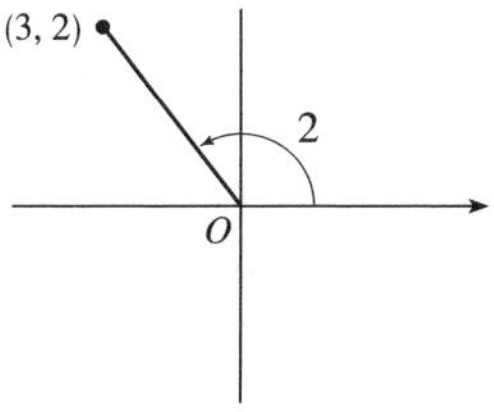

$(3, 2 + 2\pi), (-3, 2 + \pi)$

3. (a)

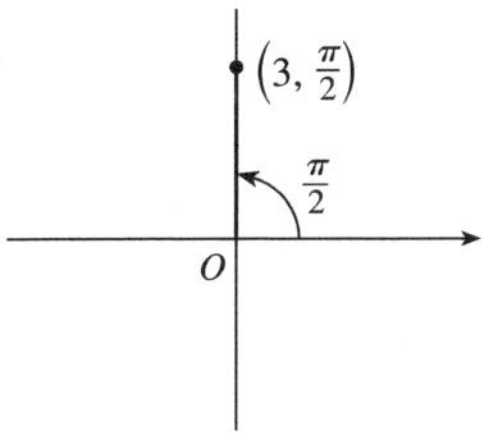

$x = 3\cos\frac{\pi}{2} = 3(0) = 0$ and $y = 3\sin\frac{\pi}{2} = 3(1) = 3$ give us the Cartesian coordinates $(0, 3)$.

(b)

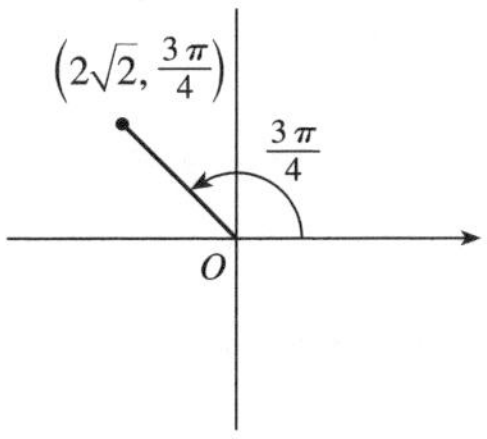

$$x = 2\sqrt{2}\cos\frac{3\pi}{4} = 2\sqrt{2}\left(-\frac{1}{\sqrt{2}}\right) = -2 \text{ and}$$

$$y = 2\sqrt{2}\sin\frac{3\pi}{4} = 2\sqrt{2}\left(\frac{1}{\sqrt{2}}\right) = 2$$

give us $(-2, 2)$.

(c)

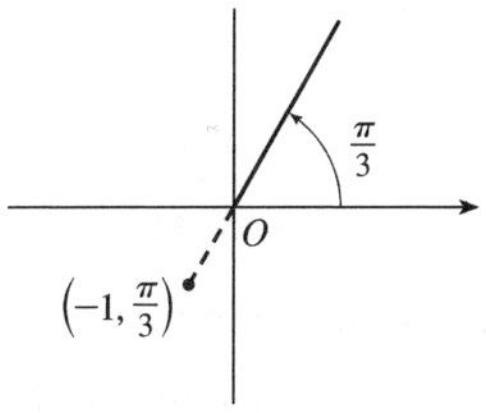

$x = -1\cos\frac{\pi}{3} = -\frac{1}{2}$ and $y = -1\sin\frac{\pi}{3} = -\frac{\sqrt{3}}{2}$ give us $\left(-\frac{1}{2}, -\frac{\sqrt{3}}{2}\right)$.

5. (a) $x = 1$ and $y = 1 \;\Rightarrow\; r = \sqrt{1^2 + 1^2} = \sqrt{2}$ and $\theta = \tan^{-1}\left(\frac{1}{1}\right) = \frac{\pi}{4}$. Since $(1, 1)$ is in the first quadrant, the polar coordinates are (i) $\left(\sqrt{2}, \frac{\pi}{4}\right)$ and (ii) $\left(-\sqrt{2}, \frac{5\pi}{4}\right)$.

(b) $x = 2\sqrt{3}$ and $y = -2 \;\Rightarrow\; r = \sqrt{\left(2\sqrt{3}\right)^2 + (-2)^2} = \sqrt{12 + 4} = \sqrt{16} = 4$ and $\theta = \tan^{-1}\left(-\frac{2}{2\sqrt{3}}\right) = \tan^{-1}\left(-\frac{1}{\sqrt{3}}\right) = -\frac{\pi}{6}$. Since $\left(2\sqrt{3}, -2\right)$ is in the fourth quadrant and $0 \le \theta \le 2\pi$, the polar coordinates are (i) $\left(4, \frac{11\pi}{6}\right)$ and (ii) $\left(-4, \frac{5\pi}{6}\right)$.

7. The curves $r = 1$ and $r = 2$ represent circles with centre O and radii 1 and 2. The region in the plane satisfying $1 \le r \le 2$ consists of both circles and the shaded region between them in the figure.

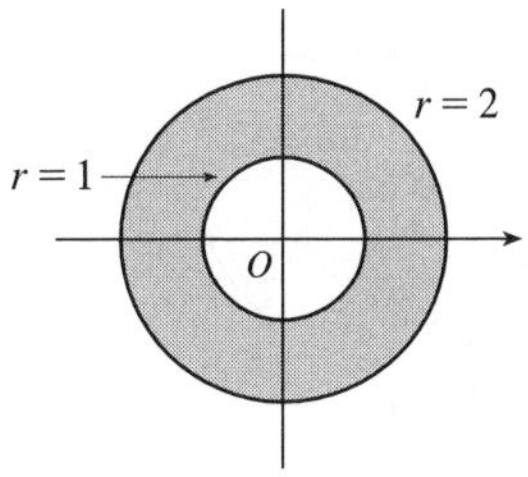

9. The region satisfying $0 \le r < 4$ and $-\pi/2 \le \theta < \pi/6$ does not include the circle $r = 4$ nor the line $\theta = \frac{\pi}{6}$.

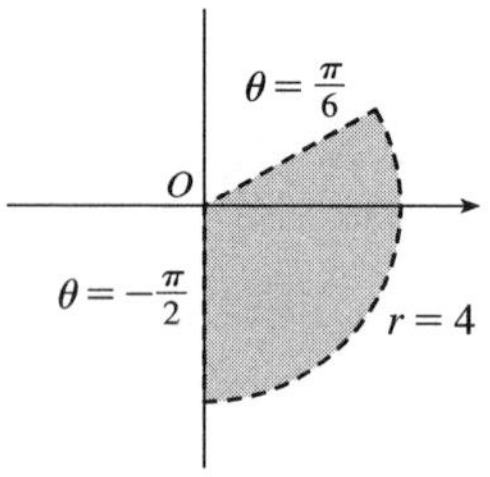

11. $2 < r < 3, \quad \frac{5\pi}{3} \le \theta \le \frac{7\pi}{3}$

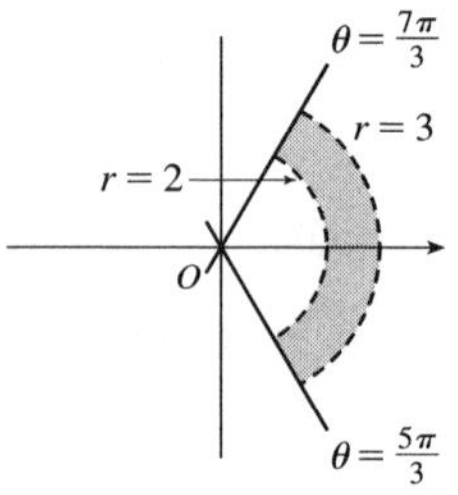

13. $r = 3\sin\theta \;\Rightarrow\; r^2 = 3r\sin\theta \;\Leftrightarrow\; x^2 + y^2 = 3y \;\Leftrightarrow\; x^2 + \left(y - \frac{3}{2}\right)^2 = \left(\frac{3}{2}\right)^2$, a circle of radius $\frac{3}{2}$ centred at $\left(0, \frac{3}{2}\right)$. The first two equations are actually equivalent since $r^2 = 3r\sin\theta \;\Rightarrow\; r(r - 3\sin\theta) = 0 \;\Rightarrow\; r = 0$ or $r = 3\sin\theta$. But $r = 3\sin\theta$ gives the point $r = 0$ (the pole) when $\theta = 0$. Thus, the single equation $r = 3\sin\theta$ is equivalent to the compound condition ($r = 0$ or $r = 3\sin\theta$).

15. $r = \csc\theta \;\Leftrightarrow\; r = \dfrac{1}{\sin\theta} \;\Leftrightarrow\; r\sin\theta = 1 \;\Leftrightarrow\; y = 1$, a horizontal line 1 unit above the x-axis.

17. $x = -y^2 \;\Leftrightarrow\; r\cos\theta = -r^2\sin^2\theta \;\Leftrightarrow\; \cos\theta = -r\sin^2\theta \;\Leftrightarrow\; r = -\dfrac{\cos\theta}{\sin^2\theta} = -\cot\theta\,\csc\theta$.

19. $x^2 + y^2 = 2cx \;\Leftrightarrow\; r^2 = 2cr\cos\theta \;\Leftrightarrow\; r^2 - 2cr\cos\theta = 0 \;\Leftrightarrow\; r(r - 2c\cos\theta) = 0 \;\Leftrightarrow\; r = 0$ or $r = 2c\cos\theta$. $r = 0$ is included in $r = 2c\cos\theta$ when $\theta = \frac{\pi}{2} + n\pi$, so the curve is represented by the single equation $r = 2c\cos\theta$.

21. (a) The description leads immediately to the polar equation $\theta = \frac{\pi}{6}$, and the Cartesian equation $y = \tan\left(\frac{\pi}{6}\right)x = \frac{1}{\sqrt{3}}\,x$ is slightly more difficult to derive.

(b) The easier description here is the Cartesian equation $x = 3$.

23. $\theta = -\pi/6$

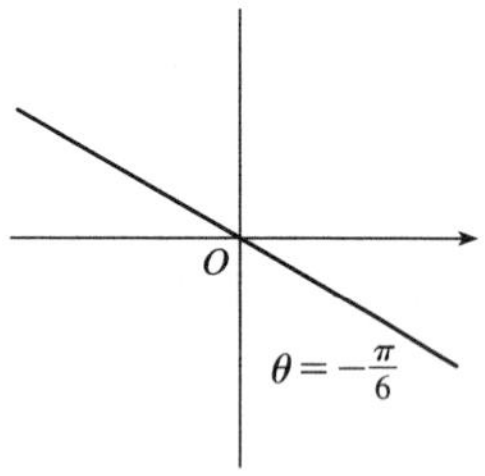

25. $r = \sin\theta \;\Leftrightarrow\; r^2 = r\sin\theta \;\Leftrightarrow\; x^2 + y^2 = y \;\Leftrightarrow\; x^2 + \left(y - \frac{1}{2}\right)^2 = \left(\frac{1}{2}\right)^2$. The reasoning here is the same as in Exercise 13. This is a circle of radius $\frac{1}{2}$ centred at $\left(0, \frac{1}{2}\right)$.

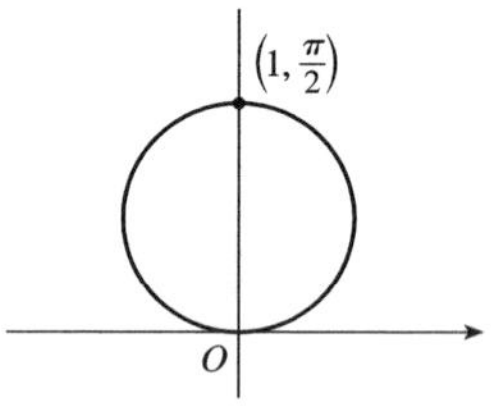

27. $r = 2(1 - \sin\theta)$. This curve is a cardioid.

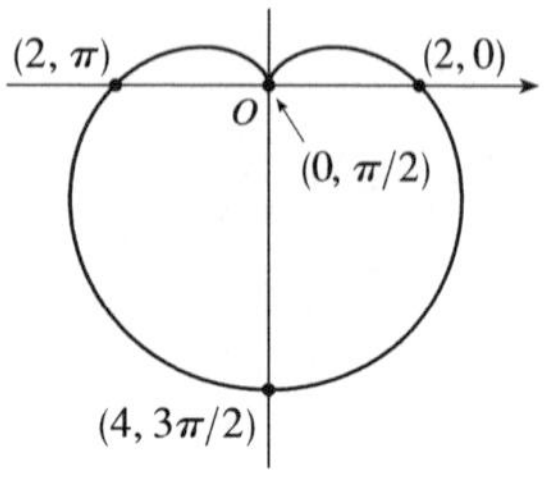

29. $r = \theta, \quad \theta \ge 0$

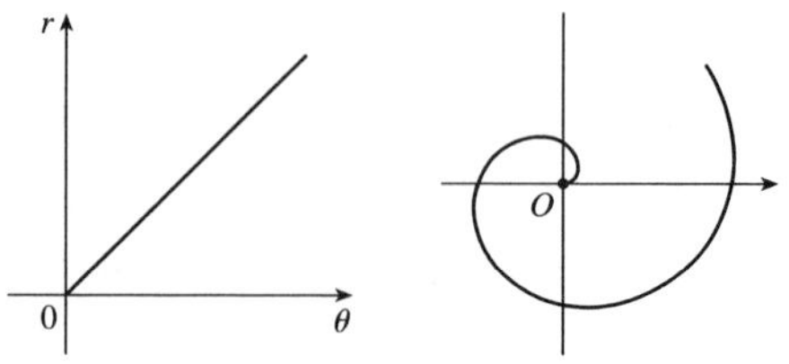

31. $r = \sin 2\theta$

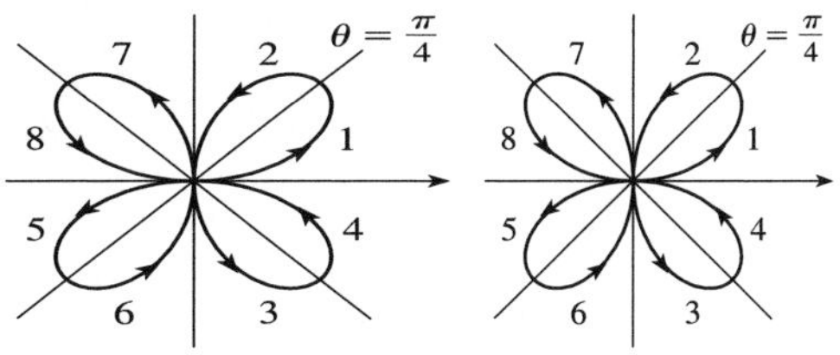

33. $r = 2\cos 4\theta$

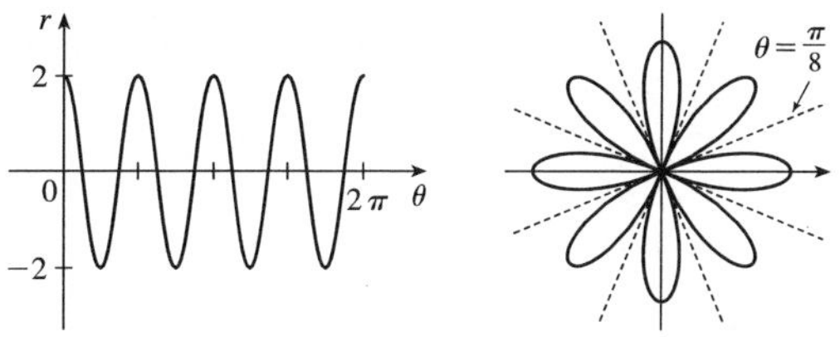

35. $r^2 = 4\cos 2\theta$

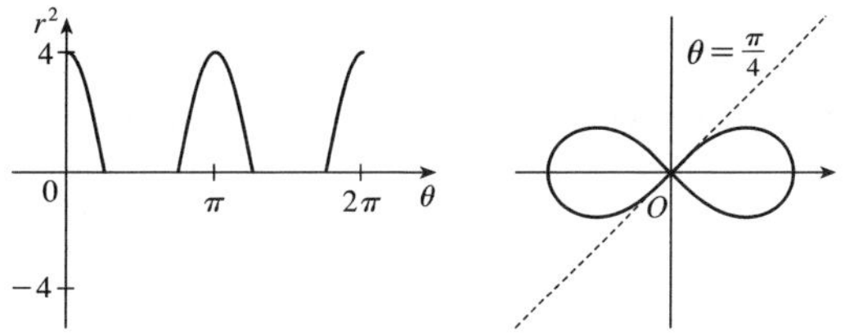

37. $r = 2\cos\left(\frac{3}{2}\theta\right)$

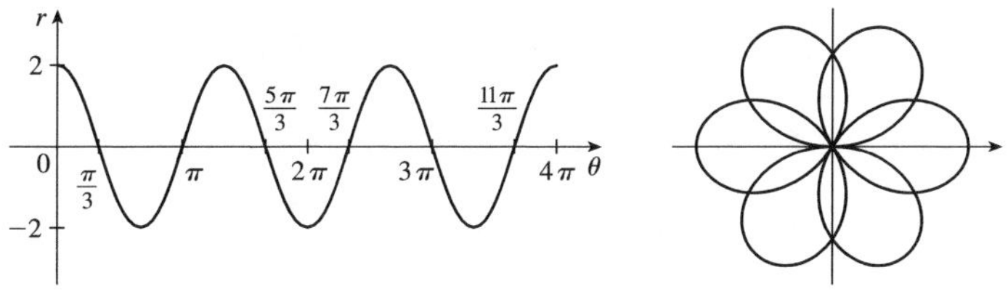

39. $r = 1 + 2\cos 2\theta$

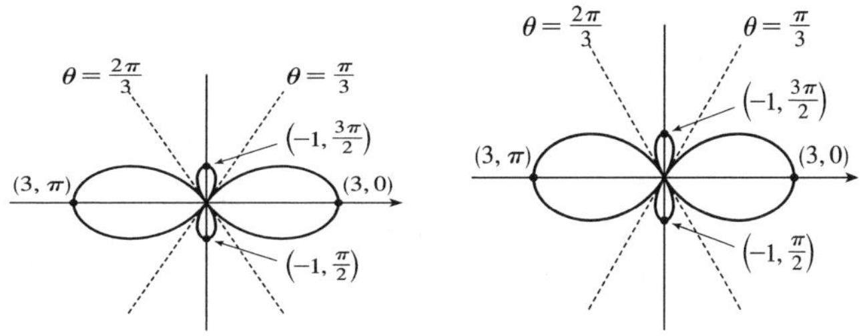

41. For $\theta = 0$, π, and 2π, r has its minimum value of about 0.5. For $\theta = \frac{\pi}{2}$ and $\frac{3\pi}{2}$, r attains its maximum value of 2. We see that the graph has a similar shape for $0 \le \theta \le \pi$ and $\pi \le \theta \le 2\pi$.

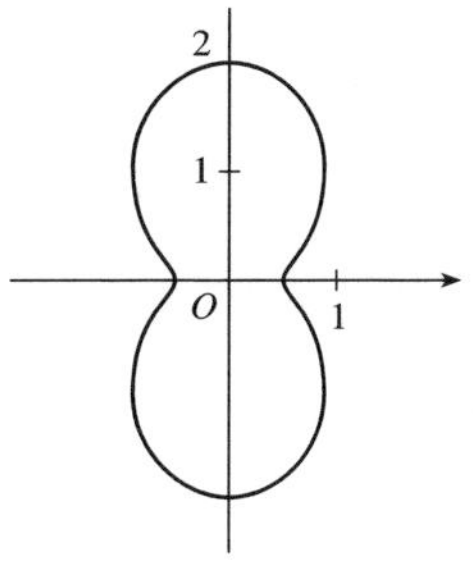

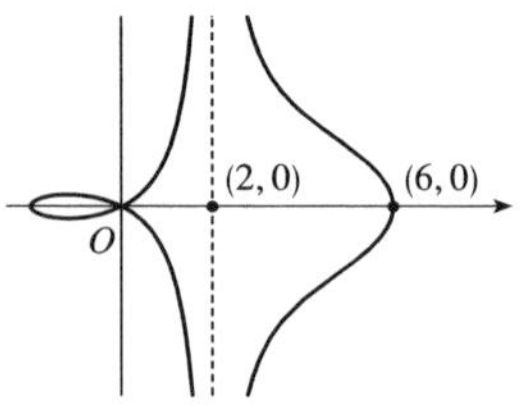

43. $x = (r)\cos\theta = (4 + 2\sec\theta)\cos\theta = 4\cos\theta + 2$. Now, $r \to \infty \Rightarrow$ $(4 + 2\sec\theta) \to \infty \Rightarrow \theta \to \left(\frac{\pi}{2}\right)^-$ or $\theta \to \left(\frac{3\pi}{2}\right)^+$ (since we need only consider $0 \le \theta < 2\pi$), so $\lim\limits_{r\to\infty} x = \lim\limits_{\theta\to\pi/2^-} (4\cos\theta + 2) = 2$. Also, $r \to -\infty \Rightarrow$ $(4 + 2\sec\theta) \to -\infty \Rightarrow \theta \to \left(\frac{\pi}{2}\right)^+$ or $\theta \to \left(\frac{3\pi}{2}\right)^-$, so $\lim\limits_{r\to-\infty} x = \lim\limits_{\theta\to\pi/2^+} (4\cos\theta + 2) = 2$. Therefore, $\lim\limits_{r\to\pm\infty} x = 2 \Rightarrow x = 2$ is a vertical asymptote.

45. (a) We see that the curve crosses itself at the origin, where $r = 0$ (in fact the inner loop corresponds to negative r-values,) so we solve the equation of the limaçon for $r = 0 \Leftrightarrow c\sin\theta = -1 \Leftrightarrow \sin\theta = -1/c$. Now if $|c| < 1$, then this equation has no solution and hence there is no inner loop. But if $c < -1$, then on the interval $(0, 2\pi)$ the equation has the two solutions $\theta = \sin^{-1}(-1/c)$ and $\theta = \pi - \sin^{-1}(-1/c)$, and if $c > 1$, the solutions are $\theta = \pi + \sin^{-1}(1/c)$ and $\theta = 2\pi - \sin^{-1}(1/c)$. In each case, $r < 0$ for θ between the two solutions, indicating a loop.

(b) For $0 < c < 1$, the dimple (if it exists) is characterized by the fact that y has a local maximum at $\theta = \frac{3\pi}{2}$. So we determine for what c-values $\dfrac{d^2y}{d\theta^2}$ is negative at $\theta = \frac{3\pi}{2}$, since by the Second Derivative Test this indicates a maximum:

$$y = r\sin\theta = \sin\theta + c\sin^2\theta \Rightarrow \frac{dy}{d\theta} = \cos\theta + 2c\sin\theta\cos\theta = \cos\theta + c\sin 2\theta \Rightarrow \frac{d^2y}{d\theta^2} = -\sin\theta + 2c\cos 2\theta.$$

At $\theta = \frac{3\pi}{2}$, this is equal to $-(-1) + 2c(-1) = 1 - 2c$, which is negative only for $c > \frac{1}{2}$. A similar argument shows that for $-1 < c < 0$, y only has a local minimum at $\theta = \frac{\pi}{2}$ (indicating a dimple) for $c < -\frac{1}{2}$.

47. $r = 2\sin\theta \Rightarrow x = r\cos\theta = 2\sin\theta\cos\theta = \sin 2\theta$, $y = r\sin\theta = 2\sin^2\theta \Rightarrow$

$$\frac{dy}{dx} = \frac{dy/d\theta}{dx/d\theta} = \frac{2 \cdot 2\sin\theta\cos\theta}{\cos 2\theta \cdot 2} = \frac{\sin 2\theta}{\cos 2\theta} = \tan 2\theta$$

When $\theta = \dfrac{\pi}{6}$, $\dfrac{dy}{dx} = \tan\!\left(2 \cdot \dfrac{\pi}{6}\right) = \tan\dfrac{\pi}{3} = \sqrt{3}$.

49. $r = 1/\theta \Rightarrow x = r\cos\theta = (\cos\theta)/\theta$, $y = r\sin\theta = (\sin\theta)/\theta \Rightarrow$

$$\frac{dy}{dx} = \frac{dy/d\theta}{dx/d\theta} = \frac{\sin\theta(-1/\theta^2) + (1/\theta)\cos\theta}{\cos\theta(-1/\theta^2) - (1/\theta)\sin\theta} \cdot \frac{\theta^2}{\theta^2} = \frac{-\sin\theta + \theta\cos\theta}{-\cos\theta - \theta\sin\theta}$$

When $\theta = \pi$, $\dfrac{dy}{dx} = \dfrac{-0 + \pi(-1)}{-(-1) - \pi(0)} = \dfrac{-\pi}{1} = -\pi$.

51. $r = 3\cos\theta \Rightarrow x = r\cos\theta = 3\cos\theta\,\cos\theta$, $y = r\sin\theta = 3\cos\theta\,\sin\theta \Rightarrow$ $dy/d\theta = -3\sin^2\theta + 3\cos^2\theta = 3\cos 2\theta = 0 \Rightarrow 2\theta = \frac{\pi}{2}$ or $\frac{3\pi}{2} \Leftrightarrow \theta = \frac{\pi}{4}$ or $\frac{3\pi}{4}$. So the tangent is horizontal at $\left(\frac{3}{\sqrt{2}}, \frac{\pi}{4}\right)$ and $\left(-\frac{3}{\sqrt{2}}, \frac{3\pi}{4}\right)$ $\left[\text{same as } \left(\frac{3}{\sqrt{2}}, -\frac{\pi}{4}\right)\right]$. $dx/d\theta = -6\sin\theta\cos\theta = -3\sin 2\theta = 0 \Rightarrow 2\theta = 0$ or $\pi \Leftrightarrow$ $\theta = 0$ or $\frac{\pi}{2}$. So the tangent is vertical at $(3, 0)$ and $\left(0, \frac{\pi}{2}\right)$.

53. $r = 1 + \cos\theta \Rightarrow x = r\cos\theta = \cos\theta(1+\cos\theta),\ y = r\sin\theta = \sin\theta(1+\cos\theta) \Rightarrow$

$dy/d\theta = (1+\cos\theta)\cos\theta - \sin^2\theta = 2\cos^2\theta + \cos\theta - 1 = (2\cos\theta - 1)(\cos\theta + 1) = 0 \Rightarrow$

$\cos\theta = \frac{1}{2}$ or $-1 \Rightarrow \theta = \frac{\pi}{3}, \pi,$ or $\frac{5\pi}{3} \Rightarrow$ horizontal tangent at $\left(\frac{3}{2}, \frac{\pi}{3}\right)$, $(0, \pi)$, and $\left(\frac{3}{2}, \frac{5\pi}{3}\right)$.

$dx/d\theta = -(1+\cos\theta)\sin\theta - \cos\theta\sin\theta = -\sin\theta(1+2\cos\theta) = 0 \Rightarrow \sin\theta = 0$ or $\cos\theta = -\frac{1}{2} \Rightarrow$

$\theta = 0, \pi, \frac{2\pi}{3},$ or $\frac{4\pi}{3} \Rightarrow$ vertical tangent at $(2, 0)$, $\left(\frac{1}{2}, \frac{2\pi}{3}\right)$, and $\left(\frac{1}{2}, \frac{4\pi}{3}\right)$.

Note that the tangent is horizontal, not vertical when $\theta = \pi$, since $\displaystyle\lim_{\theta\to\pi} \frac{dy/d\theta}{dx/d\theta} = 0$.

55. $r = a\sin\theta + b\cos\theta \Rightarrow r^2 = ar\sin\theta + br\cos\theta \Rightarrow x^2 + y^2 = ay + bx \Rightarrow$

$x^2 - bx + \left(\frac{1}{2}b\right)^2 + y^2 - ay + \left(\frac{1}{2}a\right)^2 = \left(\frac{1}{2}b\right)^2 + \left(\frac{1}{2}a\right)^2 \Rightarrow \left(x - \frac{1}{2}b\right)^2 + \left(y - \frac{1}{2}a\right)^2 = \frac{1}{4}(a^2 + b^2)$, and this is a circle with centre $\left(\frac{1}{2}b, \frac{1}{2}a\right)$ and radius $\frac{1}{2}\sqrt{a^2 + b^2}$.

Note for Exercises 57–60: Maple is able to plot polar curves using the `polarplot` command, or using the `coords=polar` option in a regular `plot` command. In Mathematica, use `PolarPlot`. In Derive, change to `Polar` under `Options State`. If your graphing device cannot plot polar equations, you must convert to parametric equations. For example, in Exercise 57, $x = r\cos\theta = \left[e^{\sin\theta} - 2\cos(4\theta)\right]\cos\theta$, $y = r\sin\theta = \left[e^{\sin\theta} - 2\cos(4\theta)\right]\sin\theta$.

57. $r = e^{\sin\theta} - 2\cos(4\theta)$. The parameter interval is $[0, 2\pi]$.

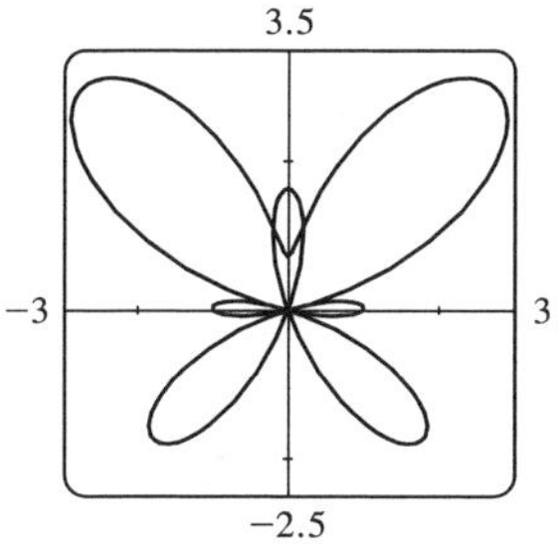

59. $r = 2 - 5\sin(\theta/6)$. The parameter interval is $[-6\pi, 6\pi]$.

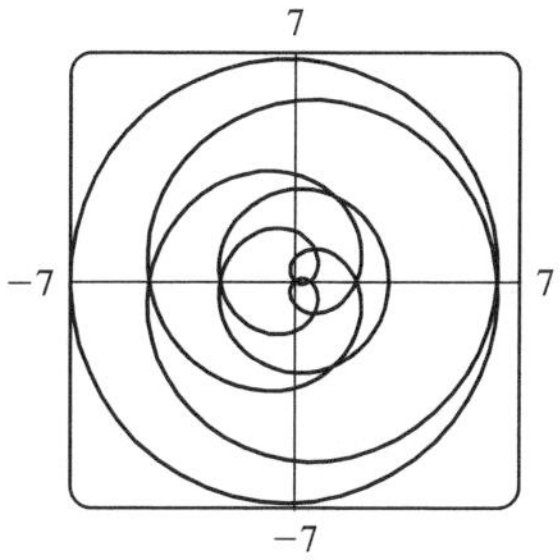

61.

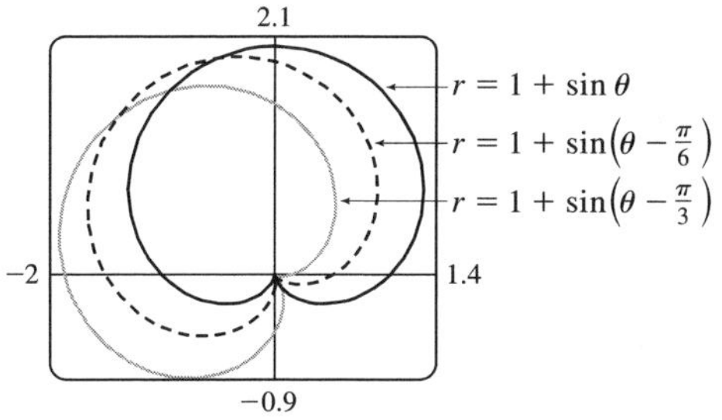

It appears that the graph of $r = 1 + \sin\left(\theta - \frac{\pi}{6}\right)$ is the same shape as the graph of $r = 1 + \sin\theta$, but rotated counterclockwise about the origin by $\frac{\pi}{6}$. Similarly, the graph of $r = 1 + \sin\left(\theta - \frac{\pi}{3}\right)$ is rotated by $\frac{\pi}{3}$. In general, the graph of $r = f(\theta - \alpha)$ is the same shape as that of $r = f(\theta)$, but rotated counterclockwise through α about the origin. That is, for any point (r_0, θ_0) on the curve $r = f(\theta)$, the point $(r_0, \theta_0 + \alpha)$ is on the curve $r = f(\theta - \alpha)$, since $r_0 = f(\theta_0) = f((\theta_0 + \alpha) - \alpha)$.

63. (a) $r = \sin n\theta$. From the graphs, it seems that when n is even, the number of loops in the curve (called a rose) is $2n$, and when n is odd, the number of loops is simply n. This is because in the case of n odd, every point on the graph is traversed twice, due to the fact that

$$r(\theta + \pi) = \sin[n(\theta + \pi)] = \sin n\theta \cos n\pi + \cos n\theta \sin n\pi = \begin{cases} \sin n\theta & \text{if } n \text{ is even} \\ -\sin n\theta & \text{if } n \text{ is odd} \end{cases}$$

$n = 2$ $\quad$ $n = 3$ $\quad$ $n = 4$ $\quad$ $n = 5$

(b) The graph of $r = |\sin n\theta|$ has $2n$ loops whether n is odd or even, since $r(\theta + \pi) = r(\theta)$.

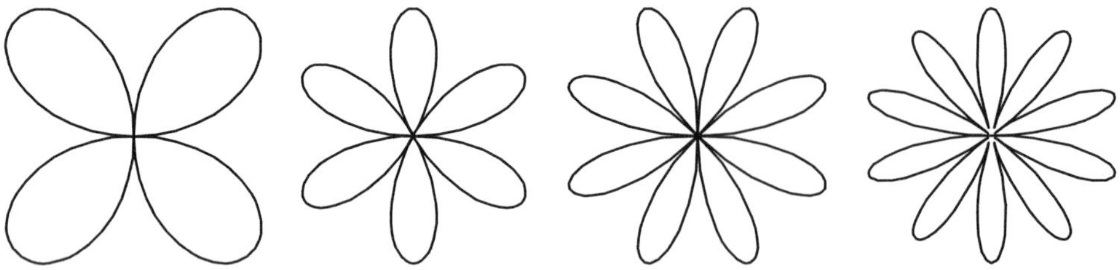

$n = 2$ $\quad$ $n = 3$ $\quad$ $n = 4$ $\quad$ $n = 5$

65. $r = \dfrac{1 - a\cos\theta}{1 + a\cos\theta}$. We start with $a = 0$, since in this case the curve is simply the circle $r = 1$.

As a increases, the graph moves to the left, and its right side becomes flattened. As a increases through about 0.4, the right side seems to grow a dimple, which upon closer investigation (with narrower θ-ranges) seems to appear at $a \approx 0.42$ (the actual value is $\sqrt{2} - 1$). As $a \to 1$, this dimple becomes more pronounced, and the curve begins to stretch out horizontally, until at $a = 1$ the denominator vanishes at $\theta = \pi$, and the dimple becomes an actual cusp. For $a > 1$ we must choose our parameter interval carefully, since $r \to \infty$ as $1 + a\cos\theta \to 0 \quad \Leftrightarrow \quad \theta \to \pm\cos^{-1}(-1/a)$. As a increases from 1, the curve splits into two parts. The left part has a loop, which grows larger as a increases, and the right part grows broader vertically, and its left tip develops a dimple when $a \approx 2.42$ (actually, $\sqrt{2} + 1$). As a increases, the dimple grows more and more pronounced.

If $a < 0$, we get the same graph as we do for the corresponding positive a-value, but with a rotation through π about the pole, as happened when c was replaced with $-c$ in Exercise 64.

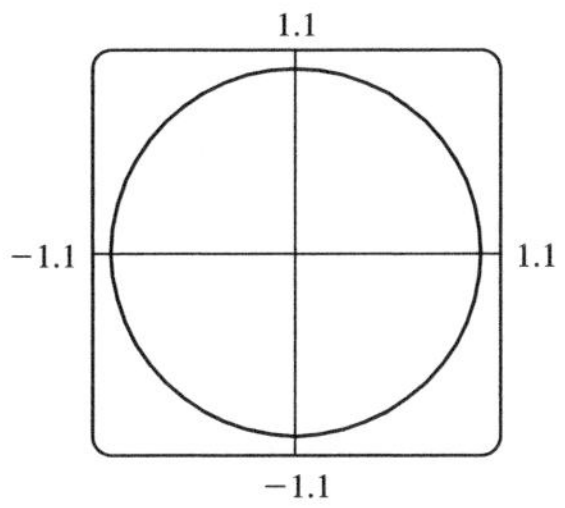

$a = 0$

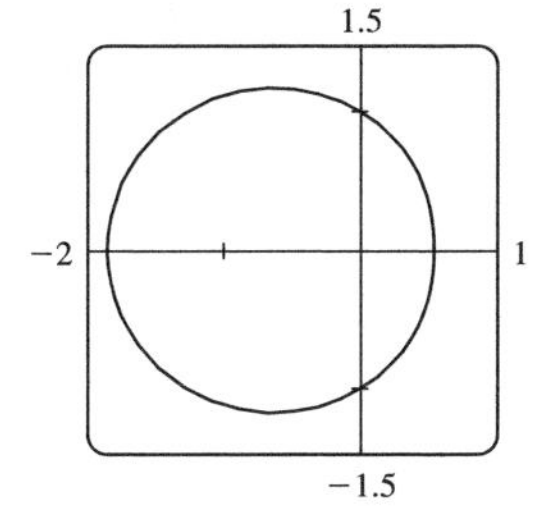

$a = 0.3$

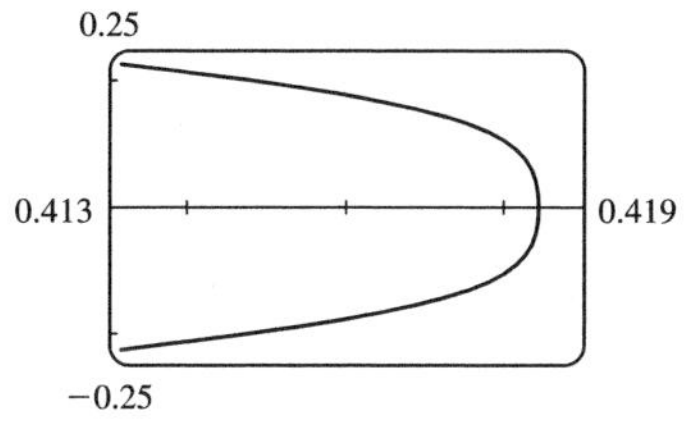

$a = 0.41, |\theta| \le 0.5$

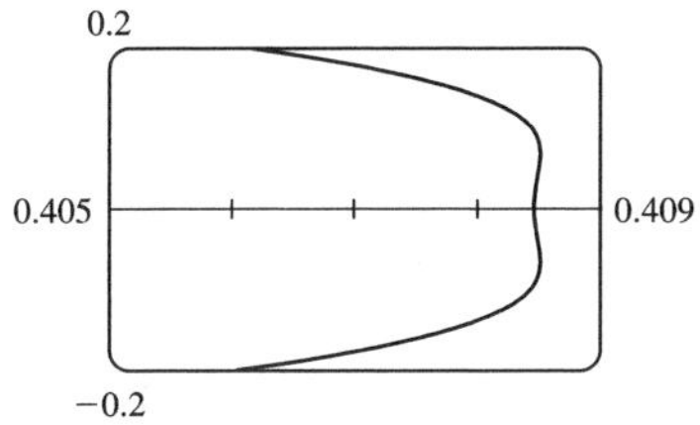

$a = 0.42, |\theta| \le 0.5$

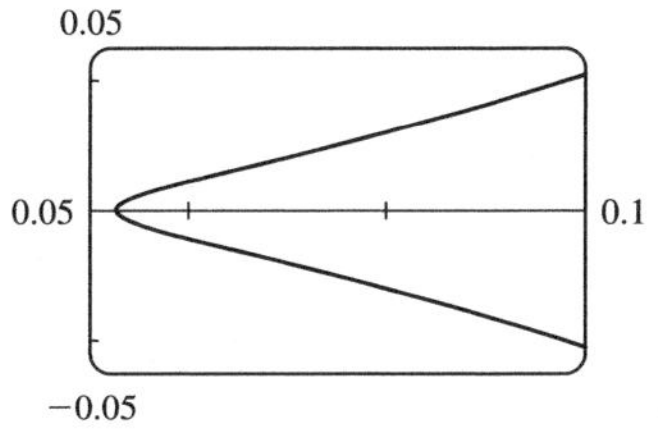

$a = 0.9, |\theta| \le 0.5$

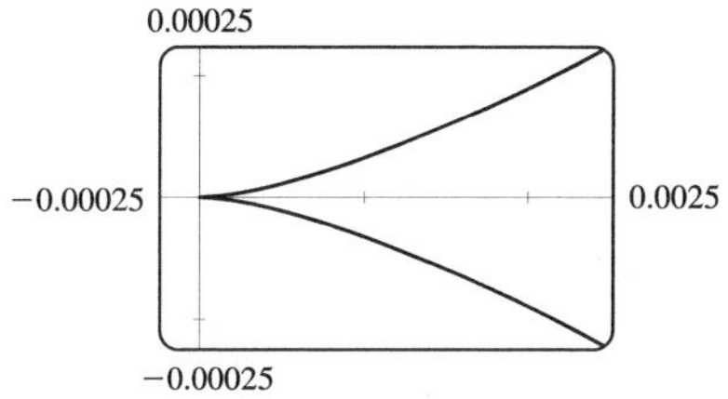

$a = 1, |\theta| \le 0.1$

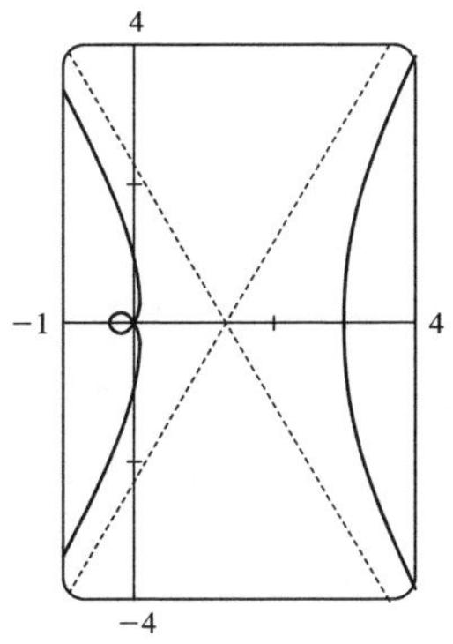

$a = 2$

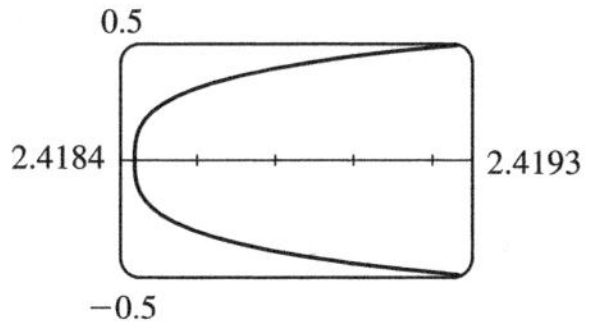

$a = 2.41, |\theta - \pi| \le 0.2$

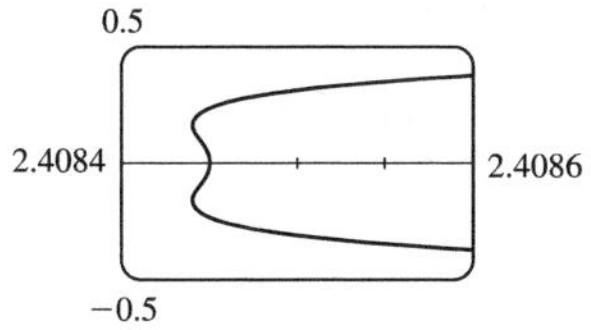

$a = 2.42, |\theta - \pi| \le 0.2$

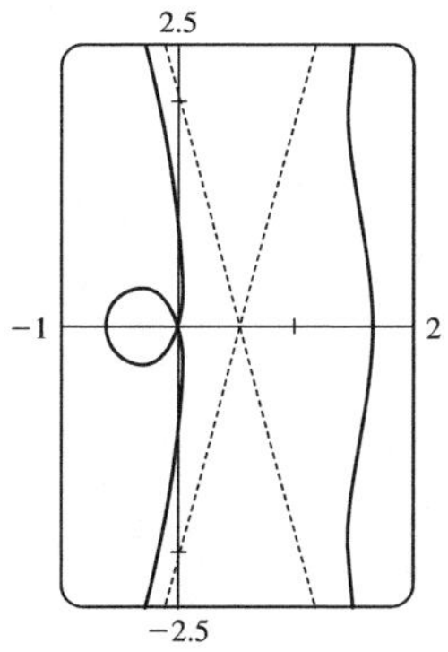

$a = 4$

67.

$$\tan\psi = \tan(\phi - \theta) = \frac{\tan\phi - \tan\theta}{1 + \tan\phi\,\tan\theta} = \frac{\dfrac{dy}{dx} - \tan\theta}{1 + \dfrac{dy}{dx}\tan\theta} = \frac{\dfrac{dy/d\theta}{dx/d\theta} - \tan\theta}{1 + \dfrac{dy/d\theta}{dx/d\theta}\tan\theta}$$

$$= \frac{\dfrac{dy}{d\theta} - \dfrac{dx}{d\theta}\tan\theta}{\dfrac{dx}{d\theta} + \dfrac{dy}{d\theta}\tan\theta} = \frac{\left(\dfrac{dr}{d\theta}\sin\theta + r\cos\theta\right) - \tan\theta\left(\dfrac{dr}{d\theta}\cos\theta - r\sin\theta\right)}{\left(\dfrac{dr}{d\theta}\cos\theta - r\sin\theta\right) + \tan\theta\left(\dfrac{dr}{d\theta}\sin\theta + r\cos\theta\right)}$$

$$= \frac{r\cos\theta + r\cdot\dfrac{\sin^2\theta}{\cos\theta}}{\dfrac{dr}{d\theta}\cos\theta + \dfrac{dr}{d\theta}\cdot\dfrac{\sin^2\theta}{\cos\theta}} = \frac{r\cos^2\theta + r\sin^2\theta}{\dfrac{dr}{d\theta}\cos^2\theta + \dfrac{dr}{d\theta}\sin^2\theta} = \frac{r}{dr/d\theta}$$

H.2 Areas and Lengths in Polar Coordinates

1. $r = \sqrt{\theta}, 0 \le \theta \le \frac{\pi}{4}$. $A = \displaystyle\int_0^{\pi/4} \tfrac{1}{2} r^2 \, d\theta = \int_0^{\pi/4} \tfrac{1}{2}\left(\sqrt{\theta}\right)^2 d\theta = \int_0^{\pi/4} \tfrac{1}{2}\theta \, d\theta = \left[\tfrac{1}{4}\theta^2\right]_0^{\pi/4} = \tfrac{1}{64}\pi^2$

3. $r = \sin\theta, \frac{\pi}{3} \le \theta \le \frac{2\pi}{3}$.

$$A = \int_{\pi/3}^{2\pi/3} \tfrac{1}{2}\sin^2\theta \, d\theta = \tfrac{1}{4}\int_{\pi/3}^{2\pi/3} (1 - \cos 2\theta)\, d\theta = \tfrac{1}{4}\left[\theta - \tfrac{1}{2}\sin 2\theta\right]_{\pi/3}^{2\pi/3}$$

$$= \tfrac{1}{4}\left[\tfrac{2\pi}{3} - \tfrac{1}{2}\sin\tfrac{4\pi}{3} - \tfrac{\pi}{3} + \tfrac{1}{2}\sin\tfrac{2\pi}{3}\right] = \tfrac{1}{4}\left[\tfrac{2\pi}{3} - \tfrac{1}{2}\left(-\tfrac{\sqrt{3}}{2}\right) - \tfrac{\pi}{3} + \tfrac{1}{2}\left(\tfrac{\sqrt{3}}{2}\right)\right] = \tfrac{1}{4}\left(\tfrac{\pi}{3} + \tfrac{\sqrt{3}}{2}\right) = \tfrac{\pi}{12} + \tfrac{\sqrt{3}}{8}$$

5. $r = \theta, 0 \le \theta \le \pi$. $A = \displaystyle\int_0^{\pi} \tfrac{1}{2}\theta^2 d\theta = \left[\tfrac{1}{6}\theta^3\right]_0^{\pi} = \tfrac{1}{6}\pi^3$

7. $r = 4 + 3\sin\theta, -\frac{\pi}{2} \le \theta \le \frac{\pi}{2}$.

$$A = \int_{-\pi/2}^{\pi/2} \tfrac{1}{2}(4 + 3\sin\theta)^2 d\theta = \tfrac{1}{2}\int_{-\pi/2}^{\pi/2} \left(16 + 24\sin\theta + 9\sin^2\theta\right) d\theta$$

$$= \tfrac{1}{2}\int_{-\pi/2}^{\pi/2} \left(16 + 9\sin^2\theta\right) d\theta \quad \text{[by Theorem 5.5.6(b)]}$$

$$= \tfrac{1}{2}\cdot 2\int_0^{\pi/2} \left[16 + 9\cdot\tfrac{1}{2}(1 - \cos 2\theta)\right] d\theta \quad \text{[by Theorem 5.5.6(a)]}$$

$$= \int_0^{\pi/2} \left(\tfrac{41}{2} - \tfrac{9}{2}\cos 2\theta\right) d\theta = \left[\tfrac{41}{2}\theta - \tfrac{9}{4}\sin 2\theta\right]_0^{\pi/2} = \left(\tfrac{41\pi}{4} - 0\right) - (0 - 0) = \tfrac{41\pi}{4}$$

9. The curve goes through the pole when $\theta = \pi/4$, so we'll find the area for $0 \le \theta \le \pi/4$ and multiply it by 4.

$A = 4\int_0^{\pi/4} \tfrac{1}{2} r^2 \, d\theta = 2\int_0^{\pi/4} (4\cos 2\theta)\, d\theta$

$= 8\int_0^{\pi/4} \cos 2\theta \, d\theta = 4\left[\sin 2\theta\right]_0^{\pi/4} = 4$

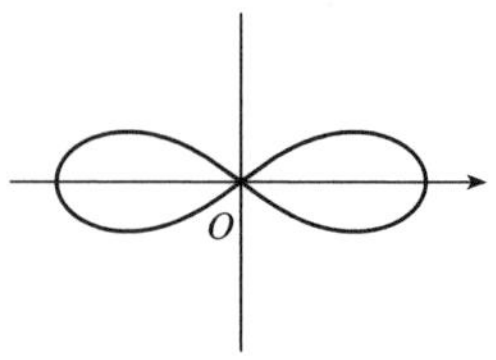

11. One-sixth of the area lies above the polar axis and is bounded by the curve

$r = 2\cos 3\theta$ for $\theta = 0$ to $\theta = \pi/6$.

$A = 6\int_0^{\pi/6} \tfrac{1}{2}(2\cos 3\theta)^2 \, d\theta = 12\int_0^{\pi/6} \cos^2 3\theta \, d\theta$

$= \tfrac{12}{2}\int_0^{\pi/6} (1 + \cos 6\theta) d\theta$

$= 6\left[\theta + \tfrac{1}{6}\sin 6\theta\right]_0^{\pi/6} = 6\left(\tfrac{\pi}{6}\right) = \pi$

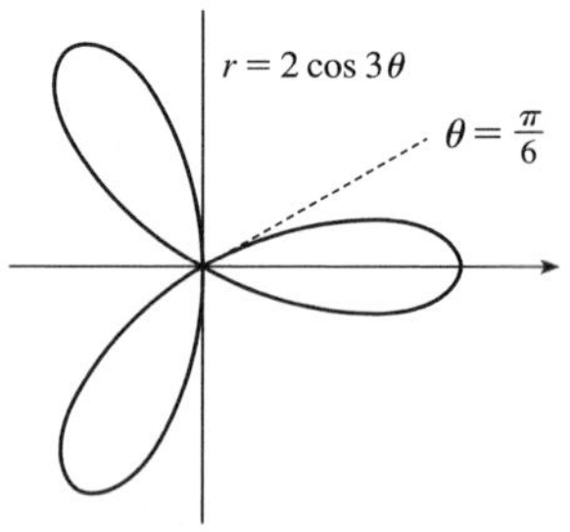

13. $A = \int_0^{2\pi} \tfrac{1}{2}(1 + 2\sin 6\theta)^2 \, d\theta = \tfrac{1}{2}\int_0^{2\pi} (1 + 4\sin 6\theta + 4\sin^2 6\theta) d\theta$

$= \tfrac{1}{2}\int_0^{2\pi} \left[1 + 4\sin 6\theta + 4\cdot\tfrac{1}{2}(1 - \cos 12\theta)\right] d\theta$

$= \tfrac{1}{2}\int_0^{2\pi} (3 + 4\sin 6\theta - 2\cos 12\theta)\, d\theta$

$= \tfrac{1}{2}\left[3\theta - \tfrac{2}{3}\cos 6\theta - \tfrac{1}{6}\sin 12\theta\right]_0^{2\pi}$

$= \tfrac{1}{2}\left[\left(6\pi - \tfrac{2}{3} - 0\right) - \left(0 - \tfrac{2}{3} - 0\right)\right] = 3\pi$

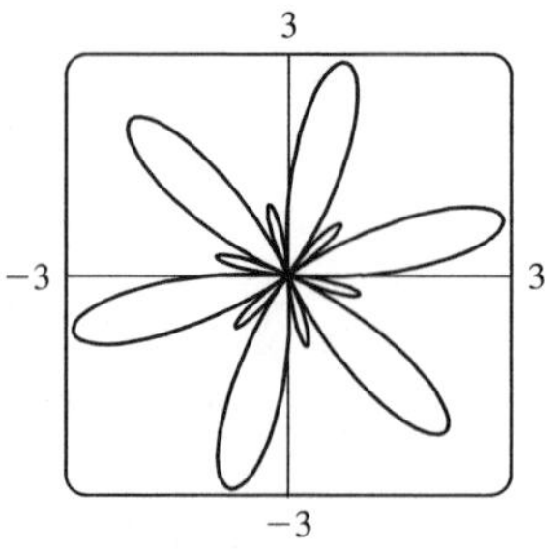

15. The shaded loop is traced out from $\theta = 0$ to $\theta = \pi/2$.

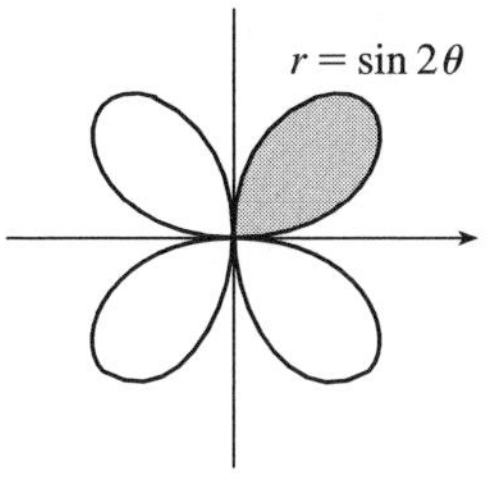

$$A = \int_0^{\pi/2} \tfrac{1}{2} r^2 \, d\theta = \tfrac{1}{2} \int_0^{\pi/2} \sin^2 2\theta \, d\theta$$

$$= \tfrac{1}{2} \int_0^{\pi/2} \tfrac{1}{2}(1 - \cos 4\theta) \, d\theta = \tfrac{1}{4}\left[\theta - \tfrac{1}{4}\sin 4\theta\right]_0^{\pi/2} = \tfrac{1}{4}\left(\tfrac{\pi}{2}\right) = \tfrac{\pi}{8}$$

17.

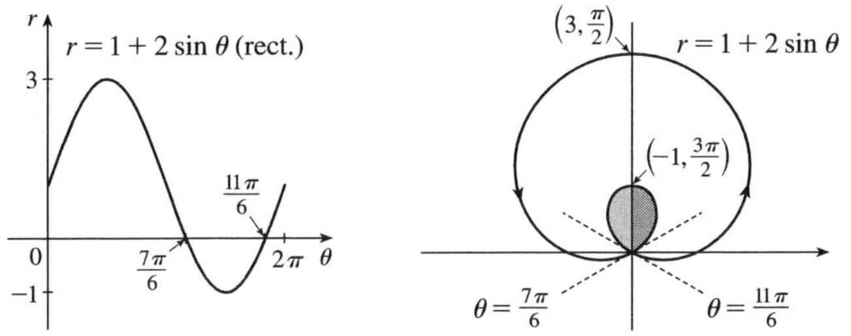

This is a limaçon, with inner loop traced out between $\theta = \frac{7\pi}{6}$ and $\frac{11\pi}{6}$ [found by solving $r = 0$].

$$A = 2\int_{7\pi/6}^{3\pi/2} \tfrac{1}{2}(1 + 2\sin\theta)^2 \, d\theta = \int_{7\pi/6}^{3\pi/2} \left(1 + 4\sin\theta + 4\sin^2\theta\right) d\theta = \int_{7\pi/6}^{3\pi/2} \left[1 + 4\sin\theta + 4 \cdot \tfrac{1}{2}(1 - \cos 2\theta)\right] d\theta$$

$$= \left[\theta - 4\cos\theta + 2\theta - \sin 2\theta\right]_{7\pi/6}^{3\pi/2} = \left(\tfrac{9\pi}{2}\right) - \left(\tfrac{7\pi}{2} + 2\sqrt{3} - \tfrac{\sqrt{3}}{2}\right) = \pi - \tfrac{3\sqrt{3}}{2}$$

19. $4\sin\theta = 2 \;\Leftrightarrow\; \sin\theta = \frac{1}{2} \;\Leftrightarrow\; \theta = \frac{\pi}{6}$ or $\frac{5\pi}{6}$ (for $0 \le \theta \le 2\pi$). We'll subtract the unshaded area from the shaded area for $\pi/6 \le \theta \le \pi/2$ and double that value.

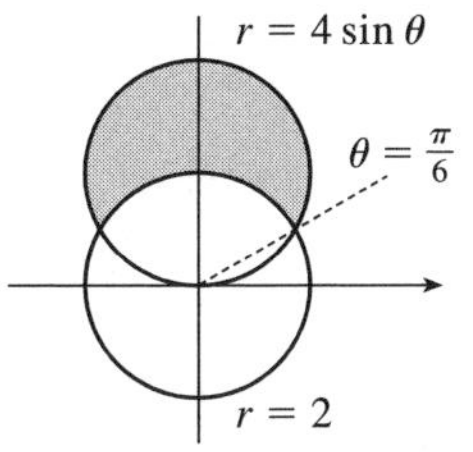

$$A = 2\int_{\pi/6}^{\pi/2} \tfrac{1}{2}(4\sin\theta)^2 \, d\theta - 2\int_{\pi/6}^{\pi/2} \tfrac{1}{2}(2)^2 \, d\theta = 2\int_{\pi/6}^{\pi/2} \tfrac{1}{2}\left[(4\sin\theta)^2 - 2^2\right] d\theta$$

$$= \int_{\pi/6}^{\pi/2} \left(16\sin^2\theta - 4\right) d\theta = \int_{\pi/6}^{\pi/2} \left[8(1 - \cos 2\theta) - 4\right] d\theta$$

$$= \int_{\pi/6}^{\pi/2} (4 - 8\cos 2\theta) \, d\theta = \left[4\theta - 4\sin 2\theta\right]_{\pi/6}^{\pi/2} = (2\pi - 0) - \left(\tfrac{2\pi}{3} - 4 \cdot \tfrac{\sqrt{3}}{2}\right) = \tfrac{4}{3}\pi + 2\sqrt{3}$$

21. $3\cos\theta = 1 + \cos\theta \;\Leftrightarrow\; \cos\theta = \frac{1}{2} \;\Rightarrow\; \theta = \frac{\pi}{3}$ or $-\frac{\pi}{3}$.

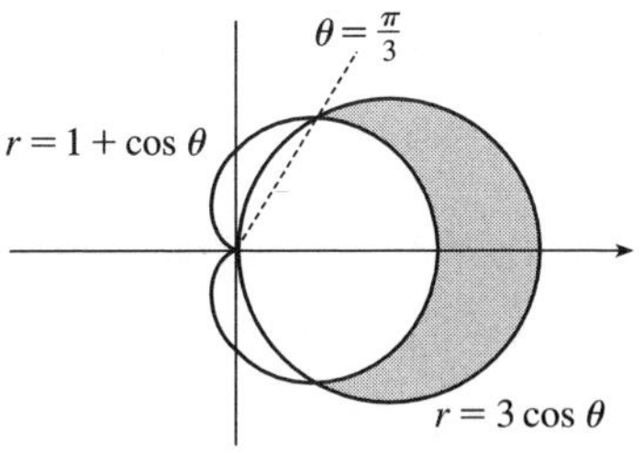

$$A = 2\int_0^{\pi/3} \tfrac{1}{2}\left[(3\cos\theta)^2 - (1 + \cos\theta)^2\right] d\theta$$

$$= \int_0^{\pi/3} \left(8\cos^2\theta - 2\cos\theta - 1\right) d\theta = \int_0^{\pi/3} \left[4(1 + \cos 2\theta) - 2\cos\theta - 1\right] d\theta$$

$$= \int_0^{\pi/3} (3 + 4\cos 2\theta - 2\cos\theta) \, d\theta = \left[3\theta + 2\sin 2\theta - 2\sin\theta\right]_0^{\pi/3}$$

$$= \pi + \sqrt{3} - \sqrt{3} = \pi$$

23. $A = 2\int_0^{\pi/4} \frac{1}{2}\sin^2\theta\,d\theta = \displaystyle\int_0^{\pi/4} \frac{1}{2}(1-\cos 2\theta)\,d\theta$

$= \frac{1}{2}\left[\theta - \frac{1}{2}\sin 2\theta\right]_0^{\pi/4} = \frac{1}{2}\left[\left(\frac{\pi}{4} - \frac{1}{2}\cdot 1\right) - (0-0)\right]$

$= \frac{1}{8}\pi - \frac{1}{4}$

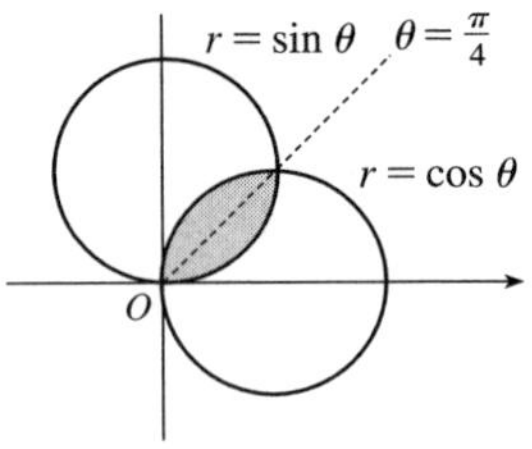

25. $\sin 2\theta = \cos 2\theta \;\Rightarrow\; \dfrac{\sin 2\theta}{\cos 2\theta} = 1 \;\Rightarrow\; \tan 2\theta = 1 \;\Rightarrow\; 2\theta = \frac{\pi}{4} \;\Rightarrow$

$\theta = \frac{\pi}{8} \;\Rightarrow$

$$A = 8\cdot 2\int_0^{\pi/8} \tfrac{1}{2}\sin^2 2\theta\,d\theta = 8\int_0^{\pi/8} \tfrac{1}{2}(1-\cos 4\theta)\,d\theta$$

$$= 4\left[\theta - \tfrac{1}{4}\sin 4\theta\right]_0^{\pi/8} = 4\left(\tfrac{\pi}{8} - \tfrac{1}{4}\cdot 1\right) = \tfrac{1}{2}\pi - 1$$

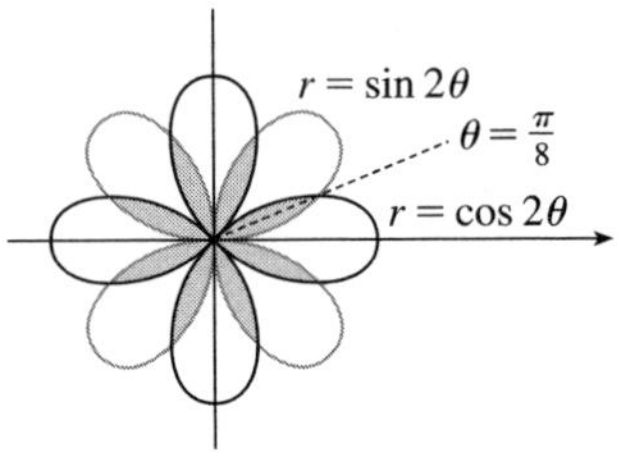

27. The darker shaded region (from $\theta = 0$ to $\theta = 2\pi/3$) represents $\frac{1}{2}$ of the desired area plus $\frac{1}{2}$ of the area of the inner loop. From this area, we'll subtract $\frac{1}{2}$ of the area of the inner loop (the lighter shaded region from $\theta = 2\pi/3$ to $\theta = \pi$), and then double that difference to obtain the desired area.

$$A = 2\left[\int_0^{2\pi/3} \tfrac{1}{2}\left(\tfrac{1}{2}+\cos\theta\right)^2 d\theta - \int_{2\pi/3}^{\pi} \tfrac{1}{2}\left(\tfrac{1}{2}+\cos\theta\right)^2 d\theta\right]$$

$$= \int_0^{2\pi/3}\left(\tfrac{1}{4}+\cos\theta+\cos^2\theta\right)d\theta - \int_{2\pi/3}^{\pi}\left(\tfrac{1}{4}+\cos\theta+\cos^2\theta\right)d\theta$$

$$= \int_0^{2\pi/3}\left[\tfrac{1}{4}+\cos\theta+\tfrac{1}{2}(1+\cos 2\theta)\right]d\theta - \int_{2\pi/3}^{\pi}\left[\tfrac{1}{4}+\cos\theta+\tfrac{1}{2}(1+\cos 2\theta)\right]d\theta$$

$$= \left[\frac{\theta}{4}+\sin\theta+\frac{\theta}{2}+\frac{\sin 2\theta}{4}\right]_0^{2\pi/3} - \left[\frac{\theta}{4}+\sin\theta+\frac{\theta}{2}+\frac{\sin 2\theta}{4}\right]_{2\pi/3}^{\pi}$$

$$= \left(\tfrac{\pi}{6}+\tfrac{\sqrt{3}}{2}+\tfrac{\pi}{3}-\tfrac{\sqrt{3}}{8}\right) - \left(\tfrac{\pi}{4}+\tfrac{\pi}{2}\right) + \left(\tfrac{\pi}{6}+\tfrac{\sqrt{3}}{2}+\tfrac{\pi}{3}-\tfrac{\sqrt{3}}{8}\right)$$

$$= \tfrac{\pi}{4}+\tfrac{3}{4}\sqrt{3} = \tfrac{1}{4}\left(\pi+3\sqrt{3}\right)$$

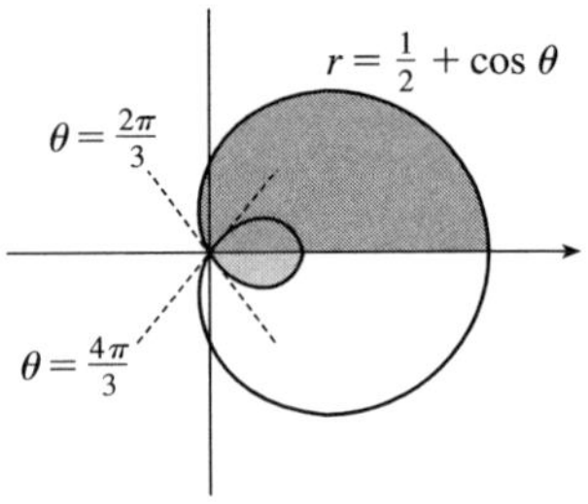

29. The curves intersect at the pole since $\left(0, \frac{\pi}{2}\right)$ satisfies $r = \cos\theta$ and $(0,0)$ satisfies $r = 1-\cos\theta$. Now $\cos\theta = 1-\cos\theta \;\Rightarrow\; 2\cos\theta = 1 \;\Rightarrow$ $\cos\theta = \frac{1}{2} \;\Rightarrow\; \theta = \frac{\pi}{3}$ or $\frac{5\pi}{3} \;\Rightarrow$ the other intersection points are $\left(\frac{1}{2}, \frac{\pi}{3}\right)$ and $\left(\frac{1}{2}, \frac{5\pi}{3}\right)$.

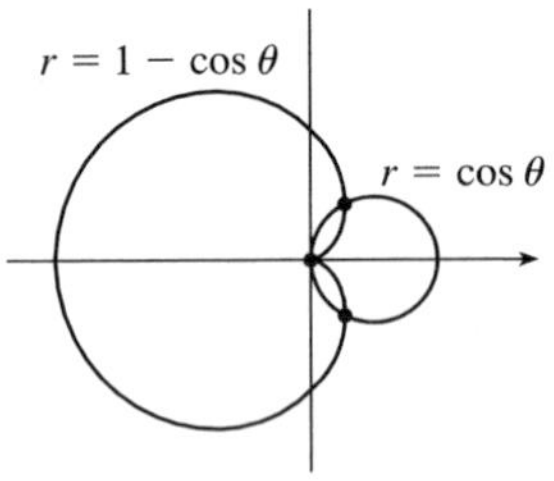

31. The pole is a point of intersection. $\sin\theta = \sin 2\theta = 2\sin\theta\,\cos\theta \;\Leftrightarrow$ $\sin\theta\,(1-2\cos\theta) = 0 \;\Leftrightarrow\; \sin\theta = 0$ or $\cos\theta = \frac{1}{2} \;\Rightarrow$ $\theta = 0, \pi, \frac{\pi}{3}$, or $-\frac{\pi}{3} \;\Rightarrow$ the other intersection points are $\left(\frac{\sqrt{3}}{2}, \frac{\pi}{3}\right)$ and $\left(\frac{\sqrt{3}}{2}, \frac{2\pi}{3}\right)$ (by symmetry).

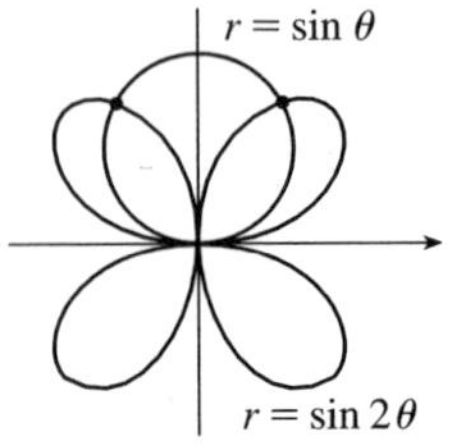

33.

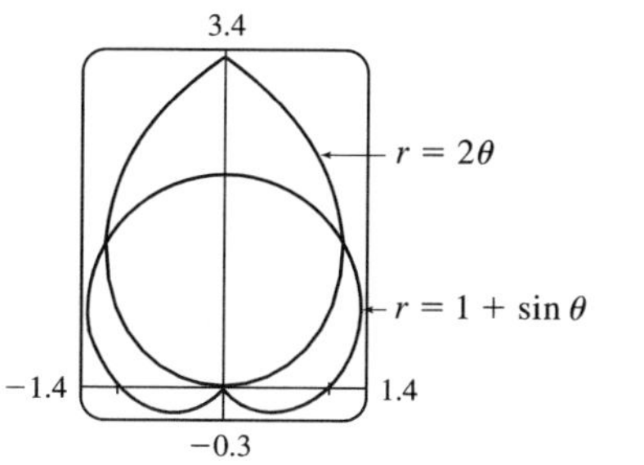

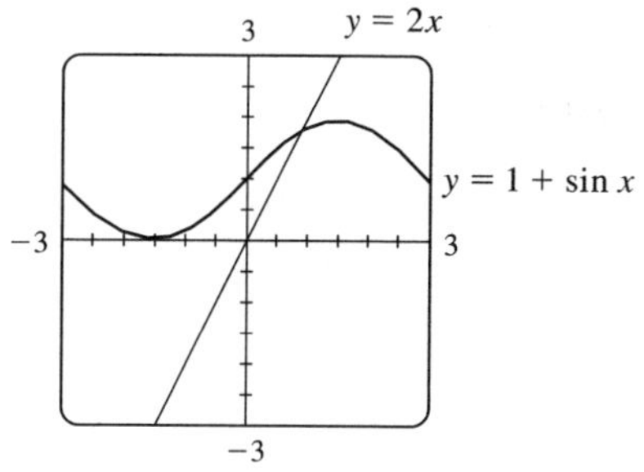

From the first graph, we see that the pole is one point of intersection. By zooming in or using the cursor, we find the θ-values of the intersection points to be $\alpha \approx 0.88786 \approx 0.89$ and $\pi - \alpha \approx 2.25$. (The first of these values may be more easily estimated by plotting $y = 1 + \sin x$ and $y = 2x$ in rectangular coordinates; see the second graph.)
By symmetry, the total area contained is twice the area contained in the first quadrant, that is,

$$A = 2\int_0^{\alpha} \tfrac{1}{2}(2\theta)^2\,d\theta + 2\int_{\alpha}^{\pi/2} \tfrac{1}{2}(1+\sin\theta)^2\,d\theta = \int_0^{\alpha} 4\theta^2\,d\theta + \int_{\alpha}^{\pi/2}\left[1 + 2\sin\theta + \tfrac{1}{2}(1-\cos 2\theta)\right]d\theta$$

$$= \left[\tfrac{4}{3}\theta^3\right]_0^{\alpha} + \left[\theta - 2\cos\theta + \left(\tfrac{1}{2}\theta - \tfrac{1}{4}\sin 2\theta\right)\right]_{\alpha}^{\pi/2}$$

$$= \tfrac{4}{3}\alpha^3 + \left[\left(\tfrac{\pi}{2} + \tfrac{\pi}{4}\right) - \left(\alpha - 2\cos\alpha + \tfrac{1}{2}\alpha - \tfrac{1}{4}\sin 2\alpha\right)\right] \approx 3.4645$$

35. $L = \int_a^b \sqrt{r^2 + (dr/d\theta)^2}\,d\theta = \int_0^{\pi/3} \sqrt{(3\sin\theta)^2 + (3\cos\theta)^2}\,d\theta = \int_0^{\pi/3} \sqrt{9(\sin^2\theta + \cos^2\theta)}\,d\theta$

$$= 3\int_0^{\pi/3} d\theta = 3\left[\,\theta\,\right]_0^{\pi/3} = 3\left(\tfrac{\pi}{3}\right) = \pi.$$

As a check, note that the circumference of a circle with radius $\frac{3}{2}$ is $2\pi\left(\frac{3}{2}\right) = 3\pi$, and since $\theta = 0$ to $\pi = \frac{\pi}{3}$ traces out $\frac{1}{3}$ of the circle (from $\theta = 0$ to $\theta = \pi$), $\frac{1}{3}(3\pi) = \pi$.

37. $L = \int_a^b \sqrt{r^2 + (dr/d\theta)^2}\,d\theta = \int_0^{2\pi} \sqrt{(\theta^2)^2 + (2\theta)^2}\,d\theta = \int_0^{2\pi} \sqrt{\theta^4 + 4\theta^2}\,d\theta$

$$= \int_0^{2\pi} \sqrt{\theta^2(\theta^2+4)}\,d\theta = \int_0^{2\pi} \theta\sqrt{\theta^2+4}\,d\theta$$

Now let $u = \theta^2 + 4$, so that $du = 2\theta\,d\theta$ $\left[\theta\,d\theta = \frac{1}{2}\,du\right]$ and

$$\int_0^{2\pi} \theta\sqrt{\theta^2+4}\,d\theta = \int_4^{4\pi^2+4} \tfrac{1}{2}\sqrt{u}\,du = \frac{1}{2}\cdot\frac{2}{3}\left[u^{3/2}\right]_4^{4(\pi^2+1)} = \frac{1}{3}\left[4^{3/2}(\pi^2+1)^{3/2} - 4^{3/2}\right] = \frac{8}{3}\left[(\pi^2+1)^{3/2} - 1\right]$$

39. The curve $r = 3\sin 2\theta$ is completely traced with $0 \le \theta \le 2\pi$. $r^2 + \left(\frac{dr}{d\theta}\right)^2 = (3\sin 2\theta)^2 + (6\cos 2\theta)^2 \Rightarrow$
$L = \int_0^{2\pi} \sqrt{9\sin^2 2\theta + 36\cos^2 2\theta}\,d\theta \approx 29.0653$

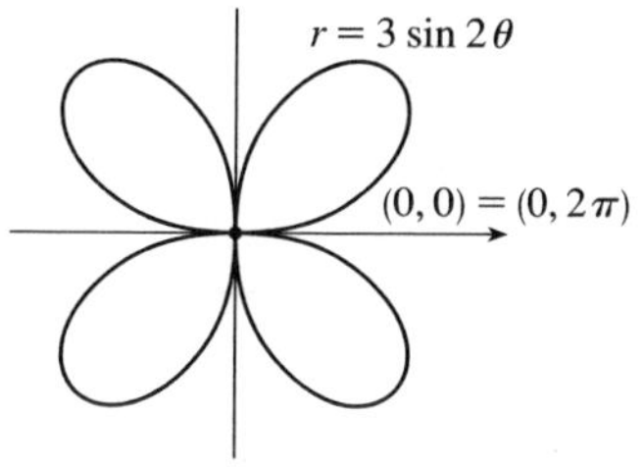

I COMPLEX NUMBERS

1. $(5-6i)+(3+2i)=(5+3)+(-6+2)i=8+(-4)i=8-4i$

3. $(2+5i)(4-i)=2(4)+2(-i)+(5i)(4)+(5i)(-i)=8-2i+20i-5i^2$
$=8+18i-5(-1)=8+18i+5=13+18i$

5. $\overline{12+7i}=12-7i$

7. $\dfrac{1+4i}{3+2i}=\dfrac{1+4i}{3+2i}\cdot\dfrac{3-2i}{3-2i}=\dfrac{3-2i+12i-8(-1)}{3^2+2^2}=\dfrac{11+10i}{13}=\dfrac{11}{13}+\dfrac{10}{13}i$

9. $\dfrac{1}{1+i}=\dfrac{1}{1+i}\cdot\dfrac{1-i}{1-i}=\dfrac{1-i}{1-(-1)}=\dfrac{1-i}{2}=\dfrac{1}{2}-\dfrac{1}{2}i$

11. $i^3=i^2\cdot i=(-1)i=-i$

13. $\sqrt{-25}=\sqrt{25}\,i=5i$

15. $\overline{12-5i}=12+15i$ and $|12-15i|=\sqrt{12^2+(-5)^2}=\sqrt{144+25}=\sqrt{169}=13$

17. $\overline{-4i}=\overline{0-4i}=0+4i=4i$ and $|-4i|=\sqrt{0^2+(-4)^2}=\sqrt{16}=4$

19. $4x^2+9=0 \;\Leftrightarrow\; 4x^2=-9 \;\Leftrightarrow\; x^2=-\frac{9}{4} \;\Leftrightarrow\; x=\pm\sqrt{-\frac{9}{4}}=\pm\sqrt{\frac{9}{4}}\,i=\pm\frac{3}{2}i.$

21. By the quadratic formula, $x^2+2x+5=0 \;\Leftrightarrow\; x=\dfrac{-2\pm\sqrt{2^2-4(1)(5)}}{2(1)}=\dfrac{-2\pm\sqrt{-16}}{2}=\dfrac{-2\pm 4i}{2}=-1\pm 2i.$

23. By the quadratic formula, $z^2+z+2=0 \;\Leftrightarrow\; z=\dfrac{-1\pm\sqrt{1^2-4(1)(2)}}{2(1)}=\dfrac{-1\pm\sqrt{-7}}{2}=-\dfrac{1}{2}\pm\dfrac{\sqrt{7}}{2}i.$

25. For $z=-3+3i$, $r=\sqrt{(-3)^2+3^2}=3\sqrt{2}$ and $\tan\theta=\frac{3}{-3}=-1 \;\Rightarrow\; \theta=\frac{3\pi}{4}$ (since z lies in the second quadrant).

Therefore, $-3+3i=3\sqrt{2}\left(\cos\frac{3\pi}{4}+i\sin\frac{3\pi}{4}\right)$.

27. For $z=3+4i$, $r=\sqrt{3^2+4^2}=5$ and $\tan\theta=\frac{4}{3} \;\Rightarrow\; \theta=\tan^{-1}\left(\frac{4}{3}\right)$ (since z lies in the first quadrant). Therefore,

$3+4i=5\left\{\cos\left[\tan^{-1}\left(\frac{4}{3}\right)\right]+i\sin\left[\tan^{-1}\left(\frac{4}{3}\right)\right]\right\}$.

29. For $z=\sqrt{3}+i$, $r=\sqrt{\left(\sqrt{3}\right)^2+1^2}=2$ and $\tan\theta=\frac{1}{\sqrt{3}} \;\Rightarrow\; \theta=\frac{\pi}{6} \;\Rightarrow\; z=2\left(\cos\frac{\pi}{6}+i\sin\frac{\pi}{6}\right)$.

For $w=1+\sqrt{3}\,i$, $r=2$ and $\tan\theta=\sqrt{3} \;\Rightarrow\; \theta=\frac{\pi}{3} \;\Rightarrow\; w=2\left(\cos\frac{\pi}{3}+i\sin\frac{\pi}{3}\right)$.

Therefore, $zw=2\cdot 2\left[\cos\left(\frac{\pi}{6}+\frac{\pi}{3}\right)+i\sin\left(\frac{\pi}{6}+\frac{\pi}{3}\right)\right]=4\left(\cos\frac{\pi}{2}+i\sin\frac{\pi}{2}\right)$,

$z/w=\frac{2}{2}\left[\cos\left(\frac{\pi}{6}-\frac{\pi}{3}\right)+i\sin\left(\frac{\pi}{6}-\frac{\pi}{3}\right)\right]=\cos\left(-\frac{\pi}{6}\right)+i\sin\left(-\frac{\pi}{6}\right)$, and $1=1+0i=1(\cos 0+i\sin 0) \;\Rightarrow$

$1/z=\frac{1}{2}\left[\cos\left(0-\frac{\pi}{6}\right)+i\sin\left(0-\frac{\pi}{6}\right)\right]=\frac{1}{2}\left[\cos\left(-\frac{\pi}{6}\right)+i\sin\left(-\frac{\pi}{6}\right)\right]$. For $1/z$, we could also use the formula that precedes Example 5 to obtain $1/z=\frac{1}{2}\left(\cos\frac{\pi}{6}-i\sin\frac{\pi}{6}\right)$.

31. For $z = 2\sqrt{3} - 2i$, $r = \sqrt{(2\sqrt{3})^2 + (-2)^2} = 4$ and $\tan\theta = \frac{-2}{2\sqrt{3}} = -\frac{1}{\sqrt{3}} \Rightarrow \theta = -\frac{\pi}{6} \Rightarrow$

$z = 4\left[\cos\left(-\frac{\pi}{6}\right) + i\sin\left(-\frac{\pi}{6}\right)\right]$. For $w = -1 + i$, $r = \sqrt{2}$, $\tan\theta = \frac{1}{-1} = -1 \Rightarrow \theta = \frac{3\pi}{4} \Rightarrow$

$w = \sqrt{2}\left(\cos\frac{3\pi}{4} + i\sin\frac{3\pi}{4}\right)$. Therefore, $zw = 4\sqrt{2}\left[\cos\left(-\frac{\pi}{6} + \frac{3\pi}{4}\right) + i\sin\left(-\frac{\pi}{6} + \frac{3\pi}{4}\right)\right] = 4\sqrt{2}\left(\cos\frac{7\pi}{12} + i\sin\frac{7\pi}{12}\right)$,

$z/w = \frac{4}{\sqrt{2}}\left[\cos\left(-\frac{\pi}{6} - \frac{3\pi}{4}\right) + i\sin\left(-\frac{\pi}{6} - \frac{3\pi}{4}\right)\right] = \frac{4}{\sqrt{2}}\left[\cos\left(-\frac{11\pi}{12}\right) + i\sin\left(-\frac{11\pi}{12}\right)\right]$

$= 2\sqrt{2}\left(\cos\frac{13\pi}{12} + i\sin\frac{13\pi}{12}\right)$, and

$1/z = \frac{1}{4}\left[\cos\left(-\frac{\pi}{6}\right) - i\sin\left(-\frac{\pi}{6}\right)\right] = \frac{1}{4}\left(\cos\frac{\pi}{6} + i\sin\frac{\pi}{6}\right)$.

33. For $z = 1 + i$, $r = \sqrt{2}$ and $\tan\theta = \frac{1}{1} = 1 \Rightarrow \theta = \frac{\pi}{4} \Rightarrow z = \sqrt{2}\left(\cos\frac{\pi}{4} + i\sin\frac{\pi}{4}\right)$. So by De Moivre's Theorem,

$$(1+i)^{20} = \left[\sqrt{2}\left(\cos\frac{\pi}{4} + i\sin\frac{\pi}{4}\right)\right]^{20} = \left(2^{1/2}\right)^{20}\left(\cos\frac{20\cdot\pi}{4} + i\sin\frac{20\cdot\pi}{4}\right)$$

$$= 2^{10}(\cos 5\pi + i\sin 5\pi) = 2^{10}[-1 + i(0)] = -2^{10} = -1024$$

35. For $z = 2\sqrt{3} + 2i$, $r = \sqrt{(2\sqrt{3})^2 + 2^2} = \sqrt{16} = 4$ and $\tan\theta = \frac{2}{2\sqrt{3}} = \frac{1}{\sqrt{3}} \Rightarrow \theta = \frac{\pi}{6} \Rightarrow z = 4\left(\cos\frac{\pi}{6} + i\sin\frac{\pi}{6}\right)$.

So by De Moivre's Theorem,

$$\left(2\sqrt{3} + 2i\right)^5 = \left[4\left(\cos\frac{\pi}{6} + i\sin\frac{\pi}{6}\right)\right]^5 = 4^5\left(\cos\frac{5\pi}{6} + i\sin\frac{5\pi}{6}\right) = 1024\left[-\frac{\sqrt{3}}{2} + \frac{1}{2}i\right] = -512\sqrt{3} + 512i$$

37. $1 = 1 + 0i = 1(\cos 0 + i\sin 0)$. Using Equation 3 with $r = 1$, $n = 8$, and $\theta = 0$, we have

$w_k = 1^{1/8}\left[\cos\left(\frac{0 + 2k\pi}{8}\right) + i\sin\left(\frac{0 + 2k\pi}{8}\right)\right] = \cos\frac{k\pi}{4} + i\sin\frac{k\pi}{4}$, where $k = 0, 1, 2, \ldots, 7$.

$w_0 = 1(\cos 0 + i\sin 0) = 1$, $w_1 = 1\left(\cos\frac{\pi}{4} + i\sin\frac{\pi}{4}\right) = \frac{1}{\sqrt{2}} + \frac{1}{\sqrt{2}}i$,

$w_2 = 1\left(\cos\frac{\pi}{2} + i\sin\frac{\pi}{2}\right) = i$, $w_3 = 1\left(\cos\frac{3\pi}{4} + i\sin\frac{3\pi}{4}\right) = -\frac{1}{\sqrt{2}} + \frac{1}{\sqrt{2}}i$,

$w_4 = 1(\cos\pi + i\sin\pi) = -1$, $w_5 = 1\left(\cos\frac{5\pi}{4} + i\sin\frac{5\pi}{4}\right) = -\frac{1}{\sqrt{2}} - \frac{1}{\sqrt{2}}i$,

$w_6 = 1\left(\cos\frac{3\pi}{2} + i\sin\frac{3\pi}{2}\right) = -i$, $w_7 = 1\left(\cos\frac{7\pi}{4} + i\sin\frac{7\pi}{4}\right) = \frac{1}{\sqrt{2}} - \frac{1}{\sqrt{2}}i$

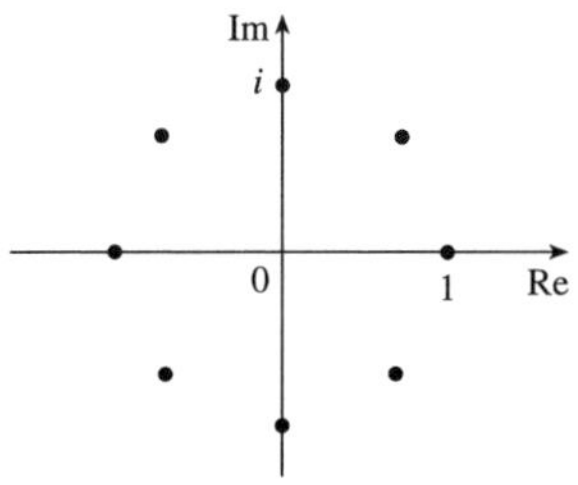

39. $i = 0 + i = 1\left(\cos\frac{\pi}{2} + i\sin\frac{\pi}{2}\right)$. Using Equation 3 with $r = 1$, $n = 3$, and $\theta = \frac{\pi}{2}$, we have

$w_k = 1^{1/3}\left[\cos\left(\frac{\frac{\pi}{2} + 2k\pi}{3}\right) + i\sin\left(\frac{\frac{\pi}{2} + 2k\pi}{3}\right)\right]$, where $k = 0, 1, 2$.

$w_0 = \left(\cos\frac{\pi}{6} + i\sin\frac{\pi}{6}\right) = \frac{\sqrt{3}}{2} + \frac{1}{2}i$

$w_1 = \left(\cos\frac{5\pi}{6} + i\sin\frac{5\pi}{6}\right) = -\frac{\sqrt{3}}{2} + \frac{1}{2}i$

$w_2 = \left(\cos\frac{9\pi}{6} + i\sin\frac{9\pi}{6}\right) = -i$

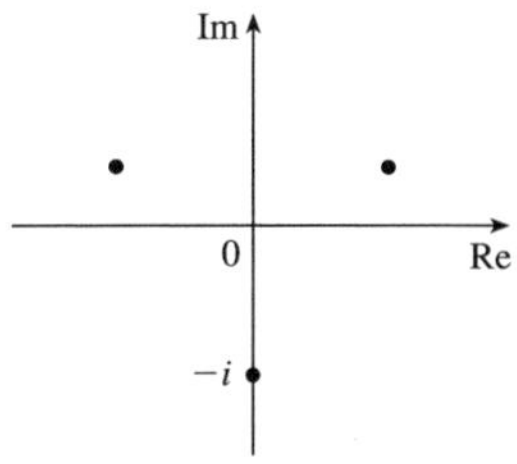

41. Using Euler's formula (6) with $y = \frac{\pi}{2}$, we have $e^{i\pi/2} = \cos\frac{\pi}{2} + i\sin\frac{\pi}{2} = 0 + 1i = i$.

43. Using Euler's formula (6) with $y = \frac{\pi}{3}$, we have $e^{i\pi/3} = \cos\frac{\pi}{3} + i\sin\frac{\pi}{3} = \frac{1}{2} + \frac{\sqrt{3}}{2}i$.

45. Using Equation 7 with $x = 2$ and $y = \pi$, we have $e^{2+i\pi} = e^2 e^{i\pi} = e^2(\cos\pi + i\sin\pi) = e^2(-1+0) = -e^2$.

47. Take $r = 1$ and $n = 3$ in De Moivre's Theorem to get

$$\begin{aligned}
[1(\cos\theta + i\sin\theta)]^3 &= 1^3(\cos 3\theta + i\sin 3\theta) \\
(\cos\theta + i\sin\theta)^3 &= \cos 3\theta + i\sin 3\theta \\
\cos^3\theta + 3(\cos^2\theta)(i\sin\theta) + 3(\cos\theta)(i\sin\theta)^2 + (i\sin\theta)^3 &= \cos 3\theta + i\sin 3\theta \\
\cos^3\theta + (3\cos^2\theta\,\sin\theta)i - 3\cos\theta\,\sin^2\theta - (\sin^3\theta)i &= \cos 3\theta + i\sin 3\theta \\
(\cos^3\theta - 3\sin^2\theta\,\cos\theta) + (3\sin\theta\,\cos^2\theta - \sin^3\theta)i &= \cos 3\theta + i\sin 3\theta
\end{aligned}$$

Equating real and imaginary parts gives

$$\cos 3\theta = \cos^3\theta - 3\sin^2\theta\,\cos\theta \quad\text{and}\quad \sin 3\theta = 3\sin\theta\,\cos^2\theta - \sin^3\theta$$

49. $F(x) = e^{rx} = e^{(a+bi)x} = e^{ax+bxi} = e^{ax}(\cos bx + i\sin bx) = e^{ax}\cos bx + i(e^{ax}\sin bx) \;\Rightarrow$

$$\begin{aligned}
F'(x) &= (e^{ax}\cos bx)' + i(e^{ax}\sin bx)' \\
&= (ae^{ax}\cos bx - be^{ax}\sin bx) + i(ae^{ax}\sin bx + be^{ax}\cos bx) \\
&= a\left[e^{ax}(\cos bx + i\sin bx)\right] + b\left[e^{ax}(-\sin bx + i\cos bx)\right] \\
&= ae^{rx} + b\left[e^{ax}(i^2\sin bx + i\cos bx)\right] \\
&= ae^{rx} + bi\left[e^{ax}(\cos bx + i\sin bx)\right] = ae^{rx} + bie^{rx} = (a+bi)e^{rx} = re^{rx}
\end{aligned}$$